Fundamental Equations of Dynamics

KINEMATICS

Particle Rectilinear Motion

Variable a
$$a = \frac{dv}{dt}$$
$$v = \frac{ds}{dt}$$
$$a\,ds = v\,dv$$

Constant $a = a_c$
$$v = v_0 + a_c t$$
$$s = s_0 + v_0 t + \tfrac{1}{2} a_c t^2$$
$$v^2 = v_0^2 + 2a_c(s - s_0)$$

Particle Curvilinear Motion

x, y, z Coordinates
$v_x = \dot{x} \quad a_x = \ddot{x}$
$v_y = \dot{y} \quad a_y = \ddot{y}$
$v_z = \dot{z} \quad a_z = \ddot{z}$

r, θ, z Coordinates
$v_r = \dot{r} \quad a_r = \ddot{r} - r\dot{\theta}^2$
$v_\theta = r\dot{\theta} \quad a_\theta = r\ddot{\theta} + 2\dot{r}\dot{\theta}$
$v_z = \dot{z} \quad a_z = \ddot{z}$

n, t, b Coordinates
$v = \dot{s}$
$a_t = \dot{v} = v\dfrac{dv}{ds}$
$a_n = \dfrac{v^2}{\rho} \quad \rho = \dfrac{[1 + (dy/dx)^2]^{3/2}}{|d^2y/dx^2|}$

Relative Motion
$\mathbf{v}_B = \mathbf{v}_A + \mathbf{v}_{B/A} \qquad \mathbf{a}_B = \mathbf{a}_A + \mathbf{a}_{B/A}$

Rigid Body Motion About a Fixed Axis

Variable α
$$\alpha = \frac{d\omega}{dt}$$
$$\omega = \frac{d\theta}{dt}$$
$$\omega\,d\omega = \alpha\,d\theta$$

Constant $\alpha = \alpha_c$
$$\omega = \omega_0 + \alpha_c t$$
$$\theta = \theta_0 + \omega_0 t + \tfrac{1}{2}\alpha_c t^2$$
$$\omega^2 = \omega_0^2 + 2\alpha_c(\theta - \theta_0)$$

For Point P
$s = \theta r \quad v = \omega r \quad a_t = \alpha r \quad a_n = \omega^2 r$

Relative General Plane Motion — Translating Axes
$\mathbf{v}_B = \mathbf{v}_A + \mathbf{v}_{B/A(\text{pin})} \qquad \mathbf{a}_B = \mathbf{a}_A + \mathbf{a}_{B/A(\text{pin})}$

Relative General Plane Motion — Trans. and Rot. Axis
$\mathbf{v}_B = \mathbf{v}_A + \mathbf{\Omega} \times \mathbf{r}_{B/A} + (\mathbf{v}_{B/A})_{xyz}$
$\mathbf{a}_B = \mathbf{a}_A + \dot{\mathbf{\Omega}} \times \mathbf{r}_{B/A} + \mathbf{\Omega} \times (\mathbf{\Omega} \times \mathbf{r}_{B/A}) + 2\mathbf{\Omega} \times (\mathbf{v}_{B/A})_{xyz} + (\mathbf{a}_{B/A})_{xyz}$

KINETICS

Mass Moment of Inertia $\quad I = \displaystyle\int r^2\,dm$

Parallel-Axis Theorem $\quad I = I_G + md^2$

Radius of Gyration $\quad k = \sqrt{\dfrac{I}{m}}$

Equations of Motion

Particle	$\Sigma \mathbf{F} = m\mathbf{a}$
Rigid Body (Plane Motion)	$\Sigma F_x = m(a_G)_x$ $\Sigma F_y = m(a_G)_y$ $\Sigma M_G = I_G \alpha$ or $\Sigma M_P = \Sigma(m_k)_P$

Principle of Work and Energy
$T_1 + \Sigma U_{1-2} = T_2$

Kinetic Energy

Particle	$T = \tfrac{1}{2} m v^2$
Rigid Body (Plane Motion)	$T = \tfrac{1}{2} m v_G^2 + \tfrac{1}{2} I_G \omega^2$

Work
Variable force $\quad U_F = \displaystyle\int F\cos\theta\,ds$
Constant force $\quad U_F = (F_c \cos\theta)\,\Delta s$
Weight $\quad U_W = -W\Delta y$
Spring $\quad U_s = -\left(\tfrac{1}{2} k s_2^2 - \tfrac{1}{2} k s_1^2\right)$
Couple moment $\quad U_M = M\Delta\theta$

Power and Efficiency
$$P = \frac{dU}{dt} = \mathbf{F}\cdot\mathbf{v} \qquad \varepsilon = \frac{P_{\text{out}}}{P_{\text{in}}} = \frac{U_{\text{out}}}{U_{\text{in}}}$$

Conservation of Energy Theorem
$T_1 + V_1 = T_2 + V_2$

Potential Energy
$V = V_g + V_e$, where $V_g = \pm Wy$, $V_e = +\tfrac{1}{2} k s^2$

Principle of Linear Impulse and Momentum

Particle	$m\mathbf{v}_1 + \Sigma \displaystyle\int \mathbf{F}\,dt = m\mathbf{v}_2$
Rigid Body	$m(\mathbf{v}_G)_1 + \Sigma \displaystyle\int \mathbf{F}\,dt = m(\mathbf{v}_G)_2$

Conservation of Linear Momentum
$\Sigma(\text{syst. } m\mathbf{v})_1 = \Sigma(\text{syst. } m\mathbf{v})_2$

Coefficient of Restitution $\quad e = \dfrac{(v_B)_2 - (v_A)_2}{(v_A)_1 - (v_B)_1}$

Principle of Angular Impulse and Momentum

Particle	$(\mathbf{H}_O)_1 + \Sigma \displaystyle\int \mathbf{M}_O\,dt = (\mathbf{H}_O)_2$ where $H_O = (d)(mv)$
Rigid Body (Plane motion)	$(\mathbf{H}_G)_1 + \Sigma \displaystyle\int \mathbf{M}_G\,dt = (\mathbf{H}_G)_2$ where $H_G = I_G \omega$ $(\mathbf{H}_O)_1 + \Sigma \displaystyle\int \mathbf{M}_O\,dt = (\mathbf{H}_O)_2$ where $H_O = I_O \omega$

Conservation of Angular Momentum
$\Sigma(\text{syst. } \mathbf{H})_1 = \Sigma(\text{syst. } \mathbf{H})_2$

ENGINEERING MECHANICS

STATICS and DYNAMICS

FIFTEENTH EDITION

ENGINEERING MECHANICS

STATICS and DYNAMICS

15TH EDITION

R.C. HIBBELER

Content Management: Dawn Murrin, Erin Ault
Content Development: Coleen Morrison
Content Production: Rose Kernan, Sandra Rodriguez, Pallavi Pandit
Product Management: Holly Stark
Product Marketing: Wayne Stevens
Rights and Permissions: Rimpy Sharma

Please contact https://support.pearson.com/getsupport/s/ with any queries on this content

Cover Image by Bozac/Alamy Stock Photo

Microsoft and/or its respective suppliers make no representations about the suitability of the information contained in the documents and related graphics published as part of the services for any purpose. All such documents and related graphics are provided "as is" without warranty of any kind. Microsoft and/or its respective suppliers hereby disclaim all warranties and conditions with regard to this information, including all warranties and conditions of merchantability, whether express, implied or statutory, fitness for a particular purpose, title and non-infringement. In no event shall Microsoft and/or its respective suppliers be liable for any special, indirect or consequential damages or any damages whatsoever resulting from loss of use, data or profits, whether in an action of contract, negligence or other tortious action, arising out of or in connection with the use or performance of information available from the services.

The documents and related graphics contained herein could include technical inaccuracies or typographical errors. Changes are periodically added to the information herein. Microsoft and/or its respective suppliers may make improvements and/or changes in the product(s) and/or the program(s) described herein at any time. Partial screen shots may be viewed in full within the software version specified.

Microsoft® and Windows® are registered trademarks of the Microsoft Corporation in the U.S.A. and other countries. This book is not sponsored or endorsed by or affiliated with the Microsoft Corporation.

Copyright © 2022, 2018, 2015, 2013, 2011 by R. C. Hibbeler. Published by Pearson Inc. or its affiliates, 221 River Street, Hoboken, NJ 07030. All Rights Reserved. Manufactured in the United States of America. This publication is protected by copyright, and permission should be obtained from the publisher prior to any prohibited reproduction, storage in a retrieval system, or transmission in any form or by any means, electronic, mechanical, photocopying, recording, or otherwise. For information regarding permissions, request forms, and the appropriate contacts within the Pearson Education Global Rights and Permissions department, please visit www.pearsoned.com/permissions/.

Acknowledgments of third-party content appear on the appropriate page within the text.

PEARSON, ALWAYS LEARNING, and MYLAB are exclusive trademarks owned by Pearson Education, Inc. or its affiliates in the U.S. and/or other countries.

Unless otherwise indicated herein, any third-party trademarks, logos, or icons that may appear in this work are the property of their respective owners, and any references to third-party trademarks, logos, icons, or other trade dress are for demonstrative or descriptive purposes only. Such references are not intended to imply any sponsorship, endorsement, authorization, or promotion of Pearson's products by the owners of such marks, or any relationship between the owner and Pearson Education, Inc., or its affiliates, authors, licensees, or distributors.

Library of Congress Cataloging-in-Publication Data
Names: Hibbeler, R. C., author.
Title: Engineering mechanics. Statics and dynamics / Russell C. Hibbeler.
Other titles: Statics and dynamics
Description: 15th edition. | Hoboken : Pearson, [2022] | Includes index. |
  Summary: "The main purpose of this book is to provide the student with a
  clear and thorough presentation of the theory and application of
  engineering mechanics"— Provided by publisher.
Identifiers: LCCN 2021011903 | ISBN 9780134780955 (hardcover)
Subjects: LCSH: Mechanics, Applied—Textbooks. | Mechanical
  engineering—Textbooks.
Classification: LCC TA645 .H48 2022 | DDC 624.1/7—dc23
LC record available at https://lccn.loc.gov/2021011903

2 2021

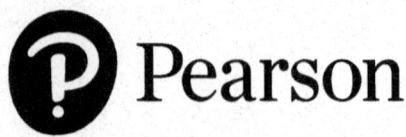

Rental
ISBN-10:   0-13-478095-7
ISBN-13: 978-0-13-478095-5

To the Student

With the hope that this work will stimulate
an interest in Engineering Mechanics
and provide an acceptable guide to its understanding.

PREFACE

The main purpose of this book is to provide the student with a clear and thorough presentation of the theory and application of engineering mechanics. To achieve this objective, this work has been shaped by the comments and suggestions of hundreds of reviewers in the teaching profession, as well as many of the author's students.

New to this Edition

Expanded Answer Section. The answer section in the back of the book now includes additional information related to the solution of select Fundamental Problems in order to offer the student some guidance in solving the problems.

Re-writing of Text Material. Further clarification of some concepts has been included in this edition, and throughout the book the accuracy has been enhanced, and important definitions are now in boldface throughout the text to highlight their importance.

Additional Fundamental Problems. Some new fundamental problems have been added along with their partial solutions which are given in the back of the book.

New Photos. The relevance of knowing the subject matter is reflected by the real-world applications depicted in the over 29 new or updated photos placed throughout the book. These photos generally are used to explain how the relevant principles apply to real-world situations and how materials behave under load.

New Problems. There are approximately 30% new problems that have been added to this edition, which involve applications to many different fields of engineering.

Hallmark Features

Besides the new features mentioned, other outstanding features that define the contents of the book include the following:

Organization and Approach. Each chapter is organized into well-defined sections that contain an explanation of specific topics, illustrative example problems, and a set of homework problems. The topics within each section are placed into subgroups defined by boldface titles. The purpose of this is to present a structured method for introducing each new definition or concept and to make the book convenient for later reference and review.

Chapter Contents. Each chapter begins with an illustration demonstrating a broad-range application of the material within the chapter. A bulleted list of the chapter contents is provided to give a general overview of the material that will be covered.

Emphasis on Free-Body Diagrams. Drawing a free-body diagram is particularly important when solving problems, and for this reason this step is strongly emphasized throughout the book. In particular, special sections and examples are devoted to show how to draw free-body diagrams. Specific homework problems have also been added to develop this practice.

Procedures for Analysis. A general procedure for analyzing any mechanics problem is presented at the end of the first chapter. Then this procedure is customized to relate to specific types of problems that are covered throughout the book. This unique feature provides the student with a logical and orderly method to follow when applying the theory. The example problems are solved using this outlined method in order to clarify its numerical application. Realize, however, that once the relevant principles have been mastered and enough confidence and judgment have been obtained, the student can then develop his or her own procedures for solving problems.

Important Points. This feature provides a review or summary of the most important concepts in a section and highlights the most significant points that should be known when applying the theory to solve problems.

Fundamental Problems. These problem sets are selectively located just after most of the example problems. They provide students with simple applications of the concepts, and therefore, the chance to develop their problem-solving skills before attempting to solve any of the standard problems that follow. In addition, they can be used for preparing for exams, and they can be used at a later time when preparing for the Fundamentals in Engineering Exam. The partial solutions are given in the back of the book.

Conceptual Understanding. Through the use of photographs placed throughout the book, the theory is applied in a simplified way in order to illustrate some of its more important conceptual features and instill the physical meaning of many of the terms used in the equations.

Homework Problems. Apart from the Fundamental and Conceptual type problems mentioned previously, other types of problems contained in the book include the following:

- **Free-Body Diagram Problems.** Some sections of the book contain introductory problems that only require drawing the free-body diagram for the specific problems within a problem set. These assignments will impress upon the student the importance of mastering this skill as a requirement for a complete solution of any equilibrium problem.

- **General Analysis and Design Problems.** The majority of problems in the book depict realistic situations encountered in engineering practice. Some of these problems come from actual products used in industry. It is hoped that this realism will both stimulate the student's interest in engineering mechanics and provide a means for developing the skill to reduce any such problem from its physical description to a model or symbolic representation to which the principles of mechanics may be applied.

Throughout the book, there is an approximate balance of problems using either SI or FPS units. Furthermore, in any set, an attempt has been made to arrange the problems in order of increasing difficulty except for the end of chapter review problems, which are presented in random order.

- **Computer Problems.** An effort has been made to include a few problems that may be solved using a numerical procedure executed on either a desktop computer or a programmable pocket calculator. The intent here is to broaden the student's capacity for using other forms of mathematical analysis without sacrificing the time needed to focus on the application of the principles of mechanics. Problems of this type, which either can or must be solved using numerical procedures, are identified by a "square" symbol (■) preceding the problem number.

The many homework problems in this edition, have been placed into two different categories. Problems that are simply indicated by a problem number have an answer and in some cases an additional numerical result given in the back of the book. An asterisk (*) before every fourth problem number indicates a problem without an answer.

Accuracy. As with the previous editions, apart from the author, the accuracy of the text and problem solutions has been thoroughly checked by Kai Beng Yap, a practicing engineer, and a team of specialists at EPAM, including Georgii Kolobov, Ekaterina Radchenko, and Artur Akberov. Thanks are also due to Keith Steuer from Snow College and Mike Freeman, Professor Emeritus at the University of Alabama.

Contents

Statics

The book is divided into 11 chapters, in which the principles are first applied to simple, then to more complicated situations. In a general sense, each principle is applied first to a particle, then a rigid body subjected to a coplanar system of forces, and finally to three-dimensional force systems acting on a rigid body.

Chapter 1 begins with an introduction to mechanics and a discussion of units. The vector properties of a concurrent force system are introduced in Chapter 2. This theory is then applied to the equilibrium of a particle in Chapter 3. Chapter 4 contains a general discussion of both concentrated and distributed force systems and the methods used to simplify them. The principles of rigid-body equilibrium are developed in Chapter 5 and then applied to specific problems involving the equilibrium of trusses, frames, and machines in Chapter 6, and to the analysis of internal forces in beams and cables in Chapter 7. Applications to problems involving frictional forces are discussed in Chapter 8, and topics related to the center of gravity and centroid are treated in Chapter 9. If time permits, sections involving more advanced topics, indicated by stars (★), may be covered. Most of these topics are included in Chapter 10 (area and mass moments of inertia) and Chapter 11 (virtual work and potential energy). Note that this material also provides a suitable reference for basic principles when it is discussed in more advanced courses. Finally, Appendix A of the Statics section provides a review and list of mathematical formulas needed to solve the problems in the book.

Alternative Coverage.
At the discretion of the instructor, some of the material may be presented in a different sequence with no loss of continuity. For example, it is possible to introduce the concept of a force and all the necessary methods of vector analysis by first covering Chapter 2 and Section 4 (the cross product). Then after covering the rest of Chapter 4 (force and moment systems), the equilibrium methods of Chapters 3 and 5 can be discussed.

Dynamics

The book is divided into 11 chapters, in which the principles are first applied to simple, then to more complicated situations.

The kinematics of a particle is discussed in Chapter 12, followed by a discussion of particle kinetics in Chapter 13 (Equation of Motion), Chapter 14 (Work and Energy), and Chapter 15 (Impulse and Momentum). The concepts of particle dynamics contained in these four chapters are then summarized in a "review" section, and the student is given the chance to identify and solve a variety of problems. A similar sequence of presentation is given for the planar motion of a rigid body: Chapter 16 (Planar Kinematics), Chapter 17 (Equations of Motion), Chapter 18 (Work and Energy), and Chapter 19 (Impulse and Momentum), followed by a summary and review set of problems for these chapters.

If time permits, some of the material involving three-dimensional rigid-body motion may be included in the course. The kinematics and kinetics of this motion are discussed in Chapters 20 and 21, respectively. Chapter 22 (Vibrations) may be included if the student has the necessary mathematical background. Sections of the book that are considered to be beyond the scope of the basic dynamics course are indicated by a star (★) and may be omitted. Note that this material also provides a suitable reference for basic principles when it is discussed in more advanced courses. Finally, Appendix A of the Dynamics section provides a list of mathematical formulas needed to solve the problems in the book, Appendix B of the Dynamics section provides a brief review of vector analysis, and Appendix C of the Dynamics section reviews application of the chain rule.

Alternative Coverage. At the discretion of the instructor, it is possible to cover Chapters 12 through 19 in the following order with no loss in continuity: Chapters 12 and 16 (Kinematics), Chapters 13 and 17 (Equations of Motion), Chapter 14 and 18 (Work and Energy), and Chapters 15 and 19 (Impulse and Momentum).

Acknowledgments

The author has endeavored to write this book so that it will appeal to both the student and instructor. Through the years, many people have helped in its development, and I will always be grateful for their valued suggestions and comments. Specifically, I wish to thank all the individuals who have sent comments to me. These include J. Aurand, J. Ari-Gur, D. Boyajian, R. Boyd, O. Byer, J. Callahan, D. Dikin, I. Elishakoff, E. Erisman, C. Heinke, R. Hendricks, F. Herrera, J. Hilton, H. Kuhlman, K. Leipold, E. Most, S. Moustafa, H. Nazeri, D. Pox, C. Roche, M. Rosengren, J. Ross, D. Rowlison, R. Scott, K. Steurer and J. Tashbar.

A long-time friend and associate, Kai Beng Yap, was of great help to me in preparing and checking problem solutions. During the production process I am thankful for the assistance of Rose Kernan, my production editor, and Marta Samsel, who worked on the cover of the book. And finally, to my wife, Conny, who helped in the proofreading of the manuscript for publication.

Lastly, many thanks are extended to all my students and to members of the teaching profession who have freely taken the time to offer their suggestions and comments. Since this list is too long to mention, it is hoped that those who have given help in this manner will accept this anonymous recognition.

I would greatly appreciate hearing from you if at any time you have any comments, suggestions, or issues related to any matters regarding this edition.

Russell Charles Hibbeler
hibbeler@bellsouth.net

Mastering Engineering

This online tutorial and assessment program allows you to integrate dynamic homework and practice problems with automated grading of exercises from the textbook. Tutorials and many end-of-section problems provide enhanced student feedback and optional hints. Mastering Engineering™ allows you to easily track the performance of your entire class on an assignment-by-assignment basis, or the detailed work of an individual student. For more information visit www.masteringengineering.com.

Resources for Instructors

Instructor's Solutions Manual This supplement provides complete solutions supported by problem statements and problem figures. The Instructor's Solutions Manual is available in PDF format on Pearson Higher Education website: www.pearson.com.

PowerPoint Slides A complete set of all the figures and tables from the textbook are available in PowerPoint format.

Resources for Students

Videos Developed by the author, three different types of videos are now available to reinforce learning the basic theory and applying the principles. The first set provides a lecture review and a self-test of the material related to the theory and concepts presented in the book. The second set provides a self-test of the example problems and the basic procedures used for their solution. And the third set provides an engagement for solving the Fundamental Problems throughout the book. For more information on how to access these videos visit www.pearson.com/hibbeler.

CONTENTS

1
General Principles 3

    Chapter Objectives   3
1.1   Mechanics   3
1.2   Fundamental Concepts   4
1.3   Units of Measurement   7
1.4   The International System of Units   9
1.5   Numerical Calculations   10
1.6   General Procedure for Analysis   12

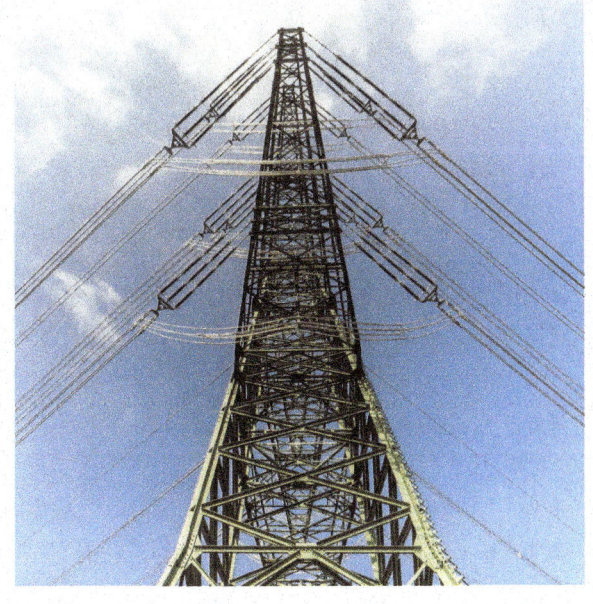

2
Force Vectors 17

    Chapter Objectives   17
2.1   Scalars and Vectors   17
2.2   Vector Operations   18
2.3   Vector Addition of Forces   20
2.4   Addition of a System of Coplanar Forces   32
2.5   Cartesian Vectors   43
2.6   Addition of Cartesian Vectors   46
2.7   Position Vectors   54
2.8   Force Vector Directed Along a Line   56
2.9   Dot Product   64

3
Equilibrium of a Particle 81

Chapter Objectives 81

3.1 Condition for the Equilibrium of a Particle 81

3.2 The Free-Body Diagram 82

3.3 Coplanar Force Systems 85

3.4 Three-Dimensional Force Systems 98

4
Force System Resultants 113

Chapter Objectives 113

4.1 Moment of a Force—Scalar Formulation 113

4.2 Principle of Moments 115

4.3 Cross Product 123

4.4 Moment of a Force—Vector Formulation 126

4.5 Moment of a Force about a Specified Axis 136

4.6 Moment of a Couple 145

4.7 Simplification of a Force and Couple System 157

4.8 Further Simplification of a Force and Couple System 168

4.9 Reduction of a Simple Distributed Loading 180

CONTENTS xv

5
Equilibrium of a Rigid Body 195

Chapter Objectives 195

5.1 Conditions for Rigid-Body Equilibrium 195

5.2 Free-Body Diagrams 197

5.3 Equations of Equilibrium 208

5.4 Two- and Three-Force Members 218

5.5 Free-Body Diagrams 231

5.6 Equations of Equilibrium 236

5.7 Constraints and Statical Determinacy 237

6
Structural Analysis 257

Chapter Objectives 257

6.1 Simple Trusses 257

6.2 The Method of Joints 260

6.3 Zero-Force Members 266

6.4 The Method of Sections 274

6.5 Space Trusses 284

6.6 Frames and Machines 288

7
Internal Forces 325

Chapter Objectives 325

7.1 Internal Loadings 325
7.2 Shear and Moment Equations and Diagrams 341
7.3 Relations among Distributed Load, Shear, and Moment 350
7.4 Cables 361

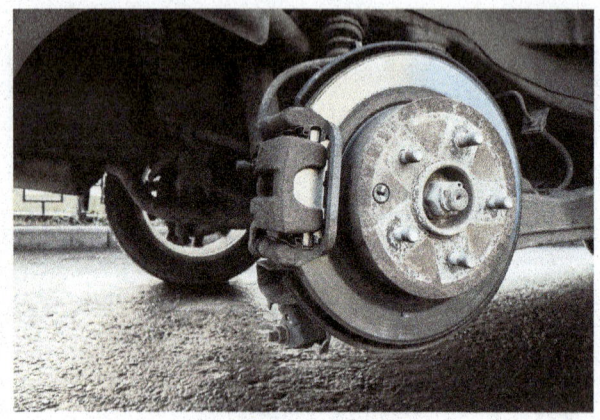

8
Friction 381

Chapter Objectives 381

8.1 Characteristics of Dry Friction 381
8.2 Problems Involving Dry Friction 386
8.3 Wedges 408
8.4 Frictional Forces on Screws 410
8.5 Frictional Forces on Flat Belts 417
8.6 Frictional Forces on Collar Bearings, Pivot Bearings, and Disks 425
8.7 Frictional Forces on Journal Bearings 428
8.8 Rolling Resistance 430

CONTENTS   XVII

9
Center of Gravity and Centroid 443

Chapter Objectives 443

9.1 Center of Gravity, Center of Mass, and the Centroid of a Body 443

9.2 Composite Bodies 466

9.3 Theorems of Pappus and Guldinus 480

9.4 Resultant of a General Distributed Loading 489

9.5 Fluid Pressure 490

10
Moments of Inertia 507

Chapter Objectives 507

10.1 Definition of Moments of Inertia for Areas 507

10.2 Parallel-Axis Theorem for an Area 508

10.3 Radius of Gyration of an Area 509

10.4 Moments of Inertia for Composite Areas 518

10.5 Product of Inertia for an Area 526

10.6 Moments of Inertia for an Area about Inclined Axes 530

10.7 Mohr's Circle for Moments of Inertia 533

10.8 Mass Moment of Inertia 541

11
Virtual Work 559

    Chapter Objectives 559

11.1 Definition of Work 559

11.2 Principle of Virtual Work 561

11.3 Principle of Virtual Work for a System of Connected Rigid Bodies 563

11.4 Conservative Forces 575

11.5 Potential Energy 576

11.6 Potential-Energy Criterion for Equilibrium 578

11.7 Stability of Equilibrium Configuration 579

Appendix

A. Mathematical Review and Formulations 594

Fundamental Problem Solutions and Answers 598

Review Problem Answers 615

Selected Answers 619

Index 631

Contents xix

12
Kinematics of a Particle 3

- Chapter Objectives 3
- **12.1** Introduction 3
- **12.2** Rectilinear Kinematics: Continuous Motion 5
- **12.3** Rectilinear Kinematics: Erratic Motion 19
- **12.4** General Curvilinear Motion 32
- **12.5** Curvilinear Motion: Rectangular Components 34
- **12.6** Motion of a Projectile 39
- **12.7** Curvilinear Motion: Normal and Tangential Components 53
- ***12.8** Curvilinear Motion: Cylindrical Components 67
- **12.9** Absolute Dependent Motion Analysis of Two Particles 81
- **12.10** Relative Motion of Two Particles Using Translating Axes 93

13
Kinetics of a Particle: Force and Acceleration 109

- Chapter Objectives 109
- **13.1** Newton's Second Law of Motion 109
- **13.2** The Equation of Motion 112
- **13.3** Equation of Motion for a System of Particles 114
- **13.4** Equations of Motion: Rectangular Coordinates 116
- **13.5** Equations of Motion: Normal and Tangential Coordinates 134
- ***13.6** Equations of Motion: Cylindrical Coordinates 148
- ***13.7** Central-Force Motion and Space Mechanics 160

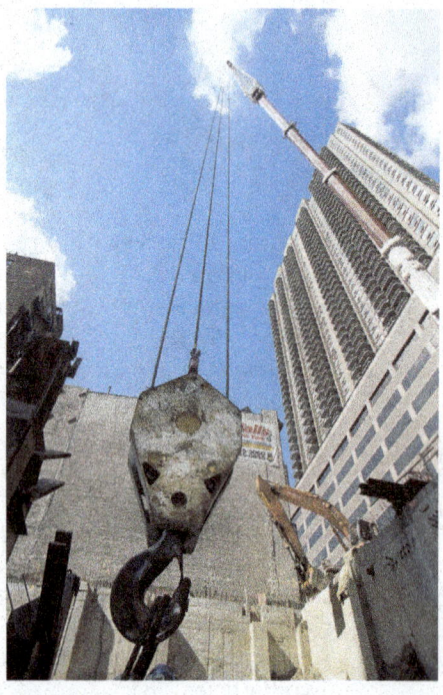

14
Kinetics of a Particle: Work and Energy 175

      Chapter Objectives 175
- 14.1 The Work of a Force 175
- 14.2 Principle of Work and Energy 180
- 14.3 Principle of Work and Energy for a System of Particles 182
- 14.4 Power and Efficiency 199
- 14.5 Conservative Forces and Potential Energy 208
- 14.6 Conservation of Energy 212

15
Kinetics of a Particle: Impulse and Momentum 231

      Chapter Objectives 231
- 15.1 Principle of Linear Impulse and Momentum 231
- 15.2 Principle of Linear Impulse and Momentum for a System of Particles 234
- 15.3 Conservation of Linear Momentum for a System of Particles 247
- 15.4 Impact 259
- 15.5 Angular Momentum 274
- 15.6 Relation Between the Moment of a Force and Angular Momentum 275
- 15.7 Principle of Angular Impulse and Momentum 278
- 15.8 Bodies Subjected to a Mass Flow 289
- 15.9 Steady Flow of a Fluid Stream 291
- 15.10 Bodies that Lose or Gain Mass 295

Contents xxi

16
Planar Kinematics of a Rigid Body 309

      Chapter Objectives 309
16.1 Planar Rigid-Body Motion 309
16.2 Translation 311
16.3 Rotation about a Fixed Axis 312
*16.4 Absolute Motion Analysis 328
16.5 Relative-Motion Analysis: Velocity 336
16.6 Instantaneous Center of Zero Velocity 349
16.7 Relative-Motion Analysis: Acceleration 361
*16.8 Relative-Motion Analysis using Rotating Axes 375

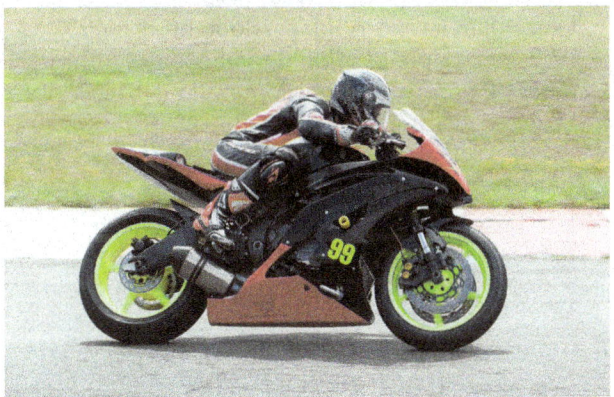

17
Planar Kinetics of a Rigid Body: Force and Acceleration 393

      Chapter Objectives 393
17.1 Mass Moment of Inertia 393
17.2 Planar Kinetic Equations of Motion 407
17.3 Equations of Motion: Translation 410
17.4 Equations of Motion: Rotation About a Fixed Axis 423
17.5 Equations of Motion: General Plane Motion 437

18
Planar Kinetics of a Rigid Body: Work and Energy 453

      Chapter Objectives 453
- **18.1** Kinetic Energy 453
- **18.2** The Work of a Force 456
- **18.3** The Work of a Couple Moment 458
- **18.4** Principle of Work and Energy 460
- **18.5** Conservation of Energy 475

19
Planar Kinetics of a Rigid Body: Impulse and Momentum 495

      Chapter Objectives 495
- **19.1** Linear and Angular Momentum 495
- **19.2** Principle of Impulse and Momentum 501
- **19.3** Conservation of Momentum 516
- ***19.4** Eccentric Impact 520

CONTENTS XXIII

20
Three-Dimensional Kinematics of a Rigid Body 535

     Chapter Objectives 535
20.1 Rotation About a Fixed Point 535
*20.2 The Time Derivative of a Vector Measured from a Fixed or Translating-Rotating System 538
20.3 General Motion 543
*20.4 Relative-Motion Analysis Using Translating and Rotating Axes 552

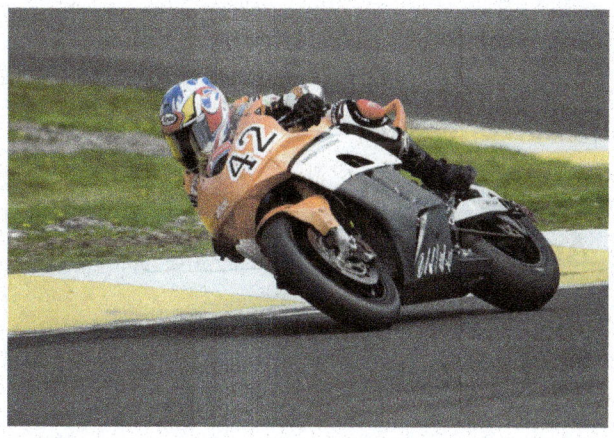

21
Three-Dimensional Kinetics of a Rigid Body 565

     Chapter Objectives 565
*21.1 Moments and Products of Inertia 565
21.2 Angular Momentum 575
21.3 Kinetic Energy 578
*21.4 Equations of Motion 586
*21.5 Gyroscopic Motion 600
21.6 Torque-Free Motion 606

22
Vibrations 617

Chapter Objectives 617
22.1 Undamped Free Vibration 617
*22.2 Energy Methods 631
*22.3 Undamped Forced Vibration 637
*22.4 Viscous Damped Free Vibration 641
*22.5 Viscous Damped Forced Vibration 644
*22.6 Electrical Circuit Analogs 647

Appendix
A. Mathematical Expressions 656
B. Vector Analysis 659
C. The Chain Rule 665

Fundamental Problems Solutions and Answers 669

Review Problem Answers 691

Answers to Selected Problems 693

Index 705

CREDITS

Cover: Bozac/Alamy Stock Photo

Chapter 1: Image Credits
002 ALAMY IMAGES: Gaertner/Alamy Stock Photo; 007 SHUTTERSTOCK: NikoNomad/Shutterstock

Chapter 2: Image Credits
016 ALAMY IMAGES: Marcus Hofmann/Alamy Stock Photo

Chapter 3: Image Credits
080 SHUTTERSTOCK: Djgis/Shutterstock

Chapter 4: Image Credits
112 SHUTTERSTOCK: Zbynek Jirousek/Shutterstock

Chapter 5: Image Credits
194 SHUTTERSTOCK: StephanScherhag/Shutterstock

Chapter 6: Image Credits
256 ALAMY IMAGES: Ray Hardinge/Alamy Stock Photo

Chapter 7: Image Credits
324 SHUTTERSTOCK: Northlight/Shutterstock

Chapter 8: Image Credits
380 ALAMY IMAGES: Eugene Sergeev/Alamy Stock Photo

Chapter 9: Image Credits
442 SHUTTERSTOCK: Ann Baldwin/Shutterstock

Chapter 10: Image Credits
506 ALAMY IMAGES: Olaf Speier/Alamy Stock Photo

Chapter 11: Image Credits
558 ALAMY IMAGES: APIX/Alamy Stock Photo

Chapter 12: Image Credits
002 GETTY IMAGES INCORPORATED: Sollina Images/Getty Images
039 SHUTTERSTOCK: NamMun Photo/Shutterstock

Chapter 13: Image Credits
108 ALAMY IMAGES: H. Mark Weidman Photography/Alamy Stock Photo
113 GETTY IMAGES INCORPORATED: Keystone/Stringer/Hulton Archive/Getty Images
138 123RF GB LIMITED: John Sandy/123RF
161 GETTY IMAGES INCORPORATED: Universal images group/Getty Images

Chapter 14: Image Credits
174 ALAMY IMAGES: Michael Doolittle/Alamy Images

Chapter 15: Image Credits
230 123RF GB LIMITED: Andrey Kekyalyaynen/123RF
248 REB Images/Image Source/Getty Images
261 SHUTTERSTOCK: NamMun Photo/Shutterstock
280 123RF GB LIMITED: Andrey Kekyalyaynen/123RF
296 NASA: © NASA

Chapter 16: Image Credits
308 SHUTTERSTOCK: Georgi Roshkov/Shutterstock

Chapter 17: Image Credits
392 SHUTTERSTOCK: Maksim Dobytko/Shutterstock

Chapter 18: Image Credits
452 SHUTTERSTOCK: Canbedone/Shutterstock

Chapter 19: Image Credits
494 ALAMY IMAGES: NASA Images/Alamy Stock Photo
520 GETTY IMAGES INCORPORATED: Mike Kemp/Rubberball/Getty Images

Chapter 20: Image Credits
534 123RF GB LIMITED: Romsvetnik/123RF

Chapter 21: Image Credits
564 ALAMY IMAGES: CW Motorsport Images/Alamy Stock Photo
568 GETTY IMAGES INCORPORATED: Ablestock/Getty Images
578 NASA: © NASA
592 SHUTTERSTOCK: F Armstrong Photography/Shutterstock
603 123RF GB LIMITED: Ruben Martinez Barricarte/123RF

Chapter 22: Image Credits
616 SHUTTERSTOCK: Wadas Jerzy/Shutterstock

ENGINEERING MECHANICS

STATICS

FIFTEENTH EDITION

CHAPTER 1

Cranes such as this one are required to lift extremely large loads. Their design is based on the basic principles of statics and dynamics, which form the subject matter of engineering mechanics.

GENERAL PRINCIPLES

CHAPTER OBJECTIVES

- To provide an introduction to the basic quantities and idealizations of mechanics.
- To state Newton's Laws of Motion and Gravitation.
- To review the principles for applying the SI system of units.
- To examine the standard procedures for performing numerical calculations.
- To present a general guide for solving problems.

1.1 MECHANICS

Mechanics is a branch of the physical sciences that is concerned with the state of rest or motion of bodies that are subjected to the action of forces. In general, this subject can be subdivided into three branches: *rigid-body mechanics, deformable-body mechanics*, and *fluid mechanics*. In this book we will study rigid-body mechanics since it is a basic requirement for the study of the mechanics of deformable bodies and the mechanics of fluids. Furthermore, rigid-body mechanics is essential for the design and analysis of many types of structural members, mechanical components, or electrical devices encountered in engineering.

Rigid-body mechanics is divided into two areas: statics and dynamics. **Statics** deals with the equilibrium of bodies, that is, those that are either at rest or move with a constant velocity; whereas **dynamics** is concerned with the accelerated motion of bodies. We can consider statics as a special case of dynamics, in which the acceleration is zero; however, statics deserves separate treatment in engineering education since many objects are designed with the intention that they remain in equilibrium.

Historical Development. The subject of statics developed at a very early time because its principles can be formulated simply from measurements of geometry and force. For example, the writings of Archimedes (287–212 B.C.) deal with the principle of the lever. Studies of the pulley, inclined plane, and wrench are also recorded in ancient writings—at times when the requirements for engineering were limited primarily to building construction.

Since the principles of dynamics depend on an accurate measurement of time, this subject developed much later. Galileo Galilei (1564–1642) was one of the first major contributors to this field. His work consisted of experiments using pendulums and falling bodies. The most significant contributions in dynamics, however, were made by Isaac Newton (1642–1727), who is noted for his formulation of the three fundamental laws of motion and the law of universal gravitational attraction. Shortly after these laws were postulated, important techniques for their application were developed by other scientists and engineers, some of whom will be mentioned throughout the book.

1.2 FUNDAMENTAL CONCEPTS

Before we begin our study of engineering mechanics, it is important to understand the meaning of certain fundamental concepts and principles.

Basic Quantities. The following four quantities are used throughout mechanics.

Length. *Length* is used to locate the position of a point in space and thereby describe the size of a physical system. Once a standard unit of length is defined, one can then use it to define distances and geometric properties of a body as multiples of this unit.

Time. *Time* is conceived as a succession of events. Although the principles of statics are time independent, this quantity plays an important role in the study of dynamics.

Mass. *Mass* is a measure of a quantity of matter that is used to compare the action of one body with that of another. This property manifests itself as a gravitational attraction between two bodies and provides a measure of the resistance of matter to a change in velocity.

Force. In general, *force* is considered as a "push" or "pull" exerted by one body on another. This interaction can occur when there is direct contact between the bodies, such as a person pushing on a wall, or it can occur through a distance when the bodies are physically separated. Examples of the latter type include gravitational, electrical, and magnetic forces. In any case, a force is completely characterized by its magnitude, direction, and point of application.

Idealizations. Models or idealizations are used in mechanics in order to simplify application of the theory. Here we will consider three important idealizations.

Particle. A ***particle*** has a mass, but a size that can be neglected. For example, the size of the earth is insignificant compared to the size of its orbit, and therefore the earth can be modeled as a particle when studying its orbital motion. When a body is idealized as a particle, the principles of mechanics reduce to a rather simplified form since the geometry of the body *will not be involved* in the analysis of the problem.

Rigid Body. A ***rigid body*** can be considered as a combination of a large number of particles in which all the particles remain at a fixed distance from one another, both before and after applying a load. This model is important because the body's shape does not change when a load is applied, and so we do not have to consider the type of material from which the body is made. In most cases the actual deformations occurring in structures, machines, mechanisms, and the like are relatively small, and the rigid-body assumption is suitable for analysis.

Concentrated Force. A ***concentrated force*** represents the effect of a loading which is assumed to act at a point on a body. We can represent a load by a concentrated force, provided the area over which the load is applied is very small compared to the overall size of the body. An example would be the contact force between a wheel and the ground.

Three forces act on the ring. Since these forces all meet at a point, then for any force analysis, we can assume the ring to be represented as a particle.

Steel is a common engineering material that does not deform very much under load. Therefore, we can consider this railroad wheel to be a rigid body acted upon by the concentrated force of the rail.

Newton's Three Laws of Motion. Engineering mechanics is formulated on the basis of Newton's three laws of motion, the validity of which is based on experimental observation. These laws apply to the motion of a particle as measured from a *nonaccelerating* reference frame. They may be briefly stated as follows.

First Law. A particle originally at rest, or moving in a straight line with constant velocity, tends to remain in this equilibrium state provided the particle is *not* subjected to an unbalanced force, Fig. 1–1a.

Equilibrium
(a)

Second Law. A particle acted upon by an *unbalanced force* **F** experiences an acceleration **a** that has the same direction as the force and a magnitude that is directly proportional to the force, Fig. 1–1b.* If the particle has a mass m, this law may be expressed mathematically as

$$\mathbf{F} = m\mathbf{a} \qquad (1\text{–}1)$$

Accelerated motion
(b)

Third Law. The mutual forces of action and reaction between two particles are equal, opposite, and collinear, Fig. 1–1c.

Action–reaction
(c)

Fig. 1–1

---

*Stated another way, the unbalanced force acting on the particle is proportional to the time rate of change of the particle's linear momentum.

Newton's Law of Gravitational Attraction. Shortly after formulating his three laws of motion, Newton postulated a law governing the gravitational attraction between any two particles. Stated mathematically,

$$F = G \frac{m_1 m_2}{r^2} \quad (1-2)$$

where

F = force of gravitation between the two particles

G = universal constant of gravitation; according to experimental evidence, $G = 66.73(10^{-12})$ m^3/(kg·s^2)

m_1, m_2 = mass of each of the two particles

r = distance between the two particles

Weight. According to Eq. 1–2, any two particles or bodies have a mutual attractive (gravitational) force acting between them. In the case of a particle located at or near the surface of the earth, however, the only gravitational force having any sizable magnitude is that between the earth, because of its very large mass, and the particle. Consequently, this force, called the **weight**, will be the only gravitational force we will consider.

From Eq. 1–2, if the particle has a mass $m_1 = m$, and we assume the earth is a nonrotating sphere of constant density and having a mass $m_2 = M_e$, then if r is the distance between the earth's center and the particle, the weight W of the particle becomes

$$W = G \frac{mM_e}{r^2}$$

If we let $g = GM_e/r^2$, we have

$$\boxed{W = mg} \quad (1-3)$$

The astronaut's weight is diminished since she is far removed from the gravitational field of the earth.

If we allow the particle to fall downward, then neglecting air resistance, the only force acting on the particle is its weight, and so Eq. 1–1 becomes $W = ma$. Comparing this result with Eq. 1–3, we see that $a = g$. In other words, g is the acceleration due to gravity. Since it depends on r, then the weight of the particle or body is *not* an absolute quantity. Instead, its magnitude depends upon the elevation where the measurement was made. For most engineering calculations, however, g is determined at sea level and at a latitude of 45°, which is considered the "standard location."

1.3 UNITS OF MEASUREMENT

The four basic quantities—length, time, mass, and force—are not all independent from one another; in fact, they are *related* by Newton's second law of motion, **F** = m**a**. Because of this, the *units* used to measure these quantities cannot *all* be selected arbitrarily. The equality **F** = m**a** is maintained only if three of the four units, called **base units**, are *defined* and the fourth unit is then *derived* from the equation.

8   CHAPTER 1   GENERAL PRINCIPLES

SI Units. The International System of units, abbreviated SI after the French "Système International d'Unités," is a modern version of the metric system which has received worldwide recognition. As shown in Table 1–1, the SI system defines length in meters (m), time in seconds (s), and mass in kilograms (kg).† The unit of force, called a **newton** (N), is *derived* from $\mathbf{F} = m\mathbf{a}$. Thus, 1 newton is equal to a force required to give 1 kilogram of mass an acceleration of 1 m/s^2 ($\text{N} = \text{kg} \cdot \text{m/s}^2$). Think of this force as the weight of a small apple.

If the weight of a body located at the "standard location" is to be determined in newtons, then Eq. 1–3 must be applied. Here measurements give $g = 9.806\ 65 \text{ m/s}^2$; however, for calculations, the value $g = 9.81 \text{ m/s}^2$ will be used. Thus,

$$W = mg \quad (g = 9.81 \text{ m/s}^2) \quad (1\text{--}4)$$

Therefore, a body of mass 1 kg has a weight of 9.81 N, a 2-kg body weighs 19.62 N, and so on, Fig. 1–2a.

U.S. Customary. In the U.S. Customary system of units (FPS) length is measured in feet (ft), time in seconds (s), and force in pounds (lb), Table 1–1. The unit of mass, called a **slug**, is *derived* from $\mathbf{F} = m\mathbf{a}$. Hence, 1 slug is equal to the amount of matter accelerated at 1 ft/s^2 when acted upon by a force of 1 lb ($\text{slug} = \text{lb} \cdot \text{s}^2/\text{ft}$).

Therefore, if the measurements are made at the "standard location," where $g = 32.2 \text{ ft/s}^2$, then from Eq. 1–3,

Fig. 1-2

$$m = \frac{W}{g} \quad (g = 32.2 \text{ ft/s}^2) \quad (1\text{--}5)$$

And so a body weighing 32.2 lb has a mass of 1 slug, a 64.4-lb body has a mass of 2 slugs, and so on, Fig. 1–2b.

TABLE 1–1	Systems of Units			
Name	Length	Time	Mass	Force
International System of Units SI	meter m	second s	kilogram kg	newton* N $\left(\dfrac{\text{kg} \cdot \text{m}}{\text{s}^2}\right)$
U.S. Customary FPS	foot ft	second s	slug* $\left(\dfrac{\text{lb} \cdot \text{s}^2}{\text{ft}}\right)$	pound lb

*Derived unit.

†Historically, the meter was defined as 1/10,000,000 the distance from the Equator to the North Pole, and the kilogram is 1/1000 of a cubic meter of water.

Conversion of Units. Table 1–2 provides a set of direct conversion factors between FPS and SI units for the basic quantities. Also, in the FPS system, recall that 1 ft = 12 in. (inches), 5280 ft = 1 mi (mile), 1000 lb = 1 kip (kilo-pound), and 2000 lb = 1 ton.

TABLE 1–2 Conversion Factors

Quantity	Unit of Measurement (FPS)	Equals	Unit of Measurement (SI)
Force	lb		4.448 N
Mass	slug		14.59 kg
Length	ft		0.3048 m

1.4 THE INTERNATIONAL SYSTEM OF UNITS

The SI system of units is used extensively in this book since it is intended to become the worldwide standard for measurement. Therefore, we will now present some of the rules for its use and some of its terminology relevant to engineering mechanics.

Prefixes. When a numerical quantity is either very large or very small, the units used to define its size may be modified by using a prefix. Some of the prefixes used in the SI system are shown in Table 1–3. Each represents a multiple or submultiple of a unit which, if applied successively, moves the decimal point of a numerical quantity to every third place.* For example, 4 000 000 N = 4000 kN (kilo-newton) = 4 MN (mega-newton), or 0.005 m = 5 mm (milli-meter). Notice that the SI system does not include the multiple deca (10) or the submultiple centi (0.01), which form part of the metric system. Except for some volume and area measurements, the use of these prefixes is generally avoided in science and engineering.

TABLE 1–3 Prefixes

	Exponential Form	Prefix	SI Symbol
Multiple			
1 000 000 000	10^9	giga	G
1 000 000	10^6	mega	M
1 000	10^3	kilo	k
Submultiple			
0.001	10^{-3}	milli	m
0.000 001	10^{-6}	micro	μ
0.000 000 001	10^{-9}	nano	n

*The kilogram is the only base unit that is defined with a prefix.

Rules for Use. Here are a few of the important rules that describe the proper use of the various SI symbols:

- Quantities defined by several units which are multiples of one another are separated by a *dot* to avoid confusion with prefix notation, as indicated by $N = kg \cdot m/s^2 = kg \cdot m \cdot s^{-2}$. Also, m·s (meter-second), whereas ms (milli-second).
- The exponential power on a unit having a prefix refers to both the unit *and* its prefix. For example, $\mu N^2 = (\mu N)^2 = \mu N \cdot \mu N$. Likewise, mm^2 represents $(mm)^2 = mm \cdot mm$.
- With the exception of the base unit the kilogram, in general avoid the use of a prefix in the denominator of composite units. For example, do not write N/mm, but rather kN/m; also, m/mg should be written as Mm/kg.
- When performing calculations, represent the numbers in terms of their *base or derived units* by converting all prefixes to powers of 10. The final result should then be expressed using a *single prefix*. Also, after calculation, it is best to keep numerical values between 0.1 and 1000; otherwise, a suitable prefix should be chosen. For example,

$$(50 \text{ kN})(60 \text{ nm}) = [50(10^3) \text{ N}][60(10^{-9}) \text{ m}]$$
$$= 3000(10^{-6}) \text{ N} \cdot \text{m} = 3(10^{-3}) \text{ N} \cdot \text{m} = 3 \text{ mN} \cdot \text{m}$$

1.5 NUMERICAL CALCULATIONS

Numerical work in engineering practice is most often performed by using handheld calculators and computers. It is important, however, that the answers to any problem be reported with justifiable accuracy using appropriate significant figures. In this section we will discuss these topics together with some other important aspects involved in all engineering calculations.

Dimensional Homogeneity. The terms of any equation used to describe a physical process must be **dimensionally homogeneous**; that is, each term must be expressed in the same units. Provided this is the case, all the terms of an equation can then be combined if numerical values are substituted for the variables. Consider, for example, the equation $s = vt + \frac{1}{2}at^2$, where, in SI units, s is the position in meters, m, t is time in seconds, s, v is velocity in m/s and a is acceleration in m/s². Regardless of how this equation is evaluated, it maintains its dimensional homogeneity. In the form stated, each of the three terms is expressed in meters $\left[m, (m/\cancel{s})\cancel{s}, (m/\cancel{s^2})\cancel{s^2} \right]$ or solving for a, $a = 2s/t^2 - 2v/t$, the terms are each expressed in units of m/s² $\left[m/s^2, m/s^2, (m/s)/s \right]$.

Keep in mind that problems in mechanics always involve the solution of dimensionally homogeneous equations, and so this fact can then be used as a partial check for algebraic manipulations of an equation.

Significant Figures. The number of significant figures contained in any number determines the accuracy of the number. For instance, the number 4981 contains four significant figures. However, if zeros occur at the end of a whole number, it may be unclear as to how many significant figures the number represents. For example, 23 400 might have three (234), four (2340), or five (23 400) significant figures. To avoid these ambiguities, we will use *engineering notation* to report a result. This requires that numbers be rounded off to the appropriate number of significant digits and then expressed in multiples of (10^3), such as (10^3), (10^6), or (10^{-9}). For instance, if 23 400 has five significant figures, it is written as $23.400(10^3)$, but if it has only three significant figures, it is written as $23.4(10^3)$.

If zeros occur at the beginning of a number that is less than one, then the zeros are not significant. For example, 0.008 21 has three significant figures. Using engineering notation, this number is expressed as $8.21(10^{-3})$. Likewise, 0.000 582 can be expressed as $0.582(10^{-3})$ or $582(10^{-6})$.

Rounding Off Numbers. Rounding off a number is necessary so that the accuracy of the result will be the same as that of the problem data. As a general rule, any numerical figure ending in a number greater than five is rounded up and a number less than five is not rounded up. The rules for rounding off numbers are best illustrated by examples. Suppose the number 3.5587 is to be rounded off to *three* significant figures. Because the fourth digit (8) is *greater than* 5, the third number is rounded up to 3.56. Likewise 0.5896 becomes 0.590 and 9.3866 becomes 9.39. If we round off 1.341 to three significant figures, because the fourth digit (1) is *less than* 5, then we get 1.34. Likewise 0.3762 becomes 0.376 and 9.871 becomes 9.87. There is a special case for any number that ends in a 5. As a general rule, if the digit preceding the 5 is an *even number*, then this digit is *not* rounded up. If the digit preceding the 5 is an *odd number*, then it is rounded up. For example, 75.25 rounded off to three significant digits becomes 75.2, 0.1275 becomes 0.128, and 0.2555 becomes 0.256.

Calculations. When a sequence of calculations is performed, it is best to store the intermediate results in the calculator. In other words, do not round off calculations until expressing the final result. This procedure maintains precision throughout the series of steps to the final solution. In this book we will generally round off the answers to *three significant figures* since most of the data in engineering mechanics, such as geometry and loads, may be reliably measured to this accuracy.

1.6 GENERAL PROCEDURE FOR ANALYSIS

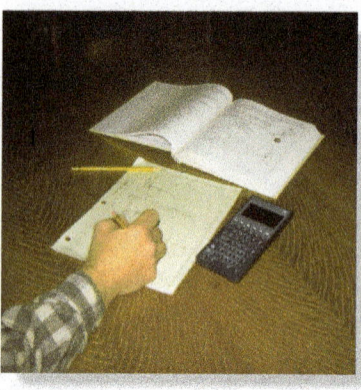

When solving problems, do the work as neatly as possible. Being neat will stimulate clear and orderly thinking, and vice versa.

Attending a lecture, reading this book, and studying the example problems helps, but **the most effective way of learning the principles of engineering mechanics is to *solve problems***. To be successful at this, it is important to always present the work in a *logical* and *orderly manner*, as suggested by the following sequence of steps:

- Read the problem carefully and try to correlate the actual physical situation with the theory studied.
- Tabulate the problem data and *draw to a large scale* any necessary diagrams.
- Apply the relevant principles, generally in mathematical form. When writing any equations, be sure they are dimensionally homogeneous.
- Solve the necessary equations, and report the answer with no more than three significant figures.
- Study the answer with technical judgment and common sense to determine whether or not it seems reasonable.

▶ IMPORTANT POINTS

- Statics is the study of bodies that are at rest or move with constant velocity.
- A particle has a mass but a size that can be neglected, and a rigid body does not deform under load.
- A force is considered as a "push" or "pull" of one body on another.
- Concentrated forces are assumed to act at a point on a body.
- Newton's three laws of motion should be memorized.
- Mass is measure of a quantity of matter that does not change from one location to another. Weight refers to the gravitational attraction of the earth on a body or quantity of mass. Its magnitude depends upon the elevation at which the mass is located.
- In the SI system the unit of force, the newton, is a derived unit. The meter, second, and kilogram are base units.
- Prefixes G, M, k, m, μ, and n are used to represent large and small numerical quantities. Their exponential size should be known, along with the rules for using the SI units.
- Perform numerical calculations with several significant figures, and then report the final answer to three significant figures.
- Algebraic manipulations of an equation can be checked in part by verifying that the equation remains dimensionally homogeneous.
- Know the rules for rounding off numbers.

EXAMPLE 1.1

Convert 2 km/h to m/s. How many ft/s is this?

SOLUTION
Since 1 km = 1000 m and 1 h = 3600 s, the factors of conversion are arranged in the following order, so that a cancellation of the units can be applied:

$$2 \text{ km/h} = \frac{2 \text{ km}}{\text{h}} \left(\frac{1000 \text{ m}}{\text{km}}\right)\left(\frac{1 \text{ h}}{3600 \text{ s}}\right)$$

$$= \frac{2000 \text{ m}}{3600 \text{ s}} = 0.556 \text{ m/s} \qquad Ans.$$

From Table 1–2, 1 ft = 0.3048 m. Thus,

$$0.556 \text{ m/s} = \left(\frac{0.556 \text{ m}}{\text{s}}\right)\left(\frac{1 \text{ ft}}{0.3048 \text{ m}}\right)$$

$$= 1.82 \text{ ft/s} \qquad Ans.$$

Note: Remember to round off the final answer to three significant figures.

EXAMPLE 1.2

Convert the quantities 300 lb · s and 52 slug/ft^3 to appropriate SI units.

SOLUTION
Using Table 1–2, 1 lb = 4.448 N.

$$300 \text{ lb} \cdot \text{s} = 300 \text{ lb} \cdot \text{s} \left(\frac{4.448 \text{ N}}{1 \text{ lb}}\right)$$

$$= 1334.4 \text{ N} \cdot \text{s} = 1.33 \text{ kN} \cdot \text{s} \qquad Ans.$$

Since 1 slug = 14.59 kg and 1 ft = 0.3048 m, then

$$52 \text{ slug/ft}^3 = \frac{52 \text{ slug}}{\text{ft}^3}\left(\frac{14.59 \text{ kg}}{1 \text{ slug}}\right)\left(\frac{1 \text{ ft}}{0.3048 \text{ m}}\right)^3$$

$$= 26.8(10^3) \text{ kg/m}^3$$

$$= 26.8 \text{ Mg/m}^3 \qquad Ans.$$

EXAMPLE 1.3

Evaluate each of the following and express with SI units having an appropriate prefix: (a) (50 mN)(6 GN), (b) (400 mm)(0.6 MN)^2, (c) 45 MN^3/900 Gg.

SOLUTION
First convert each number to base units, perform the indicated operations, then choose an appropriate prefix.

Part (a)

$$(50 \text{ mN})(6 \text{ GN}) = [50(10^{-3}) \text{ N}][6(10^9) \text{ N}]$$

$$= 300(10^6) \text{ N}^2$$

$$= 300(10^6) \text{ N}^2 \left(\frac{1 \text{ kN}}{10^3 \text{ N}}\right)\left(\frac{1 \text{ kN}}{10^3 \text{ N}}\right)$$

$$= 300 \text{ kN}^2 \qquad \textit{Ans.}$$

Note: Keep in mind the convention kN^2 = (kN)^2 = 10^6 N^2.

Part (b)

$$(400 \text{ mm})(0.6 \text{ MN})^2 = [400(10^{-3}) \text{ m}][0.6(10^6) \text{ N}]^2$$

$$= [400(10^{-3}) \text{ m}][0.36(10^{12}) \text{ N}^2]$$

$$= 144(10^9) \text{ m} \cdot \text{N}^2$$

$$= 144 \text{ Gm} \cdot \text{N}^2 \qquad \textit{Ans.}$$

We can also write

$$144(10^9) \text{ m} \cdot \text{N}^2 = 144(10^9) \text{ m} \cdot \text{N}^2 \left(\frac{1 \text{ MN}}{10^6 \text{ N}}\right)\left(\frac{1 \text{ MN}}{10^6 \text{ N}}\right)$$

$$= 0.144 \text{ m} \cdot \text{MN}^2 \qquad \textit{Ans.}$$

Part (c)

$$\frac{45 \text{ MN}^3}{900 \text{ Gg}} = \frac{45(10^6 \text{ N})^3}{900(10^6) \text{ kg}}$$

$$= 50(10^9) \text{ N}^3/\text{kg}$$

$$= 50(10^9) \text{ N}^3 \left(\frac{1 \text{ kN}}{10^3 \text{ N}}\right)^3 \frac{1}{\text{kg}}$$

$$= 50 \text{ kN}^3/\text{kg} \qquad \textit{Ans.}$$

PROBLEMS

The answers to all but every fourth problem (asterisk) are given in the back of the book.

1–1. Round off the following numbers to three significant figures: (a) 3.455 55 m, (b) 45.556 s, (c) 5555 N, (d) 4525 kg.

1–2. Represent each of the following combinations of units in the correct SI form using an appropriate prefix: (a) kN/μs, (b) Mg/mN, (c) MN/(kg·ms).

1–3. Represent each of the following combinations of units in the correct SI form using an appropriate prefix: (a) Mg/mm, (b) mN /μs, (c) μm · Mg.

*1–4. What is the weight in newtons of an object that has a mass of (a) 8 kg, (b) 0.04 mg, (c) 760 Mg?

1–5. Represent each of the following as a number between 0.1 and 1000 using an appropriate prefix: (a) 45 320 kN, (b) 568(10^5) mm, (c) 0.005 63 mg.

1–6. Round off the following numbers to three significant figures: (a) 58 342 m, (b) 68.534 s, (c) 2553 N, (d) 7555 kg.

1–7. Represent each of the following quantities in the correct SI form using an appropriate prefix: (a) 0.000 431 kg, (b) 35.3(10^3) N, (c) 0.005 32 km.

*1–8. Represent each of the following combinations of units in the correct SI form: (a) Mg/ms, (b) N/mm, (c) mN/(kg·μs).

1–9. Represent each of the following combinations of units in the correct SI form using an appropriate prefix: (a) m/ms, (b) μkm, (c) ks/mg, (d) km · μN.

1–10. Represent each of the following combinations of units in the correct SI form using an appropriate prefix: (a) GN · μm, (b) kg/μm, (c) N/ks^2, (d) kN/μs.

1–11. Represent each of the following combinations of units in the correct SI form using an appropriate prefix: (a) μMN, (b) N/μm, (c) MN/ks^2, (d) kN/ms.

*1–12. Water has a density of 1.94 slug/ft^3. What is the density expressed in SI units? Express the answer to three significant figures.

1–13. The density (mass/volume) of aluminum is 5.26 slug/ft^3. Determine its density in SI units. Use an appropriate prefix.

1–14. Evaluate each of the following to three significant figures and express each answer in SI units using an appropriate prefix: (a) $(212 \text{ mN})^2$, (b) $(52\,800 \text{ ms})^2$, (c) $[548(10^6)]^{1/2}$ms.

1–15. Using the SI system of units, show that Eq. 1–2 is a dimensionally homogeneous equation which gives F in newtons. Determine to three significant figures the gravitational force acting between two spheres that are touching each other. The mass of each sphere is 200 kg and the radius is 300 mm.

*1–16. Evaluate each of the following to three significant figures and express each answer in SI units using an appropriate prefix: (a) $(200 \text{ kN})^2$, (b) $(0.005 \text{ mm})^2$, (c) $(400 \text{ m})^3$.

1–17. If a car is traveling at 55 mi/h, determine its speed in kilometers per hour and meters per second.

1–18. Evaluate (204 mm)(0.004 57 kg)/(34.6 N) to three significant figures and express the answer in SI units using an appropriate prefix.

1–19. The specific weight (wt./vol.) of brass is 520 lb/ft^3. Determine its density (mass/vol.) in SI units. Use an appropriate prefix.

*1–20. If a man weighs 155 lb on earth, specify (a) his mass in slugs, (b) his mass in kilograms, and (c) his weight in newtons. If the man is on the moon, where the acceleration due to gravity is $g_m = 5.30$ ft/s^2, determine (d) his weight in pounds, and (e) his mass in kilograms.

1–21. Two particles have a mass of 8 kg and 12 kg, respectively. If they are 800 mm apart, determine the force of gravity acting between them. Compare this result with the weight of each particle.

CHAPTER 2

This electric transmission tower is stabilized by cables that exert forces on the tower at their points of connection. In this chapter we will show how to express these forces as Cartesian vectors, and then determine their resultant.

FORCE VECTORS

CHAPTER OBJECTIVES

- To show how to add forces and resolve them into components using the parallelogram law.
- To express force and position in Cartesian vector form and explain how to determine the vector's magnitude and direction.
- To introduce the dot product in order to use it to find the angle between two vectors or the projection of one vector onto another.

2.1 SCALARS AND VECTORS

Many physical quantities in engineering mechanics are measured using either scalars or vectors.

Scalar. A *scalar* is any positive or negative physical quantity that can be completely specified by its *magnitude*. Examples of scalar quantities include length, mass, and time.

Vector. A *vector* is any physical quantity that requires both a *magnitude* and a *direction* for its complete description. Examples of vectors encountered in statics are force, position, and moment. A vector is shown graphically by an arrow, Fig. 2–1. The length of the arrow represents the *magnitude* of the vector, and the angle θ between the vector and a fixed axis defines the *direction of its line of action*. The head or tip of the arrow indicates the *sense of direction* of the vector.

In print, vector quantities are represented by boldface letters such as **A**, and the magnitude of a vector is italicized, A. For handwritten work, it is often convenient to denote a vector quantity by simply drawing an arrow above it, $\vec{A}$.

Fig. 2–1

2.2 VECTOR OPERATIONS

Multiplication and Division of a Vector by a Scalar. If a vector is multiplied or divided by a positive scalar, its magnitude is changed by that amount. Multiplying or dividing by a negative scalar will also change the directional sense of the vector. Graphic examples of these operations are shown in Fig. 2–2.

Scalar multiplication and division

Fig. 2–2

Vector Addition. When adding two vectors together it is important to account for both their magnitudes and their directions. To do this we must use the *parallelogram law of addition.* To illustrate, the two *component vectors* **A** and **B** in Fig. 2–3a are added to form a *resultant vector* **R** = **A** + **B** using the following procedure:

- First join the tails of the components at a point to make them concurrent, Fig. 2–3b.

- From the head of **B**, draw a line parallel to **A**. Draw another line from the head of **A** that is parallel to **B**. These two lines intersect at point *P* to form the adjacent sides of a parallelogram.

- The diagonal of this parallelogram that extends to *P* forms **R**, which then represents the resultant vector **R** = **A** + **B**, Fig. 2–3c.

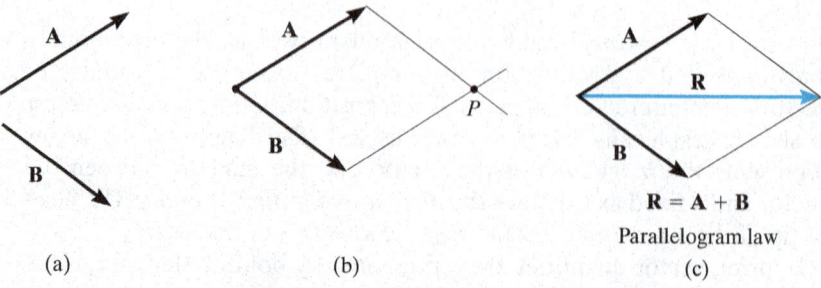

Fig. 2–3

We can also add **B** to **A**, Fig. 2–4a, using the *triangle rule*, which is a special case of the parallelogram law, whereby vector **B** is added to vector **A** in a "head-to-tail" fashion, i.e., by connecting the tail of **B** to the head of **A**, Fig. 2–4b. The resultant **R** extends from the tail of **A** to the head of **B**. In a similar manner, **R** can also be obtained by adding **A** to **B**, Fig. 2–4c. By comparison, it is seen that vector addition is commutative; in other words, the vectors can be added in either order, i.e., **R** = **A** + **B** = **B** + **A**.

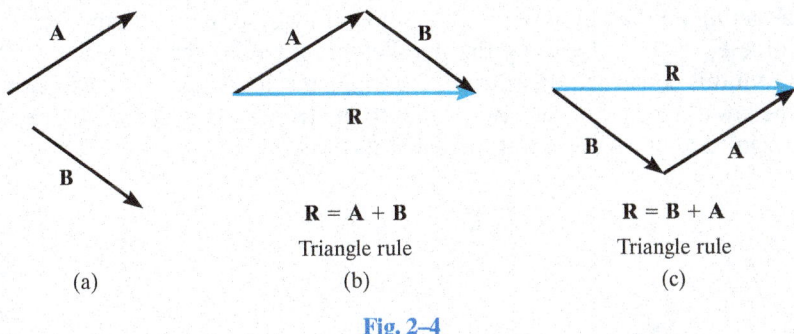

Fig. 2–4

As a special case, if the two vectors **A** and **B** are **collinear**, i.e., both have the same line of action, the parallelogram law reduces to an *algebraic* or *scalar addition* $R = A + B$, as shown in Fig. 2–5.

Vector Subtraction. The resultant of the *difference* between two vectors **A** and **B** of the same type may be expressed as

$$\mathbf{R'} = \mathbf{A} - \mathbf{B} = \mathbf{A} + (-\mathbf{B})$$

This vector sum is shown graphically in Fig. 2–6. Subtraction is therefore defined as a special case of addition, so the rules of vector addition also apply to vector subtraction.

Addition of collinear vectors

Fig. 2–5

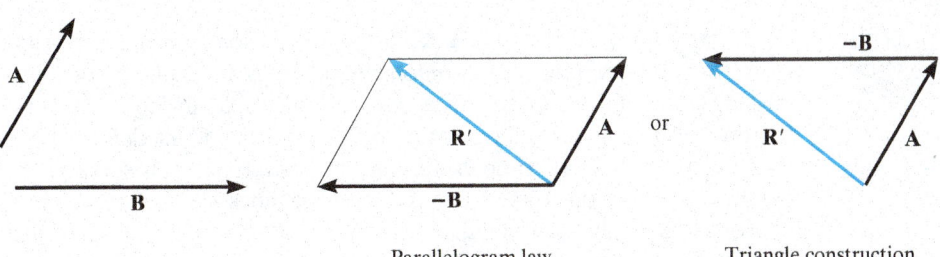

Parallelogram law

Triangle construction

Vector subtraction

Fig. 2–6

2.3 VECTOR ADDITION OF FORCES

Experimental evidence has shown that a force is a vector quantity since it has a specified magnitude, direction, and sense and it adds according to the parallelogram law. Two common problems in statics involve either finding the resultant force, knowing its components, or resolving a known force into two components. We will now describe how each of these problems is solved using the parallelogram law.

Finding a Resultant Force. The two component forces $\mathbf{F}_1$ and $\mathbf{F}_2$ acting on the pin in Fig. 2–7a are added together to form the resultant force $\mathbf{F}_R = \mathbf{F}_1 + \mathbf{F}_2$, using the parallelogram law as shown in Fig. 2–7b. From this construction, or using the triangle rule, Fig. 2–7c, we can apply the law of cosines or the law of sines to the triangle in order to obtain the magnitude of the resultant force and its direction.

The parallelogram law must be used to determine the resultant of the two forces acting on the hook.

Fig. 2–7

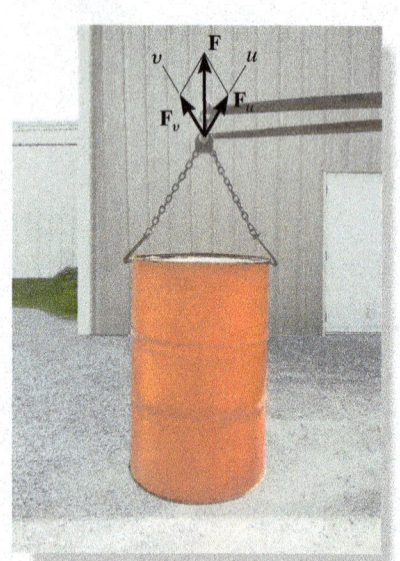

Using the parallelogram law the supporting force **F** can be resolved into components acting along the u and v axes.

Finding the Components of a Force. Sometimes it is necessary to resolve a force into two *components* in order to study its pulling or pushing effect in two specific directions. For example, in Fig. 2–8a, **F** is to be resolved into two components along the two members, defined by the u and v axes. In order to determine the magnitude of each component, a parallelogram is constructed first, by drawing lines starting from the tip of **F**, one line parallel to u, and the other line parallel to v. These lines intersect the v and u axes, forming a parallelogram. The force components $\mathbf{F}_u$ and $\mathbf{F}_v$ are established by simply joining them to the tail of **F**, to the intersection points on the u and v axes, Fig. 2–8b. This parallelogram can be reduced to a triangle, which represents the triangle rule, Fig. 2–8c. From this, the law of sines can be applied to determine the unknown magnitudes of the components.

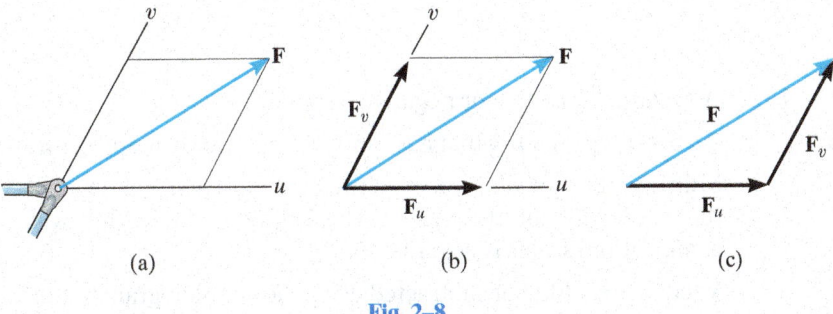

(a)　　　　　　　　(b)　　　　　　　　(c)

Fig. 2–8

Addition of Several Forces. If more than two forces are to be added, successive applications of the parallelogram law can be carried out in order to obtain the resultant force. For example, if three forces $\mathbf{F}_1$, $\mathbf{F}_2$, $\mathbf{F}_3$ act at a point O, Fig. 2–9, the resultant of any two of the forces is found, say, $\mathbf{F}_1 + \mathbf{F}_2$, and then this resultant is added to the third force, yielding the resultant of all three forces; i.e., $\mathbf{F}_R = (\mathbf{F}_1 + \mathbf{F}_2) + \mathbf{F}_3$. Using the parallelogram law to add more than two forces, as shown here, generally requires extensive geometric and trigonometric calculation to determine the magnitude and direction of the resultant. Instead, problems of this type are easily solved by using the "rectangular-component method," which is explained in the next section.

Fig. 2–9

The resultant force $\mathbf{F}_R$ on the hook requires the addition of $\mathbf{F}_1 + \mathbf{F}_2$, then this resultant is added to $\mathbf{F}_3$.

IMPORTANT POINTS

- A scalar is a positive or negative number.
- A vector is a quantity that has a magnitude, direction, and sense.
- Multiplication or division of a vector by a scalar will change the magnitude of the vector. The sense of the vector will change if the scalar is negative.
- Vectors are added or subtracted using the parallelogram law or the triangle rule.
- As a special case, if the vectors are collinear, the resultant is formed by an algebraic or scalar addition.

(a)

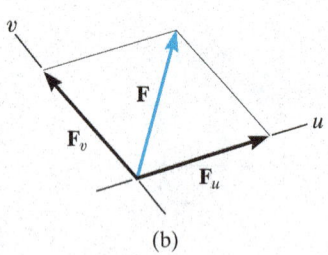

(b)

PROCEDURE FOR ANALYSIS

Problems that involve the addition of two forces can be solved as follows:

Parallelogram Law.

- Sketch the two "component" forces F_1 and F_2 added together according to the parallelogram law, yielding the *resultant* force F_R that forms the diagonal of the parallelogram, Fig. 2–10a.
- If a force F is to be resolved into *components* along two axes u and v, then start at the head of force F and construct lines parallel to the axes, thereby forming the parallelogram, Fig. 2–10b. The sides of the parallelogram represent the components, F_u and F_v.
- Label all the known and unknown force magnitudes and the angles on the sketch and identify the two unknowns as the magnitude and direction of F_R, or the magnitudes of its components.

Trigonometry.

- Redraw a half portion of the parallelogram to illustrate the triangular head-to-tail addition of the components.
- From this triangle, the magnitude of the resultant force can be determined using the law of cosines, and its direction is determined from the law of sines. The magnitudes of two force components are determined from the law of sines. The formulas are given in Fig. 2–10c.

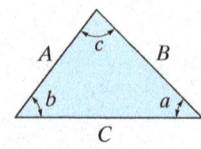

Cosine law:
$C = \sqrt{A^2 + B^2 - 2AB \cos c}$
Sine law:
$\dfrac{A}{\sin a} = \dfrac{B}{\sin b} = \dfrac{C}{\sin c}$

(c)

Fig. 2–10

EXAMPLE 2.1

The screw eye in Fig. 2–11a is subjected to two forces, $\mathbf{F}_1$ and $\mathbf{F}_2$. Determine the magnitude and direction of the resultant force.

(a)

(b)

SOLUTION

Parallelogram Law. The parallelogram is formed by drawing a line from the head of $\mathbf{F}_1$ that is parallel to $\mathbf{F}_2$, and another line from the head of $\mathbf{F}_2$ that is parallel to $\mathbf{F}_1$. The resultant force $\mathbf{F}_R$ extends to where these lines intersect at point A, Fig. 2–11b. The two unknowns are the magnitude of $\mathbf{F}_R$ and the angle θ (theta).

Trigonometry. From the parallelogram, the vector triangle is constructed, Fig. 2–11c. Using the law of cosines

$$F_R = \sqrt{(100\text{ N})^2 + (150\text{ N})^2 - 2(100\text{ N})(150\text{ N})\cos 115°}$$
$$= \sqrt{10\,000 + 22\,500 - 30\,000(-0.4226)} = 212.6\text{ N}$$
$$= 213\text{ N} \qquad \textit{Ans.}$$

Applying the law of sines to determine θ,

$$\frac{150\text{ N}}{\sin\theta} = \frac{212.6\text{ N}}{\sin 115°} \qquad \sin\theta = \frac{150\text{ N}}{212.6\text{ N}}(\sin 115°)$$

$$\theta = 39.8°$$

Thus, the direction ϕ (phi) of $\mathbf{F}_R$, measured from the horizontal, is

$$\phi = 39.8° + 15.0° = 54.8° \qquad \textit{Ans.}$$

NOTE: The results seem reasonable, since Fig. 2–11b shows $\mathbf{F}_R$ to have a magnitude larger than its components and a direction that is between them.

(c)

Fig. 2–11

EXAMPLE 2.2

Resolve the horizontal 600-lb force in Fig. 2–12a into components acting along the u and v axes and determine the magnitudes of these components.

Fig. 2–12

SOLUTION
The parallelogram is constructed by extending a line from the *head* of the 600-lb force parallel to the v axis until it intersects the u axis at point B, Fig. 2–12b. The arrow from A to B represents $\mathbf{F}_u$. Similarly, the line extended from the head of the 600-lb force drawn parallel to the u axis intersects the v axis at point C, which gives $\mathbf{F}_v$.

The vector addition using the triangle rule is shown in Fig. 2–12c. The two unknowns are the magnitudes of $\mathbf{F}_u$ and $\mathbf{F}_v$. Applying the law of sines,

$$\frac{F_u}{\sin 120°} = \frac{600 \text{ lb}}{\sin 30°}$$
$$F_u = 1039 \text{ lb} \qquad \textit{Ans.}$$

$$\frac{F_v}{\sin 30°} = \frac{600 \text{ lb}}{\sin 30°}$$
$$F_v = 600 \text{ lb} \qquad \textit{Ans.}$$

NOTE: The result for F_u shows that sometimes a component can have a greater magnitude than the resultant.

EXAMPLE 2.3

Determine the magnitude of the component force **F** in Fig. 2–13a and the magnitude of the resultant force **F**_R if **F**_R is directed along the positive y axis.

Fig. 2–13

SOLUTION
The parallelogram law of addition is shown in Fig. 2–13b, and the triangle rule is shown in Fig. 2–13c. The magnitudes of **F**_R and **F** are the two unknowns. They can be determined by applying the law of sines.

$$\frac{F}{\sin 60°} = \frac{200 \text{ lb}}{\sin 45°}$$

$$F = 245 \text{ lb} \qquad \textit{Ans.}$$

$$\frac{F_R}{\sin 75°} = \frac{200 \text{ lb}}{\sin 45°}$$

$$F_R = 273 \text{ lb} \qquad \textit{Ans.}$$

EXAMPLE 2.4

It is required that the resultant force acting on the eyebolt in Fig. 2–14a be directed along the positive x axis and that $\mathbf{F}_2$ have a *minimum* magnitude. Determine this magnitude, the angle θ, and the corresponding resultant force.

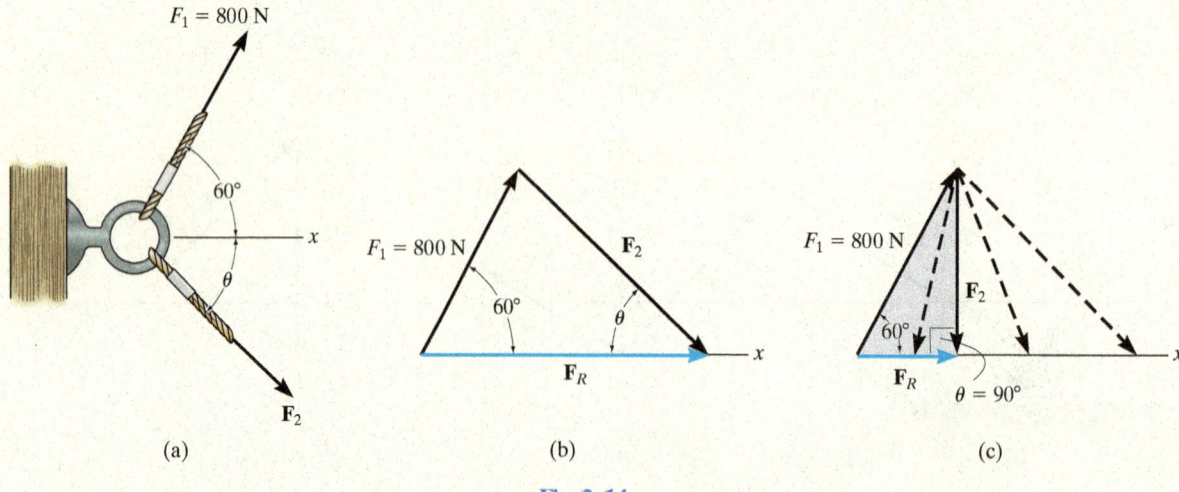

Fig. 2–14

SOLUTION

The triangle rule for $\mathbf{F}_R = \mathbf{F}_1 + \mathbf{F}_2$ is shown in Fig. 2–14b. Since the magnitudes (lengths) of $\mathbf{F}_R$ and $\mathbf{F}_2$ are not specified, then $\mathbf{F}_2$ can actually be any vector that has its head touching the line of action of $\mathbf{F}_R$, Fig. 2–14c. However, as shown, the magnitude of $\mathbf{F}_2$ is a *minimum* or the shortest length when its line of action is *perpendicular* to the line of action of $\mathbf{F}_R$, that is, when

$$\theta = 90° \qquad \textit{Ans.}$$

Since the vector addition now forms the shaded right triangle, the two unknown magnitudes can be obtained by trigonometry.

$$F_R = (800 \text{ N})\cos 60° = 400 \text{ N} \qquad \textit{Ans.}$$

$$F_2 = (800 \text{ N})\sin 60° = 693 \text{ N} \qquad \textit{Ans.}$$

FUNDAMENTAL PROBLEMS

Partial solutions and answers to all Fundamental Problems are given in the back of the book.

F2–1. Determine the magnitude of the resultant force and its direction measured clockwise from the positive x axis.

Prob. F2–1

F2–2. Two forces act on the hook. Determine the magnitude of the resultant force.

Prob. F2–2

F2–3. Determine the magnitude of the resultant force and its direction measured counterclockwise from the positive x axis.

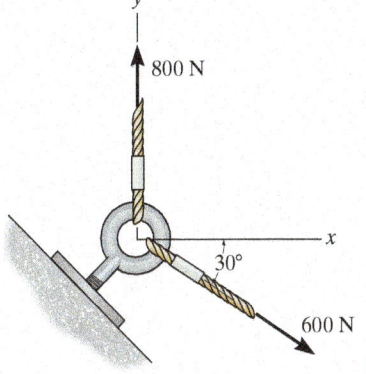

Prob. F2–3

F2–4. Resolve the 30-lb force into components along the u and v axes, and determine the magnitude of each of these components.

Prob. F2–4

F2–5. The force $F = 450$ lb acts on the frame. Resolve this force into components acting along members AB and AC, and determine the magnitude of each component.

Prob. F2–5

F2–6. If force $\mathbf{F}$ is to have a component along the u axis of $F_u = 6$ kN, determine the magnitude of $\mathbf{F}$ and the magnitude of its component $\mathbf{F}_v$ along the v axis.

Prob. F2–6

PROBLEMS

2–1. Determine the magnitude of the resultant force $F_R = F_1 + F_2$ and its orientation θ, measured counterclockwise from the positive x axis.

2–2. Determine the magnitude of the resultant force $F'_R = F_1 - F_2$ and its orientation θ, measured counterclockwise from the positive x axis.

*2–4.** Determine the magnitudes of the two components of **F** along members AB and AC. Set $F = 500$ N.

2–5. Solve Prob. 2–4 with $F = 350$ lb.

Probs. 2–1/2

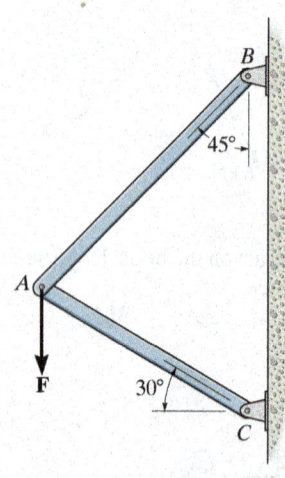

Probs. 2–4/5

2–3. Two forces are applied at the end of a screw eye in order to remove the post. Determine the angle $\theta(0° \leq \theta \leq 90°)$ and the magnitude of force **F** so that the resultant force acting on the post is directed vertically upward and has a magnitude of 750 N.

2–6. Determine the magnitude of the resultant force $F_R = F_1 + F_2$ and its direction, measured clockwise from the positive u axis.

2–7. Resolve the force F_1 into components along the u and v axes and determine the magnitudes of the components.

*2–8.** Resolve the force F_2 into components along the u and v axes and determine the magnitudes of the components.

Prob. 2–3

Probs. 2–6/7/8

2–9. Determine the magnitude of the resultant force $F_R = F_1 + F_2$ and its orientation θ, measured clockwise from the positive x axis.

2–10. Determine the magnitude of the resultant force $F_R = F_1 + F_3$ and its orientation θ, measured counterclockwise from the positive x axis.

Probs. 2–9/10

2–11. If $\theta = 60°$, determine the magnitude of the resultant force and its direction measured clockwise from the horizontal.

*2–12. Determine the angle θ for connecting member A to the plate so that the resultant force of F_A and F_B is directed horizontally to the right. Also, what is the magnitude of the resultant force?

Probs. 2–11/12

2–13. The pelvis P is connected to the femur F at A using three different muscles, which exert the forces shown on the femur. Determine the resultant force and specify its orientation θ, measured counterclockwise from the positive x axis.

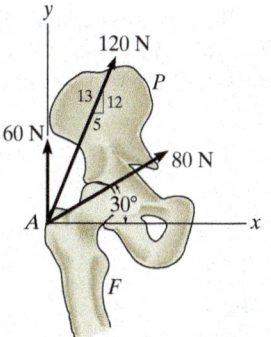

Prob. 2–13

2–14. If the resultant force acting on the support is to be 1200 lb, directed horizontally to the right, determine the force **F** in rope A and the corresponding angle θ.

Prob. 2–14

2–15. The plate is subjected to the forces acting on members A and B. If $\theta = 60°$, determine the magnitude of the resultant of these forces and its direction measured clockwise from the positive x axis.

*2–16. Determine the angle θ for connecting member B to the plate so that the resultant of F_A and F_B is directed along the positive x axis. What is the magnitude of the resultant force?

Probs. 2–15/16

2–17. Determine the magnitude and direction of the resultant $\mathbf{F}_R = \mathbf{F}_1 + \mathbf{F}_2 + \mathbf{F}_3$ of the three forces by first finding the resultant $\mathbf{F}' = \mathbf{F}_1 + \mathbf{F}_2$ and then finding $\mathbf{F}_R = \mathbf{F}' + \mathbf{F}_3$.

2–18. Determine the magnitude and direction of the resultant $\mathbf{F}_R = \mathbf{F}_1 + \mathbf{F}_2 + \mathbf{F}_3$ of the three forces by first finding the resultant $\mathbf{F}' = \mathbf{F}_2 + \mathbf{F}_3$ and then finding $\mathbf{F}_R = \mathbf{F}' + \mathbf{F}_1$.

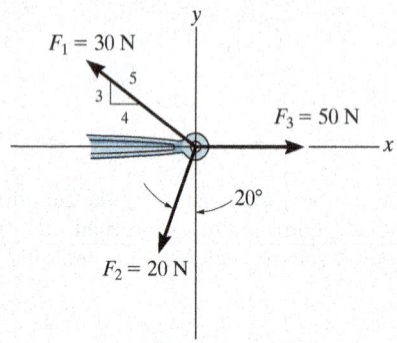

Probs. 2–17/18

2–19. The plate is subjected to the two forces at A and B. If $\theta = 60°$, determine the magnitude of the resultant force and its direction measured from the positive x axis.

*2–20. Determine the angle θ for connecting member A to the plate so that the resultant of F_A and F_B is directed along the positive x axis. Also, what is the magnitude of the resultant force?

Probs. 2–19/20

2–21. Determine the design angle θ ($0° \leq \theta \leq 90°$) for member AB so that the 400-lb horizontal force has a component of 500 lb directed from A toward C. What is the component of force acting along member AB? Take $\phi = 40°$.

2–22. Determine the design angle ϕ ($0° \leq \phi \leq 90°$) between members AB and AC so that the 400-lb horizontal force has a component of 600 lb which acts up to the right, in the direction from B toward A. Also calculate the magnitude of the force component along AC. Take $\theta = 30°$.

Probs. 2–21/22

2–23. Determine the magnitude of the two towing forces $\mathbf{F}_B$ and $\mathbf{F}_C$ if the resultant force has a magnitude $F_R = 10$ kN and is directed along the positive x axis. Set $\theta = 15°$.

*2–24. If the resultant $\mathbf{F}_R$ of the two forces acting on the jet aircraft is to be directed along the positive x axis and have a magnitude of 10 kN, determine the angle θ of the cable attached to the truck at B so that F_B is a minimum. What is the magnitude of force in each cable when this occurs?

Probs. 2–23/24

2–25. The 500-lb force is to be resolved into two components acting along the axis of the struts AB and AC. If the component of force along AC is required to be 300 lb, directed from A to C, determine the magnitude of the force acting along AB and the angle θ of the 500-lb force.

***2–28.** If $F_1 = F_2 = 30$ lb, determine the angles θ and ϕ so that the resultant force is directed along the positive x axis and has a magnitude of $F_R = 20$ lb.

Prob. 2–25

Prob. 2–28

2–26. Determine the magnitude and direction θ of $\mathbf{F}_A$ so that the resultant force is directed along the positive x axis and has a magnitude of 1250 N.

2–27. Determine the magnitude of the resultant force acting on the ring at O, if $F_A = 750$ N and $\theta = 45°$. What is its direction, measured counterclockwise from the positive x axis?

2–29. If the resultant force of the two tugboats is 3 kN, directed along the positive x axis, determine the required magnitude of force $\mathbf{F}_B$ and its direction θ.

2–30. If $F_B = 3$ kN and $\theta = 45°$, determine the magnitude of the resultant force and its direction measured clockwise from the positive x axis.

2–31. If the resultant force of the two tugboats is required to be directed toward the positive x axis, and F_B is to be a minimum, determine the magnitude of $\mathbf{F}_R$ and $\mathbf{F}_B$ and the angle θ.

Probs. 2–26/27

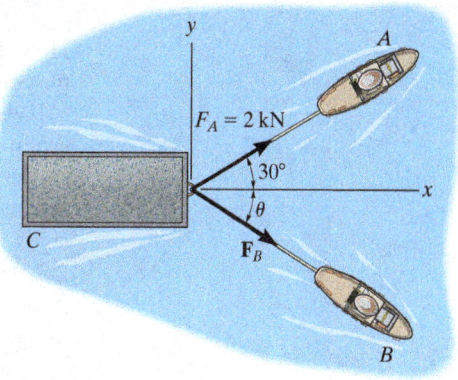

Probs. 2–29/30/31

2.4 ADDITION OF A SYSTEM OF COPLANAR FORCES

When a force is resolved into two components along the *x* and *y* axes, the components are then called **rectangular components**. For analytical work we can represent these components in one of two ways, using either scalar or Cartesian vector notation.

Scalar Notation. The rectangular components of force **F** shown in Fig. 2–15*a* are found using the parallelogram law, so that $\mathbf{F} = \mathbf{F}_x + \mathbf{F}_y$. Because these components form a right triangle, they can be determined from

$$F_x = F \cos \theta \quad \text{and} \quad F_y = F \sin \theta$$

Instead of using the angle θ, however, the direction of **F** can also be defined using a small "slope" triangle, as in the example shown in Fig. 2–15*b*. Since this triangle and the larger shaded triangle are similar, the proportional length of the sides gives

$$\frac{F_x}{F} = \frac{a}{c}$$

or

$$F_x = F\left(\frac{a}{c}\right)$$

and

$$\frac{F_y}{F} = \frac{b}{c}$$

or

$$F_y = -F\left(\frac{b}{c}\right)$$

Here the *y* component is a *negative scalar* since $\mathbf{F}_y$ is directed along the negative *y* axis.

It is important to keep in mind that this positive and negative scalar notation is to be used only for calculations, not for graphical representations in figures. Throughout the text, the *head of a vector arrow* in any figure indicates the sense of the vector *graphically*; algebraic signs are not used for this purpose. Thus, the vectors in Figs. 2–15*a* and 2–15*b* are designated by using boldface (vector) notation.* Whenever italic symbols are written near vector arrows in figures, they indicate the *magnitude* of the vector, which is *always* a *positive* quantity.

(a)

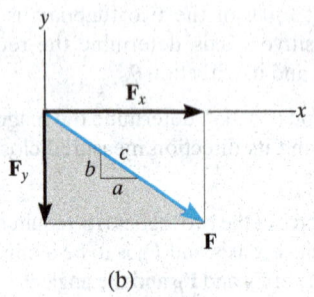

(b)

Fig. 2–15

---
*Negative signs are used only in figures with boldface notation when showing equal but opposite pairs of vectors, as in Fig. 2–2.

2.4 Addition of a System of Coplanar Forces

Cartesian Vector Notation. It is also possible to represent the *x* and *y* components of a force in terms of Cartesian unit vectors **i** and **j**. They are called unit vectors because they have a dimensionless magnitude of 1, and so they can be used to designate the *directions* of the *x* and *y* axes, respectively, Fig. 2–16.*

Since the *magnitude* of each component of **F** is *always a positive quantity*, which is represented by the (positive) scalars F_x and F_y, then we can express **F** as a *Cartesian vector*,

$$\mathbf{F} = F_x \mathbf{i} + F_y \mathbf{j}$$

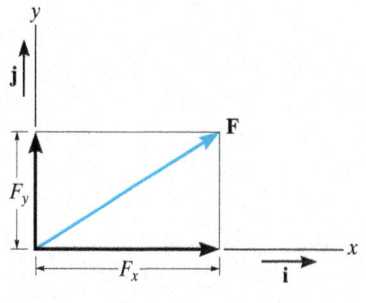

Fig. 2–16

Coplanar Force Resultants. We can use either of the two methods just described to determine the resultant of several **coplanar forces**. To do this, each force is first resolved into its *x* and *y* components, and then the respective components are added using *scalar algebra* since they are collinear. The resultant force is then formed by adding the resultant components using the parallelogram law. For example, consider the three concurrent forces in Fig. 2–17a, which have *x* and *y* components shown in Fig. 2–17b. Using Cartesian vector notation, each force is first represented as a Cartesian vector, i.e.,

$$\mathbf{F}_1 = F_{1x}\mathbf{i} + F_{1y}\mathbf{j}$$
$$\mathbf{F}_2 = -F_{2x}\mathbf{i} + F_{2y}\mathbf{j}$$
$$\mathbf{F}_3 = F_{3x}\mathbf{i} - F_{3y}\mathbf{j}$$

The vector resultant, Fig. 2–17c, is therefore

$$\mathbf{F}_R = \mathbf{F}_1 + \mathbf{F}_2 + \mathbf{F}_3$$
$$= F_{1x}\mathbf{i} + F_{1y}\mathbf{j} - F_{2x}\mathbf{i} + F_{2y}\mathbf{j} + F_{3x}\mathbf{i} - F_{3y}\mathbf{j}$$
$$= (F_{1x} - F_{2x} + F_{3x})\mathbf{i} + (F_{1y} + F_{2y} - F_{3y})\mathbf{j}$$
$$= (F_R)_x \mathbf{i} + (F_R)_y \mathbf{j}$$

Fig. 2–17

If *scalar notation* is used, then indicating the positive directions of components along the *x* and *y* axes with symbolic arrows, we have

$$\xrightarrow{+} \quad (F_R)_x = F_{1x} - F_{2x} + F_{3x}$$
$$+\uparrow \quad (F_R)_y = F_{1y} + F_{2y} - F_{3y}$$

Notice that these are the *same* results as the **i** and **j** components of $\mathbf{F}_R$ determined above.

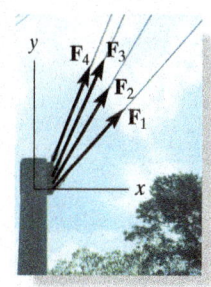

---

*For handwritten work, unit vectors are usually indicated using a circumflex, e.g., $\hat{\imath}$ and $\hat{\jmath}$. Also, realize that F_x and F_y in Fig. 2–16 represent the *magnitudes* of the components, which are *always positive scalars*. The directions are defined by **i** and **j**. If instead we used scalar notation, then F_x and F_y could be positive or negative scalars, since they would account for *both* the magnitude and direction of the components.

The resultant force of the four cable forces acting on the post can be determined by adding algebraically the separate *x* and *y* components of each cable force. This resultant $\mathbf{F}_R$ produces the *same pulling effect* on the post as all four cables.

In general then, the components of the resultant force of any number of coplanar forces can be represented by the algebraic sum of the x and y components of all the forces, i.e.,

$$(F_R)_x = \Sigma F_x$$
$$(F_R)_y = \Sigma F_y$$
(2–1)

Once these components are determined, they may be sketched along the x and y axes with their proper sense of direction, and the resultant force can be determined from vector addition, Fig. 2–17c. From this sketch, the magnitude of $\mathbf{F}_R$ is then found from the Pythagorean theorem; that is,

$$F_R = \sqrt{(F_R)_x^2 + (F_R)_y^2}$$

Also, the angle θ, which specifies the direction of the resultant force, is determined from trigonometry:

$$\theta = \tan^{-1} \left| \frac{(F_R)_y}{(F_R)_x} \right|$$

The above concepts are illustrated numerically in the examples which follow.

(c)

Fig. 2–17 (cont.)

IMPORTANT POINTS

- The resultant of several coplanar forces can easily be determined if an x, y coordinate system is established and the forces are resolved into components along the axes.

- The direction of each force is specified by the angle its line of action makes with one of the axes, or by a slope triangle.

- The orientation of the x and y axes is arbitrary, and their positive direction can be specified by the Cartesian unit vectors **i** and **j**.

- The x and y components of the *resultant force* are simply the algebraic addition of the components of all the coplanar forces.

- The magnitude of the resultant force is determined from the Pythagorean theorem, and when the resultant components are sketched on the x and y axes, Fig. 2–17c, the direction θ of the resultant can be determined from trigonometry.

EXAMPLE 2.5

Determine the x and y components of $\mathbf{F}_1$ and $\mathbf{F}_2$ acting on the boom shown in Fig. 2–18a. Express each force as a Cartesian vector.

SOLUTION

Scalar Notation. By the parallelogram law, $\mathbf{F}_1$ is resolved into x and y components, Fig. 2–18b. Since $\mathbf{F}_{1x}$ acts in the $-x$ direction, and $\mathbf{F}_{1y}$ acts in the $+y$ direction, we have

$$F_{1x} = -200 \sin 30° \text{ N} = -100 \text{ N} = 100 \text{ N} \leftarrow \qquad Ans.$$

$$F_{1y} = 200 \cos 30° \text{ N} = 173 \text{ N} = 173 \text{ N} \uparrow \qquad Ans.$$

The force $\mathbf{F}_2$ is resolved into its x and y components, as shown in Fig. 2–18c. From this "slope triangle" we could obtain the angle θ, e.g., $\theta = \tan^{-1}\left(\frac{5}{12}\right)$, and then proceed to determine the magnitudes of the components in the same manner as for $\mathbf{F}_1$. The easier method, however, consists of using proportional parts of similar triangles, i.e.,

$$\frac{F_{2x}}{260 \text{ N}} = \frac{12}{13} \qquad F_{2x} = 260 \text{ N}\left(\frac{12}{13}\right) = 240 \text{ N}$$

Similarly,

$$F_{2y} = 260 \text{ N}\left(\frac{5}{13}\right) = 100 \text{ N}$$

Notice how the magnitude of the *horizontal component*, F_{2x}, was obtained by multiplying the force magnitude by the ratio of the *horizontal leg* of the slope triangle divided by the hypotenuse; whereas the magnitude of the *vertical component*, F_{2y}, was obtained by multiplying the force magnitude by the ratio of the *vertical leg* divided by the hypotenuse. Using scalar notation to represent the components, we have

$$F_{2x} = 240 \text{ N} = 240 \text{ N} \rightarrow \qquad Ans.$$

$$F_{2y} = -100 \text{ N} = 100 \text{ N} \downarrow \qquad Ans.$$

Cartesian Vector Notation. Having determined the magnitudes and directions of the components of each force, we can express each force as a Cartesian vector.

$$\mathbf{F}_1 = \{-100\mathbf{i} + 173\mathbf{j}\} \text{ N} \qquad Ans.$$

$$\mathbf{F}_2 = \{240\mathbf{i} - 100\mathbf{j}\} \text{ N} \qquad Ans.$$

(a)

(b)

(c)

Fig. 2–18

EXAMPLE 2.6

(a)

(b)

(c)

Fig. 2–19

The link in Fig. 2–19a is subjected to two forces $\mathbf{F}_1$ and $\mathbf{F}_2$. Determine the magnitude and direction of the resultant force.

SOLUTION I

Scalar Notation. First we resolve each force into its x and y components, Fig. 2–19b, then we sum these components algebraically.

$$\xrightarrow{+} (F_R)_x = \Sigma F_x; \quad (F_R)_x = 600 \cos 30° \text{ N} - 400 \sin 45° \text{ N}$$
$$= 236.8 \text{ N} \rightarrow$$

$$+\uparrow (F_R)_y = \Sigma F_y; \quad (F_R)_y = 600 \sin 30° \text{ N} + 400 \cos 45° \text{ N}$$
$$= 582.8 \text{ N} \uparrow$$

The resultant force, shown in Fig. 2–19c, has a *magnitude* of

$$F_R = \sqrt{(236.8 \text{ N})^2 + (582.8 \text{ N})^2}$$
$$= 629 \text{ N} \qquad Ans.$$

and a direction,

$$\theta = \tan^{-1}\left(\frac{582.8 \text{ N}}{236.8 \text{ N}}\right) = 67.9° \qquad Ans.$$

SOLUTION II

Cartesian Vector Notation. From Fig. 2–19b, each force is first expressed as a Cartesian vector.

$$\mathbf{F}_1 = \{600 \cos 30° \mathbf{i} + 600 \sin 30° \mathbf{j}\} \text{ N}$$
$$\mathbf{F}_2 = \{-400 \sin 45° \mathbf{i} + 400 \cos 45° \mathbf{j}\} \text{ N}$$

Then,

$$\mathbf{F}_R = \mathbf{F}_1 + \mathbf{F}_2 = (600 \cos 30° \text{ N} - 400 \sin 45° \text{ N})\mathbf{i}$$
$$+ (600 \sin 30° \text{ N} + 400 \cos 45° \text{ N})\mathbf{j}$$
$$= \{236.8\mathbf{i} + 582.8\mathbf{j}\} \text{ N}$$

The magnitude and direction of $\mathbf{F}_R$ are determined in the same manner as before.

NOTE: Comparing the two methods of solution, notice that the use of scalar notation is more efficient since the components can be found *directly*, without first having to express each force as a Cartesian vector before adding the components. Later, however, we will show that Cartesian vector analysis is very beneficial for solving three-dimensional problems.

EXAMPLE 2.7

The end of the boom O in Fig. 2–20a is subjected to three concurrent and coplanar forces. Determine the magnitude and direction of the resultant force.

(a)

SOLUTION
Each force is resolved into its x and y components, Fig. 2–20b. Summing the x components, we have

$\xrightarrow{+} (F_R)_x = \Sigma F_x;$ $(F_R)_x = -400 \text{ N} + 250 \sin 45° \text{ N} - 200\left(\frac{4}{5}\right) \text{ N}$
$= -383.2 \text{ N} = 383.2 \text{ N} \leftarrow$

Summing the y components yields

$+\uparrow (F_R)_y = \Sigma F_y;$ $(F_R)_y = 250 \cos 45° \text{ N} + 200\left(\frac{3}{5}\right) \text{ N}$
$= 296.8 \text{ N} \uparrow$

The resultant force, shown in Fig. 2–20c, has a *magnitude* of

$F_R = \sqrt{(-383.2 \text{ N})^2 + (296.8 \text{ N})^2}$
$= 485 \text{ N}$ Ans.

From the vector addition in Fig. 2–20c, the direction angle θ is

$\theta = \tan^{-1}\left(\frac{296.8}{383.2}\right) = 37.8°$ Ans.

NOTE: Application of this method is more convenient, compared to using two applications of the parallelogram law, first to add $\mathbf{F}_1$ and $\mathbf{F}_2$, then adding $\mathbf{F}_3$ to this resultant.

(b)

(c)

Fig. 2–20

FUNDAMENTAL PROBLEMS

F2–7. Resolve each force acting on the post into its x and y components.

Prob. F2–7

F2–8. Determine the magnitude and direction of the resultant force.

Prob. F2–8

F2–9. Determine the magnitude of the resultant force acting on the corbel and its direction θ, measured counterclockwise from the x axis.

Prob. F2–9

F2–10. If the resultant force acting on the bracket is to be 750 N directed along the positive x axis, determine the magnitude of **F** and its direction θ.

Prob. F2–10

F2–11. If the magnitude of the resultant force acting on the bracket is to be 80 lb directed along the u axis, determine the magnitude of **F** and its direction θ.

Prob. F2–11

F2–12. Determine the magnitude of the resultant force and its direction θ, measured counterclockwise from the positive x axis.

Prob. F2–12

PROBLEMS

***2–32.** The three cable forces act on the eyebolt. Determine two possible magnitudes for **P** so that the resultant force has a magnitude of 800 N.

2–34. Express $\mathbf{F}_1$ and $\mathbf{F}_2$ as Cartesian vectors.

2–35. Determine the magnitude of the resultant force and its direction measured counterclockwise from the positive x axis.

Prob. 2–32

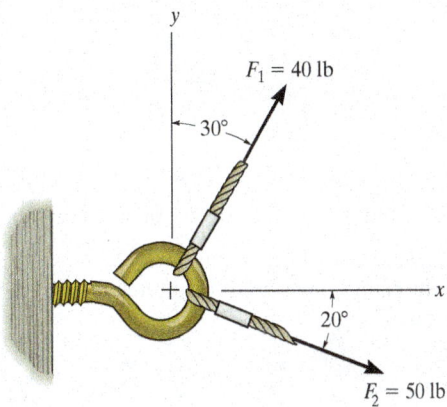

Probs. 2–34/35

2–33. Four concurrent forces act on the plate. Determine the magnitude of the resultant force and its direction measured counterclockwise from the positive x axis.

***2–36.** Resolve each force acting on the *gusset plate* into its x and y components, and express each force as a Cartesian vector.

2–37. Determine the magnitude of the resultant force acting on the *gusset plate* and its direction, measured counterclockwise from the positive x axis.

Prob. 2–33

Probs. 2–36/37

2–38. The three forces are applied to the bracket. Determine the range of values for the magnitude of force **P** so that the resultant of the three forces does not exceed 2400 N.

Prob. 2–38

2–39. Determine the x and y components of F_1 and F_2.

***2–40.** Determine the magnitude of the resultant force and its direction measured counterclockwise from the positive x axis.

Probs. 2–39/40

2–41. Determine the magnitude of the resultant force and its direction measured counterclockwise from the positive x axis.

Prob. 2–41

2–42. Express F_1, F_2, and F_3 as Cartesian vectors.

2–43. Determine the magnitude of the resultant force and its direction measured counterclockwise from the positive x axis.

Probs. 2–42/43

*2–44. The three concurrent forces acting on the post produce a zero resultant force $F_R = 0$. If $F_2 = \frac{1}{2}F_1$, and F_1 is to be 90° from F_2 as shown, determine the required magnitude of F_3 expressed in terms of F_1 and the angle θ.

Prob. 2–44

2–45. Three forces act on the ring. Determine the range of values for the magnitude of **P** so that the magnitude of the resultant force does not exceed 2500 N. Force **P** is always directed to the right.

Prob. 2–45

2–46. Determine the magnitude and orientation θ of $\mathbf{F}_B$ so that the resultant force is directed along the positive y axis and has a magnitude of 1500 N.

2–47. If $F_B = 600$ N and $\theta = 20°$, determine the magnitude of the resultant force and its direction measured counterclockwise from the positive y axis.

2–50. Express $\mathbf{F}_1$ and $\mathbf{F}_2$ as Cartesian vectors.

2–51. Determine the magnitude of the resultant force and its direction measured counterclockwise from the positive x axis.

Probs. 2–46/47

Probs. 2–50/51

*2–52. The four concentric forces act on the post. Determine the resultant force and its direction, measured counterclockwise from the positive x axis.

*2–48. Three forces act on the bracket. Determine the magnitude and direction θ of $\mathbf{F}_1$ so that the resultant force is directed along the positive x' axis and has a magnitude of 800 N.

2–49. If $F_1 = 300$ N and $\theta = 10°$, determine the magnitude of the resultant force and its direction measured counterclockwise from the positive x' axis.

Probs. 2–48/49

Prob. 2–52

2–53. Three forces act on the bracket. Determine the magnitude and direction θ of $\mathbf{F}_2$ so that the resultant force is directed along the positive u axis and has a magnitude of 50 lb.

2–54. If $F_2 = 150$ lb and $\theta = 55°$, determine the magnitude of the resultant force and its direction measured clockwise from the positive x axis.

*2–56. If the magnitude of the resultant force acting on the bracket is to be 450 N directed along the positive u axis, determine the magnitude of $\mathbf{F}_1$ and its direction ϕ.

2–57. If the resultant force acting on the bracket is required to be a minimum, determine the magnitudes of $\mathbf{F}_1$ and the resultant force. Set $\phi = 30°$.

Probs. 2–53/54

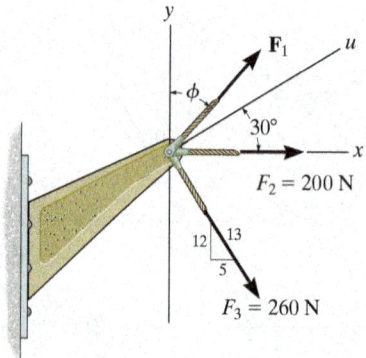

Probs. 2–56/57

2–55. Determine the resultant force acting on the hook, and its direction measured clockwise from the positive x axis.

2–58. Three forces act on the bracket. Determine the magnitude and direction θ of $\mathbf{F}$ so that the resultant force is directed along the positive x' axis and has a magnitude of 8 kN.

2–59. If $F = 5$ kN and $\theta = 30°$, determine the magnitude of the resultant force and its direction measured counterclockwise from the positive x axis.

Prob. 2–55

Probs. 2–58/59

2.5 CARTESIAN VECTORS

The operations of vector algebra, when applied to solving problems in *three dimensions*, are greatly simplified if the vectors are first represented in Cartesian vector form. In this section we will present a general method for doing this; then in the next section we will use this method for finding the resultant force of a system of concurrent forces.

Right-Handed Coordinate System. We will use a right-handed coordinate system to describe the theory of vector algebra that follows. Specifically, a rectangular coordinate system is said to be *right-handed* if the thumb of the right hand points in the direction of the positive z axis when the right-hand fingers are curled about this axis and directed from the positive x towards the positive y axis, Fig. 2–21.

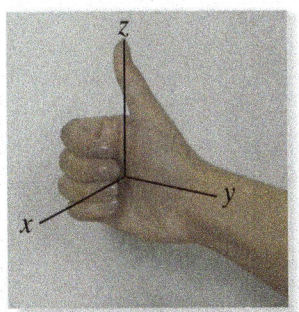

Fig. 2–21

Rectangular Components of a Vector. In general, a vector $\mathbf{A}$ may have one, two, or three rectangular components along the x, y, z coordinate axes, Fig. 2–22. These components are determined using two successive applications of the parallelogram law; that is, $\mathbf{A} = \mathbf{A}' + \mathbf{A}_z$ and then $\mathbf{A}' = \mathbf{A}_x + \mathbf{A}_y$. Combining these equations to eliminate $\mathbf{A}'$, $\mathbf{A}$ is represented by the vector sum of its *three* rectangular components,

$$\mathbf{A} = \mathbf{A}_x + \mathbf{A}_y + \mathbf{A}_z \qquad (2\text{--}2)$$

Cartesian Vector Representation. In three dimensions, the set of Cartesian unit vectors, $\mathbf{i}$, $\mathbf{j}$, $\mathbf{k}$, is used to designate the directions of the x, y, z axes, respectively, Fig. 2–23. Using these vectors the three components of $\mathbf{A}$ in Fig. 2–24 can be written in Cartesian vector form as

Fig. 2–22

Fig. 2–24

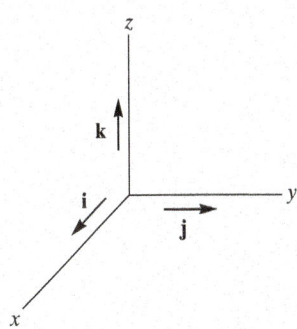

Fig. 2–23

44   CHAPTER 2   FORCE VECTORS

$$\mathbf{A} = A_x\mathbf{i} + A_y\mathbf{j} + A_z\mathbf{k} \qquad (2\text{-}3)$$

There is a distinct advantage to writing vectors in this manner. Separating the *magnitude* and *direction* of each *component vector* will simplify the operations of vector algebra, particularly in three dimensions.

Magnitude of a Cartesian Vector.

If $\mathbf{A}$ is expressed as a Cartesian vector, then its magnitude can be determined. As shown in Fig. 2–25, from the blue right triangle, $A = \sqrt{A'^2 + A_z^2}$, and from the gray right triangle, $A' = \sqrt{A_x^2 + A_y^2}$. Combining these equations to eliminate A' yields

$$A = \sqrt{A_x^2 + A_y^2 + A_z^2} \qquad (2\text{-}4)$$

Fig. 2–25

Hence, the magnitude of $\mathbf{A}$ is equal to the positive square root of the sum of the squares of the magnitudes of its components.

Coordinate Direction Angles.

We will define the *direction* of $\mathbf{A}$ by the **coordinate direction angles** α (alpha), β (beta), and γ (gamma), measured between the *tail* of $\mathbf{A}$ and the *positive* x, y, z axes, Fig. 2–26. Note that regardless of where $\mathbf{A}$ is directed, each of these angles will be between 0° and 180°.

To determine α, β, and γ, consider the projection of $\mathbf{A}$ onto the x, y, z axes, Fig. 2–27. Referring to the shaded right triangles shown in the figure, we have

Fig. 2–26

$$\cos\alpha = \frac{A_x}{A} \qquad \cos\beta = \frac{A_y}{A} \qquad \cos\gamma = \frac{A_z}{A} \qquad (2\text{-}5)$$

These numbers are known as the **direction cosines** of $\mathbf{A}$. Once they have been obtained, the coordinate direction angles α, β, γ can then be determined from the inverse cosines.

An easy way of obtaining these direction cosines is to form a unit vector $\mathbf{u}_A$ in the direction of $\mathbf{A}$, Fig. 2–26. To do this, divide $\mathbf{A}$ by its magnitude A, so that

$$\mathbf{u}_A = \frac{\mathbf{A}}{A} = \frac{A_x}{A}\mathbf{i} + \frac{A_y}{A}\mathbf{j} + \frac{A_z}{A}\mathbf{k} \qquad (2\text{-}6)$$

By comparison with Eqs. 2–5, it is seen that *the* $\mathbf{i}$, $\mathbf{j}$, $\mathbf{k}$ *components* of $\mathbf{u}_A$ *represent the direction cosines of* $\mathbf{A}$, i.e.,

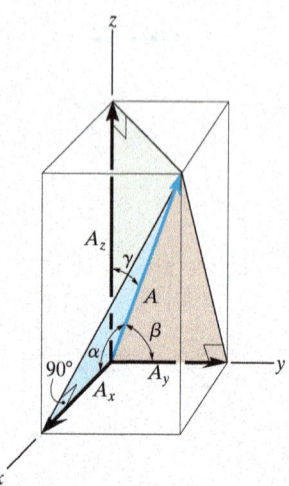

Fig. 2–27

$$\mathbf{u}_A = \cos\alpha\,\mathbf{i} + \cos\beta\,\mathbf{j} + \cos\gamma\,\mathbf{k} \qquad (2\text{-}7)$$

Since the magnitude of $\mathbf{u}_A = \cos\alpha\mathbf{i} + \cos\beta\mathbf{j} + \cos\gamma\mathbf{k}$ is one, then from this equation an important relation among the direction cosines is

$$\cos^2\alpha + \cos^2\beta + \cos^2\gamma = 1 \qquad (2\text{–}8)$$

Therefore, if only *two* of the coordinate angles are known, the third angle can be found using this equation.

Finally, if the magnitude and coordinate direction angles of **A** are known, then **A** may be expressed in Cartesian vector form as

$$\begin{aligned}\mathbf{A} &= A\mathbf{u}_A \\ &= A\cos\alpha\mathbf{i} + A\cos\beta\mathbf{j} + A\cos\gamma\mathbf{k} \\ &= A_x\mathbf{i} + A_y\mathbf{j} + A_z\mathbf{k}\end{aligned} \qquad (2\text{–}9)$$

Horizontal and Vertical Angles. Sometimes, the direction of **A** can be specified using a *horizontal angle* θ and a *vertical angle* ϕ (phi), such as shown in Fig. 2–28. The components of **A** can then be determined by applying trigonometry first to the light blue right triangle, which yields

$$A_z = A\cos\phi \quad\text{and}\quad A' = A\sin\phi$$

Now applying trigonometry to the dark blue right triangle,

$$A_x = A'\cos\theta = A\sin\phi\cos\theta$$
$$A_y = A'\sin\theta = A\sin\phi\sin\theta$$

Therefore, **A** written in Cartesian vector form becomes

$$\mathbf{A} = A\sin\phi\cos\theta\,\mathbf{i} + A\sin\phi\sin\theta\,\mathbf{j} + A\cos\phi\,\mathbf{k}$$

This equation should not be memorized; rather, it is important to understand how the components were determined using trigonometry.

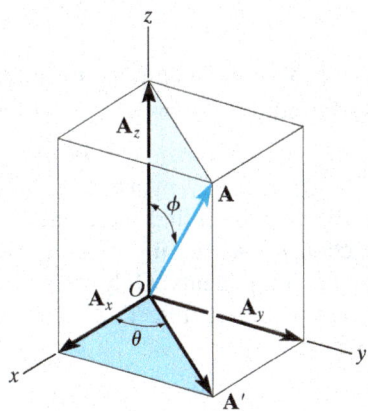

Fig. 2–28

2.6 ADDITION OF CARTESIAN VECTORS

The addition (or subtraction) of two or more vectors is greatly simplified if the vectors are expressed in terms of their Cartesian components. For example, if $\mathbf{A} = A_x\mathbf{i} + A_y\mathbf{j} + A_z\mathbf{k}$ and $\mathbf{B} = B_x\mathbf{i} + B_y\mathbf{j} + B_z\mathbf{k}$, Fig. 2–29, then the resultant vector, $\mathbf{R}$, has components which are the scalar sums of the $\mathbf{i}, \mathbf{j}, \mathbf{k}$ components of $\mathbf{A}$ and $\mathbf{B}$, i.e.,

$$\mathbf{R} = \mathbf{A} + \mathbf{B} = (A_x + B_x)\mathbf{i} + (A_y + B_y)\mathbf{j} + (A_z + B_z)\mathbf{k}$$

If this is generalized and applied to a system of several concurrent forces, then the force resultant is the vector sum of all the forces in the system and can be written as

$$\mathbf{F}_R = \Sigma\mathbf{F} = \Sigma F_x\mathbf{i} + \Sigma F_y\mathbf{j} + \Sigma F_z\mathbf{k} \qquad (2\text{–}10)$$

Here ΣF_x, ΣF_y, and ΣF_z represent the algebraic sums of the respective x, y, z or $\mathbf{i}, \mathbf{j}, \mathbf{k}$ components of each force in the system.

Fig. 2–29

IMPORTANT POINTS

- A Cartesian vector $\mathbf{A}$ has $\mathbf{i}, \mathbf{j}, \mathbf{k}$ components along the x, y, z axes. If $\mathbf{A}$ is known, its magnitude is $A = \sqrt{A_x^2 + A_y^2 + A_z^2}$.

- The direction of a Cartesian vector can be defined by the three coordinate direction angles α, β, γ, measured from the *positive x, y, z* axes to the *tail* of the vector. To find these angles, formulate a unit vector in the direction of $\mathbf{A}$, i.e., $\mathbf{u}_A = \mathbf{A}/A$, and determine the inverse cosines of its components. Only two of these angles are independent of one another; the third angle is found from $\cos^2\alpha + \cos^2\beta + \cos^2\gamma = 1$.

- The direction of a Cartesian vector can also be specified using a horizontal angle θ and vertical angle ϕ.

EXAMPLE 2.8

Express the force **F** shown in Fig. 2–30a as a Cartesian vector.

SOLUTION

The angles of 60° and 45° defining the direction of **F** are *not* coordinate direction angles. Two successive applications of the parallelogram law are needed to resolve **F** into its x, y, z components. First $\mathbf{F} = \mathbf{F}' + \mathbf{F}_z$, then $\mathbf{F}' = \mathbf{F}_x + \mathbf{F}_y$, Fig. 2–30b. By trigonometry, the magnitudes of the components are

$$F_z = 100 \sin 60° \text{ lb} = 86.6 \text{ lb}$$

$$F' = 100 \cos 60° \text{ lb} = 50 \text{ lb}$$

$$F_x = F' \cos 45° = 50 \cos 45° \text{ lb} = 35.4 \text{ lb}$$

$$F_y = F' \sin 45° = 50 \sin 45° \text{ lb} = 35.4 \text{ lb}$$

Realizing that $\mathbf{F}_y$ is in the $-\mathbf{j}$ direction, we have

$$\mathbf{F} = \{35.4\mathbf{i} - 35.4\mathbf{j} + 86.6\mathbf{k}\} \text{ lb} \qquad Ans.$$

(a)

To show that the magnitude of this vector is indeed 100 lb, apply Eq. 2–4,

$$F = \sqrt{F_x^2 + F_y^2 + F_z^2}$$

$$= \sqrt{(35.4)^2 + (35.4)^2 + (86.6)^2} = 100 \text{ lb}$$

If needed, the coordinate direction angles of **F** can be determined from the components of the unit vector acting in the direction of **F**.

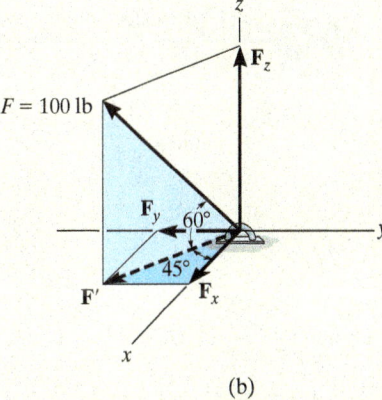

(b)

$$\mathbf{u} = \frac{\mathbf{F}}{F} = \frac{F_x}{F}\mathbf{i} + \frac{F_y}{F}\mathbf{j} + \frac{F_z}{F}\mathbf{k}$$

$$= \frac{35.4}{100}\mathbf{i} - \frac{35.4}{100}\mathbf{j} + \frac{86.6}{100}\mathbf{k}$$

$$= 0.354\mathbf{i} - 0.354\mathbf{j} + 0.866\mathbf{k}$$

so that

$$\alpha = \cos^{-1}(0.354) = 69.3°$$

$$\beta = \cos^{-1}(-0.354) = 111°$$

$$\gamma = \cos^{-1}(0.866) = 30.0°$$

These results are shown in Fig. 2–30c.

(c)

Fig. 2–30

EXAMPLE 2.9

Fig. 2–31

Two forces act on the hook shown in Fig. 2–31a. Specify the magnitude of $\mathbf{F}_2$ and its coordinate direction angles so that the resultant force $\mathbf{F}_R$ acts along the positive y axis and has a magnitude of 800 N.

SOLUTION
To solve this problem, the resultant force $\mathbf{F}_R$ and its two components, $\mathbf{F}_1$ and $\mathbf{F}_2$, will each be expressed in Cartesian vector form. Then, as shown in Fig. 2–31b, it is necessary that $\mathbf{F}_R = \mathbf{F}_1 + \mathbf{F}_2$.

Applying Eq. 2–9,

$$\mathbf{F}_1 = F_1 \cos\alpha_1 \mathbf{i} + F_1 \cos\beta_1 \mathbf{j} + F_1 \cos\gamma_1 \mathbf{k}$$
$$= 300 \cos 45°\, \mathbf{i} + 300 \cos 60°\, \mathbf{j} + 300 \cos 120°\, \mathbf{k}$$
$$= \{212.1\mathbf{i} + 150\mathbf{j} - 150\mathbf{k}\}\, \text{N}$$
$$\mathbf{F}_2 = F_{2x}\mathbf{i} + F_{2y}\mathbf{j} + F_{2z}\mathbf{k}$$

Since $\mathbf{F}_R$ has a magnitude of 800 N and acts in the $+\mathbf{j}$ direction,

$$\mathbf{F}_R = (800\, \text{N})(+\mathbf{j}) = \{800\mathbf{j}\}\, \text{N}$$

We require

$$\mathbf{F}_R = \mathbf{F}_1 + \mathbf{F}_2$$
$$800\mathbf{j} = 212.1\mathbf{i} + 150\mathbf{j} - 150\mathbf{k} + F_{2x}\mathbf{i} + F_{2y}\mathbf{j} + F_{2z}\mathbf{k}$$
$$800\mathbf{j} = (212.1 + F_{2x})\mathbf{i} + (150 + F_{2y})\mathbf{j} + (-150 + F_{2z})\mathbf{k}$$

To satisfy this equation, the $\mathbf{i}, \mathbf{j}, \mathbf{k}$ components of $\mathbf{F}_R$ must be equal to the corresponding $\mathbf{i}, \mathbf{j}, \mathbf{k}$ components of $(\mathbf{F}_1 + \mathbf{F}_2)$. Hence,

$$0 = 212.1 + F_{2x} \qquad F_{2x} = -212.1\, \text{N}$$
$$800 = 150 + F_{2y} \qquad F_{2y} = 650\, \text{N}$$
$$0 = -150 + F_{2z} \qquad F_{2z} = 150\, \text{N}$$

The magnitude of $\mathbf{F}_2$ is thus

$$F_2 = \sqrt{(-212.1\, \text{N})^2 + (650\, \text{N})^2 + (150\, \text{N})^2}$$
$$= 700\, \text{N} \qquad \textit{Ans.}$$

We can use Eq. 2–9 to determine $\alpha_2, \beta_2, \gamma_2$.

$$\cos\alpha_2 = \frac{-212.1}{700}; \qquad \alpha_2 = 108° \qquad \textit{Ans.}$$

$$\cos\beta_2 = \frac{650}{700}; \qquad \beta_2 = 21.8° \qquad \textit{Ans.}$$

$$\cos\gamma_2 = \frac{150}{700}; \qquad \gamma_2 = 77.6° \qquad \textit{Ans.}$$

These results are shown in Fig. 2–31b.

FUNDAMENTAL PROBLEMS

F2–13. Determine the coordinate direction angles of the force.

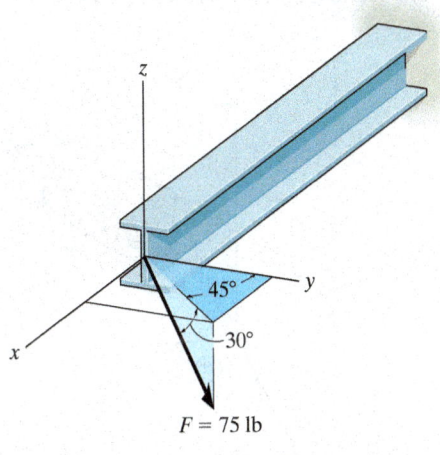

Prob. F2–13

F2–14. Express the force as a Cartesian vector.

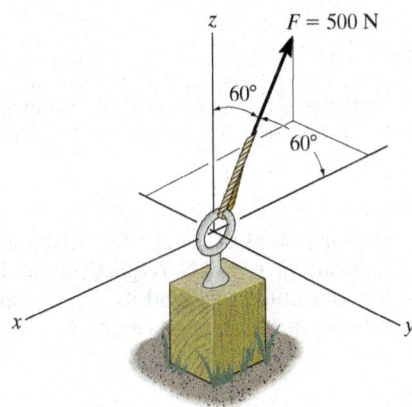

Prob. F2–14

F2–15. Express the force as a Cartesian vector.

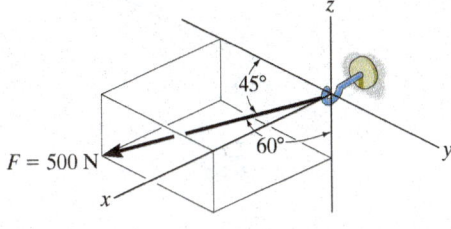

Prob. F2–15

F2–16. Express the force as a Cartesian vector.

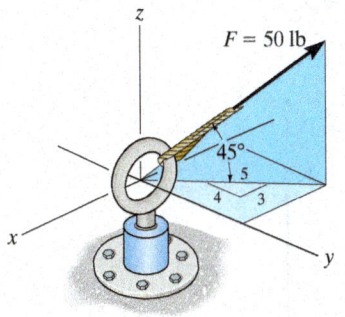

Prob. F2–16

F2–17. Express the force as a Cartesian vector.

Prob. F2–17

F2–18. Determine the resultant force acting on the hook.

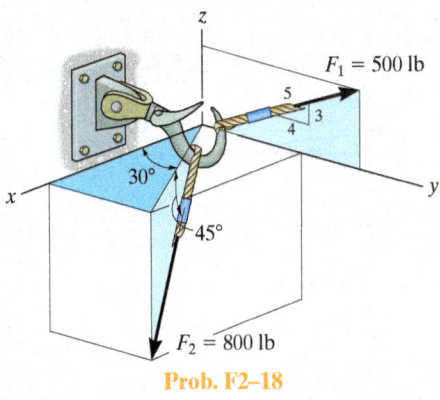

Prob. F2–18

PROBLEMS

*2–60. Determine the magnitudes of the x, y, z components of the force **F**.

2–62. The cable at the end of the beam exerts a force of 450 lb on the beam. Express **F** as a Cartesian vector.

Prob. 2–60

Prob. 2–62

2–61. The bolt is subjected to the force **F**, which has components acting along the x, y, z axes. If the magnitude of **F** is 80 N, and $\alpha = 60°$ and $\gamma = 45°$, determine the magnitudes of its components.

2–63. The force **F** acts on the bracket within the octant shown. If $F = 400$ N, $\beta = 60°$, and $\gamma = 45°$, determine the x, y, z components of **F**.

*2–64. The force **F** acts on the bracket within the octant shown. If the magnitudes of the x and z components of **F** are $F_x = 300$ N and $F_z = 600$ N, respectively, and $\beta = 60°$, determine the magnitude of **F** and its y component. Also, find the coordinate direction angles α and γ.

Prob. 2–61

Probs. 2–63/64

2–65. Express each force in Cartesian vector form and then determine the resultant force. Find the magnitude and coordinate direction angles of the resultant force.

2–66. Determine the coordinate direction angles of $\mathbf{F}_1$.

Probs. 2–65/66

2–67. The two forces $\mathbf{F}_1$ and $\mathbf{F}_2$ have a resultant force of $\mathbf{F}_R = \{-100\mathbf{k}\}$ lb. Determine the magnitude and coordinate direction angles of $\mathbf{F}_2$.

*2–68. Determine the coordinate direction angles of $\mathbf{F}_1$ and indicate them on the figure.

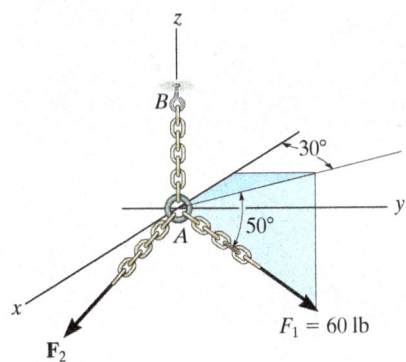

Probs. 2–67/68

2–69. The bracket is subjected to the two forces shown. Express each force in Cartesian vector form and then determine the resultant force $\mathbf{F}_R$. Find the magnitude and coordinate direction angles of the resultant force.

Prob. 2–69

2–70. Determine the magnitude and coordinate direction angles of the resultant force, and sketch this vector on the coordinate system.

Prob. 2–70

2–71. Determine the magnitude and coordinate direction angles $\alpha_1, \beta_1, \gamma_1$ of $\mathbf{F}_1$ so that the resultant of the three forces acting on the bracket is $\mathbf{F}_R = \{-350\mathbf{k}\}$ lb. Note that $\mathbf{F}_3$ lies in the x–y plane.

Prob. 2–71

*2–72. Two forces $\mathbf{F}_1$ and $\mathbf{F}_2$ act on the screw eye. If the resultant force $\mathbf{F}_R$ has a magnitude of 150 lb and the coordinate direction angles shown, determine the angle γ, and the magnitude of $\mathbf{F}_2$ and its coordinate direction angles.

2–75. The spur gear is subjected to the two forces. Express each force as a Cartesian vector.

*2–76. Determine the resultant of the two forces and express the result as a Cartesian vector.

Prob. 2–72

Probs. 2–75/76

2–73. Express each force as a Cartesian vector.

2–74. Determine the magnitude and coordinate direction angles of the resultant force, and sketch this vector on the coordinate system.

2–77. Specify the magnitude F_3 and directions $\alpha_3, \beta_3,$ and γ_3 so that the resultant force of the three forces is $\mathbf{F}_R = \{9\mathbf{j}\}$ kN.

Probs. 2–73/74

Prob. 2–77

PROBLEMS

2–78. Determine the coordinate angle γ for $\mathbf{F}_2$ and then express each force acting on the bracket as a Cartesian vector.

2–79. Determine the magnitude and coordinate direction angles of the resultant force acting on the bracket.

Probs. 2–78/79

***2–80.** Three forces act on the ring. Determine the magnitude and coordinate direction angles of the resultant force.

Prob. 2–80

2–81. If the coordinate direction angles for $\mathbf{F}_3$ are $\alpha_3 = 120°, \beta_3 = 60°$, and $\gamma_3 = 45°$, determine the magnitude and coordinate direction angles of the resultant force acting on the eyebolt.

2–82. If the coordinate direction angles for $\mathbf{F}_3$ are $\alpha_3 = 120°, \beta_3 = 45°$, and $\gamma_3 = 60°$, determine the magnitude and coordinate direction angles of the resultant force acting on the eyebolt.

2–83. If the direction of the resultant force acting on the eyebolt is defined by the unit vector $\mathbf{u}_{F_R} = \cos 30°\mathbf{j} + \sin 30°\mathbf{k}$, determine the coordinate direction angles of $\mathbf{F}_3$ and the magnitude of $\mathbf{F}_R$.

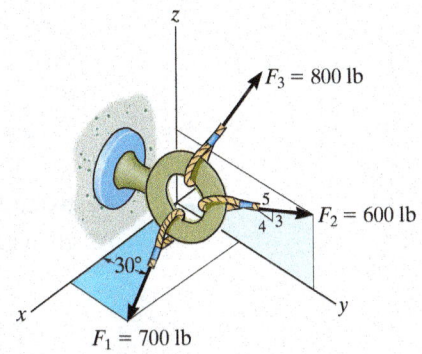

Probs. 2–81/82/83

***2–84.** The pipe is subjected to the force $\mathbf{F}$, which has components acting along the x, y, z axes. If the magnitude of $\mathbf{F}$ is 12 kN, and $\alpha = 120°$ and $\gamma = 45°$, determine the magnitudes of its three components.

2–85. The pipe is subjected to the force $\mathbf{F}$, which has components $F_x = 1.5$ kN and $F_z = 1.25$ kN. If $\beta = 75°$, determine the magnitude of $\mathbf{F}$ and $\mathbf{F}_y$.

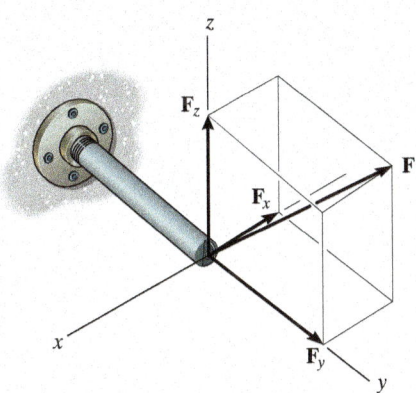

Probs. 2–84/85

2.7 POSITION VECTORS

In this section we will introduce the concept of a position vector. It will be shown that this vector is of importance in formulating a Cartesian force vector directed between two points in space.

x, y, z Coordinates. Throughout the book we will use the convention followed in many technical texts, which requires the positive z axis to be directed *upward* (the zenith direction) so that it measures the height of an object or the altitude of a point. The x, y axes then lie in the horizontal plane, Fig. 2–32.

Points in space are located relative to the origin of coordinates, O, by successive measurements along the x, y, z axes. For example, the coordinates of point A are obtained by starting at O and measuring $x_A = +4$ m along the x axis, then $y_A = +2$ m along the y axis, and finally $z_A = -6$ m along the z axis, so that $A(4\text{ m}, 2\text{ m}, -6\text{ m})$.

Fig. 2–32

Position Vector. A *position vector* **r** is defined as a fixed vector which locates a point in space relative to another point. For example, if **r** extends from the origin of coordinates, O, to point $P(x, y, z)$, Fig. 2–33a, then **r** can be expressed in Cartesian vector form as

$$\mathbf{r} = x\mathbf{i} + y\mathbf{j} + z\mathbf{k}$$

Note how the head-to-tail vector addition of the three components yields vector **r**, Fig. 2–33b. Starting at the origin O, one "travels" x in the $+\mathbf{i}$ direction, then y in the $+\mathbf{j}$ direction, and finally z in the $+\mathbf{k}$ direction to arrive at point $P(x, y, z)$.

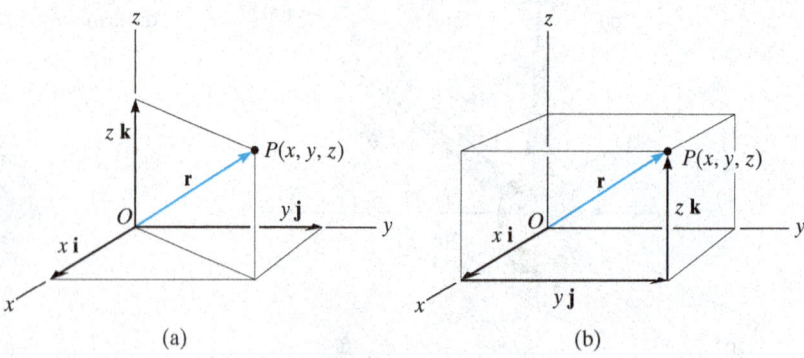

Fig. 2–33

2.7 Position Vectors

In the more general case, the position vector may be directed from point A to point B in space, Fig. 2–34a. By the head-to-tail vector addition, using the triangle rule, we require

$$\mathbf{r}_A + \mathbf{r} = \mathbf{r}_B$$

Solving for $\mathbf{r}$ and expressing $\mathbf{r}_A$ and $\mathbf{r}_B$ in Cartesian vector form yields

$$\mathbf{r} = \mathbf{r}_B - \mathbf{r}_A = (x_B\mathbf{i} + y_B\mathbf{j} + z_B\mathbf{k}) - (x_A\mathbf{i} + y_A\mathbf{j} + z_A\mathbf{k})$$

or

$$\mathbf{r} = (x_B - x_A)\mathbf{i} + (y_B - y_A)\mathbf{j} + (z_B - z_A)\mathbf{k} \qquad (2\text{–}11)$$

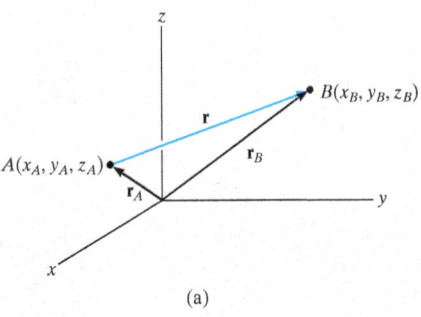

Thus, the $\mathbf{i}, \mathbf{j}, \mathbf{k}$ components of $\mathbf{r}$ may be formed by taking the coordinates of the tail of the vector $A(x_A, y_A, z_A)$ and subtracting them from the corresponding coordinates of the head $B(x_B, y_B, z_B)$. We can also form these components *directly*, Fig. 2–34b, by starting at A and moving through a distance of $(x_B - x_A)$ along the positive x axis ($+\mathbf{i}$), then $(y_B - y_A)$ along the positive y axis ($+\mathbf{j}$), and finally $(z_B - z_A)$ along the positive z axis ($+\mathbf{k}$) to get to B.

Fig. 2–34

If an x, y, z coordinate system is established, then the coordinates of two points A and B on the cable can be determined. From this the position vector $\mathbf{r}$ acting along the cable can be formulated. Its magnitude represents the distance from A to B, and its unit vector, $\mathbf{u} = \mathbf{r}/r$, gives the direction defined by α, β, γ.

2.8 FORCE VECTOR DIRECTED ALONG A LINE

Quite often in three-dimensional statics problems, the direction of a force is specified by two points through which its line of action passes. Such a situation is shown in Fig. 2–35, where the force **F** is directed along the cord AB. We can formulate **F** as a Cartesian vector by realizing that it has the *same direction* and *sense* as the position vector **r** directed from point A to point B on the cord. This common direction is specified by the *unit vector* $\mathbf{u} = \mathbf{r}/r$, and so once **u** is determined, then

$$\mathbf{F} = F\mathbf{u} = F\left(\frac{\mathbf{r}}{r}\right) = F\left(\frac{(x_B - x_A)\mathbf{i} + (y_B - y_A)\mathbf{j} + (z_B - z_A)\mathbf{k}}{\sqrt{(x_B - x_A)^2 + (y_B - y_A)^2 + (z_B - z_A)^2}}\right)$$

Although we have represented **F** symbolically in Fig. 2–36, note that it has *units of force*, unlike **r**, which has units of length.

Fig. 2–35

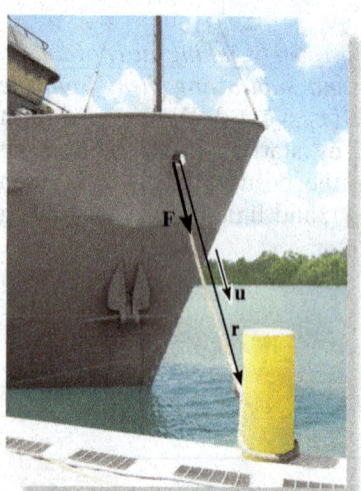

The force **F** acting along the rope can be represented as a Cartesian vector by establishing x, y, z axes and first forming a position vector **r** along the length of the rope. Then the corresponding unit vector $\mathbf{u} = \mathbf{r}/r$ that defines the direction of both the rope and the force can be determined. Finally, the magnitude of the force is combined with its direction, so that $\mathbf{F} = F\mathbf{u}$.

IMPORTANT POINTS

- A position vector locates one point in space relative to another point.

- The easiest way to formulate the components of a position vector is to determine the distance and direction that one must travel along the x, y, z directions—going from the tail to the head of the vector.

- A force **F** acting in the direction of a position vector **r** can be represented in Cartesian form if the unit vector **u** of the position vector is determined and it is multiplied by the magnitude of the force, i.e., $\mathbf{F} = F\mathbf{u} = F(\mathbf{r}/r)$.

EXAMPLE 2.10

The man shown in Fig. 2–36a pulls on the cord with a force of 70 lb. Represent this force acting on the support A as a Cartesian vector and determine its direction.

SOLUTION
Force **F** is shown in Fig. 2–36b. The *direction* of this vector, **u**, is determined from the position vector **r**, which extends from A to B. Rather than using the coordinates of the end points of the cord, **r** can be determined *directly* by noting in Fig. 2–36a that one must travel from A $\{-24\mathbf{k}\}$ ft, then $\{-8\mathbf{j}\}$ ft, and finally $\{12\mathbf{i}\}$ ft to get to B. Thus,

$$\mathbf{r} = \{12\mathbf{i} - 8\mathbf{j} - 24\mathbf{k}\} \text{ ft}$$

The magnitude of **r**, which represents the *length* of cord AB, is

$$r = \sqrt{(12 \text{ ft})^2 + (-8 \text{ ft})^2 + (-24 \text{ ft})^2} = 28 \text{ ft}$$

Forming the unit vector that defines the direction and sense of both **r** and **F**, we have

$$\mathbf{u} = \frac{\mathbf{r}}{r} = \frac{12}{28}\mathbf{i} - \frac{8}{28}\mathbf{j} - \frac{24}{28}\mathbf{k}$$

Since **F** has a *magnitude* of 70 lb and a *direction* specified by **u**, then

$$\mathbf{F} = F\mathbf{u} = 70 \text{ lb}\left(\frac{12}{28}\mathbf{i} - \frac{8}{28}\mathbf{j} - \frac{24}{28}\mathbf{k}\right)$$

$$= \{30\mathbf{i} - 20\mathbf{j} - 60\mathbf{k}\} \text{ lb} \qquad Ans.$$

The coordinate direction angles are measured between the tails of **r** (or **F**) and the *positive axes* of a localized coordinate system with origin placed at A, Fig. 2–36b. From the components of the unit vector:

$$\alpha = \cos^{-1}\left(\frac{12}{28}\right) = 64.6° \qquad Ans.$$

$$\beta = \cos^{-1}\left(\frac{-8}{28}\right) = 107° \qquad Ans.$$

$$\gamma = \cos^{-1}\left(\frac{-24}{28}\right) = 149° \qquad Ans.$$

NOTE: These results make sense when compared with the angles identified in Fig. 2–36b.

(a)

(b)

Fig. 2–36

EXAMPLE 2.11

The roof is supported by two cables as shown in the photo. If the cables exert forces $F_{AB} = 100$ N and $F_{AC} = 120$ N on the wall hook at A as shown in Fig. 2–37a, determine the resultant force acting at A. Express the result as a Cartesian vector.

SOLUTION
The resultant force $\mathbf{F}_R$ is shown graphically in Fig. 2–37b. We can express this force as a Cartesian vector by first formulating $\mathbf{F}_{AB}$ and $\mathbf{F}_{AC}$ as Cartesian vectors and then adding their components. The directions of $\mathbf{F}_{AB}$ and $\mathbf{F}_{AC}$ are specified by forming unit vectors $\mathbf{u}_{AB}$ and $\mathbf{u}_{AC}$ along the cables. These unit vectors are obtained from the associated position vectors $\mathbf{r}_{AB}$ and $\mathbf{r}_{AC}$. With reference to Fig. 2–37a, to go from A to B, we must travel $\{-4\mathbf{k}\}$ m, and then $\{4\mathbf{i}\}$ m. Thus,

$$\mathbf{r}_{AB} = \{4\mathbf{i} - 4\mathbf{k}\} \text{ m}$$

$$r_{AB} = \sqrt{(4 \text{ m})^2 + (-4 \text{ m})^2} = 5.66 \text{ m}$$

$$\mathbf{F}_{AB} = F_{AB}\left(\frac{\mathbf{r}_{AB}}{r_{AB}}\right) = (100 \text{ N})\left(\frac{4}{5.66}\mathbf{i} - \frac{4}{5.66}\mathbf{k}\right)$$

$$\mathbf{F}_{AB} = \{70.7\mathbf{i} - 70.7\mathbf{k}\} \text{ N}$$

To go from A to C, we must travel $\{-4\mathbf{k}\}$ m, then $\{2\mathbf{j}\}$ m, and finally $\{4\mathbf{i}\}$. Thus,

$$\mathbf{r}_{AC} = \{4\mathbf{i} + 2\mathbf{j} - 4\mathbf{k}\} \text{ m}$$

$$r_{AC} = \sqrt{(4 \text{ m})^2 + (2 \text{ m})^2 + (-4 \text{ m})^2} = 6 \text{ m}$$

$$\mathbf{F}_{AC} = F_{AC}\left(\frac{\mathbf{r}_{AC}}{r_{AC}}\right) = (120 \text{ N})\left(\frac{4}{6}\mathbf{i} + \frac{2}{6}\mathbf{j} - \frac{4}{6}\mathbf{k}\right)$$

$$= \{80\mathbf{i} + 40\mathbf{j} - 80\mathbf{k}\} \text{ N}$$

The resultant force is therefore

$$\mathbf{F}_R = \mathbf{F}_{AB} + \mathbf{F}_{AC} = \{70.7\mathbf{i} - 70.7\mathbf{k}\} \text{ N} + \{80\mathbf{i} + 40\mathbf{j} - 80\mathbf{k}\} \text{ N}$$

$$= \{151\mathbf{i} + 40\mathbf{j} - 151\mathbf{k}\} \text{ N} \qquad \textit{Ans.}$$

(a)

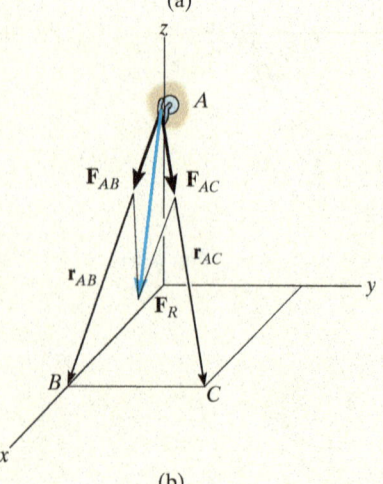

(b)

Fig. 2–37

EXAMPLE 2.12

The force in Fig. 2–38a acts on the hook. Express it as a Cartesian vector.

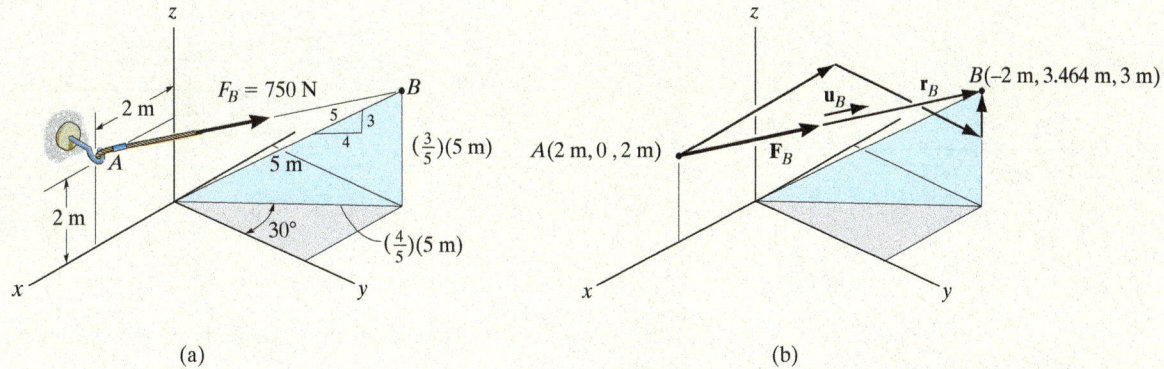

Fig. 2–38

SOLUTION

As shown in Fig. 2–38b, the coordinates for points A and B are

$$A(2 \text{ m}, 0, 2 \text{ m})$$

and

$$B\left[-\left(\frac{4}{5}\right)5 \sin 30° \text{ m}, \left(\frac{4}{5}\right)5 \cos 30° \text{ m}, \left(\frac{3}{5}\right)5 \text{ m}\right]$$

or

$$B(-2 \text{ m}, 3.464 \text{ m}, 3 \text{ m})$$

Therefore, to go from A to B, one must travel $\{-4\mathbf{i}\}$ m, then $\{3.464\mathbf{j}\}$ m, and finally $\{1\mathbf{k}\}$ m. Thus,

$$\mathbf{u}_B = \left(\frac{\mathbf{r}_B}{r_B}\right) = \frac{\{-4\mathbf{i} + 3.464\mathbf{j} + 1\mathbf{k}\} \text{ m}}{\sqrt{(-4 \text{ m})^2 + (3.464 \text{ m})^2 + (1 \text{ m})^2}}$$

$$= -0.7428\mathbf{i} + 0.6433\mathbf{j} + 0.1857\mathbf{k}$$

Force $\mathbf{F}_B$ expressed as a Cartesian vector becomes

$$\mathbf{F}_B = F_B \mathbf{u}_B = (750 \text{ N})(-0.74281\mathbf{i} + 0.6433\mathbf{j} + 0.1857\mathbf{k})$$

$$= \{-557\mathbf{i} + 482\mathbf{j} + 139\mathbf{k}\} \text{ N} \qquad \textit{Ans.}$$

FUNDAMENTAL PROBLEMS

F2–19. Express the position vector $\mathbf{r}_{AB}$ in Cartesian vector form, then determine its magnitude and coordinate direction angles.

Prob. F2–19

F2–20. Determine the length of the rod and the position vector directed from A to B. What is the angle θ?

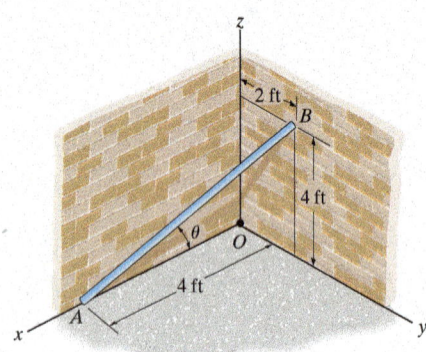

Prob. F2–20

F2–21. Express the force as a Cartesian vector.

Prob. F2–21

F2–22. Express the force as a Cartesian vector.

Prob. F2–22

F2–23. Determine the magnitude of the resultant force at A.

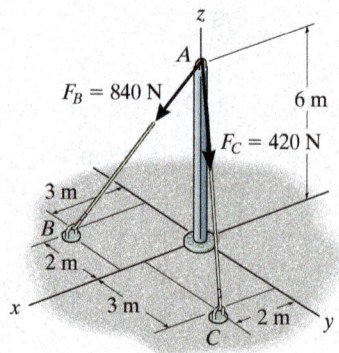

Prob. F2–23

F2–24. Determine the resultant force at A.

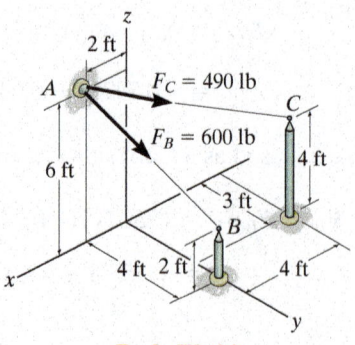

Prob. F2–24

PROBLEMS

2–86. Determine the length of the connecting rod AB by first formulating a position vector from A to B and then determining its magnitude.

*2–88. Express each force as a Cartesian vector, and then determine the magnitude and coordinate direction angles of the resultant force.

Prob. 2–86

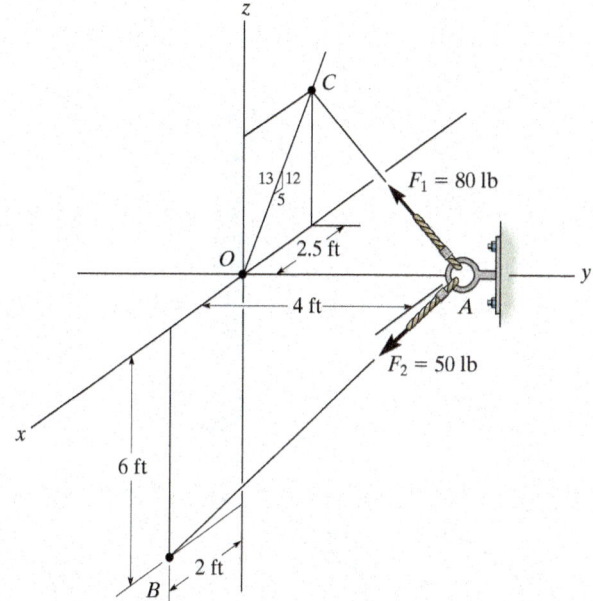

Prob. 2–88

2–87. Determine the distance between the end points A and B on the pipe assembly.

2–89. The positions of point A on the flag pole and point B on the light have been measured relative to the electronic distance meter at O. Determine the distance between A and B.

Prob. 2–87

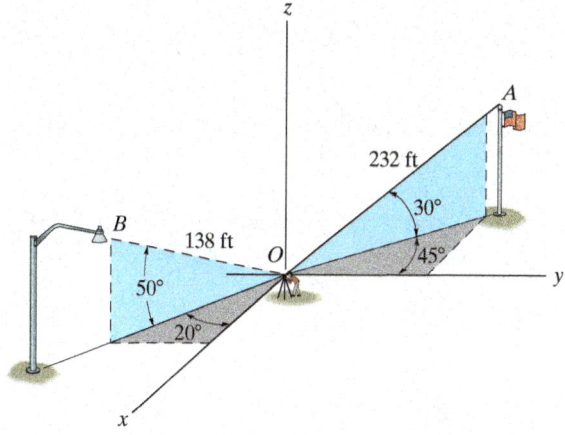

Prob. 2–89

2–90. The 8-m-long cable is anchored to the ground at A. If $x = 4$ m and $y = 2$ m, determine the coordinate z to the highest point of attachment along the column.

2–91. The 8-m-long cable is anchored to the ground at A. If $z = 5$ m, determine the location $+x$, $+y$ of the support at A. Choose a value such that $x = y$.

2–93. The 80-ft-long cable is attached to the boom at A. If $x = 60$ ft, determine the coordinates y and z to the end of the cable B. Take $z = 3y$.

2–94. The cable is attached to the boom at A and to point B, which has coordinates of $x = 80$ ft, $y = 40$ ft, $z = 40$ ft. Determine the length of the cable and the angle between the positive z axis and the cable.

Probs. 2–90/91

Probs. 2–93/94

*2–92. Determine the magnitude and coordinate direction angles of the resultant force.

2–95. At a given instant, the position of a plane at A and a train at B are measured relative to a radar antenna at O. Determine the distance d between A and B at this instant. To solve the problem, formulate a position vector, directed from A to B, and then determine its magnitude.

Prob. 2–92

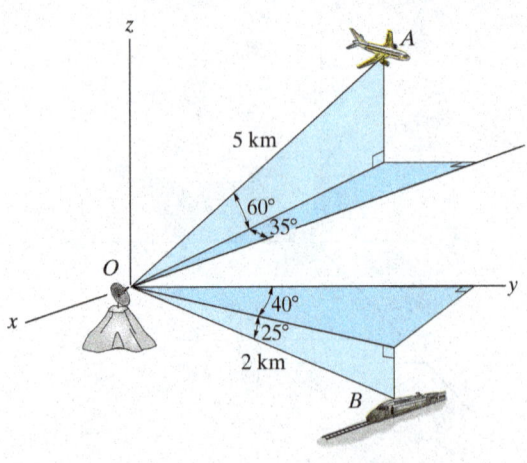

Prob. 2–95

*2–96. Represent each cable force as a Cartesian vector.

2–97. Determine the magnitude and coordinate direction angles of the resultant of the two forces acting at point A.

2–99. Position vectors along the robotic arm from O to B and B to A are $\mathbf{r}_{OB} = \{100\mathbf{i} + 300\mathbf{j} + 400\mathbf{k}\}$ mm and $\mathbf{r}_{BA} = \{350\mathbf{i} + 225\mathbf{j} - 640\mathbf{k}\}$ mm, respectively. Determine the distance from O to the grip at A.

*2–100. If $\mathbf{r}_{OA} = \{0.5\mathbf{i} + 4\mathbf{j} + 0.25\mathbf{k}\}$ m and $\mathbf{r}_{OB} = \{0.3\mathbf{i} + 2\mathbf{j} + 2\mathbf{k}\}$ m, express $\mathbf{r}_{BA}$ as a Cartesian vector.

Probs. 2–96/97

Probs. 2–99/100

2–98. The force $\mathbf{F}$ has a magnitude of 80 lb and acts at the midpoint C of the rod. Express this force as a Cartesian vector.

2–101. Determine the position $(x, y, 0)$ for fixing cable BA so that the resultant force exerted on the pole is directed along its axis, from B toward O. Also, what is the magnitude of the resultant force?

2–102. Determine the magnitude and coordinate direction angles α, β, γ of the resultant force acting on the pole. Set $x = 4$ m, $y = 2$ m.

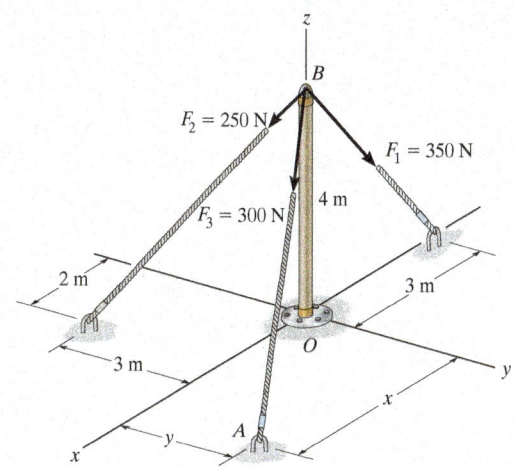

Prob. 2–98

Probs. 2–101/102

2.9 DOT PRODUCT

Occasionally in statics one has to find the angle between two lines or the components of a force parallel and perpendicular to a line. In two dimensions, these problems can readily be solved by trigonometry since the geometry is easy to visualize. In three dimensions, however, this is often difficult, and consequently vector methods should be employed for the solution. The dot product, which defines a particular method for "multiplying" two vectors, can be used to solve the above-mentioned problems.

The **dot product** of vectors **A** and **B**, written **A · B** and read "**A** dot **B**," is defined as the product of the magnitudes of **A** and **B** and the cosine of the angle θ between their tails, Fig. 2–39. Expressed in equation form,

$$\mathbf{A} \cdot \mathbf{B} = AB \cos \theta \qquad (2\text{--}12)$$

where $0° \leq \theta \leq 180°$. The dot product is often referred to as the *scalar product* of vectors since the result is a *scalar* and not a vector.

Fig. 2–39

Laws of Operation.
The following three laws of operation apply.

1. Commutative law: **A · B = B · A**

2. Multiplication by a scalar: a (**A · B**) = (a**A**) **· B** = **A ·** (a**B**)

3. Distributive law: **A · (B + D) = (A · B) + (A · D)**

Cartesian Vector Formulation.
If we apply Eq. 2–12, we can find the dot product for any two Cartesian unit vectors. For example, **i · i** = $(1)(1) \cos 0° = 1$ and **i · j** = $(1)(1) \cos 90° = 0$. If we want to find the dot product of two general vectors **A** and **B** that are expressed in Cartesian vector form, then we have

$$\begin{aligned}\mathbf{A} \cdot \mathbf{B} &= (A_x\mathbf{i} + A_y\mathbf{j} + A_z\mathbf{k}) \cdot (B_x\mathbf{i} + B_y\mathbf{j} + B_z\mathbf{k}) \\ &= A_xB_x(\mathbf{i} \cdot \mathbf{i}) + A_xB_y(\mathbf{i} \cdot \mathbf{j}) + A_xB_z(\mathbf{i} \cdot \mathbf{k}) \\ &+ A_yB_x(\mathbf{j} \cdot \mathbf{i}) + A_yB_y(\mathbf{j} \cdot \mathbf{j}) + A_yB_z(\mathbf{j} \cdot \mathbf{k}) \\ &+ A_zB_x(\mathbf{k} \cdot \mathbf{i}) + A_zB_y(\mathbf{k} \cdot \mathbf{j}) + A_zB_z(\mathbf{k} \cdot \mathbf{k})\end{aligned}$$

Carrying out the dot-product operations, the final result becomes

$$\mathbf{A} \cdot \mathbf{B} = A_xB_x + A_yB_y + A_zB_z \qquad (2\text{--}13)$$

Thus, to determine the dot product of two Cartesian vectors, *multiply their corresponding x, y, z components and sum these products algebraically*. The result will be either a positive or negative *scalar*, or it could be zero.

2.9 Dot Product

Applications. The dot product has two important applications.

- **The angle formed between two vectors or intersecting lines.** The angle θ between the tails of vectors **A** and **B** in Fig. 2–39 can be determined from Eq. 2–12 and written as

Fig. 2–39 (Repeated)

$$\theta = \cos^{-1}\left(\frac{\mathbf{A} \cdot \mathbf{B}}{AB}\right) \qquad 0° \leq \theta \leq 180°$$

Here $\mathbf{A} \cdot \mathbf{B}$ is found from Eq. 2–13. As a special case, if $\mathbf{A} \cdot \mathbf{B} = 0$, then $\theta = \cos^{-1} 0 = 90°$ so that **A** will be *perpendicular* to **B**.

- **The components of a vector parallel and perpendicular to a line.** The component of vector *A* parallel to or collinear with the line *aa* in Fig. 2–40 is defined by $A_a = A \cos \theta$. This component is sometimes referred to as the **projection** of **A** onto the line, since a *right angle* is formed in the construction. If the *direction* of the line is specified by the unit vector $\mathbf{u}_a$, then since $u_a = 1$, we can determine the magnitude of $\mathbf{A}_a$ directly from the dot product (Eq. 2–12); i.e.,

$$A_a = \mathbf{A} \cdot \mathbf{u}_a = A \cos \theta.$$

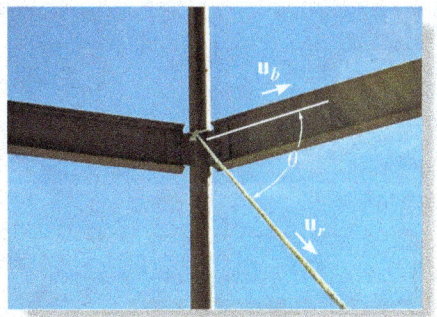

The angle θ between the rope and the beam can be determined by formulating unit vectors along the beam and rope and then using the dot product $\mathbf{u}_b \cdot \mathbf{u}_r = (1)(1) \cos \theta.$

Hence, the scalar projection of **A** along a line is determined from the dot product of **A** and the unit vector $\mathbf{u}_a$ which defines the direction of the line. Notice that if this result is positive, then $\mathbf{A}_a$ has a directional sense which is the same as $\mathbf{u}_a$, whereas if A_a is a negative scalar, then $\mathbf{A}_a$ has the opposite sense of direction to $\mathbf{u}_a$.

The component $\mathbf{A}_a$ represented as a *vector* is therefore

$$\mathbf{A}_a = A_a \mathbf{u}_a$$

The perpendicular component of **A** can also be obtained, Fig. 2–40. Since $\mathbf{A} = \mathbf{A}_a + \mathbf{A}_\perp$, then $\mathbf{A}_\perp = \mathbf{A} - \mathbf{A}_a$. There are two possible ways of obtaining $A_\perp$. One way would be to determine θ from the dot product, $\theta = \cos^{-1}(\mathbf{A} \cdot \mathbf{u}_A / A)$, then $A_\perp = A \sin \theta$. Alternatively, if A_a is known, then by the Pythagorean theorem we can also write $A_\perp = \sqrt{A^2 - A_a^2}$.

Fig. 2–40

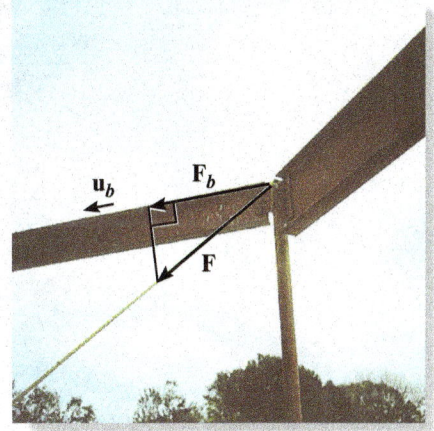

The projection of the cable force **F** along the beam can be determined by first finding the unit vector $\mathbf{u}_b$ that defines this direction. Then apply the dot product, $F_b = \mathbf{F} \cdot \mathbf{u}_b.$

IMPORTANT POINTS

- The dot product is used to determine the angle between two vectors or the projection of a vector in a specified direction.
- If vectors **A** and **B** are expressed in Cartesian vector form, the dot product is determined by multiplying the respective x, y, z scalar components and algebraically adding the results, i.e., $\mathbf{A} \cdot \mathbf{B} = A_x B_x + A_y B_y + A_z B_z$.
- From the definition of the dot product, the angle formed between the tails of vectors **A** and **B** is $\theta = \cos^{-1}(\mathbf{A} \cdot \mathbf{B}/AB)$.
- The magnitude of the projection of vector **A** along a line aa whose direction is specified by $\mathbf{u}_a$ is determined from the dot product $A_a = \mathbf{A} \cdot \mathbf{u}_a$.

EXAMPLE 2.13

Determine the magnitudes of the projections of the force **F** in Fig. 2–41a onto the u and v axes.

Projections of **F**
(a)

Components of **F**
(b)

Fig. 2–41

SOLUTION

Projections of Force. The graphical representation of the *projections* is shown in Fig. 2–41a. From this figure, the magnitudes of the projections of **F** onto the u and v axes can be obtained by trigonometry:

$$(F_u)_{\text{proj}} = (100 \text{ N})\cos 45° = 70.7 \text{ N} \qquad \textit{Ans.}$$

$$(F_v)_{\text{proj}} = (100 \text{ N})\cos 15° = 96.6 \text{ N} \qquad \textit{Ans.}$$

NOTE: These projections are not equal to the magnitudes of the components of force **F** along the u and v axes found from the parallelogram law, Fig. 2–41b. They would only be equal if the u and v axes were *perpendicular* to one another.

EXAMPLE 2.14

The frame shown in Fig. 2-42a is subjected to a horizontal force $\mathbf{F} = \{300\mathbf{j}\}$ N. Determine the magnitudes of the components of this force parallel and perpendicular to member AB.

Fig 2-42

SOLUTION

The magnitude of the projected component of $\mathbf{F}$ along AB is equal to the dot product of $\mathbf{F}$ and the unit vector $\mathbf{u}_B$, which defines the direction of AB, Fig. 2-42b. Since

$$\mathbf{u}_B = \frac{\mathbf{r}_B}{r_B} = \frac{2\mathbf{i} + 6\mathbf{j} + 3\mathbf{k}}{\sqrt{(2)^2 + (6)^2 + (3)^2}} = 0.286\mathbf{i} + 0.857\mathbf{j} + 0.429\mathbf{k}$$

then

$$F_{AB} = F\cos\theta = \mathbf{F} \cdot \mathbf{u}_B = (300\mathbf{j}) \cdot (0.286\mathbf{i} + 0.857\mathbf{j} + 0.429\mathbf{k})$$
$$= (0)(0.286) + (300)(0.857) + (0)(0.429)$$
$$= 257.1 \text{ N} \qquad\qquad Ans.$$

Since the result is a positive scalar, $\mathbf{F}_{AB}$ has the same sense of direction as $\mathbf{u}_B$, Fig. 2-42b.

Expressing $\mathbf{F}_{AB}$ in Cartesian vector form, we have

$$\mathbf{F}_{AB} = F_{AB}\mathbf{u}_B = (257.1 \text{ N})(0.286\mathbf{i} + 0.857\mathbf{j} + 0.429\mathbf{k})$$
$$= \{73.5\mathbf{i} + 220\mathbf{j} + 110\mathbf{k}\}\text{N} \qquad\qquad Ans.$$

The perpendicular component, Fig. 2-42b, is therefore

$$\mathbf{F}_\perp = \mathbf{F} - \mathbf{F}_{AB} = 300\mathbf{j} - (73.5\mathbf{i} + 220\mathbf{j} + 110\mathbf{k})$$
$$= \{-73.5\mathbf{i} + 79.6\mathbf{j} - 110\mathbf{k}\}\text{N}$$

Its magnitude can be determined either from this vector or by using the Pythagorean theorem, Fig. 2-42b:

$$F_\perp = \sqrt{F^2 - F_{AB}^2} = \sqrt{(300 \text{ N})^2 - (257.1 \text{ N})^2}$$
$$= 155 \text{ N} \qquad\qquad Ans.$$

EXAMPLE 2.15

The pipe in Fig. 2–43a is subjected to the force of $F = 80$ lb. Determine the angle θ between **F** and the pipe segment BA and the projection of **F** along this segment.

(a)

SOLUTION

Angle θ. First we will establish position vectors from B to A and B to C; Fig. 2–43b. Then we will determine the angle θ between the tails of these two vectors.

$$\mathbf{r}_{BA} = \{-2\mathbf{i} - 2\mathbf{j} + 1\mathbf{k}\} \text{ ft}, \quad r_{BA} = 3 \text{ ft}$$

$$\mathbf{r}_{BC} = \{-3\mathbf{j} + 1\mathbf{k}\} \text{ ft}, \quad r_{BC} = \sqrt{10} \text{ ft}$$

Thus,

$$\cos\theta = \frac{\mathbf{r}_{BA} \cdot \mathbf{r}_{BC}}{r_{BA} r_{BC}} = \frac{(-2)(0) + (-2)(-3) + (1)(1)}{3\sqrt{10}} = 0.7379$$

$$\theta = 42.5° \qquad \textit{Ans.}$$

Projection of F. The projected component of **F** along BA is shown in Fig. 2–43c. We must first formulate the unit vector along BA and force **F** as Cartesian vectors.

$$\mathbf{u}_{BA} = \frac{\mathbf{r}_{BA}}{r_{BA}} = \frac{(-2\mathbf{i} - 2\mathbf{j} + 1\mathbf{k})}{3} = -\frac{2}{3}\mathbf{i} - \frac{2}{3}\mathbf{j} + \frac{1}{3}\mathbf{k}$$

$$\mathbf{F} = 80 \text{ lb}\left(\frac{\mathbf{r}_{BC}}{r_{BC}}\right) = 80\left(\frac{-3\mathbf{j} + 1\mathbf{k}}{\sqrt{10}}\right) = -75.89\mathbf{j} + 25.30\mathbf{k}$$

Thus,

$$F_{BA} = \mathbf{F} \cdot \mathbf{u}_{BA} = (-75.89\mathbf{j} + 25.30\mathbf{k}) \cdot \left(-\frac{2}{3}\mathbf{i} - \frac{2}{3}\mathbf{j} + \frac{1}{3}\mathbf{k}\right)$$

$$= 0\left(-\frac{2}{3}\right) + (-75.89)\left(-\frac{2}{3}\right) + (25.30)\left(\frac{1}{3}\right)$$

$$= 59.0 \text{ lb} \qquad \textit{Ans.}$$

NOTE: Since θ has been calculated, then also,

$$F_{BA} = F\cos\theta = 80 \text{ lb} \cos 42.5° = 59.0 \text{ lb}$$

(b)

(c)

Fig 2–43

FUNDAMENTAL PROBLEMS

F2–25. Determine the angle θ between the force and the line AO.

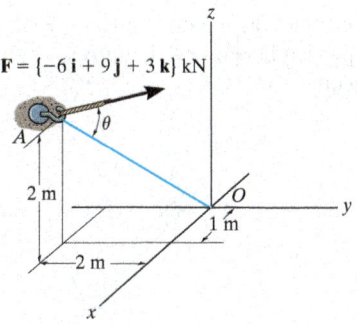

Prob. F2–25

F2–26. Determine the angle θ between the force and the line AB.

Prob. F2–26

F2–27. Determine the angle θ between the force and the line OA.

F2–28. Determine the projected component of the force along the line OA.

Probs. F2–27/28

F2–29. Find the magnitude of the projected component of the force along the pipe AO.

Prob. F2–29

F2–30. Determine the components of the force acting parallel and perpendicular to the axis of the pole.

Prob. F2–30

F2–31. Determine the magnitudes of the components of the force $F = 56$ N acting along and perpendicular to line AO.

Prob. F2–31

PROBLEMS

2–103. Given the three vectors **A**, **B**, and **D**, show that $\mathbf{A} \cdot (\mathbf{B} + \mathbf{D}) = (\mathbf{A} \cdot \mathbf{B}) + (\mathbf{A} \cdot \mathbf{D})$.

*2–104. Determine the angles θ and ϕ between the wire segments.

Prob. 2–104

2–105. Determine the magnitude of the projected component of force **F** along the pole.

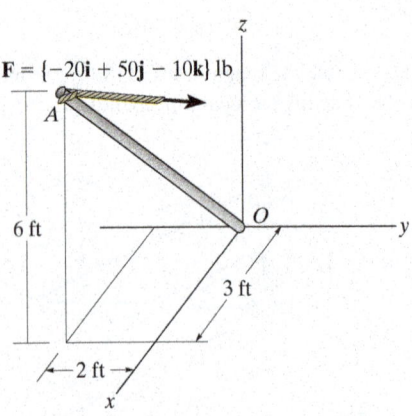

Prob. 2–105

2–106. Determine the components of **F** that act along rod AC and perpendicular to it. Point B is located at the midpoint of the rod.

2–107. Determine the components of **F** that act along rod AC and perpendicular to it. Point B is located 3 m along the rod from end C.

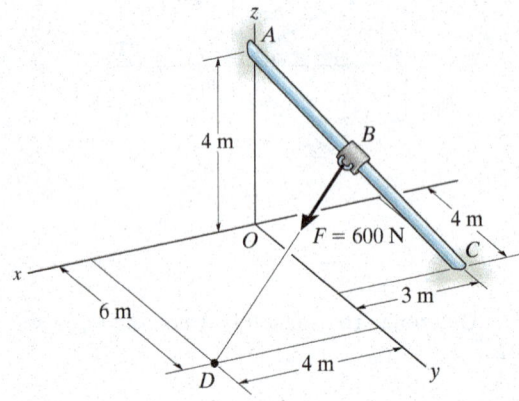

Probs. 2–106/107

*2–108. The force $\mathbf{F} = \{25\mathbf{i} - 50\mathbf{j} + 10\mathbf{k}\}$ lb acts at the end A of the pipe assembly. Determine the magnitude of the components $\mathbf{F}_1$ and $\mathbf{F}_2$, which act along AB and perpendicular to it.

Prob. 2–108

2–109. A force of $\mathbf{F} = \{-40\mathbf{k}\}$ lb acts at the end of the pipe. Determine the magnitudes of the components $\mathbf{F}_1$ and $\mathbf{F}_2$ which are directed along the pipe's axis and perpendicular to it.

2–111. Determine the angle θ between the two cables.

***2–112.** Determine the magnitude of the projection of the force $\mathbf{F}_1$ along cable AC.

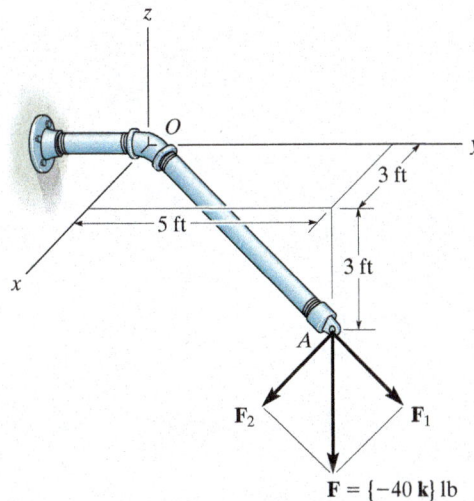

Prob. 2–109

Probs. 2–111/112

2–110. Determine the magnitudes of the components of $F = 600$ N acting along and perpendicular to segment DE of the pipe assembly.

2–113. If $\mathbf{F} = \{16\mathbf{i} + 10\mathbf{j} - 14\mathbf{k}\}$ N, determine the magnitude of the projection of $\mathbf{F}$ along the axis of the pole and perpendicular to it.

Prob. 2–110

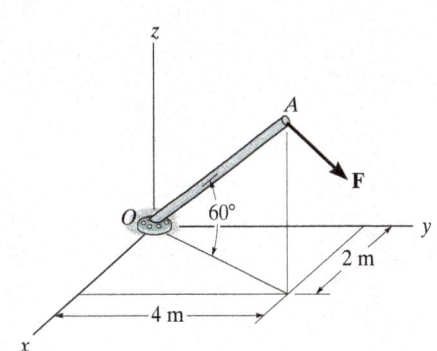

Prob. 2–113

2–114. Determine the magnitude of the projection of $\mathbf{F}_1$ along the line of action of $\mathbf{F}_2$.

2–115. Determine the angle θ between the two cables attached to the pipe.

2–117. Determine the magnitude of the projected component of the 100-lb force acting along the axis BC of the pipe.

2–118 Determine the angle θ between pipe segments BA and BC.

Probs. 2–114/115

Probs. 2–117/118

2–119. Determine the angle θ between the cables AB and AC.

*2–120. Determine the magnitude of the projection of the force $\mathbf{F} = \{400\mathbf{i} - 200\mathbf{j} + 500\mathbf{k}\}$ N acting along the cable BA.

2–121. Determine the magnitude of the projection of the force $\mathbf{F} = \{400\mathbf{i} - 200\mathbf{j} + 500\mathbf{k}\}$ N acting along the cable CA.

*2–116. Determine the magnitudes of the components of the force acting along the axis AB of the wrench handle and perpendicular to it.

Prob. 2–116

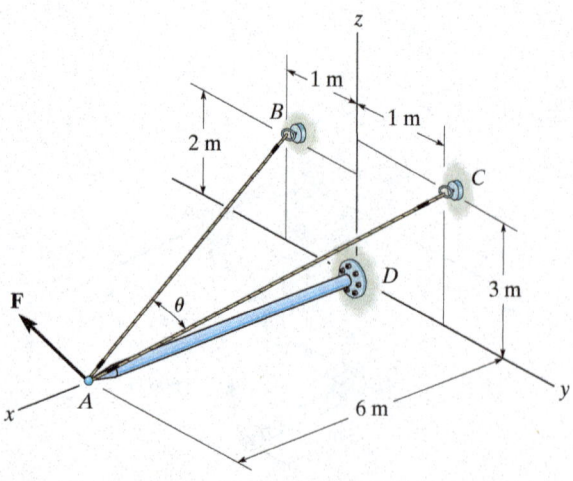

Probs. 2–119/120/121

2–122. Determine the magnitudes of the components of **F** acting along and perpendicular to segment BC of the pipe assembly.

2–123. Determine the magnitude of the projection of **F** along line AC. Express this component as a Cartesian vector.

2–125. Determine the magnitudes of the projection of the force $\mathbf{F} = \{-80\mathbf{i} + 30\mathbf{j} + 20\mathbf{k}\}$ lb in the direction of the cables AB and AC.

Probs. 2–122/123

Prob. 2–125

*2–124. Determine the angle θ between the pipe segments BA and BC.

2–126. Determine the magnitudes of the components of the force acting parallel and perpendicular to diagonal AB of the crate.

Prob. 2–124

Prob. 2–126

2–127. Determine the magnitudes of the projection of the force acting along the x and y axes.

2–129. Determine the angle θ between the two cables.

Prob. 2–127

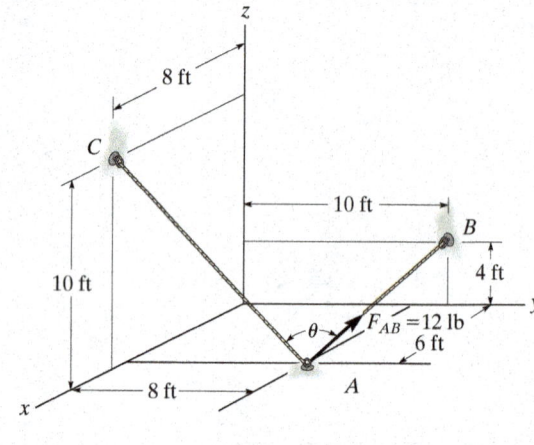

Prob. 2–129

***2–128.** Determine the magnitude of the projection of the force acting along line OA.

2–130. Determine the projection of the force acting in the direction of cable AC. Express the result as a Cartesian vector.

Prob. 2–128

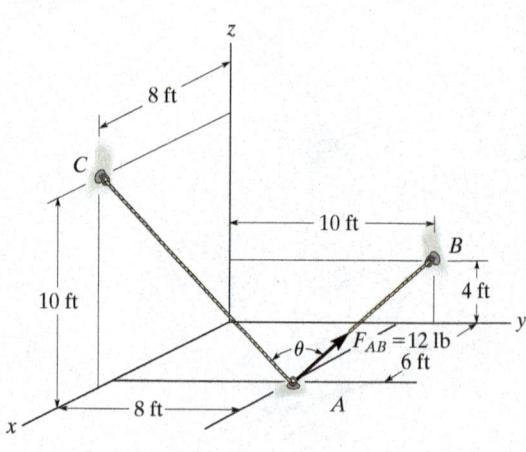

Prob. 2–130

CHAPTER REVIEW

A scalar is a positive or negative number; e.g., mass and temperature.

A vector has a magnitude and direction, where the arrowhead represents the sense of the vector.

Multiplication or division of a vector by a positive scalar will change only the magnitude of the vector. If the scalar is negative, the sense of the vector will also change, so that it acts in the opposite direction.

If vectors are collinear, the resultant is simply the algebraic or scalar addition.

$R = A + B$

Parallelogram Law

Two forces add according to the parallelogram law. The *components* form the sides of the parallelogram and the *resultant* is the diagonal.

To find the components of a force along any two axes, extend lines from the head of the force, parallel to the axes, to form the components.

The force components can be added tip-to-tail using the triangle rule, and then the law of cosines and the law of sines can be used to calculate unknown values.

$$F_R = \sqrt{F_1^2 + F_2^2 - 2 F_1 F_2 \cos \theta_R}$$

$$\frac{F_1}{\sin \theta_1} = \frac{F_2}{\sin \theta_2} = \frac{F_R}{\sin \theta_R}$$

Rectangular Components: Two Dimensions

Vectors $\mathbf{F}_x$ and $\mathbf{F}_y$ are rectangular components of $\mathbf{F}$.

The resultant force is determined from the algebraic sum of its components.

$(F_R)_x = \Sigma F_x$

$(F_R)_y = \Sigma F_y$

$F_R = \sqrt{(F_R)_x^2 + (F_R)_y^2}$

$\theta = \tan^{-1} \left| \dfrac{(F_R)_y}{(F_R)_x} \right|$

Cartesian Vectors

The unit vector $\mathbf{u}$ has a length of 1, no units, and it points in the direction of the vector $\mathbf{F}$.

A force can be resolved into its Cartesian components along the x, y, z axes so that $\mathbf{F} = F_x\mathbf{i} + F_y\mathbf{j} + F_z\mathbf{k}$.

The magnitude of $\mathbf{F}$ is determined from the positive square root of the sum of the squares of the magnitudes of its components.

The coordinate direction angles α, β, γ are determined by formulating a unit vector in the direction of $\mathbf{F}$. The x, y, z components of $\mathbf{u}$ represent $\cos \alpha, \cos \beta, \cos \gamma$.

$\mathbf{u} = \dfrac{\mathbf{F}}{F}$

$F = \sqrt{F_x^2 + F_y^2 + F_z^2}$

$\mathbf{u} = \dfrac{\mathbf{F}}{F} = \dfrac{F_x}{F}\mathbf{i} + \dfrac{F_y}{F}\mathbf{j} + \dfrac{F_z}{F}\mathbf{k}$

$\mathbf{u} = \cos \alpha \, \mathbf{i} + \cos \beta \, \mathbf{j} + \cos \gamma \, \mathbf{k}$

The coordinate direction angles are related, so that only two of the three angles are independent of one another.	$\cos^2 \alpha + \cos^2 \beta + \cos^2 \gamma = 1$	
To find the resultant of a concurrent force system, express each force as a Cartesian vector and add the **i, j, k** components of all the forces in the system.	$\mathbf{F}_R = \Sigma \mathbf{F} = \Sigma F_x \mathbf{i} + \Sigma F_y \mathbf{j} + \Sigma F_z \mathbf{k}$	
Position and Force Vectors A position vector locates one point in space relative to another. The easiest way to formulate the components of a position vector is to determine the distance and direction that one must travel along the x, y, and z directions—going from the tail to the head of the vector.	$\mathbf{r} = (x_B - x_A)\mathbf{i}$ $+ (y_B - y_A)\mathbf{j}$ $+ (z_B - z_A)\mathbf{k}$	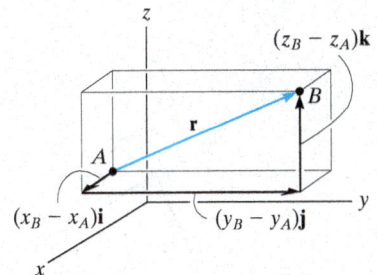
If the line of action of a force passes through points A and B, then the force acts in the same direction **u** as the position vector **r**, extending from A to B. Knowing F and **u**, the force can then be expressed as a Cartesian vector.	$\mathbf{F} = F\mathbf{u} = F\left(\dfrac{\mathbf{r}}{r}\right)$	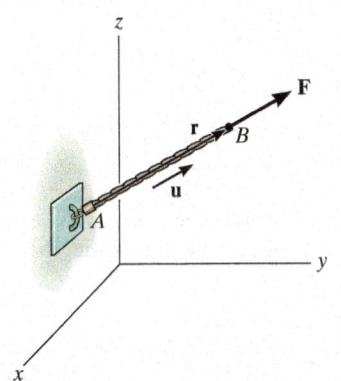
Dot Product The dot product of two vectors **A** and **B** yields a scalar. If **A** and **B** are expressed in Cartesian vector form, then the dot product is the sum of the products of their x, y, and z components.	$\mathbf{A} \cdot \mathbf{B} = AB \cos \theta$ $= A_x B_x + A_y B_y + A_z B_z$	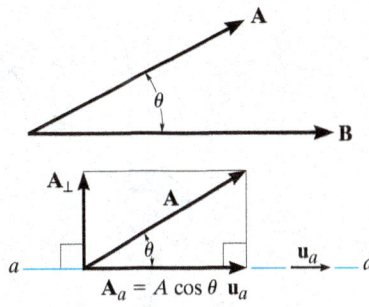
The dot product can be used to determine the angle between **A** and **B**.	$\theta = \cos^{-1}\left(\dfrac{\mathbf{A} \cdot \mathbf{B}}{AB}\right)$	
The dot product is also used to determine the projected component of a vector **A** onto an axis aa defined by its unit vector $\mathbf{u}_a$.	$\mathbf{A}_a = A \cos \theta \, \mathbf{u}_a = (\mathbf{A} \cdot \mathbf{u}_a)\mathbf{u}_a$	

REVIEW PROBLEMS

Partial solutions and answers to all Review Problems are given in the back of the book.

R2–1. Determine the magnitude of the resultant force $\mathbf{F}_R$ and its direction, measured clockwise from the positive u axis.

R2–3. Determine the magnitude of the resultant force acting on the *gusset plate*.

Prob. R2–1

Prob. R2–3

R2–2. Resolve the force into components along the u and v axes and determine the magnitudes of these components.

R2–4. The cable exerts a force of 250 lb on the crane boom. Express this force as a Cartesian vector.

Prob. R2–2

Prob. R2–4

REVIEW PROBLEMS 79

R2–5. The cable attached to the tractor at B exerts a force of 350 lb on the framework. Express this force as a Cartesian vector.

R2–7. Determine the angle θ between the edges of the sheet-metal bracket.

Prob. R2–7

Prob. R2–5

R2–8. Determine the projection of the force **F** along the pole.

R2–6. Express $\mathbf{F}_1$ and $\mathbf{F}_2$ as Cartesian vectors.

Prob. R2–6

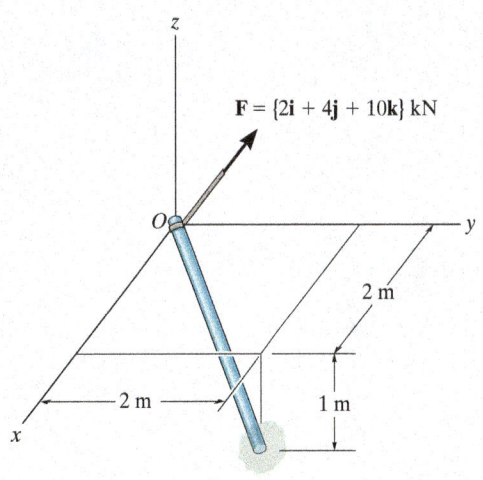

Prob. R2–8

CHAPTER 3

When this load is lifted at constant velocity, or it is just suspended, then it is in a state of equilibrium. In this chapter we will study equilibrium for a particle and show how these ideas can be used to calculate the forces in cables used to hold suspended loads.

EQUILIBRIUM OF A PARTICLE

CHAPTER OBJECTIVES

- To introduce the concept of the free-body diagram.
- To show how to apply the equations of equilibrium to solve particle equilibrium problems.

3.1 CONDITION FOR THE EQUILIBRIUM OF A PARTICLE

A particle is in *equilibrium* if it remains at rest if originally at rest, or has a constant velocity if originally in motion. To maintain equilibrium, it is *necessary* to satisfy Newton's first law of motion, which requires the *resultant force* acting on a particle to be equal to *zero*. This condition is stated by the **equation of equilibrium**,

$$\Sigma \mathbf{F} = \mathbf{0} \qquad (3\text{--}1)$$

where $\Sigma \mathbf{F}$ is the vector *sum of all the forces* acting on the particle.

Not only is Eq. 3–1 a necessary condition for equilibrium, it is also a sufficient condition. This follows from Newton's second law of motion, which can be written as $\Sigma \mathbf{F} = m\mathbf{a}$. Since the force system satisfies Eq. 3–1, then $m\mathbf{a} = \mathbf{0}$, and therefore the particle's acceleration $\mathbf{a} = \mathbf{0}$. Consequently, the particle indeed moves with constant velocity or remains at rest.

3.2 THE FREE-BODY DIAGRAM

To apply the equation of equilibrium, we must account for all the known and unknown forces ($\Sigma \mathbf{F}$) which act on the particle. The best way to do this is to make a drawing of the particle showing the particle isolated or "free" from its surroundings. When this drawing shows *all* the forces that act on the particle, it is called a ***free-body diagram*** *(FBD)*.

Before presenting a formal procedure as to how to draw a free-body diagram, we will first consider three types of supports often encountered in particle equilibrium problems.

(a)

Springs. If a ***linear elastic spring*** (or cord) having an undeformed length l_0 is subjected to a force $\mathbf{F}$, Fig. 3–1a, then the length l of the spring will change by an amount $s = (l - l_0)$ that is directly proportional to the force F. As a result, the relationship between F and s is then

$$F = ks \tag{3–2}$$

Here k is called the ***spring constant*** or ***stiffness***, measured in lb/ft or N/m. If $\mathbf{F}$ pulls on the spring then s represents an elongation, whereas if $\mathbf{F}$ pushes on the spring then s is a contraction.

Cable is in tension
(b)

Fig. 3–1

Cables and Pulleys. A cable can *only* support a tension or "pulling" force, and this force always acts in the direction of the cable. In Chapter 5, it will be shown that when a cable passes over a frictionless pulley then for any angle θ it must have the same constant tension T throughout its length in order to maintain equilibrium, Fig. 3–1b.

Smooth Contact. If an object (or particle) rests on a *smooth surface*, then the surface will exert a force $\mathbf{N}$ on the object that is *normal* to the surface at the point of contact. For example, consider the disk shown in Fig. 3–2a. The normal force $\mathbf{N}$ of the surface on the disk is directed as shown in Fig. 3–2b.

Fig. 3–2

PROCEDURE FOR DRAWING A FREE-BODY DIAGRAM

Since we must account for *all the forces acting on the particle* when applying the equations of equilibrium, the importance of first drawing a free-body diagram cannot be overemphasized. To construct a free-body diagram, the following three steps are necessary.

Draw Outlined Shape.

Imagine the particle to be *isolated* or cut "free" from its surroundings. This requires *removing* all the supports and drawing the particle's outlined shape.

Show All Forces.

Indicate on this sketch *all* the forces that act *on the particle*. These forces can be *active forces*, which tend to set the particle in motion, or they can be *reactive forces*, which are the result of the constraints or supports that tend to prevent motion. To account for all these forces, it may be helpful to trace around the particle's boundary, carefully noting each force acting on it.

Identify Each Force.

The forces that are *known* should be labeled with their proper magnitudes and directions. Letters are used to represent the magnitudes and directions of forces that are unknown.

The bucket is suspended by the cable, and instinctively we know that for equilibrium, the force in the cable must equal the weight of the bucket. By drawing a free-body diagram of the bucket we can understand why this is so. This diagram shows that there are only two forces *acting on the bucket*, namely, its weight **W** and the force **T** of the cable, and since the resultant of these forces must be equal to zero, then $T = W$.

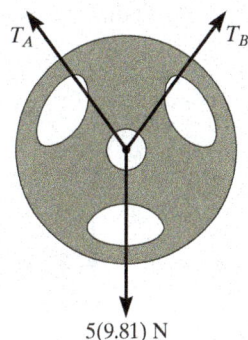

The 5-kg plate is suspended by two straps A and B. To find the force in each strap we should consider the free-body diagram of the plate. As noted, the three forces acting on it are concurrent and therefore they can be applied to a fictitious particle at the center of the plate.

EXAMPLE 3.1

The 6-kg sphere in Fig. 3–3a is supported as shown. Draw a free-body diagram of the sphere, the cord CE, and the knot at C.

SOLUTION

Sphere. There are four forces acting on the sphere, namely, its weight, 6 kg (9.81 m/s^2) = 58.9 N, the pulling or tension force of cord CE, and the two normal forces caused by the smooth inclined planes. The free-body diagram is shown in Fig. 3–3b.

Cord CE. When the cord CE is isolated, its free-body diagram shows only two forces acting on it, namely, the force of the sphere and the force of the knot at C, Fig. 3–3c. Notice that F_{CE} shown here is equal but opposite to that shown in Fig. 3–3b, a consequence of Newton's third law of action–reaction. Also, F_{CE} and F_{EC} both *pull on the cord* and keep it in *tension* so that it doesn't collapse. For equilibrium, $F_{CE} = F_{EC}$.

Knot. The knot at C is subjected to three forces, Fig. 3–3d. They are caused by the cords CBA and CE and the spring CD. As required, the free-body diagram shows all these forces labeled with their magnitudes and directions.

Fig. 3–3

3.3 COPLANAR FORCE SYSTEMS

If a particle is subjected to a system of coplanar forces that lie in the x–y plane, as in Fig. 3–4, then each force can be resolved into its **i** and **j** components. For equilibrium, these forces must sum to produce a zero force resultant, i.e.,

$$\Sigma \mathbf{F} = \mathbf{0}$$
$$\Sigma F_x \mathbf{i} + \Sigma F_y \mathbf{j} = \mathbf{0}$$

For this vector equation to be satisfied, the resultant force's x and y components must both be equal to zero. Therefore,

$$\Sigma F_x = 0$$
$$\Sigma F_y = 0 \qquad (3\text{–}3)$$

Fig. 3–4

These two equations can be solved for at most two unknowns, generally represented as angles and magnitudes of forces shown on the particle's free-body diagram.

When applying the equations of equilibrium, we must account for the sense of direction of any component by using an *algebraic sign* which corresponds to the arrowhead direction of the component along the x or y axis. When the force has an *unknown magnitude*, then the arrowhead sense of the force on the free-body diagram can be *assumed*. If the *solution* yields a *negative scalar* for this magnitude, then this will indicate that the sense of the force is opposite to that which was assumed. For example, consider the free-body diagram of the particle subjected to the two forces shown in Fig. 3–5. Here it is *assumed* that the *unknown force* **F** acts to the right to maintain equilibrium. Applying the equation of equilibrium along the x axis, we have

$$\xrightarrow{+} \Sigma F_x = 0; \qquad +F + 10 \text{ N} = 0$$

When this equation is solved, $F = -10$ N, and so the *negative sign* indicates that **F** must act to the left, Fig. 3–5. Notice that if the +x axis in Fig. 3–5 were directed to the left, both terms in the above equation would be negative, but again, after solving, $F = -10$ N, indicating that **F** would have to be directed to the left.

Fig. 3–5

IMPORTANT POINTS

The first step in solving any equilibrium problem is to draw the particle's free-body diagram. This requires *removing all the supports* and isolating or freeing the particle from its surroundings and then showing all the forces that act on the particle.

Equilibrium means the particle is at rest or moving at constant velocity. In two dimensions, the necessary and sufficient conditions for equilibrium require $\Sigma F_x = 0$ and $\Sigma F_y = 0$.

PROCEDURE FOR ANALYSIS

Coplanar force equilibrium problems for a particle can be solved using the following procedure.

Free-Body Diagram.
- Label all the known and unknown force magnitudes and directions measured from the x, y axes.
- The directional sense of a force having an unknown magnitude can be assumed.

Equations of Equilibrium.
- Apply the equations of equilibrium, $\Sigma F_x = 0$ and $\Sigma F_y = 0$. For convenience, arrows can be written alongside each equation to define the positive directions.
- Components are positive if they are directed along a positive axis, and negative if they are directed along a negative axis.
- If more than two unknowns exist and the problem involves a spring, apply $F = ks$ to relate the spring force to the displacement s of the spring.
- Since the magnitude of a force is always a positive quantity, then if the solution for a force yields a negative result, this indicates that its sense is the reverse of that shown on the free-body diagram.

The chains exert three forces on the ring at A, as shown on its free-body diagram. The ring will not move, or will move with constant velocity, provided the summation of these forces along the x and along the y axis equals zero. If one of the three forces is known, the magnitudes of the other two forces can be obtained from the two equations of equilibrium.

EXAMPLE 3.2

Determine the tension in cables BA and BC necessary to support the 60-kg cylinder in Fig. 3–6a.

(a)

(b)

SOLUTION

Free-Body Diagram. Equilibrium of the cylinder requires the tension in cable BD to be $T_{BD} = 60(9.81)$ N, Fig. 3–6b. The forces in cables BA and BC can be determined by investigating the equilibrium of ring B. Its free-body diagram is shown in Fig. 3–6c. Here the magnitudes of $\mathbf{T}_A$ and $\mathbf{T}_C$ are unknown, but their directions are known.

Equations of Equilibrium. Applying the equations of equilibrium along the x and y axes, we have

$$\xrightarrow{+} \Sigma F_x = 0; \quad T_C \cos 45° - \left(\tfrac{4}{5}\right) T_A = 0 \tag{1}$$

$$+\uparrow \Sigma F_y = 0; \quad T_C \sin 45° + \left(\tfrac{3}{5}\right) T_A - 60(9.81) \text{ N} = 0 \tag{2}$$

Equation (1) can be written as $T_A = 0.8839 T_C$. Substituting this into Eq. (2) yields

$$T_C \sin 45° + \left(\tfrac{3}{5}\right)(0.8839 T_C) - 60(9.81) \text{ N} = 0$$

so that

$$T_C = 475.66 \text{ N} = 476 \text{ N} \qquad \textit{Ans.}$$

Substituting this result into either Eq. (1) or Eq. (2), we get

$$T_A = 420 \text{ N} \qquad \textit{Ans.}$$

NOTE: It is important to realize how this problem was solved. The forces in the cables are all *internal* in Fig. 3–6a. By drawing the free-body diagram of the ring, Fig. 3–6c, we have "exposed" these forces as "external" to the ring, and thereby have been able to obtain the results by applying the conditions of equilibrium.

(c)

Fig. 3–6

EXAMPLE 3.3

Ropes AB and AC in Fig. 3–7a can each support a maximum force of 10 kN. If AB always remains horizontal, determine the smallest angle θ to which the 200-kg crate can be suspended before one of the ropes breaks.

(a)

(b)

Fig. 3–7

SOLUTION

Free-Body Diagram. We will study the equilibrium of ring A, which is subjected to three forces. As shown in Fig. 3–7b, the magnitude of $\mathbf{F}_D$ is equal to the weight of the crate, i.e., $F_D = 200(9.81)\text{N} = 1962 \text{ N} < 10 \text{ kN}$.

Equations of Equilibrium. Applying the equations of equilibrium along the x and y axes,

$$\xrightarrow{+} \Sigma F_x = 0; \qquad -F_C \cos\theta + F_B = 0; \quad F_C = \frac{F_B}{\cos\theta} \qquad (1)$$

$$+\uparrow \Sigma F_y = 0; \qquad F_C \sin\theta - 1962 \text{ N} = 0 \qquad (2)$$

From Eq. (1), F_C is always greater than F_B since $\cos\theta \leq 1$. Therefore, rope AC will reach the maximum tensile force of 10 kN *before* rope AB. Substituting $F_C = 10$ kN into Eq. (2), we get

$$[10(10^3) \text{ N}] \sin\theta - 1962 \text{ N} = 0$$

$$\theta = \sin^{-1}(0.1962) = 11.31° = 11.3° \qquad \textit{Ans.}$$

The force developed in rope AB can be obtained by substituting the values for θ and F_C into Eq. (1).

$$10(10^3) \text{ N} = \frac{F_B}{\cos 11.31°}$$

$$F_B = 9.81 \text{ kN}$$

EXAMPLE 3.4

Determine the required length of cord AC in Fig. 3–8a so that the 8-kg lamp can be suspended in the position shown. The *unstretched* length of spring AB is $l'_{AB} = 0.4$ m, and the spring has a stiffness of $k_{AB} = 300$ N/m.

(a)

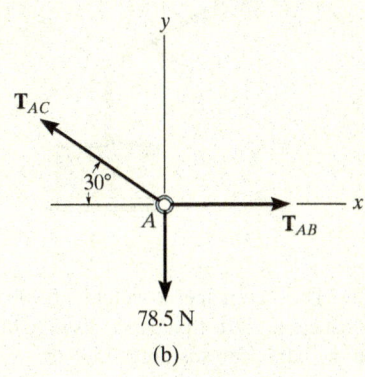

(b)

Fig. 3–8

SOLUTION

If the force in spring AB is known, then the stretch of the spring can be found using $F = ks$. From the problem geometry, it is then possible to calculate the required length of AC.

Free-Body Diagram. The lamp has a weight $W = 8(9.81) = 78.5$ N and so the free-body diagram of the ring at A is shown in Fig. 3–8b.

Equations of Equilibrium. Using the x, y axes,

$\xrightarrow{+} \Sigma F_x = 0;$ $\quad T_{AB} - T_{AC} \cos 30° = 0$
$+\uparrow \Sigma F_y = 0;$ $\quad T_{AC} \sin 30° - 78.5 \text{ N} = 0$

Solving, we obtain

$$T_{AC} = 157.0 \text{ N}$$
$$T_{AB} = 135.9 \text{ N}$$

The stretch of spring AB is therefore

$$T_{AB} = k_{AB} s_{AB}; \quad 135.9 \text{ N} = 300 \text{ N/m}(s_{AB})$$
$$s_{AB} = 0.453 \text{ m}$$

And so the stretched length is

$$l_{AB} = l'_{AB} + s_{AB}$$
$$l_{AB} = 0.4 \text{ m} + 0.453 \text{ m} = 0.853 \text{ m}$$

The horizontal distance from C to B, Fig. 3–8a, requires

$$2 \text{ m} = l_{AC} \cos 30° + 0.853 \text{ m}$$
$$l_{AC} = 1.32 \text{ m} \qquad \textit{Ans.}$$

FUNDAMENTAL PROBLEMS

All solutions must include a free-body diagram.

F3–1. The crate has a weight of 550 lb. Determine the force in each supporting cable.

Prob. F3–1

F3–2. The beam has a weight of 700 lb. Determine the shortest cable ABC that can be used to lift it if the maximum force the cable can sustain is 1500 lb.

Prob. F3–2

F3–3. If the 5-kg block is suspended from the pulley B and the sag of the cord is $d = 0.15$ m, determine the force in cord ABC. Neglect the size of the pulley.

Prob. F3–3

F3–4. The block has a mass of 5 kg and rests on the smooth plane. Determine the unstretched length of the spring.

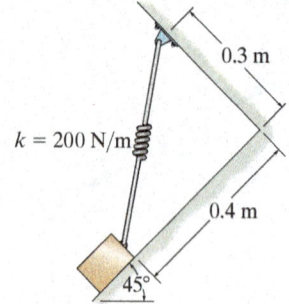

Prob. F3–4

F3–5. If the mass of cylinder C is 40 kg, determine the mass of cylinder A in order to hold the assembly in the position shown.

Prob. F3–5

F3–6. Determine the tension in cables AB, BC, and CD, necessary to support the 10-kg and 15-kg traffic lights at B and C, respectively. Also, find the angle θ.

Prob. F3–6

PROBLEMS

All solutions must include a free-body diagram.

3–1. The members of a truss are pin connected at joint O. Determine the magnitudes of $\mathbf{F}_1$ and $\mathbf{F}_2$ for equilibrium. Set $\theta = 60°$.

3–2. The members of a truss are pin connected at joint O. Determine the magnitude of $\mathbf{F}_1$ and its angle θ for equilibrium. Set $F_2 = 6$ kN.

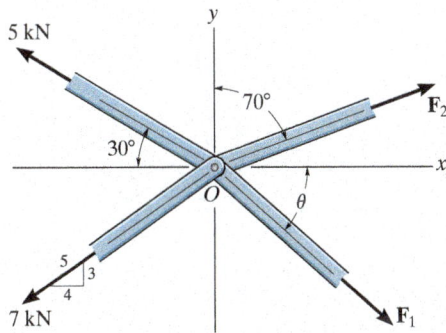

Probs. 3–1/2

3–3. Knowing the forces in members A and C, determine the forces $\mathbf{F}_B$ and $\mathbf{F}_D$ acting on members B and D that are required for equilibrium. The force system is concurrent at point O.

Prob. 3–3

*3–4. Determine the maximum weight W that can be supported in the position shown if each chain AC and AB can support a maximum tension of 600 lb before it fails.

Prob. 3–4

3–5. If the forces are concurrent at point O, determine the magnitudes of $\mathbf{F}$ and $\mathbf{T}$ for equilibrium. Take $\theta = 90°$.

3–6. Determine the force in member C and its angle θ for equilibrium. The forces are concurrent at point O. Take $F = 8$ kN.

Probs. 3–5/6

3–7. The man attempts to pull on the tree using the cable and *small* pulley arrangement shown. If the tension in AB is 60 lb, determine the tension in cable CAD and the angle θ.

3–9. The 30-kg block is supported by two springs having the stiffness shown. Determine the unstretched length of each spring after the block is removed.

Prob. 3–9

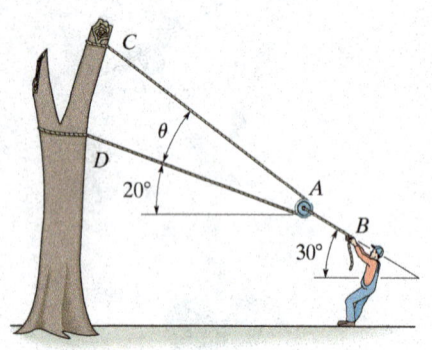

Prob. 3–7

3–10. If the spring DB has an unstretched length of 2 m, determine the stiffness of the spring to hold the 40-kg crate in the position shown.

3–11. Determine the unstretched length of DB to hold the 40-kg crate in the position shown. Take $k = 180$ N/m.

*3–8. Determine the force in cables AB and AC necessary to support the 12-kg traffic light.

Prob. 3–8

Probs. 3–10/11

*3–12. Determine the force in each cord for equilibrium of the 60-kg bucket.

3–14. Determine the stretch in each spring for equilibrium of the 2-kg block. The springs are shown in the equilibrium position.

3–15. The unstretched length of spring AB is 3 m. If the block is held in the equilibrium position, determine the mass of the block at D.

Prob. 3–12

Probs. 3–14/15

3–13. The lift sling is used to hoist a container having a mass of 500 kg. Determine the force in each of the cables AB and AC as a function of θ. If the maximum tension allowed in each cable is 5 kN, determine the shortest length of cables AB and AC that can be used for the lift. The center of gravity of the container is located at G.

*3–16. Determine the stretch of each spring for equilibrium of the 20-kg cylinder.

Prob. 3–13

Prob. 3–16

3–17. The spring has a stiffness of $k = 800$ N/m and an unstretched length of 200 mm. Determine the force in cables *BC* and *BD* when the spring is held in the position shown.

*3–20. The 10-lb weight is supported by the cord *AC* and roller and by the spring that has a stiffness of $k = 10$ lb/in. and an unstretched length of 12 in. Determine the distance *d* to maintain equilibrium.

3–21. The 10-lb weight is supported by the cord *AC* and roller and by the spring. If the spring has an unstretched length of 8 in., and the weight is in equilibrium when $d = 4$ in., determine the stiffness *k* of the spring.

Probs. 3–20/21

Prob. 3–17

3–18. The lamp has a weight of 15 lb and is supported by the six cords connected together as shown. Determine the tension in each cord and the angle θ for equilibrium. Cord *BC* is horizontal.

3–19. Each cord can sustain a maximum tension of 20 lb. Determine the largest weight of the lamp that can be supported. Also, determine θ of cord *DC* for equilibrium.

3–22. The springs *BA* and *BC* each have a stiffness of 500 N/m and an unstretched length of 3 m. Determine the horizontal force **F** applied to the cord which is attached to the small ring *B* so that the displacement of *AB* from the wall is $d = 1.5$ m.

3–23. The springs *BA* and *BC* each have a stiffness of 500 N/m and an unstretched length of 3 m. Determine the displacement *d* of the cord from the wall when a force $F = 175$ N is applied to the cord.

Probs. 3–18/19

Probs. 3–22/23

*3–24. Determine the distances x and y for equilibrium if $F_1 = 800$ N and $F_2 = 1000$ N.

3–25. Determine the magnitude of F_1 and the distance y if $x = 1.5$ m and $F_2 = 1000$ N.

*3–28. If the cords suspend the two buckets in the equilibrium position, determine the weight of bucket B. Bucket A has a weight of 60 lb.

Probs. 3–24/25

Prob. 3–28

3–26. The 30-kg pipe is supported at A by a system of five cords. Determine the force in each cord for equilibrium.

3–27. Each cord can sustain a maximum tension of 500 N. Determine the largest mass of pipe that can be supported.

3–29. If the bucket weighs 50 lb, determine the tension developed in each of the wires.

3–30. Determine the maximum weight of the bucket that can be supported so that no single wire develops a tension exceeding 100 lb.

Probs. 3–26/27

Probs. 3–29/30

3–31. Determine the forces in cables AC and AB needed to hold the 20-kg cylinder D in equilibrium. Set F = 300 N and d = 1 m.

*3–32. The cylinder D has a mass of 20 kg. If a force of F = 100 N is applied horizontally to the ring at A, determine the largest dimension d so that the force in cable AC is zero.

3–35. The ring of negligible size is subjected to a vertical force of 200 lb. Determine the required length l of cord AC such that the tension acting in AC is 160 lb. Also, what is the force in cord AB? *Hint*: Use the equilibrium condition to determine the required angle θ for attachment, then determine l using trigonometry applied to triangle ABC.

Prob. 3–35

Probs. 3–31/32

*3–36. If the bucket and its contents have a mass of 10 kg and are suspended from the cable by means of a small pulley at C, determine the location x of the pulley for equilibrium. The cable is 6 m long.

Prob. 3–36

3–33. Blocks D and F weigh 5 lb each and block E weighs 8 lb. Determine the sag s for equilibrium. Neglect the size of the pulleys.

3–34. If blocks D and F weigh 5 lb each, determine the weight of block E if the sag s = 3 ft. Neglect the size of the pulleys.

3–37. A 4-kg sphere rests on the smooth parabolic surface. Determine the normal force it exerts on the surface and the mass m_B of block B needed to hold it in the equilibrium position.

Probs. 3–33/34

Prob. 3–37

3–38. The block has a weight of 20 lb and is being hoisted at uniform velocity. Determine the angle θ for equilibrium and the force in each cord.

3–39. Determine the maximum weight W of the block that can be hoisted at constant velocity if each cord can support a maximum tension of 80 lb. Also, what is the angle θ for equilibrium?

3–41. If the mass of the block at A is 20 kg, determine the mass of the block at B and at C for equilibrium. Also, determine the angles θ and ϕ.

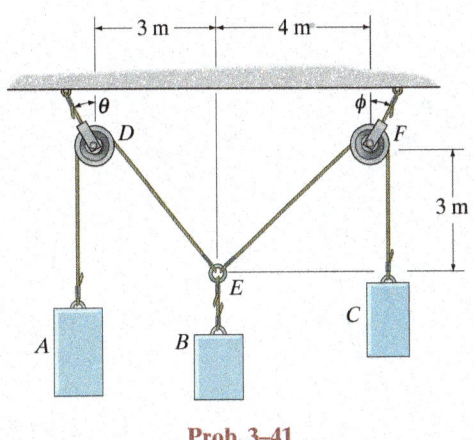

Prob. 3–41

Probs. 3–38/39

*3–40. The 200-lb uniform container is suspended by a 6-ft-long cable, which is attached to the sides of the container and passes over the small pulley located at O. If the cable can be attached at either points A and B, or C and D, determine which attachment produces the least amount of tension in the cable. What is this tension?

3–42. A "scale" is constructed with a 4-ft-long cord and the 10-lb block D. The cord is fixed to a pin at A and passes over two *small* pulleys. Determine the weight of the suspended block B if the system is in equilibrium when $s = 1.5$ ft.

Prob. 3–40

Prob. 3–42

3.4 THREE-DIMENSIONAL FORCE SYSTEMS

In Section 3.1 we stated that the necessary and sufficient condition for particle equilibrium is

$$\Sigma \mathbf{F} = \mathbf{0} \quad (3\text{–}4)$$

In the case of a three-dimensional force system, as in Fig. 3–9, we can resolve the forces into their respective $\mathbf{i}$, $\mathbf{j}$, $\mathbf{k}$ components, so that $\Sigma F_x \mathbf{i} + \Sigma F_y \mathbf{j} + \Sigma F_z \mathbf{k} = \mathbf{0}$. To satisfy this equation we require

$$\begin{aligned} \Sigma F_x &= 0 \\ \Sigma F_y &= 0 \\ \Sigma F_z &= 0 \end{aligned} \quad (3\text{–}5)$$

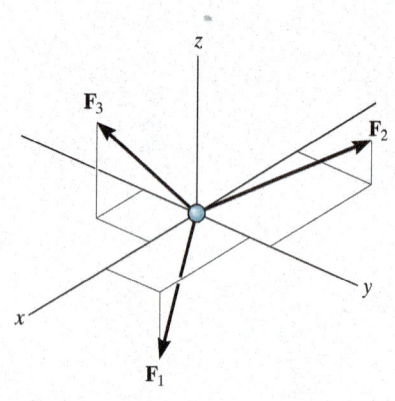

Fig. 3–9

These three equations state that the *algebraic sum* of the components of all the forces acting on the particle along each of the coordinate axes must be zero. Using them we can solve for at most three unknowns, generally represented as coordinate direction angles or magnitudes of forces shown on the particle's free-body diagram.

PROCEDURE FOR ANALYSIS

Three-dimensional force equilibrium problems for a particle can be solved using the following procedure.

Free-Body Diagram.

- Label all the known and unknown force magnitudes and directions measured from the x, y, z axes.

- The directional sense of a force having an unknown magnitude can be assumed.

Equations of Equilibrium.

- Use the scalar equations of equilibrium, $\Sigma F_x = 0$, $\Sigma F_y = 0$, $\Sigma F_z = 0$, in cases where it is easy to resolve each force into its x, y, z components.

- If the three-dimensional geometry appears difficult, then first express each force on the free-body diagram as a Cartesian vector, substitute these vectors into $\Sigma \mathbf{F} = \mathbf{0}$, and then set the $\mathbf{i}$, the $\mathbf{j}$, and the $\mathbf{k}$ components each equal to zero.

- If the solution for a force yields a negative result, this indicates that its sense is the reverse of that shown on the free-body diagram.

The joint at A is subjected to the force from the support as well as forces from each of the three chains. If the tire and any load on it have a weight W, then the force at the support will be W, and the three scalar equations of equilibrium can be applied to the free-body diagram of the joint in order to determine the chain forces, $\mathbf{F}_B$, $\mathbf{F}_C$, and $\mathbf{F}_D$.

EXAMPLE 3.5

If the 90-lb load in Fig. 3–10a is supported by two cables and a spring having a stiffness $k = 500$ lb/ft, determine the force in the cables and the stretch of the spring for equilibrium. Cable AD lies in the x–y plane and cable AC lies in the x–z plane.

SOLUTION
The stretch of the spring can be determined once the force in the spring is determined.

Free-Body Diagram. The connection at A is chosen for the equilibrium analysis since the cable forces are concurrent at this point. The free-body diagram is shown in Fig. 3–10b.

Equations of Equilibrium. By inspection, each force can easily be resolved into its x, y, z components, and so the three scalar equations of equilibrium will be used. We have

$\Sigma F_x = 0;$ $\quad F_D \sin 30° - \left(\frac{4}{5}\right) F_C = 0$ $\qquad$ (1)

$\Sigma F_y = 0;$ $\quad -F_D \cos 30° + F_B = 0$ $\qquad$ (2)

$\Sigma F_z = 0;$ $\quad \left(\frac{3}{5}\right) F_C - 90 \text{ lb} = 0$ $\qquad$ (3)

Solving Eq. (3) for F_C, then Eq. (1) for F_D, and finally Eq. (2) for F_B, yields

$$F_C = 150 \text{ lb} \qquad \textit{Ans.}$$

$$F_D = 240 \text{ lb} \qquad \textit{Ans.}$$

$$F_B = 207.8 \text{ lb} = 208 \text{ lb} \qquad \textit{Ans.}$$

The stretch of the spring is therefore

$$F_B = k s_{AB}$$

$$207.8 \text{ lb} = (500 \text{ lb/ft})(s_{AB})$$

$$s_{AB} = 0.416 \text{ ft} \qquad \textit{Ans.}$$

NOTE: Since the results for all the cable forces are positive, each cable is in tension; that is, it pulls on point A as expected, Fig. 3–10b.

(a)

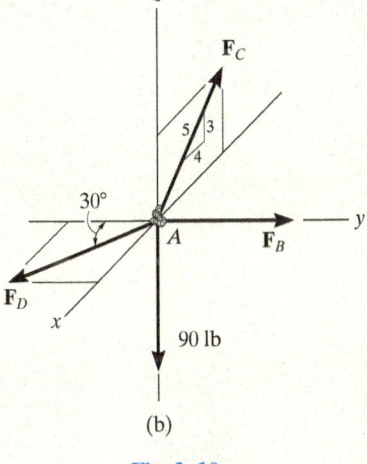

(b)

Fig. 3–10

EXAMPLE 3.6

The 10-kg lamp in Fig. 3–11a is suspended from the three equal-length cords. Determine the smallest vertical distance s from the ceiling if the force developed in any cord is not allowed to exceed 50 N.

(a)

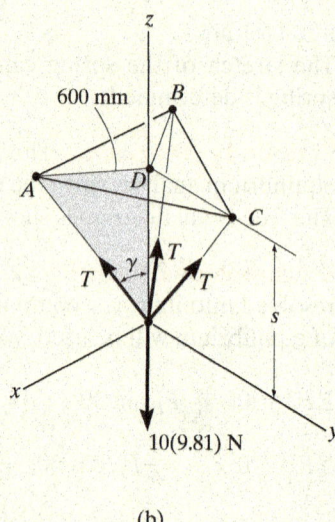
(b)

Fig. 3–11

SOLUTION

Free-Body Diagram. Due to symmetry, Fig. 3–11b, the distance $DA = DB = DC = 600$ mm, and also the tension T in each cord will be the same, and the angle between each cord and the z axis is γ.

Equation of Equilibrium. Applying the equilibrium equation along the z axis, with $T = 50$ N, we have

$\Sigma F_z = 0;$ $\qquad 3[(50 \text{ N}) \cos \gamma] - 10(9.81) \text{ N} = 0$

$$\gamma = \cos^{-1} \frac{98.1}{150} = 49.16°$$

From the shaded triangle shown in Fig. 3–11b,

$$\tan 49.16° = \frac{600 \text{ mm}}{s}$$

$$s = 519 \text{ mm} \qquad\qquad Ans.$$

EXAMPLE 3.7

Determine the force in each cable used to support the 40-lb crate shown in Fig. 3–12a.

SOLUTION

Free-Body Diagram. As shown in Fig. 3–12b, the free-body diagram of point A is considered in order to "expose" the three unknown forces in the cables.

Equations of Equilibrium. First we will express each force as a Cartesian vector. Since the coordinates of points B and C are $B(-3 \text{ ft}, -4 \text{ ft}, 8 \text{ ft})$ and $C(-3 \text{ ft}, 4 \text{ ft}, 8 \text{ ft})$, we have

$$\mathbf{F}_B = F_B \left[\frac{-3\mathbf{i} - 4\mathbf{j} + 8\mathbf{k}}{\sqrt{(-3)^2 + (-4)^2 + (8)^2}} \right]$$

$$= -0.318 F_B \mathbf{i} - 0.424 F_B \mathbf{j} + 0.848 F_B \mathbf{k}$$

$$\mathbf{F}_C = F_C \left[\frac{-3\mathbf{i} + 4\mathbf{j} + 8\mathbf{k}}{\sqrt{(-3)^2 + (4)^2 + (8)^2}} \right]$$

$$= -0.318 F_C \mathbf{i} + 0.424 F_C \mathbf{j} + 0.848 F_C \mathbf{k}$$

$$\mathbf{F}_D = F_D \mathbf{i}$$

$$\mathbf{W} = \{-40\mathbf{k}\} \text{ lb}$$

Equilibrium requires

$$\Sigma \mathbf{F} = \mathbf{0}; \qquad \mathbf{F}_B + \mathbf{F}_C + \mathbf{F}_D + \mathbf{W} = \mathbf{0}$$

$$(-0.318 F_B \mathbf{i} - 0.424 F_B \mathbf{j} + 0.848 F_B \mathbf{k}) +$$
$$(-0.318 F_C \mathbf{i} + 0.424 F_C \mathbf{j} + 0.848 F_C \mathbf{k}) + F_D \mathbf{i} - 40\mathbf{k} = \mathbf{0}$$

Equating the respective $\mathbf{i}, \mathbf{j}, \mathbf{k}$ components to zero yields

$$\Sigma F_x = 0; \qquad -0.318 F_B - 0.318 F_C + F_D = 0 \qquad (1)$$

$$\Sigma F_y = 0; \qquad -0.424 F_B + 0.424 F_C = 0 \qquad (2)$$

$$\Sigma F_z = 0; \qquad 0.848 F_B + 0.848 F_C - 40 = 0 \qquad (3)$$

Equation (2) states that $F_B = F_C$. Thus, solving Eq. (3) for F_B and F_C and substituting the result into Eq. (1) to obtain F_D, we get

$$F_B = F_C = 23.6 \text{ lb} \qquad \textit{Ans.}$$

$$F_D = 15.0 \text{ lb} \qquad \textit{Ans.}$$

(a)

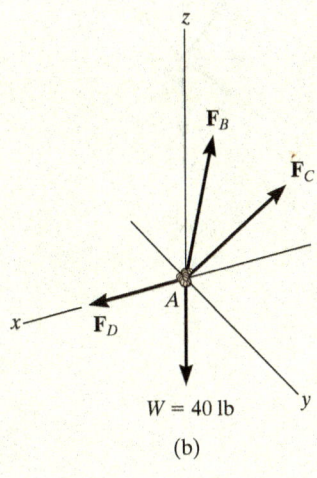

(b)

Fig. 3–12

EXAMPLE 3.8

Determine the tension in each cord used to support the 100-kg crate shown in Fig. 3–13a.

SOLUTION

Free-Body Diagram. The force in each of the cords can be determined by investigating the equilibrium of point A, Fig. 3–13b. Here the weight of the crate is $W = 100(9.81) = 981$ N.

Equations of Equilibrium. Each force on the free-body diagram will first be expressed as a Cartesian vector. Using Eq. 2–9 for $\mathbf{F}_C$ and noting point $D(-1 \text{ m}, 2 \text{ m}, 2 \text{ m})$ for $\mathbf{F}_D$, we have

$$\mathbf{F}_B = F_B \mathbf{i}$$

$$\mathbf{F}_C = F_C \cos 120° \mathbf{i} + F_C \cos 135° \mathbf{j} + F_C \cos 60° \mathbf{k}$$
$$= -0.5 F_C \mathbf{i} - 0.707 F_C \mathbf{j} + 0.5 F_C \mathbf{k}$$

$$\mathbf{F}_D = F_D \left[\frac{-1\mathbf{i} + 2\mathbf{j} + 2\mathbf{k}}{\sqrt{(-1)^2 + (2)^2 + (2)^2}} \right]$$
$$= -0.333 F_D \mathbf{i} + 0.667 F_D \mathbf{j} + 0.667 F_D \mathbf{k}$$

$$\mathbf{W} = \{-981\mathbf{k}\} \text{ N}$$

Equilibrium requires

$$\Sigma \mathbf{F} = \mathbf{0}; \qquad \mathbf{F}_B + \mathbf{F}_C + \mathbf{F}_D + \mathbf{W} = \mathbf{0}$$

$$(F_B \mathbf{i} - 0.5 F_C \mathbf{i} - 0.707 F_C \mathbf{j} + 0.5 F_C \mathbf{k}) +$$
$$(-0.333 F_D \mathbf{i} + 0.667 F_D \mathbf{j} + 0.667 F_D \mathbf{k}) - 981 \mathbf{k} = \mathbf{0}$$

Equating the respective $\mathbf{i}, \mathbf{j}, \mathbf{k}$ components to zero,

$$\Sigma F_x = 0; \qquad F_B - 0.5 F_C - 0.333 F_D = 0 \qquad (1)$$

$$\Sigma F_y = 0; \qquad -0.707 F_C + 0.667 F_D = 0 \qquad (2)$$

$$\Sigma F_z = 0; \qquad 0.5 F_C + 0.667 F_D - 981 = 0 \qquad (3)$$

Solving Eq. 2 for F_D in terms of F_C and substituting this into Eq. 3 yields F_C. F_D is then determined from Eq. 2. Finally, substituting the results into Eq. 1 gives F_B. Hence,

$$F_C = 813 \text{ N} \qquad \textit{Ans.}$$

$$F_D = 862 \text{ N} \qquad \textit{Ans.}$$

$$F_B = 694 \text{ N} \qquad \textit{Ans.}$$

(a)

(b)

Fig. 3–13

FUNDAMENTAL PROBLEMS

All solutions must include a free-body diagram.

F3–7. Determine the tension developed in cables AB, AC, and AD.

Prob. F3–7

F3–8. Determine the tension developed in cables AB, AC, and AD.

Prob. F3–8

F3–9. Determine the magnitude of forces $\mathbf{F}_1$, $\mathbf{F}_2$, $\mathbf{F}_3$, so that the particle is held in equilibrium.

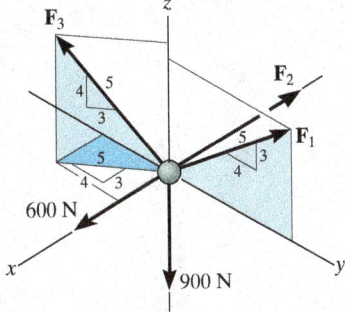

Prob. F3–9

F3–10. Determine the tension developed in cables AB, AC, and AD.

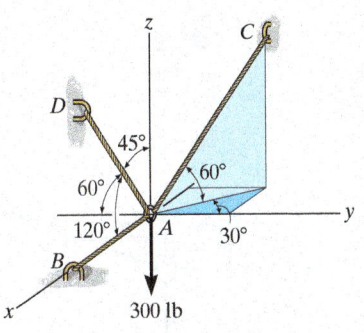

Prob. F3–10

F3–11. Determine the tension in cables AB, AC, and AD needed to support the 150-lb crate.

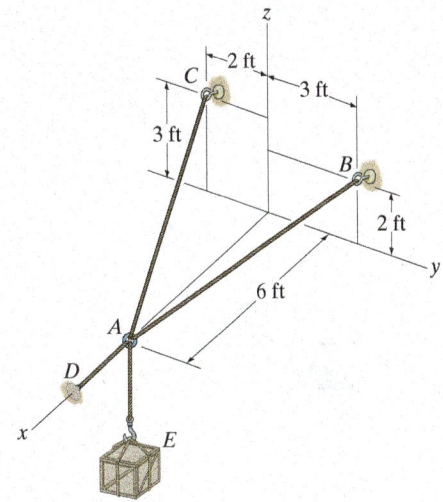

Prob. F3–11

PROBLEMS

All solutions must include a free-body diagram.

3–43. Determine the magnitude of the force **P** and the orientation of the 200-lb force required to keep the particle in equilibrium.

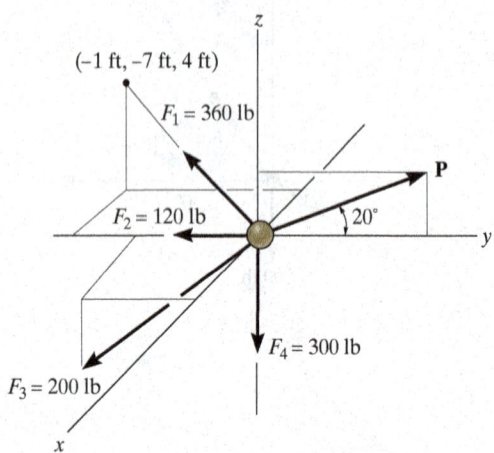

Prob. 3–43

*3–44. Determine the magnitudes of F_1, F_2, and F_3 for equilibrium of the particle.

Prob. 3–44

3–45. If a vertical force of 2.5 kN is applied to the hook at A, determine the tension in each of the three cables for equilibrium. Set $d = 1$ m.

Prob. 3–45

3–46. Determine the force in each cable needed to support the 500-lb load.

Prob. 3–46

3–47. The shear leg derrick is used to haul the 200-kg net of fish onto the dock. Determine the compressive force along each of the legs AB and CB and the tension in the winch cable DB. Assume the force in each leg acts along its axis.

3–50. Determine the force in each cable if $F = 500$ lb.

3–51. Determine the greatest force $\mathbf{F}$ that can be applied to the ring if each cable can support a maximum force of 800 lb.

Prob. 3–47

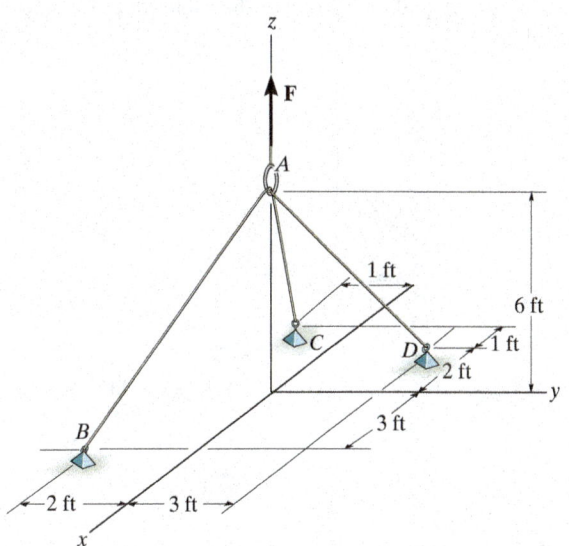

Probs. 3–50/51

*3–48.** Determine the tension in the cables in order to support the 100-kg crate.

3–49. Determine the maximum mass of the crate so that the tension developed in any cable does not exceed 3 kN.

*3–52.** Determine the tension developed in cables AB and AC and the force developed along strut AD for equilibrium of the 400-lb crate.

3–53. If the tension developed in each cable cannot exceed 300 lb, determine the largest weight of the crate that can be supported. Also, what is the force developed along strut AD?

Probs. 3–48/49

Probs. 3–52/53

3–54. Determine the tension developed in each cable for equilibrium of the 300-lb crate.

3–55. Determine the maximum weight of the crate that can be suspended from cables AB, AC, and AD so that the tension developed in any one of the cables does not exceed 250 lb.

3–58. The triangular frame ABC can be adjusted vertically between the three equal-length cords. If it remains in a horizontal plane, determine the required distance s so that the tension in each of the cords, OA, OB, and OC, equals 20 N. The lamp has a mass of 5 kg.

Probs. 3–54/55

Prob. 3–58

*3–56. The 25-kg flowerpot is supported at A by the three cords. Determine the force acting in each cord.

3–57. If each cord can sustain a maximum tension of 50 N before it fails, determine the greatest weight of the flowerpot the cords can support.

3–59. Determine the force in cables AB, AC, CD, CE, and CF needed to support the 500-lb crate. *Hint:* First analyze the equilibrium of point A, then using the result for AC, analyze the equilibrium of point C.

Probs. 3–56/57

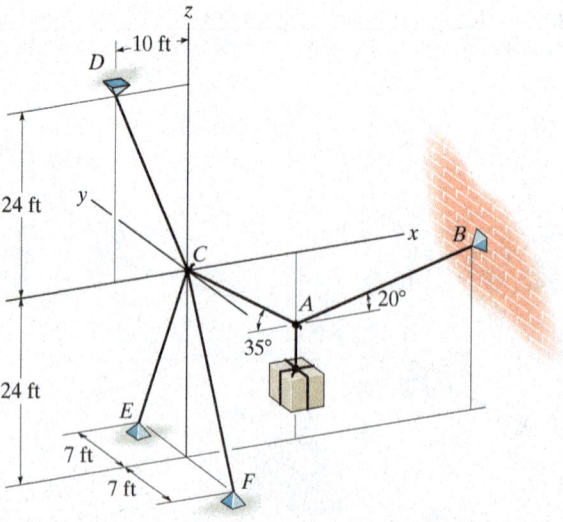

Prob. 3–59

PROBLEMS   107

*3–60. The 9500-lb pipe is supported by three cables. Determine the force in each cable for equilibrium of the hook at A.

Prob. 3–60

3–61. Determine the force in each of the three cables needed to lift the machine that has a weight of 10 kN.

Prob. 3–61

3–62. Determine the tension in cables OD and OB and the force in strut OC, required to support the 50-kg crate. The spring OA has an unstretched length of 0.8 m and a stiffness $k_{OA} = 1.2$ kN/m. The force in the strut acts along the axis of the strut.

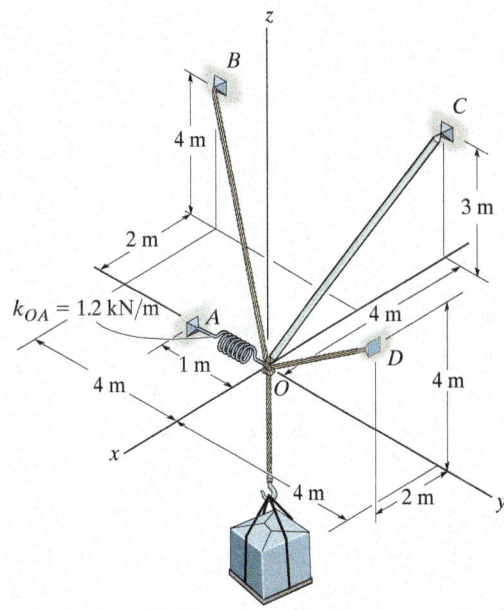

Prob. 3–62

3–63. Determine the force in each cable used to lift the 9.50-Mg surge arrester at constant velocity.

Prob. 3–63

*3–64. If cable AD is tightened by a turnbuckle and develops a tension of 1300 lb, determine the tension developed in cables AB and AC, and the force developed along the z axis of the antenna tower AE.

3–66. Determine the force in each of the three cables needed to support the tractor tread assembly, which has a mass of 8 Mg.

Prob. 3–64

Prob. 3–66

3–65. If the tension developed in either cable AB or AC cannot exceed 1000 lb, determine the maximum tension that can be developed in cable AD when it is tightened by the turnbuckle. Also, what is the force developed along the z axis of the antenna tower?

3–67. Three 10-lb balls and one 15-lb ball are suspended from the pulley-and-cable system. If the pulleys are frictionless and the centers of all of them lie in the same horizontal plane, determine the sag s for equilibrium of the system.

Prob. 3–65

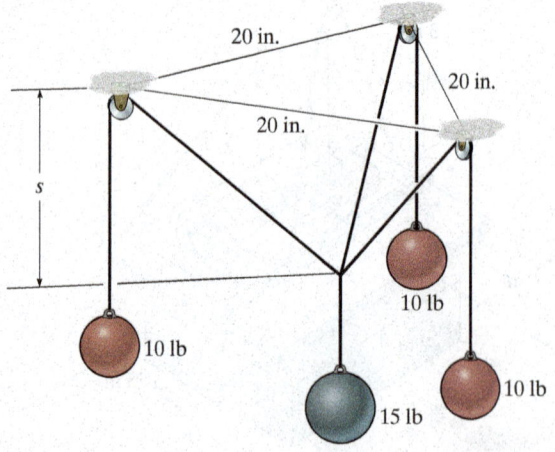

Prob. 3–67

CHAPTER REVIEW

Particle Equilibrium

A particle is in equilibrium when it is at rest or moves with constant velocity. This requires that all the forces acting on the particle produce a zero resultant force.

To account for all the forces that act on a particle, it is necessary to draw a free-body diagram. This diagram is an outlined shape of the particle that shows all the forces that act on it.

$$F_R = \Sigma F = 0$$

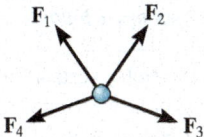

Two Dimensions

If the problem involves a linearly elastic spring, then the stretch or compression s of the spring can be related to the force applied to it.

$$F = ks$$

The force developed in a *continuous cable* that passes over a frictionless pulley must have a *constant* magnitude throughout the cable's length to keep the cable in equilibrium.

The two scalar equations of force equilibrium can be applied with reference to an established x, y coordinate system.

$$\Sigma F_x = 0$$
$$\Sigma F_y = 0$$

Three Dimensions

If the three-dimensional geometry is difficult to visualize, then the equilibrium equation should be applied using a Cartesian vector analysis. This requires first expressing each force on the free-body diagram as a Cartesian vector. When the forces are summed and set equal to zero, then the **i**, **j**, and **k** components are also zero.

$$\Sigma \mathbf{F} = 0$$

$$\Sigma F_x = 0$$
$$\Sigma F_y = 0$$
$$\Sigma F_z = 0$$

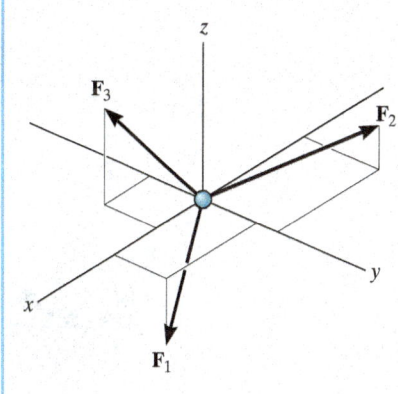

REVIEW PROBLEMS

All solutions must include a free-body diagram.

R3–1. If the bolt exerts a force of 50 lb on the pipe in the direction shown, determine the forces F_A and F_B that the smooth contacts at A and B exert on the pipe.

Prob. R3–1

R3–2. Determine the maximum weight of the engine that can be supported without exceeding a tension of 450 lb in chain AB and 480 lb in chain AC.

Prob. R3–2

R3–3. Determine the maximum weight of the flowerpot that can be supported without exceeding a cable tension of 50 lb in either cable AB or AC.

Prob. R3–3

R3–4. When y is zero, the springs sustain a force of 60 lb. Determine the magnitude of the applied vertical forces $\mathbf{F}$ and $-\mathbf{F}$ required to pull point A away from point B a distance of $y = 2$ ft. The ends of cords CAD and CBD are attached to rings at C and D.

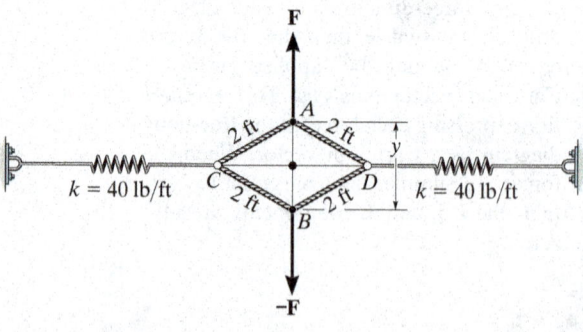

Prob. R3–4

R3–5. The joint of a space frame is subjected to four member forces. Member OA lies in the x–y plane and member OB lies in the y–z plane. Determine the force acting in each member required for equilibrium of the joint.

R3–7. Determine the force in each cable needed to support the 500-lb load.

Prob. R3–5

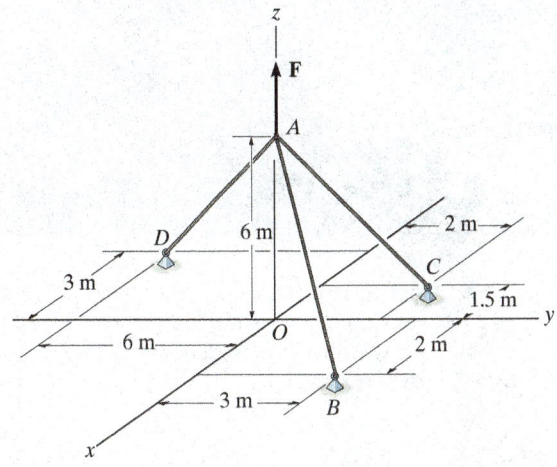

Prob. R3–7

R3–6. Determine the magnitudes of F_1, F_2, and F_3 for equilibrium of the particle.

R3–8. If cable AB is subjected to a tension of 700 N, determine the tension in cables AC and AD and the magnitude of the vertical force $\mathbf{F}$.

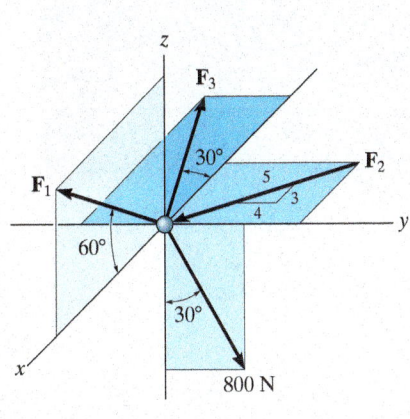

Prob. R3–6

Prob. R3–8

CHAPTER 4

Application of force on these gears will cause them to turn, thereby producing a rotation or a tendency for rotation. This effect is called a moment, and in this chapter we will study how to determine the moment of a system of loadings and calculate their resultants.

FORCE SYSTEM RESULTANTS

CHAPTER OBJECTIVES

- To discuss the concept of the moment of a force and show how to calculate it in two and three dimensions.
- To provide a method for finding the moment of a force about a specified axis.
- To define the moment of a couple.
- To show how to find the resultant effect of a nonconcurrent force system.
- To indicate how to reduce a simple distributed loading to a resultant force acting at a specified location.

(a)

(b)

4.1 MOMENT OF A FORCE—SCALAR FORMULATION

When a force is applied to a body it will produce a tendency for the body to rotate about a point that is not on the line of action of the force. This tendency to rotate is sometimes called a *torque*, but most often it is called the moment of a force or simply the *moment*. For example, consider applying a force to the handle of the wrench used to unscrew the bolt in Fig. 4–1a. The magnitude of the moment is directly proportional to the magnitude of $\mathbf{F}$ and the perpendicular distance or moment arm d. The larger the force or the longer the moment arm, the greater the moment or turning effect. If the force $\mathbf{F}$ is applied at an angle $\theta \neq 90°$, Fig. 4–1b, then it will be more difficult to turn the bolt since the moment arm $d' = d \sin \theta$ will be smaller than d. If $\mathbf{F}$ is applied along the wrench, Fig. 4–1c, its moment arm will be zero since the line of action of $\mathbf{F}$ will intersect point O (the z axis). As a result, the moment of $\mathbf{F}$ about O is also zero and no turning can occur.

(c)

Fig. 4–1

113

(a)

(b)

Fig. 4–2

In general, then, if we consider the force **F** and point O to lie in the shaded plane as shown in Fig. 4–2a, then the moment $\mathbf{M}_O$ about point O, or about an axis passing through O and perpendicular to the plane, is a *vector quantity* since it has a specified magnitude and direction.

Magnitude. The magnitude of $\mathbf{M}_O$ is

$$M_O = Fd \qquad (4\text{–}1)$$

where d is the **moment arm** or *perpendicular distance* from the axis at point O to the line of action of the force. Units of moment magnitude consist of force times distance, e.g., N · m or lb · ft.

Direction. The direction of $\mathbf{M}_O$ is defined by its **moment axis**, which is perpendicular to the plane that contains the force **F** and its moment arm d. The right-hand rule is used to establish the sense of direction of $\mathbf{M}_O$, where the natural curl of the fingers of the right hand, as they are drawn toward the palm, represent the rotation, or the tendency for rotation caused by the moment. Doing this, the thumb will give the directional sense of $\mathbf{M}_O$, Fig. 4–2a. Here the moment vector is represented three-dimensionally by a curl around an arrow. In two dimensions this vector is represented only by the curl as in Fig. 4–2b. Since this produces a counterclockwise rotation, the moment vector is actually directed out of the page.

Fig. 4–3

Resultant Moment. For two-dimensional problems, where all the forces lie within the x–y plane, Fig. 4–3, the resultant moment $(\mathbf{M}_R)_O$ about point O (the z axis) can be determined by *finding the algebraic sum* of the moments caused by all the forces in the system. As a convention, we will generally consider *positive moments* as *counterclockwise* since they are directed along the positive z axis (out of the page). *Clockwise moments* will be *negative*. Doing this, the directional sense of each moment can be represented by a *plus* or *minus* sign. Using this sign convention, with a symbolic curl to define the positive direction, the resultant moment in Fig. 4–3 is therefore

$$\zeta + (M_R)_O = \Sigma Fd; \qquad (M_R)_O = F_1 d_1 - F_2 d_2 + F_3 d_3$$

If the numerical result of this sum is a positive scalar, $(\mathbf{M}_R)_O$ will be a counterclockwise moment (out of the page); and if the result is negative, $(\mathbf{M}_R)_O$ will be a clockwise moment (into the page).

The moment of the force about point O is $M_O = Fd$. But it is easier to find this moment using $M_O = F_x(0) + F_y r = F_y r$.

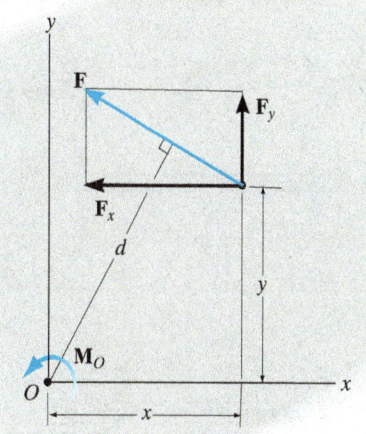

Fig. 4–4

4.2 PRINCIPLE OF MOMENTS

A concept often used in mechanics is the *principle of moments*, which is sometimes referred to as Varignon's theorem since it was originally developed by the French mathematician Pierre Varignon (1654–1722). It states that *the moment of a force about a point is equal to the sum of the moments of the components of the force about the point*. This theorem is easily proved using vector analysis as shown in Sec. 4.3. Here in Fig. 4–4 it can be demonstrated in two dimensions. The moment of **F** about point O is $M_O = Fd$, but by resolving **F** into its two rectangular components, the same moment is then

$$M_O = F_x y + F_y x$$

As noted by Example 4–1, using this method is generally easier than finding the same moment using $M_O = Fd$.

IMPORTANT POINTS

- The moment of a force creates the tendency of a body to turn about an axis passing through a specific point O.

- Using the right-hand rule, the sense of rotation is indicated by the curl of the fingers, and the thumb produces the sense of direction of the moment.

- The magnitude of the moment is determined from $M_O = Fd$, where d is called the moment arm, which represents the perpendicular or shortest distance from point O to the line of action of the force.

- In two dimensions it is often easier to use the principle of moments and find the moment of the force components about point O, rather than using $M_O = Fd$.

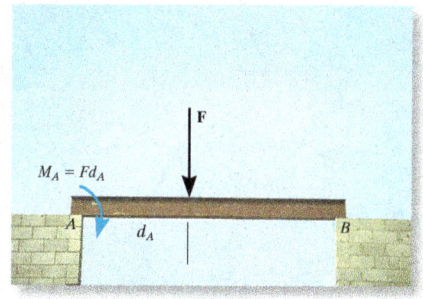

The force **F** tends to rotate the beam clockwise about its support at A with a moment $M_A = Fd_A$. The actual rotation would occur only if the support at B were removed.

EXAMPLE 4.1

For each case illustrated in Fig. 4–5, determine the moment of the force about point O.

SOLUTION (SCALAR ANALYSIS)
The line of action of each force is extended as a dashed line in order to establish the moment arm d. Also illustrated is the tendency of rotation of the member as caused by the force, and the orbit of the force about O is shown as a colored curl. Thus,

Fig. 4–4a $M_O = (100 \text{ N})(2 \text{ m}) = 200 \text{ N} \cdot \text{m} \downarrow$ Ans.
Fig. 4–4b $M_O = (50 \text{ N})(0.75 \text{ m}) = 37.5 \text{ N} \cdot \text{m} \downarrow$ Ans.
Fig. 4–4c $M_O = (40 \text{ lb})(4 \text{ ft} + 2 \cos 30° \text{ ft}) = 229 \text{ lb} \cdot \text{ft} \downarrow$ Ans.
Fig. 4–4d $M_O = (60 \text{ lb})(1 \sin 45° \text{ ft}) = 42.4 \text{ lb} \cdot \text{ft} \uparrow$ Ans.
Fig. 4–4e $M_O = (7 \text{ kN})(4 \text{ m} - 1 \text{ m}) = 21.0 \text{ kN} \cdot \text{m} \uparrow$ Ans.

(a)

(b)

(c)

(d)

(e)

Fig. 4–5

EXAMPLE 4.2

Force **F** acts at the end of the angle bracket in Fig. 4–6a. Determine the moment of the force about point O.

(a)

Fig. 4–6

(b)

SOLUTION
The force is resolved into its x and y components, Fig. 4–6b, then

$\downarrow + M_O = 400 \sin 30° \text{ N}(0.2 \text{ m}) - 400 \cos 30° \text{ N}(0.4 \text{ m})$
$= -98.6 \text{ N} \cdot \text{m} = 98.6 \text{ N} \cdot \text{m} \downarrow$ Ans.

EXAMPLE 4.3

Determine the moment of the force in Fig. 4–7a about point O.

(a)

SOLUTION I
The moment arm d in Fig. 4–7a can be found from trigonometry.

$$d = (3 \text{ m}) \sin 75° = 2.898 \text{ m}$$

Thus,

$$M_O = Fd = (5 \text{ kN})(2.898 \text{ m}) = 14.5 \text{ kN} \cdot \text{m} \;\downarrow \qquad Ans.$$

Since the force tends to rotate or orbit clockwise about point O, the moment is directed into the page.

SOLUTION II
The x and y components of the force are indicated in Fig. 4–7b. Considering counterclockwise moments as positive, and applying the principle of moments, we have

$$\zeta+ M_O = -F_x d_y - F_y d_x$$

$$= -(5 \cos 45° \text{ kN})(3 \sin 30° \text{ m}) - (5 \sin 45° \text{ kN})(3 \cos 30° \text{ m})$$

$$= -14.5 \text{ kN} \cdot \text{m} = 14.5 \text{ kN} \cdot \text{m} \;\downarrow \qquad Ans.$$

(b)

SOLUTION III
The x and y axes can be oriented parallel and perpendicular to the rod's axis, as shown in Fig. 4–7c. Here $\mathbf{F}_x$ produces no moment about point O since its line of action passes through this point. Therefore,

$$\zeta+ M_O = -F_y d_x$$

$$= -(5 \sin 75° \text{ kN})(3 \text{ m})$$

$$= -14.5 \text{ kN} \cdot \text{m} = 14.5 \text{ kN} \cdot \text{m} \;\downarrow \qquad Ans.$$

(c)

Fig. 4–7

FUNDAMENTAL PROBLEMS

F4–1. In each case, determine the moment of the force about point O.

(a)

(b)

Prob. F4–1

F4–2. In each case, determine the moment about point O.

(a)

(b)

Prob. F4–2

F4–3. In each case determine the moment about point O.

(a)

(b)

(c)

Prob. F4–3

FUNDAMENTAL PROBLEMS   119

F4–4. Determine the moment of the force about point O.

Prob. F4–4

F4–5. Determine the moment of the force about point O.

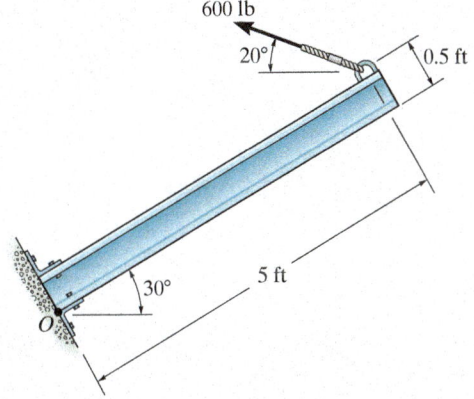

Prob. F4–5

F4–6. Determine the moment of the force about point O.

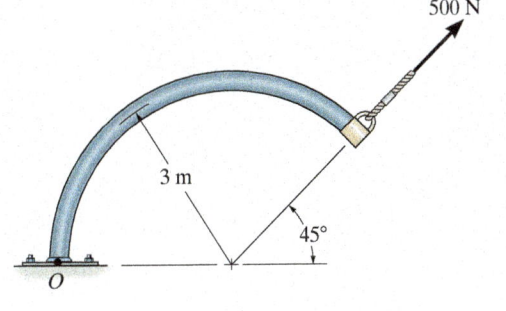

Prob. F4–6

F4–7. Determine the resultant moment produced by the forces about point O.

Prob. F4–7

F4–8. Determine the resultant moment produced by the forces about point O.

Prob. F4–8

F4–9. Determine the resultant moment produced by the forces about point O.

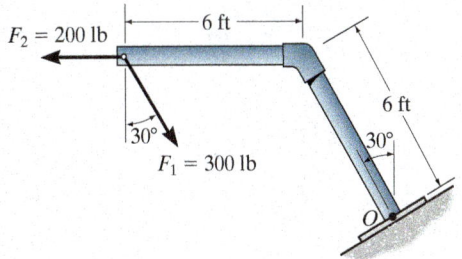

Prob. F4–9

PROBLEMS

4–1. Determine the moment about point A of each force acting on the beam.

4–2. Determine the moment about point B of each force acting on the beam.

*4–4.** Determine the moment of each of the three forces about point A.

4–5. Determine the moment of each of the three forces about point B.

Probs. 4–1/2

Probs. 4–4/5

4–3. The man exerts the two forces on the handle of the shovel. Determine the resultant moment of these forces about the blade at A.

4–6. Determine the moment of each force about the bolt at A. Take $F_B = 40$ lb, $F_C = 50$ lb.

4–7. If $F_B = 30$ lb and $F_C = 45$ lb, determine the resultant moment about the bolt at A.

Prob. 4–3

Probs. 4–6/7

*4–8. The cable exerts a force of $P = 6$ kN at the end of the 8-m-long crane boom. If $\theta = 30°$, determine the placement x of the boom at B so that this force creates a maximum moment about point O. What is this moment?

4–9. The cable exerts a force of $P = 6$ kN at the end of the 8-m-long crane boom. If $x = 10$ m, determine the angle θ of the boom so that this force creates a maximum moment about point O. What is this moment?

Probs. 4–8/9

4–10. Determine the angle θ at which the 500-N force must act at A so that the moment of this force about point B is equal to zero.

Prob. 4–10

4–11. Two forces act on the skew caster. Determine the resultant moment of these forces about point A and about point B.

Prob. 4–11

*4–12. Two men exert forces of $F = 80$ lb and $P = 50$ lb on the ropes. Determine the moment of each force about point A. Which way will the pole rotate, clockwise or counterclockwise?

4–13. If the man at B exerts a force of $P = 30$ lb on the rope, determine the magnitude of the force **F** the man at C must exert to prevent the pole from rotating, i.e., so the resultant moment about point A of both forces is zero.

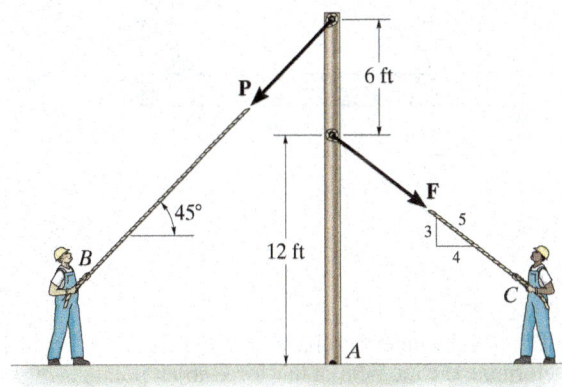

Probs. 4–12/13

4–14. A force of 80 N acts on the handle of the paper cutter at A. If $\theta = 60°$, determine the moment created by this force about the hinge at O. At what angle θ should the force be applied so that the moment it creates about point O is a maximum (clockwise)? What is this maximum moment?

Prob. 4–14

4–15. Determine the orientation θ ($0° \leq \theta \leq 180°$) of the 40-lb force **F** so that it produces (a) the maximum moment about point A and (b) no moment about point A. Find the moment in case (a).

*4–16. Determine the moment of the force **F** about point A as a function of θ. Plot the result of **M** (ordinate) versus θ (abscissa) for $0° \leq \theta \leq 180°$.

Probs. 4–15/16

4–17. The hammer is subjected to the force of $F = 20$ lb. Determine the moment of this force about point A.

4–18. In order to pull out the nail at B, the force **F** exerted on the hammer must produce a clockwise moment of 500 lb·in. about point A. Determine the required magnitude of force **F**.

Probs. 4–17/18

4–19. If the tension in the belt is 52 lb, determine the moment of each of these forces about the pin at A.

Prob. 4–19

*4–20. The tower crane is used to hoist the 2-Mg load upward at constant velocity. The 1.5-Mg jib BD, 0.5-Mg jib BC, and 6-Mg counterweight C have centers of mass at G_1, G_2, and G_3, respectively. Determine the resultant moment produced by the load and the weights of the tower crane jibs about point A and about point B.

4–21. The tower crane is used to hoist a 2-Mg load upward at constant velocity. The 1.5-Mg jib BD and 0.5-Mg jib BC have centers of mass at G_1 and G_2, respectively. Determine the required mass of the counterweight C so that the resultant moment produced by the load and the weights of the tower crane jibs about point A is zero. The center of mass for the counterweight is located at G_3.

Probs. 4–20/21

4.3 CROSS PRODUCT

The moment of a force will be formulated using Cartesian vectors in the next section. Before doing this, however, it is first necessary to expand our knowledge of vector algebra and introduce the cross-product method of vector multiplication.

The **cross product** of two vectors **A** and **B** yields the vector **C**, which is written

$$\mathbf{C} = \mathbf{A} \times \mathbf{B} \qquad (4\text{–}2)$$

and is read "**C** equals **A** cross **B**."

Magnitude. The *magnitude* of **C** is defined as the product of the magnitudes of **A** and **B** and the sine of the angle θ between their tails, where $0° \leq \theta \leq 180°$. Thus,

$$C = AB \sin \theta$$

Direction. Vector **C** has a *direction* that is perpendicular to the plane containing **A** and **B** such that the directional sense of **C** is specified by the right-hand rule; i.e., curling the fingers of the right hand from vector **A** (cross) to vector **B**, the thumb points in the direction of **C**, as shown in Fig. 4–8.

Knowing both the magnitude and direction of **C**, we can therefore write

$$\mathbf{C} = \mathbf{A} \times \mathbf{B} = (AB \sin \theta)\mathbf{u}_C \qquad (4\text{–}3)$$

The terms of Eq. 4–3 are illustrated graphically in Fig. 4–8.

Fig. 4–8

124   CHAPTER 4   FORCE SYSTEM RESULTANTS

Fig. 4–9

Laws of Operation.

The following three laws of operation apply:

- The commutative law is *not* valid; i.e., $\mathbf{A} \times \mathbf{B} \neq \mathbf{B} \times \mathbf{A}$. Rather,

$$\mathbf{A} \times \mathbf{B} = -\mathbf{B} \times \mathbf{A}$$

This is shown in Fig. 4–9 by using the right-hand rule. The cross product $\mathbf{B} \times \mathbf{A}$ yields a vector that has the same magnitude but acts in the opposite sense of direction to $\mathbf{C}$; i.e., $\mathbf{B} \times \mathbf{A} = -\mathbf{C}$.

- If the cross product is multiplied by a scalar a, it obeys the associative law;

$$a(\mathbf{A} \times \mathbf{B}) = (a\mathbf{A}) \times \mathbf{B} = \mathbf{A} \times (a\mathbf{B}) = (\mathbf{A} \times \mathbf{B})a$$

This property is easily shown since the magnitude of the resultant vector ($|a| AB \sin\theta$) and its direction are the same in each case.

- The vector cross product also obeys the distributive law of addition,*

$$\mathbf{A} \times (\mathbf{B} + \mathbf{D}) = (\mathbf{A} \times \mathbf{B}) + (\mathbf{A} \times \mathbf{D})$$

It is important to note that *proper order* of these cross products must be maintained, since they are not commutative.

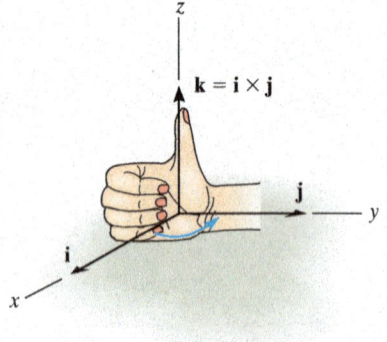

Fig. 4–10

Cartesian Vector Formulation.

Equation 4–3 may be used to find the cross product of any pair of Cartesian unit vectors. For example, to find $\mathbf{i} \times \mathbf{j}$, the magnitude of the resultant vector is $(i)(j)(\sin 90°) = (1)(1)(1) = 1$, and its direction is determined using the right-hand rule, Fig. 4–10. Here the resultant vector points in the $+\mathbf{k}$ direction, so that $\mathbf{i} \times \mathbf{j} = (1)\mathbf{k}$. In a similar manner,

$$\mathbf{i} \times \mathbf{j} = \mathbf{k} \qquad \mathbf{i} \times \mathbf{k} = -\mathbf{j} \qquad \mathbf{i} \times \mathbf{i} = 0$$
$$\mathbf{j} \times \mathbf{k} = \mathbf{i} \qquad \mathbf{j} \times \mathbf{i} = -\mathbf{k} \qquad \mathbf{j} \times \mathbf{j} = 0$$
$$\mathbf{k} \times \mathbf{i} = \mathbf{j} \qquad \mathbf{k} \times \mathbf{j} = -\mathbf{i} \qquad \mathbf{k} \times \mathbf{k} = 0$$

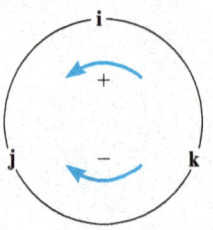

Fig. 4–11

These results should *not* be memorized; rather, it should be clearly understood how each is obtained by using the right-hand rule and the definition of the cross product. A simple scheme shown in Fig. 4–11 can sometimes be helpful for obtaining the same results when the need arises. If the circle is constructed as shown, then "crossing" two unit vectors in a *counterclockwise* fashion around the circle yields the *positive* third unit vector; e.g., $\mathbf{k} \times \mathbf{i} = \mathbf{j}$. "Crossing" *clockwise*, a *negative* unit vector is obtained; e.g., $\mathbf{i} \times \mathbf{k} = -\mathbf{j}$.

*See Prob. 4–22.

Let us now consider the cross product of two general vectors **A** and **B** which are expressed in Cartesian vector form. We have

$$\mathbf{A} \times \mathbf{B} = (A_x\mathbf{i} + A_y\mathbf{j} + A_z\mathbf{k}) \times (B_x\mathbf{i} + B_y\mathbf{j} + B_z\mathbf{k})$$

$$= A_xB_x(\mathbf{i} \times \mathbf{i}) + A_xB_y(\mathbf{i} \times \mathbf{j}) + A_xB_z(\mathbf{i} \times \mathbf{k})$$

$$+ A_yB_x(\mathbf{j} \times \mathbf{i}) + A_yB_y(\mathbf{j} \times \mathbf{j}) + A_yB_z(\mathbf{j} \times \mathbf{k})$$

$$+ A_zB_x(\mathbf{k} \times \mathbf{i}) + A_zB_y(\mathbf{k} \times \mathbf{j}) + A_zB_z(\mathbf{k} \times \mathbf{k})$$

Carrying out the cross-product operations and combining terms yields

$$\mathbf{A} \times \mathbf{B} = (A_yB_z - A_zB_y)\mathbf{i} - (A_xB_z - A_zB_x)\mathbf{j} + (A_xB_y - A_yB_x)\mathbf{k} \qquad (4\text{--}4)$$

This equation may also be written in a more compact determinant form as

$$\mathbf{A} \times \mathbf{B} = \begin{vmatrix} \mathbf{i} & \mathbf{j} & \mathbf{k} \\ A_x & A_y & A_z \\ B_x & B_y & B_z \end{vmatrix} \qquad (4\text{--}5)$$

Thus, to find the cross product of **A** and **B**, it is necessary to expand a determinant whose first row of elements consists of the unit vectors **i**, **j**, and **k** and whose second and third rows represent the x, y, z components of the two vectors **A** and **B**, respectively.*

---

*A determinant having three rows and three columns can be expanded using three minors, each of which is multiplied by one of the three terms in the first row. There are four elements in each minor, for example,

$$\begin{vmatrix} A_{11} & A_{12} \\ A_{21} & A_{22} \end{vmatrix}$$

By *definition*, this determinant notation represents the terms $(A_{11}A_{22} - A_{12}A_{21})$, which is simply the product of the two elements intersected by the arrow slanting downward to the right $(A_{11}A_{22})$ *minus* the product of the two elements intersected by the arrow slanting downward to the left $(A_{12}A_{21})$. For a 3 × 3 determinant, such as Eq. 4–5, the three minors can be generated in accordance with the following scheme:

For element **i**: $\begin{vmatrix} \mathbf{i} & \mathbf{j} & \mathbf{k} \\ A_x & A_y & A_z \\ B_x & B_y & B_z \end{vmatrix} = \mathbf{i}(A_yB_z - A_zB_y)$

Remember the negative sign

For element **j**: $\begin{vmatrix} \mathbf{i} & \mathbf{j} & \mathbf{k} \\ A_x & A_y & A_z \\ B_x & B_y & B_z \end{vmatrix} = -\mathbf{j}(A_xB_z - A_zB_x)$

For element **k**: $\begin{vmatrix} \mathbf{i} & \mathbf{j} & \mathbf{k} \\ A_x & A_y & A_z \\ B_x & B_y & B_z \end{vmatrix} = \mathbf{k}(A_xB_y - A_yB_x)$

Adding these results and noting that the **j** element *must include the minus sign* yields the expanded form of $\mathbf{A} \times \mathbf{B}$ given by Eq. 4–4.

4.4 MOMENT OF A FORCE—VECTOR FORMULATION

The moment of a force **F** about point *O*, Fig. 4–12*a*, can be expressed using the vector cross product, namely,

$$\mathbf{M}_O = \mathbf{r} \times \mathbf{F} \qquad (4\text{–}6)$$

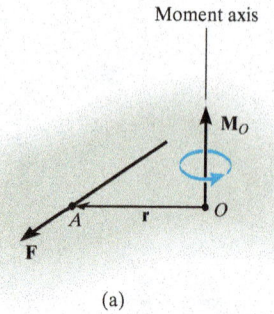

(a)

Here **r** represents a position vector directed *from O* to *any point* on the line of action of **F**. We will now show that indeed the moment $\mathbf{M}_O$, when determined by this cross product, has the proper magnitude and direction.

Magnitude. The magnitude of the cross product is defined from Eq. 4–3 as $M_O = rF \sin \theta$, where the angle θ is measured between the *tails* of **r** and **F**. To establish this angle, **r** must be treated as a ***sliding vector*** so that θ can be constructed properly, Fig. 4–12*b*. Since the moment arm $d = r \sin \theta$, then

$$M_O = rF \sin \theta = F(r \sin \theta) = Fd$$

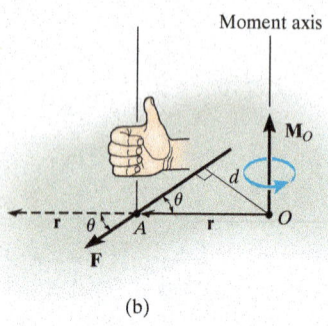

(b)

Fig. 4–12

which agrees with Eq. 4–1.

Direction. The direction and sense of $\mathbf{M}_O$ in Eq. 4–6 are determined by the right-hand rule as it applies to the cross product. Thus, sliding **r** to the dashed position and curling the right-hand fingers from **r** toward **F**, "**r** cross **F**," the thumb is directed upward or perpendicular to the plane containing **r** and **F** and this is in the *same direction* as $\mathbf{M}_O$, the moment of the force about point *O*, Fig. 4–12*b*. Remember that the cross product does not obey the commutative law, and so the order of $\mathbf{r} \times \mathbf{F}$ must be maintained to produce the correct sense of direction for $\mathbf{M}_O$.

Principle of Transmissibility. The cross product operation is often used in three dimensions since the perpendicular distance or moment arm from point *O* to the line of action of the force is not needed. In other words, we can use any position vector **r** measured from point *O* to any point on the line of action of the force **F**, Fig. 4–13. Thus,

$$\mathbf{M}_O = \mathbf{r}_1 \times \mathbf{F} = \mathbf{r}_2 \times \mathbf{F} = \mathbf{r}_3 \times \mathbf{F}$$

Fig. 4–13

Since **F** can be applied at any point along its line of action and still create this *same moment* about point *O*, then **F** can be considered a *sliding vector*. This property is called the ***principle of transmissibility*** of a force.

4.4 Moment of a Force—Vector Formulation

Cartesian Vector Formulation. If we establish x, y, z coordinate axes, then the position vector $\mathbf{r}$ and force $\mathbf{F}$ can be expressed as Cartesian vectors, Fig. 4–14a. Applying Eq. 4–5, we have

$$\mathbf{M}_O = \mathbf{r} \times \mathbf{F} = \begin{vmatrix} \mathbf{i} & \mathbf{j} & \mathbf{k} \\ r_x & r_y & r_z \\ F_x & F_y & F_z \end{vmatrix} \quad (4\text{–}7)$$

where

r_x, r_y, r_z represent the x, y, z components of the position vector drawn from point O to *any point* on the line of action of the force

F_x, F_y, F_z represent the x, y, z components of the force vector

If the determinant is expanded, then like Eq. 4–4 we have

$$\mathbf{M}_O = (r_y F_z - r_z F_y)\mathbf{i} - (r_x F_z - r_z F_x)\mathbf{j} + (r_x F_y - r_y F_x)\mathbf{k} \quad (4\text{–}8)$$

The physical meaning of these three moment components becomes evident by studying Fig. 4–14b. For example, the $\mathbf{i}$ component of $\mathbf{M}_O$ can be determined from the moments of $\mathbf{F}_x$, $\mathbf{F}_y$, and $\mathbf{F}_z$ about the x axis. The component $\mathbf{F}_x$ does *not* create a moment or tendency to cause turning about the x axis since this force is *parallel* to the x axis. The line of action of $\mathbf{F}_y$ passes through point B, and so the magnitude of the moment of $\mathbf{F}_y$ about point A on the x axis is $r_z F_y$. By the right-hand rule this component acts in the *negative* $\mathbf{i}$ direction. Likewise, $\mathbf{F}_z$ passes through point C and so it contributes a moment component of $r_y F_z \mathbf{i}$ about the x axis. Thus, $(M_O)_x = (r_y F_z - r_z F_y)$ as shown in Eq. 4–8. As an exercise, try to establish the $\mathbf{j}$ and $\mathbf{k}$ components of $\mathbf{M}_O$ in this manner and show that indeed the expanded form of the determinant, Eq. 4–8, represents the moment of $\mathbf{F}$ about point O. Once $\mathbf{M}_O$ is determined, realize that it will always be *perpendicular* to the shaded plane containing vectors $\mathbf{r}$ and $\mathbf{F}$, Fig. 4–14a.

(a)

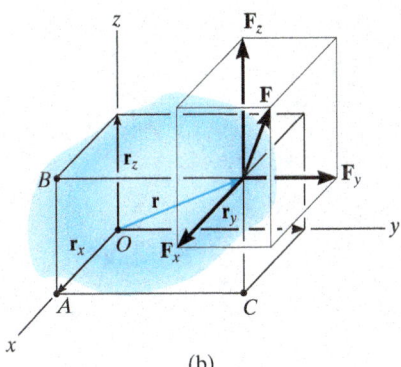

(b)

Fig. 4–14

Resultant Moment of a System of Forces. If a body is acted upon by a system of forces, Fig. 4–15, the resultant moment of the forces about point O can be determined by vector addition of the moment of each force. This resultant can be written symbolically as

$$(\mathbf{M}_R)_O = \Sigma(\mathbf{r} \times \mathbf{F}) \quad (4\text{–}9)$$

Fig. 4–15

Fig. 4–16

Principle of Moments. This concept was discussed in Sec. 4.1. Recall that it states that *the moment of a force about a point is equal to the sum of the moments of the components of the force about the point.* This theorem can be proven easily using the vector cross product since the cross product obeys the *distributive law*. For example, consider the moments of the force **F** and two of its components about point O, Fig. 4–16. Since $\mathbf{F} = \mathbf{F}_1 + \mathbf{F}_2$ we have

$$\mathbf{M}_O = \mathbf{r} \times \mathbf{F} = \mathbf{r} \times (\mathbf{F}_1 + \mathbf{F}_2) = \mathbf{r} \times \mathbf{F}_1 + \mathbf{r} \times \mathbf{F}_2$$

EXAMPLE 4.4

Fig. 4–17

Force **F** acts at the end of the angle bracket in Fig. 4–17a. Determine the moment of the force about point O.

SOLUTION
Using a Cartesian vector approach, the force and position vectors, Fig. 4–17b, are

$$\mathbf{r} = \{0.4\mathbf{i} - 0.2\mathbf{j}\} \text{ m}$$

$$\mathbf{F} = \{400 \sin 30°\mathbf{i} - 400 \cos 30°\mathbf{j}\} \text{ N}$$

$$= \{200.0\mathbf{i} - 346.4\mathbf{j}\} \text{ N}$$

The moment is therefore

$$\mathbf{M}_O = \mathbf{r} \times \mathbf{F} = \begin{vmatrix} \mathbf{i} & \mathbf{j} & \mathbf{k} \\ 0.4 & -0.2 & 0 \\ 200.0 & -346.4 & 0 \end{vmatrix}$$

$$= 0\mathbf{i} - 0\mathbf{j} + [0.4(-346.4) - (-0.2)(200.0)]\mathbf{k}$$

$$= \{-98.6\mathbf{k}\} \text{ N} \cdot \text{m} \qquad \textit{Ans.}$$

NOTE: The scalar analysis in Example 4.2 provides a more *convenient method* for analysis than this solution, since the direction of the moment and the moment arm for each component force are easy to establish. For this reason, this method is generally recommended for solving problems in two dimensions, whereas a Cartesian vector analysis is generally recommended only for solving three-dimensional problems.

EXAMPLE 4.5

Determine the moment produced by the force **F** in Fig. 4–18a about point O. Express the result as a Cartesian vector.

SOLUTION
As shown in Fig. 4–18b, either $\mathbf{r}_A$ or $\mathbf{r}_B$ can be used to determine the moment about point O. These position vectors are

$$\mathbf{r}_A = \{12\mathbf{k}\} \text{ m} \quad \mathbf{r}_B = \{4\mathbf{i} + 12\mathbf{j}\} \text{ m}$$

Force **F** expressed as a Cartesian vector is

$$\mathbf{F} = F\mathbf{u}_{AB} = 2 \text{ kN} \left[\frac{\{4\mathbf{i} + 12\mathbf{j} - 12\mathbf{k}\} \text{ m}}{\sqrt{(4 \text{ m})^2 + (12 \text{ m})^2 + (-12 \text{ m})^2}} \right]$$

$$= \{0.4588\mathbf{i} + 1.376\mathbf{j} - 1.376\mathbf{k}\} \text{ kN}$$

Thus

$$\mathbf{M}_O = \mathbf{r}_A \times \mathbf{F} = \begin{vmatrix} \mathbf{i} & \mathbf{j} & \mathbf{k} \\ 0 & 0 & 12 \\ 0.4588 & 1.376 & -1.376 \end{vmatrix}$$

$$= [0(-1.376) - 12(1.376)]\mathbf{i} - [0(-1.376) - 12(0.4588)]\mathbf{j}$$
$$+ [0(1.376) - 0(0.4588)]\mathbf{k}$$

$$= \{-16.5\mathbf{i} + 5.51\mathbf{j}\} \text{ kN} \cdot \text{m} \quad \textit{Ans.}$$

or

$$\mathbf{M}_O = \mathbf{r}_B \times \mathbf{F} = \begin{vmatrix} \mathbf{i} & \mathbf{j} & \mathbf{k} \\ 4 & 12 & 0 \\ 0.4588 & 1.376 & -1.376 \end{vmatrix}$$

$$= [12(-1.376) - 0(1.376)]\mathbf{i} - [4(-1.376) - 0(0.4588)]\mathbf{j}$$
$$+ [4(1.376) - 12(0.4588)]\mathbf{k}$$

$$= \{-16.5\mathbf{i} + 5.51\mathbf{j}\} \text{ kN} \cdot \text{m} \quad \textit{Ans.}$$

NOTE: As shown in Fig. 4–18b, $\mathbf{M}_O$ acts perpendicular to the plane that contains **F**, $\mathbf{r}_A$, and $\mathbf{r}_B$. Had this problem been worked using $M_O = Fd$, notice the difficulty that would arise in obtaining the moment arm d.

(a)

(b)

Fig. 4–18

EXAMPLE 4.6

Two forces act on the rod shown in Fig. 4–19a. Determine the resultant moment they create about the flange at O. Express the result as a Cartesian vector.

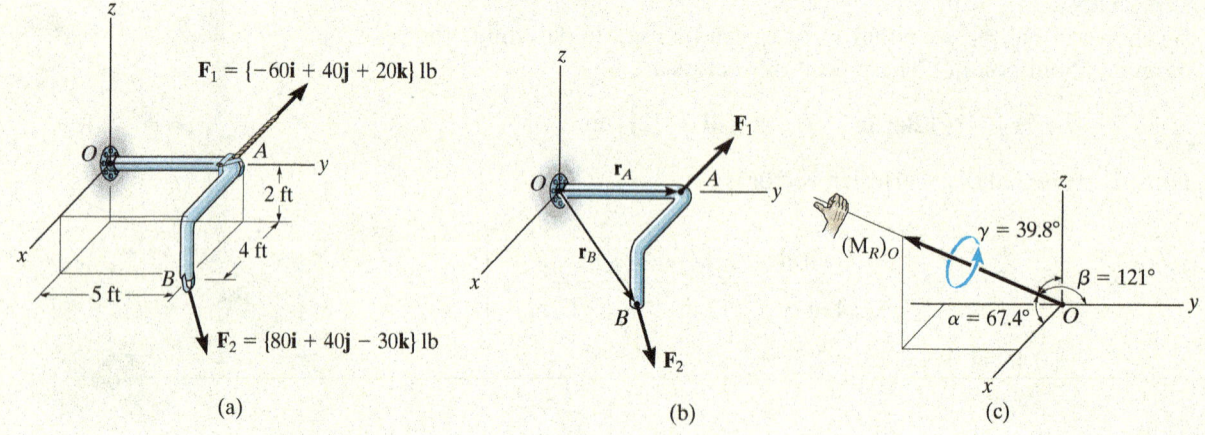

Fig. 4–19

SOLUTION
Position vectors are directed from point O to each force as shown in Fig. 4–19b. These vectors are

$$r_A = \{5j\} \text{ ft}$$
$$r_B = \{4i + 5j - 2k\} \text{ ft}$$

The resultant moment about O is therefore

$$(M_R)_O = \Sigma(r \times F)$$
$$= r_A \times F_1 + r_B \times F_2$$

$$= \begin{vmatrix} i & j & k \\ 0 & 5 & 0 \\ -60 & 40 & 20 \end{vmatrix} + \begin{vmatrix} i & j & k \\ 4 & 5 & -2 \\ 80 & 40 & -30 \end{vmatrix}$$

$$= [5(20) - 0(40)]i - [0]j + [0(40) - (5)(-60)]k$$
$$+ [5(-30) - (-2)(40)]i - [4(-30) - (-2)(80)]j + [4(40) - 5(80)]k$$
$$= \{30i - 40j + 60k\} \text{ lb} \cdot \text{ft} \qquad \textit{Ans.}$$

NOTE: This result is shown in Fig. 4–19c. The coordinate direction angles were determined from the unit vector for $(M_R)_O$. Realize that the two forces tend to cause the rod to rotate about the moment axis in the manner shown by the curl indicated on the moment vector.

FUNDAMENTAL PROBLEMS

F4–10. Determine the moment of force **F** about point O. Express the result as a Cartesian vector.

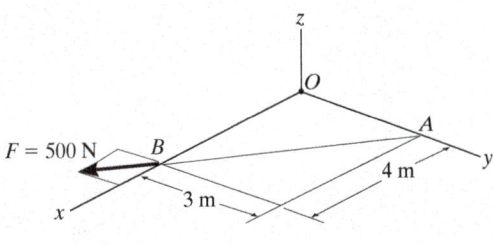

Prob. F4–10

F4–11. Determine the moment of force **F** about point O. Express the result as a Cartesian vector.

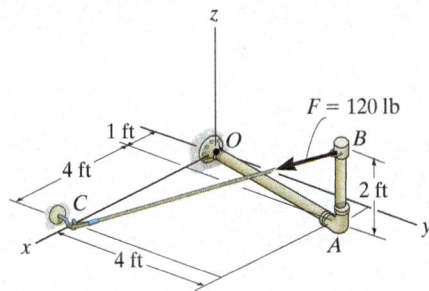

Prob. F4–11

F4–12. If $\mathbf{F}_1 = \{100\mathbf{i} - 120\mathbf{j} + 75\mathbf{k}\}$ lb and $\mathbf{F}_2 = \{-200\mathbf{i} + 250\mathbf{j} + 100\mathbf{k}\}$ lb, determine the resultant moment produced by these forces about point O. Express the result as a Cartesian vector.

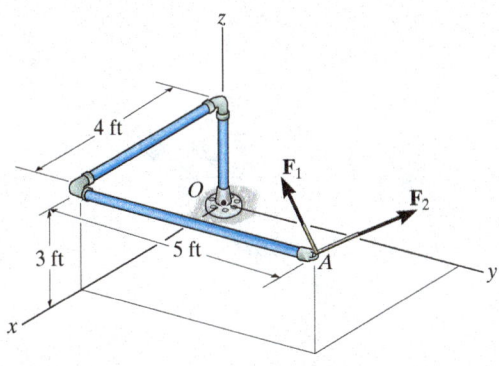

Prob. F4–12

PROBLEMS

4–22. If **A**, **B**, and **D** are given vectors, prove the distributive law for the vector cross product, i.e., $\mathbf{A} \times (\mathbf{B} + \mathbf{D}) = (\mathbf{A} \times \mathbf{B}) + (\mathbf{A} \times \mathbf{D})$.

4–23. Prove the triple scalar product identity $\mathbf{A} \cdot (\mathbf{B} \times \mathbf{C}) = (\mathbf{A} \times \mathbf{B}) \cdot \mathbf{C}$.

*4–24. Given the three nonzero vectors **A**, **B**, and **C**, show that if $\mathbf{A} \cdot (\mathbf{B} \times \mathbf{C}) = 0$, the three vectors *must* lie in the same plane.

4–25. The force $\mathbf{F} = \{400\mathbf{i} - 100\mathbf{j} - 700\mathbf{k}\}$ lb acts at the end of the beam. Determine the moment of this force about point O.

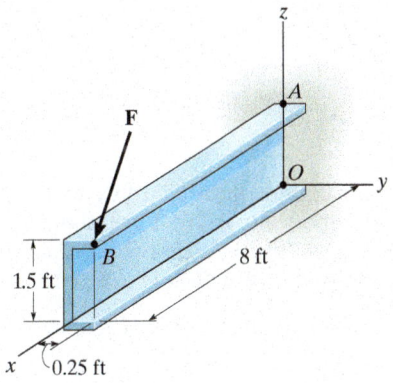

Prob. 4–25

4–26. The force $\mathbf{F} = \{400\mathbf{i} - 100\mathbf{j} - 700\mathbf{k}\}$ lb acts at the end of the beam. Determine the moment of this force about point A.

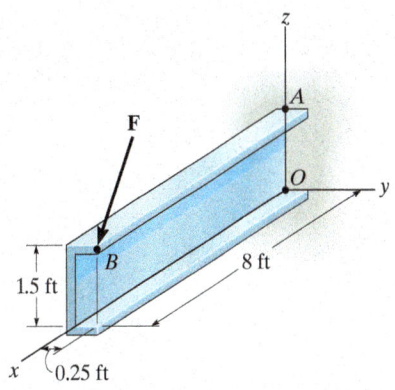

Prob. 4–26

4–27. Determine the moment of the force **F** about point O. Express the result as a Cartesian vector.

***4–28.** Determine the moment of the force **F** about point P. Express the result as a Cartesian vector.

4–31. Determine the moment of the force **F** about point P. Express the result as a Cartesian vector.

Probs. 4–27/28

Prob. 4–31

4–29. The man pulls on the rope with a force of $F = 20$ N. Determine the moment of this force about the base of the pole at O. Solve the problem two ways, i.e., by using a position vector from O to A, then O to B.

4–30. Determine the smallest force F that must be applied to the rope in order to create a moment of $M = 900$ N·m at point O.

*4–32. If $\mathbf{F} = \{50\mathbf{i} + 60\mathbf{j} + 30\mathbf{k}\}$ lb, determine the magnitude and coordinate direction angles of the moment of **F** about point A.

4–33. Determine the coordinate direction angles of the force **F** applied at the end of the pipe such that the moment of **F** about point A is zero.

Probs. 4–29/30

Probs. 4–32/33

4–34. Determine the moment of the force of $F = 600$ N about point A.

4–35. Determine the smallest force F that must be applied along the rope in order to develop a moment of $M = 1500$ N·m at A.

4–38. Determine the moment of the force **F** about the door hinge at A. Express the result as a Cartesian vector.

4–39. Determine the moment of the force **F** about the door hinge at B. Express the result as a Cartesian vector.

Probs. 4–34/35

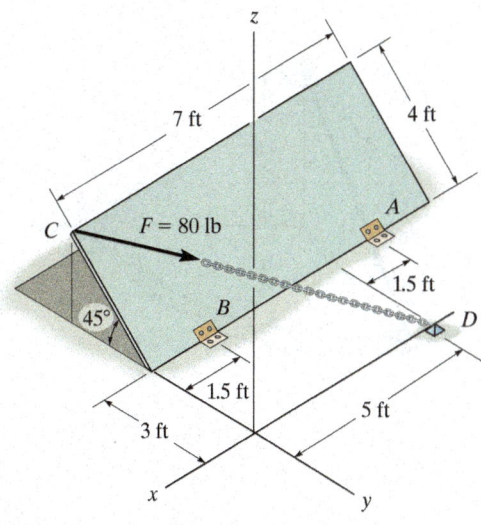

Probs. 4–38/39

*4–36. Determine the coordinate direction angles α, β, γ of force **F**, so that the moment of **F** about O is zero.

4–37. Determine the moment of force **F** about point O. The force has a magnitude of 800 N and coordinate direction angles of $\alpha = 60°$, $\beta = 120°$, $\gamma = 45°$. Express the result as a Cartesian vector.

*4–40. The curved rod has a radius of 5 ft. If a force of 60 lb acts at its end as shown, determine the moment of this force about point C.

4–41. Determine the smallest force F that must be applied along the rope in order to cause the curved rod, which has a radius of 5 ft, to fail at the support C. This requires a moment of $M = 80$ lb·ft to be developed at C.

Probs. 4–36/37

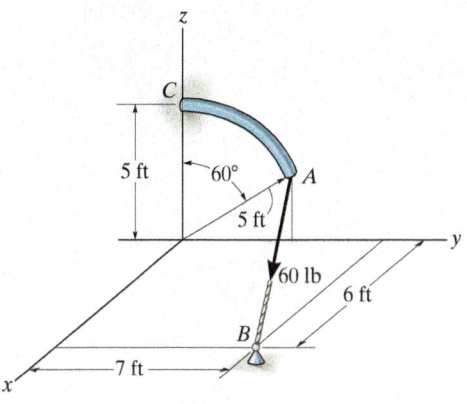

Probs. 4–40/41

4–42. The curved rod has a radius of 4 ft. If the cable AB exerts a force of 80 lb on the rod, determine the moment of this force about point C. Solve the problem by using two different position vectors.

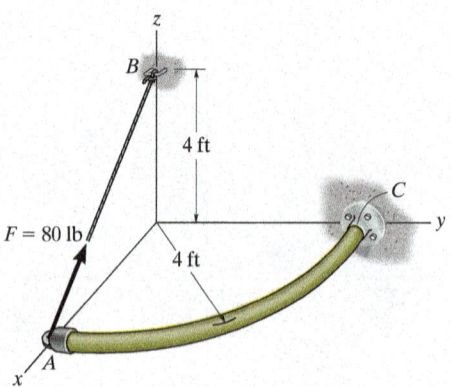

Prob. 4–42

4–43. The pipe assembly is subjected to the 80-N force. Determine the moment of this force about point A.

*4–44. The pipe assembly is subjected to the 80-N force. Determine the moment of this force about point B.

Probs. 4–43/44

4–45. A force of $\mathbf{F} = \{6\mathbf{i} - 2\mathbf{j} + 1\mathbf{k}\}$ kN produces a moment of $\mathbf{M}_O = \{4\mathbf{i} + 5\mathbf{j} - 14\mathbf{k}\}$ kN·m about the origin, point O. If the force acts at a point having an x coordinate of x = 1 m, determine the y and z coordinates. *Note:* The figure shows **F** and $\mathbf{M}_O$ in an arbitrary position.

4–46. The force $\mathbf{F} = \{6\mathbf{i} + 8\mathbf{j} + 10\mathbf{k}\}$ N creates a moment about point O of $\mathbf{M}_O = \{-14\mathbf{i} + 8\mathbf{j} + 2\mathbf{k}\}$ N·m. If the force passes through a point having an x coordinate of 1 m, determine the y and z coordinates of the point. Also, realizing that $M_O = Fd$, determine the perpendicular distance d from point O to the line of action of **F**. *Note:* The figure shows **F** and $\mathbf{M}_O$ in an arbitrary position.

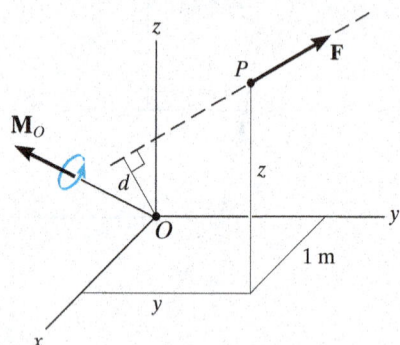

Probs. 4–45/46

4–47. The cable exerts a 140-N force on the telephone pole. Determine the moment of this force about point A. Solve the problem using two different position vectors.

Prob. 4–47

*4–48. Determine the moment about point O of each force acting on the pipe assembly. Add these moments and calculate the magnitude and coordinate direction angles of the resultant moment.

4–50. Using a ring collar the 75-N force can act in the vertical plane at various angles θ. Determine the magnitude of the moment it produces about point A, plot the results of M (ordinate) versus θ (abscissa) for $0° \leq \theta \leq 180°$, and specify the angles that give the maximum and minimum moment.

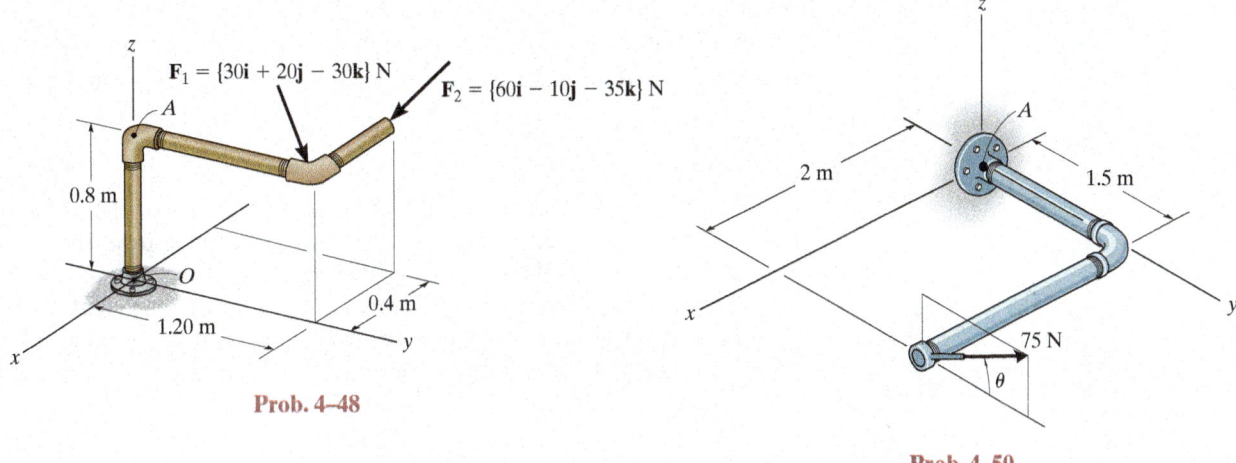

Prob. 4–48

Prob. 4–50

4–49. Determine the moment of each force about point A. Add these moments and calculate the magnitude and coordinate direction angles of the resultant moment.

4–51. The curved pipe has a radius of 5 ft. If a force of 80 lb acts at its end as shown, determine the moment of this force about point C. Solve the problem by using two different position vectors.

Prob. 4–49

Prob. 4–51

4.5 MOMENT OF A FORCE ABOUT A SPECIFIED AXIS

Sometimes, the moment produced by a force about a *specified axis* must be determined. For example, suppose the lug nut at O on the car tire in Fig. 4–20a needs to be loosened. The force applied to the wrench will create a tendency for the wrench and the nut to rotate about the *moment axis* passing through O; however, the nut can only rotate about the y axis. Therefore, to determine the turning effect, only the y component of the moment is needed, and the total moment produced is not important. To determine this component, we can use either a scalar or vector analysis.

Scalar Analysis. To use a scalar analysis in the case of the lug nut in Fig. 4–20a, the moment arm, or perpendicular distance from the axis to the line of action of the force, is $d_y = d \cos \theta$. Thus, the moment of **F** about the y axis is $M_y = F\, d_y = F(d \cos \theta)$. According to the right-hand rule, $\mathbf{M}_y$ is directed along the positive y axis as shown in the figure. In general, for any axis a, the moment is

$$M_a = F d_a \tag{4–10}$$

Realize that a force will not create a moment about an axis (or turn about an axis) if it is *parallel to the axis*, or if its *line of action passes through the axis*.

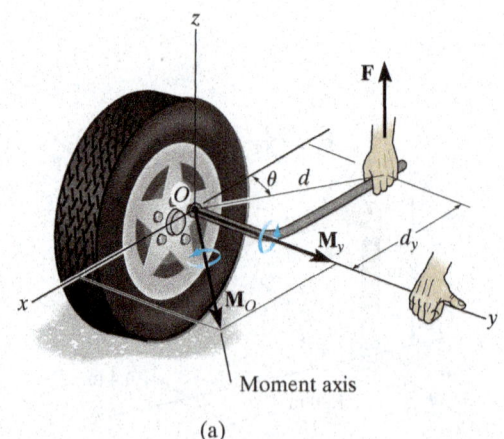

(a)

Fig. 4–20

4.5 MOMENT OF A FORCE ABOUT A SPECIFIED AXIS

Vector Analysis. To find the moment of force **F** in Fig. 4–20b about the y axis using a vector analysis, we must first determine the moment of the force about *any point O* on the y axis by applying Eq. 4–7, $\mathbf{M}_O = \mathbf{r} \times \mathbf{F}$. The component $\mathbf{M}_y$ along the y axis is the *projection* of $\mathbf{M}_O$ onto the y axis. It can be found using the *dot product* discussed in Chapter 2, so that $M_y = \mathbf{j} \cdot \mathbf{M}_O = \mathbf{j} \cdot (\mathbf{r} \times \mathbf{F})$, where **j** is the unit vector for the y axis.

We can generalize this approach by letting $\mathbf{u}_a$ be the unit vector that specifies the direction of the a axis shown in Fig. 4–21. Then the moment of **F** about a point O on the axis is $\mathbf{M}_O = \mathbf{r} \times \mathbf{F}$, and the projection of this moment onto the a axis is $M_a = \mathbf{u}_a \cdot (\mathbf{r} \times \mathbf{F})$. This combination is referred to as the *scalar triple product*. If the vectors are written in Cartesian form, we have

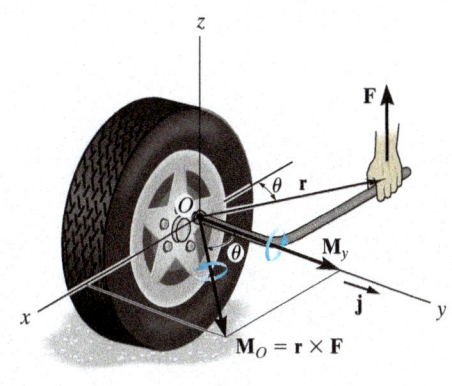

Fig. 4–20 (cont.)

$$M_a = [u_{a_x}\mathbf{i} + u_{a_y}\mathbf{j} + u_{a_z}\mathbf{k}] \cdot \begin{vmatrix} \mathbf{i} & \mathbf{j} & \mathbf{k} \\ r_x & r_y & r_z \\ F_x & F_y & F_z \end{vmatrix}$$

$$= u_{a_x}(r_y F_z - r_z F_y) - u_{a_y}(r_x F_z - r_z F_x) + u_{a_z}(r_x F_y - r_y F_x)$$

This result can also be written in the form of a determinant, making it easier to memorize.*

$$M_a = \mathbf{u}_a \cdot (\mathbf{r} \times \mathbf{F}) = \begin{vmatrix} u_{a_x} & u_{a_y} & u_{a_z} \\ r_x & r_y & r_z \\ F_x & F_y & F_z \end{vmatrix} \quad (4\text{–}11)$$

where

$u_{a_x}, u_{a_y}, u_{a_z}$ represent the x, y, z components of the unit vector defining the direction of the a axis

r_x, r_y, r_z represent the x, y, z components of the position vector extended from *any point O* on the a axis to *any point A* on the line of action of the force

F_x, F_y, F_z represent the x, y, z components of the force vector.

When M_a is evaluated from Eq. 4–11, it will yield a positive or negative scalar. The sign of this scalar indicates the sense of direction of $\mathbf{M}_a$ along the a axis. If it is positive, then $\mathbf{M}_a$ will have the same sense as $\mathbf{u}_a$, whereas if it is negative, then $\mathbf{M}_a$ will act opposite to $\mathbf{u}_a$. Once the a axis is established, point your right-hand thumb in the direction of $\mathbf{M}_a$, and the curl of your fingers will indicate the sense of twist about the axis, Fig. 4–21.

Fig. 4–21

---

*Expand this determinant to show that it will yield the above result.

Once M_a is determined, we can then express $\mathbf{M}_a$ as a Cartesian vector, namely,

$$\mathbf{M}_a = M_a \mathbf{u}_a \qquad (4\text{--}12)$$

The examples which follow illustrate numerical applications of the above concepts.

> ## IMPORTANT POINTS
>
> - The moment of a force about a specified axis can be determined provided the perpendicular distance d_a from the force line of action to the axis can be determined. Then $M_a = Fd_a$.
> - No moment is produced if the force is parallel to the axis or its line of action passes through it.
> - If vector analysis is used, then $M_a = \mathbf{u}_a \cdot (\mathbf{r} \times \mathbf{F})$, where $\mathbf{u}_a$ defines the direction of the axis and $\mathbf{r}$ is extended from *any point* on the axis to *any point* on the line of action of the force.
> - If M_a is calculated as a negative scalar, then the sense of direction of $\mathbf{M}_a$ is opposite to $\mathbf{u}_a$.
> - The moment $\mathbf{M}_a$ expressed as a Cartesian vector is determined from $\mathbf{M}_a = M_a \mathbf{u}_a$.

EXAMPLE 4.7

Determine the resultant moment of the three forces in Fig. 4–22 about the x axis, the y axis, and the z axis.

SOLUTION
A force that is *parallel* to a coordinate axis or has a line of action that passes through the axis does *not* produce any moment or tendency for turning about the axis. Defining the positive direction of the moment of a force according to the right-hand rule, as shown in the figure, we have

$$M_x = (60 \text{ lb})(2 \text{ ft}) + (50 \text{ lb})(2 \text{ ft}) + 0 = 220 \text{ lb} \cdot \text{ft} \qquad Ans.$$

$$M_y = 0 - (50 \text{ lb})(3 \text{ ft}) - (40 \text{ lb})(2 \text{ ft}) = -230 \text{ lb} \cdot \text{ft} \qquad Ans.$$

$$M_z = 0 + 0 - (40 \text{ lb})(2 \text{ ft}) = -80 \text{ lb} \cdot \text{ft} \qquad Ans.$$

The negative signs indicate that $\mathbf{M}_y$ and $\mathbf{M}_z$ act in the $-y$ and $-z$ directions, respectively.

Fig. 4–22

EXAMPLE 4.8

Determine the moment $\mathbf{M}_{AB}$ produced by the force $\mathbf{F}$ in Fig. 4–23a, which tends to rotate the rod about the AB axis.

SOLUTION

A vector analysis using $M_{AB} = \mathbf{u}_B \cdot (\mathbf{r} \times \mathbf{F})$ will be considered for the solution rather than trying to find the moment arm or perpendicular distance from the line of action of $\mathbf{F}$ to the AB axis. Each of the terms in the equation will now be identified.

Unit vector $\mathbf{u}_B$ defines the direction of the AB axis of the rod, Fig. 4–23b, where

$$\mathbf{u}_B = \frac{\mathbf{r}_B}{r_B} = \frac{\{0.4\mathbf{i} + 0.2\mathbf{j}\}\text{ m}}{\sqrt{(0.4\text{ m})^2 + (0.2\text{ m})^2}} = 0.8944\mathbf{i} + 0.4472\mathbf{j}$$

Vector $\mathbf{r}$ is directed from *any point* on the AB axis to *any point* on the line of action of the force. For example, position vectors $\mathbf{r}_C$ and $\mathbf{r}_D$ are suitable, Fig. 4–23b. (Although not shown, $\mathbf{r}_{BC}$ or $\mathbf{r}_{BD}$ can also be used.) For simplicity, we choose $\mathbf{r}_D$, where

$$\mathbf{r}_D = \{0.6\mathbf{i}\}\text{ m}$$

The force is

$$\mathbf{F} = \{-300\mathbf{k}\}\text{ N}$$

Substituting these vectors into the determinant form of the triple product and expanding, we have

$$M_{AB} = \mathbf{u}_B \cdot (\mathbf{r}_D \times \mathbf{F}) = \begin{vmatrix} 0.8944 & 0.4472 & 0 \\ 0.6 & 0 & 0 \\ 0 & 0 & -300 \end{vmatrix}$$

$$= 0.8944[0(-300) - 0(0)] - 0.4472[0.6(-300) - 0(0)]$$
$$+ 0[0.6(0) - 0(0)]$$

$$= 80.50\text{ N}\cdot\text{m}$$

Fig. 4–23

This positive result indicates that the sense of $\mathbf{M}_{AB}$ is in the same direction as $\mathbf{u}_B$, Fig. 4–23b.

Expressing $\mathbf{M}_{AB}$ in Fig. 4–23b as a Cartesian vector yields

$$\mathbf{M}_{AB} = M_{AB}\mathbf{u}_B = (80.50\text{ N}\cdot\text{m})(0.8944\mathbf{i} + 0.4472\mathbf{j})$$

$$= \{72.0\mathbf{i} + 36.0\mathbf{j}\}\text{ N}\cdot\text{m} \qquad \textit{Ans.}$$

NOTE: If the axis AB is defined using a unit vector directed from B toward A, then in the above determinant $-\mathbf{u}_B$ would have to be used. This would lead to $M_{AB} = -80.50\text{ N}\cdot\text{m}$. Consequently, $\mathbf{M}_{AB} = M_{AB}(-\mathbf{u}_B)$, and the same vector result would be obtained.

EXAMPLE 4.9

Determine the magnitude of the moment of force **F** about segment *OA* of the pipe assembly in Fig. 4–24a.

SOLUTION

The moment of **F** about the *OA* axis is determined from $M_{OA} = \mathbf{u}_{OA} \cdot (\mathbf{r} \times \mathbf{F})$, where **r** is a position vector extending from any point on the *OA* axis to any point on the line of action of **F**. As indicated in Fig. 4–24b, either $\mathbf{r}_{OD}$, $\mathbf{r}_{OC}$, $\mathbf{r}_{AD}$, or $\mathbf{r}_{AC}$ can be used. Here, $\mathbf{r}_{OD}$ will be considered since it will simplify the calculation.

The unit vector $\mathbf{u}_{OA}$, which specifies the direction of the *OA* axis, is

$$\mathbf{u}_{OA} = \frac{\mathbf{r}_{OA}}{r_{OA}} = \frac{\{0.3\mathbf{i} + 0.4\mathbf{j}\} \text{ m}}{\sqrt{(0.3 \text{ m})^2 + (0.4 \text{ m})^2}} = 0.6\mathbf{i} + 0.8\mathbf{j}$$

and the position vector $\mathbf{r}_{OD}$ is

$$\mathbf{r}_{OD} = \{0.5\mathbf{i} + 0.5\mathbf{k}\} \text{ m}$$

The force **F** expressed as a Cartesian vector is

$$\mathbf{F} = F \left(\frac{\mathbf{r}_{CD}}{r_{CD}} \right)$$

$$= (300 \text{ N}) \left[\frac{\{0.4\mathbf{i} - 0.4\mathbf{j} + 0.2\mathbf{k}\} \text{ m}}{\sqrt{(0.4 \text{ m})^2 + (-0.4 \text{ m})^2 + (0.2 \text{ m})^2}} \right]$$

$$= \{200\mathbf{i} - 200\mathbf{j} + 100\mathbf{k}\} \text{ N}$$

Therefore,

$$M_{OA} = \mathbf{u}_{OA} \cdot (\mathbf{r}_{OD} \times \mathbf{F})$$

$$= \begin{vmatrix} 0.6 & 0.8 & 0 \\ 0.5 & 0 & 0.5 \\ 200 & -200 & 100 \end{vmatrix}$$

$$= 0.6[0(100) - (0.5)(-200)] - 0.8[0.5(100) - (0.5)(200)] + 0$$

$$= 100 \text{ N} \cdot \text{m} \qquad Ans.$$

(a)

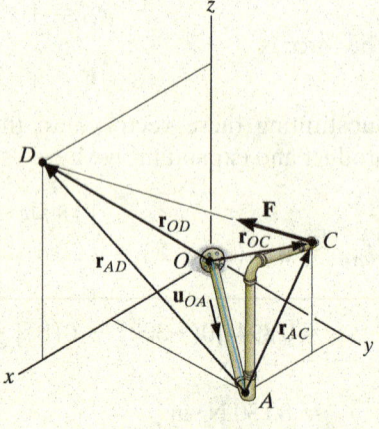

(b)

Fig. 4–24

FUNDAMENTAL PROBLEMS

F4–13. Determine the magnitude of the moment of the force about the y axis.

$\mathbf{F} = \{30\mathbf{i} - 20\mathbf{j} + 50\mathbf{k}\}$ N

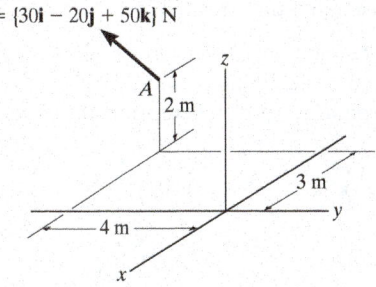

Prob. F4–13

F4–14. Determine the magnitude of the moment of the force $\mathbf{F} = \{300\mathbf{i} - 200\mathbf{j} + 150\mathbf{k}\}$ N about the x axis.

F4–15. Determine the magnitude of the moment of the force $\mathbf{F} = \{300\mathbf{i} - 200\mathbf{j} + 150\mathbf{k}\}$ N about the OA axis.

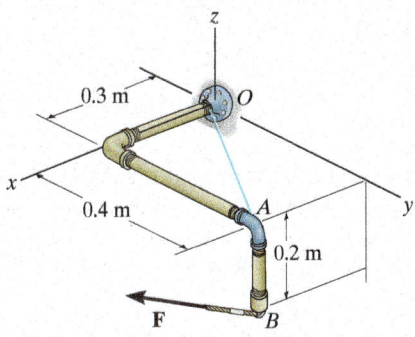

Probs. F4–14/15

F4–16. Determine the moment of the force $\mathbf{F} = \{50\mathbf{i} - 40\mathbf{j} + 20\mathbf{k}\}$ lb about the AB axis. Express the result as a Cartesian vector.

Prob. F4–16

F4–17. Determine the magnitude of the moment of the 200-N force about the x axis. Solve the problem using both a scalar and a vector analysis.

Prob. F4–17

F4–18. Determine the moment of force $\mathbf{F}$ about the x, the y, and the z axis. Solve the problem using both a scalar and a vector analysis.

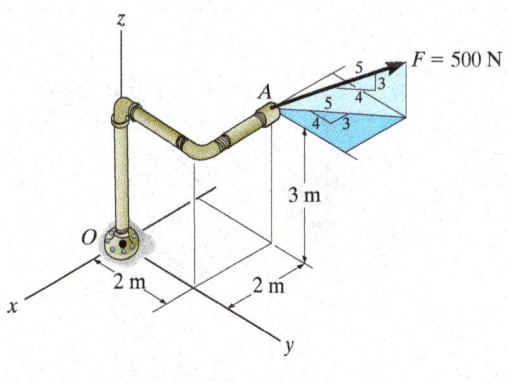

Prob. F4–18

PROBLEMS

***4–52.** Determine the moment of force **F** about the x, y, and z axes. Solve the problem (a) using a Cartesian vector approach and (b) using a scalar approach.

4–53. Determine the moment of force **F** about an axis extending between O and A. Express the result as a Cartesian vector.

***4–56.** The force of $F = 80$ lb acts along the edge DB of the tetrahedron. Determine the magnitude of the moment of this force about the edge CA.

4–57. If the moment of the force **F** about the edge CA of the tetrahedron has a magnitude of $M = 200$ lb·ft and is directed from C toward A, determine the magnitude of **F**.

Probs. 4–52/53

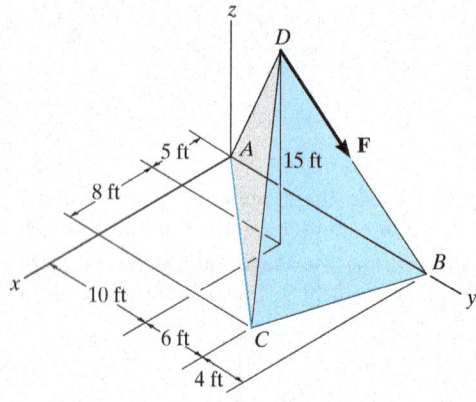

Probs. 4–56/57

4–54. Determine the moment of the force **F** about the diagonal AF of the rectangular block. Express the result as a Cartesian vector.

4–55. Determine the moment of the force **F** about the diagonal OD of the rectangular block. Express the result as a Cartesian vector.

4–58. The board is used to hold the end of the cross lug wrench in the position shown when the man applies a force of $F = 100$ N. Determine the magnitude of the moment produced by this force about the x axis. Force **F** lies in a vertical plane.

4–59. The board is used to hold the end of the cross lug wrench in the position shown. If a torque of 30 N·m about the x axis is required to tighten the nut, determine the required magnitude of the force **F** needed to turn the wrench. Force **F** lies in a vertical plane.

Probs. 4–54/55

Probs. 4–58/59

*4–60. The bevel gear is subjected to the force **F** which is caused from contact with another gear. Determine the moment of this force about the *y* axis of the gear shaft.

Prob. 4–60

4–61. Determine the magnitude of the moment of the force **F** = {50**i** − 20**j** − 80**k**} N about member *AB* of the tripod.

4–62. Determine the magnitude of the moment of the force **F** = {50**i** − 20**j** − 80**k**} N about member *BC* of the tripod.

4–63. Determine the magnitude of the moment of the force **F** = {50**i** − 20**j** − 80**k**} N about member *CA* of the tripod.

Probs. 4–61/62/63

*4–64. Determine the moment of each force acting on the handle of the wrench about the *a* axis. Take **F**₁ = {−2**i** + 4**j** − 8**k**} lb, **F**₂ = {3**i** + 2**j** − 6**k**} lb.

4–65. Determine the moment of each force acting on the handle of the wrench about the *z* axis. Take **F**₁ = {−2**i** + 4**j** − 8**k**} lb, **F**₂ = {3**i** + 2**j** − 6**k**} lb.

Probs. 4–64/65

4–66. The Rollerball skate is an in-line tandem skate that uses two large spherical wheels on each skate, rather than traditional wafer-shape wheels. During skating the two forces acting on the ball of one skate consist of a 78-lb normal force and a 13-lb friction force. Determine the moment of both of these forces about the axle *AB*.

Prob. 4–66

144   CHAPTER 4   FORCE SYSTEM RESULTANTS

4–67. Determine the resultant moment of the two forces about the Oa axis. Express the result as a Cartesian vector.

4–70. The tool is used to shut off gas valves that are difficult to access. If the force **F** is applied to the handle, determine the component of the moment created about the z axis of the valve.

Prob. 4–67

Prob. 4–70

*4–68. The pipe assembly is secured on the wall by the two brackets. If the flowerpot has a weight of 50 lb, determine the magnitude of the moment produced by the weight about the OA axis.

4–69. The pipe assembly is secured on the wall by the two brackets. If the frictional force of both brackets can resist a maximum moment of 150 lb·ft, determine the largest weight of the flowerpot that can be supported by the assembly.

4–71. The A-frame is being hoisted into an upright position by the vertical force of $F = 80$ lb. Determine the moment of this force about the y' axis passing through points A and B when the frame is in the position shown.

*4–72. The A-frame is being hoisted into an upright position by the vertical force of $F = 80$ lb. Determine the moment of this force about the x axis when the frame is in the position shown.

Probs. 4–68/69

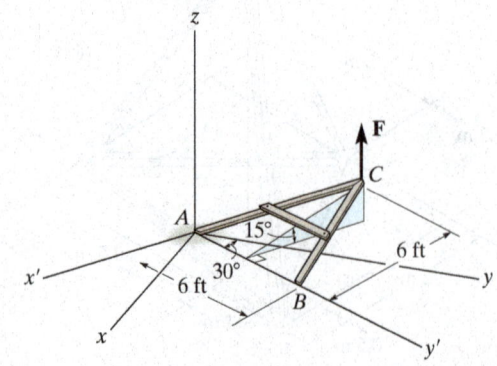

Probs. 4–71/72

4.6 MOMENT OF A COUPLE

A *couple* is defined as two parallel forces that have the same magnitude, but opposite directions, and are separated by a perpendicular distance d, Fig. 4–25. Since the resultant force is zero, the only effect of a couple is to produce a rotation, or if no movement is possible, there is a tendency for rotation.

The moment produced by a couple is called a *couple moment*. We can determine its value by finding the sum of the moments of both couple forces about *any* arbitrary point. For example, in Fig. 4–26, position vectors $\mathbf{r}_A$ and $\mathbf{r}_B$ are directed from point O to points A and B lying on the line of action of $-\mathbf{F}$ and $\mathbf{F}$. The couple moment determined about O is therefore

$$\mathbf{M} = \mathbf{r}_B \times \mathbf{F} + \mathbf{r}_A \times (-\mathbf{F}) = (\mathbf{r}_B - \mathbf{r}_A) \times \mathbf{F}$$

However, $\mathbf{r}_B = \mathbf{r}_A + \mathbf{r}$ or $\mathbf{r} = \mathbf{r}_B - \mathbf{r}_A$, so that

$$\mathbf{M} = \mathbf{r} \times \mathbf{F} \quad (4\text{--}13)$$

Fig. 4–25

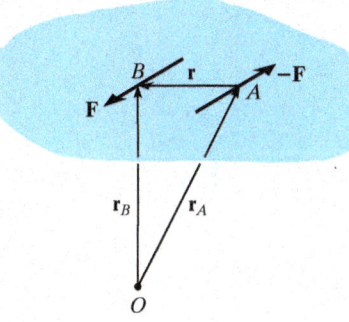

Fig. 4–26

This result indicates that a couple moment is a *free vector*, i.e., it can act at *any point* since $\mathbf{M}$ depends *only* upon the position vector $\mathbf{r}$ directed *between* the forces and *not* the position vectors $\mathbf{r}_A$ and $\mathbf{r}_B$, directed from point O to the forces.

Scalar Formulation.
The moment of a couple, Fig. 4–27, has a *magnitude* of

$$\boxed{M = Fd} \quad (4\text{--}14)$$

where F is the magnitude of one of the forces and d is the perpendicular distance or moment arm between the forces. The *direction* and sense of the couple moment are determined by the right-hand rule, where the thumb indicates this direction when the fingers are curled with the sense of rotation caused by the couple forces. In all cases, $\mathbf{M}$ will act perpendicular to the plane containing these forces.

Vector Formulation.
As noted above, the moment of a couple can also be expressed by the vector cross product using Eq. 4–13, i.e.,

$$\boxed{\mathbf{M} = \mathbf{r} \times \mathbf{F}} \quad (4\text{--}15)$$

Application of this equation is easily remembered if one thinks of taking the moments of both forces about a point lying on the line of action of one of the forces. For example, if moments are taken about point A in Fig. 4–26, the moment of $-\mathbf{F}$ is *zero* about this point, and the moment of $\mathbf{F}$ is defined from Eq. 4–15. Therefore, in the formulation $\mathbf{r}$ is crossed with the force $\mathbf{F}$ to which it is directed.

Fig. 4–27

146   CHAPTER 4   FORCE SYSTEM RESULTANTS

Fig. 4–28

(a)

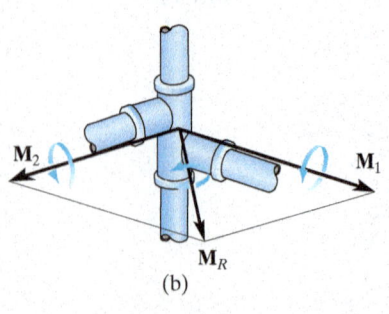

(b)

Fig. 4–29

Equivalent Couples. If two couples produce a moment with the *same magnitude and direction*, then these two couples are *equivalent*. For example, the two couples shown in Fig. 4–28 are *equivalent* because each couple moment has a magnitude of $M = 30\,\text{N}(0.4\,\text{m}) = 40\,\text{N}(0.3\,\text{m}) = 12\,\text{N}\cdot\text{m}$, and each is directed into the plane of the page. Notice that larger forces are required in the second case to create the same turning effect about the center shaft because the hands are placed closer together. Also, if the wheel was connected to the shaft at a point other than at its center, then the wheel would still be subjected to the same turning effect when each couple is applied since the 12 N · m couple is a free vector.

Resultant Couple Moment. Since couple moments are free vectors, their resultant can be determined by moving them to a single point and using vector addition. For example, to find the resultant of couple moments $\mathbf{M}_1$ and $\mathbf{M}_2$ acting on the pipe assembly in Fig. 4–29a, we can join their tails at point O and find the resultant couple moment, $\mathbf{M}_R = \mathbf{M}_1 + \mathbf{M}_2$, as shown in Fig. 4–29b.

If more than two couple moments act on the body, we may generalize this concept and write the vector resultant as

$$\mathbf{M}_R = \Sigma(\mathbf{r} \times \mathbf{F}) \qquad (4\text{--}16)$$

These concepts are illustrated numerically in the examples that follow. In general, problems projected in two dimensions should be solved using a scalar analysis since the moment arms and force components are easy to determine.

IMPORTANT POINTS

- A couple moment is produced by two noncollinear forces that are equal in magnitude but opposite in direction. Its effect is to produce pure rotation, or tendency for rotation in a specified direction.

- A couple moment is a free vector, and as a result it causes the same rotational effect on a body regardless of where the couple moment is applied to the body.

- The moment of the two couple forces can be determined about *any point*. For convenience, this point is often chosen on the line of action of one of the forces in order to eliminate the moment of this force about the point.

- In three dimensions the couple moment is often determined using the vector formulation, $\mathbf{M} = \mathbf{r} \times \mathbf{F}$, where $\mathbf{r}$ is directed from *any point* on the line of action of one of the forces to *any point* on the line of action of the other force $\mathbf{F}$.

- A resultant couple moment is simply the vector sum of all the couple moments of the system.

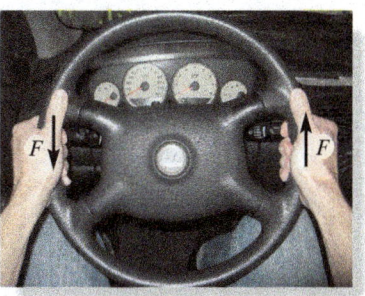

Steering wheels today have a smaller diameter than on older vehicles because power steering does not require the driver to apply a large couple moment to the wheel.

EXAMPLE 4.10

Determine the resultant couple moment of the three couples acting on the plate in Fig. 4–30.

Fig. 4–30

SOLUTION
As shown the perpendicular distances between each pair of couple forces are $d_1 = 4$ ft, $d_2 = 3$ ft, and $d_3 = 5$ ft. Considering counterclockwise couple moments as positive, we have

$\zeta+ M_R = \Sigma M;\ M_R = -F_1 d_1 + F_2 d_2 - F_3 d_3$
$= -(200\ \text{lb})(4\ \text{ft}) + (450\ \text{lb})(3\ \text{ft}) - (300\ \text{lb})(5\ \text{ft})$
$= -950\ \text{lb} \cdot \text{ft} = 950\ \text{lb} \cdot \text{ft}\ \downarrow$ *Ans.*

EXAMPLE 4.11

Determine the magnitude and direction of the couple moment acting on the gear in Fig. 4–31a.

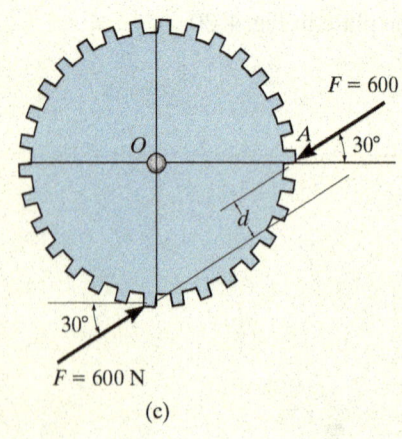

Fig. 4–31

SOLUTION

The easiest solution requires resolving each force into its components as shown in Fig. 4–31b. The couple moment can be determined by summing the moments of these force components about any point, for example, the center O of the gear or point A. If we consider counterclockwise moments as positive, we have

$\zeta+ M = \Sigma M_O;\ M = (600 \cos 30°\ N)(0.2\ m) - (600 \sin 30°\ N)(0.2\ m)$

$= 43.9\ N \cdot m \zeta$ *Ans.*

or

$\zeta+ M = \Sigma M_A;\ M = (600 \cos 30°\ N)(0.2\ m) - (600 \sin 30°\ N)(0.2\ m)$

$= 43.9\ N \cdot m \zeta$ *Ans.*

This positive result indicates that **M** has a counterclockwise rotational sense, so it is directed outward, perpendicular to the page.

NOTE: The same result can also be obtained using $M = Fd$, where d is the perpendicular distance between the lines of action of the couple forces, Fig. 4–31c. However, the computation for d is more involved. Also, realize that the couple moment is a free vector and can act at any point on the gear and produce the same turning effect about point O.

EXAMPLE 4.12

Determine the couple moment acting on the pipe shown in Fig. 4–32a. Segment AB is directed $30°$ below the x–y plane.

SOLUTION I (VECTOR ANALYSIS)

The moment of the two couple forces can be found about *any point*. If point O is considered, Fig. 4–32b, we have

$$\mathbf{M} = \mathbf{r}_A \times (-25\mathbf{k}) + \mathbf{r}_B \times (25\mathbf{k})$$
$$= (8\mathbf{j}) \times (-25\mathbf{k}) + (6\cos 30°\mathbf{i} + 8\mathbf{j} - 6\sin 30°\mathbf{k}) \times (25\mathbf{k})$$
$$= -200\mathbf{i} - 129.9\mathbf{j} + 200\mathbf{i}$$
$$= \{-130\mathbf{j}\} \text{ lb}\cdot\text{in.} \qquad Ans.$$

It is *easier* to take moments of the couple forces about a point lying on the line of action of one of the forces, e.g., point A, Fig. 4–32c. In this case the moment of the force at A is zero, so that

$$\mathbf{M} = \mathbf{r}_{AB} \times (25\mathbf{k})$$
$$= (6\cos 30°\mathbf{i} - 6\sin 30°\mathbf{k}) \times (25\mathbf{k})$$
$$= \{-130\mathbf{j}\} \text{ lb}\cdot\text{in.} \qquad Ans.$$

SOLUTION II (SCALAR ANALYSIS)

Although this problem is shown in three dimensions, the geometry is simple enough to use the scalar equation $M = Fd$, where $d = 6\cos 30° = 5.196$ in., Fig. 4–32d. Hence, taking moments of the forces about either point A or point B yields

$$M = Fd = 25 \text{ lb } (5.196 \text{ in.}) = 129.9 \text{ lb}\cdot\text{in.}$$

Applying the right-hand rule, $\mathbf{M}$ acts in the $-\mathbf{j}$ direction. Thus,

$$\mathbf{M} = \{-130\mathbf{j}\} \text{ lb}\cdot\text{in.} \qquad Ans.$$

Fig. 4–32

EXAMPLE 4.13

Replace the two couples acting on the pipe assembly in Fig. 4–33a by a resultant couple moment.

Fig. 4–33

SOLUTION (VECTOR ANALYSIS)

The couple moment $\mathbf{M}_1$, developed by the forces at A and B, can easily be determined from a scalar formulation.

$$M_1 = Fd = 150 \text{ N}(0.4 \text{ m}) = 60 \text{ N} \cdot \text{m}$$

By the right-hand rule, $\mathbf{M}_1$ acts in the $+\mathbf{i}$ direction, Fig. 4–33b. Hence,

$$\mathbf{M}_1 = \{60\mathbf{i}\} \text{ N} \cdot \text{m}$$

Vector analysis will be used to determine $\mathbf{M}_2$, caused by forces at C and D. If moments are calculated about point D, Fig. 4–33a, $\mathbf{M}_2 = \mathbf{r}_{DC} \times \mathbf{F}_C$, then

$$\mathbf{M}_2 = \mathbf{r}_{DC} \times \mathbf{F}_C = (0.3\mathbf{i}) \times \left[125\left(\tfrac{4}{5}\right)\mathbf{j} - 125\left(\tfrac{3}{5}\right)\mathbf{k}\right]$$
$$= (0.3\mathbf{i}) \times [100\mathbf{j} - 75\mathbf{k}] = 30(\mathbf{i} \times \mathbf{j}) - 22.5(\mathbf{i} \times \mathbf{k})$$
$$= \{22.5\mathbf{j} + 30\mathbf{k}\} \text{ N} \cdot \text{m}$$

Since $\mathbf{M}_1$ and $\mathbf{M}_2$ are free vectors, Fig. 4–33b, they may be moved to some arbitrary point and added vectorially, Fig. 4–33c. The resultant couple moment becomes

$$\mathbf{M}_R = \mathbf{M}_1 + \mathbf{M}_2 = \{60\mathbf{i} + 22.5\mathbf{j} + 30\mathbf{k}\} \text{ N} \cdot \text{m} \qquad Ans.$$

FUNDAMENTAL PROBLEMS

F4–19. Determine the resultant couple moment acting on the beam.

Prob. F4–19

F4–20. Determine the resultant couple moment acting on the triangular plate.

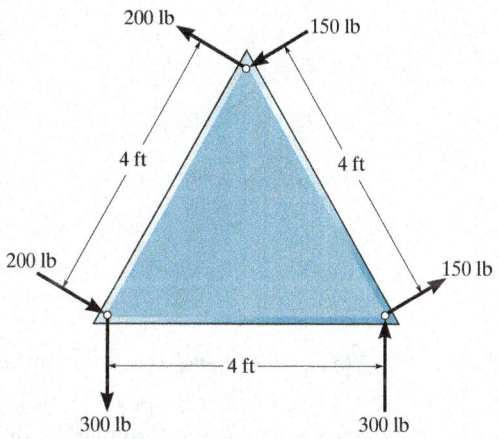

Prob. F4–20

F4–21. Determine the couple moment acting on the beam.

Prob. F4–21

F4–22. Determine the magnitude of **F** so that the resultant couple moment acting on the beam is 1.5 kN · m clockwise.

Prob. F4–22

F4–23. Determine the resultant couple moment acting on the pipe assembly.

Prob. F4–23

F4–24. Determine the couple moment acting on the pipe assembly and express the result as a Cartesian vector.

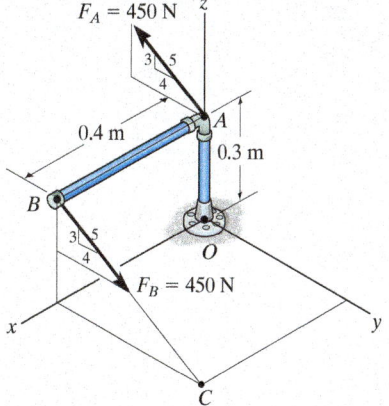

Prob. F4–24

PROBLEMS

4–73. The crossbar wrench is used to remove a lug nut from the automobile wheel. The mechanic applies a couple to the wrench such that his hands are a constant distance apart. Is it necessary that $a = b$ in order to produce the most effective turning of the nut? Explain. Also, what is the effect of changing the shaft dimension c in this regard? The forces act in the vertical plane.

4–75. Determine the magnitude of the forces **F** and **−F**, so that the resultant couple moment is 400 N · m clockwise.

Prob. 4–73

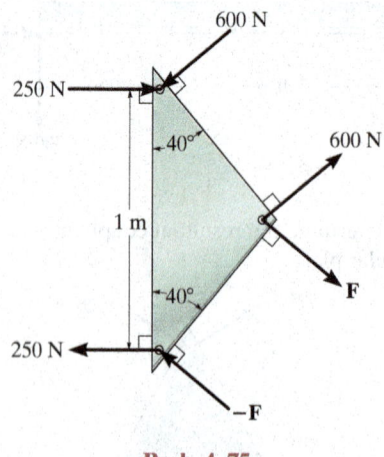

Prob. 4–75

4–74. The hanger H is used to support the end of the floor beam B. If the loading on the hanger consists of two couples, determine the magnitude of the horizontal forces **F** and **−F** so that the resultant couple moment on the hanger is zero.

Prob. 4–74

*__4–76.__ If $F = 125$ lb, determine the resultant couple moment.

4–77. Determine the magnitude of **F** so that the resultant couple moment is 450 lb · ft, counterclockwise. Where on the beam does the resultant couple moment act?

Probs. 4–76/77

4–78. The road exerts a torque of $M_A = 400$ N·m and $M_B = 200$ N·m on the brushes of the road sweeper. Determine the magnitude of the couple forces that are developed by the road on the rear wheels of the sweeper, so that the resultant couple moment on the sweeper is zero. What is the magnitude of these forces if the brush at B is turned off?

*4–80. The man tries to open the valve by applying the couple forces of $F = 75$ N to the wheel. Determine the couple moment produced.

4–81. If the valve can be opened with a couple moment of 25 N·m, determine the required magnitude of each couple force which must be applied to the wheel.

Prob. 4–78

Probs. 4–80/81

4–79. The cord passing over the two small pegs A and B on the board is subjected to a tension of 10 lb. Determine the minimum tension P and the orientation θ of the cord passing over pegs C and D, so that the resultant couple produced by both cords is 20 lb·in. clockwise.

4–82. Two couples act on the beam. If the resultant couple is to be zero, determine the magnitudes of **P** and **F**, and the distance d between A and B.

Prob. 4–79

Prob. 4–82

4–83. Three couple moments act on the pipe assembly. Determine the magnitude of the resultant couple moment if $M_2 = 50\ \text{N}\cdot\text{m}$ and $M_3 = 35\ \text{N}\cdot\text{m}$.

*4–84. Three couple moments act on the pipe assembly. Determine the magnitudes of $\mathbf{M}_2$ and $\mathbf{M}_3$ so that the resultant couple moment is zero.

*4–88. Express the moment of the couple acting on the pipe assembly as a Cartesian vector. What is the magnitude of the couple moment?

Prob. 4–88

Probs. 4–83/84

4–85. Two couples act on the frame. If the resultant couple moment is to be zero, determine the distance d between the 100-lb couple forces.

4–86. Two couples act on the frame. If $d = 4$ ft, determine the resultant couple moment by (a) summing the moments of the two couples and (b) resolving each force into x and y components and then summing the moments about A of all the force components.

4–87. Two couples act on the frame. If $d = 6$ ft, determine the resultant couple moment by (a) summing the moments of the two couples and (b) resolving each force into x and y components and then summing the moments about B of all the force components.

4–89. If $M_1 = 180\ \text{lb}\cdot\text{ft}$, $M_2 = 90\ \text{lb}\cdot\text{ft}$, and $M_3 = 120\ \text{lb}\cdot\text{ft}$, determine the magnitude and coordinate direction angles of the resultant couple moment.

4–90. Determine the magnitudes of couple moments $\mathbf{M}_1$, $\mathbf{M}_2$, and $\mathbf{M}_3$ so that the resultant couple moment is zero.

Probs. 4–85/86/87

Probs. 4–89/90

4–91. Determine the magnitude and coordinate direction angles of the resultant couple moment.

Prob. 4–91

4–94. Express the moment of the couple acting on the frame as a Cartesian vector. The forces are applied perpendicular to the frame. What is the magnitude of the couple moment? Take $F = 50$ N.

4–95. If the component of the couple moment along the x axis is $\mathbf{M}_x = \{-20\mathbf{i}\}$ N·m, determine the magnitude F of the couple forces.

Probs. 4–94/95

*4–92. If $F = 80$ N, determine the magnitude and coordinate direction angles of the couple moment. The pipe assembly lies in the x–y plane.

4–93. If the magnitude of the couple moment acting on the pipe assembly is 50 N·m, determine the magnitude of the couple forces applied to each wrench. The pipe assembly lies in the x–y plane.

*4–96. Express the moment of the couple acting on the pipe as a Cartesian vector. What is the magnitude of this couple moment? Take $F = 125$ N.

4–97. If the couple moment acting on the pipe has a magnitude of 300 N·m, determine the magnitude of the forces applied to the wrenches.

Probs. 4–92/93

Probs. 4–96/97

4–98. If $M_1 = 500$ N·m, $M_2 = 600$ N·m, and $M_3 = 450$ N·m, determine the magnitude and coordinate direction angles of the resultant couple moment.

4–99. Determine the required magnitude of couple moments M_1, M_2, and M_3 so that the resultant couple moment is $M_R = \{-300\mathbf{i} + 450\mathbf{j} - 600\mathbf{k}\}$ N·m.

4–101. If $F = 80$ N, determine the magnitude and coordinate direction angles of the couple moment. The pipe assembly lies in the x–y plane.

Probs. 4–98/99

Prob. 4–101

*4–100. If the resultant couple moment acting on the triangular block is to be zero, determine the magnitudes F and P.

4–102. If the magnitude of the couple moment acting on the pipe assembly is 50 N·m, determine the magnitude of the couple forces applied to each wrench. The pipe assembly lies in the x–y plane.

Prob. 4–100

Prob. 4–102

4.7 SIMPLIFICATION OF A FORCE AND COUPLE SYSTEM

Sometimes it is convenient to reduce a system of forces and couple moments acting on a body to a simpler form by replacing it with an *equivalent system*, consisting of a single resultant force and a resultant couple moment. A system is equivalent if the *external effects* it produces on a body are the same as those caused by the original force and couple moment system. If the body is free to move, then the external effects of a system refer to the *translating and rotating motion* of the body, or if the body is fully supported, they refer to the *reactive forces* at the supports.

For example, consider holding the stick in Fig. 4–34a, which is subjected to the force **F** at point A. If we attach a pair of equal but opposite forces **F** and −**F** at point B, which is *on the line of action* of **F**, Fig. 4–34b, we observe that −**F** at B and **F** at A will cancel each other, leaving only **F** at B, Fig. 4–34c. Force **F** has now been moved from A to B without modifying its *external effects* on the stick; i.e., the reaction at the grip remains the same. This demonstrates the ***principle of transmissibility***, which states that a force acting on a body (stick) is a **sliding vector** since it can be applied at any point along its line of action.

We can also use the above procedure to move a force to a point that is *not* on the line of action of the force. If **F** is applied perpendicular to the stick, as in Fig. 4–35a, then we can attach a pair of equal but opposite forces **F** and −**F** to B, Fig. 4–35b. Force **F** is now applied at B, and the other two forces, **F** at A and −**F** at B, form a couple that produces the couple moment $M = Fd$, Fig. 4–35c. Therefore, the force **F** can be moved from A to B provided a couple moment **M** is *added* to maintain an equivalent system. This couple moment is determined by taking the moment of **F** about B. Since **M** is actually a *free vector*, it can act at any point on the stick. In each case, the systems are equivalent, which causes a downward force **F** and clockwise couple moment $M = Fd$ to be felt at the grip.

Fig. 4–34

Fig. 4–35

System of Forces and Couple Moments.

Using this method, a system of several forces and couple moments acting on a body can be reduced to an equivalent single resultant force acting at a point O and a resultant couple moment. For example, in Fig. 4–36a, O is not on the line of action of $\mathbf{F}_1$, and so this force can be moved to point O provided a couple moment $(\mathbf{M}_O)_1 = \mathbf{r}_1 \times \mathbf{F}_1$ is added to the body. Similarly, the couple moment $(\mathbf{M}_O)_2 = \mathbf{r}_2 \times \mathbf{F}_2$ should be added to the body when we move $\mathbf{F}_2$ to point O. Finally, since the couple moment $\mathbf{M}$ is a free vector, it can just be moved to point O. By doing this, we obtain the equivalent system shown in Fig. 4–36b, which produces the same external effects on the body as that of the force and couple system shown in Fig. 4–36a. If we sum the forces and couple moments, we obtain the resultant force $\mathbf{F}_R = \mathbf{F}_1 + \mathbf{F}_2$ and the resultant couple moment $(\mathbf{M}_R)_O = \mathbf{M} + (\mathbf{M}_O)_1 + (\mathbf{M}_O)_2$, Fig. 4–36c.

Notice that $\mathbf{F}_R$ is independent of the location of point O since it is simply a summation of the forces. However, $(\mathbf{M}_R)_O$ depends upon this location since the moments $\mathbf{M}_1$ and $\mathbf{M}_2$ are determined using the position vectors $\mathbf{r}_1$ and $\mathbf{r}_2$, which extend from O to each force. Also note that $(\mathbf{M}_R)_O$ is a free vector and can act at *any point* on the body, although point O is generally chosen as its point of application.

We can now generalize the above method of reducing a force and couple system to an equivalent resultant force $\mathbf{F}_R$ acting at point O and a resultant couple moment $(\mathbf{M}_R)_O$ by using the following two equations.

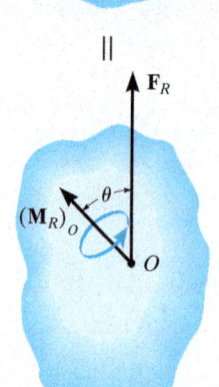

Fig. 4–36

$$\begin{aligned} \mathbf{F}_R &= \Sigma \mathbf{F} \\ (\mathbf{M}_R)_O &= \Sigma \mathbf{M}_O + \Sigma \mathbf{M} \end{aligned} \qquad (4\text{–}17)$$

The first equation states that the resultant force of the system is equivalent to the sum of all the forces; and the second equation states that the resultant couple moment of the system is equivalent to the sum of the moments of all the forces about point O, $\Sigma \mathbf{M}_O$, plus the sum of all the couple moments, $\Sigma \mathbf{M}$.

If the force system lies in the x–y plane and any couple moments are perpendicular to this plane, then the above equations reduce to the following three scalar equations.

$$\begin{aligned} (F_R)_x &= \Sigma F_x \\ (F_R)_y &= \Sigma F_y \\ (M_R)_O &= \Sigma M_O + \Sigma M \end{aligned} \qquad (4\text{–}18)$$

Here the resultant force is determined from the vector sum of its two components $(F_R)_x$ and $(F_R)_y$.

IMPORTANT POINTS

- Force is a sliding vector, since it will create the same external effects on a body when it is applied at any point P along its line of action. This is called the principle of transmissibility.

- A couple moment is a free vector since it will create the same external effects on a body when it is applied at any point P on the body.

- When a force is moved to another point P that is not on its line of action, it will create the same external effects on the body if a couple moment is also applied to the body. The couple moment is determined by taking the moment of the force about point P.

The weights of these traffic lights can be replaced by their equivalent resultant force $W_R = W_1 + W_2$ and a couple moment $(M_R)_O = W_1 d_1 + W_2 d_2$ at the support, O. In both cases the support must provide the same resistance to translation and rotation in order to keep the member in the horizontal position.

PROCEDURE FOR ANALYSIS

The following points should be kept in mind when simplifying a force and couple moment system to an equivalent resultant force and couple system.

- Establish the coordinate axes with the origin located at point O and the axes having a selected orientation.

Force Summation.

- If the force system is *coplanar*, resolve each force into its x and y components. If a component is directed along the positive x or y axis, it represents a positive scalar; whereas if it is directed along the negative x or y axis, it is a negative scalar.

- In three dimensions, represent each force as a Cartesian vector before summing the forces.

Moment Summation.

- When determining the moments of a *coplanar* force system about point O, it is generally advantageous to use the principle of moments, i.e., determine the moments of the components of each force, rather than the moment of the force itself.

- In three dimensions use the vector cross product to determine the moment of each force about point O. Here the position vectors extend from O to any point on the line of action of each force.

EXAMPLE 4.14

Replace the force and couple system shown in Fig. 4–37a by an equivalent resultant force and couple moment acting at point O.

(a) (b)

SOLUTION

Force Summation. The 3 kN and 5 kN forces are resolved into their x and y components as shown in Fig. 4–37b. We have

$\xrightarrow{+} (F_R)_x = \Sigma F_x;$ $(F_R)_x = (3 \text{ kN}) \cos 30° + \left(\frac{3}{5}\right)(5 \text{ kN}) = 5.598 \text{ kN} \rightarrow$

$+\uparrow (F_R)_y = \Sigma F_y;$ $(F_R)_y = (3 \text{ kN}) \sin 30° - \left(\frac{4}{5}\right)(5 \text{ kN}) - 4 \text{ kN} = -6.50 \text{ kN} = 6.50 \text{ kN} \downarrow$

Using the Pythagorean theorem, Fig. 4–37c, the magnitude of $\mathbf{F}_R$ is

$F_R = \sqrt{(F_R)_x^2 + (F_R)_y^2} = \sqrt{(5.598 \text{ kN})^2 + (6.50 \text{ kN})^2} = 8.58 \text{ kN}$ *Ans.*

Its direction θ is

$\theta = \tan^{-1}\left(\frac{(F_R)_y}{(F_R)_x}\right) = \tan^{-1}\left(\frac{6.50 \text{ kN}}{5.598 \text{ kN}}\right) = 49.3°$ *Ans.*

Moment Summation. Referring to Fig. 4–37b, we have

$\downarrow + (M_R)_O = \Sigma M_O;$

$(M_R)_O = (3 \text{ kN}) \sin 30°(0.2 \text{ m}) - (3 \text{ kN}) \cos 30°(0.1 \text{ m}) + \left(\frac{3}{5}\right)(5 \text{ kN})(0.1 \text{ m})$

$\qquad - \left(\frac{4}{5}\right)(5 \text{ kN})(0.5 \text{ m}) - (4 \text{ kN})(0.2 \text{ m})$

$= -2.46 \text{ kN} \cdot \text{m} = 2.46 \text{ kN} \cdot \text{m} \downarrow$ *Ans.*

This clockwise moment is shown in Fig. 4–37c.

(c)

Fig. 4–37

NOTE: Realize that the resultant force and couple moment in Fig. 4–37c will produce the same external effects or reactions at the wall as those produced by the force system, Fig. 4–37a.

EXAMPLE 4.15

Replace the force and couple system acting on the member in Fig. 4–38a by an equivalent resultant force and couple moment acting at point O.

Fig. 4–38

SOLUTION

Force Summation. Since the couple forces of 200 N are equal but opposite, they produce a zero resultant force, and so it is not necessary to consider them in the force summation. The 500-N force is resolved into its x and y components, thus,

$$\xrightarrow{+} (F_R)_x = \Sigma F_x; \quad (F_R)_x = \left(\tfrac{3}{5}\right)(500 \text{ N}) = 300 \text{ N} \rightarrow$$

$$+\uparrow (F_R)_y = \Sigma F_y; \quad (F_R)_y = (500 \text{ N})\left(\tfrac{4}{5}\right) - 750 \text{ N} = -350 \text{ N} = 350 \text{ N} \downarrow$$

From Fig. 4–38b, the magnitude of $\mathbf{F}_R$ is

$$F_R = \sqrt{(F_R)_x^2 + (F_R)_y^2}$$
$$= \sqrt{(300 \text{ N})^2 + (350 \text{ N})^2} = 461 \text{ N} \qquad Ans.$$

And the angle θ is

$$\theta = \tan^{-1}\left(\frac{(F_R)_y}{(F_R)_x}\right) = \tan^{-1}\left(\frac{350 \text{ N}}{300 \text{ N}}\right) = 49.4° \qquad Ans.$$

Moment Summation. Since the couple moment is a free vector, it can act at any point on the member. Referring to Fig. 4–38a, we have

$$\downarrow + (M_R)_O = \Sigma M_O + \Sigma M:$$

$$(M_R)_O = (500 \text{ N})\left(\tfrac{4}{5}\right)(2.5 \text{ m}) - (500 \text{ N})\left(\tfrac{3}{5}\right)(1 \text{ m})$$
$$- (750 \text{ N})(1.25 \text{ m}) + 200 \text{ N} \cdot \text{m}$$
$$= -37.5 \text{ N} \cdot \text{m} = 37.5 \text{ N} \cdot \text{m} \downarrow \qquad Ans.$$

This clockwise moment is shown in Fig. 4–38b.

EXAMPLE 4.16

The structural member is subjected to a couple moment **M** and forces $\mathbf{F}_1$ and $\mathbf{F}_2$ in Fig. 4–39a. Replace this system by an equivalent resultant force and couple moment acting at its base, point O.

SOLUTION (VECTOR ANALYSIS)
The three-dimensional aspects of the problem can be simplified by using a Cartesian vector analysis. Expressing the forces and couple moment as Cartesian vectors, we have

$$\mathbf{F}_1 = \{-800\mathbf{k}\} \text{ N}$$

$$\mathbf{F}_2 = (300 \text{ N})\mathbf{u}_{CB}$$

$$= (300 \text{ N})\left(\frac{\mathbf{r}_{CB}}{r_{CB}}\right)$$

$$= 300 \text{ N}\left[\frac{\{-0.15\mathbf{i} + 0.1\mathbf{j}\} \text{ m}}{\sqrt{(-0.15 \text{ m})^2 + (0.1 \text{ m})^2}}\right] = \{-249.6\mathbf{i} + 166.4\mathbf{j}\} \text{ N}$$

$$\mathbf{M} = -500\left(\tfrac{4}{5}\right)\mathbf{j} + 500\left(\tfrac{3}{5}\right)\mathbf{k} = \{-400\mathbf{j} + 300\mathbf{k}\} \text{ N} \cdot \text{m}$$

(b)

Fig. 4–39

Force Summation.

$$\mathbf{F}_R = \Sigma \mathbf{F}; \quad \mathbf{F}_R = \mathbf{F}_1 + \mathbf{F}_2 = -800\mathbf{k} - 249.6\mathbf{i} + 166.4\mathbf{j}$$

$$= \{-250\mathbf{i} + 166\mathbf{j} - 800\mathbf{k}\} \text{ N} \qquad \textit{Ans.}$$

Moment Summation.

$$(\mathbf{M}_R)_O = \Sigma \mathbf{M}_O + \Sigma \mathbf{M}$$

$$(\mathbf{M}_R)_O = \mathbf{r}_C \times \mathbf{F}_1 + \mathbf{r}_B \times \mathbf{F}_2 + \mathbf{M}$$

$$(\mathbf{M}_R)_O = (1\mathbf{k}) \times (-800\mathbf{k}) + \begin{vmatrix} \mathbf{i} & \mathbf{j} & \mathbf{k} \\ -0.15 & 0.1 & 1 \\ -249.6 & 166.4 & 0 \end{vmatrix} + (-400\mathbf{j} + 300\mathbf{k})$$

$$= (0) + (-166.4\mathbf{i} - 249.6\mathbf{j}) + (-400\mathbf{j} + 300\mathbf{k})$$

$$= \{-166\mathbf{i} - 650\mathbf{j} + 300\mathbf{k}\} \text{ N} \cdot \text{m} \qquad \textit{Ans.}$$

The results are shown in Fig. 4–39b.

FUNDAMENTAL PROBLEMS

F4–25. Replace the loading by an equivalent resultant force and couple moment acting at point A.

Prob. F4–25

F4–26. Replace the loading by an equivalent resultant force and couple moment acting at point A.

Prob. F4–26

F4–27. Replace the loading by an equivalent resultant force and couple moment acting at point A.

Prob. F4–27

F4–28. Replace the loading by an equivalent resultant force and couple moment acting at point A.

Prob. F4–28

F4–29. Replace the loading by an equivalent resultant force and couple moment acting at point O.

Prob. F4–29

F4–30. Replace the loading by an equivalent resultant force and couple moment acting at point O.

Prob. F4–30

PROBLEMS

4–103. Replace the force system by an equivalent resultant force and couple moment at point O.

***4–104.** Replace the force system by an equivalent resultant force and couple moment at point P.

4–107. Replace the forces acting on the gear by an equivalent resultant force and couple moment acting at point O.

Probs. 4–103/104

4–105. Replace the force system acting on the beam by an equivalent resultant force and couple moment at point B.

4–106. Replace the force system acting on the beam by an equivalent resultant force and couple moment at point A.

Prob. 4–107

***4–108.** Replace the force and couple moment system acting on the beam by an equivalent resultant force and couple moment at point A.

4–109. Replace the force and couple moment system acting on the beam by an equivalent resultant force and couple moment at point O.

Probs. 4–105/106

Probs. 4–108/109

4–110. Replace the force system acting on the beam by an equivalent resultant force, and specify its point of application on the beam measured from point O.

4–113. Replace the force system by an equivalent resultant force and couple moment at point O. Set $F = 20$ lb.

4–114. Replace the force system by an equivalent resultant force and couple moment at point O. Set $F = 15$ lb.

Prob. 4–110

Probs. 4–113/114

4–111. Determine the magnitude and orientation of θ of force $\mathbf{F}$ and its placement d on the beam so the loading system is equivalent to a resultant force of 15 kN acting vertically downward at point O and a clockwise couple moment of 60 kN·m.

***4–112.** Determine the magnitude and orientation of θ of force $\mathbf{F}$ and its placement d on the beam so the loading system is equivalent to a resultant force of 20 kN acting vertically downward at point O and a clockwise couple moment of 80 kN·m.

4–115. Replace the loading acting on the frame by an equivalent resultant force and couple moment acting at point A.

***4–116.** Replace the loading acting on the frame by an equivalent resultant force and couple moment acting at point B.

Probs. 4–111/112

Probs. 4–115/116

4–117. Replace the loading acting on the post by a resultant force and couple moment at point O.

Prob. 4–117

4–118. Determine the magnitudes of $\mathbf{F}_1$ and $\mathbf{F}_2$ and the angle θ so that the loading creates a zero-resultant force and couple moment on the handle.

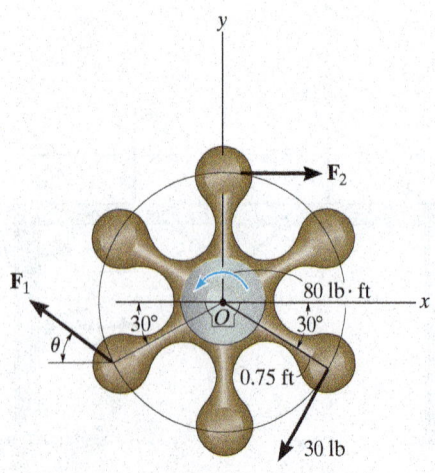

Prob. 4–118

4–119. Replace the force system acting on the truss by a resultant force and couple moment at point C.

Prob. 4–119

*4–120. A biomechanical model of the lumbar region of the human trunk is shown. The forces acting on the four muscle groups consist of $F_R = 35$ N for the rectus, $F_O = 45$ N for the oblique, $F_L = 23$ N for the lumbar latissimus dorsi, and $F_E = 32$ N for the erector spinae. These loadings are symmetric with respect to the y–z plane. Replace this system of parallel forces by an equivalent force and couple moment acting at the spine, point O. Express the results in Cartesian vector form.

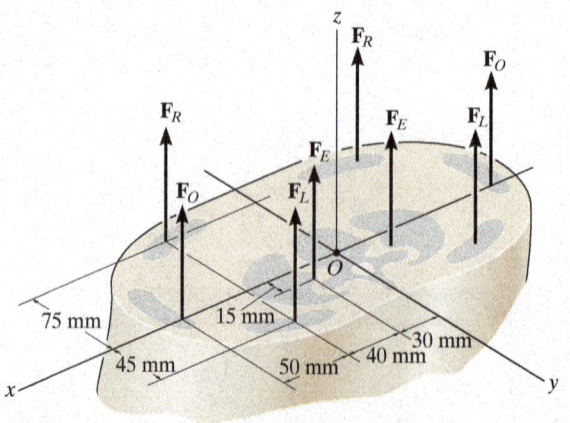

Prob. 4–120

4–121. The resultant force of the wind and the weights of the various components act on the sign. Determine the equivalent force and moment acting at its base, A.

4–123. The crate is on the ground and is to be hoisted using the three slings shown. Replace the system of forces acting on the slings by an equivalent resultant force and couple moment at point O. The force $\mathbf{F}_1$ is vertical.

Prob. 4–123

Prob. 4–121

*4–124. The forces and couple moments exerted on the toe and heel plates of a snow ski are $\mathbf{F}_t = \{-50\mathbf{i} + 80\mathbf{j} - 158\mathbf{k}\}\,\text{N}, \mathbf{M}_t = \{-6\mathbf{i} + 4\mathbf{j} + 2\mathbf{k}\}\,\text{N}\cdot\text{m}$, and $\mathbf{F}_h = \{-20\mathbf{i} + 60\mathbf{j} - 250\mathbf{k}\}\,\text{N}, \mathbf{M}_h = \{-20\mathbf{i} + 8\mathbf{j} + 3\mathbf{k}\}\,\text{N}\cdot\text{m}$, respectively. Replace this system by an equivalent force and couple moment acting at point O. Express the results in Cartesian vector form.

4–125. The forces and couple moments exerted on the toe and heel plates of a snow ski are $\mathbf{F}_t = \{-50\mathbf{i} + 80\mathbf{j} - 158\mathbf{k}\}\,\text{N}, \mathbf{M}_t = \{-6\mathbf{i} + 4\mathbf{j} + 2\mathbf{k}\}\,\text{N}\cdot\text{m}$, and $\mathbf{F}_h = \{-20\mathbf{i} + 60\mathbf{j} - 250\mathbf{k}\}\,\text{N}, \mathbf{M}_h = \{-20\mathbf{i} + 8\mathbf{j} + 3\mathbf{k}\}\,\text{N}\cdot\text{m}$, respectively. Replace this system by an equivalent force and couple moment acting at point P. Express the results in Cartesian vector form.

4–122. Replace the force system acting on the pipe assembly by a resultant force and couple moment at point O. Express the results in Cartesian vector form.

Prob. 4–122

Probs. 4–124/125

4.8 FURTHER SIMPLIFICATION OF A FORCE AND COUPLE SYSTEM

In the preceding section, we developed a way to reduce a force and couple moment system acting on a rigid body into an equivalent resultant force $\mathbf{F}_R$ acting at a specific point O and a resultant couple moment, $(\mathbf{M}_R)_O$. The force system can be further reduced to an equivalent single resultant force provided the lines of action of $\mathbf{F}_R$ and $(\mathbf{M}_R)_O$ are *perpendicular* to each other. This occurs when the force system is either concurrent, coplanar, or parallel.

Concurrent Force System. Since a *concurrent force system* is one in which the lines of action of all the forces intersect at a common point O, Fig. 4–40a, then the force system produces no moment about this point. As a result, the equivalent system can be represented by a single resultant force $\mathbf{F}_R = \Sigma \mathbf{F}$ acting at O, Fig. 4–40b.

Coplanar Force System. In the case of a *coplanar force system*, the lines of action of all the forces lie in the same plane, Fig. 4–41a, and so the resultant force $\mathbf{F}_R = \Sigma \mathbf{F}$ of this system also lies in this plane. Furthermore, the moment of each of the forces about any point O is directed perpendicular to this plane. Thus, the resultant moment $(\mathbf{M}_R)_O$ and resultant force $\mathbf{F}_R$ will be *mutually perpendicular*, Fig. 4–41b. The resultant moment can be replaced by moving the resultant force $\mathbf{F}_R$ a perpendicular or moment arm distance d away from point O such that $\mathbf{F}_R$ produces the *same moment* $(\mathbf{M}_R)_O$ about point O, Fig. 4–41c. This distance d can be determined from the scalar equation $(M_R)_O = F_R d = \Sigma M_O$ or $d = (M_R)_O / F_R$.

Fig. 4–40

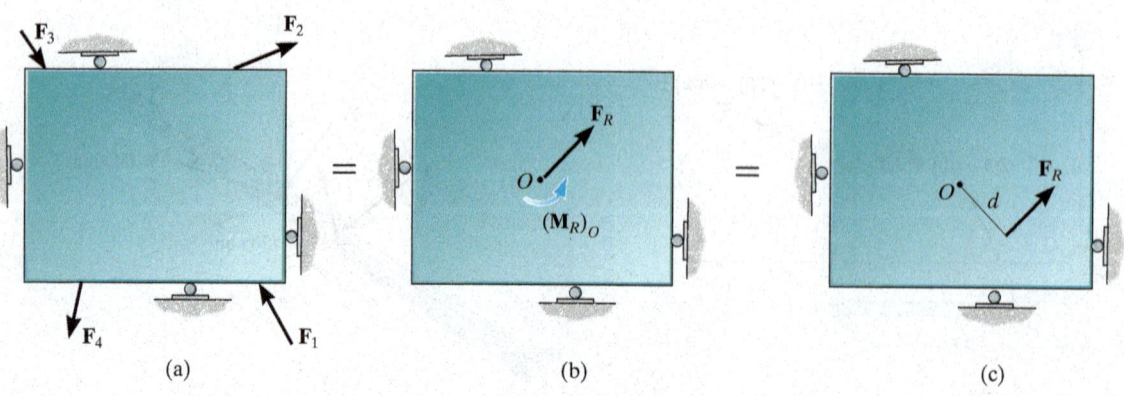

Fig. 4–41

4.8 FURTHER SIMPLIFICATION OF A FORCE AND COUPLE SYSTEM 169

Here the weights of the traffic lights are replaced by their resultant force $W_R = W_1 + W_2$, which acts at a distance $d = (W_1 d_1 + W_2 d_2)/W_R$ from O. Both systems are equivalent.

Parallel Force System. The *parallel force system* shown in Fig. 4–42a consists of forces that are all parallel to the z axis. Thus, the resultant force $\mathbf{F}_R = \Sigma\mathbf{F}$ at point O must also be parallel to this axis, Fig. 4–42b. The moment produced by each force lies in the plane of the plate, and so the resultant couple moment, $(\mathbf{M}_R)_O$, will also lie in this plane, along the moment axis a since $\mathbf{F}_R$ and $(\mathbf{M}_R)_O$ are mutually perpendicular. As a result, the force system can be further reduced to an equivalent single resultant force $\mathbf{F}_R$, acting through point P located on the perpendicular b axis, Fig. 4–42c. The distance d along this axis from point O requires $(M_R)_O = F_R d = \Sigma M_O$ or $d = \Sigma M_O / F_R$.

The four cable forces are all concurrent at point O on this bridge tower. Consequently they produce no resultant moment there, only a resultant force $\mathbf{F}_R$. Note that the designers have positioned the cables so that $\mathbf{F}_R$ is directed *along* the bridge tower directly to the support, so that it does not cause any bending of the tower.

> **IMPORTANT POINT**
>
> - A concurrent, coplanar, or parallel force system can always be reduced to a single resultant force acting at a specific point P. For any other type of force system, the simplest reduction is a wrench, which consists of a resultant force and collinear couple moment acting at a specific point P.

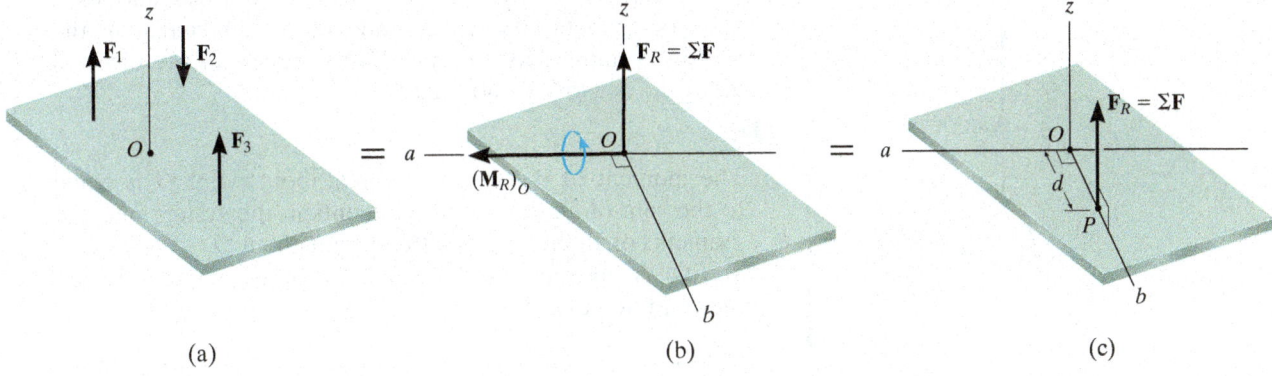

(a)   (b)   (c)

Fig. 4–42

170   CHAPTER 4   FORCE SYSTEM RESULTANTS

Reduction to a Wrench.
In general, a three-dimensional force and couple moment system will have an equivalent resultant force $\mathbf{F}_R$ acting at point O and a resultant couple moment $(\mathbf{M}_R)_O$ that are *not* perpendicular to one another, as shown in Fig. 4–43a. Although a force system such as this cannot be further reduced to an equivalent single resultant force, the resultant couple moment $(\mathbf{M}_R)_O$ can be resolved into components parallel and perpendicular to the line of action of $\mathbf{F}_R$, Fig. 4–43a. If this appears difficult to do in three dimensions, use the dot product to get $\mathbf{M}_{\|} = (\mathbf{M}_R) \cdot \mathbf{u}_{F_R}$ and then $\mathbf{M}_{\perp} = \mathbf{M}_R - \mathbf{M}_{\|}$. The perpendicular component $\mathbf{M}_{\perp}$ can be replaced if we move $\mathbf{F}_R$ to point P, a distance d from point O along the b axis, Fig. 4–43b. As shown, this axis is perpendicular to both the a axis and the line of action of $\mathbf{F}_R$. The location of P can be determined from $d = M_{\perp}/F_R$. Finally, because $\mathbf{M}_{\|}$ is a free vector, it can be moved to point P, Fig. 4–43c. This combination of a resultant force $\mathbf{F}_R$ and collinear couple moment $\mathbf{M}_{\|}$ will tend to translate and rotate the body about its axis and is referred to as a **wrench** or **screw**. A wrench is the simplest system that can represent any general force and couple moment system acting on a body.

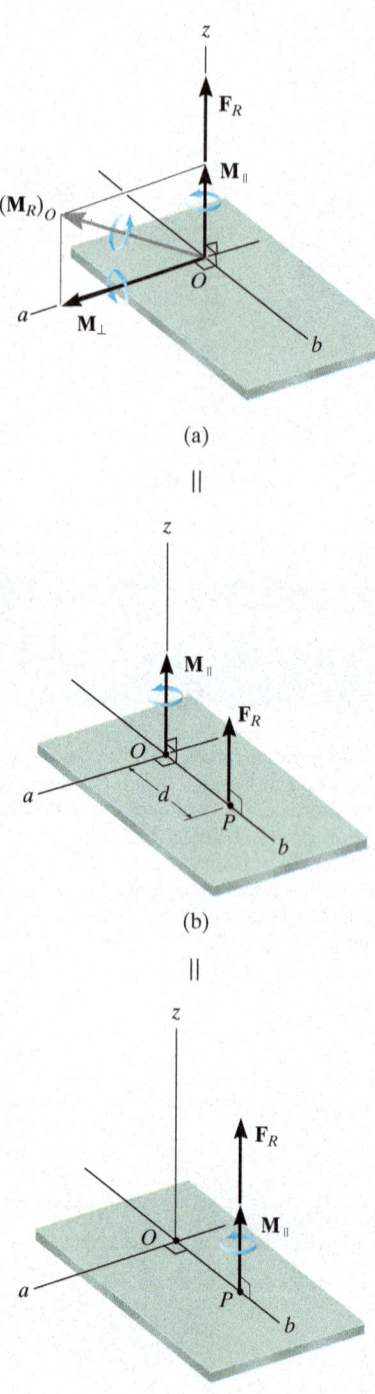

Fig. 4–43

PROCEDURE FOR ANALYSIS

The technique used to reduce a coplanar or parallel force system to a single resultant force follows a similar procedure outlined in the previous section.

- Establish the x, y, z axes and locate the resultant force $\mathbf{F}_R$ an arbitrary distance away from the origin of the coordinates.

Force Summation.

- The resultant force is equal to the sum of all the forces in the system.
- For a coplanar force system, resolve each force into its x and y components. Positive components are directed along the positive x and y axes, and negative components are directed along the negative x and y axes.

Moment Summation.

- The moment of the resultant force about point O is equal to the sum of all the couple moments in the system plus the moments of all the forces in the system about O.
- To find the location d of the resultant force from point O, use the condition that $d = \Sigma M_O / F_R$.

EXAMPLE 4.17

Replace the force and couple moment system acting on the beam in Fig. 4–44a by an equivalent resultant force, and find where its line of action intersects the beam, measured from point O.

Fig. 4–44

SOLUTION

Force Summation. Summing the force components,

$$\xrightarrow{+} (F_R)_x = \Sigma F_x; \quad (F_R)_x = 8 \text{ kN} \left(\tfrac{3}{5}\right) = 4.80 \text{ kN} \rightarrow$$

$$+\uparrow (F_R)_y = \Sigma F_y; \quad (F_R)_y = -4 \text{ kN} + 8 \text{ kN}\left(\tfrac{4}{5}\right) = 2.40 \text{ kN} \uparrow$$

From Fig. 4–44b, the magnitude of F_R is

$$F_R = \sqrt{(4.80 \text{ kN})^2 + (2.40 \text{ kN})^2} = 5.37 \text{ kN} \qquad Ans.$$

The angle θ is

$$\theta = \tan^{-1}\left(\frac{2.40 \text{ kN}}{4.80 \text{ kN}}\right) = 26.6° \qquad Ans.$$

Moment Summation. We must equate the moment of F_R about point O in Fig. 4–44b to the sum of the moments of the force and couple moment system about point O in Fig. 4–44a. Since the line of action of $(F_R)_x$ passes through point O, only $(F_R)_y$ produces a moment about this point. Thus,

$$\zeta+ (M_R)_O = \Sigma M_O; \quad 2.40 \text{ kN}(d) = -(4 \text{ kN})(1.5 \text{ m}) - 15 \text{ kN} \cdot \text{m}$$

$$- \left[8 \text{ kN}\left(\tfrac{3}{5}\right)\right](0.5 \text{ m}) + \left[8 \text{ kN}\left(\tfrac{4}{5}\right)\right](4.5 \text{ m})$$

$$d = 2.25 \text{ m} \qquad Ans.$$

EXAMPLE 4.18

Fig. 4–45

The jib crane shown in Fig. 4–45a is subjected to three coplanar forces. Replace this loading by an equivalent resultant force and specify where the resultant's line of action intersects the column AB and boom BC.

SOLUTION

Force Summation. Resolving the 250-lb force into x and y components and summing the force components yields

$\xrightarrow{+} (F_R)_x = \Sigma F_x;\quad (F_R)_x = -250\text{ lb}\left(\frac{3}{5}\right) - 175\text{ lb} = -325\text{ lb} = 325\text{ lb} \leftarrow$

$+\uparrow (F_R)_y = \Sigma F_y;\quad (F_R)_y = -250\text{ lb}\left(\frac{4}{5}\right) - 60\text{ lb} = -260\text{ lb} = 260\text{ lb} \downarrow$

As shown by the vector addition in Fig. 4–45b,

$$F_R = \sqrt{(325\text{ lb})^2 + (260\text{ lb})^2} = 416\text{ lb} \qquad Ans.$$

$$\theta = \tan^{-1}\left(\frac{260\text{ lb}}{325\text{ lb}}\right) = 38.7° \qquad Ans.$$

Moment Summation. Moments will be summed about point A. Assuming the line of action of $\mathbf{F}_R$ intersects AB at a distance y from A, Fig. 4–45b, we have

$\zeta + (M_R)_A = \Sigma M_A;\quad 325\text{ lb }(y) + 260\text{ lb }(0)$

$= 175\text{ lb }(5\text{ ft}) - 60\text{ lb }(3\text{ ft}) + 250\text{ lb}\left(\frac{3}{5}\right)(11\text{ ft}) - 250\text{ lb}\left(\frac{4}{5}\right)(8\text{ ft})$

$$y = 2.29\text{ ft} \qquad Ans.$$

By the principle of transmissibility, $\mathbf{F}_R$ can be placed at a distance x where it intersects BC, Fig. 4–45b. In this case we have

$\zeta + (M_R)_A = \Sigma M_A;\quad 325\text{ lb }(11\text{ ft}) - 260\text{ lb }(x)$

$= 175\text{ lb }(5\text{ ft}) - 60\text{ lb }(3\text{ ft}) + 250\text{ lb}\left(\frac{3}{5}\right)(11\text{ ft}) - 250\text{ lb}\left(\frac{4}{5}\right)(8\text{ ft})$

$$x = 10.9\text{ ft} \qquad Ans.$$

EXAMPLE 4.19

The slab in Fig. 4–46a is subjected to four parallel forces. Determine the magnitude and direction of a resultant force equivalent to the given force system, and locate its point of application on the slab.

Fig. 4–46

SOLUTION (SCALAR ANALYSIS)

Force Summation. From Fig. 4–46a, the resultant force is

$+\uparrow F_R = \Sigma F;$ $\quad F_R = -600\text{ N} + 100\text{ N} - 400\text{ N} - 500\text{ N}$
$\qquad = -1400\text{ N} = 1400\text{ N}\downarrow$ *Ans.*

Moment Summation. We require the moment about the x axis of the resultant force, Fig. 4–46b, to be equal to the sum of the moments about the x axis of all the forces in the system, Fig. 4–46a. The moment arms are determined from the y coordinates, since these coordinates represent the *perpendicular distances* from the x axis to the lines of action of the forces. Using the right-hand rule, we have

$(M_R)_x = \Sigma M_x;$
$-(1400\text{ N})y = 600\text{ N}(0) + 100\text{ N}(5\text{ m}) - 400\text{ N}(10\text{ m}) + 500\text{ N}(0)$
$\qquad -1400y = -3500 \qquad y = 2.50\text{ m}$ *Ans.*

In a similar manner, a moment equation can be written about the y axis using moment arms defined by the x coordinates of each force.

$(M_R)_y = \Sigma M_y;$
$(1400\text{ N})x = 600\text{ N}(8\text{ m}) - 100\text{ N}(6\text{ m}) + 400\text{ N}(0) + 500\text{ N}(0)$
$\qquad 1400x = 4200$
$\qquad x = 3\text{ m}$ *Ans.*

NOTE: A force of $F_R = 1400\text{ N}$ placed at point $P(3.00\text{ m}, 2.50\text{ m})$ on the slab, Fig. 4–46b, is therefore equivalent to the parallel force system acting on the slab in Fig. 4–46a.

EXAMPLE 4.20

(a)

(b)

Fig. 4-47

Replace the force system in Fig. 4–47a by an equivalent resultant force and specify its point of application on the pedestal.

SOLUTION

Force Summation. Here we will demonstrate a vector analysis. Summing forces,

$$F_R = \Sigma F; \quad F_R = F_A + F_B + F_C$$
$$= \{-300k\} \text{ lb} + \{-500k\} \text{ lb} + \{100k\} \text{ lb}$$
$$= \{-700k\} \text{ lb} \qquad \textit{Ans.}$$

Location. Moments will be summed about point O. The resultant force F_R is assumed to act through point $P(x, y, 0)$, Fig. 4–47b. Thus

$$(M_R)_O = \Sigma M_O;$$
$$r_P \times F_R = (r_A \times F_A) + (r_B \times F_B) + (r_C \times F_C)$$
$$(x\mathbf{i} + y\mathbf{j}) \times (-700\mathbf{k}) = [(4\mathbf{i}) \times (-300\mathbf{k})]$$
$$+ [(-4\mathbf{i} + 2\mathbf{j}) \times (-500\mathbf{k})] + [(-4\mathbf{j}) \times (100\mathbf{k})]$$
$$-700x(\mathbf{i} \times \mathbf{k}) - 700y(\mathbf{j} \times \mathbf{k}) = -1200(\mathbf{i} \times \mathbf{k}) + 2000(\mathbf{i} \times \mathbf{k})$$
$$-1000(\mathbf{j} \times \mathbf{k}) - 400(\mathbf{j} \times \mathbf{k})$$
$$700x\mathbf{j} - 700y\mathbf{i} = 1200\mathbf{j} - 2000\mathbf{j} - 1000\mathbf{i} - 400\mathbf{i}$$

Equating the **i** and **j** components,

$$-700y = -1400 \qquad (1)$$
$$y = 2 \text{ in.} \qquad \textit{Ans.}$$
$$700x = -800 \qquad (2)$$
$$x = -1.14 \text{ in.} \qquad \textit{Ans.}$$

The negative sign indicates that the x coordinate of point P is negative.

NOTE: As demonstrated in Example 4–19, it is also possible to establish Eqs. 1 and 2 directly by summing moments about the x and y axes. Using the right-hand rule, we have

$$(M_R)_x = \Sigma M_x; \quad -700y = -100 \text{ lb}(4 \text{ in.}) - 500 \text{ lb}(2 \text{ in.})$$

$$(M_R)_y = \Sigma M_y; \quad 700x = 300 \text{ lb}(4 \text{ in.}) - 500 \text{ lb}(4 \text{ in.})$$

FUNDAMENTAL PROBLEMS

F4–31. Replace the loading by an equivalent resultant force and specify where the resultant's line of action intersects the beam measured from O.

Prob. F4–31

F4–32. Replace the loading by an equivalent resultant force and specify where the resultant's line of action intersects the member measured from A.

Prob. F4–32

F4–33. Replace the loading by an equivalent resultant force and specify where the resultant's line of action intersects the horizontal segment of the member measured from A.

Prob. F4–33

F4–34. Replace the loading by an equivalent resultant force and specify where the resultant's line of action intersects the member AB measured from A.

Prob. F4–34

F4–35. Replace the loading by an equivalent single resultant force and specify the x and y coordinates of its line of action.

Prob. F4–35

F4–36. Replace the loading by an equivalent single resultant force and specify the x and y coordinates of its line of action.

Prob. F4–36

PROBLEMS

4–126. The weights of the various components of the truck are shown. Replace this system of forces by an equivalent resultant force and specify its location, measured from point B.

4–127. The weights of the various components of the truck are shown. Replace this system of forces by an equivalent resultant force and specify its location, measured from point A.

4–130. The system of four forces acts on the roof truss. Determine the equivalent resultant force and specify its location along AB, measured from point A.

Probs. 4–126/127

Prob. 4–130

*4–128. Replace the force and couple moment system by an equivalent resultant force and specify its coordinate point of application $(x, 0)$ on the x axis.

4–129. Replace the force and couple moment system by an equivalent resultant force and specify its coordinate point of application $(0, y)$ on the y axis.

4–131. Replace the loading acting on the beam by a single resultant force. Specify where the force acts, measured from end A.

*4–132. Replace the loading acting on the beam by a single resultant force. Specify where the force acts, measured from B.

Probs. 4–128/129

Probs. 4–131/132

4–133. Replace the loading acting on the post by an equivalent resultant force and couple moment at point O.

4–134. Replace the loading acting on the post by an equivalent resultant force and couple moment at point P.

4–137. Replace the force system acting on the frame by an equivalent resultant force and specify where the resultant's line of action intersects member AB, measured from point A.

4–138. Replace the force system acting on the frame by an equivalent resultant force and specify where the resultant's line of action intersects member BC, measured from point B.

Probs. 4–133/134

Probs. 4–137/138

4–135. Replace the loading on the frame by a single resultant force. Specify where its line of action intersects member AB, measured from A.

*4–136. Replace the loading on the frame by a single resultant force. Specify where its line of action intersects member CD, measured from end C.

4–139. Replace the force system acting on the frame by an equivalent resultant force, and specify where the resultant's line of action intersects member AB, measured from point A.

*4–140. Replace the force system acting on the frame by an equivalent resultant force, and specify where the resultant's line of action intersects member BC, measured from point B.

Probs. 4–135/136

Probs. 4–139/140

4–141. Replace the loading acting on the frame by an equivalent resultant force and specify where the resultant's line of action intersects member AB, measured from A.

4–142. Replace the loading acting on the frame by an equivalent resultant force and specify where the resultant's line of action intersects member BC, measured from B.

4–145. The tube supports the four parallel forces. Determine the magnitudes of forces $\mathbf{F}_C$ and $\mathbf{F}_D$ acting at C and D so that the equivalent resultant force of the force system acts through the midpoint O of the tube.

Prob. 4–145

Probs. 4–141/142

4–146. Three parallel bolting forces act on the circular plate. Determine the resultant force, and specify its location (x, z) on the plate. $F_A = 200$ lb, $F_B = 100$ lb, and $F_C = 400$ lb.

4–143. If $F_A = 7$ kN and $F_B = 5$ kN, represent the force system by a resultant force, and specify its location on the x–y plane.

*4–144. Determine the magnitudes of $\mathbf{F}_A$ and $\mathbf{F}_B$ so that the resultant force passes through point O.

4–147. The three parallel bolting forces act on the circular plate. If the force at A has a magnitude of $F_A = 200$ lb, determine the magnitudes of $\mathbf{F}_B$ and $\mathbf{F}_C$ so that the resultant force $\mathbf{F}_R$ of the system has a line of action that coincides with the y axis. *Hint:* This requires $\Sigma M_x = 0$ and $\Sigma M_z = 0$.

Probs. 4–143/144

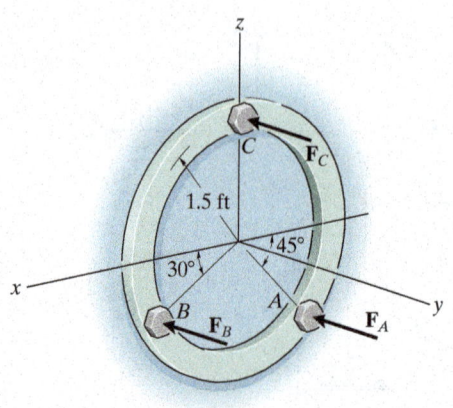

Probs. 4–146/147

*4–148. Replace the three forces acting on the plate by a wrench. Specify the force and couple moment for the wrench and the point $P(x, y)$ where its line of action intersects the plate.

4–150. The pipe assembly is subjected to the action of a wrench at B and a couple at A. Determine the magnitude F of the couple forces so that the system can be simplified to a wrench acting at point C.

Prob. 4–148

Prob. 4–150

4–149. Replace the three forces acting on the plate by a wrench. Specify the magnitude of the force and couple moment for the wrench and the point $P(y, z)$ where its line of action intersects the plate.

4–151. The pipe assembly is subjected to the action of a wrench at B and a couple at A. Simplify this system to a resultant wrench and specify the location of the wrench along the axis of pipe CD, measured from point C. Set $F = 40$ N.

Prob. 4–149

Prob. 4–151

180   CHAPTER 4   FORCE SYSTEM RESULTANTS

4.9 REDUCTION OF A SIMPLE DISTRIBUTED LOADING

Sometimes, a body may be subjected to a loading that is distributed over its surface. For example, wind on the face of a sign, water within a tank, or the weight of sand on the floor of a storage container all exert *distributed loadings*. The pressure caused by these loadings at each point on the surface represents the intensity of the loading. It is measured using pascals, Pa, (or N/m^2) in SI units, or lb/ft^2 in the U.S. Customary system.

Loading Along a Single Axis. The most common type of distributed pressure loading is represented along a single axis.* For example, consider the beam (or plate) in Fig. 4–48a that has a constant width and is subjected to a pressure loading that varies only along the x axis. This loading can be described by the function $p = p(x) \text{ N}/m^2$. Since it contains only one variable, x, we can represent it as a *coplanar distributed load*. To do so, we must multiply it by the width b m of the beam, so that $w(x) = p(x)b$ N/m, Fig. 4–48b. Using the methods of Sec. 4.8, we can replace this coplanar parallel force system with a single equivalent resultant force $\mathbf{F}_R$, Fig. 4–48c.

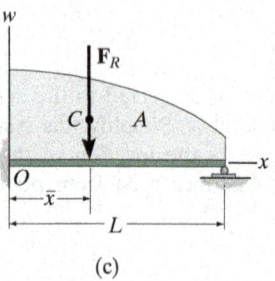

Fig. 4–48

Magnitude of Resultant Force. The magnitude of $\mathbf{F}_R$ is equivalent to the sum of all the forces in the system, $F_R = \Sigma F$. In this case integration must be used since there is an infinite number of parallel forces $d\mathbf{F}$ acting on the beam, Fig. 4–48b. Each $d\mathbf{F}$ is acting on an element of length dx, and since $w(x)$ is a force per unit length, then $dF = w(x) \, dx = dA$. For the entire length L,

$$+\downarrow F_R = \Sigma F; \qquad \boxed{F_R = \int_L w(x) \, dx = \int_A dA = A} \qquad (4\text{--}19)$$

Therefore, *the magnitude of the resultant force is equal to the area A under the loading diagram*, Fig. 4–48c.

---

*The more general case of a surface loading acting on a body is considered in Sec. 9.5.

4.9 REDUCTION OF A SIMPLE DISTRIBUTED LOADING   181

Location of Resultant Force. The location $\bar{x}$ of $\mathbf{F}_R$ can be determined by equating the moments of the force resultant and the parallel force distribution about point O. Since dF produces a moment of $x\,dF = x\,[w(x)\,dx]$ about O, Fig. 4–48b, then for the entire length L, Fig. 4–48c,

$\downarrow + (M_R)_O = \Sigma M_O;$ $-\bar{x} F_R = -\displaystyle\int_L x w(x)\, dx$

Solving for $\bar{x}$, using Eq. 4–19, we have

$$\bar{x} = \frac{\displaystyle\int_L x w(x)\, dx}{\displaystyle\int_L w(x)\, dx} = \frac{\displaystyle\int_A x\, dA}{\displaystyle\int_A dA} \quad (4\text{--}20)$$

This coordinate $\bar{x}$ locates the geometric center or **centroid** of the *area under the distributed loading. In other words, the line of action of the resultant force passes through the centroid C (geometric center) of the area under the loading diagram*, Fig. 4–48c.

When the distributed-loading diagram is in the shape of a rectangle, triangle, or some other simple geometric form, then the centroid location for such common shapes does not have to be determined from the above equation. Rather it can be obtained directly from the tabulation given on the inside back cover.

Once $\bar{x}$ is determined, $\mathbf{F}_R$ by symmetry passes through point $(\bar{x}, 0)$ on the surface of the beam in Fig. 4–48a. And so in three dimensions the resultant force has a magnitude equal to the volume under the loading curve $p = p(x)$ and a line of action which passes through the centroid (geometric center) of this volume.

(a)

(b)

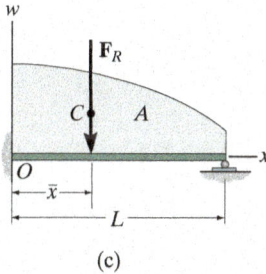

(c)

Fig. 4–48 (Repeated)

IMPORTANT POINTS

- Coplanar distributed loadings are defined by using a loading function $w = w(x)$ that indicates the intensity of the loading along the length of a member. This intensity is measured in N/m or lb/ft.

- The external effects caused by a coplanar distributed load acting on a member can be represented by a resultant force.

- This resultant is equivalent to the *area* under the loading diagram, and has a line of action that passes through the *centroid* or geometric center of this area.

The pile of bricks creates a uniform distributed loading on the board.

EXAMPLE 4.21

Determine the magnitude and location of the equivalent resultant force acting on the shaft in Fig. 4–49a.

Fig. 4–49

SOLUTION
Since $w = w(x)$ is given, this problem will be solved by integration. The differential element has an area $dA = w\,dx = 60x^2\,dx$. Applying Eq. 4–19,

$$+\downarrow F_R = \Sigma F;$$

$$F_R = \int_A dA = \int_0^{2\,m} 60x^2\,dx = 60\left(\frac{x^3}{3}\right)\Big|_0^{2\,m} = 60\left(\frac{2^3}{3} - \frac{0^3}{3}\right)$$

$$= 160\ \text{N} \hspace{4em} \textit{Ans.}$$

The location $\bar{x}$ of $\mathbf{F}_R$ measured from O, Fig. 4–49b, is determined from Eq. 4–20.

$$\bar{x} = \frac{\int_A x\,dA}{\int_A dA} = \frac{\int_0^{2\,m} x(60x^2)\,dx}{160\ \text{N}} = \frac{60\left(\frac{x^4}{4}\right)\Big|_0^{2\,m}}{160\ \text{N}} = \frac{60\left(\frac{2^4}{4} - \frac{0^4}{4}\right)}{160\ \text{N}}$$

$$= 1.5\ \text{m} \hspace{4em} \textit{Ans.}$$

NOTE: These results can be checked by using the table on the inside back cover, where it is shown that the formula for an exparabolic area of length a, height b, and shape shown in Fig. 4–49a, is

$$A = \frac{ab}{3} = \frac{2\ \text{m}(240\ \text{N/m})}{3} = 160\ \text{N and } \bar{x} = \frac{3}{4}a = \frac{3}{4}(2\ \text{m}) = 1.5\ \text{m}$$

EXAMPLE 4.22

A distributed loading of $p = (800x)$ Pa acts over the top surface of the beam shown in Fig. 4–50a. Determine the magnitude and location of the equivalent resultant force.

(a)

SOLUTION
Since the loading intensity is uniform along the width of the beam (the y axis), the loading can be viewed in two dimensions as shown in Fig. 4–50b. Here

$$w = (800x \text{ N/m}^2)(0.2 \text{ m})$$
$$= (160x) \text{ N/m}$$

At $x = 9$ m, $w = 1440$ N/m. Although we may again apply Eqs. 4–19 and 4–20 as in the previous example, it is simpler to use the table on the inside back cover.

The magnitude of the resultant force is equivalent to the area of the triangle.

$$F_R = \tfrac{1}{2}(9 \text{ m})(1440 \text{ N/m}) = 6480 \text{ N} = 6.48 \text{ kN} \qquad Ans.$$

The line of action of $\mathbf{F}_R$ passes through the *centroid C* of this triangle. Hence,

$$\bar{x} = 9 \text{ m} - \tfrac{1}{3}(9 \text{ m}) = 6 \text{ m} \qquad Ans.$$

The results are shown in Fig. 4–50c.

NOTE: We may also view the resultant $\mathbf{F}_R$ as *acting* through the *centroid* of the *volume* of the loading diagram $p = p(x)$ in Fig. 4–50a. Then $\mathbf{F}_R$ intersects the x–y plane at the point (6 m, 0). Furthermore, the magnitude of $\mathbf{F}_R$ is equal to the volume under this loading diagram; i.e.,

$$F_R = V = \tfrac{1}{2}(7200 \text{ N/m}^2)(9 \text{ m})(0.2 \text{ m}) = 6.48 \text{ kN} \qquad Ans.$$

(b)

(c)

Fig. 4–50

EXAMPLE 4.23

(a)

(b)

(c)

(d)

Fig. 4–51

The granular material exerts the distributed loading on the beam as shown in Fig. 4–51a. Determine the magnitude and location of the equivalent resultant of this load.

SOLUTION
The area of the loading diagram is a *trapezoid*, and therefore the solution can be obtained directly from the area and centroid formulas for a trapezoid listed on the inside back cover. Since these formulas are not easily remembered, instead we will solve this problem by using "composite" areas. Here we will divide the trapezoidal loading into a rectangular and a triangular loading as shown in Fig. 4–51b. The magnitude of the force represented by each of these loadings is equal to its associated *area*,

$$F_1 = \tfrac{1}{2}(9 \text{ ft})(50 \text{ lb/ft}) = 225 \text{ lb}$$

$$F_2 = (9 \text{ ft})(50 \text{ lb/ft}) = 450 \text{ lb}$$

The lines of action of these parallel forces act through the respective *centroids* of their associated areas. Using the table on the inside back cover for the triangle, we have

$$\bar{x}_1 = \tfrac{1}{3}(9 \text{ ft}) = 3 \text{ ft}$$

$$\bar{x}_2 = \tfrac{1}{2}(9 \text{ ft}) = 4.5 \text{ ft}$$

The two parallel forces $\mathbf{F}_1$ and $\mathbf{F}_2$ can be reduced to a single resultant $\mathbf{F}_R$. The magnitude of $\mathbf{F}_R$ is

$$+\downarrow F_R = \Sigma F; \qquad F_R = 225 + 450 = 675 \text{ lb} \qquad \textit{Ans.}$$

We can find the location of $\mathbf{F}_R$ with reference to point A, Figs. 4–51b and 4–51c. We require

$$\curvearrowleft + (M_R)_A = \Sigma M_A; \qquad \bar{x}(675) = 3(225) + 4.5(450)$$

$$\bar{x} = 4 \text{ ft} \qquad \textit{Ans.}$$

NOTE: The trapezoidal area in Fig. 4–51a can also be divided into two triangular areas as shown in Fig. 4–51d. In this case

$$F_3 = \tfrac{1}{2}(9 \text{ ft})(100 \text{ lb/ft}) = 450 \text{ lb}$$

$$F_4 = \tfrac{1}{2}(9 \text{ ft})(50 \text{ lb/ft}) = 225 \text{ lb}$$

and

$$\bar{x}_3 = \tfrac{1}{3}(9 \text{ ft}) = 3 \text{ ft}$$

$$\bar{x}_4 = 9 \text{ ft} - \tfrac{1}{3}(9 \text{ ft}) = 6 \text{ ft}$$

Using these results, show again that $F_R = 675$ lb and $\bar{x} = 4$ ft.

PROBLEMS

***4–152.** Replace the loading by an equivalent resultant force and couple moment acting at point O.

Prob. 4–152

4–153. Replace the distributed loading with an equivalent resultant force, and specify its location on the beam, measured from O.

Prob. 4–153

4–154. Replace the loading by an equivalent resultant force and specify its location on the beam, measured from A.

Prob. 4–154

4–155. Currently eighty-five percent of all neck injuries are caused by rear-end car collisions. To alleviate this problem, an automobile seat restraint has been developed that provides additional pressure contact with the cranium. During dynamic tests the distribution of load on the cranium has been plotted and shown to be parabolic. Determine the equivalent resultant force and its location, measured from A.

Prob. 4–155

*4–156. Determine the length b of the triangular load and its position a on the beam such that the equivalent resultant force is zero and the resultant couple moment is $8\ \text{kN}\cdot\text{m}$ clockwise.

Prob. 4–156

4–157. Replace the loading by an equivalent resultant force and specify its location measured from A.

Prob. 4–157

4–158. The distribution of soil loading on the bottom of a building slab is shown. Replace this loading by an equivalent resultant force and specify its location, measured from point O.

Prob. 4–158

4–159. Determine the magnitude of the equivalent resultant force of the distributed loading and specify its location on the beam measured from point A.

Prob. 4–159

*4–160. Replace the distributed loading by an equivalent resultant force and couple moment acting at point A.

Prob. 4–160

4–161. Determine the resultant moment of both the 100-lb force and the triangular distributed load about point O.

Prob. 4–161

4–162. If the soil exerts a trapezoidal distribution of load on the bottom of the footing, determine the intensities w_1 and w_2 of this distribution needed to support the column loadings..

Prob. 4–162

4–163. The form is used to cast a concrete wall having a width of 5 m. Determine the equivalent resultant force the wet concrete exerts on the form AB if the pressure distribution due to the concrete can be approximated as shown. Specify the location of the resultant force, measured from point B.

Prob. 4–163

PROBLEMS

*4–164. Replace the loading by an equivalent resultant force and couple moment acting at point O.

Prob. 4–164

4–165. Replace the loading on the beam by an equivalent resultant force and specify its location, measured from point A.

Prob. 4–165

4–166. Replace the loading by an equivalent resultant force and couple moment acting at point A.

4–167. Replace the loading by a single resultant force, and specify its location on the beam measured from point A.

Probs. 4–166/167

*4–168. Replace the distributed loading by an equivalent resultant force and specify where its line of action intersects a horizontal line along member AB, measured from A.

4–169. Replace the distributed loading by an equivalent resultant force and specify where its line of action intersects a vertical line along member BC, measured from C.

Probs. 4–168/169

4–170. Determine the length b of the triangular load and its position a on the beam so that the equivalent resultant force is zero and the resultant couple moment is $8\ \text{kN} \cdot \text{m}$ clockwise.

Prob. 4–170

4–171. Determine the magnitude of the equivalent resultant force of the distributed load and specify its location on the beam measured from point A.

Prob. 4–171

*4–172. If the distribution of the ground reaction on the pipe per foot of length can be approximated as shown, determine the magnitude of the resultant force due to this loading.

Prob. 4–172

4–173. Replace the loading by an equivalent resultant force and specify its location on the beam, measured from point O.

Prob. 4–173

4–174. Replace the distributed loading with an equivalent resultant force, and specify its location on the beam measured from point A.

Prob. 4–174

4–175. Replace the loading by an equivalent resultant force and couple moment acting at point O.

Prob. 4–175

*4–176. Wet concrete exerts a pressure distribution along the wall of the form. Determine the resultant force of this distribution and specify the height h where the bracing strut should be placed so that it lies through the line of action of the resultant force. The wall has a width of 5 m.

Prob. 4–176

CHAPTER REVIEW

Moment of Force—Scalar Definition

A force produces a turning effect or moment about a point O that does not lie on the force's line of action. In scalar form, the moment *magnitude* is the product of the force and the moment arm or perpendicular distance from point O to the line of action of the force.

The *direction* of the moment is defined using the right-hand rule. $\mathbf{M}_O$ always acts along an axis perpendicular to the plane containing $\mathbf{F}$ and d, and passes through point O.

$$M_O = Fd$$

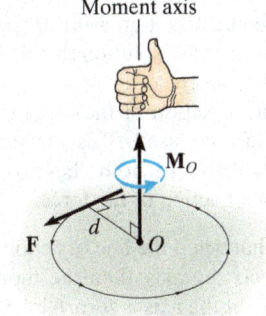

Principle of Moments

Rather than finding d, it is normally easier to resolve the force into its x and y components, determine the moment of each component about the point, and then sum the results.

$$M_O = Fd = F_x y - F_y x$$

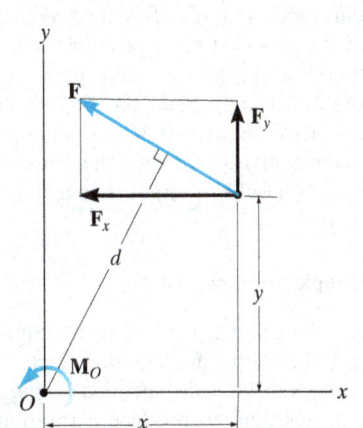

Moment of a Force—Vector Definition

Since three-dimensional geometry is generally more difficult to visualize, the vector cross product should be used to determine the moment. Here $\mathbf{M}_O = \mathbf{r} \times \mathbf{F}$, where $\mathbf{r}$ is a vector that extends from point O to any point A, B, or C on the line of action of $\mathbf{F}$.

$$\mathbf{M}_O = \mathbf{r}_A \times \mathbf{F} = \mathbf{r}_B \times \mathbf{F} = \mathbf{r}_C \times \mathbf{F}$$

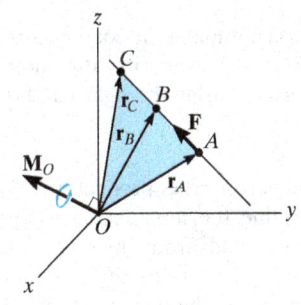

If the position vector $\mathbf{r}$ and force $\mathbf{F}$ are expressed as Cartesian vectors, then the cross product can be evaluated from the expansion of a determinant.

$$\mathbf{M}_O = \mathbf{r} \times \mathbf{F} = \begin{vmatrix} \mathbf{i} & \mathbf{j} & \mathbf{k} \\ r_x & r_y & r_z \\ F_x & F_y & F_z \end{vmatrix}$$

Moment about an Axis

If the moment of a force **F** is to be determined about an arbitrary axis a, then for a scalar solution the moment arm, or shortest distance d_a from the line of action of the force to the axis must be used. This distance is perpendicular to both the axis and the line of action of the force.

$$M_a = Fd_a$$

Note that when the line of action of **F** intersects the axis, then the moment of **F** about the axis is zero. Also, when the line of action of **F** is parallel to the axis, the moment of **F** about the axis is zero.

In three dimensions, the scalar triple product should be used. Here $\mathbf{u}_a$ is the unit vector that specifies the direction of the axis, and **r** is a position vector that is directed from any point on the axis to any point on the line of action of the force. If M_a is calculated as a negative scalar, then the sense of direction of $\mathbf{M}_a$ is opposite to $\mathbf{u}_a$.

$$M_a = \mathbf{u}_a \cdot (\mathbf{r} \times \mathbf{F}) = \begin{vmatrix} u_{a_x} & u_{a_y} & u_{a_z} \\ r_x & r_y & r_z \\ F_x & F_y & F_z \end{vmatrix}$$

Couple Moment

A couple consists of two equal but opposite forces that act a perpendicular distance d apart. Couples tend to produce a rotation without translation.

The magnitude of the couple moment is $M = Fd$, and its direction is established using the right-hand rule.

$$M = Fd$$

If the vector cross product is used to determine the moment of a couple, then **r** extends from any point on the line of action of one of the forces to any point on the line of action of the other force **F** that is used in the cross product.

$$\mathbf{M} = \mathbf{r} \times \mathbf{F}$$

Simplification of a Force and Couple System

Any system of forces and couples can be reduced to a single resultant force and resultant couple moment acting at a point. The resultant force is the sum of all the forces in the system, $\mathbf{F}_R = \Sigma \mathbf{F}$, and the resultant couple moment is equal to the sum of the moments of all the forces about the point O and the sum of the couple moments, $(\mathbf{M}_R)_O = \Sigma \mathbf{M}_O + \Sigma \mathbf{M}$.

Further simplification to a single resultant force is possible provided the force system is concurrent, coplanar, or parallel. To find the location of the resultant force from a point, it is necessary to equate the moment of the resultant force about the point O to the moment of the forces and couples in the system about the same point.

If the resultant force and couple moment at a point are not perpendicular to one another, then this system can be reduced to a wrench, which consists of the resultant force and collinear couple moment.

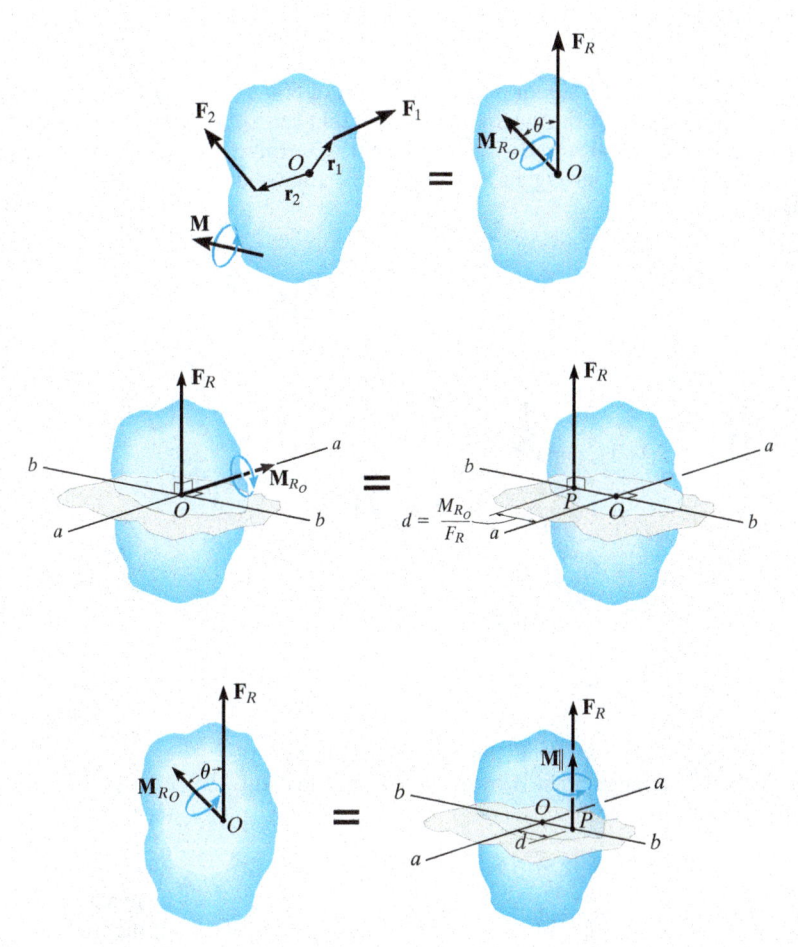

Coplanar Distributed Loading

A simple distributed loading can be represented by its resultant force, which is equivalent to the *area* under the loading curve. This resultant has a line of action that passes through the *centroid* or geometric center of the area or volume under the loading diagram.

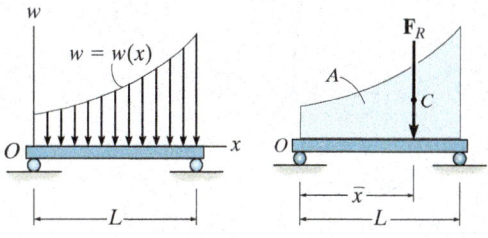

REVIEW PROBLEMS

R4–1. The boom has a length of 30 ft, a weight of 800 lb, and mass center at G. If the maximum moment that can be developed by a motor at A is $M = 20(10^3)$ lb·ft, determine the maximum load W, having a mass center at G', that can be lifted.

R4–3. The hood of the automobile is supported by the strut AB, which exerts a force of $F = 24$ lb on the hood. Determine the moment of this force about the hinged axis y.

Prob. R4–1

Prob. R4–3

R4–2. Replace the force $\mathbf{F}$ having a magnitude of $F = 50$ lb and acting at point A by an equivalent force and couple moment at point C.

R4–4. Friction on the concrete surface creates a couple moment of $M_O = 100$ N·m on the blades of the trowel. Determine the magnitude of the couple forces so that the resultant couple moment on the trowel is zero. The forces lie in a horizontal plane and act perpendicular to the handle of the trowel.

Prob. R4–2

Prob. R4–4

REVIEW PROBLEMS   193

R4–5. Replace the force and couple system by an equivalent force and couple moment at point P.

R4–7. The building slab is subjected to four parallel column loadings. Determine the equivalent resultant force and specify its location (x, y) on the slab. Take $F_1 = 30$ kN, $F_2 = 40$ kN.

Prob. R4–5

Prob. R4–7

R4–6. Replace the force system acting on the frame by a resultant force, and specify where its line of action intersects member AB, measured from point A.

R4–8. Replace the distributed loading by an equivalent resultant force, and specify its location on the beam, measured from the pin at C.

Prob. R4–6

Prob. R4–8

CHAPTER 5

It is important to be able to determine the forces in the cables used to support this boom to ensure that the boom does not fail. In this chapter we will study how to apply equilibrium methods to determine the forces acting on the supports of a rigid body such as this.

EQUILIBRIUM OF A RIGID BODY

CHAPTER OBJECTIVES

- To develop the equations of equilibrium.
- To introduce the concept of the free-body diagram.
- To show how to solve rigid-body equilibrium problems in two and three dimensions.

5.1 CONDITIONS FOR RIGID-BODY EQUILIBRIUM

In this section, we will develop both the necessary and sufficient conditions for the equilibrium of the rigid body in Fig. 5–1a. As shown, this body is subjected to an external force and couple moment system that is the result of the effects of gravitational, electrical, magnetic, or contact forces caused by supports or adjacent bodies. The internal forces caused by interactions between particles within the body are not shown in this figure, because these forces occur in equal but opposite collinear pairs and hence will cancel out, a consequence of Newton's third law.

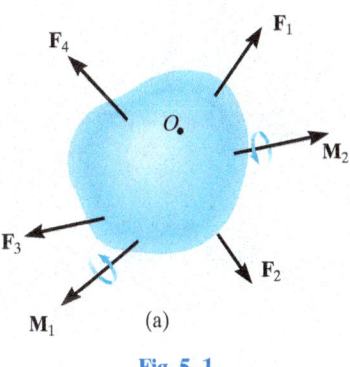

Fig. 5–1

196   Chapter 5   Equilibrium of a Rigid Body

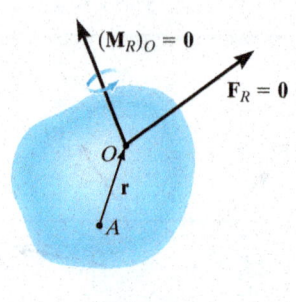

Fig. 5–1 (cont.)

Using the methods of the previous chapter, the force and couple moment system acting on a body can be reduced to an equivalent resultant force and resultant couple moment at any arbitrary point O on or off the body, Fig. 5–1b. If these two resultants are both equal to zero, then the body is said to be in *equilibrium*, which means it is at rest or will move with constant velocity. Mathematically, the equilibrium of a body is expressed as

$$\mathbf{F}_R = \Sigma \mathbf{F} = \mathbf{0} \qquad (5\text{–}1)$$

$$(\mathbf{M}_R)_O = \Sigma \mathbf{M}_O = \mathbf{0}$$

The first of these equations states that the sum of the forces acting on the body is equal to *zero*. The second equation states that the sum of the moments of all the forces in the system about point O, added to all the couple moments, is equal to *zero*. These two equations are not only necessary for equilibrium, they are also sufficient. To show this, consider summing moments about some other point, such as point A in Fig. 5–1c. We require

$$\Sigma \mathbf{M}_A = \mathbf{r} \times \mathbf{F}_R + (\mathbf{M}_R)_O = \mathbf{0}$$

Since $\mathbf{r} \neq \mathbf{0}$, this equation is satisfied if Eqs. 5–1 are satisfied, namely $\mathbf{F}_R = \mathbf{0}$ and $(\mathbf{M}_R)_O = \mathbf{0}$.

When applying the equations of equilibrium, we will assume that the body remains rigid. In reality, all bodies deform when subjected to loads; however, most engineering materials such as steel and concrete are very stiff and so their deformation is usually very small. Therefore, when applying the equations of equilibrium, we can generally assume that the body will remain *rigid* and *not deform* under the applied load without introducing any significant error. This way the direction of the applied forces and their moment arms with respect to a fixed reference remain the same both before and after a load is applied.

EQUILIBRIUM IN TWO DIMENSIONS

In the first part of the chapter, we will consider the case where the force system acting on a rigid body lies in or may be projected onto a *single* plane and, furthermore, any couple moments acting on the body are directed perpendicular to this plane. This type of force and couple system is often referred to as a two-dimensional or *coplanar* force system. For example, the airplane in Fig. 5–2 has a plane of symmetry through its center axis, and so the loads acting on the airplane are symmetrical with respect to this plane. Thus, each of the two wing tires will support the same load $\mathbf{T}$, which is represented on the side (two-dimensional) view of the plane as $2\mathbf{T}$.

Fig. 5–2

5.2 FREE-BODY DIAGRAMS

Successful application of the equations of equilibrium, which will be discussed in Sec. 5.3, requires a complete specification of *all* the known and unknown external forces that act *on* the body. The best way to account for these forces is to draw a ***free-body diagram*** of the body. This diagram is a sketch of the outlined shape of the body, which represents it as being *isolated* or "free" from its surroundings, i.e., a "free body." On this sketch it is necessary to show *all* the forces and couple moments that the supports and the surroundings exert *on the body*, so that these effects can be accounted for when the equations of equilibrium are applied. *A thorough understanding of how to draw a free-body diagram is of primary importance for solving problems in mechanics.*

Support Reactions. Before presenting a formal procedure as to how to draw a free-body diagram, we will first consider the various types of reactions that occur at supports and at points of contact between bodies subjected to coplanar force systems. As a general rule,

- A support prevents the translation of a body by exerting a force on the body.
- A support prevents the rotation of a body by exerting a couple moment on the body.

For example, let us consider three ways in which a horizontal member, such as a beam, is supported at its end. One method consists of a *roller* or cylinder, Fig. 5–3a. Since this support only prevents the beam from *translating* in the vertical direction, the roller will only exert a *force* on the beam in this direction, Fig. 5–3b.

The beam can be supported in a more restrictive manner by using a *pin*, Fig. 5–3c. The pin passes through a hole in the beam and two leaves which are fixed to the ground. Here the pin can prevent *translation* of the beam in *any direction* ϕ, Fig. 5–3d, and so the pin must exert a force **F** on the beam in the opposite direction. For purposes of analysis, it is generally easier to represent this resultant force **F** by its two rectangular components $\mathbf{F}_x$ and $\mathbf{F}_y$, Fig. 5–3e. Once these components are known, then F and ϕ can be calculated.

The most restrictive way to support the beam would be to use a *fixed support* as shown in Fig. 5–3f. This support will prevent both *translation and rotation* of the beam. As a result, a *force and couple moment* must be developed on the beam at its point of connection, Fig. 5–3g. As in the case of the pin, the force is usually represented by its rectangular components $\mathbf{F}_x$ and $\mathbf{F}_y$.

Table 5–1 lists other common types of supports for bodies subjected to coplanar force systems. (In all cases the angle θ is assumed to be known.) Carefully study each of the symbols used to represent these supports and the types of reactions they exert on their contacting members.

Fig. 5–3

TABLE 5–1 Supports for Rigid Bodies Subjected to Two-Dimensional Force Systems

Types of Connection	Reaction	Number of Unknowns
(1) cable		One unknown. The reaction is a tension force which acts away from the member in the direction of the cable.
(2) weightless link	or	One unknown. The reaction is a force which acts along the axis of the link.
(3) roller		One unknown. The reaction is a force which acts perpendicular to the surface at the point of contact.
(4) rocker		One unknown. The reaction is a force which acts perpendicular to the surface at the point of contact.
(5) smooth contacting surface		One unknown. The reaction is a force which acts perpendicular to the surface at the point of contact.
(6) roller or pin in confined smooth slot	or	One unknown. The reaction is a force which acts perpendicular to the slot.
(7) member pin connected to collar on smooth rod	or	One unknown. The reaction is a force which acts perpendicular to the rod.

continued

TABLE 5–1 Continued

Types of Connection	Reaction	Number of Unknowns
(8) smooth pin or hinge	F_y, F_x or F, ϕ	Two unknowns. The reactions are two components of force, or the magnitude and direction ϕ of the resultant force. Note that ϕ and θ are not necessarily equal [usually not, unless the rod shown is a link as in (2)].
(9) member fixed connected to collar on smooth rod	F, M	Two unknowns. The reactions are the couple moment and the force which acts perpendicular to the rod.
(10) fixed support	F_y, F_x, M or F, ϕ, M	Three unknowns. The reactions are the couple moment and the two force components, or the couple moment and the magnitude and direction ϕ of the resultant force.

Typical examples of actual supports are shown in the following sequence of photos. The numbers refer to the connection types in Table 5–1.

The cable exerts a force on the bracket in the direction of the cable. (1)

The rocker support for this bridge girder allows horizontal movement so the bridge is free to expand and contract due to a change in temperature. (4)

This concrete girder rests on the ledge that is assumed to act as a smooth contacting surface. (5)

Typical pin support for a beam. (8)

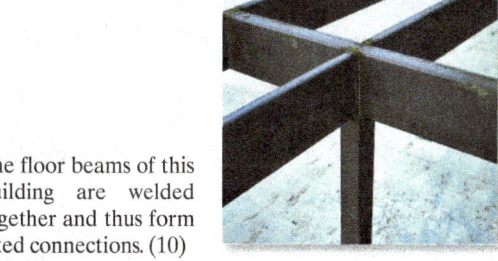

The floor beams of this building are welded together and thus form fixed connections. (10)

Internal Forces. As stated in Sec. 5.1, the internal forces that act between adjacent particles in a body always occur in collinear pairs such that they have the same magnitude and act in opposite directions (Newton's third law). Since these forces cancel each other, they will not create an *external effect* on the body. It is for this reason that the internal forces should not be included on the free-body diagram if the entire body is to be considered. For example, the engine shown in Fig. 5–4a has a free-body diagram shown in Fig. 5–4b. The internal forces between all its connected parts, such as the screws and bolts, will cancel out because they form equal and opposite collinear pairs. Only the external forces T_1 and T_2, exerted by the chains and the engine weight W, are shown on the free-body diagram.

(a) (b)

Fig. 5–4

Weight and the Center of Gravity. When a body is within a gravitational field, then each of its particles has a specified weight. It was shown in Sec. 4.8 that such a system of forces can be reduced to a single resultant force acting through a specified point. We refer to this force resultant as the *weight* W of the body and to the location of its point of application as the **center of gravity**. The methods used for its determination will be developed in Chapter 9.

In the examples and problems that follow, if the weight of the body is important for the analysis, this force will be reported in the problem statement. Also, when the body is *uniform* or made from the same material, the center of gravity will be located at the body's *geometric center* or *centroid*; however, if the body consists of a nonuniform distribution of material, or has an unusual shape, then the location of its center of gravity G will be given.

Idealized Models. When an engineer performs a force analysis of any object, he or she considers a corresponding analytical or idealized model that gives results that approximate as closely as possible the actual situation. To do this, careful choices have to be made so that selection of the type of supports, the material behavior, and the object's dimensions can be justified. This way one can feel confident that any design or analysis will yield results which can be trusted. In complex cases this process may require developing several different models of the object that must be analyzed. However, in any case, this selection process requires both skill and experience.

The following two cases illustrate what is required to develop a proper model. In Fig. 5–5a, the steel beam is to be used to support the three roof joists of a building. For a force analysis it is reasonable to assume the material (steel) is rigid since only very small deflections will occur when the beam is loaded. A bolted connection at A will allow for any slight rotation that occurs here when the load is applied, and so a *pin* can be considered for this support. At B a *roller* can be considered since this support offers no resistance to horizontal movement. Building code is used to specify the roof loading A so that the joist loads **F** can be calculated. These forces are intended to be larger than any actual loading on the beam since they account for extreme loading cases and for dynamic or vibrational effects. Finally, the weight of the beam is generally neglected when it is small compared to the load the beam supports. The idealized model of the beam is therefore shown with average dimensions a, b, c, and d in Fig. 5–5b.

As a second case, consider the lift boom in Fig. 5–6a. By inspection, it is supported by a pin at A and by the hydraulic cylinder BC, which can be approximated as a weightless link. The material can be assumed rigid, and with its density known, the weight of the boom and the location of its center of gravity G are determined. When a design loading **P** is specified, the idealized model shown in Fig. 5–6b can be used for a force analysis. Average dimensions (not shown) are used to specify the location of the loads and the supports.

Idealized models of specific objects will be given in some of the examples throughout the text. In all cases, it should be realized, however, that each case represents the reduction of a practical situation using simplified assumptions like the ones illustrated here.

(a)

(b)

Fig. 5–5

(a)

(b)

Fig. 5–6

IMPORTANT POINTS

- No equilibrium problem should be solved without *first drawing the free-body diagram*, so as to account for all the forces and couple moments that act on the body.
- If a support *prevents translation* of a body, then the support, when it is removed, exerts a *force* on the body.
- If *rotation is prevented*, then the support, when it is removed, exerts a *couple moment* on the body.
- Study Table 5–1.
- Internal forces are *never shown* on the free-body diagram since they occur in equal but opposite collinear pairs and therefore cancel out.
- The weight of a body is an external force, and its effect is represented by a single resultant force acting through the body's center of gravity *G*.
- *Couple moments* can be placed anywhere on the free-body diagram since they are *free vectors*. *Forces* can act at any point along their lines of action since they are *sliding vectors*.

PROCEDURE FOR ANALYSIS

To construct a free-body diagram for a rigid body or any group of bodies considered as a single system, the following steps should be performed:

Draw Outlined Shape.

Imagine the body to be *isolated* or cut "free" from its constraints and connections and draw (sketch) its outlined shape. Be sure to *remove all the supports* from the body.

Show All Forces and Couple Moments.

Identify all the known and unknown *external forces* and couple moments that *act on the body*. Those generally encountered are due to (1) applied loadings, (2) reactions occurring at the supports or at points of contact with other bodies (see Table 5–1), and (3) the weight of the body. To account for all these effects, it may help to trace over the boundary, carefully noting each force or couple moment acting on it.

Identify Each Loading and Give Dimensions.

The forces and couple moments that are known should be labeled with their proper magnitudes and directions. Letters are used to represent the magnitudes and direction angles of forces and couple moments that are unknown. Finally, indicate the dimensions of the body necessary for calculating the moments of forces.

EXAMPLE 5.1

Draw the free-body diagram of the uniform beam shown in Fig. 5–7a. The beam has a mass of 100 kg.

(a)

SOLUTION
The free-body diagram of the beam is shown in Fig. 5–7b. Since the support at A is fixed, the wall exerts three reactions *on the beam*, denoted as $\mathbf{A}_x$, $\mathbf{A}_y$, and $\mathbf{M}_A$. The magnitudes of these reactions are *unknown*, and their sense has been *assumed*. The weight of the beam, $W = 100(9.81)$ N $= 981$ N, acts through the beam's center of gravity G, which is 3 m from A since the beam is uniform.

(b)

Fig. 5–7

EXAMPLE 5.2

Draw the free-body diagram of the foot lever shown in Fig. 5–8a. The operator applies a vertical force to the pedal so that the spring is stretched 1.5 in. and the force on the link at B is 20 lb.

(b)

(a)

Fig. 5–8

(c)

SOLUTION

By inspection of the photo the lever is loosely bolted to the frame at A and so this bolt acts as a pin. (See (8) in Table 5–1.) Although not shown here the link at B is pinned at both ends and so it is like (2) in Table 5–1. After making the proper measurements, the idealized model of the lever is shown in Fig. 5–8b. From this, the free-body diagram is shown in Fig. 5–8c. Since the pin at A is *removed*, it exerts force components A_x and A_y on the lever. The link exerts a force of 20 lb, acting in the direction of the link. In addition the spring also exerts a horizontal force on the lever. If the stiffness is measured and found to be $k = 20$ lb/in., then since the stretch $s = 1.5$ in., using Eq. 3–2, $F_s = ks = 20$ lb/in. (1.5 in.) $= 30$ lb. Finally, the operator's shoe applies a vertical force of **F** on the pedal. The dimensions of the lever are also shown on the free-body diagram, since this information will be useful when calculating the moments of the forces. As usual, the senses of the unknown forces at A have been assumed. The correct senses will become apparent after solving the equilibrium equations.

EXAMPLE 5.3

Two smooth pipes, each having a mass of 300 kg, are supported by the forked tines of the tractor in Fig. 5–9a. Draw the free-body diagrams for each pipe and both pipes together.

SOLUTION

The idealized model from which we must draw the free-body diagrams is shown in Fig. 5–9b. Here the pipes are identified, the dimensions have been added, and the physical situation reduced to its simplest form.

Removing the surfaces of contact, the free-body diagram for pipe A is shown in Fig. 5–9c. Its weight is $W = 300(9.81)$ N $= 2943$ N. Assuming all contacting surfaces are *smooth*, the reactive forces **T**, **F**, **R** act in a direction *normal* to the tangent at their surfaces of contact.

The free-body diagram of the isolated pipe B is shown in Fig. 5–9d. Can you identify each of the three forces acting *on this pipe*? Note that **R** representing the force of A on B, Fig. 5–9d, is equal and opposite to **R** representing the force of B on A, Fig. 5–9c.

The free-body diagram of both pipes combined ("system") is shown in Fig. 5–9e. Here the contact force **R**, which acts between A and B, is considered an *internal* force and hence is not shown on the free-body diagram. That is, it represents a pair of equal but opposite collinear forces which cancel each other.

Fig. 5–9

EXAMPLE 5.4

Draw the free-body diagram of the unloaded platform that is suspended off the edge of the oil rig shown in Fig. 5–10a. The platform has a mass of 200 kg.

(a)

(b)

(c)

Fig. 5–10

SOLUTION
The idealized model of the platform will be considered in two dimensions because by observation the loading and the dimensions are all symmetrical about a vertical plane passing through its center, Fig. 5–10b. The connection at A is considered to be a pin, and the cable supports the platform at B. The direction of the cable and average dimensions of the platform are listed, and the center of gravity G has been determined. It is from this model that we have drawn the free-body diagram shown in Fig. 5–10c. The platform's weight is $200(9.81) = 1962$ N. The supports have been *removed*, and the force components $\mathbf{A}_x$ and $\mathbf{A}_y$ along with the cable force $\mathbf{T}$ represent the reactions that *both* pins and *both* cables exert on the platform, Fig. 5–10a. As a result, half their magnitudes are developed on each side of the platform.

PROBLEMS

5–1. Draw the free-body diagram for the following problems.

a) The beam in Prob. 5–10.

b) The beam in Prob. 5–11.

c) The beam in Prob. 5–12.

d) The beam in Prob. 5–14.

5–2. Draw the free-body diagram for the following problems.

a) The beam in Prob. 5–15.

b) The linkage in Prob. 5–16.

c) The frame in Prob. 5–17.

d) The beam in Prob. 5–18.

5–3. Draw the free-body diagram for the following problems.

a) The lamp in Prob. 5–19.

b) The rod in Prob. 5–20.

c) The assembly in Prob. 5–21.

d) The beam in Prob. 5–22.

*5–4. Draw the free-body diagram for the following problems.

a) The rod in Prob. 5–25.

b) The bar in Prob. 5–27.

c) The disk in Prob. 5–28.

5–5. Draw the free-body diagram for the following problems.

a) The clamp in Prob. 5–32.

b) The jib crane in Prob. 5–33.

c) The crane in Prob. 5–35.

d) The beam in Prob. 5–36.

5–6. Draw the free-body diagram for the following problems.

a) The jib crane in Prob. 5–37.

b) The bar in Prob. 5–39.

c) The davit in Prob. 5–41.

d) The boom in Prob. 5–42.

5–7. Draw the free-body diagram for the following problems.

a) The rod in Prob. 5–44.

b) The hand truck and load when it is lifted in Prob. 5–45.

c) The beam in Prob. 5–47.

d) The boom frame in Prob. 5–52.

*5–8. Draw the free-body diagram for the following problems.

a) The beam in Prob. 5–53.

b) The brake pedal in Prob. 5–54.

c) The exercise machine in Prob. 5–55.

d) The hatch door in Prob. 5–58.

5–9. Draw the free-body diagram for the following problems.

a) The crane in Prob. 5–59.

b) The smooth pipe in Prob. 5–61.

c) The bar in Prob. 5–62.

5.3 EQUATIONS OF EQUILIBRIUM

In Sec. 5.1 we developed the two equations which are both necessary and sufficient for the equilibrium of a rigid body, namely, $\Sigma \mathbf{F} = \mathbf{0}$ and $\Sigma \mathbf{M}_O = \mathbf{0}$. When the body is subjected to a system of forces, which all lie in the x–y plane, then the forces can be resolved into their x and y components. Consequently, the conditions for equilibrium in two dimensions then become

$$\Sigma F_x = 0$$
$$\Sigma F_y = 0 \qquad (5\text{--}2)$$
$$\Sigma M_O = 0$$

Here ΣF_x and ΣF_y represent, respectively, the algebraic sums of the x and y components of all the forces acting on the body, and ΣM_O represents the algebraic sum of all the couple moments and the moments of all the forces about an arbitrary point O.

Alternative Sets of Equilibrium Equations.

Although Eqs. 5–2 are *most often* used for solving coplanar equilibrium problems, two *alternative* sets of three independent equilibrium equations may also be used. One such set is

$$\Sigma F_x = 0$$
$$\Sigma M_A = 0 \qquad (5\text{--}3)$$
$$\Sigma M_B = 0$$

When using these equations it is required that a line passing through points A and B is not parallel to the y axis. To show that these equations provide the *conditions* for equilibrium, consider the free-body diagram of the plate shown in Fig. 5–11a. Using the methods of Sec. 4.7, all the forces on the free-body diagram are first replaced by an equivalent resultant force $\mathbf{F}_R = \Sigma \mathbf{F}$, and a resultant couple moment $(\mathbf{M}_R)_A = \Sigma \mathbf{M}_A$, Fig. 5–11b. If $\Sigma M_A = 0$ is satisfied, it is necessary that $(\mathbf{M}_R)_A = \mathbf{0}$. If $\Sigma F_x = 0$ is satisfied, then $\mathbf{F}_R$ must have *no component* along the x axis, and therefore it must be parallel to the y axis, Fig. 5–11c. Finally, if $\Sigma M_B = 0$, where B does not lie on the line of action of $\mathbf{F}_R$, then $\mathbf{F}_R = \mathbf{0}$ and therefore the plate in Fig. 5–11a must be in equilibrium.

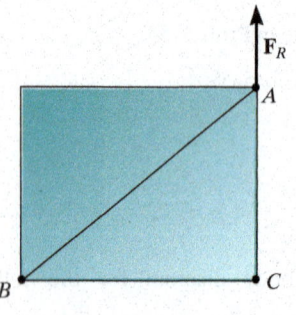

Fig. 5–11

A second alternative set of equilibrium equations is

$$\Sigma M_A = 0$$
$$\Sigma M_B = 0 \quad \quad (5\text{-}4)$$
$$\Sigma M_C = 0$$

Here it is necessary that points $A, B,$ and C do not lie on the same line. To show that these equations, when satisfied, ensure equilibrium, consider again the free-body diagram in Fig. 5–11b. If $\Sigma M_A = 0$ is to be satisfied, then $(\mathbf{M}_R)_A = \mathbf{0}$. If $\Sigma M_C = 0$ is satisfied, then the line of action of $\mathbf{F}_R$ passes through point C, Fig. 5–11c. Finally, if $\Sigma M_B = 0$ is satisfied, then $\mathbf{F}_R = \mathbf{0}$, and so the plate in Fig. 5–11a must be in equilibrium.

PROCEDURE FOR ANALYSIS

Coplanar force equilibrium problems can be solved using the following procedure.

Free-Body Diagram.

- Establish the x, y coordinate axes in any suitable orientation.

- Remove all supports and draw an outlined shape of the body.

- Show all the forces and couple moments acting on the body.

- Label all the loadings and specify their directions relative to the x or y axis. The sense of a force or couple moment having an *unknown* magnitude but known line of action can be *assumed*.

- Indicate the dimensions of the body necessary for calculating the moments of forces.

Equations of Equilibrium.

- Apply the moment equation of equilibrium, $\Sigma M_O = 0$, about a point O that lies at the intersection of the lines of action of two unknown forces. In this way, the moments of these unknowns are zero about O, and a *direct solution* for the third unknown can be determined.

- When applying the force equilibrium equations, $\Sigma F_x = 0$ and $\Sigma F_y = 0$, orient the x and y axes along lines that will provide the simplest resolution of the forces into their x and y components.

- If the solution of the equilibrium equations yields a negative scalar for a force or couple moment magnitude, this indicates that the sense is opposite to that which was assumed on the free-body diagram.

EXAMPLE 5.5

Determine the horizontal and vertical components of reaction on the beam caused by the pin at B and the rocker at A as shown in Fig. 5–12a. Neglect the weight of the beam.

Fig. 5–12

SOLUTION

Free-Body Diagram. The supports are *removed*, and the free-body diagram of the beam is shown in Fig. 5–12b. For simplicity, the 600-N force is represented by its x and y components as shown in Fig. 5–12b.

Equations of Equilibrium. Summing forces in the x direction yields

$\xrightarrow{+} \Sigma F_x = 0;$ $600 \cos 45° \text{ N} - B_x = 0$

$$B_x = 424 \text{ N} \qquad \textit{Ans.}$$

A direct solution for A_y can be obtained by applying the moment equation $\Sigma M_B = 0$ about point B.

$\zeta + \Sigma M_B = 0;$ $100 \text{ N}(2 \text{ m}) + (600 \sin 45° \text{ N})(5 \text{ m})$
$- (600 \cos 45° \text{ N})(0.2 \text{ m}) - A_y(7 \text{ m}) = 0$

$$A_y = 319 \text{ N} \qquad \textit{Ans.}$$

Summing forces in the y direction, using this result, gives

$+\uparrow \Sigma F_y = 0;$ $319 \text{ N} - 600 \sin 45° \text{ N} - 100 \text{ N} - 200 \text{ N} + B_y = 0$

$$B_y = 405 \text{ N} \qquad \textit{Ans.}$$

NOTE: The support forces in Fig. 5–12b are caused by the pins that contact the beam. The opposite forces act on the pins. For example, Fig. 5–12c shows the equilibrium of the pin at A and the rocker.

EXAMPLE 5.6

The cord shown in Fig. 5–13a supports a force of 100 lb and wraps over the frictionless pulley. Determine the tension in the cord at C and the horizontal and vertical components of reaction at pin A.

(a)

Fig. 5–13

SOLUTION

Free-Body Diagrams. The free-body diagrams of the cord and pulley are shown in Fig. 5–13b. Note that the principle of action, equal but opposite reaction must be carefully observed when drawing each of these diagrams: the cord exerts an unknown load distribution p on the pulley at the contact surface, whereas the pulley exerts an equal but opposite effect on the cord. For the solution, however, it is simpler to *combine* the free-body diagrams of the pulley and this portion of the cord, so that the distributed load becomes *internal* to this "system" and is therefore eliminated from the analysis, Fig. 5–13c.

(b)

(c)

Equations of Equilibrium. Summing moments about point A to eliminate A_x and A_y, Fig. 5–13c, we have

$\zeta + \Sigma M_A = 0;$ $100 \text{ lb } (0.5 \text{ ft}) - T(0.5 \text{ ft}) = 0$

$T = 100 \text{ lb}$ *Ans.*

Using this result,

$\xrightarrow{+} \Sigma F_x = 0;$ $-A_x + 100 \sin 30° \text{ lb} = 0$

$A_x = 50.0 \text{ lb}$ *Ans.*

$+\uparrow \Sigma F_y = 0;$ $A_y - 100 \text{ lb} - 100 \cos 30° \text{ lb} = 0$

$A_y = 187 \text{ lb}$ *Ans.*

NOTE: From the moment equation, it is seen that the tension in the cord remains *constant* as the cord passes over the pulley. (This of course is true for *any angle* θ at which the cord is directed and for *any radius* r of the pulley.)

EXAMPLE 5.7

The member shown in Fig. 5–14a is pin connected at A and rests against a smooth support at B. Determine the horizontal and vertical components of reaction at the pin A.

Fig. 5–14

SOLUTION

Free-Body Diagram. As shown in Fig. 5–14b, the supports are removed and the reaction N_B must be perpendicular to the member at B. Also, horizontal and vertical components of reaction are represented at A. The resultant of the distributed loading is $\frac{1}{2}(1.5\text{ m})(80\text{ N/m}) = 60\text{ N}$. It acts through the centroid of the triangle, 1 m from A as shown.

Equations of Equilibrium. Summing moments about A, we obtain a direct solution for N_B,

$\zeta + \Sigma M_A = 0;$ $-90\text{ N}\cdot\text{m} - 60\text{ N}(1\text{ m}) + N_B(0.75\text{ m}) = 0$

$$N_B = 200\text{ N}$$

Using this result,

$\xrightarrow{+} \Sigma F_x = 0;$ $A_x - 200\sin 30°\text{ N} = 0$

$$A_x = 100\text{ N} \qquad \textit{Ans.}$$

$+\uparrow \Sigma F_y = 0;$ $A_y - 200\cos 30°\text{ N} - 60\text{ N} = 0$

$$A_y = 233\text{ N} \qquad \textit{Ans.}$$

EXAMPLE 5.8

The box wrench in Fig. 5–15a is used to tighten the bolt at A. If the wrench does not turn when the load is applied to the handle, determine the torque or moment applied to the bolt and the force of the wrench on the bolt.

SOLUTION

Free-Body Diagram. The free-body diagram for the wrench is shown in Fig. 5–15b. Since the bolt acts as a "fixed support," when it is removed it exerts force components A_x and A_y and a moment M_A on the wrench at A.

Fig. 5–15

Equations of Equilibrium.

$\xrightarrow{+} \Sigma F_x = 0;$ $A_x - 52\left(\frac{5}{13}\right) \text{N} + 30 \cos 60° \text{ N} = 0$

$\qquad A_x = 5.00 \text{ N}$ *Ans.*

$+\uparrow \Sigma F_y = 0;$ $A_y - 52\left(\frac{12}{13}\right) \text{N} - 30 \sin 60° \text{ N} = 0$

$\qquad A_y = 74.0 \text{ N}$ *Ans.*

$\zeta + \Sigma M_A = 0;$ $M_A - \left[52\left(\frac{12}{13}\right)\text{N}\right](0.3 \text{ m}) - (30 \sin 60° \text{ N})(0.7 \text{ m}) = 0$

$\qquad M_A = 32.6 \text{ N} \cdot \text{m}$ *Ans.*

Note that M_A must be *included* in this moment summation. This couple moment is a free vector and represents the twisting resistance of the bolt on the wrench. By Newton's third law, the wrench exerts an equal but opposite moment or torque on the bolt. Furthermore, the resultant force on the wrench is

$$F_A = \sqrt{(5.00)^2 + (74.0)^2} = 74.1 \text{ N} \qquad \textit{Ans.}$$

NOTE: Although only *three* independent equilibrium equations can be written for a rigid body, it is a good practice to *check* the calculations using a fourth equilibrium equation. For example, the above computations may be verified in part by summing moments about point C:

$\zeta + \Sigma M_C = 0;$ $\left[52\left(\frac{12}{13}\right)\text{N}\right](0.4 \text{ m}) + 32.6 \text{ N} \cdot \text{m} - 74.0 \text{ N}(0.7 \text{ m}) = 0$

$\qquad 19.2 \text{ N} \cdot \text{m} + 32.6 \text{ N} \cdot \text{m} - 51.8 \text{ N} \cdot \text{m} = 0$

EXAMPLE 5.9

Determine the horizontal and vertical components of reaction on the member at the pin A, and the normal reaction at the roller B in Fig. 5–16a.

SOLUTION

Free-Body Diagram. All the supports are removed and so the free-body diagram is shown in Fig. 5–16b. The pin at A exerts two components of reaction on the member, $\mathbf{A}_x$ and $\mathbf{A}_y$.

Fig. 5–16

Equations of Equilibrium. The reaction N_B can be obtained *directly* by summing moments about point A, since $\mathbf{A}_x$ and $\mathbf{A}_y$ produce no moment about A.

$\zeta + \Sigma M_A = 0;$

$$[N_B \cos 30°](6 \text{ ft}) - [N_B \sin 30°](2 \text{ ft}) - 750 \text{ lb}(3 \text{ ft}) = 0$$

$$N_B = 536.2 \text{ lb} = 536 \text{ lb} \qquad \qquad Ans.$$

Using this result,

$\xrightarrow{+} \Sigma F_x = 0;\qquad A_x - (536.2 \text{ lb}) \sin 30° = 0$

$\qquad\qquad\qquad\quad A_x = 268 \text{ lb} \qquad\qquad\qquad Ans.$

$+\uparrow \Sigma F_y = 0;\qquad A_y + (536.2 \text{ lb}) \cos 30° - 750 \text{ lb} = 0$

$\qquad\qquad\qquad\quad A_y = 286 \text{ lb} \qquad\qquad\qquad Ans.$

Details of the equilibrium of the pin at A are shown in Fig. 5–16c.

EXAMPLE 5.10

The uniform smooth rod shown in Fig. 5–17a is subjected to a force and couple moment. If the rod is supported at A by a smooth wall and at B and C either at the top or bottom by rollers, determine the reactions at these supports. Neglect the weight of the rod.

SOLUTION

Free-Body Diagram. Removing the supports as shown in Fig. 5–17b, all the reactions act normal to the surfaces of contact since these surfaces are smooth. The reactions at B and C are shown acting in the positive y' direction. This assumes that only the rollers located on the bottom of the rod are used for support.

Equations of Equilibrium. Using the x, y coordinate system in Fig. 5–17b, we have

$$\xrightarrow{+} \Sigma F_x = 0; \quad C_{y'} \sin 30° + B_{y'} \sin 30° - A_x = 0 \quad (1)$$

$$+\uparrow \Sigma F_y = 0; \quad -300 \text{ N} + C_{y'} \cos 30° + B_{y'} \cos 30° = 0 \quad (2)$$

$$\zeta + \Sigma M_A = 0; \quad -B_{y'}(2 \text{ m}) + 4000 \text{ N} \cdot \text{m} - C_{y'}(6 \text{ m})$$
$$+ (300 \cos 30° \text{ N})(8 \text{ m}) = 0 \quad (3)$$

When writing the moment equation, it should be noted that the line of action of the force component 300 sin 30° N passes through point A, and therefore this force is not included in the moment equation.

Solving Eqs. 2 and 3 simultaneously, we obtain

$$B_{y'} = -1000.0 \text{ N} = -1 \text{ kN} \qquad Ans.$$

$$C_{y'} = 1346.4 \text{ N} = 1.35 \text{ kN} \qquad Ans.$$

Since $B_{y'}$ is a negative scalar, the sense of $\mathbf{B}_{y'}$ is opposite to that shown on the free-body diagram in Fig. 5–17b. Therefore, the top roller at B serves as the support rather than the bottom one. Retaining the negative sign for $B_{y'}$ (Why?) and substituting the results into Eq. 1, we obtain

$$1346.4 \sin 30° \text{ N} + (-1000.0 \sin 30° \text{ N}) - A_x = 0$$

$$A_x = 173 \text{ N} \qquad Ans.$$

(a)

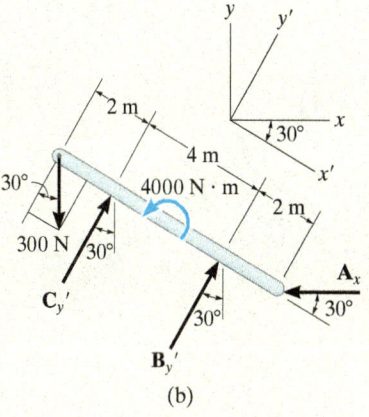

(b)

Fig. 5–17

EXAMPLE 5.11

(a)

The uniform truck ramp shown in Fig. 5–18a has a weight of 400 lb and is pinned to the body of the truck at each side and held in the position shown by the two side cables. Determine the tension in the cables.

SOLUTION
The idealized model of the ramp, which indicates all necessary dimensions and supports, is shown in Fig. 5–18b. Here the center of gravity is located at the midpoint since the ramp is considered to be uniform.

Free-Body Diagram. Removing the supports from the idealized model, the ramp's free-body diagram is shown in Fig. 5–18c.

Equations of Equilibrium. Summing moments about point A will yield a direct solution for the cable tension. Using the principle of moments, there are several ways of determining the moment of **T** about A. If we use x and y components, with **T** applied at B, we have

(b)

$\zeta + \Sigma M_A = 0;$ $-T \cos 20°(7 \sin 30° \text{ ft}) + T \sin 20°(7 \cos 30° \text{ ft})$
$+ 400 \text{ lb }(5 \cos 30° \text{ ft}) = 0$

$T = 1425 \text{ lb}$

We can also determine the moment of **T** about A by resolving it into components along and perpendicular to the ramp at B. Then the moment of the component along the ramp will be zero about A, so that

$\zeta + \Sigma M_A = 0;$ $-T \sin 10°(7 \text{ ft}) + 400 \text{ lb }(5 \cos 30° \text{ ft}) = 0$

$T = 1425 \text{ lb}$

Since there are two cables supporting the ramp,

$T' = \dfrac{T}{2} = 712 \text{ lb}$ *Ans.*

(c)

Fig. 5–18

NOTE: As an exercise, show that $A_x = 1339$ lb and $A_y = 887$ lb.

EXAMPLE 5.12

Determine the support reactions on the member in Fig. 5–19a. The collar at A is fixed to the member and can slide vertically along the vertical shaft.

(a)

(b)

Fig. 5–19

SOLUTION

Free-Body Diagram. Removing the supports, the free-body diagram of the member is shown in Fig. 5–19b. The collar exerts a horizontal force $\mathbf{A}_x$ and moment $\mathbf{M}_A$ on the member. The reaction $\mathbf{N}_B$ of the roller on the member is vertical.

Equations of Equilibrium. The forces A_x and N_B can be determined directly from the force equations of equilibrium.

$\stackrel{+}{\rightarrow} \Sigma F_x = 0;$ $A_x = 0$ *Ans.*

$+\uparrow \Sigma F_y = 0;$ $N_B - 900 \text{ N} = 0$

$N_B = 900 \text{ N}$ *Ans.*

The moment M_A can be determined by summing moments either about point A or point B.

$\zeta + \Sigma M_A = 0;$

$M_A - 900 \text{ N}(1.5 \text{ m}) - 500 \text{ N} \cdot \text{m} + 900 \text{ N}\,[3 \text{ m} + (1 \text{ m}) \cos 45°] = 0$

$M_A = -1486 \text{ N} \cdot \text{m} = 1.49 \text{ kN} \cdot \text{m} \downarrow$ *Ans.*

or

$\zeta + \Sigma M_B = 0;$ $M_A + 900 \text{ N}\,[1.5 \text{ m} + (1 \text{ m}) \cos 45°] - 500 \text{ N} \cdot \text{m} = 0$

$M_A = -1486 \text{ N} \cdot \text{m} = 1.49 \text{ kN} \cdot \text{m} \downarrow$ *Ans.*

The negative sign indicates that $\mathbf{M}_A$ has the opposite sense of rotation to that shown on the free-body diagram.

5.4 TWO- AND THREE-FORCE MEMBERS

The solutions to some equilibrium problems can be simplified by recognizing members that are subjected to only two or three forces.

Two-Force Members. As the name implies, a *two-force member* has forces applied at only two points on the member. An example of a two-force member is shown in Fig. 5–20a. To satisfy force equilibrium, F_A and F_B must be equal in magnitude, $F_A = F_B = F$, but opposite in direction ($\Sigma \mathbf{F} = \mathbf{0}$), Fig. 5–20b. Furthermore, moment equilibrium requires that F_A and F_B share the same line of action, which can only happen if they are directed along the line joining points A and B ($\Sigma \mathbf{M}_A = \mathbf{0}$ or $\Sigma \mathbf{M}_B = \mathbf{0}$), Fig. 5–20c. Therefore, for any two-force member to be in equilibrium, the two forces acting on the member *must have the same magnitude, act in opposite directions, and have the same line of action, directed along the line joining the two points where these forces act.*

The hydraulic cylinder AB is a typical example of a two-force member since it is pin connected at its ends and, provided its weight is neglected, only the resultant pin forces act on this member.

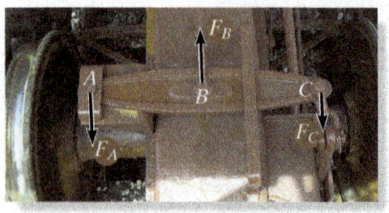

The link used for this railroad car brake is a three-force member. Since the force $\mathbf{F}_B$ in the tie rod at B and $\mathbf{F}_C$ from the link at C are parallel, then for equilibrium the resultant force $\mathbf{F}_A$ at the pin A must also be parallel with these two forces.

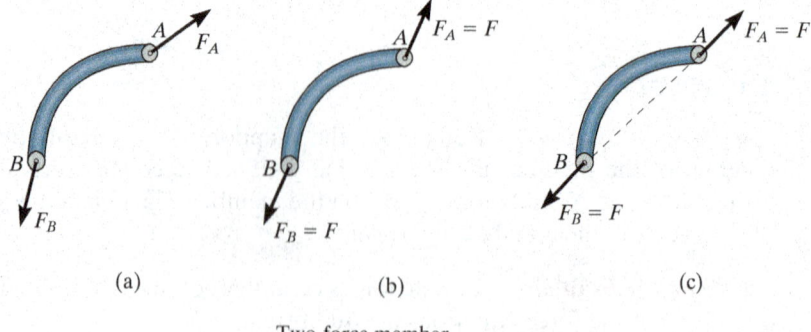

Two-force member

Fig. 5–20

Three-Force Members. If a member is subjected to only *three* forces, it is called a *three-force member*. Moment equilibrium can be satisfied only if the three forces form a *concurrent* or *parallel* force system. To illustrate, consider the member in Fig. 5–21a subjected to the three forces $\mathbf{F}_1$, $\mathbf{F}_2$, and $\mathbf{F}_3$. If the lines of action of $\mathbf{F}_1$ and $\mathbf{F}_2$ intersect at point O, then the line of action of $\mathbf{F}_3$ must *also* pass through point O so that the forces satisfy $\Sigma \mathbf{M}_O = \mathbf{0}$. As a special case, if the three forces are all parallel, Fig. 5–21b, the location of the point of intersection, O, will approach infinity.

The boom and bucket on this lift is a three-force member, provided its weight is neglected. Here the lines of action of the weight of the worker, $\mathbf{W}$, and the force of the two-force member (hydraulic cylinder) at B, $\mathbf{F}_B$, intersect at O. For moment equilibrium, the resultant force at the pin A, $\mathbf{F}_A$, must also be directed towards O.

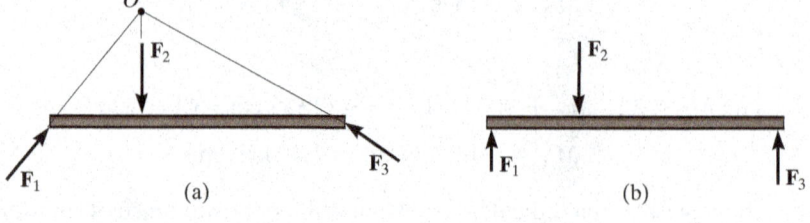

Three-force member

Fig. 5–21

EXAMPLE 5.13

The lever *ABC* is pin supported at *A* and connected to a short link *BD* as shown in Fig. 5–22a. If the weight of the members is negligible, determine the force of the pin on the lever at *A*.

SOLUTION

Free-Body Diagrams. As shown in Fig. 5–22b, the short link *BD* is a *two-force member*, so the *resultant forces* from the pins *D* and *B* must be equal, opposite, and collinear. Although the magnitude of the force is unknown, the line of action is known since it passes through *B* and *D*.

Lever *ABC* is a *three-force member*, and therefore, in order to satisfy moment equilibrium, the three nonparallel forces acting on it must be concurrent at *O*, Fig. 5–22c. Note that the force **F** on the lever at *B* is equal but opposite to the force **F** acting at *B* on the link. The distance *CO* must be 0.5 m since the line of action of **F** is known.

Equations of Equilibrium. By requiring the force system to be concurrent at *O*, since $\Sigma M_O = 0$, the angle θ which defines the line of action of $\mathbf{F}_A$ can be determined from trigonometry,

$$\theta = \tan^{-1}\left(\frac{0.7}{0.4}\right) = 60.3°$$

Using the *x*, *y* axes and applying the force equilibrium equations,

$\xrightarrow{+} \Sigma F_x = 0;$ $F_A \cos 60.3° - F \cos 45° + 400 \text{ N} = 0$

$+\uparrow \Sigma F_y = 0;$ $F_A \sin 60.3° - F \sin 45° = 0$

Solving, we get

$$F_A = 1.07 \text{ kN} \qquad \textit{Ans.}$$

$$F = 1.32 \text{ kN}$$

NOTE: We can also solve this problem by representing the force at *A* by its two components $\mathbf{A}_x$ and $\mathbf{A}_y$ and applying $\Sigma M_A = 0$ to get *F*, then $\Sigma F_x = 0, \Sigma F_y = 0$ to get A_x and A_y. Once A_x and A_y are determined, we can get F_A and θ.

Fig. 5–22

FUNDAMENTAL PROBLEMS

All solutions must include a free-body diagram.

F5–1. Determine the horizontal and vertical components of reaction at the supports. Neglect the thickness of the beam.

Prob. F5–1

F5–2. Determine the horizontal and vertical components of reaction at the pin A and the reaction on the beam at C.

Prob. F5–2

F5–3. The truss is supported by a pin at A and a roller at B. Determine the support reactions.

Prob. F5–3

F5–4. Determine the components of reaction at the fixed support A. Neglect the thickness of the beam.

Prob. F5–4

F5–5. The 25-kg bar has a center of mass at G. If it is supported by a smooth peg at C, a roller at A, and cord AB, determine the reactions at these supports.

Prob. F5–5

F5–6. Determine the reactions at the smooth contact points A, B, and C on the bar.

Prob. F5–6

PROBLEMS

All solutions must include a free-body diagram.

5–10. Determine the reactions at the supports.

Prob. 5–10

5–11. Determine the reactions at the supports.

Prob. 5–11

*5–12. Determine the horizontal and vertical components of reaction at the pin A and at the rocker B.

Prob. 5–12

5–13. Determine the reactions on the beam at A and B.

Prob. 5–13

5–14. Determine the reactions at the supports.

Prob. 5–14

5–15. Determine the reaction at fixed support A.

Prob. 5–15

*5–16. The linkage supports a force of 500 lb and rides along the top and bottom flanges of the crane rail. Determine the force of each roller on the flange.

Prob. 5–16

5–17. Determine the reactions at the roller A and pin B.

Prob. 5–17

5–18. Determine the reactions at the roller B, the rocker C, and where the beam contacts the smooth plane at A. Neglect the thickness of the beam.

Prob. 5–18

5–19. The 10-kg lamp has a center of mass at G. Determine the horizontal and vertical components of reaction at A, and the force in the cable BC.

Prob. 5–19

*5–20. A uniform glass rod having a length L is placed in the smooth hemispherical bowl having a radius r. Determine the angle of inclination θ for equilibrium.

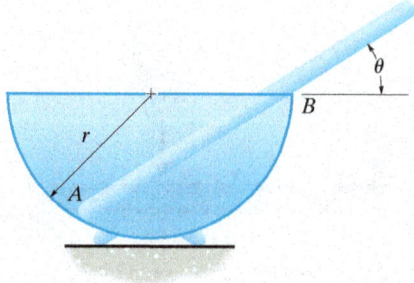

Prob. 5–20

5–21. The 2.5-Mg assembly has a center of mass at G and is hoisted with a constant velocity. Determine the force in each cable and the distance d for equilibrium.

***5–24.** The relay regulates voltage and current. Determine the force in the spring CD, which has a stiffness of $k = 120$ N/m, so that it will allow the armature to make contact at A in figure (a) with a vertical force of 0.4 N. Also, determine the force in the spring when the coil is energized and attracts the armature to E, figure (b), thereby breaking contact at A.

Prob. 5–21

(a)   (b)

Prob. 5–24

5–22. If the intensity of the distributed load acting on the beam is $w = 3$ kN/m, determine the reactions at the roller A and pin B.

5–23. If the roller at A and the pin at B can support a load up to 4 kN and 8 kN, respectively, determine the maximum intensity of the distributed load w, measured in kN/m, so that failure of the supports does not occur.

5–25. Determine the reactions on the bent rod which is supported by a smooth surface at B and by a collar at A, which is fixed to the rod and is free to slide over the fixed inclined rod.

Probs. 5–22/23

Prob. 5–25

5–26. The file cabinet contains four uniform drawers, each 2.5 ft long and weighing 30 lb/ft. If the empty cabinet (without drawers) has a weight of 40 lb and a center of mass at G, determine how many of the drawers can be fully pulled out and the extension x of the last drawer that will cause the assembly to be on the verge of tipping over.

*5–28. The disk has a mass of 10 kg. If it is suspended from a spring having an unstretched length of 400 mm, determine the angle θ for equilibrium.

Prob. 5–26

Prob. 5–28

5–27. Determine the reactions acting on the smooth uniform bar, which has a mass of 20 kg.

5–29. Determine the smallest force P needed to pull the 50-kg roller over the smooth step. Take $\theta = 30°$.

5–30. Determine the magnitude and direction θ of the minimum force P needed to pull the 50-kg roller over the smooth step.

Prob. 5–27

Probs. 5–29/30

5–31. The mechanism shown was thought by its inventor to be a perpetual-motion machine. It consists of the stand A, two smooth idler wheels B and C, and in between a uniform hollow cylindrical ring D suspended in the manner shown. The ring has a weight W and it was expected to revolve in the direction indicated by the arrow. Draw a free-body diagram of the ring and, using an appropriate equation of equilibrium, show that it will not rotate.

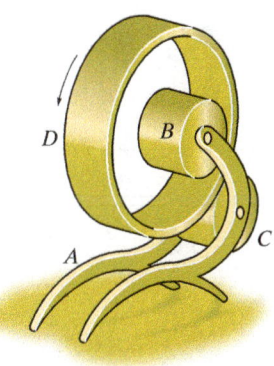

Prob. 5–31

*5–32. The hold-down clamp exerts a compressive force of 400 N on the wood block at C. If the clamp is loosely bolted to the bench using two symmetrically placed bolts B, one of which is shown, determine the force along the axis of each bolt and the vertical reaction of the bench on the clamp at A.

Prob. 5–32

5–33. If the jib crane has a mass of 800 kg and a center of mass at G, and the force at its end is $F = 15$ kN, determine the reactions at its bearings. The bearing at A is a journal bearing and supports only a horizontal force, whereas the bearing at B is a thrust bearing that supports both horizontal and vertical components.

5–34. The crane has a mass of 800 kg and a center of mass at G. The bearing at A is a journal bearing and can support a horizontal force, whereas the bearing at B is a thrust bearing that supports both horizontal and vertical components. Determine the maximum load F that can be suspended from its end if the bearings at A and B can sustain a maximum resultant load of 24 kN and 34 kN, respectively.

Probs. 5–33/34

5–35. The crane lifts the 400-kg load L. The primary boom AB has a mass of 1.20 Mg and a center of mass at G_1, whereas the secondary boom BC has a mass of 0.6 Mg and a center of mass at G_2. Determine the tension in the cable BD and the horizontal and vertical components of reaction at the pin A.

Prob. 5–35

*5–36. The beam is horizontal and the springs are unstretched when there is no load on the beam. Determine the angle of tilt of the beam when the load is applied.

Prob. 5–36

5–37. The cantilevered jib crane is used to support the load of 780 lb. If $x = 5$ ft, determine the reactions at the supports. Note that the supports are collars that allow the crane to rotate freely about the vertical axis. The collar at B supports the force in the vertical direction, whereas the one at A does not.

5–38. The cantilevered jib crane is used to support the load of 780 lb. If the trolley T can be placed anywhere between $1.5 \text{ ft} \le x \le 7.5 \text{ ft}$, determine the maximum magnitude of reaction at the supports A and B. Note that the supports are collars that allow the crane to rotate freely about the vertical axis. The collar at B supports the force in the vertical direction, whereas the one at A does not.

Probs. 5–37/38

5–39. The bar of negligible weight is supported by two springs, each having a stiffness $k = 100$ N/m. If the springs are originally unstretched, determine the angle θ the bar makes with the horizontal when the 30-N force is applied to the bar.

*5–40. Determine the stiffness k of each spring so that the 30-N force causes the bar to tip $\theta = 15°$ when the force is applied. Originally the bar is horizontal and the springs are unstretched. Neglect the weight of the bar.

Probs. 5–39/40

5–41. The davit is used to suspend the lifeboat over the side of a ship. If the boat exerts a force of 800 lb on the chain at D, determine the force acting on the hydraulic cylinder BC, and the horizontal and vertical components of reaction at the pin A.

Prob. 5–41

5–42. The boom supports the two vertical loads. Neglect the size of the collars at D and B and the thickness of the boom, and determine the horizontal and vertical components of force at the pin A and the force in cable CB. Set $F_1 = 800$ N and $F_2 = 350$ N.

5–43. The boom is intended to support two vertical loads, F_1 and F_2. If the cable CB can sustain a maximum load of 1500 N before it fails, determine the critical loads if $F_1 = 2F_2$. Also, what is the magnitude of the maximum reaction at pin A?

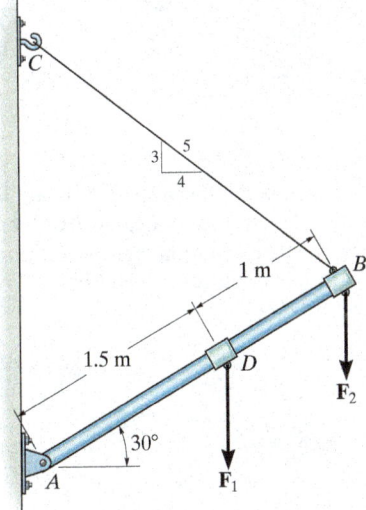

Probs. 5–42/43

∗5–44. The 10-kg uniform rod is pinned at end A. If it is subjected to a couple moment of 50 N·m, determine the smallest angle θ for equilibrium. The spring is unstretched when $\theta = 0$, and has a stiffness of $k = 60$ N/m.

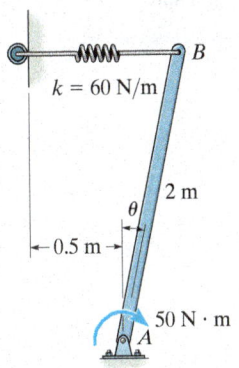

Prob. 5–44

5–45. If the hand truck and its contents have a mass of 50 kg with center of gravity at G, determine the normal reaction on both wheels and the magnitude and direction of the minimum force required at the grip B needed to lift the load.

Prob. 5–45

5–46. Three uniform books, each having a weight W and length a, are stacked as shown. Determine the maximum distance d that the top book can extend out from the bottom one so the stack does not topple over.

Prob. 5–46

5–47. Determine the reactions at the pin A and the tension in cord BC. Set $F = 40$ kN. Neglect the thickness of the beam.

∗5–48. If rope BC will fail when the tension becomes 50 kN, determine the greatest vertical load F that can be applied to the beam at B. What is the magnitude of the reaction at A for this loading? Neglect the thickness of the beam.

Probs. 5–47/48

5–49. The assembly is used to support the 120-kg container having a center of mass at G. If the spring has an unstretched length of 250 mm and stiffness of k = 300 kN/m, determine its height h and the reaction at the rollers A and B.

Prob. 5–49

5–50. The check valve is used to regulate pressure in the pipe. If the stiffness of the spring is k = 80 kN/m and its uncompressed length is 120 mm, determine the maximum pressure in the tank if the lever is to remain in the horizontal position as shown. The plug at A is circular.

Prob. 5–50

5–51. The block has a mass of 80 kg and center of mass at G. Determine the smooth reactions on the supporting platform at A, B, and C.

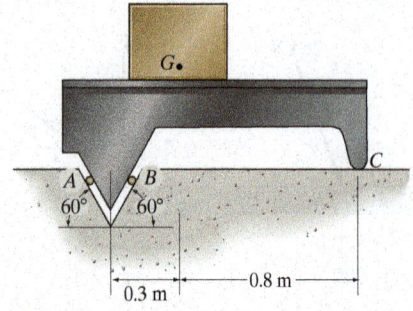

Prob. 5–51

*5–52. Calculate the tension in the cable at B required to support the load of 200 kg. The cable passes over a frictionless pulley located at C. Neglect the size of this pulley and weight of the boom.

Prob. 5–52

5–53. The uniform beam has a weight W and length l and is supported by a pin at A and a cable BC. Determine the horizontal and vertical components of reaction at A and the tension in the cable necessary to hold the beam in the position shown.

Prob. 5–53

5–54. When no force is applied to the brake pedal of the lightweight truck, the retainer spring AB keeps the pedal in contact with the smooth brake light switch at C. If the force on the switch is 3 N, determine the unstretched length of the spring if the stiffness of the spring is k = 80 N/m.

Prob. 5–54

5–55. The exercise machine consists of a rigid bar, pin connected to the frame at A. As the man pushes up on the bar it stretches the rubber band BC. If the stiffness of the rubber band is $k = 500$ lb/ft, determine the applied vertical force $\mathbf{F}$ needed to hold the bar in the position $\theta = 30°$. The rubber band is unstretched when $\theta = 0°$.

5–57. A 10-kN load is suspended from the boom at D. Determine the force in the hydraulic cylinder BC and the pin reaction at A.

Prob. 5–55

Prob. 5–57

5–58. Determine the force in the cable as a function of θ, needed to support the 100-kg hatch door.

Prob. 5–58

*5–56. The uniform rod has a length l and weight W. It is supported at one end A by a smooth wall and the other end by a cord of length s which is attached to the wall. Determine the placement h for equilibrium.

5–59. The crane supports the load of 800 kg. Determine the reactions at the supports A and B as a function of position d of the trolley. Plot the reactions vs. d for $0.5 \text{ m} \leq d \leq 3 \text{ m}$.

Prob. 5–56

Prob. 5–59

*5–60. The smooth uniform rod has a mass m and is placed on the semicircular arch and against the wall. Show that for equilibrium the angle θ must satisfy $\sin\theta = \frac{1}{r}(\sqrt{1+3\cos^2\theta})(d - l\sin\theta)$.

5–62. Determine the distance d for placement of the load $\mathbf{P}$ for equilibrium of the smooth bar when it is held in the position θ. Neglect the weight of the bar.

Prob. 5–62

Prob. 5–60

5–61. The smooth pipe rests against the wall at the points of contact A, B, and C. Determine the reactions at these points needed to support the vertical force of 45 lb. Neglect the pipe's thickness in the calculation.

5–63. The rod BC is supported by two cords, each of length a, which are attached to the pin at A. If weights W and $2W$ are suspended from the ends of the rod, determine the angle θ for equilibrium, measured from the horizontal. Express the answer in terms of a and l. Neglect the weight of the rod.

Prob. 5–61

Prob. 5–63

EQUILIBRIUM IN THREE DIMENSIONS
5.5 FREE-BODY DIAGRAMS

The first step in solving three-dimensional equilibrium problems, as in the case of two dimensions, is to draw a free-body diagram. Before we can do this, however, it is first necessary to discuss the types of reactions that can occur at the supports.

Support Reactions. The reactive forces and couple moments acting at various types of supports and connections, when the members are viewed in three dimensions, are listed in Table 5–2. It is important to recognize the symbols used to represent each of these supports and to understand clearly how the forces and couple moments are developed. As in the two-dimensional case:

- A support prevents the translation of a body by exerting a force on the body.

- A support prevents the rotation of a body by exerting a couple moment on the body.

For example, in Table 5–2, item (4), the ball-and-socket joint prevents any translation of the connecting member; therefore, a force must act on the member at the point of connection. This force has three components having unknown magnitudes, F_x, F_y, F_z. Provided these components are known, one can obtain the magnitude of force, $F = \sqrt{F_x^2 + F_y^2 + F_z^2}$, and the force's orientation defined by its coordinate direction angles α, β, γ, Eqs. 2–5.[*] Since the connecting member is allowed to rotate freely about *any* axis, no couple moment is resisted by a ball-and-socket joint.

Notice that the *single* bearing supports in items (5) through (7), the *single* pin (8), and the *single* hinge (9) are shown to resist both force and couple-moment components. If, however, these supports are used with *other* bearings, pins, or hinges to hold a rigid body in equilibrium and these supports are *properly aligned* when connected to the body, then the *force reactions* at these supports *alone* are adequate for supporting the body. In other words, the couple moments will not develop since the body is *prevented* from rotating by the other supports. The reason for this should become clear after studying the examples which follow.

---

[*] The three unknowns may also be represented as an unknown force magnitude F and two unknown coordinate direction angles. The third direction angle is obtained using the identity $\cos^2 \alpha + \cos^2 \beta + \cos^2 \gamma = 1$, Eq. 2–8.

232   CHAPTER 5   EQUILIBRIUM OF A RIGID BODY

TABLE 5–2 Supports for Rigid Bodies Subjected to Three-Dimensional Force Systems

Types of Connection	Reaction	Number of Unknowns
(1) cable	F	One unknown. The reaction is a force which acts away from the member in the known direction of the cable.
(2) smooth surface support	F	One unknown. The reaction is a force which acts perpendicular to the surface at the point of contact.
(3) roller	F	One unknown. The reaction is a force which acts perpendicular to the surface at the point of contact.
(4) ball and socket	F_x, F_y, F_z	Three unknowns. The reactions are three rectangular force components.
(5) single journal bearing	M_z, M_x, F_z, F_x	Four unknowns. The reactions are two force components, because a journal bearing allows movement along the axis of the shaft, and two couple-moment components which act perpendicular to the shaft. *Note:* The couple moments are *generally not applied* if the body is supported elsewhere. See the examples.

continued

5.5 Free-Body Diagrams

TABLE 5–2 Continued

Types of Connection	Reaction	Number of Unknowns
(6) single journal bearing with square shaft	M_z, F_z, M_y, M_x, F_x	Five unknowns. The reactions are two force and three couple-moment components. *Note*: The couple moments *are generally not applied* if the body is supported elsewhere. See the examples.
(7) single thrust bearing	M_z, F_y, F_z, M_x, F_x	Five unknowns. The reactions are three force components, because a thrust bearing prevents movement along the axis of the shaft, and two couple-moment components. *Note*: The couple moments *are generally not applied* if the body is supported elsewhere. See the examples.
(8) single smooth pin	M_z, F_z, F_y, M_y, F_x	Five unknowns. The reactions are three force and two couple-moment components. *Note*: The couple moments *are generally not applied* if the body is supported elsewhere. See the examples.
(9) single hinge	M_z, F_z, F_y, F_x, M_x	Five unknowns. The reactions are three force and two couple-moment components. *Note*: The couple moments *are generally not applied* if the body is supported elsewhere. See the examples.
(10) fixed support	M_z, F_z, F_x, M_x, F_y, M_y	Six unknowns. The reactions are three force and three couple-moment components.

Typical examples of actual supports that are referenced to Table 5–2 are shown in the following sequence of photos.

This ball-and-socket joint provides a connection for the housing of an earth grader to its frame. (4)

These journal bearings are used to support the drive shaft on a machine. (5)

The thrust bearing supports the ends of the shaft. (7)

This pin is used to support the end of the strut used on a tractor. (8)

Free-Body Diagrams. The general procedure for establishing the free-body diagram of a rigid body has been outlined in Sec. 5.2. Essentially it requires first "isolating" the body by drawing its outlined shape. This is followed by a careful *labeling* of *all* the forces and couple moments with reference to an established x, y, z coordinate system. As a general rule, show the unknown components of reaction as acting on the free-body diagram in the *positive sense*. In this way, if any negative values are obtained, they will indicate that the components act in the negative coordinate directions.

EXAMPLE 5.14

Consider the two rods and plate, along with their associated free-body diagrams, shown in Fig. 5–23. The x, y, z axes are established on the diagram and the unknown reaction components are indicated in the *positive sense*. The weight is neglected.

SOLUTION

Properly aligned journal bearings at A, B, C.

The force reactions developed by the bearings are *sufficient* for equilibrium since they prevent the shaft from rotating about each of the coordinate axes. No couple moments at each bearing are developed.

Pin at A and cable BC.

Moment components are developed by the pin on the rod to prevent rotation about the x and z axes.

Properly aligned journal bearing at A and hinge at C. Roller at B.

Only force reactions are developed by the bearing and hinge on the plate to prevent rotation about each coordinate axis. No moments are developed at the hinge.

Fig. 5–23

5.6 EQUATIONS OF EQUILIBRIUM

As stated in Sec. 5.1, the conditions for equilibrium of a rigid body subjected to a three-dimensional force system require that both the *resultant* force and *resultant* couple moment acting on the body be equal to *zero*.

Vector Equations of Equilibrium. The two conditions for equilibrium of a rigid body may be expressed mathematically in vector form as

$$\Sigma \mathbf{F} = \mathbf{0}$$
$$\Sigma \mathbf{M}_O = \mathbf{0} \qquad (5\text{-}5)$$

where $\Sigma \mathbf{F}$ is the vector sum of all the external forces acting on the body and $\Sigma \mathbf{M}_O$ is the sum of the couple moments and the moments of all the forces about any point O located either on or off the body.

Scalar Equations of Equilibrium. If all the external forces and couple moments are expressed in Cartesian vector form and substituted into Eqs. 5–5, we have

$$\Sigma \mathbf{F} = \Sigma F_x \mathbf{i} + \Sigma F_y \mathbf{j} + \Sigma F_z \mathbf{k} = \mathbf{0}$$

$$\Sigma \mathbf{M}_O = \Sigma M_x \mathbf{i} + \Sigma M_y \mathbf{j} + \Sigma M_z \mathbf{k} = \mathbf{0}$$

Since the **i, j**, and **k** components are independent from one another, then these equations are satisfied provided

$$\Sigma F_x = 0$$
$$\Sigma F_y = 0$$
$$\Sigma F_z = 0 \qquad (5\text{-}6a)$$

and

$$\Sigma M_x = 0$$
$$\Sigma M_y = 0$$
$$\Sigma M_z = 0 \qquad (5\text{-}6b)$$

These *six scalar equilibrium equations* may be used to solve for at most six unknowns shown on the free-body diagram. Equations 5–6a require the sum of the external force components acting in the x, y, z directions to be zero, and Eqs. 5–6b require the sum of the moment components about the x, y, z axes to be zero.

5.7 CONSTRAINTS AND STATICAL DETERMINACY

To ensure the equilibrium of a rigid body, it is not only necessary to satisfy the equations of equilibrium, but the body must also be properly held or constrained by its supports. Some bodies may have more supports than are necessary for equilibrium, whereas others may not have enough, or the supports may be arranged in a particular manner that could cause the body to move. Each of these cases will now be discussed.

Redundant Constraints. When a body has redundant supports, that is, more supports than are necessary to hold it in equilibrium, it becomes *statically indeterminate*. For example, the beam in Fig. 5–24a and the pipe assembly in Fig. 5–24b, shown together with their free-body diagrams, are both statically indeterminate because of the additional (or redundant) support reactions. For the beam there are five unknowns, $M_A, A_x, A_y, B_y,$ and C_y, for which only three equilibrium equations can be written ($\Sigma F_x = 0, \Sigma F_y = 0,$ and $\Sigma M_O = 0$, Eq. 5–2). The pipe assembly has eight unknowns, and here there are only six equilibrium equations, Eqs. 5–6.

The additional equations needed to solve statically indeterminate problems of the type shown in Fig. 5–24 are generally obtained from the displacement conditions at the supports. These equations involve the physical properties of the body which are studied in subjects dealing with the mechanics of deformation, such as "mechanics of materials."[*]

(a)

(b)

Fig. 5–24

---

[*] See R. C. Hibbeler, *Mechanics of Materials*, 10th edition, Pearson Education/Prentice Hall, Inc.

Improper Constraints. Having the same number of unknown reactive forces as available equations of equilibrium does not always guarantee that a body will be stable when subjected to a particular loading. For example, the pin support at A and the roller support at B for the beam in Fig. 5–25a are placed in such a way that the lines of action of the reactive forces are *concurrent* at point A. Consequently, the applied loading **P** will cause the beam to rotate slightly about A, and so the beam is improperly constrained, $\Sigma M_A \neq 0$.

In three dimensions, a body will be improperly constrained if the lines of action of all the reactive forces intersect a common axis. For example, the reactive forces at the ball-and-socket supports at A and B in Fig. 5–25b all intersect the axis passing through A and B. Since the moments of these forces about A and B are all zero, then the loading **P** will rotate the member about the AB axis, $\Sigma M_{AB} \neq 0$.

(a)

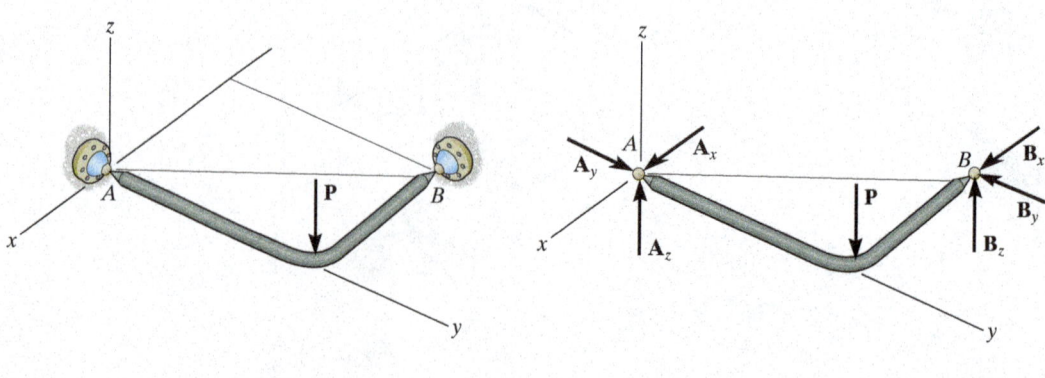

(b)

Fig. 5–25

5.7 CONSTRAINTS AND STATICAL DETERMINACY

(a)

(b)

Fig. 5–26

Another way in which improper constraining leads to instability occurs when the *reactive forces* are all *parallel*. Two- and three-dimensional examples of this are shown in Fig. 5–26. In both cases, the body is only **partially constrained**, where the summation of forces along a horizontal axis will not be zero.

To summarize these points, a body is considered ***improperly constrained*** if all the reactive forces intersect at a common point or pass through a common axis, or if all the reactive forces are parallel. In engineering practice, these situations should be avoided at all times since they will cause an unstable condition.

Stability is always an important concern when operating a crane, not only when lifting a load, but also when moving it about.

IMPORTANT POINTS

- Always draw the free-body diagram first when solving any equilibrium problem.
- If a support *prevents translation* of a body, then the support exerts a *force* on the body.
- If a support *prevents rotation*, then the support exerts a *couple moment* on the body.
- If a body is subjected to more unknown reactions than available equations of equilibrium, then the problem is *statically indeterminate*.
- A stable body requires that the lines of action of the reactive forces do not intersect a common axis and are not parallel to one another.

PROCEDURE FOR ANALYSIS

Three-dimensional equilibrium problems for a rigid body can be solved using the following procedure.

Free-Body Diagram.

- Draw an outlined shape of the body.
- Show all the forces and couple moments acting on the body.
- Establish the origin of the x, y, z axes at a convenient point and orient the axes so that they are parallel to as many of the external forces and moments as possible.
- Label all the loadings and specify their directions. In general, show all the *unknown* components having a *positive sense* along the x, y, z axes.
- Indicate the dimensions of the body necessary for calculating the moments of forces.

Equations of Equilibrium.

- If the x, y, z force and moment components seem easy to determine, then apply the six scalar equations of equilibrium; otherwise use the vector equations.
- It is not necessary that the set of axes chosen for force summation coincide with the set of axes chosen for moment summation.
- Choose the direction of an axis for moment summation such that it intersects the lines of action of as many unknown forces as possible. Realize that the moments of forces passing through points on this axis, and the moments of forces which are parallel to the axis, will then be zero.
- If the solution of the equilibrium equations yields a negative scalar for a force or couple moment magnitude, it indicates that the sense is opposite to that assumed on the free-body diagram.

EXAMPLE 5.15

The homogeneous plate shown in Fig. 5–27a has a mass of 100 kg and is subjected to a force and couple moment along its edges. If it is supported in the horizontal plane by a roller at A, a ball-and-socket joint at B, and a cord at C, determine the components of reaction at these supports.

SOLUTION (SCALAR ANALYSIS)

Free-Body Diagram. There are five unknown reactions acting on the plate, as shown in Fig. 5–27b. Each of these reactions is assumed to act in a positive coordinate direction.

Equations of Equilibrium. Since the three-dimensional geometry is rather simple, a *scalar analysis* provides a *direct solution* to this problem. A force summation along each axis yields

$\Sigma F_x = 0;$ $B_x = 0$ *Ans.*
$\Sigma F_y = 0;$ $B_y = 0$ *Ans.*
$\Sigma F_z = 0;$ $A_z + B_z + T_C - 300 \text{ N} - 981 \text{ N} = 0$ (1)

Recall that the moment of a force about an axis is equal to the product of the force magnitude and the perpendicular distance (moment arm) from the line of action of the force to the axis. Also, forces that are parallel to an axis or pass through it create no moment about the axis. Hence, summing moments about the positive x and y axes, we have

Fig. 5–27

$\Sigma M_x = 0;$ $T_C(2 \text{ m}) - 981 \text{ N}(1 \text{ m}) + B_z(2 \text{ m}) = 0$ (2)

$\Sigma M_y = 0;$
$300 \text{ N}(1.5 \text{ m}) + 981 \text{ N}(1.5 \text{ m}) - B_z(3 \text{ m}) - A_z(3 \text{ m}) - 200 \text{ N} \cdot \text{m} = 0$ (3)

The components of the force at B can be eliminated if moments are summed about the x' and y' axes. We obtain

$\Sigma M_{x'} = 0;$ $981 \text{ N}(1 \text{ m}) + 300 \text{ N}(2 \text{ m}) - A_z(2 \text{ m}) = 0$ (4)

$\Sigma M_{y'} = 0;$
$-300 \text{ N}(1.5 \text{ m}) - 981 \text{ N}(1.5 \text{ m}) - 200 \text{ N} \cdot \text{m} + T_C(3 \text{ m}) = 0$ (5)

Solving Eqs. 1 through 3 or the more convenient Eqs. 1, 4, and 5 yields

$A_z = 790 \text{ N}$ $B_z = -217 \text{ N}$ $T_C = 707 \text{ N}$ *Ans.*

The negative sign indicates that $\mathbf{B}_z$ acts downward.

NOTE: The solution of this problem does not require a summation of moments about the z axis. The plate is partially constrained since the supports cannot prevent the plate from turning about the z axis if a force is applied to it in the x–y plane.

EXAMPLE 5.16

Determine the components of reaction that the ball-and-socket joint at A, the smooth journal bearing at B, and the roller support at C exert on the rod assembly in Fig. 5–28a.

Fig. 5–28

SOLUTION (SCALAR ANALYSIS)

Free-Body Diagram. As shown on the free-body diagram, Fig. 5–28b, the reactive forces of the supports will prevent the assembly from rotating about each coordinate axis, and so the journal bearing at B only exerts reactive forces on the member. No couple moments are required.

Equations of Equilibrium. Because all the forces are either horizontal or vertical, it is convenient to use a scalar analysis. A direct solution for A_y can be obtained by summing forces along the y axis.

$\Sigma F_y = 0;$ $A_y = 0$ Ans.

The force F_C can be determined directly by summing moments about the y axis.

$\Sigma M_y = 0;$ $F_C(0.6\text{ m}) - 900\text{ N}(0.4\text{ m}) = 0$
$F_C = 600\text{ N}$ Ans.

Using this result, B_z can be determined by summing moments about the x axis.

$\Sigma M_x = 0;$ $B_z(0.8\text{ m}) + 600\text{ N}(1.2\text{ m}) - 900\text{ N}(0.4\text{ m}) = 0$
$B_z = -450\text{ N}$ Ans.

The negative sign indicates that $\mathbf{B}_z$ acts downward. The force B_x can be found by summing moments about the z axis.

$\Sigma M_z = 0;$ $-B_x(0.8\text{ m}) = 0$ $B_x = 0$ Ans.

Thus,

$\Sigma F_x = 0;$ $A_x + 0 = 0$ $A_x = 0$ Ans.

Finally, using the results of B_z and F_C.

$\Sigma F_z = 0;$ $A_z + (-450\text{ N}) + 600\text{ N} - 900\text{ N} = 0$
$A_z = 750\text{ N}$ Ans.

EXAMPLE 5.17

The boom is used to support the 75-lb flowerpot in Fig. 5–29a. Determine the tension developed in wires AB and AC.

SOLUTION (VECTOR ANALYSIS)

Free-Body Diagram. The free-body diagram of the boom is shown in Fig. 5–29b.

Equations of Equilibrium. Here the cable forces are directed at angles with the coordinate axes, so we will use a vector analysis.

$$\mathbf{F}_{AB} = F_{AB}\left(\frac{\mathbf{r}_{AB}}{r_{AB}}\right) = F_{AB}\left(\frac{\{2\mathbf{i} - 6\mathbf{j} + 3\mathbf{k}\} \text{ ft}}{\sqrt{(2 \text{ ft})^2 + (-6 \text{ ft})^2 + (3 \text{ ft})^2}}\right)$$

$$= \tfrac{2}{7}F_{AB}\mathbf{i} - \tfrac{6}{7}F_{AB}\mathbf{j} + \tfrac{3}{7}F_{AB}\mathbf{k}$$

$$\mathbf{F}_{AC} = F_{AC}\left(\frac{\mathbf{r}_{AC}}{r_{AC}}\right) = F_{AC}\left(\frac{\{-2\mathbf{i} - 6\mathbf{j} + 3\mathbf{k}\} \text{ ft}}{\sqrt{(-2 \text{ ft})^2 + (-6 \text{ ft})^2 + (3 \text{ ft})^2}}\right)$$

$$= -\tfrac{2}{7}F_{AC}\mathbf{i} - \tfrac{6}{7}F_{AC}\mathbf{j} + \tfrac{3}{7}F_{AC}\mathbf{k}$$

We can eliminate the force reaction at O by writing the moment equation of equilibrium about point O.

$$\Sigma \mathbf{M}_O = \mathbf{0}; \quad \mathbf{r}_A \times (\mathbf{F}_{AB} + \mathbf{F}_{AC} + \mathbf{W}) = \mathbf{0}$$

$$(6\mathbf{j}) \times \left[\left(\tfrac{2}{7}F_{AB}\mathbf{i} - \tfrac{6}{7}F_{AB}\mathbf{j} + \tfrac{3}{7}F_{AB}\mathbf{k}\right)\right.$$

$$\left. + \left(-\tfrac{2}{7}F_{AC}\mathbf{i} - \tfrac{6}{7}F_{AC}\mathbf{j} + \tfrac{3}{7}F_{AC}\mathbf{k}\right) + (-75\mathbf{k})\right] = \mathbf{0}$$

$$\left(\tfrac{18}{7}F_{AB} + \tfrac{18}{7}F_{AC} - 450\right)\mathbf{i} + \left(-\tfrac{12}{7}F_{AB} + \tfrac{12}{7}F_{AC}\right)\mathbf{k} = \mathbf{0}$$

$\Sigma M_x = 0;$ $\quad \tfrac{18}{7}F_{AB} + \tfrac{18}{7}F_{AC} - 450 = 0 \quad$ (1)

$\Sigma M_y = 0;$ $\quad 0 = 0$

$\Sigma M_z = 0;$ $\quad -\tfrac{12}{7}F_{AB} + \tfrac{12}{7}F_{AC} = 0 \quad$ (2)

Solving Eqs. 1 and 2 simultaneously,

$$F_{AB} = F_{AC} = 87.5 \text{ lb} \qquad \textit{Ans.}$$

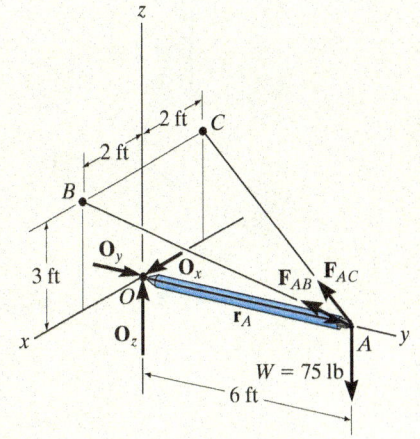

Fig. 5–29

EXAMPLE 5.18

Rod AB shown in Fig. 5–30a is subjected to the 200-N force. Determine the reactions at the ball-and-socket joint A and the tension in the cables BD and BE. The collar at C is fixed to the rod.

SOLUTION (VECTOR ANALYSIS)

Free-Body Diagram. Fig. 5–30b.

Equations of Equilibrium. Representing each force on the free-body diagram in Cartesian vector form, we have

$$\mathbf{F}_A = A_x\mathbf{i} + A_y\mathbf{j} + A_z\mathbf{k}$$
$$\mathbf{T}_E = T_E\mathbf{i}$$
$$\mathbf{T}_D = T_D\mathbf{j}$$
$$\mathbf{F} = \{-200\mathbf{k}\}\ \text{N}$$

Applying the force equation of equilibrium,

$$\Sigma \mathbf{F} = 0; \qquad \mathbf{F}_A + \mathbf{T}_E + \mathbf{T}_D + \mathbf{F} = 0$$
$$(A_x + T_E)\mathbf{i} + (A_y + T_D)\mathbf{j} + (A_z - 200)\mathbf{k} = 0$$

$\Sigma F_x = 0;\quad A_x + T_E = 0$ \hfill (1)
$\Sigma F_y = 0;\quad A_y + T_D = 0$ \hfill (2)
$\Sigma F_z = 0;\quad A_z - 200 = 0$ \hfill (3)

Summing moments about point A yields

$$\Sigma \mathbf{M}_A = 0; \qquad \mathbf{r}_C \times \mathbf{F} + \mathbf{r}_B \times (\mathbf{T}_E + \mathbf{T}_D) = 0$$

Since $\mathbf{r}_C = \frac{1}{2}\mathbf{r}_B$, then

$$(0.5\mathbf{i} + 1\mathbf{j} - 1\mathbf{k}) \times (-200\mathbf{k}) + (1\mathbf{i} + 2\mathbf{j} - 2\mathbf{k}) \times (T_E\mathbf{i} + T_D\mathbf{j}) = 0$$

Expanding and rearranging terms gives

$$(2T_D - 200)\mathbf{i} + (-2T_E + 100)\mathbf{j} + (T_D - 2T_E)\mathbf{k} = 0$$

$\Sigma M_x = 0;\quad 2T_D - 200 = 0$ \hfill (4)
$\Sigma M_y = 0;\quad -2T_E + 100 = 0$ \hfill (5)
$\Sigma M_z = 0;\quad T_D - 2T_E = 0$ \hfill (6)

Solving Eqs. 1 through 5, we get

$T_D = 100\ \text{N}$ \hfill *Ans.*
$T_E = 50\ \text{N}$ \hfill *Ans.*
$A_x = -50\ \text{N}$ \hfill *Ans.*
$A_y = -100\ \text{N}$ \hfill *Ans.*
$A_z = 200\ \text{N}$ \hfill *Ans.*

NOTE: The negative sign indicates that A_x and A_y have a sense which is opposite to that shown on the free-body diagram, Fig. 5–30b. Also, notice that Eqs. 1–6 can be set up *directly* using a scalar analysis.

Fig. 5–30

EXAMPLE 5.19

The bent rod in Fig. 5–31a is supported at A by a journal bearing, at D by a ball-and-socket joint, and at B by a cable BC. Using only *one equilibrium equation*, obtain a direct solution for the tension in cable BC. The bearing at A is capable of exerting force components only in the z and y directions since it is properly aligned on the shaft. In other words, no couple moments are required at this support.

SOLUTION (*VECTOR ANALYSIS*)

Free-Body Diagram. As shown in Fig. 5–31b, there are six unknowns.

Equations of Equilibrium. The cable tension T_B may be obtained *directly* by summing moments about an axis that passes through points D and A. Why? The direction of this axis is defined by the unit vector **u**, where

$$\mathbf{u} = \frac{\mathbf{r}_{DA}}{r_{DA}} = -\frac{1}{\sqrt{2}}\mathbf{i} - \frac{1}{\sqrt{2}}\mathbf{j}$$
$$= -0.7071\mathbf{i} - 0.7071\mathbf{j}$$

Hence, the sum of the moments about this axis is zero provided

$$\Sigma M_{DA} = \mathbf{u} \cdot \Sigma (\mathbf{r} \times \mathbf{F}) = 0$$

Here **r** represents a position vector drawn from *any point* on the axis DA to any point on the line of action of force **F** (see Eq. 4–11). With reference to Fig. 5–31b, we can therefore write

$$\mathbf{u} \cdot (\mathbf{r}_B \times \mathbf{T}_B + \mathbf{r}_E \times \mathbf{W}) = 0$$
$$(-0.7071\mathbf{i} - 0.7071\mathbf{j}) \cdot [(-1\mathbf{j}) \times (T_B\mathbf{k})$$
$$+ (-0.5\mathbf{j}) \times (-981\mathbf{k})] = 0$$
$$(-0.7071\mathbf{i} - 0.7071\mathbf{j}) \cdot [(-T_B + 490.5)\mathbf{i}] = 0$$
$$-0.7071(-T_B + 490.5) + 0 + 0 = 0$$
$$T_B = 490.5 \text{ N} \quad \quad Ans.$$

NOTE: Since the moment arms from the axis to T_B and **W** are easy to obtain, we can also determine this result using a scalar analysis. As shown in Fig. 5–31b,

$$\Sigma M_{DA} = 0; \quad T_B(1 \text{ m sin } 45°) - 981 \text{ N}(0.5 \text{ m sin } 45°) = 0$$
$$T_B = 490.5 \text{ N} \quad \quad Ans.$$

(a)

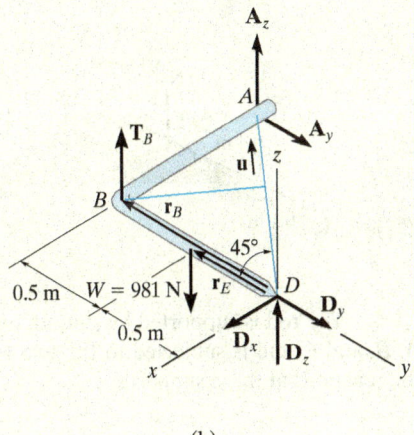

(b)

Fig. 5–31

FUNDAMENTAL PROBLEMS

All solutions must include a free-body diagram.

F5–7. The uniform plate has a weight of 500 lb. Determine the tension in each of the supporting cables.

Prob. F5–7

F5–8. Determine the reactions at the roller support A, the ball-and-socket joint D, and the tension in cable BC for the plate.

Prob. F5–8

F5–9. The rod is supported by smooth journal bearings at A, B, and C and is subjected to the two forces. Determine the reactions at these supports.

Prob. F5–9

F5–10. Determine the support reactions at the smooth journal bearings A, B, and C of the pipe assembly.

Prob. F5–10

F5–11. Determine the force developed in the short link BD, and the tension in the cords CE and CF, and the reactions of the ball-and-socket joint A on the block.

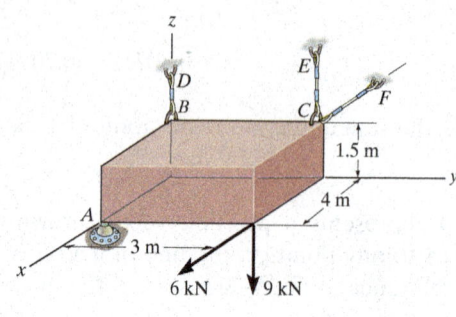

Prob. F5–11

F5–12. Determine the components of reaction that the thrust bearing A and cable BC exert on the bar.

Prob. F5–12

PROBLEMS

All solutions must include a free-body diagram.

***5–64.** The uniform load has a mass of 600 kg and is lifted using a uniform 30-kg strongback beam BAC and the four ropes. Determine the tension in each rope and the force that must be applied at A.

5–66. Determine the reaction at the ball supports B and C and the components of reaction at the ball-and-socket A (not shown) for the uniformly loaded plate.

Prob. 5–64

Prob. 5–66

5–65. The cable of the tower crane is subjected to a force of 840 N. Determine the x, y, z components of reaction at the fixed base A.

5–67. Determine the normal floor reaction on each wheel of the engine stand. The engine weighs 750 lb and has a center of gravity at G.

Prob. 5–65

Prob. 5–67

*5–68. Determine the components of reaction at the ball-and-socket B and the reaction at the rollers A and C due to the loading shown.

Prob. 5–68

5–69. Determine the x, y, z components of reaction acting on the ball-and-socket at A, the reaction at the roller B, and the tension in the cord CD required for equilibrium of the plate.

Prob. 5–69

5–70. The 100-lb door has its center of gravity at G. Determine the components of reaction at hinges A and B if hinge B resists only forces in the x and y directions and A resists forces in the x, y, z directions.

Prob. 5–70

5–71. Determine the tension in each cable and the components of reaction at D needed to support the load.

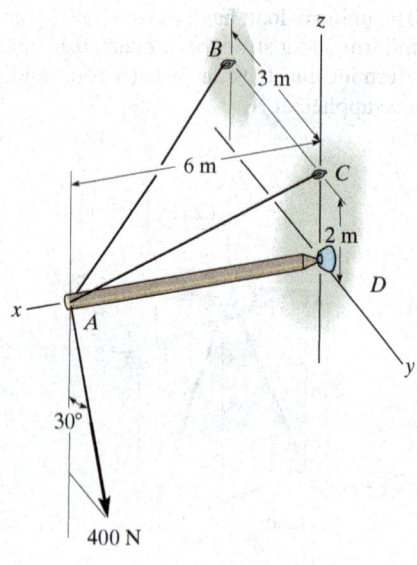

Prob. 5–71

*5–72. Determine the components of reaction acting at the ball-and-socket A, roller B, and cord CD.

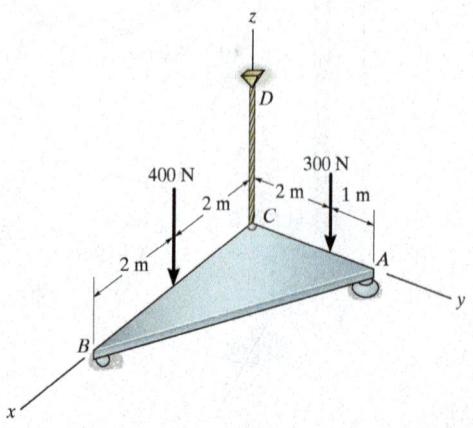

Prob. 5–72

5–73. The triangular plate, having a 90° angle at A, supports the 150-lb load. Determine the tension in cables BD and CD and the x, y, z components of reaction at the ball-and-socket joint at A.

5–75. The bent rod is supported at A, B, and C by smooth journal bearings. Determine the magnitude of F which will cause the positive x component of reaction at the bearing C to be $Cx = 50$ N. The bearings are in proper alignment and exert only force reactions on the rod.

Prob. 5–73

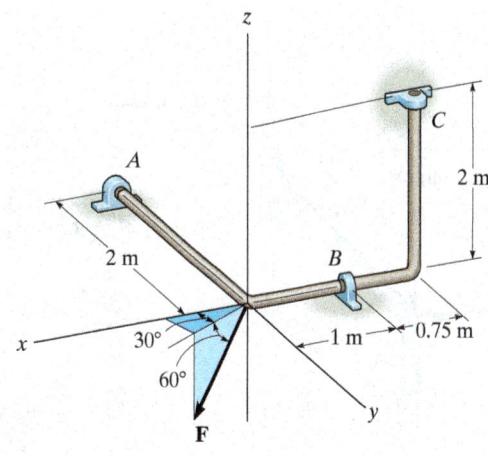

Prob. 5–75

5–74. The bent rod is supported at A, B, and C by smooth journal bearings. Determine the components of reaction at the bearings if the rod is subjected to the force $F = 800$ N. The bearings are in proper alignment and exert only force reactions on the rod.

***5–76.** The windlass supports the 50-kg bucket. Determine the horizontal force **P** needed to hold the handle in the position shown, and the x, y, z components of reaction at the ball-and-socket joint A and the smooth bearing B. The bearing at B is in proper alignment and exerts only force reactions on the windlass.

Prob. 5–74

Prob. 5–76

5–77. Member *AB* is supported by cable *BC* and at *A* by a *square* rod which fits loosely through the square hole in the collar fixed to the member as shown. Determine the components of reaction at *A* and the tension in the cable needed to hold the rod in equilibrium.

5–79. The forked rod is supported by a collar at *A*, a thrust bearing at *B*, and a cable *CD*. Determine the tension within cable *CD*, and the x, y, z components of reaction at supports *A* and *B*. The supports at *A* and *B* are in proper alignment and exert only force reactions on the rod.

Prob. 5–77

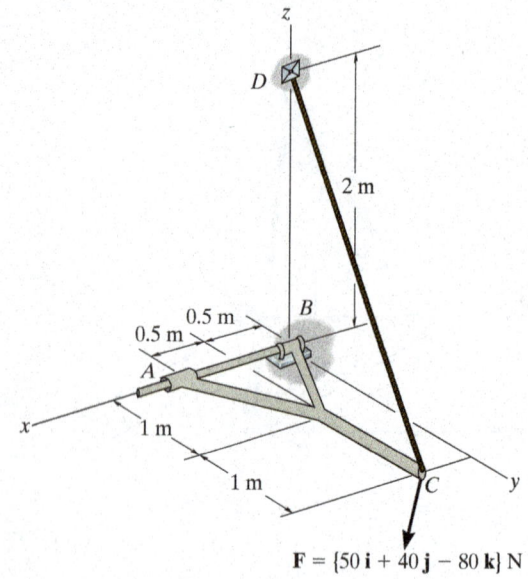

$\mathbf{F} = \{50\,\mathbf{i} + 40\,\mathbf{j} - 80\,\mathbf{k}\}$ N

Prob. 5–79

5–78. The member is supported by a pin at *A* and cable *BC*. Determine the components of reaction at these supports if the cylinder has a mass of 40 kg.

*5–80. The bent rod is supported at *A*, *B*, and *C* by journal bearings. Determine the components of reaction at the bearings if the rod is subjected to the 200-lb vertical force and the 30 lb·ft couple moment. The bearings are in proper alignment and exert only force reactions on the rod.

Prob. 5–78

Prob. 5–80

PROBLEMS 251

5–81. The bent rod is supported at A, B, and C by smooth journal bearings. Determine the x, y, z components of reaction at the bearings if the rod is subjected to forces $F_1 = 300$ lb and $F_2 = 250$ lb. $\mathbf{F}_1$ lies in the y–z plane. The bearings are in proper alignment and exert only force reactions on the rod.

5–83. The bar AB is supported by two smooth collars. At A the connection is a ball-and-socket joint and at B it is fixed to the collar. If a 50-lb load is applied to the bar, determine the x, y, z components of reaction at A and B.

Prob. 5–81

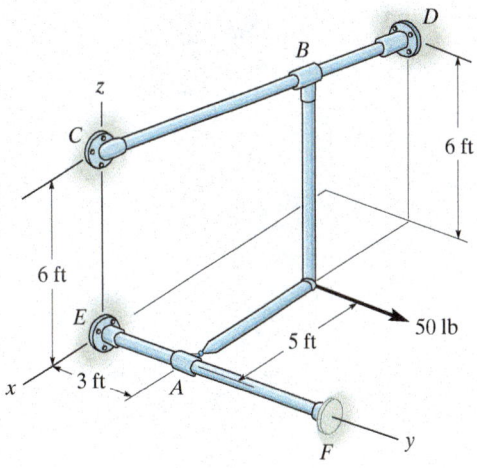

Prob. 5–83

5–82. The bent rod is supported at A, B, and C by smooth journal bearings. Determine the magnitude of $\mathbf{F}_2$ which will cause the reaction C_y at the bearing C to be equal to zero. The bearings are in proper alignment and exert only force reactions on the rod. Set $F_1 = 300$ lb.

*5–84. The rod has a weight of 6 lb/ft. If it is supported by a ball-and-socket joint at C and a journal bearing at D, determine the x, y, z components of reaction at these supports and the moment M that must be applied along the axis of the rod to hold it in the position shown.

Prob. 5–82

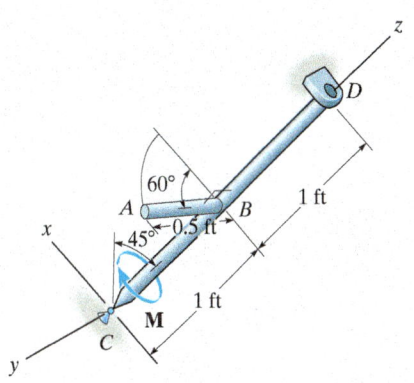

Prob. 5–84

5–85. The sign has a mass of 100 kg with center of mass at G. Determine the x, y, z components of reaction at the ball-and-socket joint A and the tension in wires BC and BD.

5–87. Both pulleys are fixed to the shaft and as the shaft turns with constant angular velocity, the power of pulley A is transmitted to pulley B. Determine the horizontal tension T in the belt on pulley B and the x, y, z components of reaction at the journal bearing C and thrust bearing D if $\theta = 45°$. The bearings are in proper alignment and exert only force reactions on the shaft.

Prob. 5–85

Prob. 5–87

5–86. Both pulleys are fixed to the shaft and as the shaft turns with constant angular velocity, the power of pulley A is transmitted to pulley B. Determine the horizontal tension T in the belt on pulley B and the x, y, z components of reaction at the journal bearing C and thrust bearing D if $\theta = 0°$. The bearings are in proper alignment and exert only force reactions on the shaft.

*5–88. Member AB is supported by a cable BC and at A it is fixed connected to a collar that passes through a *square* rod. Determine the x, y, z components of reaction at A and the tension in the cable needed to hold the 800-lb cylinder in equilibrium.

Prob. 5–86

Prob. 5–88

CHAPTER REVIEW

Equilibrium

A body in equilibrium is at rest or can translate with constant velocity.

$\Sigma F = 0$

$\Sigma M = 0$

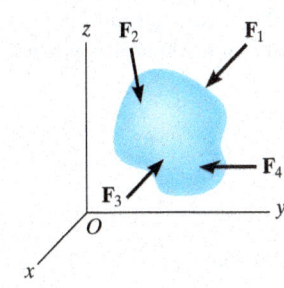

Two Dimensions

Before analyzing the equilibrium of a body, it is first necessary to draw its free-body diagram. This is an outlined shape of the body, which shows all the forces and couple moments that act on it.

Couple moments can be placed anywhere on a free-body diagram since they are free vectors. Forces can act at any point along their line of action since they are sliding vectors.

Angles used to resolve forces, and dimensions used to take moments of the forces, should also be shown on the free-body diagram.

Remember that a support will exert a force on the body in a particular direction if it prevents translation of the body in that direction, and it will exert a couple moment on the body if it prevents rotation.

roller

smooth pin or hinge

fixed support

The three scalar equations of equilibrium can be applied when solving problems in two dimensions, since the geometry is easy to visualize.

$\Sigma F_x = 0$
$\Sigma F_y = 0$
$\Sigma M_O = 0$

For the most direct solution, try to sum forces along an axis that will eliminate as many unknown forces as possible. Sum moments about a point A that passes through the line of action of as many unknown forces as possible.

$\Sigma F_x = 0;$
$A_x - P_2 = 0 \quad A_x = P_2$

$\Sigma M_A = 0;$

$P_2 d_2 + B_y d_B - P_1 d_1 = 0$

$B_y = \dfrac{P_1 d_1 - P_2 d_2}{d_B}$

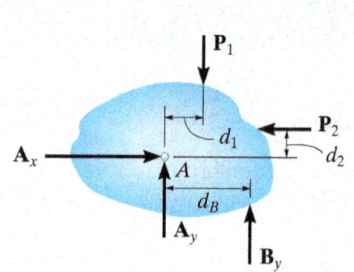

Three Dimensions

Some common types of supports and their reactions are shown here in three dimensions.

roller   ball-and-socket

fixed support

In three dimensions, it is often advantageous to use a Cartesian vector analysis when applying the equations of equilibrium. To do this, first express each known and unknown force and couple moment shown on the free-body diagram as a Cartesian vector. Then set the force summation equal to zero. Take moments about a point O that lies on the line of action of as many unknown force components as possible. From point O direct position vectors to each force, and then use the cross product to determine the moment of each force.

$\Sigma \mathbf{F} = \mathbf{0}$

$\Sigma \mathbf{M}_O = \mathbf{0}$

The six scalar equations of equilibrium are established by setting the respective **i**, **j**, and **k** components of these force and moment summations equal to zero.

$\Sigma F_x = 0 \qquad \Sigma M_x = 0$
$\Sigma F_z = 0 \qquad \Sigma M_y = 0$
$\Sigma F_y = 0 \qquad \Sigma M_z = 0$

Determinacy and Stability

If a body is supported by a minimum number of constraints to ensure equilibrium, then it is statically determinate. If it has more constraints than required, then it is statically indeterminate.

To properly constrain the body, the reactions must not all be parallel to one another or concurrent.

Statically indeterminate, five reactions, three equilibrium equations

Proper constraint, statically determinate

REVIEW PROBLEMS

All solutions must include a free-body diagram.

R5–1. Determine the reactions at the supports A and B.

Prob. R5–1

R5–2. Determine the horizontal and vertical components of reaction at the pin A and the reaction at the roller B on the lever.

Prob. R5–2

R5–3. Determine the x and z components of reaction at the journal bearing A, and the tension in cords BC and BD.

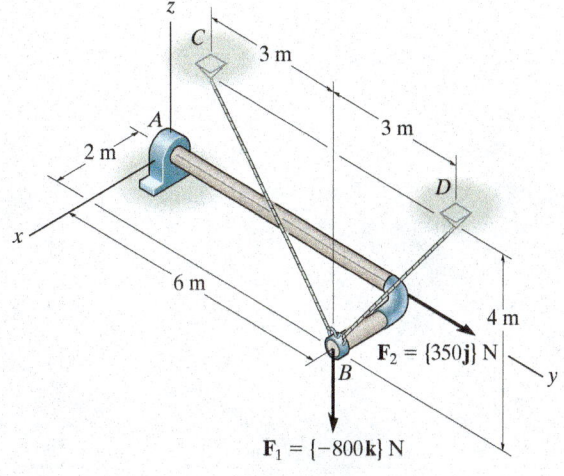

Prob. R5–3

R5–4. A vertical force of 80 lb acts on the crankshaft. Determine the horizontal equilibrium force $\mathbf{P}$ that must be applied to the handle, and the x, y, z components of reaction at the journal bearing A and thrust bearing B. The bearings are properly aligned and exert only force reactions on the shaft.

Prob. R5–4

CHAPTER 6

In order to design the many parts of this lift it is required that we first know the forces that they must support. In this chapter we will show how to analyze a machine such as this using the equations of equilibrium.

STRUCTURAL ANALYSIS

CHAPTER OBJECTIVES

- To show how to determine the forces in the members of a truss using the method of joints and the method of sections.
- To analyze the forces acting on the members of frames and machines composed of pin-connected members.

6.1 SIMPLE TRUSSES

A *truss* is a structure composed of slender members joined together at their end points. The members commonly used in construction consist of wooden struts or metal bars. In particular, *planar trusses* lie in a single plane and are often used to support roofs and bridges. The truss shown in Fig. 6–1a is an example of a typical roof-supporting truss. Here the roof load is transmitted to the truss *at the joints* by means of a series of purlins. Since this loading acts in the same plane as the truss, Fig. 6–1b, the analysis of the forces developed in the truss members will be two-dimensional.

(a)　　　　　　　　　　　　　　(b)

Fig. 6–1

258   CHAPTER 6   STRUCTURAL ANALYSIS

Stringer
Deck
Floor beam
A
(a)

Bridge truss
(b)

Fig. 6–2

In the case of a bridge, such as shown in Fig. 6–2a, the load on the deck is first transmitted to stringers, then to floor beams, and finally to the joints of the two supporting side trusses. Like the roof truss, the bridge truss loading is also coplanar, Fig. 6–2b.

When bridge or roof trusses extend over large distances, a rocker or roller is commonly used for supporting one end, for example, joint A in Figs. 6–1a and 6–2a. This type of support allows freedom for expansion or contraction of the members due to a change in temperature or application of loads.

Assumptions for Design. To design both the members and the connections of a truss, it is first necessary to determine the *force* developed in each member when the truss is subjected to a given loading. To do a force analysis we will make two important assumptions:

Gusset plate
(a)

(b)

Fig. 6–3

- *All loadings are applied at the joints.* In most situations, such as for bridge and roof trusses, this assumption is true. Frequently the weight of the members is neglected because the force supported by each member is usually much larger than its weight. However, if the weight is to be included in the analysis, it is generally satisfactory to apply it as a vertical force, with half of its magnitude applied at each end of the member.

- *The members are joined together by smooth pins.* The joint connections are usually formed by bolting or welding the ends of the members to a common plate, called a **gusset plate**, Fig. 6–3a, or by simply passing a large bolt or pin through each of the members, Fig. 6–3b. We can assume these connections act as pins provided the centerlines of the joining members are *concurrent*, as in Fig. 6–3.

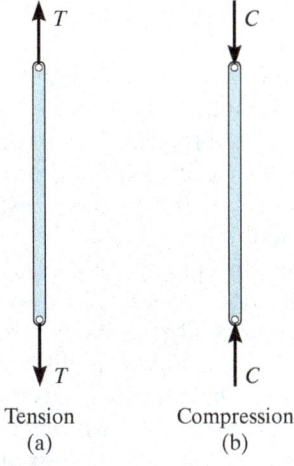

Tension　　Compression
(a)　　　　(b)

Fig. 6–4

Because of these two assumptions, *each truss member will act as a two-force member*, and therefore the force acting at each end of the member will be directed along the axis of the member. If the force tends to *elongate* the member, it is a ***tensile force*** (T), Fig. 6–4a; whereas if it tends to *shorten* the member, it is a ***compressive force*** (C), Fig. 6–4b. In the actual design of a truss it is important to state whether the force is tensile or compressive. Often, compression members must be made *thicker* than tension members because of the buckling or sudden collapse that occurs when a member is in compression.

Simple Truss. If three members are pin connected at their ends, they form a *triangular truss* that will be *rigid*, Fig. 6–5. Attaching two more members and connecting them to a new joint *D* forms a larger truss, Fig. 6–6. This procedure can be repeated as many times as desired to form an even larger truss, and by doing this one forms a ***simple truss***.

The use of metal gusset plates in the construction of these Warren trusses is clearly evident.

Fig. 6–5

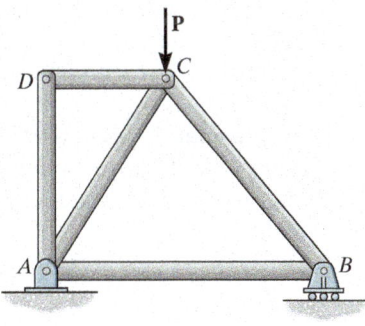

Fig. 6–6

260   CHAPTER 6   STRUCTURAL ANALYSIS

Fig. 6–7

6.2 THE METHOD OF JOINTS

One way to determine the force in each member of a truss is to use the *method of joints*. This method is based on the fact that if the entire truss is in equilibrium, then each of its joints is also in equilibrium. Therefore, if the free-body diagram of each joint is drawn, the force equilibrium equations, $\Sigma F_x = 0$ and $\Sigma F_y = 0$, can then be used to obtain the member forces acting on each joint.

For example, consider the pin at joint B of the truss in Fig. 6–7a. As shown on its free-body diagram, in two dimensions, Fig. 6–7b, three forces act on the pin, namely, the 500-N force and the forces exerted by members BA and BC. Here, $\mathbf{F}_{BA}$ is "pulling" on the pin, which means that member BA is in *tension*; whereas $\mathbf{F}_{BC}$ is "pushing" on the pin, and consequently member BC is in *compression*. These effects can also be seen by isolating the joint with small segments of the member connected to the pin, Fig. 6–7c. The pushing or pulling on these small segments indicates the effect on the members being either in compression or tension.

When using the method of joints, always start at a joint having *at least one known force and at most two unknown forces*, as in Fig. 6–7b. In this way, application of $\Sigma F_x = 0$ and $\Sigma F_y = 0$ yields two algebraic equations which can be solved for the two unknowns. When applying these equations, the correct sense of an unknown member force can be determined using one of two possible methods.

- The *correct* sense of direction of an unknown member force can, in many cases, be determined "by inspection." For example, $\mathbf{F}_{BC}$ in Fig. 6–7b must push on the pin (compression) since its horizontal component, $F_{BC} \sin 45°$, must balance the 500-N force ($\Sigma F_x = 0$). Likewise, $\mathbf{F}_{BA}$ is a tensile force since it balances the vertical component, $F_{BC} \cos 45°$ ($\Sigma F_y = 0$). In more complicated cases, the sense of an unknown member force can be *assumed*; then, after applying the equilibrium equations, the assumed sense can be verified from the numerical results. A *positive scalar* indicates that the sense is *correct*, whereas a *negative scalar* indicates that the sense shown on the free-body diagram must be *reversed*.

- *Always assume* the unknown member forces acting on the joint's free-body diagram to be in *tension*; i.e., the forces "pull" on the pin. If this is done, then numerical solution of the equilibrium equations will yield *positive scalars for members in tension and negative scalars for members in compression*. Once an unknown member force is found, use its *correct* magnitude and sense (T or C) on subsequent joint free-body diagrams.

IMPORTANT POINTS

- Simple trusses are composed of triangular elements. The members are assumed to be pin connected at their ends and the loads applied at the joints.

- If a truss is in equilibrium, then each of its joints is in equilibrium. The internal forces in the members become external forces when the free-body diagram of each joint of the truss is drawn. A force pulling on a joint is caused by tension in a member, and a force pushing on a joint is caused by compression.

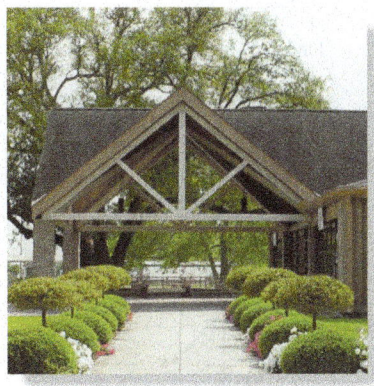

The forces in the members of this simple roof truss can be determined using the method of joints.

PROCEDURE FOR ANALYSIS

The following procedure provides a means for analyzing a truss using the method of joints.

- Draw the free-body diagram of a joint having at least one known force and at most two unknown forces. (If this joint is at one of the supports, then it may be necessary first to calculate the external reactions at the support.)

- Orient the x and y axes such that the forces on the free-body diagram can be easily resolved into their x and y components and then apply the two force equilibrium equations $\Sigma F_x = 0$ and $\Sigma F_y = 0$. Solve for the two unknown member forces and verify their correct sense.

- Using the calculated results, continue to analyze each of the other joints. Remember that a member in *compression* "pushes" on the joint and a member in *tension* "pulls" on the joint.

EXAMPLE 6.1

Fig. 6–8

Determine the force in each member of the truss shown in Fig. 6–8a and indicate whether the members are in tension or compression.

SOLUTION
Since we should have no more than two unknown forces at the joint and at least one known force acting there, we will begin our analysis at joint B.

Joint B. The free-body diagram of the joint at B is shown in Fig. 6–8b. Applying the equations of equilibrium, we have

$\xrightarrow{+} \Sigma F_x = 0;$ $500 \text{ N} - F_{BC} \sin 45° = 0;$ $F_{BC} = 707.1 \text{ N (C)}$ *Ans.*

$+\uparrow \Sigma F_y = 0;$ $F_{BC} \cos 45° - F_{BA} = 0;$ $F_{BA} = 500 \text{ N (T)}$ *Ans.*

Now that the force in member BC has been calculated, we can proceed to analyze joint C to determine the force in member CA and the support reaction at the rocker.

Joint C. From the free-body diagram, Fig. 6–8c, we have

$\xrightarrow{+} \Sigma F_x = 0;$ $-F_{CA} + 707.1 \cos 45° \text{ N} = 0;$ $F_{CA} = 500 \text{ N (T)}$ *Ans.*

$+\uparrow \Sigma F_y = 0;$ $C_y - 707.1 \sin 45° \text{ N} = 0;$ $C_y = 500 \text{ N}$ *Ans.*

Joint A. Although it is not necessary, we can determine the components of the support reactions at joint A using the results of F_{CA} and F_{BA}. From the free-body diagram, Fig. 6–8d, we have

$\xrightarrow{+} \Sigma F_x = 0;$ $500 \text{ N} - A_x = 0;$ $A_x = 500 \text{ N}$

$+\uparrow \Sigma F_y = 0;$ $500 \text{ N} - A_y = 0;$ $A_y = 500 \text{ N}$

NOTE: The results of this analysis are summarized in Fig. 6–8e. Here the free-body diagram of each joint (or pin) shows the effects of all the connected members and external forces applied to the joint, whereas the free-body diagram of each member shows only the effects of the joints on the member.

EXAMPLE 6.2

Determine the forces acting in all the members of the truss shown in Fig. 6–9a and indicate whether the members are in tension or compression.

SOLUTION
By inspection, there are more than two unknowns at each joint. As a result, the support reactions on the truss must first be determined. Show that they have been correctly calculated on the free-body diagram in Fig. 6–9b. We can now begin the analysis at joint C.

Joint C. From the free-body diagram, Fig. 6–9c,

$\xrightarrow{+} \Sigma F_x = 0;\qquad -F_{CD} \cos 30° + F_{CB} \sin 45° = 0$

$+\uparrow \Sigma F_y = 0;\qquad 1.5 \text{ kN} + F_{CD} \sin 30° - F_{CB} \cos 45° = 0$

These two equations must be solved *simultaneously* for each of the two unknowns. A more *direct solution* can be obtained by applying a force summation along an axis that is *perpendicular* to the direction of the other unknown force. For example, summing forces along the y' axis, which is perpendicular to the direction of $\mathbf{F}_{CD}$, Fig. 6–9d, yields a *direct solution* for F_{CB}.

$+\nearrow \Sigma F_{y'} = 0;\qquad 1.5 \cos 30° \text{ kN} - F_{CB} \sin 15° = 0$

$\qquad F_{CB} = 5.019 \text{ kN} = 5.02 \text{ kN (C)}\qquad Ans.$

Then,

$+\searrow \Sigma F_{x'} = 0;$

$-F_{CD} + 5.019 \cos 15° - 1.5 \sin 30° = 0;\quad F_{CD} = 4.10 \text{ kN (T)}\quad Ans.$

Joint D. We can now proceed to analyze joint D. The free-body diagram is shown in Fig. 6–9e.

$\xrightarrow{+} \Sigma F_x = 0;\qquad -F_{DA} \cos 30° + 4.10 \cos 30° \text{ kN} = 0$

$\qquad F_{DA} = 4.10 \text{ kN}\quad (T)\qquad Ans.$

$+\uparrow \Sigma F_y = 0;\qquad F_{DB} - 2(4.10 \sin 30° \text{ kN}) = 0$

$\qquad F_{DB} = 4.10 \text{ kN}\quad (T)\qquad Ans.$

NOTE: The force in the last member, BA, can be obtained from joint B or joint A. As an exercise, draw the free-body diagram of joint B, sum the forces in the horizontal direction, and show that $F_{BA} = 0.776 \text{ kN (C)}$.

(a)

(b)

(c)

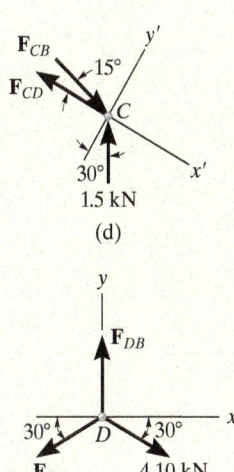
(d)

(e)

Fig. 6–9

EXAMPLE 6.3

Determine the force in each member of the truss shown in Fig. 6–10a and indicate whether the members are in tension or compression.

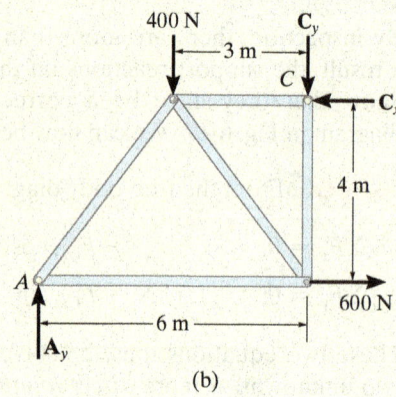

(a) (b)

SOLUTION

Support Reactions. No joint can be analyzed until the support reactions are determined, because each joint has at least three unknown forces acting on it. A free-body diagram of the entire truss is given in Fig. 6–10b. Applying the equations of equilibrium, we have

$\xrightarrow{+} \Sigma F_x = 0;$ $600 \text{ N} - C_x = 0;$ $C_x = 600 \text{ N}$

$\zeta + \Sigma M_C = 0;$ $-A_y(6 \text{ m}) + 400 \text{ N}(3 \text{ m}) + 600 \text{ N}(4 \text{ m}) = 0$

$A_y = 600 \text{ N}$

$+\uparrow \Sigma F_y = 0;$ $600 \text{ N} - 400 \text{ N} - C_y = 0;$ $C_y = 200 \text{ N}$

The analysis can now start at either joint A or C. The choice is arbitrary since there are one known and two unknown member forces acting on the pin at each of these joints.

Joint A. (Fig. 6–10c). As shown on the free-body diagram, $\mathbf{F}_{AB}$ is assumed to be compressive and $\mathbf{F}_{AD}$ is tensile. Applying the equations of equilibrium, we have

$+\uparrow \Sigma F_y = 0;$ $600 \text{ N} - \tfrac{4}{5} F_{AB} = 0;$ $F_{AB} = 750 \text{ N}$ (C)   *Ans.*

$\xrightarrow{+} \Sigma F_x = 0;$ $F_{AD} - \tfrac{3}{5}(750 \text{ N}) = 0;$ $F_{AD} = 450 \text{ N}$ (T)   *Ans.*

(c)

Fig. 6–10

Joint D. (Fig. 6–10d). Using the result for F_{AD} and summing forces in the horizontal direction, Fig. 6–10d, we have

$\xrightarrow{+} \Sigma F_x = 0;$ $-450 \text{ N} + \frac{3}{5}F_{DB} + 600 \text{ N} = 0;$ $F_{DB} = -250 \text{ N}$

The negative sign indicates that $\mathbf{F}_{DB}$ acts in the *opposite sense* to that shown in Fig. 6–10d.* Hence,

$$F_{DB} = 250 \text{ N (T)} \qquad \textit{Ans.}$$

To determine $\mathbf{F}_{DC}$, we can either correct the sense of $\mathbf{F}_{DB}$ on the free-body diagram, and then apply $\Sigma F_y = 0$, or apply this equation and retain the negative sign for F_{DB}, i.e.,

$+\uparrow \Sigma F_y = 0;$ $-F_{DC} - \frac{4}{5}(-250 \text{ N}) = 0;$ $F_{DC} = 200 \text{ N}$ (C)   *Ans.*

Joint C. (Fig. 6–10e).

$\xrightarrow{+} \Sigma F_x = 0;$ $F_{CB} - 600 \text{ N} = 0;$ $F_{CB} = 600 \text{ N}$ (C)   *Ans.*

$+\uparrow \Sigma F_y = 0;$ $200 \text{ N} - 200 \text{ N} \equiv 0$ (check)

NOTE: The analysis is summarized in Fig. 6–10f, which shows the free-body diagram for each joint and member.

(d)

(e)

(f)

Fig. 6–10 (cont.)

*The proper sense could have been determined by inspection, prior to applying $\Sigma F_x = 0$.

6.3 ZERO-FORCE MEMBERS

Truss analysis using the method of joints is greatly simplified if we can first identify those members which support *no loading*. These *zero-force members* are used to increase the stability of the truss during construction and to provide added support if the loading is changed.

The zero-force members of a truss can generally be found by inspection of each of the joints. For example, consider the truss shown in Fig. 6–11a. If a free-body diagram of the pin at joint A is drawn, Fig. 6–11b, it is seen that members AB and AF are zero-force members. (We could not come to this conclusion if we had considered the free-body diagrams of joints F or B, simply because there are five unknowns at each of these joints.) In a similar manner, consider the free-body diagram of joint D, Fig. 6–11c. Here again it is seen that DC and DE are zero-force members.

To summarize, then, *if only two non-collinear members form a truss joint and no external load or support reaction is applied to the joint, the two members must be zero-force members.* The load on the truss in Fig. 6–11a is therefore actually supported by only five members, as shown in Fig. 6–11d.

Fig. 6–11

6.3 ZERO-FORCE MEMBERS

Now consider the truss shown in Fig. 6–12a. The free-body diagram of the pin at joint D is shown in Fig. 6–12b. By orienting the y axis along members DC and DE and the x axis along member DA, it is seen that DA is a zero-force member. From the free-body diagram of joint C, Fig. 6–12c, it can be seen that this is also the case for member CA. In general then, *if three members form a truss joint for which two of the members are collinear, the third member is a zero-force member provided no external force or support reaction has a component that acts along this member.* The truss shown in Fig. 6–12d is therefore suitable for supporting the load $\mathbf{P}$.

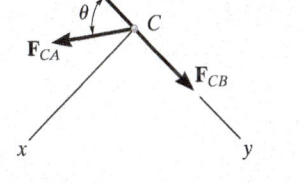

(a)

$+\swarrow \Sigma F_x = 0;\quad F_{DA} = 0$
$+\searrow \Sigma F_y = 0;\quad F_{DC} = F_{DE}$

(b)

$+\swarrow \Sigma F_x = 0;\quad F_{CA} \sin\theta = 0;\quad F_{CA} = 0 \text{ since } \sin\theta \neq 0;$
$+\searrow \Sigma F_y = 0;\quad F_{CB} = F_{CD}$

(c)

(d)

Fig. 6–12

IMPORTANT POINT

- Zero-force members support no load; however, they are necessary for stability, and are available when additional loadings are applied to the joints of the truss. These members can usually be identified by inspection. They occur at joints where only two members are connected and no external load acts along either member. Also, at joints having two collinear members, a third member will be a zero-force member if no external force components act along this member.

268   CHAPTER 6   STRUCTURAL ANALYSIS

EXAMPLE 6.4

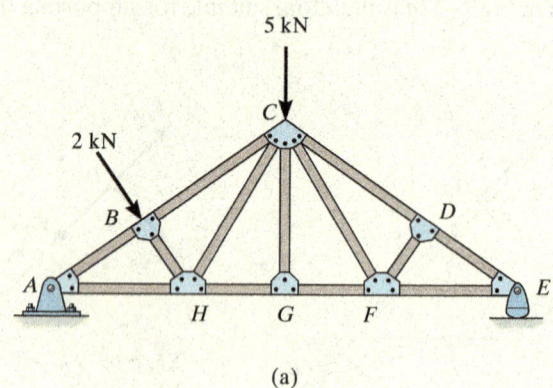

Using the method of joints, determine all the zero-force members of the *Fink roof truss* shown in Fig. 6–13a. Assume all joints are pin connected.

(b)

(c)

(d)

(e)

(f)

Fig. 6–13

SOLUTION
Look for joint geometries that have three members for which two are collinear. We have

Joint G. (Fig. 6–13b).

$+\uparrow \Sigma F_y = 0;$ $F_{GC} = 0$ Ans.

Realize that we could not conclude that *GC* is a zero-force member by considering joint *C*, where there are five unknowns. The fact that *GC* is a zero-force member means that the 5-kN load at *C* must be supported by members *CB, CH, CF,* and *CD*.

Joint D. (Fig. 6–13c).

$+\nearrow \Sigma F_x = 0;$ $F_{DF} = 0$ Ans.

Joint F. (Fig. 6–13d).

$+\uparrow \Sigma F_y = 0;$ $F_{FC} \cos \theta = 0;$ Since $\theta \neq 90°$, $F_{FC} = 0$ Ans.

NOTE: If joint *B* is analyzed, Fig. 6–13e,

$+\searrow \Sigma F_x = 0;$ $2 \text{ kN} - F_{BH} = 0;$ $F_{BH} = 2 \text{ kN}$ (C)

Also, F_{HC} must satisfy $\Sigma F_y = 0$, Fig. 6–13f, and therefore *HC* is *not* a zero-force member.

FUNDAMENTAL PROBLEMS

All solutions must include a free-body diagram.

F6–1. Determine the force in each member of the truss. State if the members are in tension or compression.

Prob. F6–1

F6–2. Determine the force in each member of the truss. State if the members are in tension or compression.

Prob. F6–2

F6–3. Determine the force in each member of the truss. State if the members are in tension or compression.

Prob. F6–3

F6–4. Determine the greatest load P that can be applied to the truss so that none of the members are subjected to a force exceeding either 2 kN in tension or 1.5 kN in compression.

Prob. F6–4

F6–5. Identify the zero-force members in the truss.

Prob. F6–5

F6–6. Determine the force in each member of the truss. State if the members are in tension or compression.

Prob. F6–6

PROBLEMS

All solutions must include a free-body diagram.

6–1. Determine the force in each member of the truss and state if the members are in tension or compression. Set $P_1 = 2$ kN and $P_2 = 1.5$ kN.

6–2. Determine the force in each member of the truss and state if the members are in tension or compression. Set $P_1 = P_2 = 4$ kN.

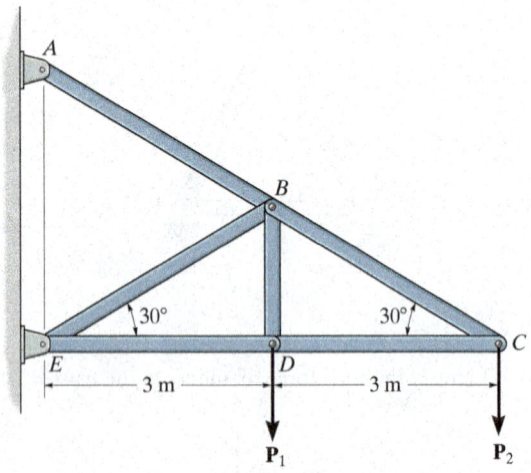

Probs. 6–1/2

6–3. Determine the force in each member of the truss and state if the members are in tension or compression. Set $P_1 = 200$ lb, $P_2 = 500$ lb.

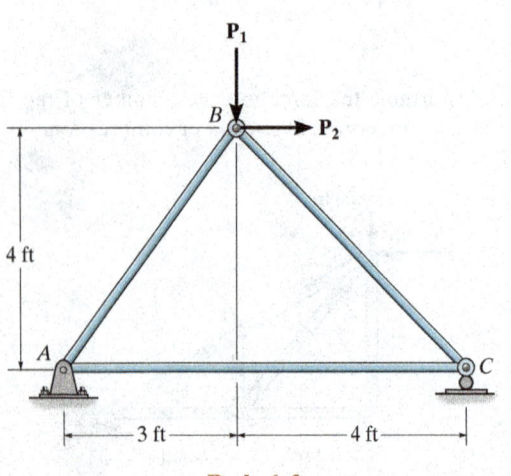

Prob. 6–3

*6–4. Determine the force in each member of the truss and state if the members are in tension or compression.

Prob. 6–4

6–5. Determine the force in each member of the truss and state if the members are in tension or compression. Set $P_1 = 7$ kN, $P_2 = 7$ kN.

6–6. Determine the force in each member of the truss and state if the members are in tension or compression. Set $P_1 = 8$ kN, $P_2 = 10$ kN.

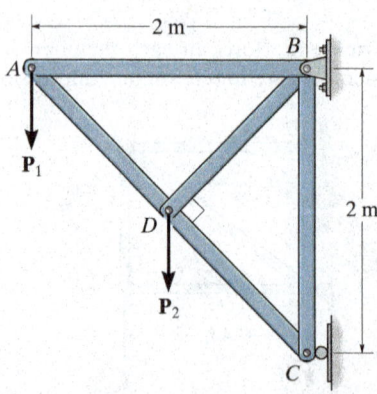

Probs. 6–5/6

6–7. Determine the force in each member of the truss and state if the members are in tension or compression.

Prob. 6–7

***6–8.** Determine the force in each member of the truss and state if the members are in tension or compression.

Prob. 6–8

6–9. Determine the force in each member of the truss in terms of the external loading and state if the members are in tension or compression.

6–10. The maximum allowable tensile force in the members of the truss is $(F_t)_{max} = 1500$ lb, and the maximum allowable compressive force is $(F_c)_{max} = 800$ lb. Determine the maximum magnitude P of the two loads that can be applied to the truss. Take $a = 8$ ft.

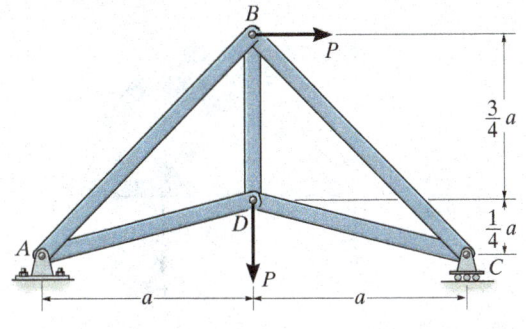

Probs. 6–9/10

6–11. Determine the zero-force members in the *Pratt roof truss*. Explain your answers using appropriate joint free-body diagrams.

Prob. 6–11

***6–12.** Determine the force in each member of the truss and state if the members are in tension or compression. Assume all members are pin connected.

Prob. 6–12

6–13. Determine the force in each member of the truss in terms of the load P and state if the members are in tension or compression.

6–14. Members AB and BC can each support a maximum compressive force of 800 lb, and members AD, DC, and BD can support a maximum tensile force of 1500 lb. If $a = 10$ ft, determine the greatest load P the truss can support.

6–15. Members AB and BC can each support a maximum compressive force of 800 lb, and members AD, DC, and BD can support a maximum tensile force of 2000 lb. If $a = 6$ ft, determine the greatest load P the truss can support.

6–18. Determine the force in each member of the truss and state if the members are in tension or compression. Set $P_1 = 10$ kN, $P_2 = 8$ kN.

6–19. Determine the force in each member of the truss and state if the members are in tension or compression. Set $P_1 = 8$ kN, $P_2 = 12$ kN.

Probs. 6–18/19

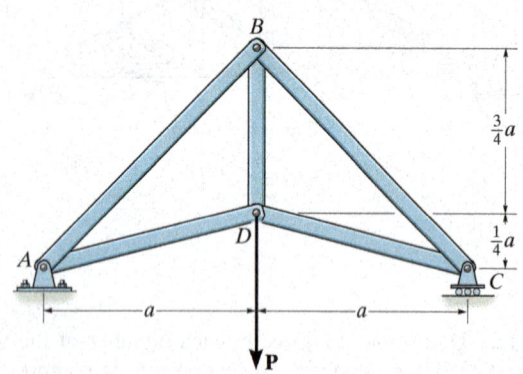

Probs. 6–13/14/15

*6–16. Determine the force in each member of the truss and state if the members are in tension or compression. Set $P = 8$ kN.

6–17. If the maximum force that any member can support is 8 kN in tension and 6 kN in compression, determine the maximum force P that can be supported at joint D.

*6–20. Determine the force in each member of the truss and state if the members are in tension or compression. Set $P_1 = 20$ kN, $P_2 = 10$ kN.

6–21. Determine the force in each member of the truss and state if the members are in tension or compression. Set $P_1 = 40$ kN, $P_2 = 20$ kN.

Probs. 6–16/17

Probs. 6–20/21

6–22. Determine the force in each member of the truss and indicate whether the members are in tension or compression. Assume that all members are pin connected.

*6–24. Determine the force in each member of the truss and state if the member is in tension or compression.

Prob. 6–22

Prob. 6–24

6–23. The *Fink truss* supports the loads shown. Determine the force in each member and state if the members are in tension or compression. Approximate each joint as a pin.

6–25. Determine the force in members BF and FD of the truss and state if the members are in tension or compression. Set $P_1 = 200$ lb, $P_2 = 150$ lb.

6–26. Determine the force in members ED, EF, DF, and DC of the truss and state if the members are in tension or compression. Set $P_1 = 800$ lb, $P_2 = 600$ lb.

Prob. 6–23

Probs. 6–25/26

6.4 THE METHOD OF SECTIONS

When we need to find the force in only a few members of a truss, we can analyze the truss using the *method of sections*. It is based on the principle that if the truss is in equilibrium, then any part of the truss is also in equilibrium. For example, consider the two truss members shown in Fig. 6–14. If the forces within the members are to be determined, then an imaginary section, indicated by the blue line, can be used to cut each member into two parts and thereby "expose" each internal force as "external" to the free-body diagrams of the parts shown on the right. Clearly, it can be seen that equilibrium requires that the member in tension (T) be subjected to a "pull," whereas the member in compression (C) is subjected to a "push."

The method of sections can also be used to "cut" or section the members of an entire truss. If the section passes through the truss and the free-body diagram of either of its two parts is drawn, we can then apply the equations of equilibrium to that part to determine the member forces at the section. Since only *three* independent equilibrium equations ($\Sigma F_x = 0$, $\Sigma F_y = 0$, $\Sigma M_O = 0$) can be applied to the free-body diagram of any part, then we should try to select a section that, in general, passes through not more than *three* members in which the forces are unknown. For example, consider the truss in Fig. 6–15a. If the forces in members BC, GC, and GF are to be determined, then section aa would be appropriate. The free-body diagrams of the two parts are shown in Figs. 6–15b and 6–15c. Notice that the direction of each member force is specified from the *geometry* of the truss, since the force in a member is along its axis. Also, the member forces acting on one part of the truss are equal but opposite to those acting on the other part—Newton's third law. Members BC and GC are assumed to be in *tension* since they are subjected to a "pull," whereas GF in *compression* since it is subjected to a "push."

The three unknown member forces $\mathbf{F}_{BC}$, $\mathbf{F}_{GC}$, and $\mathbf{F}_{GF}$ can be obtained by applying the three equilibrium equations to the free-body diagram in Fig. 6–15b. If, however, the free-body diagram in Fig. 6–15c is considered, then the three support reactions $\mathbf{D}_x$, $\mathbf{D}_y$, and $\mathbf{E}_x$ will have to be known, because only three equations of equilibrium are available. (This, of course, is done in the usual manner by considering a free-body diagram of the *entire truss*.)

Fig. 6–14

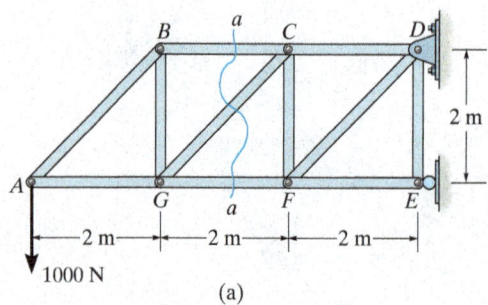

(a)

Fig. 6–15

6.4 THE METHOD OF SECTIONS   275

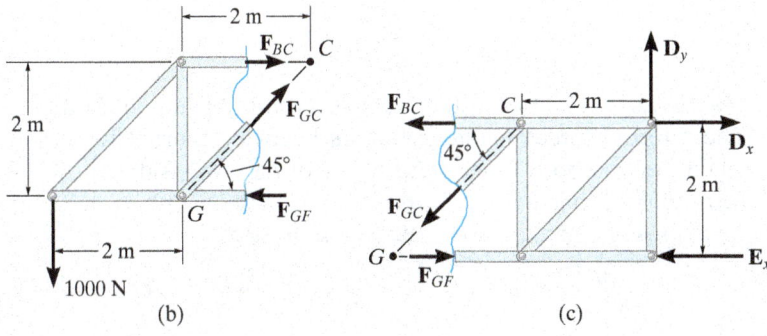

Fig. 6–15 (cont.)

When applying the equilibrium equations, we should carefully consider ways of writing the equations so as to yield a *direct solution* for each of the unknowns, rather than having to solve simultaneous equations. For example, using the part in Fig. 6–15b and summing moments about C will yield a direct solution for $\mathbf{F}_{GF}$ since $\mathbf{F}_{BC}$ and $\mathbf{F}_{GC}$ create zero moment about C. Likewise, $\mathbf{F}_{BC}$ can be directly obtained by summing moments about G. Finally, $\mathbf{F}_{GC}$ can be found directly from a force summation in the vertical direction since $\mathbf{F}_{GF}$ and $\mathbf{F}_{BC}$ have no vertical components. This ability to *determine directly* the force in a particular truss member is one of the main advantages of using the method of sections.*

As in the method of joints, there are two ways in which we can determine the correct sense of an unknown member force:

The forces in selected members of this Pratt truss can readily be determined using the method of sections.

- The correct sense can in many cases be determined "by inspection." For example, $\mathbf{F}_{BC}$ is shown as a tensile force in Fig. 6–15b since moment equilibrium about G requires that $\mathbf{F}_{BC}$ create a moment opposite to that of the 1000-N force. Also, $\mathbf{F}_{GC}$ is tensile since its vertical component must balance the 1000-N force which acts downward. In more complicated cases, the sense of an unknown force may be *assumed*. If the solution yields a *negative* scalar, it indicates that the force's sense of direction is *opposite* to that shown on the free-body diagram.

- *Always assume* that the unknown member forces at the section are *tensile* forces, i.e., "pulling" on the member. By doing this, the numerical solution of the equilibrium equations will yield *positive scalars for members in tension and negative scalars for members in compression.*

---

*If the method of joints were used to determine, say, the force in member GC, it would be necessary to analyze joints A, B, and G in sequence.

Simple trusses are often used in the construction of large cranes in order to reduce the weight of the boom and tower.

IMPORTANT POINT

- If a truss is in equilibrium, then each of its parts is in equilibrium. The internal forces in the members become external forces when the free-body diagram of a part of the truss is drawn. A force pulling on a member causes tension in the member, and a force pushing on a member causes compression.

PROCEDURE FOR ANALYSIS

The forces in the members of a truss may be determined by the method of sections using the following procedure.

Free-Body Diagram.

- Decide how to "cut" or section the truss through the members where forces are to be determined.

- Before isolating any part of the truss, it may first be necessary to determine the truss's support reactions that act on the part. Once this is done, the three equilibrium equations will be available to solve for member forces at the section.

- Draw the free-body diagram of that part of the sectioned truss which has the least number of forces acting on it.

Equations of Equilibrium.

- Moments should be summed about a point that lies at the intersection of the lines of action of two unknown forces, so that the third unknown force can be determined directly from the moment equation.

- If two of the unknown forces are *parallel*, forces may be summed *perpendicular* to their direction in order to directly determine the third unknown force.

EXAMPLE 6.5

Determine the force in members *GE*, *GC*, and *BC* of the truss shown in Fig. 6–16a. Indicate whether the members are in tension or compression.

SOLUTION

Section *aa* in Fig. 6–16a has been chosen since it cuts through the *three* members whose forces are to be determined. In order to use the method of sections, however, it is *first* necessary to determine the external reactions at *A* or *D*. A free-body diagram of the entire truss is shown in Fig. 6–16b. Applying the equations of equilibrium, we have

$\xrightarrow{+} \Sigma F_x = 0;$ $400 \text{ N} - A_x = 0;$ $A_x = 400 \text{ N}$

$\zeta + \Sigma M_A = 0;$ $-1200 \text{ N}(8 \text{ m}) - 400 \text{ N}(3 \text{ m}) + D_y(12 \text{ m}) = 0$

$$D_y = 900 \text{ N}$$

$+\uparrow \Sigma F_y = 0;$ $A_y - 1200 \text{ N} + 900 \text{ N} = 0;$ $A_y = 300 \text{ N}$

Free-Body Diagram. For the analysis the free-body diagram of the left part of the sectioned truss will be used, since it involves the least number of forces on it compared to the right part, Fig. 6–16c.

Equations of Equilibrium. Summing moments about point *G* eliminates $\mathbf{F}_{GE}$ and $\mathbf{F}_{GC}$ and yields a direct solution for F_{BC}.

$\zeta + \Sigma M_G = 0;$ $-300 \text{ N}(4 \text{ m}) - 400 \text{ N}(3 \text{ m}) + F_{BC}(3 \text{ m}) = 0$

$$F_{BC} = 800 \text{ N} \quad (T) \qquad \textit{Ans.}$$

Summing moments about point *C*, we obtain a direct solution for F_{GE}.

$\zeta + \Sigma M_C = 0;$ $-300 \text{ N}(8 \text{ m}) + F_{GE}(3 \text{ m}) = 0$

$$F_{GE} = 800 \text{ N} \quad (C) \qquad \textit{Ans.}$$

Since $\mathbf{F}_{BC}$ and $\mathbf{F}_{GE}$ have no vertical components, summing forces in the *y* direction directly yields F_{GC}, i.e.,

$+\uparrow \Sigma F_y = 0;$ $300 \text{ N} - \tfrac{3}{5} F_{GC} = 0$

$$F_{GC} = 500 \text{ N} \quad (T) \qquad \textit{Ans.}$$

Fig. 6–16

NOTE: Here it is possible to tell, by inspection, the proper direction for each unknown member force. For example, $\Sigma M_C = 0$ requires $\mathbf{F}_{GE}$ to be *compressive* because it must balance the moment of the 300-N force about *C*.

EXAMPLE 6.6

Determine the force in member *CF* of the truss shown in Fig. 6–17a. Indicate whether the member is in tension or compression. Assume each member is pin connected.

(a)

(b)

SOLUTION

Free-Body Diagram. Section *aa* in Fig. 6–17a will be used since this section will "expose" the internal force in member *CF* as "external" on the free-body diagram of either the right or left portion of the truss. It is first necessary, however, to determine the support reactions on either the left or right side. Verify the results shown on the free-body diagram in Fig. 6–17b.

The free-body diagram of the right part of the truss, which is the easiest to analyze, is shown in Fig. 6–17c. There are three unknowns, F_{FG}, F_{CF}, and F_{CD}.

Equations of Equilibrium. We will apply the moment equation about point *O* in order to eliminate the two unknowns F_{FG} and F_{CD}. The location of point *O* measured from *E* can be determined from proportional triangles, i.e., $4/(4 + x) = 6/(8 + x)$, $x = 4$ m. Or, stated in another manner, since the slope of member *GF* has a drop of 2 m to a horizontal distance of 4 m, and *FD* is 4 m, Fig. 6–17c, then from *D* to *O* the distance must be 8 m.

An easy way to determine the moment of $\mathbf{F}_{CF}$ about point *O* is to use the principle of transmissibility and slide $\mathbf{F}_{CF}$ to point *C*, and then resolve $\mathbf{F}_{CF}$ into its two rectangular components. We have

$\zeta + \Sigma M_O = 0$;

$-F_{CF} \sin 45°(12 \text{ m}) + (3 \text{ kN})(8 \text{ m}) - (4.75 \text{ kN})(4 \text{ m}) = 0$

$F_{CF} = 0.589$ kN (C) *Ans.*

(c)

Fig. 6–17

EXAMPLE 6.7

Determine the force in member *EB* of the roof truss shown in Fig. 6–18a. Indicate whether the member is in tension or compression.

SOLUTION

Free-Body Diagrams. By the method of sections, any section that passes through *EB* will also have to pass through three other members for which the forces are unknown. For example, section *aa* passes through *ED*, *EB*, *FB*, and *AB*. If a free-body diagram of the left part of this section is considered, Fig. 6–18b, it is possible to obtain F_{ED} by summing moments about *B* to eliminate the other three unknowns; however, F_{EB} cannot be determined from the remaining two equilibrium equations.

One possible way of obtaining F_{EB} is first to determine F_{ED} from section *aa*, then use this result on section *bb*, Fig. 6–18a, which is shown in Fig. 6–18c. Here the force system is concurrent and our sectioned free-body diagram is the same as the free-body diagram for the joint at *E*.

Fig. 6–18

Equations of Equilibrium. In order to determine the moment of F_{ED} about point *B*, Fig. 6–18b, we will use the principle of transmissibility and slide this force to point *C* and then resolve it into its rectangular components as shown. Therefore,

$\zeta + \Sigma M_B = 0;$ $1000 \text{ N}(4 \text{ m}) + 3000 \text{ N}(2 \text{ m}) - 4000 \text{ N}(4 \text{ m})$
$+ F_{ED} \sin 30°(4 \text{ m}) = 0$

$F_{ED} = 3000 \text{ N} \quad (C)$

Considering now the free-body diagram of section *bb*, Fig. 6–18c, we have

$\xrightarrow{+} \Sigma F_x = 0;$ $F_{EF} \cos 30° - 3000 \cos 30° \text{ N} = 0$

$F_{EF} = 3000 \text{ N} \quad (C)$

$+\uparrow \Sigma F_y = 0;$ $2(3000 \sin 30° \text{ N}) - 1000 \text{ N} - F_{EB} = 0$

$F_{EB} = 2000 \text{ N} \quad (T) \qquad \qquad Ans.$

FUNDAMENTAL PROBLEMS

All solutions must include a free-body diagram.

F6–7. Determine the force in members *BC*, *CF*, and *FE* and state if the members are in tension or compression.

Prob. F6–7

F6–8. Determine the force in members *LK*, *KC*, and *CD* of the Pratt truss and state if the members are in tension or compression.

Prob. F6–8

F6–9. Determine the force in members *KJ*, *KD*, and *CD* of the Pratt truss and state if the members are in tension or compression.

Prob. F6–9

F6–10. Determine the force in members *EF*, *CF*, and *BC* of the truss and state if the members are in tension or compression.

Prob. F6–10

F6–11. Determine the force in members *GF*, *GD*, and *CD* of the truss and state if the members are in tension or compression.

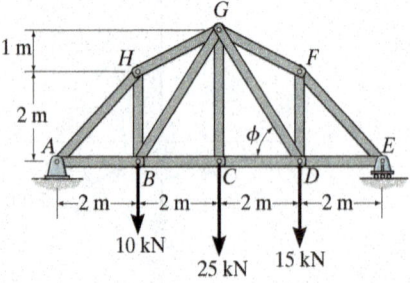

Prob. F6–11

F6–12. Determine the force in members *DC*, *HI*, and *JI* of the truss and state if the members are in tension or compression. *Suggestion:* Use the sections shown.

Prob. F6–12

PROBLEMS

All solutions must include a free-body diagram.

6–27. Determine the force in members *DC*, *HC*, and *HI* of the truss and state if the members are in tension or compression.

*6–28. Determine the force in members *ED*, *EH*, and *GH* of the truss and state if the members are in tension or compression.

6–31. Determine the force in members *CE*, *FE*, and *CD* and state if the members are in tension or compression. *Hint*: The force acting at the pin *G* is directed along member *GD*. Why?

*6–32. Determine the force in members *BC*, *FC*, and *FE* and state if the members are in tension or compression. *Hint*: The force acting at the pin *G* is directed along member *GD*. Why?

Probs. 6–31/32

Probs. 6–27/28

6–33. Determine the force in members *BC*, *HC*, and *HG* of the bridge truss and state if the members are in tension or compression.

6–29. Determine the force in members *HG*, *HE*, and *DE* of the truss and state if the members are in tension or compression.

6–30. Determine the force in members *CD*, *HI*, and *CH* of the truss and state if the members are in tension or compression.

Prob. 6–33

6–34. Determine the force in members *DE*, *DF*, and *GF* of the cantilevered truss and state if the members are in tension or compression.

Probs. 6–29/30

Prob. 6–34

6–35. Determine the force in members *GF*, *FC*, and *CD* of the bridge truss and state if the members are in tension or compression.

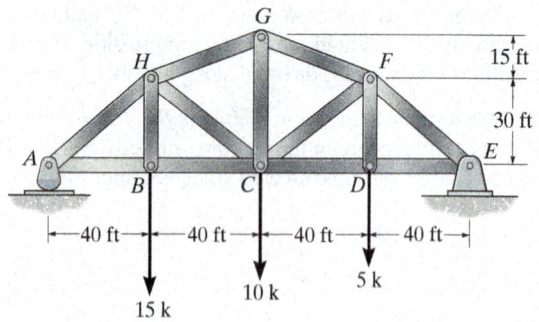

Prob. 6–35

***6–36.** For the given loading, determine the force in members *CD*, *CJ*, and *KJ* of the Howe roof truss and state if the members are in tension or compression.

Prob. 6–36

6–37. Determine the force in members *EF*, *BE*, *BC*, and *BF* of the truss and state if the members are in tension or compression. Set $P_1 = 9$ kN, $P_2 = 12$ kN, and $P_3 = 6$ kN.

6–38. Determine the force in members *BC*, *BE*, and *EF* of the truss and state if the members are in tension or compression. Set $P_1 = 6$ kN, $P_2 = 9$ kN, and $P_3 = 12$ kN.

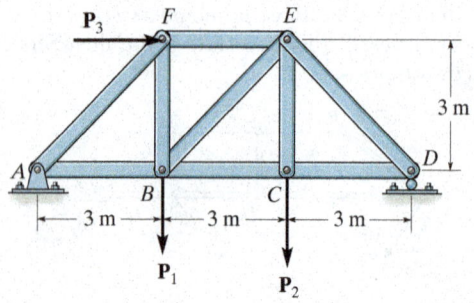

Probs. 6–37/38

6–39. Determine the force in members *FE*, *FB*, and *BC* of the truss and state if the members are in tension or compression. After sectioning the truss, solve for each force *directly* using a single equilibrium equation to obtain each force. Assume that all joints are pin connected.

Prob. 6–39

***6–40.** Determine the force in members *CF*, *FG*, and *GC* of the truss and state if the members are in tension or compression.

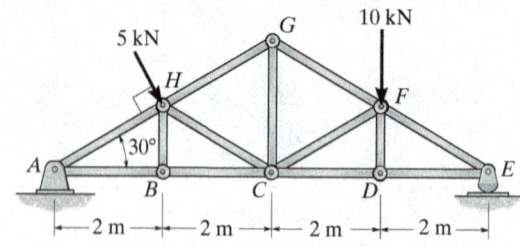

Prob. 6–40

6–41. Determine the force in members *BC*, *FC*, and *FE* and state if the members are in tension or compression.

Prob. 6–41

6–42. Determine the force in members *BC*, *HC*, and *HG* and state if the members are in tension or compression.

6–43. Determine the force in members *CD*, *CJ*, *GJ*, and *CG* and state if the members are in tension or compression.

Probs. 6–42/43

*6–44.** Determine the force in members *DE*, *JI*, and *DO* of the *K* truss and state if the members are in tension or compression. *Hint:* Use sections *aa* and *bb*.

6–45. Determine the force in members *GD* and *KJ* of the *K* truss and state if the members are in tension or compression.

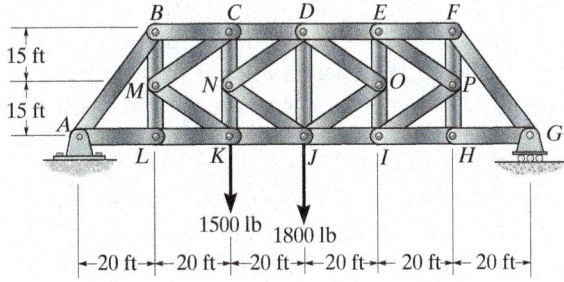

Probs. 6–44/45

6–46. The truss supports the vertical load of 600 N. Determine the force in members *BC*, *BG*, and *HG* as the dimension *L* varies. Plot the results of *F* (ordinate with tension as positive) versus *L* (abscissa) for $0 \le L \le 3$ m.

Prob. 6–46

6–47. Determine the force in members *GF*, *CF*, and *CD* of the roof truss and state if the members are in tension or compression.

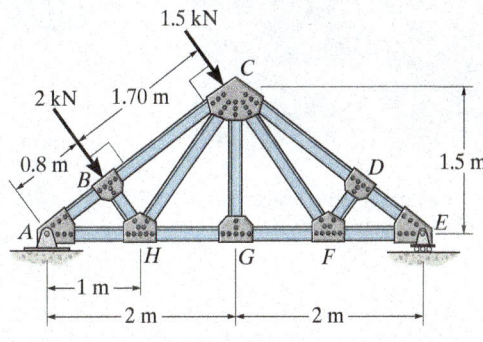

Prob. 6–47

*6–48.** Determine the force in members *JK*, *CJ*, and *CD* of the truss and state if the members are in tension or compression.

6–49. Determine the force in members *HI*, *FI*, and *EF* of the truss and state if the members are in tension or compression.

Probs. 6–48/49

*6.5 SPACE TRUSSES

A *space truss* consists of members joined together at their ends to form a stable three-dimensional structure. The simplest form of a space truss is a *tetrahedron*, constructed by connecting six members together, as shown in Fig. 6–19. Any additional members added to this basic element would be redundant in supporting the force **P**. A *simple space truss* can be built from this basic tetrahedral element by adding three additional members and a joint, and continuing in this manner to form a system of multiconnected tetrahedrons.

Fig. 6–19

Assumptions for Design. The members of a space truss may be treated as two-force members provided the external loading is applied at the joints and the joints consist of ball-and-socket connections. These assumptions are justified when the welded or bolted connections of the joined members intersect at a common point and the weight of the members can be neglected. In cases where the weight of a member is to be included in the analysis, it is generally satisfactory to apply it as a vertical force, half of its magnitude applied at each end of the member.

Typical roof-supporting space truss. Notice the use of ball-and-socket joints for the connections.

For economic reasons, large electrical transmission towers are often constructed using space trusses.

PROCEDURE FOR ANALYSIS

Either the method of joints or the method of sections can be used to determine the forces developed in the members of a simple space truss.

Method of Joints.

If the forces in *all* the members of the truss are to be determined, then the method of joints is most suitable for the analysis. Here it is necessary to apply the three equilibrium equations $\Sigma F_x = 0$, $\Sigma F_y = 0$, $\Sigma F_z = 0$ to the forces acting at each joint. The solution of many simultaneous equations can be avoided if the force analysis begins at a joint having at least one known force and at most three unknown forces. Also, if the three-dimensional geometry of the force system at the joint is hard to visualize, it is recommended that a Cartesian vector analysis be used for the solution.

Method of Sections.

If only a *few* member forces are to be determined, the method of sections can be used. When an imaginary section is passed through a truss and the truss is separated into two parts, the force system acting on one of the parts must satisfy the *six* equilibrium equations: $\Sigma F_x = 0$, $\Sigma F_y = 0$, $\Sigma F_z = 0$, $\Sigma M_x = 0$, $\Sigma M_y = 0$, $\Sigma M_z = 0$ (Eqs. 5–6). By proper choice of the section and axes for summing forces and moments, many of the unknown member forces in a space truss can be calculated *directly*, using a single equilibrium equation.

EXAMPLE 6.8

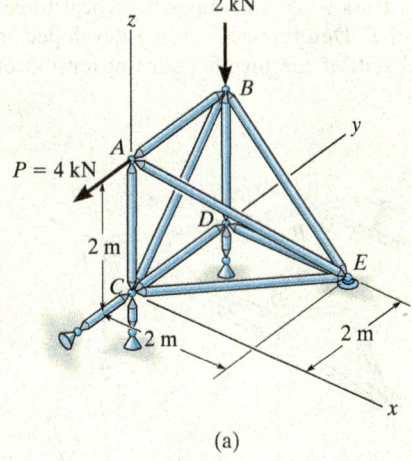

(a)

Determine the forces acting in the members of the space truss shown in Fig. 6–20a. Indicate whether the members are in tension or compression.

SOLUTION
Since there are one known force and three unknown forces acting at joint A, the analysis of the truss will begin at this joint.

Joint A. (Fig. 6–20b). Expressing each force acting on the free-body diagram of joint A as a Cartesian vector, we have

$$\mathbf{P} = \{-4\mathbf{j}\} \text{ kN} \quad \mathbf{F}_{AB} = F_{AB}\mathbf{j} \quad \mathbf{F}_{AC} = -F_{AC}\mathbf{k}$$

$$\mathbf{F}_{AE} = F_{AE}\left(\frac{\mathbf{r}_{AE}}{r_{AE}}\right) = F_{AE}(0.577\mathbf{i} + 0.577\mathbf{j} - 0.577\mathbf{k})$$

For equilibrium,

$$\Sigma \mathbf{F} = 0; \quad \mathbf{P} + \mathbf{F}_{AB} + \mathbf{F}_{AC} + \mathbf{F}_{AE} = 0$$

$$-4\mathbf{j} + F_{AB}\mathbf{j} - F_{AC}\mathbf{k} + 0.577F_{AE}\mathbf{i} + 0.577F_{AE}\mathbf{j} - 0.577F_{AE}\mathbf{k} = 0$$

$\Sigma F_x = 0; \quad 0.577 F_{AE} = 0$

$\Sigma F_y = 0; \quad -4 + F_{AB} + 0.577 F_{AE} = 0$

$\Sigma F_z = 0; \quad -F_{AC} - 0.577 F_{AE} = 0$

$F_{AC} = F_{AE} = 0$ *Ans.*

$F_{AB} = 4 \text{ kN (T)}$ *Ans.*

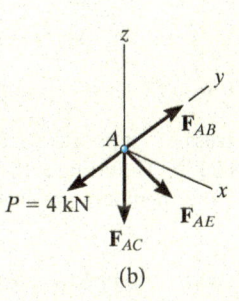

(b)

Since F_{AB} is known, joint B can be analyzed next.

Joint B. (Fig. 6–20c).

$\Sigma F_x = 0; \quad F_{BE} \dfrac{1}{\sqrt{2}} = 0$

$\Sigma F_y = 0; \quad -4 + F_{CB} \dfrac{1}{\sqrt{2}} = 0$

$\Sigma F_z = 0; \quad -2 + F_{BD} - F_{BE} \dfrac{1}{\sqrt{2}} + F_{CB} \dfrac{1}{\sqrt{2}} = 0$

$F_{BE} = 0 \quad F_{CB} = 5.65 \text{ kN (C)} \quad F_{BD} = 2 \text{ kN (T)}$ *Ans.*

(c)

Fig. 6–20

The *scalar* equations of equilibrium can now be applied to the forces acting on the free-body diagrams of joints D and C. Show that

$$F_{DE} = F_{DC} = F_{CE} = 0 \quad \textit{Ans.}$$

PROBLEMS

All solutions must include a free-body diagram.

6–50. Determine the force in each member of the three-member space truss and state if the members are in tension or compression.

Prob. 6–50

6–51. Determine the force in each member of the space truss and state if the members are in tension or compression. The truss is supported by short links at A, B, and C.

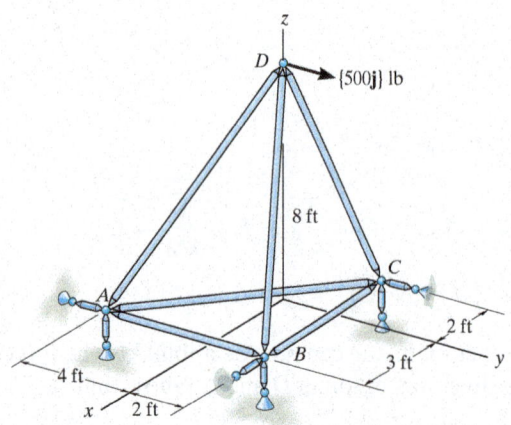

Prob. 6–51

*6–52. The space truss is used to support vertical forces at joints D, E, and F. Determine the force developed in each member and state if the members are in tension or compression.

Prob. 6–52

6–53. The space truss supports a force $\mathbf{F} = \{-500\mathbf{i} + 600\mathbf{j} + 400\mathbf{k}\}$ lb. Determine the force in each member and state if the members are in tension or compression.

6–54. The space truss supports a force $\mathbf{F} = \{600\mathbf{i} + 450\mathbf{j} - 750\mathbf{k}\}$ lb. Determine the force in each member and state if the members are in tension or compression.

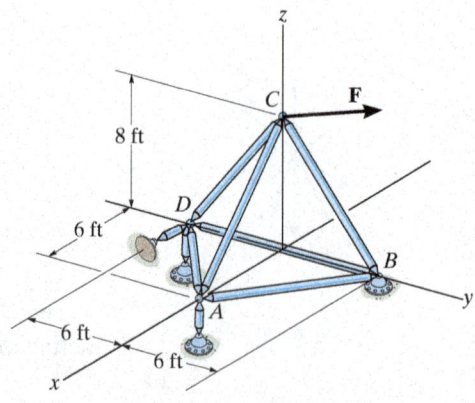

Probs. 6–53/54

6–55. Determine the force in each member of the space truss and state if the members are in tension or compression. The truss is supported by a ball-and-socket joint at A and short links at B and C.

6–57. Determine the force in members BD, AD, and AF of the space truss and state if the members are in tension or compression. The truss is supported by short links at $A, B, D,$ and F.

6–58. Determine the force in members CF, EF, and DF of the space truss and state if the members are in tension or compression. The truss is supported by short links at $A, B, D,$ and F.

Prob. 6–55

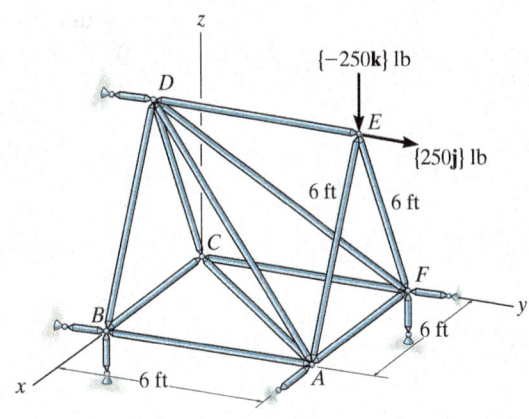

Probs. 6–57/58

6–59. Determine the force in each member of the space truss and state if the members are in tension or compression. The truss is supported by ball-and-socket joints at A, B, and E. Set $\mathbf{F} = \{800\mathbf{j}\}$ N. *Hint:* The support reaction at E acts along member EC. Why?

*6–56. Determine the force in each member of the space truss and state if the members are in tension or compression.

*6–60. Determine the force in each member of the space truss and state if the members are in tension or compression. The truss is supported by ball-and-socket joints at $A, B,$ and E. Set $\mathbf{F} = \{-200\mathbf{i} + 400\mathbf{j}\}$ N. *Hint:* The support reaction at E acts along member EC. Why?

Prob. 6–56

Probs. 6–59/60

This crane is a typical example of a framework.

6.6 FRAMES AND MACHINES

Frames and machines are two types of structures that are often composed of pin-connected **multiforce members**, i.e., members that are subjected to more than two forces. **Frames** are used to support loads, whereas **machines** contain moving parts and are designed to transmit and alter the effect of forces. Provided a frame or machine contains no more supports or members than are necessary to prevent its collapse, the forces acting at the joints and supports can be determined by applying the equations of equilibrium to each of its members. Once these forces are obtained, it is then possible to *design* the size of the members, connections, and supports using the theory of mechanics of materials and an appropriate engineering design code.

Free-Body Diagrams.
In order to determine the forces acting at the joints and supports of a frame or machine, the structure must be disassembled and the free-body diagrams of its parts must be drawn. The following important points *must* be observed:

- Isolate each part by drawing its *outlined shape*. Then show all the forces and/or couple moments that act on the part. Make sure to *label* or *identify* each known and unknown force and couple moment with reference to an established x, y coordinate system. Also, indicate any dimensions used for taking moments. As usual, the sense of an unknown force or couple moment can be assumed.

- Identify all the two-force members in the structure and represent their free-body diagrams as having two equal but opposite collinear forces acting at their points of application. (See Sec. 5.4.) By doing this we can avoid solving an unnecessary number of equilibrium equations.

- Forces common to *any* two *contacting* members act with equal magnitudes but opposite sense on the free-body diagrams of the respective members.

Common tools such as these pliers act as simple machines. Here the applied force on the handles creates a much larger force at the jaws.

The following examples graphically illustrate how to draw the free-body diagrams of a dismembered frame or machine. In all cases, the weight of the members is neglected.

EXAMPLE 6.9

For the frame shown in Fig. 6–21a, draw the free-body diagram of (a) each member, (b) the pins at B and A, and (c) the two members connected together.

(a)   (b)   (c)

SOLUTION

Part (a). By inspection, members BA and BC are *not* two-force members. Instead, as shown on the free-body diagrams, Fig. 6–21b, BC is subjected to a force from each of the pins at B and C and the external force **P**. Likewise, AB is subjected to a force from each of the pins at A and B and the external couple moment **M**. The pin forces are represented by their x and y components.

Part (b). The pin at B is subjected to only *two forces*, i.e., the force of member BC and the force of member AB. For *equilibrium* these forces (or their respective components) must be equal but opposite, Fig. 6–21c. Notice that Newton's third law is applied between the pin and its connected members, i.e., the effect of the pin on the two members, Fig. 6–21b, and the equal but opposite effect of the two members on the pin, Fig. 6–21c. In the same manner, there are three forces on pin A, Fig. 6–21d, caused by the force components of member AB and each of the two pin leafs.

Part (c). The free-body diagram of both members connected together, yet removed from the supporting pins at A and C, is shown in Fig. 6–21e. The force components B_x and B_y are *not* shown on this diagram since they are *internal* forces (Fig. 6–21b) and therefore cancel out. Also, to be consistent when later applying the equilibrium equations, the unknown force components at A and C in Fig. 6–21c must act in the *same sense* as those shown in Fig. 6–21b.

Pin A

(d)

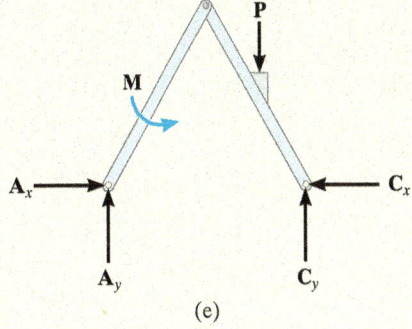

(e)

Fig. 6–21

EXAMPLE 6.10

A constant tension in the conveyor belt is maintained by using the device shown in Fig. 6–22a. Draw the free-body diagrams of the frame and the cylinder (or pulley) that the belt surrounds. The suspended block has a weight of W.

(b)

(c)

(d)

Fig. 6–22

(a)

SOLUTION

The idealized model of the device is shown in Fig. 6–22b. Here the angle θ is assumed to be known. From this model, the free-body diagrams of the pulley and frame are shown in Figs. 6–22c and 6–22d, respectively. Note that the force components $\mathbf{B}_x$ and $\mathbf{B}_y$ that the pin at B exerts on the pulley must be equal but opposite to the ones acting on the frame. See Fig. 6–21c of Example 6.9.

6.6 FRAMES AND MACHINES

EXAMPLE 6.11

For the frame shown in Fig. 6–23a, draw the free-body diagrams of (a) the entire frame including the pulleys and cords, (b) the frame without the pulleys and cords, and (c) each of the pulleys.

(a)

SOLUTION

Part (a). When the entire frame including the pulleys and cords is considered, the interactions at the points where the pulleys and cords are connected to the frame become pairs of *internal* forces which cancel each other and therefore are not shown on the free-body diagram, Fig. 6–23b.

Part (b). When the cords and pulleys are removed, their effect *on the frame* must be shown, Fig. 6–23c.

Part (c). The force components B_x, B_y, C_x, C_y of the pins on the pulleys, Fig. 6–23d, are equal but opposite to the force components exerted by the pins on the frame, Fig. 6–23c. See Example 6.9.

Fig. 6–23

EXAMPLE 6.12

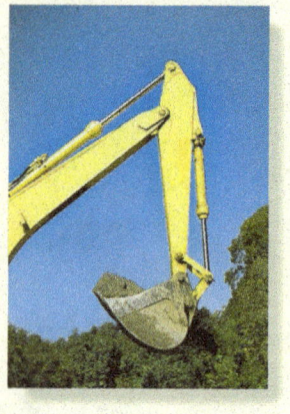

(a)

Draw the free-body diagrams of the members of the backhoe, shown in the photo, Fig. 6–24a. The bucket and its contents have a weight W.

SOLUTION
The idealized model of the assembly is shown in Fig. 6–24b. By inspection, members $AB, BC, BE,$ and HI are all two-force members since they are pin connected at their end points and no other forces act on them. The free-body diagrams of the bucket and the stick are shown in Fig. 6–24c. Note that pin C is subjected to only two forces, whereas the pin at B is subjected to three forces, Fig. 6–24d. The free-body diagram of the entire assembly is shown in Fig. 6–24e.

(b)

(c)

(d)

(e)

Fig. 6–24

EXAMPLE 6.13

Draw the free-body diagram of each part of the smooth piston and link mechanism used to crush recycled cans, Fig. 6–25a.

SOLUTION

By inspection, member AB is a two-force member. The free-body diagrams of the three parts are shown in Fig. 6–25b. Since the pins at B and D connect only two parts together, the forces there are shown as equal but opposite on the separate free-body diagrams of their connected members. In particular, four components of force act on the piston: D_x and D_y represent the effect of the pin (or lever EBD), N_w is the *resultant force* of the wall support, and **P** is the resultant compressive force caused by the can C. The directional sense of each of the unknown forces is assumed, and the correct sense will be established after the equations of equilibrium are applied.

NOTE: A free-body diagram of the entire assembly is shown in Fig. 6–25c. Here the forces between the components are internal and are not shown on the free-body diagram.

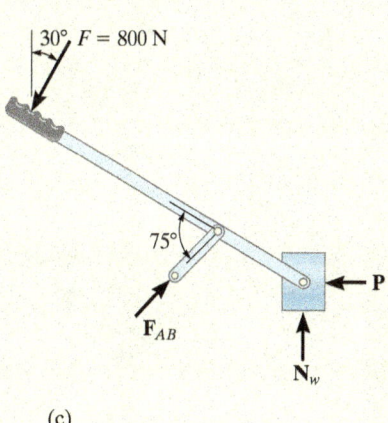

Fig. 6–25

Before proceeding, it is highly recommended that you cover the solutions of these examples and attempt to draw the requested free-body diagrams. When doing so, make sure the work is neat and that all the forces and couple moments are properly labeled.

PROCEDURE FOR ANALYSIS

The joint reactions on frames or machines (structures) composed of multiforce members can be determined using the following procedure.

Free-Body Diagram.
- Draw the free-body diagram of the entire frame or machine, a portion of it, or each of its members. The choice should be made so that it leads to the most direct solution of the problem.

- Identify the two-force members. Remember that regardless of their shape, they have equal but opposite collinear forces acting at their ends.

- When the free-body diagram of a group of members of a frame or machine is drawn, the forces between the connected parts of this group are internal forces and are not shown on the free-body diagram of the group.

- Forces common to two members which are in contact act with equal magnitude but opposite sense on the respective free-body diagrams of the members.

- In many cases it is possible to tell by inspection the proper sense of the unknown forces acting on a member; however, if this seems difficult, the sense can be assumed.

- Remember that once the free-body diagram is drawn, a couple moment is a free vector and can act at any point on the diagram. Also, a force is a sliding vector and can act at any point along its line of action.

Equations of Equilibrium.
- Count the number of unknowns and compare it to the total number of equilibrium equations that are available. In two dimensions, there are three equilibrium equations that can be written for each member.

- Sum moments about a point that lies at the intersection of the lines of action of as many of the unknown forces as possible.

- If the solution of a force or couple moment is found to be a negative scalar, it means the sense of the force is the reverse of that shown on the free-body diagram.

EXAMPLE 6.14

Determine the tension in the cables and also the force **P** required to support the 600-N force using the frictionless pulley system shown in Fig. 6–26a.

(a) (b)

Fig. 6–26

SOLUTION

Free-Body Diagram. A free-body diagram of each pulley *including* its pin and a portion of the contacting cable is shown in Fig. 6–26b. Since the cable is *continuous*, it has a *constant tension P* acting throughout its length. The link connection between pulleys B and C is a two-force member, and therefore it has an unknown tension T acting on it. Notice that the *principle of action, equal but opposite reaction* must be carefully observed for forces **P** and **T** when the *separate* free-body diagrams are drawn.

Equations of Equilibrium. The three unknowns are obtained as follows:

Pulley A
$+\uparrow \Sigma F_y = 0;$ $3P - 600 \text{ N} = 0;$ $P = 200 \text{ N}$ *Ans.*

Pulley B
$+\uparrow \Sigma F_y = 0;$ $T - 2P = 0;$ $T = 400 \text{ N}$ *Ans.*

Pulley C
$+\uparrow \Sigma F_y = 0;$ $R - 2P - T = 0;$ $R = 800 \text{ N}$ *Ans.*

EXAMPLE 6.15

The 500-kg elevator car in Fig. 6–27a is being hoisted at constant speed by motor A using the pulley system shown. Determine the force developed in the two cables.

(a)　　　　(b)

Fig. 6–27

SOLUTION

Free-Body Diagram. We can solve this problem using the free-body diagrams of the elevator car and pulley C, Fig. 6–27b. The tensile forces developed in the two cables are denoted as T_1 and T_2.

Equations of Equilibrium. For pulley C,

$+\uparrow \Sigma F_y = 0; \quad T_2 - 2T_1 = 0 \quad \text{or} \quad T_2 = 2T_1$ 　　(1)

For the elevator car,

$+\uparrow \Sigma F_y = 0; \quad 3T_1 + 2T_2 - 500(9.81) \text{ N} = 0$ 　　(2)

Substituting Eq. 1 into Eq. 2 yields

$$3T_1 + 2(2T_1) - 500(9.81) \text{ N} = 0$$

$$T_1 = 700.71 \text{ N} = 701 \text{ N} \qquad \textit{Ans.}$$

Substituting this result into Eq. 1,

$$T_2 = 2(700.71) \text{ N} = 1401 \text{ N} = 1.40 \text{ kN} \qquad \textit{Ans.}$$

EXAMPLE 6.16

Determine the horizontal and vertical components of force which the pin at C exerts on member BC of the frame in Fig. 6–28a.

SOLUTION I

Free-Body Diagrams. By inspection it can be seen that AB is a two-force member. The free-body diagrams are shown in Fig. 6–28b.

Equations of Equilibrium. The *three unknowns* can be determined by applying the three equations of equilibrium to member BC.

$\zeta + \Sigma M_C = 0$; $2000 \text{ N}(2 \text{ m}) - (F_{AB} \sin 60°)(4 \text{ m}) = 0$; $F_{AB} = 1154.7 \text{ N}$

$\xrightarrow{+} \Sigma F_x = 0$; $1154.7 \cos 60° \text{ N} - C_x = 0$; $C_x = 577 \text{ N}$ *Ans.*

$+\uparrow \Sigma F_y = 0$; $1154.7 \sin 60° \text{ N} - 2000 \text{ N} + C_y = 0$

$C_y = 1000 \text{ N}$ *Ans.*

SOLUTION II

Free-Body Diagrams. If one does not recognize that AB is a two-force member, then more work is involved in solving this problem. The free-body diagrams are shown in Fig. 6–28c.

Equations of Equilibrium. The *six unknowns* are determined by applying the three equations of equilibrium to each member.

Member AB

$\zeta + \Sigma M_A = 0$; $B_x(3 \sin 60° \text{ m}) - B_y(3 \cos 60° \text{ m}) = 0$ (1)

$\xrightarrow{+} \Sigma F_x = 0$; $A_x - B_x = 0$ (2)

$+\uparrow \Sigma F_y = 0$; $A_y - B_y = 0$ (3)

Member BC

$\zeta + \Sigma M_C = 0$; $2000 \text{ N}(2 \text{ m}) - B_y(4 \text{ m}) = 0$ (4)

$\xrightarrow{+} \Sigma F_x = 0$; $B_x - C_x = 0$ (5)

$+\uparrow \Sigma F_y = 0$; $B_y - 2000 \text{ N} + C_y = 0$ (6)

The results for C_x and C_y can be determined by solving these equations in the following sequence: 4, 1, 5, then 6. The results are

$B_y = 1000 \text{ N}$
$B_x = 577 \text{ N}$
$C_x = 577 \text{ N}$ *Ans.*
$C_y = 1000 \text{ N}$ *Ans.*

By comparison, Solution I is simpler since the requirement that F_{AB} in Fig. 6–28b be equal, opposite, and collinear at the ends of member AB automatically satisfies Eqs. 1, 2, and 3 above, and therefore eliminates the need to write these equations. *As a result, save yourself some time and effort by always identifying the two-force members before starting the analysis!*

Fig. 6–28

EXAMPLE 6.17

The compound beam shown in Fig. 6–29a is pin connected at B. Determine the components of reaction at its supports. Neglect its weight and thickness.

Fig. 6–29

SOLUTION

Free-Body Diagrams. By inspection, if we consider a free-body diagram of the *entire beam ABC*, there will be three unknown reactions at A and one at C. These four unknowns cannot all be obtained from the three available equations of equilibrium, and so for the solution it will become necessary to separate the beam into its two members as shown in Fig. 6–29b.

Equations of Equilibrium. The six unknowns are determined as follows:

Segment BC

$\xrightarrow{+} \Sigma F_x = 0;\quad B_x = 0$

$\zeta + \Sigma M_B = 0;\quad -8 \text{ kN}(1 \text{ m}) + C_y(2 \text{ m}) = 0$

$+\uparrow \Sigma F_y = 0;\quad B_y - 8 \text{ kN} + C_y = 0$

Segment AB

$\xrightarrow{+} \Sigma F_x = 0;\quad A_x - (10 \text{ kN})(\tfrac{3}{5}) + B_x = 0$

$\zeta + \Sigma M_A = 0;\quad M_A - (10 \text{ kN})(\tfrac{4}{5})(2 \text{ m}) - B_y(4 \text{ m}) = 0$

$+\uparrow \Sigma F_y = 0;\quad A_y - (10 \text{ kN})(\tfrac{4}{5}) - B_y = 0$

Solving each of these equations successively, using previously calculated results, we obtain

$A_x = 6 \text{ kN} \qquad A_y = 12 \text{ kN} \qquad M_A = 32 \text{ kN} \cdot \text{m}$ *Ans.*

$B_x = 0 \qquad B_y = 4 \text{ kN}$

$C_y = 4 \text{ kN}$ *Ans.*

EXAMPLE 6.18

The two planks in Fig. 6–30a are connected together by cable BC and a smooth spacer DE. Determine the reactions at the smooth supports A and F, and also find the force developed in the cable and spacer.

Fig. 6–30

SOLUTION

Free-Body Diagrams. The free-body diagram of each plank is shown in Fig. 6–30b. It is important to apply Newton's third law to the interaction forces F_{BC} and F_{DE} as shown.

Equations of Equilibrium. For plank AD,

$\zeta + \Sigma M_A = 0;$ $F_{DE}(6 \text{ ft}) - F_{BC}(4 \text{ ft}) - 100 \text{ lb}(2 \text{ ft}) = 0$

For plank CF,

$\zeta + \Sigma M_F = 0;$ $F_{DE}(4 \text{ ft}) - F_{BC}(6 \text{ ft}) + 200 \text{ lb}(2 \text{ ft}) = 0$

Solving simultaneously,

$F_{DE} = 140 \text{ lb}$ $F_{BC} = 160 \text{ lb}$ *Ans.*

Using these results, for plank AD,

$+\uparrow \Sigma F_y = 0;$ $N_A + 140 \text{ lb} - 160 \text{ lb} - 100 \text{ lb} = 0$

$N_A = 120 \text{ lb}$ *Ans.*

And for plank CF,

$+\uparrow \Sigma F_y = 0;$ $N_F + 160 \text{ lb} - 140 \text{ lb} - 200 \text{ lb} = 0$

$N_F = 180 \text{ lb}$ *Ans.*

NOTE: Draw the free-body diagram of the system of both planks and apply $\Sigma M_A = 0$ to determine N_F. Then use the free-body diagram of CEF to determine F_{DE} and F_{BC}.

EXAMPLE 6.19

The 75-kg man in Fig. 6–31a attempts to lift the 40-kg uniform beam off the roller support at B. Determine the tension developed in the cable attached to B and the normal reaction of the man on the beam when this is about to occur.

SOLUTION

Free-Body Diagrams. The tensile force in the cable will be denoted as T_1. The free-body diagrams of the pulley E, the man, and the beam are shown in Fig. 6–31b. Since the man must lift the beam off the roller B then $N_B = 0$. When drawing each of these diagrams, it is very important to apply Newton's third law.

Equations of Equilibrium. Using the free-body diagram of pulley E,

$+\uparrow \Sigma F_y = 0; \quad 2T_1 - T_2 = 0 \quad \text{or} \quad T_2 = 2T_1$ \quad (1)

Referring to the free-body diagram of the man using this result,

$+\uparrow \Sigma F_y = 0; \quad N_m + 2T_1 - 75(9.81) \text{ N} = 0$ \quad (2)

Summing moments about point A on the beam,

$\zeta + \Sigma M_A = 0; \quad T_1(3 \text{ m}) - N_m(0.8 \text{ m}) - [40(9.81) \text{ N}](1.5 \text{ m}) = 0$ \quad (3)

Solving Eqs. 2 and 3 simultaneously for T_1 and N_m, then using Eq. (1) for T_2, we obtain

$T_1 = 256 \text{ N} \quad N_m = 224 \text{ N} \quad T_2 = 512 \text{ N}$ \quad *Ans.*

SOLUTION II

A direct solution for T_1 can be obtained by considering the beam, the man, and pulley E as a *single system*. The free-body diagram is shown in Fig. 6–31c. Thus,

$\zeta + \Sigma M_A = 0; \quad 2T_1(0.8 \text{ m}) - [75(9.81) \text{ N}](0.8 \text{ m})$

$- [40(9.81) \text{ N}](1.5 \text{ m}) + T_1(3 \text{ m}) = 0$

$T_1 = 256 \text{ N}$ \quad *Ans.*

With this result Eqs. 1 and 2 can then be used to find N_m and T_2.

Fig. 6–31

EXAMPLE 6.20

The smooth disk shown in Fig. 6–32a is pinned at D and has a weight of 20 lb. Neglecting the weights of the other members, determine the horizontal and vertical components of reaction at pins B and D.

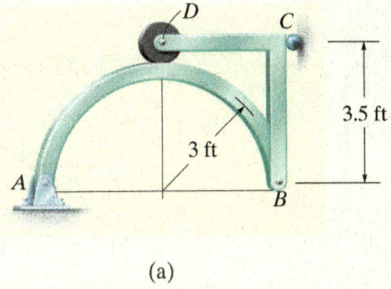

(a)

SOLUTION

Free-Body Diagrams. The free-body diagrams of the entire frame and each of its members are shown in Fig. 6–32b.

Equations of Equilibrium. The eight unknowns can of course be obtained by applying the eight equilibrium equations to each member—three to member AB, three to member BCD, and two to the disk. (Moment equilibrium is automatically satisfied for the disk.) If this is done, however, all the results can be obtained only from a simultaneous solution of some of the equations. (Try it and find out.) To avoid this situation, it is best first to determine the three support reactions on the *entire* frame; then, using these results, the remaining five equilibrium equations can be applied to two other parts in order to solve successively for the other unknowns.

Entire Frame

$\zeta + \Sigma M_A = 0;$ $-20 \text{ lb } (3 \text{ ft}) + C_x(3.5 \text{ ft}) = 0;$ $C_x = 17.1 \text{ lb}$

$\stackrel{+}{\rightarrow} \Sigma F_x = 0;$ $A_x - 17.1 \text{ lb} = 0;$ $A_x = 17.1 \text{ lb}$

$+\uparrow \Sigma F_y = 0;$ $A_y - 20 \text{ lb} = 0;$ $A_y = 20 \text{ lb}$

Member AB

$\stackrel{+}{\rightarrow} \Sigma F_x = 0;$ $17.1 \text{ lb} - B_x = 0;$ $B_x = 17.1 \text{ lb}$ *Ans.*

$\zeta + \Sigma M_B = 0;$ $-20 \text{ lb } (6 \text{ ft}) + N_D(3 \text{ ft}) = 0;$ $N_D = 40 \text{ lb}$

$+\uparrow \Sigma F_y = 0;$ $20 \text{ lb} - 40 \text{ lb} + B_y = 0;$ $B_y = 20 \text{ lb}$ *Ans.*

Disk

$\stackrel{+}{\rightarrow} \Sigma F_x = 0;$ $D_x = 0$ *Ans.*

$+\uparrow \Sigma F_y = 0;$ $40 \text{ lb} - 20 \text{ lb} - D_y = 0;$ $D_y = 20 \text{ lb}$ *Ans.*

(b)

Fig. 6–32

EXAMPLE 6.21

The frame in Fig. 6–33a supports the 50-kg cylinder. Determine the horizontal and vertical components of reaction at A and the force at C.

Fig. 6–33

SOLUTION

Free-Body Diagrams. The free-body diagram of pulley D, along with the cylinder and a portion of the cord (a system), is shown in Fig. 6–33b. Member BC is a two-force member as indicated by its free-body diagram. The free-body diagram of member ABD is also shown.

Equations of Equilibrium. We will begin by analyzing the equilibrium of the pulley. The moment equation of equilibrium is automatically satisfied with $T = 50(9.81)$ N, and so

$\xrightarrow{+} \Sigma F_x = 0; \quad D_x - 50(9.81) \text{ N} = 0; \quad D_x = 490.5 \text{ N}$

$+\uparrow \Sigma F_y = 0; \quad D_y - 50(9.81) \text{ N} = 0; \quad D_y = 490.5 \text{ N}$

Using these results, F_{BC} can be determined by summing moments about point A on member ABD.

$\zeta + \Sigma M_A = 0; \quad F_{BC}(0.6 \text{ m}) + 490.5 \text{ N}(0.9 \text{ m}) - 490.5 \text{ N}(1.20 \text{ m}) = 0$

$F_{BC} = 245.25 \text{ N}$ *Ans.*

Now A_x and A_y can be determined by summing forces.

$\xrightarrow{+} \Sigma F_x = 0; \quad A_x - 245.25 \text{ N} - 490.5 \text{ N} = 0; \quad A_x = 736 \text{ N}$ *Ans.*

$+\uparrow \Sigma F_y = 0; \quad A_y - 490.5 \text{ N} = 0; \quad A_y = 490.5 \text{ N}$ *Ans.*

EXAMPLE 6.22

Determine the force the pins at A and B exert on the two-member frame shown in Fig. 6–34a.

SOLUTION I

Free-Body Diagrams. By inspection AB and BC are two-force members. Their free-body diagrams, along with that of the pulley, are shown in Fig. 6–34b. In order to solve this problem we must also include the free-body diagram of the pin at B because this pin connects all *three* members together, Fig. 6–34c.

Equations of Equilibrium: Apply the equations of force equilibrium to pin B.

$\xrightarrow{+} \Sigma F_x = 0; \quad F_{BA} - 800 \text{ N} = 0; \quad F_{BA} = 800 \text{ N}$ *Ans.*

$+\uparrow \Sigma F_y = 0; \quad F_{BC} - 800 \text{ N} = 0; \quad F_{BC} = 800 \text{ N}$ *Ans.*

NOTE: The free-body diagram of the pin at A, Fig. 6–34d, indicates how the force F_{AB} is balanced by the force $(F_{AB}/2)$ exerted on the pin by each of the two pin leaves.

SOLUTION II

Free-Body Diagram. If we realize that AB and BC are two-force members, then the free-body diagram of the *entire frame* produces an easier solution, Fig. 6–34e. The force equations of equilibrium are the same as those above. Note that moment equilibrium will be satisfied, regardless of the radius of the pulley.

Fig. 6–34

FUNDAMENTAL PROBLEMS

All solutions must include a free-body diagram.

F6–13. Determine the force P needed to hold the 60-lb weight in equilibrium.

F6–15. If a 100-N force is applied to the handles of the pliers, determine the clamping force exerted on the smooth pipe B and the magnitude of the resultant force that one of the members exerts on pin A.

Prob. F6–13

Prob. F6–15

F6–14. Determine the horizontal and vertical components of reaction at pin C.

F6–16. Determine the horizontal and vertical components of reaction at pin C.

Prob. F6–14

Prob. F6–16

F6–17. Determine the normal force that the 100-lb plate A exerts on the 30-lb plate B.

F6–19. Determine the components of reaction at A and B.

Prob. F6–17

Prob. F6–19

F6–18. Determine the force P needed to lift the load. Also, determine the proper placement x of the hook for equilibrium. Neglect the weight of the beam.

F6–20. Determine the reactions at D.

Prob. F6–18

Prob. F6–20

F6–21. Determine the components of reaction at A and C.

F6–23. Determine the components of reaction at E.

Prob. F6–23

Prob. F6–21

F6–24. Determine the components of reaction at D and the components of reaction the pin at A exerts on member BA.

F6–22. Determine the components of reaction at C.

Prob. F6–22

Prob. F6–24

PROBLEMS

All solutions must include a free-body diagram.

6–61. Determine the force **P** required to hold the 100-lb weight in equilibrium.

Prob. 6–61

6–62. Determine the force **P** needed to suspend the 100-lb weight. Each pulley has a weight of 10 lb. Also, what are the cord reactions at *A* and *B*?

6–63. If each cord can support a maximum tension of 500 lb determine the largest weight that can be supported by the pulley system. Each pulley has a weight of 10 lb.

Probs. 6–62/63

*6–64. Determine the force **P** required to hold the 50-kg mass in equilibrium.

Prob. 6–64

6–65. Determine the horizontal and vertical components of force at pins *A* and *C* of the two-member frame.

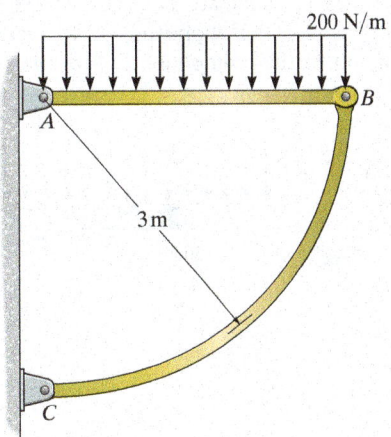

Prob. 6–65

6–66. Determine the horizontal and vertical components of force at pins A and D.

Prob. 6–66

6–67. Determine the horizontal and vertical components of force acting at the pins A, B, and C of the frame.

Prob. 6–67

***6–68.** The bridge frame consists of three parts which can be considered pinned at A, D, and E, rocker supported at C and F, and roller supported at B. Determine the horizontal and vertical components of reaction at all these supports.

Prob. 6–68

6–69. Determine the reactions at supports A and B.

Prob. 6–69

6–70. Determine the horizontal and vertical components of force at each pin. The suspended cylinder has a weight of 80 lb.

Prob. 6–70

6–71. Determine the horizontal and vertical components of force that the pins C and D exert on member ECD of the frame.

*6–72. Determine the horizontal and vertical components of force that the pins at B and D exert on member ABD of the frame.

6–74. The two ends of the spanner wrench fit loosely into the smooth slots of the bolt head. Determine the required force P on the handle in order to develop a torque of $M = 50\ \text{N} \cdot \text{m}$ on the bolt. Also, what is the resultant force on the pin at B?

6–75. The two ends of the spanner wrench fit loosely into the smooth slots of the bolt head. Determine the torque M on the bolt and the resultant force on the pin at B when a force of $P = 80\ \text{N}$ is applied to the handle.

Probs. 6–71/72

Probs. 6–74/75

*6–76. Determine the horizontal and vertical components of force which the pins at A and B exert on the frame.

6–73. Determine the compressive force exerted on the stone by a vertical load of 50 N applied to the toggle press.

Prob. 6–73

Prob. 6–76

6–77. Determine the reactions at the supports. The pin, attached to member BCD, passes through a smooth slot in member AB.

6–79. When a force of 2 lb is applied to the handles of the brad squeezer, it pulls in the smooth rod AB. Determine the force P exerted on each of the smooth brads at C and D.

Prob. 6–77

Prob. 6–79

6–78. The compound beam is pin supported at B and supported by rockers at A and C. There is a hinge (pin) at D. Determine the reactions at the supports.

***6–80.** A force of 4 lb is applied to the handle of the toggle clamp. Determine the normal force F that the bolt exerts on the clamp.

Prob. 6–78

Prob. 6–80

6–81. The hoist supports the 125-kg engine. Determine the force this load creates along member DB and along member FB, which contains the hydraulic cylinder H.

6–83. Determine the horizontal and vertical components of force that the pins exert on member ABD.

Prob. 6–81

Prob. 6–83

6–82. The compound beam is fixed at E and supported by rollers at A and B. Determine the reactions at these supports. There are pins at C and D.

*6–84. Determine the horizontal and vertical components of force acting on pins A and C.

Prob. 6–82

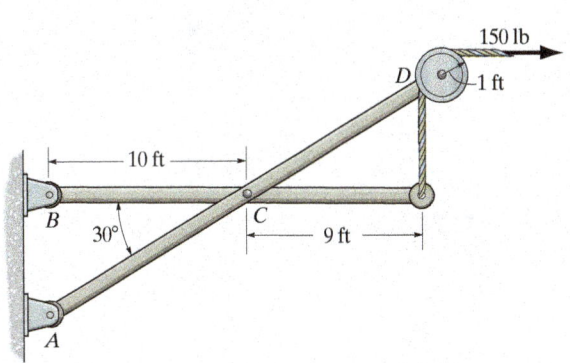

Prob. 6–84

6–85. The picture frame is glued together at its corners and held in place by the 4-corner clamp. If the tension in the adjusting screw is 14 N, determine the horizontal and vertical components of the clamping force that member AD exerts on the smooth joints at B and C. All connections are pins.

6–87. The pillar crane is subjected to the crate having a mass of 500 kg. Determine the force in the tie rod AB and the horizontal and vertical reactions at the pin support C when the boom is held in the position shown.

Prob. 6–85

Prob. 6–87

6–86. The double tree AB is used to support the loadings applied to each of the single trees. Determine the total load that must be supported by the chain EG and its placement d for AB to remain horizontal.

***6–88.** The machine is used for forming metal plates. It consists of two toggles ABC and DEF, which are operated by the hydraulic cylinder H. The toggles push the movable bar G forward, pressing the plate p into the cavity. If the force which the plate exerts on the head is $P = 12$ kN, determine the force F in the hydraulic cylinder when $\theta = 30°$.

Prob. 6–86

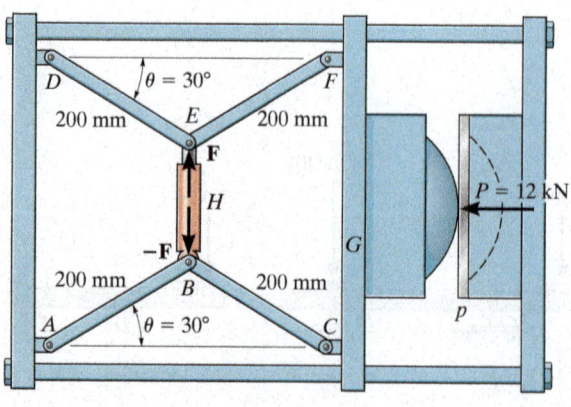

Prob. 6–88

6–89. Determine the horizontal and vertical components of force which pin C exerts on member ABC. The 600-N force is applied to the pin.

6–91. Determine the force in the hydraulic cylinders EF and AD in order to hold the shovel in the position shown. The load has a mass of 1.25 Mg and a center of gravity at G. All joints are pin connected.

Prob. 6–89

Prob. 6–91

6–90. The pipe cutter is clamped around the pipe P. If the wheel at A exerts a normal force of $F_A = 80$ N on the pipe, determine the normal forces of wheels B and C on the pipe. Also find the pin reaction on the wheel at C. The three wheels each have a radius of 7 mm and the pipe has an outer radius of 10 mm.

*6–92. The clamp is used to hold the smooth strut S in place. If the tensile force in the bolt GH is 300 N, determine the force exerted at the smooth surface at A and B.

Prob. 6–90

Prob. 6–92

6–93. The gin-pole derrick is used to lift the 300-kg stone with constant velocity. If the derrick and the block and tackle are in the position shown, determine the horizontal and vertical components of force at the pin support A and the orientation θ and tension in the guy cable BC.

Prob. 6–93

6–94. Five coins are stacked in the smooth plastic container. If each coin weighs 0.0235 lb, determine the normal reactions of the bottom coin on the container at points A and B.

Prob. 6–94

6–95. The nail cutter consists of the handle and the two cutting blades. Assuming the blades are pin connected at B and the surface at D is smooth, determine the normal force on the fingernail when a force of 1 lb is applied to the handles as shown. The pin AC slides through a smooth hole at A and is attached to the bottom member at C.

Prob. 6–95

*6–96. A man having a weight of 175 lb attempts to hold himself using one of the two methods shown. Determine the total force he must exert on bar AB in each case and the normal reaction he exerts on the platform at C. Neglect the weight of the platform. Do the results seem realistic?

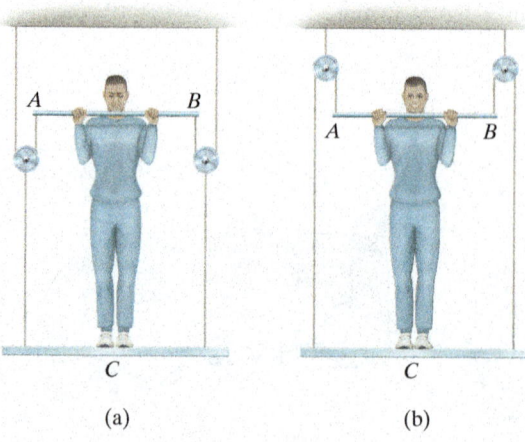

Prob. 6–96

6–97. A woman having a weight of 175 lb attempts to hold herself using one of the two methods shown. Determine the total force she must exert on bar AB in each case and the normal reaction she exerts on the platform at C. The platform has a weight of 30 lb. Do the results seem realistic?

6–99. If the 300-kg drum has a center of mass at G, determine the horizontal and vertical components of force acting at pin A and the reactions on the smooth pads C and D. The grip at B on member DAB resists both horizontal and vertical components of force at the rim of the drum.

Prob. 6–97

Prob. 6–99

*6–100.** Determine the horizontal and vertical components of force that the pins exert on member $ABCD$. The pin at C is attached to member $ABCD$ and passes through the smooth slot in member ECF.

6–98. Determine the couple moment M needed to create a force of $F = 200$ N on the slider block at C.

Prob. 6–98

Prob. 6–100

6–101. Determine the force that must be developed in the hydraulic cylinder AB in order to develop a normal force of 4 MN in the grip. Also, determine the magnitude of the force developed in the pin at C.

Prob. 6–101

6–102. If a force of $F = 350$ N is applied to the handle of the toggle clamp, determine the resulting clamping force at A.

Prob. 6–102

6–103. The double link grip is used to lift the beam. If the beam weighs 4 kN, determine the horizontal and vertical components of force acting on the pin at A and the horizontal and vertical components of force that the flange of the beam exerts on the grip at B. Assume B and F are pins.

Prob. 6–103

*6–104. The foot pump is used to fill the tire. If the force required to develop the necessary pressure in the cylinder is 70 lb, determine the vertical force **P** that must be applied to the pedal. The assembly is pin connected at A, B, and C.

Prob. 6–104

6–105. Determine the reactions at pins *A* and *B* of the frame needed to support the 200-lb load. The pulley at *C* has a radius of 0.5 ft. The pulley at *D* has a radius of 0.25 ft.

6–107. If a force of $F = 50$ lb is applied to the pads at *A* and *C*, determine the smallest dimension *d* required for equilibrium if the spring has an unstretched length of 1 ft.

Prob. 6–105

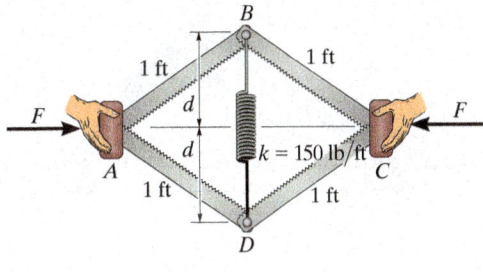

Prob. 6–107

6–106. If $d = 0.75$ ft and the spring has an unstretched length of 1 ft, determine the force *F* required for equilibrium.

*6–108. The clamp has a rated load capacity of 1500 lb. Determine the compressive force this creates along segment *AB* of the screw and the magnitude of force exerted at pin *C*. The screw is pin connected at its end *B* and passes through the pin-connected (swivel) block at *A*.

Prob. 6–106

Prob. 6–108

6–109. If the vertical force **P** is applied to the two-bar mechanism, determine the equilibrium force F on the block. Plot this magnitude as a function of θ, where $0° \leq \theta \leq 90°$.

*6–112.** The piston C moves vertically between the two smooth walls. If the spring has a stiffness of $k = 15$ lb/in., and is unstretched when $\theta = 0°$, determine the couple moment **M** that must be applied to AB to hold the mechanism in equilibrium when $\theta = 30°$.

Prob. 6–109

Prob. 6–112

6–110. The spring has an unstretched length of 0.3 m. Determine the angle θ for equilibrium if the uniform bars each have a mass of 20 kg.

6–111. The spring has an unstretched length of 0.3 m. Determine the mass m of each uniform bar if $\theta = 30°$ for equilibrium.

6–113. The platform scale consists of a combination of third and first class levers so that the load on one lever becomes the effort that moves the next lever. Through this arrangement, a small weight can balance a heavy object. If $x = 450$ mm, determine the required mass of the counterweight S required to balance the load having a mass of 90 kg.

6–114. The platform scale consists of a combination of third and first class levers so that the load on one lever becomes the effort that moves the next lever. Through this arrangement, a small weight can balance a heavy object. If $x = 450$ mm, and the mass of the counterweight S is 2 kg, determine the mass of the load L required to maintain the balance.

Probs. 6–110/111

Probs. 6–113/114

6–115. The four-member "A" frame is supported at A and E by smooth collars and at G by a pin. All the other joints are ball-and-sockets. If the pin at G will fail when the resultant force there is 800 N, determine the largest vertical force P that can be supported by the frame. Also, what are the x, y, z force components which member BD exerts on members EDC and ABC? The collars at A and E and the pin at G only exert force components on the frame.

6–117. Member AD is supported by cable AB and a roller at C, and fits through a smooth circular hole at D. Member ED is supported by a roller at D and a pole that fits in a smooth snug circular hole at E. Determine the x, y, z components of reaction at E and the tension in cable AB.

Prob. 6–115

Prob. 6–117

*6–116.** Member AB is supported by a ball-and-socket at A and smooth collar at B. Member CD is supported by a pin at C. Determine the x, y, z components of reaction at A and C.

6–118. The three pin-connected members shown in the *top view* support a downward force of 60 lb at G. If only vertical forces are supported at the connections B, C, E and at the pad supports A, D, F, determine the reactions at each pad.

Prob. 6–116

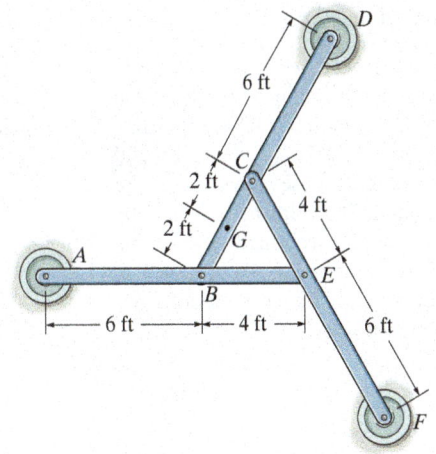

Prob. 6–118

Chapter Review

Simple Truss

A simple truss consists of triangular elements connected together by pinned joints. The forces within its members can be determined by assuming the members are all two-force members, connected concurrently at each joint. The members are either in tension or compression, or carry no force.

Roof truss

Method of Joints

The method of joints states that if a truss is in equilibrium, then each of its joints is also in equilibrium. For a plane truss, the concurrent force system at each joint must satisfy force equilibrium, $\Sigma F_x = 0$, $\Sigma F_y = 0$.

Method of Sections

The method of sections states that if a truss is in equilibrium, then each part of the truss is also in equilibrium. Pass a section through the truss and the member whose force is to be determined. Then draw the free-body diagram of the sectioned part having the least number of forces on it. The three equilibrium equations are available for solution.

Space Truss

A space truss is a three-dimensional truss built from tetrahedral elements, and is analyzed using the same methods as for plane trusses. The joints are assumed to be ball-and-socket connections.

Frames and Machines

Frames and machines are structures that contain one or more multiforce members, that is, members with three or more forces or couples acting on them. Frames are designed to support loads, and machines transmit and alter the effect of forces.

The forces acting at the joints of a frame or machine can be determined by drawing the free-body diagrams of each of its members or parts. The principle of action–reaction should be carefully observed when indicating these forces on the free-body diagram of each adjacent member or pin. For a coplanar force system, there are three equilibrium equations available for each member.

To simplify the analysis, be sure to recognize all two-force members. They have equal but opposite collinear forces at their ends.

REVIEW PROBLEMS

All solutions must include a free-body diagram.

R6–1. Determine the force in each member of the truss and state if the members are in tension or compression.

R6–3. Determine the force in member GJ and GC of the truss and state if the members are in tension or compression.

Prob. R6–1

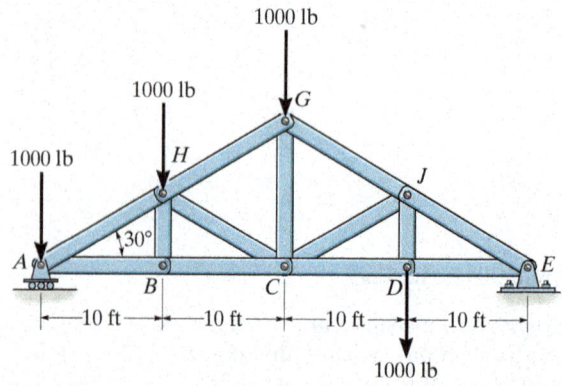

Prob. R6–3

R6–2. Determine the force in each member of the truss and state if the members are in tension or compression.

R6–4. Determine the force in members GF, FB, and BC of the *Fink truss* and state if the members are in tension or compression.

Prob. R6–2

Prob. R6–4

REVIEW PROBLEMS   323

R6–5. Determine the force in members AB, AD, and AC of the space truss and state if the members are in tension or compression.

R6–7. Determine the horizontal and vertical components of force that the pins A and C exert on the two-member frame.

Prob. R6–7

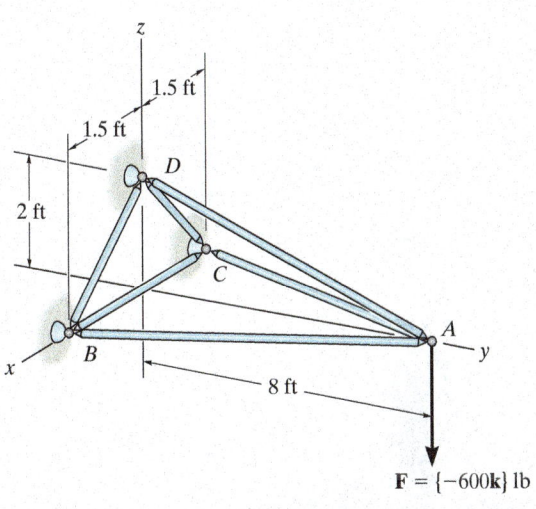

Prob. R6–5

R6–6. Determine the horizontal and vertical components of force that the pins A and B exert on the two-member frame.

R6–8. Determine the resultant forces at pins B and C on member ABC of the four-member frame.

Prob. R6–6

Prob. R6–8

CHAPTER 7

The loads within the beams and columns of this building frame must first be determined if they are to be properly designed. In this chapter we will study how to find these internal loadings.

INTERNAL FORCES

CHAPTER OBJECTIVES

- To use the method of sections to determine the internal loadings in a member at a specific point.
- To show how to obtain the internal shear and moment throughout a member and express the result graphically in the form of shear and moment diagrams.
- To analyze the forces and the shape of cables supporting various types of loadings.

7.1 INTERNAL LOADINGS

To design a structural member or mechanical element it is necessary to know the loading acting within it to be sure the material can resist this loading. Internal loadings are determined by using the ***method of sections***. To illustrate the method, consider finding the internal loading at point B of the cantilever beam shown in Fig. 7–1a. If we pass an imaginary section perpendicular to the axis of the beam through point B, and then separate the beam into two segments, then the internal loadings acting at B will be exposed and become *external* on the free-body diagram of each segment, Fig. 7–1b.

(a)

(b)

Fig. 7–1

326   CHAPTER 7   INTERNAL FORCES

In each case, the link on the backhoe is a two-force member. In the top photo it is subjected to both bending and a normal force at its center. It is more efficient to make the member straight, as in the bottom photo; then only an axial force acts within the member.

(b)

Fig. 7–1 (Repeated)

The force component N_B that acts *perpendicular* to the cross section is termed the **normal force**. The force component V_B that is tangent to the cross section is called the **shear force**, and the couple moment M_B is referred to as the **bending moment**, Fig. 7–2a. According to Newton's third law, these loadings must act in opposite directions on each segment, as shown in Fig. 7–1b. They can be determined by applying the equations of equilibrium to the free-body diagram of either segment. In this case, however, the right segment is the better choice since it does not involve the unknown support reactions at A. A direct solution for N_B is obtained by applying $\Sigma F_x = 0$, V_B is obtained from $\Sigma F_y = 0$, and M_B can be obtained by applying $\Sigma M_B = 0$, since the moments of N_B and V_B about B are zero.

In three dimensions, a general resultant internal force and couple moment resultant will act at the section. The x, y, z components of these loadings are shown in Fig. 7–2b. Here N_y is the *normal force*, and V_x and V_z are *shear force components*. M_y is a *torsional or twisting moment*, and M_x and M_z are *bending moment components*. For most applications, these *resultant loadings* will act at the geometric center or centroid (C) of the section's cross-sectional area. Although the magnitude for each loading generally will be different at various points along the axis of the member, the method of sections can always be used to determine their values.

(a)

(b)

Fig. 7–2

Sign Convention. For problems in two dimensions engineers generally use a sign convention to report the three internal loadings **N**, **V**, and **M**. Although this sign convention can be arbitrarily assigned, the one that is widely accepted will be used here, Fig. 7–3. The normal force is said to be positive if it creates *tension*, a positive shear force will cause the beam segment on which it acts to rotate clockwise, and a positive bending moment will tend to bend the segment on which it acts in a concave upward manner. Loadings that are opposite to these are considered negative.

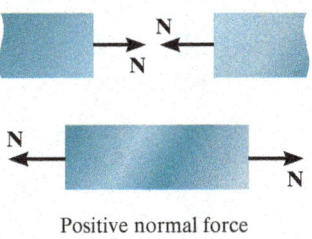

Positive normal force

IMPORTANT POINT

- There can be four types of resultant internal loads in a member. They are the normal and shear forces and the bending and torsional moments. These loadings generally vary from point to point. They can be determined using the method of sections.

Positive shear

PROCEDURE FOR ANALYSIS

The method of sections can be used to determine the internal loadings on the cross section of a member using the following procedure.

Support Reactions.

- Before the member is sectioned, it may first be necessary to determine its support reactions.

Free-Body Diagram.

- It is important that all distributed loadings, couple moments, and forces acting on the member be kept in their *exact locations*, then pass an imaginary section through the member, perpendicular to its axis at the point where the internal loadings are to be determined.

- After the section is made, draw a free-body diagram of the segment that has the least number of loads on it, and indicate the components of the internal force and couple moment resultants at the cross section acting in their positive directions in accordance with the established sign convention.

Equations of Equilibrium.

- Moments should be summed at the section. This way the normal and shear forces at the section are eliminated, and we can obtain a direct solution for the moment.

- If the solution of the equilibrium equations yields a negative scalar, the sense of the internal loading is then opposite to that shown on the free-body diagram.

Positive moment

Fig. 7–3

The designer of this shop crane realized the need for additional reinforcement around the joint at *A* in order to prevent severe internal bending of the joint when a large load is suspended from the chain hoist.

EXAMPLE 7.1

Determine the normal force, shear force, and bending moment acting just to the left, point B, and just to the right, point C, of the 6-kN force on the beam in Fig. 7–4a.

(a)

SOLUTION

Support Reactions. The free-body diagram of the beam is shown in Fig. 7–4b. When determining the *external reactions*, realize that the 9-kN·m couple moment is a free vector and therefore it can be placed *anywhere* on the free-body diagram of the entire beam. Here we will only determine A_y, since the left segments will be used for the analysis.

$\zeta + \Sigma M_D = 0;$ $\quad 9 \text{ kN} \cdot \text{m} + (6 \text{ kN})(6 \text{ m}) - A_y(9 \text{ m}) = 0$

$$A_y = 5 \text{ kN}$$

Free-Body Diagrams. The free-body diagrams of the left segments AB and AC of the beam are shown in Figs. 7–4c and 7–4d. In this case the 9-kN·m couple moment is *not included* on these diagrams, since it acts on the right segment.

Equations of Equilibrium.

Segment AB

$\xrightarrow{+} \Sigma F_x = 0;$ $\quad\quad\quad\quad\quad\quad\quad N_B = 0$ $\quad\quad$ Ans.

$+\uparrow \Sigma F_y = 0;$ $\quad\quad 5 \text{ kN} - V_B = 0;$ $\quad V_B = 5 \text{ kN}$ $\quad\quad$ Ans.

$\zeta + \Sigma M_B = 0;$ $\quad -(5 \text{ kN})(3 \text{ m}) + M_B = 0;$ $\quad M_B = 15 \text{ kN} \cdot \text{m}$ $\quad$ Ans.

Segment AC

$\xrightarrow{+} \Sigma F_x = 0;$ $\quad\quad\quad\quad\quad\quad\quad N_C = 0$ $\quad\quad$ Ans.

$+\uparrow \Sigma F_y = 0;$ $\quad 5 \text{ kN} - 6 \text{ kN} - V_C = 0;$ $\quad V_C = -1 \text{ kN}$ $\quad\quad$ Ans.

$\zeta + \Sigma M_C = 0;$ $\quad -(5 \text{ kN})(3 \text{ m}) + M_C = 0;$ $\quad M_C = 15 \text{ kN} \cdot \text{m}$ $\quad$ Ans.

NOTE: The negative sign indicates that $\mathbf{V}_C$ acts in the opposite sense to that shown on the free-body diagram. Also, the moment arm for the 5-kN force in both cases is approximately 3 m since B and C are "almost" coincident.

Fig. 7–4

EXAMPLE 7.2

Determine the normal force, shear force, and bending moment at point C of the beam in Fig. 7–5a.

(a)

(b)

SOLUTION

Free-Body Diagram. It is not necessary to find the support reactions at A since segment BC of the beam can be used to determine the internal loadings at C. The intensity of the distributed load at C is determined using similar triangles from the geometry shown in Fig. 7–5b, i.e.,

(c)

Fig. 7–5

$$w_C = (1200 \text{ N/m})\left(\frac{1.5 \text{ m}}{3 \text{ m}}\right) = 600 \text{ N/m}$$

The distributed load acting on segment BC can now be replaced by its resultant force, and its location is indicated on the free-body diagram, Fig. 7–5c.

Equations of Equilibrium.

$\xrightarrow{+} \Sigma F_x = 0;$ $\qquad N_C = 0$ *Ans.*

$+\uparrow \Sigma F_y = 0;$ $V_C - \frac{1}{2}(600 \text{ N/m})(1.5 \text{ m}) = 0$

$$V_C = 450 \text{ N} \qquad \textit{Ans.}$$

$\zeta + \Sigma M_C = 0;$ $-M_C - \frac{1}{2}(600 \text{ N/m})(1.5 \text{ m})(0.5 \text{ m}) = 0$

$$M_C = -225 \text{ N} \cdot \text{m} \qquad \textit{Ans.}$$

The negative sign indicates that $\mathbf{M}_C$ acts in the opposite sense to that shown on the free-body diagram.

EXAMPLE 7.3

Determine the normal force, shear force, and bending moment acting at point B of the two-member frame shown in Fig. 7–6a.

SOLUTION

Support Reactions. A free-body diagram of each member is shown in Fig. 7–6b. Since CD is a two-force member, the equations of equilibrium need to be applied only to member AC.

$\zeta + \Sigma M_A = 0;$ $-400 \text{ lb}(4 \text{ ft}) + \left(\frac{3}{5}\right) F_{DC}(8 \text{ ft}) = 0$ $F_{DC} = 333.3 \text{ lb}$

$\xrightarrow{+} \Sigma F_x = 0;$ $-A_x + \left(\frac{4}{5}\right)(333.3 \text{ lb}) = 0$ $A_x = 266.7 \text{ lb}$

$+\uparrow \Sigma F_y = 0;$ $A_y - 400 \text{ lb} + \left(\frac{3}{5}\right)(333.3 \text{ lb}) = 0$ $A_y = 200 \text{ lb}$

(b)

Fig. 7–6

Free-Body Diagrams. Passing an imaginary section perpendicular to the axis of member AC through point B yields the free-body diagrams of segments AB and BC shown in Fig. 7–6c. When constructing these diagrams it is important to keep the distributed loading where it is until *after the section is made*. Only then can it be replaced by a single resultant force.

Equations of Equilibrium. Applying the equations of equilibrium to segment AB, we have

$\xrightarrow{+} \Sigma F_x = 0;$ $N_B - 266.7 \text{ lb} = 0;$ $N_B = 267 \text{ lb}$ *Ans.*

$+\uparrow \Sigma F_y = 0;$ $200 \text{ lb} - 200 \text{ lb} - V_B = 0;$ $V_B = 0$ *Ans.*

$\zeta + \Sigma M_B = 0;$ $M_B - 200 \text{ lb}(4 \text{ ft}) + 200 \text{ lb}(2 \text{ ft}) = 0$

$M_B = 400 \text{ lb} \cdot \text{ft}$ *Ans.*

NOTE: As an exercise, try to obtain these same results using segment BC.

EXAMPLE 7.4

Determine the normal force, shear force, and bending moment acting at point E of the frame loaded as shown in Fig. 7–7a.

(a) (b)

SOLUTION

Support Reactions. By inspection, members AC and CD are two-force members, Fig. 7–7b. In order to determine the internal loadings at E, we must first determine the force **R** acting at the end of member AC. To obtain it, we will analyze the equilibrium of the pin at C.

Summing forces in the vertical direction on the pin, Fig. 7–7b, we have

$+\uparrow \Sigma F_y = 0;$ $R \sin 45° - 600 \text{ N} = 0;$ $R = 848.5 \text{ N}$

Free-Body Diagram. The free-body diagram of segment CE is shown in Fig. 7–7c.

Equations of Equilibrium.

$\xrightarrow{+} \Sigma F_x = 0;$ $848.5 \cos 45° \text{ N} - V_E = 0;$ $V_E = 600 \text{ N}$ *Ans.*

$+\uparrow \Sigma F_y = 0;$ $-848.5 \sin 45° \text{ N} + N_E = 0;$ $N_E = 600 \text{ N}$ *Ans.*

$\zeta + \Sigma M_E = 0;$ $848.5 \cos 45° \text{ N}(0.5 \text{ m}) - M_E = 0$

$M_E = 300 \text{ N} \cdot \text{m}$ *Ans.*

NOTE: If member AC is *straight* (from A to C), then the shear and bending moment would be *eliminated* and only a normal force of 848.5 N would act within the member.

(c)

Fig. 7–7

EXAMPLE 7.5

The uniform sign shown in Fig. 7–8a has a mass of 650 kg and is supported on the fixed column. Determine the internal loadings at A if the maximum uniform wind loading on the sign is 900 Pa.

SOLUTION
The idealized model for the sign is shown in Fig. 7–8b. Here the necessary dimensions are indicated. We will consider the free-body diagram of a section above point A since it does not involve the support reactions.

Free-Body Diagram. The sign has a weight of $W = 650(9.81)$ N $= 6.376$ kN, and the wind creates a resultant force of $F_w = 900$ N/m^2(6 m)(2.5 m) $= 13.5$ kN, which acts perpendicular to the face of the sign. These loadings are shown on the free-body diagram, Fig. 7–8c.

Equations of Equilibrium. Since the problem is three dimensional, a vector analysis will be used.

$\Sigma \mathbf{F} = \mathbf{0};$ $\mathbf{F}_A - 13.5\mathbf{i} - 6.376\mathbf{k} = \mathbf{0}$

$\mathbf{F}_A = \{13.5\mathbf{i} + 6.38\mathbf{k}\}$ kN   *Ans.*

$\Sigma \mathbf{M}_A = \mathbf{0};$ $\mathbf{M}_A + \mathbf{r} \times (\mathbf{F}_w + \mathbf{W}) = \mathbf{0}$

$$\mathbf{M}_A + \begin{vmatrix} \mathbf{i} & \mathbf{j} & \mathbf{k} \\ 0 & 3 & 5.25 \\ -13.5 & 0 & -6.376 \end{vmatrix} = \mathbf{0}$$

$\mathbf{M}_A = \{19.1\mathbf{i} + 70.9\mathbf{j} - 40.5\mathbf{k}\}$ kN·m   *Ans.*

NOTE: Here $\mathbf{F}_{A_z} = \{6.38\mathbf{k}\}$ kN represents the normal force, whereas $\mathbf{F}_{A_x} = \{13.5\mathbf{i}\}$ kN is the shear force. Also, the torsional moment is $\mathbf{M}_{A_z} = \{-40.5\mathbf{k}\}$ kN·m, and the bending moment is determined from its components $\mathbf{M}_{A_x} = \{19.1\mathbf{i}\}$ kN·m and $\mathbf{M}_{A_y} = \{70.9\mathbf{j}\}$ kN·m; i.e., $(M_b)_A = \sqrt{(M_A)_x^2 + (M_A)_y^2} = 73.4$ kN·m.

Fig. 7–8

FUNDAMENTAL PROBLEMS

All solutions must include a free-body diagram.

F7–1. Determine the normal force, shear force, and moment at point C.

Prob. F7–1

F7–2. Determine the normal force, shear force, and moment at point C.

Prob. F7–2

F7–3. Determine the normal force, shear force, and moment at point C.

Prob. F7–3

F7–4. Determine the normal force, shear force, and moment at point C.

Prob. F7–4

F7–5. Determine the normal force, shear force, and moment at point C.

Prob. F7–5

F7–6. Determine the normal force, shear force, and moment at point C. Assume A is pinned and B is a roller.

Prob. F7–6

334   CHAPTER 7   INTERNAL FORCES

PROBLEMS

All solutions must include a free-body diagram.

7–1. Determine the shear force and moment at points C and D.

Prob. 7–1

7–2. Determine the normal force, shear force, and moment in the shaft at points B and D. These points lie just to the right of the 150-lb force and the bearing at C, respectively. There is a thrust bearing at A and a journal bearing at C.

Prob. 7–2

7–3. The structural connections transmit the loads shown to the column. Determine the normal force, shear force, and moment acting in the column at a section passing horizontally through point A.

Prob. 7–3

*****7–4.** The beam-column is fixed to the floor and supports the load shown. Determine the normal force, shear force, and moment at points A and B.

Prob. 7–4

7–5. The pliers are used to grip the tube at B. If a force of 20 lb is applied to the handles, determine the shear force and moment at point C. Assume the jaws of the pliers exert only normal forces on the tube.

Prob. 7–5

7–6. Determine the distance a as a fraction of the beam's length L for locating the roller support so that the moment in the beam at B is zero.

Prob. 7–6

7–7. Determine the normal force, shear force, and moment at points C and D in the beam.

Prob. 7–7

***7–8.** Determine the shear force and moment acting at point C in the beam.

Prob. 7–8

7–9. Determine the normal force, shear force, and moment at point C. Take $P = 8$ kN.

Prob. 7–9

7–10. The cable will fail when subjected to a tension of 2 kN. Determine the largest vertical load P the frame will support and calculate the normal force, shear force, and moment at point C for this loading.

Prob. 7–10

7–11. Determine the normal force, shear force, and moment acting at point C.

Prob. 7–11

*7–12. Determine the distance a between the bearings in terms of the shaft's length L so that the moment in the *symmetric* shaft is zero at its center.

Prob. 7–12

7–13. The 9-kip force is supported by the floor panel DE, which in turn is simply supported at its ends by floor beams. These beams transmit their loads to the girder AB. Determine the shear and moment acting at point C in the girder.

Prob. 7–13

7–14. Determine the normal force, shear force, and moment at points F and E.

Prob. 7–14

7–15. Determine the normal force, shear force, and moment at point C.

Prob. 7–15

*7–16. Determine the normal force, shear force, and moment at point C of the beam.

Prob. 7–16

7–17. The cantilevered rack is used to support each end of a smooth pipe that has a total weight of 300 lb. Determine the normal force, shear force, and moment that act in the arm at its fixed support A along a vertical section.

Prob. 7–17

7–18. Determine the normal force, shear force, and moment in the beam at points C and D. The beam is supported at A and B.

Prob. 7–18

7–19. Determine the normal force, shear force, and moment at points F and E in the frame. The crate weighs 300 lb.

Prob. 7–19

*7–20. Rod AB is fixed to a smooth collar D, which slides freely along the vertical guide. Determine the normal force, shear force, and moment at point C, which is located just to the left of the 60-lb concentrated load.

Prob. 7–20

7–21. Determine the normal force, shear force, and moment at points E and F of the compound beam. Point E is located just to the left of the 800-N force.

Prob. 7–21

7–22. Determine the shear force and moment acting at point D of the beam.

Prob. 7–22

7–23. Determine the normal force, shear force, and moment at point C.

Prob. 7–23

*7–24. Determine the shear force and moment acting at point *C* of the beam. For the calculation use Simpson's rule to evaluate the integrals.

Prob. 7–24

7–25. Determine the normal force, shear force, and moment in the beam at points *D* and *E*. Point *E* is just to the right of the 3-kip load.

Prob. 7–25

7–26. Determine the distance *a* in terms of the beam's length *L* between the symmetrically placed supports *A* and *B* so that the moment at the center of the beam is zero.

Prob. 7–26

7–27. Determine the normal force, shear force, and moment at points *F* and *G* in the compound beam. Point *F* is located just to the right of the 500-lb force, while point *G* is located just to the right of the 600-lb force.

Prob. 7–27

*7–28. Determine the normal force, shear force, and moment at points *C* and *D*.

Prob. 7–28

7–29. Determine the normal force, shear force, and moment acting at point *C*.

7–30. Determine the normal force, shear force, and moment acting at point *D*.

Probs. 7–29/30

7–31. Determine the normal force, shear force, and moment in the beam at points C and D. Point D is just to the right of the 5-kip load.

Prob. 7–31

*7–32. Determine the normal force, shear force, and moment at points D and E of the frames.

Prob. 7–32

7–33. Determine the normal force, shear force, and moment at point D of the two-member frame.

7–34. Determine the normal force, shear force, and moment at point E.

Probs. 7–33/34

7–35. The man has a weight of 250 lb and a center of gravity at G. If the bench upon which he is sitting is fixed connected to the center support, determine the x, y, z components of loading at sections passing through point A and the base B.

Prob. 7–35

*7–36. Determine the normal force, shear force, and moment acting at points B and C on the curved member.

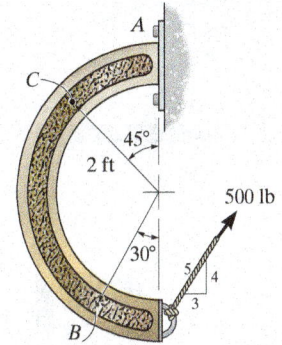

Prob. 7–36

7–37. Determine the normal force, shear force, and moment at point D of the two-member frame.

7–38. Determine the normal force, shear force, and moment at point E of the two-member frame.

Probs. 7–37/38

7-39. The distributed loading $w = w_0 \sin\theta$, measured per unit length, acts on the curved rod. Determine the normal force, shear force, and moment in the rod at $\theta = 45°$.

*7-40. Solve Prob. 7-39 for $\theta = 120°$.

7-42. Determine the x, y, z components of force and moment at point C in the pipe assembly. Neglect the weight of the pipe. The load acting at (0, 3.5 ft, 3 ft) is $\mathbf{F}_1 = \{-24\mathbf{i} - 10\mathbf{k}\}$ lb and $\mathbf{M} = \{-30\mathbf{k}\}$ lb·ft and at point (0, 3.5 ft, 0) $\mathbf{F}_2 = \{-80\mathbf{i}\}$ lb.

Probs. 7–39/40

7-41. Determine the normal force, shear force, and moment at point D of the two-member frame. Neglect the thickness of the member.

Prob. 7–42

7-43. Determine the x, y, z components of loading at a section passing through point B in the pipe assembly. Neglect the weight of the pipe. Take $\mathbf{F}_1 = \{200\mathbf{i} - 100\mathbf{j} - 400\mathbf{k}\}$ N and $\mathbf{F}_2 = \{300\mathbf{i} - 500\mathbf{k}\}$ N.

*7-44. Determine the x, y, z components of loading at a section passing through point B in the pipe assembly. Neglect the weight of the pipe. Take $\mathbf{F}_1 = \{100\mathbf{i} - 200\mathbf{j} - 300\mathbf{k}\}$ N and $\mathbf{F}_2 = \{100\mathbf{i} + 500\mathbf{j}\}$ N.

Prob. 7–41

Probs. 7–43/44

*7.2 SHEAR AND MOMENT EQUATIONS AND DIAGRAMS

Beams are structural members designed to support loadings applied perpendicular to their axes. In general, they are long and straight and have a constant cross-sectional area. They are often classified as to how they are supported. For example, a **simply supported beam** is pinned at one end and roller supported at the other, as in Fig. 7–9a, whereas a **cantilevered beam** is fixed at one end and free at the other. The actual design of a beam requires a detailed knowledge of the *variation* of the internal shear force V and bending moment M acting at *each point* along the axis of the beam.*

These *variations* of V and M along the beam's axis can be obtained by using the method of sections discussed in Sec. 7.1. In this case, however, it is necessary to section the beam at an arbitrary distance x from one end and then apply the equations of equilibrium to the segment having the length x. Doing this we can then obtain V and M as functions of x.

In general, the internal shear and bending-moment functions will be discontinuous, or their slopes will be discontinuous, at points where a distributed load changes or where concentrated forces or couple moments are applied. Because of this, these functions must be determined for *each segment* of the beam located between any two discontinuities of loading. For example, in Fig. 7–9a, x_1 is used for $0 \le x_1 < a$, x_2 is used for $a < x_2 < b$, and x_3 is used for $b < x_3 \le L$. If these functions of x are plotted, the graphs are termed the **shear diagram** and **bending moment diagram**, Fig. 7–9b and Fig. 7–9c.

To save on material and thereby produce an efficient design, these beams, also called girders, have been tapered, since the internal moment in the beam will be larger at the supports, or piers, than at the center of the span.

Fig. 7–9

---

*The internal normal force is not considered for two reasons. In most cases, the loads applied to a beam act perpendicular to the beam's axis and hence produce only an internal shear force and bending moment. And for design purposes, the beam's resistance to shear, and particularly to bending, is more important than its ability to resist a normal force.

342   CHAPTER 7   INTERNAL FORCES

Positive shear

Positive moment

Beam sign convention

Fig. 7–10

IMPORTANT POINTS

- Shear and moment diagrams for a beam provide graphical descriptions of how the internal shear and moment vary throughout the beam's length.

- To obtain these diagrams, the method of sections is used to determine V and M as functions of x. These results are then plotted. If the load on the beam suddenly changes, then regions between each load must be selected to obtain each function of x.

PROCEDURE FOR ANALYSIS

The shear and bending-moment diagrams for a beam can be constructed using the following procedure.

Support Reactions.

- Determine all the reactive forces and couple moments acting on the beam and resolve all the forces into components acting perpendicular and parallel to the beam's axis.

Shear and Moment Functions.

- Specify separate coordinates x having an origin at the beam's left end and extending to regions of the beam *between* concentrated forces and/or couple moments, or where the distributed loading is continuous.

- Section the beam at each distance x and draw the free-body diagram of one of the segments. Be sure **V** and **M** are shown acting in their *positive sense*, in accordance with the sign convention given in Fig. 7–10.

- The shear V is obtained by summing forces perpendicular to the beam's axis, and the moment M is obtained by summing moments about the sectioned end of the segment.

Shear and Moment Diagrams.

- Plot the shear diagram (V versus x) and the moment diagram (M versus x). If calculated values of the functions describing V and M are *positive*, then these values are plotted above the x axis, whereas *negative* values are plotted below the axis.

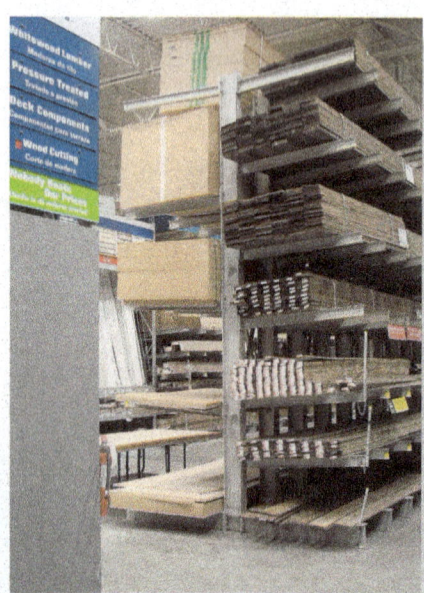
The shelving arms must be designed to resist the internal loading in the arms caused by the lumber.

EXAMPLE 7.6

Draw the shear and moment diagrams for the shaft shown in Fig. 7–11a. The support at A is a thrust bearing and the support at C is a journal bearing.

(a)

SOLUTION

Support Reactions. The support reactions are shown on the shaft's free-body diagram, Fig. 7–11d.

Shear and Moment Functions. The shaft is sectioned at an arbitrary distance x from point A, extending within the region AB, and the free-body diagram of the left segment is shown in Fig. 7–11b. The unknowns **V** and **M** are assumed to act in the *positive sense* on the right-hand face of the segment according to the established sign convention. Applying the equilibrium equations yields

(b)

$+\uparrow \Sigma F_y = 0;$ $\qquad V = 2.5 \text{ kN}$ (1)

$\zeta+\Sigma M = 0;$ $\qquad M = 2.5x \text{ kN} \cdot \text{m}$ (2)

A free-body diagram for a left segment of the shaft extending from A a distance x, within the region BC is shown in Fig. 7–11c. As always, **V** and **M** are shown acting in the positive sense. Hence,

(c)

$+\uparrow \Sigma F_y = 0;$ $\qquad 2.5 \text{ kN} - 5 \text{ kN} - V = 0$

$\qquad\qquad V = -2.5 \text{ kN}$ (3)

$\zeta+\Sigma M = 0;$ $\qquad M + 5 \text{ kN}(x - 2 \text{ m}) - 2.5 \text{ kN}(x) = 0$

$\qquad\qquad M = (10 - 2.5x) \text{ kN} \cdot \text{m}$ (4)

Shear and Moment Diagrams. When Eqs. 1 through 4 are plotted within the regions in which they are valid, the shear and moment diagrams shown in Fig. 7–11d are obtained. The shear diagram indicates that the internal shear force is always 2.5 kN (positive) within segment AB. Just to the right of point B, the shear changes sign and remains at a constant value of -2.5 kN for segment BC. The moment diagram starts at zero, increases linearly to point B at $x = 2$ m, where $M_{max} = 2.5 \text{ kN}(2 \text{ m}) = 5 \text{ kN} \cdot \text{m}$, and thereafter decreases back to zero.

NOTE: It is seen in Fig. 7–11d that the graphs of the shear and moment diagrams "jump" or change *abruptly* where the concentrated force acts, i.e., at points A, B, and C. This, of course, is the result of assuming an *idealization of a concentrated force and couple moment*. Physically, loads are always applied over a finite area, and if the actual load variation could be accounted for, the shear and moment diagrams would then be continuous over the shaft's entire length.

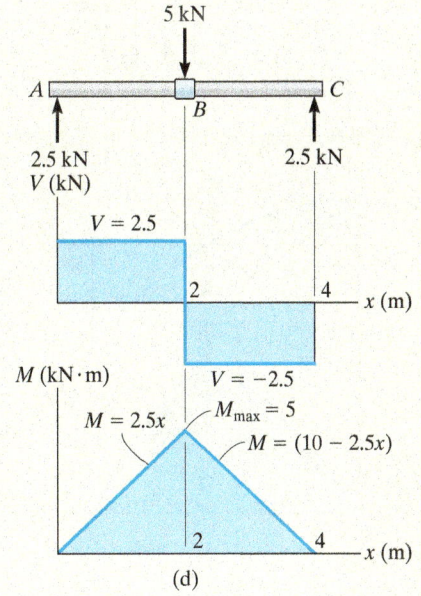
(d)

Fig. 7–11

EXAMPLE 7.7

Fig. 7–12

Draw the shear and moment diagrams for the beam shown in Fig. 7–12a.

SOLUTION

Support Reactions. The support reactions are shown on the beam's free-body diagram, Fig. 7–12c.

Shear and Moment Functions. A free-body diagram for a left segment of the beam having a length x is shown in Fig. 7–12b. The distributed loading acting at the end of this segment has an intensity which is found by proportional triangles, that is, $w/x = 6/9$ or $w = (2/3)x$. The *magnitude* of the resultant force is equal to $\frac{1}{2}(x)(\frac{2}{3}x) = \frac{1}{3}x^2$. This force *acts through the centroid* of the distributed loading area, a distance $\frac{1}{3}x$ from the right end. Applying the two equations of equilibrium yields

$$+\uparrow \Sigma F_y = 0; \qquad 9 - \frac{1}{3}x^2 - V = 0$$

$$V = \left(9 - \frac{x^2}{3}\right) \text{kN} \qquad (1)$$

$$\zeta + \Sigma M = 0; \qquad M + \frac{1}{3}x^2\left(\frac{x}{3}\right) - 9x = 0$$

$$M = \left(9x - \frac{x^3}{9}\right) \text{kN} \cdot \text{m} \qquad (2)$$

Shear and Moment Diagrams. The shear and moment diagrams shown in Fig. 7–12c are obtained by plotting Eqs. 1 and 2. The point of *zero shear* can be found using Eq. 1:

$$V = 9 - \frac{x^2}{3} = 0$$

$$x = 5.20 \text{ m}$$

NOTE: It will be shown in Sec. 7.3 that this value of x happens to represent the point on the beam where the *maximum moment* occurs. Using Eq. 2, this moment is

$$M_{max} = \left(9(5.20) - \frac{(5.20)^3}{9}\right) \text{kN} \cdot \text{m}$$

$$= 31.2 \text{ kN} \cdot \text{m}$$

FUNDAMENTAL PROBLEMS

F7–7. Determine the shear and moment as a function of x, and then draw the shear and moment diagrams.

Prob. F7–7

F7–8. Determine the shear and moment as a function of x, and then draw the shear and moment diagrams.

Prob. F7–8

F7–9. Determine the shear and moment as a function of x, and then draw the shear and moment diagrams.

Prob. F7–9

F7–10. Determine the shear and moment as a function of x, and then draw the shear and moment diagrams.

Prob. F7–10

F7–11. Determine the shear and moment as a function of x, where $0 \leq x < 3$ m and 3 m $< x \leq 6$ m, and then draw the shear and moment diagrams.

Prob. F7–11

F7–12. Determine the shear and moment as a function of x, where $0 \leq x < 3$ m and 3 m $< x \leq 6$ m, and then draw the shear and moment diagrams.

Prob. F7–12

346   CHAPTER 7   INTERNAL FORCES

PROBLEMS

All solutions must include a free-body diagram.

7–45. Draw the shear and moment diagrams for the shaft (a) in terms of the parameters shown; (b) set $P = 9$ kN, $a = 2$ m, $L = 6$ m. There is a thrust bearing at A and a journal bearing at B.

Prob. 7–45

7–46. Draw the shear and moment diagrams for the beam (a) in terms of the parameters shown; (b) set $P = 800$ lb, $a = 5$ ft, $L = 12$ ft.

Prob. 7–46

7–47. Draw the shear and moment diagrams for the beam (a) in terms of the parameters shown; (b) set $P = 600$ lb, $a = 5$ ft, $b = 7$ ft.

Prob. 7–47

*7–48. Draw the shear and moment diagrams for the cantilevered beam.

Prob. 7–48

7–49. Draw the shear and moment diagrams of the beam (a) in terms of the parameters shown; (b) set $M_0 = 500$ N·m, $L = 8$ m.

7–50. If $L = 9$ m, the beam will fail when the maximum shear force is $V_{max} = 5$ kN or the maximum bending moment is $M_{max} = 2$ kN·m. Determine the magnitude M_0 of the largest couple moments it will support.

Probs. 7–49/50

7–51. Draw the shear and moment diagrams for the beam.

Prob. 7–51

***7–52.** Draw the shear and moment diagrams for the beam.

Prob. 7–52

7–53. Draw the shear and moment diagrams for the beam.

Prob. 7–53

7–54. The shaft is supported by a smooth thrust bearing at A and a smooth journal bearing at B. Draw the shear and moment diagrams for the shaft (a) in terms of the parameters shown; (b) set $w = 500$ lb/ft, $L = 10$ ft.

Prob. 7–54

7–55. Draw the shear and moment diagrams for the compound beam. The beam is pin connected at E and F.

Prob. 7–55

***7–56.** Draw the shear and moment diagrams for the beam.

Prob. 7–56

7–57. Draw the shear and moment diagrams for beam ABC. There is a pin at B. Solve the problem (a) in terms of the parameters shown; (b) set $w = 5$ kN/m, $L = 12$ m.

Prob. 7–57

7–58. The two segments of the girder are pin connected at B. Draw the shear and moment diagrams for the girder.

Prob. 7–58

7–59. Draw the shear and moment diagrams for the beam.

Prob. 7–59

*7–60. The shaft is supported by a smooth thrust bearing at A and a smooth journal bearing at B. Draw the shear and moment diagrams for the shaft.

Prob. 7–60

7–61. Draw the shear and moment diagrams for the beam.

Prob. 7–61

7–62. Determine the placement a of the roller support B so that the maximum moment within the span AB is equivalent to the moment at the support B.

Prob. 7–62

7–63. The jib crane supports a load of 750 lb. If the boom AB has a uniform weight of 60 lb/ft, draw the shear and moment diagrams for the boom.

Prob. 7–63

*7–64. The semicircular rod is subjected to a distributed loading $w = w_0 \sin \theta$. Determine the normal force, shear force, and moment in the rod as a function of θ.

Prob. 7–64

7–65. Draw the shear and moment diagrams for the beam.

Prob. 7–65

7–66. Express the shear and moment acting in the pipe as a function of y, where $0 \le y \le 4$ ft.

***7–68.** Determine the normal force, shear force, and moment in the curved rod as a function of θ.

Prob. 7–66

Prob. 7–68

7–67. Determine the normal force, shear force, and moment in the curved rod as a function of θ. The force **P** acts at the constant angle ϕ.

7–69. Determine the distance a between the bearings in terms of the shaft's length L so that the moment in the *symmetric* shaft is zero at its center.

Prob. 7–67

Prob. 7–69

*7.3 RELATIONS AMONG DISTRIBUTED LOAD, SHEAR, AND MOMENT

If a beam is subjected to several concentrated forces, couple moments, and distributed loads, the method of constructing the shear and bending moment diagrams discussed in Sec. 7.2 may become quite tedious. In this section a simpler method for constructing these diagrams is discussed—a method based on differential relations that exist among the load, shear, and bending moment.

In order to design the beam used to support these power lines, it is important to first draw the shear and moment diagrams for the beam.

Distributed Load. Consider the beam in Fig. 7–13a, which is subjected to an arbitrary distributed load $w = w(x)$ and a concentrated force and couple moment. In the following discussion, the *distributed load* will be considered *positive* when this *load acts upward* as shown. A free-body diagram for a small segment of the beam having a length Δx and located within the region of the distributed loading is shown in Fig. 7–13b. The internal shear force and bending moment are assumed to act in the *positive sense* according to the established sign convention. Note that both the shear and moment acting on the right hand face must be increased by a small, finite amount in order to keep the segment in equilibrium. The distributed loading has been replaced by a resultant force $\Delta F = w(x)\Delta x$ that acts at a fractional distance $k(\Delta x)$ from the right end, where $0 < k < 1$ [for example, if $w(x)$ is *uniform*, $k = \tfrac{1}{2}$].

Relation between the Distributed Load and Shear. If we apply the force equation of equilibrium to the segment, then

$$+\uparrow \Sigma F_y = 0; \qquad V + w(x)\Delta x - (V + \Delta V) = 0$$
$$\Delta V = w(x)\Delta x$$

Dividing by Δx, and letting $\Delta x \to 0$, we get

$$\frac{dV}{dx} = w(x)$$

$$\frac{\text{Slope of}}{\text{shear diagram}} = \frac{\text{Distributed load}}{\text{intensity}} \tag{7–1}$$

Fig. 7–13

If we rewrite this equation in the form $dV = w(x)\,dx$ and perform an integration between any two points A and B on the beam, we see that

$$\Delta V = \int w(x)\,dx \quad (7\text{-}2)$$

Change in shear = Area under loading curve

Relation between the Shear and Moment.
If we apply the moment equation of equilibrium about point O in Fig. 7–13b, we get

$\zeta + \Sigma M_O = 0;\qquad (M + \Delta M) - [w(x)\Delta x]k\Delta x - V\Delta x - M = 0$
$\Delta M = V\Delta x + kw(x)\Delta x^2$

Dividing both sides of this equation by Δx, and letting $\Delta x \to 0$, yields

$$\frac{dM}{dx} = V \quad (7\text{-}3)$$

Slope of moment diagram = Shear

Notice that a maximum bending moment $|M|_{max}$ will occur at the point where the slope $dM/dx = 0$, and this is where the shear is equal to zero.

If Eq. 7–3 is rewritten in the form $dM = V\,dx$ and integrated between any two points A and B on the beam, we have

$$\Delta M = \int V\,dx \quad (7\text{-}4)$$

Change in moment = Area under shear diagram

These equations do not apply at points where a *concentrated* force or couple moment act since these two special cases create *discontinuities* in the shear and moment diagrams.

Force. A free-body diagram of a small segment of the beam in Fig. 7–13a, taken from under one of the forces, is shown in Fig. 7–14a. Here force equilibrium requires

$+\uparrow \Sigma F_y = 0;\qquad \Delta V = F \qquad (7\text{-}5)$

Since the *change in shear is positive*, the shear diagram will "jump" upward when **F** acts upward on the beam. Likewise, the jump in shear (ΔV) is downward when **F** acts downward.

(a)

Fig. 7–14

(b)

Fig. 7-14 (cont.)

Couple Moment. If we remove a segment of the beam in Fig. 7–13a that is located at the couple moment $\mathbf{M}_0$, the free-body diagram shown in Fig. 7–14b results. In this case letting $\Delta x \to 0$, moment equilibrium requires

$$\zeta + \Sigma M = 0; \qquad \Delta M = M_0 \qquad (7\text{–}6)$$

Thus, the *change in moment is positive*, or the moment diagram will "jump" *upward if* $\mathbf{M}_0$ *is clockwise*. Likewise, the jump ΔM is downward when $\mathbf{M}_0$ is counterclockwise.

IMPORTANT POINTS

- The slope of the shear diagram at a point is equal to the intensity of the distributed loading, where positive distributed loading is upward, i.e., $dV/dx = w(x)$.

- The change in the shear ΔV between two points is equal to *the area* under the distributed-loading curve between the points.

- If a concentrated force acts upward on the beam, the shear will jump upward by the same amount.

- The slope of the moment diagram at a point is equal to the shear, i.e., $dM/dx = V$.

- The change in the moment ΔM between two points is equal to the *area* under the shear diagram between the two points.

- If a *clockwise* couple moment acts on the beam, the shear will not be affected; however, the moment diagram will jump *upward* by the amount of the moment.

- Points of *zero shear* represent points of *maximum or minimum moment* since $dM/dx = 0$.

- Because two integrations of $w = w(x)$ are involved to first determine the change in shear, $\Delta V = \int w(x)\, dx$, then to determine the change in moment, $\Delta M = \int V\, dx$, then if the loading curve $w = w(x)$ is a polynomial of degree n, $V = V(x)$ will be a curve of degree $n + 1$, and $M = M(x)$ will be a curve of degree $n + 2$.

This concrete beam is used to support the deck. Its size and the placement of steel reinforcement within it can be determined once the shear and moment diagrams have been established.

EXAMPLE 7.8

Draw the shear and moment diagrams for the cantilever beam in Fig. 7–15a.

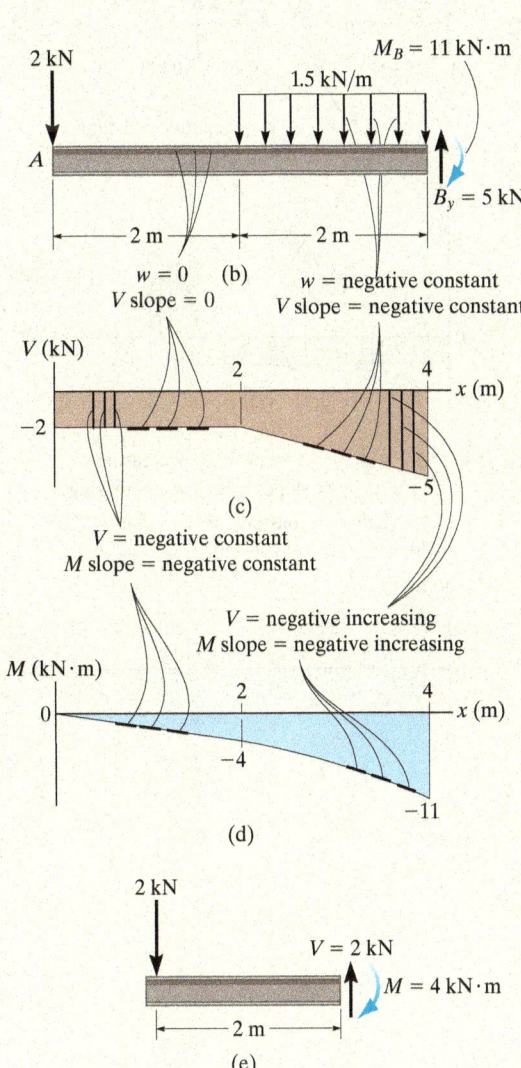

Fig. 7–15

SOLUTION
The support reactions at the fixed support B are shown in Fig. 7–15b.

Shear Diagram. The shear at end A is -2 kN. This value is plotted at $x = 0$, Fig. 7–15c. The shear at $x = 4$ m is -5 kN, the reaction on the beam. Along the beam, notice how the shear diagram is constructed by following the slopes defined by the the loading w.

Moment Diagram. At $x = 0$, $M = 0$, and at $x = 4$ m, $M = -11$ kN·m. These values are plotted in Fig. 7–15d. Notice how construction of the moment diagram is based on knowing that its slope is equal to the shear at each point. The change of moment from $x = 0$ to $x = 2$ m is determined from the area under the shear diagram, and so the moment at $x = 2$ m is

$$M|_{x=2\,\text{m}} = M|_{x=0} + \Delta M = 0 + [-2\,\text{kN}(2\,\text{m})] = -4\,\text{kN}\cdot\text{m}$$

This same value can be determined from the method of sections, Fig. 7–15e.

EXAMPLE 7.9

Draw the shear and moment diagrams for the overhang beam in Fig. 7–16a.

(a)

SOLUTION
The support reactions are shown on the free-body diagram in Fig. 7–16b.

Shear Diagram. At $x = 0$, $V = -2$ kN, and at $x = 6$ m, $V = 0$, Fig. 7–16c. The slopes are determined from the loading and from this the shear diagram is constructed. In particular, notice the positive jump of 10 kN at $x = 4$ m, which is due to the force $B_y = 10$ kN.

Moment Diagram. At $x = 0$, $M = 0$, and at $x = 6$ m, $M = 0$, Fig. 7–16d. Following the behavior of the slope found from the shear diagram, the moment diagram is constructed. The moment at $x = 4$ m is found from the area under the shear diagram.

$$M|_{x=4\text{ m}} = M|_{x=0} + \Delta M = 0 + [-2\text{ kN}(4\text{ m})] = -8\text{ kN}\cdot\text{m}$$

We can also obtain this value by using the method of sections, as shown in Fig. 7–16e.

Fig. 7–16

EXAMPLE 7.10

The shaft in Fig. 7–17a is supported by a thrust bearing at A and a journal bearing at B. Draw the shear and moment diagrams.

(a)

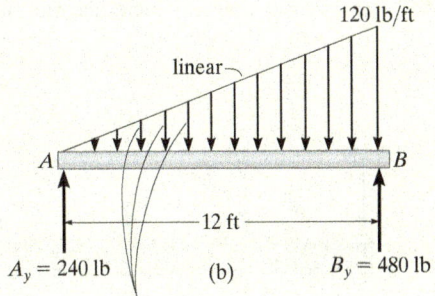

(b)

SOLUTION
The support reactions are shown in Fig. 7–17b.

Shear Diagram. At $x = 0$, $V = 240$ lb and at $x = 12$ ft, $V = -480$ lb, Fig. 7–17c. Following the slope defined by the loading, the shear diagram is constructed. Since the shear changes sign, the point where $V = 0$ must be located. To do this we will use the method of sections. The free-body diagram of the left segment of the shaft, sectioned at an arbitrary position x within the region $0 \leq x < 12$ ft, is shown in Fig. 7–17e. Here the intensity of the distributed load at x is $w = 10x$, which has been found by proportional triangles, i.e., $120/12 = w/x$.

Thus, for $V = 0$,

$+\uparrow \Sigma F_y = 0;\qquad 240 \text{ lb} - \tfrac{1}{2}(10x)x = 0$

$x = 6.93 \text{ ft}$

Moment Diagram. At $x = 0, M = 0$ and at $x = 12$ ft, $M = 0$, Fig. 7–17d. The moment diagram is constructed based on the slope as determined from the shear diagram. The maximum moment occurs at $x = 6.93$ ft, where the shear is equal to zero, since $dM/dx = V = 0$. From Fig. 7–17e,

$\zeta + \Sigma M = 0;$

$M_{max} + \tfrac{1}{2}[(10)(6.93)] 6.93 \left(\tfrac{1}{3}(6.93)\right) - 240(6.93) = 0$

$M_{max} = 1109 \text{ lb} \cdot \text{ft}$

NOTE: Integration, first of the loading w which is linear, produces a shear diagram which is parabolic, and then a moment diagram which is cubic.

(c)

(d)

(e)

Fig. 7–17

FUNDAMENTAL PROBLEMS

All solutions must include a free-body diagram.

F7–13 Draw the shear and moment diagrams for the beam.

Prob. F7–13

F7–14. Draw the shear and moment diagrams for the beam.

Prob. F7–14

F7–15. Draw the shear and moment diagrams for the beam.

Prob. F7–15

F7–16. Draw the shear and moment diagrams for the beam.

Prob. F7–16

F7–17. Draw the shear and moment diagrams for the beam.

Prob. F7–17

F7–18. Draw the shear and moment diagrams for the beam.

Prob. F7–18

PROBLEMS

All solutions must include a free-body diagram.

7–70. Draw the shear and moment diagrams for the beam.

Prob. 7–70

7–71. Draw the shear and moment diagrams for the cantilever beam.

Prob. 7–71

*7–72. Draw the shear and moment diagrams for the beam. There is a pin at C and A is a fixed support.

Prob. 7–72

7–73. Draw the shear and moment diagrams for the beam.

Prob. 7–73

7–74. Draw the shear and moment diagrams for the beam. The supports at A and B are a thrust bearing and journal bearing, respectively.

Prob. 7–74

7–75. Draw the shear and moment diagrams for the beam.

Prob. 7–75

*7–76. Draw the shear and moment diagrams for the beam.

Prob. 7–76

7–79. Draw the shear and moment diagrams for the beam.

Prob. 7–79

7–77. Draw the shear and moment diagrams for the beam.

Prob. 7–77

*7–80. Draw the shear and moment diagrams for the beam.

Prob. 7–80

7–78. Draw the shear and moment diagrams for the beam.

Prob. 7–78

7–81. Draw the shear and moment diagrams for the beam.

Prob. 7–81

7–82. Draw the shear and moment diagrams for the beam.

Prob. 7–82

7–85. Draw the shear and moment diagrams for the beam.

Prob. 7–85

7–83. Draw the shear and moment diagrams for the beam.

Prob. 7–83

7–86. Draw the shear and moment diagrams for the beam.

Prob. 7–86

*7–84. Draw the shear and moment diagrams for the beam.

7–87. Draw the shear and moment diagrams for the beam.

Prob. 7–84

Prob. 7–87

*7–88. Draw the shear and moment diagrams for the beam.

7–91. The beam consists of two segments pin connected at B. Draw the shear and moment diagrams for the beam.

Prob. 7–88

Prob. 7–91

7–89. Draw the shear and moment diagrams for the shaft. The support at A is a thrust bearing and at B it is a journal bearing.

*7–92. Draw the shear and moment diagrams for the beam.

Prob. 7–89

Prob. 7–92

7–90. Draw the shear and moment diagrams for the beam.

7–93. Draw the shear and moment diagrams for the beam.

Prob. 7–90

Prob. 7–93

*7.4 CABLES

Flexible cables and chains combine strength with lightness and often are used in structures for support and to transmit loads from one member to another. When used to support suspension bridges and trolley wheels, cables form the main load-carrying element of the structure. For these cases the weight of the cable can be neglected because the weight is often small compared to the load the cable will support. However, when cables are used as transmission lines and guys for radio antennas and derricks, the cable weight may become important and must be included in the analysis.

Three cases will be considered in what follows. In each case we will make the assumption that the cable is *perfectly flexible* and *inextensible*. Due to its flexibility, the cable offers no resistance to bending, and therefore, the tensile force acting in the cable is always tangent to the cable at points along its length. Being inextensible, the cable has a constant length both before and after the load is applied. As a result, once the load is applied, the geometry of the cable remains unchanged, and the cable or a segment of it can be treated as a rigid body.

Each of the cable segments remains approximately straight as they support the weight of these traffic lights.

Cable Subjected to Concentrated Loads.

When a cable of negligible weight supports several concentrated loads, the cable takes the form of several straight-line segments, each of which is subjected to a constant tensile force. Consider, for example, the cable shown in Fig. 7–18, where the distances h, L_1, L_2, and L_3 and the loads $\mathbf{P}_1$ and $\mathbf{P}_2$ are known. The problem here is to determine the *nine unknowns* consisting of the tension in each of the *three* segments, the *four* components of reaction at A and B, and the *two* sags y_C and y_D at points C and D. For the solution we can write *two* equations of force equilibrium at each of points A, B, C, and D. This results in a total of *eight equations*, and so to obtain the ninth equation, we need to know something about the geometry of the cable. For example, if the cable's total *length* L is specified, then the Pythagorean theorem can be used to relate each of the three segmental lengths, written in terms of h, y_C, y_D, L_1, L_2, and L_3, to the total length L. Unfortunately, this type of problem cannot be solved easily by hand. Another possibility, however, is to specify one of the sags, either y_C or y_D, instead of the cable length. By doing this, the equilibrium equations are then sufficient for obtaining the unknown forces and the remaining sag. Once the sag at each point of loading is obtained, the length of the cable can then be determined by trigonometry. The following example illustrates a procedure for performing the equilibrium analysis for a problem of this type.

Fig. 7–18

EXAMPLE 7.11

Fig. 7–19

Determine the tension in each segment of the cable shown in Fig. 7–19a.

SOLUTION
By inspection, there are four unknown external reactions ($A_x, A_y, E_x,$ and E_y) and four unknown cable tensions, one in each cable segment. These eight unknowns along with the two unknown sags y_B and y_D can be determined from *ten* available equilibrium equations. One method is to apply the force equations of equilibrium ($\Sigma F_x = 0, \Sigma F_y = 0$) to each of the five points A through E. Here, however, we will take a more direct approach.

Consider the free-body diagram for the entire cable, Fig. 7–19b. Thus,

$\xrightarrow{+} \Sigma F_x = 0;$ $\qquad -A_x + E_x = 0$

$\zeta + \Sigma M_E = 0;$

$-A_y(18 \text{ m}) + 4 \text{ kN }(15 \text{ m}) + 15 \text{ kN }(10 \text{ m}) + 3 \text{ kN }(2 \text{ m}) = 0$

$A_y = 12 \text{ kN}$

$+\uparrow \Sigma F_y = 0;$ $\qquad 12 \text{ kN} - 4 \text{ kN} - 15 \text{ kN} - 3 \text{ kN} + E_y = 0$

$E_y = 10 \text{ kN}$

Since the sag $y_C = 12$ m is known, we will now consider the leftmost segment of a section through cable BC, Fig. 7–19c.

$\zeta + \Sigma M_C = 0;$ $A_x(12 \text{ m}) - 12 \text{ kN }(8 \text{ m}) + 4 \text{ kN }(5 \text{ m}) = 0$

$A_x = E_x = 6.33 \text{ kN}$

$\xrightarrow{+} \Sigma F_x = 0;$ $\qquad T_{BC} \cos \theta_{BC} - 6.33 \text{ kN} = 0$

$+\uparrow \Sigma F_y = 0;$ $\qquad 12 \text{ kN} - 4 \text{ kN} - T_{BC} \sin \theta_{BC} = 0$

Thus,

$\theta_{BC} = 51.6°$

$T_{BC} = 10.2 \text{ kN}$ *Ans.*

Fig. 7–19 (cont.)

Proceeding now to analyze the equilibrium of points A, C, and E in sequence, we have

Point A. (Fig. 7–19d).

$\xrightarrow{+} \Sigma F_x = 0;$ $T_{AB} \cos \theta_{AB} - 6.33 \text{ kN} = 0$

$+\uparrow \Sigma F_y = 0;$ $-T_{AB} \sin \theta_{AB} + 12 \text{ kN} = 0$

$$\theta_{AB} = 62.2°$$
$$T_{AB} = 13.6 \text{ kN} \qquad Ans.$$

Point C. (Fig. 7–19e).

$\xrightarrow{+} \Sigma F_x = 0;$ $T_{CD} \cos \theta_{CD} - 10.2 \cos 51.6° \text{ kN} = 0$

$+\uparrow \Sigma F_y = 0;$ $T_{CD} \sin \theta_{CD} + 10.2 \sin 51.6° \text{ kN} - 15 \text{ kN} = 0$

$$\theta_{CD} = 47.9°$$
$$T_{CD} = 9.44 \text{ kN} \qquad Ans.$$

Point E. (Fig. 7–19f).

$\xrightarrow{+} \Sigma F_x = 0;$ $6.33 \text{ kN} - T_{ED} \cos \theta_{ED} = 0$

$+\uparrow \Sigma F_y = 0;$ $10 \text{ kN} - T_{ED} \sin \theta_{ED} = 0$

$$\theta_{ED} = 57.7°$$
$$T_{ED} = 11.8 \text{ kN} \qquad Ans.$$

NOTE: Since the slope angles that the cable segments make with the horizontal have all been determined, it is now possible to determine the sags y_B and y_D, Fig. 7–19a, using trigonometry.

364   CHAPTER 7   INTERNAL FORCES

The cable and suspenders are used to support the uniform load of a gas pipe which crosses the river.

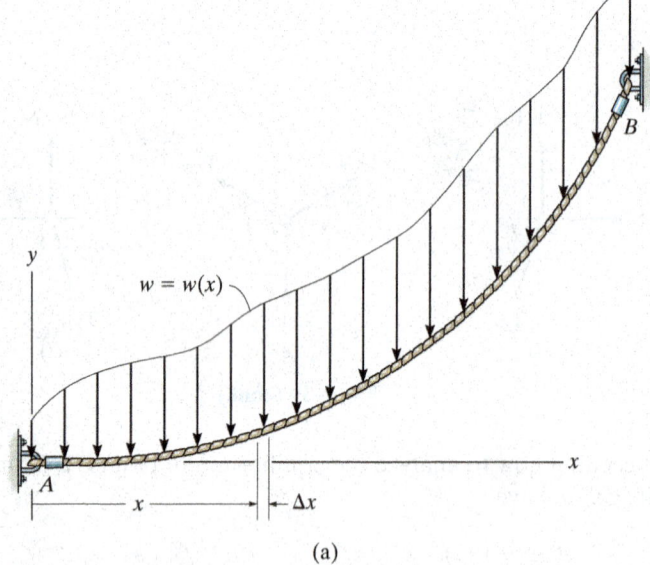

(a)

Fig. 7–20

Cable Subjected to a Distributed Load.

The weightless cable shown in Fig. 7–20a is subjected to a distributed loading $w = w(x)$ that is *measured in the x direction*. The free-body diagram of a small segment of the cable having a length Δs is shown in Fig. 7–20b. Since the tensile force changes in both magnitude and direction along the cable's length, we will denote this change on the free-body diagram by ΔT. Finally, the distributed load is represented by its resultant force $w(x)(\Delta x)$, which acts at a fractional distance $k(\Delta x)$ from point O, where $0 < k < 1$. Applying the equations of equilibrium, we have

$\xrightarrow{+} \Sigma F_x = 0;\qquad -T\cos\theta + (T + \Delta T)\cos(\theta + \Delta\theta) = 0$

$+\uparrow \Sigma F_y = 0;\qquad -T\sin\theta - w(x)(\Delta x) + (T + \Delta T)\sin(\theta + \Delta\theta) = 0$

$\downarrow + \Sigma M_O = 0;\qquad w(x)(\Delta x)k(\Delta x) - T\cos\theta\,\Delta y + T\sin\theta\,\Delta x = 0$

Dividing each of these equations by Δx and taking the limit as $\Delta x \rightarrow 0$, and therefore $\Delta y \rightarrow 0$, $\Delta\theta \rightarrow 0$, and $\Delta T \rightarrow 0$, we obtain

$$\frac{d(T\cos\theta)}{dx} = 0 \qquad (7\text{–}7)$$

$$\frac{d(T\sin\theta)}{dx} - w(x) = 0 \qquad (7\text{–}8)$$

$$\frac{dy}{dx} = \tan\theta \qquad (7\text{–}9)$$

(b)

Fig. 7–20 (cont.)

Integrating Eq. 7–7, we have

$$T \cos \theta = \text{constant} = F_H \tag{7–10}$$

where F_H represents the horizontal component of tensile force at *any point* along the cable.

Integrating Eq. 7–8 gives

$$T \sin \theta = \int w(x) \, dx \tag{7–11}$$

Dividing Eq. 7–11 by Eq. 7–10 eliminates T. Then, using Eq. 7–9, we can obtain the slope of the cable.

$$\tan \theta = \frac{dy}{dx} = \frac{1}{F_H} \int w(x) \, dx$$

Performing a second integration yields

$$\boxed{y = \frac{1}{F_H} \int \left(\int w(x) \, dx \right) dx} \tag{7–12}$$

This equation is used to determine the curve for the cable, $y = f(x)$. The horizontal force component F_H and the two additional constants, say C_1 and C_2, resulting from the integration are determined by applying the boundary conditions for the curve.

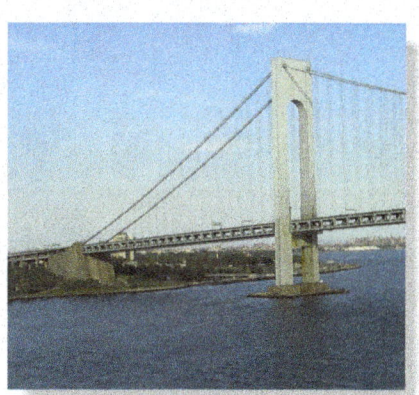

The cables of the suspension bridge exert very large forces on the tower and the foundation block, which have to be accounted for in their design.

EXAMPLE 7.12

The cable of a suspension bridge supports half of the uniform road surface between the two towers at A and B, Fig. 7–21a. If this distributed loading is w_0, determine the maximum force developed in the cable and the cable's required length. The span length L and sag h are known.

(a)

Fig. 7–21

SOLUTION

We can determine the unknowns in the problem by first finding the equation of the curve that defines the shape of the cable using Eq. 7–12. For reasons of symmetry, the origin of coordinates has been placed at the cable's center. Noting that $w(x) = w_0$, we have

$$y = \frac{1}{F_H} \int \left(\int w_0 \, dx \right) dx$$

Performing the two integrations gives

$$y = \frac{1}{F_H} \left(\frac{w_0 x^2}{2} + C_1 x + C_2 \right) \qquad (1)$$

The constants of integration are determined using the boundary conditions $y = 0$ at $x = 0$ and $dy/dx = 0$ at $x = 0$. Substituting into Eq. 1 and its derivative yields $C_1 = C_2 = 0$. The equation of the curve then becomes

$$y = \frac{w_0}{2 F_H} x^2 \qquad (2)$$

This is the equation of a *parabola*. The constant F_H is obtained using the boundary condition $y = h$ at $x = L/2$. Thus,

$$F_H = \frac{w_0 L^2}{8h} \qquad (3)$$

Therefore, Eq. 2 becomes

$$y = \frac{4h}{L^2} x^2 \qquad (4)$$

Since F_H is known, the tension in the cable may now be determined using Eq. 7–10, written as $T = F_H/\cos\theta$. For $0 \leq \theta < \pi/2$, the maximum tension will occur when θ is *maximum*, i.e., at point B, Fig. 7–21a. From Eq. 2, the slope at this point is

$$\left.\frac{dy}{dx}\right|_{x=L/2} = \tan\theta_{max} = \left.\frac{w_0}{F_H} x\right|_{x=L/2}$$

or

$$\theta_{max} = \tan^{-1}\left(\frac{w_0 L}{2 F_H}\right) \qquad (5)$$

And so

$$T_{max} = \frac{F_H}{\cos(\theta_{max})} \qquad (6)$$

Using the triangular relationship shown in Fig. 7–21b, which is based on Eq. 5, Eq. 6 may be written as

$$T_{max} = \frac{\sqrt{4 F_H^2 + w_0^2 L^2}}{2}$$

(b)

Fig. 7–21 (cont.)

Substituting Eq. 3 into this equation yields

$$T_{max} = \frac{w_0 L}{2} \sqrt{1 + \left(\frac{L}{4h}\right)^2} \qquad \textit{Ans.}$$

For a differential segment of cable length ds, we can write

$$ds = \sqrt{(dx)^2 + (dy)^2} = \sqrt{1 + \left(\frac{dy}{dx}\right)^2}\, dx$$

Hence, the total length of the cable can be determined by integration. Using Eq. 4, we have

$$\mathcal{L} = \int ds = 2 \int_0^{L/2} \sqrt{1 + \left(\frac{8h}{L^2} x\right)^2}\, dx \qquad (7)$$

Integrating yields

$$\mathcal{L} = \frac{L}{2}\left[\sqrt{1 + \left(\frac{4h}{L}\right)^2} + \frac{L}{4h}\sinh^{-1}\left(\frac{4h}{L}\right)\right] \qquad \textit{Ans.}$$

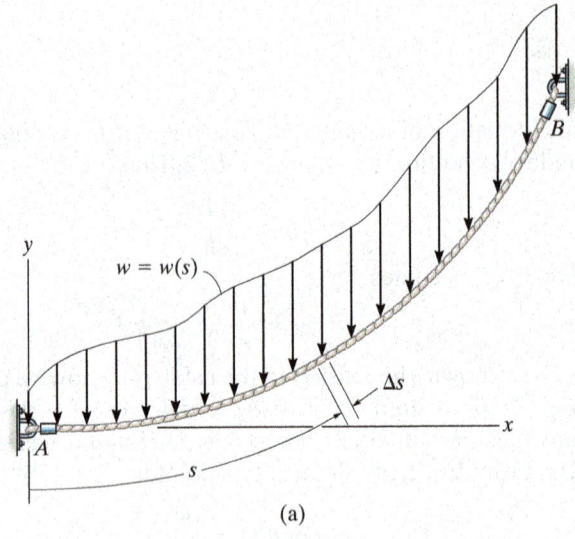

Fig. 7–22

Cable Subjected to Its Own Weight.
When the weight of a cable becomes important, the loading function w along the cable will be a function of the arc length s rather than the projected length x, that is, $w = w(s)$, as shown in Fig. 7–22a. The free-body diagram for a small segment Δs of the cable is shown in Fig. 7–22b. Applying the equilibrium equations to the force system on this diagram, one obtains relationships identical to those given by Eqs. 7–7 through 7–9, but with s replacing x in Eqs. 7–7 and 7–8. Therefore,

$$T \cos \theta = F_H$$

$$T \sin \theta = \int w(s) \, ds \tag{7–13}$$

$$\frac{dy}{dx} = \frac{1}{F_H} \int w(s) \, ds \tag{7–14}$$

To perform a direct integration of Eq. 7–14, it is necessary to replace dy/dx by ds/dx. Since

$$ds = \sqrt{dx^2 + dy^2}$$

then

$$\frac{dy}{dx} = \sqrt{\left(\frac{ds}{dx}\right)^2 - 1}$$

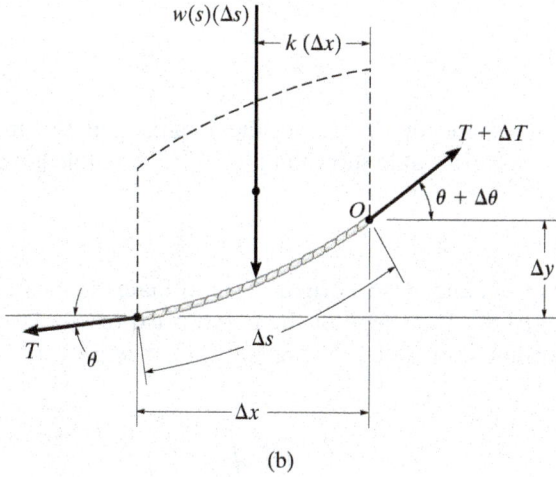

(b)

Fig. 7–22 (cont.)

Therefore,

$$\frac{ds}{dx} = \left[1 + \frac{1}{F_H^2}\left(\int w(s)\, ds\right)^2\right]^{1/2}$$

Separating the variables and integrating we obtain

$$x = \int \frac{ds}{\left[1 + \frac{1}{F_H^2}\left(\int w(s)\, ds\right)^2\right]^{1/2}} \qquad (7\text{--}15)$$

The two constants of integration, say C_1 and C_2, are found using the boundary conditions for the curve.

Electrical transmission towers must be designed to support the weight of the suspended power lines.

EXAMPLE 7.13

Determine the shape of the curve, the length, and the maximum tension in the uniform cable shown in Fig. 7–23. The cable has a weight per unit length of $w_0 = 5$ N/m.

Fig. 7–23

SOLUTION

For reasons of symmetry, the origin of coordinates is located at the center of the cable. The shape of the curve is expressed as $y = f(x)$. We can determine it by first applying Eq. 7–15, where $w(s) = w_0$.

$$x = \int \frac{ds}{\left[1 + (1/F_H^2)\left(\int w_0 \, ds\right)^2\right]^{1/2}}$$

Integrating the term under the integral sign, we have

$$x = \int \frac{ds}{[1 + (1/F_H^2)(w_0 s + C_1)^2]^{1/2}}$$

Substituting $u = (1/F_H)(w_0 s + C_1)$ so that $du = (w_0/F_H) \, ds$, a second integration yields

$$x = \frac{F_H}{w_0}(\sinh^{-1} u + C_2)$$

or

$$x = \frac{F_H}{w_0}\left\{\sinh^{-1}\left[\frac{1}{F_H}(w_0 s + C_1)\right] + C_2\right\} \quad (1)$$

To evaluate the two constants note that, from Eq. 7–14,

$$\frac{dy}{dx} = \frac{1}{F_H}\int w_0 \, ds \quad \text{or} \quad \frac{dy}{dx} = \frac{1}{F_H}(w_0 s + C_1)$$

Since $dy/dx = 0$ at $s = 0$, then $C_1 = 0$. Thus,

$$\frac{dy}{dx} = \frac{w_0 s}{F_H} \quad (2)$$

The constant C_2 may be found by using the condition $s = 0$ at $x = 0$ in Eq. 1, in which case $C_2 = 0$. To obtain the curve, solve for s in Eq. 1, which yields

$$s = \frac{F_H}{w_0}\sinh\left(\frac{w_0}{F_H}x\right) \quad (3)$$

Now substitute into Eq. 2, which gives

$$\frac{dy}{dx} = \sinh\left(\frac{w_0}{F_H}x\right)$$

Hence,
$$y = \frac{F_H}{w_0}\cosh\left(\frac{w_0}{F_H}x\right) + C_3$$

If the boundary condition $y = 0$ at $x = 0$ is applied, the constant $C_3 = -F_H/w_0$, and therefore

$$y = \frac{F_H}{w_0}\left[\cosh\left(\frac{w_0}{F_H}x\right) - 1\right] \quad (4)$$

This equation defines the shape of a *catenary curve*. The constant F_H is obtained by using the boundary condition that $y = h$ at $x = L/2$, in which case

$$h = \frac{F_H}{w_0}\left[\cosh\left(\frac{w_0 L}{2F_H}\right) - 1\right] \quad (5)$$

Since $w_0 = 5$ N/m, $h = 6$ m, and $L = 20$ m, Eqs. 4 and 5 become

$$y = \frac{F_H}{5\text{ N/m}}\left[\cosh\left(\frac{5\text{ N/m}}{F_H}x\right) - 1\right] \quad (6)$$

$$6\text{ m} = \frac{F_H}{5\text{ N/m}}\left[\cosh\left(\frac{50\text{ N}}{F_H}\right) - 1\right] \quad (7)$$

Equation 7 can be solved for F_H by using a trial-and-error procedure. The result is

$$F_H = 45.9\text{ N}$$

and therefore the deflection curve, Eq. 6, becomes

$$y = 9.19[\cosh(0.109x) - 1]\text{ m} \quad \textit{Ans.}$$

Using Eq. 3, with $x = 10$ m, the half-length of the cable is

$$\frac{\mathscr{L}}{2} = \frac{45.9\text{ N}}{5\text{ N/m}}\sinh\left[\frac{5\text{ N/m}}{45.9\text{ N}}(10\text{ m})\right] = 12.1\text{ m}$$

Hence,
$$\mathscr{L} = 24.2\text{ m} \quad \textit{Ans.}$$

Since $T = F_H/\cos\theta$, the maximum tension occurs where θ is maximum, i.e., at $s = \mathscr{L}/2 = 12.1$ m. Using Eq. 2 yields

$$\left.\frac{dy}{dx}\right|_{s=12.1\text{ m}} = \tan\theta_{max} = \frac{5\text{ N/m}(12.1\text{ m})}{45.9\text{ N}} = 1.32$$

$$\theta_{max} = 52.8°$$

And so,
$$T_{max} = \frac{F_H}{\cos\theta_{max}} = \frac{45.9\text{ N}}{\cos 52.8°} = 75.9\text{ N} \quad \textit{Ans.}$$

PROBLEMS

All solutions must include a free-body diagram.

7–94. The cable supports the three loads shown. Determine the sags y_B and y_D of B and D. Take $P_1 = 800$ N, $P_2 = 500$ N.

7–95. The cable supports the three loads shown. Determine the magnitude of P_1 if $P_2 = 600$ N and $y_B = 3$ m. Also find sag y_D.

7–97. The cable supports the loading shown. Determine the distance x_B the force at B acts from A. Set $P = 800$ N.

7–98. The cable supports the loading shown. Determine the magnitude of the horizontal force P so that $x_B = 5$ m.

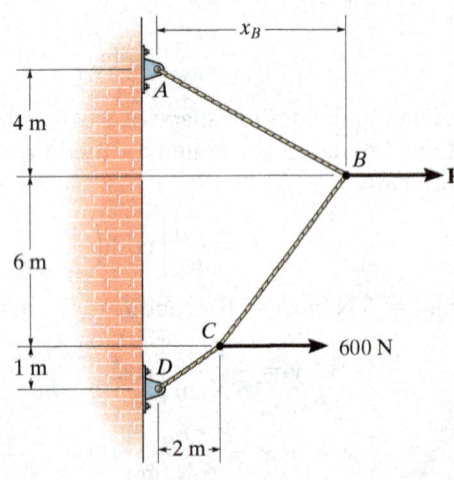

Probs. 7–94/95

Probs. 7–97/98

*__7–96.__ Determine the tension in each segment of the cable and the cable's total length.

7–99. The cable supports the three loads shown. Determine the sags y_B and y_D of points B and D. Take $P_1 = 400$ lb, $P_2 = 250$ lb.

*__7–100.__ The cable supports the three loads shown. Determine the magnitude of P_1 if $P_2 = 300$ lb and $y_B = 8$ ft. Also find the sag y_D.

Prob. 7–96

Probs. 7–99/100

7–101. Cable $ABCD$ supports the 120-kg uniform beam. Determine the maximum tension in this cable and the sag of point B.

***7–104.** The cable AB is subjected to a triangular loading. If the angles with the tangents at points A and B are $45°$ and $80°$, respectively, determine the deflection curve of the cable and the maximum tension developed in the cable.

Prob. 7–101

Prob. 7–104

7–102. Determine the maximum uniform loading w, measured in lb/ft, that the cable can support if it is capable of sustaining a maximum tension of 3000 lb before it will break.

7–103. The cable is subjected to a uniform loading of $w = 250$ lb/ft. Determine the maximum and minimum tension in the cable.

7–105. If $x = 2$ ft and the crate weighs 300 lb, which cable segment AB, BC, or CD has the greatest tension? What is this force and what is the sag y_B?

7–106. If $y_B = 1.5$ ft, determine the largest weight of the crate and its placement x so that neither cable segment AB, BC, or CD is subjected to a tension that exceeds 200 lb.

Probs. 7–102/103

Probs. 7–105/106

7–107. The cable supports a girder which weighs 850 lb/ft. Determine the tension in the cable at points A, B, and C.

Prob. 7–107

***7–108.** The cable is subjected to the parabolic loading $w = 200(1 - (x/50)^2)$ lb/ft, where x is in ft. Determine the equation $y = f(x)$ which defines the cable shape AB, and the maximum tension in the cable. The slope of the cable at A is zero.

Prob. 7–108

7–109. If the pipe has a mass per unit length of 1500 kg/m, determine the maximum tension developed in the cable.

7–110. If the pipe has a mass per unit length of 1500 kg/m, determine the minimum tension developed in the cable.

Probs. 7–109/110

7–111. The power line has a mass of 500 g/m. If it has a length of 32 m between the poles, determine the maximum tension in the line and its sag.

Prob. 7–111

***7–112.** The cable will break when the maximum tension reaches $T_{max} = 10$ kN. Determine the minimum sag h if it supports the uniform distributed load of $w = 600$ N/m.

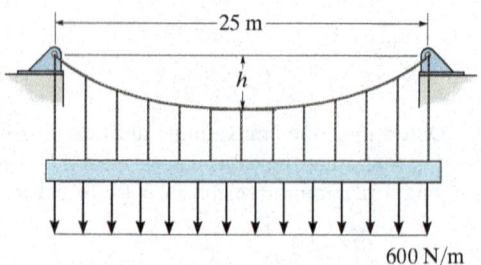

Prob. 7–112

7–113. The chain has a weight of 3 lb/ft. Determine the tension at points A, B, and C.

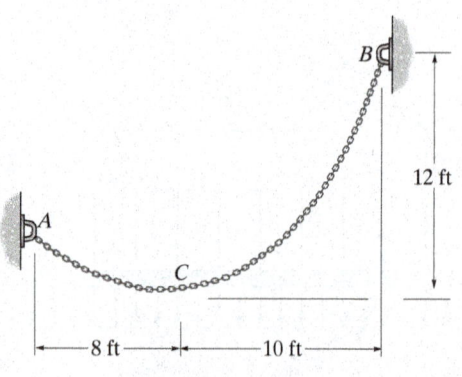

Prob. 7–113

7–114. If the side loading on the suspension bridge in Example 7.12 is $w_0 = 0.75$ kN/m and $L = 190$ m, $h = 20$ m, determine the maximum tension in the cable.

7–115. The cable will break when the maximum tension reaches $T_{max} = 12$ kN. Determine the uniform distributed load w required to develop this maximum tension. Set $h = 6$ m.

Prob. 7–115

*7–116. The uniform beam weighs 80 lb/ft and is held in the horizontal position by means of cable AB, which has a weight of 10 lb/ft. If the slope of the cable at A is 15°, determine the length of the cable.

Prob. 7–116

7–117. The cable has a weight of 8 N/m and length of 10 m. If the tension in the cable at C is 300 N, determine the distance from A to C.

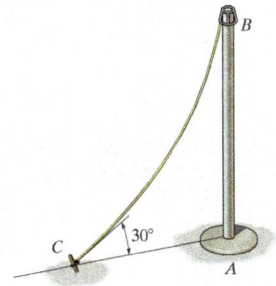

Prob. 7–117

7–118. Show that the deflection curve of the cable discussed in Example 7.13 reduces to Eq. 4 in Example 7.12 when the *hyperbolic cosine function* is expanded in terms of a series and only the first two terms are retained. (The answer indicates that the *catenary* may be replaced by a *parabola* in the analysis of problems in which the sag is small. In this case, the cable weight is assumed to be uniformly distributed along the horizontal.)

7–119. The power line is supported at A by the tower. If the cable weighs 0.75 lb/ft, and the sag $s = 3$ ft, determine the resultant horizontal force the cable exerts at A.

*7–120. The power line is supported at A by the tower. If the cable weighs 0.75 lb/ft, determine the total length of the cable, BAC. Set $s = 3$ ft.

7–121. The power line is supported at A by the tower. If the cable weighs 0.75 lb/ft, determine the required sag s so that the resultant horizontal force the cable exerts at A is zero.

Probs. 7–119/120/121

7–122. The man picks up the 52-ft chain and holds it just high enough so it is completely off the ground. The chain has points of attachment A and B that are 50 ft apart. If the chain has a weight of 3 lb/ft, and the man weighs 150 lb, determine the force he exerts on the ground. Also, how high h must he lift the chain? *Hint:* The slopes at A and B are zero.

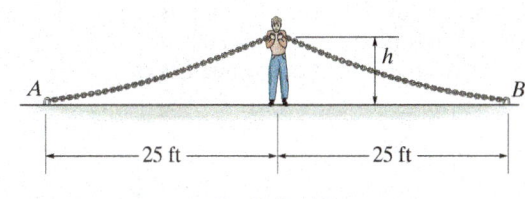

Prob. 7–122

CHAPTER REVIEW

Internal Loadings

If a coplanar force system acts on a member, then in general a resultant internal *normal force* **N**, *shear force* **V**, and *bending moment* **M** will act at any cross section along the member. For two-dimensional problems the positive directions of these loadings are shown in the figure.

The resultant internal normal force, shear force, and bending moment are determined using the method of sections. To find them, the member is sectioned at the point C where the internal loadings are to be determined. A free-body diagram of one of the sectioned parts is then drawn and the internal loadings are shown in their positive directions.

The resultant normal force is determined by summing forces normal to the cross section. The resultant shear force is found by summing forces tangent to the cross section, and the resultant bending moment is found by summing moments about the geometric center or centroid of the cross-sectional area.

$\Sigma F_x = 0$

$\Sigma F_y = 0$

$\Sigma M_C = 0$

If the member is subjected to a three-dimensional loading, then, in general, a *torsional moment* will also act on the cross section. It can be determined by summing moments about an axis that is perpendicular to the cross section and passes through its centroid.

Shear and Moment Diagrams

To construct the shear and moment diagrams for a member, it is necessary to section the member at an arbitrary point, located a distance x from the left end.

If the external loading consists of changes in the distributed load, or a series of concentrated forces and couple moments act on the member, then different expressions for V and M must be determined within regions between any load discontinuities.

The unknown shear and moment are indicated on the cross section in the positive direction according to the established sign convention, and then the internal shear and moment are determined as functions of x.

Each of the functions of the shear and moment is then plotted to create the shear and moment diagrams.

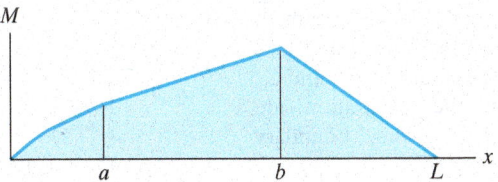

Relations between Shear and Moment

It is possible to plot the shear and moment diagrams quickly by using the differential relationships that exist between w and V and V and M.

The slope of the shear diagram is equal to the distributed loading at any point. The slope is positive if the distributed load acts upward, and vice-versa.

$$\frac{dV}{dx} = w$$

The slope of the moment diagram is equal to the shear at any point. The slope is positive if the shear is positive, or vice-versa.

$$\frac{dM}{dx} = V$$

The change in shear between any two points is equal to the area under the distributed loading curve between the points.

$$\Delta V = \int w\, dx$$

The change in the moment is equal to the area of the shear diagram between the points.

$$\Delta M = \int V\, dx$$

Cables

When a flexible and inextensible cable is subjected to a series of concentrated forces, then the analysis of the cable can be performed by using the equations of equilibrium applied to free-body diagrams of either segments of the cable or points of application of the loading.

If external distributed loads or the weight of the cable are to be considered, then the shape of the cable must be determined by first analyzing the forces on a differential segment of the cable and then integrating this result. The two constants, say C_1 and C_2, resulting from the integration are determined by applying boundary conditions for the cable.

$$y = \frac{1}{F_H} \int \left(\int w(x)\, dx \right) dx$$

Distributed load

$$x = \int \frac{ds}{\left[1 + \frac{1}{F_H^2} \left(\int w(s)\, ds \right)^2 \right]^{1/2}}$$

Cable weight

REVIEW PROBLEMS

All solutions must include a free-body diagram.

R7–1. Determine the normal force, shear force, and moment at points D and E in the frame.

R7–3. Draw the shear and moment diagrams for the beam.

Prob. R7–3

Prob. R7–1

R7–4. Draw the shear and moment diagrams for the beam.

Prob. R7–4

R7–2. Determine the normal force, shear force, and moment at points B and C in the beam.

R7–5. Draw the shear and moment diagrams for the beam.

Prob. R7–5

R7–6. A chain is suspended between points at the same elevation and spaced a distance of 60 ft apart. If it has a weight per unit length of 0.5 lb/ft and the sag is 3 ft, determine the maximum tension in the chain.

Prob. R7–2

CHAPTER 8

This brake is designed to resist the frictional forces developed by the wheel in the most efficient manner. In this chapter we will study dry friction, and show how to analyze friction forces for various types of engineering applications.

FRICTION

CHAPTER OBJECTIVES

- To introduce the concept of dry friction and show how to analyze the equilibrium of rigid bodies subjected to this force.
- To present specific applications of frictional force analysis on wedges, screws, belts, and bearings.
- To investigate the concept of rolling resistance.

8.1 CHARACTERISTICS OF DRY FRICTION

Friction is a force that resists the movement between two contacting surfaces that slide relative to one another. This force always acts *tangent* to the surface at the points of contact and is directed so as to oppose the possible or existing motion between the surfaces.

In this chapter, we will study the effects of **dry friction**, which is sometimes called *Coulomb friction* since its characteristics were studied extensively by the French physicist Charles-Augustin de Coulomb in 1781. Dry friction occurs between the contacting surfaces of bodies when there is no lubricating fluid.*

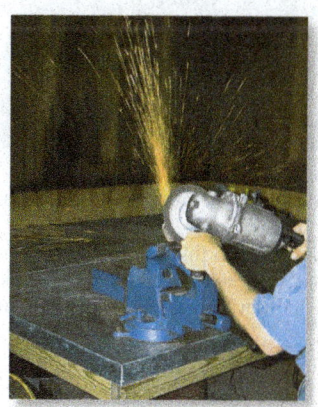

The heat generated by the abrasive action of friction can be noticed when using this grinder to sharpen a metal blade.

---

*Another type of friction, called fluid friction, is studied in fluid mechanics.

382   Chapter 8 Friction

Fig. 8–1

Regardless of the weight of the rake or shovel that is suspended, the device has been designed so that the small roller holds the handle in equilibrium due to frictional forces that develop at the points of contact, A, B, C.

Theory of Dry Friction. The theory of dry friction can be explained by considering the effects caused by pulling horizontally on a block of uniform weight **W** which is resting on a rough horizontal surface that is *nonrigid or deformable*, Fig. 8–1a. As shown on the free-body diagram of the block, Fig. 8–1b, the floor exerts an uneven *distribution* of both *normal force* ΔN_n and *frictional force* ΔF_n along the contacting surface. For equilibrium, the normal forces must act *upward* to balance the block's weight **W**, and the frictional forces act to the left to prevent the applied force **P** from moving the block to the right. Close examination of the contacting surfaces between the floor and block reveals how these frictional and normal forces develop, Fig. 8–1c. It can be seen that many microscopic irregularities exist between the two surfaces and, as a result, reactive forces ΔR_n are developed at each point of contact.* As shown, each reactive force contributes both a frictional component ΔF_n and a normal component ΔN_n.

Equilibrium. The effect of the *distributed* normal and frictional loadings is indicated by their *resultants* **N** and **F** on the free-body diagram, Fig. 8–1d. Notice that **N** acts a distance x to the right of the line of action of **W**, Fig. 8–1d. This location, which coincides with the centroid or geometric center of the normal force distribution in Fig. 8–1b, is necessary in order to balance the "tipping effect" caused by **P**. For example, if **P** is applied at a height h from the surface, Fig. 8–1d, then moment equilibrium about point O is satisfied if $Wx = Ph$ or $x = Ph/W$.

---

*This explanation based only on mechanical interactions is referred to as a classical approach. For a more complete analysis a detailed treatment of the nature of frictional forces must also include the effects of temperature, density, cleanliness, and atomic or molecular adhesion between the contacting surfaces, as well as the deformation and fracture at the points of contact. See, for example, M. Nosonovsky, *Mechanical Engineering*, March 2018.

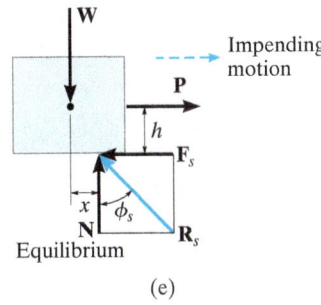

Equilibrium
(e)

Fig. 8–1 (cont.)

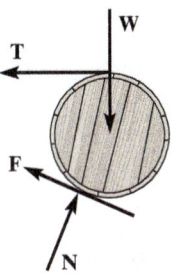

Some objects, such as this barrel, may not be on the verge of slipping, and therefore the friction force **F** must be determined strictly from the equations of equilibrium.

Impending Motion. In cases where the surfaces of contact are rather "slippery," the frictional force **F** may *not* be great enough to balance **P**, and consequently the block will tend to slip. In other words, as *P* is slowly increased, *F* correspondingly increases until it attains a certain maximum value F_s, called the **limiting static frictional force**, Fig. 8–1e. When this value is reached, the block is in *unstable equilibrium* since any further increase in *P* will cause the block to move. Experimentally, it has been determined that F_s is *directly proportional* to the resultant normal force *N*. Expressed mathematically,*

$$F_s = \mu_s N \quad (8\text{--}1)$$

where the constant of proportionality, μ_s (mu "sub" *s*), is called the **coefficient of static friction**.

Thus, when the block is on the *verge of sliding*, the normal force **N** and frictional force $\mathbf{F}_s$ combine to create a resultant $\mathbf{R}_s$, Fig. 8–1e. The angle ϕ_s (phi "sub" *s*) that $\mathbf{R}_s$ makes with **N** is called the **angle of static friction**. From the figure,

$$\phi_s = \tan^{-1}\left(\frac{F_s}{N}\right) = \tan^{-1}\left(\frac{\mu_s N}{N}\right) = \tan^{-1} \mu_s$$

Typical values for μ_s are given in Table 8–1. As indicated, these values will vary, since experimental testing was done under variable conditions of roughness and cleanliness of the contacting surfaces. For applications, therefore, it is very important that both caution and judgment be exercised when selecting a coefficient of friction for a given set of conditions. When a more accurate calculation of F_s is required, the coefficient of friction should be determined directly by an experiment that involves the two contacting materials to be used.

TABLE 8–1 Typical Values for μ_s	
Contact Materials	Coefficient of Static Friction (μ_s)
Metal on ice	0.03–0.05
Wood on wood	0.30–0.70
Leather on wood	0.20–0.50
Leather on metal	0.30–0.60
Copper on copper	0.74–1.21

*As noted in the previous footnote, this empirical result has very limited applications in modern day engineering. It is used in this chapter to illustrate how friction forces are considered for a basic equilibrium analysis.

Fig. 8–2

Motion. If the magnitude of **P** acting on the block is increased so that it becomes slightly greater than F_s, the frictional force at the contacting surface will drop to a smaller value F_k, called the **kinetic frictional force**. The block will then begin to slide with increasing speed, Fig. 8–2a. As this occurs, the block will "ride" on top of these peaks at the points of contact, as shown in Fig. 8–2b. The continued breakdown of the nonrigid surface is the dominant mechanism creating kinetic friction.

Experiments with sliding blocks indicate that the magnitude of the kinetic friction force is directly proportional to the magnitude of the resultant normal force, expressed mathematically as

$$F_k = \mu_k N \qquad (8\text{–}2)$$

Here the constant of proportionality, μ_k, is called the **coefficient of kinetic friction**. Typical values for μ_k are approximately 25 percent *smaller* than those listed in Table 8–1 for μ_s.

As shown in Fig. 8–2a, in this case, the resultant force at the surface of contact, $\mathbf{R}_k$, has a line of action defined by ϕ_k. This angle is referred to as the **angle of kinetic friction**, where

$$\phi_k = \tan^{-1}\left(\frac{F_k}{N}\right) = \tan^{-1}\left(\frac{\mu_k N}{N}\right) = \tan^{-1} \mu_k$$

By comparison, $\phi_s \geq \phi_k$.

The above effects regarding friction can be summarized by referring to the graph in Fig. 8–3, which shows the variation of the frictional force F versus the applied load P. Here the frictional force is categorized in three different ways:

- F is a *static frictional force* if equilibrium is maintained.

- F is a *limiting static frictional force* F_s when it reaches a maximum value needed to maintain equilibrium.

- F is a *kinetic frictional force* F_k when sliding occurs at the contacting surface.

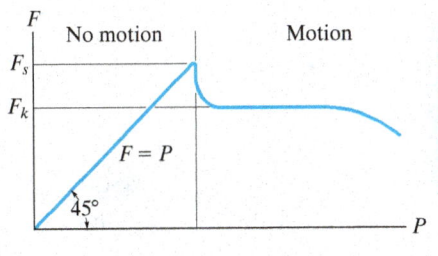

Fig. 8–3

Notice also from the graph that for very large values of P or for high speeds, aerodynamic effects will cause F_k and likewise μ_k to begin to decrease.

Characteristics of Dry Friction.
As a result of *experiments* that pertain to the foregoing discussion, we can state the following rules which apply to bodies subjected to dry friction.

- The frictional force acts *tangent* to the contacting surfaces in a direction *opposed* to the *motion* or tendency for motion of one surface relative to another.

- The maximum static frictional force F_s that can be developed is independent of the area of contact between the surfaces, provided the normal pressure is not very low nor great enough to severely deform or crush the surfaces between the bodies.

- The maximum static frictional force is generally greater than the kinetic frictional force for any two surfaces of contact. However, if one of the bodies is moving with a *very low velocity* over the surface of another, F_k becomes approximately equal to F_s, i.e., $\mu_s \approx \mu_k$.

- When *slipping* at the surface of contact is *about to occur*, the maximum static frictional force is proportional to the normal force, such that $F_s = \mu_s N$.

- When *slipping* at the surface of contact is *occurring*, the kinetic frictional force is proportional to the normal force, such that $F_k = \mu_k N$.

8.2 PROBLEMS INVOLVING DRY FRICTION

If a rigid body is in equilibrium when it is subjected to a system of forces that includes the effect of friction, the force system must satisfy not only the equations of equilibrium but *also* the laws that govern the frictional forces.

(a)

(b)

Fig. 8–4

Types of Friction Problems.
In general, there are three types of static problems involving dry friction. They can easily be classified once free-body diagrams are drawn and the total number of unknowns are identified and compared with the total number of available equilibrium equations.

No Apparent Impending Motion.
Problems in this category are strictly equilibrium problems, which require the number of unknowns to be *equal* to the number of available equilibrium equations. Once the frictional forces are determined from the solution, however, their numerical values must be checked to be sure they satisfy the inequality $F \le \mu_s N$; otherwise, slipping will occur and the body will not remain in equilibrium. A problem of this type is shown in Fig. 8–4a. Here we must determine the frictional forces at A and C to check if the equilibrium position of the two-member frame can be maintained. If the members are uniform and have known weights of 100 N each, then the free-body diagrams are as shown in Fig. 8–4b. There are six unknown force components which can be determined *strictly* from the six equilibrium equations (three for each member). Once F_A, N_A, F_C, and N_C are determined, then the members will remain in equilibrium provided $F_A \le 0.3 N_A$ and $F_C \le 0.5 N_C$ are satisfied.

(a)

(b)

Fig. 8–5

Impending Motion at All Points of Contact.
In this case the total number of unknowns will *equal* the total number of available equilibrium equations *plus* the total number of available frictional equations, $F = \mu N$. When *motion is impending* at the points of contact, then $F_s = \mu_s N$; whereas if the body is *slipping*, then $F_k = \mu_k N$. For example, consider the problem of finding the smallest angle θ at which the 100-N bar in Fig. 8–5a can be placed against the wall without slipping. The free-body diagram of the bar is shown in Fig. 8–5b. Here the *five* unknowns are determined from the *three* equilibrium equations and *two* static frictional equations which apply at *both* points of contact, so that $F_A = 0.3 N_A$ and $F_B = 0.4 N_B$.

Impending Motion at Some Points of Contact. For these types of problems, the number of unknowns will be *less* than the number of available equilibrium equations plus the number of available frictional equations or conditional equations for tipping. As a result, several possibilities for motion or impending motion will exist and the problem will involve a determination of the kind of motion which actually occurs. For example, consider the two-member frame in Fig. 8–6a. In this problem we wish to determine the horizontal force P needed to cause movement. If each member has a weight of 100 N, then the free-body diagrams are as shown in Fig. 8–6b. There are *seven* unknowns. For a unique solution we must satisfy the *six* equilibrium equations (three for each member) and only *one* of two possible static frictional equations. This means that as P increases it will either cause slipping at A and no slipping at C, so that $F_A = 0.3 N_A$ and $F_C \leq 0.5 N_C$; or slipping will occur at C and no slipping at A, in which case $F_C = 0.5 N_C$ and $F_A \leq 0.3 N_A$. The actual situation can be determined by calculating P for each case, and then choosing the case for which P is *smaller*. If in both cases the *same value* for P is calculated, which would be highly improbable, then slipping at both points occurs simultaneously; i.e., the *seven unknowns* would have to satisfy *eight equations*.

Fig. 8–6

Equilibrium versus Frictional Equations. Whenever we solve a problem such as the one in Fig. 8–4, where the friction force F is to be an "equilibrium force" and satisfies the inequality $F < \mu_s N$, then we can assume the sense of direction of F on the free-body diagram. The correct sense is made known *after* solving the equations of equilibrium for F. If F is a negative scalar the sense of **F** is the reverse of that which was assumed. This convenience of *assuming* the sense of **F** is possible because the equilibrium equations equate to zero the *components of vectors* acting in the *same direction*. However, in cases where the frictional equation $F = \mu N$ is used in the solution of a problem, then this convenience of *assuming* the sense of **F** is *lost*, since the frictional equation relates only the *magnitudes* of two *perpendicular* vectors. Consequently, **F** must always be shown acting with its *correct sense* on the free-body diagram, *whenever* the frictional equation is used for the solution of a problem.

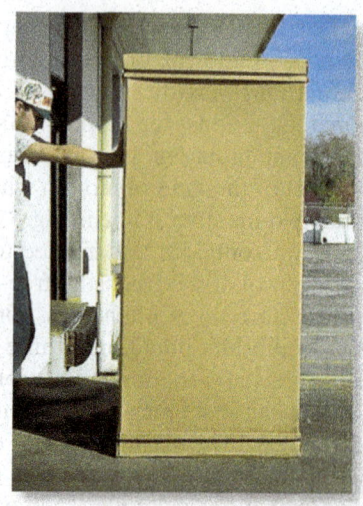

Depending upon where the man pushes on the crate, it will either tip or slip.

IMPORTANT POINTS

- Friction is a tangential force that resists the movement of one surface relative to another.

- If no sliding occurs, the maximum value for the friction force is equal to the product of the coefficient of static friction and the normal force at the surface, $F_s = \mu_s N$.

- If sliding occurs, then the friction force is the product of the coefficient of kinetic friction and the normal force at the surface, $F_k = \mu_k N$.

- There are three types of static friction problems. Each of these problems is analyzed by first drawing the necessary free-body diagrams, and then applying the equations of equilibrium, while satisfying either the conditions of friction or the possibility of tipping.

PROCEDURE FOR ANALYSIS

Equilibrium problems involving dry friction can be solved using the following procedure.

Free-Body Diagrams.
- Draw the necessary free-body diagrams, and unless it is stated in the problem that impending motion or slipping occurs, *always* show the frictional forces as unknowns (i.e., *do not assume* $F = \mu N$).

- Determine the number of unknowns and compare this with the number of available equilibrium equations.

- If there are more unknowns than equations of equilibrium, it will be necessary to apply the frictional equation at some, if not all, points of contact to obtain the extra equations needed for a complete solution.

- If the equation $F = \mu N$ is to be used, it will be necessary to show **F** acting in the correct sense of direction on the free-body diagram.

Equations of Equilibrium and Friction.
- Apply the equations of equilibrium and the necessary frictional equations (or conditional equations if tipping is possible) and solve for the unknowns.

- If the problem involves a three-dimensional force system such that it becomes difficult to obtain the force components or the necessary moment arms, apply the equations of equilibrium using Cartesian vectors.

EXAMPLE 8.1

The uniform crate shown in Fig. 8–7a has a mass of 20 kg. If a force $P = 80$ N is applied to the crate, determine if it remains in equilibrium. The coefficient of static friction is $\mu_s = 0.3$.

(a)

SOLUTION

Free-Body Diagram. As shown in Fig. 8–7b, the *resultant* normal force $\mathbf{N}_C$ must act a distance x from the crate's centerline in order to counteract the tipping effect caused by $\mathbf{P}$. There are *three unknowns*, F, N_C, and x, which can be determined strictly from the *three* equations of equilibrium.

Equations of Equilibrium.

$\xrightarrow{+} \Sigma F_x = 0;$ $\qquad 80 \cos 30° \text{ N} - F = 0$

$+\uparrow \Sigma F_y = 0;$ $\qquad -80 \sin 30° \text{ N} + N_C - 196.2 \text{ N} = 0$

$\zeta + \Sigma M_O = 0;$
$$80 \sin 30° \text{ N}(0.4 \text{ m}) - 80 \cos 30° \text{ N}(0.2 \text{ m}) + N_C(x) = 0$$

Solving,

$$F = 69.3 \text{ N}$$
$$N_C = 236.2 \text{ N}$$
$$x = -0.00908 \text{ m} = -9.08 \text{ mm}$$

Since x is negative it indicates the *resultant* normal force acts (slightly) to the *left* of the crate's centerline. Also, the *maximum* frictional force which can be developed at the surface of contact is $F_{\max} = \mu_s N_C = 0.3(236.2 \text{ N}) = 70.9 \text{ N}$. Since $F = 69.3 \text{ N} < 70.9 \text{ N}$, the crate will *not slip*, although it is very close to doing so.

(b)

Fig. 8–7

EXAMPLE 8.2

It is observed that when the bed of the dump truck is raised to an angle of $\theta = 25°$ the vending machines will begin to slide off the bed, Fig. 8–8a. Determine the static coefficient of static friction between a vending machine and the surface of the truck bed.

SOLUTION
An idealized model of a vending machine resting on the truck bed is shown in Fig. 8–8b. The dimensions have been measured and the center of gravity has been located. We will assume that the vending machine weighs W.

(a)

Free-Body Diagram. As shown in Fig. 8–8c, the dimension x is used to locate the position of the resultant normal force **N**. There are four unknowns, N, F, μ_s, and x.

Equations of Equilibrium.

$+\searrow\Sigma F_x = 0;$ $\qquad W \sin 25° - F = 0 \qquad (1)$

$+\nearrow\Sigma F_y = 0;$ $\qquad N - W \cos 25° = 0 \qquad (2)$

$\downarrow+\Sigma M_O = 0;$ $\qquad -W \sin 25°(2.5 \text{ ft}) + W \cos 25°(x) = 0 \qquad (3)$

Since slipping impends at $\theta = 25°$, using Eqs. 1 and 2, we have

$F_s = \mu_s N;$ $\qquad W \sin 25° = \mu_s(W \cos 25°)$

$\qquad \mu_s = \tan 25° = 0.466 \qquad$ *Ans.*

The angle of $\theta = 25°$ is referred to as the **angle of repose**, and by comparison, it is equal to the angle of static friction, $\theta = \phi_s$. Since θ is independent of the weight of the vending machine, then knowing θ provides a convenient method for determining the coefficient of static friction.

NOTE: From Eq. 3, we find $x = 1.17$ ft. Since 1.17 ft < 1.5 ft, indeed the vending machine will slip down the truck bed and not tip over.

(b)

(c)

Fig. 8–8

EXAMPLE 8.3

The uniform 10-kg ladder in Fig. 8–9a rests against the smooth wall at B, and the end A rests on the rough horizontal plane for which the coefficient of static friction is $\mu_s = 0.3$. Determine the angle of inclination θ of the ladder and the normal reaction at B if the ladder is on the verge of slipping.

(a)

SOLUTION

Free-Body Diagram. As shown on the free-body diagram, Fig. 8–9b, the frictional force $\mathbf{F}_A$ must act to the right since impending motion at A is to the left.

Equations of Equilibrium and Friction. Since the ladder is on the verge of slipping, then $F_A = \mu_s N_A = 0.3 N_A$. By inspection, N_A can be obtained directly.

$+\uparrow \Sigma F_y = 0;$ $\qquad N_A - 10(9.81) \text{ N} = 0 \quad N_A = 98.1 \text{ N}$

Using this result, $F_A = 0.3(98.1 \text{ N}) = 29.43 \text{ N}$. Now N_B can be found.

$\xrightarrow{+} \Sigma F_x = 0;$ $\qquad 29.43 \text{ N} - N_B = 0$

$\qquad N_B = 29.43 \text{ N} = 29.4 \text{ N}$ $\qquad$ *Ans.*

Finally, the angle θ can be determined by summing moments about point A.

$\zeta + \Sigma M_A = 0;$ $\quad (29.43 \text{ N})(4 \text{ m}) \sin \theta - [10(9.81) \text{ N}](2 \text{ m}) \cos \theta = 0$

$$\frac{\sin \theta}{\cos \theta} = \tan \theta = 1.6667$$

$\theta = 59.04° = 59.0°$ $\qquad$ *Ans.*

(b)

Fig. 8–9

EXAMPLE 8.4

Beam AB is subjected to a uniform load of 200 N/m and is supported at B by post BC, Fig. 8–10a. If the coefficients of static friction at B and C are $\mu_B = 0.2$ and $\mu_C = 0.5$, determine the force **P** needed to pull the post out from under the beam. Neglect the weight of the members and the thickness of the beam.

(a)

SOLUTION

Free-Body Diagrams. The free-body diagram of the beam is shown in Fig. 8–10b. Applying $\Sigma M_A = 0$, we obtain $N_B = 400$ N. This result is shown on the free-body diagram of the post, Fig. 8–10c. Referring to this member, the *four* unknowns F_B, P, F_C, and N_C are determined from the *three* equations of equilibrium and *one* frictional equation applied either at B or C.

(b)

Equations of Equilibrium and Friction.

$\xrightarrow{+} \Sigma F_x = 0;$ $P - F_B - F_C = 0$ (1)

$+\uparrow \Sigma F_y = 0;$ $N_C - 400 \text{ N} = 0$ (2)

$\zeta + \Sigma M_C = 0;$ $-P(0.25 \text{ m}) + F_B(1 \text{ m}) = 0$ (3)

(Post Slips at B and Rotates about C.) This requires $F_C \leq \mu_C N_C$ and

$F_B = \mu_B N_B;$ $F_B = 0.2(400 \text{ N}) = 80$ N

Using this result and solving Eqs. 1 through 3, we obtain

$P = 320$ N
$F_C = 240$ N
$N_C = 400$ N

(c)

Fig. 8–10

Since $F_C = 240 \text{ N} > \mu_C N_C = 0.5(400 \text{ N}) = 200$ N, slipping at C occurs. Thus the other case of movement must be investigated.

(Post Slips at C and Rotates about B.) Here $F_B \leq \mu_B N_B$ and

$F_C = \mu_C N_C;$ $F_C = 0.5 N_C$ (4)

Solving Eqs. 1 through 4 yields

$P = 267$ N   *Ans.*
$N_C = 400$ N
$F_C = 200$ N
$F_B = 66.7$ N

Obviously, this case occurs first since it requires a *smaller* value for P.

EXAMPLE 8.5

(a)

(b)

Fig. 8–11

Blocks A and B have a mass of 3 kg and 9 kg, respectively, and are connected to the weightless links shown in Fig. 8–11a. Determine the largest vertical force P that can be applied at the pin C without causing any movement. The coefficient of static friction between the blocks and the contacting surfaces is $\mu_s = 0.3$.

SOLUTION

Free-Body Diagrams. The links are two-force members and so the free-body diagrams of pin C and blocks A and B are shown in Fig. 8–11b. Since the horizontal component of $\mathbf{F}_{AC}$ tends to move block A to the left, $\mathbf{F}_A$ must act to the right. Similarly, $\mathbf{F}_B$ must act to the left to oppose the tendency of motion of block B to the right, caused by $\mathbf{F}_{BC}$. There are seven unknowns and six available force equilibrium equations, two for the pin and two for each block, so that *only one* frictional equation is needed.

Equations of Equilibrium and Friction. The force in links AC and BC can be related to P by considering the equilibrium of pin C.

$+\uparrow \Sigma F_y = 0;$ $F_{AC} \cos 30° - P = 0;$ $F_{AC} = 1.155P$

$\xrightarrow{+} \Sigma F_x = 0;$ $1.155P \sin 30° - F_{BC} = 0;$ $F_{BC} = 0.5774P$

Using the result for F_{AC}, for block A,

$\xrightarrow{+} \Sigma F_x = 0;$ $F_A - 1.155P \sin 30° = 0;$ $F_A = 0.5774P$ (1)

$+\uparrow \Sigma F_y = 0;$ $N_A - 1.155P \cos 30° - 3(9.81 \text{ N}) = 0;$

$N_A = P + 29.43 \text{ N}$ (2)

Using the result for F_{BC}, for block B,

$\xrightarrow{+} \Sigma F_x = 0;$ $(0.5774P) - F_B = 0;$ $F_B = 0.5774P$ (3)

$+\uparrow \Sigma F_y = 0;$ $N_B - 9(9.81 \text{ N}) = 0;$ $N_B = 88.29 \text{ N}$

Movement of the system may be caused by the initial slipping of *either* block A or block B. If we *assume* that block A slips first, then

$F_A = \mu_s N_A = 0.3 N_A$ (4)

Substituting Eqs. 1 and 2 into Eq. 4,

$0.5774P = 0.3(P + 29.43)$

$P = 31.8 \text{ N}$ *Ans.*

Substituting this result into Eq. 3, we obtain $F_B = 18.4$ N. Since the maximum static frictional force at B is $(F_B)_{max} = \mu_s N_B = 0.3(88.29 \text{ N}) = 26.5 \text{ N} > F_B$, block B will not slip. Thus, the above assumption is correct. Notice that if the inequality were not satisfied, we would have to assume slipping of block B ($F_B = 0.3 N_B$) and then solve for P.

FUNDAMENTAL PROBLEMS

All solutions must include a free-body diagram.

F8–1. Determine the friction developed between the 50-kg crate and the ground if (a) $P = 200$ N, and (b) $P = 400$ N. The coefficients of static and kinetic friction between the crate and the ground are $\mu_s = 0.3$ and $\mu_k = 0.2$.

Prob. F8–1

F8–2. Determine the minimum force P to prevent the 30-kg rod AB from sliding. The contact surface at B is smooth, whereas the coefficient of static friction between the rod and the wall at A is $\mu_s = 0.2$.

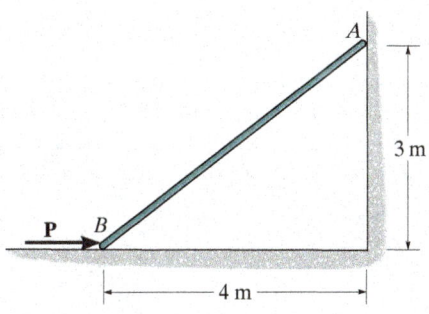

Prob. F8–2

F8–3. Determine the maximum force P that can be applied without causing the two 50-kg crates to move. The coefficient of static friction between each crate and the ground is $\mu_s = 0.25$.

Prob. F8–3

F8–4. If the coefficient of static friction at contact points A and B is $\mu_s = 0.3$, determine the maximum force P that can be applied without causing the 100-kg spool to move.

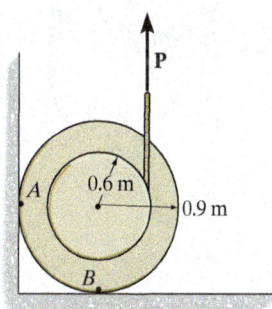

Prob. F8–4

F8–5. Determine the maximum force P that can be applied without causing movement of the 250-lb crate that has a center of gravity at G. The coefficient of static friction at the floor is $\mu_s = 0.4$.

Prob. F8–5

F8–6. Determine the minimum coefficient of static friction between the uniform 50-kg spool and the wall so that the spool does not slip.

F8–8. If the coefficient of static friction at all contacting surfaces is μ_s, determine the inclination θ at which the identical blocks, each of weight W, begin to slide.

Prob. F8–6

Prob. F8–8

F8–7. Blocks A, B, and C have weights of 50 N, 25 N, and 15 N, respectively. Determine the smallest horizontal force P that will cause impending motion. The coefficient of static friction between A and B is $\mu_s = 0.3$, between B and C, $\mu_s' = 0.4$, and between block C and the ground, $\mu_s'' = 0.35$.

F8–9. Blocks A and B have a mass of 7 kg and 10 kg, respectively. Using the coefficients of static friction indicated, determine the largest force P which can be applied to the cord without causing motion. There are pulleys at C and D.

Prob. F8–7

Prob. F8–9

PROBLEMS

All solutions must include a free-body diagram.

8–1. Determine the maximum force P the connection can support so that no slipping occurs between the plates. There are four bolts used for the connection and each is tightened so that it is subjected to a tension of 4 kN. The coefficient of static friction between the plates is $\mu_s = 0.4$.

Prob. 8–1

8–2. If the coefficient of static friction between the incline and the 200-lb crate is $\mu_s = 0.3$, determine the force P that must be applied to the rope to begin moving the crate up the incline.

8–3. The mine car and its contents have a total mass of 6 Mg and a center of gravity at G. If the coefficient of static friction between the wheels and the tracks is $\mu_s = 0.4$ when the wheels are locked, find the normal force acting on the front wheels at B and the rear wheels at A when the brakes at both A and B are locked. Does the car move?

Prob. 8–3

*8–4.** The winch on the truck is used to hoist the garbage bin onto the bed of the truck. If the loaded bin has a weight of 8500 lb and center of gravity at G, determine the force in the cable needed to begin the lift. The coefficients of static friction at A and B are $\mu_A = 0.3$ and $\mu_B = 0.2$, respectively. Neglect the height of the support at A.

Prob. 8–2

Prob. 8–4

8–5. The automobile has a mass of 2 Mg and center of mass at G. Determine the towing force F required to move the car if the back brakes are locked, and the front wheels are free to roll. Take $\mu_s = 0.3$.

8–6. The automobile has a mass of 2 Mg and center of mass at G. Determine the towing force F required to move the car. Both the front and rear brakes are locked. Take $\mu_s = 0.3$.

*8–8.** The bracket can move freely along the pole until a force **P** is placed on it. If the coefficient of static friction at A and B is $\mu_s = 0.6$, determine the smallest distance d at which the force can be applied so that the bracket does not slip. Neglect the weight of the bracket.

Probs. 8–5/6

Prob. 8–8

8–7. The 100-lb cylinder rests between the two inclined planes. When $P = 15$ lb, the cylinder is on the verge of impending motion. Determine the coefficient of static friction between the surfaces of contact and the cylinder.

8–9. The pipe of weight W is to be pulled up the inclined plane of slope α using a force **P**. If **P** acts at an angle ϕ, show that for slipping $P = W \sin(\alpha + \theta)/\cos(\phi + \theta)$, where θ is the angle of static friction; $\theta = \tan^{-1}\mu_s$.

8–10. Determine the angle ϕ at which the applied force **P** should act on the pipe so that P is as small as possible when pulling the pipe up the incline. What is the corresponding value of P? The pipe weighs W and the slope α is known. Express the answer in terms of the angle of kinetic friction, $\theta = \tan^{-1}\mu_k$.

Prob. 8–7

Probs. 8–9/10

8–11. The man has a mass of 40 kg. He plans to scale the vertical crevice using the method shown. If the coefficient of static friction between his shoes and the rock is $\mu_s = 0.4$ and between his backside and the rock, $\mu_b = 0.3$, determine the smallest horizontal force his body must exert on the rocks in order to do this.

8–13. The brake is to be designed to be self locking, that is, it will not rotate when no load **P** is applied to it when the disk is subjected to a clockwise couple moment $\mathbf{M}_0$. Determine the distance d of the lever that will allow this to happen. The coefficient of static friction at B is $\mu_s = 0.5$.

Prob. 8–11

Prob. 8–13

*8–12. The block brake is used to stop the wheel from rotating when the wheel is subjected to a couple moment $\mathbf{M}_0$. If the coefficient of static friction between the wheel and the block is μ_s, determine the smallest force P that should be applied.

8–14. The car has a mass of 1.6 Mg and center of mass at G. If the coefficient of static friction between the shoulder of the road and the tires is $\mu_s = 0.4$, determine the greatest slope θ the shoulder can have without causing the car to slip or tip over if the car travels along the shoulder at constant velocity.

Prob. 8–12

Prob. 8–14

8–15. The log has a coefficient of static friction of $\mu_s = 0.3$ with the ground and a weight of 40 lb/ft. If a man can pull on the rope with a maximum force of 80 lb, determine the greatest length l of log he can drag.

Prob. 8–15

8–18. The spool of wire having a weight of 300 lb rests on the ground at B and against the wall at A. Determine the force P required to begin pulling the wire horizontally off the spool. The coefficient of static friction between the spool and its points of contact is $\mu_s = 0.25$.

8–19. The spool of wire having a weight of 300 lb rests on the ground at B and against the wall at A. Determine the normal force acting on the spool at A if $P = 300$ lb. The coefficient of static friction between the spool and the ground at B is $\mu_s = 0.35$. The wall at A is smooth.

Probs. 8–18/19

__*8–16.__ Determine the torque M that must be applied by the engine to the rear wheels of the tractor to cause them to slip. The front wheels are free to roll. The tractor weighs 3500 lb and has a center of gravity at G. The coefficient of static friction between the rear wheels and the ground is $\mu_s = 0.5$.

8–17. If the coefficient of static friction between the rear wheels of the tractor and the ground is $\mu_s = 0.6$, determine if the rear wheels slip or the front wheels lift off the ground as the engine provides a torque M to the rear wheels. What is the torque needed to cause the motion? The front wheels are free to roll. The tractor weighs 2500 lb and has a center of gravity at G.

__*8–20.__ The ring has a mass of 0.5 kg and is resting on the surface of the table. To move the ring, a normal force **P** from the finger is exerted on it. Determine its magnitude when the ring is on the verge of slipping at A. The coefficient of static friction at A is $\mu_A = 0.2$ and at B, $\mu_B = 0.3$.

Probs. 8–16/17

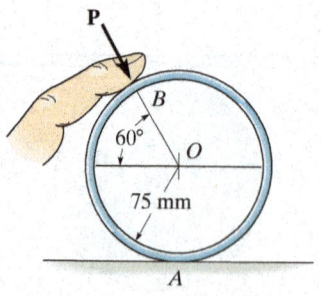

Prob. 8–20

8–21. The fork lift has a weight of 2400 lb and a center of gravity at G. If the rear wheels are powered, whereas the front wheels are free to roll, determine the maximum number of 300-lb crates the fork lift can push forward. The coefficient of static friction between the wheels and the ground is $\mu_s = 0.4$, and between each crate and the ground $\mu_s' = 0.35$.

8–23. Determine the minimum force P needed to push the tube E up the incline. The tube has a mass of 75 kg and the roller D has a mass of 100 kg. The force acts parallel to the plane, and the coefficients of static friction at the contacting surfaces are $\mu_A = 0.3$, $\mu_B = 0.25$, and $\mu_C = 0.4$. The roller and tube each have a radius of 150 mm.

Prob. 8–23

Prob. 8–21

*8–24. The uniform thin pole has a weight of 30 lb and a length of 26 ft. If it is placed against the smooth wall and on the rough floor in the position $d = 10$ ft, will it remain in this position when it is released? The coefficient of static friction is $\mu_s = 0.3$.

8–22. The cylinder is confined by the brake, where $\mu_s = 0.4$. Determine the required compression in the spring in order to resist a torque of 800 N · m on the cylinder. The spring has a stiffness of $k = 3$ MN/m and an unstretched length of 60 mm. End C of member ABC slides freely along the smooth vertical guide.

8–25. The uniform pole has a weight of 30 lb and a length of 26 ft. Determine the maximum distance d it can be placed from the smooth wall and not slip. The coefficient of static friction between the floor and the pole is $\mu_s = 0.3$.

Prob. 8–22

Probs. 8–24/25

8–26. The truck has a mass of 1.25 Mg and a center of mass at G. Determine the greatest load it can move if (a) the truck has rear-wheel drive while the front wheels are free to roll, and (b) the truck has four-wheel drive. The coefficient of static friction between the wheels and the ground is $\mu_w = 0.5$, and between the crate and the ground, it is $\mu_c = 0.4$.

8–27. Solve Prob. 8–26 if the truck and crate are resting on an upward 10° incline.

8–29. The friction pawl is pinned at A and rests against the wheel at B. It allows freedom of movement when the wheel is rotating counterclockwise about C. Clockwise rotation is prevented due to friction of the pawl which tends to bind the wheel. If $(\mu_s)_B = 0.6$, determine the design angle θ which will prevent clockwise motion for any value of applied moment M. *Hint:* Neglect the weight of the pawl so that it becomes a two-force member.

Probs. 8–26/27

Prob. 8–29

*__8–28.__ The uniform crate resting on the dolly has a mass of 500 kg and mass center at G. If the front casters contact a high step, and the coefficient of static friction between the crate and the dolly is $\mu_s = 0.45$, determine the greatest force P that can be applied without causing motion of the crate. The dolly does not move.

8–30. Two blocks A and B have a weight of 10 lb and 6 lb, respectively. They are resting on the incline for which the coefficients of static friction are $\mu_A = 0.15$ and $\mu_B = 0.25$. Determine the incline angle θ for which both blocks begin to slide. Also find the required stretch or compression in the connecting spring for this to occur. The spring has a stiffness of $k = 2$ lb/ft.

8–31. Two blocks A and B have a weight of 10 lb and 6 lb, respectively. They are resting on the incline for which the coefficients of static friction are $\mu_A = 0.15$ and $\mu_B = 0.25$. Determine the angle θ which will cause motion of one of the blocks. What is the friction force under each of the blocks when this occurs? The spring has a stiffness of $k = 2$ lb/ft and is originally unstretched.

Prob. 8–28

Probs. 8–30/31

*8–32. Determine the smallest force P that must be applied to begin moving the 150-lb uniform crate. The coefficient of static friction between the crate and the floor is $\mu_s = 0.5$.

8–33. The man having a weight of 200 lb pushes horizontally on the crate. If the coefficient of static friction between the 450-lb crate and the floor is $\mu_s = 0.3$ and between his shoes and the floor $\mu'_s = 0.6$, determine if he can move the crate.

*8–36. Determine the minimum force P needed to push the tube E up the incline. The coefficients of static friction at the contacting surfaces are $\mu_A = 0.2$, $\mu_B = 0.3$, and $\mu_C = 0.4$. The 100-kg roller and 40-kg tube each have a radius of 150 mm.

Prob. 8–36

Probs. 8–32/33

8–34. The uniform hoop of weight W is subjected to the horizontal force **P**. Determine the coefficient of static friction between the hoop and the surface at A and B if the hoop is on the verge of rotating.

8–35. Determine the maximum horizontal force P that can be applied to the 30-lb hoop without causing it to rotate. The coefficient of static friction between the hoop and the surfaces A and B is $\mu_s = 0.2$. Take $r = 300$ mm.

8–37. The double-block brake mechanism is used to prevent the wheel from turning when the wheel is subjected to the torque of $M = 5$ N·m. If the coefficient of static friction between the blocks and the wheel is $\mu_s = 0.8$, determine the smallest vertical force P applied to the handle needed to stop the wheel.

8–38. Solve Prob. 8–37 if the torque is $M = 5$ N·m clockwise.

Probs. 8–34/35

Probs. 8–37/38

8–39. The car has a weight of 4000 lb and a center of gravity at G. If it pulls off the side of the road, determine the greatest angle of tilt θ it can have without slipping or tripping over. The coefficient of static friction between its wheels and the ground is $\mu_s = 0.4$.

Prob. 8–39

8–41. If the coefficient of static friction at A and B is $\mu_s = 0.6$, determine the maximum angle θ so that the frame remains in equilbrium, regardless of the mass of the cylinder. Neglect the mass of the rods.

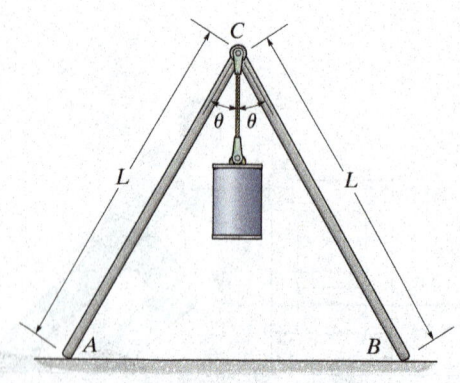

Prob. 8–41

*8–40. The 800-lb concrete pipe is being lowered from the truck bed when it is in the position shown. If the coefficient of static friction at the points of support A and B is $\mu_s = 0.4$, determine where it begins to slip first: at A or B, or at both A and B.

8–42. The 100-kg disk rests on a surface for which $\mu_B = 0.2$. Determine the smallest vertical force **P** that can be applied tangentially to the disk which will cause motion to impend.

Prob. 8–40

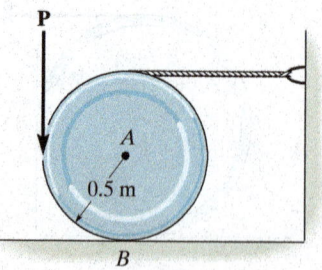

Prob. 8–42

8–43. The tongs are used to suspend the 80-kg plate. Determine the smallest coefficient of static friction to prevent slipping of the plate.

8–45. The man has a mass of 60 kg and the crate has a mass of 100 kg. If the coefficient of static friction between his shoes and the ground is $\mu_s = 0.4$ and between the crate and the ground is $\mu_c = 0.3$, determine if the man is able to move the crate using the rope-and-pulley system shown.

Prob. 8–43

Prob. 8–45

***8–44.** The 50-lb board is placed across the channel and a 100-lb girl attempts to walk across. If the coefficient of static friction at A and B is $\mu_s = 0.4$, determine if she can make the crossing; and if not, how far will she get from A before the board slips?

8–46. The uniform cylindrical tank has a mass of 250 kg. If the coefficient of static friction is $\mu_s = 0.15$, determine the force P applied to the rope needed to move the tank.

Prob. 8–44

Prob. 8–46

8–47. A chain having a length of 20 ft and weight of 8 lb/ft rests on a surface for which the coefficient of static friction is $\mu_s = 0.2$. If a crane is used to hoist the chain, determine the force P it applies to the chain if the length of chain remaining on the ground begins to slip when the horizontal component of **P** becomes 10 lb. What length of chain remains on the ground?

Prob. 8–47

*8–48. Determine the smallest couple moment which can be applied to the 20-lb wheel that will cause impending motion. The cord is attached to the 30-lb block, and the coefficients of static friction are $\mu_B = 0.2$ and $\mu_D = 0.3$.

Prob. 8–48

8–49. The crate has a weight of 200 lb and a center of gravity at G. Determine the horizontal force P required to tow it. Also, find the location of the resultant normal force measured from A. Take $h = 4$ ft and $\mu_s = 0.4$.

8–50. The crate has a weight of 200 lb and a center of gravity at G. Determine the height h of the tow rope so that the crate slips and tips at the same time. What horizontal force P is required to do this? Take $\mu_s = 0.4$.

Probs. 8–49/50

8–51. Beam AB has a negligible mass and thickness, and supports the 200-kg uniform block. It is pinned at A and rests on the top of a post, having a mass of 20 kg and negligible thickness. Determine the minimum force P needed to move the post. The coefficients of static friction at B and C are $\mu_B = 0.4$ and $\mu_C = 0.2$, respectively.

*8–52. Beam AB has a negligible mass and thickness, and supports the 200-kg uniform block. It is pinned at A and rests on the top of a post, having a mass of 20 kg and negligible thickness. Determine the two coefficients of static friction at B and at C so that when the magnitude of the applied force is increased to $P = 300$ N, the post slips at both B and C simultaneously.

Probs. 8–51/52

8–53. Determine the smallest couple moment that can be applied to the 150-lb wheel that will cause impending motion. The uniform concrete block has a weight of 300 lb. The coefficients of static friction are $\mu_A = 0.2$, $\mu_B = 0.3$, and between the concrete block and the floor, $\mu = 0.4$.

Prob. 8–53

8–54. Two blocks A and B, each having a mass of 6 kg, are connected by the linkage shown. If the coefficients of static friction at the contacting surfaces are $\mu_B = 0.8$ and $\mu_A = 0.2$, determine the largest vertical force P that may be applied to pin C without causing the blocks to slip. Neglect the weight of the links.

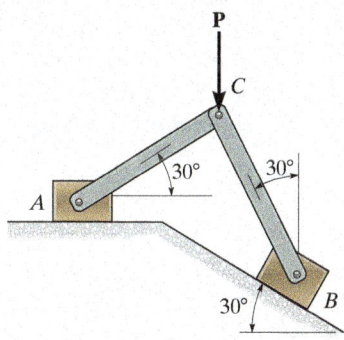

Prob. 8–54

8–55. Two blocks A and B, each having a mass of 6 kg, are connected by the linkage shown. If the coefficient of static friction at the contacting surfaces is $\mu_s = 0.5$, determine the largest vertical force P that can be applied to pin C without causing the blocks to move. Neglect the weight of the links.

Prob. 8–55

*8–56. The uniform 5-kg rod rests on the ground at B and against the wall at A. If the rod does not slip at B, determine the smallest coefficient of static friction at A which will prevent slipping at A.

Prob. 8–56

8–57. The uniform pole has a weight W and is lowered slowly from a vertical position $\theta = 90°$ toward the horizontal using cable AB. If the coefficient of static friction is $\mu_s = 0.3$ at C, determine the angle θ at which the pole will start to slip.

Prob. 8–57

8.3 WEDGES

A *wedge* is a simple machine that is often used to transform an applied force into much larger forces. Wedges can also be used to slightly move or adjust heavy loads.

To find the forces on a wedge consider the one shown in Fig. 8–12a, which is used to *lift* the block. On the free-body diagrams in Fig. 8–12b, we have excluded the weight of the wedge since it is usually *small* compared to the weight **W** of the block. Since the force **P** causes impending motion to the right, the frictional forces $\mathbf{F}_1$ and $\mathbf{F}_2$ must oppose this motion, and the frictional force $\mathbf{F}_3$ of the wall on the block must act downward. The locations of the resultant normal forces are not important here, since neither the block nor wedge will "tip." Hence the moment equilibrium equations will not be considered. There are seven unknowns, consisting of the applied force **P**, needed to cause the impending motion, and six normal and frictional forces. To solve, the seven available equations consist of four force equilibrium equations, $\Sigma F_x = 0$, $\Sigma F_y = 0$, applied to the wedge and block, and three frictional equations, $F = \mu N$, applied at each surface of contact.

If the block is to be *lowered*, then the frictional forces will all act in a sense opposite to that shown in Fig. 8–12b. For example, if the coefficient of friction is very *small* or the wedge angle θ is *large*, then the applied force **P** must act to the right to keep the block from moving downward. For large coefficients of friction or small angles θ, then **P** will have a reverse sense of direction in order to *pull* on the wedge to remove it and lower the block. If **P** is *not applied* and friction forces hold the block in place, then the wedge is referred to as *self-locking*.

Wedges are often used to adjust the elevation of structural or mechanical parts. Also, they provide stability for objects such as this pipe.

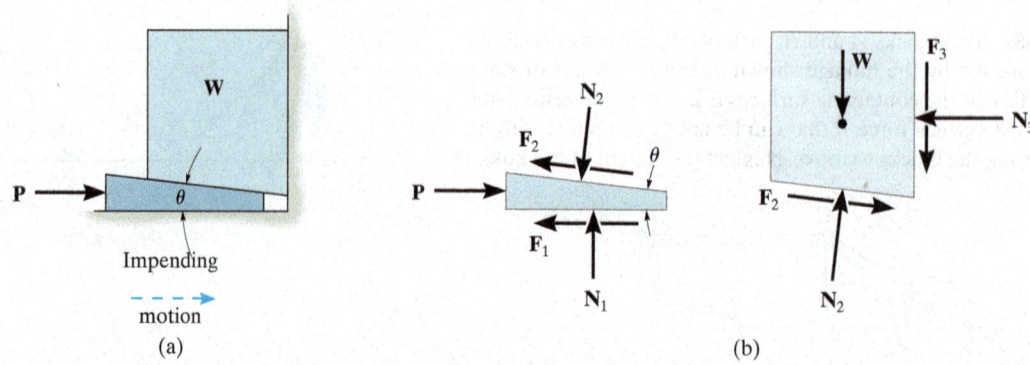

Fig. 8–12

EXAMPLE 8.6

The uniform stone in Fig. 8–13a has a mass of 500 kg and is held in the horizontal position using a wedge at B. If the coefficient of static friction is $\mu_s = 0.3$ at the surfaces of contact, determine the minimum force P needed to remove the wedge. Assume that the stone does not slip at A.

Fig. 8–13

SOLUTION

The minimum force P requires $F = \mu_s N$ at the surfaces of contact with the wedge. The free-body diagrams of the stone and wedge are shown in Fig. 8–13b. On the wedge the friction force opposes the impending motion, and on the stone at A, $F_A \leq \mu_s N_A$, since slipping does not occur there. The five unknowns are determined from the three equilibrium equations for the stone and two for the wedge. From the free-body diagram of the stone,

$\zeta + \Sigma M_A = 0;$ $-4905 \text{ N}(0.5 \text{ m}) + (N_B \cos 7° \text{ N})(1 \text{ m})$
$\qquad + (0.3 N_B \sin 7° \text{ N})(1 \text{ m}) = 0$
$\qquad N_B = 2383.1 \text{ N}$

Using this result for the wedge, we have

$+\uparrow \Sigma F_y = 0;$ $N_C - 2383.1 \cos 7° \text{ N} - 0.3(2383.1 \sin 7° \text{ N}) = 0$
$\qquad N_C = 2452.5 \text{ N}$

$\xrightarrow{+} \Sigma F_x = 0;$ $2383.1 \sin 7° \text{ N} - 0.3(2383.1 \cos 7° \text{ N}) +$
$\qquad\qquad P - 0.3(2452.5 \text{ N}) = 0$

$\qquad P = 1154.9 \text{ N} = 1.15 \text{ kN}$ *Ans.*

NOTE: Since P is positive, indeed the wedge must be pulled out. If P were zero, the wedge would remain in place (self-locking) and the frictional forces developed at B and C would satisfy $F_B < \mu_s N_B$ and $F_C < \mu_s N_C$.

8.4 FRICTIONAL FORCES ON SCREWS

In most cases, screws are used as fasteners; however, in many types of machines they are incorporated to transmit power or motion from one part of the machine to another. A *square-threaded screw* is commonly used for this purpose, especially when large forces are applied along its axis. In this section, we will analyze the forces acting on square-threaded screws. The analysis of other types of screws, such as those that have a V-thread, is based on these same principles.

For analysis, a square-threaded screw, as in Fig. 8–14a, can be considered a cylinder having an inclined square ridge or *thread* wrapped around it. If we unwind the thread by one revolution, as shown in Fig. 8–14b, the slope or the *lead angle* θ is determined from $\theta = \tan^{-1}(l/2\pi r)$. Here l and $2\pi r$ are the vertical and horizontal distances between A and B, where r is the mean radius of the thread. The distance l is called the *lead* of the screw and it is equivalent to the distance the screw advances when it turns one revolution.

Upward Impending Motion. Consider the case of the square-threaded screw jack in Fig. 8–15 that is subjected to upward impending motion caused by the applied torsional moment **M**.* A free-body diagram of the *entire unraveled thread h* in contact with the jack can be represented as a block, as shown in Fig. 8–16a. The force **W** is the vertical force acting on the thread or the axial force applied to the shaft, Fig. 8–15, and M/r is the resultant horizontal force produced by the moment M about the axis of the shaft. The reaction **R** of the groove on the thread has both frictional and normal components, where $F = \mu_s N$. The angle of static friction is $\phi_s = \tan^{-1}(F/N) = \tan^{-1}\mu_s$. Applying the force equations of equilibrium, we have

$$\xrightarrow{+} \Sigma F_x = 0; \quad M/r - R\sin(\theta + \phi_s) = 0$$

$$+\uparrow \Sigma F_y = 0; \quad R\cos(\theta + \phi_s) - W = 0$$

Eliminating R from these equations, we obtain

$$M = rW\tan(\theta + \phi_s) \qquad (8\text{–}3)$$

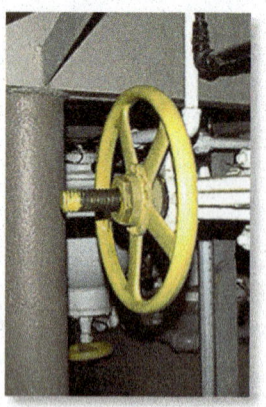

Square-threaded screws find applications on valves, jacks, and vises, where particularly large forces must be developed along the axis of the screw.

(a)

(b)

Fig. 8–14

*For applications, **M** can be developed by applying a horizontal force **P** at a right angle to the end of a lever that would be fixed to the screw.

8.4 Frictional Forces on Screws

Fig. 8-15

Upward screw motion
(a)

Self-locking screw ($\theta = \phi_s$)
(on the verge of rotating downward)
(b)

Self-Locking Screw. A screw is said to be *self-locking* if it remains in place under any axial load **W** when the moment **M** is removed. For this to occur, the direction of the frictional force must be reversed so that **R** acts on the other side of **N**. Here the angle of static friction ϕ_s becomes greater than or equal to θ, Fig. 8–16d. If $\phi_s = \theta$, Fig. 8–16b, then **R** will act vertically to balance **W**, and the screw will be on the verge of winding downward.

Downward Impending Motion, ($\theta > \phi_s$). If the screw is not self-locking, it is necessary to apply a moment **M'** to prevent the screw from winding downward and so a horizontal force M'/r must push against the thread, Fig. 8–16c. Using the same procedure as before, the magnitude M' required to prevent this unwinding is

$$M' = rW \tan(\theta - \phi_s) \qquad (8\text{-}4)$$

Downward Impending Motion, ($\phi_s > \theta$). If a screw is self-locking, a moment M'' must be applied to the screw in the opposite direction to wind the screw downward ($\phi_s > \theta$). This causes a reverse horizontal force M''/r as indicated in Fig. 8–16d. In this case, we obtain

$$M'' = rW \tan(\phi_s - \theta) \qquad (8\text{-}5)$$

If *motion of the screw* occurs, Eqs. 8-3, 8-4, and 8-5 can be applied by simply replacing ϕ_s with ϕ_k.

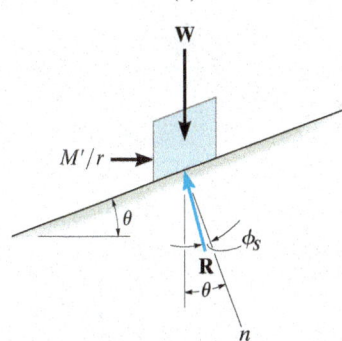

Downward screw motion ($\theta > \phi_s$)
(c)

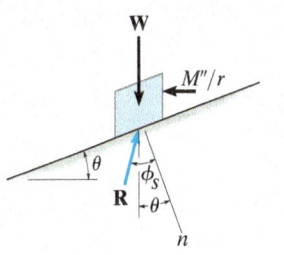

Downward screw motion ($\theta < \phi_s$)
(d)

Fig. 8-16

EXAMPLE 8.7

The turnbuckle shown in Fig. 8–17 has a square thread with a mean radius of 5 mm and a lead of 2 mm. If the coefficient of static friction between the screw and the turnbuckle is $\mu_s = 0.25$, determine the moment **M** that must be applied to draw the end screws closer together.

Fig. 8–17

SOLUTION

The moment can be obtained by applying Eq. 8–3. Since friction at *two screws* must be overcome, this requires

$$M = 2[rW \tan(\theta + \phi_s)] \quad (1)$$

Here $W = 2000$ N, $\phi_s = \tan^{-1}\mu_s = \tan^{-1}(0.25) = 14.04°$, $r = 5$ mm, and $\theta = \tan^{-1}(l/2\pi r) = \tan^{-1}(2 \text{ mm}/[2\pi(5 \text{ mm})]) = 3.64°$.

Substituting these values into Eq. 1 and solving gives

$$M = 2[(2000 \text{ N})(5 \text{ mm}) \tan(14.04° + 3.64°)]$$

$$= 6374.7 \text{ N} \cdot \text{mm} = 6.37 \text{ N} \cdot \text{m} \qquad Ans.$$

NOTE: When the moment is *removed*, the turnbuckle will be self-locking; i.e., it will not unscrew since $\phi_s > \theta$.

PROBLEMS

All solutions must include a free-body diagram.

8–58. The blocks each have a weight of 50 lb. If the coefficient of static friction at A is $\mu_s = 0.2$ and between each block $\mu'_s = 0.4$, determine how many blocks can be stacked as shown before they begin to topple.

Prob. 8–58

8–59. If the coefficient of static friction between the axe and the wood is $\mu_s = 0.2$, determine the smallest angle θ of the blade which will cause the axe to be self-locking. Neglect the weight of the axe.

Prob. 8–59

*8–60. Each block has a weight of 400 lb. Determine how far the force **P** can compress the spring until block B slips on block A. What is P for this to occur?

Prob. 8–60

8–61. If the coefficient of static friction between all the surfaces of contact is μ_s, determine the force P that must be applied to the wedge in order to lift the brace that supports the load F. Neglect friction between the wall and the brace.

Prob. 8–61

8–62. If $P = 250$ N, determine the required minimum compression in the spring so that the wedge will not move to the right. Neglect the weight of A and B. The coefficient of static friction for all contacting surfaces is $\mu_s = 0.35$. Neglect friction at the rollers.

8–63. Determine the minimum applied force P required to move wedge A to the right. The spring is compressed a distance of 175 mm. Neglect the weight of A and B. The coefficient of static friction for all contacting surfaces is $\mu_s = 0.35$. Neglect friction at the rollers.

Probs. 8–62/63

*8–64. The wedge has a negligible weight and a coefficient of static friction $\mu_s = 0.35$ with all contacting surfaces. Determine the angle θ so that it is self-locking. This requires no slipping for any magnitude of the force P applied to the joint.

Prob. 8–64

8–65. If it takes a vertical force of 60,000 lb to separate the two parts at A and C, determine the force P the hydraulic cylinder must exert on the wedge in order to push it forward. The coefficient of static friction between the wedge and members AB and CD is $\mu_s = 0.2$.

Prob. 8–65

8–66. The load bar of negligible weight is screwed in place between the walls of the truck bed in order to keep loads from shifting. Determine the torsional moment M that must be applied to the center of the bar so that it produces a compression in the bar of 80 lb. The square-threaded screw at A has a lead of 0.2 in. and a radius of 0.5 in., and the coefficient of static friction between the bar and the screw is $\mu_s = 0.3$. Also, what amount of horizontal force F must be applied to the center of the bar to pull it out? The coefficient of static friction between the walls of the truck and the end pads is $\mu_s = 0.7$.

Prob. 8–66

8–67. If the clamping force at G is 900 N, determine the horizontal force $\mathbf{F}$ that must be applied perpendicular to the handle of the lever at E. The mean diameter and lead of both single square-threaded screws at C and D are 25 mm and 5 mm, respectively. The coefficient of static friction is $\mu_s = 0.3$.

*8–68. If a horizontal force of $F = 50$ N is applied perpendicular to the handle of the lever at E, determine the clamping force developed at G. The mean diameter and lead of the single square-threaded screw at C and D are 25 mm and 5 mm, respectively. The coefficient of static friction is $\mu_s = 0.3$.

Probs. 8–67/68

8–69. The hand clamp is constructed using a square-threaded screw having a mean diameter of 36 mm, a lead of 4 mm, and a coefficient of static friction at the screw of $\mu_s = 0.3$. To tighten the screw, a force of $F = 20$ N is applied perpendicular to the handle. Determine the clamping force in the board AB.

8–70. The hand clamp is constructed using a square-threaded screw having a mean diameter of 36 mm, a lead of 4 mm, and a coefficient of static friction at the screw of $\mu_s = 0.3$. If the clamping force in the board AB is 300 N, determine the reversed force $-\mathbf{F}$ that must be applied perpendicular to the handle in order to loosen the screw.

Probs. 8–69/70

8–71. If couple forces of $F = 10$ lb are applied perpendicular to the lever of the clamp at A and B, determine the clamping force on the boards. The single square-threaded screw of the clamp has a mean diameter of 1 in. and a lead of 0.25 in. The coefficient of static friction is $\mu_s = 0.3$.

*8–72. If the clamping force on the boards is 600 lb, determine the required magnitude of the couple forces that must be applied perpendicular to the lever AB of the clamp at A and B in order to loosen the screw. The single square-threaded screw has a mean diameter of 1 in. and a lead of 0.25 in. The coefficient of static friction is $\mu_s = 0.3$.

Probs. 8–71/72

8–73. Prove that the lead l must be less than $2\pi r \mu_s$ for the jack screw shown in Fig. 8–15 to be self-locking.

8–74. The square-threaded bolt is used to join two plates together. If the bolt has a mean diameter of $d = 20$ mm and a lead of $l = 3$ mm, determine the smallest torque M required to loosen the bolt if the tension in the bolt is $T = 40$ kN. The coefficient of static friction between the threads and the bolt is $\mu_s = 0.15$.

Prob. 8–74

8–75. The shaft has a square-threaded screw with a lead of 9 mm and a mean radius of 15 mm. If it is in contact with a plate gear having a mean radius of 20 mm, determine the resisting torque M on the gear which can be overcome if a torque of 7 N · m is applied to the shaft. The coefficient of static friction between the gear and the screw is $\mu_s = 0.2$. Neglect friction at the bearings located at A and B.

Prob. 8–75

*8–76. If couple forces of $F = 35$ N are applied to the handle of the machinist's vise, determine the compressive force developed in the block. Neglect friction at the bearing A. The guide at B is smooth. The single square-threaded screw has a mean radius of 6 mm and a lead of 8 mm, and the coefficient of static friction is $\mu_s = 0.27$.

Prob. 8–76

8–77. Determine the horizontal force P applied perpendicular to the handle of the jack screw necessary to start lifting the 3-kN load. The square-threaded screw has a lead of 5 mm and a mean diameter of 60 mm. The coefficient of static friction for the screw is $\mu_s = 0.2$.

Prob. 8–77

8–78. The automobile jack is subjected to a vertical load of $F = 8$ kN. If a square-threaded screw, having a lead of 5 mm and a mean diameter of 10 mm, is used in the jack, determine the force that must be applied perpendicular to the handle to (a) raise the load, and (b) lower the load; $\mu_s = 0.2$. The supporting plate exerts only vertical forces at A and B.

Prob. 8–78

8–79. If the required clamping force at the board A is to be 50 N, determine the force F that must be applied to the handle to tighten the clamp. The single square-threaded screw has a mean radius of 10 mm, a lead of 3 mm, and the coefficient of static friction is $\mu_s = 0.35$.

*8–80. Determine the clamping force on the board A if the screw of the hold-down clamp is tightened with a force of $F = 400$ N. The single square-threaded screw has a mean radius of 8 mm, a lead of 2 mm, and the coefficient of static friction is $\mu_s = 0.3$.

Probs. 8–79/80

8–81. If a horizontal force of $P = 100$ N is applied perpendicular to the handle of the lever at A, determine the compressive force F exerted on the material. Each single square-threaded screw has a mean diameter of 25 mm and a lead of 7.5 mm. The coefficient of static friction at all contacting surfaces of the wedges is $\mu_s = 0.2$, and the coefficient of static friction at the screw is $\mu_s' = 0.15$.

8–82. Determine the horizontal force P that must be applied perpendicular to the handle of the lever at A in order to develop a compressive force of 12 kN on the material. Each single square-threaded screw has a mean diameter of 25 mm and a lead of 7.5 mm. The coefficient of static friction at all contacting surfaces of the wedges is $\mu_s = 0.2$, and the coefficient of static friction at the screw is $\mu_s' = 0.15$.

Probs. 8–81/82

8.5 FRICTIONAL FORCES ON FLAT BELTS

Whenever belt drives or band brakes are designed, it is necessary to determine the frictional forces developed between the belt and its contacting surface. In this section we will analyze the frictional forces acting on a flat belt, although the analysis of other types of belts, such as the V-belt, is based on similar principles.

Consider the flat belt shown in Fig. 8–18a, which passes over a fixed curved surface. The total angle of belt-to-surface contact in radians is β, and the coefficient of friction between the two surfaces is μ. We wish to determine the tension T_2 in the belt, which is needed to pull the belt counterclockwise over the surface, and thereby overcome both the frictional forces at the surface of contact and the tension T_1 in the other end of the belt. Obviously, $T_2 > T_1$.

Frictional Analysis. A free-body diagram of the belt segment in contact with the surface is shown in Fig. 8–18b. The normal and frictional forces, acting at different points along the belt, will vary in both magnitude and direction. Due to this *unknown* distribution, the analysis of the problem will begin by considering a differential element of the belt.

A free-body diagram of this element having a length ds is shown in Fig. 8–18c. Assuming either impending motion or motion of the belt, the magnitude of the frictional force $dF = \mu\, dN$. This force opposes the sliding motion of the belt, and so it will increase the magnitude of the tensile force acting in the belt by dT. Applying the two force equations of equilibrium, we have

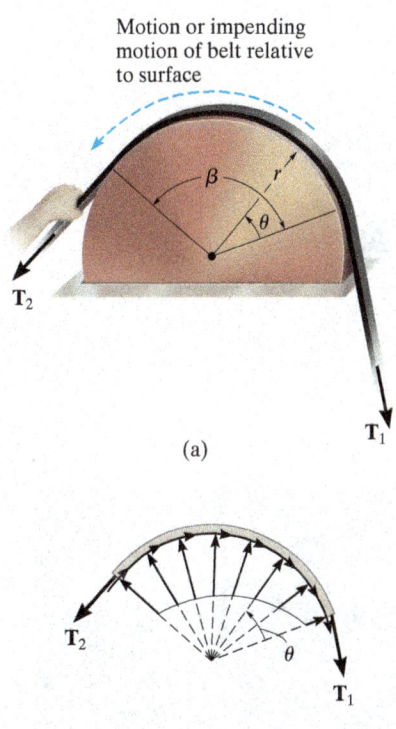

(a)

(b)

$$\searrow + \Sigma F_x = 0; \quad T\cos\left(\frac{d\theta}{2}\right) + \mu\, dN - (T + dT)\cos\left(\frac{d\theta}{2}\right) = 0$$

$$+\nearrow \Sigma F_y = 0; \quad dN - (T + dT)\sin\left(\frac{d\theta}{2}\right) - T\sin\left(\frac{d\theta}{2}\right) = 0$$

Since $d\theta$ is of *infinitesimal size*, $\sin(d\theta/2) = d\theta/2$ and $\cos(d\theta/2) = 1$. Also, the *product* of the two infinitesimals dT and $d\theta/2$ may be neglected when compared to infinitesimals of the first order. As a result, these two equations become

$$\mu\, dN = dT$$

and

$$dN = T\, d\theta$$

Eliminating dN yields

$$\frac{dT}{T} = \mu\, d\theta$$

(c)

Fig. 8–18

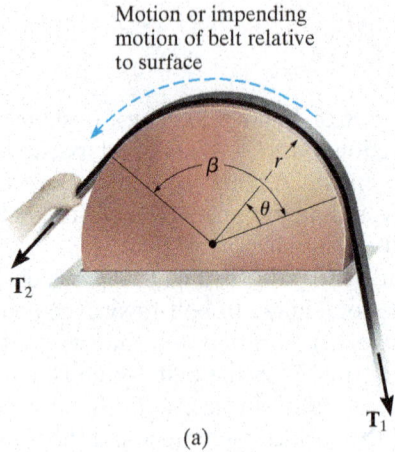

Fig. 8–18 (Repeated)

Integrating this equation between all points of contact that the belt makes with the surface, and noting that $T = T_1$ at $\theta = 0$ and $T = T_2$ at $\theta = \beta$, yields

$$\int_{T_1}^{T_2} \frac{dT}{T} = \mu \int_{0}^{\beta} d\theta$$

$$\ln \frac{T_2}{T_1} = \mu \beta$$

Solving for T_2, we obtain

$$T_2 = T_1 e^{\mu \beta} \tag{8-6}$$

where

$T_2, T_1 =$ belt tensions; T_1 opposes the direction of motion (or impending motion) of the belt measured relative to the surface, while T_2 acts in the direction of the relative belt motion (or impending motion); because of friction, $T_2 > T_1$

$\mu =$ coefficient of static or kinetic friction between the belt and the surface of contact

$\beta =$ angle of belt-to-surface contact, measured in radians

$e = 2.718\ldots$, base of the natural logarithm

Note that T_2 is *independent* of the *radius r* of the surface, and instead it is a function of the angle of belt to surface contact, β. As a result, this equation is valid for flat belts passing over any curved contacting surface.

Flat or V-belts are often used to transmit the torque developed by a motor to a wheel attached to a pump, fan, or blower.

EXAMPLE 8.8

The maximum tension that can be developed in the cord shown in Fig. 8–19a is 500 N. If the pulley at A is free to rotate and the coefficient of static friction at the fixed drums B and C is $\mu_s = 0.25$, determine the largest mass of the cylinder that can be lifted by the cord.

(a)

SOLUTION

Lifting the cylinder, which has a weight $W = mg$, causes the cord to move counterclockwise over the drums, and so the maximum tension T_2 in the cord occurs at D. Thus, $F = T_2 = 500$ N. A section of the cord passing over the drum at B is shown in Fig. 8–19b. Since $180° = \pi$ rad the angle of contact between the drum and the cord is $\beta = (135°/180°)\pi = 3\pi/4$ rad. Using Eq. 8–6, we have

$$T_2 = T_1 e^{\mu_s \beta}; \qquad 500 \text{ N} = T_1 e^{0.25[(3/4)\pi]}$$

Hence,

$$T_1 = \frac{500 \text{ N}}{e^{0.25[(3/4)\pi]}} = \frac{500 \text{ N}}{1.80} = 277.4 \text{ N}$$

Since the pulley at A is free to rotate, equilibrium requires that the tension in the cord remains the *same* on both sides of the pulley.

The section of the cord passing over the drum at C is shown in Fig. 8–19c. The weight $W < 277.4$ N. Applying Eq. 8–6, we obtain

$$T_2 = T_1 e^{\mu_s \beta}; \qquad 277.4 \text{ N} = W e^{0.25[(3/4)\pi]}$$

$$W = 153.9 \text{ N}$$

so that

$$m = \frac{W}{g} = \frac{153.9 \text{ N}}{9.81 \text{ m/s}^2}$$

$$= 15.7 \text{ kg} \qquad \textit{Ans.}$$

(b)

(c)

Fig. 8–19

420   CHAPTER 8   FRICTION

PROBLEMS

All solutions must include a free-body diagram.

8–83. Determine the *minimum* tension in the rope at points A and B that is necessary to maintain equilibrium. Take $\mu_s = 0.3$ between the rope and the fixed post D.

***8–84.** Determine the *maximum* tension in the rope at points A and B that is necessary to maintain equilibrium. Take $\mu_s = 0.3$ between the rope and the fixed post D.

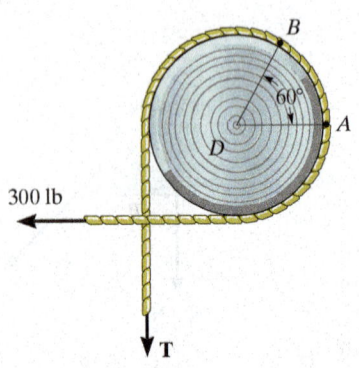

Probs. 8–83/84

8–85. A 180-lb farmer tries to restrain the cow from escaping by wrapping the rope two turns around the tree trunk. If the cow exerts a force of 250 lb on the rope, determine if the farmer can successfully restrain the cow. The coefficient of static friction between the rope and the tree trunk is $\mu_s = 0.15$, and between the farmer's shoes and the ground $\mu_s' = 0.3$.

Prob. 8–85

8–86. The 100-lb boy at A is suspended from the rope that passes over the quarter circular cliff rock. Determine if it is possible for the 185-lb woman to hoist him up; and if this is possible, what smallest force must she exert on the rope? The coefficient of static friction between the rope and the rock is $\mu_s = 0.2$, and between the shoes of the woman and the ground $\mu_s' = 0.8$.

Prob. 8–86

8–87. The 100-lb boy at A is suspended from the rope that passes over the quarter circular cliff rock. What force must the woman exert on the rope in order to let the boy descend at constant velocity? The coefficients of static and kinetic friction between the rope and the rock are $\mu_s = 0.4$ and $\mu_k = 0.35$, respectively.

Prob. 8–87

*8–88. If the force T_1 is applied to the rope at A, determine the force T_2 at B needed to pull the rope over the two fixed drums having the angles of contact and the coefficients of static friction shown.

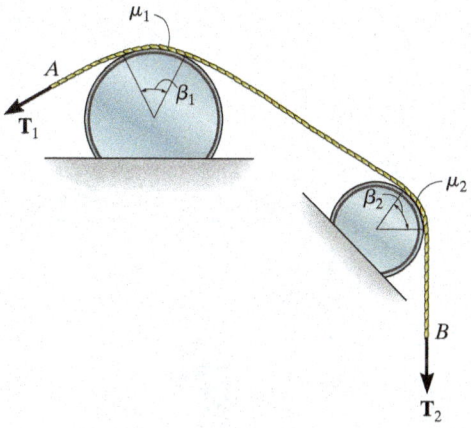

Prob. 8–88

8–89. Determine the *minimum* tension in the rope at points A and B which is necessary to maintain equilibrium. The coefficient of static friction between the rope and the fixed post D is $\mu_s = 0.3$. The rope is wrapped only once around the post.

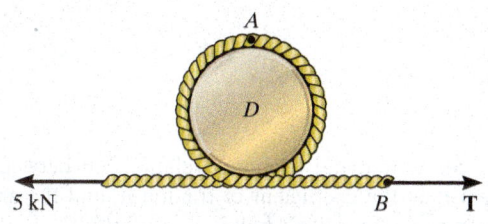

Prob. 8–89

8–90. The choker sling is used to lift the smooth pipe that has a mass of 600 kg. If the coefficient of static friction between the loop at the end A of the sling and the rope is $\mu_s = 0.3$, determine the angle θ at the connection.

Prob. 8–90

8–91. Two 8-kg blocks are attached to a cord that passes over two fixed drums. If $\mu_s = 0.3$ at the drums, determine the equilibrium angle θ the cord makes with the horizontal when a vertical force $P = 200$ N is applied to the cord to pull it downward.

Prob. 8–91

*8–92. Determine the force P that must be applied to the handle of the lever so that the wheel is on the verge of turning if $M = 300$ N·m. The coefficient of static friction between the belt and the wheel is $\mu_s = 0.3$.

8–93. If a force of $P = 30$ N is applied to the handle of the lever, determine the largest couple moment M that can be resisted so that the wheel does not turn. The coefficient of static friction between the belt and the wheel is $\mu_s = 0.3$.

Probs. 8–92/93

8–94. The 50-lb cylinder is attached to a cord which passes over the fixed drum. If the coefficient of static friction at the drum is $\mu_s = 0.3$, determine the maximum force P that can be applied to the roller when $\theta = 30°$ without causing the cylinder to move.

8–95. The 50-lb cylinder is attached to a cord which passes over the fixed drum. If the coefficient of static friction between the cord and the drum is $\mu_s = 0.3$, determine the angle θ of the cord if the maximum vertical load that can be applied to the roller without causing the cylinder to move is $P = 100$ lb.

Probs. 8–94/95

*8–96. The load can be supported by two boys when the cord is suspended over the pipe (a half-turn). If each boy can pull with a force of 125 lb, determine the maximum weight of the load. Can one boy support the same load when the cord is wrapped one-and-a-half times around the pipe? The coefficient of static friction is $\mu_s = 0.3$.

Prob. 8–96

8–97. A girl weighing 100 lb attempts to pull herself up a tree using a rope which is draped over the limb at B. If $\mu_s = 0.4$, determine the smallest force at which she must pull on the rope to lift herself. Do you think this is possible?

Prob. 8–97

8–98. Show that the frictional relationship between the belt tensions, the coefficient of friction μ, and the angular contacts α and β for the V-belt is $T_2 = T_1 e^{\mu\beta/\sin(\alpha/2)}$.

Prob. 8–98

8–99. The wheel is subjected to a torque of $M = 50$ N·m. If the coefficient of kinetic friction between the band brake and the rim of the wheel is $\mu_k = 0.3$, determine the smallest horizontal force P that must be applied to the lever to stop the wheel from rotating.

8–101. The uniform bar AB is supported by a rope that passes over a frictionless pulley at C and a fixed peg at D. If the coefficient of static friction between the rope and the peg is $\mu_D = 0.3$, determine the smallest distance x from the end of the bar at which a 20-N force may be placed and not cause the bar to move.

Prob. 8–99

Prob. 8–101

*8–100. Blocks A and B have a mass of 7 kg and 10 kg, respectively. Using the coefficients of static friction indicated, determine the largest vertical force P which can be applied to the cord without causing motion.

8–102. The belt on the portable dryer wraps around the drum D, idler pulley A, and motor pulley B. If the motor can develop a maximum torque of $M = 0.80$ N·m, determine the smallest spring tension required to prevent the belt from slipping. The coefficient of static friction between the belt and the drum and motor pulley is $\mu_s = 0.3$.

Prob. 8–100

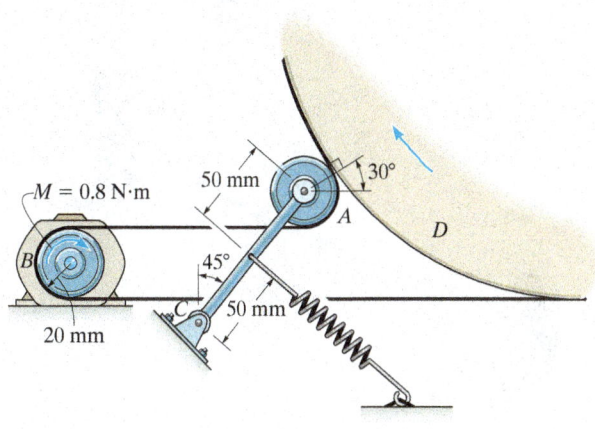

Prob. 8–102

8–103. Two boys have a tug-of-war using a rope. In the process the rope rubs against a post such that it is deflected 3° on each side of the post. If boy A weighs 60 lb and boy B weighs 50 lb, determine if it possible for one of them to pull the rope toward him. The coefficient of static friction between the rope and the post is $\mu_p = 0.3$, and the coefficient of static friction between the boys' shoes and the ground is $\mu_s = 0.4$.

8–105. A 10-kg cylinder D, which is attached to a small pulley B, is placed on the cord as shown. Determine the largest angles θ so that the cord does not slip over the peg at C. The cylinder at E also has a mass of 10 kg, and the coefficient of static friction between the cord and the peg is $\mu_s = 0.1$.

Prob. 8–103

Prob. 8–105

*8–104. Two boys have a tug-of-war using a rope. In the process the rope rubs against a post such that it is deflected 3° on each side of the post. If boy A weighs 60 lb and boy B weighs 50 lb, determine if it is possible for one of them to pull the rope toward him. The coefficient of static friction between the rope and the post is $\mu_p = 0.3$, and the coefficient of static friction between the shoes of the boy at A and the ground is $\mu_A = 0.4$, whereas for boy B it is $\mu_B = 0.5$.

8–106. A conveyer belt is used to transfer granular material and the frictional resistance on the top of the belt is $F = 500$ N. Determine the smallest stretch of the spring attached to the moveable axle of the idle pulley B so that the belt does not slip at the drive pulley A when the torque **M** is applied. What minimum torque M is required to keep the belt moving? The coefficient of static friction between the belt and the wheel at A is $\mu_s = 0.2$.

Prob. 8–104

Prob. 8–106

*8.6 FRICTIONAL FORCES ON COLLAR BEARINGS, PIVOT BEARINGS, AND DISKS

Pivot and *collar bearings* are commonly used in machines to support an *axial load* on a rotating shaft. Typical examples are shown in Fig. 8–20. Provided these bearings are not lubricated, or are only partially lubricated, the laws of dry friction may be applied to determine the moment needed to turn the shaft when it supports an axial force.

Pivot bearing
(a)

Collar bearing
(b)

Fig. 8–20

Frictional Analysis. The collar bearing on the shaft shown in Fig. 8–21 is subjected to an axial force $\mathbf{P}$ and has a total bearing or contact area $\pi(R_2^2 - R_1^2)$. Provided the bearing is new and evenly supported, then the normal pressure p on the bearing will be *uniformly distributed* over this area. Since $\Sigma F_z = 0$, then p, measured as a force per unit area, is $p = P/\pi(R_2^2 - R_1^2)$.

The moment needed to cause impending rotation of the shaft can be determined from moment equilibrium about the z axis. A differential area element $dA = (r\, d\theta)(dr)$, shown in Fig. 8–21, is subjected to both a normal force $dN = p\, dA$ and an associated frictional force,

$$dF = \mu_s\, dN = \mu_s p\, dA = \frac{\mu_s P}{\pi(R_2^2 - R_1^2)} dA$$

Fig. 8–21

Fig. 8–21(repeated)

The normal force does not create a moment about the axis of the shaft; however, the frictional force does; namely, $dM_z = r\,dF$. Integration is needed to determine the applied moment **M** required to overcome all the frictional forces. Therefore, for impending rotational motion,

$$\Sigma M_z = 0; \qquad M - \int_A r\,dF = 0$$

Substituting for dF and dA and integrating over the entire bearing area, we have

$$M = \int_{R_1}^{R_2}\int_0^{2\pi} r\left[\frac{\mu_s P}{\pi(R_2^2 - R_1^2)}\right](r\,d\theta\,dr) = \frac{\mu_s P}{\pi(R_2^2 - R_1^2)}\int_{R_1}^{R_2} r^2\,dr \int_0^{2\pi} d\theta$$

or

$$M = \frac{2}{3}\mu_s P\left(\frac{R_2^3 - R_1^3}{R_2^2 - R_1^2}\right) \qquad (8\text{–}7)$$

The moment developed at the end of the shaft, when it is *rotating* at constant speed, can be found by substituting μ_k for μ_s.

In the case of a pivot bearing, Fig. 8–20a, then $R_2 = R$ and $R_1 = 0$, and Eq. 8–7 reduces to

$$M = \frac{2}{3}\mu_s PR \qquad (8\text{–}8)$$

Equations 8–7 and 8–8 apply only for bearing surfaces subjected to *constant pressure*. If the pressure is not uniform, a variation of the pressure as a function of the bearing area must be determined before integrating to obtain the moment. The following example illustrates this situation.

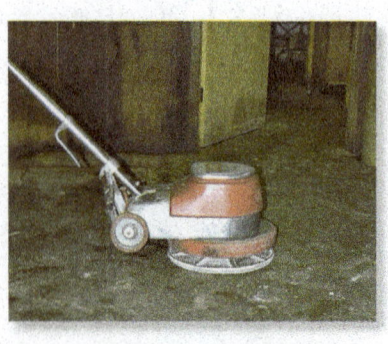

The motor that turns the disk of this sanding machine develops a torque that must overcome the frictional forces acting on the disk.

EXAMPLE 8.9

The uniform bar shown in Fig. 8–22a has a weight of 4 lb. If it is assumed that the normal pressure acting at the contacting surface varies linearly along the length of the bar as shown, determine the couple moment **M** required to rotate the bar. Assume that the bar's width is negligible in comparison to its length. The coefficient of static friction is $\mu_s = 0.3$.

SOLUTION

A free-body diagram of the bar is shown in Fig. 8–22b. The intensity w_0 of the distributed load at the center ($x = 0$) is determined from vertical force equilibrium, Fig. 8–22a.

$$+\uparrow \Sigma F_z = 0; \quad -4 \text{ lb} + 2\left[\frac{1}{2}(2 \text{ ft})w_0\right] = 0 \quad w_0 = 2 \text{ lb/ft}$$

Since $w = 0$ at $x = 2$ ft, the distributed load expressed as a function of x is

$$w = (2 \text{ lb/ft})\left(1 - \frac{x}{2 \text{ ft}}\right) = 2 - x$$

The magnitude of the normal force acting on a differential segment having a length dx is therefore

$$dN = w\, dx = (2 - x)\, dx$$

The magnitude of the frictional force acting on the same element is

$$dF = \mu_s\, dN = 0.3(2 - x)\, dx$$

Hence, the moment created by this force about the z axis is

$$dM = x\, dF = 0.3(2x - x^2)\, dx$$

The summation of moments about the z axis of the bar is determined by integration, which yields

$$\Sigma M_z = 0; \quad M - 2\int_0^2 (0.3)(2x - x^2)\, dx = 0$$

$$M = 0.6\left(x^2 - \frac{x^3}{3}\right)\Big|_0^2$$

$$M = 0.8 \text{ lb} \cdot \text{ft} \quad \textit{Ans.}$$

(a)

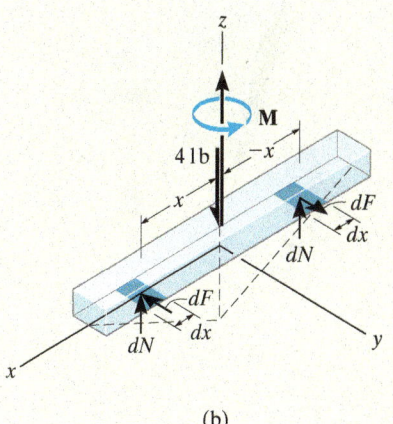

(b)

Fig. 8–22

8.7 FRICTIONAL FORCES ON JOURNAL BEARINGS

When a shaft or axle is subjected to lateral loads, a *journal bearing* is commonly used for support. Provided the bearing is not lubricated, or is only partially lubricated, a reasonable analysis of the frictional resistance on the bearing can be based on the laws of dry friction.

Frictional Analysis. A typical journal-bearing support is shown in Fig. 8–23a. As the shaft rotates, the contact point A moves up the wall of the bearing where slipping occurs. If the vertical load acting at the end of the shaft is **P**, then the bearing reactive force **R** acting at A will be equal and opposite to **P**, Fig. 8–23b. The moment needed to maintain constant rotation of the shaft can be found by summing moments about the axis of the shaft; i.e.,

$$\Sigma M_z = 0; \qquad M - (R \sin \phi_k)r = 0$$

or

$$M = Rr \sin \phi_k \qquad (8\text{–}9)$$

where ϕ_k is the angle of kinetic friction defined by $\tan \phi_k = F/N = \mu_k N/N = \mu_k$. In Fig. 8–23c, it is seen that $r \sin \phi_k = r_f$. The dashed circle with this radius is called the *friction circle*, and as the shaft rotates, the reaction **R** will always be tangent to it. If the bearing is partially lubricated, μ_k is small, and therefore $\sin \phi_k \approx \tan \phi_k \approx \mu_k$. Under these conditions, a reasonable *approximation* to the moment needed to overcome the frictional resistance becomes

$$M \approx Rr\mu_k \qquad (8\text{–}10)$$

This equation implies that M can be minimized by making the bearing radius r as small as possible. For large bearings, however, this type of journal bearing is not suitable for long service, since friction between the shaft and bearing will eventually wear down the surfaces. Instead, designers will incorporate "ball bearings" or "rollers" in journal bearings to minimize frictional losses.

Unwinding the cable from this spool requires overcoming friction from the supporting shaft.

(a)

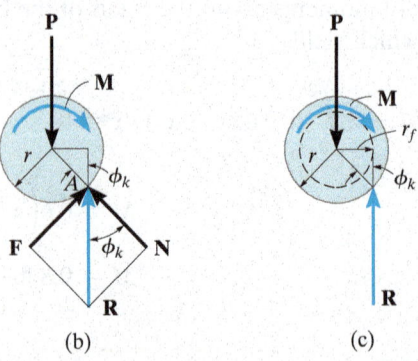

(b)   (c)

Fig. 8–23

EXAMPLE 8.10

The 100-mm-diameter pulley shown in Fig. 8–24a fits loosely on a 10-mm-diameter shaft for which the coefficient of static friction is $\mu_s = 0.4$. Determine the minimum tension T in the belt needed to (a) raise the 100-kg block and (b) lower the block. Assume that no slipping occurs between the belt and pulley, and neglect the weight of the pulley.

SOLUTION

Part (a). A free-body diagram of the pulley is shown in Fig. 8–24b. When $T = 981$ N, the pulley will make contact with the shaft at point P_1. As the tension T is *increased*, the contact point will move around the shaft to point P_2 before motion impends. From the figure, the friction circle has a radius $r_f = r \sin \phi_s$. Using the simplification that $\sin \phi_s \approx \tan \phi_s \approx \mu_s$ then $r_f \approx r\mu_s = (5 \text{ mm})(0.4) = 2$ mm, and so summing moments about P_2 gives

$\zeta + \Sigma M_{P_2} = 0;$ $981 \text{ N}(52 \text{ mm}) - T(48 \text{ mm}) = 0$

$T = 1063 \text{ N} = 1.06 \text{ kN}$ *Ans.*

If a more exact analysis is used, then $\phi_s = \tan^{-1} 0.4 = 21.8°$. Thus, the radius of the friction circle would be $r_f = r \sin \phi_s = 5 \sin 21.8° = 1.86$ mm. Therefore,

$\zeta + \Sigma M_{P_2} = 0;$

$981 \text{ N}(50 \text{ mm} + 1.86 \text{ mm}) - T(50 \text{ mm} - 1.86 \text{ mm}) = 0$

$T = 1057 \text{ N} = 1.06 \text{ kN}$ *Ans.*

Part (b). When the block is *lowered*, the resultant force **R** acts at point P_3 on the shaft as shown in Fig. 8–24c. Summing moments about this point yields

$\zeta + \Sigma M_{P_3} = 0;$ $981 \text{ N}(48 \text{ mm}) - T(52 \text{ mm}) = 0$

$T = 906 \text{ N}$ *Ans.*

The difference between raising and lowering the block is thus 157 N.

Fig. 8–24

430   CHAPTER 8   FRICTION

*8.8 ROLLING RESISTANCE

When a *rigid* cylinder rolls at constant velocity along a *rigid* surface, the normal force exerted by the surface on the cylinder acts perpendicular to the tangent at the point of contact, Fig. 8–25a. Actually, however, no materials are perfectly rigid, and therefore the reaction of the surface on the cylinder consists of a distribution of normal pressure. For example, consider the cylinder to be made of a very hard material, and the surface upon which it rolls to be relatively soft. As the cylinder of weight **W** rolls, the surface material in front of the cylinder *retards* the motion since it is being *deformed*, Fig. 8–25b, whereas the material in the rear is *restored* from the deformed state, and therefore tends to *push* the cylinder forward. The normal pressures acting on the cylinder in this manner are represented in Fig. 8–25b by their resultant forces $\mathbf{N}_d$ and $\mathbf{N}_r$. The magnitude of the force of *deformation*, $\mathbf{N}_d$, and its horizontal component is *always greater* than that of *restoration*, $\mathbf{N}_r$, and consequently a horizontal driving force **P** must be applied to the cylinder to maintain the motion, Fig. 8–25b.*

Rigid surface of contact

(a)

Soft surface of contact

(b)

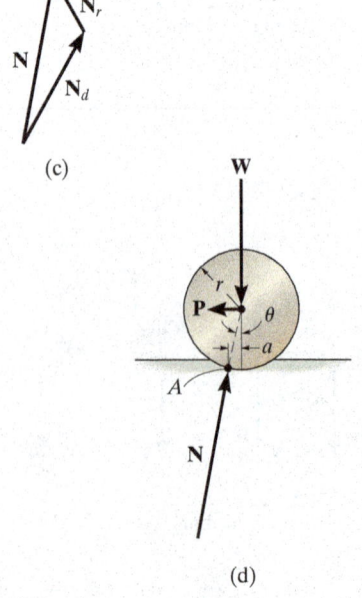

(c)

(d)

Fig. 8–25

Rolling resistance is caused primarily by this effect, although it is also, to a lesser degree, the result of surface adhesion and relative microsliding between the surfaces of contact. Because the actual force **P** needed to overcome these effects is difficult to determine, a simplified method will be developed here to explain one way engineers have analyzed this phenomenon. To do this, we will consider the resultant of the *entire* normal pressure, $\mathbf{N} = \mathbf{N}_d + \mathbf{N}_r$, acting on the cylinder, Fig. 8–25c. As shown in Fig. 8–25d, this force acts at an angle θ with the vertical. To keep the cylinder in equilibrium, i.e., rolling at a constant rate, it is necessary that **N** be *concurrent* with the driving force **P** and the weight **W**. Summing moments about point A gives $Wa = P(r \cos \theta)$. Since the deformations are generally very small in relation to the cylinder's radius, $\cos \theta \approx 1$; hence,

$$Wa \approx Pr$$

or

$$P \approx \frac{Wa}{r} \qquad (8\text{–}11)$$

The distance a is termed the ***coefficient of rolling resistance***, which has the dimension of length. For instance, $a \approx 0.5$ mm for a wheel rolling on a rail, both of which are made of mild steel. For hardened steel ball

---

*Actually, the deformation force $\mathbf{N}_d$ causes *energy* to be stored in the material as its magnitude is increased, whereas the restoration force $\mathbf{N}_r$, as its magnitude is decreased, allows some of this energy to be released. The remaining energy is *lost* since it is used to heat up the surface, and if the cylinder's weight is very large, it accounts for permanent deformation of the surface. Work must be done by the horizontal force **P** to make up for this loss.

bearings on steel, $a \approx 0.1$ mm. Experimentally, though, this factor is difficult to measure, since it depends on such parameters as the rate of rotation of the cylinder, the elastic properties of the contacting surfaces, and the surface finish. For this reason, little reliance is placed on the data for determining a. The analysis presented here does, however, indicate why a heavy load (W) offers greater resistance to motion (P) than a light load under the same conditions. Furthermore, since Wa/r is generally very small compared to $\mu_k W$, the force needed to *roll* a cylinder over the surface will be much less than that needed to *slide* it across the surface. It is for this reason that a roller or ball bearings are often used to minimize the frictional resistance between moving parts.

Rolling resistance of railroad wheels on the rails is small since steel is very stiff. By comparison, the rolling resistance of the wheels of a tractor in a wet field is very large.

EXAMPLE 8.11

A 10-kg steel wheel shown in Fig. 8–26a has a radius of 100 mm and rests on an inclined plane made of soft wood. If θ is increased so that the wheel begins to roll down the incline with constant velocity when $\theta = 1.2°$, determine the coefficient of rolling resistance.

(a)

(b)

Fig. 8–26

SOLUTION

As shown on the free-body diagram, Fig. 8–26b, when the wheel has impending motion, the normal reaction **N** acts at point A defined by the dimension a. Resolving the weight into components parallel and perpendicular to the incline, and summing moments about point A, yields

$\zeta + \Sigma M_A = 0$;

$-(98.1 \cos 1.2° \text{ N})(a) + (98.1 \sin 1.2° \text{ N})(100 \cos 1.2° \text{ mm}) = 0$

Solving, we obtain

$a = 2.09$ mm   *Ans.*

432   CHAPTER 8   FRICTION

PROBLEMS

All solutions must include a free-body diagram.

8–107. The collar bearing uniformly supports an axial force of $P = 5$ kN. If the coefficient of static friction is $\mu_s = 0.3$, determine the smallest torque M required to overcome friction.

*8–108. The collar bearing uniformly supports an axial force of $P = 8$ kN. If a torque of $M = 200$ N·m is applied to the shaft and causes it to rotate at constant velocity, determine the coefficient of kinetic friction at the surface of contact.

Probs. 8–107/108

8–109. The floor-polishing machine rotates at a constant angular velocity. If it has a weight of 80 lb, determine the couple forces F the operator must apply to the handles to hold the machine stationary. The coefficient of kinetic friction between the floor and brush is $\mu_k = 0.3$. Assume the brush exerts a uniform pressure on the floor.

Prob. 8–109

8–110. The *double-collar bearing* is subjected to an axial force $P = 4$ kN. Assuming that collar A supports $0.75P$ and collar B supports $0.25P$, both with a uniform distribution of pressure, determine the maximum frictional moment M that may be resisted by the bearing. Take $\mu_s = 0.2$ for both collars.

Prob. 8–110

8–111. The *double-collar bearing* is subjected to an axial force $P = 16$ kN. Assuming that collar A supports $0.75P$ and collar B supports $0.25P$, both with a uniform distribution of pressure, determine the smallest torque M that must be applied to overcome friction. Take $\mu_s = 0.2$ for both collars.

Prob. 8–111

*8–112. The pivot bearing is subjected to a pressure distribution at its surface of contact which varies as shown. If the coefficient of static friction is μ, determine the torque M required to overcome friction if the shaft supports an axial force **P**. For the solution it is necessary to determine the peak pressure p_0 in terms of P and the bearing radius R.

8–114. A tube has a total weight of 200 lb, length $l = 8$ ft, and radius $= 0.75$ ft. If it rests in sand for which the coefficient of static friction is $\mu_s = 0.23$, determine the torque M needed to turn it. Assume that the pressure distribution along the length of the tube is defined by $p = p_0 \sin \theta$. For the solution it is necessary to determine p_0, the peak pressure, in terms of the weight and tube dimensions.

Prob. 8–112

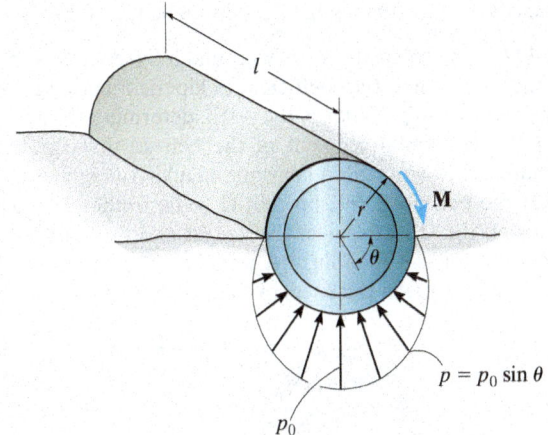

Prob. 8–114

8–113. Because of wearing at the edges, the pivot bearing is subjected to a conical pressure distribution at its surface of contact. Determine the torque M required to overcome friction and turn the shaft, which supports an axial force **P**. The coefficient of static friction is μ_s. For the solution it is necessary to determine the peak pressure p_0 in terms of P and the bearing radius R.

8–115. The corkscrew is used to remove the 15-mm-diameter cork from the bottle. Determine the smallest vertical force P that must be applied to the handle if the gauge pressure in the bottle is $p = 175$ kPa and the cork pushes against the sides of the bottle's neck with a uniform pressure of 90 kPa. The coefficient of static friction between the bottle and the cork is $\mu_s = 0.4$. *Hint:* The force exerted on the bottom of the cork is $F = pA$, where A is the surface area of the cork's bottom and p is the gauge pressure.

Prob. 8–113

Prob. 8–115

*8–116. The collar fits *loosely* around a fixed shaft that has a radius of 2 in. If the coefficient of kinetic friction between the shaft and the collar is $\mu_k = 0.3$, determine the force P on the horizontal segment of the belt so that the collar rotates *counterclockwise* with a constant angular velocity. Assume that the belt does not slip on the collar; rather, the collar slips on the shaft. Neglect the weight and thickness of the belt and collar. The radius, measured from the center of the collar to the mean thickness of the belt, is 2.25 in.

8–117. The collar fits *loosely* around a fixed shaft that has a radius of 2 in. If the coefficient of kinetic friction between the shaft and the collar is $\mu_k = 0.3$, determine the force P on the horizontal segment of the belt so that the collar rotates *clockwise* with a constant angular velocity. Assume that the belt does not slip on the collar; rather, the collar slips on the shaft. Neglect the weight and thickness of the belt and collar. The radius, measured from the center of the collar to the mean thickness of the belt, is 2.25 in.

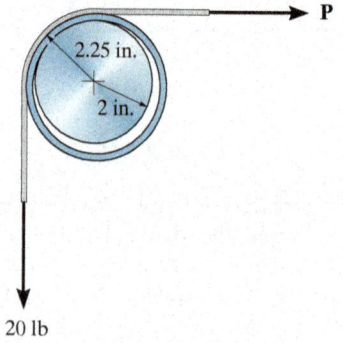

Probs. 8–116/117

8–118. If the coefficient of static friction at the cone clutch is μ_s, determine the smallest force P that should be applied to the handle in order to transmit the torque **M**.

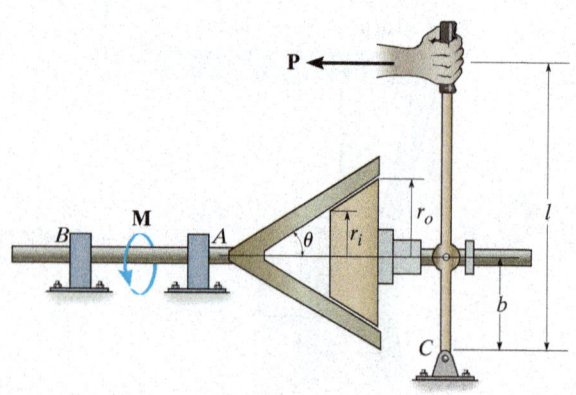

Prob. 8–118

8–119. A disk having an outer diameter of 120 mm fits loosely over a fixed shaft having a diameter of 30 mm. If the coefficient of static friction between the disk and the shaft is $\mu_s = 0.15$ and the disk has a mass of 50 kg, determine the smallest vertical force F acting on the rim which must be applied to the disk to cause it to slip over the shaft.

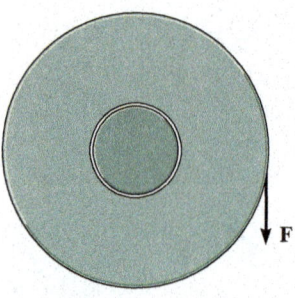

Prob. 8–119

*8–120. The 4-lb pulley has a diameter of 1 ft and the axle has a diameter of 1 in. If the coefficient of kinetic friction between the axle and the pulley is $\mu_k = 0.20$, determine the vertical force P on the rope required to lift the 20-lb block at constant velocity.

8–121. Solve Prob. 8–120 if the force **P** is applied horizontally to the left.

Probs. 8–120/121

8–122. Determine the tension T in the belt needed to overcome the tension of 200 lb created on the other side. Also, what are the normal and frictional components of force developed on the collar bushing? The coefficient of static friction is $\mu_s = 0.21$.

8–123. If a tension $T = 215$ lb is required to pull the 200-lb force around the collar bushing, determine the coefficient of static friction at the contacting surface. The belt does not slip on the collar.

Probs. 8–122/123

*8–124. If the smallest tension required to pull the belt downward at A over the shaft S is $T_A = 500$ N, determine the coefficient of kinetic friction between the loosely fitting collar bushing B and the shaft. Assume that the belt does not slip on the collar; rather, the collar slips on the shaft.

Prob. 8–124

8–125. The 5-kg skateboard rolls down the 5° slope at constant speed. If the coefficient of kinetic friction between the 12.5-mm-diameter axles and the wheels is $\mu_k = 0.3$, determine the radius of the wheels. Neglect rolling resistance of the wheels on the surface. The center of mass for the skateboard is at G.

Prob. 8–125

8–126. The bell crank fits loosely into a 0.5-in.-diameter pin. Determine the required force P which is just sufficient to rotate the bell crank clockwise. The coefficient of static friction between the pin and the bell crank is $\mu_s = 0.3$.

Prob. 8–126

8–127. The bell crank fits loosely into a 0.5-in.-diameter pin. If $\mathbf{P} = 41$ lb, the bell crank is then on the verge of rotating counterclockwise. Determine the coefficient of static friction between the pin and the bell crank.

Prob. 8–127

436   CHAPTER 8   FRICTION

*8–128. The lawn roller has a mass of 80 kg. If the arm BA is held at an angle of 30° from the horizontal and the coefficient of rolling resistance for the roller is 25 mm, determine the force P needed to push the roller at constant speed. Neglect friction developed at the axle, A, and assume that the resultant force **P** acting on the handle is applied along arm BA.

8–130. The cylinder is subjected to a load that has a weight W. If the coefficients of rolling resistance for the cylinder's top and bottom surfaces are a_A and a_B, respectively, show that a horizontal force having a magnitude of $P = [W(a_A + a_B)]/2r$ is required to move the load and thereby roll the cylinder forward. Neglect the weight of the cylinder.

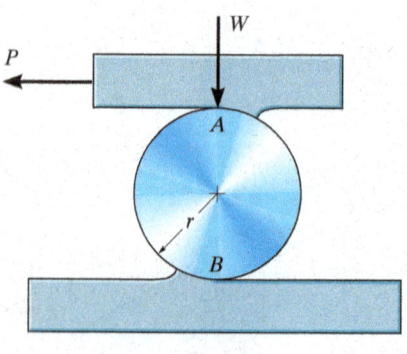

Prob. 8–130

Prob. 8–128

8–129. The tractor has a weight of 16,000 lb and the coefficient of rolling resistance is $a = 2$ in. Determine the force P needed to overcome rolling resistance at all four wheels and push it forward.

8–131. A large stone having a mass of 500 kg is moved along the incline using a series of 150-mm-diameter rollers for which the coefficient of rolling resistance is 3 mm at the ground and 4 mm at the bottom surface of the stone. Determine the force T that will cause the stone to descend the plane at a constant speed.

Prob. 8–129

Prob. 8–131

CHAPTER REVIEW

Dry Friction

Frictional forces exist between two rough surfaces of contact. These forces act on a body so as to oppose its motion or tendency of motion.

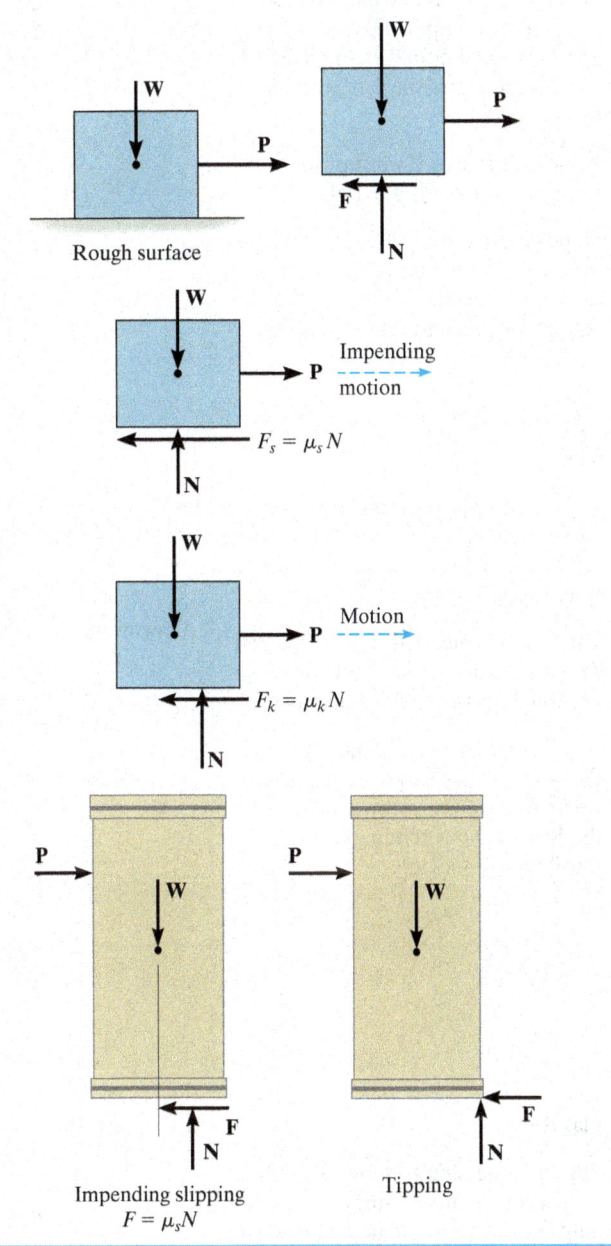

A static frictional force has a maximum value of $F_s = \mu_s N$, where μ_s is the *coefficient of static friction*. In this case, motion between the contacting surfaces is *impending*.

If slipping occurs, then the friction force remains essentially constant and equal to $F_k = \mu_k N$. Here μ_k is the *coefficient of kinetic friction*.

The solution of a problem involving friction requires first drawing the free-body diagram of the body. If the unknowns cannot be determined strictly from the equations of equilibrium, and the possibility of slipping occurs, then the friction equation should be applied at the appropriate points of contact in order to complete the solution.

It may also be possible for slender objects, like crates, to tip over, and this situation should also be investigated.

Wedges

Wedges are inclined planes used to increase the application of a force. The two force equilibrium equations are used to relate the forces acting on the wedge.

An applied force **P** must push on the wedge to move it to the right.

If the coefficients of friction between the surfaces are large enough, then **P** can be removed, and the wedge will be self-locking and remain in place.

$\Sigma F_x = 0$
$\Sigma F_y = 0$

Screws

Square-threaded screws are used to move heavy loads. They represent an inclined plane, wrapped around a cylinder.

The moment needed to turn a screw depends upon the coefficient of friction and the screw's lead angle θ.

If the coefficient of friction between the surfaces is large enough, then when $M = 0$, the screw will support the load without tending to turn, i.e., it will be self-locking.

$M = rW \tan(\theta + \phi_s)$

Upward Impending Screw Motion

$M' = rW \tan(\theta - \phi_s)$

Downward Impending Screw Motion

$\theta > \phi_s$

$M'' = rW \tan(\phi_s - \theta)$

Downward Screw Motion

$\phi_s > \theta$

Flat Belts

The force needed to move a flat belt over a rough curved surface depends only on the angle of belt contact, β, and the coefficient of friction.

$T_2 = T_1 e^{\mu\beta}$

$T_2 > T_1$

Collar Bearings and Disks

The frictional analysis of a collar bearing or disk requires looking at a differential element of the contact area. The normal force acting on this element is determined from force equilibrium along the shaft, and the moment needed to turn the shaft at a constant rate is determined from moment equilibrium about the axis of the shaft.

If the pressure on the surface of a collar bearing is uniform, then integration gives the result shown.

$$M = \frac{2}{3} \mu_s P \left(\frac{R_2^3 - R_1^3}{R_2^2 - R_1^2} \right)$$

Journal Bearings

When a moment is applied to a shaft in a nonlubricated or partially lubricated journal bearing, the shaft will tend to roll up the side of the bearing until slipping occurs. This defines the radius of a friction circle, and from it the moment needed to turn the shaft can be determined.

$$M = Rr \sin \phi_k$$

Rolling Resistance

The resistance of a wheel to rolling over a surface is caused by localized *deformation* of the two materials in contact. This causes the resultant normal force acting on the rolling body to be inclined so that it provides a component that acts in the opposite direction of the applied force **P** causing the motion. One way to analyze this effect is to use a *coefficient of rolling resistance, a,* which is determined from experiment.

$$P \approx \frac{Wa}{r}$$

REVIEW PROBLEMS

All solutions must include a free-body diagram.

R8–1. The uniform 20-lb ladder rests on the rough floor for which the coefficient of static friction is $\mu_s = 0.4$ and against the smooth wall at B. Determine the horizontal force P the man must exert on the ladder in order to cause it to move.

R8–3. A 35-kg disk rests on an inclined surface for which $\mu_s = 0.2$. Determine the maximum vertical force P that may be applied to bar AB without causing the disk to slip at C. Neglect the mass of the bar.

Prob. R8–1

Prob. R8–3

R8–2. The uniform 60-kg crate C rests uniformly on a 10-kg dolly D. If the front casters of the dolly at A are locked to prevent rolling while the casters at B are free to roll, determine the maximum force P that may be applied without causing motion of the crate. The coefficient of static friction between the casters and the floor is $\mu_f = 0.35$ and between the dolly and the crate, $\mu_d = 0.5$.

R8–4. The cam is subjected to a couple moment of $5 \text{ N} \cdot \text{m}$. Determine the minimum force P that should be applied to the follower in order to hold the cam in the position shown. The coefficient of static friction between the cam and the follower is $\mu = 0.4$. The guide at A is smooth.

Prob. R8–2

Prob. R8–4

R8–5. The cam is subjected to a couple moment of 5 N·m. Determine the minimum force **P** that should be applied to the follower in order to hold the cam in the position shown. The coefficient of static friction between the cam and the follower is $\mu = 0.4$. The guide at **A** is smooth.

R8–7. The uniform 50-lb beam is supported by the rope that is attached to the end of the beam, wraps over the rough peg, and is then connected to the 100-lb block. If the coefficient of static friction between the beam and the block, and between the rope and the peg, is $\mu_s = 0.4$, determine the maximum distance that the block can be placed from A and still remain in equilibrium. Assume the block will not tip.

Prob. R8–5

Prob. R8–7

R8–6. The jacking mechanism consists of a link that has a square-threaded screw with a mean diameter of 0.5 in. and a lead of 0.20 in., and the coefficient of static friction is $\mu_s = 0.4$. Determine the torque M that should be applied to the screw to start lifting the 6000-lb load acting at the end of member ABC.

R8–8. The hand cart has wheels with a diameter of 80 mm. If a crate having a mass of 500 kg is placed on the cart so that each wheel carries an equal load, determine the horizontal force P that must be applied to the handle to overcome rolling resistance. The coefficient of rolling resistance is 2 mm. Neglect the mass of the cart.

Prob. R8–6

Prob. R8–8

CHAPTER 9

When a tank of any shape is designed, it is important to be able to determine its center of gravity, calculate its volume and surface area, and determine the forces caused by the liquid it contains. These topics will be covered in this chapter.

CENTER OF GRAVITY AND CENTROID

CHAPTER OBJECTIVES

- To show how to determine the location of the center of gravity and centroid for a body of arbitrary shape and one composed of composite parts.
- To use the theorems of Pappus and Guldinus for finding the surface area and volume for a body having axial symmetry.
- To present a method for finding the resultant of a general distributed loading and to show how it applies to finding the resultant force of a pressure loading caused by a fluid.

9.1 CENTER OF GRAVITY, CENTER OF MASS, AND THE CENTROID OF A BODY

Knowing the resultant or total weight of a body and its location is important when considering the effect this force produces on the body. The point of location is called the center of gravity, and in this section we will show how to find it for an irregularly shaped body. We will then extend this method to show how to find the body's center of mass, and its geometric center or centroid.

Center of Gravity. A body is composed of an infinite number of particles of differential size, and so if the body is located within a gravitational field, then each of these particles will have a weight dW. These weights will form a parallel force system, and the resultant of this system is the total weight of the body, which passes through a single point called the *center of gravity*, G.*

---

*In a strict sense this is true as long as the gravity field is assumed to have the same magnitude and direction everywhere. Although the actual force of gravity is directed toward the center of the earth, and this force varies with its distance from the center, for most engineering applications we can assume the gravity field has the same magnitude and direction everywhere.

444   CHAPTER 9   CENTER OF GRAVITY AND CENTROID

(a)

(b)

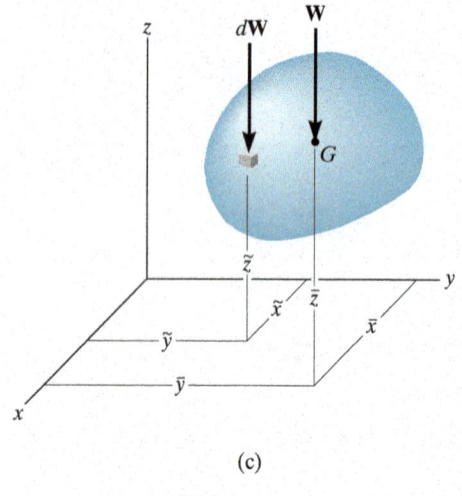

(c)

Fig. 9–1

To show how to determine the location of the center of gravity, consider the rod in Fig. 9–1a, where the segment having the weight dW is located at the arbitrary position $\tilde{x}$. Using the methods outlined in Sec. 4.8, the total weight of the rod is the sum of the weights of all of its particles, that is

$$+\downarrow F_R = \Sigma F_z; \qquad W = \int dW$$

The location of the center of gravity, measured from the y axis, is determined by equating the moment of W about the y axis, Fig. 9–1b, to the sum of the moments of the weights of all its particles about this same axis. Therefore,

$$(M_R)_y = \Sigma M_y; \qquad \bar{x}W = \int \tilde{x}\, dW$$

$$\bar{x} = \frac{\int \tilde{x}\, dW}{\int dW}$$

In a similar manner, if the body represents a plate, Fig. 9–1b, then a moment balance about the x and y axes would be required to determine the location $(\bar{x}, \bar{y})$ of point G. Finally we can generalize this idea to a three-dimensional body, Fig. 9–1c, and perform a moment balance about each of these axes to locate G for any rotated position of the axes. This results in the following equations.

$$\bar{x} = \frac{\int \tilde{x}\, dW}{\int dW} \qquad \bar{y} = \frac{\int \tilde{y}\, dW}{\int dW} \qquad \bar{z} = \frac{\int \tilde{z}\, dW}{\int dW} \qquad (9\text{–}1)$$

where

$\bar{x}, \bar{y}, \bar{z}$ are the coordinates of the center of gravity G.
$\tilde{x}, \tilde{y}, \tilde{z}$ are the coordinates of an arbitrary particle in the body.

9.1 CENTER OF GRAVITY, CENTER OF MASS, AND THE CENTROID OF A BODY

Center of Mass of a Body. In order to study the *dynamic response* or accelerated motion of a body, it becomes important to locate the body's *center of mass* C_m, Fig. 9–2. This location can be determined by substituting $dW = g\, dm$ into Eqs. 9–1. Provided g is constant, it cancels out, and so

$$\bar{x} = \frac{\int \tilde{x}\, dm}{\int dm} \qquad \bar{y} = \frac{\int \tilde{y}\, dm}{\int dm} \qquad \bar{z} = \frac{\int \tilde{z}\, dm}{\int dm} \qquad (9\text{–}2)$$

Fig. 9–2

Centroid of a Volume. If the body in Fig. 9–3 is made from a *homogeneous material*, then its density ρ (rho) will be *constant*. Therefore, a differential element of volume dV has a mass $dm = \rho\, dV$. Substituting this into Eqs. 9–2 and canceling out ρ, we obtain formulas that locate the *centroid C* or geometric center of the body; namely

$$\bar{x} = \frac{\int_V \tilde{x}\, dV}{\int_V dV} \qquad \bar{y} = \frac{\int_V \tilde{y}\, dV}{\int_V dV} \qquad \bar{z} = \frac{\int_V \tilde{z}\, dV}{\int_V dV} \qquad (9\text{–}3)$$

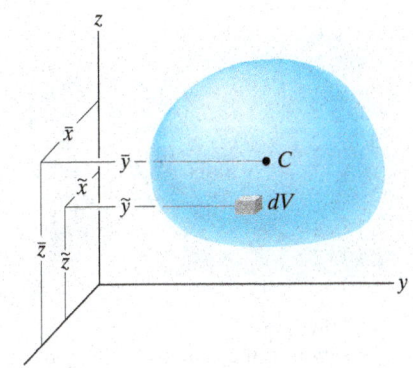

Fig. 9–3

Since these equations represent a balance of the moments of the volume of the body, then, if the volume possesses two planes of symmetry, its centroid will lie along the line of intersection of these two planes. For example, the cone in Fig. 9–4 has a centroid that lies on the y axis so that $\bar{x} = \bar{z} = 0$. To find the location $\bar{y}$ we can use the second of Eqs. 9–3. Here, a single integration is possible if we choose a differential element represented by a *thin disk* having a thickness dy and radius $r = z$. Its volume is $dV = \pi r^2 dy = \pi z^2 dy$ and its centroid is at $\tilde{x} = 0, \tilde{y} = y, \tilde{z} = 0$.

Fig. 9–4

Fig. 9–5

Integration must be used to determine the location of the center of gravity of this lamp post due to the curvature of the member.

Centroid of an Area.
If an area lies in the x–y plane and is bounded by the curve $y = f(x)$, as shown in Fig. 9–5a, then its centroid will be in this plane and can be determined from integrals similar to Eqs. 9–3, namely,

$$\bar{x} = \frac{\int_A \tilde{x}\, dA}{\int_A dA} \qquad \bar{y} = \frac{\int_A \tilde{y}\, dA}{\int_A dA} \qquad (9\text{–}4)$$

These integrals can be evaluated by performing a *single integration* if we use a *rectangular strip* for the differential area element. For example, if a vertical strip is used, Fig. 9–5b, the area of the element is $dA = y\, dx$, and its centroid is located at $\tilde{x} = x$ and $\tilde{y} = y/2$. If we consider a horizontal strip, Fig. 9–5c, then $dA = x\, dy$, and its centroid is located at $\tilde{x} = x/2$ and $\tilde{y} = y$.

Centroid of a Line.
If a line segment (or rod) lies within the x–y plane and it can be described by a curve $y = f(x)$, Fig. 9–6a, then its centroid is determined from

$$\bar{x} = \frac{\int_L \tilde{x}\, dL}{\int_L dL} \qquad \bar{y} = \frac{\int_L \tilde{y}\, dL}{\int_L dL} \qquad (9\text{–}5)$$

Here, the length of the differential element is given by the Pythagorean theorem, $dL = \sqrt{(dx)^2 + (dy)^2}$, which can also be written in the form

$$dL = \sqrt{\left(\frac{dx}{dx}\right)^2 dx^2 + \left(\frac{dy}{dx}\right)^2 dx^2}$$

$$= \left(\sqrt{1 + \left(\frac{dy}{dx}\right)^2}\right) dx$$

or

$$dL = \sqrt{\left(\frac{dx}{dy}\right)^2 dy^2 + \left(\frac{dy}{dy}\right)^2 dy^2}$$

$$= \left(\sqrt{\left(\frac{dx}{dy}\right)^2 + 1}\right) dy$$

Either one of these expressions can be used; however, for application, the one that will result in a simpler integration should be selected. For example, consider the rod in Fig. 9–6b, defined by $y = 2x^2$. The length of the element is $dL = \sqrt{1 + (dy/dx)^2}\, dx$, and since $dy/dx = 4x$, then $dL = \sqrt{1 + (4x)^2}\, dx$. The centroid for this element is located at $\tilde{x} = x$ and $\tilde{y} = y$.

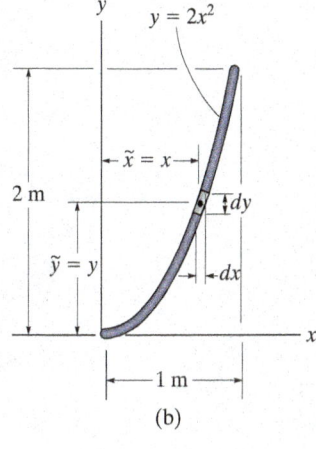

Fig. 9–6

IMPORTANT POINTS

- The centroid represents the geometric center of a body. This point coincides with the center of mass or the center of gravity only if the material composing the body is uniform or homogeneous.

- Formulas used to locate the center of gravity or the centroid simply represent a balance between the sum of moments of all the parts of the system and the moment of the "resultant" for the system.

- In some cases the centroid is located at a point that is not on the object, as in the case of a ring, where the centroid is at its center. Also, this point will lie on any axis of symmetry for the body, Fig. 9–7.

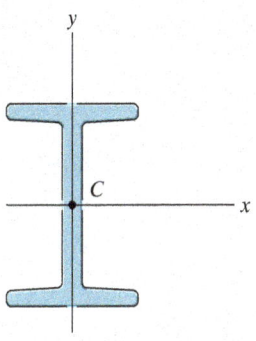

Fig. 9–7

PROCEDURE FOR ANALYSIS

The center of gravity or centroid of an object or shape can be determined by single integrations using the following procedure.

Differential Element.
- Select an appropriate coordinate system, specify the coordinate axes, and then choose a differential element for integration.
- For lines the element is represented by a differential line segment of length dL.
- For areas the element is generally a rectangle of area dA, having a finite length and differential width.
- For volumes the element can be a circular disk of volume dV, having a finite radius and differential thickness.
- Locate the element so that it touches the arbitrary point (x, y, z) on the curve that defines the boundary of the shape.

Size and Moment Arms.
- Express the length dL, area dA, or volume dV of the element in terms of the coordinates describing the curve.
- Express the moment arms $\tilde{x}, \tilde{y}, \tilde{z}$ for the centroid or center of gravity of the element in terms of the coordinates describing the curve.

Integrations.
- Substitute the formulations for $\tilde{x}, \tilde{y}, \tilde{z}$ and $dL, dA,$ or dV into the appropriate equations (Eqs. 9–1 through 9–5).
- Express the function in the integrand in terms of the same variable as the differential thickness of the element.
- The limits of the integral are defined from the two extreme locations of the element's differential thickness, so that when the elements are "summed" or the integration performed, the entire region is covered.

EXAMPLE 9.1

Locate the centroid of the rod bent into the shape of a parabolic arc as shown in Fig. 9–8a.

SOLUTION

Differential Element. The differential element is shown in Fig. 9–8a. It is located on the curve at the *arbitrary point* (x, y).

Area and Moment Arms. The differential element of length dL can be expressed in terms of the differentials dx and dy, Fig. 9–8b, using the Pythagorean theorem.

$$dL = \sqrt{(dx)^2 + (dy)^2} = \sqrt{\left(\frac{dx}{dy}\right)^2 + 1}\, dy$$

Since $x = y^2$, then $dx/dy = 2y$. Therefore, expressing dL in terms of y and dy, we have

$$dL = \sqrt{(2y)^2 + 1}\, dy$$

As shown in Fig. 9–8a, the centroid of the element is located at $\tilde{x} = x$, $\tilde{y} = y$.

Integrations. Applying Eq. 9–5 and using the integration formula to evaluate the integrals, we get

$$\bar{x} = \frac{\int_L \tilde{x}\, dL}{\int_L dL} = \frac{\int_0^{1\,m} x\sqrt{4y^2 + 1}\, dy}{\int_0^{1\,m} \sqrt{4y^2 + 1}\, dy} = \frac{\int_0^{1\,m} y^2\sqrt{4y^2 + 1}\, dy}{\int_0^{1\,m} \sqrt{4y^2 + 1}\, dy}$$

$$= \frac{0.6063}{1.479} = 0.410 \text{ m} \qquad Ans.$$

$$\bar{y} = \frac{\int_L \tilde{y}\, dL}{\int_L dL} = \frac{\int_0^{1\,m} y\sqrt{4y^2 + 1}\, dy}{\int_0^{1\,m} \sqrt{4y^2 + 1}\, dy} = \frac{0.8484}{1.479} = 0.574 \text{ m} \qquad Ans.$$

NOTE: These results for C seem reasonable when they are plotted on Fig. 9–8a.

Fig. 9–8

EXAMPLE 9.2

Locate the centroid of the circular wire segment shown in Fig. 9–9.

Fig. 9–9

SOLUTION
Polar coordinates will be used to solve this problem since the arc is circular.

Differential Element. A differential circular arc is selected as shown in the figure. This element lies on the curve at (R, θ).

Length and Moment Arm. The length of the differential element is $dL = R\, d\theta$, and its centroid is located at $\tilde{x} = R\cos\theta$ and $\tilde{y} = R\sin\theta$.

Integrations. Applying Eqs. 9–5 and integrating with respect to θ, we obtain

$$\bar{x} = \frac{\int_L \tilde{x}\, dL}{\int_L dL} = \frac{\int_0^{\pi/2} (R\cos\theta) R\, d\theta}{\int_0^{\pi/2} R\, d\theta} = \frac{R^2 \int_0^{\pi/2} \cos\theta\, d\theta}{R \int_0^{\pi/2} d\theta} = \frac{2R}{\pi} \quad Ans.$$

$$\bar{y} = \frac{\int_L \tilde{y}\, dL}{\int_L dL} = \frac{\int_0^{\pi/2} (R\sin\theta) R\, d\theta}{\int_0^{\pi/2} R\, d\theta} = \frac{R^2 \int_0^{\pi/2} \sin\theta\, d\theta}{R \int_0^{\pi/2} d\theta} = \frac{2R}{\pi} \quad Ans.$$

NOTE: As expected, the two coordinates are numerically the same due to the symmetry of the wire.

EXAMPLE 9.3

Determine the distance $\bar{y}$ measured from the x axis to the centroid of the area of the triangle shown in Fig. 9–10.

Fig. 9–10

SOLUTION

Differential Element. Consider a rectangular element having a thickness dy, and located in an arbitrary position so that it intersects the boundary at (x, y), Fig. 9–10.

Area and Moment Arms. The area of the element is $dA = x\,dy = \frac{b}{h}(h - y)\,dy$, and its centroid is located a distance $\tilde{y} = y$ from the x axis.

Integration. Applying the second of Eqs. 9–4 and integrating with respect to y yields

$$\bar{y} = \frac{\int_A \tilde{y}\,dA}{\int_A dA} = \frac{\int_0^h y\left[\frac{b}{h}(h - y)\,dy\right]}{\int_0^h \frac{b}{h}(h - y)\,dy} = \frac{\frac{1}{6}bh^2}{\frac{1}{2}bh}$$

$$= \frac{h}{3} \qquad \textit{Ans.}$$

NOTE: This result is valid for any shape of triangle. It states that the centroid is located at one-third the height, measured from the base of the triangle.

EXAMPLE 9.4

Locate the centroid for the area of a quarter circle shown in Fig. 9–11.

Fig. 9–11

SOLUTION

Differential Element. Polar coordinates will be used, since the boundary is circular. We choose the element in the shape of a *triangle*, Fig. 9–11. (Actually the shape is a circular sector; however, neglecting higher-order differentials, the element becomes triangular.) The element intersects the curve at point (R, θ).

Area and Moment Arms. The area of the element is

$$dA = \tfrac{1}{2}(R)(R\, d\theta) = \frac{R^2}{2} d\theta$$

and using the results of Example 9.3, the centroid of the (triangular) element is located at $\tilde{x} = \tfrac{2}{3} R \cos\theta$, $\tilde{y} = \tfrac{2}{3} R \sin\theta$.

Integrations. Applying Eqs. 9–4 and integrating with respect to θ, we obtain

$$\bar{x} = \frac{\int_A \tilde{x}\, dA}{\int_A dA} = \frac{\int_0^{\pi/2} \left(\tfrac{2}{3} R \cos\theta\right) \dfrac{R^2}{2} d\theta}{\int_0^{\pi/2} \dfrac{R^2}{2} d\theta} = \frac{\left(\tfrac{2}{3} R\right) \int_0^{\pi/2} \cos\theta\, d\theta}{\int_0^{\pi/2} d\theta} = \frac{4R}{3\pi} \quad \text{Ans.}$$

$$\bar{y} = \frac{\int_A \tilde{y}\, dA}{\int_A dA} = \frac{\int_0^{\pi/2} \left(\tfrac{2}{3} R \sin\theta\right) \dfrac{R^2}{2} d\theta}{\int_0^{\pi/2} \dfrac{R^2}{2} d\theta} = \frac{\left(\tfrac{2}{3} R\right) \int_0^{\pi/2} \sin\theta\, d\theta}{\int_0^{\pi/2} d\theta} = \frac{4R}{3\pi} \quad \text{Ans.}$$

EXAMPLE 9.5

Locate the centroid of the area shown in Fig. 9–12a.

SOLUTION I

Differential Element. A differential element of thickness dx is shown in Fig. 9–12a. The element intersects the curve at the *arbitrary point* (x, y), and so it has a height y.

Area and Moment Arms. The area of the element is $dA = y\, dx$, and its centroid is located at $\tilde{x} = x$, $\tilde{y} = y/2$.

Integrations. Applying Eqs. 9-4 and integrating with respect to x yields

$$\bar{x} = \frac{\int_A \tilde{x}\, dA}{\int_A dA} = \frac{\int_0^{1\,m} xy\, dx}{\int_0^{1\,m} y\, dx} = \frac{\int_0^{1\,m} x^3\, dx}{\int_0^{1\,m} x^2\, dx} = \frac{0.250}{0.333} = 0.75\text{ m} \quad Ans.$$

$$\bar{y} = \frac{\int_A \tilde{y}\, dA}{\int_A dA} = \frac{\int_0^{1\,m} (y/2)y\, dx}{\int_0^{1\,m} y\, dx} = \frac{\int_0^{1\,m} (x^2/2)x^2\, dx}{\int_0^{1\,m} x^2\, dx} = \frac{0.100}{0.333} = 0.3\text{ m} \quad Ans.$$

SOLUTION II

Differential Element. The differential element of thickness dy is shown in Fig. 9–12b. The element intersects the curve at the *arbitrary point* (x, y), and so it has a length $(1 - x)$.

Area and Moment Arms. The area of the element is $dA = (1-x)\, dy$, and its centroid is located at

$$\tilde{x} = x + \left(\frac{1-x}{2}\right) = \frac{1+x}{2}, \quad \tilde{y} = y$$

Integrations. Applying Eqs. 9-4 and integrating with respect to y, we obtain

$$\bar{x} = \frac{\int_A \tilde{x}\, dA}{\int_A dA} = \frac{\int_0^{1\,m} [(1+x)/2](1-x)\, dy}{\int_0^{1\,m} (1-x)\, dy} = \frac{\frac{1}{2}\int_0^{1\,m} (1-y)\, dy}{\int_0^{1\,m} (1-\sqrt{y})\, dy} = \frac{0.250}{0.333} = 0.75\text{ m} \quad Ans.$$

$$\bar{y} = \frac{\int_A \tilde{y}\, dA}{\int_A dA} = \frac{\int_0^{1\,m} y(1-x)\, dy}{\int_0^{1\,m} (1-x)\, dy} = \frac{\int_0^{1\,m} (y - y^{3/2})\, dy}{\int_0^{1\,m} (1-\sqrt{y})\, dy} = \frac{0.100}{0.333} = 0.3\text{ m} \quad Ans.$$

NOTE: Plot these results and notice that they seem reasonable. Also, for this problem, elements of thickness dx offer a simpler solution.

Fig. 9–12

454   Chapter 9   Center of Gravity and Centroid

EXAMPLE 9.6

Locate the centroid of the semi-elliptical area shown in Fig. 9–13a.

Fig. 9–13

SOLUTION I

Differential Element. The rectangular differential element parallel to the y axis shown shaded in Fig. 9–13a will be considered. This element has a thickness of dx and a height of y.

Area and Moment Arms. Thus, the area is $dA = y\,dx$, and its centroid is located at $\tilde{x} = x$ and $\tilde{y} = y/2$.

Integration. Since the area is symmetrical about the y axis,

$$\bar{x} = 0 \qquad \text{Ans.}$$

Applying the second of Eqs. 9–4 with $y = \sqrt{1 - \dfrac{x^2}{4}}$, we have

$$\bar{y} = \frac{\int_A \tilde{y}\,dA}{\int_A dA} = \frac{\int_{-2\,\text{ft}}^{2\,\text{ft}} \dfrac{y}{2}(y\,dx)}{\int_{-2\,\text{ft}}^{2\,\text{ft}} y\,dx} = \frac{\dfrac{1}{2}\int_{-2\,\text{ft}}^{2\,\text{ft}}\left(1 - \dfrac{x^2}{4}\right)dx}{\int_{-2\,\text{ft}}^{2\,\text{ft}} \sqrt{1 - \dfrac{x^2}{4}}\,dx} = \frac{4/3}{\pi} = 0.424\text{ ft} \qquad \text{Ans.}$$

SOLUTION II

Differential Element. The shaded rectangular differential element of thickness dy and width $2x$, parallel to the x axis, will be considered, Fig. 9–13b.

Area and Moment Arms. The area is $dA = 2x\,dy$, and its centroid is at $\tilde{x} = 0$ and $\tilde{y} = y$.

Integration. Applying the second of Eqs. 9–4, with $x = 2\sqrt{1 - y^2}$, we have

$$\bar{y} = \frac{\int_A \tilde{y}\,dA}{\int_A dA} = \frac{\int_0^{1\,\text{ft}} y(2x\,dy)}{\int_0^{1\,\text{ft}} 2x\,dy} = \frac{\int_0^{1\,\text{ft}} 4y\sqrt{1 - y^2}\,dy}{\int_0^{1\,\text{ft}} 4\sqrt{1 - y^2}\,dy} = \frac{4/3}{\pi}\text{ ft} = 0.424\text{ ft} \qquad \text{Ans.}$$

EXAMPLE 9.7

Locate the $\bar{y}$ centroid for the paraboloid of revolution, shown in Fig. 9–14.

Fig. 9–14

SOLUTION

Differential Element. An element having the shape of a *thin disk* is chosen. This element has a thickness dy, it intersects the generating curve at the *arbitrary point* $(0, y, z)$, and so its radius is $r = z$.

Volume and Moment Arm. The volume of the element is $dV = (\pi z^2)\, dy$, and its centroid is located at $\tilde{y} = y$.

Integration. Applying the second of Eqs. 9–3 and integrating with respect to y yields

$$\bar{y} = \frac{\int_V \tilde{y}\, dV}{\int_V dV} = \frac{\int_0^{100\,\text{mm}} y(\pi z^2)\, dy}{\int_0^{100\,\text{mm}} (\pi z^2)\, dy} = \frac{100\pi \int_0^{100\,\text{mm}} y^2\, dy}{100\pi \int_0^{100\,\text{mm}} y\, dy} = 66.7\ \text{mm} \quad Ans.$$

EXAMPLE 9.8

Determine the location of the center of mass of the cylinder shown in Fig. 9–15 if its density varies directly with the distance from its base, i.e., $\rho = 200z$ kg/m^3.

Fig. 9–15

SOLUTION
For reasons of material symmetry,

$$\bar{x} = \bar{y} = 0 \qquad \text{Ans.}$$

Differential Element. A disk element of radius 0.5 m and thickness dz is chosen for integration, Fig. 9–15, since the *density of the entire element is constant* for a given value of z. The element is located along the z axis at the *arbitrary point* $(0, 0, z)$.

Volume and Moment Arm. The volume of the element is $dV = \pi(0.5)^2\, dz$, and its centroid is located at $\tilde{z} = z$.

Integrations. Using the third of Eqs. 9–2 with $dm = \rho\, dV$ and integrating with respect to z, noting that $\rho = 200z$, we have

$$\bar{z} = \frac{\int_V \tilde{z} \rho\, dV}{\int_V \rho\, dV} = \frac{\int_0^{1\,\text{m}} z(200z)[\pi(0.5)^2\, dz]}{\int_0^{1\,\text{m}} (200z)\pi(0.5)^2\, dz}$$

$$= \frac{\int_0^{1\,\text{m}} z^2\, dz}{\int_0^{1\,\text{m}} z\, dz} = 0.667 \text{ m} \qquad \text{Ans.}$$

FUNDAMENTAL PROBLEMS

F9–1. Determine the centroid $(\bar{x}, \bar{y})$ of the area.

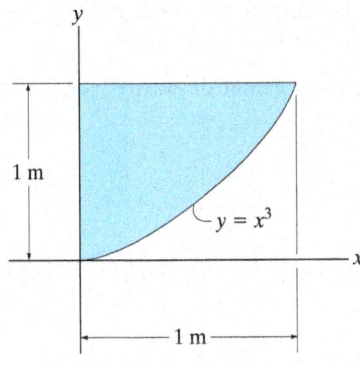

Prob. F9–1

F9–2. Determine the centroid $(\bar{x}, \bar{y})$ of the area.

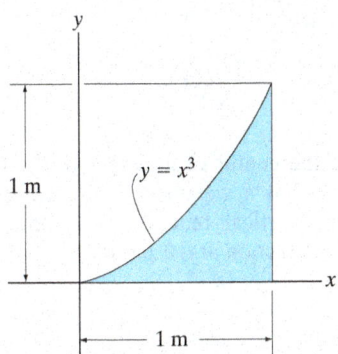

Prob. F9–2

F9–3. Determine the centroid $\bar{y}$ of the area.

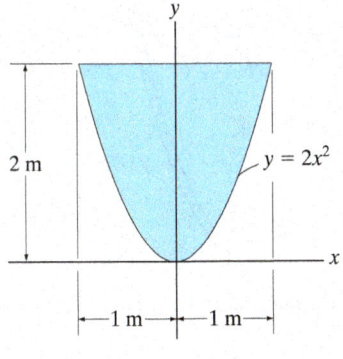

Prob. F9–3

F9–4. Locate the center of mass $\bar{x}$ of the straight rod if its mass per unit length is given by $m = 0.5(1+x^2)$ kg.

Prob. F9–4

F9–5. Locate the centroid $\bar{y}$ of the homogeneous solid formed by revolving the shaded area about the y axis.

Prob. F9–5

F9–6. Locate the centroid $\bar{z}$ of the homogeneous solid formed by revolving the shaded area about the z axis.

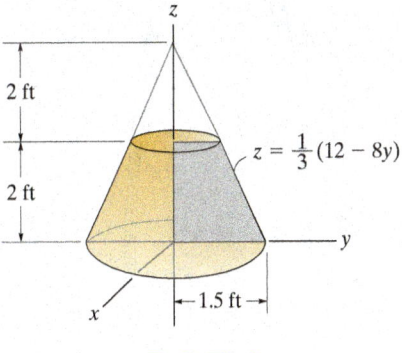

Prob. F9–6

PROBLEMS

9–1. Determine the distance $\bar{x}$ to the center of gravity of the rod bent in the form of a parabola. If the rod has a weight per unit length of 0.5 lb/ft, determine the components of reaction at the fixed support A.

9–3. Locate the center of gravity $\bar{x}$ of the homogeneous rod. If the rod has a weight per unit length of 100 N/m, determine the vertical reaction at A and the x and y components of reaction at the pin B.

*9–4. Locate the center of gravity $\bar{y}$ of the homogeneous rod.

Prob. 9–1

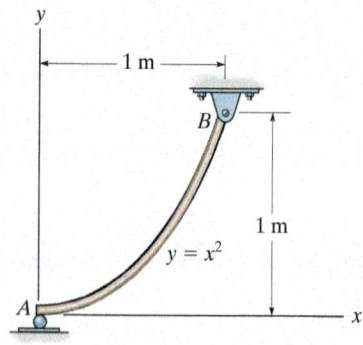

Probs. 9–3/4

9–2. Locate the center of gravity $(\bar{x}, \bar{y})$ of the homogeneous rod. What is its weight if it has a mass per the unit length of 3 kg/m? Solve the problem by evaluating the integrals using Simpson's rule.

9–5. Locate the center of gravity $\bar{x}$ of the homogeneous rod. If the rod has a weight per unit length of 0.5 lb/ft, determine the vertical reaction at A and the x and y components of reaction at the pin B.

Prob. 9–2

Prob. 9–5

9–6. Locate the centroid $\bar{y}$ of the area.

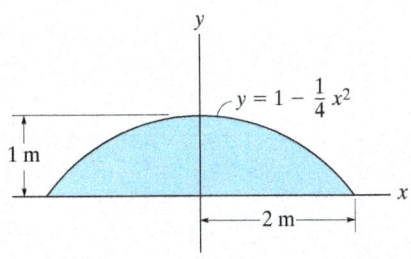

Prob. 9–6

9–7. Locate the centroid $\bar{x}$ of the area.

Prob. 9–7

*9–8. Determine the location $(\bar{x}, \bar{y})$ of the centroid of the area.

Prob. 9–8

9–9. Locate the centroid $\bar{x}$ of the area.

9–10. Locate the centroid $\bar{y}$ of the area.

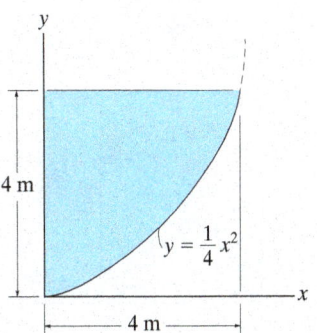

Probs. 9–9/10

9–11. Locate the centroid $\bar{x}$ of the area.

*9–12. Locate the centroid $\bar{y}$ of the area.

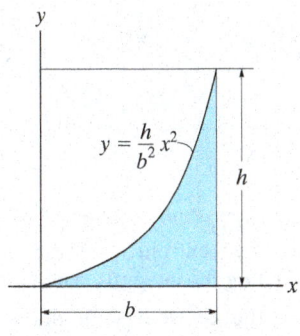

Probs. 9–11/12

9–13. Locate the centroid $\bar{x}$ of the area.

9–14. Locate the centroid $\bar{y}$ of the area.

Probs. 9–13/14

9–15. Locate the centroid $\bar{x}$ of the area. Solve the problem by evaluating the integrals using Simpson's rule.

*9–16. Locate the centroid $\bar{y}$ of the area. Solve the problem by evaluating the integrals using Simpson's rule.

Probs. 9–15/16

9–17. Locate the centroid of the area.

Prob. 9–17

9–18. Locate the centroid $\bar{x}$ of the area. Solve the problem by evaluating the integrals using Simpson's rule.

9–19. Locate the centroid $\bar{y}$ of the area. Solve the problem by evaluating the integrals using Simpson's rule.

Probs. 9–18/19

*9–20. Locate the centroid $(\bar{x}, \bar{y})$ of the exparabolic segment of area.

Prob. 9–20

9–21. Locate the centroid $\bar{x}$ of the area.

9–22. Locate the centroid $\bar{y}$ of the area.

Probs. 9–21/22

9–23. Locate the centroid $\bar{x}$ of the area.

*9–24. Locate the centroid $\bar{y}$ of the area.

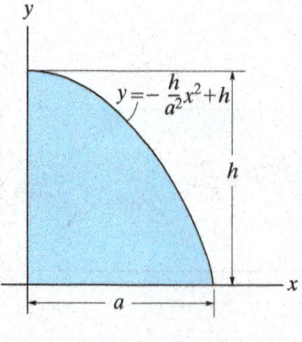

Probs. 9–23/24

9–25. The plate has a thickness of 0.25 ft and a specific weight of $\gamma = 180$ lb/ft^3. Determine the location of its center of gravity. Also, find the tension in each of the cords used to support it.

*9–28. Locate the centroid $\bar{x}$ of the area.

9–29. Locate the centroid $\bar{y}$ of the area.

Prob. 9–25

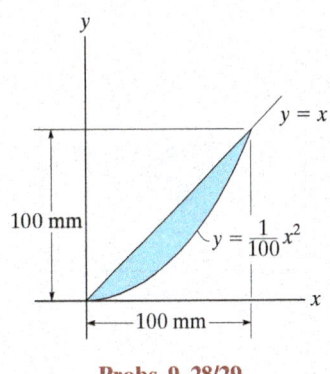

Probs. 9–28/29

9–26. Locate the centroid $\bar{x}$ of the area.

9–27. Locate the centroid $\bar{y}$ of the area.

9–30. Locate the centroid $\bar{x}$ of the area.

9–31. Locate the centroid $\bar{y}$ of the area.

Probs. 9–26/27

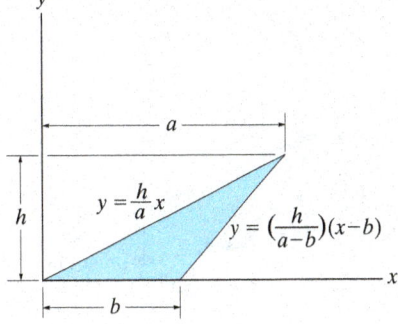

Probs. 9–30/31

*9–32. Locate the centroid $\bar{x}$ of the area.

9–33. Locate the centroid $\bar{y}$ of the area.

9–35. Locate the centroid $\bar{x}$ of the area.

*9–36. Locate the centroid $\bar{y}$ of the area.

Probs. 9–32/33

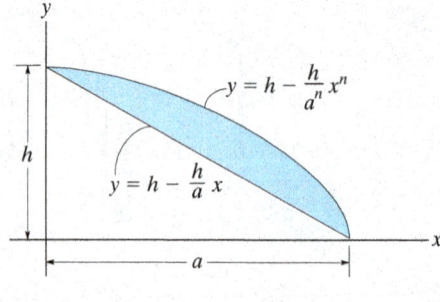

Probs. 9–35/36

9–34. Locate the center of gravity of the homogeneous cantilever beam and determine the reactions at the fixed support. The material has a density of 8 Mg/m³.

9–37. Determine the distance $\bar{y}$ to the centroid of the cone.

9–38. Determine the distance $\bar{y}$ to the center of mass of the cone. The density of the material varies linearly from zero at the origin to ρ_0 at $x = h$.

Prob. 9–34

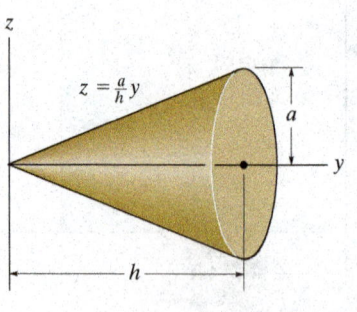

Probs. 9–37/38

9–39. Locate the centroid of the volume formed by rotating the shaded area about the *aa* axis.

9–41. Locate the centroid of the solid.

Prob. 9–39

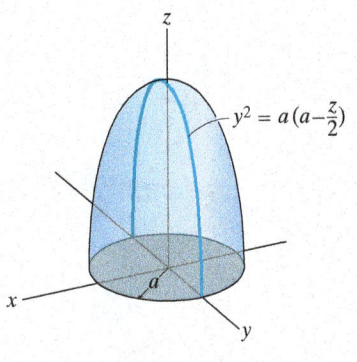

Prob. 9–41

9–42. Locate the centroid $\bar{x}$ of the solid.

*9–40. Locate the centroid $\bar{y}$ of the paraboloid.

Prob. 9–40

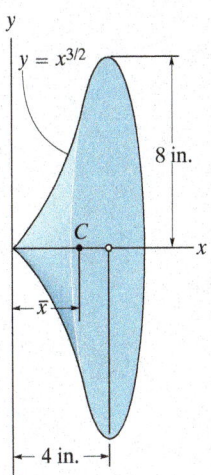

Prob. 9–42

9–43. Locate the center of gravity for the homogeneous half-cone.

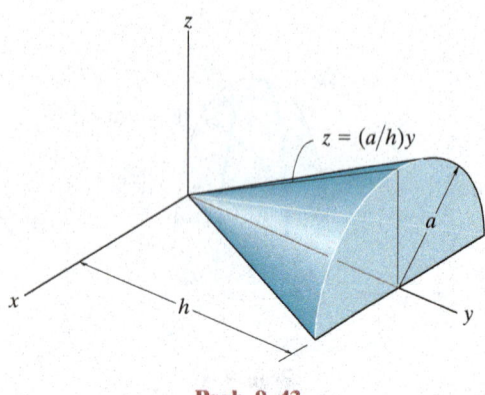

Prob. 9–43

*9–44. The king's chamber of the Great Pyramid of Giza is located at its centroid. Assuming the pyramid to be a solid, prove that this point is at $\bar{z} = \frac{1}{4}h$. *Suggestion*: Use a rectangular differential plate element having a thickness dz and area $(2x)(2y)$.

9–45. Determine the distance $\bar{y}$ to the centroid of the bell-shaped volume.

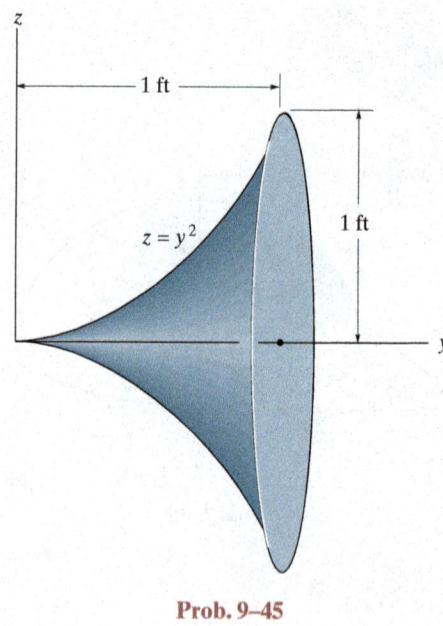

Prob. 9–45

9–46. Determine the distance $\bar{y}$ to the centroid of the semi-ellipsoid.

Prob. 9–44

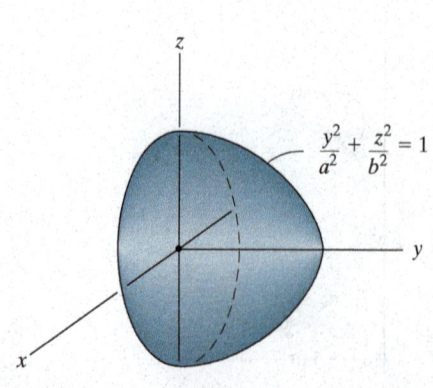

Prob. 9–46

9–47. Locate the center of mass $\bar{x}$ of the hemisphere. The density of the material varies linearly from zero at the origin O to ρ_0 at the surface. *Hint:* Choose a hemispherical shell element for integration.

9–49. Locate the center of gravity $\bar{y}$ of the volume. The material is homogeneous.

Prob. 9–49

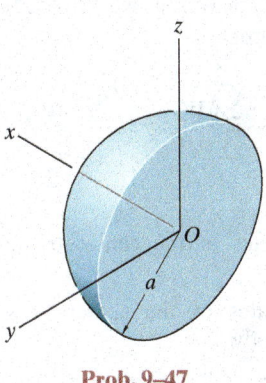

Prob. 9–47

*9–48.** Determine the distance $\bar{y}$ to the center of gravity of the volume. The material is homogeneous.

9–50. Determine the location $\bar{z}$ of the centroid for the tetrahedron. *Suggestion:* Use a triangular "plate" element parallel to the x–y plane and of thickness dz.

Prob. 9–48

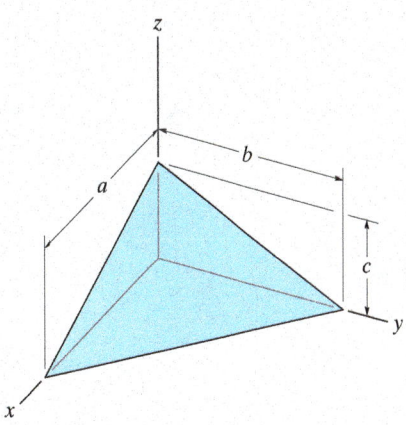

Prob. 9–50

9.2 COMPOSITE BODIES

A *composite body* consists of a series of connected "simpler" shaped bodies, which may be rectangular, triangular, semicircular, etc. Such a body can often be sectioned or divided into its composite parts and, provided the *weight* and location of the center of gravity of each of these parts are known, we can then eliminate the need for integration to determine the center of gravity for the body. The method for doing this follows the same procedure outlined in Sec. 9.1, and so formulas analogous to Eqs. 9–1 result. Here, however, we have a finite number of weights and so the equations become

$$\bar{x} = \frac{\Sigma \tilde{x} W}{\Sigma W} \quad \bar{y} = \frac{\Sigma \tilde{y} W}{\Sigma W} \quad \bar{z} = \frac{\Sigma \tilde{z} W}{\Sigma W} \tag{9-6}$$

Here

$\bar{x}, \bar{y}, \bar{z}$ represent the coordinates of the center of gravity G of the composite body.

$\tilde{x}, \tilde{y}, \tilde{z}$ represent the coordinates of the center of gravity of each composite part of the body.

ΣW is the sum of the weights of all the composite parts of the body, or simply the total weight of the body.

When the body has a *constant density or specific weight*, the center of gravity *coincides* with the centroid of the body. The centroid for composite lines, areas, and volumes can then be found using relations analogous to Eqs. 9–6; however, the W's are replaced by L's, A's, and V's, respectively. Centroids for common shapes of lines, areas, shells, and volumes that often make up a composite body are given in the table on the inside back cover.

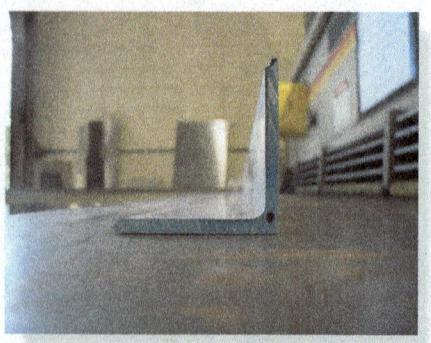

A stress analysis of this angle requires that the centroid of its cross-sectional area be located.

In order to determine the force required to tip over this concrete barrier, it is first necessary to determine the location of its center of gravity G. This point will lie on the vertical axis of symmetry.

PROCEDURE FOR ANALYSIS

The location of the center of gravity of a body or the centroid of a composite geometrical object represented by a line, area, or volume can be determined using the following procedure.

Composite Parts.

- Using a sketch, divide the body or object into a finite number of composite parts that have simpler shapes.

- If a composite body has a *hole*, or a geometric region having no material, then consider the composite body without the hole and consider the hole as an *additional* composite part having *negative* weight or size.

Moment Arms.

- Establish the coordinate axes on the sketch and determine the coordinates $\tilde{x}, \tilde{y}, \tilde{z}$ of the center of gravity or centroid of each part.

Summations.

- Determine $\bar{x}, \bar{y}, \bar{z}$ by applying the center of gravity equations, Eqs. 9–6, or the analogous centroid equations.

- If an object is *symmetrical* about an axis, the centroid of the object lies on this axis.

If desired, the calculations can be arranged in tabular form, as indicated in the following three examples.

The center of gravity of this water tank can be determined by dividing it into composite parts and applying Eqs. 9–6.

EXAMPLE 9.9

Locate the centroid of the wire shown in Fig. 9–16a.

SOLUTION

Composite Parts. The wire is divided into three segments as shown in Fig. 9–16b.

Moment Arms. The location of the centroid for each segment is determined and indicated in the figure. In particular, the centroid of segment ① is determined either by integration or by using the table on the inside back cover.

Summations. For convenience, the calculations can be tabulated as follows:

Segment	L (mm)	$\tilde{x}$ (mm)	$\tilde{y}$ (mm)	$\tilde{z}$ (mm)	$\tilde{x}L$ (mm²)	$\tilde{y}L$ (mm²)	$\tilde{z}L$ (mm²)
1	$\pi(60) = 188.5$	60	−38.2	0	11 310	−7200	0
2	40	0	20	0	0	800	0
3	20	0	40	−10	0	800	−200
	$\Sigma L = 248.5$				$\Sigma \tilde{x}L = 11\ 310$	$\Sigma \tilde{y}L = -5600$	$\Sigma \tilde{z}L = -200$

Thus,

$$\bar{x} = \frac{\Sigma \tilde{x}L}{\Sigma L} = \frac{11\ 310}{248.5} = 45.5 \text{ mm} \qquad \textit{Ans.}$$

$$\bar{y} = \frac{\Sigma \tilde{y}L}{\Sigma L} = \frac{-5600}{248.5} = -22.5 \text{ mm} \qquad \textit{Ans.}$$

$$\bar{z} = \frac{\Sigma \tilde{z}L}{\Sigma L} = \frac{-200}{248.5} = -0.805 \text{ mm} \qquad \textit{Ans.}$$

Fig. 9–16

EXAMPLE 9.10

Locate the centroid of the plate area shown in Fig. 9–17a.

(a)

SOLUTION

Composite Parts. The plate is divided into three segments as shown in Fig. 9–17b. Here the area of the small rectangle ③ is considered "negative" since it must be subtracted from the larger area ②.

Moment Arms. The location of the centroid of each segment is shown in the figure. Note that the $\tilde{x}$ coordinates of ② and ③ are *negative*.

Summations. Taking the data from Fig. 9–17b, the calculations are tabulated as follows:

Segment	A (ft^2)	$\tilde{x}$ (ft)	$\tilde{y}$ (ft)	$\tilde{x}A$ (ft^3)	$\tilde{y}A$ (ft^3)
1	$\frac{1}{2}(3)(3) = 4.5$	1	1	4.5	4.5
2	$(3)(3) = 9$	-1.5	1.5	-13.5	13.5
3	$-(2)(1) = -2$	-2.5	2	5	-4
	$\Sigma A = 11.5$			$\Sigma \tilde{x}A = -4$	$\Sigma \tilde{y}A = 14$

(b)

Fig. 9–17

Thus,

$$\bar{x} = \frac{\Sigma \tilde{x}A}{\Sigma A} = \frac{-4}{11.5} = -0.348 \text{ ft} \quad Ans.$$

$$\bar{y} = \frac{\Sigma \tilde{y}A}{\Sigma A} = \frac{14}{11.5} = 1.22 \text{ ft} \quad Ans.$$

NOTE: If these results are plotted in Fig. 9–17a, the location of point C seems reasonable.

EXAMPLE 9.11

(a)

Locate the center of mass of the assembly shown in Fig. 9–18a. The conical frustum has a density of $\rho_c = 8 \text{ Mg/m}^3$, and the hemisphere has a density of $\rho_h = 4 \text{ Mg/m}^3$. There is a 25-mm-radius cylindrical hole in the center of the frustum.

SOLUTION

Composite Parts. The assembly can be thought of as consisting of four segments as shown in Fig. 9–18b. For the calculations, ③ and ④ must be considered as "negative" segments in order that the four segments, when added together, yield the total composite shape shown in Fig. 9–18a.

Moment Arm. Using the table on the inside back cover, the calculations for the centroid $\tilde{z}$ of each piece are shown in the figure.

Summations. Because of *symmetry*, note that

$$\bar{x} = \bar{y} = 0 \qquad \text{Ans.}$$

Since $W = mg$, and g is constant, the third of Eqs. 9–6 becomes $\bar{z} = \Sigma \tilde{z}m / \Sigma m$. The mass of each piece can be calculated from $m = \rho V$. Also, $1 \text{ Mg/m}^3 = 10^{-6} \text{ kg/mm}^3$, so that

Segment	m (kg)	$\tilde{z}$ (mm)	$\tilde{z}m$ (kg·mm)
1	$8(10^{-6})(\frac{1}{3})\pi(50)^2(200) = 4.189$	50	209.440
2	$4(10^{-6})(\frac{2}{3})\pi(50)^3 = 1.047$	-18.75	-19.635
3	$-8(10^{-6})(\frac{1}{3})\pi(25)^2(100) = -0.524$	$100 + 25 = 125$	-65.450
4	$-8(10^{-6})\pi(25)^2(100) = -1.571$	50	-78.540
	$\Sigma m = 3.142$		$\Sigma \tilde{z}m = 45.815$

Thus,

$$\bar{z} = \frac{\Sigma \tilde{z}m}{\Sigma m} = \frac{45.815}{3.142} = 14.6 \text{ mm} \qquad \text{Ans.}$$

(b)

Fig. 9–18

FUNDAMENTAL PROBLEMS

F9–7. Locate the centroid ($\bar{x}, \bar{y}, \bar{z}$) of the wire bent in the shape shown.

Prob. F9–7

F9–8. Locate the centroid $\bar{y}$ of the beam's cross-sectional area.

Prob. F9–8

F9–9. Locate the centroid $\bar{y}$ of the beam's cross-sectional area.

Prob. F9–9

F9–10. Locate the centroid ($\bar{x}, \bar{y}$) of the cross-sectional area.

Prob. F9–10

F9–11. Locate the center of mass ($\bar{x}, \bar{y}, \bar{z}$) of the homogeneous solid block.

Prob. F9–11

F9–12. Locate the center of mass ($\bar{x}, \bar{y}, \bar{z}$) of the homogeneous solid block.

Prob. F9–12

PROBLEMS

9–51. The truss is made from five members, each having a length of 4 m and a mass of 7 kg/m. If the mass of the gusset plates at the joints and the thickness of the members can be neglected, determine the distance d to where the hoisting cable must be attached, so that the truss does not tip (rotate) when it is lifted.

9–53. Determine the location $\bar{y}$ of the centroid C of the "roll-formed" member. Neglect the thickness of the material and any slight bends at the corners.

Prob. 9–51

Prob. 9–53

*__9–52.__ Determine the location $(\bar{x}, \bar{y}, \bar{z})$ of the centroid of the homogeneous rod.

9–54. Locate the centroid $(\bar{x}, \bar{y})$ of the metal cross section. Neglect the thickness of the material and slight bends at the corners.

Prob. 9–52

Prob. 9–54

9–55. Determine the location $(\bar{x}, \bar{y})$ of the centroid of the sheet metal cross section. Neglect the thickness of the material and slight bends at the corners.

9–57. Determine the location $(\bar{x}, \bar{y})$ of the centroid of the rod. Neglect the rod's thickness.

Prob. 9–55

Prob. 9–57

*9–56. The steel and aluminum plate assembly is bolted together and fastened to the wall. Each plate has a constant width in the z direction of 200 mm and thickness of 20 mm. If the density of A and B is $\rho_s = 7.85$ Mg/m^3, and for C, $\rho_{al} = 2.71$ Mg/m^3, determine the location $\bar{x}$ of the center of mass. Neglect the size of the bolts.

9–58. The rectangular horn antenna is used for receiving microwaves. Determine the location $\bar{x}$ of its center of gravity G. The horn is made of plates having a constant thickness and density and is open at each end.

Prob. 9–56

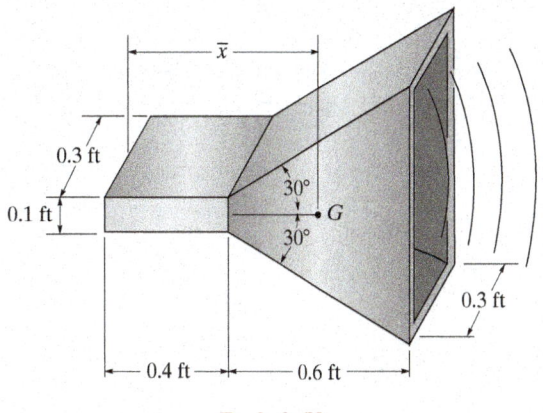

Prob. 9–58

9–59. Determine the distance $\bar{y}$ to the centroid of the trapezoidal area in terms of the dimensions shown.

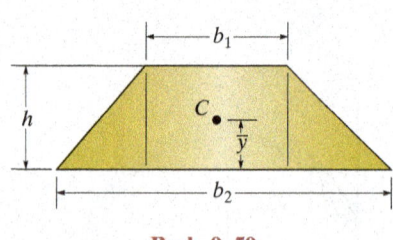

Prob. 9–59

***9–60.** Locate the centroid $\bar{y}$ for the beam's cross-sectional area.

Prob. 9–60

9–61. Determine the location y of the centroid for the beam's cross-sectional area.

Prob. 9–61

9–62. Locate the centroid $(\bar{x}, \bar{y})$ of the area.

Prob. 9–62

9–63. Locate the centroid $\bar{y}$ of the cross-sectional area of the beam.

Prob. 9–63

***9–64.** Determine the location $(\bar{x}, \bar{y})$ of the centroid of the area.

Prob. 9–64

9–65. Determine the location $\bar{y}$ of the centroid of the area.

Prob. 9–65

9–66. Determine the location $\bar{y}$ of the centroid for the cross-sectional area.

Prob. 9–66

9–67. Locate the centroid $\bar{y}$ of the cross-sectional area of the beam constructed from a channel and a plate. Assume all corners are square and neglect the size of the weld at A.

Prob. 9–67

*9–68. Determine the distance $\bar{y}$ to the centroid of the area.

Prob. 9–68

9–69. Determine the distance $\bar{y}$ to the centroid of the beam's cross-sectional area.

Prob. 9–69

9–70. Determine the location $\bar{x}$ of the centroid of the solid made from a hemisphere, cylinder, and cone.

Prob. 9–70

9–71. Determine the location ($\bar{x}, \bar{y}$) of the centroid of the area.

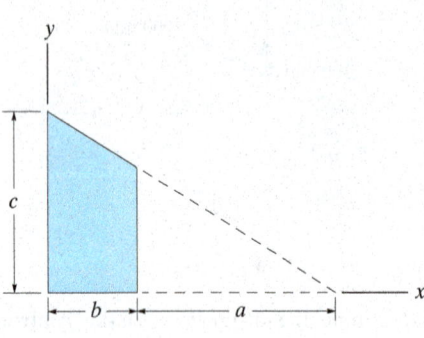

Prob. 9–71

*9–72. A toy skyrocket consists of a solid conical top, $\rho_i = 600 \text{ kg/m}^3$, a hollow cylinder, $\rho_c = 400 \text{ kg/m}^3$, and a stick having a circular cross section, $\rho_s = 300 \text{ kg/m}^3$. Determine the length of the stick, x, so that the center of gravity G of the skyrocket is located along line aa.

Prob. 9–72

9–73. Determine the location ($\bar{x}, \bar{y}$) of the centroid of the area.

Prob. 9–73

9–74. Determine the distance $\bar{x}$ to the center of gravity of the gas dehydrator assembly. The weight and the center of gravity of each of the various components are indicated. What are the vertical reactions at A and B needed to support the assembly?

Prob. 9–74

9–75. Determine the distance $\bar{y}$ to the center of mass of the assembly, which has a hole bored through its center. The material has a density of 3 Mg/m³.

Prob. 9–75

*9–76. The sheet metal part has the dimensions shown. Determine the location ($\bar{x}, \bar{y}, \bar{z}$) of its centroid.

9–77. The sheet metal part has a weight per unit area of 2 lb/ft² and is supported by the smooth rod and the cord at C. If the cord is cut, the part will rotate about the y axis until it reaches equilibrium. Determine the equilibrium angle of tilt, measured downward from the negative x axis, that AD makes with the −x axis.

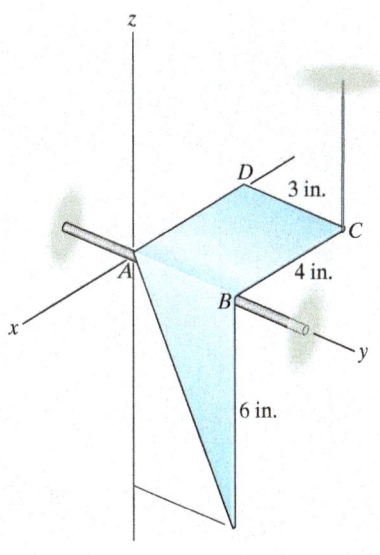

Probs. 9–76/77

9–78. Determine the distance $\bar{x}$ to the center of gravity of the generator assembly. The weight and the center of gravity of each of the various components are indicated. What are the vertical reactions at blocks A and B needed to support the assembly?

Prob. 9–78

9–79. The buoy is made from two homogeneous cones each having a radius of 1.5 ft. If $h = 1.2$ ft, find the distance $\bar{z}$ to the buoy's center of gravity G.

*9–80. The buoy is made from two homogeneous cones each having a radius of 1.5 ft. If it is required that the buoy's center of gravity G be located at $\bar{z} = 0.5$ ft, determine the height h of the top cone.

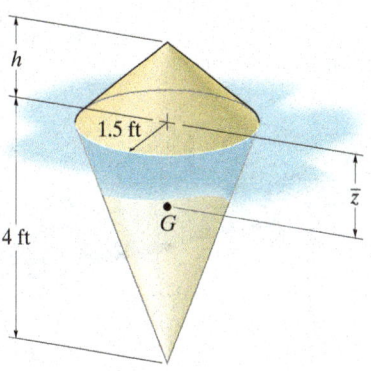

Probs. 9–79/80

9–81. The assembly is made from a steel hemisphere, $\rho_{st} = 7.80\ \text{Mg/m}^3$, and an aluminum cylinder, $\rho_{al} = 2.70\ \text{Mg/m}^3$. Determine the mass center of the assembly if the height of the cylinder is $h = 200$ mm.

9–82. The assembly is made from a steel hemisphere, $\rho_{st} = 7.80\ \text{Mg/m}^3$, and an aluminum cylinder, $\rho_{al} = 2.70\ \text{Mg/m}^3$. Determine the height h of the cylinder so that the mass center of the assembly is located at $\bar{z} = 160$ mm.

Probs. 9–81/82

9–83. The car rests on four scales and in this position the scale readings of both the front and rear tires are shown by F_A and F_B. When the rear wheels are elevated to a height of 3 ft above the front scales, the new readings of the front wheels are also recorded. Use this data to calculate the location $\bar{x}$ and $\bar{y}$ to the center of gravity G of the car. The tires each have a diameter of 1.98 ft.

Prob. 9–83

*9–84. The solid is formed by boring a conical hole into the cylinder. Determine the distance to the center of gravity.

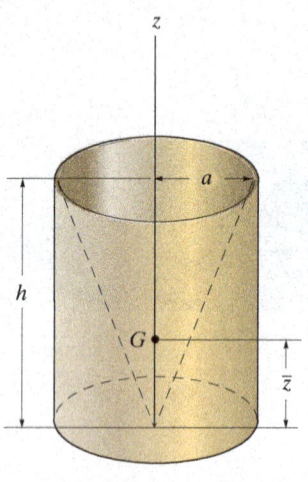

Prob. 9–84

9–85. Determine the distance $\bar{z}$ to the center of mass of the casting that is formed from a hollow cylinder having a density of 8 Mg/m³ and a hemisphere having a density of 3 Mg/m³.

Prob. 9–85

9–86. Locate the center of mass $\bar{z}$ of the assembly. The material has a density of $\rho = $ Mg/m^3. There is a 30-mm-diameter hole bored through the center.

***9–88.** Determine the dimension h of the block so that the centroid C of the assembly lies at the base of the cylinder as shown.

Prob. 9–86

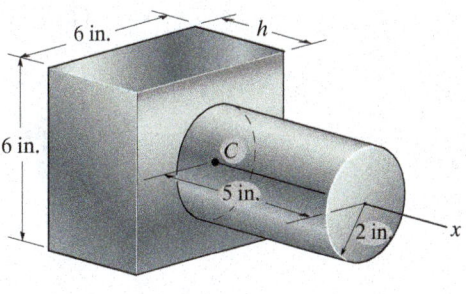

Prob. 9–88

9–89. The composite plate is made from both steel (A) and brass (B) segments. Determine the mass and location ($\bar{x}, \bar{y}, \bar{z}$) of its mass center G. Take $\rho_{st} = 7.85$ Mg/m^3 and $\rho_{br} = 8.74$ Mg/m^3.

9–87. Major floor loadings in a shop are caused by the weights of the objects shown. Each force acts through its respective center of gravity G. Locate the center of gravity ($\bar{x}, \bar{y}$) of all these objects.

Prob. 9–87

Prob. 9–89

*9.3 THEOREMS OF PAPPUS AND GULDINUS

The two **theorems of Pappus and Guldinus** are used to find the surface area and volume of any body of revolution. They were first developed by Pappus of Alexandria during the fourth century A.D. and then restated at a much later time by the Swiss mathematician Paul Guldin or Guldinus (1577–1643).

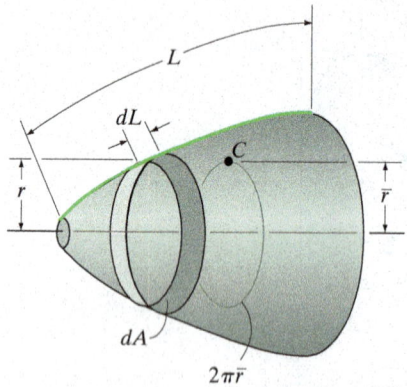

Fig. 9–19

Surface Area. If we revolve a *plane curve* about an axis that does not intersect the curve we will generate a *surface area of revolution*. For example, the surface area in Fig. 9–19 is formed by revolving the curve of length L about the horizontal axis. To determine this surface area, consider the differential element of length dL. If this element is revolved 2π radians about the axis, a ring having a surface area of $dA = 2\pi r\, dL$ will be generated. Thus, the surface area of the entire body is $A = 2\pi \int r\, dL$. However, $\int r\, dL = \bar{r}L$ (Eq. 9-5), and so $A = 2\pi \bar{r}L$. If instead the curve is revolved only through an angle θ (radians), then

$$A = \theta \bar{r} L \qquad (9\text{--}7)$$

where

A = surface area of revolution
θ = angle of revolution measured in radians, $\theta \leq 2\pi$
$\bar{r}$ = perpendicular distance from the axis of revolution to the centroid of the generating curve
L = length of the generating curve

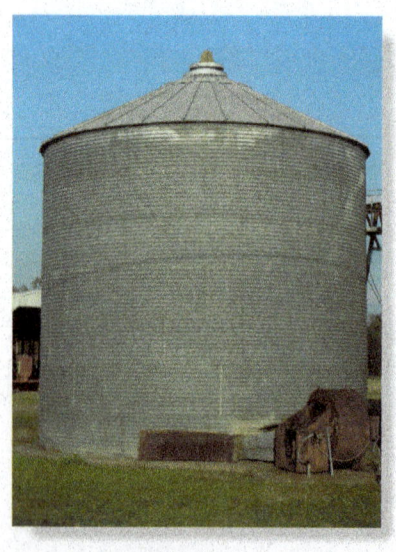

The amount of material used on this storage building can be estimated by using the first theorem of Pappus and Guldinus to determine its surface area.

Therefore the first theorem of Pappus and Guldinus states that *the area of a surface of revolution equals the product of the length L of the generating curve and the distance $\theta \bar{r}$ traveled by the centroid of the curve in generating the surface area*, Fig. 9–19.

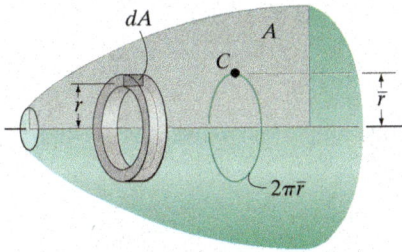

Fig. 9–20

Volume. A *volume* can be generated by revolving a *plane area* about an axis that does not intersect the area. For example, if we revolve the shaded area A in Fig. 9–20 about the horizontal axis, it generates the volume shown. To determine this volume, consider revolving the differential element of area dA 2π radians about the axis. It produces a ring having the volume $dV = 2\pi r\, dA$. The entire volume is then $V = 2\pi \int r\, dA$. However, $\int r\, dA = \bar{r} A$, Eq. 9–4, so that $V = 2\pi \bar{r} A$. If instead the area is only revolved through an angle θ (radians), then

$$V = \theta \bar{r} A \tag{9-8}$$

where

V = volume of revolution
θ = angle of revolution measured in radians, $\theta \leq 2\pi$
$\bar{r}$ = perpendicular distance from the axis of revolution to the centroid of the generating area
A = generating area

Therefore the second theorem of Pappus and Guldinus states that *the volume of a body of revolution equals the product of the generating area A and the distance $\theta \bar{r}$ traveled by the centroid of the area in generating the volume.*

Composite Shapes. We may also apply the above two theorems to lines or areas that are composed of a series of composite parts. In this case the total surface area or volume generated is the addition of the surface areas or volumes generated by each of these parts. If the perpendicular distance from the axis of revolution to the centroid of each part is $\tilde{r}$, then

$$A = \theta \Sigma(\tilde{r} L) \tag{9-9}$$

and

$$V = \theta \Sigma(\tilde{r} A) \tag{9-10}$$

Application of the above theorems is illustrated numerically in the following examples.

The volume of fertilizer contained within this silo can be determined using the second theorem of Pappus and Guldinus.

EXAMPLE 9.12

Show that the surface area of a sphere is $A = 4\pi R^2$ and its volume is $V = \frac{4}{3}\pi R^3$.

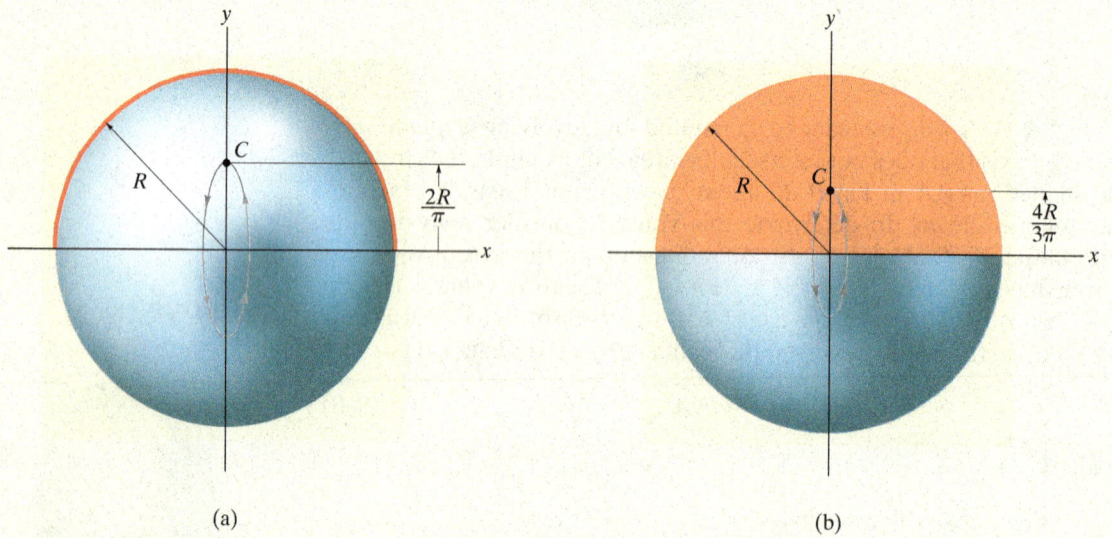

(a)   (b)

Fig. 9–21

SOLUTION

Surface Area. The surface area of the sphere in Fig. 9–21a is generated by revolving a semicircular arc about the x axis. Using the table on the inside back cover, the centroid of this arc is located at a distance $\bar{r} = 2R/\pi$ from the axis of revolution (x axis). Since the centroid moves through an angle of $\theta = 2\pi$ rad to generate the sphere, then applying Eq. 9–7 we have

$$A = \theta \bar{r} L; \qquad A = 2\pi \left(\frac{2R}{\pi}\right) \pi R = 4\pi R^2 \qquad \textit{Ans.}$$

Volume. The volume of the sphere is generated by revolving the semicircular *area* in Fig. 9–21b about the x axis. Using the table on the inside back cover to locate the centroid of the area, i.e., $\bar{r} = 4R/3\pi$, and applying Eq. 9–8, we have

$$V = \theta \bar{r} A; \qquad V = 2\pi \left(\frac{4R}{3\pi}\right)\left(\frac{1}{2}\pi R^2\right) = \frac{4}{3}\pi R^3 \qquad \textit{Ans.}$$

EXAMPLE 9.13

Determine the surface area and volume of the full solid in Fig. 9–22a.

(a)   (b)

SOLUTION

Surface Area. The surface area is generated by revolving the four line segments shown in Fig. 9–22b 2π radians about the z axis. The distances from the centroid of each segment to the z axis are also shown in the figure. Applying Eq. 9–7 yields

$$A = 2\pi\Sigma \bar{r}L = 2\pi[(2.5 \text{ in.})(2 \text{ in.}) + (3 \text{ in.})\left(\sqrt{(1 \text{ in.})^2 + (1 \text{ in.})^2}\right)$$

$$+ (3.5 \text{ in.})(3 \text{ in.}) + (3 \text{ in.})(1 \text{ in.})]$$

$$= 143 \text{ in}^2 \qquad \qquad Ans.$$

Volume. The volume of the solid is generated by revolving the two area segments shown in Fig. 9–22c 2π radians about the z axis. The distances from the centroid of each segment to the z axis are also shown in the figure. Applying Eq. 9–10, we have

$$V = 2\pi\Sigma\bar{r}A = 2\pi\left\{(3.1667 \text{ in.})\left[\frac{1}{2}(1 \text{ in.})(1 \text{ in.})\right] + (3 \text{ in.})[(2 \text{ in.})(1 \text{ in.})]\right\}$$

$$= 47.6 \text{ in}^3 \qquad \qquad Ans.$$

(c)

Fig. 9–22

FUNDAMENTAL PROBLEMS

F9–13. Determine the surface area and volume of the solid formed by revolving the shaded area 360° about the z axis.

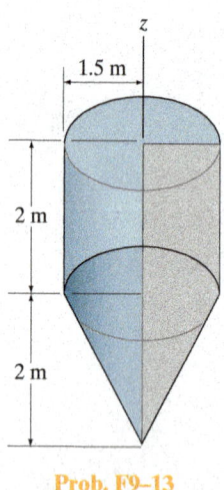

Prob. F9–13

F9–14. Determine the surface area and volume of the solid formed by revolving the shaded area 360° about the z axis.

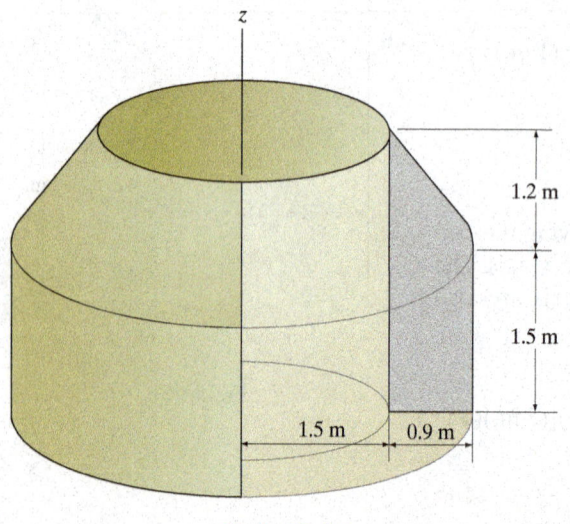

Prob. F9–14

F9–15. Determine the surface area and volume of the solid formed by revolving the shaded area 360° about the z axis.

Prob. F9–15

F9–16. Determine the surface area and volume of the solid formed by revolving the shaded area 360° about the z axis.

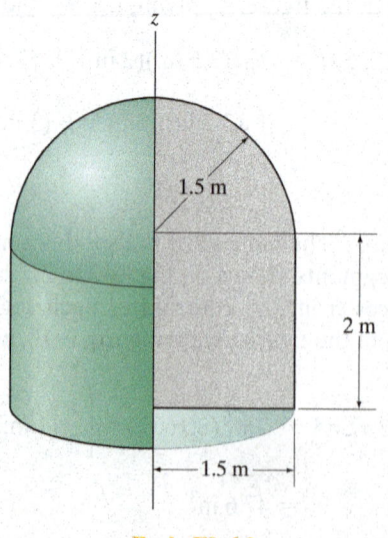

Prob. F9–16

PROBLEMS

9–90. The starter for an electric motor is a full cylinder and has the cross-sectional areas shown. If copper wiring has a density of $\rho_{cu} = 8.90 \, \text{Mg/m}^3$ and the steel frame has a density of $\rho_{st} = 7.80 \, \text{Mg/m}^3$, estimate the total mass of the starter. Neglect the size of the copper wire.

9–93. The Gates Manufacturing Co. produces pulley wheels such as the one shown. Determine the weight of the wheel if it is made from steel having a specific weight of 490 lb/ft^3.

9–94. The Gates Manufacturing Co. produces pulley wheels such as the one shown. Determine the total surface area of the wheel in order to estimate the amount of paint needed to protect its surface from rust.

Prob. 9–90

Probs. 9–93/94

9–91. Determine the outside surface area of the storage tank.

*9–92. Determine the volume of the storage tank.

9–95. A ring is generated by rotating the quarter circular area about the x axis. Determine its volume.

*9–96. A ring is generated by rotating the quarter circular area about the x axis. Determine its surface area.

Probs. 9–91/92

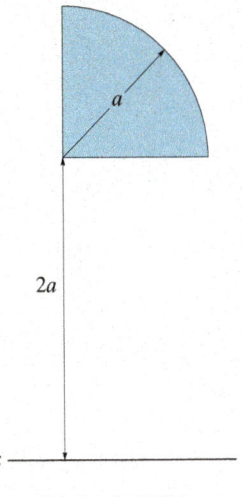

Probs. 9–95/96

9–97. Determine the volume of concrete needed to construct the curb.

9–98. Determine the surface area of the curb. Do not include the area of the ends in the calculation.

Probs. 9–97/98

9–99. Determine the approximate amount of aluminum necessary to make the funnel. It consists of a full circular part having a thickness of 2 mm.

*9–100. Determine the approximate outer surface area of the funnel. It consists of a full circular part of negligible thickness.

Probs. 9–99/100

9–101. The elevated water storage tank has a conical top and hemispherical bottom and is fabricated using thin steel plate. Determine how many square feet of plate is needed to fabricate the tank.

9–102. The elevated water storage tank has a conical top and hemispherical bottom and is fabricated using thin steel plate. Determine the volume within the tank.

Probs. 9–101/102

9–103. Determine the surface area and the volume of the ring formed by rotating the square about the vertical axis.

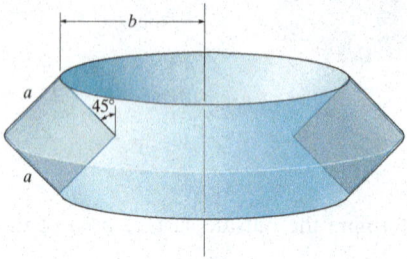

Prob. 9–103

*9–104. Using integration, determine both the area and the distance $\bar{y}$ to the centroid of the shaded area. Then using the second theorem of Pappus-Guldinus, determine the volume of the solid generated by revolving the shaded area about the x axis.

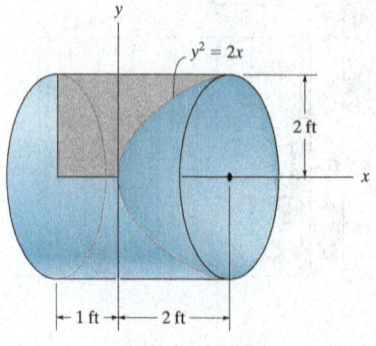

Prob. 9–104

9–105. The heat exchanger radiates thermal energy at the rate of 2500 kJ/h for each square meter of its surface area. Determine how many joules (J) are radiated within a 5-hour period.

Prob. 9–105

9–106. Using integration, determine the area and the centroidal distance $\bar{y}$ of the shaded area. Then, using the second theorem of Pappus-Guldinus, determine the volume of a paraboloid formed by revolving the area about the x axis.

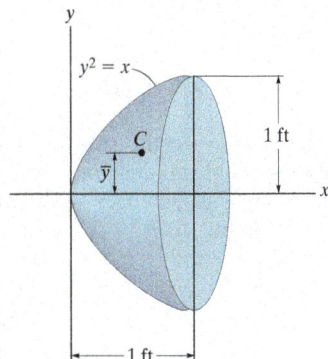

Prob. 9–106

9–107. The suspension bunker is made from plates which are curved to the natural shape which a completely flexible membrane would take if subjected to a full load of coal. This curve may be approximated by a parabola, $y = 0.2x^2$. Determine the weight of coal which the bunker would contain when completely filled. Coal has a specific weight of $\gamma = 50$ lb/ft^3, and assume there is a 20% loss in volume due to air voids. Solve the problem by integration to determine the cross-sectional area of ABC; then use the second theorem of Pappus–Guldinus to find the volume.

Prob. 9–107

*****9–108.** Sand is piled between two walls as shown. Assume the pile to be a quarter section of a cone and that 26 percent of this volume is voids (air space). Use the second theorem of Pappus-Guldinus to determine the volume of sand.

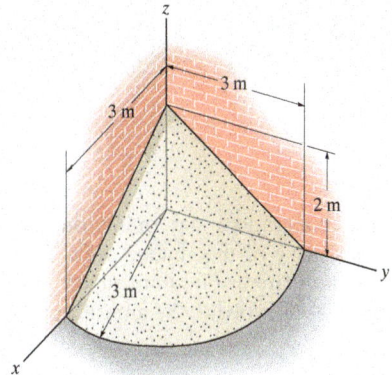

Prob. 9–108

9–109. Determine the surface area and volume of the wheel formed by revolving the cross-sectional area 360° about the z axis.

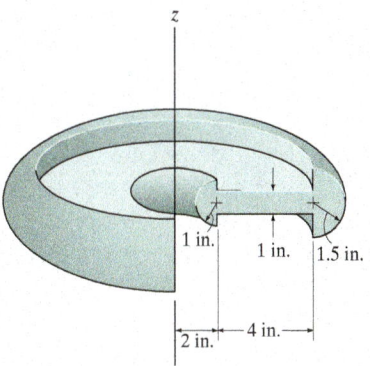

Prob. 9–109

9–110. Determine the surface area of the silo which consists of a cylinder and hemispherical cap. Neglect the thickness of the plates.

Prob. 9–110

9–111. Using integration, determine the area and the centroidal distance $\bar{y}$ of the shaded area. Then, using the second theorem of Pappus-Guldinus, determine the volume of a solid formed by revolving the area about the x axis.

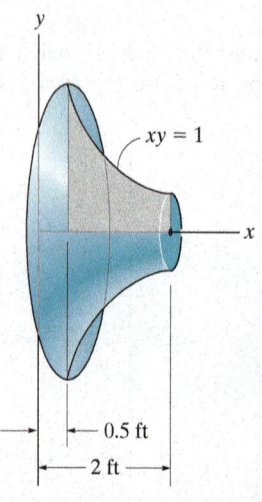

Prob. 9–111

*9–112. Using integration, determine the area and the centroidal distance $\bar{y}$ of the shaded area. Then, using the second theorem of Pappus-Guldinus, determine the volume of a paraboloid formed by revolving the area about the x axis.

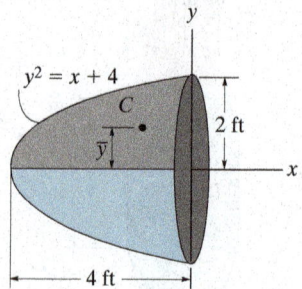

Prob. 9–112

9–113. Determine the volume of material needed to make the casting.

Prob. 9–113

9–114. Determine the weight of the wedge which is formed by rotating a right triangle of base 6 in. and height 3 in. through an angle of 30°. The specific weight of the material is $\gamma = 0.22$ lb/in^3.

Prob. 9–114

*9.4 RESULTANT OF A GENERAL DISTRIBUTED LOADING

In Sec. 4.9, we discussed the method used to simplify a two-dimensional distributed loading to a single resultant force acting at a specific point. In this section we will generalize this method to include flat surfaces that have an arbitrary shape and are subjected to a variable load distribution. To show how to do this, we will consider the flat plate in Fig. 9–23a, which is subjected to the loading defined by $p = p(x, y)$ Pa, where 1 Pa (pascal) = 1 N/m². Knowing this function, we can determine the resultant force $\mathbf{F}_R$ acting on the plate and its location $(\bar{x}, \bar{y})$, Fig. 9–23b.

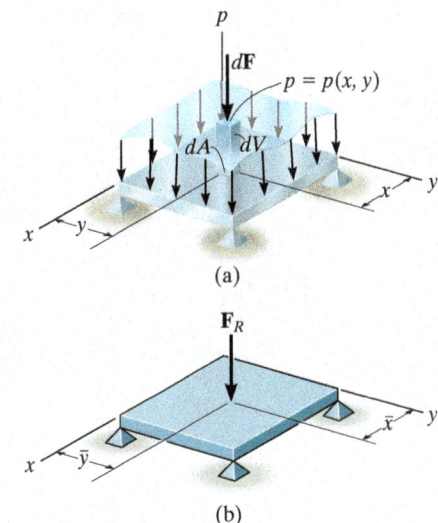

(a)

(b)

Fig. 9–23

Magnitude of Resultant Force. The force $d\mathbf{F}$ acting on the differential area dA m² of the plate, located at the arbitrary point (x, y), has a magnitude of $dF = [p(x, y) \text{ N/m}^2](dA \text{ m}^2) = [p(x, y) \, dA]$ N. This result is equivalent to the differential *volume element* dV shown in Fig. 9–23a. The *magnitude* of $\mathbf{F}_R$ is the sum of these differential forces acting over the plate's *entire surface area* A. Thus:

$$F_R = \Sigma F; \qquad \boxed{F_R = \int_A p(x, y) \, dA = \int_V dV = V} \qquad (9\text{–}11)$$

Therefore, *the magnitude of the resultant force is equal to the total volume under the distributed-loading diagram.*

Location of Resultant Force. The location $(\bar{x}, \bar{y})$ of $\mathbf{F}_R$ is determined by setting the moment of $\mathbf{F}_R$ equal to the moments of all the differential forces $d\mathbf{F}$ about the y and the x axis. From Figs. 9–23a and 9–23b, using Eq. 9–11, we get

The resultant of a wind loading that is distributed on the front or side walls of this building must be calculated using integration in order to design the framework that holds the building together.

$$\boxed{\bar{x} = \dfrac{\int_A x p(x, y) \, dA}{\int_A p(x, y) \, dA} = \dfrac{\int_V x \, dV}{\int_V dV} \qquad \bar{y} = \dfrac{\int_A y p(x, y) \, dA}{\int_A p(x, y) \, dA} = \dfrac{\int_V y \, dV}{\int_V dV}} \qquad (9\text{–}12)$$

Hence, *the line of action of the resultant force passes through the geometric center or centroid of the volume under the distributed-loading diagram.*

*9.5 FLUID PRESSURE

According to Pascal's law, a fluid at rest creates a pressure p at a point that is the *same* in *all* directions. Also, the magnitude of p, measured as a force per unit area, depends on the specific weight γ or mass density ρ of the fluid, and the depth z of the point from the fluid surface. The relationship can be expressed mathematically as

$$p = \gamma z = \rho g z \qquad (9\text{-}13)$$

where g is the acceleration due to gravity.*

To illustrate how Eq. 9–13 is applied, consider the submerged plate shown in Fig. 9–24. Three points on the plate have been specified. Since point B is at depth z_1 from the liquid surface, the *pressure* at this point has a magnitude $p_1 = \gamma z_1$. Likewise, points C and D are both at depth z_2; hence, $p_2 = \gamma z_2$. In all cases, the pressure acts *normal* to the surface area dA located at the specified point.

Since Eq. 9–13 defines how the pressure will vary over a surface, then using the results of Sec. 9.4, we will be able to determine the resultant force caused by a liquid and specify its location on the surface of a submerged plate. Three different shapes of plates will now be considered.

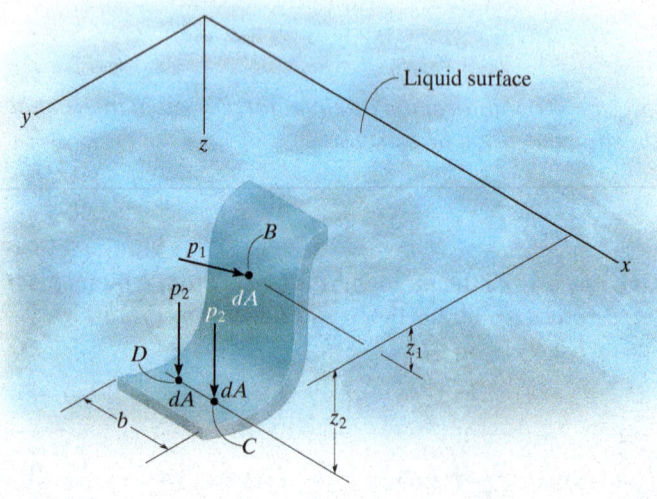

Fig. 9–24

---

*This equation is valid only for fluids that are assumed *incompressible*, as in the case of most liquids. Gases are compressible fluids, and since their density changes significantly with both pressure and temperature, Eq. 9–13 should not be used. See R.C. Hibbeler, *Fluid Mechanics*, 2e, Pearson Inc.

9.5 FLUID PRESSURE

Flat Plate of Constant Width. If the plate is flat and has a rectangular shape of constant width, Fig. 9–25a, then the distribution of pressure over the plate's surface is represented by a trapezoidal volume having an intensity of $p_1 = \gamma z_1$ at depth z_1 and $p_2 = \gamma z_2$ at depth z_2. As noted in Sec. 9.4, the magnitude of the *resultant force* $\mathbf{F}_R$ is equal to the *volume* of this loading diagram, and $\mathbf{F}_R$ has a *line of action* that passes through this volume's *centroid C*. As a result, $\mathbf{F}_R$ acts on the plate at point *P*, called the **center of pressure**.

Since the plate has a *constant width*, the loading distribution may also be viewed in two dimensions, Fig. 9–25b. Here the loading intensity is measured as a force/length and varies linearly from $w_1 = bp_1 = b\gamma z_1$ to $w_2 = bp_2 = b\gamma z_2$. The magnitude of $\mathbf{F}_R$ in this case equals the trapezoidal area, and $\mathbf{F}_R$ has a *line of action* that passes through the area's *centroid C*.

The walls of the tank must be designed to support the pressure loading of the liquid that is contained within it.

(a)

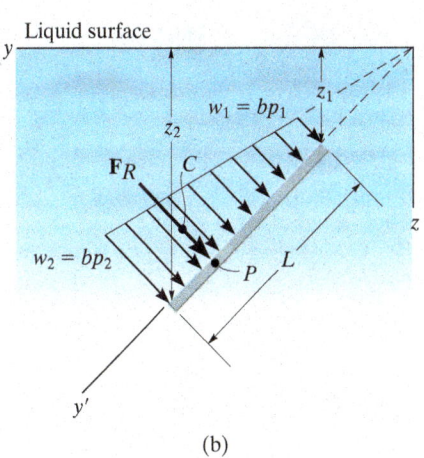

(b)

Fig. 9–25

492　CHAPTER 9　CENTER OF GRAVITY AND CENTROID

Fig. 9–26

Curved Plate of Constant Width. When a submerged plate of constant width is curved, the pressure acting normal to the plate continually changes both its magnitude and direction, and therefore the calculation for the magnitude of $\mathbf{F}_R$ and its location P is more difficult than for a flat plate. Three- and two-dimensional views of the loading distribution are shown in Figs. 9–26a and 9–26b, respectively. Although integration can be used to solve this problem, a simpler method exists. This method requires separate calculations for the horizontal and vertical *components* of $\mathbf{F}_R$.

For the plate in Fig. 9–26b, the distributed loading is represented by the *equivalent loading* shown in Fig. 9–26c. Here the plate supports the weight of liquid W_f contained within the block BDA. This force has a magnitude $W_f = (\gamma b)(\text{area}_{BDA})$ and acts through the centroid of BDA. In addition, there are the pressure distributions caused by the liquid acting along the vertical and horizontal sides of this block. Along the vertical side AD, the force $\mathbf{F}_{AD}$ has a magnitude equal to the area of the trapezoid. It acts through the centroid C_{AD} of this area. The distributed loading along the horizontal side AB is constant since all points lying in this plane are at the same depth from the surface of the liquid. The magnitude of $\mathbf{F}_{AB}$ is simply the area of the rectangle. This force acts through the centroid C_{AB} or at the midpoint of AB. Summing these three forces yields $\mathbf{F}_R = \Sigma \mathbf{F} = \mathbf{F}_{AD} + \mathbf{F}_{AB} + \mathbf{W}_f$. Finally, the location of the center of pressure P on the plate is determined by applying $M_R = \Sigma M$, which states that the moment of the resultant force about a convenient reference point such as D or B in Fig. 9–26b is equal to the sum of the moments of the three forces in Fig. 9–26c about this same point.

9.5 FLUID PRESSURE

Flat Plate of Variable Width. The pressure distribution acting on the surface of a flat submerged plate having a variable width is shown in Fig. 9–27. The pressure acting on the differential area strip dA, parallel to the x axis and at a depth z, is $p = \gamma z$. Therefore the force on this element is $dF = p\, dA$, and so $dF = (\gamma z) dA$. The resultant force is

$$F_R = \int dF = \gamma \int z\, dA$$

If the depth to the centroid C' of the area is $\bar{z}$, Fig. 9–27, then $\int z\, dA = \bar{z} A$. Substituting, we have

$$F_R = \gamma \bar{z} A \qquad (9\text{--}14)$$

In other words, *the magnitude of the resultant force acting on any flat plate is equal to the product of the area A of the plate and the pressure $p = \gamma \bar{z}$ at the depth of the area's centroid C'*. As discussed in Sec. 9.4, this force is also equivalent to the volume under the pressure distribution. The line of action of the resultant force passes through the centroid C of this *volume* and intersects the plate at the center of pressure P, Fig. 9–27.

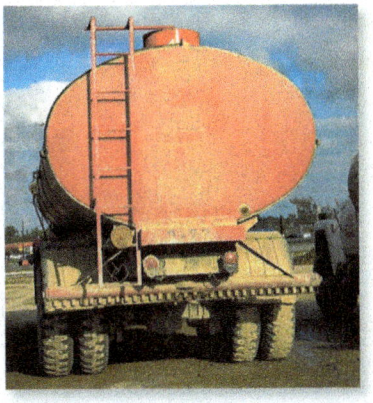

The resultant force of the water pressure and its location on the elliptical back plate of this tank truck must be determined by integration.

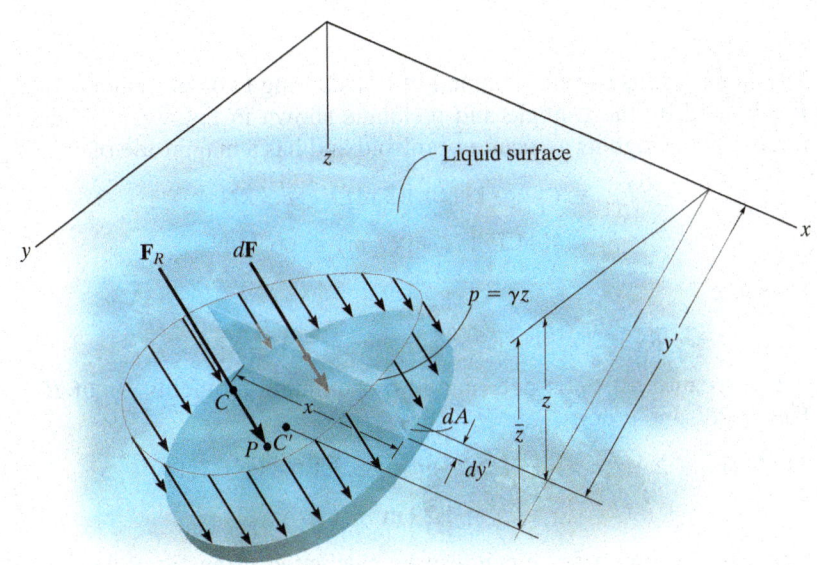

Fig. 9–27

EXAMPLE 9.14

Determine the magnitude and location of the resultant hydrostatic force acting on the submerged rectangular plate AB shown in Fig. 9–28a. The plate has a width of 1.5 m; $\rho_w = 1000$ kg/m^3.

SOLUTION I
The water pressures at depths A and B are

$$p_A = \rho_w g z_A = (1000 \text{ kg/m}^3)(9.81 \text{ m/s}^2)(2 \text{ m}) = 19.62 \text{ kPa}$$

$$p_B = \rho_w g z_B = (1000 \text{ kg/m}^3)(9.81 \text{ m/s}^2)(5 \text{ m}) = 49.05 \text{ kPa}$$

Since the plate has a constant width, the pressure loading can be viewed in two dimensions, as shown in Fig. 9–28b. The intensities of the load at A and B are

$$w_A = b p_A = (1.5 \text{ m})(19.62 \text{ kPa}) = 29.43 \text{ kN/m}$$

$$w_B = b p_B = (1.5 \text{ m})(49.05 \text{ kPa}) = 73.58 \text{ kN/m}$$

From the table on the inside back cover, the magnitude of the resultant force $\mathbf{F}_R$ created by this distributed load is

$$F_R = \text{area of a trapezoid} = \tfrac{1}{2}(3)(29.4 + 73.6) = 154.5 \text{ kN} \qquad \textit{Ans.}$$

This force acts through the centroid of this area,

$$h = \frac{1}{3}\left(\frac{2(29.43) + 73.58}{29.43 + 73.58}\right)(3) = 1.29 \text{ m} \qquad \textit{Ans.}$$

measured upward from B, Fig. 9–28b.

SOLUTION II
The same results can be obtained by considering two components of $\mathbf{F}_R$, defined by the triangle and rectangle shown in Fig. 9–28c. Each force acts through its associated centroid and has a magnitude of

$$F_{Re} = (29.43 \text{ kN/m})(3 \text{ m}) = 88.3 \text{ kN}$$

$$F_t = \tfrac{1}{2}(44.15 \text{ kN/m})(3 \text{ m}) = 66.2 \text{ kN}$$

Hence,

$$F_R = F_{Re} + F_t = 88.3 + 66.2 = 154.5 \text{ kN} \qquad \textit{Ans.}$$

The location of $\mathbf{F}_R$ is determined by summing moments about B, Figs. 9–28b and c, i.e.,

$$\zeta + (M_R)_B = \Sigma M_B; \qquad (154.5)h = 88.3(1.5) + 66.2(1)$$

$$h = 1.29 \text{ m} \qquad \textit{Ans.}$$

NOTE: The resultant force can also be calculated using Eq. 9–14, i.e.,
$F_R = \gamma \bar{z} A = (9810 \text{ N/m}^3)(3.5 \text{ m})(3 \text{ m})(1.5 \text{ m}) = 154.5 \text{ kN}.$

(a)

(b)

(c)

Fig. 9–28

EXAMPLE 9.15

Determine the magnitude of the resultant hydrostatic force acting on the surface of a seawall shaped in the form of a parabola, as shown in Fig. 9–29a. The wall is 5 m long; $\rho_w = 1020 \text{ kg/m}^3$.

Fig. 9–29

SOLUTION

The horizontal and vertical components of the resultant force will be calculated, Fig. 9–29b. Since

$$p_B = \rho_w g z_B = (1020 \text{ kg/m}^3)(9.81 \text{ m/s}^2)(3 \text{ m}) = 30.02 \text{ kPa}$$

then

$$w_B = b p_B = 5 \text{ m}(30.02 \text{ kPa}) = 150.1 \text{ kN/m}$$

Thus,

$$F_h = \tfrac{1}{2}(3 \text{ m})(150.1 \text{ kN/m}) = 225.1 \text{ kN}$$

The area of the parabolic section ABC can be determined using the formula for a parabolic area listed on the inside back cover, $A = \tfrac{1}{3}ab$. Hence, the weight of water within this 5-m-long region is

$$F_v = (\rho_w g b)(\text{area}_{ABC})$$

$$= (1020 \text{ kg/m}^3)(9.81 \text{ m/s}^2)(5 \text{ m})\left[\tfrac{1}{3}(1 \text{ m})(3 \text{ m})\right] = 50.0 \text{ kN}$$

The resultant force is therefore

$$F_R = \sqrt{F_h^2 + F_v^2} = \sqrt{(225.1 \text{ kN})^2 + (50.0 \text{ kN})^2}$$

$$= 231 \text{ kN} \qquad \textit{Ans.}$$

EXAMPLE 9.16

Determine the magnitude and location of the resultant hydrostatic force acting on the triangular end plates of the water trough shown in Fig. 9–30a; $\rho_w = 1000$ kg/m^3.

(a)

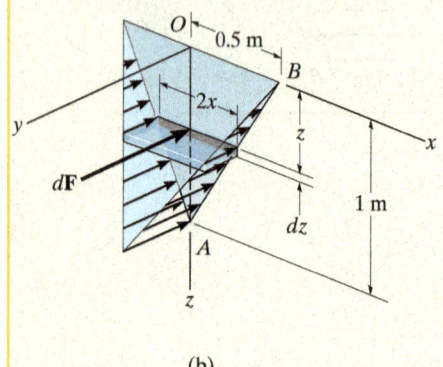

(b)

Fig. 9–30

SOLUTION

The pressure distribution acting on the end plate E is shown in Fig. 9–30b. The magnitude of the resultant force is equal to the volume of this loading distribution. We will solve the problem by integration. Choosing the horizontal volume element shown in the figure, we have

$$dF = dV = p\, dA = \rho_w g z (2x\, dz) = 19\,620 z x\, dz$$

The equation of line AB is

$$x = 0.5(1 - z)$$

Therefore, substituting and integrating with respect to z from $z = 0$ to $z = 1$ m yields

$$F = V = \int_V dV = \int_0^{1\,\text{m}} (19\,620)z[0.5(1-z)]\, dz$$

$$= 9810 \int_0^{1\,\text{m}} (z - z^2)\, dz = 1635\text{ N} = 1.64\text{ kN} \qquad Ans.$$

This resultant passes through the *centroid of the volume*. Because of symmetry,

$$\bar{x} = 0 \qquad Ans.$$

Since $\tilde{z} = z$ for the volume element, then

$$\bar{z} = \dfrac{\int_V \tilde{z}\, dV}{\int_V dV} = \dfrac{\int_0^{1\,\text{m}} z(19\,620)z[0.5(1-z)]\, dz}{1635} = \dfrac{9810 \int_0^{1\,\text{m}} (z^2 - z^3)\, dz}{1635}$$

$$= 0.5\text{ m} \qquad Ans.$$

NOTE: We can also determine the resultant force by applying Eq. 9–14,

$$F_R = \gamma \bar{z} A = (9810\text{ N/m}^3)\left(\tfrac{1}{3}\right)(1\text{ m})\left[\tfrac{1}{2}(1\text{ m})(1\text{ m})\right] = 1.64\text{ kN}.$$

FUNDAMENTAL PROBLEMS

F9–17. Determine the magnitude of the hydrostatic force acting per meter width of the wall. Water has a density of $\rho = 1 \text{ Mg/m}^3$.

Prob. F9–17

F9–18. Determine the magnitude of the hydrostatic force acting on gate AB, which has a width of 4 ft. The specific weight of water is $\gamma = 62.4 \text{ lb/ft}^3$.

Prob. F9–18

F9–19. Determine the magnitude of the hydrostatic force acting on gate AB, which has a width of 1.5 m. Water has a density of $\rho = 1 \text{ Mg/m}^3$.

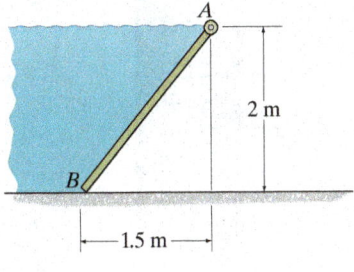

Prob. F9–19

F9–20. Determine the magnitude of the hydrostatic force acting on gate AB, which has a width of 2 m. Water has a density of $\rho = 1 \text{ Mg/m}^3$.

Prob. F9–20

F9–21. Determine the magnitude of the hydrostatic force acting on gate AB, which has a width of 2 ft. The specific weight of water is $\gamma = 62.4 \text{ lb/ft}^3$.

Prob. F9–21

PROBLEMS

9–115. The pressure loading on the plate is described by the function $p = [-240/(x + 1) + 340]$ Pa. Determine the magnitude of the resultant force and the coordinates $(\bar{x}, \bar{y})$ of the point where the line of action of the source intersects the plate.

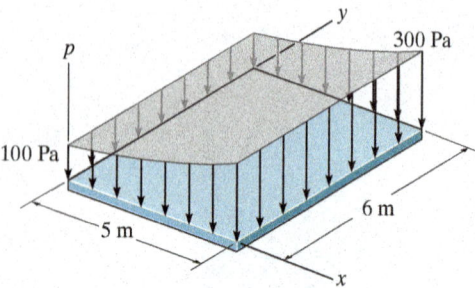

Prob. 9–115

*9–116. The load over the plate varies linearly along the sides of the plate such that $p = (12 - 6x + 4y)$ kPa. Determine the magnitude of the resultant force and the coordinates $(\bar{x}, \bar{y})$ of the point where the line of action of the force intersects the plate.

Prob. 9–116

9–117. Determine the magnitude and location of the resultant force of the parabolic loading acting on the plate.

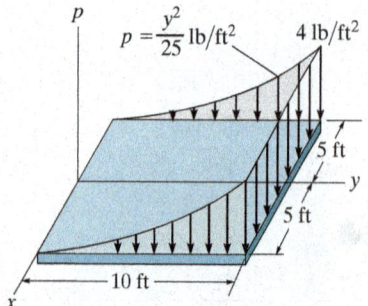

Prob. 9–117

9–118. The rectangular plate is subjected to a distributed load over its *entire surface*. The load is defined by the expression $p = p_0 \sin(\pi x/a) \sin(\pi y/b)$, where p_0 represents the pressure acting at the center of the plate. Determine the magnitude and location of the resultant force acting on the plate.

Prob. 9–118

9–119. A wind loading creates a positive pressure on one side of the chimney and a negative (suction) pressure on the other side, as shown. If this pressure loading acts uniformly along the chimney's length, determine the magnitude of the resultant force created by the wind.

Prob. 9–119

*9–120. When the tide water A subsides, the tide gate automatically swings open to drain the marsh B. For the condition of high tide shown, determine the horizontal reactions developed at the hinge C and stop block D. The length of the gate is 6 m and its height is 4 m. $\rho_w = 1.0 \text{ Mg/m}^3$.

Prob. 9–120

9–121. The tank is filled with water to a depth of $d = 4$ m. Determine the resultant force the water exerts on side A and side B of the tank. If oil instead of water is placed in the tank, to what depth d should it reach so that it creates the same resultant forces? $\rho_o = 900 \text{ kg/m}^3$ and $\rho_w = 1000 \text{ kg/m}^3$.

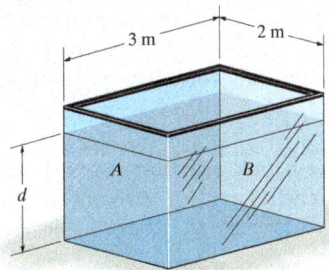

Prob. 9–121

9–122. Determine the magnitude and location of the resultant hydrostatic force acting on each of the cover plates A and B. $\rho_w = 1.0 \text{ Mg/m}^3$.

Prob. 9–122

9–123. The tank is used to store a liquid having a specific weight of $\gamma = 80 \text{ lb/ft}^3$. If it is filled to the top, determine the magnitude of the resultant force the liquid exerts on side $ABEF$.

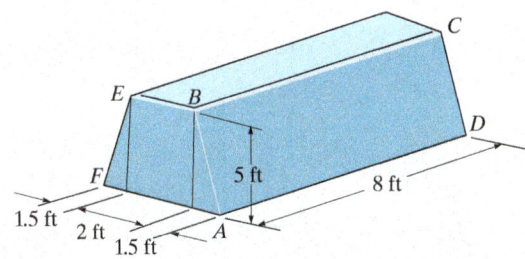

Prob. 9–123

*9–124. The tank is used to store a liquid having a specific weight of $\gamma = 80 \text{ lb/ft}^3$. If it is filled to the top, determine the magnitude of the resultant force the liquid exerts on side $ABCD$.

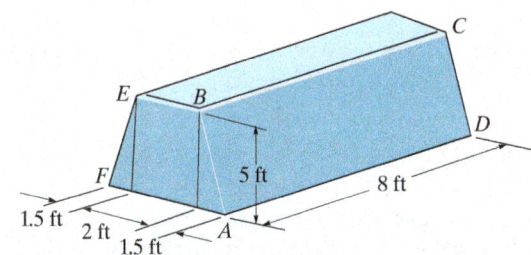

Prob. 9–124

9–125. The trapezoidal back plate of the boat has the dimensions shown. If the water on the outside of the boat is 3 in. from the top, determine the resultant force on the plate and its location, measured from the top of the boat. $\gamma_w = 62.4 \text{ lb/ft}^3$.

Prob. 9–125

9–126. The quarter circular "gravity" dam is held in place by its own weight. Determine the smallest possible density of the material composing the dam so that it will be prevented from overturning about its end A. $\rho_w = 1.0 \, \text{Mg/m}^3$.

Prob. 9–126

9–127. The 2-m-wide rectangular gate is pinned at its center A and is prevented from rotating by the block at B. Determine the reactions at these supports due to hydrostatic pressure. $\rho_w = 1.0 \, \text{Mg/m}^3$.

Prob. 9–127

*9–128. The tank is filled with a liquid that has a density of $900 \, \text{kg/m}^3$. Determine the resultant force that it exerts on the elliptical end plate, and the location of the center of pressure, measured from the x axis.

Prob. 9–128

9–129. The structure is used for temporary storage of oil at sea for later loading into ships. When it is empty the water level is at A (sea level). As oil is poured into it, the water is displaced through exit ports at D. If the riser EC is filled with oil, i.e., to a depth of C, determine the height h to B of the oil level above sea level. $\rho_o = 900 \, \text{kg/m}^3$ and $\rho_w = 1020 \, \text{kg/m}^3$.

9–130. If the structure in Prob. 9–129 is totally filled with oil, i.e., until it reaches a depth of 58 m below sea level, how high h will the oil level extend above sea level?

Probs. 9–129/130

CHAPTER REVIEW

Center of Gravity and Centroid

The *center of gravity* G represents a point where the weight of the body can be considered concentrated. The distance from an axis to this point can be determined from a balance of moments, which requires that the moment of the weight of all the particles of the body about this axis must equal the moment of the entire weight of the body about the axis.

$$\bar{x} = \frac{\int \tilde{x}\, dW}{\int dW}$$

$$\bar{y} = \frac{\int \tilde{y}\, dW}{\int dW}$$

$$\bar{z} = \frac{\int \tilde{z}\, dW}{\int dW}$$

The center of mass will coincide with the center of gravity provided the acceleration of gravity is constant.

$$\bar{x} = \frac{\int_L \tilde{x}\, dL}{\int_L dL} \quad \bar{y} = \frac{\int_L \tilde{y}\, dL}{\int_L dL} \quad \bar{z} = \frac{\int_L \tilde{z}\, dL}{\int_L dL}$$

$$\bar{x} = \frac{\int_A \tilde{x}\, dA}{\int_A dA} \quad \bar{y} = \frac{\int_A \tilde{y}\, dA}{\int_A dA} \quad \bar{z} = \frac{\int_A \tilde{z}\, dA}{\int_A dA}$$

The *centroid* is the location of the geometric center for the body. It is determined in a similar manner, using a moment balance of geometric elements such as line, area, or volume segments. For bodies having an arbitrary shape, moments are summed (integrated) using differential elements.

$$\bar{x} = \frac{\int_V \tilde{x}\, dV}{\int_V dV} \quad \bar{y} = \frac{\int_V \tilde{y}\, dV}{\int_A dV} \quad \bar{z} = \frac{\int_V \tilde{z}\, dV}{\int_V dV}$$

The center of mass will coincide with the centroid provided the material is homogeneous, i.e., the density of the material is the same throughout. The centroid will always lie on an axis of symmetry.

Composite Body

If the body is a composite of several shapes, each having a known location for its center of gravity or centroid, then the location of the center of gravity or centroid of the body can be determined from a discrete summation using its composite parts.

$$\bar{x} = \frac{\Sigma \tilde{x} W}{\Sigma W}$$

$$\bar{y} = \frac{\Sigma \tilde{y} W}{\Sigma W}$$

$$\bar{z} = \frac{\Sigma \tilde{z} W}{\Sigma W}$$

Theorems of Pappus and Guldinus

The theorems of Pappus and Guldinus can be used to determine the surface area and volume of a body of revolution.

The *surface area* equals the product of the length L of the generating curve and the distance traveled by the centroid of the curve needed to generate the area $\theta \bar{r}$.

$$A = \theta \bar{r} L$$

The *volume* of the body equals the product of the generating area A and the distance traveled by the centroid of this area needed to generate the volume $\theta \bar{r}$.

$$V = \theta \bar{r} A$$

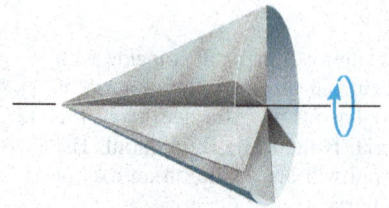

CHAPTER REVIEW 503

General Distributed Loading

The magnitude of the resultant force is equal to the volume under the distributed-loading diagram. The line of action of the resultant force passes through the geometric center or centroid of this volume.

$$F_R = \int_A p(x, y)\, dA = \int_V dV$$

$$\bar{x} = \frac{\int_V x\, dV}{\int_V dV}$$

$$\bar{y} = \frac{\int_V y\, dV}{\int_V dV}$$

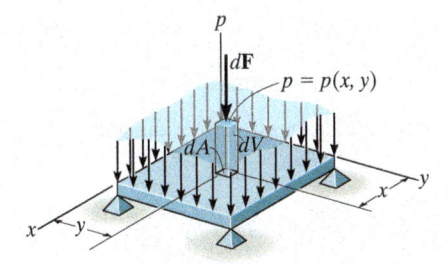

Fluid Pressure

The pressure of a liquid at a point on a submerged surface depends upon the depth of the point and the density of the liquid in accordance with $p = \rho g h = \gamma h$. This pressure will create a *linear distribution* of loading on a flat vertical or inclined surface.

If the surface is horizontal, then the loading will be *uniform*.

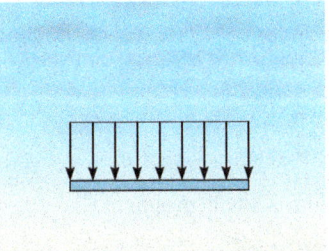

The resultant force of a hydrostatic loading can be determined by finding the volume under the loading curve or using $F_R = \gamma \bar{z} A$, where $\bar{z}$ is the depth to the centroid of the plate's surface. The line of action of the resultant force passes through the centroid of the volume of the loading diagram and acts at a point P on the plate called the center of pressure.

REVIEW PROBLEMS

R9–1. Locate the centroid $\bar{x}$ of the area.

R9–2. Locate the centroid $\bar{y}$ of the area.

R9–4. Locate the centroid of the rod.

Probs. R9–1/2

R9–3. Locate the centroid $\bar{z}$ of the hemisphere.

Prob. R9–4

R9–5. Locate the centroid $\bar{y}$ of the beam's cross-sectional area.

Prob. R9–3

Prob. R9–5

Review Problems

R9–6. A circular V-belt has an inner radius of 600 mm and a cross-sectional area as shown. Determine the surface area of the belt.

R9–7. A circular V-belt has an inner radius of 600 mm and a cross-sectional area as shown. Determine the volume of material required to make the belt.

R9–9. The gate AB is 8 m wide. Determine the horizontal and vertical components of force acting on the pin at B and the vertical reaction at the smooth support A; $\rho_w = 1.0 \text{ Mg/m}^3$.

Probs. R9–6/7

Prob. R9–9

R9–8. The rectangular bin is filled with coal, which creates a pressure distribution along wall A that varies as shown, i.e., $p = 4z^{1/3}$ lb/ft^2, where z is in feet. Determine the resultant force created by the coal, and its location, measured from the top surface of the coal.

R9–10. Determine the magnitude of the resultant hydrostatic force acting per foot of length on the seawall; $\gamma_w = 62.4 \text{ lb/ft}^3$.

Prob. R9–8

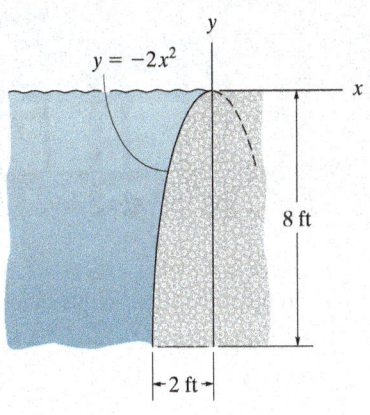

Prob. R9–10

CHAPTER 10

The design of these structural members requires calculating their cross-sectional moments of inertia. In this chapter we will discuss how this is done.

MOMENTS OF INERTIA

CHAPTER OBJECTIVES

- To show how to determine the moment of inertia for an area by integration and by using formulas.
- To introduce the product of inertia and show how to find the maximum and minimum moments of inertia for an area.
- To discuss the mass moment of inertia.

10.1 DEFINITION OF MOMENTS OF INERTIA FOR AREAS

Whenever a distributed load acts perpendicular to an area and its intensity varies linearly, the calculation of the moment of the loading about an axis will involve an integral of the form $\int y^2 dA$. For example, consider the plate in Fig. 10–1, which is submerged in a fluid and subjected to the pressure p. As discussed in Sec. 9.5, this pressure varies linearly with depth, such that $p = \gamma y$, where γ is the specific weight of the fluid. As a result, the force acting on the differential area dA of the plate is $dF = p\, dA = (\gamma y)\, dA$, and so the *moment* of this force about the x axis is $dM = y\, dF = \gamma y^2\, dA$. Integrating dM over the entire area of the plate yields $M = \gamma \int y^2\, dA$.

The integral $\int y^2\, dA$ is sometimes referred to as the "second moment" of the area about an axis (the x axis), but more often it is called the ***moment of inertia of the area***. The word "inertia" is used here since the formulation is similar to the mass moment of inertia, $\int y^2 dm$, which is a dynamical property described in Sec. 10.8. Although for an area this integral has no physical meaning, it often arises in formulas used in fluid mechanics, mechanics of materials, structural mechanics, and mechanical design, and so an engineer needs to be familiar with the methods used to determine its value.

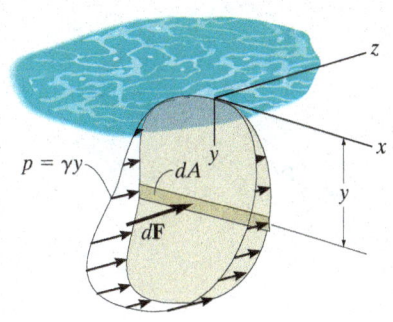

Fig. 10–1

508   CHAPTER 10   MOMENTS OF INERTIA

Moment of Inertia. By definition, the moments of inertia of a differential area dA about the x and y axes are $dI_x = y^2\,dA$ and $dI_y = x^2\,dA$, respectively, Fig. 10–2. For the entire area A the **moments of inertia** are determined by integration; i.e.,

$$I_x = \int_A y^2\,dA$$
$$I_y = \int_A x^2\,dA \qquad (10\text{–}1)$$

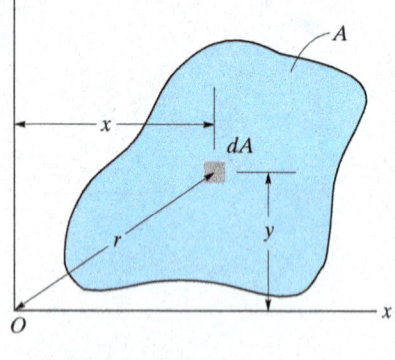

Fig. 10–2

We can also formulate this quantity for dA about the "pole" O or z axis, Fig. 10–2. This is referred to as the **polar moment of inertia**. It is defined as $dJ_O = r^2\,dA$, where r is the perpendicular distance from the pole (z axis) to the element dA. For the entire area the *polar moment of inertia* is

$$J_O = \int_A r^2\,dA = I_x + I_y \qquad (10\text{–}2)$$

This relation between J_O and I_x, I_y is possible since $r^2 = x^2 + y^2$, Fig. 10–2.

From the above formulations it is seen that I_x, I_y, and J_O will *always* be *positive* since they involve the product of distance squared and area. Furthermore, the units involve length raised to the fourth power, e.g., m⁴, mm⁴, or ft⁴, in⁴.

10.2 PARALLEL-AXIS THEOREM FOR AN AREA

If the moment of inertia is known about an axis passing through the centroid of an area, then the **parallel-axis theorem** can be used to find the moment of inertia of the area about *any axis* that is parallel to the centroidal axis. To develop this theorem, consider finding the moment of inertia of the shaded area shown in Fig. 10–3 about the x axis. If we choose a differential element dA located at an arbitrary distance y' from the *centroidal* x' axis, then the distance between the parallel x and x' axis is d_y, and so the moment of inertia of dA about the x axis is $dI_x = (y' + d_y)^2\,dA$. For the entire area,

$$I_x = \int_A (y' + d_y)^2\,dA$$
$$= \int_A y'^2\,dA + 2d_y \int_A y'\,dA + d_y^2 \int_A dA$$

Fig. 10–3

The first integral represents the moment of inertia of the area about the centroidal axis, $\bar{I}_{x'}$. The second integral is zero since the x' axis passes through the area's centroid C; i.e., $\int y'\, dA = \bar{y}' \int dA = 0$ since $\bar{y}' = 0$. Since the third integral represents the area A, the final result is therefore

$$I_x = \bar{I}_{x'} + A d_y^2 \qquad (10\text{–}3)$$

A similar expression can be written for I_y; i.e.,

$$I_y = \bar{I}_{y'} + A d_x^2 \qquad (10\text{–}4)$$

And finally, for the polar moment of inertia, since $\bar{J}_C = \bar{I}_{x'} + \bar{I}_{y'}$ and $d^2 = d_x^2 + d_y^2$, we have

$$J_O = \bar{J}_C + A d^2 \qquad (10\text{–}5)$$

The form of each of these three equations states that *the moment of inertia for an area about an axis is equal to its moment of inertia about a parallel axis passing through the area's centroid plus the product of the area and the square of the perpendicular distance between the axes.*

In order to predict the strength and deflection of this beam, it is necessary to calculate the moment of inertia of the beam's cross-sectional area.

10.3 RADIUS OF GYRATION OF AN AREA

The **radius of gyration** of an area about an axis has units of length and is a quantity that is often used for the design of columns in structural mechanics. Provided the areas and moments of inertia are *known*, the radii of gyration are determined from the formulas

$$k_x = \sqrt{\frac{I_x}{A}}$$

$$k_y = \sqrt{\frac{I_y}{A}} \qquad (10\text{–}6)$$

$$k_O = \sqrt{\frac{J_O}{A}}$$

The form of these equations is easily remembered since it is similar to that for finding the moment of inertia for a differential area about an axis. For example, $I_x = k_x^2 A$; whereas for a differential area, $dI_x = y^2\, dA$.

IMPORTANT POINTS

- The moment of inertia is a geometric property of an area that is used to determine the strength of a structural member or the location of a resultant pressure force acting on a plate submerged in a fluid. It is sometimes referred to as the second moment of the area about an axis, because the distance from the axis to each area element is squared.

- If the moment of inertia of an area is known about its centroidal axis, then the moment of inertia about a corresponding parallel axis can be determined using the parallel-axis theorem.

PROCEDURE FOR ANALYSIS

In most cases the moment of inertia can be determined using a single integration. The following procedure shows two ways in which this can be done.

- If the curve defining the boundary of the area is expressed as $y = f(x)$, then select a rectangular differential element such that it has a finite length and differential width.

- The element should be located so that it intersects the curve at the *arbitrary point* (x, y).

Case 1.

- Orient the element so that its length is *parallel* to the axis about which the moment of inertia is calculated. This situation occurs when the rectangular element shown in Fig. 10–4a is used to determine I_x for the area. Here the entire element is at a distance y from the x axis since it has a thickness dy. Thus $I_x = \int y^2 dA$. To find I_y, the element is oriented as shown in Fig. 10–4b. This element lies at the *same* distance x from the y axis so that $I_y = \int x^2 dA$.

Case 2.

- The length of the element can be oriented *perpendicular* to the axis about which the moment of inertia is calculated; however, Eq. 10–1 *does not apply* since all points on the element will *not* lie at the same moment-arm distance from the axis. For example, if the rectangular element in Fig. 10–4a is used to determine I_y, it will first be necessary to calculate the moment of inertia of the *element* about an axis parallel to the y axis that passes through the element's centroid, and then determine the moment of inertia of the *element* about the y axis using the parallel-axis theorem. Integration of this result will yield I_y. See Examples 10.2 and 10.3.

(a)

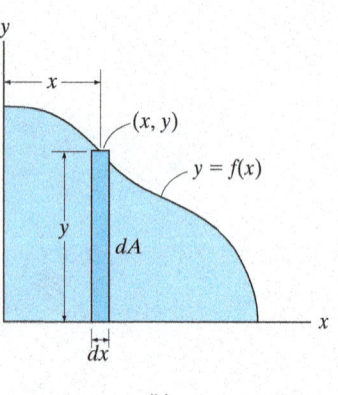

(b)

Fig. 10–4

EXAMPLE 10.1

Determine the moment of inertia for the area shown in Fig. 10–5 with respect to (a) the centroidal x' axis, (b) the axis x_b passing through the base of the rectangle, and (c) the pole or z' axis perpendicular to the x'–y' plane and passing through the centroid C.

SOLUTION (CASE 1)

Part (a). The horizontal element shown in Fig. 10–5 is chosen for integration. Because of its location and orientation, the *entire element* is at a distance y' from the x' axis. Here it is necessary to integrate from $y' = -h/2$ to $y' = h/2$. Since $dA = b\,dy'$, then

$$\bar{I}_{x'} = \int_A y'^2\, dA = \int_{-h/2}^{h/2} y'^2 (b\,dy') = b \int_{-h/2}^{h/2} y'^2\, dy'$$

$$\bar{I}_{x'} = \frac{1}{12}bh^3 \quad \text{Ans.}$$

Fig. 10–5

Part (b). The moment of inertia about an axis passing through the base of the rectangle can be obtained by using the above result and applying the parallel-axis theorem, Eq. 10–3.

$$I_{x_b} = \bar{I}_{x'} + A d_y^2$$

$$= \frac{1}{12}bh^3 + bh\left(\frac{h}{2}\right)^2 = \frac{1}{3}bh^3 \quad \text{Ans.}$$

Part (c). To obtain the polar moment of inertia about point C, we must first obtain $\bar{I}_{y'}$, which is simply found by interchanging the dimensions b and h in the result of part (a), i.e.,

$$\bar{I}_{y'} = \frac{1}{12}hb^3$$

Using Eq. 10–2, the polar moment of inertia about C is therefore

$$\bar{J}_C = \bar{I}_{x'} + \bar{I}_{y'} = \frac{1}{12}bh(h^2 + b^2) \quad \text{Ans.}$$

EXAMPLE 10.2

Fig. 10–6

Determine the moment of inertia for the shaded area shown in Fig. 10–6a about the x axis.

SOLUTION I (CASE 1)

A differential element that is *parallel* to the x axis, Fig. 10–6a, is chosen for integration. It intersects the curve at the *arbitrary point* (x, y). Since this element has a thickness dy, its area is $dA = (100 - x) \, dy$. Furthermore, the element lies at the same distance y from the x axis. Hence, integrating with respect to y, from $y = 0$ to $y = 200$ mm, we have

$$I_x = \int_A y^2 \, dA = \int_0^{200 \text{ mm}} y^2 (100 - x) \, dy$$

$$= \int_0^{200 \text{ mm}} y^2 \left(100 - \frac{y^2}{400}\right) dy = \int_0^{200 \text{ mm}} \left(100 y^2 - \frac{y^4}{400}\right) dy$$

$$= 107(10^6) \text{ mm}^4 \qquad \qquad \textit{Ans.}$$

SOLUTION II (CASE 2)

A differential element *parallel* to the y axis, as shown in Fig. 10–6b, is chosen for integration. It intersects the curve at the *arbitrary point* (x, y). In this case, all points of the element do *not* lie at the same distance from the x axis, and therefore the parallel-axis theorem must be used to determine the *moment of inertia of the element* with respect to this axis. For a rectangle having a base b and height h, the moment of inertia about its centroidal axis has been determined in part (a) of Example 10.1. There it was found that $\bar{I}_{x'} = \frac{1}{12} b h^3$. For the differential element shown in Fig. 10–6b, $b = dx$ and $h = y$, and so $d\bar{I}_{x'} = \frac{1}{12} dx \, y^3$. Since the centroid of the element is $\tilde{y} = y/2$ from the x axis, the moment of inertia of the element about this axis is

$$dI_x = d\bar{I}_{x'} + dA \, \tilde{y}^2 = \frac{1}{12} dx \, y^3 + y \, dx \left(\frac{y}{2}\right)^2 = \frac{1}{3} y^3 \, dx$$

(This result can also be concluded from part (b) of Example 10.1.) Integrating with respect to x, from $x = 0$ to $x = 100$ mm, yields

$$I_x = \int dI_x = \int_0^{100 \text{ mm}} \frac{1}{3} y^3 \, dx = \int_0^{100 \text{ mm}} \frac{1}{3} (400x)^{3/2} \, dx$$

$$= 107(10^6) \text{ mm}^4 \qquad \qquad \textit{Ans.}$$

EXAMPLE 10.3

Determine the moment of inertia with respect to the x axis for the circular area shown in Fig. 10–7a.

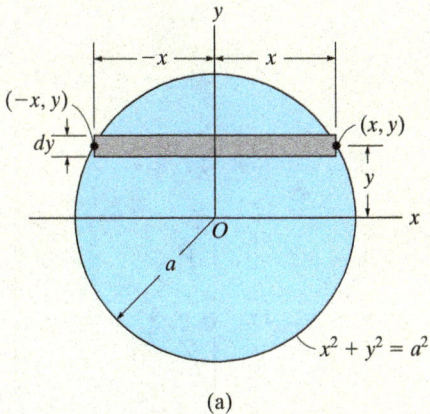

(a)

SOLUTION I (CASE 1)

Using the differential element shown in Fig. 10–7a, since $dA = 2x\,dy$, we have

$$I_x = \int_A y^2\,dA = \int_A y^2(2x)\,dy$$

$$= \int_{-a}^{a} y^2\left(2\sqrt{a^2 - y^2}\right) dy = \frac{\pi a^4}{4} \qquad \text{Ans.}$$

SOLUTION II (CASE 2)

When the differential element shown in Fig. 10–7b is chosen, the centroid for the element happens to lie on the x axis, and since $\bar{I}_{x'} = \frac{1}{12}bh^3$ for a rectangle, we have

$$dI_x = \frac{1}{12} dx (2y)^3$$

$$= \frac{2}{3} y^3 \, dx$$

Integrating with respect to x yields

$$I_x = \int_{-a}^{a} \frac{2}{3}(a^2 - x^2)^{3/2}\,dx = \frac{\pi a^4}{4} \qquad \text{Ans.}$$

NOTE: By comparison, Solution I requires less calculation. Therefore, if an integral using a particular element appears difficult to evaluate, try solving the problem using an element oriented in the other direction.

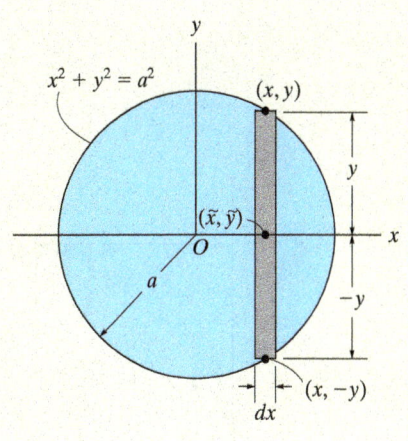

(b)

Fig. 10–7

514   CHAPTER 10   MOMENTS OF INERTIA

FUNDAMENTAL PROBLEMS

F10–1. Determine the moment of inertia of the area about the x axis.

F10–3. Determine the moment of inertia of the area about the y axis.

Prob. F10–1

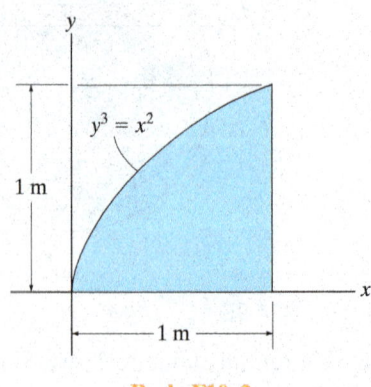

Prob. F10–3

F10–2. Determine the moment of inertia of the area about the x axis.

F10–4. Determine the moment of inertia of the area about the y axis.

Prob. F10–2

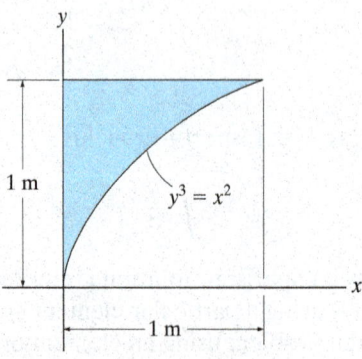

Prob. F10–4

PROBLEMS

10–1. Determine the moment of inertia of the area about the x axis.

10–2. Determine the moment of inertia of the area about the y axis.

10–5. Determine the moment of inertia of the area about the x axis.

10–6. Determine the moment of inertia of the area about the y axis.

Probs. 10–1/2

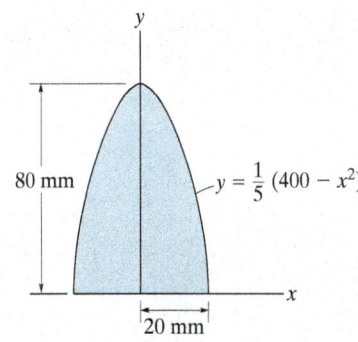

Probs. 10–5/6

10–3. Determine the moment of inertia of the triangular area about the x axis.

*10–4. Determine the moment of inertia of the triangular area about the y axis.

10–7. Determine the moment of inertia of the area about the x axis.

*10–8. Determine the moment of inertia of the area about the y axis.

Probs. 10–3/4

Probs. 10–7/8

10–9. Determine the moment of inertia of the area about the *x* axis. Solve the problem in two ways, using rectangular differential elements: (a) having a thickness dx and (b) having a thickness of dy.

Prob. 10–9

10–10. Determine the radius of gyration k_y of the parabolic area.

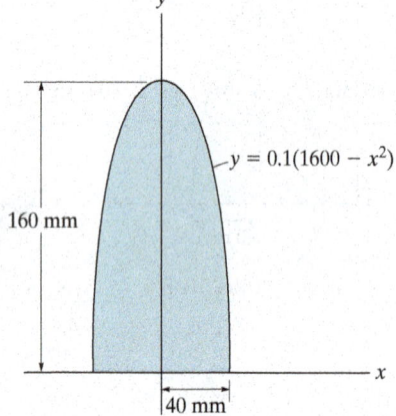

Prob. 10–10

10–11. Determine the moment of inertia for the area about the *x* axis.

***10–12.** Determine the moment of inertia for the area about the *y* axis.

Probs. 10–11/12

10–13. Determine the moment of inertia for the area about the *x* axis.

10–14. Determine the moment of inertia for the area about the *y* axis.

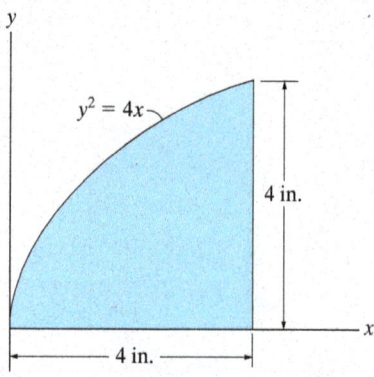

Probs. 10–13/14

10–15. Determine the moment of inertia for the area about the *x* axis.

Prob. 10–15

*10–16. Determine the moment of inertia for the area about the *y* axis.

Prob. 10–16

10–17. Determine the moment of inertia of the quarter circular area about the x axis.

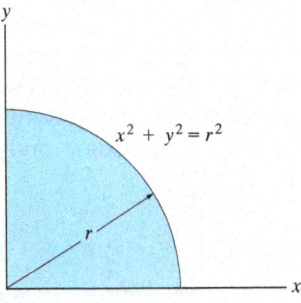

Prob. 10–17

10–18. Determine the moment of inertia for the area about the y axis.

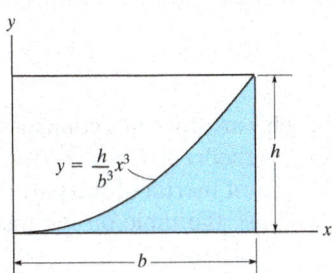

Prob. 10–18

10–19. Determine the moment of inertia of the area about the x axis.

*10–20. Determine the moment of inertia of the area about the y axis.

Probs. 10–19/20

10–21. Determine the moment of inertia of the area about the x axis.

10–22. Determine the moment of inertia of the area about the y axis.

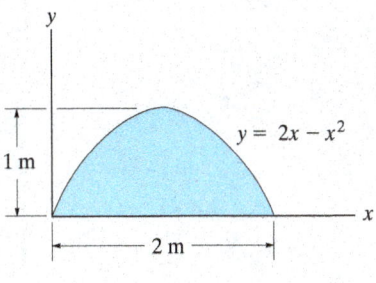

Probs. 10–21/22

10–23. Determine the moment of inertia for the area about the x axis.

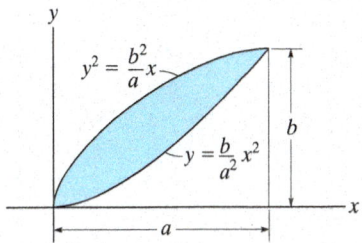

Prob. 10–23

*10–24. Determine the moment of inertia of the equilateral triangle about the x' axis passing through its centroid.

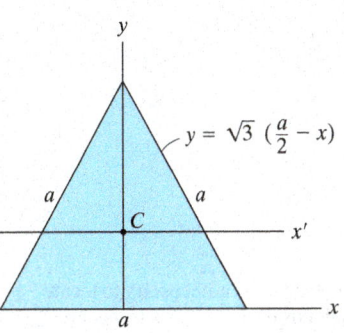

Prob. 10–24

10.4 MOMENTS OF INERTIA FOR COMPOSITE AREAS

The moment of inertia for a composite area that consists of a series of connected "simpler" parts or shapes can be determined about any axis provided the moment of inertia of each of its parts is known or can be determined about the axis. The following procedure outlines a method for doing this.

PROCEDURE FOR ANALYSIS

Composite Parts.
- Using a sketch, divide the area into its composite parts and indicate the perpendicular distance from the centroid of each part to the axis.

Parallel-Axis Theorem.
- If the centroidal axis for each part does not coincide with the reference axis, the parallel-axis theorem, $I = \bar{I} + Ad^2$, must be used to determine the moment of inertia of the part about the axis. For the calculation of $\bar{I}$, use the table on the inside back cover.

Summation.
- The moment of inertia of the entire area about the axis is determined by summing the results of its composite parts about this axis.

- If a composite part has an empty region or hole, its moment of inertia is found by subtracting the moment of inertia of the hole from the moment of inertia of the entire part including the hole.

To design this T-beam, it is necessary to locate the centroidal axis of its cross-sectional area, and then find the moment of inertia of this area about the axis.

EXAMPLE 10.4

Determine the moment of inertia of the area shown in Fig. 10–8a about the x axis.

Fig. 10–8

SOLUTION

Composite Parts. The area can be obtained by *subtracting* the circle from the rectangle shown in Fig. 10–8b. The centroid of each area is located in the figure.

Parallel-Axis Theorem. The moments of inertia about the x axis are determined using the parallel-axis theorem and the moment of inertia formulae for circular and rectangular areas $I_x = \frac{1}{4}\pi r^4$ and $I_x = \frac{1}{12}bh^3$, found on the inside back cover.

Circle

$$I_x = \bar{I}_{x'} + Ad_y^2$$

$$= \frac{1}{4}\pi(25)^4 + \pi(25)^2(75)^2 = 11.4(10^6) \text{ mm}^4$$

Rectangle

$$I_x = \bar{I}_{x'} + Ad_y^2$$

$$= \frac{1}{12}(100)(150)^3 + (100)(150)(75)^2 = 112.5(10^6) \text{ mm}^4$$

Summation. The moment of inertia for the area is therefore

$$I_x = -11.4(10^6) + 112.5(10^6)$$

$$= 101(10^6) \text{ mm}^4 \qquad \qquad \text{Ans.}$$

EXAMPLE 10.5

Fig. 10–9

Determine the moments of inertia for the cross-sectional area of the member shown in Fig. 10–9a about the x and y centroidal axes.

SOLUTION

Composite Parts. The cross section can be subdivided into the three rectangular areas A, B, and D shown in Fig. 10–9b. For the calculation, the centroid of each of these rectangles is located in the figure.

Parallel-Axis Theorem. From the table on the inside back cover, or Example 10.1, the moment of inertia of a rectangle about its centroidal axis is $\bar{I} = \frac{1}{12}bh^3$. Hence, using the parallel-axis theorem for rectangles A and D, the calculations are as follows:

Rectangles A and D

$$I_x = \bar{I}_{x'} + Ad_y^2 = \frac{1}{12}(100)(300)^3 + (100)(300)(200)^2$$
$$= 1.425(10^9) \text{ mm}^4$$

$$I_y = \bar{I}_{y'} + Ad_x^2 = \frac{1}{12}(300)(100)^3 + (100)(300)(250)^2$$
$$= 1.90(10^9) \text{ mm}^4$$

Rectangle B

$$I_x = \frac{1}{12}(600)(100)^3 = 0.05(10^9) \text{ mm}^4$$

$$I_y = \frac{1}{12}(100)(600)^3 = 1.80(10^9) \text{ mm}^4$$

Summation. The moments of inertia for the entire cross section are thus

$$I_x = 2[1.425(10^9)] + 0.05(10^9)$$
$$= 2.90(10^9) \text{ mm}^4 \qquad Ans.$$

$$I_y = 2[1.90(10^9)] + 1.80(10^9)$$
$$= 5.60(10^9) \text{ mm}^4 \qquad Ans.$$

FUNDAMENTAL PROBLEMS

F10–5. Determine the moment of inertia of the cross-sectional area of the beam about the centroidal x and y axes.

F10–7. Determine the moment of inertia of the cross-sectional area of the channel with respect to the y axis.

Prob. F10–5

Prob. F10–7

F10–6. Determine the moment of inertia of the cross-sectional area of the beam about the centroidal x and y axes.

F10–8. Determine the moment of inertia of the cross-sectional area of the T-beam with respect to the x' axis passing through the centroid of the cross section.

Prob. F10–6

Prob. F10–8

PROBLEMS

10–25. Determine the moment of inertia of the composite area about the x axis.

10–26. Determine the moment of inertia of the composite area about the y axis.

*10–28. Determine the moment of inertia of the beam's cross-sectional area with respect to the x and y centroidal axes.

Probs. 10–25/26

Prob. 10–28

10–27. Determine the moment of inertia of the "Z" section with respect to the x and y centroidal axes.

10–29. Determine the location $\bar{y}$ of the centroid of the channel's cross-sectional area and then calculate the moment of inertia of the area about this axis.

Prob. 10–27

Prob. 10–29

10–30. Determine the moment of inertia of the cross-sectional area of the beam about the x axis.

10–31. Determine the moment of inertia of the cross-sectional area of the beam about the y axis.

10–34. Determine the moment of inertia of the cross-sectional area of the beam about the y axis.

10–35. Determine $\bar{y}$, which locates the centroidal axis x' for the cross-sectional area of the T-beam, and then find the moment of inertia about the x' axis.

Probs. 10–30/31

Probs. 10–34/35

*10–32. Determine the moment of inertia I_x of the area about the x axis.

10–33. Determine the moment of inertia I_y of the area about the y axis.

*10–36. Determine the distance $\bar{x}$ to the centroid for the beam's cross-sectional area, then find $\bar{I}_y$.

10–37. Determine the moment of inertia of the beam's cross-sectional area about the x axis.

Probs. 10–32/33

Probs. 10–36/37

10–38. Determine the moment of inertia of the area about the x axis.

10–39. Determine the moment of inertia of the area about the y axis.

10–42. Determine the moment of inertia of the beam's cross-sectional area about the x axis.

10–43. Determine the moment of inertia of the beam's cross-sectional area about the y axis.

*10–44. Determine the distance $\bar{y}$ to the centroid C of the beam's cross-sectional area and then find the moment of inertia $\bar{I}_{x'}$ about the x' axis.

10–45. Determine the distance $\bar{x}$ to the centroid C of the beam's cross-sectional area and then find the moment of inertia $\bar{I}_{y'}$ about the y' axis.

Probs. 10–38/39

Probs. 10–42/43/44/45

*10–40. Determine the distance $\bar{x}$ to the centroid for the beam's cross-sectional area, then find $\bar{I}_y$.

10–41. Determine the moment of inertia of the beam's cross-sectional area about the x axis.

10–46. Determine $\bar{y}$ which locates the centroid C of the cross-sectional area of the wing channel, and then determine the moment of inertia $\bar{I}_{x'}$ about the centroidal x' axis. Neglect the effect of rounded corners. The material has a uniform thickness of 0.5 in.

10–47. Determine the moment of inertia of the cross-sectional area of the wing channel about the y axis.

Probs. 10–40/41

Probs. 10–46/47

PROBLEMS   525

***10–48.** The assembly consists of two pipes and a plate. Determine the inner radii r of the pipes so that the moment of inertia of the assembly about the x axis is $I_x = 527$ in^4.

Prob. 10–48

10–49. Determine the polar moment of inertia of the shaft's cross-sectional area about the center O.

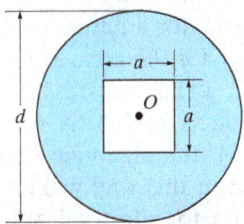

Prob. 10–49

10–50. Locate the centroid $\bar{y}$ of the cross section and determine the moment of inertia of the section about the x' axis.

Prob. 10–50

10–51. Determine the moment of inertia for the beam's cross-sectional area about the x' axis that passes through the centroid C of the cross section.

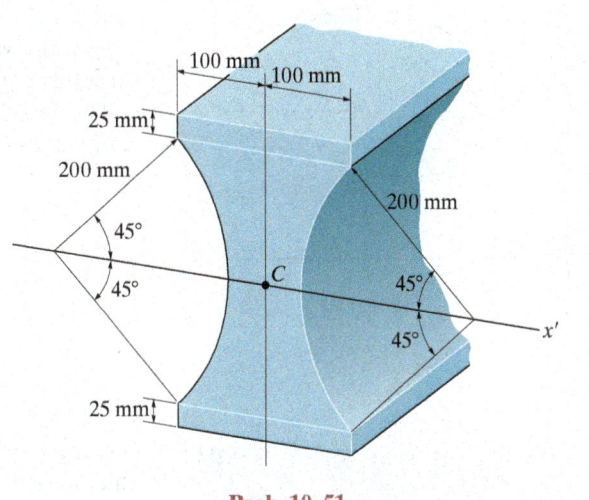

Prob. 10–51

*10–52. Determine the moment of inertia of the area about the x axis.

10–53. Determine the moment of inertia of the area about the y axis.

Probs. 10–52/53

*10.5 PRODUCT OF INERTIA FOR AN AREA

It will be shown in the next section that the property of an area, called the *product of inertia*, is required in order to determine the *maximum* and *minimum* moments of inertia for the area. These maximum and minimum values are important since they are needed for designing structural and mechanical members such as beams, columns, and shafts.

The **product of inertia** of the area in Fig. 10–10 with respect to the x and y axes is defined as

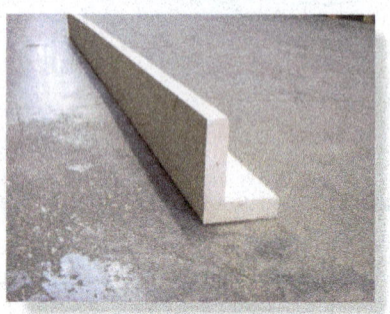

Fig. 10–10

$$I_{xy} = \int_A xy \, dA \qquad (10\text{–}7)$$

Example 10.6 shows how we can evaluate this integral using a single integration by choosing an element dA having a differential size or thickness in only one direction.

Like the moment of inertia, the product of inertia has units of length raised to the fourth power, e.g., m^4, mm^4 or ft^4, in^4. However, since x or y may be negative, the product of inertia may either be positive, negative, or zero, depending on the location and orientation of the coordinate axes. For example, the product of inertia I_{xy} for an area will be *zero* if either the x or y axis is an axis of *symmetry* for the area, as in Fig. 10–11. Here every element dA located at point (x, y) has a corresponding element dA located at $(x, -y)$. Since the products of inertia for these elements are, respectively, $xy \, dA$ and $-xy \, dA$, the algebraic sum or integration of all the elements that are chosen in this way will cancel each other. Consequently, the product of inertia for the total area becomes zero. It also follows from the definition of I_{xy} that the "sign" of this quantity depends on the quadrant where the area is located. As shown in Fig. 10–12, if the area is rotated from one quadrant to another, the sign of I_{xy} will change.

The effectiveness of this beam to resist bending can be determined once its moments of inertia and its product of inertia are known.

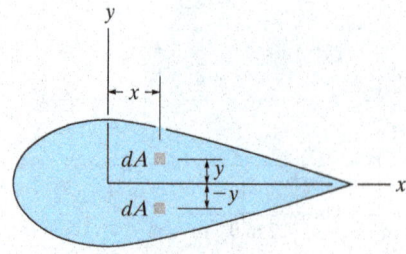

Fig. 10–11

10.5 Product of Inertia for an Area   527

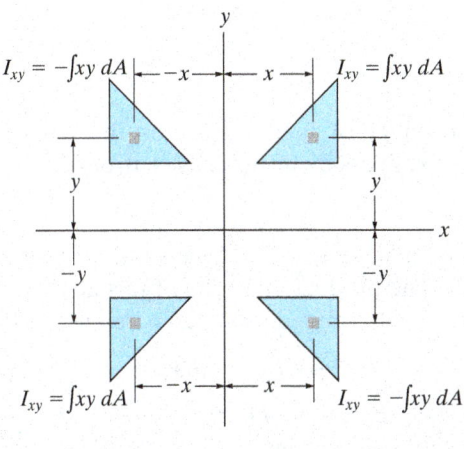

Fig. 10–12

Parallel-Axis Theorem. Consider the shaded area shown in Fig. 10–13, where x' and y' represent a set of axes passing through the *centroid* of the area, and x and y represent a corresponding set of parallel axes. Since the product of inertia of dA with respect to the x and y axes is $dI_{xy} = (x' + d_x)(y' + d_y)\, dA$, then for the entire area,

$$I_{xy} = \int_A (x' + d_x)(y' + d_y)\, dA$$

$$= \int_A x'y'\, dA + d_x \int_A y'\, dA + d_y \int_A x'\, dA + d_x d_y \int_A dA$$

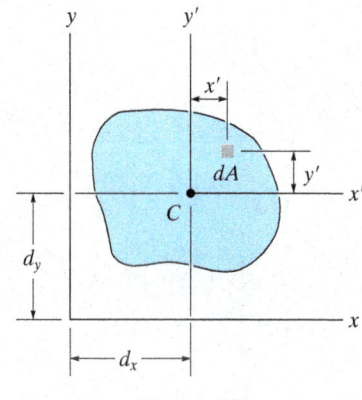

Fig. 10–13

The first term on the right represents the product of inertia for the area with respect to the centroidal axes, $\bar{I}_{x'y'}$. The integrals in the second and third terms are zero since the moments of the area are taken about the centroidal axis. Realizing that the fourth integral represents the entire area A, the parallel-axis theorem for the product of inertia then becomes

$$\boxed{I_{xy} = \bar{I}_{x'y'} + A d_x d_y} \qquad (10\text{–}8)$$

It is important that the *algebraic signs* for d_x and d_y be maintained when applying this equation.

EXAMPLE 10.6

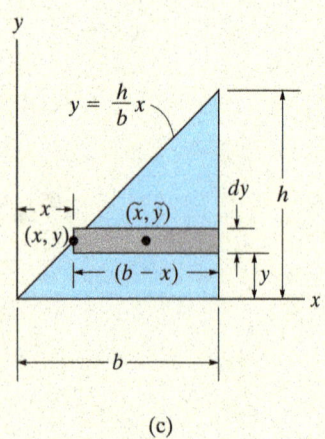

Fig. 10–14

Determine the product of inertia $\bar{I}_{x'y'}$ for the triangle shown in Fig. 10–14a about axes passing through its centroid C.

SOLUTION I

The differential element in Fig. 10–14b has a thickness dx, and so its area is $dA = y\,dx$. The product of inertia of this element with respect to the x and y axes is determined using the parallel-axis theorem.

$$dI_{xy} = d\bar{I}_{x'y'} + dA\,\tilde{x}\,\tilde{y}$$

where $\tilde{x}$ and $\tilde{y}$ locate the *centroid* of the element or the origin of the x', y' axes. (See Fig. 10–13.) Since $d\bar{I}_{x'y'} = 0$, due to symmetry, and $\tilde{x} = x$, $\tilde{y} = y/2$, then

$$dI_{xy} = 0 + (y\,dx)x\left(\frac{y}{2}\right) = \left(\frac{h}{b}x\,dx\right)x\left(\frac{h}{2b}x\right) = \frac{h^2}{2b^2}x^3\,dx$$

Integrating with respect to x from $x = 0$ to $x = b$ yields

$$I_{xy} = \frac{h^2}{2b^2}\int_0^b x^3\,dx = \frac{b^2h^2}{8} \qquad \textit{Ans.}$$

SOLUTION II

The differential element in Fig. 10–14c has a thickness dy, and so its area is $dA = (b - x)\,dy$. The *centroid* is located at point $\tilde{x} = x + (b - x)/2 = (b + x)/2$, $\tilde{y} = y$. The product of inertia of the element is then

$$dI_{xy} = d\bar{I}_{x'y'} + dA\,\tilde{x}\,\tilde{y} = 0 + (b - x)\,dy\left(\frac{b + x}{2}\right)y$$

$$= \left(b - \frac{b}{h}y\right)dy\left[\frac{b + (b/h)y}{2}\right]y = \frac{1}{2}y\left(b^2 - \frac{b^2}{h^2}y^2\right)dy$$

Integrating with respect to y from $y = 0$ to $y = h$ yields

$$I_{xy} = \frac{1}{2}\int_0^h y\left(b^2 - \frac{b^2}{h^2}y^2\right)dy = \frac{b^2h^2}{8}$$

Therefore,

$$I_{xy} = \bar{I}_{x'y'} + A\,d_x\,d_y$$

$$\frac{b^2h^2}{8} = \bar{I}_{x'y'} + \left(\frac{1}{2}bh\right)\left(\frac{2b}{3}\right)\left(\frac{h}{3}\right)$$

$$\bar{I}_{x'y'} = -\frac{b^2h^2}{72} \qquad \textit{Ans.}$$

EXAMPLE 10.7

Determine the product of inertia for the cross-sectional area of the member shown in Fig. 10–15a, about the x and y centroidal axes.

(a)   (b)

Fig. 10–15

SOLUTION

As in Example 10.5, the cross section can be subdivided into three composite rectangular areas A, B, and D, Fig. 10–15b. The coordinates for the centroid of each of these rectangles are shown in the figure. Due to symmetry, the product of inertia of *each rectangle* is *zero* about a set of x', y' axes that passes through the centroid of each rectangle. Applying the parallel-axis theorem, we have

Rectangle A

$$I_{xy} = \bar{I}_{x'y'} + A d_x d_y$$
$$= 0 + (300)(100)(-250)(200) = -1.50(10^9) \text{ mm}^4$$

Rectangle B

$$I_{xy} = \bar{I}_{x'y'} + A d_x d_y$$
$$= 0 + 0 = 0$$

Rectangle D

$$I_{xy} = \bar{I}_{x'y'} + A d_x d_y$$
$$= 0 + (300)(100)(250)(-200) = -1.50(10^9) \text{ mm}^4$$

The product of inertia for the entire cross section is therefore

$$I_{xy} = -1.50(10^9) + 0 - 1.50(10^9) = -3.00(10^9) \text{ mm}^4 \quad Ans.$$

NOTE: This negative result occurred because rectangles A and D have centroids located with negative x and negative y coordinates, respectively.

*10.6 MOMENTS OF INERTIA FOR AN AREA ABOUT INCLINED AXES

In structural and mechanical design, it is sometimes necessary to calculate the moments and product of inertia I_u, I_v, and I_{uv} for an area with respect to a set of inclined u and v axes when the values for θ, I_x, I_y, and I_{xy} are known, Fig. 10–16. To do this we will use *transformation equations* which relate the x, y and u, v coordinates. From the figure, these equations are

$$u = x \cos \theta + y \sin \theta$$
$$v = y \cos \theta - x \sin \theta$$

Fig. 10–16

Therefore, the moments and product of inertia of dA about the u and v axes become

$$dI_u = v^2\, dA = (y \cos \theta - x \sin \theta)^2\, dA$$
$$dI_v = u^2\, dA = (x \cos \theta + y \sin \theta)^2\, dA$$
$$dI_{uv} = uv\, dA = (x \cos \theta + y \sin \theta)(y \cos \theta - x \sin \theta)\, dA$$

Expanding each expression and integrating, realizing that $I_x = \int y^2\, dA$, $I_y = \int x^2\, dA$, and $I_{xy} = \int xy\, dA$, we obtain

$$I_u = I_x \cos^2 \theta + I_y \sin^2 \theta - 2I_{xy} \sin \theta \cos \theta$$
$$I_v = I_x \sin^2 \theta + I_y \cos^2 \theta + 2I_{xy} \sin \theta \cos \theta$$
$$I_{uv} = I_x \sin \theta \cos \theta - I_y \sin \theta \cos \theta + I_{xy}(\cos^2 \theta - \sin^2 \theta)$$

Using the trigonometric identities $\sin 2\theta = 2 \sin \theta \cos \theta$ and $\cos 2\theta = \cos^2 \theta - \sin^2 \theta$, we can simplify these expressions, in which case

$$\boxed{\begin{aligned} I_u &= \frac{I_x + I_y}{2} + \frac{I_x - I_y}{2} \cos 2\theta - I_{xy} \sin 2\theta \\ I_v &= \frac{I_x + I_y}{2} - \frac{I_x - I_y}{2} \cos 2\theta + I_{xy} \sin 2\theta \\ I_{uv} &= \frac{I_x - I_y}{2} \sin 2\theta + I_{xy} \cos 2\theta \end{aligned}} \quad (10\text{–}9)$$

If the first and second equations are added together, we can show that the polar moment of inertia about the z axis passing through point O is, as expected, *independent* of the orientation of the u and v axes; i.e.,

$$J_O = I_u + I_v = I_x + I_y$$

Principal Moments of Inertia.

Equations 10–9 indicate that I_u, I_v, and I_{uv} depend on the angle of inclination, θ, of the u, v axes, and here it is important to determine the orientation of these axes about which the moments of inertia for the area are maximum and minimum. This particular set of axes is called the *principal axes* and the corresponding moments of inertia with respect to these axes are called the *principal moments of inertia*. In general, there is a set of principal axes for every location of point O; however, for structural and mechanical design, this point is to be located at the centroid of the area.

The angle which defines the orientation of the principal axes can be found by differentiating the first of Eqs. 10–9 with respect to θ and setting the result equal to zero.

$$\frac{dI_u}{d\theta} = -2\left(\frac{I_x - I_y}{2}\right)\sin 2\theta - 2I_{xy}\cos 2\theta = 0$$

Therefore, at $\theta = \theta_p$,

$$\tan 2\theta_p = \frac{-I_{xy}}{(I_x - I_y)/2} \qquad (10\text{–}10)$$

The two roots θ_{p_1} and θ_{p_2} are $90°$ apart, and so they each specify the inclination of one of the principal axes. We can find the sine and cosine of $2\theta_{p_1}$ and $2\theta_{p_2}$ using the ratios from the triangles shown in Fig. 10–17, which are based on Eq. 10–10. Substituting each ratio into the first or second of Eqs. 10–9 and simplifying, we obtain

$$I_{\substack{\max \\ \min}} = \frac{I_x + I_y}{2} \pm \sqrt{\left(\frac{I_x - I_y}{2}\right)^2 + I_{xy}^2} \qquad (10\text{–}11)$$

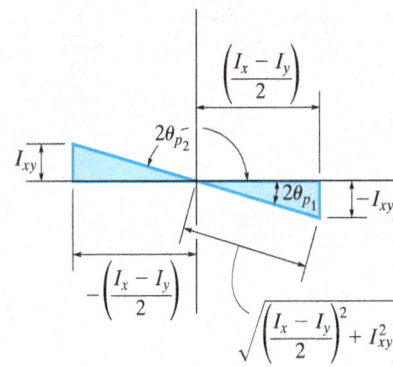

Fig. 10–17

Depending on the sign chosen, this result gives the maximum or minimum moment of inertia for the area. Furthermore, if the trigonometric ratios are substituted into the third of Eqs. 10–9, it can be shown that $I_{uv} = 0$; that is, the *product of inertia with respect to the principal axes is zero.* Since this is the case for any symmetrical axis, it therefore follows that *any symmetrical axis represents a principal axis of inertia for the area.*

EXAMPLE 10.8

Fig. 10–18

Determine the principal moments of inertia and the orientation of the principal axes of inertia for the cross-sectional area of the member shown in Fig. 10–18a with respect to an axis passing through the centroid.

SOLUTION

The moments and product of inertia of the cross section with respect to the x, y axes have been determined in Examples 10.5 and 10.7. The results are

$$I_x = 2.90(10^9) \text{ mm}^4 \quad I_y = 5.60(10^9) \text{ mm}^4 \quad I_{xy} = -3.00(10^9) \text{ mm}^4$$

Using Eq. 10–10, the angles of inclination of the principal axes u and v, Fig. 10-18b, are

$$\tan 2\theta_p = \frac{-I_{xy}}{(I_x - I_y)/2} = \frac{-[-3.00(10^9)]}{[2.90(10^9) - 5.60(10^9)]/2} = -2.22$$

$$2\theta_p = -65.8° \text{ and } 114.2°$$

Thus,

$$\theta_{p_2} = -32.9° \quad \text{and} \quad \theta_{p_1} = 57.1° \quad \textit{Ans.}$$

The principal moments of inertia with respect to these axes are determined from Eq. 10–11. Hence,

$$I_{\substack{\max \\ \min}} = \frac{I_x + I_y}{2} \pm \sqrt{\left(\frac{I_x - I_y}{2}\right)^2 + I_{xy}^2}$$

$$= \frac{2.90(10^9) + 5.60(10^9)}{2}$$

$$\pm \sqrt{\left[\frac{2.90(10^9) - 5.60(10^9)}{2}\right]^2 + [-3.00(10^9)]^2}$$

$$I_{\substack{\max \\ \min}} = 4.25(10^9) \pm 3.29(10^9)$$

or

$$I_{\max} = 7.54(10^9) \text{ mm}^4 \quad I_{\min} = 0.960(10^9) \text{ mm}^4 \quad \textit{Ans.}$$

The maximum moment of inertia, $I_{\max} = 7.54(10^9) \text{ mm}^4$, occurs with respect to the u axis since by inspection most of the cross-sectional area is farthest away from this axis. Also, we can show this by substituting the data with $\theta = 57.1°$ into the first of Eqs. 10–9 and solving for I_u, which will give $I_{\max}$.

*10.7 MOHR'S CIRCLE FOR MOMENTS OF INERTIA

Equations 10–9 to 10–11 have a graphical solution that is convenient to use and generally easy to remember. Squaring the first and third of Eqs. 10–9 and adding, it is found that

$$\left(I_u - \frac{I_x + I_y}{2}\right)^2 + I_{uv}^2 = \left(\frac{I_x - I_y}{2}\right)^2 + I_{xy}^2$$

Here I_x, I_y, and I_{xy} are *known constants*. Thus, the above equation may be written in compact form as

$$(I_u - a)^2 + I_{uv}^2 = R^2$$

When this equation is plotted on a set of axes that represent the respective moment of inertia and the product of inertia, as shown in Fig. 10–19, the resulting graph represents a *circle* of radius

$$R = \sqrt{\left(\frac{I_x - I_y}{2}\right)^2 + I_{xy}^2}$$

and having its center located at point $(a, 0)$, where $a = (I_x + I_y)/2$. The circle so constructed is called **Mohr's circle**, named after the German engineer Otto Mohr (1835–1918).

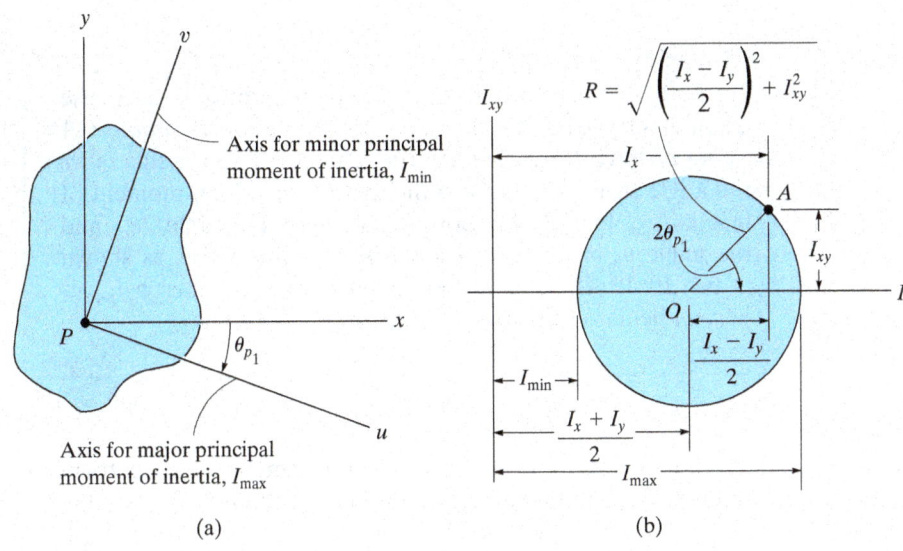

Fig. 10–19

534   CHAPTER 10   MOMENTS OF INERTIA

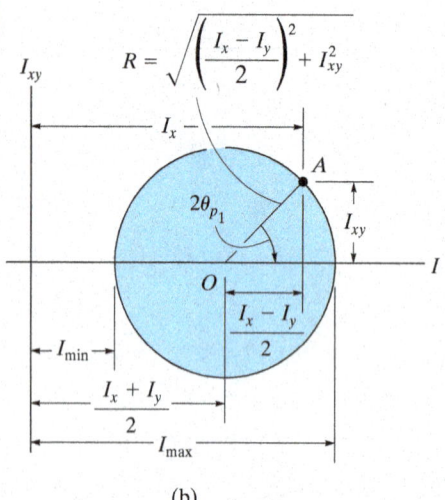

Fig. 10–19 (Repeated)

PROCEDURE FOR ANALYSIS

The main purpose in using Mohr's circle here is to have a convenient means for finding the principal moments of inertia for an area. The following procedure provides a method for doing this.

Determine I_x, I_y, and I_{xy}.

- Establish the x, y axes and determine I_x, I_y, and I_{xy}, Fig. 10–19a.

Construct the Circle.

- Construct a rectangular coordinate system such that the horizontal axis represents the moment of inertia I, and the vertical axis represents the product of inertia I_{xy}, Fig. 10–19b.

- Determine the center of the circle, O, which is located at a distance $(I_x + I_y)/2$ from the origin, and plot the reference point A having coordinates (I_x, I_{xy}). Remember, I_x is always positive, whereas I_{xy} can be either positive or negative.

- Connect the reference point A with the center of the circle and determine the distance OA by trigonometry. This distance represents the radius of the circle, Fig. 10–19b. Finally, draw the circle.

Principal Moments of Inertia.

- The points where the circle intersects the I axis give the values of the principal moments of inertia I_{min} and I_{max}. Notice that, as expected, the *product of inertia will be zero at these points*, Fig. 10–19b.

Principal Axes.

- To find the orientation of the major principal axis, use trigonometry to find the angle $2\theta_{p_1}$, *measured from the radius OA to the positive I axis*, Fig. 10–19b. This angle represents *twice* the angle from the x axis to the axis of maximum moment of inertia I_{max}, Fig. 10–19a. Both the angle on the circle, $2\theta_{p_1}$, and the angle θ_{p_1} must be measured in the same sense, as shown in Fig. 10–19. The axis for minimum moment of inertia I_{min} is perpendicular to the axis for I_{max}.

Using trigonometry, the above procedure can be verified to be in accordance with the equations developed in Sec. 10.6.

EXAMPLE 10.9

Using Mohr's circle, determine the principal moments of inertia and the orientation of the major principal axes of inertia for the cross-sectional area of the member shown in Fig. 10–20a, with respect to an axis passing through the centroid.

SOLUTION

Determine I_x, I_y, I_{xy}. The moments and product of inertia have been determined in Examples 10.5 and 10.7 with respect to the x, y axes shown in Fig. 10–20a. The results are $I_x = 2.90(10^9)$ mm^4, $I_y = 5.60(10^9)$ mm^4, and $I_{xy} = -3.00(10^9)$ mm^4.

Construct the Circle. The I and I_{xy} axes are shown in Fig. 10–20b. The center of the circle, O, lies at a distance $(I_x + I_y)/2 = (2.90 + 5.60)/2 = 4.25$ from the origin. When the reference point $A(I_x, I_{xy})$ or $A(2.90, -3.00)$ is connected to point O, the radius OA is determined from the triangle OBA using the Pythagorean theorem.

$$OA = \sqrt{(1.35)^2 + (-3.00)^2} = 3.29$$

The circle is constructed in Fig. 10–20c.

Principal Moments of Inertia. The circle intersects the I axis at points $(7.54, 0)$ and $(0.960, 0)$. Hence,

$$I_{max} = (4.25 + 3.29)10^9 = 7.54(10^9) \text{ mm}^4 \quad Ans.$$
$$I_{min} = (4.25 - 3.29)10^9 = 0.960(10^9) \text{ mm}^4 \quad Ans.$$

Principal Axes. As shown in Fig. 10–20c, the angle $2\theta_{p_1}$ is determined from the circle by measuring counterclockwise from OA to the positive I axis. Hence,

$$2\theta_{p_1} = 180° - \sin^{-1}\left(\frac{|BA|}{|OA|}\right) = 180° - \sin^{-1}\left(\frac{3.00}{3.29}\right) = 114.2°$$

The principal axis for $I_{max} = 7.54(10^9)$ mm^4 is therefore oriented at an angle $\theta_{p_1} = 57.1°$, measured *counterclockwise*, from the *positive* x axis to the *positive* u axis. The v axis is perpendicular to this axis. The results are shown in Fig. 10–20d.

Fig. 10–20

PROBLEMS

10–54. Determine the product of inertia of the thin strip of area with respect to the x and y axes. The strip is oriented at an angle θ from the x axis. Assume that $t \ll l$.

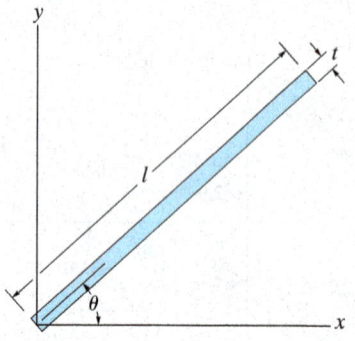

Prob. 10–54

10–55. Determine the product of inertia of the area with respect to the x and y axes.

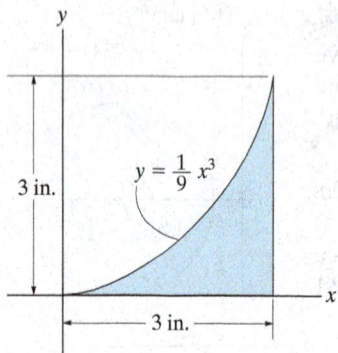

Prob. 10–55

*10–56. Determine the product of inertia for the parabola with respect to the x and y axes.

Prob. 10–56

10–57. Determine the product of inertia of the area with respect to the x and y axes, and then use the parallel-axis theorem to find the product of inertia of the area with respect to the centroidal x' and y' axes.

Prob. 10–57

10–58. Determine the product of inertia for the parabolic area with respect to the x and y axes.

Prob. 10–58

10–59. Determine the product of inertia of the area with respect to the x and y axes.

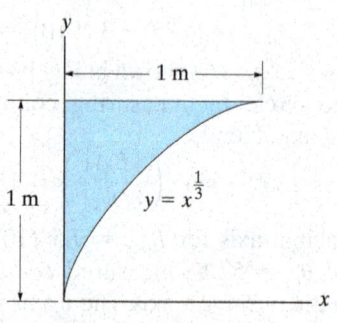

Prob. 10–59

*10–60. Determine the product of inertia of the area with respect to the x and y axes.

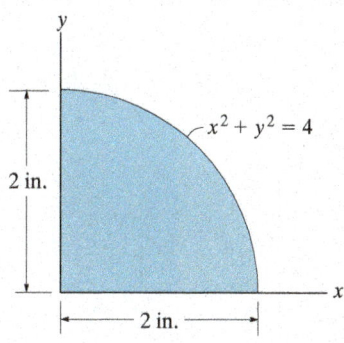

Prob. 10–60

10–61. Determine the product of inertia of the area with respect to the x and y axes.

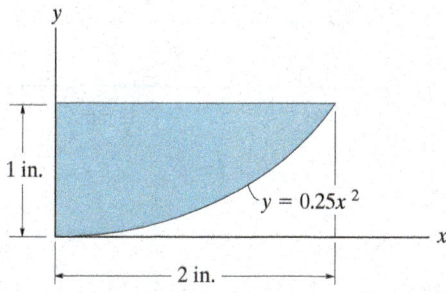

Prob. 10–61

10–62. Determine the product of inertia of the beam's cross-sectional area with respect to the x and y axes.

Prob. 10–62

10–63. Determine the moments of inertia of the beam's cross-sectional area with respect to the u and v axes.

Prob. 10–63

*10–64. Locate the centroid $(\bar{x}, \bar{y})$ of the beam's cross-sectional area, and then determine the product of inertia of this area with respect to the centroidal x' and y' axes.

Prob. 10–64

10–65. Determine the product of inertia for the area with respect to the x and y axes.

Prob. 10–65

10–66. Determine the product of inertia of the cross-sectional area of the member with respect to the x and y axes.

Prob. 10–66

10–67. Determine the location ($\bar{x}, \bar{y}$) to the centroid C of the angle's cross-sectional area, and then determine the product of inertia with respect to the x' and y' axes.

Prob. 10–67

*10–68. Locate the centroid, $\bar{y}$, and determine the orientation of the principal centroidal axes for the composite area. What are the moments of inertia with respect to these axes?

Prob. 10–68

10–69. Determine the principal moments of the composite area with respect to a set of principal axes that have their origin located at the centroid C. Use the equations developed in Sec. 10.7. $I_{xy} = -15.84(10^6)$ mm⁴.

*10–72. Determine the moments of inertia I_u and I_v and the product of inertia I_{uv} for the rectangular area. The u and v axes pass through the centroid C. Take $\theta = 30°$.

10–73. Determine the moments of inertia I_u and I_v and the product of inertia I_{uv} for the rectangular area. The u and v axes pass through the centroid C. Take $\theta = 20°$.

Prob. 10–69

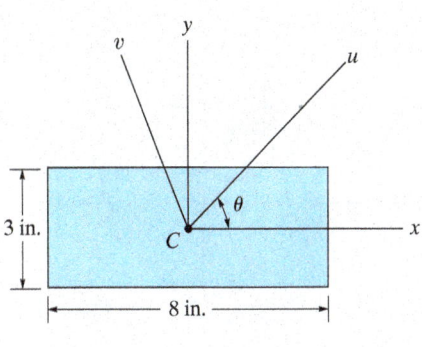

Probs. 10–72/73

10–70. Locate the position $\bar{x}, \bar{y}$ for the centroid C of the beam's cross-sectional area, and then determine the product of inertia with respect to the x' and y' axes.

10–71. Determine the product of inertia of the beam's cross-sectional area with respect to the x and y axes.

10–74. Determine the principal moments of inertia of the composite area with respect to a set of principal axes that have their origin located at the centroid C. Use the equations developed in Sec.10.7. $I_{xy} = 15.84(10^6)$ mm⁴.

10–75. Solve Prob. 10–74 using Mohr's circle.

Probs. 10–70/71

Probs. 10–74/75

*10–76. Determine the moments of inertia and the product of inertia of the beam's cross-sectional area with respect to the u and v axes.

10–77. Solve Prob. 10–76 using Mohr's circle.

*10–80. Determine the directions of the principal axes with origin located at point O, and the principal moments of inertia for the area about these axes.

10–81. Solve Prob. 10–80 using Mohr's circle.

Probs. 10–76/77

Probs. 10–80/81

10–78. The cross section of an airplane wing has the moments of inertia and product of inertia about the x and y axes passing through the centroid C of $\bar{I}_x = 450$ in^4, $\bar{I}_y = 1730$ in^4, $\bar{I}_{xy} = 138$ in^4. Determine the orientation of the principal axes and the principal moments of inertia.

10–79. Solve Prob. 10–78 using Mohr's circle.

10–82. Determine the directions of the principal axes with origin located at point O and the principal moments of inertia of the area about these axes.

Probs. 10–78/79

Prob. 10–82

10–83. Solve Prob. 10–82 using Mohr's circle.

10.8 MASS MOMENT OF INERTIA

The mass moment of inertia of a body is a measure of the body's resistance to angular acceleration. Since it is used in dynamics to study rotational motion, methods for its calculation will now be discussed.

Consider the rigid body shown in Fig. 10–21. We define the *mass moment of inertia* of the body about the z axis as

$$I = \int_m r^2 \, dm \qquad (10\text{–}12)$$

Here r is the perpendicular distance from the axis to the arbitrary element dm. Since the formulation involves r, the value of I is *unique* for each axis about which it is determined. The axis which is generally chosen, however, passes through the body's mass center G. Common units used for its measurement are $kg \cdot m^2$ or $slug \cdot ft^2$.

If the body consists of material having a density ρ, then $dm = \rho \, dV$, Fig. 10–22a. Substituting this into Eq. 10–12, the body's moment of inertia is then determined using *volume elements* for integration; i.e.,

$$I = \int_V r^2 \rho \, dV \qquad (10\text{–}13)$$

For most applications, ρ will be a *constant*, and so this term may be factored out of the integral, and the integration is then purely a function of geometry.

$$I = \rho \int_V r^2 \, dV \qquad (10\text{–}14)$$

Fig. 10–21

(a)

Fig. 10–22

Fig. 10–22 (cont.)

PROCEDURE FOR ANALYSIS

If a body is symmetrical with respect to an axis, as in Fig. 10–22, then its mass moment of inertia about the axis can be determined by using a single integration. Shell and disk elements are used for this purpose.

Shell Element.

- If a *shell element* having a height z, radius y, and thickness dy is chosen for integration, Fig. 10–22b, then its volume is $dV = (2\pi y)(z)\, dy$.

- This element can be used in Eq. 10–13 or 10–14 for determining the moment of inertia I_z since the *entire element*, due to its "thinness," lies at the *same* perpendicular distance $r = y$ from the z axis (see Example 10.10).

Disk Element.

- If a disk element having a radius y and a thickness dz is chosen for integration, Fig. 10–22c, then its volume is $dV = (\pi y^2)\, dz$.

- In this case the element is *finite* in the radial direction, and consequently its points *do not* all lie at the *same radial distance* r from the z axis. As a result, Eqs. 10–13 or 10–14 *cannot* be used to determine I_z. Instead, to perform the integration using this element, it is first necessary to determine the moment of inertia *of the element* about the z axis and then integrate this result (see Example 10.11).

EXAMPLE 10.10

Determine the mass moment of inertia of the cylinder shown in Fig. 10–23a about the z axis. The density of the material is ρ.

(a)   (b)

Fig. 10–23

SOLUTION

Shell Element. This problem will be solved using the *shell element* in Fig. 10–23b and thus only a single integration is required. The volume of the element is $dV = (2\pi r)(h)\,dr$, and so its mass is $dm = \rho\,dV = \rho(2\pi h r\,dr)$. Since the *entire element* lies at the same distance r from the z axis, the moment of inertia of the element is

$$dI_z = r^2\,dm = \rho 2\pi h r^3\,dr$$

Integrating over the entire cylinder yields

$$I_z = \int_m r^2\,dm = \rho 2\pi h \int_0^R r^3\,dr = \frac{\rho \pi}{2} R^4 h$$

Since the mass of the cylinder is

$$m = \int_m dm = \rho 2\pi h \int_0^R r\,dr = \rho \pi h R^2$$

then

$$I_z = \frac{1}{2} m R^2 \qquad\qquad Ans.$$

EXAMPLE 10.11

Determine the mass moment of inertia of the solid in Fig. 10–24a about the y axis. Take $\rho = 5$ slug/ft³.

Fig. 10–24

SOLUTION

Disk Element. The moment of inertia will be determined using a *disk element*, as shown in Fig. 10–24b. Here the element intersects the curve at the arbitrary point (x, y) and has a mass

$$dm = \rho \, dV = \rho(\pi x^2) \, dy$$

Although all points on the element are *not* located at the same distance from the y axis, it is still possible to determine the moment of inertia dI_y *of the element* about the y axis. In the previous example it was shown that the moment of inertia of a homogeneous cylinder about its longitudinal axis is $I = \frac{1}{2}mR^2$, where m and R are the mass and radius of the cylinder. Since the height of the cylinder is not involved in this formula, we can also use this result for a disk. Thus, for the disk element in Fig. 10–24b, we have

$$dI_y = \frac{1}{2}(dm)x^2 = \frac{1}{2}[\rho(\pi x^2) \, dy]x^2$$

Substituting $x = y^2$, $\rho = 5$ slug/ft³, and integrating with respect to y, from $y = 0$ to $y = 1$ ft, yields the moment of inertia for the entire solid.

$$I_y = \frac{5\pi}{2} \int_0^{1 \text{ ft}} x^4 \, dy = \frac{5\pi}{2} \int_0^{1 \text{ ft}} y^8 \, dy = 0.873 \text{ slug} \cdot \text{ft}^2 \quad \textit{Ans.}$$

Fig. 10–25

Parallel-Axis Theorem. If the moment of inertia of a body about an axis passing through the body's mass center is known, then the moment of inertia about any other *parallel axis* can be determined by using the **parallel-axis theorem**. To derive this theorem, consider the body shown in Fig. 10–25. The z' axis passes through the mass center G, whereas the corresponding *parallel z axis* lies a distance d away. Selecting the differential element of mass dm, which is located at point (x', y'), and using the Pythagorean theorem, $r^2 = (d + x')^2 + y'^2$, the moment of inertia of the body about the z axis is

$$I = \int_m r^2 \, dm = \int_m [(d + x')^2 + y'^2] \, dm$$

$$= \int_m (x'^2 + y'^2) \, dm + 2d \int_m x' \, dm + d^2 \int_m dm$$

Since $r'^2 = x'^2 + y'^2$, the first integral represents I_G. The second integral is equal to *zero*, since the z' axis passes through the body's mass center, i.e., $\int x' \, dm = \bar{x} \int dm = 0$ since $\bar{x} = 0$. Finally, the third integral is the total mass m of the body. Therefore, the moment of inertia about the z axis becomes

$$\boxed{I = I_G + md^2} \qquad (10\text{–}15)$$

where

I_G = moment of inertia about the z' axis passing through the mass center G

m = mass of the body

d = distance between the parallel axes

Radius of Gyration.
Occasionally, the moment of inertia of a body about a specified axis is reported in handbooks using the *radius of gyration*, k. This value has units of length, and when it and the body's mass m are known, the moment of inertia can be determined from the equation

$$I = mk^2 \quad \text{or} \quad k = \sqrt{\frac{I}{m}} \qquad (10\text{--}16)$$

Note the *similarity* between the definition of k in this formula and r in the equation $dI = r^2\, dm$, which defines the moment of inertia of a differential element of mass dm of the body about an axis.

Composite Bodies.
If a body is constructed from a number of simple shapes such as disks, spheres, and rods, the moment of inertia of the body about any axis z can be determined by adding algebraically the moments of inertia of all the composite shapes calculated about the same axis. Algebraic addition is necessary since a composite part must be considered as a negative quantity if it has already been included within another part—as in the case of a "hole" subtracted from a solid plate. Also, the parallel-axis theorem is needed for the calculations if the center of mass of each composite part does not lie on the z axis. For calculations, a table of some simple shapes is given on the inside back cover.

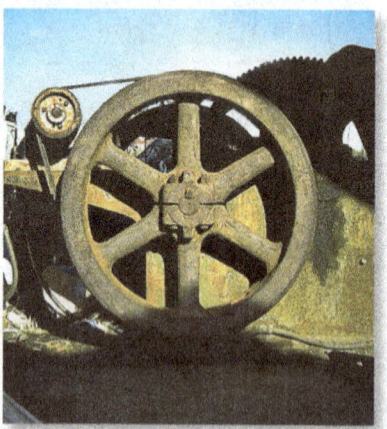

This flywheel, which operates a metal cutter, has a large moment of inertia about its center. Once it begins rotating it is difficult to stop it and therefore a uniform motion can be effectively transferred to the cutting blade.

EXAMPLE 10.12

If the plate shown in Fig. 10–26a has a density of 8000 kg/m³ and a thickness of 10 mm, determine its mass moment of inertia about an axis perpendicular to the page and passing through point O.

(a) (b)

Fig. 10–26

SOLUTION

The plate consists of two composite parts, the 0.25-m-radius disk *minus* a 0.125-m-radius disk, Fig. 10–26b. The moment of inertia about O can be determined by finding the moment of inertia of each of these parts about O and then *algebraically* adding the results. The calculations are performed by using the parallel-axis theorem.

Disk. From the inside back cover, the moment of inertia of a disk about an axis perpendicular to the plane of the disk and passing through G is $I_G = \frac{1}{2}mr^2$, and since the mass center is 0.25 m from point O, we have

$m_d = \rho_d V_d = 8000 \text{ kg/m}^3 \, [\pi(0.25 \text{ m})^2(0.01 \text{ m})] = 15.71 \text{ kg}$

$(I_O)_d = \frac{1}{2}m_d r_d^2 + m_d d^2$

$\quad = \frac{1}{2}(15.71 \text{ kg})(0.25 \text{ m})^2 + (15.71 \text{ kg})(0.25 \text{ m})^2$

$\quad = 1.473 \text{ kg} \cdot \text{m}^2$

Hole. For the smaller disk (hole), we have

$m_h = \rho_h V_h = 8000 \text{ kg/m}^3 \, [\pi(0.125 \text{ m})^2(0.01 \text{ m})] = 3.93 \text{ kg}$

$(I_O)_h = \frac{1}{2}m_h r_h^2 + m_h d^2$

$\quad = \frac{1}{2}(3.93 \text{ kg})(0.125 \text{ m})^2 + (3.93 \text{ kg})(0.25 \text{ m})^2$

$\quad = 0.276 \text{ kg} \cdot \text{m}^2$

The moment of inertia of the plate about O is therefore

$I_O = (I_O)_d - (I_O)_h$

$\quad = 1.473 \text{ kg} \cdot \text{m}^2 - 0.276 \text{ kg} \cdot \text{m}^2$

$\quad = 1.20 \text{ kg} \cdot \text{m}^2$ *Ans.*

EXAMPLE 10.13

Fig. 10–27

The pendulum in Fig. 10–27 consists of two thin rods each having a weight of 10 lb. Determine the pendulum's mass moment of inertia about an axis passing through (a) the pin at O, and (b) the mass center G of the pendulum.

SOLUTION

Part (a). Using the table on the inside back cover, the moment of inertia of rod OA about an axis perpendicular to the page and passing through the end point O of the rod is $I_O = \frac{1}{3}ml^2$. Hence,

$$(I_{OA})_O = \frac{1}{3}ml^2 = \frac{1}{3}\left(\frac{10 \text{ lb}}{32.2 \text{ ft/s}^2}\right)(2 \text{ ft})^2 = 0.414 \text{ slug} \cdot \text{ft}^2$$

This same value can also be determined using $I_G = \frac{1}{12}ml^2$ and the parallel-axis theorem; i.e.,

$$(I_{OA})_O = \frac{1}{12}ml^2 + md^2 = \frac{1}{12}\left(\frac{10 \text{ lb}}{32.2 \text{ ft/s}^2}\right)(2 \text{ ft})^2 + \frac{10 \text{ lb}}{32.2 \text{ ft/s}^2}(1 \text{ ft})^2$$

$$= 0.414 \text{ slug} \cdot \text{ft}^2$$

For rod BC we have

$$(I_{BC})_O = \frac{1}{12}ml^2 + md^2 = \frac{1}{12}\left(\frac{10 \text{ lb}}{32.2 \text{ ft/s}^2}\right)(2 \text{ ft})^2 + \frac{10 \text{ lb}}{32.2 \text{ ft/s}^2}(2 \text{ ft})^2$$

$$= 1.346 \text{ slug} \cdot \text{ft}^2$$

The moment of inertia of the pendulum about O is therefore

$$I_O = 0.414 + 1.346 = 1.76 \text{ slug} \cdot \text{ft}^2 \qquad \textit{Ans.}$$

Part (b). The mass center G will be located relative to the pin at O. Assuming this distance to be $\bar{y}$, Fig. 10–27, and using the formula for determining the mass center, we have

$$\bar{y} = \frac{\Sigma \tilde{y} m}{\Sigma m} = \frac{1(10/32.2) + 2(10/32.2)}{(10/32.2) + (10/32.2)} = 1.50 \text{ ft}$$

The moment of inertia I_G may be calculated in the same manner as I_O, which requires successive applications of the parallel-axis theorem in order to transfer the moments of inertia of rods OA and BC to G. A more direct solution, however, involves applying the parallel-axis theorem using the result for I_O determined above; i.e.,

$$I_O = I_G + md^2; \qquad 1.76 \text{ slug} \cdot \text{ft}^2 = I_G + \left(\frac{20 \text{ lb}}{32.2 \text{ ft/s}^2}\right)(1.50 \text{ ft})^2$$

$$I_G = 0.362 \text{ slug} \cdot \text{ft}^2 \qquad \textit{Ans.}$$

PROBLEMS

*10–84. Determine the moment of inertia of the thin ring about the z axis. The ring has a mass m.

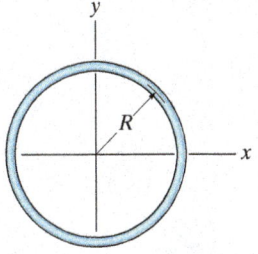

Prob. 10–84

10–85. Determine the moment of inertia of the ellipsoid with respect to the x axis and express the result in terms of the mass m of the ellipsoid. The material has a constant density ρ.

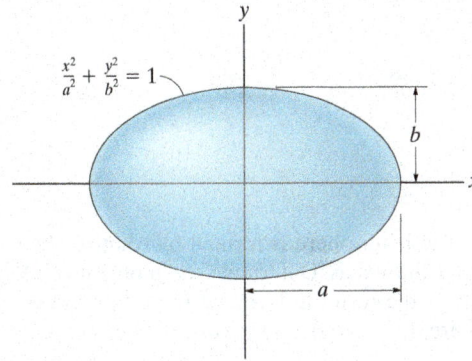

Prob. 10–85

10–86. The paraboloid is formed by revolving the shaded area around the x axis. Determine the radius of gyration k_x of the paraboloid. The density of the material is $\rho = 5 \text{ Mg/m}^3$.

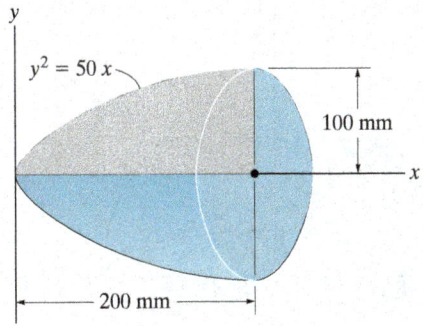

Prob. 10–86

10–87. The paraboloid is formed by revolving the shaded area around the x axis. Determine the moment of inertia about the x axis and express the result in terms of the mass m of the paraboloid. The material has a constant density ρ.

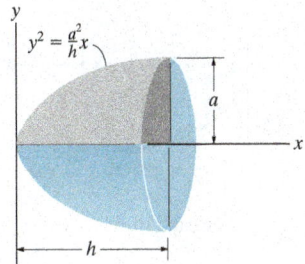

Prob. 10–87

*10–88. Determine the moment of inertia of the homogeneous triangular prism with respect to the y axis. Express the result in terms of the mass m of the prism. *Hint*: For integration, use thin plate elements parallel to the x–y plane having a thickness of dz.

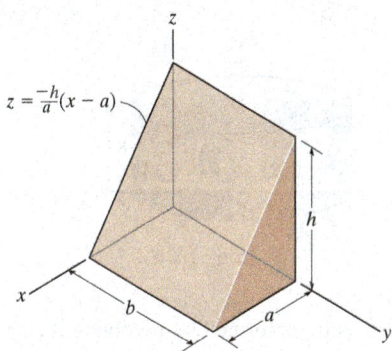

Prob. 10–88

10–89. Determine the moment of inertia of the semi-ellipsoid with respect to the x axis and express the result in terms of the mass m of the semi-ellipsoid. The material has a constant density ρ.

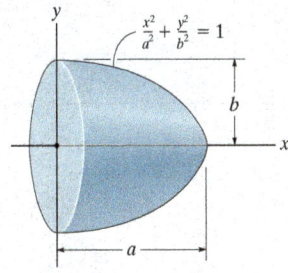

Prob. 10–89

10–90. Determine the radius of gyration k_x of the solid formed by revolving the shaded area about the x axis. The density of the material is ρ.

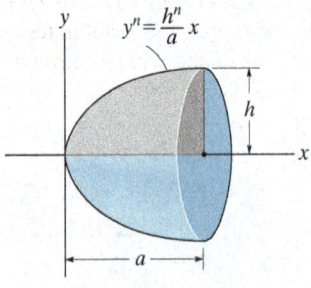

Prob. 10–90

10–91. The concrete shape is formed by rotating the shaded area about the y axis. Determine the moment of inertia I_y. The specific weight of concrete is $\gamma = 150 \text{ lb/ft}^3$.

Prob. 10–91

*10–92. The solid is formed by revolving the shaded area around the x axis. Determine the radius of gyration k_x. The density of the material is $\rho = 5 \text{ Mg/m}^3$.

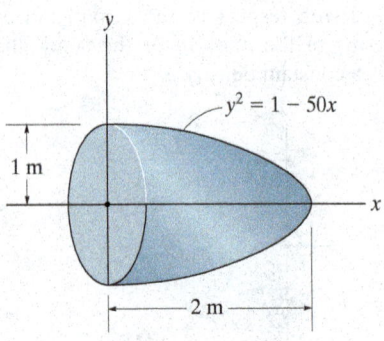

Prob. 10–92

10–93. The right circular cone is formed by revolving the shaded area around the x axis. Determine the moment of inertia I_x and express the result in terms of the total mass m of the cone. The cone has a constant density ρ.

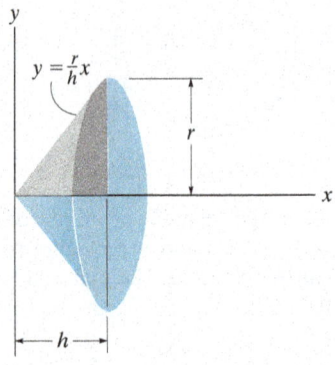

Prob. 10–93

10–94. The hemisphere is formed by rotating the shaded area about the y axis. Determine the moment of inertia I_y and express the result in terms of the total mass m of the hemisphere. The material has a constant density ρ.

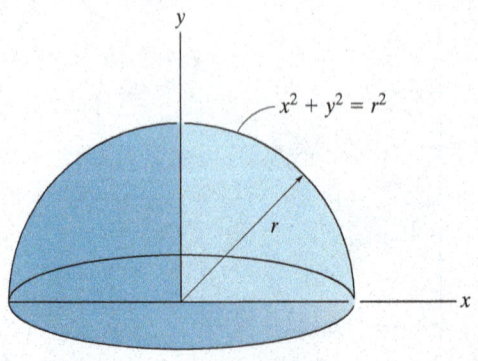

Prob. 10–94

10–95. Determine the moment of inertia of the wire triangle about an axis perpendicular to the page and passing through point O. Also, locate the mass center G and determine the moment of inertia about an axis perpendicular to the page and passing through point G. The wire has a mass of 0.3 kg/m. Neglect the size of the ring at O.

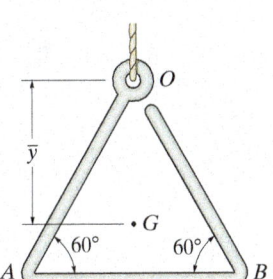

Prob. 10–95

*10–96. The pendulum consists of an 8-kg circular disk A, a 2-kg circular disk B, and a 4-kg slender rod. Determine the radius of gyration of the pendulum about an axis perpendicular to the page and passing through point O.

Prob. 10–96

10–97. Determine the moment of inertia I_z of the frustum of the cone which has a conical depression. The material has a density of $\rho = 200$ kg/m^3.

Prob. 10–97

10–98. The slender rods have a weight of 3 lb/ft. Determine the moment of inertia of the assembly about an axis perpendicular to the page and passing through point A.

Prob. 10–98

10–99. Determine the moment of inertia of the thin plate about an axis perpendicular to the page and passing through the pin at O. The plate has a hole in its center. Its thickness is 50 mm, and the material has a density of $\rho = 50$ kg/m^3.

Prob. 10–99

*10–100. The pendulum consists of a plate having a weight of 12 lb and a slender rod having a weight of 4 lb. Determine the radius of gyration of the pendulum about an axis perpendicular to the page and passing through point O.

Prob. 10–100

10–101. If the large ring, small ring and each of the spokes weigh 100 lb, 15 lb, and 20 lb, respectively, determine the mass moment of inertia of the wheel about an axis perpendicular to the page and passing through point A.

10–102. Determine the moment of inertia of the crank about the x-axis. The material is steel having a specific weight of $\gamma_{st} = 490\ \text{lb/ft}^3$.

Prob. 10–102

10–103. Each of the three slender rods has a mass m. Determine the moment of inertia of the assembly about an axis that is perpendicular to the page and passes through the center point O.

Prob. 10–101

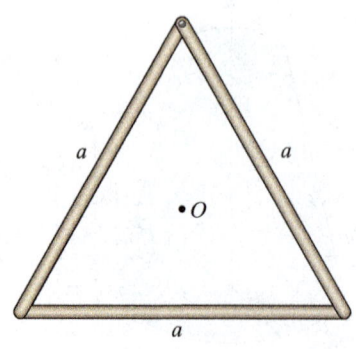

Prob. 10–103

PROBLEMS 553

*10–104. The thin plate has a mass per unit area of 10 kg/m². Determine its mass moment of inertia about the *y* axis.

10–105. The thin plate has a mass per unit area of 10 kg/m². Determine its mass moment of inertia about the *z* axis.

*10–108. The pendulum consists of two slender rods *AB* and *OC* which have a mass of 3 kg/m. The thin plate has a mass of 12 kg/m². Determine the location $\bar{y}$ of the center of mass *G* of the pendulum, then calculate the moment of inertia of the pendulum about an axis perpendicular to the page and passing through *G*.

Probs. 10–104/105

Prob. 10–108

10–106. Determine the moment of inertia of the wheel about the *x* axis that passes through the center of mass *G*. The material has a specific weight of $\gamma = 90$ lb/ft³.

10–107. Determine the moment of inertia of the wheel about the *x'* axis that passes through point *O*. The material has a specific weight of $\gamma = 90$ lb/ft³.

10–109. Determine the moment of inertia I_z of the frustum of the cone which has a conical depression. The material has a density of $\rho = 200$ kg/m³.

Probs. 10–106/107

Prob. 10–109

CHAPTER REVIEW

Area Moment of Inertia

The *area moment of inertia* represents the second moment of the area about an axis. It is frequently used in formulas related to the strength and stability of structural members or mechanical elements.

If the area shape is irregular but can be described mathematically, then a differential element must be selected and integration over the entire area must be performed to determine the moment of inertia.

$$I_y = \int_A x^2 \, dA$$

Parallel-Axis Theorem

If the moment of inertia for an area is known about its centroidal axis, then its moment of inertia about a parallel axis can be determined using the parallel-axis theorem.

$$I = \bar{I} + Ad^2$$

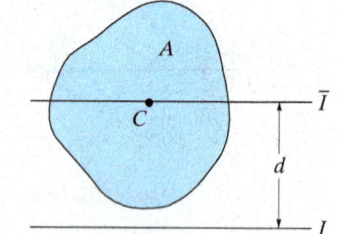

Composite Area

If an area is a composite of common shapes, as found on the inside back cover, then its moment of inertia is equal to the algebraic sum of the moments of inertia of each of its parts.

 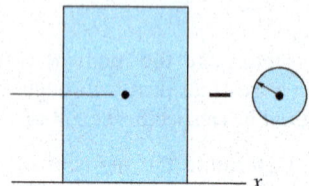

Product of Inertia

The *product of inertia* of an area is used in formulas to determine the orientation of an axis about which the moment of inertia for the area is a maximum or minimum.

$$I_{xy} = \int_A xy \, dA$$

If the product of inertia for an area is known with respect to its centroidal x', y' axes, then its value can be determined with respect to any x, y axes using the parallel-axis theorem for the product of inertia.

$$I_{xy} = \bar{I}_{x'y'} + A d_x d_y$$

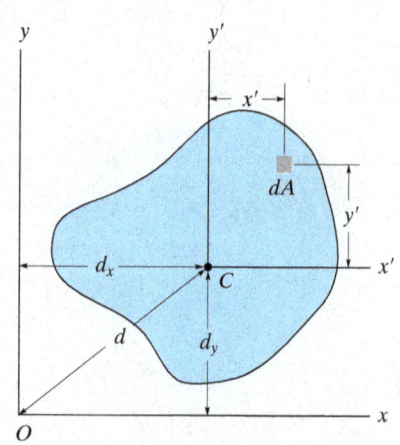

Principal Moments of Inertia

Provided the moments of inertia, I_x and I_y, and the product of inertia, I_{xy}, are known, then the transformation formulas, or Mohr's circle, can be used to determine the maximum and minimum or *principal moments of inertia* for the area, as well as finding the orientation of the principal axes of inertia.

$$I_{\substack{max \\ min}} = \frac{I_x + I_y}{2} \pm \sqrt{\left(\frac{I_x - I_y}{2}\right)^2 + I_{xy}^2}$$

$$\tan 2\theta_p = \frac{-I_{xy}}{(I_x - I_y)/2}$$

Mass Moment of Inertia

The *mass moment of inertia* is a property of a body that measures its resistance to a change in its rotation. It is defined as the "second moment" of the mass elements of the body about an axis.

$$I = \int_m r^2\, dm$$

For homogeneous bodies having axial symmetry, the mass moment of inertia can be determined by a single integration, using a disk or shell element.

$$I = \rho \int_V r^2\, dV$$

The mass moment of inertia of a composite body is determined by using tabular values of its composite shapes, found on the inside back cover, along with the parallel-axis theorem.

$$I = I_G + md^2$$

REVIEW PROBLEMS

R10–1. Determine the moment of inertia for the area about the x axis.

Prob. R10–1

R10–2. Determine the moment of inertia for the area about the x axis.

Prob. R10–2

R10–3. Determine the moment of inertia of the area about the y axis.

Prob. R10–3

R10–4. Determine the moment of inertia of the area about the x axis. Use the parallel-axis theorem to find the area moment of inertia about the x' axis that passes through the centroid C of the area. $\bar{y} = 120$ mm.

Prob. R10–4

R10–5. Determine the area moment of inertia of the triangular area about (a) the x axis, and (b) the centroidal x' axis.

R10–7. Determine the moment of inertia of the beam's cross-sectional area about the x axis which passes through the centroid C.

Prob. R10–5

Prob. R10–7

R10–6. Determine the product of inertia of the area with respect to the x and y axes.

R10–8. Determine the mass moment of inertia I_x of the body and express the result in terms of the mass m of the body. The density of the material is ρ.

Prob. R10–6

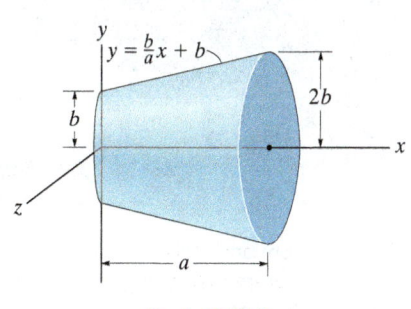

Prob. R10–8

CHAPTER 11

Equilibrium and stability of this scissors lift as a function of its position can be determined using the methods of work and energy, which are explained in this chapter.

VIRTUAL WORK

CHAPTER OBJECTIVES

- To introduce the principle of virtual work and show how it applies to finding the equilibrium configuration of a system of pin-connected members.
- To establish the potential-energy function and use the potential-energy method to investigate the type of equilibrium or stability of a rigid body or system of pin-connected members.

11.1 DEFINITION OF WORK

The principle of virtual work was proposed by the Swiss mathematician Jean Bernoulli in the eighteenth century. It provides an alternative method for solving problems involving the equilibrium of a particle, a rigid body, or a system of connected rigid bodies. Before we discuss this principle, however, we must first define the work produced by a force and by a couple moment.

Work of a Force. A force does work when it undergoes a displacement. For example, the force **F** in Fig. 11–1a is given a differential displacement $d\mathbf{r}$, and if θ is the angle between the force and this

560   CHAPTER 11   VIRTUAL WORK

(a)

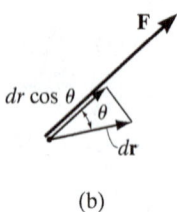

(b)

Fig. 11–1

displacement, then the component of **F** in the direction of the displacement is $F \cos \theta$. The **work** produced by **F** is then defined as

$$dU = F\, dr \cos \theta$$

This expression is also the product of the force F and the component of displacement in the direction of the force, $dr \cos \theta$, Fig. 11–1b. If we use the definition of the dot product (Eq. 2–11), the work can also be written as

$$dU = \mathbf{F} \cdot d\mathbf{r}$$

Notice that work is a **scalar**, and like other scalar quantities, it has a magnitude that can either be *positive* or *negative*.

In the SI system, the unit of work is a **joule** (J), where $1\text{ J} = 1\text{ N}\cdot\text{m}$. In the FPS system, it is the foot-pound (ft·lb). The moment of a force has this same combination of units; however, the concepts of moment and work are in no way related. A moment is a vector quantity, whereas work is a scalar.

Work of a Couple Moment.
The rotation of a couple moment also produces work. Consider the rigid body in Fig. 11–2, which is acted upon by the couple forces **F** and $-$**F** that produce a couple moment **M** having a magnitude $M = Fr$. When the body undergoes the differential displacement shown, points A and B move $d\mathbf{r}_A$ and $d\mathbf{r}_B$ to their final positions A' and B', respectively. Since $d\mathbf{r}_B = d\mathbf{r}_A + d\mathbf{r}'$, this movement can be thought of as a *translation* $d\mathbf{r}_A$, where A and B move to A' and B'', and a *rotation* about A', where the body rotates through the angle $d\theta$ about A. The couple forces do no work during the translation $d\mathbf{r}_A$ because each force undergoes the same amount of displacement in opposite directions, thus canceling out the work. During rotation, however, **F** is displaced $d\mathbf{r}' = r\, d\theta$, and so it does work $dU = F\, dr' = Fr\, d\theta$. Since $M = Fr$, the work of the couple moment **M** is therefore

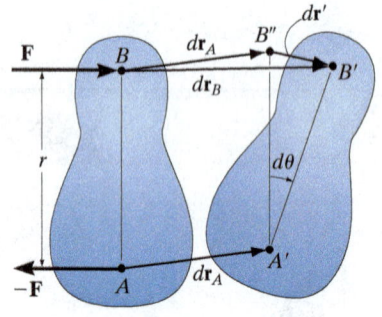

Fig. 11–2

$$dU = M\, d\theta$$

If **M** and $d\theta$ have the same sense, the work is *positive*; however, if they have the opposite sense, the work will be *negative*.

Virtual Work. The definitions of the work of a force and a couple have been presented in terms of *actual movements* expressed by differential displacements having magnitudes of dr and $d\theta$. Consider now an *imaginary* or ***virtual movement*** of a body in static equilibrium. Here the displacement or rotation is assumed to occur, but *does not actually exist*. These movements are first-order differential quantities and will be denoted by the symbols δr and $\delta \theta$ (delta r and delta θ), respectively. The ***virtual work*** done by a force having this virtual displacement is therefore

$$\delta U = F \cos \theta \, \delta r \qquad (11\text{-}1)$$

Similarly, when a couple undergoes a virtual rotation in the plane of the couple forces, the ***virtual work*** is

$$\delta U = M \, \delta \theta \qquad (11\text{-}2)$$

11.2 PRINCIPLE OF VIRTUAL WORK

The ***principle of virtual work*** states that if a body is in equilibrium, then the algebraic sum of the virtual work done by all the forces and couple moments acting on the body is zero for any virtual displacement of the body. Thus,

$$\delta U = 0 \qquad (11\text{-}3)$$

For example, consider the free-body diagram of the particle (ball) that rests on the floor, Fig. 11–3. If we "imagine" the ball to be displaced downward a virtual amount δy, then the weight does positive virtual work, $W \, \delta y$, and the normal force does negative virtual work, $-N \, \delta y$. For equilibrium the total virtual work must be zero, so that $\delta U = W \, \delta y - N \, \delta y = (W - N) \, \delta y = 0$. Since $\delta y \neq 0$, then $N = W$ as required by applying $\Sigma F_y = 0$.

Fig. 11–3

In a similar manner, we can also apply the virtual-work equation $\delta U = 0$ to a rigid body subjected to a coplanar force system. Here, separate virtual translations in the x and y directions, and a virtual rotation about an axis perpendicular to the x–y plane that passes through an arbitrary point O, will correspond to the three equilibrium equations, $\Sigma F_x = 0$, $\Sigma F_y = 0$, and $\Sigma M_O = 0$. When writing these equations, it is *not necessary* to include the work done by the *internal forces* acting within the body since a rigid body *does not deform* when subjected to an external loading, and furthermore, when the body moves through any virtual displacement, the internal forces occur in equal but opposite collinear pairs, so that the corresponding work done by each pair of forces will cancel.

To demonstrate an application, consider the simply supported beam in Fig. 11–4a. When the beam is given a virtual rotation $\delta\theta$ about point B, Fig. 11–4b, the only forces that do work are $\mathbf{P}$ and $\mathbf{A}_y$. Since $\delta y = l\,\delta\theta$ and $\delta y' = (l/2)\,\delta\theta$, the virtual-work equation for this case is $\delta U = A_y(l\,\delta\theta) - P(l/2)\,\delta\theta = (A_y l - Pl/2)\,\delta\theta = 0$. Since $\delta\theta \neq 0$, then $A_y = P/2$. Excluding $\delta\theta$, notice that the terms in parentheses actually represent the application of $\Sigma M_B = 0$.

Fig. 11–4

11.3 PRINCIPLE OF VIRTUAL WORK FOR A SYSTEM OF CONNECTED RIGID BODIES

The method of virtual work is particularly effective for solving equilibrium problems that involve a system of several *connected* rigid bodies, such as the ones shown in Fig. 11–5. Each of these systems is said to have only one degree of freedom since the arrangement of the links can be completely specified using only one coordinate $(q =)\theta$. In other words, with this single coordinate and the length of the members, we can, using trigonometry, locate the position of the forces **F** and **P**.

In this book, we will only consider the application of the principle of virtual work to systems containing one degree of freedom. Because they are less complicated, they will serve as a way to approach the solution of more complex problems involving systems with many degrees of freedom. The procedure for solving problems involving a system of frictionless connected rigid bodies follows.*

Fig. 11–5

IMPORTANT POINTS

- A force does work when it moves through a displacement in the direction of the force. A couple moment does work when it moves through a collinear rotation. Specifically, positive work is done when the force or couple moment and its displacement have the same sense of direction.

- The principle of virtual work is generally used to determine the equilibrium configuration for a system of multiple connected members.

- A virtual displacement is imaginary; i.e., it does not really happen. It is a differential displacement that is given in the positive direction of a position coordinate.

- Forces or couple moments that do not virtually displace do no virtual work.

This scissors lift has one degree of freedom. Without the need for dismembering the mechanism, the force in the hydraulic cylinder AB required to provide the lift can be determined *directly* by using the principle of virtual work.

---

*This method of applying the principle of virtual work is sometimes called the ***method of virtual displacements*** because a virtual displacement is applied in order to determine an unknown force. Although it is not used here, we can also apply the principle of virtual work as a ***method of virtual forces***. This method is often used to apply a virtual force and then determine the displacements of points on a deformable body. See R. C. Hibbeler, *Mechanics of Materials*, Pearson, Inc.

Procedure for Analysis

Free-Body Diagram.

- Draw the free-body diagram of the entire system of connected bodies and define the *coordinate q*.

- Sketch the "deflected position" of the system on the free-body diagram when the system undergoes a *positive virtual* displacement δq.

Virtual Displacements.

- Indicate *position coordinates s*, each measured from a *fixed point* on the free-body diagram. These coordinates are only directed to the forces that do work.

- Each of these coordinate axes should be *along the line of action* of the force to which it is directed, so that the virtual work in the coordinate direction can be calculated.

- Relate each of the position coordinates s to the coordinate q; then *differentiate* these expressions in order to express each virtual displacement δs in terms of δq.

Virtual-Work Equation.

- Write the *virtual-work equation* for the system assuming that, whether possible or not, each position coordinate s undergoes a *positive* virtual displacement δs. If a force or couple moment is in the same direction as the positive virtual displacement, the work is positive. Otherwise, it is negative.

- Express the work of *each* force and couple moment in the equation in terms of δq.

- Factor out this common displacement from all the terms, and solve for the unknown force, couple moment, or equilibrium position q.

EXAMPLE 11.1

Determine the angle θ for equilibrium of the two-member linkage shown in Fig. 11–6a. Each member has a mass of 10 kg.

SOLUTION

Free-Body Diagram. The system has only one degree of freedom since the location of both links can be specified by the single coordinate, $(q =) \theta$. As shown on the free-body diagram in Fig. 11–6b, when θ is given a *positive* (clockwise) virtual rotation $\delta\theta$, only the force **F** and the two 98.1-N weights do work. (The reactive forces $\mathbf{D}_x$ and $\mathbf{D}_y$ are fixed, and $\mathbf{B}_y$ does not displace along its line of action.)

Virtual Displacements. If the origin of coordinates is established at the *fixed* pin support D, then the position of **F** and **W** can be specified by the *position coordinates* x_B and y_w. In order to determine the work, note that, as required, these coordinates are along the lines of action of their associated forces. Expressing these position coordinates in terms of θ and taking the derivatives yields

$$x_B = 2(1 \cos \theta) \text{ m} \quad \delta x_B = -2 \sin \theta \, \delta\theta \text{ m} \quad (1)$$
$$y_w = \tfrac{1}{2}(1 \sin \theta) \text{ m} \quad \delta y_w = 0.5 \cos \theta \, \delta\theta \text{ m} \quad (2)$$

It is seen by the *signs* of these equations, and indicated in Fig. 11–6b, that an *increase* in θ (i.e., $\delta\theta$) causes a *decrease* in x_B and an *increase* in y_w.

Virtual-Work Equation. If the virtual displacements δx_B and δy_w are *both positive*, then the forces **W** and **F** will do positive work since the forces and their corresponding displacements have the same sense. Hence, the virtual-work equation for the displacement $\delta\theta$ is

$$\delta U = 0; \qquad W \, \delta y_w + W \, \delta y_w + F \, \delta x_B = 0 \quad (3)$$

Substituting Eqs. 1 and 2 into Eq. 3 in order to relate the virtual displacements to the common virtual displacement $\delta\theta$ yields

$$98.1(0.5 \cos \theta \, \delta\theta) + 98.1(0.5 \cos \theta \, \delta\theta) + 25(-2 \sin \theta \, \delta\theta) = 0$$

Notice that the "negative work" done by **F** (force in the opposite sense to displacement) has actually been *accounted for* in the above equation by the "negative sign" of Eq. 1. Factoring out the *common displacement* $\delta\theta$ and solving for θ, noting that $\delta\theta \neq 0$, yields

$$(98.1 \cos \theta - 50 \sin \theta) \, \delta\theta = 0$$

$$\theta = \tan^{-1} \frac{98.1}{50} = 63.0° \qquad \textit{Ans.}$$

NOTE: If this problem had been solved using the equations of equilibrium, it would be necessary to dismember the links and apply three scalar equations to *each* link. The principle of virtual work, by means of calculus, has eliminated this task so that the answer is obtained directly.

(a)

(b)

Fig. 11–6

EXAMPLE 11.2

Determine the required force P in Fig. 11–7a needed to maintain equilibrium of the scissors linkage when $\theta = 60°$. The spring is unstretched when $\theta = 30°$. Neglect the mass of the links.

SOLUTION

Free-Body Diagram. Only $\mathbf{F}_s$ and $\mathbf{P}$ do work when θ undergoes a *positive* virtual displacement $\delta\theta$, Fig. 11–7b. For the arbitrary position θ, the spring is stretched $(0.3 \text{ m}) \sin\theta - (0.3 \text{ m}) \sin 30°$, so that

$$F_s = ks = 5000 \text{ N/m}\,[(0.3 \text{ m}) \sin\theta - (0.3 \text{ m}) \sin 30°]$$

$$= (1500 \sin\theta - 750) \text{ N}$$

Virtual Displacements. The position coordinates, x_B and x_D, measured from the *fixed point A*, are used to locate $\mathbf{F}_s$ and $\mathbf{P}$. These coordinates are along or parallel to the line of action of their corresponding forces. Expressing x_B and x_D in terms of the angle θ using trigonometry,

$$x_B = (0.3 \text{ m}) \sin\theta$$

$$x_D = 3[(0.3 \text{ m}) \sin\theta] = (0.9 \text{ m}) \sin\theta$$

Differentiating, we obtain the virtual displacements of points B and D.

$$\delta x_B = 0.3 \cos\theta\, \delta\theta \qquad (1)$$

$$\delta x_D = 0.9 \cos\theta\, \delta\theta \qquad (2)$$

(a)

(b)

Fig. 11–7

Virtual-Work Equation. Force $\mathbf{P}$ does positive work since it acts in the same direction as its positive virtual displacement. The spring force $\mathbf{F}_s$ does negative work since it acts opposite to its positive virtual displacement. Thus, the virtual-work equation becomes

$$\delta U = 0; \qquad -F_s\, \delta x_B + P\, \delta x_D = 0$$

$$-[1500 \sin\theta - 750](0.3 \cos\theta\, \delta\theta) + P(0.9 \cos\theta\, \delta\theta) = 0$$

$$[0.9P + 225 - 450 \sin\theta] \cos\theta\, \delta\theta = 0$$

Since $\cos\theta\, \delta\theta \neq 0$, then this equation requires

$$P = 500 \sin\theta - 250$$

When $\theta = 60°$,

$$P = 500 \sin 60° - 250 = 183 \text{ N} \qquad \textit{Ans.}$$

EXAMPLE 11.3

If the box in Fig. 11–8a has a mass of 10 kg, determine the couple moment M needed to maintain equilibrium when $\theta = 60°$. Neglect the mass of the members.

Fig. 11–8

SOLUTION

Free-Body Diagram. When θ undergoes a positive virtual displacement $\delta\theta$, only the couple moment **M** and the weight of the box do work, Fig. 11–8b.

Virtual Displacements. The position coordinate y_E, measured from the *fixed point B*, locates the weight, 10(9.81) N. Here,

$$y_E = (0.45 \text{ m}) \sin\theta + b$$

where b is a constant distance. Differentiating this equation, we obtain

$$\delta y_E = 0.45 \text{ m} \cos\theta \, \delta\theta \qquad (1)$$

Virtual-Work Equation. For positive virtual displacements $\delta\theta$ and δy_E the virtual-work equation becomes

$\delta U = 0;$ $\qquad M \, \delta\theta - [10(9.81) \text{ N}]\delta y_E = 0$

Substituting Eq. 1 into this equation

$$M \, \delta\theta - 10(9.81) \text{ N}(0.45 \text{ m} \cos\theta \, \delta\theta) = 0$$
$$\delta\theta(M - 44.145 \cos\theta) = 0$$

Since $\delta\theta \neq 0$, then

$$M - 44.145 \cos\theta = 0$$

It is required that $\theta = 60°$, so that

$$M = 44.145 \cos 60° = 22.1 \text{ N}\cdot\text{m} \qquad Ans.$$

EXAMPLE 11.4

(b)

Fig. 11–9

The mechanism in Fig. 11–9a supports the 50-lb cylinder. Determine the angle θ for equilibrium if the spring has an unstretched length of 2 ft when $\theta = 0°$. Neglect the mass of the members.

SOLUTION

Free-Body Diagram. When the mechanism undergoes a positive virtual displacement $\delta\theta$, Fig. 11–9b, only $\mathbf{F}_s$ and the 50-lb force do work. Since the final length of the spring is $2(1 \text{ ft} \cos \theta)$, then

$$F_s = ks = (200 \text{ lb/ft})(2 \text{ ft} - 2 \text{ ft} \cos \theta) = (400 - 400 \cos \theta) \text{ lb}$$

Virtual Displacements. The position coordinates x_D and x_E are established from the *fixed point A* to locate F_s at D and at E. The coordinate y_B, also measured from A, specifies the position of the 50-lb force at B. The coordinates can be expressed in terms of θ using trigonometry.

$$x_D = (1 \text{ ft}) \cos \theta$$
$$x_E = 3[(1 \text{ ft}) \cos \theta] = (3 \text{ ft}) \cos \theta$$
$$y_B = (2 \text{ ft}) \sin \theta$$

Differentiating, we obtain the virtual displacements of points D, E, and B as

$$\delta x_D = -1 \sin \theta \, \delta\theta \qquad (1)$$
$$\delta x_E = -3 \sin \theta \, \delta\theta \qquad (2)$$
$$\delta y_B = 2 \cos \theta \, \delta\theta \qquad (3)$$

Virtual-Work Equation. The virtual-work equation is written as if all virtual displacements are positive, so that

$$\delta U = 0; \qquad F_s \, \delta x_E + 50 \, \delta y_B - F_s \, \delta x_D = 0$$

$$(400 - 400 \cos \theta)(-3 \sin \theta \, \delta\theta) + 50(2 \cos \theta \, \delta\theta)$$
$$- (400 - 400 \cos \theta)(-1 \sin \theta \, \delta\theta) = 0$$

$$\delta\theta(800 \sin \theta \cos \theta - 800 \sin \theta + 100 \cos \theta) = 0$$

Since $\delta\theta \neq 0$, then

$$800 \sin \theta \cos \theta - 800 \sin \theta + 100 \cos \theta = 0$$

Solving by trial and error,

$$\theta = 34.9° \qquad \textit{Ans.}$$

FUNDAMENTAL PROBLEMS

F11–1. Determine the required magnitude of force **P** to maintain equilibrium of the linkage at $\theta = 60°$. Each link has a mass of 20 kg.

Prob. F11–1

F11–2. Determine the magnitude of force **P** required to hold the 50-kg smooth rod in equilibrium at $\theta = 60°$.

Prob. F11–2

F11–3. The linkage is subjected to a force of $P = 2$ kN. Determine the angle θ for equilibrium. The spring is unstretched when $\theta = 0°$. Neglect the mass of the links.

Prob. F11–3

F11–4. The linkage is subjected to a force of $P = 6$ kN. Determine the angle θ for equilibrium. The spring is unstretched at $\theta = 60°$. Neglect the mass of the links.

Prob. F11–4

F11–5. Determine the angle θ where the 50-kg bar is in equilibrium. The spring is unstretched at $\theta = 60°$.

Prob. F11–5

F11–6. The scissors linkage is subjected to a force of $P = 150$ N. Determine the angle θ for equilibrium. The spring is unstretched at $\theta = 0°$. Neglect the mass of the links.

Prob. F11–6

PROBLEMS

11–1. Determine the tension in cable AC. The lamp weighs 10 lb.

11–3. The disk has a weight of 10 lb and is subjected to a vertical force $P = 8$ lb and a couple moment $M = 8$ lb·ft. Determine the disk's rotation θ if the end of the spring wraps around the periphery of the disk as the disk turns. The spring is originally unstretched.

Prob. 11–1

Prob. 11–3

11–2. The scissors jack supports a load **P**. Determine the axial force in the screw necessary for equilibrium when the jack is in the position θ. Each of the four links has a length L and is pin connected at its center. Points B and D can move horizontally.

*11–4. If a force of $P = 5$ lb is applied to the handle of the mechanism, determine the force the screw exerts on the cork of the bottle. The screw is attached to the pin at A and passes through the collar that rests on the bottle neck at B.

Prob. 11–2

Prob. 11–4

11-5. A 5-kg uniform serving table is supported on each side by pairs of two identical links, *AB* and *CD*, and springs *CE*. If the bowl has a mass of 1 kg, determine the angle θ where the table is in equilibrium. The springs each have a stiffness of $k = 200$ N/m and are unstretched when $\theta = 90°$. Neglect the mass of the links.

11-6. A 5-kg uniform serving table is supported on each side by pairs of two identical links, *AB* and *CD*, and springs *CE*. If the bowl has a mass of 1 kg and is in equilibrium when $\theta = 45°$, determine the stiffness k of each spring. The springs are unstretched when $\theta = 90°$. Neglect the mass of the links.

Probs. 11–5/6

11-7. When $\theta = 20°$, the 50-lb uniform block compresses the two vertical springs 4 in. If the uniform links *AB* and *CD* each weigh 10 lb, determine the magnitude of the applied couple moments **M** needed to maintain equilibrium when $\theta = 20°$.

Prob. 11–7

***11-8.** The bar is supported by the spring and smooth collar that allows the spring to be always perpendicular to the bar for any angle θ. If the unstretched length of the spring is l_0, determine the force *P* needed to hold the bar in the equilibrium position θ. Neglect the weight of the bar.

Prob. 11–8

11-9. Determine the force **P** that must be applied perpendicular to the handle in order to hold the mechanism in equilibrium for any angle θ of rod *CD*. There is a couple moment **M** applied to the link *BA*. A smooth collar attached at *A* slips along rod *CD*.

Prob. 11–9

11–10. The thin rod of weight W rests against the smooth wall and floor. Determine the magnitude of force **P** needed to hold it in equilibrium at the angle θ.

*11–12. If the disk is subjected to a couple moment M, determine the disk's rotation θ required for equilibrium. The end of the spring wraps around the periphery of the disk as the disk turns. The spring is originally unstretched.

Prob. 11–12

11–13. Determine the horizontal force F required to maintain equilibrium of the slider mechanism when $\theta = 60°$. Set $M = 6\,\text{N}\cdot\text{m}$.

11–11. If each of the three links of the mechanism has a mass of 4 kg, determine the angle θ for equilibrium. The spring, which always remains vertical, is unstretched when $\theta = 0°$.

11–14. Determine the moment M that must be applied to the slider mechanism in order to maintain the equilibrium position $\theta = 60°$ when the horizontal force $F = 100\,\text{N}$ is applied at D.

Prob. 11–11

Probs. 11–13/14

11–15. The spring is unstretched when $\theta = 0°$. If $P = 8$ lb, determine the angle θ for equilibrium. Due to the roller guide, the spring always remains vertical. Neglect the weight of the links.

11–17. If a force $P = 40$ N is applied perpendicular to the handle of the toggle press, determine the vertical compressive force developed at C; $\theta = 30°$.

Prob. 11–15

Prob. 11–17

*__11–16.__ If each of the three links of the mechanism has a weight of 20 lb, determine the angle θ for equilibrium. The spring, which always remains vertical due to the roller guide, is unstretched when $\theta = 0°$. Set $\mathbf{P} = \mathbf{0}$.

11–18. The "Nuremberg scissors" is subjected to a horizontal force of $P = 600$ N. Determine the angle θ for equilibrium. The spring has a stiffness of $k = 15$ kN/m and is unstretched when $\theta = 15°$.

11–19. The "Nuremberg scissors" is subjected to a horizontal force of $P = 600$ N. Determine the stiffness k of the spring for equilibrium when $\theta = 60°$. The spring is unstretched when $\theta = 15°$.

Prob. 11–16

Probs. 11–18/19

*11–20. The spring has an unstretched length of 0.3 m. Determine the angle θ for equilibrium if the uniform links each have a mass of 5 kg. Set $P = 0$.

11–21. The spring has an unstretched length of 0.3 m. Determine the angle θ for equilibrium if $P = 50$ N. Neglect the weight of the links.

11–23. If vertical forces $P_1 = 40$ lb and $P_2 = 20$ lb act at C and E as shown, determine the angle θ for equilibrium. The spring is unstretched when $\theta = 45°$. Neglect the weight of the members.

*11–24. If vertical forces $P_1 = P_2 = 30$ lb act at C and E as shown, determine the angle θ for equilibrium. The spring is unstretched when $\theta = 45°$. Neglect the weight of the members.

Probs. 2–20/21

Probs. 11–23/24

11–22. The dumpster has a weight W and a center of gravity at G. Determine the force in the hydraulic cylinder needed to hold it in the general position θ.

11–25. The members of the mechanism are pin connected. If a vertical force of 800 N acts at A, determine the angle θ for equilibrium. The spring is unstretched when $\theta = 0°$. Neglect the mass of the links.

Prob. 11–22

Prob. 11–25

*11.4 CONSERVATIVE FORCES

When a force does work that depends only upon the initial and final positions of the force, and it is *independent* of the path it travels, then the force is referred to as a ***conservative force***.

Weight. Consider a block of weight **W** that travels along the path in Fig. 11–10a. When it is displaced up the path by an amount d**r**, then the work is $dU = \mathbf{W} \cdot d\mathbf{r}$ or $dU = -W(dr \cos \theta) = -W dy$, as shown in Fig. 11–10b. The work is *negative* since **W** acts in the opposite sense of dy. Thus, if the block moves up from A to B, through the vertical displacement h, the work is

$$U = -\int_0^h W\, dy = -Wh$$

Fig. 11–10

The weight of a body is therefore a conservative force, since the work done by the weight depends only on the *vertical displacement* of the body, and is independent of the path along which the body travels.

Spring Force. Now consider the linearly elastic spring in Fig. 11–11, which undergoes a displacement ds. The work done by the spring force on the block is $dU = -F_s\, ds = -ks\, ds$. The work of $\mathbf{F}_s$ when the block is displaced from $s = s_1$ to $s = s_2$ is therefore

$$U = -\int_{s_1}^{s_2} ks\, ds = -\left(\tfrac{1}{2}ks_2^2 - \tfrac{1}{2}ks_1^2\right)$$

Here the work depends only on the spring's initial and final positions, s_1 and s_2, measured from the spring's unstretched position. Since this result is independent of the path taken by the block as it moves from s_1 to s_2, then a spring force is also a *conservative force*.

Fig. 11–11

576   CHAPTER 11   VIRTUAL WORK

Friction. In contrast to a conservative force, consider the force of *friction* exerted on a sliding body by a fixed surface. The work done by the frictional force depends on the path; the longer the path, the greater the work. Consequently, frictional forces are **nonconservative**, and some of the work done by them is dissipated from the body in the form of heat.

*11.5 POTENTIAL ENERGY

A conservative force can give the body the capacity to do work. This capacity, measured as **potential energy**, depends on the location or "position" of the body measured relative to a fixed reference position or datum.

Gravitational Potential Energy. If a body is located a distance y *above* a fixed horizontal reference or datum as in Fig. 11–12, the weight of the body has *positive* gravitational potential energy V_g since the weight has the capacity of doing positive work when the body is moved back down to the datum. Likewise, if the body is located a distance y *below* the datum, V_g is *negative* since the weight does negative work when the body is moved back up to the datum. At the datum, $V_g = 0$.

Measuring y as *positive upward*, the gravitational potential energy of the body's weight is therefore

$$V_g = Wy \qquad (11\text{–}4)$$

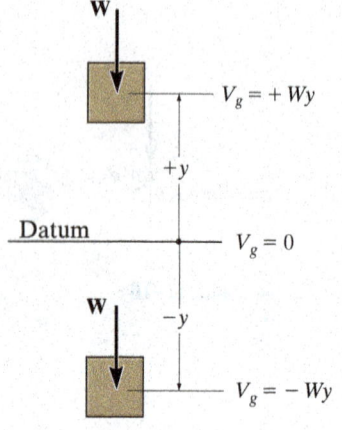

Fig. 11–12

Elastic Potential Energy. When a spring is either elongated or compressed by an amount s from its unstretched position (the datum), the energy stored in the spring is called *elastic potential energy*. It is determined from

$$V_e = \tfrac{1}{2} k s^2 \qquad (11\text{–}5)$$

This energy is always a positive quantity since the spring force acting on the attached body does *positive* work on the body as the force returns the body to the spring's unstretched position, Fig. 11–13.

$$V_e = +\tfrac{1}{2} k s^2$$

Fig. 11–13

11.5 POTENTIAL ENERGY

Potential Function. In the general case, if a body is subjected to *both* gravitational and elastic forces, the *potential energy* or **potential function** V of the body can be expressed as the algebraic sum

$$V = V_g + V_e \qquad (11\text{–}6)$$

where measurement of V depends on the location of the body with respect to a selected datum in accordance with Eqs. 11–4 and 11–5.

If a *system* of frictionless connected rigid bodies has a *single degree of freedom* that can be defined by the coordinate q, then the potential function for the system becomes $V = V(q)$. The work done by all the weight and spring forces acting on the system as it moves from q_1 to q_2, is measured by the *difference* in V; i.e.,

$$U_{1-2} = V(q_1) - V(q_2) \qquad (11\text{–}7)$$

For example, the potential function for a system consisting of a block supported by a spring, as in Fig. 11–14, can be expressed in terms of the coordinate $(q =\ )\, y$, measured from a fixed datum located at the unstretched length of the spring. Here

$$V = V_g + V_e$$
$$= -Wy + \tfrac{1}{2}ky^2 \qquad (11\text{–}8)$$

If the block moves from y_1 to y_2, then the work of $\mathbf{W}$ and $\mathbf{F}_s$ is

$$U_{1-2} = V(y_1) - V(y_2) = -W(y_1 - y_2) + \tfrac{1}{2}ky_1^2 - \tfrac{1}{2}ky_2^2$$

(a)

Fig. 11–14

*11.6 POTENTIAL-ENERGY CRITERION FOR EQUILIBRIUM

If a frictionless connected system has one degree of freedom, and its position is defined by the coordinate q, then if it displaces from q to $q + dq$, Eq. 11–7 becomes

$$dU = V(q) - V(q + dq)$$

or

$$dU = -dV$$

The counterweight at A balances the weight of the deck B of this simple lift bridge. By applying the method of potential energy we can analyze the equilibrium state of the bridge.

If the system is in equilibrium and undergoes a *virtual displacement* δq, rather than an actual displacement dq, then the above equation becomes $\delta U = -\delta V$. However, the principle of virtual work requires that $\delta U = 0$, and therefore, $\delta V = 0$, and so we can write $\delta V = (dV/dq)\delta q = 0$. Since $\delta q \neq 0$, this expression becomes

$$\boxed{\frac{dV}{dq} = 0} \qquad (11\text{–}9)$$

Hence, *when a frictionless connected system of rigid bodies is in equilibrium, the first derivative of its potential function is equal to zero.* For example, using Eq. 11–8 we can determine the equilibrium position for the spring and block in Fig. 11–14a. We have

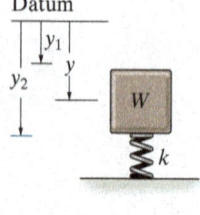

(a)

$$\frac{dV}{dy} = -W + ky = 0$$

Hence, the equilibrium position $y = y_{eq}$ is

$$y_{eq} = \frac{W}{k}$$

(b)

Fig. 11–14 (cont.)

Of course, this *same result* can be obtained by applying $\Sigma F_y = 0$ to the forces acting on the free-body diagram of the block, Fig. 11–14b.

*11.7 STABILITY OF EQUILIBRIUM CONFIGURATION

The potential function V of a system can also be used to investigate the stability of the equilibrium configuration, which is classified as *stable*, *neutral*, or *unstable*.

Stable Equilibrium. A system is said to be in **stable equilibrium** if the system has a tendency to return to its original position when it is given a small displacement. The potential energy of the system in this case is at its *minimum*. A simple example is shown in Fig. 11–15a. When the disk is displaced and released, its center of gravity G will always move (rotate) back to its equilibrium position, which is at the *lowest point* of its path. This is where the potential energy of the disk is at its *minimum*.

Neutral Equilibrium. A system is said to be in **neutral equilibrium** if the system still remains in equilibrium when the system is given a small displacement away from its original position. In this case, the potential energy of the system is *constant*. Neutral equilibrium is shown in Fig. 11–15b, where a disk is pinned at G. Each time the disk is rotated, a new equilibrium position is established and the potential energy remains unchanged.

Unstable Equilibrium. A system is said to be in **unstable equilibrium** if it has a tendency to be *displaced farther away* from its original equilibrium position when it is given a small displacement. The potential energy of this system is a *maximum*. An unstable equilibrium position of the disk is shown in Fig. 11–15c. Here the disk will rotate away from its equilibrium position when its center of gravity is slightly displaced. At this *highest point*, its potential energy is at a *maximum*.

During high winds and when going around a curve, these sugar-cane trucks can become unstable and tip over since their center of gravity is high off the road when they are fully loaded.

Stable equilibrium Neutral equilibrium Unstable equilibrium
(a) (b) (c)

Fig. 11–15

One-Degree-of-Freedom System. If a system has only one degree of freedom, and its position is defined by the coordinate q, then the potential function V for the system in terms of q can be plotted, Fig. 11–16. Provided the system is in *equilibrium*, then dV/dq, which represents the slope of this function, must be equal to zero. An investigation of stability at the equilibrium configuration therefore requires that the second derivative of the potential function be evaluated.

If d^2V/dq^2 is greater than zero, Fig. 11–16a, the potential energy of the system will be a *minimum*. This indicates that the equilibrium configuration is *stable*. Thus,

$$\frac{dV}{dq} = 0, \qquad \frac{d^2V}{dq^2} > 0 \qquad \text{stable equilibrium} \qquad (11\text{–}10)$$

If d^2V/dq^2 is less than zero, Fig. 11–16b, the potential energy of the system will be a *maximum*. This indicates *unstable* equilibrium.

$$\frac{dV}{dq} = 0, \qquad \frac{d^2V}{dq^2} < 0 \qquad \text{unstable equilibrium} \qquad (11\text{–}11)$$

Finally, if d^2V/dq^2 is equal to zero, it will be necessary to investigate the higher order derivatives to determine the stability. The equilibrium configuration will be *stable* if the first non-zero derivative is of an *even* order and it is *positive*. Likewise, the equilibrium will be *unstable* if this first non-zero derivative is an *odd* order or if it is an even order and negative. If all the higher order derivatives are *zero*, the system is said to be in *neutral equilibrium*, Fig. 11–16c. Thus,

$$\frac{dV}{dq} = \frac{d^2V}{dq^2} = \frac{d^3V}{dq^3} = \cdots = 0 \qquad \text{neutral equilibrium} \qquad (11\text{–}12)$$

This condition occurs only if the potential-energy function for the system is constant at or around the neighborhood of q_{eq}.

Stable equilibrium
(a)

Unstable equilibrium
(b)

Neutral equilibrium
(c)

Fig. 11–16

> ### IMPORTANT POINTS
>
> - A conservative force does work that is independent of the path through which the force moves. Examples include the weight and the spring force.
> - Potential energy provides the body with the capacity to do work when the body moves relative to a fixed position or datum. Gravitational potential energy will be positive when the body is above a datum, and negative when the body is below the datum. Spring or elastic potential energy is always positive. It depends upon the stretch or compression of the spring.
> - The sum of these two forms of potential energy represents the potential function. Equilibrium requires that the first derivative of the potential function be equal to zero. Stability at the equilibrium position is determined from the second or higher order derivatives of the potential function.

PROCEDURE FOR ANALYSIS

Using potential-energy methods, the equilibrium positions and the stability of a body or a system of connected bodies having a single degree of freedom can be obtained by applying the following procedure.

Potential Function.

- Sketch the system so that it is in the *arbitrary position* specified by the coordinate q.

- Establish a horizontal *datum* through a *fixed point** and express the *gravitational potential energy* V_g in terms of the weight W of each member and its vertical distance y from the datum, $V_g = Wy$.

- Express the elastic potential energy V_e of the system in terms of the stretch or compression, s, of any connecting spring, $V_e = \frac{1}{2}ks^2$.

- Formulate the potential function $V = V_g + V_e$ and express the *position coordinates* y and s in terms of the single coordinate q.

Equilibrium Position.

- The equilibrium position of the system is determined by taking the first derivative of V and setting it equal to zero, $dV/dq = 0$.

Stability.

- Stability at the equilibrium position is determined by evaluating the second or higher-order derivatives of V.

- If the second derivative is greater than zero, the system is stable; if all derivatives are equal to zero, the system is in neutral equilibrium; and if the second derivative is less than zero, the system is unstable.

---

*The location of the datum is *arbitrary*, since only the *changes* or differentials of V are required for investigation of the equilibrium position and its stability.

EXAMPLE 11.5

Fig. 11-17

The uniform link shown in Fig. 11–17a has a mass of 10 kg. If the spring is unstretched when $\theta = 0°$, determine the angle θ for equilibrium and investigate the stability at the equilibrium position.

SOLUTION

Potential Function. The datum is established at the bottom of the link, Fig. 11–17b. When the link is located in the arbitrary position θ, the spring increases its potential energy and the weight decreases its potential energy. Hence,

$$V = V_e + V_g = \frac{1}{2}ks^2 + Wy$$

Since $l = s + l\cos\theta$ or $s = l(1 - \cos\theta)$, and $y = (l/2)\cos\theta$, then

$$V = \frac{1}{2}kl^2(1 - \cos\theta)^2 + W\left(\frac{l}{2}\cos\theta\right)$$

Equilibrium Position. The first derivative of V is

$$\frac{dV}{d\theta} = kl^2(1 - \cos\theta)\sin\theta - \frac{Wl}{2}\sin\theta = 0$$

or

$$l\left[kl(1 - \cos\theta) - \frac{W}{2}\right]\sin\theta = 0$$

This equation is satisfied provided

$$\sin\theta = 0 \qquad \theta = 0° \qquad \textit{Ans.}$$

or

$$\theta = \cos^{-1}\left(1 - \frac{W}{2kl}\right) = \cos^{-1}\left[1 - \frac{10(9.81)}{2(200)(0.6)}\right] = 53.8° \qquad \textit{Ans.}$$

Stability. The second derivative of V is

$$\frac{d^2V}{d\theta^2} = kl^2(1 - \cos\theta)\cos\theta + kl^2\sin\theta\sin\theta - \frac{Wl}{2}\cos\theta$$

$$= kl^2(\cos\theta - \cos 2\theta) - \frac{Wl}{2}\cos\theta$$

Substituting values for the constants, with $\theta = 0°$ and $\theta = 53.8°$, yields

$$\left.\frac{d^2V}{d\theta^2}\right|_{\theta=0°} = 200(0.6)^2(\cos 0° - \cos 0°) - \frac{10(9.81)(0.6)}{2}\cos 0°$$

$$= -29.4 < 0 \qquad \text{(unstable equilibrium)} \qquad \textit{Ans.}$$

$$\left.\frac{d^2V}{d\theta^2}\right|_{\theta=53.8°} = 200(0.6)^2(\cos 53.8° - \cos 107.6°) - \frac{10(9.81)(0.6)}{2}\cos 53.8°$$

$$= 46.9 > 0 \qquad \text{(stable equilibrium)} \qquad \textit{Ans.}$$

EXAMPLE 11.6

If the spring AD in Fig. 11–18a has a stiffness of 18 kN/m and is unstretched when $\theta = 60°$, determine the angle θ for equilibrium. The load has a mass of 1.5 Mg. Investigate the stability at the equilibrium position.

SOLUTION

Potential Energy. The gravitational potential energy for the load with respect to the fixed datum, shown in Fig. 11–18b, is

$$V_g = mgy = 1500(9.81) \text{ N}[(4 \text{ m}) \sin \theta + h] = 58\,860 \sin \theta + 14\,715h$$

where h is a constant distance. From the geometry of the system, the stretch of the spring when the load is on the platform is $s = (4 \text{ m}) \cos \theta - (4 \text{ m}) \cos 60° = (4 \text{ m}) \cos \theta - 2 \text{ m}$.
Thus, the elastic potential energy of the system is

$$V_e = \tfrac{1}{2}ks^2 = \tfrac{1}{2}(18\,000 \text{ N/m})(4 \text{ m} \cos \theta - 2 \text{ m})^2 = 9000(4 \cos \theta - 2)^2$$

The potential energy function for the system is therefore

$$V = V_g + V_e = 58\,860 \sin \theta + 14\,715h + 9000(4 \cos \theta - 2)^2 \quad (1)$$

Equilibrium. When the system is in equilibrium,

$$\frac{dV}{d\theta} = 58\,860 \cos \theta + 18\,000(4 \cos \theta - 2)(-4 \sin \theta) = 0$$

$$58\,860 \cos \theta - 288\,000 \sin \theta \cos \theta + 144\,000 \sin \theta = 0$$

Since $\sin 2\theta = 2 \sin \theta \cos \theta$,

$$58\,860 \cos \theta - 144\,000 \sin 2\theta + 144\,000 \sin \theta = 0$$

Solving by trial and error,

$\theta = 28.18°$ and $\theta = 45.51°$ *Ans.*

Stability. Taking the second derivative of Eq. 1,

$$\frac{d^2V}{d\theta^2} = -58\,860 \sin \theta - 288\,000 \cos 2\theta + 144\,000 \cos \theta$$

Substituting $\theta = 28.18°$ yields

$$\frac{d^2V}{d\theta^2} = -60\,402 < 0 \quad \text{(Unstable equilibrium)} \quad Ans.$$

And for $\theta = 45.51°$,

$$\frac{d^2V}{d\theta^2} = 64\,073 > 0 \quad \text{(Stable equilibrium)} \quad Ans.$$

(a)

(b)

Fig. 11–18

EXAMPLE 11.7

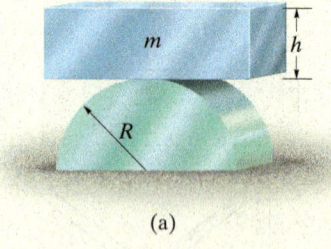

(a)

The uniform block having a mass m rests on the top surface of the half cylinder, Fig. 11–19a. Show that this is a condition of unstable equilibrium if $h > 2R$.

SOLUTION

Potential Function. The datum is established at the base of the cylinder, Fig. 11–19b. If the block is displaced by an amount θ from the equilibrium position, the potential function is

$$V = V_e + V_g$$
$$= 0 + mgy$$

From Fig. 11–19b,

$$y = \left(R + \frac{h}{2}\right)\cos\theta + R\theta\sin\theta$$

Thus,

$$V = mg\left[\left(R + \frac{h}{2}\right)\cos\theta + R\theta\sin\theta\right]$$

(b)

Fig. 11–19

Equilibrium Position.

$$\frac{dV}{d\theta} = mg\left[-\left(R + \frac{h}{2}\right)\sin\theta + R\sin\theta + R\theta\cos\theta\right] = 0$$

$$= mg\left(-\frac{h}{2}\sin\theta + R\theta\cos\theta\right) = 0$$

Note that $\theta = 0°$ satisfies this equation.

Stability. Taking the second derivative of V yields

$$\frac{d^2V}{d\theta^2} = mg\left(-\frac{h}{2}\cos\theta + R\cos\theta - R\theta\sin\theta\right)$$

At $\theta = 0°$,

$$\left.\frac{d^2V}{d\theta^2}\right|_{\theta=0°} = -mg\left(\frac{h}{2} - R\right)$$

Since all the constants are positive, the block is in unstable equilibrium provided $h > 2R$, because then $d^2V/d\theta^2 < 0$.

PROBLEMS

11–26. The potential energy of a one-degree-of-freedom system is defined by $V = (20x^3 - 10x^2 - 25x - 10)$ ft·lb, where x is in ft. Determine the equilibrium positions and investigate the stability for each of these positions.

11–27. If the potential function for a one-degree-of-freedom system is $V = (12 \sin 2\theta + 15 \cos \theta)$ J, where $0° < \theta < 180°$, determine the positions for equilibrium and investigate the stability at each of these positions.

*11–28. If the potential function for a one-degree-of-freedom system is $V = (8x^3 - 2x^2 - 10)$ J, where x is given in meters, determine the positions for equilibrium and investigate the stability at each of these positions.

11–29. If the potential function for a one-degree-of-freedom system is $V = (10 \cos 2\theta + 25 \sin \theta)$ J, where $0° < \theta < 180°$, determine the positions for equilibrium and investigate the stability at each of these positions.

11–30. If the potential energy for a one-degree-of-freedom system is expressed by the relation $V = (4x^3 - x^2 - 3x + 10)$ ft·lb, where x is given in feet, determine the equilibrium positions and investigate the stability at each position.

11–31. The uniform link AB has a mass of 3 kg. The rod BD, having negligible weight, passes through a swivel block at C. If the spring has a stiffness of $k = 100$ N/m and is unstretched when $\theta = 0°$, determine the angle θ for equilibrium and investigate the stability at the equilibrium position. Neglect the size of the swivel block.

*11–32. The spring of the scale has an unstretched length of a. Determine the angle θ for equilibrium when a weight W is supported on the platform. Neglect the weight of the members. What value W would be required to keep the scale in neutral equilibrium when $\theta = 0°$?

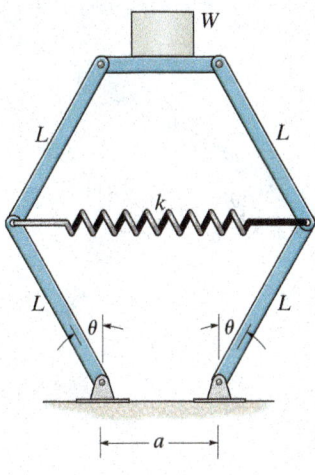

Prob. 11–32

11–33. The uniform bar has a mass of 80 kg. Determine the angle θ for equilibrium and investigate the stability of the bar when it is in this position. The spring has an unstretched length when $\theta = 90°$.

Prob. 11–33

Prob. 11–31

11–34. Determine the force in the spring required to keep the 6-kg rod in equilibrium when $\theta = 30°$. The spring remains horizontal due to the roller guide.

*11–36. Each of the three links of the mechanism has a weight of 10 lb. Determine the angle θ for equilibrium. The spring is unstretched when $\theta = 0°$. Investigate the stability at the equilibrium position.

Prob. 11–34

Prob. 11–36

11–35. The uniform beam has a weight W. If the contacting surfaces are smooth, determine the angle θ for equilibrium. The spring is uncompressed when $\theta = 90°$.

11–37. The door has a uniform weight of 50 lb. It is hinged at A and is held open by the 30-lb weight and the pulley. Determine the angle θ for equilibrium.

Prob. 11–35

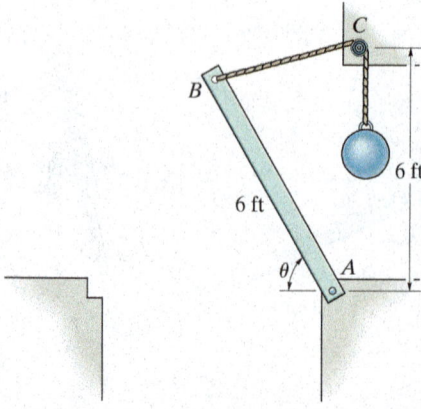

Prob. 11–37

11–38. The uniform rod OA weighs 20 lb, and when the rod is in the vertical position, the spring is unstretched. Determine the position θ for equilibrium. Investigate the stability at the equilibrium position.

*11–40. A homogeneous cone rests on top of the cylindrical surface. Determine a relationship between the radius r of the cylinder and the height h of the cone for neutral equilibrium. *Hint:* Establish the potential function for a *small* angle θ of tilt of the cone, i.e., approximate $\sin\theta \approx \theta$ and $\cos\theta \approx 1 - \theta^2/2$.

Prob. 11–38

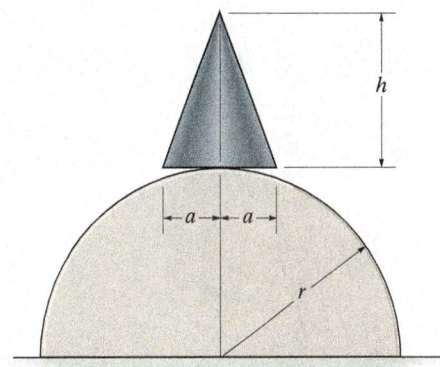

Prob. 11–40

11–39. A spring with a torsional stiffness k is attached to the hinge at B. It is unstretched when the rod assembly is in the vertical position. Determine the weight W of the block that results in neutral equilibrium. *Hint:* Establish the potential energy function for a small angle θ, i.e., approximate $\sin\theta \approx 0$, and $\cos\theta \approx 1 - \theta^2/2$.

11–41. The uniform rod has a mass of 100 kg. If the spring is unstretched when $\theta = 60°$, determine the angle θ for equilibrium and investigate the stability at the equilibrium position. The spring is always in the horizontal position due to the roller guide at B.

Prob. 11–39

Prob. 11–41

11–42. Each bar has a mass per length of m_0. Determine the angles θ and ϕ for equilibrium. The contact at A is smooth, and both are pin connected at B.

Prob. 11–42

11–43. The truck has a mass of 20 Mg and a mass center at G. Determine the steepest grade θ along which it can park without overturning and investigate the stability in this position.

Prob. 11–43

*__11–44.__ The hemisphere has a conical cavity cut into it as shown. Determine the depth d of the cavity in terms of r so that the hemisphere balances on the pivot and remains in neutral equilibrium.

Prob. 11–44

11–45. A 3-lb weight is attached to the end of rod ABC. If the rod is supported by a smooth slider block at C and rod BD, determine the angle θ for equilibrium. Neglect the weight of the rods and the slider.

Prob. 11–45

11–46. A cone is attached to the hemisphere. If both pieces have the same density, determine the height h of the cone if the configuration is to be in neutral equilibrium.

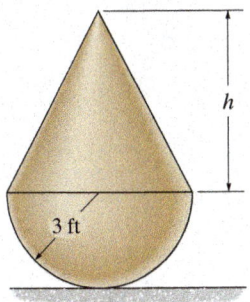

Prob. 11–46

11–47. The homogeneous block has a mass of 10 kg and rests on the smooth corners of two ledges. Determine the angle θ for placement that will cause the block to be stable.

Prob. 11–47

*11–48. The block weighs W and is supported by links AB and BC. Determine the necessary spring stiffness k required to hold the system in neutral equilibrium. The springs are subjected to an initial compression F_0 when the links are vertical as shown.

Prob. 11–48

11–49. The 60-lb hemisphere supports a cylinder having a specific weight of $\gamma = 311$ lb/ft^3. Determine the height h of the cylinder which will produce neutral equilibrium in the position shown.

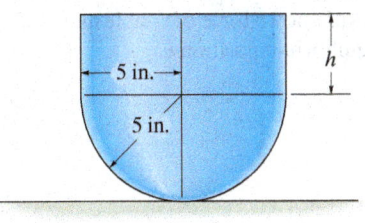

Prob. 11–49

CHAPTER REVIEW

Principle of Virtual Work

The forces on a body will do *virtual work* when the body undergoes an *imaginary* differential displacement or rotation.

δy, $\delta y'$ – virtual displacements

$\delta \theta$ – virtual rotation

For equilibrium, the sum of the virtual work done by all the forces acting on the body must be equal to zero for any virtual displacement. This is referred to as the *principle of virtual work*, and it is useful for finding the equilibrium configuration for a mechanism or a reactive force acting on a series of connected members.

$\delta U = 0$

If the system of connected members has one degree of freedom, then its position can be specified by one independent coordinate, such as θ.

To apply the principle of virtual work, it is first necessary to use *position coordinates* to locate all the forces and moments on the mechanism that will do work when the mechanism undergoes a virtual movement $\delta \theta$.

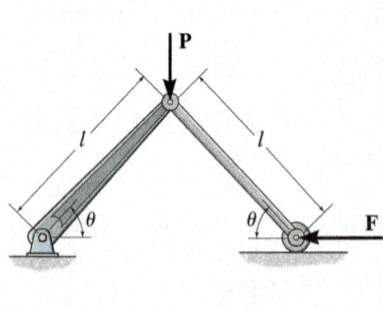

The coordinates are related to the independent coordinate θ and then these expressions are differentiated in order to relate the *virtual* coordinate displacements to the virtual displacement $\delta \theta$.

Finally, the equation of virtual work is written for the mechanism in terms of the common virtual displacement $\delta \theta$, and then it is set equal to zero. By factoring $\delta \theta$ out of the equation, it is then possible to determine either the unknown force or couple moment, or the equilibrium position θ.

Potential-Energy Criterion for Equilibrium

When a system is subjected only to conservative forces, such as weight and spring forces, then the equilibrium configuration can be determined using the *potential-energy function V* for the system.

$$V = V_g + V_e = -Wy + \tfrac{1}{2}ky^2$$

The potential-energy function is established by expressing the weight and spring potential energy for the system in terms of the independent coordinate q.

Once the potential-energy function is formulated, its first derivative is set equal to zero. The solution yields the equilibrium position q_{eq} for the system.

$$\frac{dV}{dq} = 0$$

The stability of the system can be investigated by taking the second derivative of V.

$$\frac{dV}{dq} = 0, \quad \frac{d^2V}{dq^2} > 0 \quad \text{stable equilibrium}$$

$$\frac{dV}{dq} = 0, \quad \frac{d^2V}{dq^2} < 0 \quad \text{unstable equilibrium}$$

$$\frac{dV}{dq} = \frac{d^2V}{dq^2} = \frac{d^3V}{dq^3} = \cdots = 0 \quad \text{neutral equilibrium}$$

REVIEW PROBLEMS

R11–1. The toggle joint is subjected to the load P. Determine the compressive force F it creates on the cylinder at A as a function of θ.

R11–3. If a torque of $M = 50\ \text{N} \cdot \text{m}$ is applied to the flywheel, determine the force F applied to the ram to hold it in the position $\theta = 60°$.

Prob. R11–1

Prob. R11–3

R11–2. The uniform links AB and BC each weigh 2 lb and the cylinder weighs 20 lb. Determine the horizontal force P required to hold the mechanism at $\theta = 45°$. The spring has an unstretched length of 6 in.

R11–4. The uniform bar AB weighs 10 lb. If the attached spring is unstretched when $\theta = 90°$, use the method of virtual work and determine the angle θ for equilibrium. Note that the spring always remains in the vertical position due to the roller guide.

Prob. R11–2

Prob. R11–4

R11–5. The spring has an unstretched length of 0.3 m. Use the method of virtual work and determine the angle θ for equilibrium if the uniform links each have a mass of 5 kg.

R11–7. The uniform bar AB weighs 100 lb. If both springs DE and BC are unstretched when $\theta = 90°$, determine the angle θ for equilibrium. Investigate the stability at the equilibrium position. Both springs always act in the horizontal position because of the roller guides at C and E.

Prob. R11–5

Prob. R11–7

R11–6. The block A has a mass of 40 kg. Determine the angle θ for equilibrium and investigate the stability of the mechanism in this position. The spring has a stiffness of $k = 1.5$ kN/m and is unstretched when $\theta = 90°$. Neglect the mass of the links.

R11–8. The spring attached to the mechanism has an unstretched length when $\theta = 90°$. Determine the position θ for equilibrium and investigate the stability of the mechanism at this position. Disk A is pin connected to the frame at B and has a weight of 20 lb. Neglect the weight of the bars.

Prob. R11–6

Prob. R11–8

APPENDIX A

MATHEMATICAL REVIEW AND FORMULATIONS

Geometry and Trigonometry Review

The angles θ in Fig. A–1 are equal between the transverse and two parallel lines.

Fig. A–1

For a line and its normal, the angles θ in Fig. A–2 are equal.

Fig. A–2

For the circle in Fig. A–3, $s = \theta r$, so that when $\theta = 360° = 2\pi$ rad then the circumference is $s = 2\pi r$. Also, since $180° = \pi$ rad, then θ (rad) $= (\pi/180°)\theta°$. The area of the circle is $A = \pi r^2$.

Fig. A–3

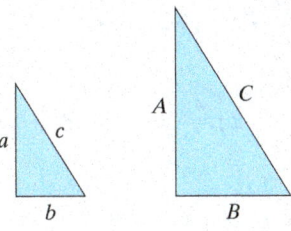

Fig. A–4

The sides of a similar triangle can be obtained by proportion as in Fig. A–4, where $\dfrac{a}{A} = \dfrac{b}{B} = \dfrac{c}{C}$.

For the right triangle in Fig. A–5, the Pythagorean theorem is

$$h = \sqrt{(o)^2 + (a)^2}$$

The trigonometric functions are

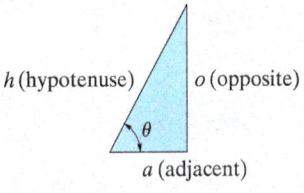

Fig. A–5

$$\sin\theta = \frac{o}{h}$$

$$\cos\theta = \frac{a}{h}$$

$$\tan\theta = \frac{o}{a}$$

This is easily remembered as "soh, cah, toa", i.e., the sine is the opposite over the hypotenuse, etc. The other trigonometric functions follow from this.

$$\csc\theta = \frac{1}{\sin\theta} = \frac{h}{o}$$

$$\sec\theta = \frac{1}{\cos\theta} = \frac{h}{a}$$

$$\cot\theta = \frac{1}{\tan\theta} = \frac{a}{o}$$

Trigonometric Identities

$\sin^2 \theta + \cos^2 \theta = 1$

$\sin(\theta \pm \phi) = \sin \theta \cos \phi \pm \cos \theta \sin \phi$

$\sin 2\theta = 2 \sin \theta \cos \theta$

$\cos(\theta \pm \phi) = \cos \theta \cos \phi \mp \sin \theta \sin \phi$

$\cos 2\theta = \cos^2 \theta - \sin^2 \theta$

$\cos \theta = \pm \sqrt{\dfrac{1 + \cos 2\theta}{2}}, \quad \sin \theta = \pm \sqrt{\dfrac{1 - \cos 2\theta}{2}}$

$\tan \theta = \dfrac{\sin \theta}{\cos \theta}$

$1 + \tan^2 \theta = \sec^2 \theta \qquad 1 + \cot^2 \theta = \csc^2 \theta$

Quadratic Formula

If $ax^2 + bx + c = 0$, then $x = \dfrac{-b \pm \sqrt{b^2 - 4ac}}{2a}$

Hyperbolic Functions

$\sinh x = \dfrac{e^x - e^{-x}}{2},$

$\cosh x = \dfrac{e^x + e^{-x}}{2},$

$\tanh x = \dfrac{\sinh x}{\cosh x}$

Power-Series Expansions

$\sin x = x - \dfrac{x^3}{3!} + \cdots, \quad \cos x = 1 - \dfrac{x^2}{2!} + \cdots$

$\sinh x = x + \dfrac{x^3}{3!} + \cdots, \quad \cosh x = 1 + \dfrac{x^2}{2!} + \cdots$

Derivatives

$\dfrac{d}{dx}(u^n) = nu^{n-1}\dfrac{du}{dx} \qquad \dfrac{d}{dx}(\sin u) = \cos u \dfrac{du}{dx}$

$\dfrac{d}{dx}(uv) = u\dfrac{dv}{dx} + v\dfrac{du}{dx} \qquad \dfrac{d}{dx}(\cos u) = -\sin u \dfrac{du}{dx}$

$\dfrac{d}{dx}\left(\dfrac{u}{v}\right) = \dfrac{v\dfrac{du}{dx} - u\dfrac{dv}{dx}}{v^2} \qquad \dfrac{d}{dx}(\tan u) = \sec^2 u \dfrac{du}{dx}$

$\dfrac{d}{dx}(\cot u) = -\csc^2 u \dfrac{du}{dx} \qquad \dfrac{d}{dx}(\sinh u) = \cosh u \dfrac{du}{dx}$

$\dfrac{d}{dx}(\sec u) = \tan u \sec u \dfrac{du}{dx} \qquad \dfrac{d}{dx}(\cosh u) = \sinh u \dfrac{du}{dx}$

$\dfrac{d}{dx}(\csc u) = -\csc u \cot u \dfrac{du}{dx}$

Integrals

$$\int x^n \, dx = \frac{x^{n+1}}{n+1} + C, n \neq -1$$

$$\int \frac{dx}{a + bx} = \frac{1}{b} \ln(a + bx) + C$$

$$\int \frac{dx}{a + bx^2} = \frac{1}{2\sqrt{-ab}} \ln\left[\frac{a + x\sqrt{-ab}}{a - x\sqrt{-ab}}\right] + C, \quad ab < 0$$

$$\int \frac{x \, dx}{a + bx^2} = \frac{1}{2b} \ln(bx^2 + a) + C$$

$$\int \frac{x^2 \, dx}{a + bx^2} = \frac{x}{b} - \frac{a}{b\sqrt{ab}} \tan^{-1}\frac{x\sqrt{ab}}{a} + C, \, ab > 0$$

$$\int \sqrt{a + bx} \, dx = \frac{2}{3b} \sqrt{(a + bx)^3} + C$$

$$\int x\sqrt{a + bx} \, dx = \frac{-2(2a - 3bx)\sqrt{(a + bx)^3}}{15b^2} + C$$

$$\int x^2 \sqrt{a + bx} \, dx = $$
$$\frac{2(8a^2 - 12abx + 15b^2x^2)\sqrt{(a + bx)^3}}{105b^3} + C$$

$$\int \sqrt{a^2 - x^2} \, dx = \frac{1}{2}\left[x\sqrt{a^2 - x^2} + a^2 \sin^{-1}\frac{x}{a}\right] + C, \quad a > 0$$

$$\int x\sqrt{a^2 - x^2} \, dx = -\frac{1}{3}\sqrt{(a^2 - x^2)^3} + C$$

$$\int x^2 \sqrt{a^2 - x^2} \, dx = -\frac{x}{4}\sqrt{(a^2 - x^2)^3}$$
$$+ \frac{a^2}{8}\left(x\sqrt{a^2 - x^2} + a^2 \sin^{-1}\frac{x}{a}\right) + C, a > 0$$

$$\int \sqrt{x^2 \pm a^2} \, dx = $$
$$\frac{1}{2}\left[x\sqrt{x^2 \pm a^2} \pm a^2 \ln\left(x + \sqrt{x^2 \pm a^2}\right)\right] + C$$

$$\int x\sqrt{x^2 \pm a^2} \, dx = \frac{1}{3}\sqrt{(x^2 \pm a^2)^3} + C$$

$$\int x^2 \sqrt{x^2 \pm a^2} \, dx = \frac{x}{4}\sqrt{(x^2 \pm a^2)^3}$$
$$\mp \frac{a^2}{8} x\sqrt{x^2 \pm a^2} - \frac{a^4}{8}\ln\left(x + \sqrt{x^2 \pm a^2}\right) + C$$

$$\int \frac{dx}{\sqrt{a + bx}} = \frac{2\sqrt{a + bx}}{b} + C$$

$$\int \frac{x \, dx}{\sqrt{x^2 \pm a^2}} = \sqrt{x^2 \pm a^2} + C$$

$$\int \frac{dx}{\sqrt{a + bx + cx^2}} = \frac{1}{\sqrt{c}} \ln\left[\sqrt{a + bx + cx^2} + x\sqrt{c} + \frac{b}{2\sqrt{c}}\right] + C, c > 0$$

$$= \frac{1}{\sqrt{-c}} \sin^{-1}\left(\frac{-2cx - b}{\sqrt{b^2 - 4ac}}\right) + C, c < 0$$

$$\int \sin x \, dx = -\cos x + C$$

$$\int \cos x \, dx = \sin x + C$$

$$\int x \cos(ax) \, dx = \frac{1}{a^2}\cos(ax) + \frac{x}{a}\sin(ax) + C$$

$$\int x^2 \cos(ax) \, dx = \frac{2x}{a^2}\cos(ax) + \frac{a^2x^2 - 2}{a^3}\sin(ax) + C$$

$$\int e^{ax} \, dx = \frac{1}{a} e^{ax} + C$$

$$\int xe^{ax} \, dx = \frac{e^{ax}}{a^2}(ax - 1) + C$$

$$\int \sinh x \, dx = \cosh x + C$$

$$\int \cosh x \, dx = \sinh x + C$$

Fundamental Problem Solutions and Answers

Chapter 2

F2–1.
$F_R = \sqrt{(2 \text{ kN})^2 + (6 \text{ kN})^2 - 2(2 \text{ kN})(6 \text{ kN}) \cos 105°}$
$= 6.798 \text{ kN} = 6.80 \text{ kN}$ *Ans.*
$\dfrac{\sin \phi}{6 \text{ kN}} = \dfrac{\sin 105°}{6.798 \text{ kN}}, \quad \phi = 58.49°$
$\theta = 45° + \phi = 45° + 58.49° = 103°$ *Ans.*

F2–2. $F_R = \sqrt{200^2 + 500^2 - 2(200)(500) \cos 140°}$
$= 666 \text{ N}$ *Ans.*

F2–3. $F_R = \sqrt{600^2 + 800^2 - 2(600)(800) \cos 60°}$
$= 721.11 \text{ N} = 721 \text{ N}$ *Ans.*
$\dfrac{\sin \alpha}{800} = \dfrac{\sin 60°}{721.11}; \quad \alpha = 73.90°$
$\phi = \alpha - 30° = 73.90° - 30° = 43.9°$ *Ans.*

F2–4. $\dfrac{F_u}{\sin 45°} = \dfrac{30}{\sin 105°}; \quad F_u = 22.0 \text{ lb}$ *Ans.*
$\dfrac{F_v}{\sin 30°} = \dfrac{30}{\sin 105°}; \quad F_v = 15.5 \text{ lb}$ *Ans.*

F2–5. $\dfrac{F_{AB}}{\sin 105°} = \dfrac{450}{\sin 30°}$
$F_{AB} = 869 \text{ lb}$ *Ans.*
$\dfrac{F_{AC}}{\sin 45°} = \dfrac{450}{\sin 30°}$
$F_{AC} = 636 \text{ lb}$ *Ans.*

F2–6. $\dfrac{F}{\sin 30°} = \dfrac{6}{\sin 105°} \quad F = 3.11 \text{ kN}$ *Ans.*
$\dfrac{F_v}{\sin 45°} = \dfrac{6}{\sin 105°} \quad F_v = 4.39 \text{ kN}$ *Ans.*

F2–7. $(F_1)_x = 0 \quad (F_1)_y = 300 \text{ N}$ *Ans.*
$(F_2)_x = -(450 \text{ N}) \cos 45° = -318 \text{ N}$ *Ans.*
$(F_2)_y = (450 \text{ N}) \sin 45° = 318 \text{ N}$ *Ans.*
$(F_3)_x = \left(\tfrac{3}{5}\right) 600 \text{ N} = 360 \text{ N}$ *Ans.*
$(F_3)_y = \left(\tfrac{4}{5}\right) 600 \text{ N} = 480 \text{ N}$ *Ans.*

F2–8. $F_{Rx} = 300 + 400 \cos 30° - 250\left(\tfrac{4}{5}\right) = 446.4 \text{ N}$
$F_{Ry} = 400 \sin 30° + 250\left(\tfrac{3}{5}\right) = 350 \text{ N}$
$F_R = \sqrt{(446.4)^2 + 350^2} = 567 \text{ N}$ *Ans.*
$\theta = \tan^{-1} \dfrac{350}{446.4} = 38.1° \measuredangle$ *Ans.*

F2–9.
$\xrightarrow{+}(F_R)_x = \Sigma F_x;$
$(F_R)_x = -(700 \text{ lb}) \cos 30° + 0 + \left(\tfrac{3}{5}\right)(600 \text{ lb})$
$= -246.22 \text{ lb}$
$+\uparrow (F_R)_y = \Sigma F_y;$
$(F_R)_y = -(700 \text{ lb}) \sin 30° - 400 \text{ lb} - \left(\tfrac{4}{5}\right)(600 \text{ lb})$
$= -1230 \text{ lb}$
$F_R = \sqrt{(246.22 \text{ lb})^2 + (1230 \text{ lb})^2} = 1254 \text{ lb} = 1.25 \text{ k}$ *Ans.*
$\phi = \tan^{-1}\left(\tfrac{1230 \text{ lb}}{246.22 \text{ lb}}\right) = 78.68°$
$\theta = 180° + \phi = 180° + 78.68° = 259°$ *Ans.*

F2–10. $\xrightarrow{+}(F_R)_x = \Sigma F_x;$
$750 \text{ N} = F \cos \theta + \left(\tfrac{5}{13}\right)(325 \text{ N}) + (600 \text{ N}) \cos 45°$
$+\uparrow (F_R)_y = \Sigma F_y;$
$0 = F \sin \theta + \left(\tfrac{12}{13}\right)(325 \text{ N}) - (600 \text{ N}) \sin 45°$
$\tan \theta = 0.6190 \quad \theta = 31.76° = 31.8° \measuredangle$ *Ans.*
$F = 236 \text{ N}$ *Ans.*

F2–11. $\xrightarrow{+}(F_R)_x = \Sigma F_x;$
$(80 \text{ lb}) \cos 45° = F \cos \theta + 50 \text{ lb} - \left(\tfrac{3}{5}\right) 90 \text{ lb}$
$+\uparrow (F_R)_y = \Sigma F_y;$
$-(80 \text{ lb}) \sin 45° = F \sin \theta - \left(\tfrac{4}{5}\right)(90 \text{ lb})$
$\tan \theta = 0.2547 \quad \theta = 14.29° = 14.3° \measuredangle$ *Ans.*
$F = 62.5 \text{ lb}$ *Ans.*

F2–12. $(F_R)_x = 15\left(\tfrac{4}{5}\right) + 0 + 15\left(\tfrac{4}{5}\right) = 24 \text{ kN} \to$
$(F_R)_y = 15\left(\tfrac{3}{5}\right) + 20 - 15\left(\tfrac{3}{5}\right) = 20 \text{ kN} \uparrow$
$F_R = 31.2 \text{ kN}$ *Ans.*
$\theta = 39.8°$ *Ans.*

F2–13. $F_x = 75 \cos 30° \sin 45° = 45.93 \text{ lb}$
$F_y = 75 \cos 30° \cos 45° = 45.93 \text{ lb}$
$F_z = -75 \sin 30° = -37.5 \text{ lb}$
$\alpha = \cos^{-1}\left(\tfrac{45.93}{75}\right) = 52.2°$ *Ans.*
$\beta = \cos^{-1}\left(\tfrac{45.93}{75}\right) = 52.2°$ *Ans.*
$\gamma = \cos^{-1}\left(\tfrac{-37.5}{75}\right) = 120°$ *Ans.*

598

F2–14. $\cos\beta = \sqrt{1 - \cos^2 120° - \cos^2 60°} = \pm 0.7071$
Require $\beta = 135°$.
$\mathbf{F} = F\mathbf{u}_F = (500\text{ N})(-0.5\mathbf{i} - 0.7071\mathbf{j} + 0.5\mathbf{k})$
$= \{-250\mathbf{i} - 354\mathbf{j} + 250\mathbf{k}\}\text{ N}$ *Ans.*

F2–15. $\cos^2\alpha + \cos^2 135° + \cos^2 120° = 1$
$\alpha = 60°$
$\mathbf{F} = F\mathbf{u}_F = (500\text{ N})(0.5\mathbf{i} - 0.7071\mathbf{j} - 0.5\mathbf{k})$
$= \{250\mathbf{i} - 354\mathbf{j} - 250\mathbf{k}\}\text{ N}$ *Ans.*

F2–16. $F_z = (50\text{ lb})\sin 45° = 35.36\text{ lb}$
$F' = (50\text{ lb})\cos 45° = 35.36\text{ lb}$
$F_x = \left(\frac{3}{5}\right)(35.36\text{ lb}) = 21.21\text{ lb}$
$F_y = \left(\frac{4}{5}\right)(35.36\text{ lb}) = 28.28\text{ lb}$
$\mathbf{F} = \{-21.2\mathbf{i} + 28.3\mathbf{j} + 35.4\mathbf{k}\}\text{ lb}$ *Ans.*

F2–17. $F_z = (750\text{ N})\sin 45° = 530.33\text{ N}$
$F' = (750\text{ N})\cos 45° = 530.33\text{ N}$
$F_x = (530.33\text{ N})\cos 60° = 265.2\text{ N}$
$F_y = (530.33\text{ N})\sin 60° = 459.3\text{ N}$
$\mathbf{F} = \{265\mathbf{i} - 459\mathbf{j} + 530\mathbf{k}\}\text{ N}$ *Ans.*

F2–18. $\mathbf{F}_1 = \left(\frac{4}{5}\right)(500\text{ lb})\mathbf{j} + \left(\frac{3}{5}\right)(500\text{ lb})\mathbf{k}$
$= \{400\mathbf{j} + 300\mathbf{k}\}\text{ lb}$
$\mathbf{F}_2 = [(800\text{ lb})\cos 45°]\cos 30°\,\mathbf{i}$
$+ [(800\text{ lb})\cos 45°]\sin 30°\,\mathbf{j}$
$+ (800\text{ lb})\sin 45°\,(-\mathbf{k})$
$= \{489.90\mathbf{i} + 282.84\mathbf{j} - 565.69\mathbf{k}\}\text{ lb}$
$\mathbf{F}_R = \mathbf{F}_1 + \mathbf{F}_2 = \{490\mathbf{i} + 683\mathbf{j} - 266\mathbf{k}\}\text{ lb}$ *Ans.*

F2–19. $\mathbf{r}_{AB} = \{-6\mathbf{i} + 6\mathbf{j} + 3\mathbf{k}\}\text{ m}$ *Ans.*
$r_{AB} = \sqrt{(-6\text{ m})^2 + (6\text{ m})^2 + (3\text{ m})^2} = 9\text{ m}$ *Ans.*
$\alpha = 132°,\quad \beta = 48.2°,\quad \gamma = 70.5°$ *Ans.*

F2–20. $\mathbf{r}_{AB} = \{-4\mathbf{i} + 2\mathbf{j} + 4\mathbf{k}\}\text{ ft}$ *Ans.*
$r_{AB} = \sqrt{(-4\text{ ft})^2 + (2\text{ ft})^2 + (4\text{ ft})^2} = 6\text{ ft}$ *Ans.*
$\alpha = \cos^{-1}\left(\frac{-4\text{ ft}}{6\text{ ft}}\right) = 131.8°$
$\theta = 180° - 131.8° = 48.2°$ *Ans.*

F2–21. $\mathbf{r}_{AB} = \{2\mathbf{i} + 3\mathbf{j} - 6\mathbf{k}\}\text{ m}$
$\mathbf{F}_{AB} = F_{AB}\mathbf{u}_{AB}$
$= (630\text{ N})\left(\frac{2}{7}\mathbf{i} + \frac{3}{7}\mathbf{j} - \frac{6}{7}\mathbf{k}\right)$
$= \{180\mathbf{i} + 270\mathbf{j} - 540\mathbf{k}\}\text{ N}$ *Ans.*

F2–22. $\mathbf{F} = F\mathbf{u}_{AB} = 900\text{ N}\left(-\frac{4}{9}\mathbf{i} + \frac{7}{9}\mathbf{j} - \frac{4}{9}\mathbf{k}\right)$
$= \{-400\mathbf{i} + 700\mathbf{j} - 400\mathbf{k}\}\text{ N}$ *Ans.*

F2–23. $\mathbf{F}_B = F_B\mathbf{u}_B$
$= (840\text{ N})\left(\frac{3}{7}\mathbf{i} - \frac{2}{7}\mathbf{j} - \frac{6}{7}\mathbf{k}\right)$
$= \{360\mathbf{i} - 240\mathbf{j} - 720\mathbf{k}\}\text{ N}$
$\mathbf{F}_C = F_C\mathbf{u}_C$
$= (420\text{ N})\left(\frac{2}{7}\mathbf{i} + \frac{3}{7}\mathbf{j} - \frac{6}{7}\mathbf{k}\right)$
$= \{120\mathbf{i} + 180\mathbf{j} - 360\mathbf{k}\}\text{ N}$
$F_R = \sqrt{(480\text{ N})^2 + (-60\text{ N})^2 + (-1080\text{ N})^2}$
$= 1.18\text{ kN}$ *Ans.*

F2–24. $\mathbf{F}_B = F_B\mathbf{u}_B$
$= (600\text{ lb})\left(-\frac{1}{3}\mathbf{i} + \frac{2}{3}\mathbf{j} - \frac{2}{3}\mathbf{k}\right)$
$= \{-200\mathbf{i} + 400\mathbf{j} - 400\mathbf{k}\}\text{ lb}$
$\mathbf{F}_C = F_C\mathbf{u}_C$
$= (490\text{ lb})\left(-\frac{6}{7}\mathbf{i} + \frac{3}{7}\mathbf{j} - \frac{2}{7}\mathbf{k}\right)$
$= \{-420\mathbf{i} + 210\mathbf{j} - 140\mathbf{k}\}\text{ lb}$
$\mathbf{F}_R = \mathbf{F}_B + \mathbf{F}_C = \{-620\mathbf{i} + 610\mathbf{j} - 540\mathbf{k}\}\text{ lb}$ *Ans.*

F2–25. $\mathbf{u}_{AO} = -\frac{1}{3}\mathbf{i} + \frac{2}{3}\mathbf{j} - \frac{2}{3}\mathbf{k}$
$\mathbf{u}_F = -0.5345\mathbf{i} + 0.8018\mathbf{j} + 0.2673\mathbf{k}$
$\theta = \cos^{-1}(\mathbf{u}_{AO}\cdot\mathbf{u}_F) = 57.7°$ *Ans.*

F2–26. $\mathbf{u}_{AB} = -\frac{3}{5}\mathbf{j} + \frac{4}{5}\mathbf{k}$
$\mathbf{u}_F = \frac{4}{5}\mathbf{i} - \frac{3}{5}\mathbf{j}$
$\theta = \cos^{-1}(\mathbf{u}_{AB}\cdot\mathbf{u}_F) = 68.9°$ *Ans.*

F2–27. $\mathbf{u}_{OA} = \frac{12}{13}\mathbf{i} + \frac{5}{13}\mathbf{j}$
$\mathbf{u}_{OA}\cdot\mathbf{j} = u_{OA}(1)\cos\theta$
$\cos\theta = \frac{5}{13};\quad \theta = 67.4°$ *Ans.*

F2–28. $\mathbf{u}_{OA} = \frac{12}{13}\mathbf{i} + \frac{5}{13}\mathbf{j}$
$\mathbf{F} = F\mathbf{u}_F = [650\mathbf{j}]\text{ N}$
$F_{OA} = \mathbf{F}\cdot\mathbf{u}_{OA} = 250\text{ N}$
$\mathbf{F}_{OA} = F_{OA}\mathbf{u}_{OA} = \{231\mathbf{i} + 96.2\mathbf{j}\}\text{ N}$ *Ans.*

F2–29.
$$\mathbf{F} = (400 \text{ N})\frac{\{4\mathbf{i} + 1\mathbf{j} - 6\mathbf{k}\}\text{m}}{\sqrt{(4 \text{ m})^2 + (1 \text{ m})^2 + (-6 \text{ m})^2}}$$
$$= \{219.78\mathbf{i} + 54.94\mathbf{j} - 329.67\mathbf{k}\} \text{ N}$$
$$\mathbf{u}_{AO} = \frac{\{-4\mathbf{j} - 6\mathbf{k}\}\text{m}}{\sqrt{(-4 \text{ m})^2 + (-6 \text{ m})^2}}$$
$$= -0.5547\mathbf{j} - 0.8321\mathbf{k}$$
$(F_{AO})_{\text{proj}} = \mathbf{F} \cdot \mathbf{u}_{AO} = 244 \text{ N}$ *Ans.*

F2–30.
$\mathbf{F} = [(-600 \text{ lb})\cos 60°]\sin 30° \mathbf{i}$
$+ [(600 \text{ lb})\cos 60°]\cos 30° \mathbf{j}$
$+ [(600 \text{ lb})\sin 60°]\mathbf{k}$
$= \{-150\mathbf{i} + 259.81\mathbf{j} + 519.62\mathbf{k}\} \text{ lb}$
$\mathbf{u}_A = -\frac{2}{3}\mathbf{i} + \frac{2}{3}\mathbf{j} + \frac{1}{3}\mathbf{k}$
$(F_A)_{\text{par}} = \mathbf{F} \cdot \mathbf{u}_A = 446.41 \text{ lb} = 446 \text{ lb}$ *Ans.*
$(F_A)_{\text{per}} = \sqrt{(600 \text{ lb})^2 - (446.41 \text{ lb})^2}$
$= 401 \text{ lb}$ *Ans.*

F2–31.
$\mathbf{F} = 56 \text{ N}(\frac{3}{7}\mathbf{i} - \frac{6}{7}\mathbf{j} + \frac{2}{7}\mathbf{k})$
$= \{24\mathbf{i} - 48\mathbf{j} + 16\mathbf{k}\} \text{ N}$
$(F_{AO})_\| = \mathbf{F} \cdot \mathbf{u}_{AO} = (24\mathbf{i} - 48\mathbf{j} + 16\mathbf{k}) \cdot (\frac{3}{7}\mathbf{i} - \frac{6}{7}\mathbf{j} - \frac{2}{7}\mathbf{k})$
$= 46.86 \text{ N} = 46.9 \text{ N}$ *Ans.*
$(F_{AO})_\perp = \sqrt{F^2 - (F_{AO})_\|^2} = \sqrt{(56)^2 - (46.86)^2}$
$= 30.7 \text{ N}$ *Ans.*

Chapter 3

F3–1. $\overset{+}{\rightarrow}\Sigma F_x = 0; \frac{4}{5}F_{AC} - F_{AB}\cos 30° = 0$
$+\uparrow\Sigma F_y = 0; \frac{3}{5}F_{AC} + F_{AB}\sin 30° - 550 = 0$
$F_{AB} = 478 \text{ lb}$ *Ans.*
$F_{AC} = 518 \text{ lb}$ *Ans.*

F3–2. $+\uparrow\Sigma F_y = 0; -2(1500)\sin\theta + 700 = 0$
$\theta = 13.5°$
$L_{ABC} = 2\left(\frac{5 \text{ ft}}{\cos 13.5°}\right) = 10.3 \text{ ft}$ *Ans.*

F3–3. $\overset{+}{\rightarrow}\Sigma F_x = 0; \quad T\cos\theta - T\cos\phi = 0$
$\phi = \theta$
$+\uparrow\Sigma F_y = 0; \quad 2T\sin\theta - 49.05 \text{ N} = 0$
$\theta = \tan^{-1}\left(\frac{0.15 \text{ m}}{(0.4 \text{ m}/2)}\right) = 36.87°$
$T = 40.9 \text{ N}$ *Ans.*

F3–4. $+\nearrow\Sigma F_x = 0; \frac{4}{5}(F_{sp}) - 5(9.81)\sin 45° = 0$
$F_{sp} = 43.35 \text{ N}$
$F_{sp} = k(l - l_0); 43.35 = 200(0.5 - l_0)$
$l_0 = 0.283 \text{ m}$ *Ans.*

F3–5. $+\uparrow\Sigma F_y = 0; \quad (392.4 \text{ N})\sin 30° - m_A(9.81) = 0$
$m_A = 20 \text{ kg}$ *Ans.*

F3–6. $+\uparrow\Sigma F_y = 0; \quad T_{AB}\sin 15° - 10(9.81) \text{ N} = 0$
$T_{AB} = 379.03 \text{ N} = 379 \text{ N}$ *Ans.*
$\overset{+}{\rightarrow}\Sigma F_x = 0; \quad T_{BC} - 379.03 \text{ N} \cos 15° = 0$
$T_{BC} = 366.11 \text{ N} = 366 \text{ N}$ *Ans.*
$\overset{+}{\rightarrow}\Sigma F_x = 0; \quad T_{CD}\cos\theta - 366.11 \text{ N} = 0$
$+\uparrow\Sigma F_y = 0; \quad T_{CD}\sin\theta - 15(9.81) \text{ N} = 0$
$T_{CD} = 395 \text{ N}$ *Ans.*
$\theta = 21.9°$ *Ans.*

F3–7. $\mathbf{F}_{AD} = F_{AD}\left(\frac{\mathbf{r}_{AD}}{r_{AD}}\right) = \frac{1}{3}F_{AD}\mathbf{i} - \frac{2}{3}F_{AD}\mathbf{j} + \frac{2}{3}F_{AD}\mathbf{k}$
$\Sigma F_z = 0; \quad \frac{2}{3}F_{AD} - 600 = 0$
$F_{AD} = 900 \text{ N}$ *Ans.*
$\Sigma F_y = 0; \quad F_{AB}\cos 30° - \frac{2}{3}(900) = 0$
$F_{AB} = 692.82 \text{ N} = 693 \text{ N}$ *Ans.*
$\Sigma F_x = 0; \quad \frac{1}{3}(900) + 692.82\sin 30° - F_{AC} = 0$
$F_{AC} = 646.41 \text{ N} = 646 \text{ N}$ *Ans.*

F3–8. $\Sigma F_z = 0; \quad F_{AD}(\frac{4}{5}) - 900 = 0$
$F_{AD} = 1125 \text{ N} = 1.125 \text{ kN}$ *Ans.*
$\Sigma F_y = 0; \quad F_{AC}(\frac{4}{5}) - 1125(\frac{3}{5}) = 0$
$F_{AC} = 843.75 \text{ N} = 844 \text{ N}$ *Ans.*
$\Sigma F_x = 0; \quad F_{AB} - 843.75(\frac{3}{5}) = 0$
$F_{AB} = 506.25 \text{ N} = 506 \text{ N}$ *Ans.*

F3–9. $\Sigma F_x = 0; \quad [(\frac{3}{5})F_3](\frac{3}{5}) + 600 \text{ N} - F_2 = 0$ (1)
$\Sigma F_y = 0; \quad (\frac{4}{5})F_1 - [(\frac{3}{5})F_3](\frac{4}{5}) = 0$ (2)
$\Sigma F_z = 0; \quad (\frac{4}{5})F_3 + (\frac{3}{5})F_1 - 900 \text{ N} = 0$ (3)
$F_3 = 776 \text{ N}$ *Ans.*
$F_1 = 466 \text{ N}$ *Ans.*
$F_2 = 879 \text{ N}$ *Ans.*

F3–10. $\mathbf{F}_{AC} = F_{AC}\{-\cos 60° \sin 30° \mathbf{i}$
$+ \cos 60° \cos 30° \mathbf{j} + \sin 60° \mathbf{k}\}$
$= -0.25F_{AC}\mathbf{i} + 0.4330F_{AC}\mathbf{j} + 0.8660F_{AC}\mathbf{k}$
$\mathbf{F}_{AD} = F_{AD}\{\cos 120° \mathbf{i} + \cos 120° \mathbf{j} + \cos 45° \mathbf{k}\}$
$= -0.5F_{AD}\mathbf{i} - 0.5F_{AD}\mathbf{j} + 0.7071F_{AD}\mathbf{k}$
$\Sigma F_y = 0; \quad 0.4330F_{AC} - 0.5F_{AD} = 0$
$\Sigma F_z = 0; \quad 0.8660F_{AC} + 0.7071F_{AD} - 300 = 0$
$F_{AD} = 175.74 \text{ lb} = 176 \text{ lb}$ *Ans.*
$F_{AC} = 202.92 \text{ lb} = 203 \text{ lb}$ *Ans.*
$\Sigma F_x = 0; \quad F_{AB} - 0.25(202.92) - 0.5(175.74) = 0$
$F_{AB} = 138.60 \text{ lb} = 139 \text{ lb}$ *Ans.*

F3–11. $\mathbf{F}_B = F_B\left(\dfrac{\mathbf{r}_{AB}}{r_{AB}}\right)$

$= F_B\left[\dfrac{\{-6\mathbf{i} + 3\mathbf{j} + 2\mathbf{k}\}\text{ ft}}{\sqrt{(-6\text{ ft})^2 + (3\text{ ft})^2 + (2\text{ ft})^2}}\right]$

$= -\tfrac{6}{7}F_B\mathbf{i} + \tfrac{3}{7}F_B\mathbf{j} + \tfrac{2}{7}F_B\mathbf{k}$

$\mathbf{F}_C = F_C\left(\dfrac{\mathbf{r}_{AC}}{r_{AC}}\right)$

$= F_C\left[\dfrac{\{-6\mathbf{i} - 2\mathbf{j} + 3\mathbf{k}\}\text{ ft}}{\sqrt{(-6\text{ ft})^2 + (-2\text{ ft})^2 + (3\text{ft})^2}}\right]$

$= -\tfrac{6}{7}F_C\mathbf{i} - \tfrac{2}{7}F_C\mathbf{j} + \tfrac{3}{7}F_C\mathbf{k}$

$\mathbf{F}_D = F_D\mathbf{i}$

$\mathbf{W} = \{-150\mathbf{k}\}\text{ lb}$

$\Sigma F_x = 0;\ -\tfrac{6}{7}F_B - \tfrac{6}{7}F_C + F_D = 0$ (1)

$\Sigma F_y = 0;\ \tfrac{3}{7}F_B - \tfrac{2}{7}F_C = 0$ (2)

$\Sigma F_z = 0;\ \tfrac{2}{7}F_B + \tfrac{3}{7}F_C - 150 = 0$ (3)

$F_B = 162$ lb *Ans.*

$F_C = 1.5(162\text{ lb}) = 242$ lb *Ans.*

$F_D = 346.15$ lb $= 346$ lb *Ans.*

Chapter 4

F4–1. a) $M_O = -\left(\tfrac{3}{5}\right)(500\text{ N})(2\text{ m}) = 600\text{ N}\cdot\text{m}\downarrow$
b) $M_O = \left(\tfrac{4}{5}\right)(500\text{ N})(3\text{ m}) = 1200\text{ N}\cdot\text{m}\uparrow$

F4–2. a) $M_O = -\left(\tfrac{3}{5}\right)(500\text{ N})(3\text{ m} - 1\text{ m})$
$+ \left(\tfrac{4}{5}\right)(500\text{ N})(1\text{ m}) = 200\text{ N}\cdot\text{m}\downarrow$

b) $M_O = \left(\tfrac{3}{5}\right)(500\text{ N})(1\text{ m}) - \left(\tfrac{4}{5}\right)(500\text{ N})(3\text{ m})$
$= 900\text{ N}\cdot\text{m}\downarrow$

F4–3. a) $\downarrow + M_O = -\left(\tfrac{4}{5}\right)(100\text{ N})(2\text{ m}) - \left(\tfrac{3}{5}\right)(100\text{ N})(5\text{ m})$
$= -460\text{ N}\cdot\text{m} = 460\text{ N}\cdot\text{m}\downarrow$ *Ans.*

b) $\downarrow + M_O = [(300\text{ N})\sin 30°][0.4\text{ m} + (0.3\text{ m})\cos 45°]$
$- [(300\text{ N})\cos 30°][(0.3\text{ m})\sin 45°]$
$= 36.7\text{ N}\cdot\text{m}$ *Ans.*

c) $\downarrow + M_O = (600\text{ lb})(4\text{ ft} + (3\text{ ft})\cos 45° - 1\text{ ft})$
$= 3.07\text{ kip}\cdot\text{ft}$ *Ans.*

F4–4. $\downarrow + M_O = -50\sin 60°(0.1 + 0.2\cos 45° + 0.1)$
$+ 50\cos 60°(0.2\sin 45°)$
$= 11.2\text{ N}\cdot\text{m}$ *Ans.*

F4–5. $\downarrow + M_O = 600\sin 50°(5) + 600\cos 50°(0.5)$
$= -2.49\text{ kip}\cdot\text{ft}$ *Ans.*

F4–6. $\downarrow + M_O = 500\sin 45°(3 + 3\cos 45°)$
$- 500\cos 45°(3\sin 45°)$
$= 1.06\text{ kN}\cdot\text{m}$ *Ans.*

F4–7. $\downarrow + (M_R)_O = \Sigma Fd;$
$(M_R)_O = -(600\text{ N})(1\text{ m})$
$+ (500\text{ N})[3\text{ m} + (2.5\text{ m})\cos 45°]$
$- (300\text{ N})[(2.5\text{ m})\sin 45°]$
$= 1254\text{ N}\cdot\text{m} = 1.25\text{ kN}\cdot\text{m}$ *Ans.*

F4–8. $\downarrow + (M_R)_O = \Sigma Fd;$
$(M_R)_O = \left[\left(\tfrac{3}{5}\right)500\text{ N}\right](0.425\text{ m})$
$- \left[\left(\tfrac{4}{5}\right)500\text{ N}\right](0.25\text{ m})$
$- [(600\text{ N})\cos 60°](0.25\text{ m})$
$- [(600\text{ N})\sin 60°](0.425\text{ m})$
$= -268\text{ N}\cdot\text{m} = 268\text{ N}\cdot\text{m}\downarrow$ *Ans.*

F4–9. $\downarrow + (M_R)_O = \Sigma Fd;$
$(M_R)_O = (300\cos 30°\text{ lb})(6\text{ ft} + 6\sin 30°\text{ ft})$
$- (300\sin 30°\text{ lb})(6\cos 30°\text{ ft})$
$+ (200\text{ lb})(6\cos 30°\text{ ft})$
$= 2.60\text{ kip}\cdot\text{ft}$ *Ans.*

F4–10. $\mathbf{F} = F\mathbf{u}_{AB} = 500\text{ N}\left(\tfrac{4}{5}\mathbf{i} - \tfrac{3}{5}\mathbf{j}\right) = \{400\mathbf{i} - 300\mathbf{j}\}\text{ N}$

$\mathbf{M}_O = \mathbf{r}_{OA} \times \mathbf{F} = \{3\mathbf{j}\}\text{ m} \times \{400\mathbf{i} - 300\mathbf{j}\}\text{ N}$
$= \{-1200\mathbf{k}\}\text{ N}\cdot\text{m}$ *Ans.*

or

$\mathbf{M}_O = \mathbf{r}_{OB} \times \mathbf{F} = \{4\mathbf{i}\}\text{ m} \times \{400\mathbf{i} - 300\mathbf{j}\}\text{ N}$
$= \{-1200\mathbf{k}\}\text{ N}\cdot\text{m}$ *Ans.*

F4–11. $\mathbf{F} = F\mathbf{u}_{BC}$

$= 120\text{ lb}\left[\dfrac{\{4\mathbf{i} - 4\mathbf{j} - 2\mathbf{k}\}\text{ ft}}{\sqrt{(4\text{ ft})^2 + (-4\text{ ft})^2 + (-2\text{ ft})^2}}\right]$

$= \{80\mathbf{i} - 80\mathbf{j} - 40\mathbf{k}\}\text{ lb}$

$\mathbf{M}_O = \mathbf{r}_C \times \mathbf{F} = \begin{vmatrix} \mathbf{i} & \mathbf{j} & \mathbf{k} \\ 5 & 0 & 0 \\ 80 & -80 & -40 \end{vmatrix}$

$= \{200\mathbf{j} - 400\mathbf{k}\}\text{ lb}\cdot\text{ft}$ *Ans.*

or

$\mathbf{M}_O = \mathbf{r}_B \times \mathbf{F} = \begin{vmatrix} \mathbf{i} & \mathbf{j} & \mathbf{k} \\ 1 & 4 & 2 \\ 80 & -80 & -40 \end{vmatrix}$

$= \{200\mathbf{j} - 400\mathbf{k}\}\text{ lb}\cdot\text{ft}$ *Ans.*

F4–12. $\mathbf{F}_R = \mathbf{F}_1 + \mathbf{F}_2$
$= \{(100 - 200)\mathbf{i} + (-120 + 250)\mathbf{j}$
$+ (75 + 100)\mathbf{k}\}\text{ lb}$
$= \{-100\mathbf{i} + 130\mathbf{j} + 175\mathbf{k}\}\text{ lb}$

$$(\mathbf{M}_R)_O = \mathbf{r}_A \times \mathbf{F}_R = \begin{vmatrix} \mathbf{i} & \mathbf{j} & \mathbf{k} \\ 4 & 5 & 3 \\ -100 & 130 & 175 \end{vmatrix}$$

$= \{485\mathbf{i} - 1000\mathbf{j} + 1020\mathbf{k}\}$ lb·ft   Ans.

F4–13. $M_y = \mathbf{j} \cdot (\mathbf{r}_A \times \mathbf{F}) = \begin{vmatrix} 0 & 1 & 0 \\ -3 & -4 & 2 \\ 30 & -20 & 50 \end{vmatrix}$

$= 210$ N·m   Ans.

F4–14. $M_x = \mathbf{i} \cdot (\mathbf{r}_{OB} \times \mathbf{F}) = \begin{vmatrix} 1 & 0 & 0 \\ 0.3 & 0.4 & -0.2 \\ 300 & -200 & 150 \end{vmatrix}$

$= 20$ N·m   Ans.

F4–15. $\mathbf{u}_{OA} = \dfrac{\mathbf{r}_A}{r_A} = \dfrac{\{0.3\mathbf{i} + 0.4\mathbf{j}\}\text{ m}}{\sqrt{(0.3\text{ m})^2 + (0.4\text{ m})^2}} = 0.6\mathbf{i} + 0.8\mathbf{j}$

$M_{OA} = \mathbf{u}_{OA} \cdot (\mathbf{r}_{AB} \times \mathbf{F}) = \begin{vmatrix} 0.6 & 0.8 & 0 \\ 0 & 0 & -0.2 \\ 300 & -200 & 150 \end{vmatrix}$

$= -72$ N·m   Ans.

$|M_{OA}| = 72$ N·m

F4–16. $\mathbf{u}_{AB} = \dfrac{\mathbf{r}_{AB}}{r_{AB}} = \dfrac{\{-4\mathbf{i} + 3\mathbf{j}\}\text{ ft}}{\sqrt{(-4\text{ ft})^2 + (3\text{ ft})^2}} = -0.8\mathbf{i} + 0.6\mathbf{j}$

$M_{AB} = \mathbf{u}_{AB} \cdot (\mathbf{r}_{AC} \times \mathbf{F})$

$= \begin{vmatrix} -0.8 & 0.6 & 0 \\ 0 & 0 & 2 \\ 50 & -40 & 20 \end{vmatrix} = -4$ lb·ft

$\mathbf{M}_{AB} = M_{AB}\mathbf{u}_{AB} = \{3.2\mathbf{i} - 2.4\mathbf{j}\}$ lb·ft   Ans.

F4–17. Scalar Analysis
The magnitudes of the force components are
$F_x = |200\cos 120°| = 100$ N
$F_y = 200\cos 60° = 100$ N
$F_z = 200\cos 45° = 141.42$ N

$M_x = -F_y(z) + F_z(y)$
$= -(100\text{ N})(0.25\text{ m}) + (141.42\text{ N})(0.3\text{ m})$
$= 17.4$ N·m   Ans.

Vector Analysis

$M_x = \begin{vmatrix} 1 & 0 & 0 \\ 0 & 0.3 & 0.25 \\ -100 & 100 & 141.42 \end{vmatrix} = 17.4$ N·m   Ans.

F4–18. Scalar Analysis
The magnitudes of the force components are
$F_x = \left(\frac{3}{5}\right)\left[\frac{4}{5}(500)\right] = 240$ N
$F_y = \frac{4}{5}\left[\frac{4}{5}(500)\right] = 320$ N
$F_z = \frac{3}{5}(500) = 300$ N
$M_x = -320(3) + 300(2) = -360$ N·m   Ans.
$M_y = -240(3) - 300(-2) = -120$ N·m   Ans.
$M_z = 240(2) - 320(2) = -160$ N·m   Ans.

Vector Analysis

$\mathbf{F} = \{-240\mathbf{i} + 320\mathbf{j} + 300\mathbf{k}\}$ N
$\mathbf{r}_{OA} = \{-2\mathbf{i} + 2\mathbf{j} + 3\mathbf{k}\}$ m
$M_x = \mathbf{i} \cdot (\mathbf{r}_{OA} \times \mathbf{F}) = -360$ N·m
$M_y = \mathbf{j} \cdot (\mathbf{r}_{OA} \times \mathbf{F}) = -120$ N·m
$M_z = \mathbf{k} \cdot (\mathbf{r}_{OA} \times \mathbf{F}) = -160$ N·m

F4–19. $\zeta + M_{C_R} = \Sigma M_A = -400(3) + 400(5) - 300(5)$
$\quad - 200(0.2) = -740$ N·m   Ans.

Also,

$\zeta + M_{C_R} = -300(5) + 400(2) - 200(0.2)$
$= -740$ N·m   Ans.

F4–20. $\zeta + M_{C_R} = 300(4) + 200(4) + 150(4)$
$= 2600$ lb·ft   Ans.

F4–21. $\zeta + M_C = 10\left(\frac{3}{5}\right)(2) - 10\left(\frac{4}{5}\right)(4) = -20$ kN·m
$= 20$ kN·m$\downarrow$ Ans.

F4–22. $\zeta + (M_B)_R = \Sigma M_B$
-1.5 kN·m $= (2\text{ kN})(0.3\text{ m}) - F(0.9\text{ m})$
$F = 2.33$ kN   Ans.

F4–23. $\mathbf{u}_1 = \dfrac{\mathbf{r}_1}{r_1} = \dfrac{\{-2\mathbf{i} + 2\mathbf{j} + 3.5\mathbf{k}\}\text{ ft}}{\sqrt{(-2\text{ ft})^2 + (2\text{ ft})^2 + (3.5\text{ ft})^2}}$

$= -\frac{2}{4.5}\mathbf{i} + \frac{2}{4.5}\mathbf{j} + \frac{3.5}{4.5}\mathbf{k}$

$\mathbf{u}_2 = -\mathbf{k}$

$\mathbf{u}_3 = \frac{1.5}{2.5}\mathbf{i} - \frac{2}{2.5}\mathbf{j}$

$(\mathbf{M}_c)_1 = (M_c)_1\mathbf{u}_1$
$= (450\text{ lb·ft})\left(-\frac{2}{4.5}\mathbf{i} + \frac{2}{4.5}\mathbf{j} + \frac{3.5}{4.5}\mathbf{k}\right)$
$= \{-200\mathbf{i} + 200\mathbf{j} + 350\mathbf{k}\}$ lb·ft

$(\mathbf{M}_c)_2 = (M_c)_2\mathbf{u}_2 = (250\text{ lb·ft})(-\mathbf{k})$
$= \{-250\mathbf{k}\}$ lb·ft

$(\mathbf{M}_c)_3 = (M_c)_3\,\mathbf{u}_3 = (300\text{ lb·ft})\left(\frac{1.5}{2.5}\mathbf{i} - \frac{2}{2.5}\mathbf{j}\right)$
$= \{180\mathbf{i} - 240\mathbf{j}\}$ lb·ft

$(\mathbf{M}_c)_R = \Sigma M_c;$

$(\mathbf{M}_c)_R = \{-20\mathbf{i} - 40\mathbf{j} + 100\mathbf{k}\}$ lb·ft Ans.

F4–24. $\mathbf{F}_B = \left(\frac{4}{5}\right)(450 \text{ N})\mathbf{j} - \left(\frac{3}{5}\right)(450 \text{ N})\mathbf{k}$
$= \{360\mathbf{j} - 270\mathbf{k}\}$ N

$\mathbf{M}_c = \mathbf{r}_{AB} \times \mathbf{F}_B = \begin{vmatrix} \mathbf{i} & \mathbf{j} & \mathbf{k} \\ 0.4 & 0 & 0 \\ 0 & 360 & -270 \end{vmatrix}$

$= \{108\mathbf{j} + 144\mathbf{k}\}$ N·m Ans.

Also,

$\mathbf{M}_c = (\mathbf{r}_A \times \mathbf{F}_A) + (\mathbf{r}_B \times \mathbf{F}_B)$

$= \begin{vmatrix} \mathbf{i} & \mathbf{j} & \mathbf{k} \\ 0 & 0 & 0.3 \\ 0 & -360 & 270 \end{vmatrix} + \begin{vmatrix} \mathbf{i} & \mathbf{j} & \mathbf{k} \\ 0.4 & 0 & 0.3 \\ 0 & 360 & -270 \end{vmatrix}$

$= \{108\mathbf{j} + 144\mathbf{k}\}$ N·m Ans.

F4–25. $\xrightarrow{+} F_{Rx} = \Sigma F_x; \quad F_{Rx} = -200 + \frac{3}{5}(100) = -140$ lb
$+\uparrow F_{Ry} = \Sigma F_y; \quad F_{Ry} = -150 + \frac{4}{5}(100) = -70$ lb
$F_R = \sqrt{140^2 + 70^2} = 157$ lb Ans.
$\theta = \tan^{-1}\left(\frac{70}{140}\right) = 26.6°$ Ans.

$\downarrow + M_{A_R} = \Sigma M_A;$
$M_{A_R} = -\frac{3}{5}(100)(4) + \frac{4}{5}(100)(6) - 150(3)$
$M_{R_A} = -210$ lb·ft Ans.

F4–26. $\xrightarrow{+} F_{Rx} = \Sigma F_x; \quad F_{Rx} = \frac{4}{5}(50) = 40$ N
$+\uparrow F_{Ry} = \Sigma F_y; \quad F_{Ry} = -40 - 30 - \frac{3}{5}(50)$
$= -100$ N
$F_R = \sqrt{(40)^2 + (100)^2} = 108$ N Ans.
$\theta = \tan^{-1}\left(\frac{100}{40}\right) = 68.2°$ Ans.

$\downarrow + M_{A_R} = \Sigma M_A;$
$M_{A_R} = -30(3) - \frac{3}{5}(50)(6) - 200$
$= -470$ N·m Ans.

F4–27. $\xrightarrow{+} (F_R)_x = \Sigma F_x;$
$(F_R)_x = 900 \sin 30° = 450$ N →
$+\uparrow (F_R)_y = \Sigma F_y;$
$(F_R)_y = -900 \cos 30° - 300$
$= -1079.42$ N $= 1079.42$ N ↓
$F_R = \sqrt{450^2 + 1079.42^2}$
$= 1169.47$ N $= 1.17$ kN Ans.
$\theta = \tan^{-1}\left(\frac{1079.42}{450}\right) = 67.4°$ Ans.

$\downarrow + (M_R)_A = \Sigma M_A;$
$(M_R)_A = 300 - 900 \cos 30°(0.75) - 300(2.25)$
$= -959.57$ N·m
$= 960$ N·m ↓ Ans.

F4–28. $\xrightarrow{+} (F_R)_x = \Sigma F_x;$
$(F_R)_x = 150\left(\frac{3}{5}\right) + 50 - 100\left(\frac{4}{5}\right) = 60$ lb →
$+\uparrow (F_R)_y = \Sigma F_y;$
$(F_R)_y = -150\left(\frac{4}{5}\right) - 100\left(\frac{3}{5}\right)$
$= -180$ lb $= 180$ lb ↓
$F_R = \sqrt{60^2 + 180^2} = 189.74$ lb $= 190$ lb Ans.
$\theta = \tan^{-1}\left(\frac{180}{60}\right) = 71.6°$ Ans.

$\downarrow + (M_R)_A = \Sigma M_A;$
$(M_R)_A = 100\left(\frac{4}{5}\right)(1) - 100\left(\frac{3}{5}\right)(6) - 150\left(\frac{4}{5}\right)(3)$
$= -640 = 640$ lb·ft ↓ Ans.

F4–29. $\mathbf{F}_R = \Sigma \mathbf{F};$
$\mathbf{F}_R = \mathbf{F}_1 + \mathbf{F}_2$
$= (-300\mathbf{i} + 150\mathbf{j} + 200\mathbf{k}) + (-450\mathbf{k})$
$= \{-300\mathbf{i} + 150\mathbf{j} - 250\mathbf{k}\}$ N Ans.

$\mathbf{r}_{OA} = (2-0)\mathbf{j} = \{2\mathbf{j}\}$ m
$\mathbf{r}_{OB} = (-1.5-0)\mathbf{i} + (2-0)\mathbf{j} + (1-0)\mathbf{k}$
$= \{-1.5\mathbf{i} + 2\mathbf{j} + 1\mathbf{k}\}$ m

$(\mathbf{M}_R)_O = \Sigma \mathbf{M};$
$(\mathbf{M}_R)_O = \mathbf{r}_{OB} \times \mathbf{F}_1 + \mathbf{r}_{OA} \times \mathbf{F}_2$

$= \begin{vmatrix} \mathbf{i} & \mathbf{j} & \mathbf{k} \\ -1.5 & 2 & 1 \\ -300 & 150 & 200 \end{vmatrix} + \begin{vmatrix} \mathbf{i} & \mathbf{j} & \mathbf{k} \\ 0 & 2 & 0 \\ 0 & 0 & -450 \end{vmatrix}$

$= \{-650\mathbf{i} + 375\mathbf{k}\}$ N·m Ans.

F4–30. $\mathbf{F}_1 = \{-100\mathbf{j}\}$ N

$\mathbf{F}_2 = (200 \text{ N}) \left[\frac{\{-0.4\mathbf{i} - 0.3\mathbf{k}\} \text{ m}}{\sqrt{(-0.4 \text{ m})^2 + (-0.3 \text{ m})^2}} \right]$

$= \{-160\mathbf{i} - 120\mathbf{k}\}$ N

$\mathbf{M}_c = \{-75\mathbf{i}\}$ N·m
$\mathbf{F}_R = \{-160\mathbf{i} - 100\mathbf{j} - 120\mathbf{k}\}$ N Ans.
$(\mathbf{M}_R)_O = (0.3\mathbf{k}) \times (-100\mathbf{j})$

$+ \begin{vmatrix} \mathbf{i} & \mathbf{j} & \mathbf{k} \\ 0 & 0.5 & 0.3 \\ -160 & 0 & -120 \end{vmatrix} + (-75\mathbf{i})$

$= \{-105\mathbf{i} - 48\mathbf{j} + 80\mathbf{k}\}$ N·m Ans.

F4–31. $+\uparrow F_R = \Sigma F_y; \quad F_R = -500 - 250 - 500$
$= -1250$ lb Ans.

$\downarrow + F_R x = \Sigma M_O;$
$-1250(x) = -500(3) - 250(6) - 500(9)$
$x = 6$ ft Ans.

F4–32. $\xrightarrow{+} (F_R)_x = \Sigma F_x;$
$(F_R)_x = 100(\frac{3}{5}) + 50 \sin 30° = 85 \text{ lb} \rightarrow$
$+\uparrow (F_R)_y = \Sigma F_y;$
$(F_R)_y = 200 + 50 \cos 30° - 100(\frac{4}{5})$
$= 163.30 \text{ lb} \uparrow$
$F_R = \sqrt{85^2 + 163.30^2} = 184 \text{ lb}$
$\theta = \tan^{-1}(\frac{163.30}{85}) = 62.5°$ Ans.
$\downarrow + (M_R)_A = \Sigma M_A;$
$163.30(d) = 200(3) - 100(\frac{4}{5})(6) + 50 \cos 30°(9)$
$d = 3.12 \text{ ft}$ Ans.

F4–33. $\xrightarrow{+} (F_R)_x = \Sigma F_x;$
$(F_R)_x = 15(\frac{4}{5}) = 12 \text{ kN} \rightarrow$
$+\uparrow (F_R)_y = \Sigma F_y;$
$(F_R)_y = -20 + 15(\frac{3}{5}) = -11 \text{ kN} = 11 \text{ kN} \downarrow$
$F_R = \sqrt{12^2 + 11^2} = 16.3 \text{ kN}$ Ans.
$\theta = \tan^{-1}(\frac{11}{12}) = 42.5°$ Ans.
$\downarrow + (M_R)_A = \Sigma M_A;$
$-11(d) = -20(2) - 15(\frac{4}{5})(2) + 15(\frac{3}{5})(6)$
$d = 0.909 \text{ m}$ Ans.

F4–34. $\xrightarrow{+} (F_R)_x = \Sigma F_x;$
$(F_R)_x = (\frac{3}{5}) 5 \text{ kN} - 8 \text{ kN}$
$= -5 \text{ kN} = 5 \text{ kN} \leftarrow$
$+\uparrow (F_R)_y = \Sigma F_y;$
$(F_R)_y = -6 \text{ kN} - (\frac{4}{5}) 5 \text{ kN}$
$= -10 \text{ kN} = 10 \text{ kN} \downarrow$
$F_R = \sqrt{5^2 + 10^2} = 11.2 \text{ kN}$ Ans.
$\theta = \tan^{-1}(\frac{10 \text{ kN}}{5 \text{ kN}}) = 63.4°$ Ans.
$\downarrow + (M_R)_A = \Sigma M_A;$
$5 \text{ kN}(d) = 8 \text{ kN}(3 \text{ m}) - 6 \text{ kN}(0.5 \text{ m})$
$- [(\frac{4}{5}) 5 \text{ kN}](2 \text{ m})$
$- [(\frac{3}{5}) 5 \text{ kN}](4 \text{ m})$
$d = 0.2 \text{ m}$ Ans.

F4–35. $+\uparrow F_R = \Sigma F_z;$ $F_R = -400 - 500 + 100$
$= -800 \text{ N}$ Ans.
$M_{Rx} = \Sigma M_x;$ $-800y = -400(4) - 500(4)$
$y = 4.50 \text{ m}$ Ans.
$M_{Ry} = \Sigma M_y;$ $800x = 500(4) - 100(3)$
$x = 2.125 \text{ m}$ Ans.

F4–36. $+\uparrow F_R = \Sigma F_z;$
$F_R = -200 - 200 - 100 - 100$
$= -600 \text{ N}$ Ans.
$M_{Rx} = \Sigma M_x;$
$-600y = 200(1) + 200(1) + 100(3) - 100(3)$
$y = -0.667 \text{ m}$ Ans.
$\uparrow + M_{Ry} = \Sigma M_y;$
$600x = 100(3) + 100(3) + 200(2) - 200(3)$
$x = 0.667 \text{ m}$ Ans.

Chapter 5

F5–1. $\xrightarrow{+} \Sigma F_x = 0;$ $-A_x + 500(\frac{3}{5}) = 0$
$A_x = 300 \text{ lb}$ Ans.
$\downarrow + \Sigma M_A = 0;$ $B_y(10) - 500(\frac{4}{5})(5) - 600 = 0$
$B_y = 260 \text{ lb}$ Ans.
$+\uparrow \Sigma F_y = 0;$ $A_y + 260 - 500(\frac{4}{5}) = 0$
$A_y = 140 \text{ lb}$ Ans.

F5–2. $\downarrow + \Sigma M_A = 0;$
$F_{CD} \sin 45°(1.5 \text{ m}) - 4 \text{ kN}(3 \text{ m}) = 0$
$F_{CD} = 11.31 \text{ kN} = 11.3 \text{ kN}$ Ans.
$\xrightarrow{+} \Sigma F_x = 0;$ $A_x + (11.31 \text{ kN}) \cos 45° = 0$
$A_x = -8 \text{ kN} = 8 \text{ kN} \leftarrow$ Ans.
$+\uparrow \Sigma F_y = 0;$
$A_y + (11.31 \text{ kN}) \sin 45° - 4 \text{ kN} = 0$
$A_y = -4 \text{ kN} = 4 \text{ kN} \downarrow$ Ans.

F5–3. $\downarrow + \Sigma M_A = 0;$
$N_B[6 \text{ m} + (6 \text{ m}) \cos 45°]$
$- 10 \text{ kN}[2 \text{ m} + (6 \text{ m}) \cos 45°]$
$- 5 \text{ kN}(4 \text{ m}) = 0$
$N_B = 8.047 \text{ kN} = 8.05 \text{ kN}$ Ans.
$\xrightarrow{+} \Sigma F_x = 0;$
$(5 \text{ kN}) \cos 45° - A_x = 0$
$A_x = 3.54 \text{ kN}$ Ans.
$+\uparrow \Sigma F_y = 0;$
$A_y + 8.047 \text{ kN} - (5 \text{ kN}) \sin 45° - 10 \text{ kN} = 0$
$A_y = 5.49 \text{ kN}$ Ans.

F5–4. $\xrightarrow{+} \Sigma F_x = 0;$ $-A_x + 400 \cos 30° = 0$
$A_x = 346 \text{ N}$ Ans.
$+\uparrow \Sigma F_y = 0;$
$A_y - 200 - 200 - 200 - 400 \sin 30° = 0$
$A_y = -800 \text{ N}$ Ans.
$\downarrow + \Sigma M_A = 0;$

FUNDAMENTAL PROBLEMS

$M_A - 200(2.5) - 200(3.5) - 200(4.5)$
$\quad - 400 \sin 30°(4.5) - 400 \cos 30°(3 \sin 60°) = 0$
$\quad\quad\quad M_A = 3.90 \text{ kN} \cdot \text{m}$ Ans.

F5–5. $\zeta + \Sigma M_A = 0;$
$N_C(0.7 \text{ m}) - [25(9.81) \text{ N}](0.5 \text{ m}) \cos 30° = 0$
$N_C = 151.71 \text{ N} = 152 \text{ N}$ Ans.
$\xrightarrow{+} \Sigma F_x = 0;$
$T_{AB} \cos 15° - (151.71 \text{ N}) \sin 30° = 0$
$\quad T_{AB} = 78.53 \text{ N} = 78.5 \text{ N}$ Ans.
$+ \uparrow \Sigma F_y = 0;$
$F_A + (78.53 \text{ N}) \sin 15°$
$\quad + (151.71 \text{ N}) \cos 30° - 25(9.81) \text{ N} = 0$
$\quad\quad F_A = 93.5 \text{ N}$ Ans.

F5–6. $\xrightarrow{+} \Sigma F_x = 0;$
$N_C \sin 30° - (250 \text{ N}) \sin 60° = 0$
$\quad N_C = 433.0 \text{ N} = 433 \text{ N}$ Ans.
$\zeta + \Sigma M_B = 0;$
$-N_A \sin 30°(0.15 \text{ m}) - 433.0 \text{ N}(0.2 \text{ m})$
$\quad + [(250 \text{ N}) \cos 30°](0.6 \text{ m}) = 0$
$\quad\quad N_A = 577.4 \text{ N} = 577 \text{ N}$ Ans.
$+ \uparrow \Sigma F_y = 0;$
$N_B - 577.4 \text{ N} + (433.0 \text{ N}) \cos 30°$
$\quad - (250 \text{ N}) \cos 60° = 0$
$\quad\quad N_B = 327 \text{ N}$ Ans.

F5–7. $\Sigma F_z = 0;$
$T_A + T_B + T_C - 200 - 500 = 0$
$\Sigma M_x = 0;$
$T_A(3) + T_C(3) - 500(1.5) - 200(3) = 0$
$\Sigma M_y = 0;$
$-T_B(4) - T_C(4) + 500(2) + 200(2) = 0$
$T_A = 350 \text{ lb}, T_B = 250 \text{ lb}, T_C = 100 \text{ lb}$ Ans.

F5–8. $\Sigma M_y = 0;$
$600 \text{ N}(0.2 \text{ m}) + 900 \text{ N}(0.6 \text{ m}) - A_z(1 \text{ m}) = 0$
$\quad\quad A_x = 660 \text{ N}$ Ans.
$\Sigma M_x = 0;$
$D_z(0.8 \text{ m}) - 600 \text{ N}(0.5 \text{ m}) - 900 \text{ N}(0.1 \text{ m}) = 0$
$\quad\quad D_z = 487.5 \text{ N}$ Ans.
$\Sigma F_x = 0; \quad\quad D_x = 0$ Ans.
$\Sigma F_y = 0; \quad\quad D_y = 0$ Ans.
$\Sigma F_z = 0;$
$T_{BC} + 660 \text{ N} + 487.5 \text{ N} - 900 \text{ N} - 600 \text{ N} = 0$
$\quad\quad T_{BC} = 352.5 \text{ N}$ Ans.

F5–9. $\Sigma F_y = 0; \quad 400 \text{ N} + C_y = 0;$
$\quad\quad C_y = -400 \text{ N}$ Ans.
$\Sigma M_y = 0; \quad -C_x(0.4 \text{ m}) - 600 \text{ N}(0.6 \text{ m}) = 0$
$\quad\quad C_x = -900 \text{ N}$ Ans.
$\Sigma M_x = 0; \quad B_z(0.6 \text{ m}) + 600 \text{ N}(1.2 \text{ m})$
$\quad + (-400 \text{ N})(0.4 \text{ m}) = 0$
$\quad\quad B_z = -933.3 \text{ N}$ Ans.
$\Sigma M_z = 0;$
$-B_x(0.6 \text{ m}) - (-900 \text{ N})(1.2 \text{ m})$
$\quad + (-400 \text{ N})(0.6 \text{ m}) = 0$
$\quad\quad B_x = 1400 \text{ N}$ Ans.
$\Sigma F_x = 0; \quad 1400 \text{ N} + (-900 \text{ N}) + A_x = 0$
$\quad\quad A_x = -500 \text{ N}$ Ans.
$\Sigma F_z = 0; \quad A_z - 933.3 \text{ N} + 600 \text{ N} = 0$
$\quad\quad A_z = 333.3 \text{ N}$ Ans.

F5–10. $\Sigma F_x = 0; \quad\quad B_x = 0$ Ans.
$\Sigma M_z = 0;$
$C_y(0.4 \text{ m} + 0.6 \text{ m}) = 0 \quad C_y = 0$ Ans.
$\Sigma F_y = 0; \quad A_y + 0 = 0 \quad A_y = 0$ Ans.
$\Sigma M_x = 0; \; C_z(0.6 \text{ m} + 0.6 \text{ m}) + B_z(0.6 \text{ m})$
$\quad - 450 \text{ N}(0.6 \text{ m} + 0.6 \text{ m}) = 0$
$1.2 C_z + 0.6 B_z - 540 = 0$
$\Sigma M_y = 0; \; -C_z(0.6 \text{ m} + 0.4 \text{ m})$
$\quad - B_z(0.6 \text{ m}) + 450 \text{ N}(0.6 \text{ m}) = 0$
$-C_z - 0.6 B_z + 270 = 0$
$C_z = 1350 \text{ N} \quad B_z = -1800 \text{ N}$ Ans.
$\Sigma F_z = 0;$
$A_z + 1350 \text{ N} + (-1800 \text{ N}) - 450 \text{ N} = 0$
$\quad\quad A_z = 900 \text{ N}$ Ans.

F5–11. $\Sigma F_y = 0; \quad A_y = 0$ Ans.
$\Sigma M_x = 0; \quad -9(3) + T_{CE}(3) = 0$
$\quad\quad T_{CE} = 9 \text{ kN}$ Ans.
$\Sigma M_z = 0; \quad T_{CF}(3) - 6(3) = 0$
$\quad\quad T_{CF} = 6 \text{ kN}$ Ans.
$\Sigma M_y = 0; \quad 9(4) - A_z(4) - 6(1.5) = 0$
$\quad\quad A_z = 6.75 \text{ kN}$ Ans.
$\Sigma F_x = 0; \quad A_x + 6 - 6 = 0 \quad A_x = 0$ Ans.
$\Sigma F_z = 0; \quad T_{BD} + 9 - 9 + 6.75 = 0$
$\quad\quad T_{BD} = -6.75 \text{ kN}$ Ans.

F5–12. $\Sigma F_x = 0;$ $A_x = 0$ Ans.
$\Sigma F_y = 0;$ $A_y = 0$ Ans.
$\Sigma F_z = 0;$ $A_z + T_{BC} - 80 = 0$
$\Sigma M_x = 0; (M_A)_x + 6T_{BC} - 80(6) = 0$
$\Sigma M_y = 0; 3F_{BC} - 80(1.5) = 0$ $T_{BC} = 40$ lb   Ans.
$\Sigma M_z = 0; (M_A)_z = 0$ Ans.
$A_z = 40$ lb $(M_A)_x = 240$ lb · ft   Ans.

Chapter 6

F6–1. Joint A.
$+\uparrow\Sigma F_y = 0;$ 225 lb $- F_{AD} \sin 45° = 0$
$F_{AD} = 318.20$ lb $= 318$ lb (C)   Ans.
$\xrightarrow{+}\Sigma F_x = 0;$ $F_{AB} - (318.20 \text{ lb}) \cos 45° = 0$
$F_{AB} = 225$ lb (T)   Ans.
Joint B.
$\xrightarrow{+}\Sigma F_x = 0;$ $F_{BC} - 225$ lb $= 0$
$F_{BC} = 225$ lb (T)   Ans.
$+\uparrow\Sigma F_y = 0;$ $F_{BD} = 0$ Ans.
Joint D.
$\xrightarrow{+}\Sigma F_x = 0;$
$F_{CD} \cos 45° + (318.20 \text{ lb}) \cos 45° - 450$ lb $= 0$
$F_{CD} = 318.20$ lb $= 318$ lb (T)   Ans.

F6–2. Joint D.
$+\uparrow\Sigma F_y = 0; \frac{3}{5}F_{CD} - 300 = 0;$
$F_{CD} = 500$ lb (T)   Ans.
$\xrightarrow{+}\Sigma F_x = 0; -F_{AD} + \frac{4}{5}(500) = 0$
$F_{AD} = 400$ lb (C)   Ans.
Joint C.
$F_{BC} = 500$ lb (T), $F_{AC} = F_{AB} = 0$ Ans.

F6–3. $D_x = 800$ lb, $D_y = 1400$ lb, $B_y = 800$ lb
Joint B.
$\xrightarrow{+}\Sigma F_x = 0; F_{BA} = 0$ Ans.
$+\uparrow\Sigma F_y = 0; -600 + F_{BC} = 0; F_{BC} = 600$ lb (T)   Ans.
Joint C.
$\uparrow\Sigma F_y = 0; \frac{3}{5}F_{CA} - 600 = 0; F_{CA} = 1000$ lb (C)   Ans.
$\xrightarrow{+}\Sigma F_x = 0; -F_{CD} + \frac{4}{5}(1000) = 0; F_{CD} = 800$ lb (C)   Ans.
Joint A.
$+\uparrow\Sigma F_y = 0; -800 - \frac{3}{5}(1000) + F_{AD} = 0$ Ans.
$F_{AD} = 1400$ lb (T)   Ans.

F6–4. Joint C.
$+\uparrow\Sigma F_y = 0;$ $2F \cos 30° - P = 0$
$F_{AC} = F_{BC} = F = \frac{P}{2 \cos 30°} = 0.5774P$ (C)
Joint B.
$\xrightarrow{+}\Sigma F_x = 0; 0.5774P \cos 60° - F_{AB} = 0$
$F_{AB} = 0.2887P$ (T)
$F_{AB} = 0.2887P = 2$ kN
$P = 6.928$ kN
$F_{AC} = F_{BC} = 0.5774P = 1.5$ kN
$P = 2.598$ kN
The *smaller value* of P is chosen,
$P = 2.598$ kN $= 2.60$ kN   Ans.

F6–5. $F_{CB} = 0$ Ans.
$F_{CD} = 0$ Ans.
$F_{AE} = 0$ Ans.
$F_{DE} = 0$ Ans.

F6–6. Joint C.
$+\uparrow\Sigma F_y = 0;$ 259.81 lb $- F_{CD} \sin 30° = 0$
$F_{CD} = 519.62$ lb $= 520$ lb (C)   Ans.
$\xrightarrow{+}\Sigma F_x = 0;$ $(519.62 \text{ lb}) \cos 30° - F_{BC} = 0$
$F_{BC} = 450$ lb (T)   Ans.
Joint D.
$+\nearrow\Sigma F_{y'} = 0;$ $F_{BD} \cos 30° = 0$ $F_{BD} = 0$ Ans.
$+\searrow\Sigma F_{x'} = 0;$ $F_{DE} - 519.62$ lb $= 0$
$F_{DE} = 519.62$ lb $= 520$ lb (C)   Ans.
Joint B.
$\uparrow\Sigma F_y = 0;$ $F_{BE} \sin\phi = 0$ $F_{BE} = 0$ Ans.
$\xrightarrow{+}\Sigma F_x = 0;$ 450 lb $- F_{AB} = 0$
$F_{AB} = 450$ lb (T)   Ans.
Joint A.
$+\uparrow\Sigma F_y = 0;$ 340.19 lb $- F_{AE} = 0$
$F_{AE} = 340$ lb (C)   Ans.

F6–7. $+\uparrow\Sigma F_y = 0; F_{CF} \sin 45° - 600 - 800 = 0$
$F_{CF} = 1980$ lb (T)   Ans.
$\downarrow+\Sigma M_C = 0; F_{FE}(4) - 800(4) = 0$
$F_{FE} = 800$ lb (T)   Ans.
$\downarrow+\Sigma M_F = 0; F_{BC}(4) - 600(4) - 800(8) = 0$
$F_{BC} = 2200$ lb (C)   Ans.

F6–8. $\zeta+\Sigma M_A = 0;\quad G_y(12\text{ m}) - 20\text{ kN}(2\text{ m})$
$- 30\text{ kN}(4\text{ m}) - 40\text{ kN}(6\text{ m}) = 0$
$G_y = 33.33\text{ kN}$
$+\uparrow\Sigma F_y = 0;\ F_{KC} + 33.33\text{ kN} - 40\text{ kN} = 0$
$F_{KC} = 6.67\text{ kN (C)}$ Ans.
$\zeta+\Sigma M_K = 0;$
$33.33\text{ kN}(8\text{ m}) - 40\text{ kN}(2\text{ m}) - F_{CD}(3\text{ m}) = 0$
$F_{CD} = 62.22\text{ kN} = 62.2\text{ kN (T)}$ Ans.
$\xrightarrow{+}\Sigma F_x = 0;\ F_{LK} - 62.22\text{ kN} = 0$
$F_{LK} = 62.2\text{ kN (C)}$ Ans.

F6–9. From the geometry of the truss,
$\phi = \tan^{-1}(3\text{ m}/2\text{ m}) = 56.31°.$
$\zeta+\Sigma M_K = 0;$
$33.33\text{ kN}(8\text{ m}) - 40\text{ kN}(2\text{ m}) - F_{CD}(3\text{ m}) = 0$
$F_{CD} = 62.2\text{ kN (T)}$ Ans.
$\zeta+\Sigma M_D = 0;\ 33.33\text{ kN}(6\text{ m}) - F_{KJ}(3\text{ m}) = 0$
$F_{KJ} = 66.7\text{ kN (C)}$ Ans.
$+\uparrow\Sigma F_y = 0;$
$33.33\text{ kN} - 40\text{ kN} + F_{KD}\sin 56.31° = 0$
$F_{KD} = 8.01\text{ kN (T)}$ Ans.

F6–10. From the geometry of the truss,
$\tan\phi = \dfrac{(9\text{ ft})\tan 30°}{3\text{ ft}} = 1.732\quad \phi = 60°$
$\zeta+\Sigma M_C = 0;$
$F_{EF}\sin 30°(6\text{ ft}) + 300\text{ lb}(6\text{ ft}) = 0$
$F_{EF} = -600\text{ lb} = 600\text{ lb (C)}$ Ans.
$\zeta+\Sigma M_D = 0;$
$300\text{ lb}(6\text{ ft}) - F_{CF}\sin 60°(6\text{ ft}) = 0$
$F_{CF} = 346.41\text{ lb} = 346\text{ lb (T)}$ Ans.
$\zeta+\Sigma M_F = 0;$
$300\text{ lb}(9\text{ ft}) - 300\text{ lb}(3\text{ ft}) - F_{BC}(9\text{ ft})\tan 30° = 0$
$F_{BC} = 346.41\text{ lb} = 346\text{ lb (T)}$ Ans.

F6–11. From the geometry of the truss,
$\theta = \tan^{-1}(1\text{ m}/2\text{ m}) = 26.57°$
$\phi = \tan^{-1}(3\text{ m}/2\text{ m}) = 56.31°.$

The location of O can be found using similar triangles.
$\dfrac{1\text{ m}}{2\text{ m}} = \dfrac{2\text{ m}}{2\text{ m} + x}$
$4\text{ m} = 2\text{ m} + x$
$x = 2\text{ m}$

$\zeta+\Sigma M_G = 0;$
$26.25\text{ kN}(4\text{ m}) - 15\text{ kN}(2\text{ m}) - F_{CD}(3\text{ m}) = 0$
$F_{CD} = 25\text{ kN (T)}$ Ans.
$\zeta+\Sigma M_D = 0;$
$26.25\text{ kN}(2\text{ m}) - F_{GF}\cos 26.57°(2\text{ m}) = 0$
$F_{GF} = 29.3\text{ kN (C)}$ Ans.
$\zeta+\Sigma M_O = 0;\ 15\text{ kN}(4\text{ m}) - 26.25\text{ kN}(2\text{ m})$
$- F_{GD}\sin 56.31°(4\text{ m}) = 0$
$F_{GD} = 2.253\text{ kN} = 2.25\text{ kN (T)}$ Ans.

F6–12. $\zeta+\Sigma M_H = 0;$
$F_{DC}(12\text{ ft}) + 1200\text{ lb}(9\text{ ft}) - 1600\text{ lb}(21\text{ ft}) = 0$
$F_{DC} = 1900\text{ lb (C)}$ Ans.
$\zeta+\Sigma M_D = 0;$
$1200\text{ lb}(21\text{ ft}) - 1600\text{ lb}(9\text{ ft}) - F_{HI}(12\text{ ft}) = 0$
$F_{HI} = 900\text{ lb (C)}$ Ans.
$\zeta+\Sigma M_C = 0;\ F_{JI}\cos 45°(12\text{ ft}) + 1200\text{ lb}(21\text{ ft})$
$- 900\text{ lb}(12\text{ ft}) - 1600\text{ lb}(9\text{ ft}) = 0$
$F_{JI} = 0$ Ans.

F6–13. $+\Sigma F_y = 0;$
$3P - 60 = 0$
$P = 20\text{ lb}$ Ans.

F6–14. $\zeta+\Sigma M_C = 0;$
$-\left(\tfrac{4}{5}\right)(F_{AB})(9) + 400(6) + 500(3) = 0$
$F_{AB} = 541.67\text{ lb}$
$\xrightarrow{+}\Sigma F_x = 0;\ -C_x + \tfrac{3}{5}(541.67) = 0$
$C_x = 325\text{ lb}$ Ans.
$+\uparrow\Sigma F_y = 0;\ C_y + \tfrac{4}{5}(541.67) - 400 - 500 = 0$
$C_y = 467\text{ lb}$ Ans.

F6–15. $\zeta+\Sigma M_A = 0;\ 100\text{ N}(250\text{ mm}) - N_B(50\text{ mm}) = 0$
$N_B = 500\text{ N}$ Ans.
$\xrightarrow{+}\Sigma F_x = 0;\ (500\text{ N})\sin 45° - A_x = 0$
$A_x = 353.55\text{ N}$
$+\uparrow\Sigma F_y = 0;\ A_y - 100\text{ N} - (500\text{ N})\cos 45° = 0$
$A_y = 453.55\text{ N}$
$F_A = \sqrt{(353.55\text{ N})^2 + (453.55\text{ N})^2}$
$= 575\text{ N}$ Ans.

F6–16. $\zeta +\Sigma M_C = 0;$
$$400(2) + 800 - F_{BA}\left(\tfrac{3}{\sqrt{10}}\right)(1)$$
$$- F_{BA}\left(\tfrac{1}{\sqrt{10}}\right)(3) = 0$$
$$F_{BA} = 843.27 \text{ N}$$
$\xrightarrow{+}\Sigma F_x = 0;\ C_x - 843.27\left(\tfrac{3}{\sqrt{10}}\right) = 0$
$$C_x = 800 \text{ N} \qquad Ans.$$
$+\uparrow\Sigma F_y = 0;\ C_y + 843.27\left(\tfrac{1}{\sqrt{10}}\right) - 400 = 0$
$$C_y = 133 \text{ N} \qquad Ans.$$

F6–17. Plate A:
$+\uparrow\Sigma F_y = 0;\ 2T + N_{AB} - 100 = 0$
Plate B:
$+\uparrow\Sigma F_y = 0;\ 2T - N_{AB} - 30 = 0$
$$T = 32.5 \text{ lb},\ N_{AB} = 35 \text{ lb} \qquad Ans.$$

F6–18. Pulley C:
$+\uparrow\Sigma F_y = 0;\ T - 2P = 0;\ T = 2P$
Beam:
$+\uparrow\Sigma F_y = 0;\ 2P + P - 6 = 0$
$$P = 2 \text{ kN} \qquad Ans.$$
$\zeta +\Sigma M_A = 0;\ 2(0.8) - 6(x) = 0$
$$x = 0.333 \text{ m} \qquad Ans.$$

F6–19. Member CD
$\zeta +\Sigma M_D = 0;\ 600(1.5) - N_C(3) = 0$
$$N_C = 300 \text{ N}$$
Member ABC
$\zeta +\Sigma M_A = 0;\ -800 + B_y(2) - (300\sin 45°)4 = 0$
$$B_y = 824.26 = 824 \text{ N} \qquad Ans.$$
$\xrightarrow{+}\Sigma F_x = 0;\ A_x - 300\cos 45° = 0;$
$$A_x = 212 \text{ N} \qquad Ans.$$
$+\uparrow\Sigma F_y = 0;\ -A_y + 824.26 - 300\sin 45° = 0;$
$$A_y = 612 \text{ N} \qquad Ans.$$

F6–20. AB is a two-force member.
Member BC
$\zeta +\Sigma M_c = 0;\ 15(3) + 10(6) - F_{BA}\left(\tfrac{4}{5}\right)(9) = 0$
$$F_{BA} = 14.58 \text{ kN}$$
$\xrightarrow{+}\Sigma F_x = 0;\ (14.58)\left(\tfrac{3}{5}\right) - C_x = 0;$
$$C_x = 8.75 \text{ kN}$$
$+\uparrow\Sigma F_y = 0;\ (14.58)\left(\tfrac{4}{5}\right) - 10 - 15 + C_y = 0;$
$$C_y = 13.3 \text{ kN}$$
Member CD
$\xrightarrow{+}\Sigma F_x = 0;\quad 8.75 - D_x = 0;\ D_x = 8.75 \text{ kN} \qquad Ans.$
$+\uparrow\Sigma F_y = 0;\quad -13.3 + D_y = 0;\ D_y = 13.3 \text{ kN} \qquad Ans.$
$\zeta +\Sigma M_D = 0;\ -8.75(4) + M_D = 0;\ M_D = 35 \text{ kN}\cdot\text{m} \quad Ans.$

F6–21. Entire frame
$\zeta +\Sigma M_A = 0;\ -600(3) - [400(3)](1.5) + C_y(3) = 0$
$$C_y = 1200 \text{ N} \qquad Ans.$$
$+\uparrow\Sigma F_y = 0;\ A_y - 400(3) + 1200 = 0$
$$A_y = 0 \qquad Ans.$$
$\xrightarrow{+}\Sigma F_x = 0;\ 600 - A_x - C_x = 0$
Member AB
$\zeta +\Sigma M_B = 0;\ 400(1.5)(0.75) - A_x(3) = 0$
$$A_x = 150 \text{ N} \qquad Ans.$$
$$C_x = 450 \text{ N} \qquad Ans.$$
These same results can be obtained by considering members AB and BC.

F6–22. Entire frame
$\zeta +\Sigma M_E = 0;\ 250(6) - A_y(6) = 0$
$$A_y = 250 \text{ N}$$
$\xrightarrow{+}\Sigma F_x = 0;\ E_x = 0$
$+\uparrow\Sigma F_y = 0;\ 250 - 250 + E_y = 0;\ E_y = 0$
Member BD
$\zeta +\Sigma M_D = 0;\ 250(4.5) - B_y(3) = 0;$
$$B_y = 375 \text{ N}$$
Member ABC
$\zeta +\Sigma M_C = 0;\ -250(3) + 375(1.5) + B_x(2) = 0$
$$B_x = 93.75 \text{ N}$$
$\xrightarrow{+}\Sigma F_x = 0;\ -C_x + B_x = 0;\ C_x = 93.75 \text{ N} \quad Ans.$
$+\uparrow\Sigma F_y = 0;\ 250 - 375 + C_y = 0;\ C_y = 125 \text{ N} \quad Ans.$

F6–23. AD, CB are two-force members.
Member AB
$\zeta +\Sigma M_A = 0;\ -[\tfrac{1}{2}(3)(4)](1.5) + B_y(3) = 0$
$$B_y = 3 \text{ kN}$$
Since BC is a two-force member $C_y = B_y = 3$ kN and $C_x = 0\ (\Sigma M_B = 0)$.
Member EDC
$\zeta +\Sigma M_E = 0;\ F_{DA}\left(\tfrac{4}{5}\right)(1.5) - 5(3) - 3(3) = 0$
$$F_{DA} = 20 \text{ kN}$$
$\xrightarrow{+}\Sigma F_x = 0;\ E_x - 20\left(\tfrac{3}{5}\right) = 0;\ E_x = 12 \text{ kN} \quad Ans.$
$+\uparrow\Sigma F_y = 0;\ -E_y + 20\left(\tfrac{4}{5}\right) - 5 - 3 = 0;$
$$E_y = 8 \text{ kN} \qquad Ans.$$

F6–24. *AC* and *DC* are two-force members.
Member *BC*
$\zeta + \Sigma M_C = 0$; $[\frac{1}{2}(3)(8)](1) - B_y(3) = 0$
$B_y = 4$ kN
Member *BA*
$\zeta + \Sigma M_B = 0$; $6(2) - A_x(4) = 0$
$A_x = 3$ kN Ans.
$+\uparrow \Sigma F_y = 0$; -4 kN $+ A_y = 0$; $A_y = 4$ kN Ans.
Entire Frame
$\zeta + \Sigma M_A = 0$; $-6(2) - [\frac{1}{2}(3)(8)](2) + D_y(3) = 0$
$D_y = 12$ kN Ans.
Since *DC* is a two-force member ($\Sigma M_C = 0$) then
$D_x = 0$ Ans.

Chapter 7

F7–1. $\zeta + \Sigma M_A = 0$; $B_y(6) - 10(1.5) - 15(4.5) = 0$
$B_y = 13.75$ kN
$\xrightarrow{+} \Sigma F_x = 0$; $N_C = 0$ Ans.
$+\uparrow \Sigma F_y = 0$; $V_C + 13.75 - 15 = 0$
$V_C = 1.25$ kN Ans.
$\zeta + \Sigma M_C = 0$; $13.75(3) - 15(1.5) - M_C = 0$
$M_C = 18.75$ kN·m Ans.

F7–2. $\zeta + \Sigma M_B = 0$; $30 - 10(1.5) - A_y(3) = 0$
$A_y = 5$ kN
$\xrightarrow{+} \Sigma F_x = 0$; $N_C = 0$ Ans.
$+\uparrow \Sigma F_y = 0$; $5 - V_C = 0$
$V_C = 5$ kN Ans.
$\zeta + \Sigma M_C = 0$; $M_C + 30 - 5(1.5) = 0$
$M_C = -22.5$ kN·m Ans.

F7–3. $\xrightarrow{+} \Sigma F_x = 0$; $B_x = 0$
$\zeta + \Sigma M_A = 0$; $3(6)(3) - B_y(9) = 0$
$B_y = 6$ kip
$\xrightarrow{+} \Sigma F_x = 0$; $N_C = 0$ Ans.
$\zeta + \uparrow \Sigma F_y = 0$; $V_C - 6 = 0$
$V_C = 6$ kip Ans.
$\zeta + \Sigma M_C = 0$; $-M_C - 6(4.5) = 0$
$M_C = -27$ kip·ft Ans.

F7–4. $\zeta + \Sigma M_A = 0$; $B_y(6) - 12(1.5) - 9(3)(4.5) = 0$
$B_y = 23.25$ kN
$\xrightarrow{+} \Sigma F_x = 0$; $N_C = 0$ Ans.
$+\uparrow \Sigma F_y = 0$; $V_C + 23.25 - 9(1.5) = 0$
$V_C = -9.75$ kN Ans.
$\zeta + \Sigma M_C = 0$;
$23.25(1.5) - 9(1.5)(0.75) - M_C = 0$
$M_C = 24.75$ kN·m Ans.

F7–5. $\zeta + \Sigma M_A = 0$; $B_y(6) - \frac{1}{2}(9)(6)(3) = 0$
$B_y = 13.5$ kN
$\xrightarrow{+} \Sigma F_x = 0$; $N_C = 0$ Ans.
$+\uparrow \Sigma F_y = 0$; $V_C + 13.5 - \frac{1}{2}(9)(3) = 0$
$V_C = 0$ Ans.
$\zeta + \Sigma M_C = 0$; $13.5(3) - \frac{1}{2}(9)(3)(1) - M_C = 0$
$M_C = 27$ kN·m Ans.

F7–6. $\zeta + \Sigma M_A = 0$;
$B_y(6) - \frac{1}{2}(6)(3)(2) - 6(3)(4.5) = 0$
$B_y = 16.5$ kN
$\xrightarrow{+} \Sigma F_x = 0$; $N_C = 0$ Ans.
$+\uparrow \Sigma F_y = 0$; $V_C + 16.5 - 6(3) = 0$
$V_C = 1.50$ kN Ans.
$\zeta + \Sigma M_C = 0$; $16.5(3) - 6(3)(1.5) - M_C = 0$
$M_C = 22.5$ kN·m Ans.

F7–7. $+\uparrow \Sigma F_y = 0$; $6 - V = 0$ $V = 6$ kN
$\zeta + \Sigma M_O = 0$; $M + 18 - 6x = 0$
$M = (6x - 18)$ kN·m

Fig. F7–7

F7–8. $+\uparrow \Sigma F_y = 0$; $-V - 2x = 0$
$V = (-2x)$ kN
$\zeta + \Sigma M_O = 0$; $M + 2x(\frac{x}{2}) - 15 = 0$
$M = (15 - x^2)$ kN·m

Fig. F7–8

F7–9. $+\uparrow \Sigma F_y = 0;\quad -V - \tfrac{1}{2}(2x)(x) = 0$
$V = -(x^2)\text{ kN}$
$\zeta + \Sigma M_O = 0;\quad M + \tfrac{1}{2}(2x)(x)(\tfrac{x}{3}) = 0$
$M = -(\tfrac{1}{3}x^3)\text{ kN·m}$

Fig. F7–9

F7–10. $+\uparrow \Sigma F_y = 0;\quad -V - 2 = 0$
$V = -2\text{ kN}$
$\zeta + \Sigma M_O = 0;\quad M + 2x = 0$
$M = (-2x)\text{ kN·m}$

Fig. F7–10

F7–11. Region $3\text{ m} \le x < 3\text{ m}$
$+\uparrow \Sigma F_y = 0;\quad -V - 5 = 0\quad V = -5\text{ kN}$
$\zeta + \Sigma M_O = 0;\quad M + 5x = 0$
$M = (-5x)\text{ kN·m}$
Region $0 < x \le 6\text{ m}$
$+\uparrow \Sigma F_y = 0;\quad V + 5 = 0\quad V = -5\text{ kN}$
$\zeta + \Sigma M_O = 0;\quad 5(6 - x) - M = 0$
$M = (5(6 - x))\text{ kN·m}$

Fig. F7–11

F7–12. Region $0 \le x < 3\text{ m}$
$+\uparrow \Sigma F_y = 0;\quad V = 0$
$\zeta + \Sigma M_O = 0;\quad M - 12 = 0$
$M = 12\text{ kN·m}$
Region $3\text{ m} < x \le 6\text{ m}$
$+\uparrow \Sigma F_y = 0;\quad V + 4 = 0\quad V = -4\text{ kN}$
$\zeta + \Sigma M_O = 0;\quad 4(6 - x) - M = 0$
$M = (4(6 - x))\text{ kN·m}$

Fig. F7–12

F7–13.

Fig. F7–13

F7–14.

Fig. F7–14

F7–15.

Fig. F7–15

F7–16.

Fig. F7–16

F7–17.

Fig. F7–17

F7–18.

Fig. F7–18

Chapter 8

F8–1. a) $+\uparrow \Sigma F_y = 0$; $N - 50(9.81) - 200(\frac{3}{5}) = 0$
$N = 610.5$ N
$\xrightarrow{+} \Sigma F_x = 0$; $F - 200(\frac{4}{5}) = 0$
$F = 160$ N
$F < F_{max} = \mu_s N = 0.3(610.5) = 183.15$ N,
therefore $F = 160$ N Ans.

b) $+\uparrow \Sigma F_y = 0$; $N - 50(9.81) - 400(\frac{3}{5}) = 0$
$N = 730.5$ N
$\xrightarrow{+} \Sigma F_x = 0$; $F - 400(\frac{4}{5}) = 0$
$F = 320$ N
$F > F_{max} = \mu_s N = 0.3(730.5) = 219.15$ N
Block slips
$F = \mu_s N = 0.2(730.5) = 146$ N Ans.

F8–2. $\zeta + \Sigma M_B = 0$;
$N_A(3) + 0.2N_A(4) - 30(9.81)(2) = 0$
$N_A = 154.89$ N
$\xrightarrow{+} \Sigma F_x = 0$; $P - 154.89 = 0$
$P = 154.89$ N $= 155$ N Ans.

F8–3. Crate A
$+\uparrow \Sigma F_y = 0$; $N_A - 50(9.81) = 0$
$N_A = 490.5$ N
$\xrightarrow{+} \Sigma F_x = 0$; $T - 0.25(490.5) = 0$
$T = 122.62$ N
Crate B
$+\uparrow \Sigma F_y = 0$; $N_B + P \sin 30° - 50(9.81) = 0$
$N_B = 490.5 - 0.5P$
$\xrightarrow{+} \Sigma F_x = 0$;
$P \cos 30° - 0.25(490.5 - 0.5 P) - 122.62 = 0$
$P = 247$ N Ans.

F8–4. $\xrightarrow{+} \Sigma F_x = 0$; $N_A - 0.3N_B = 0$
$+\uparrow \Sigma F_y = 0$;
$N_B + 0.3N_A + P - 100(9.81) = 0$
$\zeta + \Sigma M_O = 0$;
$P(0.6) - 0.3N_B(0.9) - 0.3 N_A(0.9) = 0$
$N_A = 175.70$ N $N_B = 585.67$ N
$P = 343$ N Ans.

F8–5. If slipping occurs:
$+\uparrow \Sigma F_y = 0$; $N_c - 250$ lb $= 0$; $N_c = 250$ lb
$\xrightarrow{+} \Sigma F_x = 0$; $P - 0.4(250) = 0$; $P = 100$ lb
If tipping occurs: ($\Sigma = 1.5$ m)
$\zeta + \Sigma M_A = 0$; $-P(4.5) + 250(1.5) = 0$
$P = 83.3$ lb Ans.

F8–6.
$\zeta + \Sigma M_A = 0$; $490.5(0.6) - T\cos 60°(0.3 \cos 60° + 0.6)$
$- T\sin 60° (0.3 \sin 60°) = 0$
$T = 490.5$ N
$\xrightarrow{+} \Sigma F_x = 0$; $490.5 \sin 60° - N_A = 0$; $N_A = 424.8$ N
$+\uparrow \Sigma F_y = 0$; $\mu_s(424.8) + 490.5 \cos 60° - 490.5 = 0$
$\mu_s = 0.577$ Ans.

F8–7. A will not move. Assume B is about to slip on C and A, and C is stationary.
$\stackrel{+}{\rightarrow}\Sigma F_x = 0$; $P - 0.3(50) - 0.4(75)$; $P = 45$ N
Assume C is about to slip and B does not slip on C, but is about to slip at A.
$\stackrel{+}{\rightarrow}\Sigma F_x = 0$; $P - 0.3(50) - 0.35(90) = 0$
$P = 46.5$ N > 45 N
$P = 45$ N Ans.

F8–8. A is about to move down the plane and B moves upward.
Block A
$+\nwarrow\Sigma F_y = 0$; $N = W\cos\theta$
$+\nearrow\Sigma F_x = 0$; $T + \mu_s(W\cos\theta) - W\sin\theta = 0$
$T = W\sin\theta - \mu_s W\cos\theta$ (1)
Block B
$+\nwarrow\Sigma F_y = 0$; $N' = 2W\cos\theta$
$+\nearrow\Sigma F_x = 0$; $2T - \mu_s W\cos\theta - \mu_s(2W\cos\theta) - W\sin\theta = 0$
Substituting Eq.(1) for T in the above equation and simplifying,
$\theta = \tan^{-1} 5\mu_s$ Ans.

F8–9. Assume B is about to slip on A, $F_B = 0.3\,N_B$.
$\stackrel{+}{\rightarrow}\Sigma F_x = 0$; $P - 0.3(10)(9.81) = 0$
$P = 29.4$ N
Assume B is about to tip on A, $x = 0$.
$\zeta+\Sigma M_O = 0$; $10(9.81)(0.15) - P(0.4) = 0$
$P = 36.8$ N
Assume A is about to slip, $F_A = 0.1\,N_A$.
$+\uparrow\Sigma F_y = 0$ $N - 10(9.81) - 7(9.81) = 0$
$\stackrel{+}{\rightarrow}\Sigma F_x = 0$ $P - 0.1[7(9.81) + 10(9.81)] = 0$
$P = 16.7$ N
Choose the smallest result. $P = 16.7$ N Ans.

Chapter 9

F9–1. $\bar{x} = \dfrac{\int_A \tilde{x}\, dA}{\int_A dA} = \dfrac{\dfrac{1}{2}\int_0^{1\,m} y^{2/3}\, dy}{\int_0^{1\,m} y^{1/3}\, dy} = 0.4$ m Ans.

$\bar{y} = \dfrac{\int_A \tilde{y}\, dA}{\int_A dA} = \dfrac{\int_0^{1\,m} y^{4/3}\, dy}{\int_0^{1\,m} y^{1/3}\, dy} = 0.571$ m Ans.

F9–2. $\bar{x} = \dfrac{\int_A \tilde{x}\, dA}{\int_A dA} = \dfrac{\int_0^{1\,m} x(x^3\, dx)}{\int_0^{1\,m} x^3\, dx}$
$= 0.8$ m Ans.

$\bar{y} = \dfrac{\int_A \tilde{y}\, dA}{\int_A dA} = \dfrac{\int_0^{1\,m} \dfrac{1}{2} x^3(x^3\, dx)}{\int_0^{1\,m} x^3\, dx}$
$= 0.286$ m Ans.

F9–3. $\bar{y} = \dfrac{\int_A \tilde{y}\, dA}{\int_A dA} = \dfrac{\int_0^{2\,m} y\left(2\left(\dfrac{y^{1/2}}{\sqrt{2}}\right)\right) dy}{\int_0^{2\,m} 2\left(\dfrac{y^{1/2}}{\sqrt{2}}\right) dy}$
$= 1.2$ m Ans.

F9–4. $\bar{x} = \dfrac{\int_m \tilde{x}\, dm}{\int_m dm} = \dfrac{\int_0^1 x\left[0.5\left(1 + \dfrac{x^2}{L^2}\right) dx\right]}{\int_0^1 0.5\left(1 + \dfrac{x^2}{L^2}\right) dx}$
$= \dfrac{9}{16}$ Ans.

F9–5. $\bar{y} = \dfrac{\int_V \tilde{y}\, dV}{\int_V dV} = \dfrac{\int_0^{1\,m} y\left(\dfrac{\pi}{4} y\, dy\right)}{\int_0^{1\,m} \dfrac{\pi}{4} y\, dy}$
$= 0.667$ m Ans.

F9–6. $\bar{z} = \dfrac{\int_V \tilde{z}\, dV}{\int_V dV} = \dfrac{\int_0^{2\,ft} z\left[\dfrac{9\pi}{64}(4-z)^2\, dz\right]}{\int_0^{2\,ft} \dfrac{9\pi}{64}(4-z)^2\, dz}$
$= 0.786$ ft Ans.

F9–7. $\bar{x} = \dfrac{\Sigma \tilde{x} L}{\Sigma L} = \dfrac{150(300) + 300(600) + 300(400)}{300 + 600 + 400}$
$= 265$ mm Ans.

$\bar{y} = \dfrac{\Sigma \tilde{y} L}{\Sigma L} = \dfrac{0(300) + 300(600) + 600(400)}{300 + 600 + 400}$
$= 323$ mm Ans.

$\bar{z} = \dfrac{\Sigma \tilde{z} L}{\Sigma L} = \dfrac{0(300) + 0(600) + (-200)(400)}{300 + 600 + 400}$
$= -61.5$ mm Ans.

F9–8. $\bar{y} = \dfrac{\Sigma \tilde{y} A}{\Sigma A} = \dfrac{150[300(50)] + 325[50(300)]}{300(50) + 50(300)}$

$= 237.5$ mm *Ans.*

F9–9. $\bar{y} = \dfrac{\Sigma \tilde{y} A}{\Sigma A} = \dfrac{100[2(200)(50)] + 225[50(400)]}{2(200)(50) + 50(400)}$

$= 162.5$ mm *Ans.*

F9–10. $\bar{x} = \dfrac{\Sigma \tilde{x} A}{\Sigma A} = \dfrac{0.25[4(0.5)] + 1.75[0.5(2.5)]}{4(0.5) + 0.5(2.5)}$

$= 0.827$ in. *Ans.*

$\bar{y} = \dfrac{\Sigma \tilde{y} A}{\Sigma A} = \dfrac{2[4(0.5)] + 0.25[(0.5)(2.5)]}{4(0.5) + (0.5)(2.5)}$

$= 1.33$ in. *Ans.*

F9–11. $\bar{x} = \dfrac{\Sigma \tilde{x} V}{\Sigma V} = \dfrac{1[2(7)(6)] + 4[4(2)(3)]}{2(7)(6) + 4(2)(3)}$

$= 1.67$ ft *Ans.*

$\bar{y} = \dfrac{\Sigma \tilde{y} V}{\Sigma V} = \dfrac{3.5[2(7)(6)] + 1[4(2)(3)]}{2(7)(6) + 4(2)(3)}$

$= 2.94$ ft *Ans.*

$\bar{z} = \dfrac{\Sigma \tilde{z} V}{\Sigma V} = \dfrac{3[2(7)(6)] + 1.5[4(2)(3)]}{2(7)(6) + 4(2)(3)}$

$= 2.67$ ft *Ans.*

F9–12. $\bar{x} = \dfrac{\Sigma \tilde{x} V}{\Sigma V}$

$= \dfrac{0.25[0.5(2.5)(1.8)] + 0.25\left[\frac{1}{2}(1.5)(1.8)(0.5)\right] + (1.0)\left[\frac{1}{2}(1.5)(1.8)(0.5)\right]}{0.5(2.5)(1.8) + \frac{1}{2}(1.5)(1.8)(0.5) + \frac{1}{2}(1.5)(1.8)(0.5)}$

$= 0.391$ m *Ans.*

$\bar{y} = \dfrac{\Sigma \tilde{y} V}{\Sigma V} = \dfrac{5.00625}{3.6} = 1.39$ m *Ans.*

$\bar{z} = \dfrac{\Sigma \tilde{z} V}{\Sigma V} = \dfrac{2.835}{3.6} = 0.7875$ m *Ans.*

F9–13. $A = 2\pi \Sigma \tilde{r} L$

$= 2\pi\left[0.75(1.5) + 1.5(2) + 0.75\sqrt{(1.5)^2 + (2)^2}\right]$

$= 37.7$ m² *Ans.*

$V = 2\pi \Sigma \tilde{r} A$

$= 2\pi\left[0.75(1.5)(2) + 0.5\left(\frac{1}{2}\right)(1.5)(2)\right]$

$= 18.8$ m³ *Ans.*

F9–14. $A = 2\pi \Sigma \tilde{r} L$

$= 2\pi\left[1.95\sqrt{(0.9)^2 + (1.2)^2} + 2.4(1.5) + 1.95(0.9) + 1.5(2.7)\right]$

$= 77.5$ m² *Ans.*

$V = 2\pi \Sigma \tilde{r} A$

$= 2\pi\left[1.8\left(\frac{1}{2}\right)(0.9)(1.2) + 1.95(0.9)(1.5)\right]$

$= 22.6$ m³ *Ans.*

F9–15. $A = 2\pi \Sigma \tilde{r} L$

$= 2\pi\left[7.5(15) + 15(18) + 22.5\sqrt{15^2 + 20^2} + 15(30)\right]$

$= 8765$ in.² *Ans.*

$V = 2\pi \Sigma \tilde{r} A$

$= 2\pi\left[7.5(15)(38) + 20\left(\frac{1}{2}\right)(15)(20)\right]$

$= 45\,710$ in.³ *Ans.*

F9–16. $A = 2\pi \Sigma \tilde{r} L$

$= 2\pi\left[\dfrac{2(1.5)}{\pi}\left(\dfrac{\pi(1.5)}{2}\right) + 1.5(2) + 0.75(1.5)\right]$

$= 40.1$ m² *Ans.*

$V = 2\pi \Sigma \tilde{r} A$

$= 2\pi\left[\dfrac{4(1.5)}{3\pi}\left(\dfrac{\pi(1.5^2)}{4}\right) + 0.75(1.5)(2)\right]$

$= 21.2$ m³ *Ans.*

F9–17. $w_b = \rho_w g h b = 1000(9.81)(6)(1)$

$= 58.86$ kN/m

$F_R = \frac{1}{2}(58.76)(6) = 176.58$ kN $= 177$ kN *Ans.*

F9–18. $w_b = \gamma_w h b = 62.4(4)(4) = 998.4$ lb/ft

$F_R = 998.4(3) = 3.00$ kip *Ans.*

F9–19. $w_b = \rho_w g h_B b = 1000(9.81)(2)(1.5)$

$= 29.43$ kN/m

$F_R = \frac{1}{2}(29.43)\left(\sqrt{(1.5)^2 + (2)^2}\right)$

$= 36.8$ kN *Ans.*

F9–20. $w_A = \rho_w g h_A b = 1000(9.81)(3)(2)$

$= 58.86$ kN/m

$w_B = \rho_w g h_B b = 1000(9.81)(5)(2)$

$= 98.1$ kN/m

$F_R = \frac{1}{2}(58.86 + 98.1)(2) = 157$ kN *Ans.*

F9–21. $w_A = \gamma_w h_A b = 62.4(6)(2) = 748.8$ lb/ft

$w_B = \gamma_w h_B b = 62.4(10)(2) = 1248$ lb/ft

$F_R = \frac{1}{2}(748.8 + 1248)\left(\sqrt{(3)^2 + (4)^2}\right)$

$= 4.99$ kip *Ans.*

Chapter 10

F10–1.

$$I_x = \int_A y^2\, dA = \int_0^{1\,\text{m}} y^2\left[(1 - y^{3/2})\,dy\right] = 0.111 \text{ m}^4 \quad \textit{Ans.}$$

F10–2.

$$I_x = \int_A y^2\, dA = \int_0^{1\,\text{m}} y^2\left(y^{3/2}\,dy\right) = 0.222 \text{ m}^4 \quad \textit{Ans.}$$

F10–3.

$$I_y = \int_A x^2\, dA = \int_0^{1\,\text{m}} x^2\left(x^{2/3}\right) dx = 0.273 \text{ m}^4 \quad \textit{Ans.}$$

F10–4.

$$I_y = \int_A x^2\, dA = \int_0^{1\,\text{m}} x^2\left[(1 - x^{2/3})\, dx\right] = 0.0606 \text{ m}^4 \quad \textit{Ans.}$$

F10–5. $I_x = \left[\frac{1}{12}(50)(450^3) + 0\right] + \left[\frac{1}{12}(300)(50^3) + 0\right]$
$\qquad = 383(10^6) \text{ mm}^4 \qquad \textit{Ans.}$

$I_y = \left[\frac{1}{12}(450)(50^3) + 0\right]$
$\qquad + 2\left[\frac{1}{12}(50)(150^3) + (150)(50)(100)^2\right]$
$\qquad = 183(10^6) \text{ mm}^4 \qquad \textit{Ans.}$

F10–6. $I_x = \frac{1}{12}(360)(200^3) - \frac{1}{12}(300)(140^3)$
$\qquad = 171(10^6) \text{ mm}^4 \qquad \textit{Ans.}$

$I_y = \frac{1}{12}(200)(360^3) - \frac{1}{12}(140)(300^3)$
$\qquad = 463(10^6) \text{ mm}^4 \qquad \textit{Ans.}$

F10–7. $I_y = 2\left[\frac{1}{12}(50)(200^3) + 0\right]$
$\qquad + \left[\frac{1}{12}(300)(50^3) + 0\right]$
$\qquad = 69.8 (10^6) \text{ mm}^4 \qquad \textit{Ans.}$

F10–8.

$\bar{y} = \dfrac{\Sigma \tilde{y} A}{\Sigma A} = \dfrac{15(150)(30) + 105(30)(150)}{150(30) + 30(150)} = 60 \text{ mm}$

$\bar{I}_{x'} = \Sigma(\bar{I} + Ad^2)$
$\qquad = \left[\frac{1}{12}(150)(30)^3 + (150)(30)(60 - 15)^2\right]$
$\qquad + \left[\frac{1}{12}(30)(150)^3 + 30(150)(105 - 60)^2\right]$
$\qquad = 27.0\,(10^6) \text{ mm}^4 \qquad \textit{Ans.}$

Chapter 11

F11–1. $y_G = 0.75 \sin \theta \qquad \delta y_G = 0.75 \cos \theta\, \delta\theta$
$\qquad x_C = 2(1.5) \cos \theta \qquad \delta x_C = -3 \sin \theta\, \delta\theta$
$\qquad \delta U = 0; \quad 2W\delta y_G + P\delta x_C = 0$
$\qquad (294.3 \cos \theta - 3P \sin \theta)\delta\theta = 0$
$\qquad P = 98.1 \cot \theta|_{\theta=60°} = 56.6 \text{ N} \qquad \textit{Ans.}$

F11–2. $x_A = 5 \cos \theta \qquad \delta x_A = -5 \sin \theta\, \delta\theta$
$\qquad y_G = 2.5 \sin \theta \qquad \delta y_G = 2.5 \cos \theta\, \delta\theta$
$\qquad \delta U = 0; \qquad -P\delta x_A + (-W\delta y_G) = 0$
$\qquad (5P \sin \theta - 1226.25 \cos \theta)\delta\theta = 0$
$\qquad P = 245.25 \cot \theta|_{\theta=60°} = 142 \text{ N} \qquad \textit{Ans.}$

F11–3. $x_B = 0.6 \sin \theta \qquad \delta x_B = 0.6 \cos \theta\, \delta\theta$
$\qquad y_C = 0.6 \cos \theta \qquad \delta y_C = -0.6 \sin \theta\, \delta\theta$
$\qquad \delta U = 0; \qquad -F_{sp}\delta x_B + (-P\delta y_C) = 0$
$\qquad -9(10^3) \sin \theta\, (0.6 \cos \theta\, \delta\theta)$
$\qquad - 2000(-0.6 \sin \theta\, \delta\theta) = 0$
$\qquad \sin \theta = 0 \qquad \theta = 0° \qquad \textit{Ans.}$
$\qquad\qquad -5400 \cos \theta + 1200 = 0$
$\qquad \theta = 77.16° = 77.2° \qquad \textit{Ans.}$

F11–4. $x_B = 0.9 \cos \theta \qquad \delta x_B = -0.9 \sin \theta\, \delta\theta$
$\qquad x_C = 2(0.9 \cos \theta) \quad \delta x_C = -1.8 \sin \theta\, \delta\theta$
$\qquad \delta U = 0; \quad P\delta x_B + (-F_{sp}\, \delta x_C) = 0$
$\qquad 6(10^3)(-0.9 \sin \theta\, \delta\theta)$
$\qquad - 36(10^3)(\cos \theta - 0.5)(-1.8 \sin \theta\, \delta\theta) = 0$
$\qquad \sin \theta\, (64\,800 \cos \theta - 37\,800)\delta\theta = 0$
$\qquad \sin \theta = 0 \qquad \theta = 0° \qquad \textit{Ans.}$
$\qquad\qquad 64\,800 \cos \theta - 37\,800 = 0$
$\qquad\qquad \theta = 54.31° = 54.3° \qquad \textit{Ans.}$

F11–5. $y_G = 2.5 \sin \theta \quad \delta y_G = 2.5 \cos \theta\, \delta\theta$
$\qquad x_A = 5 \cos \theta \quad \delta x_C = -5 \sin \theta\, \delta\theta$
$\qquad \delta U = 0; \qquad (-F_{sp}\delta x_A) - W\delta y_G = 0$
$\qquad (15\,000 \sin \theta \cos \theta - 7500 \sin \theta$
$\qquad\qquad - 1226.25 \cos \theta)\delta\theta = 0$
$\qquad \theta = 56.33° = 56.3° \qquad \textit{Ans.}$
$\qquad \text{or } \theta = 9.545° = 9.55° \qquad \textit{Ans.}$

F11–6. $F_{sp} = 15\,000(0.6 - 0.6 \cos \theta)$
$\qquad x_C = 3[0.3 \sin \theta] \qquad \delta x_C = 0.9 \cos \theta\, \delta\theta$
$\qquad y_B = 2[0.3 \cos \theta] \qquad \delta y_B = -0.6 \sin \theta\, \delta\theta$
$\qquad \delta U = 0; \qquad P\delta x_C + F_{sp}\delta y_B = 0$
$\qquad (135 \cos \theta - 5400 \sin \theta + 5400 \sin \theta \cos \theta)\delta\theta = 0$
$\qquad \theta = 20.9° \qquad \textit{Ans.}$

Review Problem Answers

Chapter 2

R2–1. $605.1 = 605$ N Ans.
$\phi = 55.40° + 30° = 85.4°$ Ans.

R2–2. $F_{1v} = 129$ N Ans.
$F_{1u} = 183$ N Ans.

R2–3. $F_R = 0$ Ans.

R2–4. $\{217\mathbf{i} + 85.5\mathbf{j} - 91.2\mathbf{k}\}$ lb Ans.

R2–5. $\mathbf{F} = F\mathbf{u} = \{98.1\mathbf{i} + 269\mathbf{j} - 201\mathbf{k}\}$ lb Ans.

R2–6. $\{415.69\mathbf{i} - 240\mathbf{j} + 360\mathbf{k}\}$ N Ans.
$\{318.20\mathbf{j} + 318.20\mathbf{k}\}$ N Ans.

R2–7. $82.0°$ Ans.

R2–8. $F_{\text{Proj}} = 0.667$ kN Ans.

Chapter 3

R3–1. $F_A = 34.6$ lb $F_B = 57.3$ lb Ans.
R3–2. $W = 240$ lb Ans.
R3–3. $F_{AB} = 41.67$ lb
$W = 76.6$ lb Ans.
R3–4. $F = 40.8$ lb Ans.
R3–5. $F_1 = 0$ Ans.
$F_2 = 311$ lb Ans.
$F_3 = 238$ lb Ans.
R3–6. $F_1 = 800$ N $F_2 = 147$ N $F_3 = 564$ N Ans.
R3–7. $F_{CD} = 625$ lb $F_{CA} = F_{CB} = 198$ lb Ans.
R3–8. $F = 1060$ N $= 1.06$ kN Ans.

Chapter 4

R4–1. $W = 319$ lb Ans.

R4–2. $\mathbf{F}_R = \{14.3\mathbf{i} + 21.4\mathbf{j} - 42.9\mathbf{k}\}$ lb Ans.
$(\mathbf{M}_R)_C = \{-1929\mathbf{i} + 428.8\mathbf{j} - 428.8\mathbf{k}\}$ lb·ft Ans.

R4–3. $\mathbf{M}_y = \{-78.4\mathbf{j}\}$ lb·ft Ans.
R4–4. $F = 133$ N Ans.
R4–5. $F_R = 2.10$ kN Ans.
$\theta = 81.6°$ Ans.
$M_P = 16.8$ kN·m Ans.
R4–6. $F_R = 462$ lb Ans.
$39.1°$ Ans.
$\theta = 3.07$ ft Ans.
R4–7. $F_R = 140$ kN↓ Ans.
$y = 7.14$ m Ans.
$x = 5.71$ m Ans.
R4–8. $F_R = 18.0$ kip Ans.
$x = 11.7$ ft Ans.

Chapter 5

R5–1. $N_B = 957$ N Ans.
$A_y = 743$ N
$A_x = 0$ Ans.
R5–2. $F_B = 67.6$ lb Ans.
$A_x = 25$ lb Ans.
$A_y = 111$ lb Ans.
R5–3. $A_x = 0$ Ans.
$F_{BD} = 208$ N Ans.
$F_{BC} = 792$ N Ans.
$A_z = 0$ Ans.
$(M_A)_x = 0$ Ans.
$(M_A)_z = 700$ N·m Ans.
R5–4.
$P = 100$ lb $B_z = 40$ lb $B_x = -35.7$ lb Ans.
$A_x = 136$ lb $B_y = 0$ $A_z = 40$ lb Ans.

Chapter 6

R6–1. Joint B:

$F_{BC} = 3$ kN (C) $F_{BA} = 8$ kN (C)   Ans.

Joint A:
$F_{AC} = 1.46$ kN (C) $F_{AF} = 4.17$ kN (T)   Ans.

Joint C:
$F_{CD} = 4.17$ kN (C) $F_{CF} = 3.12$ kN (C)   Ans.

Joint E:
$F_{EF} = 0$ $F_{ED} = 13.1$ kN (C)   Ans.

Joint D:
$F_{DF} = 5.21$ kN (T)   Ans.

R6–2. Joint A:
$F_{AG} = 471$ lb (C) $F_{AB} = 333$ lb (T)   Ans.

Joint B:
$F_{BC} = 333$ lb (T) $F_{GB} = 0$ Ans.

Joint D:
$F_{DE} = 943$ lb (C) $F_{DC} = 667$ lb (T)   Ans.

Joint E:
$F_{EC} = 667$ lb (T) $F_{EG} = 667$ lb (C)   Ans.

Joint C:
$F_{GC} = 471$ lb (T)   Ans.

R6–3. $F_{GJ} = 2.00$ kip (C) $F_{GC} = 1.00$ kip (T)   Ans.

R6–4.

$A_y = 1100$ lb $A_x = 0$

$F_{GF} = 1.80$ kip (C) $F_{FB} = 693$ lb (T)   Ans.

$F_{BC} = 1.21$ kip (T)   Ans.

R6–5. Joint A:

$F_{AD} = 2.47$ kip (T)   Ans.

$F_{AC} = F_{AB} = 1220.91$ lb (C) = 1.22 kip (C)   Ans.

R6–6. CB is a two force member.

Member AC:

$A_x = 300$ N $A_y = 80.4$ N   Ans.

$B_x = B_y = 220$ N   Ans.

R6–7. Member AB:

$B_y = 500$ N

Member BC:

$B_x = 1400$ N $A_y = 250$ N   Ans.

Member AB:

$A_x = 1.40$ kN   Ans.

Member BC:

$C_x = 500$ N $C_y = 1.70$ kN   Ans.

R6–8. $F_{BE} = 1.53$ kip $F_{CD} = 350$ lb   Ans.

Chapter 7

R7–1. $F_{CD} = 86.60$ lb

Since member CF is a two-force member,

$V_D = M_D = 0$ $N_D = F_{CD} = 86.6$ lb   Ans.

$B_y = 57.735$ lb

$N_E = 0$ $V_E = 28.9$ lb $M_E = 86.6$ lb·ft   Ans.

R7–2. Segment DC

$N_C = 0$ $V_C = 9.00$ kN $M_C = -62.5$ kN·m   Ans.

Segment DB

$N_B = 0$ $V_B = 27.5$ kN $M_B = -184.5$ kN·m   Ans.

R7–3.

R7–4.

R7–5.

R7–6.

$T_{max} = 76.7$ lb *Ans.*

Chapter 8

R8–1. Assume that the ladder slips at A:

$F_A = 0.4 N_A$ $N_A = 20$ lb $F_A = 8$ lb
$P = 1$ lb   *Ans.*

$N_B = 7$ lb > 0 **OK**
The ladder will remain in contact with the wall.

R8–2. *Friction:* Assuming the crate slips on dolly, then
$F_d = \mu_{sd} N_d = 0.5(588.6) = 294.3$ N. Solving Eqs. (1) and (2)

$P = 294.3$ N $x = 0.400$ m

Since $x > 0.3$ m, the crate tips on the dolly. If this is the case $x = 0.3$ m. Solving Eqs. (1) and (2) with $x = 0.3$ m yields
$P = 220.725$ N $F_d = 220.725$ N

Assuming the dolly slips at A, then $F_A = \mu_{sf} N_A = 0.35 N_A$. Substituting this value into Eqs. (3), (4), and (5) and solving, we have

$N_A = 559$ N $N_B = 128$ N

$P = 195.6$ N $= 196$ N *(Controls)*   *Ans.*

R8–3. *Friction:* If the disk is on the verge of moving, slipping would have to occur at point C. Hence, $F_C = \mu_s N_C = 0.2 N_C$. Substituting this into Eqs. (1) and (2) and solving, we have

$P = 182$ N   *Ans.*
$N_C = 606.60$ N

R8–4.

$N_B = 147.06$ N $P = 147$ N   *Ans.*

R8–5. All blocks slip at the same time:
$P = 625$ lb   *Ans.*

R8–6. $\alpha = \tan^{-1}\left(\dfrac{10}{25}\right) = 21.80°$
$M = 1745$ lb·in $= 145$ lb·ft   **Ans.**

R8–7. $d = 4.60$ ft   *Ans.*
R8–8. $P = 245$ N   *Ans.*

Chapter 9

R9–1. Using an element of thickness dx,
$$\bar{x} = \dfrac{b-a}{\ln \dfrac{b}{a}}$$ *Ans.*

R9–2. Using an element of thickness dx,
$$\bar{y} = \dfrac{c^2(b-a)}{2ab \ln \dfrac{b}{a}}$$ *Ans.*

R9–3. $\tilde{z} = \dfrac{3}{8}a$ *Ans.*

R9–4. $\bar{x} = 1.22$ ft $\bar{y} = 0.778$ ft $\bar{z} = 0.778$ ft   *Ans.*

R9–5. $\bar{y} = 87.5$ mm   *Ans.*

R9–6. $A = 1.25$ m²   *Ans.*

R9–7. $V = 0.0227$ m³   *Ans.*

R9–8. $F = 144$ lb   *Ans.*

$\bar{z} = 4.57$ ft   *Ans.*

R9–9. $A_y = 2.51$ MN $B_x = 2.20$ MN $B_y = 859$ kN   *Ans.*

R9–10. $F_N = 2.02$ kip   *Ans.*

Chapter 10

R10–1. $I_x = 1.07 \text{ in}^4$ Ans.

R10–2. $I_x = 0.610 \text{ ft}^4$ Ans.

R10–3. $I_y = 2.13 \text{ ft}^4$ Ans.

R10–4. $I_{xy} = 0.1875 \text{ m}^4$ Ans.

R10–5.

(a) $I_x = \dfrac{bh^3}{12}$ Ans.

(b) $I_x = \dfrac{bh^3}{36}$ Ans.

R10–6. $I_{xy} = 0.1875 \text{ m}^4$ Ans.

R10–7. $I_y = 0.0954 d^4$ Ans.

R10–8. $I_x = \dfrac{93}{70} mb^2$ Ans.

Chapter 11

R11–1. $F = \dfrac{P}{2 \tan \theta}$ Ans.

R11–2. $P = 5.28 \text{ lb}$ Ans.

R11–3. $F = 512 \text{ N}$ Ans.

R11–4. $\theta = 90°$ $\theta = 30°$ Ans.

R11–5. $\theta = 15.5°$ and $\theta = 85.4°$ Ans.

R11–6. $\cos \theta = 0$ $\theta = 90°$ Ans.

$\theta = 35.5°$ Ans.

$127.17 > 0$ Stable   Ans.

$-201.10 < 0$ Unstable   Ans.

R11–7. $\left. \dfrac{d^2V}{d\theta^2} \right|_{\theta = 90°} = 1524 > 0$ Ans.

$\left. \dfrac{d^2V}{d\theta^2} \right|_{\theta = 9.467°} = 1774.7 < 0$

Thus, the system is in unstable equilibrium at $\theta = 9.47°$.   Ans.

R11–8. $\left. \dfrac{d^2V}{d\theta^2} \right|_{\theta = 37.77°} = 125.7 > 0$

Thus, the system is in stable equilibrium at $\theta = 37.8°$ Ans.

Answers to Selected Problems

Chapter 1

- **1-1** a) 3.46 m, b) 45.6 s, c) 5.56 kN, d) 4.52 Mg
- **1-2** a) GN/s, b) Gg/N, c) GN/(kg·s)
- **1-3** a) Gg/m, b) kN/s, c) mm·kg
- **1-5** a) 45.3 MN, b) 56.8 km, c) 5.63 μg
- **1-6** a) 58.3 km, b) 68.5 s, c) 2.55 kN, d) 7.56 mg
- **1-7** a) 0.431 g, b) 35.3 kN, c) 5.32 m
- **1-9** a) km/s, b) mm, c) Gs/kg, d) mm·N
- **1-10** a) kN·m, b) Gg/m, c) μN/s^2, d) GN/s
- **1-11** a) N, b) MN/m, c) N/s^2, d) MN/s
- **1-13** 2.71 Mg/m^3
- **1-14** a) 44.9(10)$^{-3}$ N^2, b) 2.79(10)^3 s^2, c) 23.4 s
- **1-15** 7.41 μN
- **1-17** 88.5 km/h, 24.6 m/s
- **1-18** 26.9 μm·kg/N
- **1-19** 8.33 Mg/m^3
- **1-21** F = 10.0 nN, W_1 = 78.5 N, W_2 = 118 N

Chapter 2

- **2-1** F_R = 320 lb, θ = 176°
- **2-2** F_R = 474 lb, θ = 75.4°
- **2-3** θ = 18.6°, F = 319 N
- **2-5** F_{AB} = 314 lb, F_{AC} = 256 lb
- **2-6** F_R = 8.03 kN, ϕ = 1.22°
- **2-7** $(F_1)_v$ = 2.93 kN, $(F_1)_u$ = 2.07 kN
- **2-9** F_R = 464 N, θ = 78.6°
- **2-10** F_R = 222 N, θ = 4.26°
- **2-11** F_R = 10.8 kN, ϕ = 3.16°
- **2-13** F_R = 240 N, θ = 61.3°
- **2-14** F = 616 lb, θ = 46.9°
- **2-15** F_R = 458 lb, α = 10.9°
- **2-17** F_R = 19.2 N, θ = 2.37° ↘
- **2-18** F_R = 19.2 N, θ = 2.37° ↘
- **2-19** F_R = 10.8 kN, ϕ = 3.16°
- **2-21** θ = 53.5°, F_{AB} = 621 lb
- **2-22** ϕ = 38.3°; F_{AC} = 323 lb
- **2-23** F_C = 4.51 kN, F_B = 5.96 kN
- **2-25** ϕ = 35.4°, F_{AB} = 485 lb
- **2-26** θ = 54.3°, F_A = 686 N
- **2-27** F_R = 1.23 kN, θ = 6.08°
- **2-29** F_B = 1.61 kN, θ = 38.3°
- **2-30** F_R = 4.01 kN, ϕ = 16.2°
- **2-31** θ = 90°, F_B = 1 kN, F_R = 1.73 kN
- **2-33** F_R = 84.3 lb; θ = 37.2°
- **2-34** $\mathbf{F}_1$ = {20**i** + 34.6**j**} lb, $\mathbf{F}_2$ = {47.0**i** − 17.1**j**} lb
- **2-35** F_R = 69.2 lb, θ = 14.7° ∠θ
- **2-37** F_R = 1.96 kN, θ = 4.12°
- **2-38** 1222.6 N ≤ P ≤ 3173.5 N
- **2-39** F_{1x} = 141 N, F_{1y} = 141 N, F_{2x} = −130 N, F_{2y} = 75 N
- **2-41** F_R = 12.5 kN, θ = 64.1°
- **2-42** $\mathbf{F}_1$ = {680**i** − 510**j**} N, $\mathbf{F}_2$ = {−312**i** − 541**j**} N, $\mathbf{F}_3$ = {−530**i** + 530**j**} N
- **2-43** F_R = 546 N, θ = 253°
- **2-45** 0 ≤ P ≤ 1.62 kN
- **2-46** θ = 68.6°, F_B = 960 N
- **2-47** F_R = 839 N, θ = 14.8°
- **2-49** F_R = 389 N, ϕ' = 42.7°
- **2-50** $\mathbf{F}_1$ = {−52.0**i** + 30.0**j**} lb; $\mathbf{F}_2$ = {−47.0**i** − 17.1**j**} lb
- **2-51** F_R = 99.8 lb; θ = 173°
- **2-53** θ = 103°, F_2 = 88.1 lb
- **2-54** F_R = 161 lb, θ = 38.3°
- **2-55** F_R = 759 N, θ = 106°
- **2-57** F_R = 380 N, F_1 = 57.8 N
- **2-58** θ = 86.0°, F = 1.97 kN
- **2-59** F_R = 11.1 kN, θ = 47.7°
- **2-61** F_x = 40 N, F_y = 40 N, F_z = 56.6 N
- **2-62** $\mathbf{F}$ = {289**i** + 225**j** − 261**k**} lb
- **2-63** F_x = −200 N, F_y = 200 N, F_z = 283 N
- **2-65** $\mathbf{F}_1$ = {−106**i** + 106**j** + 260**k**} N, $\mathbf{F}_2$ = {250**i** + 354**j** − 250**k**} N, $\mathbf{F}_R$ = {144**i** + 460**j** + 9.81**k**} N, F_R = 482 N, α = 72.6°, β = 17.4°, γ = 88.8°
- **2-66** α_1 = 111°, β_1 = 69.3°, γ_1 = 30.0°
- **2-67** F_2 = 66.4 lb, α = 59.8°, β = 107°, γ = 144°
- **2-69** $\mathbf{F}_1$ = {86.5**i** + 186**j** − 143**k**} N, $\mathbf{F}_2$ = {−200**i** + 283**j** + 200**k**} N, $\mathbf{F}_R$ = {−113**i** + 468**j** + 56.6**k**} N, F_R = 485 N, α = 104°, β = 15.1°, γ = 83.3°
- **2-70** F_R = 384 N, α = 14.8°, β = 88.9° γ = 105°
- **2-71** F_1 = 429 lb, α_1 = 62.2°, β_1 = 110°, γ_1 = 145°
- **2-73** $\mathbf{F}_1$ = {72.0**i** + 54.0**j**} N, $\mathbf{F}_2$ = {53.0**i** + 53.0**j** + 130**k**} N, $\mathbf{F}_3$ = {200 **k**}
- **2-74** F_R = 407 N, α = 72.1°, β = 82.5°, γ = 19.5°
- **2-75** $\mathbf{F}_1$ = {14.0**j** − 48.0**k**} lb, $\mathbf{F}_2$ = {90**i** − 127**j** + 90**k**} lb
- **2-77** F_3 = 9.58 kN, α_3 = 15.5°, β_3 = 98.4°, γ_3 = 77.0°
- **2-78** $\mathbf{F}_1$ = {−159**i** + 276**j** + 318**k**}N, $\mathbf{F}_2$ = {424**i** + 300**j** + 30**k**}N
- **2-79** F_R = 634 N, α = 65.3°, β = 24.8°, γ = 88.4°
- **2-81** F_R = 1.55 kip, α = 82.4°, β = 37.6°, γ = 53.4°
- **2-82** F_R = 1.60 kip, α = 82.6°, β = 29.4°, γ = 61.7°

Answers to Selected Problems

2-83 $\alpha_3 = 139°$;
$\beta_3 = 128°, \gamma_3 = 102°$, if $F_R = 387$ N
$\beta_3 = 60.7°, \gamma_3 = 64.4°$, if $F_R = 1.41$ kN
2-85 $F = 2.02$ kN, $F_y = 0.523$ kN
2-86 $r_{AB} = 397$ mm
2-87 $r = 6.40$ ft
2-89 $r_{AB} = 284$ ft
2-90 6.63 m
2-91 4.42 m
2-93 $y = 16.7$ ft, $z = 50.2$ ft
2-94 $r_{AB} = 98.0$ ft, $\gamma = 114°$
2-95 $d = 6.71$ km
2-97 $F_R = 757$ N, $\alpha = 149°$, $\beta = 90.0°$, $\gamma = 59.0°$
2-98 $F = \{-34.3\mathbf{i} + 22.9\mathbf{j} - 68.6\mathbf{k}\}$ lb
2-99 $r_{OA} = 732$ mm
2-101 $x = 1.96$ m, $y = 2.34$ m, $F_R = 704$ N
2-102 $\alpha = 82.9°$
$\beta = 93.4°$
$\gamma = 172°$
2-105 $|\text{Proj } F| = 31.4$ lb
2-106 $F_\parallel = 99.1$ N, $F_\perp = 592$ N
2-107 $F_\parallel = 82.4$ N, $F_\perp = 594$ N
2-109 $F_1 = 18.3$ lb, $F_2 = 35.6$ lb
2-110 $(F_{ED})_\parallel = 334$ N, $(F_{ED})_\perp = 498$ N
2-111 $\theta = 36.4°$
2-113 $F = 23.49$ N, $F_\perp = 23.1$ N
2-114 $(F_1)_{F_2} = 5.44$ lb
2-115 $\theta = 100°$
2-117 $F_P = 10.5$ lb
2-118 $\theta = 143°$
2-119 $\theta = 19.2°$
2-121 $F_{CA} = 162$ N
2-122 $(F_{BC})_\parallel = 28.3$ lb, $(F_{BC})_\perp = 68.0$ lb
2-123 $F_{AC} = 25.9$ lb, $F_{AC} = \{-18.0\mathbf{i} - 15.4\mathbf{j} + 10.3\mathbf{k}\}$ lb
2-125 $F_{AC} = 26.5$ lb, $F_{BA} = 66.4$ lb
2-126 $[(F)_{AB}]_\parallel = 63.2$ lb, $[(F)_{AB}]_\perp = 64.1$ lb
2-127 $F_x = 75$ N, $F_y = 260$ N
2-129 $\theta = 82.9°$
2-130 Proj $\mathbf{F}_{AB} = \{0.229\mathbf{i} - 0.916\mathbf{j} + 1.15\mathbf{k}\}$ lb

Chapter 3

3-1 $F_2 = 9.60$ kN, $F_1 = 1.83$ kN
3-2 $\theta = 4.69°$, $F_1 = 4.31$ kN
3-3 $F_D = 4.90$ kN, $F_B = 3.46$ kN
3-5 $T = 7.20$ kN, $F = 5.40$ kN
3-6 $T = 7.66$ kN, $\theta = 70.1°$
3-7 $\theta = 20°$, $T = 30.5$ lb
3-9 $l'_{AB} = 0.452$ m, $l'_{AC} = 0.658$ m
3-10 $k = 176$ N/m
3-11 $l_0 = 2.03$ m
3-13 $F_{AC} = F_{AB} = \{2.45 \cos\theta\}$ kN, $l = 1.72$ m
3-14 $x_{AD} = 0.491$ m, $x_{AC} = 0.793$ m, $x_{AB} = 0.467$ m
3-15 $m = 8.56$ kg
3-17 $F_{BD} = 171$ N, $F_{BC} = 145$ N
3-18 $T_{AB} = 11.0$ lb, $T_{AC} = 7.76$ lb, $T_{BC} = 11.0$ lb, $T_{BE} = 19.0$ lb, $T_{CD} = 17.4$ lb, $\theta = 18.4°$
3-19 $\theta = 18.4°$, $W = 15.8$ lb
3-21 $k = 6.80$ lb/in.
3-22 $F = 158$ N
3-23 $d = 1.56$ m
3-25 $y = 2$ m, $F_1 = 833$ N
3-26 $T_{HA} = 294$ N, $T_{AB} = 340$ N, $T_{AE} = 170$ N, $T_{BD} = 490$ N, $T_{BC} = 562$ N
3-27 $m = 26.7$ kg
3-29 $F_{ED} = 30.2$ lb, $F_{EB} = 43.6$ lb, $F_{BC} = 69.8$ lb, $F_{BA} = 86.6$ lb
3-30 $W = 57.7$ lb
3-31 $F_{AC} = 267$ N, $F_{AB} = 98.6$ N
3-33 $s = 5.33$ ft
3-34 $W = 6$ lb
3-35 $F_{AB} = 175$ lb, $l = 2.34$ ft, or $F_{AB} = 82.4$ lb, $l = 1.40$ ft
3-37 $m_B = 3.58$ kg, $N = 19.7$ N
3-38 $\theta = 60°$, $T_{AB} = 34.6$ lb, $\theta = 60°$, $T_{AB} = 34.6$ lb
3-39 $\theta = 60°$, $W = 46.2$ lb
3-41 $\theta = 22.5°$, $m_C = 17.7$ kg, $m_B = 24.7$ kg, $\phi = 26.6°$
3-42 $W_B = 18.3$ lb
3-43 $\alpha = 77.2°$, $\beta = 148°$, $\gamma = 119°$, $P = 639$ lb
3-45 $F_{AB} = 0.980$ kN, $F_{AC} = 0.463$ kN, $F_{AD} = 1.55$ kN
3-46 $F_{AB} = 250$ lb, $F_{AC} = 389$ lb, $F_{BD} = 194$ lb
3-47 $F_{AB} = 2.52$ kN, $F_{CB} = 2.52$ kN, $F_{BD} = 3.64$ kN
3-49 $m = 102$ kg
3-50 $F_{AC} = 113$ lb, $F_{AB} = 257$ lb, $F_{AD} = 210$ lb
3-51 $F = 1558$ lb
3-53 $F_{AD} = 557$ lb, $W = 407$ lb
3-54 $F_{AB} = 79.2$ lb, $F_{AC} = 119$ lb, $F_{AD} = 283$ lb
3-55 $W_C = 265$ lb
3-57 $W = 55.8$ N
3-58 $s = 410$ mm
3-59 $F_{AC} = 574$ lb, $F_{AB} = 500$ lb, $F_{CD} = 1.22$ kip, $F_{CE} = 416$ lb
3-61 $F_{AB} = 2.12$ kN, $F_{AC} = 2.12$ kN, $F_{AD} = 7.03$ kN
3-62 $F_{OB} = 120$ N, $F_{OC} = 150$ N, $F_{OD} = 480$ N
3-63 $F_B = 28.1$ kN, $F_C = 39.8$ kN, $F_D = 28.1$ kN
3-65 $F_{AE} = 2.91$ kip, $F = 1.61$ kip
3-66 $F_{AB} = 16.6$ kN, $F_{AD} = 55.2$ kN
3-67 $s = 6.67$ in.

Chapter 4

- **4-1** $(M_{F_1})_A = 3.00$ kip·ft ↻
  $(M_{F_2})_A = 5.60$ kip·ft ↻
  $(M_{F_3})_A = 2.59$ kip·ft ↻
- **4-2** $(M_{F_1})_B = 4.125$ kip·ft ↺
  $(M_{F_2})_B = 2.00$ kip·ft ↺
  $(M_{F_3})_B = 40.0$ lb·ft ↺
- **4-3** $M_{RA} = 1.90$ N·m ↺
- **4-5** $M_B = 150$ N·m ↺, $M_B = 600$ N·m ↺, $M_B = 0$
- **4-6** $M_B = 90.6$ lb·ft ↻, $M_C = 141$ lb·ft ↺
- **4-7** $M_A = 195$ lb·ft ↺
- **4-9** $(M_O)_{max} = 48.0$ kN·m ↺, $\theta = 31.5°$
- **4-10** $\theta = 8.53°$
- **4-11** $M_A = 60.8$ lb·in ↺, $M_B = 79.9$ lb·in ↻
- **4-13** $F = 39.8$ lb
- **4-14** $M_O = 28.1$ N·m ↻, $\theta = 88.6°$, $(M_O)_{max} = 32.0$ N·m ↻
- **4-15** $\theta_a = 76.0°$; $M_{A,max} = 330$ lb·ft; $\theta_b = 166°$
- **4-17** $M_A = 362$ lb·in ↻
- **4-18** $F = 27.6$ lb
- **4-19** $M_{A1} = 542$ lb·in; ↻, $M_{A2} = 10.0$ lb·in ↻
- **4-21** $M_C = 4.97$ Mg
- **4-25** $\mathbf{M}_O = \{-25\mathbf{i} + 6200\mathbf{j} - 900\mathbf{k}\}$ lb·ft
- **4-26** $\mathbf{M}_A = \{-175\mathbf{i} + 5600\mathbf{j} - 900\mathbf{k}\}$ lb·ft
- **4-27** $\mathbf{M}_O = \{-40\mathbf{i} - 44\mathbf{j} - 8\mathbf{k}\}$ kN·m
- **4-29** $\mathbf{M}_O = \{61.2\mathbf{i} + 81.6\mathbf{j}\}$ N·m
- **4-30** $F = 176$ N·m
- **4-31** $\mathbf{M}_P = \{-24\mathbf{i} + 24\mathbf{j} + 8\mathbf{k}\}$ kN·m
- **4-33** $\alpha = 77.8°, \beta = 29.6°, \gamma = 63.4°$
- **4-34** $\mathbf{M}_A = \{574\mathbf{i} + 350\mathbf{j} + 1385\mathbf{k}\}$ N·m
- **4-35** $F = 585$ N
- **4-37** $\mathbf{M}_O = \{163\mathbf{i} - 346\mathbf{j} - 360\mathbf{k}\}$ N·m
- **4-38** $\mathbf{M}_A = \{-82.9\mathbf{i} + 41.5\mathbf{j} + 232\mathbf{k}\}$ lb·ft
- **4-39** $\mathbf{M}_B = \{-82.9\mathbf{i} - 96.8\mathbf{j} - 52.8\mathbf{k}\}$ lb·ft
- **4-41** $F = 18.6$ lb
- **4-42** $\mathbf{M}_C = \{-226\mathbf{i} - 226\mathbf{j} - 226\mathbf{k}\}$ lb·ft
- **4-43** $\mathbf{M}_A = \{-5.39\mathbf{i} + 13.1\mathbf{j} + 11.4\mathbf{k}\}$ N·m
- **4-45** $y = 2$ m, $z = 1$ m
- **4-46** $y = 1$ m, $z = 3$ m, $d = 1.15$ m
- **4-47** $\mathbf{M}_A = \{360\mathbf{i} + 240\mathbf{j}\}$ N·m
- **4-49** $\mathbf{M}_1 = \{-36\mathbf{i} - 36\mathbf{k}\}$ N·m,
  $\mathbf{M}_2 = \{-42\mathbf{i} + 14\mathbf{j} - 68\mathbf{k}\}$ N·m,
  $(M_R)_A = 131$ N·m, $\alpha = 127°, \beta = 96.1°, \gamma = 143°$
- **4-50** $\theta = 0°, 90°, 180°$
- **4-51** $\mathbf{M}_C = \{343\mathbf{i} + 263\mathbf{j} + 198\mathbf{k}\}$ lb·ft
- **4-53** $\mathbf{M}_{OA} = \{-17.3\mathbf{i} - 13.0\mathbf{j}\}$ N·m
- **4-54** $\mathbf{M}_{AF} = \{9.33\mathbf{i} + 9.33\mathbf{j} - 4.67\mathbf{k}\}$ N·m
- **4-55** $\mathbf{M}_{OD} = \{11.3\mathbf{i} + 11.3\mathbf{j} - 5.67\mathbf{k}\}$ N·m
- **4-57** $F = 19.8$ lb
- **4-58** $M_x = 21.7$ N·m
- **4-59** $F = 139$ N
- **4-61** $M_{AB} = 136$ N·m
- **4-62** $M_{BC} = 165$ N·m
- **4-63** $M_{CA} = 226$ N·m
- **4-65** $(M_a)_1 = 38.2$ lb·in., $(M_a)_2 = 14.1$ lb·in.
- **4-66** $M_{AB} = 0$
- **4-67** $(\mathbf{M}_R)_{Oa} = \{26.1\mathbf{i} - 15.1\mathbf{j}\}$ lb·ft
- **4-69** $W = 56.8$ lb
- **4-70** $M_z = 15.5$ N·m
- **4-71** $M_{y'} = 464$ lb·ft
- **4-73** No., Changing c has no effect on turning nut.
- **4-74** $F = 67.5$ lb
- **4-75** $F = 830$ N
- **4-77** $F = 139$ lb
- **4-78** $F = 133$ N, $F = 267$ N
- **4-79** $\theta = 45°$, $P = 8.99$ lb
- **4-81** $F = 83.3$ N
- **4-82** $F = 300$ N, $P = 500$ N, $d = 3.96$ m
- **4-83** $M_R = 39.7$ N·m
- **4-85** $d = 5.54$ ft
- **4-86** $M_R = 134$ lb·ft ↻
- **4-87** $M_R = 39.6$ lb·ft ↺
- **4-89** $M_R = 96.0$ lb·ft, $\alpha = 47.4°, \beta = 74.9°, \gamma = 133°$
- **4-90** $M_3 = 318$ lb·ft, $M_1 = M_2 = 287$ lb·ft
- **4-91** $M_R = 64.0$ lb·ft, $\alpha = 94.7°, \beta = 13.2°, \gamma = 102°$
- **4-93** $F = 98.1$ N
- **4-94** $\mathbf{M}_C = \{-65.0\mathbf{i} - 37.5\mathbf{k}\}$ N·m
- **4-95** $F = 15.4$ N
- **4-97** $F = 832$ N
- **4-98** $(M_c)_R = 1.04$ kN·m, $\alpha = 120°, \beta = 61.3°, \gamma = 136°$
- **4-99** $M_2 = 346$ N·m, $M_1 = 450$ N·m, $M_3 = 427$ N·m
- **4-101** $M_c = 40.8$ N·m $\alpha = 11.3° \beta = 101° \gamma = 90°$
- **4-102** $F = 98.1$ N
- **4-103** $F_R = 365$ N, $\theta = 70.8°$ ↗, $(M_R)_O = 2364$ N·m ↺
- **4-105** $F_R = 464$ lb, $\theta = 45.1°$ ↖, $M_{RB} = 1600$ lb·ft ↻
- **4-106** $F_R = 464$ lb, $\theta = 45.1°$ ↖, $M_{RA} = 832$ lb·ft ↻
- **4-107** $F_R = 1.08$ kN, $\theta = 68.2°$ ↗, $M_{Ro} = 0.901$ kN·m ↺
- **4-109** $F_R = 342$ N, $\theta = 43.0°$ ↗, $M_{Ro} = 99.5$ N·m ↻
- **4-110** $F_R = 73.8$ lb, $\theta = 59.0°$ ↖, $d = 5.66$ ft
- **4-111** $F = 8.94$ kN, $d = 4.5$ m
- **4-113** $F_R = 29.9$ lb, $\theta = 78.4°$ ↙, $M_{Ro} = 214$ lb·in. ↺

4-114 $F_R = 26.4$ lb, $\theta = 85.7°$ ⦨, $M_{Ro} = 205$ lb·in. ↶
4-115 $F_R = 219$ lb, $\theta = 79.5°$ ⦨,
$M_{R_A} = 1.19$ kip·ft ↶
4-117 $F_R = 416$ lb, $\theta = 35.2°$ ⦨,
$(M_R)_A = 1.48$ kip·ft ↷
4-118 $F_2 = 41.1$ lb, $\theta = 44.9°$, $F_1 = 36.8$ lb
4-119 $(M_R)_C = 6.40$ kip·ft, $F_R = 850$ lb; $\theta = 61.9°$ ↷
4-121 $\mathbf{F}_R = \{-700\mathbf{j} - 2200\mathbf{k}\}$ lb;
$\mathbf{M}_{RA} = \{19.6\mathbf{i} - 21.6\mathbf{j} + 8.40\mathbf{k}\}$ kip·ft
4-122 $\mathbf{F}_R = [-30\mathbf{i} + 15\mathbf{j} + 45\mathbf{k}]$ lb;
$(\mathbf{M}_R)_o = [80\mathbf{i} - 87.5\mathbf{j} + 102.5\mathbf{k}]$ lb·ft
4-123 $\mathbf{F}_R = \{506\mathbf{i} - 23.2\mathbf{j} + 1236\mathbf{k}\}$ N,
$\mathbf{M}_{Ro} = \{-17.7\mathbf{i} - 600\mathbf{j} + 26.5\mathbf{k}\}$ N·m
4-125 $\mathbf{F}_R = \{-70\mathbf{i} + 140\mathbf{j} - 408\mathbf{k}\}$ N,
$\mathbf{M}_{RP} = \{-26\mathbf{i} + 357\mathbf{j} + 127\mathbf{k}\}$ N·m
4-126 $F_R = 10.75$ kip ↓, $d = 13.7$ ft
4-127 $F_R = 10.75$ kip ↓, $d = 9.26$ ft
4-129 $F_R = 58.3$ lb; $\theta = 31.0°$ ⦨, $y = -3.00$ ft
4-130 $F_R = 904$ lb, $\phi = 23.6°$, $d = 6.10$ ft
4-131 $F = 1302$ N, $\theta = 84.5°$ ⦨, $x = 7.36$ m ↷
4-133 $F_R = 78.1$ lb, $\theta = 87.1°$ ⦨, $M_{Ro} = 100$ lb·ft ↷
4-134 $F_R = 78.1$ lb, $\theta = 87.1°$ ⦨, $M_{R_P} = 124$ lb·ft ↷
4-135 $F_R = 991$ N, $\theta = 63.0°$ ⦨, $y = 1.78$ m
4-137 $F_R = 65.9$ lb, $\theta = 49.8°$ ⦨, $d = 2.10$ ft
4-138 $F_R = 65.9$ lb, $\theta = 49.8°$ ⦨, $d = 4.62$ ft
4-139 $F_R = 98.5$ lb; $\theta = 24.0°$ ⦨; $d = 4.33$ ft
4-141 $F_R = 197$ lb, $\theta = 42.6°$ ⦨, $d = 5.24$ ft
4-142 $F_R = 197$ lb, $\theta = 42.6°$ ⦨, $d = 0.824$ ft
4-143 $F_R = 26$ kN, $y = 82.7$ mm, $x = 3.85$ mm
4-145 $F_C = 600$ N, $F_D = 500$ N
4-146 $F_R = 700$ lb, $z = 0.447$ ft, $x = -0.117$ ft
4-147 $F_B = 163$ lb; $F_C = 223$ lb
4-149 $F_R = 495$ N, $M_R = 1.54$ kN·m, $z = 1.16$ m,
$y = 2.06$ m
4-150 $F = 53.3$ N, $\mathbf{F}_{RW} = \{-40\mathbf{i}\}$ N,
$\mathbf{M}_{RW} = \{-30\mathbf{i}\}$ N·m
4-151 $y = 0.20$ m, $\mathbf{F}_{RW} = \{-40\mathbf{i}\}$ N,
$\mathbf{M}_{RW} = \{-30\mathbf{i}\}$ N·m
4-153 $F_R = 6.75$ kN, $\bar{x} = 2.5$ m
4-154 $F_R = 21.0$ kN, $d = 3.43$ m
4-155 $F_R = 7$ lb, $\bar{x} = 0.268$ ft
4-157 $F_R = 162$ lb ↓; $d = 6.80$ ft
4-158 $F_R = 3.90$ kip, $d = 11.3$ ft
4-159 $F_R = 13.5$ kip ↓; $d = 6.0$ ft
4-161 $M_{RO} = 1.95$ kip·ft ↷
4-162 $F_R = 27$ kN, $d = 2.11$ m
4-163 $F_R = 107$ kN, $\bar{z} = 2.40$ m
4-165 $F_R = 6$ kN, $d = 1.88$ m
4-166 $F_R = 1.80$ kN, $(M_R)_A = 4.20$ kN·m ↷
4-167 $F_R = 1.80$ kN, $d = 2.33$ m
4-169 $F_R = 12.0$ kN, $\theta = 48.4°$ ⦨
4-170 $b = 5.625$ m, $a = 1.54$ m
4-171 $F_R = 1.87$ kip, $\tilde{x} = 3.66$ ft
4-173 $F_R = 30$ kN, $d = 4.10$ m
4-174 $F_R = 2.96$ kip, $\bar{x} = 9.21$ ft
4-175 $F_R = \dfrac{2Lw_0}{\pi}, (M_R)_O = \left(\dfrac{2\pi - 4}{\pi^2}\right)w_0L^2$ ↷

Chapter 5

5-10 $B_y = 400$ N, $A_x = 0$, $A_y = 200$ N
5-11 $N_A = 750$ N, $B_y = 600$ N, $B_x = 450$ N
5-14 $N_A = 3.33$ kN, $B_X = 2.40$ kN, $B_Y = 133$ N
5-15 $A_x = 0$, $A_y = 6.75$ kN, $M_A = 46.25$ kN·m
5-17 $A_y = 390$ N, $B_x = 0$, $B_y = 60.0$ N
5-18 $R_A = 667$ N, $R_B = 220$ N, $R_C = 440$ N
5-19 $F_{BC} = 204$ N, $A_x = 91.1$ N, $A_y = 84.1$ N
5-21 $d = 0$, $F_{AC} = 17.5$ kN, $F_{AB} = 19.8$ kN
5-22 $N_A = 3.71$ kN, $B_x = 1.86$ kN, $B_y = 8.78$ kN
5-23 $w = 2.67$ kN/m
5-25 $N_A = 39.7$ lb, $N_B = 82.5$ lb, $M_A = 106$ lb·ft
5-26 $x = 0.8$ ft
5-27 $N_B = 98.1$ N, $A_x = 85.0$ N, $A_y = 147$ N
5-29 $P = 272$ N
5-30 $\theta = 33.6°$, $P_{min} = 271$ N
5-31   The ring will not turn.
5-33 $A_x = 25.4$ kN, $B_y = 22.8$ kN, $B_x = 25.4$ kN
5-34 $F = 14.0$ kN
5-35 $T_{BD} = 76.8$ kN, $A_y = 65.6$ kN, $A_x = 62.9$ kN
5-37 $N_A = 975$ lb, $B_x = 975$ lb, $B_y = 780$ lb
5-38 $A_x = 1.46$ kip, $F_B = 1.66$ kip
5-39 $\theta = 17.5°$
5-41 $F_{BC} = 410$ lb, $A_x = 71.2$ lb, $A_y = 396$ lb
5-42 $F_{CB} = 782$ N, $A_x = 625$ N, $A_y = 681$ N
5-43 $F_2 = 724$ N, $F_1 = 1.45$ kN, $F_A = 1.75$ kN
5-45 $P = 660$ N, $N_A = 442$ N, $\theta = 48.0°$ ⦨
5-46 $d = \dfrac{3a}{4}$
5-47 $F_{BC} = 80$ kN, $A_x = 54$ kN, $A_y = 16$ kN
5-49 $N_A = N_B = 1.77$ kN and $h = 246.08$ mm
5-50 $P = 611$ kPa
5-51 $N_C = 214$ N, $N_A = N_B = 571$ N
5-53 $T = \dfrac{W \cos \theta}{2 \sin(\phi - \theta)}$

$A_x = \dfrac{W \cos \phi \cos \theta}{2 \sin(\phi - \theta)}$

$A_y = \dfrac{W (\sin \phi \cos \theta - 2 \cos \phi \sin \theta)}{2 \sin (\phi - \theta)}$

5-54 $l_o = 70.3$ mm
5-55 $F = 6.70$ lb
5-57 $F_{BC} = 100$ kN; $A_x = 71.0$ kN; $A_y = 61.0$ kN

Answers to Selected Problems

5-58 $p = \dfrac{491 \cdot \cos(\theta)}{\sin(\theta/2 + 45°)}$

5-59 $N_B = \{1.57d\}$ kN, $F_A = \left\{\sqrt{2.46d^2 + 61.6}\right\}$ kN

5-61 $R_C = 63.9$ lb, $R_B = 11.9$ lb, $R_A = 26.0$ lb

5-62 $d = \dfrac{a}{\cos^3 \theta}$

5-63 $\theta = \tan^{-1}\left(\dfrac{l}{6\sqrt{a^2 - \frac{l^2}{4}}}\right)$

5-65 $M_{Ax} = 7.20$ kN·m, $M_{Ay} = -10.8$ kN·m, $M_{Az} = 0$

5-66 $A_x = 0$, $A_y = 0$, $A_z = B_z = C_z = 5.33$ lb

5-67 $N_C = 375$ lb, $N_A = N_B = 187.5$ lb

5-69 $B_z = 80$ lb, $A_z = 120$ lb, $T_{CD} = 0$, $A_x = 0$, $A_y = 0$

5-70 $B_y = -37.5$ lb, $B_x = 0$, $A_z = 100$ lb, $A_x = 0$, $A_y = 37.5$ lb

5-71 $F_{AB} = 467$ N; $F_{AC} = 674$ N; $D_x = 1.04$ kN; $D_y = D_z = 0$

5-73 $F_{BD} = F_{CD} = 70.2$ lb; $A_x = A_z = 37.5$ lb; $Ay = 0$

5-74 $C_y = 800$ N, $B_z = 107$ N, $B_y = 600$ N, $C_x = 53.6$ N, $A_x = -400$ N, $A_z = 800$ N

5-75 $F = 746$ N

5-77 $T_{BC} = 1.40$ kN, $A_y = 800$ N, $A_x = 1.20$ kN, $(M_A)_x = 600$ N·m, $(M_A)_y = 1.20$ kN·m, $(M_A)_z = 2.40$ kN·m

5-78 $F_{CB} = 1.37$ kN, $(M_A)_x = 785$ N·m, $(M_A)_z = 589$ N·m, $A_x = 1.18$ kN, $A_y = 589$ N, $A_z = 0$

5-79 $F_{CD} = 115$ N; $A_y = 80.0$ N; $A_z = 0$; $B_x = -30.0$ N, $B_y = -40.0$ N, $B_z = 0$

5-81 $A_x = 633$ lb, $A_y = -141$ lb, $B_x = -721$ lb $B_z = 895$ lb, $C_y = 200$ lb, $C_z = -506$ lb

5-82 $F_2 = 674$ lb, $A_x = 357$ lb, $A_y = -200$ lb, $B_x = -595$ lb, $B_z = 974$ lb, $C_z = -286$ lb

5-83 $B_y = -50$ lb, $M_{B_z} = 0$, $M_{B_x} = -300$ lb·ft, $B_x = -16.7$ lb, $A_x = 16.7$ lb

5-85 $F_{BD} = 294$ N, $F_{BC} = 589$ N, $A_x = 0$, $A_y = 589$ N, $A_z = 490.5$ N

5-86 $T_{BC} = 131$ lb; $T_{BD} = 510$ lb; $A_x = A_y = 0$; $A_z = 589$ lb

5-87 $T = 58.0$ N, $C_z = 77.6$ N, $C_y = 24.9$ N, $D_y = 68.5$ N, $D_z = 32.1$ N

Chapter 6

6-1 $F_{CB} = 3.00$ kN (T), $F_{CD} = 2.60$ kN (C), $F_{DE} = 2.60$ kN (C), $F_{DB} = 2.00$ kN (T), $F_{BE} = 2.00$ kN (C), $F_{BA} = 5.00$ kN (T)

6-2 $F_{CB} = 8.00$ kN (T), $F_{CD} = 6.93$ kN (C), $F_{DE} = 6.93$ kN (C), $F_{DB} = 4.00$ kN (T), $F_{BE} = 4.00$ kN (C), $F_{BA} = 12.0$ kN (T)

6-3 $F_{BA} = 214$ lb (T), $F_{BC} = 525$ lb (C), $F_{CA} = 371$ lb (T)

6-5 $F_{AD} = 9.90$ kN (C), $F_{AB} = 7$ kN (T), $F_{DB} = 4.95$ kN (T), $F_{DC} = 14.8$ kN (C), $F_{CB} = 10.5$ kN (T)

6-6 $F_{AD} = 11.3$ kN (C), $F_{AB} = 8$ kN (T), $F_{DB} = 7.07$ kN (T), $F_{DC} = 18.4$ kN (C), $F_{CB} = 13.0$ kN (T)

6-7 $F_{DE} = 16.3$ kN (C), $F_{DC} = 8.40$ kN (T), $F_{EA} = 8.85$ kN (C), $F_{EC} = 6.20$ kN (C), $F_{CF} = 8.77$ kN (T), $F_{CB} = 2.20$ kN (T), $F_{BA} = 3.11$ kN (T), $F_{BF} = 6.20$ kN (C), $F_{FA} = 6.20$ kN (T)

6-9 $F_{BC} = 1.89P$ (C), $F_{CD} = 1.37P$ (T), $F_{AB} = 0.471P$ (C), $F_{BD} = 1.67P$ (T), $F_{AD} = 1.37P$ (T)

6-10 $P = 424$ lb

6-11 Members LB, BK, KC, HF, FI, IE, and EJ are zero force members.

6-13 $F_{CD} = F_{AD} = 0.687P$ (T), $F_{CB} = F_{AB} = 0.943P$ (C), $F_{DB} = 1.33P$ (T)

6-14 $P_{max} = 849$ lb

6-15 $P_{max} = 849$ lb

6-17 $P = 5.20$ kN

6-18 $F_{DE} = 8.94$ kN (T), $F_{DC} = 4.00$ kN (C), $F_{CB} = 4.00$ kN (C), $F_{CE} = 0$, $F_{EB} = 11.3$ kN (C), $F_{EF} = 12.0$ kN (T), $F_{BA} = 12.0$ kN (C), $F_{BF} = 18.0$ kN (T), $F_{FA} = 20.1$ kN (C), $F_{FG} = 21.0$ kN (T)

6-19 $F_{DE} = 13.4$ kN (T), $F_{DC} = 6.00$ kN (C), $F_{CB} = 6.00$ kN (C), $F_{CE} = 0$, $F_{EB} = 17.0$ kN (C), $F_{EF} = 18.0$ kN (T), $F_{BA} = 18.0$ kN (C), $F_{BF} = 20.0$ kN (T), $F_{FA} = 22.4$ kN (C), $F_{FG} = 28.0$ kN (T)

6-21 $F_{AB} = 43.8$ kN (C), $F_{AG} = 26.2$ kN (T), $F_{BC} = 26.2$ kN (C), $F_{BG} = 35.0$ kN (T), $F_{CG} = 6.24$ kN (T), $F_{FG} = 22.4$ kN (T), $F_{CF} = 6.24$ kN (C), $F_{CD} = 18.76$ kN (C), $F_{DE} = 31.2$ kN (C), $F_{DF} = 25$ kN (T), $F_{EF} = 18.76$ kN (T)

6-22 $F_{AB} = 10.6$ kN (C), $F_{AH} = 7.5$ kN (T), $F_{EF} = 10.6$ kN (T), $F_{ED} = 7.5$ kN (C), $F_{BC} = 7.50$ kN (C), $F_{BH} = 7.50$ kN (T), $F_{FG} = 7.50$ kN (T), $F_{FD} = 2.50$ kN (C)

6-23 $F_{AB} = 3.75$ kip (C), $F_{AH} = 3$ kip (T), $F_{BC} = 3.75$ kip (C), $F_{BH} = 1$ kip (C), $F_{HC} = 1$ kip (T), $F_{GH} = 2$ kip (T), $F_{EF} = 3$ kip (T), $F_{ED} = 3.75$ kip (C), $F_{DC} = 3.75$ kip (C), $F_{DF} = 1$ kip (C), $F_{CF} = 1$ kip (T), $F_{CG} = 2$ kip (C), $F_{FG} = 2$ kip (T)

6-25 $F_{BF} = 155$ lb (C), $F_{DF} = 0$
6-26 $F_{ED} = 2.38$ kip (C), $F_{EF} = 1.94$ kip (T), $F_{DF} = 0$, $F_{DC} = 2.38$ kip (C)
6-27 $F_{HI} = 42.5$ kN (T), $F_{HC} = 100$ kN (T), $F_{DC} = 125$ kN (C)
6-29 $F_{HG} = 1125$ lb (T), $F_{DE} = 3375$ lb (C), $F_{EH} = 3750$ lb (T)
6-30 $F_{CD} = 3375$ lb (C), $F_{HI} = 6750$ lb (T), $F_{CH} = 5625$ lb (C)
6-31 $F_{CD} = 2.5$ kip (T), $F_{FE} = 3.50$ kip (T), $F_{CE} = 2.50$ kip (C)
6-33 $F_{HG} = 29.0$ kN (C), $F_{BC} = 20.5$ kN (T), $F_{HC} = 12.0$ kN (T)
6-34 $F_{DF} = 2.0$ kip (C), $F_{GF} = 5.70$ kip (C), $F_{DE} = 6.40$ kip (T)
6-35 $F_{FC} = 1.39$ kip (T), $F_{GF} = 19.0$ kip (C), $F_{CD} = 16.7$ kip (T)
6-37 $F_{EF} = 14.0$ kN (C), $F_{BC} = 13.0$ kN (T), $F_{BE} = 1.41$ kN (T), $F_{BF} = 8.00$ kN (T)
6-38 $F_{EF} = 15.0$ kN (C), $F_{BC} = 12.0$ kN (T), $F_{BE} = 4.24$ kN (T)
6-39 $F_{FE} = 1.0$ kip (T), $F_{BC} = 2.24$ kip (C), $F_{FB} = 0$
6-41 $F_{FE} = 8.49$ kN (T), $F_{BC} = 8.49$ kN (C), $F_{FC} = 0$
6-42 $F_{HG} = 17.6$ kN (C), $F_{HC} = 5.41$ kN (C), $F_{BC} = 19.1$ kN (T)
6-43 $F_{GJ} = 17.6$ kN (C), $F_{CJ} = 8.11$ kN (C), $F_{CD} = 21.4$ kN (T), $F_{CG} = 7.50$ kN (T)
6-45 $F_{CD} = 2.53$ kip (C), $F_{KJ} = 2.53$ kip (T)
6-46 $F_{BC} = -200\,L$; $F_{HG} = 400\,L$; $F_{BG} = -200\sqrt{L^2 + 9}$
6-47 $F_{GF} = 1.78$ kN (T), $F_{CD} = 2.23$ kN (C), $F_{CF} = 0$
6-49 $F_{EF} = 12.9$ kN (T), $F_{FI} = 7.21$ kN (T), $F_{HI} = 21.1$ kN (C)
6-50 $F_{AD} = 300$ lb (C), $F_{BD} = 450$ lb (C), $F_{CD} = 568$ lb (C)
6-51 $F_{BD} = 731$ lb (C), $F_{BC} = 250$ lb (T), $F_{BA} = 167$ lb (T), $F_{AD} = 786$ lb (T), $F_{AC} = 391$ lb (C), $F_{CD} = 0$
6-53 $F_{CA} = 833$ lb (T), $F_{CB} = 667$ lb (C), $F_{CD} = 333$ lb (T), $F_{AD} = F_{AB} = 354$ lb (C), $F_{DB} = 50$ lb (T)
6-54 $F_{CA} = 1000$ lb (C), $F_{CD} = 406$ lb (T), $F_{CB} = 344$ lb (C), $F_{AB} = F_{AD} = 424$ lb (T), $F_{DB} = 544$ lb (C)
6-55 $F_{BE} = 900$ lb (T), $F_{BC} = 0$, $F_{AB} = 600$ lb (C), $F_{DE} = F_{CD} = F_{AD} = F_{CE} = F_{AC} = 0$, $F_{AE} = 671$ lb (C)
6-57 $F_{BD} = 144$ lb (C), $F_{AD} = 204$ lb (T), $F_{AF} = 72.2$ lb (T)
6-58 $F_{EF} = 144$ lb (C), $F_{DF} = 0$, $F_{CF} = 72.2$ lb (T)

6-59 $F_{AD} = 686$ N (T), $F_{BD} = 0$, $F_{CD} = 615$ N (C), $F_{BC} = 229$ N (T), $F_{AC} = 343$ N (T), $F_{EC} = 457$ N (C)
6-61 $P = 12.5$ lb
6-62 $P = 25$ lb, $R_A = 25$ lb, $R_B = 60$ lb
6-63 $W_{max} = 980$ lb
6-65 $A_x = 300$ N, $A_y = 300$ N, $C_x = 300$ N, $C_y = 300$ N
6-66 $A_x = 24.0$ kN, $A_y = 12.0$ kN, $D_x = 18.0$ kN, $D_y = 24.0$ kN
6-67 $B_y = -24.0$ lb, $B_x = 74.0$ lb, $A_x = 74.0$ lb, $A_y = 124$ lb, $C_x = 74.0$ lb, $C_y = 76.0$ lb
6-69 $B_x = 2.80$ kip, $B_y = 1.05$ kip, $A_x = 2.80$ kip, $A_y = 5.10$ kip, $M_A = 43.2$ kip·ft
6-70 $C_x = 490.5$ N, $C_y = 184$ N, $B_y = 920$ kN, $B_x = 1.23$ kN
6-71 $C_x = 1600$ lb, $C_y = 800$ lb, $D_x = 800$ lb, $D_y = 800$ lb
6-73 $F_{spec} = 111$ N
6-74 $F_B = 907$ N, $P = 156$ N
6-75 $F_B = 464$ N, $M = 25.6$ N·m
6-77 $A_x = 300$ lb, $A_y = 400$ lb, $C_y = 1200$ lb, $D_x = 300$ lb, $D_y = 800$ lb
6-78 $N_C = 3.00$ kN, $N_A = 3.00$ kN, $B_y = 18.0$ kN, $B_x = 0$
6-79 $N_C = N_D = 2$ lb
6-81 $F_{FB} = 1.94$ kN, $F_{BD} = 2.60$ kN
6-82 $B_y = 1.34$ kip, $A_y = 1.02$ kip, $E_x = 0$, $E_y = 240$ lb, $M_E = 1.20$ kip·ft
6-83 $D_x = 50$ lb, $D_y = 50$ lb, $B_x = 112$ lb, $A_x = 62.5$ lb, $B_y = 117$ lb, $A_y = 167$ lb
6-85 $N_B = 7$ N, $N_C = 9.33$ N
6-86 $F_{EG} = 1.40$ kN, $d = 0.857$ m
6-87 $F_{CB} = 14.4$ kN, $F_{AB} = 9.70$ kN, $C_x = 11.5$ kN, $C_y = 8.65$ kN
6-89 $C_x = -650$ N; $C_y = 0$
6-90 $N_B = N_C = 49.5$ N
6-91 $F_{EF} = 8.18$ kN (T), $F_{AD} = 158$ kN (C)
6-93 $T = 5.10$ kN, $\theta = 30°$, $A_x = 4.33$ kN, $A_y = 5.15$ kN
6-94 $N_B = 0.1175$ lb, $N_A = 0.0705$ lb
6-95 $F_N = 5.25$ lb
6-97 a) $F = 205$ lb, $N_C = 380$ lb, b) $F = 102$ lb, $N_C = 72.5$ lb
6-98 $M = 26.7$ N·m
6-99 $N_C = 12.7$ kN, $A_x = 12.7$ kN, $A_y = 2.94$ kN, $N_D = 1.05$ kN
6-101 $F_{AB} = 8.60$ MN, $F_C = 9.78$ MN
6-102 $N_A = 284$ N
6-103 $A_x = 6.00$ kN; $A_y = 0$; $B_x = 4.00$ kN; $B_y = 2.00$ kN

6-105 $F_A = 258$ lbs, $F_B = 142$ lbs
6-106 $F = 66.1$ lb
6-107 $d = 0.638$ ft
6-109 $F = \left(\dfrac{\sqrt{16 - 9\sin^2\theta}}{\tan\theta\ \sqrt{16 - 9\sin^2\theta} + 3\sin\theta}\right)P$
6-110 $\theta = 23.7°$
6-111 $m = 26.0$ kg
6-113 $m_S = 1.71$ kg
6-114 $m_L = 106$ kg
6-115 $P = 283$ N, $B_x = D_x = 42.5$ N, $B_y = D_y = 283$ N, $B_z = D_z = 283$ N
6-117 $E_y = 0$, $E_x = 0$, $M_{Ex} = 0.5$ kN·m, $M_{Ey} = 0$, $F_{AB} = 1.56$ kN
6-118 $F_D = 20.8$ lb, $F_F = 14.7$ lb, $F_A = 24.5$ lb

Chapter 7

7-1 $N_C = 0$, $V_C = -386$ lb, $M_C = -857$ lb·ft, $N_D = 0$, $V_D = 300$ lb, $M_D = -600$ lb·ft
7-2 $N_B = 0$, $V_B = -100$ lb, $M_B = 750$ lb·in, $N_D = 0$, $V_D = 75$ lb, $M_D = -750$ lb·in
7-3 $N_A = -39$ kN, $V_A = 0$, $M_A = -2.425$ kN·m
7-5 $V_C = -133$ lb, $M_C = 133$ lb·in.
7-6 $a = \dfrac{L}{3}$
7-7 $N_D = 0$, $N_C = 0$, $V_D = 900$ lb, $M_D = -7.35$ kip·ft, $V_C = -2.11$ kip, $M_C = -3.72$ kip·ft
7-9 $N_C = -30$ kN, $V_C = -8$ kN, $M_C = 6$ kN·m
7-10 $P = 0.533$ kN, $N_C = -2$ kN, $V_C = -0.533$ kN, $M_C = 0.400$ kN·m
7-11 $N_C = 0$, $V_C = 11.8$ kN, $M_C = -35.4$ kN·m
7-13 $V_C = 3.75$ kip, $M_C = 30$ kip·ft
7-14 $N_F = 0$, $V_F = -150$ lb, $M_F = -600$ lb·ft, $N_E = 0$, $V_E = 230$ lb, $M_E = -387$ lb·ft
7-15 $N_C = 0$, $V_C = 12$ kN, $M_C = -24$ kN·m
7-17 $N_A = 86.6$ lb, $V_A = 150$ lb, $M_A = 1.80$ kip·in.
7-18 $N_C = -4$ kip, $V_C = 1.0$ kip, $M_C = -14.0$ kip·ft, $N_D = -4$ kip, $V_D = 1.75$ kip, $M_D = -5.0$ kip·ft
7-19 $N_E = -300$ lb, $V_E = 300$ lb, $M_E = -450$ lb·ft, $N_F = 150$ lb, $V_F = -300$ lb, $M_F = -450$ lb·ft
7-21 $N_E = 720$ N, $V_E = 1.12$ kN, $M_E = -320$ N·m, $N_F = 0$, $V_F = -1.24$ kN, $M_F = -1.41$ kN·m
7-22 $N_D = 0$, $V_D = 5$ kip, $M_D = 46$ kip·ft
7-23 $N_C = -280$ lb, $V_C = 50$ lb, $M_C = 150$ lb·ft
7-25 $N_D = 0$, $V_D = 0.75$ kip, $M_D = 13.5$ kip·ft, $N_E = 0$, $V_E = -9$ kip, $M_E = -24.0$ kip·ft
7-26 $a = \dfrac{2}{3}L$

7-27 $N_F = 0$, $V_F = 20$ lb, $M_F = 1040$ lb·ft, $N_G = 0$, $V_G = -580$ lb, $M_G = 1160$ lb·ft
7-29 $N_C = -406$ lb, $V_C = 903$ lb, $M_C = 1.35$ kip·ft
7-30 $N_D = -464$ lb, $V_D = -203$ lb, $M_D = 2.61$ kip·ft
7-31 $N_C = 0$, $V_C = 2.75$ kip, $M_C = 25.5$ kip·ft, $N_D = 0$, $V_D = -5.25$ kip, $M_D = 31.5$ kip·ft
7-33 $N_D = -2.25$ kN, $V_D = 1.25$ kN, $M_D = -1.88$ kN·m
7-34 $N_E = 1.25$ kN, $V_E = 0$, $M_B = 1.69$ kN·m
7-35 $V_{A_x} = 0$, $N_{A_y} = 0$, $V_{A_z} = 250$ lb, $M_{Ax} = 187.5$ lb·ft, $M_{Ay} = 187.5$ lb·ft, $M_{Az} = 0$, $V_{B_x} = 0$, $V_{B_y} = 0$, $N_{B_z} = 250$ lb, $M_{Bx} = 437.5$ lb·ft, $M_{By} = 187.5$ lb·ft, $M_{Bz} = 0$
7-37 $N_D = 1.26$ kN, $V_D = 0$, $M_D = 500$ N·m
7-38 $N_E = -1.48$ kN, $V_E = 500$ N, $M_E = 1000$ N·m
7-39 $V = 0.278\,w_0 r$, $N = 0.0759\,w_0 r$, $M = 0.0759\,w_0 r^2$
7-41 $V_D = 135$ N, $N_D = 180$ N, $M_D = 270$ N·m
7-42 $N_C = 0$, $(V_C)_x = 104$ lb, $(V_C)_z = 10$ lb, $(M_C)_x = 20$ lb·ft, $(M_C)_y = 72$ lb·ft, $(M_C)_z = -178$ lb·ft
7-43 $N_x = -500$ N, $V_y = 100$ N, $V_z = 900$ N, $M_x = 600$ N·m, $M_y = -900$ N·m, $M_z = 400$ N·m
7-45   a) $0 \le x < a$: $V = \left(1 - \dfrac{a}{L}\right)P$, $M = \left(1 - \dfrac{a}{L}\right)Px$, $a < x \le L$: $V = -\left(\dfrac{a}{L}\right)P$, b) $M = P\left(a - \dfrac{a}{L}x\right)$
      $0 \le x < 2$ m: $V = 6$ kN, $M = \{6x\}$ kN·m
      2 m $< x \le 6$ m: $V = -3$ kN, $M = \{18 - 3x\}$ kN·m
7-46   a) For $0 \le x < a$, $V = P$, $M = Px$,
      For $a < x < L - a$, $V = 0$, $M = Pa$,
      For $L - a < x \le L$, $V = -P$, $M = P(L - x)$
      b) For $0 \le x < 5$ ft, $V = 800$ lb, $M = 800x$ lb·ft,
      For 5 ft $< x < 7$ ft, $V = 0$, $M = 4000$ lb·ft,
      For 7 ft $< x \le 12$ ft, $V = -800$ lb, $M = (9600 - 800x)$ lb·ft
7-47   a) For $0 \le x < a$, $V = \dfrac{Pb}{a + b}$, $M = \dfrac{Pb}{a + b}x$,
      For $a < x \le a + b$, $V = -\dfrac{Pa}{a + b}$,
      $M = Pa - \dfrac{Pa}{a + b}x$,
      b) For $0 \le x < 5$ ft, $V = 350$ lb, $M = 350x$ lb·ft,
      For 5 ft $< x \le 12$ ft, $V = -250$ lb, $M = \{3000 - 250x\}$ lb·ft

626 ANSWERS TO SELECTED PROBLEMS

7-49 a) $0 \le x < \dfrac{L}{3}$: $V = 0$, $M = 0$, $\dfrac{L}{3} < x < \dfrac{2L}{3}$:

$V = 0$, $M = M_0$, $\dfrac{2L}{3} < x \le L$: $V = 0$, $M = 0$,

b) $0 \le x < \dfrac{8}{3}$ m: $V = 0$, $M = 0$,

$\dfrac{8}{3}$ m $< x < \dfrac{16}{3}$ m:

$V = 0$, $M = 500$ N·m, $\dfrac{16}{3}$ m $< x \le 8$ m: $V = 0$,

$M = 0$

7-50 $M_{max} = 2$ kN·m
7-51 $V = 28 - 3x$, $M = 28x - 1.5x^2 - 114$
7-53 For $0 \le x \le 8$, $V = 133.75 - 40x$,
$M = 133.75x - 20x^2$, For $8 \le x \le 11$,
$V = 20$, $M = 20x - 370$

7-54 a) $V = \dfrac{w}{2}(L - 2x)$, $M = \dfrac{w}{2}(Lx - x^2)$

b) $V = (2500 - 500x)$ lb,
$M = (2500x - 250x^2)$ lb·ft

7-55 For $0 \le x < L$, $V = \dfrac{w}{18}(7L - 18x)$,

$M = \dfrac{w}{18}(7Lx - 9x^2)$, For $L < x < 2L$,

$V = \dfrac{w}{2}(3L - 2x)$, $M = \dfrac{w}{18}(27Lx - 20L^2 - 9x^2)$,

For $2L < x \le 3L$, $V = \dfrac{w}{18}(47L - 18x)$,

$M = \dfrac{w}{18}(47Lx - 9x^2 - 60L^2)$

7-57 (a) $V = \dfrac{w}{8}(7L - 8x)$,

$M = -\dfrac{w}{8}(4x^2 - 7Lx + 3L^2)$,

(b) $V = 5(10.5 - x)$, $M = -2.5(x^2 - 21x + 108)$
7-58 $V_{max} = -11.89$ kp, $M_{max} = 106.7$ kip·ft
7-59 $V_{max} = -540$ lb, $M_{max} = 1.36$ kip·ft

7-62 $V = [(L + X)^3 - L^3]\dfrac{\pi\gamma_0^2}{(3L^2)}$

$M = [(L + X)^4 - L^3(4x + L)]\dfrac{\pi\gamma_0^2}{(12L^2)}$

7-63 $V = \{1350 - 60x\}$ lb,
$M = \{1350x - 30x^2 - 8250\}$ lb·ft,
$V = \{600 - 60x\}$ lb,
$M = \{600x - 30x^2 - 3000\}$ lb·ft

7-65 For $0 \le x < 3$ m, $V = \{21.0 - 2x^2\}$ kN,

$M = \left\{21.0x - \dfrac{2}{3}x^3\right\}$ kN·m, For 3 m $< x \le 6$ m,

$V = \{39.0 - 12x\}$ kN,
$M = \{-6x^2 + 39x - 18\}$ kN·m
7-66 $V = \{16 - 4y\}$ lb, $M = \{-2y^2 + 16y - 40\}$ lb·ft
7-67 $N = P\sin(\theta + \phi)$, $V = -P\cos(\theta + \phi)$,
$M = Pr[\sin(\theta + \phi) - \sin\phi]$
7-69 $a = 0.366 L$
7-70 $V_{max} = 1.20$ kip, $M_{max} = 19.2$ kip·ft
7-71 $V_{max} = 50$ lb, $M_{max} = -660$ lb·ft
7-73 $V_{max} = -550$ lb, $M_{max} = 4200$ lb·ft
7-74 $x = 0.5^+$, $V = 450$ N, $M = -150$ N·m, $x = 1.5^-$,
$V = -750$ N, $M = -300$ N·m
7-75 $x = 2^+$, $V = -375$ N, $M = 750$ N·m
7-77 $x = 10^+$, $V = 20.0$ kip, $M = -50.0$ kip·ft
7-78 $V_{max} = 45$ kN, $M_{max} = -67.5$ kN·m
7-79 $V_{max} = -510$ lb, $M_{max} = 2401$ lb·ft
7-81 $V_{max} = 800$ lb, $M_{max} = -1200$ lb·ft
7-82 $V_{max} = 9$ kN, $M_{max} = -19.5$ kN·m
7-83 $V_{max} = 9$ kip; $M_{max} = -36$ kip·ft
7-85 $V_{max} = -110$ lb; $M_{max} = -180$ lb·ft
7-86 $V_{max} = 3$ kN, $M_{max} = 3$ kN·m
7-87 $x = 3$, $V = -2.25$ kN, $M = 20.25$ kN·m
7-89 $V_{max} = 1.98$ kN, $M_{max} = 0.395$ kN·m
7-90 $V_{max} = 13.5$ kN, $M_{max} = -9.5$ kN·m
7-91 $V_{max} = 1017$ lb, $M_{max} = -9.40$ lb·ft
7-93 $V_{max} = 3.75$ kip, $M_{max} = 37.5$ kip·ft
7-94 $y_B = 2.22$ m, $y_D = 1.55$ m
7-95 $P_1 = 320$ N, $y_D = 2.33$ m
7-97 $x_B = 5.39$ m
7-98 $P = 700$ N
7-99 $y_B = 8.67$ ft, $y_D = 7.04$ ft
7-101 $T = 858$ N, $y_B = 1.67$ m
7-102 $w = 51.9$ lb/ft
7-103 $T_{max} = 14.4$ kip, $T_{min} = 13.0$ kip
7-105 $T_{AB} = T_{CD} = 212$ lb (max), $y_B = 2$ ft
7-106 $x = 2.57$ ft, $W = 247$ lb
7-107 $T_A = 61.7$ kip, $T_B = 36.5$ kip, $T_C = 50.7$ kip
7-109 $T_{max} = 594$ kN
7-110 $T_{min} = 552$ kN
7-111 $T_{max} = 121$ N, $n = 5.91$ m
7-113 $T_A = 37.3$ lb, $T_B = 52.4$ lb, $T_C = 16.4$ lb
7-114 $T_{max} = 184$ kN
7-115 $w = 1.36$ kN/m
7-117 $x = 8.07$ m
7-119 $F_{Rx} = 292$ lb
7-121 $s = 5.14$ ft
7-122 $h = 6.21$ ft, $N_m = 306$ lb

Chapter 8

8-1	$P = 12.8$ kN
8-2	$P = 72.0$ lb
8-3	$N_A = 16.5$ kN, $N_B = 42.3$ kN, It does not move.
8-5	$F = 2.76$ kN
8-6	$F = 5.79$ kN
8-7	$\mu = 0.127$
8-10	$\phi = \theta$, $P = W \sin(\alpha + \theta)$
8-11	$F = 561$ N
8-13	$d = 3$ ft
8-14	car slips before it tips, $\theta = 21.8°$
8-15	$l = 26.7$ ft
8-17	$M = 2.21$ kip · ft
8-18	$P = 1350$ lb
8-19	$N_A = 200$ lb
8-21	$n = 6$ crates
8-22	$\delta = 5.60$ mm
8-23	$P = 1.17$ kN
8-25	$d = 13.4$ ft
8-26	$W_a = 6.97$ kN, $W_b = 15.3$ kN
8-27	$W_a = 1.25$ kN, $W_b = 6.89$ kN
8-29	$\theta = 11.0°$
8-30	$\theta = 10.6°$, $x = 0.184$ ft
8-31	$\theta = 8.53°$, $F_A = 1.48$ lb, $F_B = 0.890$ lb
8-33	No
8-34	$\mu_s = 0.333$
8-35	$P = 8.18$ lb
8-37	$P = 0.962$ N
8-38	$P = 0.962$ N
8-39	$\theta = 21.8°$
8-41	$\theta = 31.0°$
8-42	$P = 654$ N
8-43	$\mu_s = 0.531$
8-45	He can not move the crate.
8-46	$P = 612$ N
8-47	$P = 110$ lb, $L = 6.25$ ft
8-49	$P = 80$ lb, $d = 0.4$ ft
8-50	$P = 80$ lb, $h = 5$ ft
8-51	$P = 408$ N
8-53	$M = 55.2$ lb · ft
8-54	$P = 28.1$ N
8-55	$P = 23.9$ N
8-57	$\theta = 65.2°$
8-58	Any number of blocks can be stacked.
8-59	$\theta = 22.6°$
8-61	$P = F\left(\dfrac{(1 - \mu_s^2) \tan \alpha + 2\mu_s}{1 - 2\mu_s \tan \alpha - \mu_s^2}\right)$
8-62	$x = 18.3$ mm
8-63	$P = 2.39$ kN
8-65	$P = 32.3$ kip
8-66	$M = 29.7$ lb.in, $F = 112$ lb
8-67	$F = 66.7$ N
8-69	$F_{AB} = 164$ N
8-70	$F = 28.3$ N
8-71	$P = 617$ lb
8-74	$M = 40.6$ N · m
8-75	$M = 31.0$ N · m
8-77	$P = 51.2$ N
8-78	$P = 96.4$ N raising, $P = 10.3$ N lowering
8-79	$F = 1.35$ N
8-81	$T = 4.02$ kN, $F = 11.6$ kN
8-82	$P = 104$ N
8-83	$T_A = 187$ lb, $T_B = 137$ lb
8-85	He will successfully restrain the cow.
8-86	Yes, it is possible. $F = 137$ lb
8-87	$T_1 = 57.7$ N
8-89	$T_A = 1.95$ kN, $T_B = 0.759$ kN
8-90	$\theta = 99.2°$
8-91	$\theta = 40.1°$
8-93	$M = 458$ N · m
8-94	$P = 46.9$ lb
8-95	$\theta = 63.5°$
8-97	$T_2 = 77.8$ lb
8-99	$P = 53.6$ N
8-101	$x = 0.384$ m
8-102	$F_s = 85.4$ N
8-103	Therefore, it is possible for boy A to pull boy B towards him
8-105	$\theta_{max} = 38.8°$
8-106	$M = 50.0$ N · m, $x = 286$ mm
8-107	$M = 132$ N · m
8-109	$F = 10.7$ lb
8-110	$M = 16.1$ N · m
8-111	$M = 237$ N · m
8-113	$M = \dfrac{\mu_s PR}{2}$
8-114	$M = 43.9$ lb · ft
8-115	$P = 2.00$ N
8-117	$P = 29.0$ lb
8-118	$P = \dfrac{3Mb \sin \theta (r_o^2 - r_i^2)}{2\mu_s l(r_o^3 - r_i^3)}$
8-119	$F = 18.9$ N
8-121	$P = 20.5$ lb
8-122	$T = 289$ lb, $N = 479$ lb, $F = 101$ lb
8-123	$\mu_s = 0.0407$
8-125	$r = 20.6$ mm
8-126	$P = 42.2$ lb
8-127	$\mu_s = 0.411$
8-129	$P = 1333$ lb
8-131	$T = 2.25$ kN

Chapter 9

9-1 $\bar{x} = 0.620$ ft, $A_x = 0$, $A_y = 1.16$ lb, $M_A = 0.720$ lb·ft

9-2 $W = 247$ N, $\bar{x} = 1$ m, $\bar{y} = 2.06$ m

9-3 $\bar{x} = 0.574$ m, $B_x = 0$, $B_y = 84.8$ N, $A_y = 63.1$ N

9-5 $\bar{x} = 0.620$ ft, $A_y = 0.442$ lb, $B_x = 0$, $B_y = 0.720$ lb

9-6 $\bar{y} = \dfrac{2}{5}$ m

9-7 $\bar{x} = \dfrac{n+1}{2(n+2)}a$

9-9 $\bar{x} = \dfrac{3}{2}$ m

9-10 $\bar{y} = \dfrac{12}{5}$ m

9-11 $\bar{x} = \dfrac{3}{4}b$

9-13 $\bar{x} = 6$ m

9-14 $\bar{y} = 2.8$ m

9-15 $\bar{x} = 0.587$ m

9-17 $\bar{y} = \dfrac{\alpha \pi}{8}, \bar{x} = \dfrac{L}{2}$

9-18 $\bar{x} = 0.398$ m

9-19 $\bar{y} = 1.00$ m

9-21 $\bar{x} = 1\dfrac{3}{5}$ ft

9-22 $\bar{y} = 4.15$ ft

9-23 $\bar{x} = \dfrac{3}{8}a$

9-25 $\bar{x} = 3.20$ ft, $\bar{y} = 3.20$ ft, $T_A = 384$ lb, $T_C = 384$ lb, $T_B = 1.15$ kip

9-26 $\bar{x} = 3$ ft

9-27 $\bar{y} = \dfrac{6}{5}$ ft

9-29 $\bar{y} = 40.0$ mm

9-30 $\bar{x} = \dfrac{1}{3}(a+b)$

9-31 $\bar{y} = \dfrac{h}{3}$

9-33 $\bar{y} = \dfrac{\pi a}{8}$

9-34 $\bar{x} = -3$ m, $\bar{y} = 0.25$ m, $\bar{z} = -0.3$ m, $A_x = A_y = 0$, $A_z = 52.3$ kN·m, $M_A = 52.3$ kN·m

9-35 $\bar{x} = \left[\dfrac{2(n+1)}{3(n+2)}\right]a$

9-37 $\tilde{y} = \dfrac{3}{4}h$

9-38 $\tilde{y} = \dfrac{4}{5}h$

9-39 $\bar{x} = 0, \bar{y} = 3$ m, $\bar{z} = 0.694$ m

9-41 $\bar{x} = \bar{y} = 0, \bar{z} = \dfrac{2}{3}a$

9-42 $\bar{x} = 3.20$ in.

9-43 $\bar{x} = 0, \bar{y} = \dfrac{3}{4}h, \bar{z} = \dfrac{a}{\pi}$

9-45 $\tilde{y} = 0.833$ ft

9-46 $\tilde{y} = \dfrac{3}{8}a$

9-47 $\bar{x} = 0.4a$

9-49 $\bar{y} = 84.7$ mm

9-50 $\bar{z} = \dfrac{c}{4}$

9-51 $d = 3$ m

9-53 $\bar{y} = 67.9$ mm

9-54 $\bar{x} = 0, \bar{y} = 58.3$ mm

9-55 $\bar{x} = 2.85$ in., $\bar{y} = 5.33$ in.

9-57 $y = 0, \bar{x} = 124$ mm

9-58 $\bar{x} = 0.607$ ft

9-59 $\bar{y} = \left(\dfrac{h}{3}\right)\dfrac{(b_2 + 2b_1)}{(b_2 + b_1)}$

9-61 $\bar{y} = 238$ mm

9-62 $\bar{x} = -1.00$ in., $\bar{y} = 4.625$ in.

9-63 $\bar{y} = 11.9$ in.

9-65 $\bar{y} = 0.385\,a$

9-66 $\bar{y} = 37.5$ mm

9-67 $\bar{y} = 272$ mm

9-69 $\bar{y} = 87.5$ mm

9-70 $\bar{x} = 105$ mm

9-71 $\bar{x} = \dfrac{b(b+3a)}{3(2a+b)}, \bar{y} = \dfrac{b^2c + 3abc + 3a^2c}{3(a+b)(2a+b)}$

9-73 $\bar{x} = 2.5$ in., $\bar{y} = 4$ in.

9-74 $\bar{x} = 7.18$ ft, $A_y = 667$ lb, $B_y = 2633$ lb

9-75 $\bar{y} = 58.1$ mm, $\tilde{x} = \tilde{z} = 0$

9-77 $\theta = 53.1°$

9-78 $\bar{x} = 4.28$ ft, $B_y = 620$ lb, $A_y = 830$ lb

9-79 $\bar{z} = 0.70$ ft

9-81 $\bar{z} = 122$ mm

9-82 $h = 385$ mm

9-83 $\bar{x} = 5.07$ ft, $\bar{y} = 3.80$ ft

9-85 $\bar{x} = \bar{y} = 0, \bar{z} = 52.5$ mm
9-86 $\bar{z} = 58.1$ mm
9-87 $\bar{x} = 19.0$ ft, $\bar{y} = 11.0$ ft
9-89 $m = 16.4$ kg, $\bar{x} = 153$ mm, $\bar{y} = -15$ mm, $\bar{z} = 111$ mm
9-90 $m = 39.5$ kg
9-91 $A = 3.56\,(10^3)$ ft^2
9-93 $W = 3.01$ lb
9-94 $A = 70.0$ in^2
9-95 $V = \dfrac{\pi(6\pi + 4)}{6}a^3$
9-97 $V = 0.114$ m^3
9-98 $A = 2.25$ m^2
9-99 $V = 143(10)^{-6}$ m^3
9-101 $A = 1156$ ft^2
9-102 $V = 3485$ ft^3
9-103 $A = 8\pi ba$, $V = 2\pi ba^2$
9-105 $Q = 205$ MJ
9-106 $A = 0.667$ ft^2, $\bar{y} = 0.375$ ft, $V = 1.57$ ft^3
9-107 $W = 126$ kip
9-109 $A = 522$ in^2, $V = 263$ in^3
9-110 $A = 5.65(10^3)$ ft^2
9-111 $A = 1.39$ ft^2, $\bar{y} = 0.541$ ft, $V = 4.71$ ft^3
9-113 $V = 1.40(10^3)$ in^3
9-114 $W = 1.04$ lb
9-115 $F = 7.62$ kN, $\bar{x} = 2.74$ m, $\bar{y} = 3$ m
9-117 $F_R = 133$ lb, $\bar{y} = 7.5$ ft, $\bar{x} = 0$
9-118 $F_R = \dfrac{4ab}{\pi^2}p_0$, $\bar{x} = \dfrac{a}{2}, \bar{y} = \dfrac{b}{2}$
9-119 $F_R = \pi l r p_0$
9-121 For water: $F_{R_A} = 157$ kN, $F_{R_B} = 235$ kN
For oil: $d = 4.22$ m
9-122 $F_{R_A} = 2.45$ kN, $d' = 1.02$ m, $F_{R_B} = 9.75$ kN, Located at the center of plate B.
9-123 $F = 4$ kip
9-125 $F_R = 117$ lb, $d = 13.1$ in.
9-126 $p = 369$ kg/m^3
9-127 $F_B = 29.4$ kN, $F_A = 235$ kN
9-129 $h = 6.67$ m
9-130 $h = 7.73$ m

Chapter 10

10-1 $I_x = 0.167$ in^4
10-2 $I_y = 0.3$ in^4
10-3 $I_x = \frac{1}{12}bh^3$
10-5 $I_x = 3.12\,(10^6)$ mm^4
10-6 $I_y = 171\,(10^3)$ mm^4
10-7 $I_x = 9.05$ in^4
10-9 $I_x = 23.8$ ft^4
10-10 $k_y = 17.9$ mm
10-11 $I_x = 4.27$ in^4
10-13 $I_x = 34.1$ in^4
10-14 $I_y = 73.1$ in^4
10-15 $I_x = 205$ in^4
10-17 $I_x = \dfrac{\pi r^4}{16}$
10-18 $I_y = \dfrac{b^3 h}{6}$
10-19 $I_x = 333$ in^4
10-21 $I_x = 0.305$ m^4
10-22 $I_y = 1.6$ m^4
10-23 $I_x = \dfrac{3ab^3}{35}$
10-25 $I_x = 209$ in^4
10-26 $I_y = 533$ in^4
10-27 $I_x = 246$ in^4, $I_y = 61.5$ in^4
10-29 $\bar{y} = 91.7$ mm, $I_{x'} = 216(10^6)$ mm^4
10-30 $I_x = 182$ in^4
10-31 $I_y = 966$ in^4
10-33 $I_y = 2.03(10^9)$ mm^4
10-34 $I_y = 115\,(10^6)$ mm^4
10-35 $\bar{y} = 207$ mm, $\bar{I}_{x'} = 222\,(10^6)$ mm^4
10-37 $I_x = 54.7$ in^4
10-38 $I_x = 1971$ in^4
10-39 $I_y = 2376$ in^4
10-41 $I_x = 91.7(10^6)$ mm^4
10-42 $I_x = 154(10^6)$ mm^4
10-43 $I_y = 91.3(10^6)$ mm^4
10-45 $\bar{x} = 61.6$ mm, $\bar{I}_{y'} = 41.2(10^6)$ mm^4
10-46 $\bar{y} = 1.29$ in., $\bar{I}_{x'} = 6.74$ in^4
10-47 $I_y = 18.3$ in^4
10-49 $J_o = \dfrac{1}{32}\pi d^4 - \dfrac{1}{6}a^4$
10-50 $\bar{y} = 0.181$ m, $\bar{I}_{x'} = 4.23(10^{-3})$ m^4
10-51 $\bar{I}_{x'} = 520(10^6)$ mm^4
10-53 $I_y = 365$ in^4
10-54 $I_{xy} = \dfrac{1}{3}tl^3\sin 2\theta$
10-55 $I_{xy} = 5.06$ in^4
10-57 $I_{xy} = 10.7$ m^4, $\bar{I}_{x'y'} = 1.07$ m^4
10-58 $I_{xy} = \dfrac{1}{6}a^2b^2$
10-59 $I_{xy} = 0.0625$ m^4
10-61 $I_{xy} = 0.667$ in^4
10-62 $I_{xy} = 97.8$ in^4
10-63 $I_u = 15.75$ in^4, $I_v = 25.75$ in^4
10-65 $I_{xy} = 119$ in^4

10-66 $I_{xy} = 98.4(10^6)$ mm^4
10-67 $\bar{x} = \bar{y} = 44.1$ mm, $I_{x'y'} = -6.26(10^6)$ mm^4
10-69 $\theta = 10.5°$, $I_{max} = 117(10^6)$ mm^4,
$I_{min} = 28.8(10^6)$ mm^4
10-70 $\bar{x} = 85.0$ mm, $\bar{y} = 35.0$ mm,
$I_{x'y'} = -7.50(10^6)$ mm^4
10-71 $I_{xy} = 29.7(10^6)$ mm^4
10-73 $I_u = 195$ in^4, $I_v = 650$ in^4, $I_{uv} = -191$ in^4
10-74 $\theta_P = 10.5°$;
$I_{max} = 117(10)^6$ mm^4, $I_{min} = 28.8(10)^6$ mm^4
10-75 $I_{max} = 117(10)^6$ mm^4, $I_{min} = 28.8(10)^6$ mm^4,
$\theta = 10.5°$
10-77 $I_u = 43.9$ in^4, $I_v = 23.6$ in^4, $I_{uv} = 17.5$ in^4
10-78 $\theta = 6.08°$, $I_{max} = 1.74(10^3)$ in^4, $I_{min} = 435$ in^4
10-79 $\theta = 6.08°$, $I_{max} = 1.74(10^3)$ in^4, $I_{min} = 435$ in^4
10-81 $I_{max} = 24.6(10)^6$ mm^4, $I_{min} = 2.56(10)^6$ mm^4,
$\theta = -27.3°$
10-82 $\theta_{p1} = -45°$, $\theta_{p2} = 45°$, $I_{max} = 638(10^{-6})$ m^4,
$I_{min} = 91.1(10^{-6})$ m^4
10-83 $I_{max} = 638(10^{-6})$ m^4, $I_{min} = 91.1(10^{-6})$ m^4,
$\theta_{p1} = -45°$, $\theta_{p2} = 45°$
10-85 $I_x = \frac{2}{5}mb^2$
10-86 $k_x = 57.7$ mm
10-87 $I_x = \frac{1}{3}ma^2$
10-89 $I_x = \frac{2}{5}mb^2$
10-90 $k_x = \sqrt{\frac{n+2}{2(n+4)}}h$
10-91 $I_y = 2.25$ slug $\cdot$ ft^2
10-93 $I_x = \frac{3}{10}mr^2$
10-94 $I_y = \frac{2}{5}mr^2$
10-95 $I_O = 0.450(10^{-3})$ kg $\cdot$ m^2, $\bar{y} = 57.7$ mm,
$I_G = 0.150(10^{-3})$ kg $\cdot$ m^2
10-97 $I_z = 1.53$ kg $\cdot$ m^2
10-98 $I_A = 1.58$ slug $\cdot$ ft^2
10-99 $I_O = 6.23$ kg $\cdot$ m^2
10-101 $I_A = 222$ slug $\cdot$ ft^2
10-102 $I_x = 0.402$ slug $\cdot$ ft^2
10-103 $I_O = \frac{1}{2}ma^2$
10-105 $I_z = 0.113$ kg $\cdot$ m^2
10-106 $I_G = 118$ slug $\cdot$ ft^2
10-107 $I_O = 293$ slug $\cdot$ ft^2
10-109 $I_z = 34.2$ kg $\cdot$ m^2

Chapter 11
11-1 $F_{AC} = 7.32$ lb
11-2 $F = 2P \cot \theta$
11-3 $\theta = 42.4°$
11-5 $\theta = 90°$, $\theta = 36.1°$
11-6 $k = 166$ N/m
11-7 $M = 52.0$ lb $\cdot$ ft
11-9 $P = \dfrac{2M}{L}$
11-10 $P = \dfrac{W}{2} \cot \theta$
11-11 $\theta = 23.8°$, $\theta = 72.3°$
11-13 $F = 13.9$ N
11-14 $M = 43.3$ N $\cdot$ m
11-15 $\theta = 9.21°$
11-17 $F_C = 427$ N
11-18 $\theta = 41.2°$
11-19 $k = 9.88$ kN/m
11-21 $\theta = 90°$, $\theta = 33.4°$
11-22 $F = \dfrac{W(a+b-d\tan\theta)}{ac}\sqrt{a^2+c^2+2ac\sin\theta}$
11-23 $\theta = 16.6°$
11-25 $\theta = 23.6°$
11-26 $x = -0.5$ ft unstable, $x = 0.833$ ft stable
11-27 Unstable at $\theta = 34.6°$, stable at $\theta = 145°$
11-29 $\theta = 38.7°$ unstable, $\theta = 90°$ stable,
$\theta = 141°$ unstable
11-30 $x = -0.424$ ft unstable, $x = 0.590$ ft stable
11-31 $\theta = 20.2°$, stable
11-33 Unstable at $\theta = 90°$, stable at $\theta = 49.0°$
11-34 $F_s = 211$ N
11-35 $\theta = \sin^{-1}\left(\dfrac{W}{2kL}\right)$
11-37 $\theta = 16.3°$
11-38 $\theta = 0°$, unstable
$\theta = 64.8°$, stable
11-39 $W = \dfrac{8k}{3L}$
11-41 Stable at $\theta = 24.6°$
11-42 $\phi = 17.4°$, $\theta = 9.18°$
11-43 Unstable at $\theta = 23.2°$
11-45 $\theta = 0°$, $\theta = 33.0°$
11-46 $h = 5.20$ ft
11-47 Stable at $\theta = 45°$
11-49 $h = 3.99$ in.

Index

A

Acceleration, dynamics and, 3
Active force, 83
Angles, 34, 44–45, 64–66, 68, 76–77, 383–385, 410, 594–595
   Cartesian force vectors, 44–45
   coordinate direction, 44, 76–77
   dot product used for, 64–66, 68, 77
   dry friction and, 383–385, 410
   formed between intersecting lines, 68
   horizontal (θ), 45
   impending motion and, 383, 385
   kinetic friction (θ_k), 384–385
   lead, 410
   mathematical review of, 594–595
   projection, parallel and perpendicular, 65
   Pythagorean's theorem and, 34, 65, 595
   resultant forces from, 34
   screws, 410
   static friction (θ_s), 383, 385
   vectors and, 34, 44–45, 64–66, 68, 76–77
   vertical (ϕ), 45
Applied force (**P**), 382–385, 408–409, 437–438
Area (A), 446, 448, 452–454, 480–483, 501–502, 508–513, 518–520, 526–535, 554–555
   axial symmetry and rotation, 480–483, 502, 526–529, 555
   centroid (C) of an, 446, 448, 452–454, 480–483, 501–502, 554
   centroidal axis of, 508–509
   composite bodies (shapes), 481, 502, 518–520, 554–555
   inclined axis, about, 530–532
   integration for, 446, 452–454, 501, 507–508
   Mohr's circle for, 533–535
   moments of inertia (I) for, 508–513, 518–520, 526–535, 554–555
   Pappus and Guldinus, theorems of, 480–483, 502
   parallel-axis theorem for, 508–510, 518, 527, 554
   plane, volume generated revolution of, 481
   polar moment of inertia, 508–509
   principal moments of inertia, 531–532, 555
   procedures for analysis of, 448, 510, 518
   product of inertia for, 526–529, 531, 554
   radius of gyration of, 509
   surface of revolution, 480, 482–483, 502
   transformation equations for, 530, 555
   volume of revolution, 481–483, 502
Associative law, 124
Axes, 136–140, 180, 190, 507–513, 518–520, 526–529, 530–535, 541–548, 554–555
   area moments of inertia for, 507–513, 526–529, 530–532, 554
   centroidal axis of, 508–509, 554
   composite bodies, 518–520, 546–548, 554–555
   distributed load reduction, 180
   inclined, area about, 530–532
   line of action for, 136–138, 190
   mass moments of inertia for, 541–548, 555
   Mohr's circle for, 533–535
   moment of a force about specified, 136–140, 190
   moments of inertia (I), 507–513, 518–520, 530–535, 541–548, 554–555
   parallel-axis theorem for, 508–510, 518, 527, 545, 554
   principal, 531–535, 555
   procedures for analysis of, 510, 534, 542
   product of inertia and, 526–529, 531, 554
   radius of gyration for, 509, 546
   resultant forces and, 136–140, 180, 190
   right-hand rule for, 136–138
   scalar analysis, 136, 190
   transformation equations for, 530
   vector analysis, 137–140, 190
Axial loads, friction analysis of, 425–427
Axial revolution, 480–483, 502
Axial symmetry, 480–483, 502–503
   axial revolution and, 480–483, 503
   centroid (C) and, 480–483, 502
   composite bodies, 481
   Pappus and Guldinus, theorems of, 480–483, 503
   rotation and, 480–483, 503
   surface area and, 480, 482–483, 503
   volume and, 481–483, 503
Axis of symmetry, 447, 466–467, 502, 526–529, 531, 554
   area (A) of, 526–529
   centroid (C) and, 447, 466–467, 502, 554
   parallel-axis theorem for, 527, 554
   principal axes, 531
   product of inertia, 526–529, 531, 554

B

Ball and socket connections, 231–232, 234
Base units, 7, 10
Beams, 325–360, 376–378
   bending moments (**M**) and, 326–327, 351–355, 376
   cantilevered, 325–326, 341
   centroid (C), 326
   couple moment (**M**) and, 352
   distributed loads and, 350–355, 378
   force equilibrium, 350–351
   free-body diagrams, 325–327, 376
   internal forces, 325–360, 376–378
   internal loads of, 325–332, 350–355
   method of sections for, 325–332, 342
   moments, 326–327, 350–355, 376
   normal force (**N**) and, 326–327, 376
   procedures for analysis of, 327, 342
   resultant loadings, 326, 376
   shear and moment diagrams, 341–344, 350–355, 377–378
   shear force (**V**) and, 326–327, 350–355, 376
   sign convention for, 327, 377

632 INDEX

Beams (*Continued*)
  simply supported, 341
  torsional (twisting) moment, 326, 376
Bearings, 231–235, 425–429, 439
  axial loads, 425–427
  collar, 425–427, 439
  free-body diagrams, 231–235
  frictional analysis of, 425–429, 439
  journal, 232–234, 428–429, 439
  lateral loads, 428–429
  pivot, 425–427, 439
  rigid-body support reactions, 231–235
  thrust, 233, 234
Belts (flat), frictional analysis of, 417–419, 438
Bending moment diagrams, 341–344. *See also* Shear and moment diagrams
Bending moments (**M**), 326–327, 351–355, 376–378
  distributed loads and, 351–355, 378
  internal forces and, 326–327, 350–355, 376, 378
  method of sections for, 326–327
  shear (**V**) and, 351
  shear and moment diagrams, 350–355, 377–378
Body at rest (zero), 196
By inspection, determination of forces, 260, 266–267

C

Cables, 82, 87, 109, 198, 199, 232, 361–375, 378
  concentrated loads, 361–363, 378
  connections, 198, 199
  continuous, 109
  distributed loads, 364–367, 378
  equilibrium of, 82, 87, 109
  flexibility of, 361
  free-body diagram for, 82, 87, 109, 198, 232
  inextensible, 361
  internal forces of, 87, 361–375, 378
  sagging, 361
  support reactions, 198, 232
  weight of as force, 368–371, 378
Calculations, engineering importance of, 10–11
Cantilevered beam, 325–326, 341
Cartesian coordinate system, 33–36, 43–48, 54–59, 64, 76–77, 123–130, 189
  addition of vectors, 33–36, 46
  concurrent force resultants, 33, 43–48, 77
  coordinate direction angles, 44–45, 76–77
  coplanar force resultants, 33–36
  cross product for, 123–125
  direction and, 44–46, 76–77, 123, 126
  dot product in, 64, 77
  force vector directed across a line, 56–59
  horizontal angles (θ), 45

  magnitude of, 33, 44–46, 76, 123, 126
  moment of a force, calculations by, 126–130, 189
  position vectors (**r**), 54–55, 57–58, 77
  rectangular components, 33–36, 43–48, 76
  right-hand rule, 43, 123–124, 126
  sign convention for, 124
  three-dimensional systems, 43–48
  two-dimensional systems, 33–36
  unit vectors, 33, 43–44, 56, 76
  vector formulation, 124–130, 189
  vector representation, 43–44, 76–77
  vertical angles (ϕ), 45
Cartesian vector notation, 33
Center of gravity (*G*), 7, 200, 442–505
  center of mass (C_m) and, 445, 502
  centroid (*C*) as, 200, 442–505
  composite bodies, 466–470, 503
  constant density and, 466
  coplanar forces, 200
  free-body diagrams of, 200
  location of, 443–444, 447–456, 502
  Newton's law of gravitational attraction and, 7
  procedure for analysis of, 448, 467
  rigid-body equilibrium and, 200
  specific weight and, 466
  weight (*W*) and, 7, 200, 443–444, 466, 502
Center of mass (C_m), 445, 448, 456, 501
Center of pressure (*P*), 491, 503
Centroid (*C*), 181, 200, 326, 442–505
  area in *x–y* plane, 446, 448, 452–454, 501
  axis of symmetry, 447, 466–467, 501
  axial symmetry, 480–483, 501–502
  beam cross-section location, 326
  center of gravity (*G*) as, 200, 442–505
  center of mass (C_m) of a body, 445, 448, 456, 501
  composite bodies, 466–470, 502
  composite shapes, 481
  coplanar forces, 200
  distributed loads and, 489–496, 503
  distributed loads, 181
  flat surfaces, 489
  fluid pressure and, 490–496, 503
  free-body diagrams for, 200
  integration for determination of, 445–456, 501
  line in *x–y* plane, 446–447, 448–451, 501
  line of action for, 181, 491, 498, 503
  location of, 181, 445–456, 501
  mass of a body (C_m), 445, 447, 456
  method of sections and, 326
  Pappus and Guldinus, theorems of, 480–483, 502
  plates, 489–496
  procedure for analysis of, 448, 467

Pythagorean's theorem for, 447
resultant forces and, 181, 326, 489, 490–496, 503
rigid-body equilibrium and, 200
rotation of an axis, 480–483, 502
surface area and, 480, 482–483, 502
volume, of a, 445, 448, 481–483, 502
Centroidal axis, 508–509, 554
Coefficient of kinetic friction (μ_k), 384–385
Coefficient of rolling resistance, 430–431
Coefficient of static friction (μ_s), 383, 385, 425–426
Collar bearings, frictional analysis of, 425–427, 439
Collinear couple moment, 170
Collinear vectors, 19, 75
Commutative law, 19, 64, 124
Component vectors of a force, 18, 20–26, 75
Composite bodies, 466–470, 481, 502, 518–520, 546–548, 554–555
area (A) of, 481, 518–520, 554
axial symmetry and, 466–467, 481
center of gravity (G), 466–470, 502
centroid (C) of, 466–470, 481, 502
constant density and, 466
mass moments of inertia, 546–548, 555
moments of inertia (I), 518–520, 546–548, 554
procedure for analysis of, 467, 518
theorem of Pappus and Guldinus for parts of, 481
specific weight and, 466
weight (W) and, 466, 502
Compressive forces (C), 258–259, 260–261, 274–275
method of joints and, 260–261
method of sections and, 274–275
truss members, 258–259
Concentrated force, 5
Concentrated loads, 350–351, 361–363, 377–378
cables subjected to, 361–363, 378
distributed loads, 350–351
sagging from, 361
shear and moment discontinuities from, 351, 377
Concurrent forces, 18, 33, 43–48, 83, 98–102, 109, 168–170, 218–219, 238
addition of vectors, 18, 43–48
Cartesian coordinate system for, 33, 43–48
constraints and, 238
equilibrium of, 86, 98–102, 109, 238
equivalent systems of, 168–170
free-body diagrams, 86, 98–102, 109
force and couple systems, simplification of, 168–170
lines of action for, 166
procedure for analysis of, 170
resultant couple moment, 168
statical determinacy and, 238
three-dimensional systems, 43–48, 98–102, 168–170

three-force members, 218–219
two-dimensional coplanar resultants, 33
Connections, free-body diagrams of, 197–199. *See also* Joints; Support reactions
Conservative forces, 575–576
friction as nonconservative, 576
spring force, 575
virtual work (U) and, 575–576
weight, 575
Constant density, center of gravity (G) and, 466
Constraints, 237–245, 254
improper, 238–239, 254
procedure of analysis of, 240
redundant, 237
statical determinacy and, 237–245, 254
support reactions and, 237–245
rigid-body equilibrium and, 237–245, 254
Continuous cables, 109
Conversion of units, 9
Coordinate direction angles, 44–45, 76–77
Coordinates, 43–48, 54–58, 76–77, 563–568, 578, 590. *See also* Cartesian coordinate system
Cartesian, 43–48, 54–55, 57–58, 76–77
frictionless systems, 578
position, 54–55, 57–58, 563–568, 578, 590
potential energy and, 578
right-hand rule for, 43
vector representation, 43–48, 54–56
virtual work for rigid-body connections, 563–568, 578, 590
x, y, z positions, 43–44, 54, 76–77
Coplanar distributed loads, 180–184
Coplanar forces, 32–37, 76, 85–89, 109, 157–162, 168–174, 180–184, 196–230, 253–254
addition of systems of, 32–37, 85
Cartesian vector notation, 33
center of gravity, 200
centroid (geometric center), 200
couple moments of, 157–162, 168–174
direct solution for unknowns, 208–217, 254
direction of, 32–33, 85
distributed load reduction, 180–184
equations of equilibrium, 85, 109, 208–217, 253
equilibrium of, 85–89, 109, 196–230, 253–254
equivalent systems of, 157–162, 168–174
free-body diagrams, 85–89, 197–206, 253
idealized models of, 200–201
internal forces and, 200
lines of action, 157, 168–174
magnitude of, 33, 34, 85
particles subjected to, 85–89, 109
procedure for analysis of, 86, 159, 170, 202, 209
rectangular components, 32–37, 76

Coplanar forces (*Continued*)
   resultant couple moment, 168
   resultants, 33–37, 157–162, 168–174, 180–184
   rigid bodies, 196–230, 253–254
   scalar notation, 32, 33
   support reactions, 197–199, 253
   system components, 32–37
   systems, simplification of, 157–162, 168–174
   two- and three-force members, 218–219
   vectors for, 32–37, 76
   weight and, 200
Cosine functions, 595
Cosine law, 20, 22, 75
Cosines, direction of, 44–45
Coulomb friction, 381. *See also* Dry friction
Couple, 145
Couple moments (M_O), 145–150, 157–162, 168–174, 190–191, 195–196, 197, 352, 560–561
   collinear, 170
   concurrent force system simplification, 168–170, 191
   coplanar force system simplification, 157–162, 168–174, 190–191
   distributed loading, 157–162, 168–174, 190–191, 352
   equivalent couples, 146
   equivalent systems, 157–162, 168–174, 190–191
   force systems and, 145–150, 195–196
   free vectors, 145
   internal forces and, 352
   parallel force system simplification, 169–170, 191
   procedure for analysis of, 159, 170
   resultants, 146–147, 168–174
   right-hand rule for, 145
   rigid bodies, equilibrium of, 195–196
   rotation of, 197, 560–561
   scalar formulation of, 145
   shear and moment diagrams, 352
   shear load (**V**) relationships, 352
   support reactions and, 197
   systems, simplification of, 157–162, 168–174, 190–191
   three-dimensional systems, 157–162, 168–174, 191
   translation of, 197, 560
   vector formulation of, 145–150
   virtual work of, 561
   work of, 560
   wrench, reduction of forces to, 170, 191
Cross product, 123–125
   Cartesian vector formulation, 124–125
   direction for, 123
   laws of operation, 124
   magnitude for, 123
   right-hand rule for, 123–124
   vector multiplication using, 123–125
Curved plates, fluid pressure and, 492
Cylinders, rolling resistance of, 430–431

D

Deformation, rolling resistance and, 430, 439
Derivatives, 596
Derived units, 7–8, 10
Dimensional homogeneity, 10
Direct solution for unknowns, 208–217, 254
Direction, 17, 33–34, 44–46, 54–56, 65–68, 75–77, 83, 85, 114, 123, 126, 136–138, 145, 189, 197, 385, 387, 408, 410–412, 438
   axis, moment of a force about, 136–138
   Cartesian coordinate vectors, 44–46, 54–56, 76
   Cartesian vector notation, 33
   coordinate direction angles, 44–45, 76
   coplanar force systems, 33–34, 85
   cross product and, 123
   dot product applications, 65–68
   equilibrium and, 83, 85, 197
   force vector along a line, 56
   free-body diagrams, 83, 85, 197
   frictional forces, 385, 387, 408, 410–412, 438
   horizontal angle θ, 34
   impending motion and, 410–412, 438
   line of action, 17–18, 56, 77, 126, 136–138
   moment of a couple, 145
   moment of a force (M_O), 114, 126, 136–138, 189
   position vectors, 54–55, 77
   right-hand rule for, 43, 123, 126, 136–138, 145, 189
   screws, impending motion of, 410–412, 438
   three-dimensional systems, 44–46
   translation, 197
   vector sense of, 17, 33–34, 75
   vertical angle ϕ, 34
Direction cosines, 44–45
Disks, 425–427, 439, 542, 544, 555
   frictional analysis of, 425–427, 439
   mass moments of inertia, 542, 544, 555
Displacement (δ), 561–568, 578, 590
   frictionless systems, 578
   potential energy and, 578
   principle of virtual work and, 561–568, 590
   procedure for analysis of, 564
   rigid bodies, connected systems of, 563–568
   virtual work (U) and, 561–568, 578, 590
   virtual work equations for, 561–562, 564
Distributed loads, 180–184, 191, 350–355, 364–367, 377–378, 489–496, 503
   axis (single) loading, 180
   beams subjected to, 350–355, 377–378
   bending moment (**M**) relationships, 350–355, 378

cables subjected to, 364–367, 378
center of pressure (P), 490, 503
centroid (C) of, 181, 191, 489–496, 503
concentrated loads and, 350–351, 377–378
coplanar, 180, 191
couple moment (M_0) relationships, 352
fluid pressure from, 490–496, 503
force equilibrium, 350–352
force system resultants, 180–184, 191
incompressible fluids, 490
internal forces, 350–355, 364–367, 377–378
linearly, 491, 503
line of action of, 181, 191
loading curve for, 181, 191
magnitude and, 180, 489, 503
reduction of forces, 180–184, 191
resultant forces of, 180–184, 191, 489, 503
shear and moment diagrams, 350–355, 377–378
shear force (V) relationships, 350–355, 378
uniform, 350, 503
Distributive law, 64, 124, 128
Dot notation, 10
Dot product, 64–68, 77, 137
   angles between intersecting lines, 65, 77
   applications of, 65–68
   Cartesian vector formulation, 64, 77
   laws of operation, 64
   moment about a specified axis, 137
   projections, parallel and perpendicular, 65–66, 77
   unit vectors and, 64–65, 77
   vector angles and direction from, 64–68, 77
Dry friction, 381–441
   angles (θ) of, 383–384
   applied force (**P**) and, 382–385, 437–439
   bearings, analysis of, 425–429, 439
   belts (flat), analysis of, 417–419, 438
   collar and pivot bearings, analysis of, 425–427, 439
   characteristics of, 381–385, 437
   coefficients of (μ), 383–385, 437
   direction of force, 385, 387
   disks, analysis of, 425–427, 439
   equations for friction versus equilibrium, 387–394
   equilibrium and, 382–383, 387
   frictional force, 382–385, 428
   impending motion, 383, 386–394, 410–412, 437–438
   journal bearings, analysis of, 428–429, 439
   kinetic force (F_k), 384–385, 437
   motion and, 383–385, 386–394, 410–412, 425–431
   problems involving, 386–394
   procedure for analysis of, 389
   rolling resistance and, 430–431, 439
   screws, forces on, 410–412, 438
   sliding and, 384–385, 386–394, 437
   slipping and, 383, 385, 386–387, 437
   static force (F_s), 383, 385, 437
   theory of, 382
   tipping effect, balance of, 382, 437
   wedges and, 408–409, 438
Dynamics, study of, 2–4

E

Elastic potential energy(V_e), 576
Engineering notation, 11
Equations of equilibrium, 81, 85–86, 98–102, 196, 208–217, 236, 253–254, 387–394
   alternative sets for, 208–209
   body at rest (zero), 196
   coplanar force systems, 85–86, 208–217, 253–254
   direct solution, 208–217, 254
   direction and, 85
   frictional equations and, 387–394
   magnitude and, 85
   particles, 81, 85–86, 98–102
   procedure for analysis using, 86, 98, 209, 389
   rigid bodies, 196, 208–217, 253–254
   scalar form, 236, 253–254
   three-dimensional force systems, 98–102, 236, 254
   two- and three-force members, 218–219
   vector form, 236, 254
Equilibrium, 3, 80–111, 194–255, 350–351, 382–383, 387, 578–584, 591
   concurrent forces, 83, 98–102, 109, 238
   conditions for, 81, 195–196, 208
   constraints, 237–245
   coplanar force systems, 85–89, 109, 196–230, 253–254
   direction and, 83, 85, 197
   distributed load relationships, 350–351
   free-body diagrams, 82–89, 98–102, 197–206, 231–235, 253–254
   friction and, 382–383, 387
   frictionless systems, 578
   idealized models for, 200–201
   impending motion and, 387
   improper constraints and, 238–239, 254
   neutral, 579–580, 591
   one (single) degree-of-freedom system, 580–581
   particles, 80–111
   potential-energy (V) criterion for, 578, 591
   procedures for analysis of, 86, 98, 202, 209, 240, 581
   redundant constraints and, 237
   rigid bodies, 194–255
   shear and moment diagrams, 350–351
   stability of systems, 238–239, 254, 579–584, 591
   stable, 579–581, 591
   statical determinacy and, 237–245, 254

Equilibrium (*Continued*)
   statics and, 3
   support reactions, 197–199, 231–235, 237–245, 253–254
   three-dimensional force systems, 98–102, 109, 231–245, 254
   tipping effect, balance of, 382, 437
   two- and three-force members, 218–219
   two-dimensional force systems, 85–89, 109
   unstable, 580–581, 591
   virtual work (U) and, 578–584, 591
   zero condition, 81, 85, 109, 196
Equivalent couples, 146
Equivalent systems, 157–162, 168–174
   concurrent force systems, 168–170
   coplanar force systems, 157–162, 168–174
   external effects of, 157
   force and couple moment simplification, 157–162, 168–174
   lines of action of, 157, 168–174
   parallel force systems, 169–170
   principle of transmissibility for, 157
   procedures for analysis, 159, 170
   system of force and couple moments, 158
   three-dimensional systems, 157–162, 168–174
   wrench, reduction to, 170
Exponential notation, 10
External effects for equivalent systems, 157
External forces, 195, 261, 274

F

Fixed supports, 197, 199, 233
Flat belts, frictional analysis of, 417–419, 438
Flat plates, 489, 491, 493–494, 503
   constant width, 491
   distributed loads on, 489, 503
   fluid pressure and, 491, 493–494, 503
   variable width, 493
Floor beams, truss analysis and, 258
Fluid pressure, 490–496, 503
   acceleration due to gravity (g), 491
   center of pressure (P), 491
   centroid (C), 490–496, 503
   curved plate of constant width, 492
   flat plate of constant width, 491
   flat plate of variable width, 493
   incompressible fluids, 490
   line of action, 491
   Pascal's law, 490
   plates, 490–496, 503
   resultant forces and, 490–496, 503
Force, 3, 4, 5–9, 16–79, 80–111, 112–193, 195–196, 200, 218–219, 257–283, 288–319, 324–379, 380–441, 489, 490–496, 503, 559–561, 563–568, 575–576
   active, 83
   addition of vectors, 18–19, 20–26, 32–37, 46–48
   applied (**P**), 382–385, 408–409, 437–438
   axis, about a specified, 136–140, 180, 190
   basic quantity of mechanics, 4
   beams, 325–360, 376–377
   bending moments (M) and, 326–327, 350–355, 376, 378
   by inspection, 260, 266–267
   cables, 82, 87, 361–375, 378
   Cartesian vector notation for, 33, 54–59, 127
   components of, 20–26, 32–37, 75
   compressive (C), 258–259, 260–261, 274–275
   concentrated, 5, 350–351, 361–363, 377–378
   concurrent, 18, 33, 43–48, 77, 168–170, 238
   conservative, 575–576
   coplanar, 32–37, 76, 85–89, 109, 157–162, 168–174, 180–184, 191
   couple moments and, 145–150, 157–162, 168–174, 190–191, 195–196
   cross product, 123–125
   directed along a line, 56–59
   displacements from, 563–568
   distributed loads, 180–184, 191, 350–355, 364–367, 378, 489, 490–496
   dot product, 64–68, 77
   equilibrium and, 80–111, 195–196, 218–219, 350–351
   equivalent systems, reduction to, 157–162, 168–174
   external, 195, 261, 274
   fluid pressure, 490–496, 503
   frames, 288–303
   free-body diagrams, 82–89, 98–102, 109, 200, 288–319, 325–327
   friction as, 380–441, 576
   frictional, 382–385, 437
   gravitational, 7
   idealized models for, 200–201
   internal, 200, 274–276, 324–379
   kinetic frictional (F_k), 384–385, 437
   line of action, 17–18, 56, 77, 126–127, 136–138, 157, 168–174, 181, 190
   machines, 288–303
   mechanics of, 3
   method of joints and, 260–268
   method of sections for, 274–279, 325–332, 342
   moment (M) of, 113–117, 126–130, 136–140, 145–150, 157–162, 168–177, 189–190, 326–327, 376, 378
   motion and, 383–385
   multiforce members, 288
   Newton's laws, 6–7
   nonconservative, 576
   normal (**N**), 82, 326–327, 376, 382–384
   parallel systems, 169–170
   parallelogram law for, 18, 20–22, 75

particles subjected to, 80–111
position vectors and, 54–55, 57–58, 77
principle of moments, 115–117, 128
principle of transmissibility, 126, 157
procedures for analysis of, 22, 83, 86, 159, 170, 209, 294, 327
pulleys, 82, 88
reactive, 83
rectangular components, 32–48, 76–77
resultant, 18, 20–26, 33–37, 75, 77, 112–193, 326–327, 489, 490–496, 503
rigid bodies, equilibrium of, 195–196, 200–201
scalar notation for, 32, 33
scalar formulation, 17, 18, 64, 75, 113–114, 136, 145, 189
shear (**V**), 326–327, 350–355, 376, 378
simplification of systems, 157–162, 168–174, 191
smooth surface contact, 82
spring ($\mathbf{F}_s$), 82, 575
springs, 82, 89
static frictional (F_s), 383, 385, 437
structural analysis and, 257–283, 288–319, 325–332
structural members, 218–219, 257–259, 260–268, 274–279, 325–360
systems of, 32–37, 112–193
tensile (T), 258–259, 260–261, 274–275, 417–419
three-dimensional systems, 43–48, 54–58, 98–102, 109, 157–162, 168–174, 190
trusses, 257–283, 320–321
two- and three-force members, 218–219
unbalanced, 6
units of, 8–9
unknown, 260–265, 274–279
virtual work (U) and, 559–561, 563–568, 575–576
weight, 7, 200, 368–371, 378, 575
work (W) of, 559–561
wrench, reduction to, 170
vector formulation, 16–79, 80–111, 123–130, 137–140, 145–150, 189
Frames, 288–303, 321
  free-body diagrams for, 288–294, 321
  multiforce members of, 288, 321
  procedure for analysis of, 294
  structural analysis of, 288–303, 321
Free-body diagrams, 82–94, 98–102, 109, 197–206, 208–217, 231–235, 237–245, 253–254, 274–279, 288–294, 320–321, 325–332, 376, 563–568
  beams, 325–332, 376
  cables, 82, 87
  center of gravity, 200
  centroid (geometric center), 200, 326, 376
  concurrent forces, 98–102
  coplanar force systems, 85–89, 109, 197–206, 208–217, 253–254
  constraints, 237–245, 254

direction and, 83, 85, 197
equilibrium and, 82–94, 98–102, 109, 197–206, 208–217, 231–235, 237–245, 253–254
external forces and, 274–275
frames, 288–294, 321
idealized models of, 200–201
internal forces and, 200, 274–275, 325–332, 376
machines, 288–294, 321
method of sections using, 274–279, 325–332
particle equilibrium, 82–84
procedures for analysis using, 83, 86, 98, 202, 209, 240, 276, 294
pulleys, 82, 88
rigid bodies, 197–206, 208–217, 231–235, 237–245, 253–254
smooth surface contact, 82
springs, 82, 89
statical determinacy and, 237–245, 254
structural analysis using, 274–279, 288–294, 320–321
support reactions, 197–199, 231–235, 237–245, 253–254
three-dimensional systems, 98–102, 109, 231–235, 254
trusses, 274–279
virtual work, 563–568
weight and, 200
Free vector, 145, 202
Friction (F), 380–441, 576
  angles (θ) of, 383–384
  applied force (**P**), 382–385, 408–409, 437–438
  axial loads and, 425–427
  bearings, analysis of, 425–429, 439
  belts (flat), forces on, 417–419, 438
  characteristics of, 381–385, 437
  coefficients of (μ), 383–385, 425–426, 437
  collar bearings, analysis of, 425–427, 439
  Coulomb, 381
  disks, analysis of, 425–427, 439
  dry, 381–441
  equations for friction and equilibrium, 387–394
  equilibrium and, 382–383, 387
  impending motion, 383, 386–394, 410–412, 437–438
  journal bearings, analysis of, 428–429, 439
  kinetic force (F_k), 384–385, 437
  lateral loads and, 428–429
  nonconservative force, as a, 576
  point of contact, 381–387, 430–431, 437
  pivot bearings, analysis of, 425–427, 439
  procedure for analysis of, 389
  rolling resistance and, 430–431, 439
  screws, forces of, 410–412, 438
  shaft rotation and, 425–429, 439
  sliding and, 384–385, 386–394, 437
  slipping and, 383, 385, 386–387, 437
  static force (F_s), 383, 385, 437

638 INDEX

Friction (*F*) (*Continued*)
   virtual work (*U*) and, 576
   wedges and, 408–409, 438
Frictional circle, 428
Frictional force, 382–385, 437
Frictionless systems, 578

G

Geometric center, 181, 200, 326. *See also* Centroid (*C*)
Gravitational attraction, Newton's law of, 7
Gravitational potential energy (V_g), 576
Gravity, *see* Center of gravity (*G*)
Gusset plate, 258–259

H

Hinge connections, 199, 231, 233–234
Hyperbolic functions, 596

I

Idealizations (models) for mechanics, 5, 200–201
Impending motion, 383, 386–394, 410–412, 437–438
   all points of contact, 386
   angle of static friction for, 383
   coefficient of static friction (μ_s) for, 383, 437
   downward, 411, 438
   dry friction problems due to, 386–394
   equilibrium and frictional equations for, 387–394
   friction and, 383, 386–394, 410–412, 437–438
   no apparent, 386
   points of contact, 383, 386–387
   procedure for analysis of, 389
   screws and, 410–412, 438
   slipping, verge of, 383
   tipping and, 387
   upward, 410–411, 438
Inclined axes, moment of inertia for area about, 530–532
Incompressible fluids, 490
Inertia, *see* Moments of inertia
Integrals, 507, 597
Integration, 445–456, 489, 493, 503, 507–513, 541, 554–555
   area (*A*), centroid of, 446, 452–454, 507–513
   center of mass (C_m), determination of using, 445, 456
   centroid (*C*), determination of using, 445–456, 489, 493, 503, 554
   distributed loads, 489, 493, 503
   fluid pressure distribution from, 493, 503
   line, centroid of, 446–447, 449–451
   mass moments of inertia, determination of using, 541, 555
   moments of inertia, determination of using, 507–513, 554
   parallel-axis theorem, 508–509, 554
   procedure for analysis using, 448, 510
   resultant forces determined by, 489, 493
   volume (*V*), centroid of, 445, 455

   volume elements for, 541
Internal forces, 200, 274–275, 324–379
   beams subjected to, 325–360, 376–377
   bending moments (**M**) and, 326–327, 350–355, 376, 378
   cables subjected to, 361–375, 378
   compressive (C), 274
   concentrated loads, 350–351, 361–363, 377–378
   couple moment (M_0) and, 352
   distributed loads, 350–355, 364–368, 377–378
   force equilibrium, 350–351
   free-body diagrams, 200, 274–275, 325–332, 376
   method of sections and, 274–275, 325–332, 342
   moments (*M*) and, 326–327, 350–355, 376–378
   normal force (**N**) and, 326–327, 376
   procedures for analysis of, 327, 342
   resultant loadings, 326–327, 376
   rigid-body equilibrium and, 200
   shear and moment diagrams, 341–344, 350–355, 377–378
   shear force (**V**) and, 326–327, 350–355, 376, 378
   sign convention for, 327, 377
   structural members with, 274–275, 325–332, 376
   tensile (T), 274
   torsional (twisting) moment, 326, 376
   weight, 368–371, 378
International System (SI) of units, 8, 9–10

J

Joints, 257–268, 260–268. *See also* Method of joints
   equilibrium of, 260–261
   loadings at, 258–259
   pin connections, 258–259
   procedure for analysis of, 261
   truss analysis and, 257–259, 260–268
   unknown forces, 260–265
   zero-force members, 266–268
Joules (J), unit of, 560
Journal bearings, 232–234, 428–429, 439
   free-body diagrams, 232–234
   frictional analysis of, 428–429, 439
   support reactions, 232–234

K

Kinetic frictional force (F_k), 384–385, 437

L

Lateral loads, friction analysis of, 428–429
Lead of a screw, 410
Lead angle, 410
Length, 4, 8–9, 446–448, 449–451, 501
   basic quantity of mechanics, 4
   centroid (*C*) of lines, 446–448, 449–451, 501
   integration for, 446–447, 449–451, 501
   procedure for analysis of, 448

Pythagorean theorem for, 447
units of, 8–9
Line of action, 17–18, 56, 77, 126–127, 136–138, 157, 168–174, 190, 489, 491, 503
  centroid (C) location from, 181, 489
  collinear vectors, 18
  distributed loads, 489, 491, 503
  fluid pressure and, 491
  force and couple system simplification, 157, 168–174
  force vector directed along, 56, 77
  moment force-vector formulation, 126–127
  moment of a force about an axis, 136–138, 190
  perpendicular to force resultants, 168–174
  principle of transmissibility, 126, 157
  resultant force, 181, 489
  vector representation of, 17–18, 56, 77, 137–138
Linear elastic behavior, 82
Linear load distribution, 491, 503
Lines, centroid (C) of, 446–451. *See also* Length
  integration for, 446–447, 449–451
  procedure for analysis of, 448
Loading curve, 181, 191
Loads, 180–184, 257–259, 325–332, 350–355, 361–375, 376–378, 425–429, 489–496, 503.
  See also Distributed loads
  axial, 425–427
  beams, 325–332, 350–355, 376–377
  cables, 361–375, 378
  concentrated, 350–351, 361–363, 377–378
  distributed, 180–184, 350–355, 364–367, 377–378, 489–496
  fluid pressure, 490–496
  friction (F) and, 425–429
  internal, 325–332
  lateral, 428–429
  linear distribution of, 491, 503
  moment (M) relations with, 351–355, 378
  plates, 489–496, 503
  resultant forces, 180–184, 489–496
  reduction of distributed, 180–184
  shaft rotation and, 425–429
  shear (V), 350–355, 376, 378
  single axis representation, 180
  structural analysis and, 257–259
  three-dimensional, 326, 376
  truss joints, 257–259
  uniform, 503
  units of, 180
  weight, 368–371, 378

M

Machines, 288–303, 321
  free-body diagrams for, 288–294, 321
  multiforce members of, 288, 321
  procedure for analysis of, 294
  structural analysis of, 288–303, 321
Magnitude, 17, 20–26, 32–34, 44–45, 83, 85, 109, 114, 123, 126 145, 180–181, 189, 489, 503
  Cartesian vectors, 33, 44–46
  coplanar force systems, 32–34, 85
  constant, 109
  couple moments, 145
  cross product and, 123
  distributed load reduction and, 180–181, 489, 503
  equilibrium and, 83, 85
  force components, 20, 22, 32–34, 83
  free-body diagrams, 83, 85
  integration for, 489, 503
  moments and, 114, 123, 126, 189
  Pythagorean theorem for, 34
  resultant forces, 20, 22, 180–181, 489, 503
  right-hand rule for, 126
  sine and cosine laws for, 20, 22
  vector force addition and, 20, 22–26
  vector representation of, 17, 20, 22, 32–34, 44
  units of, 114
Mass, 4, 8–9, 445, 448, 456, 501
  basic quantity of mechanics, 4
  center of (C_m), 445, 448, 456, 501
  integration of, 445, 456, 501
  units of, 8–9
Mass moments of inertia, 541–548, 555
  axis systems, 541–548, 555
  composite bodies, 546–548, 555
  disk elements, 542, 544, 555
  integration for, 541, 555
  parallel-axis theorem for, 545
  procedure for analysis of, 542
  Pythagorean theorem for, 545
  radius of gyration for, 546
  shell elements, 542–543, 555
  volume elements for integration, 541
Mathematical expressions, 594–597
Mechanics, study of, 3
Members, 218–219, 257–268, 274–279, 288–303, 321, 325–332, 376. *See also* Beams
  compressive force (C), 258–259, 260, 274–275
  equilibrium of forces, 218–219
  frame analysis, 288–303, 321
  internal loads(forces) in, 274, 325–332, 376
  joint connections, 257–259, 260–268
  machine analysis, 288–303, 321
  method of sections for, 274–279
  multiforce, 288, 321
  pin connections, 258–259

640 INDEX

Members (*Continued*)
  procedure for analysis of, 327
  tensile force (T), 258–259
  three-force, 218–219
  truss analysis, 257–268, 274–279
  two-force, 218–219
  unknown forces, 260–265, 274–279
  zero-force, 266–268
Method of joints, 260–268, 284–285, 320
  compressive forces, 260–261
  procedures for analysis using, 261, 284
  space truss analysis, 284–285
  structural analysis using, 260–268, 284–285, 320
  tensile forces, 260–261
  truss analysis, 260–268, 284–285, 320
  unknown forces, 260–265
  zero-force members for, 266–268
Method of sections, 274–279, 284, 320, 325–332, 342
  beam analysis using, 325–332, 342
  compressive forces (C), 274–275
  external forces and, 274–275
  internal forces and, 274–275, 325–332
  free-body diagrams for, 274–279, 325–332
  procedures for analysis using, 276, 284, 327
  shear and moment diagrams from, 342
  space truss analysis, 284
  structural analysis using, 274–279, 284, 320, 325–332
  tensile forces (T), 274–275
  truss analysis, 274–279, 284, 320
  unknown member forces, 274–279
Models (idealizations), 5, 200–201
Mohr's circle, 533–535
Moment arm (perpendicular distance), 113–114, 115, 136, 190
Moment axis, 114, 126, 136–140, 190
  direction and, 114, 126
  force about a, 136–140, 190
  force-vector formulation and, 126
  right-hand rule for, 114, 126, 136–138
  scalar analysis of, 136
  vector analysis of, 137–140
Moments (*M*), 112–193, 326–327, 350–355, 376, 378. *See also* Couple moments
  bending (**M**), 326–327, 350–355, 376, 378
  concentrated load discontinuities, 351
  couple (M_0), 145–150, 157–162, 168–174, 190–191, 352
  cross product for, 123–125
  direction and, 114, 123, 126, 189
  distributed loads and, 180–184, 191, 350–355, 378
  equivalent systems, reduction to, 157–162, 168–174
  force, of, 112–193
  force-vector formulation, 126–130
  free vector, 145
  internal forces and, 326–327, 350–355, 376, 378

  magnitude and, 114, 123, 126, 189
  normal force (**N**) and, 326–327
  parallel force systems, 169–170
  perpendicular to force resultants, 168–174
  principle of moments, 115–117, 128, 189
  principle of transmissibility, 126, 157
  procedures for analysis of, 159, 170
  right-hand rule for, 114–115
  resultant forces and, 114, 127, 146–147, 180–184
  scalar formulation of, 113–114, 136, 145, 189
  shear loads (**V**) and, 326–327, 350–355, 378
  sign convention for, 114, 124
  system simplification of, 157–162, 168–174, 190–191
  torque, 114
  torsional (twisting), 326, 376
  Varignon's theorem, 115–117
  vector formulation of, 126–130, 137–140, 145–150, 189
  wrench, reduction of force and couple to, 170
Moments of inertia (*I*), 507–557
  algebraic sum of, 518
  area (*A*), 508–513, 518–520, 526–532, 554
  axis of symmetry, 526–529, 541–548, 574
  axis systems, 507–513, 518–520, 541–548
  composite bodies, 518–520, 546–548, 554–555
  disk elements, 542, 544
  inclined axis, area about, 530–532
  integrals, 507
  integration and, 507–513
  mass, 541–548, 555
  Mohr's circle for, 533–535
  parallel-axis theorem for, 508–510, 518, 527, 545, 554
  polar, 508–509
  principle, 531–532, 534, 555
  procedures for analysis of, 510, 518, 534, 542
  product of inertia and, 526–529, 533–534, 554
  radius of gyration for, 509, 546
  shell elements, 542–543, 555
  transformation equations for, 530–531, 555
Motion, 6, 383–394, 408–412, 417–419, 425–431, 437–439. *See also* Revolution; Shaft rotation
  bearings, 425–429, 439
  belt drives, 417–419, 438
  coefficients of friction (μ) and, 383–385, 430–431, 437
  downward, 411, 438
  equilibrium and frictional equations for, 387–394
  friction and, 383–394, 408–412, 417–419, 425–431, 437–439
  impending, 383, 386–394, 410–412, 437–438
  kinetic frictional force (F_k), 384–385, 437
  Newton's laws of, 6
  points of contact, 383–387
  procedure for analysis of, 389
  rolling resistance and, 430–431, 439
  screws and, 410–412, 438

self-locking mechanisms, 408, 411
shaft rotation, 425–429, 439
sliding, 384–385, 386–394, 437
slipping (impending), 383, 386–387, 437
static frictional force (F_s), 383, 385, 437
upward, 410–411, 438
verge of sliding, 383
wedges, 408–409, 438
Movement, virtual, 561
Multiforce members, 288. *See also* Frames; Machines

N

Neutral equilibrium, 579–580
Newton, unit of, 8
Newton's laws, 4, 6–7
dynamics and, 4
gravitational attraction, 7
motion, 6
Nonconservative force, friction as a, 576
Normal force (**N**), 326–327, 376, 382–384
dry friction and, 382–384
equilibrium and, 382
impending motion and, 383–384
internal forces as, 326–327, 376
method of sections for, 326–327, 376
Numerical calculations, importance of, 10–11

P

Pappus and Guldinus, theorems of, 480–483, 502
axial revolution and symmetry, 480–483, 502
centroid (*C*) and, 480–483, 502
composite shapes, 481
surface area and, 480, 482–483, 502
volume and, 481–483, 502
Parallel-axis theorem, 508–510, 518, 527, 545, 554
area (*A*) and, 508–510, 518, 527, 554
centroidal axis for, 508–510, 554
composite parts, 518
mass moments of inertia determined by, 545
moments of inertia determined by, 508–510, 518, 545, 554
product of inertia determined by, 527, 554
Parallel force systems, 169–170, 218
equilibrium of, 218, 238–239
improper constraints, 238–239
reactive, 238–239
three-force members, 218
simplification of, 169–170
Parallelogram law, 18, 20–22, 75
Particles, 5–7, 80–111
coplanar force systems, 85–89, 109
defined, 5
equations of equilibrium, 81, 85–86, 98–102
equilibrium of, 80–111

free-body diagrams, 82–84
gravitational attraction, 7
idealized model of, 5
Newton's laws applied to, 6–7
nonaccelerating reference of motion, 6
procedures for analysis of, 83, 98
three-dimensional force systems, 98–102, 109
two-dimensional force systems, 85–89, 109
zero condition, 81, 109
Pascal's law, 490
Perpendicular distance (moment arm), 113–114, 115, 136, 190
Pin connections, 197–199, 201, 231, 233–234, 258–259
concurrent forces of, 258–259
coplanar systems, 197–199, 201
free-body diagrams of, 197–199, 231, 233–234
three-dimensional systems, 231, 233–234
truss member analysis, 258–259
Pivot bearings, frictional analysis of, 425–427
Planar truss, 257
Plates, 489–496, 503
flat of constant width, 491
distributed loads on, 489, 503
centroid (*C*), 489–496, 503
curved of constant width, 492
flat of constant width, 491
flat of variable width, 493
fluid pressure and, 490–496, 503
linear distribution on, 491, 503
resultant forces acting on, 489–496, 503
Point of contact, 381–387, 430–431, 437
friction and, 381–387, 430–431, 437
impending motion (slipping), 383, 386–387
kinetic friction and, 384–385
motion (sliding), 384–385
rolling resistance and, 430–431
static friction and, 383, 385
Polar moments of inertia, 508–509
Position coordinates, 563–568, 578, 590
Position vectors (**r**), 54–55, 57–58, 77
Cartesian vector form, 54–55, 57–58, 77
head-to-tail addition, 54–55
x, y, z coordinates, 54, 77
Potential energy (*V*), 576–584, 591
elastic (V_e), 576
equilibrium, criterion for, 578, 591
equilibrium configurations, 579–584
frictionless systems, 578
gravitational (V_g), 576
position coordinates for, 578
potential function equations, 577
procedure for analysis of, 581
single (one) degree-of-freedom systems, 577, 580–581

Potential energy (V) (*Continued*)
  stability of systems and, 579–584, 591
  virtual work (V) and, 576–584, 591
Power-series expansions, 596
Pressure, *see* Fluid pressure
Principal axes, 531–535, 554
  Mohr's circle for, 533–535
  principal moment of inertia and, 531–535
  procedure for analysis of, 534
  product of inertia for, 531, 554
Principle moments of inertia, 531–532, 534, 555
  Mohr's circle for, 533–535
  principal axes, 531–532, 534, 555
  procedure for analysis of, 534
  transformation equations for, 531, 555
Principle of moments, 115–117, 128, 189
Principle of transmissibility, 126, 157
Principle of virtual work, 559, 561–568, 590
Product of inertia, 526–529, 531, 533–534, 554
  axis of symmetry for, 526–529
  centroid for, 527, 554
  Mohr's circle and, 533–534
  parallel-axis theorem for, 527, 554
  procedure for analysis of, 534
  principal axes and, 531
Procedure for analysis, 12–14
Projections, parallel and perpendicular, 65–66, 77, 137
Pulleys, free-body diagram of, 82, 88, 109
Purlins, 257
Pythagorean theorem, 34, 65, 447, 545, 595

Q
Quadratic formula, 596

R
Radius of gyration, 509, 546
Reactive force, 83, 238–239
Rectangular components, 32–48, 76–77
  coplanar force systems, 32–37, 76–77
  force vectors of, 32–48, 76–77
  resultant force, 33–34, 76
Resultant forces, 18, 20–26, 33–37, 43–49, 75–76, 112–193, 326–327, 376, 489–496, 503
  axis, moment of force about, 136–140, 180, 190
  beams, 326–327
  Cartesian vector components, 43–48
  Cartesian vector notation for, 33
  centroid (C) and, 181, 326, 489–496, 503
  concurrent forces, 33, 43–48, 77, 168–170
  coplanar forces, 33–34, 76, 157–162, 168–174
  couple moments, 145–150, 157–162, 168–174, 191
  cross product for, 124–125
  direction of, 124, 189
  distributed loads, 489–496, 503
  fluid pressure and, 490–496, 503
  force components and, 18, 20–22, 75
  force system, 33–37, 77, 112–193
  integration for, 489, 503
  internal forces, 326–267, 376
  lines of action, 136–138, 157, 168–174, 181, 489, 491
  magnitude of, 124, 180, 189, 489, 503
  method of sections for, 326–327
  moments of a force, 113–117, 127, 136–140, 189, 326–327
  parallel force systems, 169–170
  parallelogram law for, 18, 20–22, 75
  perpendicular to moments, 168–174
  plates, 489–496, 503
  principle of moments, 115–117, 189
  procedure for analysis of, 22, 170
  reduction of distributed loads, 180–184, 191
  scalar formulation of, 113–114, 136, 145, 189
  scalar notation for, 33
  system reduction for, 157–162, 168–174, 191
  vector addition for, 18–22, 33–37
  vector formulation of, 20–26, 75, 126–130, 145–150, 189–190
  vector subtraction for, 19
  wrench, reduction to, 170, 191
Revolution, 480–483, 502
  axial symmetry and, 480–483
  centroid (C) and, 480–483, 502
  composite shapes, 481
  Pappus and Guldinus, theorems of, 480–483, 502
  plane area, 481
  surface area, 480, 482–483, 502
  volume from, 481–483, 502
Right-hand rule, 43, 114–115, 123–124, 126, 136–138, 145
  axis, moment of a force about, 136–138
  cross product direction, 123–124
  force-vector formulation, 124, 126
  moment of a couple, 145
  moment of a force, 114–115, 136–138
  three-dimensional coordinate systems, 43
Rigid bodies, 3, 5, 194–255, 563–568, 578–584, 590
  center of gravity, 200
  centroid (geometric center), 200
  conditions for, 195–196
  connected systems of, 563–568, 590
  constraints of, 237–245
  coplanar force systems, 196–230, 253–254
  defined, 5
  displacement (δ) and, 563–568, 578, 590
  equations of equilibrium for, 196, 208–217, 253–254
  equilibrium of, 194–255, 578–584
  external forces and, 195
  force and couple systems acting on, 195–196
  free-body diagrams, 197–206, 231–235, 237–245, 253–254

frictionless systems, 578
idealized models of, 5, 200–201
internal forces and, 200
improper constraints for, 238–239, 254
mechanics, study of, 3
position coordinates for, 563–568, 578, 590
potential energy and, 578–584
procedures for analysis of, 202, 209, 240, 564, 581
redundant constraints for, 237
statical determinacy and, 237–245, 254
support reactions, 197–199, 231–235, 237–245, 253–254
three-dimensional systems, 231–245, 254
two- and three-force members, 218–219
uniform, 200
virtual work (V) for, 563–568, 578–584, 590
weight and, 200
Rocker connections, 198, 199
Roller connections, 197–198, 201, 232, 234
Rolling resistance, frictional forces and, 430–431, 439
Roof truss, 257–258, 320
Rotation, 197, 425–426, 428, 439, 560–561. *See also* Revolution; Shaft rotation
Rounding off numbers, 11

S

Scalar notation, 32, 33
Scalar product, 64
Scalar triple product, 137
Scalars, 17, 18, 32, 64, 85, 113–114, 136, 145, 189, 236, 253–254, 560
   axis, moment of force about, 136
   couple moments, formulation by, 145
   division of vectors by, 18
   dot product and, 64
   equations of equilibrium, 236, 253–254
   magnitude of, 17
   moment of a force, formulation by, 113–114, 136, 189
   multiplication of vectors by, 18, 64
   negative, 32, 85
   torque, 113
   vectors and, 17, 18, 64
   work of a couple moment, 560
Screws, frictional forces on, 410–412, 438
Self-locking mechanisms, 408, 411
Sense of direction, 17
Shaft rotation, 425–429, 439
   axial loads, 425–427
   collar and pivot bearings, 425–427, 438
   disks, 425–427, 438
   frictional analysis of, 425–429, 438
   frictional circle, 428
   journal bearings, 428–429, 438
   lateral loads, 428–429

Shear and moment diagrams, 341–344, 350–355, 377–378
   beam analysis using, 341–344, 350–355, 377
   couple moment (M_0) and, 352
   discontinuities in, 351
   distributed load relations and, 350–355, 378
   internal forces and, 341–344, 350–355, 377–378
   method of sections for, 342
   moment (M) relations in, 351–355, 377–378
   procedure for analysis of, 342
   shear force (V) relations in, 350–355, 377–378
Shear force (**V**), 326–327, 350–355, 376–378
   beams, 326–327, 350–355, 376
   bending moments (**M**) and, 326–327, 350–355, 376, 378
   concentrated load discontinuities, 351
   couple moment (M_0) and, 352
   distributed load relations, 350–355, 378
   internal forces, 326–327, 350–355, 376–378
   method of sections for, 326–327
   shear and moment diagrams, 350–355, 377–378
Shell elements, mass moments of inertia, 542–543, 555
Significant figures, 11
Simple trusses, 257–259, 320
Simply supported beam, 341
Sine functions, 595
Sine law, 20, 22, 75
Single degree-of-freedom systems, 577, 580–581
Sliding, 383–394, 437–438
   friction and, 383–394
   kinetic frictional force (F_k), 384–385, 437
   motion of, 384–394
   problems involving, 386–394
   verge of, 383
Sliding vector, 126, 157, 202
Slipping, 383, 385, 386–394, 437
   friction and, 383, 385–394, 437
   impending motion of, 383, 386–387, 437
   points of contact, 383, 386–387
   problems involving, 386–394
   static frictional force (F_s), 383, 385, 437
Slug, unit of, 8
Smooth surface contact (support), 82, 232
Solving problems, procedure for, 12–14
Space trusses, structural analysis of, 284–285, 321
Specific weight, center of gravity (G) and, 466
Spring constant (k), 82
Spring force (**F**$_s$), virtual work and, 575
Springs, free-body diagram of, 82, 89, 109
Stable equilibrium, 579–581, 591
Stability of a system, 238–239, 254, 579–584, 591. *See also* Equilibrium
   equilibrium configurations for, 579–584, 591
   free-body diagrams for, 238–239, 254
   improper constraints and, 238–239

Stability of a system (*Continued*)
  neutral equilibrium, 579–580, 591
  one (single) degree-of-freedom system, 580–581
  potential energy and, 579–584
  procedure for analysis of, 581
  reactive parallel forces, 238–239
  rigid-body equilibrium and, 238–239, 254
  stable equilibrium, 579–581, 591
  statical determinacy and, 238–239, 254
  unstable equilibrium, 579–581, 591
  virtual work and, 579–584, 591
Static frictional force (F_s), 383, 385, 437
Statical determinacy, 237–245, 254
  equilibrium and, 237–245
  procedure for analysis of, 240
  improper constraints and, 238–239
  indeterminacy, 237, 254
  redundant constraints and, 237
  rigid-body equilibrium and, 237–245, 254
  stability and, 238–239, 254
Statically indeterminate bodies, 237, 254
Statics, 2–15
  basic quantities, 4
  concentrated force, 5
  equilibrium and, 3
  force, 4, 5–9
  gravitational attraction, 7
  historical development of, 4
  idealizations, 5
  length, 4, 8–9
  mass, 4, 8–9
  mechanics study of, 2–3
  motion, 6
  Newton's laws, 6–7
  numerical calculations for, 10–11
  particles, 5
  procedure for analysis of, 12–14
  rigid bodies, 5
  study of, 2–15
  time, 4, 8
  units of measurement, 7–10
  weight, 7
Stiffness factor (k), 82
Stringers, 258
Structural analysis, 256–323, 325–360
  beams, 325–360
  compressive forces (C), 258–259, 260–261, 274–275
  frames, 288–303, 321
  free-body diagrams, 274–279, 288–294, 320–321
  internal forces and, 274–275, 325–360
  machines, 288–303, 321
  method of joints, 260–268, 284–285, 320
  method of sections, 274–279, 284, 320, 325–332, 342
  multiforce members, 288, 321
  procedures for analysis of, 261, 276, 284, 294, 327, 342
  shear and moment diagrams for, 341–344
  space trusses, 284–285, 321
  tensile forces (T), 258–259, 260–261, 274–275
  trusses, 257–287, 320–321
  unknown forces, 260–265, 274–279
  zero-force members, 266–268
Structural members, *see* Members
Support reactions, 197–199, 201, 231–235, 237–245, 253–254
  coplanar force systems, 197–199, 201, 253
  free-body diagrams, 197–199, 201, 231–235, 237–245, 253–254
  improper constraints, 238–239
  procedure for analysis of, 240
  redundant constraints, 237
  rigid-body equilibrium and, 197–199, 201, 231–235, 237–245
  statical determinacy and, 237–245, 254
  three-dimensional force systems, 231–235, 237–245, 254
Surface area, centroid (C) and, 480, 482–483, 502
Symmetry, *see* Axial symmetry; Axis of symmetry
System simplification, 157–162, 168–174
  concurrent force system, 168–170
  coplanar force systems, 157–162, 168–174
  equivalent system, reduction to, 157–162, 168–174
  lines of action and, 157, 168–174
  parallel force systems, 169–170
  procedures for analysis, 159, 170
  reduction to a wrench, 170
  system of force and couple moments, 158
  three-dimensional systems, 157–162, 168–174

T

Tangent functions, 595
Tensile forces (*T*), 258–259, 260–261, 274–275, 417–419
  flat belts, 417–419
  method of joints and, 260–261
  method of sections and, 274–275
  truss members, 258–259, 274–275
Tetrahedron form, 284
Thread of a screw, 410
Three-dimensional systems, 43–48, 54–59, 64–68, 76–77, 98–102, 109, 157–162, 168–174, 231–245, 254.
See also Concurrent forces
  addition of vectors, 46
  Cartesian coordinate system for, 43–48, 76–77
  Cartesian unit vectors, 43–44, 56, 76–77
  Cartesian vector representation, 43–44
  concurrent forces, 43–48, 77, 98–102, 109, 168–170, 254
  constraints for, 237–245, 254
  coordinate direction angles, 44–45, 76–77
  direction angles for, 44–45

dot product for, 64–68, 77
equations of equilibrium, 98, 236, 254
equilibrium of, 98–102, 109, 231–245, 254
equivalent systems, 157–162, 168–174
force and couple moment system simplification, 157–162, 168–174
force vectors, 43–48, 56–59, 77
free-body diagrams, 98–102, 231–235, 254
magnitude of, 44
parallel system simplification, 169–170
particles, 98–102, 109
position vectors, 54–55, 57–58, 77
procedure for analysis of, 98
reactive parallel forces, 239
rectangular components, 43–48, 76–77
resultants, 43–48
right-hand rule, 43
rigid bodies, 231–245, 254
statical determinacy and, 237–245, 254
support reactions for, 231–235, 237–245, 254
x, y, z position coordinates, 43–44, 54, 76–77
Three-force member equilibrium, 218–219
Thrust bearing connections, 233, 234
Time, 4, 8
  basic quantity of mechanics, 4
  units of, 8
Tipping effect, balance of, 382, 437
Torque, 113. *See also* Moments (M)
Torsional (twisting) moment, 326, 376
Transformation equations, moments of inertia (I) and, 530–531, 555
Translation, 197, 560
Trapezoid, distributed loading of, 184
Triangle rule, 19, 75
Triangular truss, 259
Trigonometric identities, 596
Trusses, 257–287, 320–321
  assumptions for design, 258–259, 284
  bridges, 257–258
  compressive force (C) and, 258–259, 260–261, 274–275
  floor beams, 258
  gusset plate for, 258–259
  joints, 257–259, 260–268
  method of joints, 260–268, 284–285, 320
  method of sections, 274–279, 284, 320
  planar, 257
  procedures for analysis of, 261, 276, 284
  purlins, 257
  roof, 257–258, 320
  simple, 257–259, 320
  space trusses, 284–285, 321
  stringers, 258

structural analysis for, 257–287, 320–321
tensile force (T) and, 258–259, 260–261, 274–275
triangular, 259
zero-force members, 266–268
Two-dimensional systems, 32–37, 76, 85–89, 196–230. *See also* Coplanar forces
  Cartesian unit vectors, 33, 76
  coplanar force vectors, 32–37, 76
  free-body diagrams for, 85–89
  particle equilibrium, 85–89
  procedure for analysis of, 86, 202, 209
  rigid-body equilibrium, 196–230
  scalar notation for, 32
Two-force member equilibrium, 218–219

U
U.S. Customary (FPS) system of units, 8
Unbalanced force, 6
Uniform distributed load, 350, 503
Uniform rigid bodies, 200
Unit vector (**u**), 33–34, 43–44, 56, 64–65, 76–77. *See also* Cartesian coordinates
  Cartesian vectors, 33–34, 43–44, 56, 76
  dot product and, 64–65, 77
  three-dimensional, 43–44, 56, 76–77
  force components, 33–34
  force vectors, 56, 77
Units of measurement, 7–10
  base, 7, 10
  conversion of, 9
  derived, 7–8, 10
  International System (SI) of, 8, 9–10
  prefixes, 9
  rules for use, 10
  U.S. Customary (FPS) system of, 8
Unknown member forces, 260–265, 274–279
Unstable equilibrium, 579–581, 591

V
Varignon's theorem, 115–117
Vectors, 16–79, 123–130, 137–140, 145–150, 189, 236, 254
  addition of, 18–19, 20–26, 32–37, 46–48
  addition of forces, 20–26, 32–37
  axis, moment of a force about, 137–140
  Cartesian coordinate system, 33–36, 43–48, 54–59, 76, 123–130, 189
  Cartesian notation for, 33
  components of a force, 18, 20–26, 75
  concurrent forces, 18, 33, 43–48, 77
  coplanar force systems, 32–37
  cross product method of multiplication, 123–125
  collinear, 19, 75

Vectors (*Continued*)
  couple moments, formulation by, 145–150
  direction and, 17, 33–34, 44–46, 123, 126
  division by scalars, 18, 75
  dot product, 64–68, 77, 147
  equations of equilibrium, 236, 254
  force directed along a line, 56–59
  forces and, 16–79
  free, 145, 202
  line of action, 17–18, 56, 77, 126–127, 137–138
  magnitude and, 17, 20–26, 32–34, 44–46, 123, 126
  moments of a force, formulation by, 126–130, 137–140, 189
  multiplication by scalars, 18, 64, 75
  operations, 18–19
  parallelogram law for, 18, 20–22, 75
  physical quantity requirements, 17
  position (**r**), 54–55, 57–58, 77
  principle of transmissibility, 126
  procedure for analysis of, 22
  projections, parallel and perpendicular, 65–66, 147
  rectangular components, 32–48, 76–77
  resultant couple moment, 146–147
  resultant of a force, 18, 20–26, 75, 127
  rigid-body equilibrium and, 236, 254
  scalar notation for, 32
  scalar triple product, 137
  scalars and, 17, 18, 64, 75
  sliding, 126, 202
  subtraction of forces, 19
  systems of coplanar forces, 32–37
  three-dimensional systems, 43–48, 54–59, 64–68, 76–77, 254
  triangle rule for, 19, 75
  two-dimensional systems, 32–37, 76
  unit, 33–34, 43–44, 56, 64–65, 76–77
Virtual movement, 561
Virtual work (*U*), 558–593
  conservative forces and, 575–576
  couple moment, work of, 560–561
  displacement (δ) and, 561–568, 578, 590
  equations for, 561–562, 564
  equilibrium and, 578–584, 591
  force (**F**) and, 559–560, 563–568, 575–576, 590
  friction and, 576
  frictionless systems, 578
  movement as, 561
  position coordinates for, 563–568, 578, 590
  potential energy (*V*) and, 576–584, 591
  principle of, 559, 561–568, 590
  procedures for analysis using, 564, 581
  rigid-bodies, connected systems of, 563–568
  single (one) degree-of-freedom systems, 577, 580–581
  spring force (**F**$_s$) and, 575
  stability of a system, 579–584, 591
  weight (**W**) and, 575
  work (**W**) of a force, 559–561
Volume (*V*), 445, 448, 455, 481–483, 501–502
  axial rotation and symmetry, 481–483, 502
  centroid of (*C*), 445, 448, 455, 481–483, 501–502
  integration of, 445, 455, 501
  Pappus and Guldinus, theorems of, 481–483, 502
  plane area revolution and, 481
  procedure for analysis of, 448

W

Wedges, 408–409, 438
Weight (*W*), 7, 200, 368–371, 378, 443–444, 466, 501–502, 575
  cables subjected to own, 368–371, 378
  center of gravity (*G*) and, 200, 443–444, 466, 501–502
  composite body parts, 466, 502
  conservative force of, 575
  gravitational attraction and, 7
  internal force of, 368–371, 378
  rigid-body equilibrium and, 200
  virtual work (*U*) and, 575
Weightless link, support reactions of, 198
Work (**W**) of a force, 559–561. *See also* Virtual work
  couple moment, of a, 560
  force, of a, 559–560
  virtual movement and, 561
Wrench, reduction of force and moment to, 170, 191

X

x, y, z position coordinates, 43–44, 54, 76–77

Z

Zero condition of equilibrium, 81, 109, 196
Zero-force members, method of joints and, 266–268

ENGINEERING MECHANICS

DYNAMICS

FIFTEENTH EDITION

CHAPTER 12

Although these jet planes are rather large, from a distance their motion can be analyzed as if each were a particle.

KINEMATICS OF A PARTICLE

CHAPTER OBJECTIVES

- To introduce the concepts of position, displacement, velocity, and acceleration.
- To study particle motion along a straight line and represent this motion graphically.
- To investigate particle motion along a curved path using different coordinate systems.
- To present an analysis of dependent motion of two particles.
- To examine the principles of relative motion of two particles using translating axes.

12.1 INTRODUCTION

Engineering mechanics is the study of the state of rest or motion of bodies subjected to the action of forces. It is divided into two areas, namely, statics and dynamics. **Statics** is concerned with the equilibrium of a body that is either at rest or moves with constant velocity. Here we will consider *dynamics*, which deals with the accelerated motion of a body. This subject will be presented in two parts: *kinematics*, which treats only the geometric aspects of the motion, and *kinetics*, which is the analysis of the forces causing the motion. To develop these principles, the dynamics of a particle will be discussed first, followed by topics in rigid-body dynamics in two and then three dimensions.

Historically, the principles of dynamics developed when it was possible to make an accurate measurement of time. Galileo Galilei (1564–1642) was one of the first major contributors to this field. His work consisted of experiments using pendulums and falling bodies. The most significant contributions to dynamics, however, were made by Isaac Newton (1642–1727), who is noted for his formulation of the three fundamental laws of motion and the law of universal gravitational attraction. Shortly after these laws were postulated, important techniques for their application were developed by Euler, D'Alembert, Lagrange, and others.

There are many problems in engineering whose solutions require application of the principles of dynamics. For example, bridges and frames are subjected to moving loads and natural forces caused by wind and earthquakes. The structural design of any vehicle, such as an automobile or airplane, requires consideration of the motion to which it is subjected. This is also true for many mechanical devices, such as motors, pumps, movable tools, industrial manipulators, and machinery. Furthermore, predictions of the motions of artificial satellites, projectiles, and spacecraft are based on the theory of dynamics. With further advances in technology, there will be an even greater need for knowing how to apply the principles of this subject.

Problem Solving. Dynamics is considered to be more involved than statics since both the forces applied to a body and its motion must be taken into account. Also, many applications require using calculus, rather than just algebra and trigonometry. In any case, the most effective way of learning the principles of dynamics is *to solve problems*. To be successful at this, it is necessary to present the work in a logical and orderly manner as suggested by the following sequence of steps:

1. Read the problem carefully and try to correlate the actual physical situation with the theory you have studied.

2. Draw any necessary diagrams and tabulate the problem data.

3. Establish a coordinate system and apply the relevant principles, generally in mathematical form.

4. Solve the necessary equations using a consistent set of units, and report the answer with no more than three significant figures, which is generally the accuracy of the given data.

5. Study the answer using technical judgment and common sense to determine whether or not it seems reasonable.

In applying this general procedure, do the work as neatly as possible. Being neat generally stimulates clear and orderly thinking, and vice versa. If you are having trouble developing your problem-solving skills, consider watching the videos available at www.pearson.com/hibbeler.

12.2 RECTILINEAR KINEMATICS: CONTINUOUS MOTION

We will begin our study of dynamics by discussing the kinematics of a particle that moves along a straight path. Recall that a ***particle*** has a mass but negligible size and shape, so we will limit application to those objects that have dimensions that are of no consequence in the analysis of the motion. For example, a rocket, projectile, or a vehicle can be considered as a particle, as long as its motion is characterized by the motion of its mass center, and any rotation of the body is neglected.

Rectilinear Kinematics. The kinematics of a particle is characterized by specifying, at any given instant, the particle's position, velocity, and acceleration.

Position. The rectilinear or straight-line path of a particle will be defined using a single coordinate axis s, Fig. 12–1a. The origin O on the path is a fixed point, and from this point the ***position coordinate*** s is used to specify the location of the particle at any given instant. The magnitude of s is the distance from O to the particle, usually measured in meters (m) or feet (ft), and the sense of direction is defined by the algebraic sign of s. Although the choice is arbitrary, here s will be positive when the particle is located to the right of the origin, and it will be negative if the particle is located to the left of O. Position is actually a vector quantity since it has both magnitude and direction; however, it is being represented by the algebraic scalar s, rather than in boldface $\mathbf{s}$, since the direction always remains along the coordinate axis.

Position

(a)

Displacement. The ***displacement*** of the particle is defined as the *change in its position*. For example, if the particle moves from one point to another, Fig. 12–1b, the displacement is

$$\Delta s = s' - s$$

In this case Δs is *positive* since the particle's final position is to the *right* of its initial position, i.e., $s' > s$. Displacement is also a *vector quantity*, and it should be distinguished from the distance the particle travels. Specifically, the *distance traveled* is a *positive scalar* that represents the total length of path over which the particle travels.

Displacement

(b)

Fig. 12–1

Velocity. If the particle moves through a displacement Δs during the time interval Δt, the *average velocity* of the particle is

$$v_{\text{avg}} = \frac{\Delta s}{\Delta t}$$

If we take smaller and smaller values of Δt, the magnitude of Δs becomes smaller and smaller. Consequently, the *instantaneous velocity* is a vector defined as $v = \lim_{\Delta t \to 0} (\Delta s / \Delta t)$, or

$$(\xrightarrow{+}) \qquad v = \frac{ds}{dt} \qquad (12\text{–}1)$$

Since Δt or dt is always positive, the sign used to define the *sense* of the velocity is the same as that of Δs or ds. For example, if the particle is moving to the *right*, Fig. 12–1c, the velocity is *positive*; whereas if it is moving to the *left*, the velocity is *negative*. (This is emphasized here by the arrow written at the left of Eq. 12–1.) The *magnitude* of the velocity is known as the *speed*, and it is generally expressed in units of m/s or ft/s.

Occasionally, the term "average speed" is used. The *average speed* is always a positive scalar and is defined as the total distance traveled by a particle, s_T, divided by the elapsed time Δt; i.e.,

$$(v_{\text{avg}})_{\text{sp}} = \frac{s_T}{\Delta t}$$

For example, the particle in Fig. 12–1d travels along the path of length s_T in time Δt, so its average speed is $(v_{\text{avg}})_{\text{sp}} = s_T / \Delta t$, but its average velocity is $v_{\text{avg}} = -\Delta s / \Delta t$.

Velocity

(c)

Average velocity and
Average speed

(d)

Fig. 12–1 (cont.)

12.2 RECTILINEAR KINEMATICS: CONTINUOUS MOTION

Acceleration. If the velocity of the particle is known at two points, then the *average acceleration* of the particle during the time interval Δt is defined as

$$a_{\text{avg}} = \frac{\Delta v}{\Delta t}$$

Here Δv represents the difference in the velocity during the time interval Δt, i.e., $\Delta v = v' - v$, Fig. 12–1e.

The *instantaneous acceleration* at time t is a *vector* that is found by taking smaller and smaller values of Δt and corresponding smaller and smaller values of Δv, so that $a = \lim_{\Delta t \to 0}(\Delta v / \Delta t)$, or

Acceleration
(e)

$(\xrightarrow{+})$
$$a = \frac{dv}{dt} \quad (12\text{–}2)$$

Substituting Eq. 12–1 into this result, we can also write

$(\xrightarrow{+})$
$$a = \frac{d^2s}{dt^2}$$

Both the average and instantaneous acceleration can be either positive or negative. In particular, when the particle is *slowing down*, or its speed is decreasing, the particle is said to be *decelerating*. In this case, v' in Fig. 12–1f is *less* than v, and so $\Delta v = v' - v$ will be negative. Consequently, a will also be negative, and therefore it will act to the *left*, in the *opposite sense* to v. Also, notice that if the particle is originally at rest, then it can have an acceleration if a moment later it has a velocity v'. Units commonly used to express the magnitude of acceleration are m/s^2 or ft/s^2.

Finally, an important differential relation involving the displacement, velocity, and acceleration along the path may be obtained by eliminating the time differential dt between Eqs. 12–1 and 12–2. We have

$$dt = \frac{ds}{v} = \frac{dv}{a}$$

or

$(\xrightarrow{+})$
$$a\,ds = v\,dv \quad (12\text{–}3)$$

Deceleration
(f)

Fig. 12–1 (cont.)

Although we have now produced three important kinematic equations, realize that the above equation is not independent of Eqs. 12–1 and 12–2.

Constant Acceleration, $a = a_c$. When the acceleration is constant, each of the three kinematic equations $a_c = dv/dt$, $v = ds/dt$, and $a_c \, ds = v \, dv$ can be integrated to obtain formulas that relate a_c, v, s, and t.

Velocity as a Function of Time. Integrating $a_c = dv/dt$, assuming that initially $v = v_0$ when $t = 0$, we get

$$\int_{v_0}^{v} dv = \int_{0}^{t} a_c \, dt$$

(↑)

$$\boxed{v = v_0 + a_c t} \quad (12\text{–}4)$$
Constant Acceleration

Position as a Function of Time. Integrating $v = ds/dt = v_0 + a_c t$, assuming that initially $s = s_0$ when $t = 0$, yields

$$\int_{s_0}^{s} ds = \int_{0}^{t} (v_0 + a_c t) \, dt$$

(↑)

$$\boxed{s = s_0 + v_0 t + \tfrac{1}{2} a_c t^2} \quad (12\text{–}5)$$
Constant Acceleration

Velocity as a Function of Position. If we solve for t in Eq. 12–4 and substitute it into Eq. 12–5, or integrate $v \, dv = a_c \, ds$, assuming that initially $v = v_0$ at $s = s_0$, we get

$$\int_{v_0}^{v} v \, dv = \int_{s_0}^{s} a_c \, ds$$

(↑)

$$\boxed{v^2 = v_0^2 + 2a_c(s - s_0)} \quad (12\text{–}6)$$
Constant Acceleration

The algebraic signs of s_0, v_0, and a_c, used in these equations, are determined from the positive direction of the s axis as indicated by the arrow written at the left of each equation. It is important to remember that these equations are useful *only when the acceleration is constant and when $t = 0$, $s = s_0$, $v = v_0$*. A typical example of constant accelerated motion occurs when a body falls freely toward the earth. If air resistance is neglected and the distance of fall is short, then the constant *downward* acceleration of the body when it is close to the earth is approximately 9.81 m/s² or 32.2 ft/s².

IMPORTANT POINTS

- Dynamics is the study of bodies that have accelerated motion.
- Kinematics is a study of the geometry of the motion.
- Kinetics is a study of the forces that cause the motion.
- Rectilinear kinematics refers to straight-line motion.
- Speed refers to the magnitude of velocity.
- Average speed is the total distance traveled divided by the total time. This is different from the average velocity, which is the displacement divided by the time.
- A particle that is slowing down is decelerating.
- A particle can have an acceleration and yet have zero velocity.
- The relationship $a\,ds = v\,dv$ is derived from $a = dv/dt$ and $v = ds/dt$, by eliminating dt.

During the time this rocket undergoes rectilinear motion, its altitude as a function of time can be measured and expressed as $s = s(t)$. Its velocity can then be found using $v = ds/dt$, and its acceleration can be determined from $a = dv/dt$.

PROCEDURE FOR ANALYSIS

Coordinate System.
- Establish a position coordinate s along the path and specify its *fixed origin* and positive direction.
- Since motion is along a straight line, the vector quantities position, velocity, and acceleration can be represented as algebraic scalars. For analytical work the sense of s, v, and a is then defined by their *algebraic signs*.
- The positive sense for each of these scalars can be indicated by an arrow shown alongside each kinematic equation as it is applied.

Kinematic Equations.
- If a relation is known between any *two* of the four variables a, v, s, and t, then a third variable can be obtained by using one of the kinematic equations, $a = dv/dt$, $v = ds/dt$ or $a\,ds = v\,dv$, since each equation relates all three variables.*
- Whenever integration is performed, it is important that the position and velocity be known at a given instant in order to evaluate either the constant of integration if an indefinite integral is used, or the limits of integration if a definite integral is used.
- Remember that Eqs. 12–4 through 12–6 have limited use. These equations apply *only* when the *acceleration is constant* and the initial conditions are $s = s_0$ and $v = v_0$ when $t = 0$.

* Some standard differentiation and integration formulas are given in Appendix A.

EXAMPLE 12.1

The car in Fig. 12–2 moves in a straight line such that for a short time its velocity is defined by $v = (3t^2 + 2t)$ ft/s, where t is in seconds. Determine its position and acceleration when $t = 3$ s. When $t = 0$, $s = 0$.

Fig. 12–2

SOLUTION

Coordinate System. The position coordinate extends from the fixed origin O to the car, positive to the right.

Position. Since $v = f(t)$, the car's position can be determined from $v = ds/dt$, since this equation relates v, s, and t. Noting that $s = 0$ when $t = 0$, we have*

$$(\xrightarrow{+}) \qquad v = \frac{ds}{dt} = (3t^2 + 2t)$$

$$\int_0^s ds = \int_0^t (3t^2 + 2t)\,dt$$

$$s\Big|_0^s = t^3 + t^2 \Big|_0^t$$

$$s = t^3 + t^2$$

When $t = 3$ s,

$$s = (3)^3 + (3)^2 = 36 \text{ ft} \qquad \textit{Ans.}$$

Acceleration. Since $v = f(t)$, the acceleration is determined from $a = dv/dt$, since this equation relates a, v, and t.

$$(\xrightarrow{+}) \qquad a = \frac{dv}{dt} = \frac{d}{dt}(3t^2 + 2t) = 6t + 2$$

When $t = 3$ s,

$$a = 6(3) + 2 = 20 \text{ ft/s}^2 \rightarrow \qquad \textit{Ans.}$$

NOTE: The formulas for constant acceleration *cannot* be used to solve this problem, because the acceleration is a function of time.

---

*The *same result* can be obtained by evaluating a constant of integration C rather than using definite limits on the integral. For example, integrating $ds = (3t^2 + 2t)dt$ yields $s = t^3 + t^2 + C$. Using the condition that at $t = 0$, $s = 0$, then $C = 0$.

EXAMPLE 12.2

A small projectile is fired vertically *downward* into a fluid with an initial velocity of 60 m/s. Due to the drag resistance of the fluid the projectile experiences a deceleration of $a = (-0.4v^3)\,\text{m/s}^2$, where v is in m/s. Determine the projectile's velocity and position 4 s after it is fired.

SOLUTION

Coordinate System. Since the motion is downward, the position coordinate is positive downward, with origin located at O, Fig. 12–3.

Velocity. Here $a = f(v)$ and so we must determine the velocity as a function of time using $a = dv/dt$, since this equation relates v, a, and t. (Why not use $v = v_0 + a_c t$?) Separating the variables and integrating, with $v_0 = 60$ m/s when $t = 0$, yields*

$(+\downarrow)$
$$a = \frac{dv}{dt} = -0.4v^3$$

$$\int_{60\,\text{m/s}}^{v} \frac{dv}{-0.4v^3} = \int_{0}^{t} dt$$

$$\frac{1}{-0.4}\left(\frac{1}{-2}\right)\frac{1}{v^2}\bigg|_{60}^{v} = t - 0$$

$$\frac{1}{0.8}\left[\frac{1}{v^2} - \frac{1}{(60)^2}\right] = t$$

$$v = \left\{\left[\frac{1}{(60)^2} + 0.8t\right]^{-1/2}\right\} \text{m/s}$$

Fig. 12–3

Here the positive root is taken, since the projectile will continue to move downward. When $t = 4$ s,

$$v = 0.559\,\text{m/s} \downarrow \qquad\qquad \textit{Ans.}$$

Position. Knowing $v = f(t)$, we can obtain the projectile's position from $v = ds/dt$, since this equation relates s, v, and t. Using the initial condition $s = 0$, when $t = 0$, we have

$(+\downarrow)$
$$v = \frac{ds}{dt} = \left[\frac{1}{(60)^2} + 0.8t\right]^{-1/2}$$

$$\int_0^s ds = \int_0^t \left[\frac{1}{(60)^2} + 0.8t\right]^{-1/2} dt$$

$$s = \frac{2}{0.8}\left[\frac{1}{(60)^2} + 0.8t\right]^{1/2}\bigg|_0^t$$

$$s = \frac{1}{0.4}\left\{\left[\frac{1}{(60)^2} + 0.8t\right]^{1/2} - \frac{1}{60}\right\} \text{m}$$

When $t = 4$ s,

$$s = 4.43\,\text{m} \qquad\qquad \textit{Ans.}$$

*The *same result* can be obtained by evaluating a constant of integration C rather than using definite limits on the integral.

EXAMPLE 12.3

During a test the rocket in Fig. 12–4 travels upward at 75 m/s, and when it is 40 m from the ground its engine fails. Determine the maximum height s_B reached by the rocket and its speed just before it hits the ground. While in motion the rocket is subjected to a constant downward acceleration of 9.81 m/s^2 due to gravity. Neglect the effect of air resistance.

SOLUTION

Coordinate System. The origin O for the position coordinate s is taken at ground level with positive upward, Fig. 12–4.

Maximum Height. Since the rocket is traveling *upward*, $v_A = +75$ m/s when $t = 0$. At the maximum height $s = s_B$ the velocity $v_B = 0$. For the entire motion, the acceleration is $a_c = -9.81$ m/s^2 (negative since it acts in the *opposite* sense to positive velocity or positive displacement). Since a_c is *constant* the rocket's position may be related to its velocity at the two points A and B on the path by using Eq. 12–6, namely,

$(+\uparrow)$ $\qquad v_B^2 = v_A^2 + 2a_c(s_B - s_A)$

$0 = (75 \text{ m/s})^2 + (-9.81 \text{ m/s}^2)(s_B - 40 \text{ m})$

$s_B = 327$ m *Ans.*

Velocity. To obtain the velocity of the rocket just before it hits the ground, we can apply Eq. 12–6 between points B and C, Fig. 12–4.

$(+\uparrow)$ $\qquad v_C^2 = v_B^2 + 2a_c(s_C - s_B)$

$= 0 + 2(-9.81 \text{ m/s}^2)(0 - 327 \text{ m})$

$v_C = -80.1$ m/s $= 80.1$ m/s $\downarrow$ *Ans.*

The negative root was chosen since the rocket is moving downward. Similarly, Eq. 12–6 may also be applied between points A and C, i.e.,

$(+\uparrow)$ $\qquad v_C^2 = v_A^2 + 2a_c(s_C - s_A)$

$= (75 \text{ m/s})^2 + 2(-9.81 \text{ m/s}^2)(0 - 40 \text{ m})$

$v_C = -80.1$ m/s $= 80.1$ m/s $\downarrow$ *Ans.*

Fig. 12–4

NOTE: It should be realized that the rocket is subjected to a *deceleration* from A to B of 9.81 m/s^2, and then from B to C it is *accelerated* at this rate. Furthermore, even though the rocket momentarily comes to *rest* at B ($v_B = 0$) the acceleration at B is still 9.81 m/s^2 downward!

EXAMPLE 12.4

A metallic particle is subjected to the influence of a magnetic field as it travels downward from plate A to plate B, Fig. 12–5. If the particle is released from rest at the midpoint C, $s = 100$ mm, and the acceleration is $a = (4s)$ m/s^2, where s is in meters, determine the velocity of the particle when it reaches plate B, $s = 200$ mm, and the time it takes to travel from C to B.

SOLUTION

Coordinate System. As shown in Fig. 12–5, s is positive downward, measured from plate A.

Velocity. Since $a = f(s)$, the velocity as a function of position can be obtained by using $v\, dv = a\, ds$. Realizing that $v = 0$ at $s = 0.1$ m, we have

$(+\downarrow)$
$$v\, dv = a\, ds$$
$$\int_0^v v\, dv = \int_{0.1\,\text{m}}^s 4s\, ds$$
$$\frac{1}{2}v^2\Big|_0^v = \frac{4}{2}s^2\Big|_{0.1\,\text{m}}^s$$
$$v = 2(s^2 - 0.01)^{1/2} \text{ m/s} \qquad (1)$$

At $s = 200$ mm $= 0.2$ m,
$$v_B = 0.346 \text{ m/s} = 346 \text{ mm/s} \downarrow \qquad Ans.$$

Time. The time for the particle to travel from C to B can be obtained using $v = ds/dt$ and Eq. 1, where $s = 0.1$ m when $t = 0$. From Appendix A,

$(+\downarrow)$
$$ds = v\, dt$$
$$= 2(s^2 - 0.01)^{1/2} dt$$
$$\int_{0.1}^s \frac{ds}{(s^2 - 0.01)^{1/2}} = \int_0^t 2\, dt$$
$$\ln\left(\sqrt{s^2 - 0.01} + s\right)\Big|_{0.1}^s = 2t\Big|_0^t$$
$$\ln\left(\sqrt{s^2 - 0.01} + s\right) + 2.303 = 2t$$

Fig. 12–5

At $s = 0.2$ m,
$$t = \frac{\ln\left(\sqrt{(0.2)^2 - 0.01} + 0.2\right) + 2.303}{2} = 0.658 \text{ s} \qquad Ans.$$

NOTE: The formulas for constant acceleration cannot be used here because the acceleration changes with position, i.e., $a = 4s$.

EXAMPLE 12.5

A particle moves along a horizontal path with a velocity of $v = (3t^2 - 6t)$ m/s, where t is in seconds. If it is initially located at the origin O, determine the distance traveled in 3.5 s, and the particle's average velocity and average speed during the time interval.

SOLUTION

Coordinate System. Here positive motion is to the right, measured from the origin O, Fig. 12–6a.

Distance Traveled. Since $v = f(t)$, the position as a function of time may be found by integrating $v = ds/dt$ with $t = 0, s = 0$.

$(\xrightarrow{+})$
$$ds = v\, dt$$
$$= (3t^2 - 6t)\, dt$$
$$\int_0^s ds = \int_0^t (3t^2 - 6t)\, dt$$
$$s = (t^3 - 3t^2) \text{ m} \qquad (1)$$

Fig. 12–6

In order to determine the distance traveled in 3.5 s, it is necessary to investigate the path of motion. If we graph the velocity function, Fig. 12–6b, then it shows that for $0 < t < 2$ s the velocity is *negative*, which means the particle is traveling to the *left*, and for $t > 2$ s the velocity is *positive*, and hence the particle is traveling to the *right*. Also, note that $v = 0$ when $t = 2$ s. The particle's position when $t = 0, t = 2$ s, and $t = 3.5$ s can be determined from Eq. 1. This yields

$$s|_{t=0} = 0 \quad s|_{t=2\text{s}} = -4.0 \text{ m} \quad s|_{t=3.5\text{s}} = 6.125 \text{ m}$$

The path is shown in Fig. 12–6a. Hence, the distance traveled in 3.5 s is

$$s_T = 4.0 + 4.0 + 6.125 = 14.125 \text{ m} = 14.1 \text{ m} \qquad Ans.$$

Velocity. The *displacement* from $t = 0$ to $t = 3.5$ s is

$$\Delta s = s|_{t=3.5\text{s}} - s|_{t=0} = 6.125 \text{ m} - 0 = 6.125 \text{ m}$$

and so the average velocity is

$$v_{\text{avg}} = \frac{\Delta s}{\Delta t} = \frac{6.125 \text{ m}}{3.5 \text{ s} - 0} = 1.75 \text{ m/s} \rightarrow \qquad Ans.$$

The average speed is defined in terms of the *total distance traveled* s_T. This positive scalar is

$$(v_{\text{avg}})_{\text{sp}} = \frac{s_T}{\Delta t} = \frac{14.125 \text{ m}}{3.5 \text{ s} - 0} = 4.04 \text{ m/s} \qquad Ans.$$

NOTE: In this problem, the acceleration is $a = dv/dt = (6t - 6)$ m/s^2, which is not constant.

FUNDAMENTAL PROBLEMS

F12–1. Initially, the car travels along a straight road with a speed of 35 m/s. If the brakes are applied and the speed of the car is reduced to 10 m/s in 15 s, determine the constant deceleration of the car.

Prob. F12–1

F12–2. A ball is thrown vertically upward with a speed of 15 m/s. Determine the time of flight when it returns to its original position.

Prob. F12–2

F12–3. A particle travels along a straight line with a velocity of $v = (4t - 3t^2)$ m/s, where t is in seconds. Determine the position of the particle when $t = 4$ s. $s = 0$ when $t = 0$.

F12–4. A particle travels along a straight line with a speed $v = (0.5t^3 - 8t)$ m/s, where t is in seconds. Determine the acceleration of the particle when $t = 2$ s.

F12–5. The position of the particle is $s = (2t^2 - 8t + 6)$ m, where t is in seconds. Determine the time when the velocity of the particle is zero, and the total distance traveled by the particle when $t = 3$ s.

Prob. F12–5

F12–6. A particle travels along a straight line with an acceleration of $a = (10 - 0.2s)$ m/s^2, where s is measured in meters. Determine the velocity of the particle when $s = 10$ m if $v = 5$ m/s at $s = 0$.

Prob. F12–6

F12–7. A particle moves along a straight line such that its acceleration is $a = (4t^2 - 2)$ m/s^2, where t is in seconds. When $t = 0$, the particle is located 2 m to the left of the origin, and when $t = 2$ s, it is 20 m to the left of the origin. Determine the position of the particle when $t = 4$ s.

F12–8. A particle travels along a straight line with a velocity of $v = (20 - 0.05s^2)$ m/s, where s is in meters. Determine the acceleration of the particle at $s = 15$ m.

PROBLEMS

12–1. Starting from rest, a particle moving in a straight line has an acceleration of $a = (2t - 6)$ m/s^2, where t is in seconds. What is the particle's velocity when $t = 6$ s, and what is its position when $t = 11$ s?

12–2. If a particle has an initial velocity of $v_0 = 12$ ft/s to the right, at $s_0 = 0$, determine its position when $t = 10$ s, if $a = 2$ ft/s^2 to the left.

12–3. A particle travels along a straight line with a velocity $v = (12 - 3t^2)$ m/s, where t is in seconds. When $t = 1$ s, the particle is located 10 m to the left of the origin. Determine the acceleration when $t = 4$ s, the displacement from $t = 0$ to $t = 10$ s, and the distance the particle travels during this time period.

*12–4. A particle travels along a straight line with a constant acceleration. When $s = 4$ ft, $v = 3$ ft/s and when $s = 10$ ft, $v = 8$ ft/s. Determine the velocity as a function of position.

12–5. The velocity of a particle traveling in a straight line is given by $v = (6t - 3t^2)$ m/s, where t is in seconds. If $s = 0$ when $t = 0$, determine the particle's deceleration and position when $t = 3$ s. How far has the particle traveled during the 3-s time interval, and what is its average speed?

12–6. A particle moving along a straight line is subjected to a deceleration $a = (-2v^3)$ m/s^2, where v is in m/s. If it has a velocity $v = 8$ m/s and a position $s = 10$ m when $t = 0$, determine its velocity and position when $t = 4$ s.

12–7. A particle moves along a straight line such that its position is defined by $s = (2t^3 + 3t^2 - 12t - 10)$ m. Determine the velocity, average velocity, and the average speed of the particle when $t = 3$ s.

*12–8. A particle is moving along a straight line such that its position is defined by $s = (10t^2 + 20)$ mm, where t is in seconds. Determine (a) the displacement of the particle during the time interval from $t = 1$ s to $t = 5$ s, (b) the average velocity of the particle during this time interval, and (c) the acceleration when $t = 1$ s.

12–9. A particle moves along a straight path with an acceleration of $a = (5/s)$ m/s^2, where s is in meters. Determine the particle's velocity when $s = 2$ m, if it is released from rest when $s = 1$ m.

12–10. A particle moves along a straight line with an acceleration of $a = 5/(3s^{1/3} + s^{5/2})$ m/s^2, where s is in meters. Determine the particle's velocity when $s = 2$ m, if it starts from rest when $s = 1$ m. Use a numerical method to evaluate the integral.

12–11. A particle travels along a straight-line path such that in 4 s it moves from an initial position $s_A = -8$ m to a position $s_B = +3$ m. Then in another 5 s it moves from s_B to $s_C = -6$ m. Determine the particle's average velocity and average speed during the 9-s time interval.

*12–12. The speed of a particle traveling along a straight line within a liquid is measured as a function of its position as $v = (100 - s)$ mm/s, where s is in millimeters. Determine (a) the particle's deceleration when it is located at point A, where $s_A = 75$ mm, (b) the distance the particle travels before it stops, and (c) the time needed to stop the particle.

12–13. Tests reveal that a driver takes about 0.75 s before he or she can *react* to a situation to avoid a collision. It takes about 3 s for a driver having 0.1% alcohol in his system to do the same. If such drivers are traveling on a straight road at 30 mph (44 ft/s) and their cars can decelerate at 2 ft/s^2, determine the shortest stopping distance d for each from the moment they see the pedestrians. *Moral:* If you must drink, please don't drive!

Prob. 12–13

12–14. The sports car travels along the straight road such that $v = 3\sqrt{100 - s}$ m/s, where s is in meters. Determine the time for the car to reach $s = 60$ m. How much time does it take to stop?

12–18. Car A starts from rest at $t = 0$ and travels along a straight road with a constant acceleration of 6 ft/s² until it reaches a speed of 80 ft/s. Afterwards it maintains this speed. Also, when $t = 0$, car B located 6000 ft down the road is traveling towards A at a constant speed of 60 ft/s. Determine the distance traveled by car A when they pass each other.

Prob. 12–14

Prob. 12–18

12–15. A ball is released from the bottom of an elevator, which is traveling upward with a velocity of 6 ft/s. If the ball strikes the bottom of the elevator shaft in 3 s, determine the height of the elevator from the bottom of the shaft at the instant the ball is released. Also, find the velocity of the ball when it strikes the bottom of the shaft.

*12–16. A particle is moving along a straight line with an initial velocity of 6 m/s when it is subjected to a deceleration of $a = (-1.5v^{1/2})$ m/s², where v is in m/s. Determine how far it travels before it stops. How much time does this take?

12–17. Car B is traveling a distance d ahead of car A. Both cars are traveling at 60 ft/s when the driver of B suddenly applies the brakes, causing his car to decelerate at 12 ft/s². It takes the driver of car A 0.75 s to react (this is the normal reaction time for drivers). When he applies his brakes, he decelerates at 15 ft/s². Determine the minimum distance d between the cars so as to avoid a collision.

12–19. A particle moves along a straight path with an acceleration of $a = (kt^3 + 4)$ ft/s², where t is in seconds. Determine the constant k, knowing that $v = 12$ ft/s when $t = 1$ s, and that $v = -2$ ft/s when $t = 2$ s.

*12–20. The velocity of a particle traveling along a straight line is $v = (3t^3 - 6t)$ ft/s, where t is in seconds. If $s = 4$ ft when $t = 0$, determine the position of the particle when $t = 4$ s. What is the total distance traveled during the time interval $t = 0$ to $t = 4$ s? Also, what is the acceleration when $t = 2$ s?

12–21. A freight train travels at $v = 60(1 - e^{-t})$ ft/s, where t is the elapsed time in seconds. Determine the distance traveled in three seconds, and the acceleration at this time.

Prob. 12–17

Prob. 12–21

12–22. The acceleration of the boat is defined by $a = (1.5\,v^{1/2})$ m/s. Determine its speed when $t = 4$ s if it has a speed of 3 m/s when $t = 0$.

Prob. 12–22

12–23. A particle is moving along a straight line such that its acceleration is defined as $a = (-2v)$ m/s^2, where v is in meters per second. If $v = 20$ m/s when $s = 0$ and $t = 0$, determine the particle's position, velocity, and acceleration as functions of time.

*12–24. When a particle is projected vertically upward with an initial velocity of v_0, it experiences an acceleration $a = -(g + kv^2)$, where g is the acceleration due to gravity, k is a constant, and v is the velocity of the particle. Determine the maximum height reached by the particle.

12–25. If the effects of atmospheric resistance are accounted for, a freely falling body has an acceleration defined by the equation $a = 9.81[1 - v^2(10^{-4})]$ m/s^2, where v is in m/s and the positive direction is downward. If the body is released from rest at a *very high altitude*, determine (a) the velocity when $t = 5$ s, and (b) the body's terminal or maximum attainable velocity (as $t \to \infty$).

12–26. A ball is thrown with an upward velocity of 5 m/s from the top of a 10-m-high building. One second later another ball is thrown upward from the ground with a velocity of 10 m/s. Determine the height from the ground where the two balls pass each other.

12–27. When a particle falls through the air, its initial acceleration $a = g$ diminishes until it is zero, and thereafter it falls at a constant or terminal velocity v_f. If this variation of the acceleration can be expressed as $a = (g/v_f^2)(v_f^2 - v^2)$, determine the time needed for the velocity to become $v = v_f/2$. Initially the particle falls from rest.

*12–28. A train is initially traveling along a straight track at a speed of 90 km/h. For 6 s it is subjected to a constant deceleration of 0.5 m/s^2, and then for the next 5 s it has a constant deceleration a_c. Determine a_c so that the train stops at the end of the 11-s time period.

12–29. Two cars A and B start from rest at a stop line. Car A has a constant acceleration of $a_A = 8$ m/s^2, while car B has an acceleration of $a_B = (2t^{3/2})$ m/s^2, where t is in seconds. Determine the distance between the cars when A reaches a velocity of $v_A = 120$ km/h.

12–30. A sphere is fired downward into a medium with an initial speed of 27 m/s. If it experiences a deceleration of $a = (-6t)$ m/s^2, where t is in seconds, determine the distance traveled before it stops.

12–31. A particle is moving along a straight line such that its position is given by $s = (4t - t^2)$ ft, where t is in seconds. Determine the distance traveled from $t = 0$ to $t = 5$ s, the average velocity, and the average speed of the particle during this time interval.

*12–32. Ball A is thrown vertically upward from the top of a 30-m-high building with an initial velocity of 5 m/s. At the same instant another ball B is thrown upward from the ground with an initial velocity of 20 m/s. Determine the height from the ground and the time at which they pass.

12–33. As a body is projected to a high altitude above the earth's surface, the variation of the acceleration of gravity with respect to altitude y must be taken into account. Neglecting air resistance, this acceleration is determined from the formula $a = -g_0[R^2/(R + y)^2]$, where g_0 is the constant gravitational acceleration at sea level, R is the radius of the earth, and the positive direction is measured upward. If $g_0 = 9.81$ m/s^2 and $R = 6356$ km, determine the minimum initial velocity (escape velocity) at which a projectile should be shot vertically from the earth's surface so that it does not fall back to the earth. *Hint:* This requires that $v = 0$ as $y \to \infty$.

12–34. Accounting for the variation of gravitational acceleration a with respect to altitude y (see Prob. 12–33), derive an equation that relates the velocity of a freely falling particle to its altitude. Assume that the particle is released from rest at an altitude y_0 from the earth's surface. With what velocity does the particle strike the earth if it is released from rest at an altitude $y_0 = 500$ km? Use the numerical data in Prob. 12–33.

12.3 RECTILINEAR KINEMATICS: ERRATIC MOTION

When a particle has erratic or changing motion, then its position, velocity, and acceleration *cannot* be described by a single continuous mathematical function along the entire path. Instead, a series of functions will be required to specify the motion at different intervals. For this reason, it is convenient to represent the motion as a graph. If this graph relates any two of the variables s, v, a, t, then it can be used to construct subsequent graphs relating two other variables since the variables are related by the differential relationships $v = ds/dt$, $a = dv/dt$, or $a\,ds = v\,dv$. Several situations are possible.

The s–t, v–t, and a–t Graphs. To construct the v–t graph given the s–t graph, Fig. 12–7a, the equation $v = ds/dt$ should be used, since it relates the variables s and t to v. This equation states that

$$\frac{ds}{dt} = v$$

$$\frac{\text{slope of}}{s\text{–}t\text{ graph}} = \text{velocity}$$

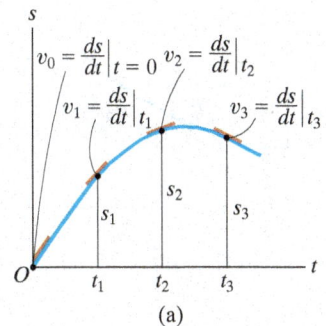

For example, by measuring the slope on the s–t graph when $t = t_1$, the velocity is v_1, Fig. 12–7a. The v–t graph can be constructed by plotting this and other values at each instant, Fig. 12–7b.

The a–t graph can be constructed from the v–t graph in a similar manner, since

$$\frac{dv}{dt} = a$$

$$\frac{\text{slope of}}{v\text{–}t\text{ graph}} = \text{acceleration}$$

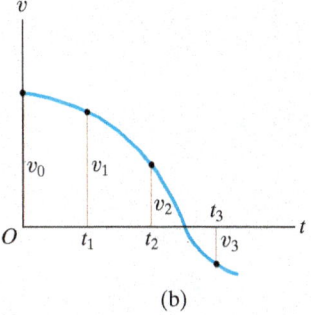

Fig. 12–7

Examples of various measurements are shown in Fig. 12–8a and plotted in Fig. 12–8b.

If the s–t curve for each interval of motion can be expressed by a mathematical function $s = s(t)$, then the equation of the v–t and a–t graph for the same interval can be obtained from succesive derivatives of this function with respect to time since $v = ds/dt$ and $a = dv/dt$. Since differentiation reduces a polynomial of degree n to that of degree $n - 1$, then if the s–t graph is parabolic (a second-degree curve), the v–t graph will be a sloping line (a first-degree curve), and the a–t graph will be a constant or a horizontal line (a zero-degree curve).

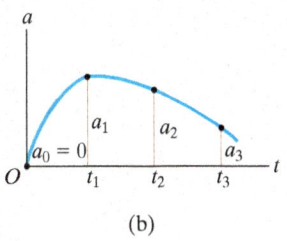

Fig. 12–8

If the *a–t* graph is given, Fig. 12–9a, the *v–t* graph may be constructed using $a = dv/dt$, written as

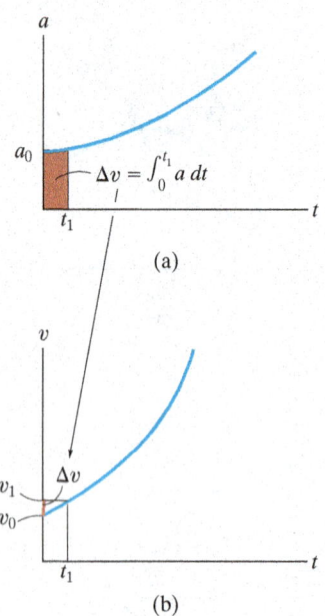

$$\Delta v = \int a \, dt$$

change in velocity = area under *a–t* graph

Therefore, to construct the *v–t* graph, we begin with the particle's initial velocity v_0 and then add to this small increments of area (Δv) determined from the *a–t* graph. In this manner successive points, $v_1 = v_0 + \Delta v$, etc., are determined, Fig. 12–9b. When doing this, an algebraic addition of the area increments of the *a–t* graph is necessary, since areas lying above the *t* axis correspond to an increase in *v* ("positive" area), whereas those lying below the axis indicate a decrease in *v* ("negative" area).

Similarly, if the *v–t* graph is given, Fig. 12–10a, it is possible to determine the *s–t* graph using $v = ds/dt$, written as

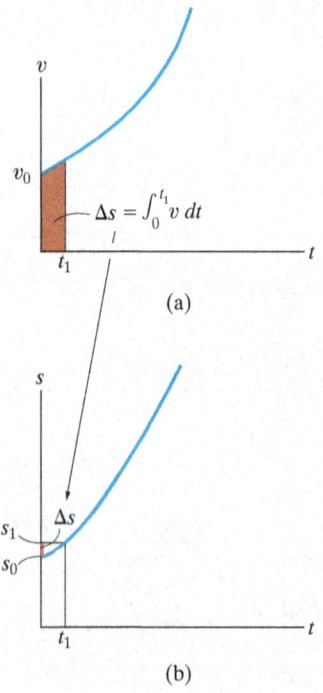

Fig. 12–9

$$\Delta s = \int v \, dt$$

displacement = area under *v–t* graph

Here we begin with the particle's initial position s_0 and add (algebraically) to this small area increments Δs determined from the *v–t* graph, Fig. 12–10b.

Due to the integration, if *segments* of the *a–t* graph can be described by a series of equations, then each of these equations can be successively *integrated* to yield equations describing the corresponding segments of the *v–t* and *s–t* graphs. As a result, if the *a–t* graph is linear (a first-degree curve), integration will yield a *v–t* graph that is parabolic (a second-degree curve) and an *s–t* graph that is cubic (third-degree curve).

Fig. 12–10

12.3 RECTILINEAR KINEMATICS: ERRATIC MOTION

The v–s and a–s Graphs. If the *a–s* graph can be constructed, then points on the *v–s* graph can be determined by using $v\,dv = a\,ds$. Integrating this equation between the limits $v = v_0$ at $s = s_0$ and $v = v_1$ at $s = s_1$, we have,

$$\tfrac{1}{2}(v_1^2 - v_0^2) = \int_{s_0}^{s_1} a\,ds$$

area under
a–s graph

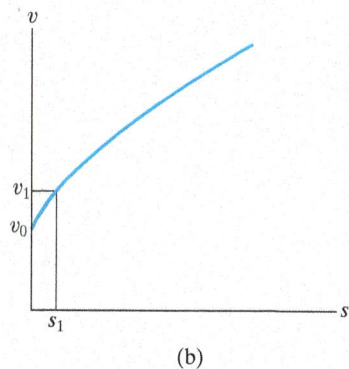

For example, if the red area in Fig. 12–11a is determined, and the initial velocity v_0 at $s_0 = 0$ is known, then $v_1 = \left(2\int_0^{s_1} a\,ds + v_0^2\right)^{1/2}$, Fig. 12–11b. Other points on the *v–s* graph can be determined in this same manner.

If the *v–s* graph is known, the acceleration a at any position s can be determined using $a\,ds = v\,dv$, written as

Fig. 12–11

$$a = v\left(\frac{dv}{ds}\right)$$

velocity times
acceleration = slope of
v–s graph

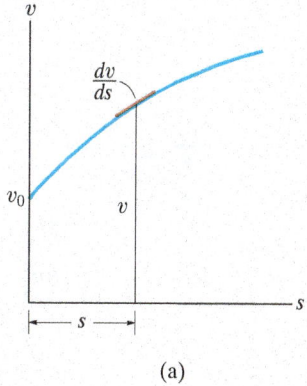

For example, at point (s, v) in Fig. 12–12a, the slope dv/ds of the *v–s* graph is measured. Then with v and dv/ds known, the value of a can be calculated, Fig. 12–12b.

The *v–s* graph can also be constructed from the *a–s* graph, or vice versa, by approximating the known graph in various intervals with mathematical functions, $v = f(s)$ or $a = g(s)$, and then using $a\,ds = v\,dv$ to obtain the other graph.

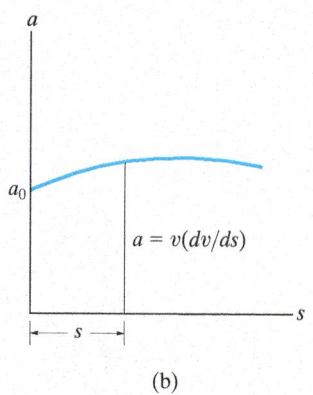

Fig. 12–12

EXAMPLE 12.6

A bicycle moves along a straight road such that its position is described by the graph shown in Fig. 12–13a. Construct the v–t and a–t graphs for $0 \le t \le 30$ s.

(a)

SOLUTION

v–t Graph. Since $v = ds/dt$, the v–t graph can be determined by differentiating the equations defining the s–t graph, Fig. 12–13a. We have

$0 \le t < 10$ s; $s = (t^2)$ ft $v = \dfrac{ds}{dt} = (2t)$ ft/s

10 s $< t \le 30$ s; $s = (20t - 100)$ ft $v = \dfrac{ds}{dt} = 20$ ft/s

These results are plotted in Fig. 12–13b. We can also obtain specific values of v by measuring the *slope* of the s–t graph at a given instant. For example, at $t = 20$ s, the slope of the s–t graph is determined from the straight line from 10 s to 30 s, i.e.,

$t = 20$ s; $v = \dfrac{\Delta s}{\Delta t} = \dfrac{500 \text{ ft} - 100 \text{ ft}}{30 \text{ s} - 10 \text{ s}} = 20$ ft/s

a–t Graph. Since $a = dv/dt$, the a–t graph can be determined by differentiating the equations defining the lines of the v–t graph. This yields

$0 \le t < 10$ s; $v = (2t)$ ft/s $a = \dfrac{dv}{dt} = 2$ ft/s^2

$10 < t \le 30$ s; $v = 20$ ft/s $a = \dfrac{dv}{dt} = 0$

These results are plotted in Fig. 12–13c.

NOTE: The sudden change in v and a at $t = 10$ s represents a discontinuity, but actually this change must occur during a short, but finite time.

(b)

(c)

Fig. 12–13

EXAMPLE 12.7

The car in Fig. 12–14a starts from rest and travels along a straight track such that it accelerates at 10 m/s² for 10 s, and then decelerates at 2 m/s². Draw the v–t graph and determine the time t' needed to stop the car.

SOLUTION

(a)

v–t Graph. Since $dv = a\,dt$, the v–t graph is determined by integrating the straight-line segments of the a–t graph. Using the *initial condition* $v = 0$ when $t = 0$, we have

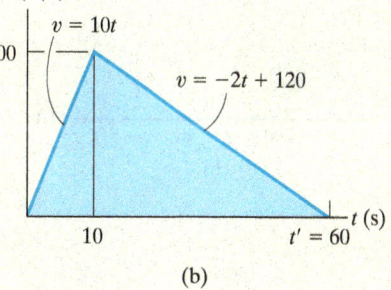

$0 \leq t < 10$ s; $a = (10)$ m/s²; $\int_0^v dv = \int_0^t 10\,dt$, $v = 10t$

When $t = 10$ s, $v = 10(10) = 100$ m/s. Using this as the *initial condition* for the next time period, we have

10 s $< t \leq t'$; $a = (-2)$ m/s²; $\int_{100\,\text{m/s}}^v dv = \int_{10\,\text{s}}^t -2\,dt$, $v = (-2t + 120)$ m/s

(b)

Fig. 12–14

When $t = t'$ we require $v = 0$. This yields, Fig. 12–14b,

$$t' = 60 \text{ s} \qquad \textit{Ans.}$$

A direct solution for t' is also possible by realizing that the area under the a–t graph is equal to the change in the car's velocity. We require $\Delta v = 0 = A_1 + A_2$, Fig. 12–14a. Thus

$$0 = 10 \text{ m/s}^2(10 \text{ s}) + (-2 \text{ m/s}^2)(t' - 10 \text{ s})$$

$$t' = 60 \text{ s} \qquad \textit{Ans.}$$

EXAMPLE 12.8

The v–s graph describing the motion of a motorcycle is shown in Fig. 12–15a. Construct the a–s graph of the motion and determine the time needed for the motorcycle to reach the position $s = 400$ ft.

SOLUTION

a–s Graph. Since the equations for segments of the v–s graph are given, the a–s graph can be determined using $a\,ds = v\,dv$.

$0 \leq s < 200$ ft; $v = (0.2s + 10)$ ft/s

$$a = v\frac{dv}{ds} = (0.2s + 10)\frac{d}{ds}(0.2s + 10) = 0.04s + 2$$

200 ft $< s \leq 400$ ft; $v = 50$ ft/s

$$a = v\frac{dv}{ds} = (50)\frac{d}{ds}(50) = 0$$

The results are plotted in Fig. 12–15b.

Time. The time can be obtained using the v–s graph and $v = ds/dt$, because this equation relates v, s, and t. For the first segment of motion, $s = 0$ when $t = 0$, so

$0 \leq s < 200$ ft; $v = (0.2s + 10)$ ft/s; $dt = \dfrac{ds}{v} = \dfrac{ds}{0.2s + 10}$

$$\int_0^t dt = \int_0^s \frac{ds}{0.2s + 10}$$

$$t = (5\ln(0.2s + 10) - 5\ln 10)\text{ s}$$

At $s = 200$ ft, $t = 5\ln[0.2(200) + 10] - 5\ln 10 = 8.05$ s. Therefore, using these initial conditions for the second segment of motion,

200 ft $< s \leq 400$ ft; $v = 50$ ft/s; $dt = \dfrac{ds}{v} = \dfrac{ds}{50}$

$$\int_{8.05\text{ s}}^t dt = \int_{200\text{ m}}^s \frac{ds}{50}$$

$$t - 8.05 = \frac{s}{50} - 4;$$

$$t = \left(\frac{s}{50} + 4.05\right)\text{ s}$$

Therefore, at $s = 400$ ft,

$$t = \frac{400}{50} + 4.05 = 12.0\text{ s} \qquad\qquad Ans.$$

NOTE: The graphical results can be checked in part by calculating slopes. For example, at $s = 0$, $a = v(dv/ds) = 10(50 - 10)/200 = 2$ m/s^2. Also, the results can be checked in part by inspection. The v–s graph indicates the initial increase in velocity (acceleration) followed by constant velocity ($a = 0$).

Fig. 12–15

FUNDAMENTAL PROBLEMS

F12–9. Due to an external force, the particle travels along a straight track such that its position is described by the s–t graph. Construct the v–t graph for the same time interval. Take $v = 0$, $a = 0$ when $t = 0$.

Prob. F12–9

F12–10. A van travels along a straight road with a velocity described by the graph. Construct the s–t and a–t graphs during the same period. Take $s = 0$ when $t = 0$.

Prob. F12–10

F12–11. The rider begins to apply a force to the rear wheel of his bicycle, thereby initiating an acceleration. If his velocity is described by the v–s graph, construct the a–s graph for the same interval.

Prob. F12–11

F12–12. The sports car starts from rest and travels along a straight road. Its initial increasing acceleration is caused by the rear wheels of the car as shown on the graph. Construct the v–s graph. What is the velocity of the car when $s = 10$ m and $s = 15$ m?

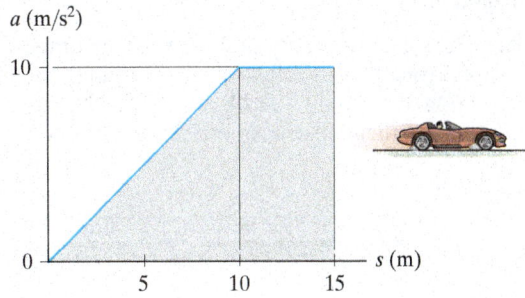

Prob. F12–12

F12–13. The dragster starts from rest and has an acceleration described by the graph. Construct the v–t graph for the time interval $0 \leq t \leq t'$, where t' is the time for the car to come to rest.

Prob. F12–13

F12–14. The dragster starts from rest and has a velocity described by the graph. Construct the s–t graph during the time interval $0 \leq t \leq 15$ s. Also, determine the total distance traveled during this time interval.

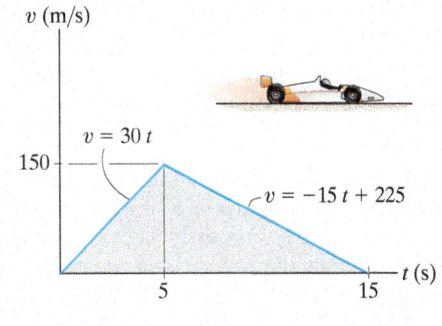

Prob. F12–14

PROBLEMS

12–35. A freight train starts from rest and travels with a constant acceleration of 0.5 ft/s². After a time t' it maintains a constant speed so that when $t = 160$ s it has traveled 2000 ft. Determine the time t' and draw the v–t graph for the motion.

*12–36. The s–t graph for a train has been experimentally determined. From the data, construct the v–t and a–t graphs for the motion; $0 \le t \le 40$ s. For $0 \le t < 30$ s, the curve is $s = (0.4t^2)$ m, and then it becomes straight for $t > 30$ s.

Prob. 12–36

12–37. Two rockets start from rest at the same elevation. Rocket A accelerates vertically at 20 m/s² for 12 s and then maintains a constant speed. Rocket B accelerates at 15 m/s² until reaching a constant speed of 150 m/s. Construct the a–t, v–t, and s–t graphs for each rocket until $t = 20$ s. What is the distance between the rockets when $t = 20$ s?

12–38. A particle starts from $s = 0$ and travels along a straight line with a velocity $v = (t^2 - 4t + 3)$ m/s, where t is in seconds. Construct the v–t and a–t graphs for the time interval $0 \le t \le 4$ s.

12–39. If the position of a particle is defined by $s = [2 \sin (\pi/5)t + 4]$ m, where t is in seconds, construct the s–t, v–t, and a–t graphs for $0 \le t \le 10$ s.

*12–40. An airplane starts from rest, travels 5000 ft down a runway, and after uniform acceleration, takes off with a speed of 162 mi/h. It then climbs in a straight line with a uniform acceleration of 3 ft/s² until it reaches a constant speed of 220 mi/h. Draw the s–t, v–t, and a–t graphs that describe the motion.

12–41. The elevator starts from rest at the first floor of the building. It can accelerate at 5 ft/s² and then decelerate at 2 ft/s². Determine the shortest time it takes to reach a floor 40 ft above the ground. The elevator starts from rest and then stops. Draw the a–t, v–t, and s–t graphs for the motion.

Prob. 12–41

12–42. A car starting from rest moves along a straight track with an acceleration as shown. Determine the time t for the car to reach a speed of 50 m/s and construct the v–t graph that describes the motion until the time t.

Prob. 12–42

12–43. The motion of a jet plane just after landing on a runway is described by the a–t graph. Determine the time t' when the jet plane stops. Construct the v–t and s–t graphs for the motion. Here $s = 0$, and $v = 300$ ft/s when $t = 0$.

Prob. 12–43

*12–44. The v–t graph for a particle moving through an electric field from one plate to another has the shape shown in the figure. The acceleration and deceleration that occur are constant and both have a magnitude of 4 m/s^2. If the plates are spaced 200 mm apart, determine the maximum velocity v_{max} and the time t' for the particle to travel from one plate to the other. Also draw the s–t graph. When $t = t'/2$ the particle is at $s = 100$ mm.

12–45. The v–t graph for a particle moving through an electric field from one plate to another has the shape shown in the figure, where $t' = 0.2$ s and $v_{max} = 10$ m/s. Draw the s–t and a–t graphs for the particle. When $t = t'/2$ the particle is at $s = 0.5$ m.

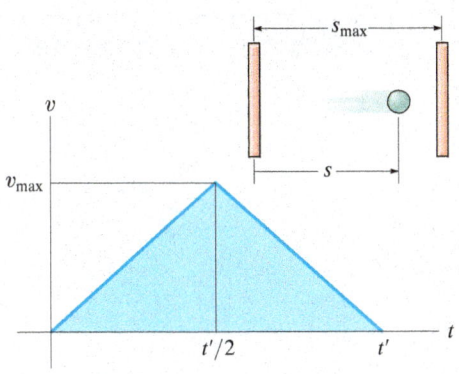

Probs. 12–44/45

12–46. The a–s graph for a rocket moving along a straight track has been experimentally determined. If the rocket starts at $s = 0$ when $v = 0$, determine its speed when it is at $s = 75$ ft, and 125 ft, respectively. Use Simpson's rule with $n = 100$ to evaluate v at $s = 125$ ft.

Prob. 12–46

12–47. The rocket has an acceleration described by the graph. If it starts from rest, construct the v–t and s–t graphs for the motion for the time interval $0 \leq t \leq 14$ s.

Prob. 12–47

***12–48.** The *s–t* graph for a train has been determined experimentally. From the data, construct the *v–t* and *a–t* graphs for the motion.

Prob. 12–48

12–49. The jet car is originally traveling at a velocity of 10 m/s when it is subjected to the acceleration shown. Determine the car's maximum velocity and the time t' when it stops. When $t = 0, s = 0$.

Prob. 12–49

12–50. From experimental data, the motion of a jet plane while traveling along a runway is defined by the *v–t* graph shown. Construct the *s–t* and *a–t* graphs for the motion.

Prob. 12–50

12–51. The *v–t* graph for a train has been experimentally determined. From the data, construct the *s–t* and *a–t* graphs for the motion for $0 \le t \le 180$ s. When $t = 0, s = 0$.

Prob. 12–51

*12–52. A motorcycle starts from rest at $s = 0$ and travels along a straight road with the speed shown by the v–t graph. Determine the total distance the motorcycle travels until it stops when $t = 15$ s. Also plot the a–t and s–t graphs.

12–53. A motorcycle starts from rest at $s = 0$ and travels along a straight road with the speed shown by the v–t graph. Determine the motorcycle's acceleration and position when $t = 8$ s and $t = 12$ s.

Probs. 12–52/53

12–54. The v–t graph for the motion of a car as it moves along a straight road is shown. Draw the s–t and a–t graphs. Also determine the average speed and the distance traveled for the 15-s time interval. When $t = 0$, $s = 0$.

Prob. 12–54

12–55. An airplane lands on the straight runway, originally traveling at 110 ft/s when $s = 0$. If it is subjected to the decelerations shown, determine the time t' needed to stop the plane and construct the s–t graph for the motion.

Prob. 12–55

*12–56. Starting from rest at $s = 0$, a boat travels in a straight line with the acceleration shown by the a–s graph. Determine the boat's speed when $s = 50$ ft, 100 ft, and 150 ft.

12–57. Starting from rest at $s = 0$, a boat travels in a straight line with the acceleration shown by the a–s graph. Construct the v–s graph.

Probs. 12–56/57

12–58. A motorcyclist starting from rest travels along a straight road and for 10 s has an acceleration as shown. Draw the v–t graph that describes the motion and find the distance traveled in 10 s.

12–62. If the position of a particle is defined as $s = (5t - 3t^2)$ ft, where t is in seconds, construct the s–t, v–t, and a–t graphs for $0 \leq t \leq 2.5$ s.

12–63. The jet bike is moving along a straight road with the speed described by the v–s graph. Construct the a–s graph.

Prob. 12–58

Prob. 12–63

12–59. The speed of a train during the first minute has been recorded as follows:

t (s)	0	20	40	60
v (m/s)	0	16	21	24

Plot the v–t graph, approximating the curve as straight-line segments between the given points. Determine the total distance traveled.

*12–60. A man riding upward in a freight elevator accidentally drops a package off the elevator when it is 100 ft from the ground. If the elevator maintains a constant upward speed of 4 ft/s, determine how high the elevator is from the ground the instant the package hits the ground. Draw the v–t curve for the package during the time it is in motion. Assume that the package was released with the same upward speed as the elevator.

12–61. Two cars start from rest side by side and travel along a straight road. Car A accelerates at 4 m/s² for 10 s and then maintains a constant speed. Car B accelerates at 5 m/s² until reaching a constant speed of 25 m/s and then maintains this speed. Construct the a–t, v–t, and s–t graphs for each car until $t = 15$ s. What is the distance between the two cars when $t = 15$ s?

*12–64. The boat is originally traveling at a speed of 8 m/s when it is subjected to the acceleration shown in the graph. Determine the boat's maximum speed and the time t when it stops.

Prob. 12–64

12–65. The jet plane starts from rest at $s = 0$ and is subjected to the acceleration shown. Determine the speed of the plane when it has traveled 1000 ft. Also, how much time is required for it to travel 1000 ft?

Prob. 12–65

12–67. The v–s graph of a cyclist traveling along a straight road is shown. Construct the a–s graph.

Prob. 12–67

12–66. The a–s graph for a freight train is given for the first 200 m of its motion. Plot the v–s graph. The train starts from rest.

Prob. 12–66

*12–68. The jet plane starts from rest at $s = 0$ and is subjected to the acceleration shown. Construct the v–s graph and determine the time needed to travel 500 ft down the runway.

Prob. 12–68

12.4 GENERAL CURVILINEAR MOTION

Curvilinear motion occurs when a particle moves along a curved path. Since this path is often described in three dimensions, vector analysis will be used to formulate the particle's position, velocity, and acceleration.* In this section the general aspects of curvilinear motion are discussed, and in subsequent sections we will consider three types of coordinate systems often used to analyze this motion.

Position. Consider a particle located at a point on a space curve defined by the path function $s(t)$, Fig. 12–16a. The position of the particle, measured from a fixed point O, will be designated by the *position vector* $\mathbf{r} = \mathbf{r}(t)$.

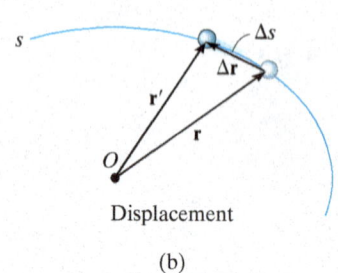

Position
Path
(a)

Displacement. If the particle moves a distance Δs along the curve to a new position, defined by $\mathbf{r}' = \mathbf{r} + \Delta \mathbf{r}$, Fig. 12–16b, then the *displacement* $\Delta \mathbf{r}$ represents the change in the particle's position and is determined by vector subtraction; i.e., $\Delta \mathbf{r} = \mathbf{r}' - \mathbf{r}$.

Velocity. If $\Delta \mathbf{r}$ occurs during the time Δt, then the *average velocity* of the particle is

$$\mathbf{v}_{avg} = \frac{\Delta \mathbf{r}}{\Delta t}$$

The *instantaneous velocity* is determined from this equation by letting $\Delta t \to 0$, and consequently the direction of $\Delta \mathbf{r}$ approaches the *tangent* to the curve. Hence, $\mathbf{v} = \lim_{\Delta t \to 0}(\Delta \mathbf{r}/\Delta t)$ or

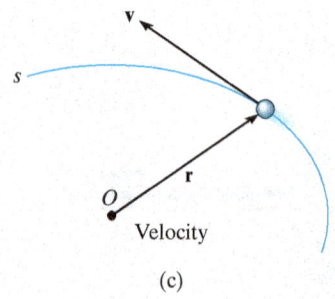

Displacement
(b)

$$\mathbf{v} = \frac{d\mathbf{r}}{dt} \qquad (12\text{–}7)$$

Since $d\mathbf{r}$ will be tangent to the curve, the *direction* of $\mathbf{v}$ is also *tangent to the curve*, Fig. 12–16c. The *magnitude* of $\mathbf{v}$, which is called the *speed*, is obtained by realizing that the length of the straight-line segment $\Delta \mathbf{r}$ in Fig. 12–16b approaches the arc length Δs as $\Delta t \to 0$, and so we have $v = \lim_{\Delta t \to 0}(\Delta r/\Delta t) = \lim_{\Delta t \to 0}(\Delta s/\Delta t)$, or

Velocity
(c)

Fig. 12–16

$$v = \frac{ds}{dt} \qquad (12\text{–}8)$$

Thus, the *speed* can be obtained by differentiating the path function s with respect to time.

---

*A summary of some of the important concepts of vector analysis is given in Appendix B.

Acceleration. If the particle has a velocity **v** at time t and a velocity $\mathbf{v}' = \mathbf{v} + \Delta\mathbf{v}$ at $t + \Delta t$, Fig. 12–16d, then the *average acceleration* of the particle during the time interval Δt is

$$\mathbf{a}_{avg} = \frac{\Delta \mathbf{v}}{\Delta t}$$

(d)

where $\Delta\mathbf{v} = \mathbf{v}' - \mathbf{v}$. To study this time rate of change, the two velocity vectors in Fig. 12–16d are plotted in Fig. 12–16e such that their tails are located at the fixed point O' and their arrowheads touch points on a curve. This curve is called a **hodograph**, and when constructed, it describes the locus of points for the arrowhead of the velocity vector in the same manner as the *path* describes the locus of points for the arrowhead of the position vector, Fig. 12–16a.

To obtain the *instantaneous acceleration*, let $\Delta t \to 0$, and so $\mathbf{a} = \lim_{\Delta t \to 0}(\Delta\mathbf{v}/\Delta t)$, or

$$\mathbf{a} = \frac{d\mathbf{v}}{dt} \qquad (12\text{–}9)$$

Substituting Eq. 12–7 into this result, we can also write

$$\mathbf{a} = \frac{d^2\mathbf{r}}{dt^2}$$

By definition of the derivative, **a** acts *tangent to the hodograph*, Fig. 12–16f, and, *in general it is not tangent to the path of motion*, Fig. 12–16g.

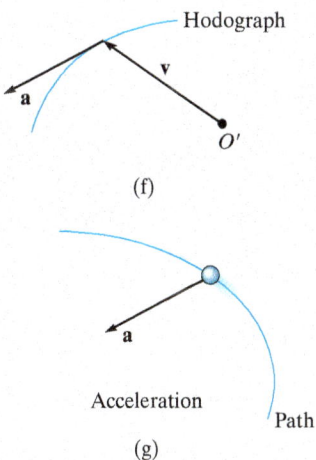

Fig. 12–16

12.5 CURVILINEAR MOTION: RECTANGULAR COMPONENTS

Occasionally the motion of a particle can best be described along a path that is expressed in terms of its x, y, z coordinates.

Position. If the particle is at point (x, y, z) on the path shown in Fig. 12–17a, then its location is defined by the *position vector*

$$\mathbf{r} = x\mathbf{i} + y\mathbf{j} + z\mathbf{k} \qquad (12\text{–}10)$$

When the particle moves, the x, y, z components of $\mathbf{r}$ will be functions of time; i.e., $x = x(t)$, $y = y(t)$, $z = z(t)$, so that $\mathbf{r} = \mathbf{r}(t)$.

At any instant the *magnitude* of $\mathbf{r}$ is determined from Eq. B–3 in Appendix B as

$$r = \sqrt{x^2 + y^2 + z^2}$$

And the *direction* of $\mathbf{r}$ is specified by the unit vector $\mathbf{u}_r = \mathbf{r}/r$.

Velocity. The time derivative of $\mathbf{r}$ yields the velocity of the particle. Hence,

$$\mathbf{v} = \frac{d\mathbf{r}}{dt} = \frac{d}{dt}(x\mathbf{i}) + \frac{d}{dt}(y\mathbf{j}) + \frac{d}{dt}(z\mathbf{k})$$

When taking this derivative, it is necessary to account for changes in *both* the magnitude and direction of each of the vector's components. For example, the derivative of the $\mathbf{i}$ component of $\mathbf{r}$ is

$$\frac{d}{dt}(x\mathbf{i}) = \frac{dx}{dt}\mathbf{i} + x\frac{d\mathbf{i}}{dt}$$

The last term is zero, because the x, y, z reference frame is *fixed*, and therefore the *direction* (and the *magnitude*) of $\mathbf{i}$ does not change with time. Differentiation of the $\mathbf{j}$ and $\mathbf{k}$ components are carried out in a similar manner, and so the final result is

$$\mathbf{v} = \frac{d\mathbf{r}}{dt} = v_x\mathbf{i} + v_y\mathbf{j} + v_z\mathbf{k} \qquad (12\text{–}11)$$

where

$$v_x = \dot{x} \quad v_y = \dot{y} \quad v_z = \dot{z} \qquad (12\text{–}12)$$

The "dot" notation $\dot{x}$, $\dot{y}$, $\dot{z}$ represents the first time derivatives of $x = x(t)$, $y = y(t)$, $z = z(t)$, respectively.

Position
(a)

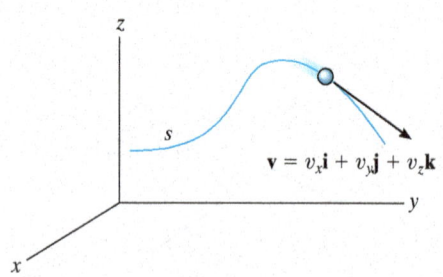

Velocity
(b)

Fig. 12–17

12.5 CURVILINEAR MOTION: RECTANGULAR COMPONENTS

The velocity has a *magnitude* that is found from

$$v = \sqrt{v_x^2 + v_y^2 + v_z^2}$$

and a *direction* that is specified by the unit vector $\mathbf{u}_v = \mathbf{v}/v$. As discussed in Sec. 12.4, this direction is *always tangent to the path*, as shown in Fig. 12–17b.

Acceleration. The acceleration of the particle is obtained by taking the time derivative of Eq. 12–11 (or the second time derivative of Eq. 12–10). We have

$$\mathbf{a} = \frac{d\mathbf{v}}{dt} = a_x\mathbf{i} + a_y\mathbf{j} + a_z\mathbf{k} \qquad (12\text{--}13)$$

where

$$\begin{aligned} a_x &= \dot{v}_x = \ddot{x} \\ a_y &= \dot{v}_y = \ddot{y} \\ a_z &= \dot{v}_z = \ddot{z} \end{aligned} \qquad (12\text{--}14)$$

Acceleration

(c)

Fig. 12–17

Here a_x, a_y, a_z represent the time derivatives of $v_x = v_x(t)$, $v_y = v_y(t)$, $v_z = v_z(t)$, or the second time derivatives of $x = x(t)$, $y = y(t)$, $z = z(t)$.

The acceleration has a *magnitude*

$$a = \sqrt{a_x^2 + a_y^2 + a_z^2}$$

and a *direction* specified by the unit vector $\mathbf{u}_a = \mathbf{a}/a$. Since $\mathbf{a}$ represents the time rate of *change* in both the magnitude and direction of the velocity, in general $\mathbf{a}$ will *not* be tangent to the path, Fig. 12–17c.

IMPORTANT POINTS

- Curvilinear motion will generally cause changes in *both* the magnitude and direction of the particle's position, velocity, and acceleration.

- The velocity is always directed *tangent* to the path.

- In general, the acceleration is *not* tangent to the path, but rather, it is tangent to the hodograph.

- If the motion is described using rectangular coordinates, then the components along the x, y, z axes do not change direction, only their magnitude and sense (algebraic sign) will change.

PROCEDURE FOR ANALYSIS

Coordinate System.
- A rectangular coordinate system should be used to solve problems in cases where the motion can conveniently be expressed in terms of its x, y, z components.

Kinematic Quantities.
- Since *rectilinear or straight-line motion* occurs along *each coordinate axis*, then $v = ds/dt$ and $a = dv/dt$; or in cases where the motion is not expressed as a function of time, the equation $a\,ds = v\,dv$ can be used.

- In two dimensions, the equation of the path $y = f(x)$ can be used to *relate* the x and y components of velocity and acceleration by applying the chain rule of calculus. A review of this concept is given in Appendix C.

- Once the x, y, z components of **v** and **a** have been determined, the magnitudes of these vectors are found from the Pythagorean theorem, Eq. B–3, and their coordinate direction angles from the components of their unit vectors, Eqs. B–4 and B–5.

EXAMPLE 12.9

At any instant the horizontal position of the weather balloon in Fig. 12–18a is defined by $x = (8t)$ ft, where t is in seconds. If the equation of the path is $y = x^2/10$, determine the magnitude and direction of the balloon's velocity and the acceleration when $t = 2$ s.

SOLUTION

Velocity. The velocity component in the x direction is

$$v_x = \dot{x} = \frac{d}{dt}(8t) = 8 \text{ ft/s} \rightarrow$$

To find the relationship between the velocity components we will use the chain rule of calculus. When $t = 2$ s, $x = 8(2) = 16$ ft, Fig. 12–18a, and so

$$v_y = \dot{y} = \frac{d}{dt}(x^2/10) = 2x\dot{x}/10 = 2(16)(8)/10 = 25.6 \text{ ft/s} \uparrow$$

When $t = 2$ s, the magnitude of velocity is therefore

$$v = \sqrt{(8 \text{ ft/s})^2 + (25.6 \text{ ft/s})^2} = 26.8 \text{ ft/s} \qquad Ans.$$

The velocity is tangent to the path, Fig. 12–18b, where

$$\theta_v = \tan^{-1}\frac{v_y}{v_x} = \tan^{-1}\frac{25.6}{8} = 72.6° \qquad Ans.$$

Acceleration. The relationship between the acceleration components is determined using the chain rule. (See Appendix C.) We have

$$a_x = \dot{v}_x = \frac{d}{dt}(8) = 0$$

$$a_y = \dot{v}_y = \frac{d}{dt}(2x\dot{x}/10) = 2(\dot{x})\dot{x}/10 + 2x(\ddot{x})/10$$
$$= 2(8)^2/10 + 2(16)(0)/10 = 12.8 \text{ ft/s}^2 \uparrow$$

Thus,

$$a = \sqrt{(0)^2 + (12.8)^2} = 12.8 \text{ ft/s}^2 \qquad Ans.$$

The direction of **a**, as shown in Fig. 12–18c, is

$$\theta_a = \tan^{-1}\frac{12.8}{0} = 90° \qquad Ans.$$

NOTE: It is also possible to obtain v_y and a_y by first expressing $y = f(t) = (8t)^2/10 = 6.4t^2$ and then taking successive time derivatives.

(a)

(b)

(c)

Fig. 12–18

EXAMPLE 12.10

For a short time, the path of the plane in Fig. 12–19a is described by $y = (0.001x^2)$ m. If the plane is rising with a constant upward velocity of 10 m/s, determine the magnitudes of the velocity and acceleration of the plane when it reaches an altitude of $y = 100$ m.

SOLUTION

When $y = 100$ m, then $100 = 0.001x^2$ or $x = 316.2$ m. Also, due to constant velocity $v_y = 10$ m/s, and so the time is

$$y = v_y t; \qquad 100 \text{ m} = (10 \text{ m/s})\, t \qquad t = 10 \text{ s}$$

Velocity. Using the chain rule (see Appendix C, Eq. C–1) to find the relationship between the velocity components, we have

$$y = 0.001x^2$$

$$v_y = \dot{y} = \frac{dy}{dx}\dot{x} = \frac{d}{dt}(0.001x^2) = (0.002x)\dot{x} = 0.002x v_x \quad (1)$$

Thus

$$10 \text{ m/s} = 0.002(316.2 \text{ m})(v_x)$$
$$v_x = 15.81 \text{ m/s}$$

The magnitude of the velocity is therefore

$$v = \sqrt{v_x^2 + v_y^2} = \sqrt{(15.81 \text{ m/s})^2 + (10 \text{ m/s})^2} = 18.7 \text{ m/s} \qquad Ans.$$

Acceleration. Using the chain rule, or Eq. C–2, the time derivative of Eq. 1 gives the relation between the acceleration components.

$$a_y = \dot{v}_y = (0.002\dot{x})\dot{x} + 0.002x(\ddot{x}) = 0.002(v_x^2 + x a_x)$$

When $x = 316.2$ m, $v_x = 15.81$ m/s, $\dot{v}_y = a_y = 0$, so that

$$0 = 0.002[(15.81 \text{ m/s})^2 + 316.2 \text{ m}(a_x)]$$
$$a_x = -0.791 \text{ m/s}^2$$

The magnitude of the plane's acceleration is therefore

$$a = \sqrt{a_x^2 + a_y^2} = \sqrt{(-0.791 \text{ m/s}^2)^2 + (0 \text{ m/s}^2)^2}$$
$$= 0.791 \text{ m/s}^2 \qquad Ans.$$

These results are shown in Fig. 12–19b.

Fig. 12–19

12.6 MOTION OF A PROJECTILE

It is convenient to analyze the free-flight motion of a projectile in terms of its rectangular components. To illustrate, consider a projectile launched at point (x_0, y_0), with an initial velocity of $\mathbf{v}_0$, having components $(\mathbf{v}_0)_x$ and $(\mathbf{v}_0)_y$, Fig. 12–20. When air resistance is neglected, the only force acting on the projectile is its weight, and this causes the projectile to have a *constant downward acceleration* of $a_c = g = 9.81 \text{ m/s}^2$ or $g = 32.2 \text{ ft/s}^2$.*

Each picture in this sequence is taken after the same time period. In the horizontal direction the distance between the ball is the same because its velocity component is constant. The vertical distance between the ball is different because of the vertical deceleration-acceleration due to gravity.

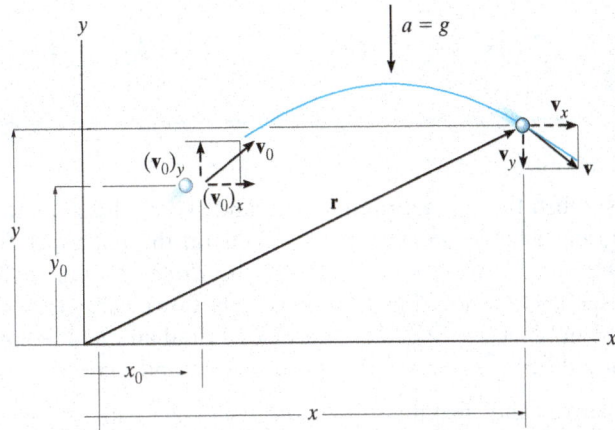

Fig. 12–20

Horizontal Motion. Since $a_x = 0$, application of the constant acceleration equations, 12–4 to 12–6, yields

$(\xrightarrow{+})$ $\qquad v = v_0 + a_c t; \qquad\qquad v_x = (v_0)_x$

$(\xrightarrow{+})$ $\qquad x = x_0 + v_0 t + \tfrac{1}{2} a_c t^2; \qquad x = x_0 + (v_0)_x t$

$(\xrightarrow{+})$ $\qquad v^2 = v_0^2 + 2a_c(x - x_0); \qquad v_x = (v_0)_x$

The first and last equations simply indicate that *the horizontal component of velocity always remains constant during the motion.*

Vertical Motion. Since $a_y = -g$, then applying Eqs. 12–4 to 12–6, we get

$(+\uparrow)$ $\qquad v = v_0 + a_c t; \qquad\qquad v_y = (v_0)_y - gt$

$(+\uparrow)$ $\qquad y = y_0 + v_0 t + \tfrac{1}{2} a_c t^2; \qquad y = y_0 + (v_0)_y t - \tfrac{1}{2} g t^2$

$(+\uparrow)$ $\qquad v^2 = v_0^2 + 2a_c(y - y_0); \qquad v_y^2 = (v_0)_y^2 - 2g(y - y_0)$

Since the last equation can be formulated on the basis of eliminating the time t from the first two equations, then *only two of the above three equations are independent of one another.*

*This assumes that the earth's gravitational field does not vary with altitude.

Once thrown, the basketball follows a parabolic trajectory.

To summarize, problems involving the motion of a projectile can have at most three unknowns since only three independent equations can be written; that is, *one equation in the horizontal direction* and *two in the vertical direction*. Once $\mathbf{v}_x$ and $\mathbf{v}_y$ are obtained, the resultant velocity $\mathbf{v}$, which is *always tangent* to the path, can be determined by the *vector sum* of $\mathbf{v}_x$ and $\mathbf{v}_y$, as shown in Fig. 12–20.

PROCEDURE FOR ANALYSIS

Coordinate System.

- Establish the *x, y* coordinate axes and sketch the trajectory of the particle. Between any *two points* on the path specify the given problem data and identify the *three unknowns*. In all cases the acceleration of gravity acts downward and equals 9.81 m/s² or 32.2 ft/s². The particle's initial and final velocities should be represented in terms of their *x* and *y* components.
- Positive and negative position, velocity, and acceleration components always act in accordance with their associated coordinate directions.

Kinematic Equations.

- Depending upon the known data and what is to be determined, a choice should be made as to which three of the following four equations should be applied between the two points on the path to obtain the most direct solution to the problem.

Horizontal Motion.

- The *velocity* in the horizontal or *x* direction is *constant*, i.e., $v_x = (v_0)_x$, and

$$x = x_0 + (v_0)_x t$$

Vertical Motion.

- In the vertical or *y* direction *only two* of the following three equations can be used for the solution.

$$v_y = (v_0)_y + a_c t$$

$$y = y_0 + (v_0)_y t + \tfrac{1}{2} a_c t^2$$

$$v_y^2 = (v_0)_y^2 + 2 a_c (y - y_0)$$

EXAMPLE 12.11

A sack slides off the ramp, shown in Fig. 12–21, with a horizontal velocity of 12 m/s. If the height of the ramp is 6 m from the floor, determine the time needed for the sack to strike the floor and the range R where the sack strikes the ground.

Fig. 12–21

SOLUTION

Coordinate System. The origin of coordinates is established at the beginning of the path, point A, Fig. 12–21. The initial velocity of the sack has components $(v_A)_x = 12$ m/s and $(v_A)_y = 0$. Also, between points A and B the acceleration is $a_y = -9.81$ m/s². Since $(v_B)_x = (v_A)_x = 12$ m/s, the three unknowns are $(v_B)_y$, R, and the time of flight t_{AB}. Here we do not need to determine $(v_B)_y$.

Vertical Motion. The vertical distance from A to B is known, and therefore we can obtain a direct solution for t_{AB} by using the equation

$(+\uparrow)$ $\qquad y_B = y_A + (v_A)_y t_{AB} + \tfrac{1}{2} a_c t_{AB}^2$

$\qquad -6 \text{ m} = 0 + 0 + \tfrac{1}{2}(-9.81 \text{ m/s}^2)t_{AB}^2$

$\qquad t_{AB} = 1.11 \text{ s}$ *Ans.*

Horizontal Motion. Since t_{AB} has been calculated, R is determined as follows:

$(\xrightarrow{+})$ $\qquad x_B = x_A + (v_A)_x t_{AB}$

$\qquad R = 0 + 12 \text{ m/s } (1.11 \text{ s})$

$\qquad R = 13.3 \text{ m}$ *Ans.*

NOTE: The calculation for t_{AB} also indicates that if the sack was released *from rest* at A, it would take the same amount of time to strike the floor at C, Fig. 12–21.

EXAMPLE 12.12

The chipping machine is designed to eject wood chips at $v_O = 25$ ft/s as shown in Fig. 12–22. If the tube is oriented at 30° from the horizontal, determine how high, h, the chips strike the pile if they land on the pile 20 ft from the tube.

Fig. 12–22

SOLUTION

Coordinate System. The three unknowns are the height h, time of flight t_{OA}, and vertical component of velocity $(v_A)_y$. [Note that $(v_A)_x = (v_O)_x$.] With the origin of coordinates at O, Fig. 12–22, the initial velocity of a chip has components of

$$(v_O)_x = (25 \cos 30°) \text{ ft/s} = 21.65 \text{ ft/s} \rightarrow$$
$$(v_O)_y = (25 \sin 30°) \text{ ft/s} = 12.5 \text{ ft/s} \uparrow$$

Also, $(v_A)_x = (v_O)_x = 21.65$ ft/s and $a_y = -32.2$ ft/s². Since we do not need to determine $(v_A)_y$, we have

Horizontal Motion.

$(\xrightarrow{+})$
$$x_A = x_O + (v_O)_x t_{OA}$$
$$20 \text{ ft} = 0 + (21.65 \text{ ft/s}) t_{OA}$$
$$t_{OA} = 0.9238 \text{ s}$$

Vertical Motion. Relating t_{OA} to the initial and final elevations of a chip, we have

$(+\uparrow)$ $\quad y_A = y_O + (v_O)_y t_{OA} + \tfrac{1}{2} a_c t_{OA}^2$
$$(h - 4 \text{ ft}) = 0 + (12.5 \text{ ft/s})(0.9238 \text{ s}) + \tfrac{1}{2}(-32.2 \text{ ft/s}^2)(0.9238 \text{ s})^2$$
$$h = 1.81 \text{ ft} \qquad \textit{Ans.}$$

EXAMPLE 12.13

The track for this racing event was designed so that riders jump off the slope at 30°, from a height of 1 m. During a race it was observed that the rider shown in Fig. 12–23a remained in mid air for 1.5 s. Determine the speed at which he was traveling off the ramp, the horizontal distance he travels before striking the ground, and the maximum height he attains. Neglect the size of the bike and rider.

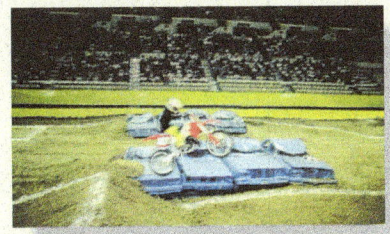

(a)

SOLUTION

Coordinate System. As shown in Fig. 12–23b, the origin of the coordinates is established at A. Between the end points of the path AB the three unknowns are the initial speed v_A, range R, and the vertical component of velocity $(v_B)_y$.

Vertical Motion. Since the time of flight and the vertical distance between the ends of the path are known, we can determine v_A.

$(+\uparrow)$ $\qquad y_B = y_A + (v_A)_y t_{AB} + \tfrac{1}{2} a_c t_{AB}^2$

$\qquad -1 \text{ m} = 0 + v_A \sin 30°(1.5 \text{ s}) + \tfrac{1}{2}(-9.81 \text{ m/s}^2)(1.5 \text{ s})^2$

$\qquad v_A = 13.38 \text{ m/s} = 13.4 \text{ m/s}$ *Ans.*

Horizontal Motion. The range R can now be determined.

$(\xrightarrow{+})$ $\qquad x_B = x_A + (v_A)_x t_{AB}$

$\qquad R = 0 + 13.38 \cos 30° \text{ m/s}(1.5 \text{ s})$

$\qquad\quad = 17.4 \text{ m}$ *Ans.*

In order to find the maximum height h we will consider the path AC, Fig. 12–23b. Here the three unknowns are the time of flight t_{AC}, the horizontal distance from A to C, and the height h. At the maximum height $(v_C)_y = 0$, and since v_A is known, we can determine h *directly* without considering t_{AC} using the following equation.

$\qquad (v_C)_y^2 = (v_A)_y^2 + 2a_c[y_C - y_A]$

$\qquad 0^2 = (13.38 \sin 30° \text{ m/s})^2 + 2(-9.81 \text{ m/s}^2)[(h - 1 \text{ m}) - 0]$

$\qquad h = 3.28 \text{ m}$ *Ans.*

NOTE: Show that the bike will strike the ground at B with a velocity having components of

$\qquad (v_B)_x = 11.6 \text{ m/s} \rightarrow, \quad (v_B)_y = 8.02 \text{ m/s} \downarrow$

(b)

Fig. 12–23

FUNDAMENTAL PROBLEMS

F12–15. If the x and y components of a particle's velocity are $v_x = (32t)$ m/s and $v_y = 8$ m/s, determine the equation of the path $y = f(x)$, if $x = 0$ and $y = 0$ when $t = 0$.

F12–16. A particle is traveling along the straight path. If its position along the x axis is $x = (8t)$ m, where t is in seconds, determine its speed when $t = 2$ s.

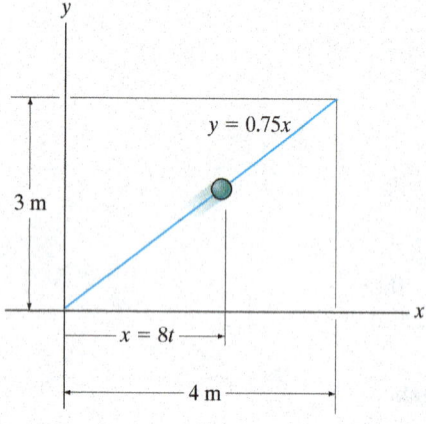

Prob. F12–16

F12–17. A particle is constrained to travel along the path. If $x = (4t^4)$ m, where t is in seconds, determine the magnitudes of the particle's velocity and acceleration when $t = 0.5$ s.

Prob. F12–17

F12–18. A particle travels along a straight-line path $y = 0.5x$. If the x component of the particle's velocity is $v_x = (2t^2)$ m/s, where t is in seconds, determine the magnitudes of the particle's velocity and acceleration when $t = 4$ s.

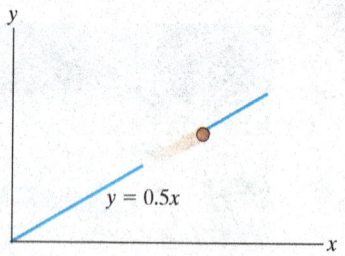

Prob. F12–18

F12–19. A particle is traveling along the parabolic path $y = 0.25x^2$. If $x = 8$ m, $v_x = 8$ m/s, and $a_x = 4$ m/s^2 when $t = 2$ s, determine the magnitudes of the particle's velocity and acceleration at this instant.

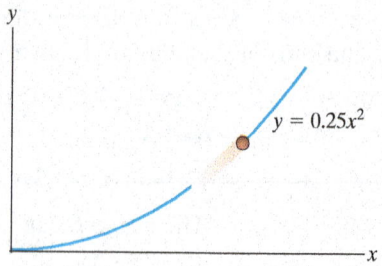

Prob. F12–19

F12–20. The box slides down the path described by the equation $y = (0.05x^2)$ m, where x is in meters. If the box has x components of velocity and acceleration of $v_x = -3$ m/s and $a_x = -1.5$ m/s^2 at $x = 5$ m, determine the y components of the velocity and the acceleration of the box at this instant.

Prob. F12–20

F12–21. The ball is kicked from point A with the initial velocity $v_A = 10$ m/s. Determine the maximum height h it reaches.

F12–22. The ball is kicked from point A with the initial velocity $v_A = 10$ m/s. Determine the range R, and the speed when the ball strikes the ground.

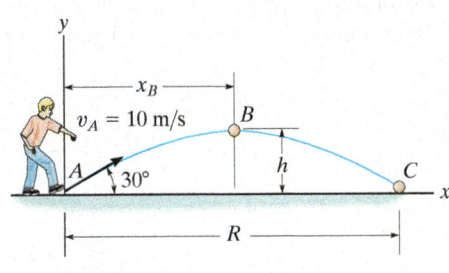

Probs. F12–21/22

F12–23. Determine the speed at which the basketball at A must be thrown at the angle of 30° so that it makes it to the basket at B.

Prob. F12–23

F12–24. Water is sprayed at an angle of 90° from the slope at 20 m/s. Determine the range R.

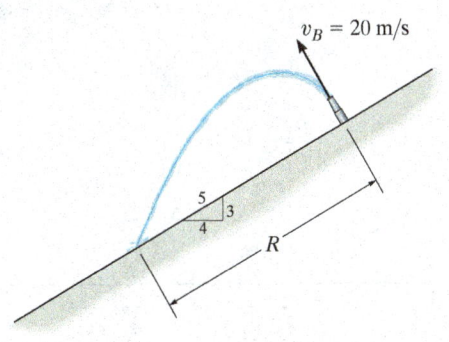

Prob. F12–24

F12–25. A ball is thrown from A. If it is required to clear the wall at B, determine the minimum magnitude of its initial velocity v_A.

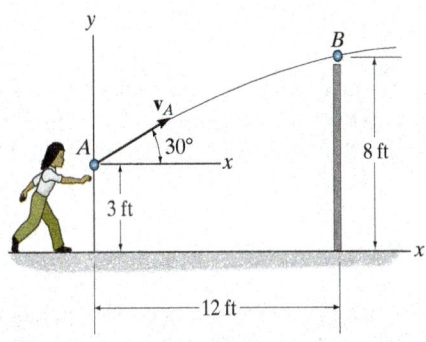

Prob. F12–25

F12–26. A projectile is fired with an initial velocity of $v_A = 150$ m/s off the roof of the building. Determine the range R where it strikes the ground at B.

Prob. F12–26

PROBLEMS

12–69. If the velocity of a particle is defined as $\mathbf{v}(t) = \{0.8t^2\mathbf{i} + 12t^{1/2}\mathbf{j} + 5\mathbf{k}\}$ m/s, determine the magnitude and coordinate direction angles α, β, γ of the particle's acceleration when $t = 2$ s.

12–70. A particle moves along the curve $y = e^{2x}$ such that its velocity has a constant magnitude of $v = 4$ ft/s. Determine the x and y components of velocity when the particle is at $y = 5$ ft.

12–71. A particle travels along the parabolic path $y = bx^2$. If its component of velocity along the y axis is $v_y = ct^2$, determine the x and y components of the particle's acceleration. Here b and c are constants.

*12–72. A particle travels along the circular path $x^2 + y^2 = r^2$. If the y component of the particle's velocity is $v_y = 2r \cos 2t$, determine the x and y components of its acceleration at any instant.

12-73. The water sprinkler, positioned at the base of a hill, releases a stream of water with a velocity of 15 ft/s as shown. Determine the point $B(x, y)$ where the water strikes the ground on the hill. Assume that the hill is defined by the equation $y = (0.05x^2)$ ft and neglect the size of the sprinkler.

Prob. 12–73

12–74. A particle moves along the curve $y = x - (x^2/400)$, where x and y are in ft. If the velocity component in the x direction is $v_x = 2$ ft/s and remains *constant*, determine the magnitudes of the velocity and acceleration when $x = 20$ ft.

12–75. A particle travels along the curve from A to B in 2 s. It takes 4 s for it to go from B to C and then 3 s to go from C to D. Determine its average speed when it goes from A to D.

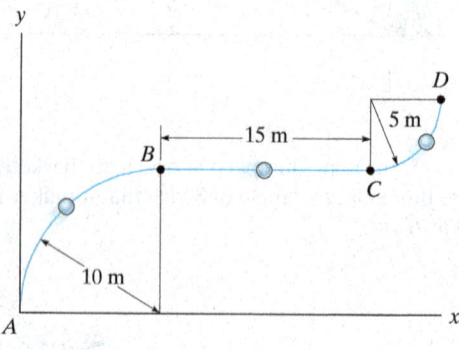

Prob. 12–75

*12-76. A particle travels along the curve from A to B in 5 s. It takes 8 s for it to go from B to C and then 10 s to go from C to A. Determine its average speed when it goes around the closed path.

Prob. 12–76

12–77. Show that if a projectile is fired at an angle θ from the horizontal with an initial velocity $\mathbf{v}_0$, the *maximum* range the projectile can travel is given by $R_{max} = v_0^2/g$, where g is the acceleration of gravity. What is the angle θ for this condition?

12–78. A rocket is fired from rest at $x = 0$ and travels along a parabolic trajectory described by $y^2 = [120(10^3)x]$ m. If the x component of acceleration is $a_x = (\frac{1}{4}t^2)$ m/s^2, where t is in seconds, determine the magnitudes of the rocket's velocity and acceleration when $t = 10$ s.

12–79. The particle travels along the path defined by the parabola $y = 0.5x^2$. If the component of velocity along the x axis is $v_x = (5t)$ ft/s, where t is in seconds, determine the particle's distance from the origin O and the magnitude of its acceleration when $t = 1$ s. When $t = 0$, $x = 0$, $y = 0$.

12–81. A car traveling along the road has the velocities indicated in the figure when it arrives at points A, B, and C. If it takes 10 s to go from A to B, and then 15 s to go from B to C, determine the average acceleration between points A and B and between points A and C.

Prob. 12–81

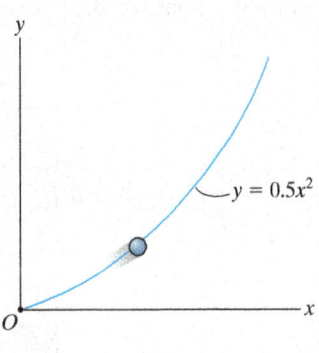

Prob. 12–79

12–82. The roller coaster car travels down the helical path at constant speed such that the parametric equations that define its position are $x = c \sin kt$, $y = c \cos kt$, $z = h - bt$, where c, h, and b are constants. Determine the magnitudes of its velocity and acceleration.

*12–80. The motorcycle travels with constant speed v_0 along the path that, for a short distance, takes the form of a sine curve. Determine the x and y components of its velocity at any instant on the curve.

Prob. 12–80

Prob. 12–82

12–83. Pegs A and B are restricted to move in the elliptical slots due to the motion of the slotted link. If the link moves with a constant speed of 10 m/s, determine the magnitudes of the velocity and acceleration of peg A when $x = 1$ m.

Prob. 12–83

*12–84. The van travels over the hill described by $y = (-1.5(10^{-3}) x^2 + 15)$ ft. If it has a constant speed of 75 ft/s, determine the x and y components of the van's velocity and acceleration when $x = 50$ ft.

Prob. 12–84

12–85. The flight path of the helicopter as it takes off from A is defined by the parametric equations $x = (2t^2)$ m and $y = (0.04t^3)$ m, where t is the time in seconds. Determine the distance the helicopter is from point A and the magnitudes of its velocity and acceleration when $t = 10$ s.

Prob. 12–85

12–86. Determine the minimum initial velocity v_0 and the corresponding angle θ_0 at which the ball must be kicked in order for it to just cross over the 3-m-high fence.

Prob. 12–86

12–87. The catapult is used to launch a ball such that it strikes the wall of the building at the maximum height of its trajectory. If it takes 1.5 s to travel from A to B, determine the velocity v_A at which it was launched, the angle of release θ, and the height h.

Prob. 12–87

*12–88. Neglecting the size of the ball, determine the magnitude v_A of the basketball's initial velocity and its velocity when it passes through the basket.

Prob. 12–88

12–89. The girl at A can throw a ball at $v_A = 10$ m/s. Calculate the maximum possible range $R = R_{max}$ and the associated angle θ at which it should be thrown. Assume the ball is caught at B at the same elevation from which it is thrown.

12–90. Show that the girl at A can throw the ball to the boy at B by launching it at equal angles measured up or down from a 45° inclination. If $v_A = 10$ m/s, determine the range R if this value is 15°, i.e., $\theta_1 = 45° - 15° = 30°$ and $\theta_2 = 45° + 15° = 60°$. Assume the ball is caught at the same elevation from which it is thrown.

12–93. A golf ball is struck with a velocity of 80 ft/s as shown. Determine the distance d to where it will land.

12–94. A golf ball is struck with a velocity of 80 ft/s as shown. Determine the speed at which it strikes the ground at B and the time of flight from A to B.

Probs. 12–93/94

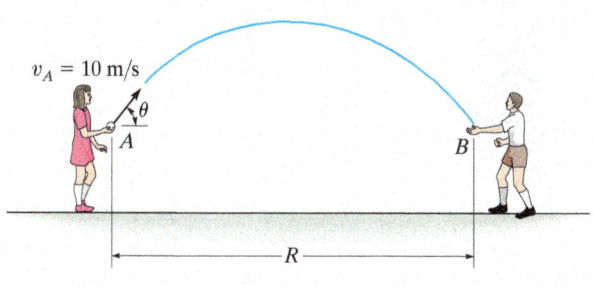

Probs. 12–89/90

12–91. The ball at A is kicked at $v_A = 16$ ft/s and at an angle $\theta_A = 30°$. Determine the point $(x, -y)$ where it strikes the ground. Assume the ground has the shape of a parabola as shown.

*12–92. The ball at A is kicked such that $\theta_A = 30°$. If it strikes the ground at B having coordinates $x = 15$ ft, $y = -9$ ft, determine the speed at which it is kicked and the speed at which it strikes the ground.

12–95. The pitching machine is adjusted so that the baseball is launched with a speed of $v_A = 30$ m/s. If the ball strikes the ground at B, determine the two possible angles θ_A at which it was launched.

Probs. 12–91/92

Prob. 12–95

*12–96. The firefighter at A uses the hose which ejects a steady stream of water at 10 m/s. Calculate the maximum range $R = R_{max}$ to which the water particles can be "thrown" and the angle θ at which he must hold the nozzle.

12–97. Show that the firefighter at A can "throw" water particles on B by aiming the nozzle at equal angles, measured up at $\theta_1 = 45° - 15° = 30°$ and $\theta_2 = 45° + 15° = 60°$. The water has a speed of 10 m/s at the nozzle.

12–99. A projectile is fired from the platform at B. The shooter fires his gun from point A at an angle of 30°. Determine the muzzle speed of the bullet if it hits the projectile at C.

Prob. 12–99

Probs. 12–96/97

12–98. The skateboard rider leaves the ramp at A with an initial velocity v_A at a 30° angle. If he strikes the ground at B, determine v_A and the time of flight.

*12–100. If the motorcycle leaves ramp A traveling at 110 ft/s, determine the height h ramp B must have so that the motorcycle lands safely.

Prob. 12–98

Prob. 12–100

12–101. The drinking fountain is designed such that the nozzle is located a distance away from the edge of the basin as shown. Determine the maximum and minimum speed at which water can be ejected from the nozzle so that it does not splash over the sides of the basin at B and C.

*12–104. The man stands 60 ft from the wall and throws a ball at it with a speed of $v_0 = 50$ ft/s. Determine the angle θ at which he should release the ball so that it strikes the wall at the highest point possible. What is this height? The room has a ceiling height of 20 ft.

Prob. 12–101

Prob. 12–104

12–102. The hose is held at an angle $\theta = 30°$ with the horizontal, and the water is discharged at A with a speed of $v_A = 40$ ft/s. If the water stream strikes the building at B, determine the two possible distances s from the building.

12–103. Water is discharged from the hose with a speed of 40 ft/s. Determine the two possible angles θ the firefighter can hold the hose so that the water strikes the building at B. Take $s = 20$ ft.

12–105. The velocity of a water jet discharging from an orifice can be obtained from $v = \sqrt{2gh}$, where $h = 2$ m is the depth of the orifice from the free water surface. Determine the time for a particle of water leaving the orifice to reach a point B and the horizontal distance x where it hits the surface.

Prob. 12–105

Probs. 12–102/103

12–106. The snowmobile is traveling at 10 m/s when it leaves the embankment at A. Determine the time of flight from A to B and the range R of the trajectory.

***12–108.** The buckets on the conveyor travel with a speed of 15 ft/s. Each bucket contains a block which falls out of the bucket when $\theta = 120°$. Determine the distance s to where the block strikes the conveyor. Neglect the size of the block.

Prob. 12–106

Prob. 12–108

12–107. The firefighter wishes to direct the flow of water from his hose to the fire at B. Determine two possible angles θ_1 and θ_2 at which this can be done. Water flows from the hose at $v_A = 80$ ft/s.

12–109. The catapult is used to launch a ball such that it strikes the wall of the building at the maximum height of its trajectory. If it takes 1.5 s to travel from A to B, determine the velocity $\mathbf{v}_A$ at which it was launched, the angle of release θ, and the height h.

Prob. 12–107

Prob. 12–109

12.7 CURVILINEAR MOTION: NORMAL AND TANGENTIAL COMPONENTS

When the path along which a particle travels is *known*, then it is often convenient to describe the motion using n and t axes that are normal and tangent to the path, respectively, and at the instant considered have their origin located at the particle.

Planar Motion. To establish these axes, consider the particle in Fig. 12–24a, which is at position s, measured from point O. The t axis is *tangent* to the curve at the particle and is positive in the direction of *increasing s*. We will designate this positive direction with the unit vector $\mathbf{u}_t$. A unique choice for the *normal axis* can be made by noting that geometrically the curve is constructed from a series of differential arc segments ds, Fig. 12–24b. Each segment ds is formed from the arc of an associated circle having a *radius of curvature* ρ (rho) and *center of curvature* O'. The normal axis n is perpendicular to the t axis and its positive direction is *towards* the center of curvature O', Fig. 12–24a. This direction, which is *always* on the concave side of the curve, will be designated by the unit vector $\mathbf{u}_n$. The plane which contains the n and t axes is referred to as the embracing or **osculating plane**, and in this case it is fixed in the plane of motion.

Velocity. As indicated in Sec. 12.4, the particle's velocity $\mathbf{v}$ has a *direction* that is *always tangent to the path*, Fig. 12–24c, and a *magnitude* that is determined by taking the time derivative of the path function $s = s(t)$, i.e., $v = ds/dt$ (Eq. 12–8). Hence,

$$\boxed{\mathbf{v} = v\mathbf{u}_t} \qquad (12\text{–}15)$$

where

$$\boxed{v = \dot{s}} \qquad (12\text{–}16)$$

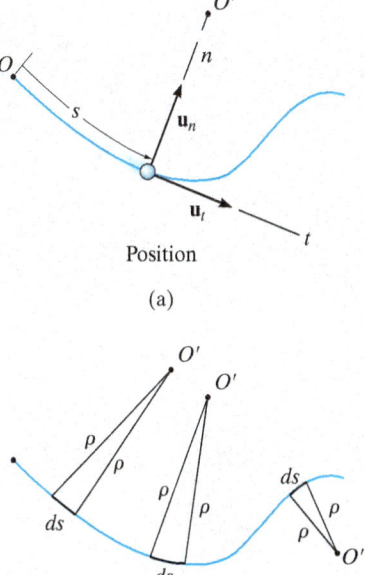

Position
(a)

Radius of curvature
(b)

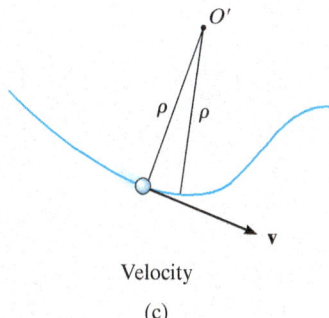

Velocity
(c)

Fig. 12–24

54   CHAPTER 12   KINEMATICS OF A PARTICLE

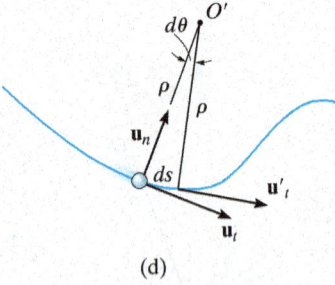

(d)

Acceleration. The acceleration of the particle is the time rate of change of the velocity. Therefore,

$$\mathbf{a} = \dot{\mathbf{v}} = \dot{v}\mathbf{u}_t + v\dot{\mathbf{u}}_t \qquad (12\text{–}17)$$

To determine the time derivative $\dot{\mathbf{u}}_t$, note that as the particle moves along the arc ds in time dt, $\mathbf{u}_t$ preserves its magnitude of unity; however, its direction changes, and becomes $\mathbf{u}'_t$, Fig. 12–24d. As shown in Fig. 12–24e, we require $\mathbf{u}'_t = \mathbf{u}_t + d\mathbf{u}_t$, where $d\mathbf{u}_t$ acts between the arrowheads of $\mathbf{u}_t$ and $\mathbf{u}'_t$. Since $u_t = u'_t = 1$, then $d\mathbf{u}_t$ has a *magnitude* of $du_t = (1)\,d\theta$, and its *direction* is defined by $\mathbf{u}_n$. Consequently, $d\mathbf{u}_t = d\theta\,\mathbf{u}_n$, and therefore the time derivative becomes $\dot{\mathbf{u}}_t = \dot{\theta}\mathbf{u}_n$. Since $ds = \rho\,d\theta$, Fig. 12–24d, then $\dot{\theta} = \dot{s}/\rho$, and therefore

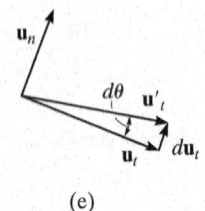

(e)

$$\dot{\mathbf{u}}_t = \dot{\theta}\mathbf{u}_n = \frac{\dot{s}}{\rho}\mathbf{u}_n = \frac{v}{\rho}\mathbf{u}_n$$

Substituting this into Eq. 12–17, **a** can be written as the sum of its two components,

$$\mathbf{a} = a_t\mathbf{u}_t + a_n\mathbf{u}_n \qquad (12\text{–}18)$$

where

$$a_t = \dot{v} \quad \text{or} \quad a_t\,ds = v\,dv \qquad (12\text{–}19)$$

and

$$a_n = \frac{v^2}{\rho} \qquad (12\text{–}20)$$

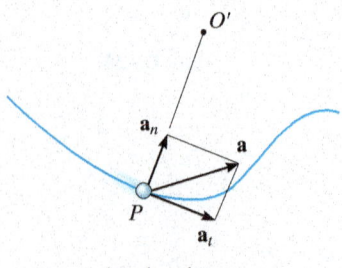

Acceleration
(f)

Fig. 12–24 (cont.)

These two mutually perpendicular components are shown in Fig. 12–24f. Therefore, the *magnitude* of acceleration is

$$a = \sqrt{a_t^2 + a_n^2} \qquad (12\text{–}21)$$

12.7 CURVILINEAR MOTION: NORMAL AND TANGENTIAL COMPONENTS

To better understand these results, consider the following two special cases of motion.

1. If the particle moves along a straight line, then $\rho \to \infty$ and from Eq. 12–20, $a_n = 0$. Therefore, $a = a_t = \dot{v}$, and so the *tangential component of acceleration represents the time rate of change in the magnitude of the velocity.*

2. If the particle moves along a curve with a constant speed, then $a_t = \dot{v} = 0$ and $a = a_n = v^2/\rho$. Therefore, the *normal component of acceleration represents the time rate of change in the direction of the velocity.* Since $\mathbf{a}_n$ always acts towards the center of curvature, this component is sometimes referred to as the *centripetal* (or center seeking) *acceleration*.

As the boy swings upward with a velocity **v**, his motion can be analyzed using *n*–*t* coordinates. As he rises, the magnitude of his velocity (speed) is decreasing, and so a_t will be negative. The rate at which the direction of his velocity changes is a_n, which is always positive, that is, towards the center of rotation.

As a result of these interpretations, a particle moving along the curved path in Fig. 12–25 will have an acceleration directed as shown.

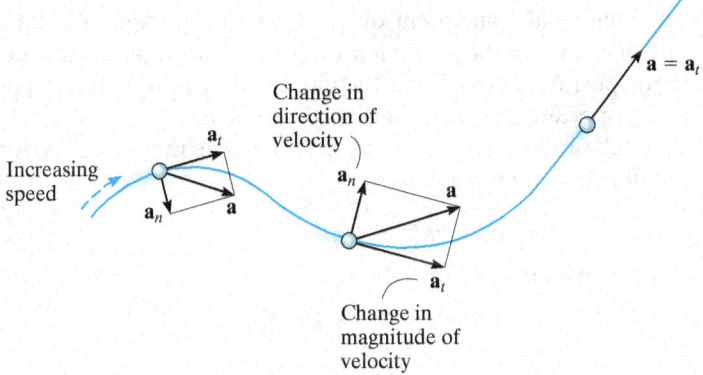

Fig. 12–25

Three-Dimensional Motion.
If the particle moves along a space curve, Fig. 12–26, then at a given instant the *t* axis is uniquely specified; however, an infinite number of straight lines can be constructed normal to the tangent axis. As in the case of planar motion, however, we will choose the positive *n* axis directed toward the path's center of curvature O'. This axis is referred to as the *principal normal* to the curve. With the *n* and *t* axes so defined, Eqs. 12–15 through 12–21 can then be used to determine **v** and **a**. Since $\mathbf{u}_t$ and $\mathbf{u}_n$ are always perpendicular to one another and lie in the osculating plane, for spatial motion a third unit vector, $\mathbf{u}_b$, defines the *binormal axis b* which is perpendicular to $\mathbf{u}_t$ and $\mathbf{u}_n$, Fig. 12–26.

Since the three unit vectors are related to one another by the vector cross product, e.g., $\mathbf{u}_b = \mathbf{u}_t \times \mathbf{u}_n$, Fig. 12–26, it may be possible to use this relation to establish the direction of one of the axes if the directions of the other two are known. For example, no motion occurs in the $\mathbf{u}_b$ direction, and if this direction and $\mathbf{u}_t$ are known, then $\mathbf{u}_n$ can be determined from $\mathbf{u}_n = \mathbf{u}_b \times \mathbf{u}_t$, Fig. 12–26.

Fig. 12–26

Once the rotation is constant, the riders will then have only a normal component of acceleration.

PROCEDURE FOR ANALYSIS

Coordinate System.
- Provided the *path* of the particle is *known*, we can establish a set of *n* and *t* coordinates having a *fixed origin*, which is coincident with the particle at the instant considered.
- The positive tangent axis acts in the direction of motion and the positive normal axis is directed toward the path's center of curvature.

Velocity.
- The particle's *velocity* is always tangent to the path.
- The magnitude of velocity is found from the time derivative of the path function.

$$v = \dot{s}$$

Tangential Acceleration.
- The tangential component of acceleration represents the time rate of change in the *magnitude* of velocity. This component acts in the positive *s* direction if the particle's speed is increasing or in the opposite direction if the speed is decreasing.
- The relations between a_t, v, t, and s are the same as for rectilinear motion, namely,

$$a_t = \dot{v} \quad a_t\, ds = v\, dv$$

- If a_t is constant, $a_t = (a_t)_c$, then

$$s = s_0 + v_0 t + \tfrac{1}{2}(a_t)_c t^2$$
$$v = v_0 + (a_t)_c t$$
$$v^2 = v_0^2 + 2(a_t)_c(s - s_0)$$

Normal Acceleration.
- The normal component of acceleration represents the time rate of change in the *direction* of the velocity. This component is *always* directed toward the center of curvature of the path, i.e., along the positive *n* axis.
- The magnitude of this component is determined from

$$a_n = \frac{v^2}{\rho}$$

- If the path is expressed as $y = f(x)$, the radius of curvature ρ at any point on the path is determined from the equation

$$\rho = \frac{[1 + (dy/dx)^2]^{3/2}}{|d^2y/dx^2|}$$

The derivation of this result is given in any standard calculus text.

Motorists traveling along this cloverleaf interchange experience a normal acceleration due to the change in direction of their velocity. A tangential component of acceleration occurs when the cars' speed is increased or decreased.

EXAMPLE 12.14

The skier has a speed of 6 m/s at point A, Fig. 12–27a, which is increasing at 2 m/s². Determine the direction of the velocity and the direction and magnitude of acceleration at this instant. Neglect the size of the skier in the calculation.

SOLUTION

Coordinate System. Although the path has been expressed in terms of its x and y coordinates, we can still establish the origin of the n, t axes at the fixed point A on the path and determine the components of $\mathbf{v}$ and $\mathbf{a}$ along these axes, Fig. 12–27a.

Velocity. By definition, the velocity is always tangent to the path. Since $y = \frac{1}{20}x^2$, $dy/dx = \frac{1}{10}x$, then at $x = 10$ m, $dy/dx = 1$. Hence, at A, $\mathbf{v}$ makes an angle of $\theta = \tan^{-1} 1 = 45°$ with the x axis, Fig. 12–27b. Therefore,

$$v_A = 6 \text{ m/s} \quad 45° \nearrow \qquad \textit{Ans.}$$

The acceleration is determined from $\mathbf{a} = \dot{v}\mathbf{u}_t + (v^2/\rho)\mathbf{u}_n$. However, we must first determine the radius of curvature of the path at A (10 m, 5 m). Since $d^2y/dx^2 = \frac{1}{10}$, then

$$\rho = \frac{[1 + (dy/dx)^2]^{3/2}}{|d^2y/dx^2|} = \frac{[1 + (\frac{1}{10}x)^2]^{3/2}}{|\frac{1}{10}|}\bigg|_{x=10 \text{ m}} = 28.28 \text{ m}$$

The acceleration is therefore

$$\mathbf{a}_A = \dot{v}\mathbf{u}_t + \frac{v^2}{\rho}\mathbf{u}_n$$

$$= 2\mathbf{u}_t + \frac{(6 \text{ m/s})^2}{28.28 \text{ m}}\mathbf{u}_n$$

$$= \{2\mathbf{u}_t + 1.273\mathbf{u}_n\} \text{ m/s}^2$$

As shown in Fig. 12–27b,

$$a = \sqrt{(2 \text{ m/s}^2)^2 + (1.273 \text{ m/s}^2)^2} = 2.37 \text{ m/s}^2$$

$$\phi = \tan^{-1}\frac{2}{1.273} = 57.5°$$

Thus, $45° + 90° + 57.5° - 180° = 12.5°$, so that

$$a = 2.37 \text{ m/s}^2 \quad 12.5° \nearrow \qquad \textit{Ans.}$$

(a)

(b)

Fig. 12–27

EXAMPLE 12.15

A race car C travels around the horizontal circular track that has a radius of 300 ft, Fig. 12–28. If the car increases its speed at a constant rate of 7 ft/s^2, starting from rest, determine the time needed for it to reach an acceleration of 8 ft/s^2. What is its speed at this instant?

Fig. 12–28

SOLUTION

Coordinate System. The origin of the n and t axes is coincident with the car at the instant considered. The t axis is in the direction of motion, and the positive n axis is directed toward the center of the circle. This coordinate system is selected since the path is known.

Acceleration. The magnitude of acceleration can be related to its components using $a = \sqrt{a_t^2 + a_n^2}$. Here $a_t = 7$ ft/s^2. Since $a_n = v^2/\rho$, the velocity as a function of time must be determined first.

$$v = v_0 + (a_t)_c t$$
$$v = 0 + 7t$$

Thus

$$a_n = \frac{v^2}{\rho} = \frac{(7t)^2}{300} = 0.163 t^2 \text{ ft/s}^2$$

The time needed for the acceleration to reach 8 ft/s^2 is therefore

$$a = \sqrt{a_t^2 + a_n^2}$$
$$8 \text{ ft/s}^2 = \sqrt{(7 \text{ ft/s}^2)^2 + (0.163 t^2)^2}$$

Solving for the positive value of t yields

$$0.163 t^2 = \sqrt{(8 \text{ ft/s}^2)^2 - (7 \text{ ft/s}^2)^2}$$
$$t = 4.87 \text{ s} \quad \textit{Ans.}$$

Velocity. The speed at time $t = 4.87$ s is

$$v = 7t = 7(4.87) = 34.1 \text{ ft/s} \quad \textit{Ans.}$$

NOTE: Remember the velocity will always be tangent to the path, whereas the acceleration will be directed within the curvature of the path.

EXAMPLE 12.16

If the box in Fig. 12–29a starts from rest at A and increases its speed, such that $a_t = (0.2t)$ m/s^2, where t is in seconds, determine the time needed for the box to reach B and the magnitude of its acceleration when it arrives at this point.

SOLUTION

Coordinate System. The position of the box at any instant is defined from the fixed point A using the position or path coordinate s, Fig. 12–29a. The acceleration is to be determined at B, so the origin of the n, t axes is at this point.

Time. Since $v_A = 0$ when $t = 0$, then

$$a_t = \dot{v} = 0.2t \quad (1)$$

$$\int_0^v dv = \int_0^t 0.2t\, dt$$

$$v = 0.1t^2 \quad (2)$$

Here $s_B = 3 + 2\pi(2)/4 = 6.142$ m, Fig. 12–29a, and since $s_A = 0$ when $t = 0$ we have

$$v = \frac{ds}{dt} = 0.1t^2$$

$$\int_0^{6.142 \text{ m}} ds = \int_0^{t_B} 0.1t^2\, dt$$

$$6.142 \text{ m} = 0.0333 t_B^3$$

$$t_B = 5.690 \text{ s} \quad \text{Ans.}$$

Acceleration. Substituting t_B into Eqs. 1 and 2 yields

$$(a_B)_t = \dot{v}_B = 0.2(5.690) = 1.138 \text{ m/s}^2$$
$$v_B = 0.1(5.69)^2 = 3.238 \text{ m/s}$$

At B, $\rho_B = 2$ m, so that

$$(a_B)_n = \frac{v_B^2}{\rho_B} = \frac{(3.238 \text{ m/s})^2}{2 \text{ m}} = 5.242 \text{ m/s}^2$$

The magnitude of $\mathbf{a}_B$, Fig. 12–29b, is therefore

$$a_B = \sqrt{(1.138 \text{ m/s}^2)^2 + (5.242 \text{ m/s}^2)^2} = 5.36 \text{ m/s}^2 \quad \text{Ans.}$$

(a)

(b)

Fig. 12–29

FUNDAMENTAL PROBLEMS

F12–27. The boat is traveling along the circular path with a speed of $v = (0.0625t^2)$ m/s, where t is in seconds. Determine the magnitude of its acceleration when $t = 10$ s.

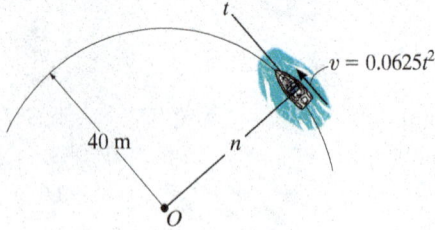

Prob. F12–27

F12–28. The car is traveling along the road with a speed of $v = (2s)$ m/s, where s is in meters. Determine the magnitude of its acceleration when $s = 10$ m.

Prob. F12–28

F12–29. If the car decelerates uniformly along the curved road from 25 m/s at A to 15 m/s at C, determine the acceleration of the car at B.

Prob. F12–29

F12–30. When $x = 10$ ft, the crate has a speed of 20 ft/s which is increasing at 6 ft/s². Determine the direction of the crate's velocity and the magnitude of its acceleration at this instant.

Prob. F12–30

F12–31. If the motorcycle has a deceleration of $a_t = -(0.001s)$ m/s² and its speed at position A is 25 m/s, determine the magnitude of its acceleration when it passes point B.

Prob. F12–31

F12–32. The car travels up the hill with a speed of $v = (0.2s)$ m/s, where s is in meters, measured from A. Determine the magnitude of its acceleration when it is at $s = 50$ m, where $\rho = 500$ m.

Prob. F12–32

PROBLEMS

12–110. A train travels along a circular track having a radius of 2000 ft. If the speed of the train is uniformly increased from 6 mi/h to 8 mi/h in 1 min, determine the magnitude of the acceleration when the speed of the train is 7 mi/h. 1 mi = 5280 ft.

12–111. A boat has an initial speed of 16 ft/s. If it then increases its speed along a circular path of radius $\rho = 80$ ft at the rate of $\dot{v} = (1.5s)$ ft/s^2, where s is in feet, determine the normal and tangential components of the boat's acceleration at $s = 16$ ft.

*12–112. A particle moves along the curve $y = 1 + \cos x$ with a constant speed $v = 4$ ft/s. Determine the normal and tangential components of its velocity and acceleration at any instant.

12–113. The position of a particle is defined by $\mathbf{r} = \{4(t - \sin t)\mathbf{i} + (2t^2 - 3)\mathbf{j}\}$ m, where t is in seconds and the argument for the sine is in radians. Determine the speed of the particle and its normal and tangential components of acceleration when $t = 1$ s.

12–114. The automobile has a speed of 80 ft/s at point A and an acceleration $\mathbf{a}$ having a magnitude of 10 ft/s^2, acting in the direction shown. Determine the radius of curvature of the path at point A and the tangential component of acceleration.

12–115. When the car reaches point A it has a speed of 25 m/s. If the brakes are applied, its speed is reduced by $a_t = (-\frac{1}{4}t^{1/2})$ m/s^2. Determine the magnitude of acceleration of the car just before it reaches point C.

*12–116. When the car reaches point A, it has a speed of 25 m/s. If the brakes are applied, its speed is reduced by $a_t = (0.001s - 1)$ m/s^2. Determine the magnitude of acceleration of the car just before it reaches point C.

Probs. 12–115/116

12–117. Starting from rest, the motorboat travels around the circular path, $\rho = 50$ m, at a speed $v = (0.8t)$ m/s, where t is in seconds. Determine the magnitudes of the boat's velocity and acceleration when it has traveled 20 m.

12–118. Starting from rest, the motorboat travels around the circular path, $\rho = 50$ m, at a speed $v = (0.2t^2)$ m/s, where t is in seconds. Determine the magnitudes of the boat's velocity and acceleration at the instant $t = 3$ s.

Prob. 12–114

Probs. 12–117/118

12–119. The satellite S travels around the earth in a circular path with a constant speed of 20 Mm/h. If the acceleration is 2.5 m/s^2, determine the altitude h. Assume the earth's diameter to be 12 713 km.

12–121. The car passes point A with a speed of 25 m/s after which its speed is defined by $v = (25 - 0.15s)$ m/s, where s is in meters. Determine the magnitude of the car's acceleration when it reaches point B, where $s = 51.5$ m and $x = 50$ m.

12–122. If the car passes point A with a speed of 20 m/s and begins to increase its speed at a constant rate of $a_t = 0.5$ m/s^2, determine the magnitude of the car's acceleration when $s = 101.68$ m and $x = 0$.

Probs. 12–121/122

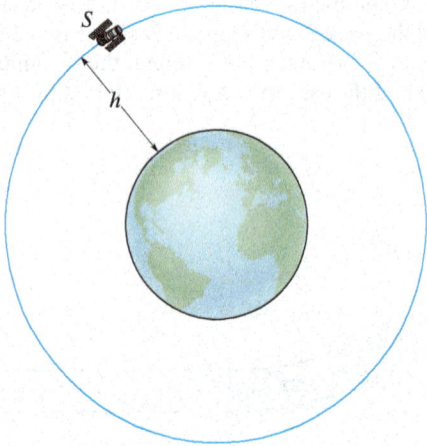

Prob. 12–119

12–123. The motorcycle travels up the hill at a constant speed of 15 m/s. Determine the magnitude of its acceleration as a function of x.

Prob. 12–123

*12–120. The motorcyclist travels along the curve at a constant speed of 30 ft/s. Determine his acceleration when he is located at point A. Neglect the size of the motorcycle and rider for the calculation.

*12–124. When the bicycle passes point A, it has a speed of 6 m/s, which is increasing at the rate of $\dot{v} = 0.5$ m/s^2. Determine the magnitude of its acceleration when it is at point A.

Prob. 12–120

Prob. 12–124

12–125. A racing car travels with a constant speed of 240 km/h around the elliptical race track. Determine the acceleration experienced by the driver at A.

12–126. The racing car travels with a constant speed of 240 km/h around the elliptical race track. Determine the acceleration experienced by the driver at B.

***12–128.** The car has an initial speed $v_0 = 20$ m/s at $s = 0$. If it increases its speed along the circular track at $a_t = (0.8s)$ m/s^2, where s is in meters, determine the time needed for the car to travel $s = 25$ m.

12–129. The car starts from rest at $s = 0$ and increases its speed at $a_t = 4$ m/s^2. Determine the time when the magnitude of acceleration becomes 20 m/s^2. At what position s does this occur?

Probs. 12–125/126

12–127. At a given instant the train engine at E has a speed of 20 m/s and an acceleration of 14 m/s^2 acting in the direction shown. Determine the rate of increase in the train's speed and the radius of curvature ρ of the path.

Probs. 12–128/129

12–130. A boat is traveling along a circular curve having a radius of 100 ft. If its speed at $t = 0$ is 15 ft/s and is increasing at $\dot{v} = (0.8t)$ ft/s^2, determine the magnitude of its acceleration at the instant $t = 5$ s.

12–131. A boat is traveling along a circular path having a radius of 20 m. Determine the magnitude of the boat's acceleration when the speed is $v = 5$ m/s and the rate of increase in the speed is $\dot{v} = 2$ m/s^2.

*12–132. Starting from rest, a bicyclist travels around a horizontal circular path, $\rho = 10$ m, at a speed of $v = (0.09t^2 + 0.1t)$ m/s, where t is in seconds. Determine the magnitudes of her velocity and acceleration when she has traveled $s = 3$ m.

12–133. A particle travels around a circular path having a radius of 50 m. If it is initially traveling with a speed of 10 m/s and its speed then increases at a rate of $\dot{v} = (0.05\,v)$ m/s^2, determine the magnitude of the particle's acceleration four seconds later.

Prob. 12–127

12–134. The motorcycle is traveling at a constant speed of 60 km/h. Determine the magnitude of its acceleration when it is at point A.

Prob. 12–134

12–135. When $t = 0$, the train has a speed of 8 m/s, which is increasing at 0.5 m/s^2. Determine the magnitude of the acceleration of the engine when it reaches point A, at $t = 20$ s. Here the radius of curvature of the tracks is $\rho_A = 400$ m.

Prob. 12–135

*12–136. A train is traveling with a constant speed of 14 m/s along the curved path. Determine the magnitude of the acceleration of the front of the train, B, at the instant it reaches point A ($y = 0$).

Prob. 12–136

12–137. The ball is ejected horizontally from the tube with a speed of 8 m/s. Find the equation of the path, $y = f(x)$, and then find the ball's velocity and the normal and tangential components of acceleration when $t = 0.25$ s.

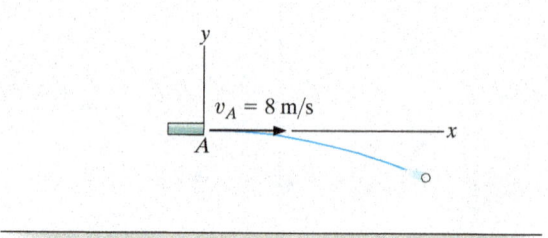

Prob. 12–137

12–138. Two cyclists, A and B, are traveling counterclockwise around a circular track at a constant speed of 8 ft/s at the instant shown. If the speed of A is increased at $(a_t)_A = (s_A)$ ft/s^2, where s_A is in feet, determine the distance measured counterclockwise along the track from B to A between the cyclists when $t = 1$ s. What is the magnitude of the acceleration of each cyclist at this instant?

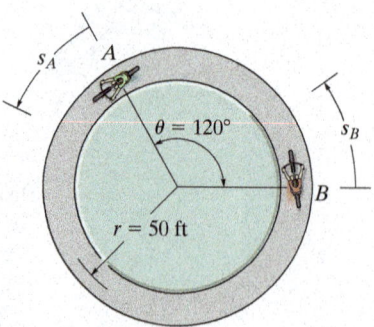

Prob. 12–138

12–139. The car is traveling at a constant speed of 30 m/s. The driver then applies the brakes at A and thereby reduces the car's speed at the rate of $a_t = (-0.08v)$ m/s², where v is in m/s. Determine the magnitude of the acceleration of the car just before it reaches point C on the circular curve. It takes 15 s for the car to travel from A to C.

*12–140. The car is traveling at a speed of 30 m/s. The driver then applies the brakes at A and thereby reduces the speed at the rate of $a_t = (-\frac{1}{8}t)$ m/s², where t is in seconds. Determine the magnitude of the acceleration of the car just before it reaches point C on the circular curve. It takes 15 s for the car to travel from A to C.

Probs. 12–139/140

12–141. A spiral transition curve is used on railroads to connect a straight portion of the track with a curved portion. If the spiral is defined by the equation $y = (10^{-6})x^3$, where x and y are in feet, determine the magnitude of the acceleration of a train engine moving with a constant speed of 40 ft/s when it is at the point where $x = 600$ ft.

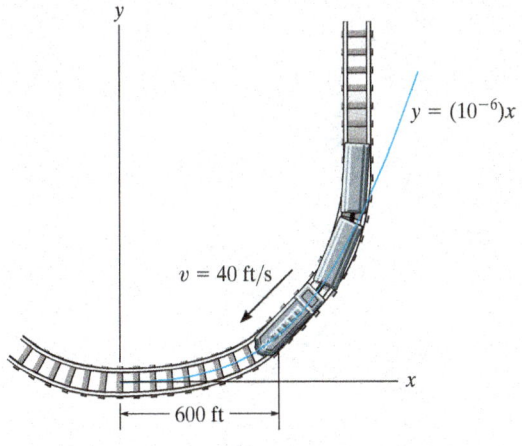

Prob. 12–141

12–142. The race car has an initial speed $v_A = 15$ m/s at A. If it increases its speed along the circular track at the rate $a_t = (0.4s)$ m/s², where s is in meters, determine the time needed for the car to travel 20 m. Take $\rho = 150$ m.

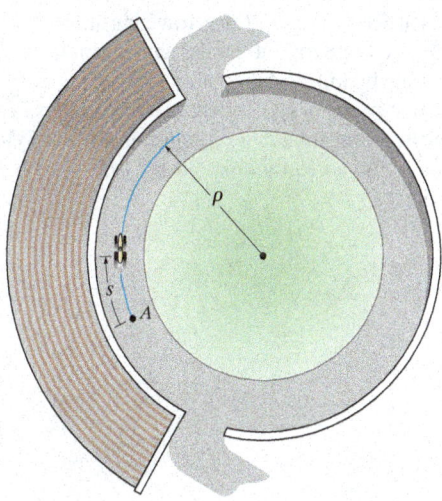

Prob. 12–142

12–143. Determine the magnitude of acceleration of the airplane during the turn. It flies along the horizontal circular path AB in 40 s, while maintaining a constant speed of 300 ft/s.

*12–144. The airplane flies along the horizontal circular path AB in 60 s. If its speed at point A is 400 ft/s, which decreases at a rate of $a_t = (-0.1t)$ ft/s², where t is in seconds, determine the magnitude of the plane's acceleration when it reaches point B.

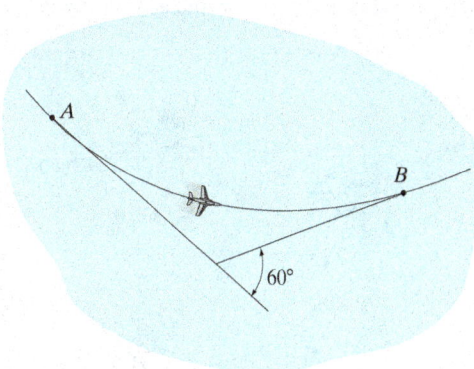

Probs. 12–143/144

12–145. Particles A and B are traveling counterclockwise around a circular track at a constant speed of 8 m/s. If at the instant shown the speed of A begins to increase by $(a_t)_A = (0.4s_A)$ m/s², where s_A is in meters, determine the distance measured counterclockwise along the track from B to A when $t = 1$ s. What is the magnitude of the acceleration of each particle at this instant?

12–146. Particles A and B are traveling around a circular track at a speed of 8 m/s at the instant shown. If the speed of B is increasing by $(a_t)_B = 4$ m/s², and at the same instant A has an increase in speed of $(a_t)_A = 0.8t$ m/s², where t is in seconds, determine how long it takes for a collision to occur. What is the magnitude of the acceleration of each particle just before the collision occurs?

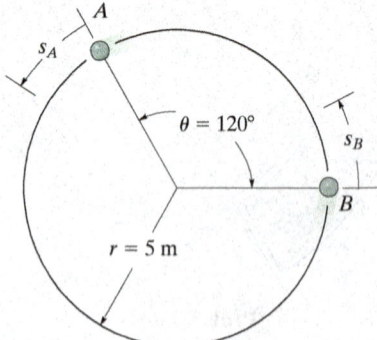

Probs. 12–145/146

12–147. When the motorcyclist is at A, he increases his speed along the vertical circular path at the rate of $\dot{v} = (0.3t)$ ft/s², where t is in seconds and $t = 0$ at A. If he starts from rest at A, determine the magnitudes of his velocity and acceleration when he reaches B.

*12–148. When the motorcyclist is at A, he increases his speed along the vertical circular path at the rate of $\dot{v} = (0.04s)$ ft/s², where s is in ft and $s = 0$ at A. If he starts at $v_A = 2$ ft/s, determine the magnitude of his velocity when he reaches B. Also, what is his initial acceleration?

Probs. 12–147/148

12–149. The train passes point B with a speed of 20 m/s which is decreasing at $a_t = -0.5$ m/s². Determine the magnitude of acceleration of the train at this point.

12–150. The train passes point A with a speed of 30 m/s and begins to decrease its speed at a constant rate of $a_t = -0.25$ m/s². Determine the magnitude of the acceleration of the train when it reaches point B, where $s_{AB} = 412$ m.

Probs. 12–149/150

12–151. The particle travels with a constant speed of 300 mm/s along the curve. Determine the particle's acceleration when it is located at point (200 mm, 100 mm) and sketch this vector on the curve.

Prob. 12–151

***12–152.** A particle P travels along an elliptical spiral path such that its position vector $\mathbf{r}$ is defined by $\mathbf{r} = \{2\cos(0.1t)\mathbf{i} + 1.5\sin(0.1t)\mathbf{j} + (2t)\mathbf{k}\}$ m, where t is in seconds and the arguments for the sine and cosine are given in radians. When $t = 8$ s, determine the coordinate direction angles α, β, and γ, which the binormal axis to the osculating plane makes with the x, y, and z axes. *Hint:* Solve for the velocity $\mathbf{v}_P$ and acceleration $\mathbf{a}_P$ of the particle in terms of their $\mathbf{i}$, $\mathbf{j}$, $\mathbf{k}$ components. The binormal is parallel to $\mathbf{v}_P \times \mathbf{a}_P$. Why?

Prob. 12–152

■**12–153.** A car travels around a circular track having a radius of $r = 300$ m such that when it is at point A it has a velocity of 5 m/s, which is increasing at the rate of $\dot{v} = (0.06t)$ m/s^2, where t is in seconds. Determine the magnitudes of its velocity and acceleration when it has traveled one-third the way around the track.

Prob. 12–153

12–154. If the speed of the crate at A is 15 ft/s, which is increasing at a rate $\dot{v} = 3$ ft/s^2, determine the magnitude of the acceleration of the crate at this instant.

Prob. 12–154

*12.8 CURVILINEAR MOTION: CYLINDRICAL COMPONENTS

Sometimes the motion of the particle is constrained on a path that is best described using cylindrical coordinates. If motion is restricted to the plane, then polar coordinates are used.

Polar Coordinates. We can specify the location of the particle shown in Fig. 12–30a using a ***radial coordinate*** r, which extends outward from the fixed origin O to the particle, and a ***transverse coordinate*** θ, which is the counterclockwise angle between a fixed reference line and the r axis. This angle is generally measured in degrees or radians, where 1 rad $= 180°/\pi$. The positive directions of the r and θ coordinates are defined by the unit vectors $\mathbf{u}_r$ and $\mathbf{u}_\theta$, respectively. Here $\mathbf{u}_r$ is in the direction of increasing r when θ is held fixed, and $\mathbf{u}_\theta$ is in a direction of increasing θ when r is held fixed.

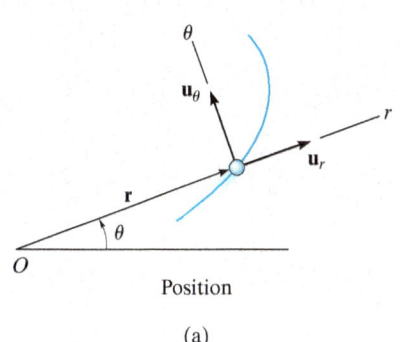

Position
(a)

Fig. 12–30

68   CHAPTER 12   KINEMATICS OF A PARTICLE

Position
(a)

(b)

Position. At any instant the position of the particle, Fig. 12–30a, is defined by the position vector

$$\mathbf{r} = r\mathbf{u}_r \qquad (12\text{–}22)$$

Velocity. The velocity $\mathbf{v}$ is obtained by taking the time derivative of $\mathbf{r}$. Using a dot to represent the time derivative, we have

$$\mathbf{v} = \dot{\mathbf{r}} = \dot{r}\mathbf{u}_r + r\dot{\mathbf{u}}_r$$

To evaluate $\dot{\mathbf{u}}_r$, notice that $\mathbf{u}_r$ only changes its direction with respect to time, since the magnitude of this vector is always one unit. During the time Δt, a change Δr will not cause a change in the direction of $\mathbf{u}_r$; however, a change $\Delta\theta$ will cause $\mathbf{u}_r$ to become $\mathbf{u}'_r$, where $\mathbf{u}'_r = \mathbf{u}_r + \Delta\mathbf{u}_r$, Fig. 12–30b. The time change in $\mathbf{u}_r$ is then $\Delta\mathbf{u}_r$. For small angles $\Delta\theta$ this vector has a magnitude $\Delta u_r \approx 1(\Delta\theta)$ and acts in the $\mathbf{u}_\theta$ direction. Therefore, $\Delta\mathbf{u}_r = \Delta\theta\mathbf{u}_\theta$, and so

$$\dot{\mathbf{u}}_r = \lim_{\Delta t \to 0} \frac{\Delta\mathbf{u}_r}{\Delta t} = \left(\lim_{\Delta t \to 0} \frac{\Delta\theta}{\Delta t}\right)\mathbf{u}_\theta$$

$$\dot{\mathbf{u}}_r = \dot{\theta}\mathbf{u}_\theta \qquad (12\text{–}23)$$

Substituting this result into the above equation, the velocity can then be written in component form as

$$\boxed{\mathbf{v} = v_r\mathbf{u}_r + v_\theta\mathbf{u}_\theta} \qquad (12\text{–}24)$$

where

$$\boxed{\begin{array}{l} v_r = \dot{r} \\ v_\theta = r\dot{\theta} \end{array}} \qquad (12\text{–}25)$$

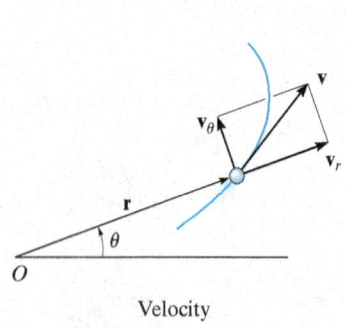

Velocity
(c)

Fig. 12–30 (cont.)

These components are shown graphically in Fig. 12–30c. The *radial component* $\mathbf{v}_r$ is a measure of the rate of increase or decrease in the length of the radial coordinate, i.e., $\dot{r}$; whereas the *transverse component* $\mathbf{v}_\theta$ can be interpreted as the velocity along the circumference of a circle having a radius r. In particular, the term $\dot{\theta} = d\theta/dt$ is called the **angular velocity**, since it indicates the time rate of change of the angle θ. Common units used for this measurement are rad/s.

Since $\mathbf{v}_r$ and $\mathbf{v}_\theta$ are mutually perpendicular, the *magnitude* of velocity or speed is simply

$$v = \sqrt{(\dot{r})^2 + (r\dot{\theta})^2} \qquad (12\text{–}26)$$

and the *direction* of $\mathbf{v}$ is, of course, tangent to the path, Fig. 12–30c.

12.8 CURVILINEAR MOTION: CYLINDRICAL COMPONENTS

Acceleration. Taking the time derivatives of Eq. 12–24, using Eqs. 12–25, we obtain the particle's acceleration,

$$\mathbf{a} = \dot{\mathbf{v}} = \ddot{r}\mathbf{u}_r + \dot{r}\dot{\mathbf{u}}_r + \dot{r}\dot{\theta}\mathbf{u}_\theta + r\ddot{\theta}\mathbf{u}_\theta + r\dot{\theta}\dot{\mathbf{u}}_\theta$$

To evaluate $\dot{\mathbf{u}}_\theta$, it is necessary only to find the change in the direction of $\mathbf{u}_\theta$ since its magnitude is always unity. During the time Δt, a change Δr will not change the direction of $\mathbf{u}_\theta$; however, a change $\Delta\theta$ will cause $\mathbf{u}_\theta$ to become $\mathbf{u}'_\theta$, where $\mathbf{u}'_\theta = \mathbf{u}_\theta + \Delta\mathbf{u}_\theta$, Fig. 12–30d. The time change in $\mathbf{u}_\theta$ is thus $\Delta\mathbf{u}_\theta$. For small angles this vector has a magnitude $\Delta u_\theta \approx 1(\Delta\theta)$ and acts in the $-\mathbf{u}_r$ direction; i.e., $\Delta\mathbf{u}_\theta = -\Delta\theta\mathbf{u}_r$. Thus,

$$\dot{\mathbf{u}}_\theta = \lim_{\Delta t \to 0} \frac{\Delta\mathbf{u}_\theta}{\Delta t} = -\left(\lim_{\Delta t \to 0} \frac{\Delta\theta}{\Delta t}\right)\mathbf{u}_r$$

$$\dot{\mathbf{u}}_\theta = -\dot{\theta}\mathbf{u}_r \quad (12\text{–}27)$$

(d)

Substituting this result and Eq. 12–23 into the above equation for **a**, we can write the acceleration in component form as

$$\mathbf{a} = a_r\mathbf{u}_r + a_\theta\mathbf{u}_\theta \quad (12\text{–}28)$$

where

$$a_r = \ddot{r} - r\dot{\theta}^2$$
$$a_\theta = r\ddot{\theta} + 2\dot{r}\dot{\theta} \quad (12\text{–}29)$$

The term $\ddot{\theta} = d^2\theta/dt^2 = d/dt(d\theta/dt)$ is called the **angular acceleration** since it measures the change made in the angular velocity during an instant of time. Units for this measurement are rad/s^2.

Since $\mathbf{a}_r$ and $\mathbf{a}_\theta$ are perpendicular to one another, the *magnitude* of acceleration is

$$a = \sqrt{(\ddot{r} - r\dot{\theta}^2)^2 + (r\ddot{\theta} + 2\dot{r}\dot{\theta})^2} \quad (12\text{–}30)$$

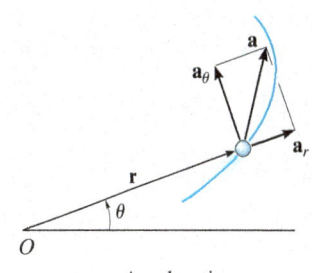

Acceleration

(e)

The *direction* is determined from the vector addition of its two components, and as shown in Fig. 12–30e, in general **a** will *not* be tangent to the path.

70   CHAPTER 12   KINEMATICS OF A PARTICLE

Fig. 12–31

Cylindrical Coordinates. If the particle moves along a space curve, Fig. 12–31, then its location can be specified by the three *cylindrical coordinates*, r, θ, z, where the z coordinate is identical to that used for rectangular coordinates. Since the unit vector defining its direction, $\mathbf{u}_z$, is constant, the time derivatives of this vector are zero, and therefore the position, velocity, and acceleration of the particle can be written in terms of its cylindrical coordinates as

$$\mathbf{r}_P = r\mathbf{u}_r + z\mathbf{u}_z$$
$$\mathbf{v} = \dot{r}\mathbf{u}_r + r\dot{\theta}\mathbf{u}_\theta + \dot{z}\mathbf{u}_z \tag{12-31}$$
$$\mathbf{a} = (\ddot{r} - r\dot{\theta}^2)\mathbf{u}_r + (r\ddot{\theta} + 2\dot{r}\dot{\theta})\mathbf{u}_\theta + \ddot{z}\mathbf{u}_z \tag{12-32}$$

Time Derivatives. The above equations require that we obtain the time derivatives $\dot{r}$, $\ddot{r}$, $\dot{\theta}$, and $\ddot{\theta}$ in order to determine the r and θ components of $\mathbf{v}$ and $\mathbf{a}$. Two types of problems generally occur:

1. If the polar coordinates are specified as time parametric equations, $r = r(t)$ and $\theta = \theta(t)$, then the time derivatives can be found directly.

2. If the time parametric equations are not given, then the path $r = f(\theta)$ must be known. Using the chain rule of calculus we can then find the relation between $\dot{r}$ and $\dot{\theta}$, and between $\ddot{r}$ and $\ddot{\theta}$.

The spiral motion of this girl can be followed by using cylindrical components. Here the radial coordinate r is constant, the transverse coordinate θ will increase with time as the girl rotates about the vertical, and her altitude z will decrease with time.

PROCEDURE FOR ANALYSIS

Coordinate System.
- Polar coordinates are a suitable choice for solving problems where the angular motion of the radial coordinate r is given to describe the particle's motion. Also, some paths of motion can conveniently be described in terms of these coordinates.
- To use polar coordinates, the origin is established at a fixed point, and the radial line r is directed to the particle.
- The transverse coordinate θ is measured from a fixed reference line to the radial line.

Velocity and Acceleration.
- Once r and the four time derivatives $\dot{r}$, $\ddot{r}$, $\dot{\theta}$, and $\ddot{\theta}$ have been evaluated at the instant considered, their values are substituted into Eqs. 12–25 and 12–29 to obtain the radial and transverse components of $\mathbf{v}$ and $\mathbf{a}$.
- If it is necessary to take the time derivatives of $r = f(\theta)$, then the chain rule must be used. See Appendix C.
- Motion in three dimensions requires a simple extension of the above procedure to include $\dot{z}$ and $\ddot{z}$.

EXAMPLE 12.17

The amusement park ride shown in Fig. 12–32a consists of a chair that is rotating in a horizontal circular path of radius r such that the arm OB has an angular velocity $\dot{\theta}$ and angular acceleration $\ddot{\theta}$. Determine the radial and transverse components of velocity and acceleration of the car. Neglect its size in the calculation.

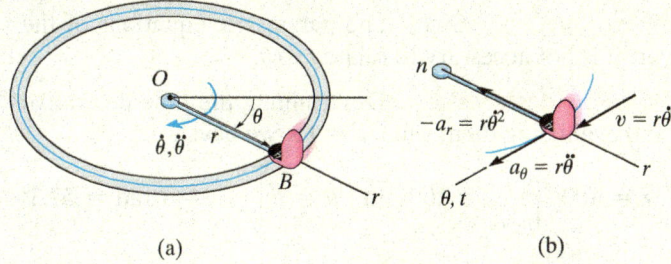

Fig. 12–32

SOLUTION

Coordinate System. Since the angular motion of the arm is known, polar coordinates will be used for the solution, Fig. 12–32a.

Velocity and Acceleration. It is first necessary to specify the first and second time derivatives of r and θ. Since r is *constant*, we have

$$r = r \quad \dot{r} = 0 \quad \ddot{r} = 0$$

Thus,

$$v_r = \dot{r} = 0 \quad \text{Ans.}$$

$$v_\theta = r\dot{\theta} \quad \text{Ans.}$$

$$a_r = \ddot{r} - r\dot{\theta}^2 = -r\dot{\theta}^2 \quad \text{Ans.}$$

$$a_\theta = r\ddot{\theta} + 2\dot{r}\dot{\theta} = r\ddot{\theta} \quad \text{Ans.}$$

These results are shown in Fig. 12–32b.

NOTE: The n, t axes are also shown in Fig. 12–32b, which in this special case of circular motion happen to be *collinear* with the r and θ axes, respectively. Since $v = v_\theta = v_t = r\dot{\theta}$, then by comparison,

$$-a_r = a_n = \frac{v^2}{\rho} = \frac{(r\dot{\theta})^2}{r} = r\dot{\theta}^2$$

$$a_\theta = a_t = \frac{dv}{dt} = \frac{d}{dt}(r\dot{\theta}) = \frac{dr}{dt}\dot{\theta} + r\frac{d\dot{\theta}}{dt} = 0 + r\ddot{\theta}$$

EXAMPLE 12.18

The rod OA in Fig. 12–33a rotates in the horizontal plane such that $\theta = (t^3)$ rad. At the same time, the collar B is sliding outward along OA so that $r = (100t^2)$ mm. If t is in seconds, determine the velocity and acceleration of the collar when $t = 1$ s.

SOLUTION

Coordinate System. Since time parametric equations of the path are given, it is not necessary to relate r to θ.

Velocity and Acceleration. Determining the time derivatives of r and θ and evaluating them when $t = 1$ s, we have

$$r = 100t^2 \Big|_{t=1\,s} = 100 \text{ mm} \qquad \theta = t^3 \Big|_{t=1\,s} = 1 \text{ rad} = 57.3°$$

$$\dot{r} = 200t \Big|_{t=1\,s} = 200 \text{ mm/s} \qquad \dot{\theta} = 3t^2 \Big|_{t=1\,s} = 3 \text{ rad/s}$$

$$\ddot{r} = 200 \Big|_{t=1\,s} = 200 \text{ mm/s}^2 \qquad \ddot{\theta} = 6t \Big|_{t=1\,s} = 6 \text{ rad/s}^2$$

(a)

(b)

As shown in Fig. 12–33b,

$$\mathbf{v} = \dot{r}\mathbf{u}_r + r\dot{\theta}\mathbf{u}_\theta$$
$$= 200\mathbf{u}_r + 100(3)\mathbf{u}_\theta = \{200\mathbf{u}_r + 300\mathbf{u}_\theta\} \text{ mm/s}$$

The magnitude of $\mathbf{v}$ is

$$v = \sqrt{(200)^2 + (300)^2} = 361 \text{ mm/s} \qquad \text{Ans.}$$

$$\delta = \tan^{-1}\left(\frac{300}{200}\right) = 56.3° \qquad \delta + 57.3° = 114° \qquad \text{Ans.}$$

As shown in Fig. 12–33c,

$$\mathbf{a} = (\ddot{r} - r\dot{\theta}^2)\mathbf{u}_r + (r\ddot{\theta} + 2\dot{r}\dot{\theta})\mathbf{u}_\theta$$
$$= [200 - 100(3)^2]\mathbf{u}_r + [100(6) + 2(200)3]\mathbf{u}_\theta$$
$$= \{-700\mathbf{u}_r + 1800\mathbf{u}_\theta\} \text{ mm/s}^2$$

(c)

Fig. 12–33

The magnitude of $\mathbf{a}$ is

$$a = \sqrt{(-700)^2 + (1800)^2} = 1930 \text{ mm/s}^2 \qquad \text{Ans.}$$

$$\phi = \tan^{-1}\left(\frac{1800}{700}\right) = 68.7° \qquad (180° - \phi) + 57.3° = 169° \qquad \text{Ans.}$$

NOTE: The velocity is tangent to the path; however, the acceleration is directed within the curvature of the path, as expected.

EXAMPLE 12.19

The searchlight in Fig. 12–34a casts a spot of light along the wall that is located 100 m from the searchlight. Determine the magnitudes of the velocity and acceleration at which the light appears to travel across the wall at the instant $\theta = 45°$. The searchlight rotates at a constant rate of $\dot{\theta} = 4$ rad/s.

(a)

SOLUTION

Coordinate System. Polar coordinates will be used to solve this problem since the angular rate of the searchlight is given. To find the necessary time derivatives it is first necessary to relate r to θ. From Fig. 12–34a,

$$r = 100/\cos\theta = 100\sec\theta$$

Velocity and Acceleration. Using the chain rule, noting that $d(\sec\theta) = \sec\theta\tan\theta\, d\theta$, and $d(\tan\theta) = \sec^2\theta\, d\theta$, we have

$$\dot{r} = 100(\sec\theta\tan\theta)\dot{\theta}$$
$$\ddot{r} = 100(\sec\theta\tan\theta)\dot{\theta}(\tan\theta)\dot{\theta} + 100\sec\theta(\sec^2\theta)\dot{\theta}(\dot{\theta})$$
$$\quad + 100\sec\theta\tan\theta(\ddot{\theta})$$
$$= 100\sec\theta\tan^2\theta\,(\dot{\theta})^2 + 100\sec^3\theta\,(\dot{\theta})^2 + 100(\sec\theta\tan\theta)\ddot{\theta}$$

Since $\dot{\theta} = 4$ rad/s = constant, then $\ddot{\theta} = 0$, and the above equations, when $\theta = 45°$, become

$$r = 100\sec 45° = 141.4$$
$$\dot{r} = 400\sec 45°\tan 45° = 565.7$$
$$\ddot{r} = 1600\,(\sec 45°\tan^2 45° + \sec^3 45°) = 6788.2$$

(b)

As shown in Fig. 12–34b,

$$\mathbf{v} = \dot{r}\mathbf{u}_r + r\dot{\theta}\mathbf{u}_\theta$$
$$= 565.7\mathbf{u}_r + 141.4(4)\mathbf{u}_\theta$$
$$= \{565.7\mathbf{u}_r + 565.7\mathbf{u}_\theta\}\text{ m/s}$$
$$v = \sqrt{v_r^2 + v_\theta^2} = \sqrt{(565.7)^2 + (565.7)^2}$$
$$= 800 \text{ m/s} \qquad Ans.$$

As shown in Fig. 12–34c,

$$\mathbf{a} = (\ddot{r} - r\dot{\theta}^2)\mathbf{u}_r + (r\ddot{\theta} + 2\dot{r}\dot{\theta})\mathbf{u}_\theta$$
$$= [6788.2 - 141.4(4)^2]\mathbf{u}_r + [141.4(0) + 2(565.7)4]\mathbf{u}_\theta$$
$$= \{4525.5\mathbf{u}_r + 4525.5\mathbf{u}_\theta\}\text{ m/s}^2$$
$$a = \sqrt{a_r^2 + a_\theta^2} = \sqrt{(4525.5)^2 + (4525.5)^2}$$
$$= 6400 \text{ m/s}^2 \qquad Ans.$$

(c)

(d)

Fig. 12–34

NOTE: It is also possible to find a without having to calculate $\ddot{r}$ (or a_r). As shown in Fig. 12–34d, since $a_\theta = 4525.5$ m/s^2, then by vector resolution, $a = 4525.5/\cos 45° = 6400$ m/s^2.

EXAMPLE 12.20

Due to the rotation of the forked rod, the ball in Fig. 12–35a travels around the slotted path, a portion of which is in the shape of a cardioid, $r = 0.5(1 - \cos\theta)$ ft, where θ is in radians. If the ball's velocity is $v = 4$ ft/s and its acceleration is $a = 30$ ft/s² at the instant $\theta = 180°$, determine the angular velocity $\dot\theta$ and angular acceleration $\ddot\theta$ of the fork.

SOLUTION

Coordinate System. This path is most unusual, and mathematically it is best expressed using polar coordinates, as done here, rather than using rectangular coordinates. Also, since $\dot\theta$ and $\ddot\theta$ must be determined, then r, θ coordinates are an obvious choice.

Velocity and Acceleration. The time derivatives of r and θ can be determined using the chain rule.

$r = 0.5(1 - \cos\theta)$ ft

$$r = 0.5(1 - \cos\theta)$$
$$\dot r = 0.5(\sin\theta)\dot\theta$$
$$\ddot r = 0.5(\cos\theta)(\dot\theta)^2 + 0.5(\sin\theta)\ddot\theta$$

Evaluating these results at $\theta = 180°$, we have

$$r = 1 \text{ ft} \quad \dot r = 0 \quad \ddot r = -0.5\dot\theta^2$$

Since $v = 4$ ft/s, using Eq. 12–26 to determine $\dot\theta$ yields

$$v = \sqrt{(\dot r)^2 + (r\dot\theta)^2}$$
$$4 = \sqrt{(0)^2 + (1\dot\theta)^2}$$
$$\dot\theta = 4 \text{ rad/s} \qquad Ans.$$

In a similar manner, $\ddot\theta$ can be found using Eq. 12–30.

$v = 4$ ft/s
$a = 30$ ft/s²

(b)

Fig. 12–35

$$a = \sqrt{(\ddot r - r\dot\theta^2)^2 + (r\ddot\theta + 2\dot r\dot\theta)^2}$$
$$30 = \sqrt{[-0.5(4)^2 - 1(4)^2]^2 + [1\ddot\theta + 2(0)(4)]^2}$$
$$(30)^2 = (-24)^2 + \ddot\theta^2$$
$$\ddot\theta = 18 \text{ rad/s}^2 \qquad Ans.$$

Vectors **a** and **v** are shown in Fig. 12–35b.

FUNDAMENTAL PROBLEMS

F12–33. The car has a speed of 55 ft/s. Determine the angular velocity $\dot{\theta}$ of the radial line OA at this instant.

Prob. F12–33

F12–34. The platform is rotating about the vertical axis such that at any instant its angular position is $\theta = (4t^{3/2})$ rad, where t is in seconds. A ball rolls outward along the radial groove so that its position is $r = (0.1t^3)$ m, where t is in seconds. Determine the magnitudes of the velocity and acceleration of the ball when $t = 1.5$ s.

Prob. F12–34

F12–35. Peg P is driven by the fork link OA along the curved path described by $r = (2\theta)$ ft. At the instant $\theta = \pi/4$ rad, the angular velocity and angular acceleration of the link are $\dot{\theta} = 3$ rad/s and $\ddot{\theta} = 1$ rad/s^2. Determine the magnitude of the peg's acceleration at this instant.

Prob. F12–35

F12–36. Peg P is driven by the forked link OA along the path described by $r = e^\theta$, where r is in meters. When $\theta = \frac{\pi}{4}$ rad, the link has an angular velocity and angular acceleration of $\dot{\theta} = 2$ rad/s and $\ddot{\theta} = 4$ rad/s^2. Determine the radial and transverse components of the peg's acceleration at this instant.

Prob. F12–36

F12–37. The collars are pin connected at B and are free to move along rod OA and the curved guide OC having the shape of a cardioid, $r = [0.2(1 + \cos\theta)]$ m. At $\theta = 30°$, the angular velocity of OA is $\dot{\theta} = 3$ rad/s. Determine the magnitude of the velocity of the collars at this point.

Prob. F12–37

F12–38. At the instant $\theta = 45°$, the athlete is running with a constant speed of 2 m/s. Determine the angular velocity at which the camera must turn in order to follow the motion.

Prob. F12–38

PROBLEMS

12–155. If a particle's position is described by the polar coordinates $r = 4(1 + \sin t)$ m and $\theta = (2e^{-t})$ rad, where t is in seconds and the argument for the sine is in radians, determine the radial and transverse components of the particle's velocity and acceleration when $t = 2$ s.

*12–156. A particle moves along a path defined by polar coordinates $r = (2e^t)$ ft and $\theta = (8t^2)$ rad, where t is in seconds. Determine the components of its velocity and acceleration when $t = 1$ s.

12–157. A particle travels around a limaçon, defined by the equation $r = b - a \cos \theta$, where a and b are constants. Determine the particle's radial and transverse components of velocity and acceleration as a funcion of θ and its time derivatives.

12–158. An airplane is flying in a straight line with a velocity of 200 mi/h and an acceleration of 3 mi/h². If the propeller has a diameter of 6 ft and is rotating at a constant angular rate of 120 rad/s, determine the magnitudes of velocity and acceleration of a particle located on the tip of the propeller.

12–159. The small washer is sliding down the cord OA. When it is at the midpoint, its speed is 28 m/s and its acceleration is 7 m/s². Express the velocity and acceleration of the washer at this point in terms of its cylindrical components.

*12–160. A car is traveling along the circular curve of radius $r = 300$ ft. At the instant shown, its angular rate of rotation is $\dot\theta = 0.4$ rad/s, which is increasing at the rate of $\ddot\theta = 0.2$ rad/s². Determine the magnitudes of the car's velocity and acceleration at this instant.

Prob. 12–160

12–161. If a particle moves along a path such that $r = (2 \cos t)$ ft and $\theta = (t/2)$ rad, where t is in seconds, plot the path $r = f(\theta)$ and determine the particle's radial and transverse components of velocity and acceleration.

12–162. If a particle's position is described by the polar coordinates $r = (2 \sin 2\theta)$ m and $\theta = (4t)$ rad, where t is in seconds, determine the radial and transverse components of its velocity and acceleration when $t = 1$ s.

12–163. The driver of the car maintains a constant speed of 40 m/s. Determine the angular velocity of the camera tracking the car when $\theta = 15°$.

*12–164. When $\theta = 15°$, the car has a speed of 50 m/s which is increasing at 6 m/s². Determine the angular velocity of the camera tracking the car at this instant.

Prob. 12–159

Probs. 12–163/164

12–165. The time rate of change of acceleration is referred to as the *jerk*, which is often used as a means of measuring passenger discomfort. Calculate this vector, $\dot{\mathbf{a}}$, in terms of its cylindrical components, using Eq. 12–32.

12–166. The position of a particle is described by $r = (300e^{-0.5t})$ mm and $\theta = (0.3t^2)$ rad, where t is in seconds. Determine the magnitudes of the particle's velocity and acceleration at the instant $t = 1.5$ s.

12–167. The pin P is constrained to move along the curve defined by the lemniscate $r = (4 \sin 2\theta)$ ft. If the angular position of the slotted arm OA is defined by $\theta = (3t^{3/2})$ rad, where t is in seconds, determine the magnitudes of the velocity and acceleration of P when $\theta = 60°$.

Prob. 12–167

*12–168. A particle travels along the portion of the "four-leaf rose" defined by the equation $r = (5 \cos 2\theta)$ m. If the angular velocity of the radial coordinate line is $\dot{\theta} = (3t^2)$ rad/s, where t is in seconds, determine the radial and transverse components of the particle's velocity and acceleration at the instant $\theta = 30°$. When $t = 0$, $\theta = 0$.

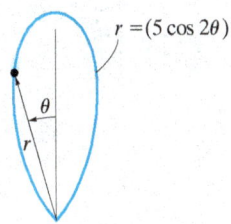

Prob. 12–168

12–169. A particle P moves along the spiral path $r = (10/\theta)$ ft, where θ is in radians. If it maintains a constant speed of $v = 20$ ft/s, determine v_r and v_θ as functions of θ and evaluate each at $\theta = 1$ rad.

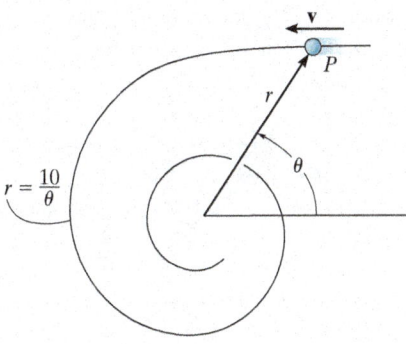

Prob. 12–169

12–170. A particle moves in the x–y plane such that its position is defined by $\mathbf{r} = \{2t\mathbf{i} + 4t^2\mathbf{j}\}$ ft, where t is in seconds. Determine the radial and transverse components of the particle's velocity and acceleration when $t = 2$ s.

12–171. The mechanism of a machine is constructed so that the roller at A follows the surface of the cam described by the equation $r = (0.3 + 0.2 \cos \theta)$ m. If $\dot{\theta} = 0.5$ rad/s and $\ddot{\theta} = 0$, determine the magnitudes of the roller's velocity and acceleration when $\theta = 30°$. Neglect the size of the roller. Also determine the velocity components $(\mathbf{v}_A)_x$ and $(\mathbf{v}_A)_y$ of the roller at this instant. The rod to which the roller is attached remains vertical and can slide up or down along the guides while the guides translate horizontally to the left.

Prob. 12–171

*12–172. The rod OA rotates clockwise with a constant angular velocity of 6 rad/s. Two pin-connected slider blocks, located at B, move freely on OA and the curved rod whose shape is a limaçon described by the equation $r = 200(2 - \cos\theta)$ mm. Determine the speed of the slider blocks at the instant $\theta = 150°$.

12–173. Determine the magnitude of the acceleration of the slider blocks in Prob. 12–172 when $\theta = 150°$.

Probs. 12–172/173

12–174. The link is pinned at O, and as a result of its rotation it drives the peg P along the vertical guide. Calculate the magnitudes of the velocity and acceleration of P if $\theta = ct$ rad, where c is a constant.

Prob. 12–174

12–175. The mechanism within a machine is constructed so that the pin follows the path described by the equation $r = (300 + 200\cos\theta)$ mm. If $\dot\theta = 0.5$ rad/s and $\ddot\theta = 0$, determine the magnitudes of the pin's velocity and acceleration when $\theta = 30°$. Neglect the size of the pin. Also, calculate the velocity components $(v_A)_x$ and $(v_A)_y$ of the pin at this instant.

Prob. 12–175

*12–176. The box slides down the helical ramp with a constant speed of $v = 2$ m/s. Determine the magnitude of its acceleration. The ramp descends a vertical distance of 1 m for every full revolution. The mean radius of the ramp is $r = 0.5$ m.

12–177. The box slides down the helical ramp such that $r = 0.5$ m, $\theta = (0.5t^3)$ rad, and $z = (2-0.2t^2)$ m, where t is in seconds. Determine the magnitudes of the velocity and acceleration of the box at the instant $\theta = 2\pi$ rad.

Probs. 12–176/177

12–178. The motion of the pin P is controlled by the rotation of the grooved link OA. If the link is rotating at a constant angular rate of $\dot\theta = 6$ rad/s, determine the magnitudes of the velocity and acceleration of P at the instant $\theta = \pi/2$ rad. The spiral path is defined by the equation $r = (40\,\theta)$ mm, where θ is in radians.

Prob. 12–178

12–179. If the circular plate rotates clockwise with a constant angular velocity of $\dot{\theta} = 1.5$ rad/s, determine the magnitudes of the velocity and acceleration of the follower rod AB when $\theta = (2/3\pi)$ rad.

*12–180. When $\theta = (2/3\pi)$ rad, the angular velocity and angular acceleration of the circular plate are $\dot{\theta} = 1.5$ rad/s and $\ddot{\theta} = 3$ rad/s², respectively. Determine the magnitudes of the velocity and acceleration of the rod AB at this instant.

*12–184. For a short time the jet plane moves along a path in the shape of a lemniscate, $r^2 = (2500 \cos 2\theta)$ km². At the instant $\theta = 30°$, the radar tracking device is rotating at $\dot{\theta} = 5(10^{-3})$ rad/s with $\ddot{\theta} = 2(10^{-3})$ rad/s². Determine the radial and transverse components of velocity and acceleration of the plane at this instant.

Prob. 12–184

$r = (10 + 50\, \theta^{1/2})$ mm

Probs. 12–179/180

12–181. The motion of peg P is constrained by the lemniscate curved slot in OB and by the slotted arm OA. If OA rotates counterclockwise with a constant angular velocity of $\dot{\theta} = 3$ rad/s, determine the magnitudes of the velocity and acceleration of peg P at $\theta = 30°$.

12–182. The motion of peg P is constrained by the lemniscate curved slot in OB and by the slotted arm OA. If OA rotates counterclockwise with an angular velocity of $\dot{\theta} = (3t^{3/2})$ rad/s, where t is in seconds, determine the magnitudes of the velocity and acceleration of peg P at $\theta = 30°$. When $t = 0, \theta = 0°$.

12–185. The rod OA rotates counterclockwise with a constant angular velocity of $\dot{\theta} = 5$ rad/s. Two pin-connected slider blocks, located at B, move freely on OA and the curved rod whose shape is a limaçon described by the equation $r = 100(2 - \cos \theta)$ mm. Determine the speed of the slider blocks at the instant $\theta = 120°$.

12–186. Determine the magnitude of the acceleration of the slider blocks in Prob. 12–185 when $\theta = 120°$.

$r^2 = (4 \cos 2\theta)$ m²

Probs. 12–181/182

12–183. A car travels along a road, which for a short distance is defined by $r = (200/\theta)$ ft, where θ is in radians. If it maintains a constant speed of $v = 35$ ft/s, determine the radial and transverse components of its velocity when $\theta = \pi/3$ rad.

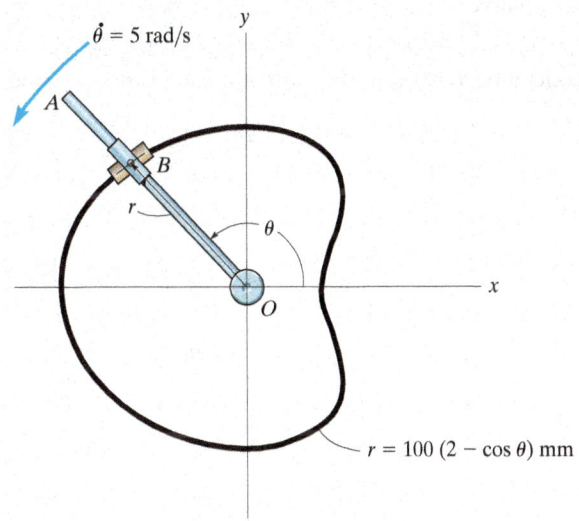

$r = 100\,(2 - \cos \theta)$ mm

Probs. 12–185/186

12–187. If the cam rotates clockwise with a constant angular velocity of $\dot{\theta} = 5$ rad/s, determine the magnitudes of the velocity and acceleration of the follower rod AB at the instant $\theta = 30°$. The surface of the cam has the shape of a limaçon defined by $r = (200 + 100 \cos \theta)$ mm.

*12–188. At the instant $\theta = 30°$, the cam rotates with a clockwise angular velocity of $\dot{\theta} = 5$ rad/s and angular acceleration of $\ddot{\theta} = 6$ rad/s². Determine the magnitudes of the velocity and acceleration of the follower rod AB at this instant. The surface of the cam has the shape of a limaçon defined by $r = (200 + 100 \cos \theta)$ mm.

12–191. The arm of the robot moves so that $r = 3$ ft is constant, and its grip A moves along the path $z = (3 \sin 4\theta)$ ft, where θ is in radians. If $\theta = (0.5t)$ rad, where t is in seconds, determine the magnitudes of the grip's velocity and acceleration when $t = 3$ s.

*12–192. For a short time the arm of the robot is extending such that $\dot{r} = 1.5$ ft/s when $r = 3$ ft, $z = (4t^2)$ ft, and $\theta = 0.5t$ rad, where t is in seconds. Determine the magnitudes of the velocity and acceleration of the grip A when $t = 3$ s.

Probs. 12–191/192

Probs. 12–187/188

12–189. A particle moves along an Archimedean spiral $r = (8\theta)$ ft, where θ is in radians. If $\dot{\theta} = 4$ rad/s (constant), determine the radial and transverse components of the particle's velocity and acceleration at the instant $\theta = \pi/2$ rad. Sketch the curve and show the components on the curve.

12–190. Solve Prob. 12–189 if the particle has an angular acceleration $\ddot{\theta} = 5$ rad/s² when $\dot{\theta} = 4$ rad/s at $\theta = \pi/2$ rad.

12–193. The slotted arm OA rotates counterclockwise about O with a constant angular velocity of $\dot{\theta}$. The motion of pin B is constrained such that it moves on the fixed circular surface and along the slot in OA. Determine the magnitudes of the velocity and acceleration of pin B as a function of θ.

12–194. The slotted arm OA rotates counterclockwise about O such that when $\theta = \pi/4$, arm OA is rotating with an angular velocity of $\dot{\theta}$ and an angular acceleration of $\ddot{\theta}$. Determine the magnitudes of the velocity and acceleration of pin B at this instant. The motion of pin B is constrained such that it moves on the fixed circular surface and along the slot in OA.

Probs. 12–189/190

Probs. 12–193/194

12.9 ABSOLUTE DEPENDENT MOTION ANALYSIS OF TWO PARTICLES

In some types of problems the motion of one particle will *depend* on the corresponding motion of another particle. This dependency commonly occurs if the particles, here represented by blocks, are interconnected by inextensible cords which are wrapped around pulleys. For example, the movement of block *A* downward along the inclined plane in Fig. 12–36 will cause a corresponding movement of block *B* up the other incline.

Although this may seem obvious, we can show this mathematically by first specifying the location of the blocks using **position coordinates** s_A and s_B, where each of the coordinate axes is (1) measured from a *fixed point* (*O*) or *fixed datum line*, (2) measured along each inclined plane *in the direction of motion* of each block, and (3) has a positive sense from the fixed datums to *A* and to *B*. If the total cord length is l_T, then the two position coordinates can be related by the equation

$$s_A + l_{CD} + s_B = l_T$$

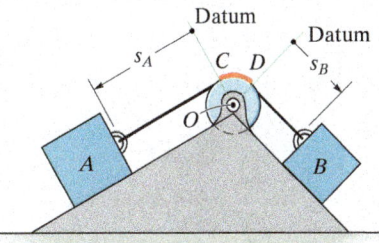

Fig. 12–36

Here l_{CD} is the length of the cord passing over arc *CD*. Taking the time derivative of this expression, realizing that l_{CD} and l_T remain constant, while s_A and s_B measure the segments of the cord that change in length, we have

$$\frac{ds_A}{dt} + \frac{ds_B}{dt} = 0 \quad \text{or} \quad v_B = -v_A$$

The negative sign indicates that when block *A* has a velocity downward, i.e., in the direction of positive s_A, it causes a corresponding upward velocity of block *B*; i.e., *B* moves in the negative s_B direction.

In a similar manner, time differentiation of the velocities yields the relation between the accelerations, i.e.,

$$a_B = -a_A$$

A more complicated example is shown in Fig. 12–37a. In this case, the position of block *A* is specified by s_A, and the position of the *end* of the cord from which block *B* is suspended is defined by s_B. As above, we have chosen position coordinates which (1) have their origin at fixed points or datums, (2) are measured in the direction of motion of each block, and (3) from the fixed datums are positive to the right for s_A and positive downward for s_B. During the motion, the length of the red colored segments of the cord in Fig. 12–37a remains constant. If *l* represents the total length of cord minus these segments, then the two position coordinates can be related by the equation

$$2s_B + h + s_A = l$$

Since *l* and *h* are constant during the motion, the two time derivatives yield

$$2v_B = -v_A \qquad 2a_B = -a_A$$

Hence, when *B* moves downward ($+s_B$), *A* moves to the left ($-s_A$) with twice the motion.

(a)

Fig. 12–37

This example can also be worked by defining the position of block B from the center of the bottom pulley (a fixed point), Fig. 12–37b. In this case

$$2(h - s_B) + h + s_A = l$$

Time differentiation yields

$$2v_B = v_A \qquad 2a_B = a_A$$

Here the signs are the same because when A moves to the right $(+s_A)$, B moves up $(+s_B)$.

Fig. 12–37 (cont.)

The cable is wrapped around several pulleys on this crane in order to reduce the required force needed to hoist a load.

PROCEDURE FOR ANALYSIS

Position-Coordinate Equation.
- Establish each position coordinate with an origin located at a *fixed* point or datum.
- It is *not necessary* that the *origin* be the *same* for each of the coordinates; however, it is *important* that each coordinate be directed along the *path of motion* of the particle.
- Using geometry or trigonometry, relate the position coordinates to the total length of the cord, l_T, or to that portion of cord, l, which *excludes* the segments that do not change length as the particles move—such as arc segments wrapped over pulleys.
- If two or more cords are wrapped around pulleys, then the position of a point on one cord must be related to the position of a point on another cord using the above procedure. Separate equations are written for each cord and the positions of the two particles are then related by these equations (see Examples 12.22 and 12.23).

Time Derivatives.
- Two successive time derivatives of the position-coordinate equations will give the velocity and acceleration equations which relate the motions of the particles.
- The signs of the terms in these equations will be consistent with those that specify the positive and negative sense of the position coordinates.

EXAMPLE 12.21

Determine the speed of block A in Fig. 12–38 if block B is moving upward at 6 ft/s.

Fig. 12–38

SOLUTION

Position-Coordinate Equation. There is *one cord* in this system having segments which change length. Position coordinates s_A and s_B will be used since each is measured from a fixed point (C or D) and extends along each block's *path of motion*. Here, s_B is directed to point E since motion of B and E is the *same*.

The four red colored segments of the cord in Fig. 12–38 maintain a constant length and do not have to be considered as the blocks move. The remaining length of cord, l, is also constant and is related to the changing position coordinates s_A and s_B by the equation

$$s_A + 3s_B = l$$

Time Derivative. Taking the time derivative,

$$v_A + 3v_B = 0$$

so that when $v_B = -6$ ft/s (upward),

$$v_A = 18 \text{ ft/s} \downarrow \qquad \qquad Ans.$$

EXAMPLE 12.22

Determine the speed of block A in Fig. 12–39 if block B has an upward speed of 6 ft/s.

Fig. 12–39

SOLUTION

Position-Coordinate Equation. As shown, the positions of A and B are defined using coordinates s_A and s_B. Since the system has *two cords* with segments that change length, it will be necessary to use a third coordinate, s_C, in order to relate s_A to s_B. In other words, the length of one of the cords can be expressed in terms of s_A and s_C, and the length of the other cord can be expressed in terms of s_B and s_C.

The red colored segments of the cords in Fig. 12–39 do not have to be considered in the analysis. Why? For the remaining cord lengths, say l_1 and l_2, we have

$$s_A + 2s_C = l_1 \qquad s_B + (s_B - s_C) = l_2$$

Time Derivative. Taking the time derivative of these equations yields

$$v_A + 2v_C = 0 \qquad 2v_B - v_C = 0$$

Eliminating v_C produces the relationship between the motions of each cylinder.

$$v_A + 4v_B = 0$$

so that when $v_B = -6$ ft/s (upward),

$$v_A = +24 \text{ ft/s} = 24 \text{ ft/s} \downarrow \qquad \qquad Ans.$$

EXAMPLE 12.23

Determine the speed of block B in Fig. 12–40 if the end of the cord at A is pulled down with a speed of 2 m/s.

Fig. 12–40

SOLUTION

Position-Coordinate Equation. The position of A is defined by s_A, and the position of block B is specified by s_B since point E on the pulley will have the *same motion* as the block. Both coordinates are measured from a horizontal datum passing through the *fixed* pin at pulley D. Since the system consists of *two* cords, the coordinates s_A and s_B cannot be directly related. Instead, by establishing a third position coordinate, s_C, we can now express the length of one of the cords in terms of s_B and s_C, and the length of the other cord in terms of s_A, s_B, and s_C.

Excluding the four red colored segments of the cords in Fig. 12–40, the remaining constant cord lengths l_1 and l_2 (along with the hook and link dimensions) can be expressed as

$$s_C + s_B = l_1$$
$$(s_A - s_C) + (s_B - s_C) + s_B = l_2$$

Time Derivative. The time derivative of each equation gives

$$v_C + v_B = 0$$
$$v_A - 2v_C + 2v_B = 0$$

Eliminating v_C, we obtain

$$v_A + 4v_B = 0$$

so that when $v_A = 2$ m/s (downward),

$$v_B = -0.5 \text{ m/s} = 0.5 \text{ m/s} \uparrow \qquad Ans.$$

EXAMPLE 12.24

The person at A is hoisting a safe S as shown in Fig. 12-41 by walking to the right with a constant velocity $v_A = 0.5$ m/s. Determine the velocity and acceleration of the safe when it reaches an elevation of 10 m. The rope is 30 m long and passes over a small pulley at D.

SOLUTION

Position-Coordinate Equation. This problem is unlike the previous examples since rope segment DA changes *both direction and magnitude*. However, the ends of the rope, A and C, move along the x and y axes, respectively.

These coordinates may be related since the rope has a fixed length $l = 30$ m, which at all times is equal to the length of segment DA plus CD. Using the Pythagorean theorem to determine l_{DA}, we have $l_{DA} = \sqrt{(15)^2 + x^2}$. Also, $l_{CD} = 15 - y$, and so

$$l = l_{DA} + l_{CD}$$
$$30 = \sqrt{(15)^2 + x^2} + (15 - y)$$
$$y = \sqrt{225 + x^2} - 15 \quad (1)$$

Time Derivatives. Taking the time derivative, using the chain rule (see Appendix C), where $v_S = dy/dt$ and $v_A = dx/dt$, we have

$$v_S = \frac{dy}{dt} = \left[\frac{1}{2}\frac{2x}{\sqrt{225 + x^2}}\right]\frac{dx}{dt}$$
$$= \frac{x}{\sqrt{225 + x^2}}v_A \quad (2)$$

At $y = 10$ m, x is determined from Eq. 1, i.e., $x = 20$ m. Hence, from Eq. 2 with $v_A = 0.5$ m/s,

$$v_S = \frac{20}{\sqrt{225 + (20)^2}}(0.5) = 0.4 \text{ m/s} = 400 \text{ mm/s} \uparrow \quad Ans.$$

The acceleration is determined by taking the time derivative of Eq. 2. Since v_A is constant, then $a_A = dv_A/dt = 0$, and so

$$a_S = \frac{d^2y}{dt^2} = \left[\frac{-x(dx/dt)}{(225 + x^2)^{3/2}}\right]xv_A + \left[\frac{1}{\sqrt{225 + x^2}}\right]\left(\frac{dx}{dt}\right)v_A + \left[\frac{1}{\sqrt{225 + x^2}}\right]x\frac{dv_A}{dt} = \frac{225v_A^2}{(225 + x^2)^{3/2}}$$

At $x = 20$ m, with $v_A = 0.5$ m/s, the acceleration becomes

$$a_S = \frac{225(0.5 \text{ m/s})^2}{[225 + (20 \text{ m})^2]^{3/2}} = 0.00360 \text{ m/s}^2 = 3.60 \text{ mm/s}^2 \uparrow \quad Ans.$$

NOTE: The constant velocity at A causes the other end C of the rope to have an acceleration since $\mathbf{v}_A$ causes segment DA to change its direction as well as its length.

Fig. 12-41

FUNDAMENTAL PROBLEMS

F12–39. Determine the velocity of block *D* if end *A* of the rope is pulled down with a speed of $v_A = 3$ m/s.

Prob. F12–39

F12–40. Determine the velocity of block *A* if end *B* of the rope is pulled down with a speed of 6 m/s.

Prob. F12–40

F12–41. Determine the velocity of block *A* if end *B* of the rope is pulled down with a speed of 1.5 m/s.

Prob. F12–41

F12–42. Determine the velocity of block *A* if end *F* of the rope is pulled down with a speed of $v_F = 3$ m/s.

Prob. F12–42

F12–43. Determine the velocity of car *A*. The motor *M* winds in the cable such that $v_P = 4$ m/s.

Prob. F12–43

F12–44. Determine the velocity of block *B* if block *A* moves downward with a speed of $v_A = 4$ ft/s.

Prob. F12–44

PROBLEMS

12–195. If the end of the cable at A is pulled down with a speed of 2 m/s, determine the speed at which block B rises.

Prob. 12–195

12–197. Determine the displacement of the block at B if A is pulled down 4 ft.

Prob. 12–197

*12–196. The motor at C pulls in the cable with an acceleration $a_C = (3t^2)$ m/s^2, where t is in seconds. The motor at D draws in its cable at $a_D = 5$ m/s^2. If both motors start at the same instant from rest when $d = 3$ m, determine (a) the time needed for $d = 0$, and (b) the velocities of block A relative to block B when this occurs.

Prob. 12–196

12–198. The motor draws in the cable at C with a constant velocity of $v_C = 4$ m/s. The motor draws in the cable at D with a constant acceleration of $a_D = 8$ m/s^2. If $v_D = 0$ when $t = 0$, determine (a) the time needed for block A to rise 3 m, and (b) the relative velocity of block A with respect to block B when this occurs.

Prob. 12–198

12–199. Determine the displacement of the log if the truck at C pulls the cable 4 ft to the right.

12–202. If the end A of the cable is moving at $v_A = 3$ m/s, determine the speed of block B.

Prob. 12–199

Prob. 12–202

*12–200. Determine the constant speed at which the cable at A must be drawn in by the motor in order to hoist the load 6 m in 1.5 s.

12–201. Starting from rest, the cable can be wound onto the drum of the motor at a rate of $v_A = (3t^2)$ m/s, where t is in seconds. Determine the time needed to lift the load 7 m.

12–203. Determine the time needed for the load at B to attain a speed of 10 m/s, starting from rest, if the cable is drawn into the motor with an acceleration of 3 m/s².

*12–204. The cable at A is being drawn toward the motor at $v_A = 8$ m/s. Determine the velocity of the block.

Probs. 12–200/201

Probs. 12–203/204

12–205. Determine the speed of B if A is moving downward with a speed of $v_A = 4$ m/s at the instant shown.

Prob. 12–205

12–206. Determine the speed of the block at B.

Prob. 12–206

12–207. Determine the constant speed at which the cable at A must be drawn in by the motor in order to hoist the load at B 15 ft in 5 s.

Prob. 12–207

*12–208. The hoist is used to lift the load at D. If the end A of the chain is traveling downward at $v_A = 5$ ft/s and the end B is traveling upward at $v_B = 2$ ft/s, determine the velocity of the load at D.

Prob. 12–208

12–209. The cylinder C is being lifted using the cable and pulley system shown. If point A on the cable is being drawn toward the drum with a speed of 2 m/s, determine the speed of the cylinder.

12–210. The cylinder C can be lifted with a maximum acceleration of $a_C = 3$ m/s^2 without causing the cables to fail. Determine the speed at which point A is moving toward the drum when $s = 4$ m if the cylinder is lifted from rest in the shortest time possible.

*12–212. The roller at A is moving upward with a velocity of $v_A = 3$ ft/s and has an acceleration of $a_A = 4$ ft/s^2 when $s_A = 4$ ft. Determine the velocity and acceleration of block B at this instant.

Prob. 12–212

Probs. 12–209/210

12–211. The roller at A is moving with a velocity of $v_A = 4$ m/s and has an acceleration of $a_A = 2$ m/s^2 when $x_A = 3$ m. Determine the velocity and acceleration of block B at this instant.

12–213. If the hydraulic cylinder H draws in rod BC at 2 ft/s, determine the speed of slider A.

Prob. 12–211

Prob. 12–213

12–214. At the instant shown, the car at A is traveling at 10 m/s around the curve while increasing its speed at 5 m/s^2. The car at B is traveling at 18.5 m/s along the straightaway and increasing its speed at 2 m/s^2. Determine the relative velocity and relative acceleration of A with respect to B at this instant.

*12–216. If block B is moving down with a velocity v_B and has an acceleration a_B, determine the velocity and acceleration of block A in terms of the parameters shown.

Prob. 12–216

Prob. 12–214

12–217. The block B is suspended from a cable that is attached to the block at E, wraps around three pulleys, and is tied to the back of a truck. If the truck starts from rest when x_D is zero, and moves forward with a constant acceleration of $a_D = 0.5$ m/s^2, determine the speed of the block at the instant $x_D = 2$ m. Neglect the size of the pulleys in the calculation. When $x_D = 0$, $y_C = 5$ m, so that points C and D are at the same elevation.

12–215. The motor draws in the cord at B with an acceleration of $a_B = 2$ m/s^2. When $s_A = 1.5$ m, $v_B = 6$ m/s. Determine the velocity and acceleration of the collar at this instant.

Prob. 12–215

Prob. 12–217

12.10 RELATIVE MOTION OF TWO PARTICLES USING TRANSLATING AXES

Throughout this chapter the absolute motion of a particle has been determined using a single fixed reference frame. There are many cases, however, where the path of motion for a particle is complicated, so that it may be easier to analyze the motion in parts by using two or more frames of reference. For example, the motion of a particle located at the tip of an airplane propeller, while the plane is in flight, is more easily described if one observes first the motion of the airplane from a fixed reference and then superimposes (vectorially) the circular motion of the particle measured from a reference attached to the airplane. In this section *translating frames of reference* will be considered for the analysis.

Position. Consider particles A and B, which move along the arbitrary paths shown in Fig. 12–42. The *absolute position* of each particle, $\mathbf{r}_A$ and $\mathbf{r}_B$, is measured from the common origin O of the *fixed* x, y, z reference frame. The origin of a translating frame of reference x', y', z' is attached to and moves with particle A. The position of B measured relative to A is denoted by the *relative-position vector* $\mathbf{r}_{B/A}$. The three vectors shown in Fig. 12–42 are related by vector addition.

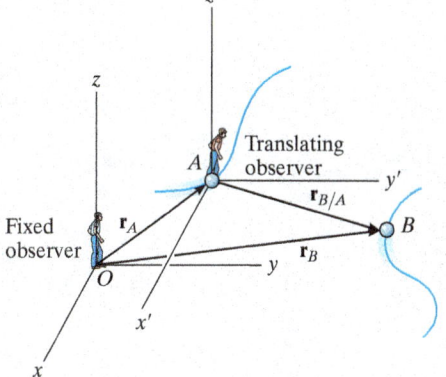

Fig. 12–42

$$\mathbf{r}_B = \mathbf{r}_A + \mathbf{r}_{B/A} \qquad (12\text{–}33)$$

Velocity. The equation that relates the velocities of the particles is determined by taking the time derivative of this equation; i.e.,

$$\mathbf{v}_B = \mathbf{v}_A + \mathbf{v}_{B/A} \qquad (12\text{–}34)$$

Here $\mathbf{v}_B = d\mathbf{r}_B/dt$ and $\mathbf{v}_A = d\mathbf{r}_A/dt$ refer to *absolute velocities*, since they are observed from the fixed frame; whereas the *relative velocity* $\mathbf{v}_{B/A} = d\mathbf{r}_{B/A}/dt$ is observed from the translating frame. It is important to note that since the x', y', z' axes translate, the *components* of $\mathbf{r}_{B/A}$ will *not* change direction and therefore the time derivative of these components will only have to account for the change in their magnitudes. Equation 12–34 therefore states that the velocity of B is equal to the velocity of A plus (vectorially) the velocity of "B with respect to A," as measured by the *translating observer* fixed to the x', y', z' reference frame.

Acceleration. The time derivative of the velocity equation yields a similar relationship between the *absolute* and *relative accelerations* of particles A and B.

$$\mathbf{a}_B = \mathbf{a}_A + \mathbf{a}_{B/A} \tag{12-35}$$

Here $\mathbf{a}_{B/A}$ is the acceleration of B as seen by the observer located at A and translating with the x', y', z' reference frame.*

PROCEDURE FOR ANALYSIS

- Specify the particle (A) that is the origin for the translating x', y', z' axes. Usually this point has a *known* velocity or acceleration.

- Since vector addition forms a triangle, there can be at most *two unknowns*, represented by the magnitudes and/or directions of the vector quantities.

- These unknowns can be solved for either graphically, using trigonometry (law of sines, law of cosines), or, as we will show, by resolving each of the three vectors into rectangular or Cartesian components, thereby generating a set of scalar equations.

The pilots of these close-flying planes must be aware of their relative positions and velocities at all times in order to avoid a collision.

---

*An easy way to remember the setup of these equations is to note the "cancellation" of the subscript A between the two terms, e.g., $\mathbf{a}_B = \mathbf{a}_{\cancel{A}} + \mathbf{a}_{B/\cancel{A}}$.

EXAMPLE 12.25

A train travels at a constant speed of 60 mi/h and crosses over a road as shown in Fig. 12–43a. If the automobile A is traveling at 45 mi/h along the road, determine the magnitude and direction of the velocity of the train relative to the automobile.

SOLUTION I

Vector Analysis. The relative velocity $\mathbf{v}_{T/A}$ is measured from the translating x', y' axes attached to the automobile, Fig. 12–43a. It is determined from $\mathbf{v}_T = \mathbf{v}_A + \mathbf{v}_{T/A}$. Since $\mathbf{v}_T$ and $\mathbf{v}_A$ are known in *both* magnitude and direction, the unknowns become the x and y components of $\mathbf{v}_{T/A}$. Using the x, y axes in Fig. 12–43a, we have

$$\mathbf{v}_T = \mathbf{v}_A + \mathbf{v}_{T/A}$$
$$60\mathbf{i} = (45\cos 45°\mathbf{i} + 45\sin 45°\mathbf{j}) + \mathbf{v}_{T/A}$$
$$\mathbf{v}_{T/A} = \{28.2\mathbf{i} - 31.8\mathbf{j}\} \text{ mi/h}$$

The magnitude of $\mathbf{v}_{T/A}$ is thus

$$v_{T/A} = \sqrt{(28.2)^2 + (-31.8)^2} = 42.5 \text{ mi/h} \qquad Ans.$$

From the direction of each component, Fig. 12–43b, the direction of $\mathbf{v}_{T/A}$ is

$$\tan \theta = \frac{(v_{T/A})_y}{(v_{T/A})_x} = \frac{31.8}{28.2}$$
$$\theta = 48.5° \searrow \qquad Ans.$$

SOLUTION II

Scalar Analysis. The unknown components of $\mathbf{v}_{T/A}$ can also be determined by applying a scalar analysis. We will assume these components act in the *positive* x and y directions. Thus,

$$\mathbf{v}_T = \mathbf{v}_A + \mathbf{v}_{T/A}$$

$$\begin{bmatrix} 60 \text{ mi/h} \\ \rightarrow \end{bmatrix} = \begin{bmatrix} 45 \text{ mi/h} \\ \nearrow 45° \end{bmatrix} + \begin{bmatrix} (v_{T/A})_x \\ \rightarrow \end{bmatrix} + \begin{bmatrix} (v_{T/A})_y \\ \uparrow \end{bmatrix}$$

Resolving each vector into its x and y components yields

$(\xrightarrow{+})$ $\qquad 60 = 45\cos 45° + (v_{T/A})_x + 0$
$(+\uparrow)$ $\qquad 0 = 45\sin 45° + 0 + (v_{T/A})_y$

Solving, we obtain the previous results,

$(v_{T/A})_x = 28.2 \text{ mi/h} = 28.2 \text{ mi/h} \rightarrow$
$(v_{T/A})_y = -31.8 \text{ mi/h} = 31.8 \text{ mi/h} \downarrow$

(a)

(b)

(c)

Fig. 12–43

EXAMPLE 12.26

Plane A in Fig. 12–44a is flying along a straight-line path, whereas plane B is flying along a circular path having a radius of curvature of $\rho_B = 400$ km. Determine the velocity and acceleration of B as measured by the pilot of A.

SOLUTION I

Velocity. Since the motion relative to plane A is to be determined, the *translating frame of reference* x', y' is attached to it, Fig. 12–44a. Applying the relative-velocity equation in scalar form, since the velocity vectors of both planes are parallel at the instant shown, we have

$$(+\uparrow) \quad v_B = v_A + v_{B/A}$$

$$600 \text{ km/h} = 700 \text{ km/h} + v_{B/A}$$

$$v_{B/A} = -100 \text{ km/h} = 100 \text{ km/h} \downarrow \quad \textit{Ans.}$$

The vector addition is shown in Fig. 12–44b.

Acceleration. Plane B has both tangential and normal components of acceleration since it is flying along a *curved path*. From Eq. 12–20, the magnitude of the normal component is

$$(a_B)_n = \frac{v_B^2}{\rho} = \frac{(600 \text{ km/h})^2}{400 \text{ km}} = 900 \text{ km/h}^2$$

Applying the relative-acceleration equation, we have

$$\mathbf{a}_B = \mathbf{a}_A + \mathbf{a}_{B/A}$$

$$900\mathbf{i} - 100\mathbf{j} = 50\mathbf{j} + \mathbf{a}_{B/A}$$

Thus,

$$\mathbf{a}_{B/A} = \{900\mathbf{i} - 150\mathbf{j}\} \text{ km/h}^2$$

From Fig. 12–44c, the magnitude and direction of $\mathbf{a}_{B/A}$ are therefore

$$a_{B/A} = 912 \text{ km/h}^2 \quad \theta = \tan^{-1}\frac{150}{900} = 9.46° \quad \textit{Ans.}$$

NOTE: The solution to this problem was possible using a translating frame of reference, since the pilot in plane A is "translating." Observation of the motion of plane A with respect to the pilot of plane B, however, must be obtained using a *rotating* set of axes attached to plane B. (This assumes, of course, that the pilot of B is fixed in the rotating frame, so she does not turn her eyes to follow the motion of A.) The analysis for this case is more difficult and is given in Example 16.21.

Fig. 12–44

EXAMPLE 12.27

At the instant shown in Fig. 12–45a, cars A and B are traveling with speeds of 18 m/s and 12 m/s, respectively. Also at this instant, A has a decrease in speed of 2 m/s^2, and B has an increase in speed of 3 m/s^2. Determine the velocity and acceleration of B with respect to A.

SOLUTION

Velocity. The fixed x, y axes are established at an arbitrary point on the ground, and the translating x', y' axes are attached to car A, Fig. 12–45a. Why? The relative velocity is determined from $\mathbf{v}_B = \mathbf{v}_A + \mathbf{v}_{B/A}$. What are the two unknowns? Using a Cartesian vector analysis, we have

$$\mathbf{v}_B = \mathbf{v}_A + \mathbf{v}_{B/A}$$

$$-12\mathbf{j} = (-18\cos 60°\mathbf{i} - 18\sin 60°\mathbf{j}) + \mathbf{v}_{B/A}$$

$$\mathbf{v}_{B/A} = \{9\mathbf{i} + 3.588\mathbf{j}\} \text{ m/s}$$

Thus,

$$v_{B/A} = \sqrt{(9)^2 + (3.588)^2} = 9.69 \text{ m/s} \qquad Ans.$$

Noting that $\mathbf{v}_{B/A}$ has $+\mathbf{i}$ and $+\mathbf{j}$ components, Fig. 12–45b, its direction is

$$\tan\theta = \frac{(v_{B/A})_y}{(v_{B/A})_x} = \frac{3.588}{9}$$

$$\theta = 21.7° \qquad Ans.$$

Acceleration. Car B has both tangential and normal components of acceleration. Why? The magnitude of the normal component is

$$(a_B)_n = \frac{v_B^2}{\rho} = \frac{(12 \text{ m/s})^2}{100 \text{ m}} = 1.440 \text{ m/s}^2$$

Applying the equation for relative acceleration yields

$$\mathbf{a}_B = \mathbf{a}_A + \mathbf{a}_{B/A}$$

$$(-1.440\mathbf{i} - 3\mathbf{j}) = (2\cos 60°\mathbf{i} + 2\sin 60°\mathbf{j}) + \mathbf{a}_{B/A}$$

$$\mathbf{a}_{B/A} = \{-2.440\mathbf{i} - 4.732\mathbf{j}\} \text{ m/s}^2$$

Here $\mathbf{a}_{B/A}$ has $-\mathbf{i}$ and $-\mathbf{j}$ components. Thus, from Fig. 12–45c,

$$a_{B/A} = \sqrt{(2.440)^2 + (4.732)^2} = 5.32 \text{ m/s}^2 \qquad Ans.$$

$$\tan\phi = \frac{(a_{B/A})_y}{(a_{B/A})_x} = \frac{4.732}{2.440}$$

$$\phi = 62.7° \qquad Ans.$$

NOTE: Is it possible to obtain the relative acceleration of $\mathbf{a}_{A/B}$ using this method? Refer to the comment made at the end of Example 12–26.

(a)

(b)

(c)

Fig. 12–45

FUNDAMENTAL PROBLEMS

F12–45. Car A is traveling with a constant speed of 80 km/h due north, while car B is traveling with a constant speed of 100 km/h due east. Determine the velocity of car B relative to car A.

Prob. F12–45

F12–46. Two planes A and B are traveling with the constant velocities shown. Determine the magnitude and direction of the velocity of B relative to A.

Prob. F12–46

F12–47. The boats A and B travel with constant speeds of $v_A = 15$ m/s and $v_B = 10$ m/s when they leave the pier at O at the same time. Determine the distance between them when $t = 4$ s.

Prob. F12–47

F12–48. Cars A and B are traveling at the speeds shown. If B is accelerating at 1200 km/h^2 while A maintains a constant speed, determine the velocity and acceleration of A with respect to B.

Prob. F12–48

PROBLEMS

12–218. Two planes, A and B, are flying at the same altitude. If their velocities are $v_A = 500$ km/h and $v_B = 700$ km/h such that the angle between their straight-line courses is $\theta = 60°$, determine the velocity of plane B with respect to plane A.

Prob. 12–218

12–219. At the instant shown, cars A and B are traveling at speeds of 55 mi/h and 40 mi/h, respectively. If B is increasing its speed by 1200 mi/h², while A maintains a constant speed, determine the velocity and acceleration of B with respect to A. Car B moves along a curve having a radius of curvature of 0.5 mi.

Prob. 12–219

*12–220. The boat can travel with a speed of 16 km/h in still water. The point of destination is located along the dashed line. If the water is moving at 4 km/h, determine the bearing angle θ at which the boat must travel to stay on course.

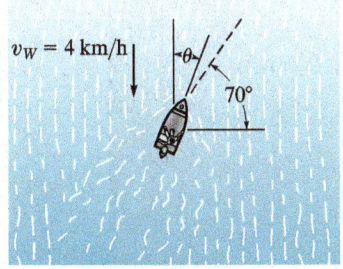

Prob. 12–220

12–221. Two boats leave the pier P at the same time and travel in the directions shown. If $v_A = 40$ ft/s and $v_B = 30$ ft/s, determine the magnitude of the velocity of boat A relative to boat B. How long after leaving the pier will the boats be 1500 ft apart?

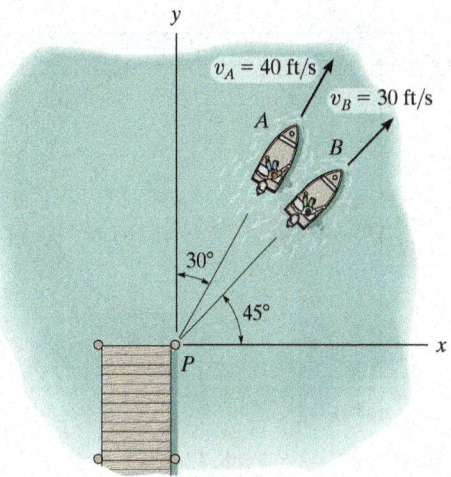

Prob. 12–221

12–222. Car A travels along a straight road at a speed of 25 m/s while accelerating at 1.5 m/s². At this same instant car C is traveling along the straight road with a speed of 30 m/s while decelerating at 3 m/s². Determine the velocity and acceleration of car A relative to car C.

12–223. Car B is traveling along the curved road with a speed of 15 m/s while decreasing its speed at 2 m/s². At this same instant car C is traveling along the straight road with a speed of 30 m/s while decelerating at 3 m/s². Determine the velocity and acceleration of car B relative to car C.

Probs. 12–222/223

*12–224. At the instant shown, car A has a speed of 20 km/h, which is being increased at the rate of 300 km/h^2 as the car enters the expressway. At the same instant, car B is decelerating at 250 km/h^2 while traveling forward at 100 km/h. Determine the velocity and acceleration of A with respect to B.

12–226. At the instant shown, the bicyclist at A is traveling at 7 m/s around the curve on the race track while increasing the bicycle's speed at 0.5 m/s^2. The bicyclist at B is traveling at 8 m/s along the straightaway and increasing the bicycle's speed at 0.7 m/s^2. Determine the relative velocity and relative acceleration of A with respect to B at this istant.

Prob. 12–224

Prob. 12–226

12–225. Cars A and B are traveling around the circular race track. At the instant shown, A has a speed of 90 ft/s and is increasing its speed at the rate of 15 ft/s^2, whereas B has a speed of 105 ft/s and is decreasing its speed at 25 ft/s^2. Determine the relative velocity and relative acceleration of car A with respect to car B at this instant.

12–227. At the instant shown, cars A and B are traveling at velocities of 40 m/s and 30 m/s, respectively. If B is increasing its velocity by 2 m/s^2, while A maintains a constant velocity, determine the velocity and acceleration of B with respect to A. The radius of curvature at B is $\rho_B = 200$ m.

*12–228. At the instant shown, cars A and B are traveling at velocities of 40 m/s and 30 m/s, respectively. If A is increasing its speed at 4 m/s^2, whereas the speed of B is decreasing at 3 m/s^2, determine the velocity and acceleration of B with respect to A. The radius of curvature at B is $\rho_B = 200$ m.

Prob. 12–225

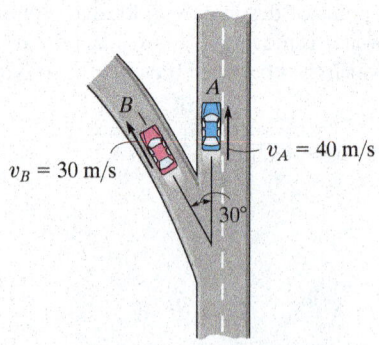

Probs. 12–227/228

12–229. A man can row a boat at 5 m/s in still water. He wishes to cross a 50-m-wide river to point B, 50 m downstream. If the river flows with a velocity of 2 m/s, determine the speed of the boat and the time needed to make the crossing.

12–231. At the instant shown car A is traveling with a velocity of 30 m/s and has an acceleration of 2 m/s^2 along the highway. At the same instant B is traveling on the trumpet interchange curve with a speed of 15 m/s, which is decreasing at 0.8 m/s^2. Determine the relative velocity and relative acceleration of B with respect to A at this instant.

Prob. 12–229

Prob. 12–231

12–230. A man can swim at 4 ft/s in still water. He wishes to cross the 40-ft-wide river to point B, 30 ft downstream. If the river flows with a velocity of 2 ft/s, determine the speed of the man and the time needed to make the crossing. *Note:* While in the water he must not direct himself toward point B to reach this point. Why?

***12–232.** The football player at A throws the ball in the y–z plane at a speed $v_A = 50$ ft/s and an angle $\theta_A = 60°$ with the horizontal. At the instant the ball is thrown, the player is at B and is running with constant speed along the line BC in order to catch it. Determine this speed, v_B, so that he makes the catch at the same elevation from which the ball was thrown.

Prob. 12–230

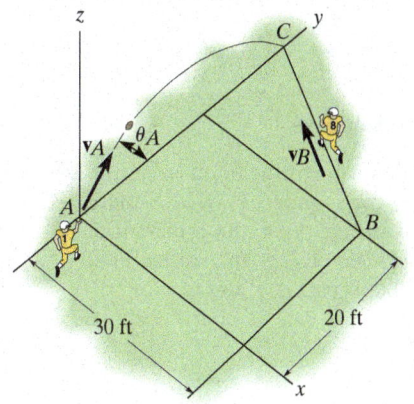

Prob. 12–232

CHAPTER REVIEW

Rectilinear Kinematics

Rectilinear kinematics refers to motion along a straight line. A position coordinate s specifies the location of a particle on the line, and the displacement Δs is the change in its position.

The average velocity is a vector quantity, defined as the displacement divided by the time interval.

$$v_{avg} = \frac{\Delta s}{\Delta t}$$

The average speed is a scalar, and is the total distance traveled divided by the time of travel.

$$(v_{sp})_{avg} = \frac{s_T}{\Delta t}$$

The time, position, velocity, and acceleration are related by three differential equations.

$$a = \frac{dv}{dt}, \quad v = \frac{ds}{dt}, \quad a\,ds = v\,dv$$

If the acceleration is known to be *constant*, then the differential equations relating time, position, velocity, and acceleration can be integrated.

$$v = v_0 + a_c t$$
$$s = s_0 + v_0 t + \tfrac{1}{2} a_c t^2$$
$$v^2 = v_0^2 + 2a_c(s - s_0)$$

Graphical Solutions

If the motion is erratic, then it can be described by a graph. If one of these graphs is given, then the others can be established using the differential relations between a, v, s, and t.

$$a = \frac{dv}{dt},$$
$$v = \frac{ds}{dt},$$
$$a\,ds = v\,dv$$

Curvilinear Motion, x, y, z

Curvilinear motion along the path can be resolved into rectilinear motion along the x, the y, and the z axes. The equation of the path is used to relate the motion along each axis.

$v_x = \dot{x} \quad a_x = \dot{v}_x$

$v_y = \dot{y} \quad a_y = \dot{v}_y$

$v_z = \dot{z} \quad a_z = \dot{v}_z$

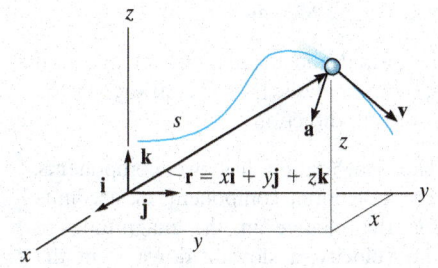

Projectile Motion

Free-flight motion of a projectile follows a parabolic path. It has a constant velocity in the horizontal direction, and a constant downward acceleration of $g = 9.81$ m/s² or 32.2 ft/s² in the vertical direction. Any two of the three equations for constant acceleration apply in the vertical direction, and in the horizontal direction only one equation applies.

$(+\uparrow) \quad v_y = (v_0)_y + a_c t$

$(+\uparrow) \quad y = y_0 + (v_0)_y t + \tfrac{1}{2} a_c t^2$

$(+\uparrow) \quad v_y^2 = (v_0)_y^2 + 2 a_c (y - y_0)$

$(\xrightarrow{+}) \quad x = x_0 + (v_0)_x t$

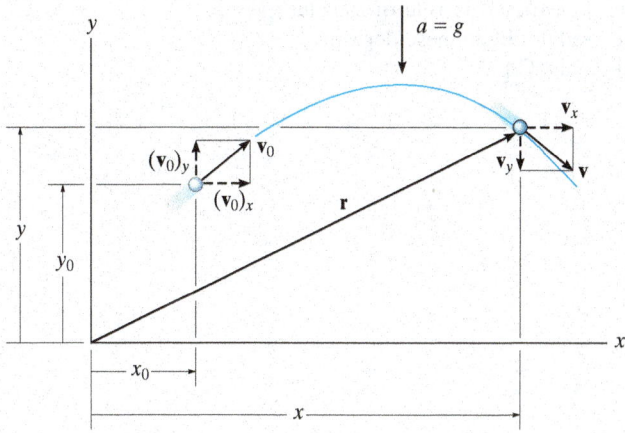

Curvilinear Motion n, t

If normal and tangential axes are used for the analysis, then **v** is always in the positive t direction.

The acceleration has two components. The tangential component, $\mathbf{a}_t$, accounts for the change in the magnitude of the velocity; a slowing down is in the negative t direction, and a speeding up is in the positive t direction. The normal component $\mathbf{a}_n$ accounts for the change in the direction of the velocity. This component is always in the positive n direction.

$$a_t = \dot{v} \quad \text{or} \quad a_t\, ds = v\, dv$$

$$a_n = \frac{v^2}{\rho}$$

Curvilinear Motion r, θ

If the path of motion is expressed in polar coordinates, then the velocity and acceleration components can be related to the time derivatives of r and θ.

To determine the components, it is necessary to know $r, \dot{r}, \ddot{r}, \theta, \dot{\theta}$ at the instant considered. If the path $r = f(\theta)$ is given, then the chain rule of calculus must be used to obtain these derivatives. (See Appendix C.)

$$v_r = \dot{r}$$
$$v_\theta = r\dot{\theta}$$

$$a_r = \ddot{r} - r\dot{\theta}^2$$
$$a_\theta = r\ddot{\theta} + 2\dot{r}\dot{\theta}$$

Velocity

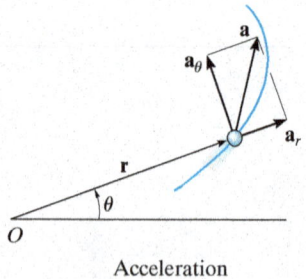

Acceleration

Absolute Dependent Motion of Two Particles

The dependent motion of blocks that are suspended from pulleys and cables can be related by the geometry of the system. This is done by first establishing position coordinates, measured from a fixed origin and directed along the line of motion of each block.

Using geometry and/or trigonometry, the coordinates are then related to the cable length in order to formulate a position coordinate equation.

The time derivatives of this equation give relationships between the velocities and the accelerations of the blocks.

$$2s_B + h + s_A = l$$
$$2v_B = -v_A$$
$$2a_B = -a_A$$

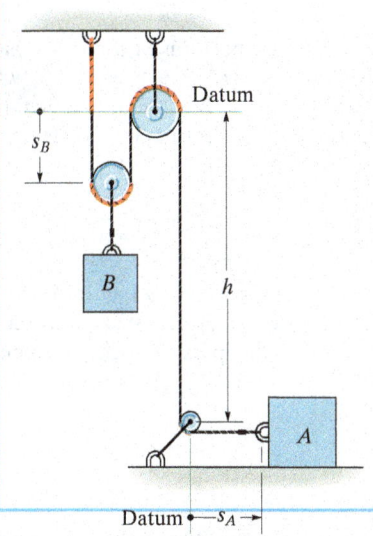

Relative-Motion Analysis Using Translating Axes

If two particles A and B undergo independent motions, then these motions can be related to their relative motion using a *translating set of axes* attached to one of the particles (A).

For planar motion, each vector equation produces two scalar component equations, one in the x, and the other in the y direction. For solution, the vectors can be expressed in Cartesian form, or the x and y scalar components can be written directly.

$$\mathbf{r}_B = \mathbf{r}_A + \mathbf{r}_{B/A}$$
$$\mathbf{v}_B = \mathbf{v}_A + \mathbf{v}_{B/A}$$
$$\mathbf{a}_B = \mathbf{a}_A + \mathbf{a}_{B/A}$$

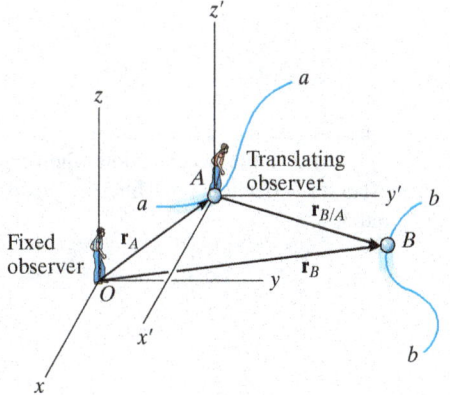

REVIEW PROBLEMS

R12–1. The position of a particle along a straight line is given by $s = (t^3 - 9t^2 + 15t)$ ft, where t is in seconds. Determine its maximum acceleration and maximum velocity during the time interval $0 \le t \le 10$ s.

R12–2. If a particle has an initial velocity $v_0 = 12$ ft/s to the right, and a constant acceleration of 2 ft/s² to the left, determine the particle's displacement in 10 s. Originally $s_0 = 0$.

R12–3. A projectile, initially at the origin, moves along a straight-line path through a fluid medium such that its velocity is $v = 1800(1 - e^{-0.3t})$ mm/s where t is in seconds. Determine the displacement of the projectile during the first 3 s.

R12–4. The v–t graph of a car while traveling along a road is shown. Determine the acceleration when $t = 2.5$ s, 10 s, and 25 s. Also if $s = 0$ when $t = 0$, find the position when $t = 5$ s, 20 s, and 30 s.

R12–5. A car traveling along the straight portions of the road has the velocities indicated in the figure when it arrives at points A, B, and C. If it takes 3 s to go from A to B, and then 5 s to go from B to C, determine the average acceleration between points A and B and between points A and C.

Prob. R12–5

R12–6. From a videotape, it was observed that a player kicked a football 126 ft during a measured time of 3.6 seconds. Determine the initial speed of the ball and the angle θ at which it was kicked.

Prob. R12–4

Prob. R12–6

R12–7. The truck travels in a circular path having a radius of 50 m at a speed of $v = 4$ m/s. For a short distance from $s = 0$, its speed is increased by $\dot{v} = (0.05s)$ m/s^2, where s is in meters. Determine its speed and the magnitude of its acceleration when it has moved $s = 10$ m.

Prob. R12–7

R12–8. Car B turns such that its speed is increased by $(a_t)_B = (0.5e^t)$ m/s^2, where t is in seconds. If the car starts from rest when $\theta = 0°$, determine the magnitudes of its velocity and acceleration when $t = 2$ s. Neglect the size of the car.

Prob. R12–8

R12–9. A particle is moving along a circular path of 2-m radius such that its position as a function of time is given by $\theta = (5t^2)$ rad, where t is in seconds. Determine the magnitude of the particle's acceleration when $\theta = 30°$. The particle starts from rest when $\theta = 0°$.

R12–10. Determine the time needed for the load at B to attain a speed of 8 m/s, starting from rest, if the cable is drawn into the motor with an acceleration of 0.2 m/s^2.

Prob. R12–10

R12–11. Two planes, A and B, are flying at the same altitude. If their velocities are $v_A = 600$ km/h and $v_B = 500$ km/h such that the angle between their straight-line courses is $\theta = 75°$, determine the velocity of plane B with respect to plane A.

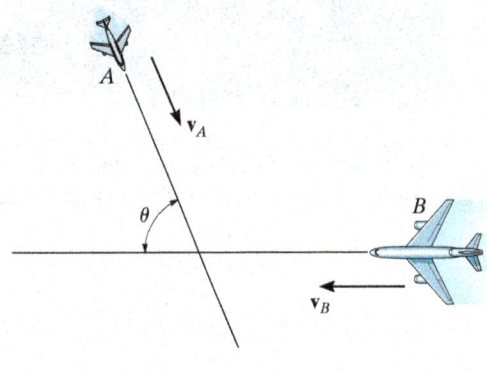

Prob. R12–11

CHAPTER 13

The design of conveyors for this bakery requires knowing the forces that act on these loaves of bread and their speed as they move along.

KINETICS OF A PARTICLE: FORCE AND ACCELERATION

CHAPTER OBJECTIVES

- To state Newton's Second Law of Motion and to define mass and weight.
- To analyze the accelerated motion of a particle using the equation of motion with different coordinate systems.
- To investigate central-force motion and apply it to problems involving space mechanics.

13.1 NEWTON'S SECOND LAW OF MOTION

Kinetics is a branch of dynamics that deals with the relationship between the change in motion of a body and the forces that cause this change. The basis for kinetics is Newton's second law, which states that when an *unbalanced force* acts on a particle, the particle will *accelerate* in the direction of the force with a magnitude that is proportional to the force.

We can determine the constant of proportionality by applying a known force **F** to a particle, and then measuring its acceleration **a**. Since the force and acceleration are directly proportional, then the constant of proportionality, m, is determined from the ratio $m = F/a$. This positive scalar m is called the ***mass*** of the particle. Since it is constant, m provides a quantitative measure of the resistance of the particle to a change in its velocity, that is, its ***inertia***.

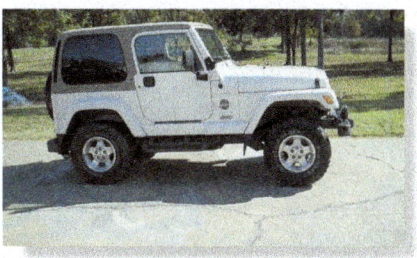

The jeep leans backward due to its inertia, which resists its forward acceleration.

For a particle of mass m, Newton's second law of motion may therefore be written in mathematical form as

$$\mathbf{F} = m\mathbf{a}$$

This equation is referred to as the **equation of motion**, and it is one of the most important formulations in mechanics.* In 1905, however, Albert Einstein developed the theory of relativity and placed limitations on the use of Newton's second law, especially when the particle's speed approaches the speed of light (0.3 Gm/s). Developments of the theory of quantum mechanics by Erwin Schrödinger and others indicate further that conclusions drawn from using this equation are also invalid when particles approach the size of an atom and move close to one another. For the most part, however, these requirements regarding particle speed and size are not encountered in engineering problems, so their effects will not be considered in this book.

Newton's Law of Gravitational Attraction.
Shortly after formulating his three laws of motion, Newton postulated a law governing the mutual attraction between any two particles. In mathematical form this law can be expressed as

$$F = G\frac{m_1 m_2}{r^2} \tag{13–1}$$

where

F = force of attraction between the two particles
G = universal constant of gravitation; according to experimental evidence $G = 66.73(10^{-12})$ m^3/(kg·s^2)
m_1, m_2 = mass of each of the two particles
r = distance between the centers of the two particles

---

*Since m is constant, we can also write the equation of motion in the form $\mathbf{F} = d(m\mathbf{v})/dt$, where $m\mathbf{v}$ is the particle's linear momentum. In other words, we can state that the unbalanced force acting on the particle is proportional to the time rate of change of the particle's linear momentum.

In the case of a particle located at or near the surface of the earth, the only gravitational force having any sizable magnitude is that between the earth and the particle. This force is termed the "weight" and, for our purpose, it will be the only gravitational force considered.

From Eq. 13–1, we can develop a general expression for finding the weight W of a particle having a mass $m_1 = m$. Let $m_2 = M_e$ be the mass of the earth and r the distance between the earth's center and the particle. Then, if we let $g = GM_e/r^2$, we have

$$W = mg$$

By comparison with $F = ma$, we see that g represents the acceleration due to gravity. For most engineering calculations we assume g is measured at a point on the surface of the earth at sea level, and at a latitude of 45°, which is considered the "standard location." Here $g = 9.81 \text{ m/s}^2 = 32.2 \text{ ft/s}^2$.

In the SI system the mass of the body is specified in kilograms, and so the weight must be calculated using the above equation, Fig. 13–1a. Thus,

$$W = mg \text{ (N)} \quad (g = 9.81 \text{ m/s}^2) \qquad (13\text{–}2)$$

SI system
(a)

Thus, a 1-kg body weighs 9.81 N; a 2-kg body weighs 19.62 N; and so on.

In the FPS system the weight of the body is specified in pounds, and the mass is measured in slugs, a term derived from "sluggish", which refers to the body's inertia. It must be calculated, Fig. 13–1b, using

$$m = \frac{W}{g} \text{ (slug)} \quad (g = 32.2 \text{ ft/s}^2) \qquad (13\text{–}3)$$

FPS system
(b)

Fig. 13–1

Therefore, a 32.2-lb body has a mass of 1 slug; a 64.4-lb body has a mass of 2 slugs; and so on.

112   CHAPTER 13   KINETICS OF A PARTICLE: FORCE AND ACCELERATION

(a)

Free-body diagram   Kinetic diagram

(b)   (c)

Fig. 13–2

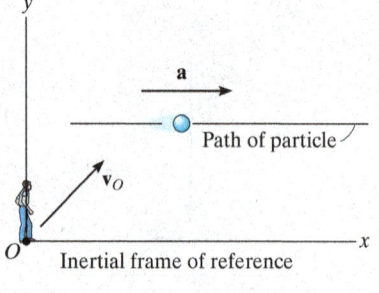

Inertial frame of reference

Fig. 13–3

13.2 THE EQUATION OF MOTION

When more than one force acts on a particle, the resultant force is determined by a vector summation of all the forces; i.e., $\mathbf{F}_R = \Sigma\mathbf{F}$. For this more general case, the equation of motion may be written as

$$\Sigma\mathbf{F} = m\mathbf{a} \qquad (13\text{–}4)$$

For example, consider the particle shown in Fig. 13–2a, which has a mass m and is subjected to the action of two forces, $\mathbf{F}_1$ and $\mathbf{F}_2$. We can graphically account for the magnitude and direction of each force acting on the particle by drawing the particle's *free-body diagram*, Fig. 13–2b. Since the *resultant* of these forces *produces* the vector $m\mathbf{a}$, its magnitude and direction can be represented graphically on the **kinetic diagram**, shown in Fig. 13–2c. The equal sign written between the diagrams symbolizes the *graphical* equivalency between the free-body diagram and the kinetic diagram; i.e., $\Sigma\mathbf{F} = m\mathbf{a}$.*

Inertial Reference Frame. When applying the equation of motion, it is important that the acceleration of the particle be measured from a reference frame that is *either fixed or translates with a constant velocity*. In this way, the observer will not accelerate and measurements of the particle's acceleration will be the *same* from *any reference* of this type. Such a frame of reference is commonly known as a **Newtonian** or *inertial reference frame*, Fig. 13–3.

When studying the motions of rockets and satellites, it is justifiable to consider the inertial reference frame as fixed to the stars, whereas dynamics problems concerned with motions on or near the surface of the earth may be solved by using an inertial frame which is assumed fixed to the earth. Even though the earth both rotates about its own axis and revolves about the sun, the accelerations created by these rotations are relatively small and so they can be neglected for most applications.

---

*The equation of motion can also be rewritten in the form $\Sigma\mathbf{F} - m\mathbf{a} = \mathbf{0}$. The vector $-m\mathbf{a}$ is referred to as the *inertia force vector*. If it is treated in the same way as a "force vector," then the state of "equilibrium" created is referred to as **dynamic equilibrium**. This method of application, which will not be used in this book, is often referred to as the *D'Alembert principle*, named after the French mathematician Jean le Rond d'Alembert.

We are all familiar with the sensation one feels when sitting in a car that is subjected to a forward acceleration. Often people think this is caused by a "force" which acts on them and tends to push them back in their seats; however, this is not the case. Instead, this sensation occurs due to their inertia or the resistance of their mass to a change in velocity.

Consider the passenger who is strapped to the seat of a rocket sled. Provided the sled is at rest or is moving with constant velocity, then no force is exerted on his back as shown on his free-body diagram.

When the thrust of the rocket engine causes the sled to accelerate, then the seat upon which he is sitting exerts a force **F** on him which pushes him forward with the sled. In the photo, notice that the inertia of his head resists this change in motion (acceleration), and so his head moves back against the seat and his face, which is nonrigid, tends to distort backward.

Upon deceleration the force of the seatbelt **F'** tends to pull his body to a stop, but his head leaves contact with the back of the seat and his face distorts forward, again due to his inertia or tendency to continue to move forward. No force is pulling him forward, although this is the sensation he receives.

13.3 EQUATION OF MOTION FOR A SYSTEM OF PARTICLES

The equation of motion will now be extended to include a system of particles enclosed within a region in space, as shown in Fig. 13–4a. At the instant considered, the arbitrary *i*-th particle, having a mass m_i, is subjected to both a resultant internal force, $\mathbf{f}_i$, and a resultant external force, $\mathbf{F}_i$. The *internal force* is the resultant of all the forces the other particles exert on the *i*th particle. The *resultant external force* represents, for example, the effect of gravitational, electrical, magnetic, or contact forces between the *i*th particle and adjacent bodies or particles not included within the system.

The free-body and kinetic diagrams for the *i*th particle are shown in Fig. 13–4b. Applying the equation of motion,

$$\Sigma \mathbf{F} = m\mathbf{a}; \qquad \mathbf{F}_i + \mathbf{f}_i = m_i \mathbf{a}_i$$

When the equation of motion is applied to each of the other particles of the system, similar equations will result, and if all these equations are added together, we obtain

$$\Sigma \mathbf{F}_i + \Sigma \mathbf{f}_i = \Sigma m_i \mathbf{a}_i$$

Inertial coordinate system
(a)

Free-body diagram

Kinetic diagram

(b)

Fig. 13–4

The summation of the internal forces, if carried out, will equal zero, since internal forces between any two particles occur in equal but opposite collinear pairs. Therefore,

$$\Sigma \mathbf{F}_i = \Sigma m_i \mathbf{a}_i \qquad (13\text{--}5)$$

If $\mathbf{r}_G$ is the position vector which locates the center of mass G of the particles, Fig. 13–4a, then by definition of the center of mass, $m\mathbf{r}_G = \Sigma m_i \mathbf{r}_i$, where $m = \Sigma m_i$ is the total mass of all the particles. Differentiating this equation twice with respect to time, assuming that no mass is entering or leaving the system, we get $m\mathbf{a}_G = \Sigma m_i \mathbf{a}_i$. Substituting this result into Eq. 13–5, we obtain

$$\boxed{\Sigma \mathbf{F} = m\mathbf{a}_G} \qquad (13\text{--}6)$$

Considering a body composed of this system of particles, we can therefore state that the sum of the external forces acting on the body is equal to the mass of the body times the acceleration of its center of mass G.

IMPORTANT POINTS

- The equation of motion is based on experimental evidence and is valid only when applied within an inertial frame of reference.
- The equation of motion states that the *unbalanced force* on a particle causes it to *accelerate*.
- An inertial frame of reference does not rotate, rather its axes either translate with constant velocity or are at rest.
- Mass is a property of matter that provides a quantitative measure of its resistance to a change in velocity. It is an absolute quantity and so it does not change from one location to another.
- Weight is a force that is caused by the earth's gravitation. It is not absolute; rather it depends on the altitude of the mass from the earth's surface.

13.4 EQUATIONS OF MOTION: RECTANGULAR COORDINATES

When a particle moves relative to an inertial x, y, z frame of reference, the forces acting on the particle, as well as its acceleration, can be expressed in terms of their **i, j, k** components, Fig. 13–5. Applying the equation of motion, we have

$$\Sigma \mathbf{F} = m\mathbf{a}; \qquad \Sigma F_x \mathbf{i} + \Sigma F_y \mathbf{j} + \Sigma F_z \mathbf{k} = m(a_x \mathbf{i} + a_y \mathbf{j} + a_z \mathbf{k})$$

To satisfy this equation, the respective **i, j, k** components on the left side must equal the corresponding components on the right side. As a result, we may write the following three scalar equations:

$$\begin{aligned} \Sigma F_x &= ma_x \\ \Sigma F_y &= ma_y \\ \Sigma F_z &= ma_z \end{aligned} \qquad (13\text{--}7)$$

If the particle is constrained to move only in the x–y plane, then only the first two of these equations are used to specify the motion.

Fig. 13–5

PROCEDURE FOR ANALYSIS

Application of the equation of motion requires the following steps.

Free-Body Diagram.

- Select the inertial coordinate system. Choose rectangular or x, y, z coordinates if the particle has *rectilinear motion*.

- Draw the particle's free-body diagram in order to account for *all the forces* ($\Sigma \mathbf{F}$) which act on the particle.

- The direction and sense of the particle's acceleration **a** should also be established. If the sense is unknown, for mathematical convenience assume its components act along the *positive* coordinate axes.

- The acceleration can be represented as the *ma* vector on the kinetic diagram.

- Identify the unknowns in the problem.

Equations of Motion.

- If the forces can be resolved directly from the free-body diagram, apply the equations of motion in their scalar component form.

- If the geometry of the problem appears complicated, which often occurs in three dimensions, Cartesian vector analysis can be used for the solution.

- *Friction.* If the particle slides on a rough surface, then use the *friction equation*, which relates the friction and normal forces with the coefficient of kinetic friction, i.e., $F_f = \mu_k N$.

- *Spring.* If the particle is connected to an *elastic spring* having negligible mass, the spring force F_s can be related to the stretch or compression s of the spring by the equation $F_s = ks$. Here k is the spring's stiffness measured as a force per unit length.

Kinematics.

- Once the particle's acceleration is known, its velocity or position can be found, using a kinematic equation.

- If *acceleration* is a function of time, integrate $a = dv/dt$ and $v = ds/dt$ to obtain $v = v(t)$ and $s = s(t)$.

- If *acceleration* is a function of displacement, integrate $a\,ds = v\,dv$ to obtain $v = f(s)$.

- If *acceleration is constant*, use $v = v_0 + a_c t$, $s = s_0 + v_0 t + \tfrac{1}{2} a_c t^2$, $v^2 = v_0^2 + 2a_c(s - s_0)$ to determine v and s.

- If the problem involves the dependent motion of several particles, use the method outlined in Sec. 12.9 to relate their accelerations.

EXAMPLE 13.1

(a)

(b)

The 50-kg crate shown in Fig. 13–6a rests on a horizontal surface for which the coefficient of kinetic friction is $\mu_k = 0.3$. If the crate is subjected to the 400-N towing force as shown, determine the velocity of the crate in 3 s starting from rest.

SOLUTION

Free-Body Diagram. The weight of the crate is $W = mg = 50 \text{ kg} (9.81 \text{ m/s}^2) = 490.5$ N. As shown in Fig. 13–6b, the frictional force has a magnitude $F = \mu_k N_C$ and acts to the left, since it opposes the motion of the crate. The acceleration $\mathbf{a}$ is assumed to act horizontally, in the positive x direction. There are two unknowns, namely, N_C and a.

Equations of Motion. From the data shown on the free-body diagram, we have

$$\xrightarrow{+} \Sigma F_x = ma_x; \quad 400 \cos 30° - 0.3 N_C = 50a \quad (1)$$
$$+\uparrow \Sigma F_y = ma_y; \quad N_C - 490.5 + 400 \sin 30° = 0 \quad (2)$$

Solving Eq. 2 for N_C, substituting the result into Eq. 1, and solving for a yields

$$N_C = 290.5 \text{ N}$$
$$a = 5.185 \text{ m/s}^2$$

Kinematics. The acceleration is *constant*, since the applied force **P** is constant. Since the initial velocity is zero, the velocity of the crate in 3 s is

$$(\xrightarrow{+}) \quad v = v_0 + a_c t$$
$$= 0 + 5.185(3)$$
$$= 15.6 \text{ m/s} \rightarrow \quad \textit{Ans.}$$

(c)

Fig. 13–6

NOTE: An alternative procedure would be to draw the crate's free-body *and* kinetic diagrams, Fig. 13–6c, prior to applying the equations of motion.

EXAMPLE 13.2

A 10-kg projectile is fired vertically upward from the ground, with an initial velocity of 50 m/s, Fig. 13–7a. Determine the maximum height to which it will travel if (a) atmospheric resistance is neglected; and (b) atmospheric resistance is measured as $F_D = (0.01v^2)$ N, where v is the speed of the projectile at any instant, measured in m/s.

SOLUTION

Part (a) Free-Body Diagram. As shown in Fig. 13–7b, the projectile's weight is $W = mg = 10(9.81) = 98.1$ N. We will assume the unknown acceleration **a** acts upward in the *positive z direction*.

Equation of Motion.
$$+\uparrow \Sigma F_z = ma_z; \qquad -98.1 = 10a, \qquad a = -9.81 \text{ m/s}^2$$

The result indicates that the projectile, like every object having free-flight motion near the earth's surface, is subjected to a *constant* downward acceleration of 9.81 m/s².

Kinematics. Initially, $z_0 = 0$ and $v_0 = 50$ m/s, and at the maximum height $z = h$, $v = 0$. Since the acceleration is *constant*, then

$$(+\uparrow) \qquad v^2 = v_0^2 + 2a_c(z - z_0)$$
$$0 = (50)^2 + 2(-9.81)(h - 0)$$
$$h = 127 \text{ m} \qquad \qquad Ans.$$

Part (b) Free-Body Diagram. Since the force $F_D = (0.01v^2)$ N tends to retard the upward motion of the projectile, it acts downward as shown on the free-body diagram, Fig. 13–7c.

Equation of Motion.
$$+\uparrow \Sigma F_z = ma_z; \qquad -0.01v^2 - 98.1 = 10a, \qquad a = -(0.001v^2 + 9.81)$$

Kinematics. Here the acceleration is *not constant* since F_D depends on the velocity. Since the acceleration is a function of velocity, then the position can be determined from

$$(+\uparrow) \; a \, dz = v \, dv; \qquad -(0.001v^2 + 9.81) \, dz = v \, dv$$

Separating the variables and integrating, realizing that initially $z_0 = 0$, $v_0 = 50$ m/s (positive upward), and at $z = h$, $v = 0$, we have

$$\int_0^h dz = -\int_{50 \text{ m/s}}^0 \frac{v \, dv}{0.001v^2 + 9.81}$$

$$h = -500 \ln(v^2 + 9810) \Big|_{50 \text{ m/s}}^0$$

$$h = 114 \text{ m} \qquad \qquad Ans.$$

NOTE: The answer indicates a lower elevation than that obtained in part (a) due to atmospheric resistance or drag.

(a)

(b)

(c)

Fig. 13–7

EXAMPLE 13.3

The baggage truck A shown in the photo has a weight of 900 lb and tows a 550-lb cart B and a 325-lb cart C. For a short time the driving frictional force developed at the wheels of the truck is $F_A = (40t)$ lb, where t is in seconds. If the truck starts from rest, determine its speed in 2 seconds. Also, what is the horizontal force acting on the coupling between the truck and cart B at this instant? Neglect the size of the truck and carts.

(a)

SOLUTION

Free-Body Diagram. As shown in Fig. 13–8a, it is the frictional driving force that gives both the truck and carts an acceleration. Here we have considered all three vehicles as a single system.

Equation of Motion. Only motion in the horizontal direction has to be considered.

$$\overset{+}{\leftarrow} \Sigma F_x = ma_x; \qquad 40t = \left(\frac{900 + 550 + 325}{32.2}\right)a$$

$$a = 0.7256t$$

Kinematics Since the acceleration is a function of time, the velocity of the truck is obtained using $a = dv/dt$ with the initial condition that $v_0 = 0$ at $t = 0$. We have

$$\int_0^v dv = \int_0^{2\,s} 0.7256t\, dt; \qquad v = 0.3628t^2 \Big|_0^{2\,s} = 1.45 \text{ ft/s} \qquad Ans.$$

Free-Body Diagram. In order to determine the force between the truck and cart B, we will consider a free-body diagram of the truck so that we can "expose" the coupling force $\mathbf{T}$ as external to the free-body diagram, Fig. 13–8b.

(b)

Fig. 13–8

Equation of Motion. When $t = 2$ s, then

$$\overset{+}{\leftarrow} \Sigma F_x = ma_x; \qquad 40(2) - T = \left(\frac{900}{32.2}\right)[0.7256(2)]$$

$$T = 39.4 \text{ lb} \qquad Ans.$$

NOTE: Try and obtain this same result by considering a free-body diagram of carts B and C as a single system.

EXAMPLE 13.4

A smooth 2-kg collar, shown in Fig. 13–9a, is attached to a spring having a stiffness $k = 3$ N/m and an unstretched length of 0.75 m. If the collar is released from rest at A, determine its acceleration and the normal force of the rod on the collar at the instant $y = 1$ m.

SOLUTION
Free-Body Diagram. The free-body diagram of the collar when it is located at the arbitrary position y is shown in Fig. 13–9b. Furthermore, the collar is *assumed* to be accelerating so that "**a**" acts downward in the *positive y* direction. There are four unknowns, namely, N_C, F_s, a, and θ.

Equations of Motion.

$$\xrightarrow{+} \Sigma F_x = ma_x; \quad -N_C + F_s \cos\theta = 0 \quad (1)$$
$$+\downarrow \Sigma F_y = ma_y; \quad 19.62 - F_s \sin\theta = 2a \quad (2)$$

From Eq. 2 it is seen that the acceleration depends on the magnitude and direction of the spring force. Solution for N_C and a is possible once F_s and θ are known.

The magnitude of the spring force is a function of the stretch s of the spring; i.e., $F_s = ks$. Here the unstretched length is $AB = 0.75$ m, Fig. 13–9a; therefore, $s = CB - AB = \sqrt{y^2 + (0.75)^2} - 0.75$. Since $k = 3$ N/m, then

Fig. 13–9

$$F_s = ks = 3(\sqrt{y^2 + (0.75)^2} - 0.75) \quad (3)$$

From Fig. 13–9a, the angle θ is related to y by trigonometry.

$$\tan\theta = \frac{y}{0.75}$$

Substituting $y = 1$ m into Eqs. 3 and 4 yields $F_s = 1.50$ N and $\theta = 53.1°$. Substituting these results into Eqs. 1 and 2, we obtain

$$N_C = 0.900 \text{ N} \quad \quad Ans.$$
$$a = 9.21 \text{ m/s}^2 \downarrow \quad \quad Ans.$$

NOTE: This is not a case of constant acceleration, since the spring force changes both its magnitude and direction as the collar moves downward.

EXAMPLE 13.5

(a)

(b)

(c)

(d)

Fig. 13–10

The 100-kg block A shown in Fig. 13–10a is released from rest. If the masses of the pulleys and the cord are neglected, determine the acceleration of the 20-kg block B.

SOLUTION
Free-Body Diagrams. Since the mass of pulley C is *neglected*, ($m_C = 0$) then $T' = 2T$ as shown in Fig. 13–10b. The free-body diagrams for blocks A and B are shown in Figs. 13–10c and d, respectively. If A were to remain stationary, then $T = 490.5$ N; whereas for B to remain stationary, $T = 196.2$ N. Hence A will move down while B moves up. Although this is the case, we will *assume* both blocks accelerate downward, in the direction of $+s_A$ and $+s_B$. The three unknowns are T, a_A, and a_B.

Equations of Motion. Block A,
$$+\downarrow \Sigma F_y = ma_y; \qquad 981 - 2T = 100 a_A \quad (1)$$
Block B,
$$+\downarrow \Sigma F_y = ma_y; \qquad 196.2 - T = 20 a_B \quad (2)$$

Kinematics. The necessary third equation is obtained by relating a_A to a_B using a dependent motion analysis, discussed in Sec. 12.9. The coordinates s_A and s_B in Fig. 13–10a measure the positions of A and B from the fixed datum. It is seen that
$$2s_A + s_B = l$$
where l is constant and represents the total vertical length of cord. Differentiating this expression twice with respect to time yields
$$2a_A = -a_B \quad (3)$$

Equations 1 to 3 have all been written so that the *positive direction was always assumed downward*. This is very important since we are seeking a simultaneous solution of equations. The results are
$$T = 327.0 \text{ N}$$
$$a_A = 3.27 \text{ m/s}^2$$
$$a_B = -6.54 \text{ m/s}^2 \qquad \textit{Ans.}$$

Hence when block A accelerates *downward*, block B accelerates *upward* as expected.

FUNDAMENTAL PROBLEMS

F13–1. The motor winds in the cable with a constant acceleration, such that the 20-kg crate moves a distance $s = 6$ m in 3 s, starting from rest. Determine the tension developed in the cable. The coefficient of kinetic friction between the crate and the plane is $\mu_k = 0.3$.

Prob. F13–1

F13–2. If motor M exerts a force of $F = (10t^2 + 100)$ N on the cable, where t is in seconds, determine the velocity of the 25-kg crate when $t = 4$ s. The coefficients of static and kinetic friction between the crate and the plane are $\mu_s = 0.3$ and $\mu_k = 0.25$, respectively. The crate is initially at rest.

Prob. F13–2

F13–3. A spring of stiffness $k = 500$ N/m is mounted against the 10-kg block. If the block is subjected to the force of $F = 500$ N, determine its velocity at $s = 0.5$ m. When $s = 0$, the block is at rest and the spring is uncompressed. The contact surface is smooth.

Prob. F13–3

F13–4. The 2-Mg car is being towed by a winch. If the winch exerts a force of $T = 100(s + 1)$ N on the cable, where s is the displacement of the car in meters, determine the speed of the car when $s = 10$ m, starting from rest. Neglect rolling resistance of the car.

Prob. F13–4

F13–5. The spring has a stiffness $k = 200$ N/m and is unstretched when the 25-kg block is at A. Determine the acceleration of the block when $s = 0.4$ m. The contact surface between the block and the plane is smooth.

Prob. F13–5

F13–6. Block B rests upon a smooth surface. If the coefficients of static and kinetic friction between A and B are $\mu_s = 0.4$ and $\mu_k = 0.3$, respectively, determine the acceleration of each block if $P = 6$ lb.

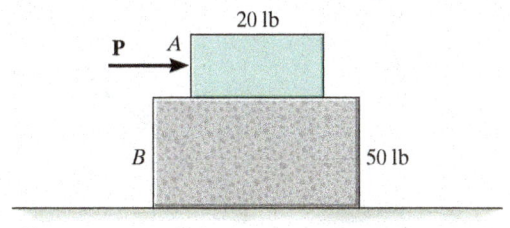

Prob. F13–6

PROBLEMS

All solutions must include a free-body diagram.

13–1. The 6-lb particle is subjected to the action of its weight and forces $\mathbf{F}_1 = \{2\mathbf{i} + 6\mathbf{j} - 2t\mathbf{k}\}$ lb, $\mathbf{F}_2 = \{t^2\mathbf{i} - 4t\mathbf{j} - 1\mathbf{k}\}$ lb, and $\mathbf{F}_3 = \{-2t\mathbf{i}\}$ lb, where t is in seconds. Determine the distance the ball is from the origin 2 s after being released from rest.

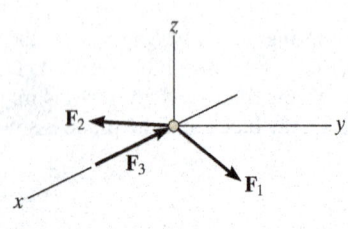

Prob. 13–1

*13–4.** The jet belt and man have a total weight of 210 lb. If the engines provide a constant thrust having x and y components of $F_x = 35$ lb and $F_y = 300$ lb, determine the total distance the man travels when $t = 5$ s after takeoff. What is his velocity at this instant? Neglect air resistance and the loss of fuel.

Prob. 13–4

13–2. If the coefficient of kinetic friction between the 50-kg crate and the ground is $\mu_k = 0.3$, determine the distance the crate travels and its velocity when $t = 3$ s. The crate starts from rest and $P = 200$ N.

13–3. If the 50-kg crate starts from rest and achieves a velocity of $v = 4$ m/s when it travels a distance of 5 m to the right, determine the magnitude of force $\mathbf{P}$ acting on the crate. The coefficient of kinetic friction between the crate and the ground is $\mu_k = 0.3$.

Probs. 13–2/3

13–5. If blocks A and B of mass 10 kg and 6 kg, respectively, are placed on the inclined plane and released, determine the force developed in the link. The coefficients of kinetic friction between the blocks and the inclined plane are $\mu_A = 0.1$ and $\mu_B = 0.3$. Neglect the mass of the link.

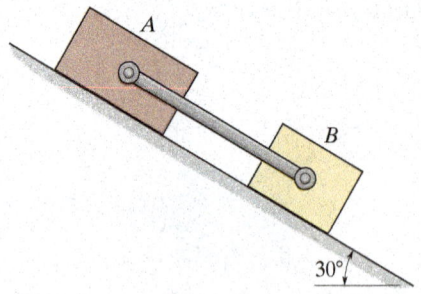

Prob. 13–5

13–6. If $P = 400$ N and the coefficient of kinetic friction between the 50-kg crate and the inclined plane is $\mu_k = 0.25$, determine the velocity of the crate after it travels 6 m up the plane. The crate starts from rest.

13–7. If the 50-kg crate starts from rest and travels a distance of 6 m up the plane in 4 s, determine the magnitude of force **P** acting on the crate. The coefficient of kinetic friction between the crate and the ground is $\mu_k = 0.25$.

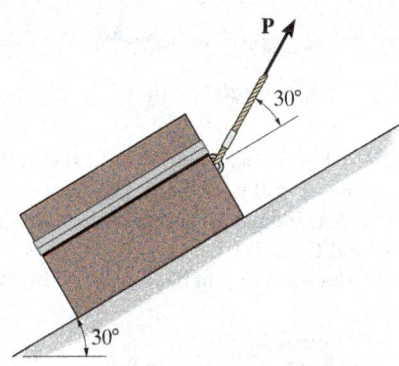

Probs. 13–6/7

*13–8. In the event of an emergency, a nuclear reactor is shut down by dropping the control rod R, having a weight of 20 lb, into the reactor core C. If the rod is immersed in "heavy" water, which offers a drag resistance to downward motion of $F_D = (0.25v)$ lb, where v is the velocity of the rod in ft/s, determine the distance s it must descend into the reactor core when it attains a speed of 15 ft/s. The rod is released from rest when $s = 0$. Neglect the effects of buoyancy.

Prob. 13–8

13–9. The 10-lb block has a speed of 4 ft/s when the force of $F = (8t^2)$ lb is applied. Determine the velocity of the block when $t = 2$ s. The coefficient of kinetic friction at the surface is $\mu_k = 0.2$.

13–10. The 10-lb block has a speed of 4 ft/s when the force of $F = (8t^2)$ lb is applied. Determine the velocity of the block when it moves $s = 30$ ft. The coefficient of kinetic friction at the surface is $\mu_k = 0.2$.

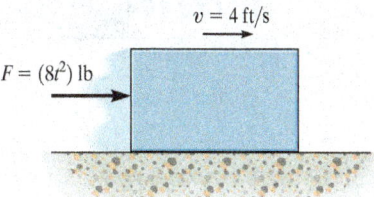

Probs. 13–9/10

13–11. Determine the time needed to pull the cord at B down 4 ft starting from rest when a force of 10 lb is applied to the cord. Block A weighs 20 lb. Neglect the mass of the pulleys and cords.

Prob. 13–11

*13–12. The conveyor belt C is moving at 6 m/s. If the coefficient of static friction between the conveyor belt and the 10-kg box B is $\mu_s = 0.2$, determine the shortest time the conveyor can stop so that the box does not slip or move on the belt.

Prob. 13–12

13–13. Block A has a weight of 8 lb and block B has a weight of 6 lb. They rest on a surface for which the coefficient of kinetic friction is $\mu_k = 0.2$. If the spring has a stiffness of $k = 20$ lb/ft, and it is compressed 0.2 ft, determine the acceleration of each block just after they are released.

Prob. 13–13

13–14. The 2-Mg truck is traveling at 15 m/s when the brakes on all its wheels are applied, causing it to skid for a distance of 10 m before coming to rest. Determine the constant horizontal force developed in the coupling C, and the frictional force developed between the tires of the truck and the road during this time. The total mass of the boat and trailer is 1 Mg.

Prob. 13–14

13–15. The motor lifts the 50-kg crate with an acceleration of 6 m/s². Determine the components of force reaction and the couple moment at the fixed support A.

Prob. 13–15

*13–16. Team A and team B consist of four members of total mass 320 kg and 330 kg, respectively. If the average coefficients of static and kinetic friction for team A are $\mu_s = 0.52$ and $\mu_k = 0.5$, and for team B, $\mu_s = 0.48$ and $\mu_k = 0.46$, which team will win the tug of war? Also, what is the acceleration of the losing team if the winning team develops its maximum static friction force?

Prob. 13–16

13–17. Block A has a mass of 60 kg and rests on block B, which has a mass of 30 kg. If the coefficients of static and kinetic friction are indicated in the figure, determine the largest horizontal force **P** which can be applied to block B so that block A does not slip on block B while block B slides.

Prob. 13–17

13–18. A 40-lb suitcase slides from rest 20 ft down the smooth ramp. Determine the point where it strikes the ground at C. How long does it take to go from A to C?

13–19. Solve Prob. 13–18 if the suitcase has an initial velocity down the ramp of $v_A = 10$ ft/s and the coefficient of kinetic friction along AB is $\mu_k = 0.2$.

Probs. 13–18/19

*13–20. The 400-kg mine car is hoisted up the incline using the cable and motor M. For a short time, the force in the cable is $F = (3200t^2)$ N, where t is in seconds. If the car has an initial velocity $v_1 = 2$ m/s when $t = 0$, determine its velocity when $t = 2$ s.

13–21. The 400-kg mine car is hoisted up the incline using the cable and motor M. For a short time, the force in the cable is $F = (3200t^2)$ N, where t is in seconds. If the car has an initial velocity $v_1 = 2$ m/s when $t = 0$, determine the distance it moves up the plane when $t = 2$ s.

13–23. Determine the acceleration of each block if A and B have a mass m and C has a mass $2\,m$. Neglect the mass of the cords and the pulleys.

Prob. 13–23

Probs. 13–20/21

*13–24. The conveyor belt delivers each 12-kg crate to the ramp at A such that the crate's speed is $v_A = 2.5$ m/s, directed down *along* the ramp. If the coefficient of kinetic friction between each crate and the ramp is $\mu_k = 0.3$, determine the speed at which each crate slides off the ramp at B. Assume that no tipping occurs. Take $\theta = 30°$.

13–25. The conveyor belt delivers each 12-kg crate to the ramp at A such that the crate's speed is $v_A = 2.5$ m/s, directed down *along* the ramp. If the coefficient of kinetic friction between each crate and the ramp is $\mu_k = 0.3$, determine the smallest incline θ of the ramp so that the crates will slide off and fall into the cart.

13–22. The 50-kg block A is released from rest. Determine the velocity of the 15-kg block B in 2 s.

Prob. 13–22

Probs. 13–24/25

13–26. At a given instant the 10-lb block A is moving downward with a speed of 6 ft/s. Determine its speed 2 s later if block B has a weight of 4 lb, and the coefficient of kinetic friction between it and the inclined plane is $\mu_k = 0.2$. Neglect the mass of the pulleys and cord.

Prob. 13–26

13–27. The 20-kg block A rests on the 60-kg plate B in the position shown. Neglecting the mass of the rope and pulley, and using the coefficients of kinetic friction indicated, determine the time needed for block A to slide 0.5 m *on the plate* when the system is released from rest.

Prob. 13–27

*13–28. At the instant shown the 100-lb block A is moving down the plane at 5 ft/s while being attached to the 50-lb block B. If the coefficient of kinetic friction between the block and the incline is $\mu_k = 0.2$, determine the acceleration of A and the distance A slides before it stops. Neglect the mass of the pulleys and cables.

Prob. 13–28

13–29. The force exerted by the motor M on the cable is shown in the graph. Determine the velocity of the 200-lb crate A when $t = 2.5$ s.

Prob. 13–29

13–30. The four 2-kg metal plates are stacked on top of one another. If the coefficient of static friction between any two plates is $\mu_s = 0.3$, determine the smallest horizontal force **F** that can be applied to the bottom plate which will cause any one of the plates to slip relative to another. The floor is smooth.

13–33. If a horizontal force $P = 35$ lb is applied to block A, determine the acceleration of block B. Neglect friction. Block A weighs 15 lb, and block B weighs 8 lb. *Hint*: Show that $a_A = a_B \tan 15°$.

Prob. 13–30

Prob. 13–33

13–31. The tractor is used to lift the 150-kg load B with the 24-m-long rope, boom, and pulley system. If the tractor travels to the right at a constant speed of 4 m/s, determine the tension in the rope when $s_A = 5$ m. When $s_A = 0$, $s_B = 0$.

*13–32. The tractor is used to lift the 150-kg load B with the 24-m-long rope, boom, and pulley system. If the tractor travels to the right with an acceleration of 3 m/s^2 and has a velocity of 4 m/s at the instant $s_A = 5$ m, determine the tension in the rope at this instant. When $s_A = 0$, $s_B = 0$.

13–34. The 3-kg crate rests on the 10-kg cart where the coefficients of static and kinetic friction are $\mu_s = 0.25$ and $\mu_k = 0.2$, respectively. Determine the smallest constant force **P** needed to cause the crate to slip. How much time does it take for the crate to slip off the cart?

Probs. 13–31/32

Prob. 13–34

13–35. The coefficient of static friction between the 200-kg crate and the flat bed of the truck is $\mu_s = 0.3$. Determine the shortest time for the truck to reach a speed of 60 km/h, starting from rest with constant acceleration, so that the crate does not slip.

Prob. 13–35

*13–36. Determine the velocity of the cylinder A when $t = 4$ s if the system is released from rest. Cylinder A has a mass of 8 kg and cylinder B has a mass of 6 kg. Neglect the mass of the pulleys and cords.

Prob. 13–36

13–37. The safe S has a weight of 200 lb and is supported by the rope and pulley arrangement shown. If the end of the rope is given to a boy B of weight 90 lb, determine his acceleration if in the confusion he doesn't let go of the rope. Neglect the mass of the pulleys and rope.

Prob. 13–37

13–38. Block A has a mass of 20 kg and block B has a mass of 50 kg. If the spring is stretched 0.2 m at the instant shown, determine the acceleration of block B at this instant if it is (a) originally at rest, (b) moving downward with a speed of 3 m/s. Neglect friction.

Prob. 13–38

13–39. An electron of mass m is discharged with an initial horizontal velocity of $\mathbf{v}_0$. If it is subjected to two fields of force for which $F_x = F_0$ and $F_y = 0.3F_0$, where F_0 is constant, determine the equation of the path, and the speed of the electron at any time t.

13–41. The 2-kg collar C is free to slide along the smooth shaft AB. Determine the acceleration of collar C if (a) the shaft is fixed from moving, (b) collar A, which is fixed to shaft AB, moves downward at constant velocity along the vertical rod, and (c) collar A is subjected to a downward acceleration of 2 m/s². In all cases, the collar moves in the plane.

Prob. 13–39

Prob. 13–41

*** 13–40.** The 400-lb cylinder at A is hoisted using the motor and the pulley system shown. If the speed of point B on the cable is increased at a constant rate from zero to $v_B = 10$ ft/s in $t = 5$ s, determine the tension in the cable at B to cause the motion.

13–42. The 2-kg collar C is free to slide along the smooth shaft AB. Determine the acceleration of collar C if collar A is subjected to an upward acceleration of 4 m/s².

Prob. 13–40

Prob. 13–42

13–43. A parachutist having a mass m opens his parachute from an at-rest position at a very high altitude. If the atmospheric drag resistance is $F_D = kv^2$, where k is a constant, determine his velocity when he has fallen for a time t. What is his velocity when he lands on the ground? This velocity is referred to as the *terminal velocity*, which is found by letting the time of fall $t \to \infty$.

Prob. 13–43

*13–44. Determine the tension developed in the cords attached to each block and the accelerations of the blocks. Neglect the mass of the pulleys and cords.

Prob. 13–44

13–45. A projectile of mass m is fired into a liquid at an angle θ_0 with an initial velocity $\mathbf{v}_0$ as shown. If the liquid develops a frictional or drag resistance on the projectile which is proportional to its velocity, i.e., $F = kv$, where k is a constant, determine the x and y components of its position at any instant. Also, what is the maximum distance x_{max} that it travels?

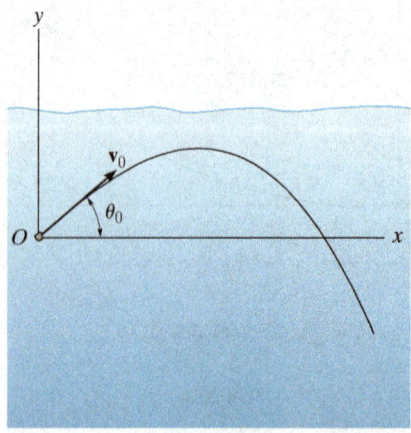

Prob. 13–45

13–46. Blocks A and B each have a mass m. Determine the largest horizontal force P which can be applied to B so that A will not move relative to B. All surfaces are smooth.

13–47. Blocks A and B each have a mass m. Determine the largest horizontal force P which can be applied to B so that A will not slip on B. The coefficient of static friction between A and B is μ_s. Neglect any friction between B and C.

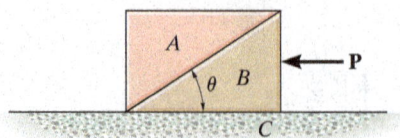

Probs. 13–46/47

*13–48. The smooth block B of negligible size has a mass m and rests on the horizontal plane. If the board AC pushes on the block at an angle θ with a constant acceleration $\mathbf{a}_0$, determine the velocity of the block along the board and the distance s the block moves along the board as a function of time t. The block starts from rest when $s = 0, t = 0$.

Prob. 13–48

13–49. The 30-lb crate is being hoisted upward with a constant acceleration of $6 \text{ ft}/\text{s}^2$. If the uniform beam AB has a weight of 200 lb, determine the components of reaction at A. Neglect the size and mass of the pulley at B. *Hint*: First find the tension in the cable, then analyze the forces in the beam using statics.

Prob. 13–49

13–50. The rock is dropped from rest within the thick liquid. If the drag force acting on it is $F = cv$, where c is a constant, determine the velocity and position of the rock as a function of time. Plot the velocity and position time graphs.

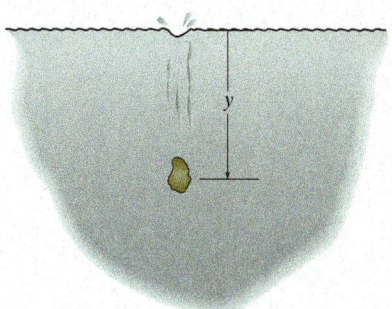

Prob. 13–50

13–51. The block A has a mass m_A and rests on the pan B, which has a mass m_B. Both are supported by a spring having a stiffness k that is attached to the bottom of the pan and to the ground. Determine the distance d the pan should be pushed down from the equilibrium position and then released from rest so that separation of the block will take place from the surface of the pan at the instant the spring becomes unstretched.

Prob. 13–51

13.5 EQUATIONS OF MOTION: NORMAL AND TANGENTIAL COORDINATES

When a particle moves along a curved path which is known, the equation of motion for the particle may be expressed in terms of components along the tangential, normal, and binormal directions, Fig. 13–11. There is no motion of the particle in the binormal direction, since the particle is constrained to move along the path. Therefore

$$\Sigma \mathbf{F} = m\mathbf{a}$$

$$\Sigma F_t \mathbf{u}_t + \Sigma F_n \mathbf{u}_n + \Sigma F_b \mathbf{u}_b = m\mathbf{a}_t + m\mathbf{a}_n$$

This equation is satisfied provided

$$\begin{aligned} \Sigma F_t &= ma_t \\ \Sigma F_n &= ma_n \\ \Sigma F_b &= 0 \end{aligned} \qquad (13\text{--}8)$$

Inertial coordinate system

Fig. 13–11

Recall that $a_t\,(= dv/dt)$ represents the time rate of change in the magnitude of velocity. So if the resultant of $\Sigma \mathbf{F}_t$ acts in the direction of motion, the particle's speed will increase; whereas if it acts in the opposite direction, the particle will slow down. Likewise, $a_n\,(= v^2/\rho)$ represents the time rate of change in the velocity's direction. It is caused by the centripetal force or the resultant of $\Sigma \mathbf{F}_n$, which *always* acts in the positive *n* direction, i.e., toward the path's center of curvature.

As a roller coaster falls downward along the track, the cars have both normal and tangential components of acceleration.

13.5 EQUATIONS OF MOTION: NORMAL AND TANGENTIAL COORDINATES

PROCEDURE FOR ANALYSIS

When a problem involves the motion of a particle along a *known curved path*, normal and tangential coordinates should be considered for the analysis since the acceleration components can be readily formulated. The method for applying the equations of motion, which relate the forces to the acceleration, has been outlined in the procedure given in Sec. 13-4. Specifically, for t, n, b coordinates it may be stated as follows.

Free-Body Diagram.

- Establish the inertial t, n, b coordinate system at the particle and draw the particle's free-body diagram.
- The particle's normal acceleration $\mathbf{a}_n$ *always* acts in the positive n direction.
- If the tangential acceleration $\mathbf{a}_t$ is unknown, assume it acts in the positive t direction.
- There is no acceleration in the b direction.
- Identify the unknowns in the problem.

Equations of Motion.

- Apply the equations of motion, Eq. 13–8.

Kinematics.

- Formulate the tangential and normal components of acceleration; i.e., $a_t = dv/dt$ or $a_t = v\,dv/ds$ and $a_n = v^2/\rho$.
- If the path is defined as $y = f(x)$, the radius of curvature at the point where the particle is located can be obtained from $\rho = [1 + (dy/dx)^2]^{3/2}/|d^2y/dx^2|$.

EXAMPLE 13.6

Determine the banking angle θ for the race track so that the wheels of the racing cars shown in Fig. 13–12a will not have to depend upon friction to prevent any car from sliding up or down the track. Assume the cars have negligible size, a mass m, and travel around the curve of radius ρ with a constant speed v.

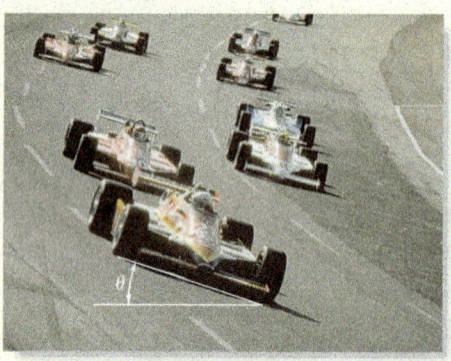

(a)

SOLUTION
Before looking at the following solution, give some thought as to why it should be solved using t, n, b coordinates.

Free-Body Diagram. As shown in Fig. 13–12b, and as stated in the problem, no frictional force acts on the car. Here N_C represents the *resultant* of the ground on all four wheels. Since a_n can be calculated, the unknowns are N_C and θ.

(b)

Fig. 13–12

Equations of Motion. Using the n, b axes shown,

$$\overset{\pm}{\rightarrow} \Sigma F_n = ma_n; \qquad N_C \sin\theta = m\frac{v^2}{\rho} \qquad (1)$$

$$+\uparrow \Sigma F_b = 0; \qquad N_C \cos\theta - mg = 0 \qquad (2)$$

Eliminating N_C and m from these equations by dividing Eq. 1 by Eq. 2, we obtain

$$\tan\theta = \frac{v^2}{g\rho}$$

$$\theta = \tan^{-1}\left(\frac{v^2}{g\rho}\right) \qquad Ans.$$

NOTE: The result is independent of the mass of the car. Also, a force summation in the tangential direction is of no consequence to the solution. If it were considered, then $a_t = dv/dt = 0$, since the car moves with *constant speed*. A further analysis of this problem is discussed in Prob. 21–53.

EXAMPLE 13.7

The 3-kg disk D is attached to the end of a cord as shown in Fig. 13–13a. The other end of the cord is attached to a ball-and-socket joint located at the center of a platform. If the platform rotates rapidly, and the disk is placed on it and released from rest, determine the time it takes for the disk to reach a speed great enough to break the cord. The maximum tension the cord can sustain is 100 N, and the coefficient of kinetic friction between the disk and the platform is $\mu_k = 0.1$.

(a)

SOLUTION

Free-Body Diagram. The frictional force has a magnitude $F = \mu_k N_D = 0.1 N_D$ and a sense of direction that opposes the *relative motion* of the disk with respect to the platform. It is this force that gives the disk a tangential component of acceleration causing v to increase, thereby causing T to increase until it reaches 100 N. The weight of the disk is $W = 3(9.81) = 29.43$ N. Since a_n can be related to v, the unknowns are N_D, a_t, and v.

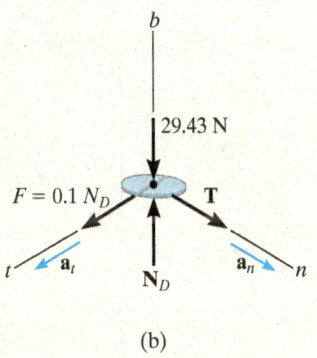

(b)

Fig. 13–13

Equations of Motion.

$\Sigma F_n = m a_n;$ $\quad T = 3 \left(\dfrac{v^2}{1} \right) \qquad (1)$

$\Sigma F_t = m a_t;$ $\quad 0.1 N_D = 3 a_t \qquad (2)$

$\Sigma F_b = 0;$ $\quad N_D - 29.43 = 0 \qquad (3)$

Setting $T = 100$ N, Eq. 1 can be solved for the critical speed v_{cr} of the disk needed to break the cord. Solving all the equations, we obtain

$$N_D = 29.43 \text{ N}$$
$$a_t = 0.981 \text{ m/s}^2$$
$$v_{cr} = 5.77 \text{ m/s}$$

Kinematics. Since a_t is *constant*, the time needed to break the cord is

$$v_{cr} = v_0 + a_t t$$
$$5.77 = 0 + (0.981)t$$
$$t = 5.89 \text{ s} \qquad \textit{Ans.}$$

EXAMPLE 13.8

If the ski jump can be approximated by the parabola shown in Fig. 13–14a, determine the normal force on the 150-lb skier the instant she arrives at the end of the jump, point A, where her velocity is 65 ft/s. Also, what is her acceleration at this point?

SOLUTION
Why consider using n, t coordinates to solve this problem?

Free-Body Diagram. Since $dy/dx = x/100|_{x=0} = 0$, the slope at A is horizontal. Therefore, a free-body diagram of the skier when she is at A is shown in Fig. 13–14b. Since the path is *curved*, there are two components of acceleration, $\mathbf{a}_n$ and $\mathbf{a}_t$.

Here we can calculate a_n and so the unknowns are a_t and N_A.

Equations of Motion.

$$+\uparrow \Sigma F_n = ma_n; \qquad N_A - 150 = \frac{150}{32.2}\left(\frac{(65)^2}{\rho}\right) \qquad (1)$$

$$\pm \Sigma F_t = ma_t; \qquad 0 = \frac{150}{32.2} a_t \qquad (2)$$

The radius of curvature ρ for the path must be determined at point $A(0, -200 \text{ ft})$. Here $y = \frac{1}{200}x^2 - 200$, $dy/dx = \frac{1}{100}x$, $d^2y/dx^2 = \frac{1}{100}$, so that at $x = 0$,

$$\rho = \frac{[1 + (dy/dx)^2]^{3/2}}{|d^2y/dx^2|}\bigg|_{x=0} = \frac{[1 + (0)^2]^{3/2}}{|\frac{1}{100}|} = 100 \text{ ft}$$

Substituting this into Eq. 1 and solving for N_A, we obtain

$$N_A = 347 \text{ lb} \qquad \qquad Ans.$$

Kinematics. From Eq. 2,

$$a_t = 0$$

Thus,

$$a_n = \frac{v^2}{\rho} = \frac{(65)^2}{100} = 42.2 \text{ ft/s}^2$$

$$a_A = a_n = 42.2 \text{ ft/s}^2 \uparrow \qquad \qquad Ans.$$

NOTE: Apply the equation of motion in the y direction and show that when the skier is in mid air, her downward acceleration is 32.2 ft/s².

Fig. 13–14

EXAMPLE 13.9

The 60-kg skateboarder in Fig. 13–15a coasts down the circular track. If he starts from rest when $\theta = 0°$, determine the magnitude of the normal reaction the track exerts on him when $\theta = 60°$. Neglect his size for the calculation.

SOLUTION
Free-Body Diagram. The free-body diagram of the skateboarder when he is at an *arbitrary position* θ is shown in Fig. 13–15b.

At $\theta = 60°$ there are three unknowns, N_s, a_t, and a_n (or v).

(a)

Equations of Motion.

$$+\nearrow \Sigma F_n = ma_n; \quad N_s - [60(9.81)\text{ N}]\sin\theta = (60\text{ kg})\left(\frac{v^2}{4\text{ m}}\right) \quad (1)$$

$$+\searrow \Sigma F_t = ma_t; \quad [60(9.81)\text{ N}]\cos\theta = (60\text{ kg})\, a_t$$

$$a_t = 9.81\cos\theta$$

Kinematics. Since the increase in speed, a_t, is expressed in terms of θ, the equation $v\, dv = a_t\, ds$ must be used to determine the speed of the skateboarder when $\theta = 60°$. Using the geometric relation $s = \theta r$, where $ds = r\, d\theta = (4\text{ m})\, d\theta$, Fig. 13–15c, and the initial condition $v = 0$ at $\theta = 0°$, we have,

$$v\, dv = a_t\, ds$$

$$\int_0^v v\, dv = \int_0^{60°} 9.81\cos\theta(4\, d\theta)$$

$$\left.\frac{v^2}{2}\right|_0^v = 39.24\sin\theta\Big|_0^{60°}$$

$$\frac{v^2}{2} - 0 = 39.24(\sin 60° - 0)$$

$$v^2 = 67.97\text{ m}^2/\text{s}^2$$

Substituting this result and $\theta = 60°$ into Eq. 1 yields

$$N_s = 1529.23\text{ N} = 1.53\text{ kN} \qquad \textit{Ans.}$$

(b)

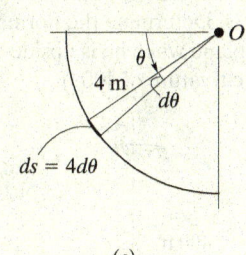

(c)

Fig. 13–15

FUNDAMENTAL PROBLEMS

F13–7. The block rests at a distance of 2 m from the center of the platform. If the coefficient of static friction between the block and the platform is $\mu_s = 0.3$, determine the maximum speed which the block can attain before it begins to slip. Assume the angular motion of the disk is slowly increasing.

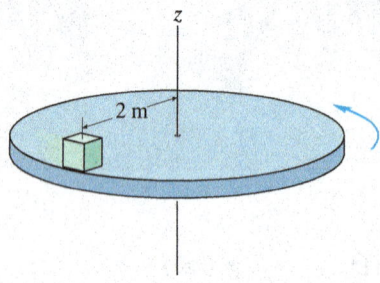

Prob. F13–7

F13–8. Determine the maximum speed that the jeep can travel over the crest of the hill and not lose contact with the road.

Prob. F13–8

F13–9. A pilot weighs 150 lb and is traveling at a constant speed of 120 ft/s. Determine the normal force he exerts on the seat of the plane when he is upside down at A. The loop has a radius of curvature of 400 ft.

Prob. F13–9

F13–10. The sports car is traveling along a 30° banked road having a radius of curvature of $\rho = 500$ ft. If the coefficient of static friction between the tires and the road is $\mu_s = 0.2$, determine the maximum speed so no slipping occurs. Neglect the size of the car.

Prob. F13–10

F13–11. If the 10-kg ball has a velocity of 3 m/s when it is at the position A, along the vertical path, determine the tension in the cord and the increase in the speed of the ball at this position.

Prob. F13–11

F13–12. The motorcycle has a mass of 0.5 Mg and a negligible size. It passes point A traveling at a speed of 15 m/s, which is increasing at a constant rate of 1.5 m/s². Determine the resultant frictional force exerted by the road on the tires at this instant.

Prob. F13–12

PROBLEMS

All solutions must include a free-body diagram.

*13–52. The 4-Mg helicopter maneuvers a horizontal turn having a radius of curvature $\rho = 400$ m. Determine the lift force F_L required and the angle of the bank θ when it is flying horizontally with a constant speed of $v = 40$ m/s. Show that θ increases if v increases by also calculating θ when $v = 60$ m/s.

Prob. 13–52

13–53. The 2-kg block B and 15-kg cylinder A are connected to a light cord that passes through a hole in the center of the smooth table. If the block is given a speed of $v = 10$ m/s, determine the radius r of the circular path along which it travels.

13–54. The 2-kg block B and 15-kg cylinder A are connected to a light cord that passes through a hole in the center of the smooth table. If the block travels along a circular path of radius $r = 1.5$ m, determine the speed of the block.

Probs. 13–53/54

13–55. Determine the maximum constant speed at which the pilot can travel around the vertical curve having a radius of curvature $\rho = 800$ m, so that he experiences a maximum acceleration $a_n = 8g = 78.5$ m/s^2. If he has a mass of 70 kg, determine the normal force he exerts on the seat of the airplane when the plane is traveling at this speed and is at its lowest point.

Prob. 13–55

*13–56. Cartons having a mass of 5 kg are required to move along the assembly line at a constant speed of 8 m/s. Determine the smallest radius of curvature ρ for the conveyor so the cartons do not slip. The coefficients of static and kinetic friction between a carton and the conveyor are $\mu_s = 0.7$ and $\mu_k = 0.5$, respectively.

Prob. 13–56

13–57. Calculate the constant speed of the cars on the amusement-park ride if it is observed that the cables are directed at 30° from the vertical. Each car, including its passengers, has a mass of 550 kg. Also, what are the components of force in the n, t, and z directions which a 60-kg passenger exerts on the car during the motion?

*13–60. At the instant $\theta = 60°$, the boy's center of mass G has a downward speed $v_G = 15$ ft/s. Determine the rate of increase in his speed and the tension in each of the two supporting cords of the swing at this instant. The boy has a weight of 60 lb. Neglect his size and the mass of the seat and cords.

13–61. At the instant $\theta = 60°$, the boy's center of mass G is momentarily at rest. Determine his speed and the tension in each of the two supporting cords of the swing when $\theta = 90°$. The boy has a weight of 60 lb. Neglect his size and the mass of the seat and cords.

Prob. 13–57

Probs. 13–60/61

13–58. The 2-kg spool S fits loosely on the inclined rod for which the coefficient of static friction is $\mu_s = 0.2$. If the spool is located 0.25 m from A, determine the minimum constant speed the spool can have so that it does not slip down the rod.

13–59. The 2-kg spool S fits loosely on the inclined rod for which the coefficient of static friction is $\mu_s = 0.2$. If the spool is located 0.25 m from A, determine the maximum constant speed the spool can have so that it does not slip up the rod.

13–62. A girl having a mass of 25 kg sits at the edge of the merry-go-round so her center of mass G is at a distance of 1.5 m from the axis of rotation. If the angular motion of the platform is *slowly* increased so that the girl's tangential component of acceleration can be neglected, determine the maximum speed which she can have before she begins to slip off the merry-go-round. The coefficient of static friction between the girl and the merry-go-round is $\mu_s = 0.3$.

Probs. 13–58/59

Prob. 13–62

13–63. The plane is traveling at a constant speed of 800 ft/s along the curve $y = 20(10^{-6})x^2 + 5000$, where x and y are in feet. If the pilot has a weight of 180 lb, determine the normal and tangential components of the force the seat exerts on the pilot when the plane is at its lowest point.

*13–64. The jet plane is traveling at a constant speed of 1000 ft/s along the curve $y = 20(10^{-6})x^2 + 5000$, where x and y are in feet. If the pilot has a weight of 180 lb, determine the normal and tangential components of the force the seat exerts on the pilot when $y = 10\,000$ ft.

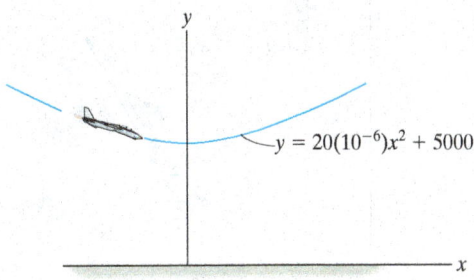

Probs. 13–63/64

13–65. The 5-lb packages ride on the surface of the conveyor belt. If the belt starts from rest and its speed increases to 2 ft/s in 2 s, determine the maximum angle θ so that none of the packages slip on the inclined surface AB of the belt. The coefficient of static friction between the belt and a package is $\mu_s = 0.3$. At what angle ϕ do the packages first begin to slip off the surface of the belt after the belt is moving at its constant speed of 2 ft/s? Neglect the size of the packages.

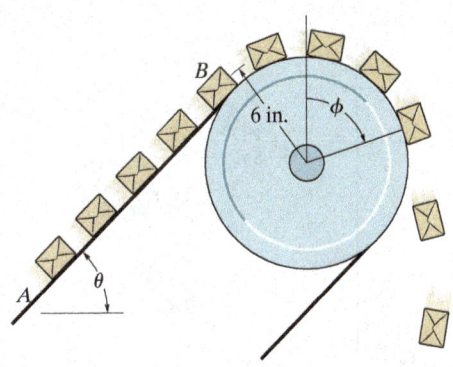

Prob. 13–65

13–66. A motorcyclist in a circus rides his motorcycle within the confines of the hollow sphere. If the coefficient of static friction between the wheels of the motorcycle and the sphere is $\mu_s = 0.4$, determine the minimum speed at which he must travel if he is to ride along the wall when $\theta = 90°$. The mass of the motorcycle and rider is 250 kg, and the radius of curvature to the center of gravity is $\rho = 20$ m. Neglect the size of the motorcycle for the calculation.

Prob. 13–66

13–67. A sled and rider, weighing 30 lb and 150 lb, respectively, travel down a smooth slope that has the shape shown. If the rider does not hold on to the sides of the sled, determine the normal force which he exerts on the sled when the sled is (a) at point B, where it has a velocity of 15 ft/s, and (b) at some point C in mid air.

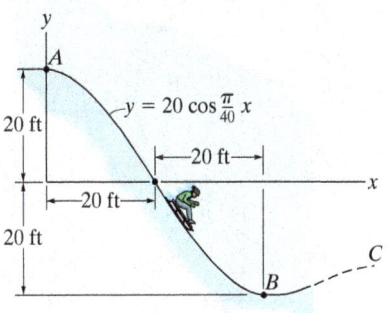

Prob. 13–67

*13–68. The collar has a mass of 2 kg and is free to slide along the smooth rod, which rotates in the *horizontal plane*. The attached spring has a stiffness $k=1$ kN/m, and when no force is applied to the spring its unstretched length is $d=0.5$ m. Determine the force which the spring exerts on the collar when the collar is at rest with respect to the rod, yet due to the rod's rotation it is moving with a constant speed $v=6$ m/s.

13–69. The collar has a mass of 2 kg and is free to slide along the smooth rod, which rotates in the *horizontal plane*. The attached spring has a stiffness $k=1$ kN/m, and when no force is applied to the spring its unstretched length is $d=0.5$ m. Determine the length d if the collar is at rest with respect to the rod, yet due to the rod's rotation it is moving with a constant speed $v=6$ m/s.

Probs. 13–68/69

13–70. A small, 6-lb collar slides along the vertical, circular, smooth rod. If the collar is released from rest when $\theta = 10°$, determine the normal force which it exerts on the rod when it arrives at points A and B.

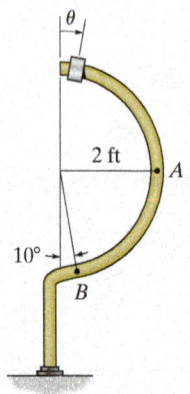

Prob. 13–70

13–71. The 150-lb man lies against the cushion for which the coefficient of static friction is $\mu_s = 0.5$. Determine the resultant normal and frictional forces the cushion exerts on him due to rotation about the z axis. He has a constant speed $v = 20$ ft/s. Neglect the size of the man and take $\theta = 60°$.

Prob. 13–71

*13–72. The 150-lb man lies against the cushion for which the coefficient of static friction is $\mu_s = 0.5$. If he rotates about the z axis with a constant speed $v = 30$ ft/s, determine the smallest angle θ of the cushion at which he will begin to slip off.

Prob. 13–72

13–73. The 52-kg skier is coasting freely down the hill with a speed of 4 m/s at $x = 10$ m. Determine the normal reaction on the ground and the rate of increase in speed at the instant shown. Neglect friction and the size of the skier.

13–75. The 2-lb collar is released from rest at A and slides down along the smooth rod. If the attached spring has a stiffness $k = 2$ lb/ft, determine its unstretched length so that it does not allow the collar to leave contact with the top surface of the rod until $\theta = 60°$.

Prob. 13–73

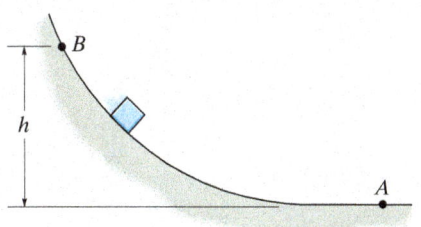

Prob. 13–75

13–74. If the bicycle and rider have a total weight of 180 lb, determine the resultant normal force acting on the bicycle when it is at point A while it is freely coasting at $v_A = 6$ ft/s. Also, calculate the increase in the bicyclist's speed at this point. Neglect the resistance due to the wind and the size of the bicycle and rider.

***13–76.** Prove that if the block is released from rest at point B on a smooth path of *arbitrary shape*, the speed it attains when it reaches point A is equal to the speed it attains when it falls freely through a distance h; i.e., $v = \sqrt{2gh}$.

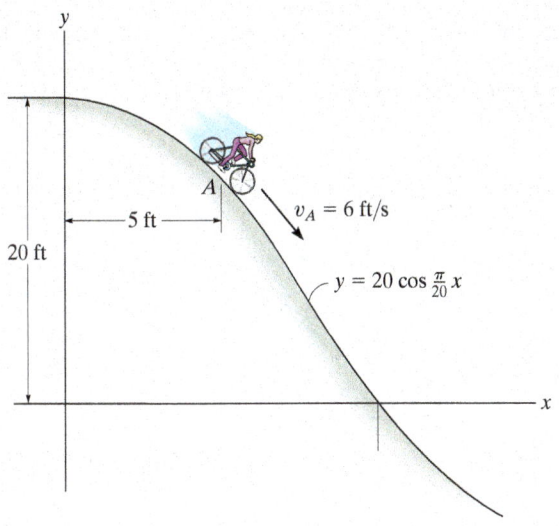

Prob. 13–74

Prob. 13–76

13–77. The rotational speed of the disk is controlled by a 30-g smooth contact arm AB which is spring mounted on the disk. When the disk is at rest, the center of mass G of the arm is located 150 mm from the center O, and the preset compression in the spring is 20 mm. If the initial gap between B and the contact at C is 10 mm, determine the (controlling) speed v_G of the arm's mass center, G, which will close the gap. The disk rotates in the horizontal plane. The spring has a stiffness of $k = 50$ N/m, and its ends are attached to the contact arm at D and to the disk at E.

13–79. The airplane, traveling at a constant speed of 50 m/s, is executing a horizontal turn. If the plane is banked at $\theta = 15°$, when the pilot experiences only a normal force on the seat of the plane, determine the radius of curvature ρ of the turn. Also, what is the normal force of the seat on the pilot if he has a mass of 70 kg?

Prob. 13–79

Prob. 13–77

13–78. The collar has a mass of 5 kg and is confined to move along the smooth circular rod which lies in the horizontal plane. The attached spring has an unstretched length of 200 mm. If at the instant $\theta = 30°$ the collar has a speed $v = 2$ m/s, determine the magnitudes of the normal force of the rod on the collar and the collar's acceleration.

*13–80. The 2-kg pendulum bob moves in the vertical plane with a velocity of 8 m/s when $\theta = 0°$. Determine the initial tension in the cord and also at the instant the bob reaches $\theta = 30°$. Neglect the size of the bob.

Prob. 13–78

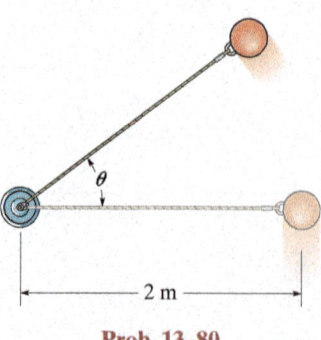

Prob. 13–80

13–81. The 2-kg pendulum bob moves in the vertical plane with a velocity of 6 m/s when $\theta = 0°$. Determine the angle θ where the tension in the cord becomes zero.

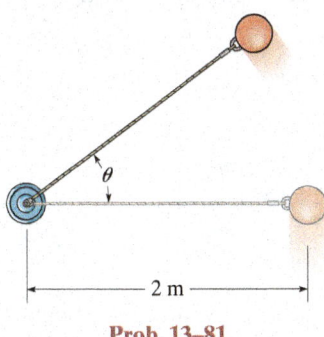

Prob. 13–81

13–82. The cyclist is coasting freely down the hill with a speed of 15 m/s at $y = 0.2$ m. Determine the resultant normal reaction on the bicycle and the rate of increase in speed at the instant shown. The bicycle and rider have a total mass of 80 kg. Neglect friction, the mass of the wheels, and the size of the bicycle.

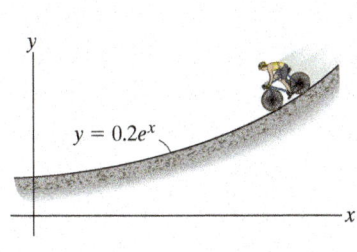

Prob. 13–82

13–83. The "eggbeater" wind turbine is stopped by using spoilers which consist of 20-lb blocks that slide out along the blades when the blades are turning. As the spoilers travel along the blade, they create a drag on the blade that slows and eventually stops the blade. If it is observed that the spoilers reach point A on the blade, determine the required minimum speed of the spoilers needed to maintain this position, and also the normal force they exert on the blade when in this position. The blade has the shape of a parabola, as shown in the figure. The coefficient of static friction between the spoilers and the blades is $\mu_s = 0.3$.

Prob. 13–83

*13–84. The 600-kg wrecking ball is suspended from the crane by a cable having a negligible mass. If the ball has a speed $v = 8$ m/s at the instant it is at its lowest point, $\theta = 0°$, determine the tension in the cable at this instant. Also, determine the angle θ to which the ball swings before it stops.

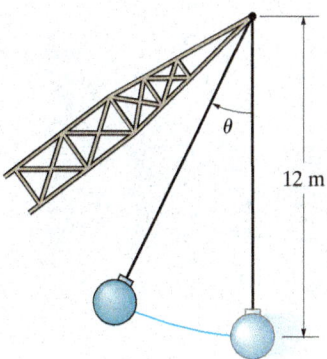

Prob. 13–84

148   CHAPTER 13   KINETICS OF A PARTICLE: FORCE AND ACCELERATION

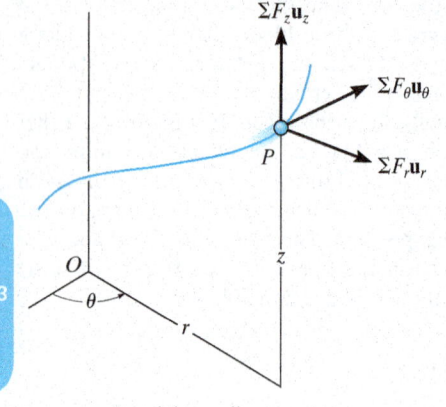

Inertial coordinate system

Fig. 13–16

*13.6 EQUATIONS OF MOTION: CYLINDRICAL COORDINATES

When all the forces acting on a particle are resolved into cylindrical components, i.e., along the unit-vector directions $\mathbf{u}_r$, $\mathbf{u}_\theta$, $\mathbf{u}_z$, Fig. 13–16, the equation of motion can be expressed as

$$\Sigma \mathbf{F} = m\mathbf{a}$$
$$\Sigma F_r \mathbf{u}_r + \Sigma F_\theta \mathbf{u}_\theta + \Sigma F_z \mathbf{u}_z = ma_r \mathbf{u}_r + ma_\theta \mathbf{u}_\theta + ma_z \mathbf{u}_z$$

To satisfy this equation, we require

$$\begin{aligned} \Sigma F_r &= ma_r \\ \Sigma F_\theta &= ma_\theta \\ \Sigma F_z &= ma_z \end{aligned} \tag{13-9}$$

If the particle is constrained to move only in the r–θ plane, then only the first two of Eq. 13–9 are used to specify the motion.

Tangential and Normal Forces.
When a particle is constrained to move along a path, as in Fig. 13–17a, then the free-body diagram for the particle must include a normal force **N** which the path exerts on the particle. Also, if the path is rough, then a friction force **F** must be included. As shown, these forces act normal to the tangent to the path and along the tangent in the opposite direction of motion. When the path is described as $r = f(\theta)$, then the *directions* of **N** and **F** can be specified relative to the radial coordinate by using the angle ψ (psi), Fig. 13–17b, which is defined between the *extended* radial line and the tangent to the curve.

Motion of the roller coaster along this spiral can be studied using cylindrical coordinates.

(a)

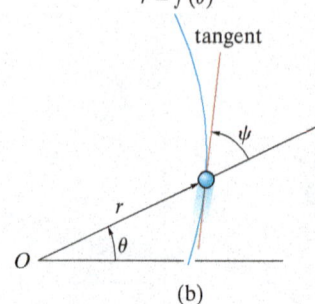

(b)

Fig. 13–17

This angle can be obtained by noting that when the particle is displaced a distance ds along the path, Fig. 13–17c, the component of displacement in the radial direction is dr and the component of displacement in the transverse direction is $r\,d\theta$. Since these two components are mutually perpendicular, the angle ψ can be determined from $\tan \psi = r\,d\theta/dr$, or

$$\tan \psi = \frac{r}{dr/d\theta} \qquad (13\text{–}10)$$

If ψ is calculated as a positive quantity, it is measured from the *extended radial line* to the tangent in a counterclockwise sense or in the positive direction of θ, Fig.13–17c. If it is negative, it is measured in the opposite direction to positive θ. For example, consider the cardioid $r = a(1 + \cos \theta)$, shown in Fig. 13–18. Because $dr/d\theta = -a \sin \theta$, then when $\theta = 30°$, $\tan \psi = a(1 + \cos 30°)/(-a \sin 30°) = -3.732$, or $\psi = -75°$, measured clockwise, opposite to $+\theta$ as shown in the figure.

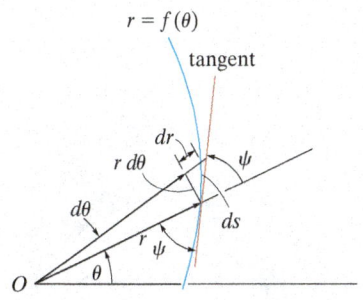

(c)

Fig. 13–17(cont.)

PROCEDURE FOR ANALYSIS

Cylindrical or polar coordinates are a suitable choice for the analysis of a problem for which data regarding the angular motion of the radial line r are given, or in cases where the path can be conveniently expressed in terms of these coordinates.

Free-Body Diagram.

- Establish the r, θ, z inertial coordinate system and draw the particle's free-body diagram.

- Assume that $\mathbf{a}_r$, $\mathbf{a}_\theta$, $\mathbf{a}_z$ act in the positive directions of r, θ, z if they are unknown.

- Identify the unknowns in the problem.

Equations of Motion.

- Apply the equations of motion, Eq. 13–9.

Kinematics.

- Use the methods of Sec. 12.8 to determine r and the time derivatives $\dot{r}$, $\ddot{r}$, $\dot{\theta}$, $\ddot{\theta}$, $\ddot{z}$, and then evaluate the acceleration components $a_r = \ddot{r} - r\dot{\theta}^2$, $a_\theta = r\ddot{\theta} + 2\dot{r}\dot{\theta}$, $a_z = \ddot{z}$.

- When taking the time derivatives of $r = f(\theta)$, it is very important to use the chain rule, which is discussed in Appendix C.

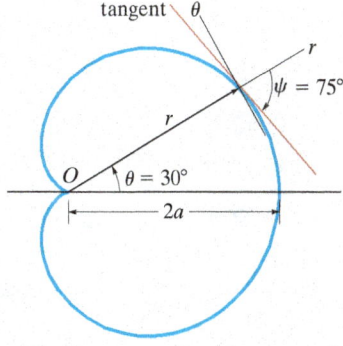

Fig. 13–18

EXAMPLE 13.10

The smooth 0.5-kg double-collar in Fig. 13–19a can freely slide on arm AB and the circular guide rod. If the arm rotates with a constant angular velocity of $\dot{\theta} = 3$ rad/s, determine the force the arm exerts on the collar at the instant $\theta = 45°$. Motion is in the horizontal plane.

SOLUTION

Free-Body Diagram. Only the normal reaction $\mathbf{N}_C$ of the guide rod and the force $\mathbf{F}$ of arm AB act on the collar, Fig. 13–19b. Here $\mathbf{F}$ acts perpendicular to arm AB, that is, in the direction of the θ axis, while $\mathbf{N}_C$ acts perpendicular to the tangent of the circular path at $\theta = 45°$. The acceleration components $\mathbf{a}_r$ and $\mathbf{a}_\theta$ are assumed to act in the positive r and θ directions. The four unknowns are N_C, F, a_r, a_θ.

Equations of Motion.

$$+\nearrow \Sigma F_r = ma_r; \qquad -N_C \cos 45° = (0.5 \text{ kg}) a_r \qquad (1)$$

$$+\nwarrow \Sigma F_\theta = ma_\theta; \qquad F - N_C \sin 45° = (0.5 \text{ kg}) a_\theta \qquad (2)$$

Kinematics. Using the chain rule (see Appendix C), the first and second time derivatives of r when $\theta = 45°$, $\dot{\theta} = 3$ rad/s, $\ddot{\theta} = 0$, are

$$r = 0.8 \cos \theta = 0.8 \cos 45° = 0.5657 \text{ m}$$

$$\dot{r} = -0.8 \sin \theta \, \dot{\theta} = -0.8 \sin 45°(3) = -1.6971 \text{ m/s}$$

$$\ddot{r} = -0.8\left[\sin \theta \, \ddot{\theta} + \cos \theta \, \dot{\theta}^2\right]$$

$$= -0.8[\sin 45°(0) + \cos 45°(3^2)] = -5.091 \text{ m/s}^2$$

Fig. 13–19

We have

$$a_r = \ddot{r} - r\dot{\theta}^2 = -5.091 \text{ m/s}^2 - (0.5657 \text{ m})(3 \text{ rad/s})^2 = -10.18 \text{ m/s}^2$$

$$a_\theta = r\ddot{\theta} + 2\dot{r}\dot{\theta} = (0.5657 \text{ m})(0) + 2(-1.6971 \text{ m/s})(3 \text{ rad/s})$$

$$= -10.18 \text{ m/s}^2$$

Substituting these results into Eqs. 1 and 2 and solving, we get

$$N_C = 7.20 \text{ N}$$

$$F = 0 \qquad \qquad Ans.$$

EXAMPLE 13.11

The smooth 2-kg cylinder C in Fig. 13–20a has a pin P through its center which passes through the slot in arm OA. If the arm is forced to rotate in the *vertical plane* at a constant rate $\dot{\theta} = 0.5$ rad/s, determine the force that the arm exerts on the peg at the instant $\theta = 60°$.

SOLUTION
Why is it a good idea to use polar coordinates to solve this problem?

Free-Body Diagram. The free-body diagram for the cylinder is shown in Fig. 13–20b. The force on the peg, F_P, acts perpendicular to the slot in the arm. As usual, $\mathbf{a}_r$ and $\mathbf{a}_\theta$ are assumed to act in the directions of *positive* r and θ, respectively. Identify the four unknowns.

Equations of Motion. Using the data in Fig. 13–20b, we have

$+\swarrow \Sigma F_r = ma_r;$ $\quad 19.62 \sin \theta - N_C \sin \theta = 2a_r \quad (1)$
$+\nwarrow \Sigma F_\theta = ma_\theta;$ $\quad 19.62 \cos \theta + F_P - N_C \cos \theta = 2a_\theta \quad (2)$

Kinematics. From Fig. 13–20a, r can be related to θ by the equation

$$r = \frac{0.4}{\sin \theta} = 0.4 \csc \theta$$

Since $d(\csc \theta) = -(\csc \theta \cot \theta)\, d\theta$ and $d(\cot \theta) = -(\csc^2 \theta)\, d\theta$, then r and the necessary time derivatives become

$\dot{\theta} = 0.5 \quad\quad r = 0.4 \csc \theta$
$\ddot{\theta} = 0 \quad\quad\quad \dot{r} = -0.4(\csc \theta \cot \theta)\dot{\theta}$
$\quad\quad\quad\quad\quad\;\; = -0.2 \csc \theta \cot \theta$
$\quad\quad\quad\quad\;\; \ddot{r} = -0.2(-\csc \theta \cot \theta)(\dot{\theta}) \cot \theta - 0.2 \csc \theta (-\csc^2 \theta)\dot{\theta}$
$\quad\quad\quad\quad\quad\;\; = 0.1 \csc \theta (\cot^2 \theta + \csc^2 \theta)$

Evaluating these formulas at $\theta = 60°$, we get

$\dot{\theta} = 0.5 \quad\quad r = 0.462$
$\ddot{\theta} = 0 \quad\quad\quad \dot{r} = -0.133$
$\quad\quad\quad\quad\; \ddot{r} = 0.192$

$a_r = \ddot{r} - r\dot{\theta}^2 = 0.192 - 0.462(0.5)^2 = 0.0770$
$a_\theta = r\ddot{\theta} + 2\dot{r}\dot{\theta} = 0 + 2(-0.133)(0.5) = -0.133$

Substituting these results into Eqs. 1 and 2 with $\theta = 60°$ and solving yields

$\quad\quad N_C = 19.4$ N $\quad\quad F_P = -0.356$ N $\quad\quad$ *Ans.*

The negative sign indicates that $\mathbf{F}_P$ acts opposite to the direction shown in Fig. 13–20b.

Fig. 13–20

EXAMPLE 13.12

A can C, having a mass of 0.5 kg, moves along a grooved horizontal slot shown in Fig. 13–21a. The slot is in the form of a spiral, which is defined by the equation $r = (0.1\theta)$ m, where θ is in radians. If the arm OA rotates at a constant rate $\dot{\theta} = 4$ rad/s, determine the force it exerts on the can at the instant $\theta = \pi$ rad. Neglect friction and the size of the can.

SOLUTION

Free-Body Diagram. The driving force $\mathbf{F}_C$ acts perpendicular to the arm OA, whereas the normal force of the slot on the can, $\mathbf{N}_C$, acts perpendicular to the tangent to the curve at $\theta = \pi$ rad, Fig. 13–21b. As usual, $\mathbf{a}_r$ and $\mathbf{a}_\theta$ are assumed to act in the *positive directions* of r and θ. Since the path is specified, the angle ψ which the extended radial line r makes with the tangent, Fig. 13–21c, can be determined from Eq. 13–10. We have $r = 0.1\theta$, so that $dr/d\theta = 0.1$, and therefore

$$\tan \psi = \frac{r}{dr/d\theta} = \frac{0.1\theta}{0.1} = \theta$$

When $\theta = \pi$, $\psi = \tan^{-1}\pi = 72.3°$, so that $\phi = 90° - \psi = 17.7°$, as shown in Fig. 13–21c. Identify the four unknowns in Fig. 13–21b.

Equations of Motion. Using $\phi = 17.7°$ and the data shown in Fig. 13–21b, we have

$\xrightarrow{+} \Sigma F_r = ma_r;$ $\qquad N_C \cos 17.7° = 0.5 a_r$ (1)

$+\downarrow \Sigma F_\theta = ma_\theta;$ $\qquad F_C - N_C \sin 17.7° = 0.5 a_\theta$ (2)

Kinematics. The time derivatives of r and θ are

$\dot{\theta} = 4$ rad/s $\qquad r = 0.1\theta$

$\ddot{\theta} = 0$ $\qquad \dot{r} = 0.1\dot{\theta} = 0.1(4) = 0.4$ m/s

$\qquad\qquad \ddot{r} = 0.1\ddot{\theta} = 0$

At the instant $\theta = \pi$ rad,

$$a_r = \ddot{r} - r\dot{\theta}^2 = 0 - 0.1(\pi)(4)^2 = -5.03 \text{ m/s}^2$$

$$a_\theta = r\ddot{\theta} + 2\dot{r}\dot{\theta} = 0 + 2(0.4)(4) = 3.20 \text{ m/s}^2$$

Substituting these results into Eqs. 1 and 2 and solving yields

$$N_C = -2.64 \text{ N}$$

$$F_C = 0.800 \text{ N} \qquad\qquad Ans.$$

What does the negative sign for N_C indicate?

Fig. 13–21

FUNDAMENTAL PROBLEMS

F13–13. Determine the constant angular velocity $\dot{\theta}$ of the vertical shaft of the amusement ride if $\phi = 45°$. Neglect the mass of the cables and the size of the passengers.

Prob. F13–13

F13–14. The 0.2-kg ball is blown through the smooth vertical circular tube whose shape is defined by $r = (0.6 \sin \theta)$ m, where θ is in radians. If $\theta = (\pi t^2)$ rad, where t is in seconds, determine the magnitude of force $\mathbf{F}$ exerted by the blower on the ball when $t = 0.5$ s.

Prob. F13–14

F13–15. The 2-Mg car is traveling along the curved road described by $r = (50e^{2\theta})$ m, where θ is in radians. If a camera is located at A and it rotates with an angular velocity of $\dot{\theta} = 0.05$ rad/s and an angular acceleration of $\ddot{\theta} = 0.01$ rad/s² at the instant $\theta = \frac{\pi}{6}$ rad, determine the resultant friction force developed between the tires and the road at this instant.

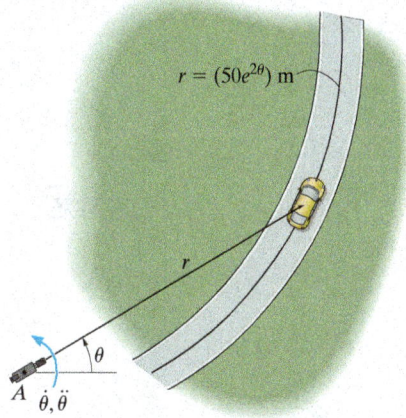

Prob. F13–15

F13–16. The 0.2-kg pin P is constrained to move in the smooth curved slot, which is defined by the lemniscate $r = (0.6 \cos 2\theta)$ m. Its motion is controlled by the rotation of the slotted arm OA, which has a constant clockwise angular velocity of $\dot{\theta} = -3$ rad/s. Determine the force arm OA exerts on the pin P when $\theta = 0°$. Motion is in the vertical plane.

Prob. F13–16

PROBLEMS

All solutions must include a free-body diagram.

13–85. The 2-kg rod AB moves up and down as its end slides on the smooth contoured surface of the cam, where $r = 0.1$ m and $z = (0.02 \sin 2\theta)$ m. If the cam is rotating at a constant rate of 5 rad/s, determine the maximum and minimum force the cam exerts on the rod.

*13–88. The tube rotates in the horizontal plane at a constant rate of $\dot{\theta} = 4$ rad/s. If a 0.2-kg ball B starts at the origin O with an initial radial velocity of $\dot{r} = 1.5$ m/s and moves outward through the tube, determine the radial and transverse components of the ball's velocity at the instant it leaves the outer end at C, $r = 0.5$ m. *Hint:* Show that the equation of motion in the r direction is $\ddot{r} - 16r = 0$. The solution is of the form $r = Ae^{-4t} + Be^{4t}$. Evaluate the integration constants A and B, and determine the time t when $r = 0.5$ m. Proceed to obtain v_r and v_θ.

Prob. 13–85

Prob. 13–88

13–86. Determine the magnitude of the resultant force acting on a 5-kg particle at the instant $t = 2$ s, if the particle is moving along a horizontal path defined by the equations $r = (2t + 10)$ m and $\theta = (1.5t^2 - 6t)$ rad, where t is in seconds.

13–87. The path of motion of a 5-lb particle in the horizontal plane is described in terms of polar coordinates as $r = (2t + 1)$ ft and $\theta = (0.5t^2 - t)$ rad, where t is in seconds. Determine the magnitude of the unbalanced force acting on the particle when $t = 2$ s.

13–89. The boy of mass 40 kg is sliding down the spiral slide at a constant speed such that his position, measured from the top of the chute, has components $r = 1.5$ m, $\theta = (0.7t)$ rad, and $z = (-0.5t)$ m, where t is in seconds. Determine the components of force $\mathbf{F}_r$, $\mathbf{F}_\theta$, and $\mathbf{F}_z$ which the slide exerts on him at the instant $t = 2$ s. Neglect the size of the boy.

Prob. 13–89

13–90. The 40-kg boy is sliding down the smooth spiral slide such that $z = -2$ m/s when he rotates $\theta = 360°$ and his speed is 2 m/s. Determine the r, θ, z components of force the slide exerts on him at this instant. Neglect the size of the boy.

Prob. 13–90

13–91. Using a forked rod, a 0.5-kg smooth peg P is forced to move along the *vertical slotted* path $r = (0.5\theta)$ m, where θ is in radians. If the angular position of the arm is $\theta = (\frac{\pi}{8}t^2)$ rad, where t is in seconds, determine the force of the rod on the peg and the normal force of the slot on the peg at the instant $t = 2$ s. The peg is in contact with only *one edge* of the rod and slot at any instant.

Prob. 13–91

*13–92. A boy standing firmly spins the girl sitting on a circular "dish" or sled in a circular path of radius $r_0 = 3$ m such that her angular velocity is $\dot\theta_0 = 0.1$ rad/s. If the attached cable OC is drawn inward such that the radial coordinate r changes with a constant speed of $\dot r = -0.5$ m/s, determine the tension it exerts on the sled at the instant $r = 2$ m. The sled and girl have a total mass of 50 kg. Neglect the size of the girl and sled and the effects of friction between the sled and ice. *Hint:* First show that the equation of motion in the θ direction yields $a_\theta = r\ddot\theta + 2\dot r\dot\theta = (1/r)\, d/dt\,(r^2\dot\theta) = 0$. When integrated, $r^2\dot\theta = C$, where the constant C is determined from the problem data.

Prob. 13–92

13–93. The collar, which has a weight of 3 lb, slides along the smooth rod lying in the *horizontal plane* and having the shape of a parabola, $r = [4/(1-\cos\theta)]$ ft, where θ is in radians. If the collar's angular rate is constant and equals $\dot\theta = 4$ rad/s, determine the tangential force P needed to cause the motion, and the normal force component that the collar exerts on the rod at the instant $\theta = 90°$.

Prob. 13–93

13–94. The smooth pin P has a mass of 80 g. It is attached to an elastic cord extending from O to P and, due to the slotted arm guide, moves along the *horizontal* circular path $r = (0.8 \sin \theta)$ m. If the cord has a stiffness $k = 30$ kN/m and an unstretched length of 0.25 m, determine the force of the guide and the normal force of the circular path on the pin when $\theta = 60°$. The guide has a constant angular velocity $\dot{\theta} = 5$ rad/s.

*13–96. If the coefficient of static friction between the conical surface and the block of mass m is $\mu_s = 0.2$, determine the minimum constant angular velocity $\dot{\theta}$ so that the block does not slide downwards.

13–97. If the coefficient of static friction between the conical surface and the block is $\mu_s = 0.2$, determine the maximum constant angular velocity $\dot{\theta}$ without causing the block to slide upwards.

Prob. 13–94

Probs. 13–96/97

13–95. Using air pressure, the 0.5-kg ball is forced to move through the tube lying in the horizontal plane and having the shape of a logarithmic spiral. If the tangential force exerted on the ball due to air pressure is 6 N, determine the rate of increase in the ball's speed at the instant $\theta = \pi/2$. Also, what is the angle ψ from the extended radial coordinate r to the line of action of the 6-N force?

13–98. The 2-lb spool slides along the smooth *horizontal* spiral rod, $r = (2\,\theta)$ ft, where θ is in radians. If its angular rate of rotation is constant and equals $\dot{\theta} = 4$ rad/s, determine the horizontal tangential force P needed to cause the motion, and the horizontal normal force component that the spool exerts on the rod at the instant $\theta = 90°$.

Prob. 13–95

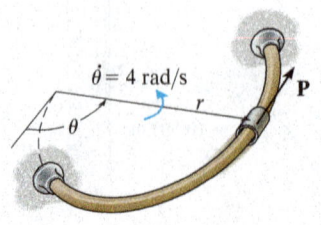

Prob. 13–98

PROBLEMS   157

13–99. If the position of the 3-kg collar C on the smooth rod AB is held at $r = 720$ mm, determine the constant angular velocity $\dot{\theta}$ at which the mechanism is rotating about the vertical axis. The spring has an unstretched length of 400 mm. Neglect the mass of the rod and the size of the collar.

13–102. Using air pressure, the 0.5-kg ball is forced to move through the tube lying in the *horizontal plane* and having the shape of a logarithmic spiral. If the tangential force exerted on the ball due to the air is 6 N, determine the rate of increase in the ball's speed at the instant $\theta = \pi/2$. What direction does the tangential force act in?

Prob. 13–99

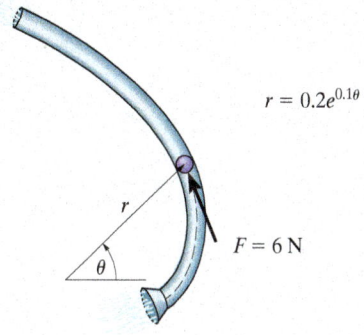

Prob. 13–102

*13–100. For a short time, the 250-kg roller-coaster car with passengers is traveling along the spiral track at a constant speed of $v = 8$ m/s. If the track descends $d = 12$ m for every full revolution, $\theta = 2\pi$ rad, determine the magnitudes of the components of force which the track exerts on the car in the $r, \theta,$ and z directions. Neglect the size of the car.

13–101. For a short time, the 250-kg roller-coaster car with passengers is traveling along the spiral track at a constant speed such that its position measured from the top of the track has components $r = 10$ m, $\theta = (0.2t)$ rad, and $z = (-0.3t)$ m, where t is in seconds. Determine the magnitudes of the components of force which the track exerts on the car in the $r, \theta,$ and z directions at the instant $t = 2$ s. Neglect the size of the car.

13–103. The amusement park ride rotates with a constant angular velocity of $\dot{\theta} = 0.8$ rad/s. If the path of the ride is defined by $r = (3 \sin \theta + 5)$ m and $z = (3 \cos \theta)$ m, determine the $r, \theta,$ and z components of force exerted by the seat on the 20-kg boy when $\theta = 120°$.

Probs. 13–100/101

Prob. 13–103

*13–104. The collar has a mass of 2 kg and travels along the smooth horizontal rod defined by the equiangular spiral $r = (e^\theta)$ m, where θ is in radians. Determine the tangential force F and the normal force N acting on the collar when $\theta = 45°$, if the force F maintains a constant angular motion $\dot{\theta} = 2$ rad/s.

Prob. 13–104

13–105. The smooth surface of the vertical cam is defined in part by the curve $r = (0.2 \cos\theta + 0.3)$ m. The forked rod is rotating with an angular acceleration of $\ddot{\theta} = 2$ rad/s^2, and when $\theta = 45°$, the angular velocity is $\dot{\theta} = 6$ rad/s. Determine the force the cam and the rod exert on the 2-kg roller at this instant. The attached spring has a stiffnesss $k = 100$ N/m and an unstretched length of 0.1 m.

Prob. 13–105

13–106. The smooth surface of the vertical cam is defined in part by the curve $r = (0.2 \cos\theta + 0.3)$ m. If the forked rod is rotating with a constant angular velocity of $\dot{\theta} = 4$ rad/s, determine the force the cam and the rod exert on the 2-kg roller when $\theta = 30°$. The attached spring has a stiffnesss $k = 30$ N/m and an unstretched length of 0.1 m.

Prob. 13–106

13–107. The forked rod is used to move the smooth 2-lb particle around the horizontal path in the shape of a limaçon, $r = (2 + \cos\theta)$ ft. If $\theta = (0.5t^2)$ rad, where t is in seconds, determine the force which the rod exerts on the particle at the instant $t = 1$ s. The fork and path contact the particle on only one side.

Prob. 13–107

*13–108. The spool, which has a mass of 2 kg, slides along the smooth *horizontal* spiral rod, $r = (0.40\theta)$ m, where θ is in radians. If its angular rate of rotation is constant and equals $\dot{\theta} = 6$ rad/s, determine the horizontal tangential force P needed to cause the motion, and the horizontal normal force component that the spool exerts on the rod at the instant $\theta = 45°$.

Prob. 13–108

13–109. Rod OA rotates counterclockwise at a constant angular rate $\dot{\theta} = 4$ rad/s. The double collar B is pin connected together such that one collar slides over the rotating rod and the other collar slides over the circular rod described by the equation $r = (1.6 \cos \theta)$ m. If *both* collars have a mass of 0.5 kg, determine the force which the circular rod exerts on one of the collars and the force that OA exerts on the other collar at the instant $\theta = 45°$. Motion is in the horizontal plane.

13–110. Solve Prob. 13–109 if motion is in the vertical plane.

Probs. 13–109/110

13–111. A 0.2-kg spool slides down along a smooth rod. If the rod has a constant angular rate of rotation $\dot{\theta} = 2$ rad/s in the vertical plane, show that the equations of motion for the spool are $\ddot{r} - 4r - 9.81 \sin \theta = 0$ and $0.8\dot{r} + N_s - 1.962 \cos \theta = 0$, where N_s is the magnitude of the normal force of the rod on the spool. Using the methods of differential equations, it can be shown that the solution of the first of these equations is $r = C_1 e^{-2t} + C_2 e^{2t} - (9.81/8) \sin 2t$. If r, $\dot{r}$, and θ are zero when $t = 0$, evaluate the constants C_1 and C_2 and determine r at the instant $\theta = \pi/4$ rad.

Prob. 13–111

*13–112. The airplane executes the vertical loop defined by $r^2 = [810(10^3) \cos 2\theta]$ m^2. If the pilot maintains a constant speed $v = 120$ m/s along the path, determine the normal force the seat exerts on him at the instant $\theta = 0°$. The pilot has a mass of 75 kg.

Prob. 13–112

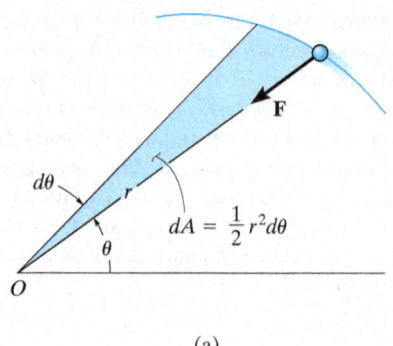

*13.7 CENTRAL-FORCE MOTION AND SPACE MECHANICS

If a particle is moving only under the influence of a force **F** which is always directed toward a fixed point, as in Fig. 13–22a, the motion is called *central-force motion*. This type of motion is commonly caused by electrostatic and gravitational forces.

For this case, the free-body diagram for the particle is shown in Fig. 13–22b. Using polar coordinates (r, θ), the equations of motion, Eq. 13–9, become

$$\Sigma F_r = ma_r; \qquad -F = m\left[\frac{d^2r}{dt^2} - r\left(\frac{d\theta}{dt}\right)^2\right] \qquad (13\text{–}11)$$

$$\Sigma F_\theta = ma_\theta; \qquad 0 = m\left(r\frac{d^2\theta}{dt^2} + 2\frac{dr}{dt}\frac{d\theta}{dt}\right)$$

The second of these equations may be written in the form

$$\frac{1}{r}\left[\frac{d}{dt}\left(r^2\frac{d\theta}{dt}\right)\right] = 0$$

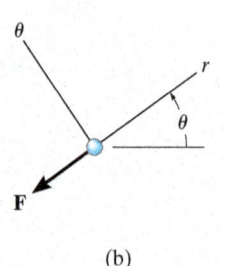

(b)

Fig. 13–22

so that integrating yields

$$r^2\frac{d\theta}{dt} = h \qquad (13\text{–}12)$$

Here h is the constant of integration.

From Fig. 13–22a notice that the shaded area described by the radius r, as r moves through an angle $d\theta$, is a triangle having two sides r, and, neglecting higher order derivatives, a base of $rd\theta$. The area of this triangle is therefore $dA = \frac{1}{2}r^2\,d\theta$. If the **areal velocity** is defined as

$$\frac{dA}{dt} = \frac{1}{2}r^2\frac{d\theta}{dt} = \frac{h}{2} \qquad (13\text{–}13)$$

then it is seen that this term is *constant*. In other words, the particle will sweep out equal segments of area per unit of time as it travels along the path.

To obtain the *path of motion*, $r = f(\theta)$, the independent variable t must be eliminated from Eq. 13–11. Using the chain rule, the time derivative d^2r/dt^2 may be replaced by

$$\frac{dr}{dt} = \frac{dr}{d\theta}\frac{d\theta}{dt} = \frac{h}{r^2}\frac{dr}{d\theta}$$

$$\frac{d^2r}{dt^2} = \frac{d}{dt}\left(\frac{h}{r^2}\frac{dr}{d\theta}\right) = \frac{d}{d\theta}\left(\frac{h}{r^2}\frac{dr}{d\theta}\right)\frac{d\theta}{dt} = \left[\frac{d}{d\theta}\left(\frac{h}{r^2}\frac{dr}{d\theta}\right)\right]\frac{h}{r^2}$$

Substituting a new dependent variable (xi) $\xi = 1/r$ into this equation, we have

$$\frac{d^2r}{dt^2} = -h^2\xi^2\frac{d^2\xi}{d\theta^2}$$

Also, the square of Eq. 13–12 becomes

$$\left(\frac{d\theta}{dt}\right)^2 = h^2\xi^4$$

Finally, substituting these two equations into Eq. 13–11 we get

$$-h^2\xi^2\frac{d^2\xi}{d\theta^2} - h^2\xi^3 = -\frac{F}{m}$$

or

$$\frac{d^2\xi}{d\theta^2} + \xi = \frac{F}{mh^2\xi^2} \qquad (13\text{--}14)$$

This differential equation defines the path over which the particle travels when it is subjected to the central force **F**.

To illustrate a common application, consider the trajectory of a space satellite or space vehicle launched into free-flight orbit with an initial velocity $\mathbf{v}_0$, Fig. 13–23. Here, it will be assumed that this velocity is initially *parallel* to the tangent at the surface of the earth, as shown in the figure.* Also, just after the satellite is released, the only gravitational force of any sizable magnitude will be the gravitational force of the earth. According to Newton's law of gravitation, this force will always act between the mass centers of the earth and the satellite, Fig. 13–23, and from Eq. 13–1, this force has a magnitude of

$$F = G\frac{M_e m}{r^2}$$

where M_e and m represent the mass of the earth and the satellite, respectively, G is the gravitational constant, and r is the distance between

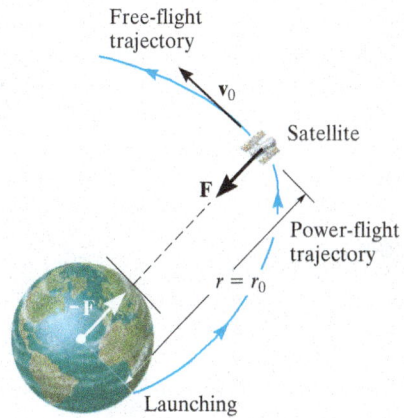

This satellite is subjected to a central force and its orbital motion can be closely predicted using the equations developed in this section. (Universal ImagesGroup/Getty Images)

Fig. 13–23

---

*The case where $\mathbf{v}_0$ acts at some initial angle θ to the tangent is best described using the conservation of angular momentum. See Prob. 15–113.

the mass centers. To obtain the orbital path, we set $\xi = 1/r$ in the foregoing equation and substitute the result into Eq. 13–14. This yields

$$\frac{d^2\xi}{d\theta^2} + \xi = \frac{GM_e}{h^2} \qquad (13\text{–}15)$$

The solution of this second-order differential equation is the sum of its complementary and particular solutions given by

$$\xi = \frac{1}{r} = C\cos(\theta - \phi) + \frac{GM_e}{h^2} \qquad (13\text{–}16)$$

where C and ϕ are constants of integration.

When graphed it represents the equation of a conic section, expressed in terms of polar coordinates. As shown in Fig. 13–24, a conic section is defined as the locus of a point P that moves in such a way that the ratio of its distance FP from a *focus*, or fixed point F, to its perpendicular distance PA measured to the fixed line DD called the **directrix**, is constant. This constant ratio will be denoted as e and is called the **eccentricity**. Therefore,

$$e = \frac{FP}{PA}$$

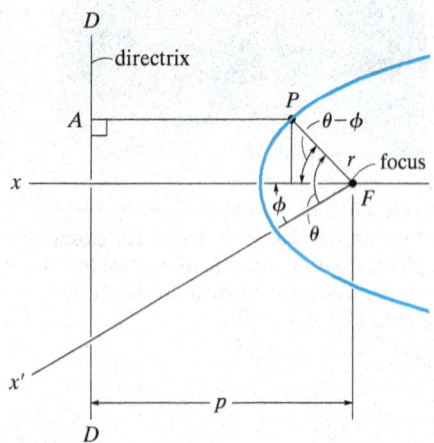

Fig. 13–24

From Fig. 13–24,

$$FP = r = e(PA) = e[\,p - r\cos(\theta - \phi)\,]$$

or

$$\frac{1}{r} = \frac{1}{p}\cos(\theta - \phi) + \frac{1}{ep}$$

Comparing this equation with Eq. 13–16, it is seen that the fixed distance from the focus to the directrix is

$$p = \frac{1}{C} \qquad (13\text{–}17)$$

And the eccentricity of the conic section is

$$\boxed{e = \frac{Ch^2}{GM_e}} \qquad (13\text{–}18)$$

When the polar angle θ in Fig. 13–24 is measured from the x axis rather than the x' axis, then the angle ϕ is zero, and therefore Eq. 13–16 reduces to

$$\frac{1}{r} = C \cos \theta + \frac{GM_e}{h^2} \qquad (13\text{--}19)$$

The constants h and C can be determined from the initial values of the position and velocity of the satellite just after the end of its *power-flight trajectory*. For example, if the initial height or distance to the space vehicle is r_0, measured from the center of the earth, and its initial speed is v_0 at the beginning of its free flight, Fig. 13–25, then the constant h may be obtained from Eq. 13–12. When $\theta = \phi = 0°$, the velocity $\mathbf{v}_0$ has no radial component; therefore, from Eq. 12–25, $v_0 = r_0(d\theta/dt)$, so that

$$h = r_0^2 \frac{d\theta}{dt}$$

or

$$h = r_0 v_0 \qquad (13\text{--}20)$$

To determine C, use Eq. 13–19 with $\theta = 0°, r = r_0$, and substitute Eq. 13–20 for h. We get

$$C = \frac{1}{r_0}\left(1 - \frac{GM_e}{r_0 v_0^2}\right) \qquad (13\text{--}21)$$

The equation for the free-flight trajectory therefore becomes

$$\frac{1}{r} = \frac{1}{r_0}\left(1 - \frac{GM_e}{r_0 v_0^2}\right)\cos \theta + \frac{GM_e}{r_0^2 v_0^2} \qquad (13\text{--}22)$$

The type of path traveled by the satellite is determined from the value of the eccentricity of the conic section as given by Eq. 13–18. If

$$\begin{array}{ll} e = 0 & \text{free-flight trajectory is a circle} \\ e = 1 & \text{free-flight trajectory is a parabola} \\ e < 1 & \text{free-flight trajectory is an ellipse} \\ e > 1 & \text{free-flight trajectory is a hyperbola} \end{array} \qquad (13\text{--}23)$$

Each of these trajectories is shown in Fig. 13–25.

Parabolic Path. It is seen that when the satellite follows a hyperbolic orbit, $e > 1$, it will never return to its starting point. When $e = 1$, the orbit is parabolic, and it is "on the border" of never returning. This launch velocity, $\mathbf{v}_0$, is called the *escape velocity*, v_e. It can be determined by using the second of Eqs. 13–23, $e = 1$, with Eqs. 13–18, 13–20, and 13–21. It is left as an exercise to show that

$$v_e = \sqrt{\frac{2GM_e}{r_0}} \qquad (13\text{–}24)$$

Circular Orbit. The speed v_c required to launch a satellite into a circular orbit can be found using the first of Eqs. 13–23, $e = 0$. Since e is related to h and C, Eq. 13–18, C must be zero to satisfy this equation (from Eq. 13–20, h cannot be zero); and therefore, using Eq. 13–21, we have

$$v_c = \sqrt{\frac{GM_e}{r_0}} \qquad (13\text{–}25)$$

Any speed at launch which is less than v_c will cause the satellite to reenter the earth's atmosphere and either burn up or crash, Fig. 13–25.

Fig. 13–25

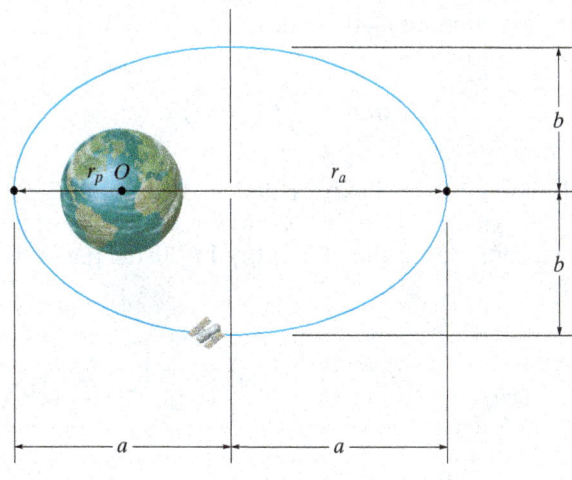

Fig. 13–26

Elliptical Orbit. All the trajectories attained by planets and most satellites are elliptical, Fig. 13–26. For a satellite, the *minimum distance* from the orbit to the center of the earth O (which is located at one of the foci of the ellipse) is r_p and can be found using Eq. 13–22 with $\theta = 0°$. Therefore

$$r_p = r_0 \qquad (13\text{–}26)$$

This distance is called the **perigee** of the orbit. The **apogee** or maximum distance r_a can be found using Eq. 13–22 with $\theta = 180°$.* Thus,

$$r_a = \frac{r_0}{(2GM_e/r_0 v_0^2) - 1} \qquad (13\text{–}27)$$

From Fig. 13–26, the half-length of the major axis of the ellipse is

$$a = \frac{r_p + r_a}{2} \qquad (13\text{–}28)$$

Using analytical geometry, it can be shown that the half-length of the minor axis is determined from the equation

$$b = \sqrt{r_p r_a} \qquad (13\text{–}29)$$

---

* Actually, the terminology perigee and apogee pertains only to orbits about the *earth*. If any other heavenly body is located at the focus of its elliptical orbit, the minimum and maximum distances are referred to respectively as the **periapsis** and **apoapsis** of the orbit.

Finally, by direct integration, the area of an ellipse is

$$A = \pi ab = \frac{\pi}{2}(r_p + r_a)\sqrt{r_p r_a} \qquad (13\text{--}30)$$

and since the areal velocity has been defined by Eq. 13–13, $dA/dt = h/2$, then integrating yields $A = hT/2$, where T is the **period** of time required to make one orbital revolution. From Eq. 13–30, the period is therefore

$$T = \frac{\pi}{h}(r_p + r_a)\sqrt{r_p r_a} \qquad (13\text{--}31)$$

In addition to predicting the orbital trajectory of earth satellites, the theory developed in this section is also valid, to a surprisingly close approximation, at predicting the actual motion of the planets orbiting the sun. In this case the mass of the sun, M_s, should be substituted for M_e when the appropriate formulas are used.

The fact that the planets do indeed follow elliptic orbits about the sun was discovered by the German astronomer Johannes Kepler in the early seventeenth century. Since his discovery was made *before* Newton had developed the laws of motion and the law of gravitation, it provided important proof as to the validity of these laws.

Kepler's three laws, developed after 20 years of planetary observation, are summarized as follows:

1. Every planet travels in its orbit such that the radial line measured to it from the center of the sun sweeps over equal areas in equal intervals of time.

2. The orbit of every planet is an ellipse with the sun placed at one of its foci.

3. The square of the period of any planet is directly proportional to the cube of the major axis of its orbit.

A mathematical statement of the first and second laws is given by Eqs. 13–13 and 13–22, respectively. The third law can be shown from Eq. 13–31 using Eqs. 13–19, 13–28, and 13–29. (See Prob. 13–117.)

EXAMPLE 13.13

A satellite is launched 600 km from the surface of the earth, with an initial velocity of 30 Mm/h acting parallel to the tangent at the surface of the earth, Fig. 13–27. Assuming that the radius of the earth is 6378 km and that its mass is $5.976(10^{24})$ kg, determine (a) the eccentricity of the orbital path, and (b) the velocity of the satellite at apogee.

SOLUTION

Part (a). The eccentricity of the orbit is obtained using Eq. 13–18. The constants h and C are first determined from Eq. 13–20 and 13–21. Since

$$r_p = r_0 = 6378 \text{ km} + 600 \text{ km} = 6.978(10^6) \text{ m}$$

$$v_0 = 30 \text{ Mm/h} = 8333.3 \text{ m/s}$$

then

$$h = r_p v_0 = 6.978(10^6)(8333.3) = 58.15(10^9) \text{ m}^2/\text{s}$$

$$C = \frac{1}{r_p}\left(1 - \frac{GM_e}{r_p v_0^2}\right)$$

$$= \frac{1}{6.978(10^6)}\left[1 - \frac{66.73(10^{-12})[5.976(10^{24})]}{6.978(10^6)(8333.3)^2}\right] = 25.4(10^{-19}) \text{ m}^{-1}$$

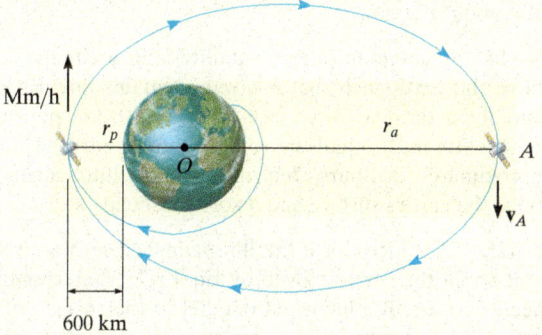

Fig. 13–27

Hence,

$$e = \frac{Ch^2}{GM_e} = \frac{2.54(10^{-8})[58.15(10^9)]^2}{66.73(10^{-12})[5.976(10^{24})]} = 0.125 < 1 \quad Ans.$$

From Eqs. 13–23, observe that the orbit is an *ellipse*.

Part (b). If the satellite were launched at the apogee A shown in Fig. 13–27, with a velocity v_A, the same orbit would be maintained provided

$$h = r_p v_0 = r_a v_A = 58.15(10^9) \text{ m}^2/\text{s}$$

Using Eq. 13–27, we have

$$r_a = \frac{r_p}{\dfrac{2GM_e}{r_p v_0^2} - 1} = \frac{6.978(10^6)}{\dfrac{2[66.73(10^{-12})][5.976(10^{24})]}{6.978(10^6)(8333.3)^2} - 1} = 10.804(10^6) \text{ m}$$

Thus,

$$v_A = \frac{58.15(10^9)}{10.804(10^6)} = 5382.2 \text{ m/s} = 19.4 \text{ Mm/h} \quad Ans.$$

NOTE: The farther the satellite is from the earth, the slower it moves, which is to be expected since h is constant.

PROBLEMS

In the following problems, except where otherwise indicated, assume that the radius of the earth is 6378 km, the earth's mass is $5.976(10^{24})$ kg, the mass of the sun is $1.99(10^{30})$ kg, and the gravitational constant is $G = 66.73(10^{-12})$ m^3/(kg·s^2).

13–113. The earth has an orbit with eccentricity 0.0167 around the sun. Knowing that the earth's minimum distance from the sun is $146(10^6)$ km, find the speed at which the earth travels when it is at this distance. Determine the equation in polar coordinates which describes the earth's orbit about the sun.

13–114. A communications satellite is in a circular orbit above the earth such that it always remains directly over a point on the earth's surface. As a result, the period of the satellite must equal the rotation of the earth, which is approximately 24 hours. Determine the satellite's altitude h above the earth's surface and its orbital speed.

13–115. The speed of a satellite launched into a circular orbit about the earth is given by Eq. 13–25. Determine the speed of a satellite launched parallel to the surface of the earth so that it travels in a circular orbit 800 km from the earth's surface.

*13–116. The rocket is docked next to a satellite located 18 Mm above the earth's surface. If the satellite is traveling in a circular orbit, determine the speed tangent to the earth's surface which must suddenly be given to the rocket, relative to the satellite, such that it travels in free flight away from the satellite along a parabolic trajectory as shown.

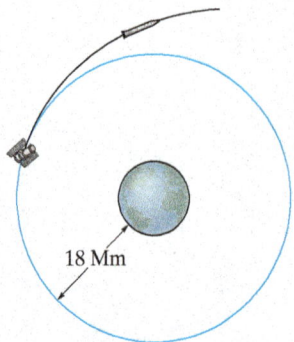

Prob. 13–116

13–117. Prove Kepler's third law of motion. *Hint*: Use Eqs. 13–19, 13–28, 13–29, and 13–31.

13–118. The satellite is in an elliptic orbit around the earth as shown. Determine its velocity at perigee P and apogee A, and the period of the satellite.

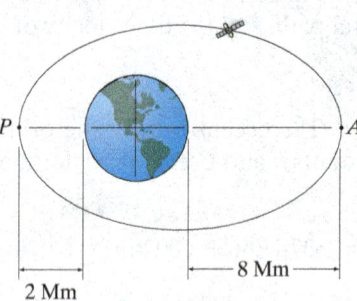

Prob. 13–118

13–119. The rocket is traveling in free flight along the elliptical orbit. The planet has no atmosphere, and its mass is 0.60 times that of the earth. If the rocket has the orbit shown, determine the rocket's speed when it is at A and at B.

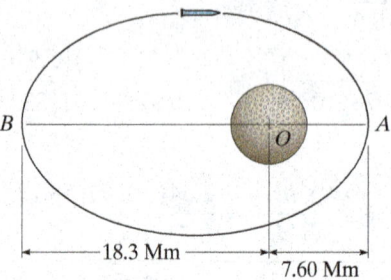

Prob. 13–119

*13–120. A satellite is to be placed into an elliptical orbit about the earth such that at the perigee of its orbit it has an *altitude* of 800 km, and at apogee its *altitude* is 2400 km. Determine its required launch velocity tangent to the earth's surface at perigee, and the period of its orbit.

Prob. 13–120

13–121. The rocket is traveling in free flight along an elliptical trajectory $A'A$. The planet has a mass 0.60 times that of the earth. If the rocket has an apoapsis and periapsis as shown in the figure, determine the speed of the rocket when it is at point A.

Prob. 13–121

13–122. The rocket shown is originally in a circular orbit 6 Mm above the surface of the earth. It is required that it travel in another circular orbit having an altitude of 14 Mm. To do this, the rocket is given a short pulse of power at A so that it travels in free flight along the dashed elliptical path from the first orbit to the second orbit. Determine the necessary speed it must have at A just after the power pulse, and the time required to get to the outer orbit along the path AA'. What adjustment in speed must be made at A' to maintain the second circular orbit?

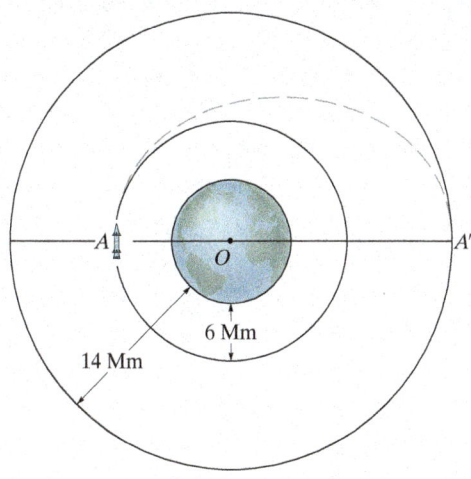

Prob. 13–122

13–123. The rocket is initially in free-flight circular orbit around the earth. Determine the speed of the rocket at A. What change in the speed at A is required so that it can move in an elliptical orbit to reach point A'?

*13–124. The rocket is in free-flight circular orbit around the earth. Determine the time needed for the rocket to travel from the inner orbit at A to the outer orbit at A'.

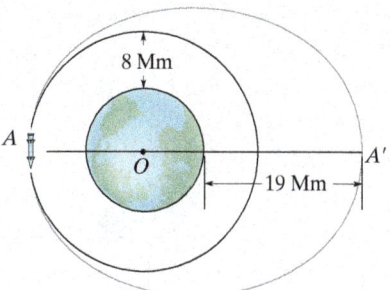

Probs. 13–123/124

13–125. The rocket is in free flight along an elliptical trajectory $A'A$. The planet has no atmosphere, and its mass is 0.60 times that of the earth. If the orbit has the apoapsis and periapsis shown, determine the rocket's velocity when it is at point A. Take $G = 34.4(10^{-9})(\text{lb} \cdot \text{ft}^2)/\text{slug}^2$, $M_e = 409(10^{21})$ slug, 1 mi = 5280 ft.

13–126. If the rocket is to land on the surface of the planet, determine the required free-flight speed it must have at A' so that the landing occurs at B. How long does it take for the rocket to land, in going from A' to B? The planet has no atmosphere, and its mass is 0.6 times that of the earth. Take $G = 34.4(10^{-9})(\text{lb} \cdot \text{ft}^2)/\text{slug}^2$, $M_e = 409(10^{21})$ slug, 1 mi = 5280 ft.

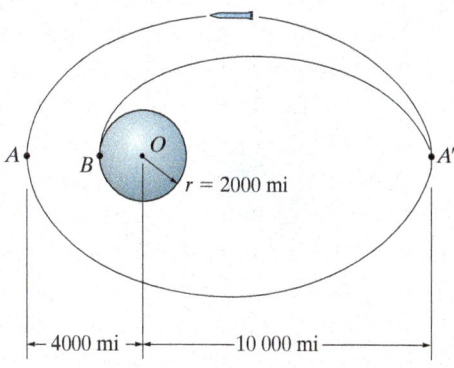

Probs. 13–125/126

13–127. The rocket is in a free-flight elliptical orbit about the earth such that the eccentricity of its orbit is e and its perigee is r_0. Determine the minimum increment of speed it should have in order to escape the earth's gravitational field when it is at this point along its orbit.

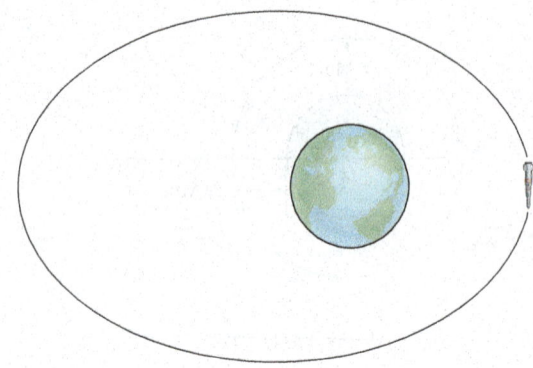

Prob. 13–127

*13–128.** The space shuttle is launched with a velocity of 17 500 mi/h parallel to the tangent of the earth's surface at point P and then travels around the elliptical orbit. When it reaches point A, its engines are turned on and its velocity is suddenly increased. Determine the required increase in velocity so that it enters the second elliptical orbit. Take $G = 34.4 \ (10^{-9}) \ \text{ft}^4/\text{lb} \cdot \text{s}^4$, $M_e = 409(10^{21})$ slug, and $r_e = 3960$ mi, where 5280 ft = mi.

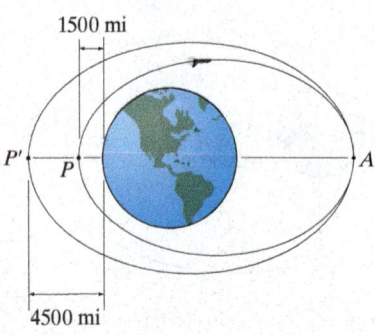

Prob. 13–128

13–129. The rocket is traveling in free flight along an elliptical trajectory $A'A$. The planet has no atmosphere, and its mass is 0.60 times that of the earth. If the rocket has the orbit shown, determine the rocket's velocity when it is at point A.

13–130. If the rocket is to land on the surface of the planet, determine the required free-flight speed it must have at A' so that the landing occurs at B. How long does it take for the rocket to land, going from A' to B? The planet has no atmosphere, and its mass is 0.6 times that of the earth.

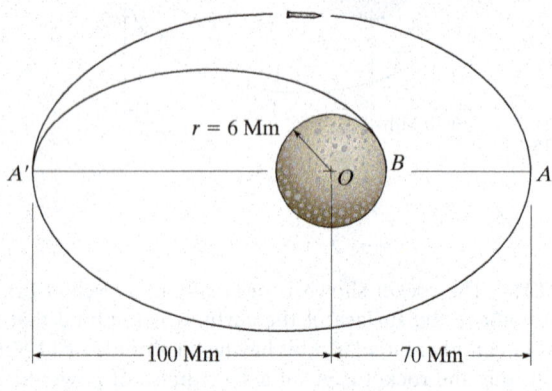

Probs. 13–129/130

13–131. The rocket is traveling around the earth in free flight along an elliptical orbit AC. If the rocket has the orbit shown, determine the rocket's velocity when it is at point A.

*13–132.** The rocket is traveling around the earth in free flight along the elliptical orbit AC. Determine its change in speed when it reaches A so that it travels along the elliptical orbit AB.

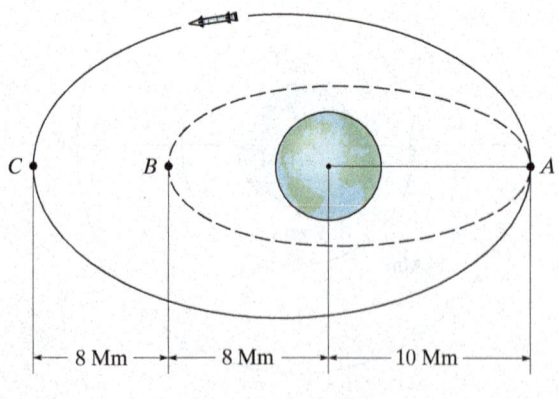

Probs. 13–131/132

CHAPTER REVIEW

Kinetics

Kinetics is the study of the relationship between forces and the acceleration they cause. This relationship is based on Newton's second law of motion, expressed mathematically as $\Sigma \mathbf{F} = m\mathbf{a}$.

Before applying the equation of motion, it is important to first draw the particle's *free-body diagram* in order to account for all of the forces that act on the particle. Graphically, this diagram is equal to the *kinetic diagram*, which shows the result of the forces, that is, the $m\mathbf{a}$ vector.

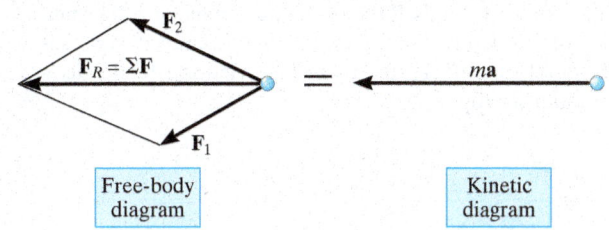

Free-body diagram   Kinetic diagram

Inertial Coordinate System

When applying the equation of motion, it is important to measure the acceleration from an inertial coordinate system. This system has axes that do not rotate but are either fixed or translate with a constant velocity. Various types of inertial coordinate systems can be used to apply $\Sigma \mathbf{F} = m\mathbf{a}$ in component form.

Inertial frame of reference

Rectangular x, y, z axes are used to describe the motion along each of the straight-line axes.

$$\Sigma F_x = ma_x \qquad \Sigma F_y = ma_y \qquad \Sigma F_z = ma_z$$

Normal, tangential, and binormal axes n, t, b are often used when the path is known. The normal component of acceleration, $\mathbf{a}_n$, is always directed in the $+n$ direction. It indicates the change in the velocity direction. The tangential component, $\mathbf{a}_t$, is tangent to the path. It indicates the change in the velocity magnitude.

$$\Sigma F_t = ma_t \qquad \Sigma F_n = ma_n \qquad \Sigma F_b = 0$$
$$a_t = dv/dt \quad \text{or} \quad a_t = v\, dv/ds$$
$$a_n = v^2/\rho \quad \text{where} \quad \rho = \frac{[1 + (dy/dx)^2]^{3/2}}{|d^2y/dx^2|}$$

Cylindrical coordinates are useful when angular motion of the radial line r is known or when the path can conveniently be described with these coordinates.

$$\Sigma F_r = m(\ddot{r} - r\dot{\theta}^2)$$
$$\Sigma F_\theta = m(r\ddot{\theta} + 2\dot{r}\dot{\theta})$$
$$\Sigma F_z = m\ddot{z}$$

Central-Force Motion

When a single force acts upon a particle, such as during the free-flight trajectory of a satellite in a gravitational field, then the motion is referred to as central-force motion. The orbit depends upon its eccentricity e; and as a result, the trajectory can either be circular, parabolic, elliptical, or hyperbolic.

REVIEW PROBLEMS

R13–1. The van is traveling at 20 km/h when the coupling of the trailer at A fails. If the trailer has a mass of 250 kg and coasts 45 m before coming to rest, determine the constant horizontal force F created by rolling friction which causes the trailer to stop.

R13–3. Block B rests on a smooth surface. If the coefficients of friction between A and B are $\mu_s = 0.4$ and $\mu_k = 0.3$, determine the acceleration of each block if $F = 50$ lb.

Prob. R13–1

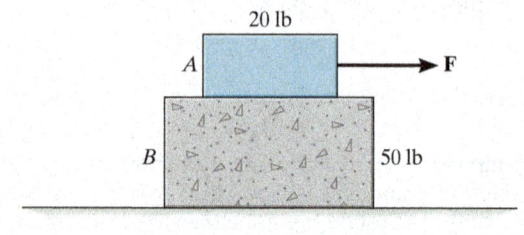

Prob. R13–3

R13–2. The motor M pulls in its attached rope with an acceleration $a_p = 6$ m/s². Determine the towing force exerted by M on the rope in order to move the 50-kg crate up the inclined plane. The coefficient of kinetic friction between the crate and the plane is $\mu_k = 0.3$. Neglect the mass of the pulleys and rope.

R13–4. If the motor draws in the cable at a rate of $v = (0.05s^{3/2})$ m/s, where s is in meters, determine the tension developed in the cable when $s = 10$ m. The crate has a mass of 20 kg, and the coefficient of kinetic friction between the crate and the ground is $\mu_k = 0.2$.

Prob. R13–2

Prob. R13–4

Review Problems

R13–5. The ball has a mass of 30 kg and a speed $v = 4$ m/s at the instant it is at its lowest point, $\theta = 0°$. Determine the tension in the cord and the rate at which the ball's speed is decreasing at the instant $\theta = 20°$. Neglect the size of the ball.

R13–7. The 10-lb suitcase slides down the curved ramp for which the coefficient of kinetic friction is $\mu_k = 0.2$. If at the instant it reaches point A it has a speed of 5 ft/s, determine the normal force on the suitcase and the rate of increase of its speed.

Prob. R13–7

Prob. R13–5

R13–8. The spool, which has a mass of 4 kg, slides along the rotating rod. At the instant shown, the angular rate of rotation of the rod is $\dot{\theta} = 6$ rad/s and this rotation is increasing at $\ddot{\theta} = 2$ rad/s^2. At this same instant, the spool has a velocity of 3 m/s and an acceleration of 1 m/s^2, both measured relative to the rod and directed away from the center O when $r = 0.5$ m. Determine the radial frictional force and the normal force, both exerted by the rod on the spool at this instant.

R13–6. The bottle rests at a distance of 3 ft from the center of the horizontal platform. If the coefficient of static friction between the bottle and the platform is $\mu_s = 0.3$, determine the maximum speed that the bottle can attain before slipping. Assume the angular motion of the platform is slowly increasing.

Prob. R13–8

Prob. R13–6

CHAPTER 14

The power requirements of this crane and the motion of this hook involve work and energy principles that can be analyzed using the methods outlined in this chapter.

KINETICS OF A PARTICLE: WORK AND ENERGY

CHAPTER OBJECTIVES

- To develop the principle of work and energy and apply it to solve kinetic problems.
- To study problems that involve power and efficiency.
- To introduce the concept of a conservative force and apply the theorem of conservation of energy to solve kinetic problems.

14.1 THE WORK OF A FORCE

In this chapter, we will analyze the motion of a particle using the concepts of work and energy. Before we do this, however, we must first define the work of a force. Specifically, a force will do *work* on a particle only when the particle undergoes a *displacement in the direction of the force*. For example, if the force $\mathbf{F}$ in Fig. 14–1a causes the particle to move along the path s from position $\mathbf{r}$ to a new position $\mathbf{r}'$, the displacement is then $d\mathbf{r} = \mathbf{r}' - \mathbf{r}$. The magnitude of $d\mathbf{r}$ becomes ds, the length of the differential segment along the path. If the angle between the tails of $d\mathbf{r}$ and $\mathbf{F}$ is θ, Fig. 14–1a, then the work dU done by $\mathbf{F}$ is a *scalar quantity*, defined by the product of the force magnitude F times the component of displacement in the direction of the force, $ds \cos \theta$, i.e.,

$$dU = F(ds \cos \theta)$$

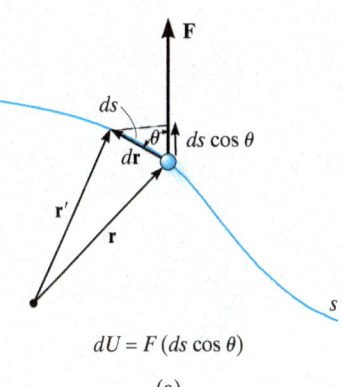

$dU = F(ds \cos \theta)$

(a)

176   CHAPTER 14   KINETICS OF A PARTICLE: WORK AND ENERGY

Fig. 14–1

We can also interpret the work as the product of displacement ds and the component of force, $F \cos \theta$, in the direction of displacement, $dU = (F \cos \theta \, ds)$. Note that if $0° \leq \theta < 90°$, then the force component and the displacement have the *same sense* so that the work is *positive*; whereas if $90° < \theta \leq 180°$, these vectors will have *opposite sense*, and therefore the work is *negative*. Also, $dU = 0$ if the force is *perpendicular* to displacement, since $\cos 90° = 0$, or if the force is applied at a *fixed point*, in which case the displacement is zero.

By definition of the dot product (see Eq. B–14) either one of the above equations can also be written as

$$dU = \mathbf{F} \cdot d\mathbf{r}$$

The unit of work in SI units is the **joule** (J), which is the amount of work done by a one-newton force when it moves through a distance of one meter in the direction of the force ($1 \text{ J} = 1 \text{ N} \cdot \text{m}$). In the FPS system, work is measured in units of foot-pounds (ft · lb), which is the work done by a one-pound force acting through a distance of one foot in the direction of the force.*

Work of a Variable Force. If the particle acted upon by the force **F** undergoes a finite displacement along its path from $\mathbf{r}_1$ to $\mathbf{r}_2$ or s_1 to s_2, Fig. 14–2a, the work of force **F** is determined by integration. Provided **F** and θ can be expressed as a function of position, then

$$U_{1-2} = \int_{\mathbf{r}_1}^{\mathbf{r}_2} \mathbf{F} \cdot d\mathbf{r} = \int_{s_1}^{s_2} F \cos \theta \, ds \quad (14\text{–}1)$$

Sometimes, this relation may be obtained by using experimental data to plot a graph of $F \cos \theta$ vs. s. Then the area under this graph bounded by s_1 and s_2 represents the total work, Fig. 14–2b.

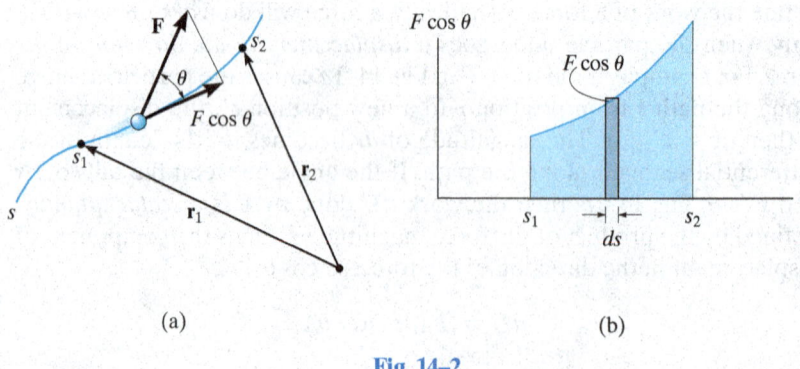

Fig. 14–2

---

*By convention, the units for the moment of a force or torque are written as lb · ft, to distinguish them from those used to signify work, ft · lb.

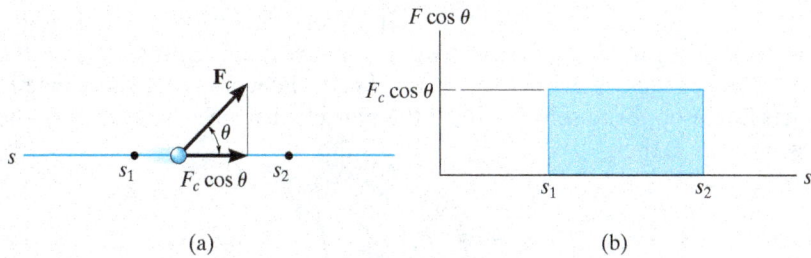

(a) (b)

Fig. 14–3

Work of a Constant Force Moving Along a Straight Line.
If the force $\mathbf{F}_c$ has a constant magnitude and acts at a constant angle θ from its straight-line path, Fig. 14–3a, then the component of $\mathbf{F}_c$ in the direction of displacement is always $F_c \cos\theta$. The work done by $\mathbf{F}_c$ when the particle is displaced from s_1 to s_2 is determined from Eq. 14–1, in which case

$$U_{1-2} = F_c \cos\theta \int_{s_1}^{s_2} ds$$

or

$$\boxed{U_{1-2} = F_c \cos\theta(s_2 - s_1)} \tag{14-2}$$

The crane must do work in order to hoist the weight of the pipe.

Here the work of $\mathbf{F}_c$ represents the *area of the rectangle* in Fig. 14–3b.

Work of a Weight.
Consider a particle of weight $\mathbf{W}$, which moves up along the path s shown in Fig. 14–4 from position s_1 to position s_2. At an intermediate point, the displacement $d\mathbf{r} = dx\mathbf{i} + dy\mathbf{j} + dz\mathbf{k}$. Since $\mathbf{W} = -W\mathbf{j}$, applying Eq. 14–1 we have

$$U_{1-2} = \int \mathbf{F} \cdot d\mathbf{r} = \int_{\mathbf{r}_1}^{\mathbf{r}_2}(-W\mathbf{j})\cdot(dx\mathbf{i} + dy\mathbf{j} + dz\mathbf{k})$$

$$= \int_{y_1}^{y_2} -W\,dy = -W(y_2 - y_1)$$

or

$$\boxed{U_{1-2} = -W\,\Delta y} \tag{14-3}$$

Here the work is independent of the path and is equal to the magnitude of the particle's weight times its vertical displacement. In the case shown, the work is *negative*, since W is downward and Δy is upward. If the particle is displaced *downward* $(-\Delta y)$, then work of the weight is *positive*.

Fig. 14–4

178   CHAPTER 14   KINETICS OF A PARTICLE: WORK AND ENERGY

Work of a Spring Force. If an elastic spring is elongated a distance ds, Fig. 14–5a, then the work done by the force that acts on the attached particle is $dU = -F_s\,ds = -ks\,ds$. The work is *negative* since F_s acts in the opposite sense to ds. If the particle displaces from s_1 to s_2, the work of $\mathbf{F}_s$ is then

$$U_{1-2} = \int_{s_1}^{s_2} F_s\,ds = \int_{s_1}^{s_2} -ks\,ds$$

$$\boxed{U_{1-2} = -\left(\tfrac{1}{2}ks_2^2 - \tfrac{1}{2}ks_1^2\right)} \qquad (14\text{–}4)$$

This work represents the trapezoidal area under the line $F_s = ks$, Fig. 14–5b.

A mistake in sign can be avoided when applying this equation if one simply notes the direction of the spring force acting on the particle and compares it with the sense of direction of displacement of the particle — if both are in the *same sense, positive work* results; if they are *opposite* to one another, the *work is negative*.

(a)

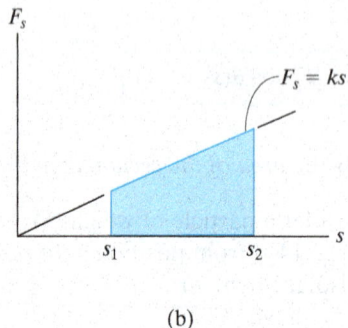

(b)

Fig. 14–5

The forces acting on the cart, as it is pulled a distance s up the incline, are shown on its free-body diagram. The constant towing force $\mathbf{T}$ does positive work of $U_T = (T\cos\phi)s$, the weight does negative work of $U_W = -(W\sin\theta)s$, and the normal force $\mathbf{N}$ does no work since there is no displacement of this force along its line of action.

EXAMPLE 14.1

The 10-kg block in Fig. 14–6a rests on the smooth incline. If the spring is originally stretched 0.5 m, determine the total work done by all the forces acting on the block when a horizontal force $P = 400$ N pushes the block up the plane $s = 2$ m.

SOLUTION
First the free-body diagram of the block is drawn in order to account for all the forces that act on the block, Fig. 14–6b.

Horizontal Force P. Since this force is *constant*, the work is determined using Eq. 14–2. The result can be calculated as the force times the component of displacement in the direction of the force; i.e.,

$$U_P = 400 \text{ N } (2 \text{ m cos } 30°) = 692.8 \text{ J}$$

or the displacement times the component of force in the direction of displacement, i.e.,

$$U_P = 400 \text{ N cos } 30°(2 \text{ m}) = 692.8 \text{ J}$$

Spring Force F_s. In the initial position the spring is stretched $s_1 = 0.5$ m and in the final position it is stretched $s_2 = 0.5$ m $+ 2$ m $= 2.5$ m. We require the work to be negative since the force and displacement are opposite to each other. The work of the spring force is thus

$$U_s = -\left[\tfrac{1}{2}(30 \text{ N/m})(2.5 \text{ m})^2 - \tfrac{1}{2}(30 \text{ N/m})(0.5 \text{ m})^2\right] = -90 \text{ J}$$

Weight W. Since the weight acts in the opposite sense to its vertical displacement, the work is negative; i.e.,

$$U_W = -(98.1 \text{ N}) (2 \text{ m sin } 30°) = -98.1 \text{ J}$$

Note that it is also possible to consider the component of weight in the direction of displacement; i.e.,

$$U_W = -(98.1 \sin 30° \text{ N}) (2 \text{ m}) = -98.1 \text{ J}$$

Normal Force N_B. This force does *no work* since it is *always* perpendicular to the displacement.

Total Work. The work of all the forces when the block is displaced 2 m is therefore

$$U_T = 692.8 \text{ J} - 90 \text{ J} - 98.1 \text{ J} = 505 \text{ J} \qquad Ans.$$

Fig 14–6

14.2 PRINCIPLE OF WORK AND ENERGY

If the particle in Fig. 14–7 has a mass m and is subjected to a system of external forces represented by the resultant $\mathbf{F}_R = \Sigma\mathbf{F}$, then the equation of motion for the particle in the tangential direction is $\Sigma F_t = ma_t$. Substituting the kinematic equation $a_t = v\,dv/ds$ and integrating both sides, assuming initially that the particle has a position $s = s_1$ and a speed $v = v_1$, and later at $s = s_2$, $v = v_2$, we have

$$\Sigma \int_{s_1}^{s_2} F_t\, ds = \int_{v_1}^{v_2} mv\, dv$$

$$\Sigma \int_{s_1}^{s_2} F_t\, ds = \tfrac{1}{2}mv_2^2 - \tfrac{1}{2}mv_1^2 \qquad (14\text{-}5)$$

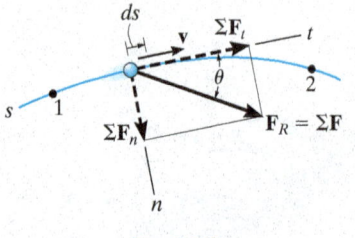

Fig. 14–7

From Fig. 14–7, note that $\Sigma F_t = \Sigma F \cos\theta$, and since all work is defined from Eq. 14–1, the final result can be written as

$$\Sigma U_{1-2} = \tfrac{1}{2}mv_2^2 - \tfrac{1}{2}mv_1^2 \qquad (14\text{-}6)$$

This equation is called the ***principle of work and energy*** for the particle. The term on the left is the sum of the work done by *all* the forces acting on the particle as the particle moves from point 1 to point 2. The two terms on the right side, which are of the form $T = \tfrac{1}{2}mv^2$, define the particle's final and initial **kinetic energy**. Like work, kinetic energy is a *scalar* and has units of joules (J) and ft · lb. However, unlike work, which can be either positive or negative, the kinetic energy is *always positive*, regardless of the direction of motion of the particle. Remember though that it is important that the velocities are measured from an inertial coordinate system.

When Eq. 14–6 is applied, it is often expressed in the form

$$\boxed{T_1 + \Sigma U_{1-2} = T_2} \qquad (14\text{-}7)$$

which states that the particle's initial kinetic energy plus the work done by all the forces acting on the particle as it moves from its initial to its final position is equal to the particle's final kinetic energy.

PROCEDURE FOR ANALYSIS

The principle of work and energy is used to solve problems involving force, displacement, and velocity, since these terms are involved in the formulation. For application the following procedure is suggested.

Work (Free-Body Diagram).

- Establish the inertial coordinate system and draw a free-body diagram of the particle in order to account for all the forces that do work on the particle as it moves along its path.

Principle of Work and Energy.

- Apply the principle of work and energy, $T_1 + \Sigma U_{1-2} = T_2$.

- The kinetic energy at the initial and final points is *always positive*, since it involves the speed squared $\left(T = \frac{1}{2}mv^2\right)$.

- A force does work when it moves through a displacement in the direction of the force.

- Work is *positive* when the force component is in the *same sense of direction* as its displacement, otherwise it is negative.

- Forces that are functions of displacement must be integrated to obtain the work. Graphically, the work is equal to the area under the force-displacement curve.

- The work of a weight is the product of the weight and the vertical displacement, $U_W = \pm Wy$. It is positive when the weight moves downwards.

- The work of a spring is of the form $U_s = \frac{1}{2}ks^2$, where k is the spring stiffness and s is the stretch or compression of the spring.

Numerical application of this procedure is illustrated in the examples following Sec. 14.3.

If an oncoming car strikes these crash barrels, the car's kinetic energy will be transformed into work, which causes the barrels, and to some extent the car, to be deformed. By knowing the amount of energy that can be absorbed by each barrel it is possible to design a crash barrier such as this.

14.3 PRINCIPLE OF WORK AND ENERGY FOR A SYSTEM OF PARTICLES

The principle of work and energy can be extended to include a system of particles isolated within an enclosed region of space as shown in Fig. 14–8. Here the arbitrary ith particle, having a mass m_i, is subjected to a resultant external force $\mathbf{F}_i$ and a resultant internal force $\mathbf{f}_i$ which all the other particles exert on the ith particle. If we apply the principle of work and energy to this and each of the other particles in the system, then since work and energy are scalar quantities, the equations can be summed algebraically, which gives

$$\Sigma T_1 + \Sigma U_{1-2} = \Sigma T_2 \qquad (14\text{--}8)$$

In this case, the initial kinetic energy of the system plus the work done by all the external and internal forces acting on the system is equal to the final kinetic energy of the system.

If the system represents a *translating rigid body*, or a series of connected translating bodies, then all the particles in each body will undergo the *same displacement*. Therefore, the work of all the internal forces will occur in equal but opposite collinear pairs and so it will cancel out. On the other hand, if the body is assumed to be *nonrigid*, the particles of the body may be displaced along *different paths*, and some of the energy due to force interactions would be given off and lost as heat or stored in the body if permanent deformations occur. We will discuss these effects briefly in what follows, and in Sec. 15.4. Throughout this book, however, the principle of work and energy will be applied to problems where direct accountability of such energy losses does not have to be considered.

Inertial coordinate system

Fig. 14–8

14.3 PRINCIPLE OF WORK AND ENERGY FOR A SYSTEM OF PARTICLES

Work of Friction Caused by Sliding. A special class of problems will now be investigated which requires a careful application of Eq. 14–8. These problems involve cases where friction occurs as a body slides over the surface of another body. Consider, for example, a block which is translating a distance s over the rough surface shown in Fig. 14–9a. If the applied force **P** just balances the *resultant* frictional force $\mu_k N$, Fig. 14–9b, then due to equilibrium a constant velocity **v** is maintained, and one would expect Eq. 14–8 to be applied as follows:

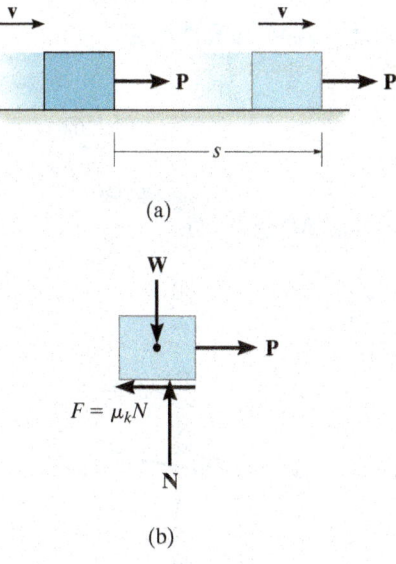

$$\tfrac{1}{2}mv^2 + Ps - \mu_k Ns = \tfrac{1}{2}mv^2$$

Indeed this equation is satisfied if $P = \mu_k N$; however, as one realizes from experience, the sliding motion will *generate heat*, a form of energy which seems not to be accounted for in the work-energy equation. In order to explain this paradox and thereby more closely represent the nature of friction, we should actually model the block so that the surfaces of contact are *deformable* (nonrigid).* Recall that the rough portions at the bottom of the block act as "teeth," and when the block slides these teeth *deform slightly* and either break off or vibrate as they pull away from "teeth" at the contacting surface, Fig. 14–9c. As a result, frictional forces that act on the block at these points are displaced slightly, due to the localized deformations, and later they are replaced by other frictional forces as other points of contact are made. At any instant, the *resultant* **F** of all these frictional forces remains essentially constant, i.e., $\mu_k N$; however, due to the many *localized deformations*, the actual displacement s' of $\mu_k N$ is *not* the same as the displacement s of the applied force **P**. Instead, s' will be *less* than s ($s' < s$), and therefore the *external work* done by the resultant frictional force will be $\mu_k N s'$ and not $\mu_k N s$. The remaining amount of work, $\mu_k N (s - s')$, manifests itself as an increase in *internal energy*, which in fact causes the block's temperature to rise.

In summary then, Eq. 14–8 can be applied to problems involving sliding friction; however, it should be fully realized that the work of the resultant frictional force is not represented by $\mu_k N s$; instead, this term represents *both* the external work of friction ($\mu_k N s'$) *and* internal work $[\mu_k N (s - s')]$ which is converted into various forms of internal energy, such as heat.†

Fig. 14–9

---

* See Chapter 8 of *Engineering Mechanics: Statics.*
† See B. A. Sherwood and W. H. Bernard, "Work and Heat Transfer in the Presence of Sliding Friction," *Am. J. Phys.* 52, 1001 (1984).

EXAMPLE 14.2

Fig. 14–10

The 3500-lb automobile shown in Fig. 14–10a travels down the 10° inclined road at a speed of 20 ft/s. If the driver jams on the brakes, causing the wheels to lock, determine how far s the tires skid on the road. The coefficient of kinetic friction between the wheels and the road is $\mu_k = 0.5$.

SOLUTION
This problem can be solved using the principle of work and energy, since it involves force, velocity, and displacement.

Work (Free-Body Diagram). As shown in Fig. 14–10b, the normal force $\mathbf{N}_A$ does no work since it never undergoes displacement along its line of action. The weight, 3500 lb, is displaced $s \sin 10°$ and does positive work. Why? The frictional force $\mathbf{F}_A$ does both external and internal work when it undergoes a displacement s. This work is negative since it is in the opposite sense of direction to the displacement. Applying the equation of equilibrium normal to the road, we have

$$+\nwarrow \Sigma F_n = 0; \quad N_A - 3500 \cos 10° \text{ lb} = 0 \quad N_A = 3446.8 \text{ lb}$$

Thus,

$$F_A = \mu_k N_A = 0.5\,(3446.8 \text{ lb}) = 1723.4 \text{ lb}$$

Principle of Work and Energy.

$$T_1 + \Sigma U_{1-2} = T_2$$

$$\frac{1}{2}\left(\frac{3500 \text{ lb}}{32.2 \text{ ft/s}^2}\right)(20 \text{ ft/s})^2 + 3500 \text{ lb}(s \sin 10°) - (1723.4 \text{ lb})s = 0$$

Solving for s yields

$$s = 19.5 \text{ ft} \qquad \qquad Ans.$$

NOTE: If this problem is solved by using the equation of motion, *two steps* are involved. First, from the free-body diagram, Fig. 14–10b, the equation of motion is applied along the incline to obtain the deceleration. This yields

$$+\swarrow \Sigma F_s = ma_s; \quad 3500 \sin 10° \text{ lb} - 1723.4 \text{ lb} = \frac{3500 \text{ lb}}{32.2 \text{ ft/s}^2} a$$

$$a = -10.3 \text{ ft/s}^2$$

Then, since a is constant, we have

$$(+\swarrow) \quad v^2 = v_0^2 + 2a_c(s - s_0);$$

$$(0)^2 = (20 \text{ ft/s})^2 + 2(-10.3 \text{ ft/s}^2)(s - 0)$$

$$s = 19.5 \text{ ft} \qquad \qquad Ans.$$

EXAMPLE 14.3

For a short time the crane in Fig. 14–11a lifts a 2.50-Mg beam with a force of $F = (28 + 3s^2)$ kN. Determine the speed of the beam when it has risen $s = 3$ m. Also, how much time does it take to attain this height starting from rest?

SOLUTION
We can solve part of this problem using the principle of work and energy since it involves force, velocity, and displacement. Kinematics must be used to determine the time. Note that at $s = 0$, $F = 28(10^3)$ N $> W = 2.50(10^3)(9.81)$ N, so motion will occur.

Work (Free-Body Diagram). As shown on the free-body diagram, Fig. 14–11b, the lifting force **F** does positive work, which must be determined by integration since this force is a variable. Also, the weight is constant and will do negative work since the displacement is upward.

Principle of Work and Energy.

$$T_1 + \Sigma U_{1-2} = T_2$$

$$0 + \int_0^s (28 + 3s^2)(10^3)\, ds - (2.50)(10^3)(9.81)s = \tfrac{1}{2}(2.50)(10^3)v^2$$

$$28(10^3)s + (10^3)s^3 - 24.525(10^3)s = 1.25(10^3)v^2$$

$$v = (2.78s + 0.8s^3)^{\frac{1}{2}} \qquad (1)$$

When $s = 3$ m,

$$v = 5.47 \text{ m/s} \qquad \textit{Ans.}$$

Kinematics. Since we were able to express the velocity as a function of displacement, Eq. 1, the time can be determined using $v = ds/dt$. In this case,

$$(2.78s + 0.8s^3)^{\frac{1}{2}} = \frac{ds}{dt}$$

$$t = \int_0^3 \frac{ds}{(2.78s + 0.8s^3)^{\frac{1}{2}}}$$

The integration can be performed numerically using a pocket calculator. The result is

$$t = 1.79 \text{ s} \qquad \textit{Ans.}$$

NOTE: The acceleration of the beam can be determined by integrating Eq. 1 using $v\, dv = a\, ds$, or more directly, by applying the equation of motion, $\Sigma F_y = ma_y$.

(a)

(b)

Fig. 14–11

EXAMPLE 14.4

The platform P, shown in Fig. 14–12a, has negligible mass and is tied down so that the 0.4-m-long cords keep a 1-m-long spring compressed 0.6 m when *nothing* is on the platform. If a 2-kg block is placed on the platform and released from rest after the platform is pushed down 0.1 m, Fig. 14–12b, determine the maximum height h the block rises in the air, measured from the ground.

(a)   (b)

SOLUTION
Work (Free-Body Diagram). Since the block is released from rest and later reaches its maximum height, the initial and final velocities are zero. The free-body diagram of the block when it is still in contact with the platform is shown in Fig. 14–12c. Note that the weight does negative work and the spring force does positive work. Why? In particular, the *initial compression* in the spring is $s_1 = 0.6$ m $+ 0.1$ m $= 0.7$ m. Due to the cords, the spring's *final compression* is $s_2 = 0.6$ m (after the block leaves the platform). The bottom of the block rises from a height of $(0.4$ m $- 0.1$ m$) = 0.3$ m to a final height h.

Principle of Work and Energy.

$$T_1 + \Sigma U_{1-2} = T_2$$
$$\tfrac{1}{2}mv_1^2 + \left[-\left(\tfrac{1}{2}ks_2^2 - \tfrac{1}{2}ks_1^2\right) - W\,\Delta y\right] = \tfrac{1}{2}mv_2^2$$

Note that here $s_1 = 0.7$ m $> s_2 = 0.6$ m and so the work of the spring as determined from Eq. 14–4 will indeed be positive once the calculation is made. Thus,

$$0 + \left\{-\left[\tfrac{1}{2}(200\text{ N/m})(0.6\text{ m})^2 - \tfrac{1}{2}(200\text{ N/m})(0.7\text{ m})^2\right]\right.$$
$$\left. - (19.62\text{ N})[h - (0.3\text{ m})]\right\} = 0$$

Solving yields

$$h = 0.963 \text{ m} \qquad \textit{Ans.}$$

(c)

Fig. 14–12

EXAMPLE 14.5

The 40-kg boy in Fig. 14–13a slides down the smooth water slide. If he starts from rest at A, determine his speed when he reaches B and the normal reaction the slide exerts on the boy at this position.

(a)

SOLUTION

Work (Free-Body Diagram). As shown on the free-body diagram, Fig. 14–13b, there are two forces acting on the boy as he goes down the slide. Note that the normal force does no work.

Principle of Work and Energy.

$$T_A + \Sigma U_{A-B} = T_B$$

$$0 + (40(9.81) \text{ N})(7.5 \text{ m}) = \tfrac{1}{2}(40 \text{ kg})v_B^2$$

$$v_B = 12.13 \text{ m/s} = 12.1 \text{ m/s} \qquad Ans.$$

(b)

Equation of Motion. Referring to the free-body diagram of the boy when he is at B, Fig. 14–13c, the normal reaction $\mathbf{N}_B$ can now be obtained by applying the equation of motion along the n axis. Here the radius of curvature of the path is

$$\rho_B = \frac{\left[1 + \left(\dfrac{dy}{dx}\right)^2\right]^{3/2}}{|d^2y/dx^2|} = \left.\frac{[1 + (0.15x)^2]^{3/2}}{|0.15|}\right|_{x=0} = 6.667 \text{ m}$$

Thus,

$$+\uparrow \Sigma F_n = ma_n; \quad N_B - 40(9.81) \text{ N} = 40 \text{ kg}\left(\frac{(12.13 \text{ m/s})^2}{6.667 \text{ m}}\right)$$

$$N_B = 1275.3 \text{ N} = 1.28 \text{ kN} \qquad Ans.$$

(c)

Fig. 14–13

EXAMPLE 14.6

Fig. 14–14

Blocks A and B shown in Fig. 14–14a have a mass of 10 kg and 100 kg, respectively. Determine the distance B travels so that its speed becomes 2 m/s after being released from rest.

SOLUTION
This problem may be solved by considering the blocks separately and applying the principle of work and energy to each block. However, it is easier to eliminate the work of the cable tension from the analysis by considering blocks A and B together as a *single system*.

Work (Free-Body Diagram). As shown on the free-body diagram of the system, Fig. 14–14b, the cable force $\mathbf{T}$ and reactions $\mathbf{R}_1$ and $\mathbf{R}_2$ do *no work*, since these forces represent the reactions at the supports and consequently they do not move while the blocks are displaced. The weights both do positive work if we *assume* both move downward, in the positive sense of direction of s_A and s_B, Fig. 14–14a.

Principle of Work and Energy. Realizing the blocks are released from rest, we have

$$\Sigma T_1 + \Sigma U_{1-2} = \Sigma T_2$$

$$\left[\tfrac{1}{2}m_A(v_A)_1^2 + \tfrac{1}{2}m_B(v_B)_1^2\right] + [W_A \,\Delta s_A + W_B\, \Delta s_B] =$$
$$\left[\tfrac{1}{2}m_A(v_A)_2^2 + \tfrac{1}{2}m_B(v_B)_2^2\right]$$

$$[0 + 0] + [98.1 \text{ N } (\Delta s_A) + 981 \text{ N } (\Delta s_B)] =$$
$$\left[\tfrac{1}{2}(10 \text{ kg})(v_A)_2^2 + \tfrac{1}{2}(100 \text{ kg})(2 \text{ m/s})^2\right] \quad (1)$$

Kinematics. Using methods of kinematics, discussed in Sec. 12.9, the total length l of all the vertical segments of cable in Fig. 14–14a can be expressed in terms of the position coordinates s_A and s_B as

$$s_A + 4s_B = l$$

Hence, a change in position yields the displacement equation

$$\Delta s_A + 4\,\Delta s_B = 0$$
$$\Delta s_A = -4\,\Delta s_B \quad (2)$$

Here a downward displacement of one block produces an upward displacement of the other block. Taking the time derivative of Eq. 2 yields

$$v_A = -4v_B = -4(2 \text{ m/s}) = -8 \text{ m/s} \quad (3)$$

Retaining the negative sign in Eq. 2 and substituting into Eq. 1 yields

$$\Delta s_B = 0.883 \text{ m} \downarrow \qquad \textit{Ans.}$$

NOTE: Since we had to solve Eqs. 1–3 *simultaneously*, it is important to write these equations using the *same* coordinate directions for s_A and s_B.

FUNDAMENTAL PROBLEMS

F14–1. The spring is placed between the wall and the 10-kg block. If the block is subjected to a force of $F = 500$ N, determine its velocity when $s = 0.5$ m. When $s = 0$, the block is at rest and the spring is uncompressed. The contact surface is smooth.

Prob. F14–1

F14–2. If the motor exerts a constant force of 300 N on the cable, determine the speed of the 20-kg crate when it travels $s = 10$ m up the plane, starting from rest. The coefficient of kinetic friction between the crate and the plane is $\mu_k = 0.3$.

Prob. F14–2

F14–3. If the motor exerts a force of $F = (600 + 2s^2)$ N on the cable, determine the speed of the 100-kg crate when it rises to $s = 15$ m. The crate is initially at rest on the ground.

Prob. F14–3

F14–4. The 1.8-Mg dragster is traveling at 125 m/s when the engine is shut off and the parachute is released. If the drag force of the parachute can be approximated by the graph, determine the speed of the dragster when it has traveled 400 m.

Prob. F14–4

F14–5. When $s = 0.6$ m, the spring is unstretched and the 10-kg block has a speed of 5 m/s down the smooth plane. Determine the distance s when the block stops.

Prob. F14–5

F14–6. The 5-lb collar is pulled by a cord that passes around a small peg at C. If the cord is subjected to a constant force of $F = 10$ lb, and the collar is at rest when it is at A, determine its speed the instant before it reaches B. Neglect friction.

Prob. F14–6

PROBLEMS

All solutions must include a free-body diagram.

14–1. The car having a mass of 2 Mg is originally traveling at 2 m/s. Determine the distance it must be towed by a force $F = 4$ kN in order to attain a speed of 5 m/s. Neglect friction and the mass of the wheels.

Prob. 14–1

14–2. For protection, the barrel barrier is placed in front of the bridge pier. If the relation between the force and deflection of the barrier is $F = (90(10^3)x^{1/2})$ lb, where x is in ft, determine the car's maximum penetration in the barrier. The car has a weight of 4000 lb and it is traveling with a speed of 75 ft/s just before it hits the barrier.

Prob. 14–2

14–3. The collar has mass of 5 kg and is moving at 8 m/s when $x = 0$ and a force of $F = 60$ N is applied to it. The direction u of this force varies such that $u = 10x$, where x is in meters and u is clockwise, measured in degrees. Determine the speed of the collar when $x = 3$ m. The coefficient of kinetic friction between the collar and the rod is $\mu_k = 0.3$.

Prob. 14–3

*14–4. The 100-kg crate is subjected to the forces shown. If it is originally at rest, determine the distance it slides in order to attain a speed of $v = 8$ m/s. The coefficient of kinetic friction between the crate and the surface is $\mu_k = 0.2$.

Prob. 14–4

14–5. Determine the required height h of the roller coaster so that when it is essentially at rest at the crest of the hill A it will reach a speed of 100 km/h when it comes to the bottom B. Also, what should be the minimum radius of curvature ρ for the track at B so that the passengers do not experience a normal force greater than $4mg = (39.24m)$ N? Neglect the size of the car.

14–7. As indicated by the derivation, the principle of work and energy is valid for observers in *any* inertial reference frame. Show that this is so, by considering the 10-kg block which rests on the smooth surface and is subjected to a horizontal force of 6 N. If observer A is in a *fixed* frame x, determine the final speed of the block if it has an initial speed of 5 m/s and travels 10 m, both directed to the right and measured from the fixed frame. Compare the result with that obtained by an observer B, attached to the x' axis and moving at a constant velocity of 2 m/s relative to A. *Hint*: The distance the block travels will first have to be calculated for observer B before applying the principle of work and energy.

Prob. 14–5

Prob. 14–7

14–6. When the driver applies the brakes of a light truck traveling 40 km/h, it skids 30 m before stopping. How far will the truck skid if it is traveling 80 km/h when the brakes are applied?

*****14–8.** If the 50-kg crate is subjected to a force of $P = 200$ N, determine its speed when it has traveled 15 m starting from rest. The coefficient of kinetic friction between the crate and the ground is $\mu_k = 0.3$.

14–9. If the 50-kg crate starts from rest and attains a speed of 6 m/s when it has traveled a distance of 15 m, determine the force P acting on the crate. The coefficient of kinetic friction between the crate and the ground is $\mu_k = 0.3$.

Prob. 14–6

Probs. 14–8/9

14–10. The force **F**, acting in a constant direction on the 20-kg block, has a magnitude which varies with the positions s of the block. Determine how far the block must slide before its velocity becomes 15 m/s. When $s = 0$ the block is moving to the right at $v = 6$ m/s. The coefficient of kinetic friction between the block and surface is $\mu_k = 0.3$.

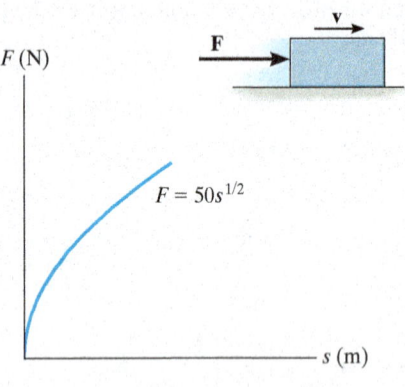

Prob. 14–10

14–11. The 20-lb block B rests on the surface of a table for which the coefficient of kinetic friction is $\mu_k = 0.1$. Determine the speed of the 10-lb block A after it has moved downward 2 ft from rest. Neglect the mass of the pulleys and cords.

Prob. 14–11

*14–12. The 1.5-kg block slides along a smooth plane and strikes a *nonlinear spring* with a speed of $v = 4$ m/s. The spring is termed "nonlinear" because it has a resistance of $F_s = ks^2$, where $k = 900$ N/m². Determine the speed of the block after it has compressed the spring $s = 0.2$ m.

Prob. 14–12

14–13. The 2-lb brick slides down a smooth roof, such that when it is at A it has a velocity of 5 ft/s. Determine the speed of the brick just before it leaves the surface at B, the distance d from the wall to where it strikes the ground, and the speed at which it hits the ground.

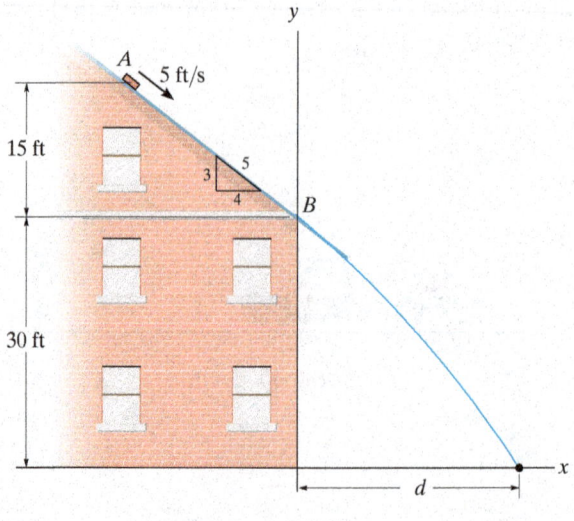

Prob. 14–13

14–14. The motion of a truck is arrested using a bed of loose stones AB and a set of crash barrels BC. If experiments show that the stones provide a rolling resistance of 160 lb per wheel and the crash barrels provide a resistance as shown in the graph, determine the distance x the 4500-lb truck penetrates the barrels if the truck is coasting at 60 ft/s when it approaches A. Take $s = 50$ ft and neglect the size of the truck.

*14–16.** A small box of mass m is given a speed of $v = \sqrt{\frac{1}{4}gr}$ at the top of the smooth half cylinder. Determine the angle θ at which the box leaves the surface.

Prob. 14–14

Prob. 14–16

14–17. The sack of mass m has a velocity of $\mathbf{v}_0$ when it is at a height h above the unstretched spring. Determine the maximum compression of the spring if it has a stiffness k.

14–15. The two blocks A and B have a weight $W_A = 60$ lb and $W_B = 10$ lb. If the coefficient of kinetic friction between the incline and block A is $\mu_k = 0.2$, determine the speed of A after it moves 3 ft down the plane starting from rest. Neglect the mass of the cord and pulleys.

Prob. 14–15

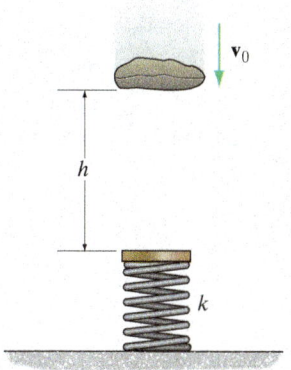

Prob. 14–17

14–18. When the 12-lb block A is released from rest it lifts the two 15-lb weights B and C. Determine the maximum distance A will fall before its motion is momentarily stopped. Neglect the weight of the cord and the size of the pulleys.

*14–20. The crash cushion for a highway barrier consists of a nest of barrels filled with an impact-absorbing material. The barrier stopping force is measured versus the vehicle penetration into the barrier. Determine the distance a car having a weight of 4000 lb will penetrate the barrier if it is originally traveling at 55 ft/s when it strikes the first barrel.

Prob. 14–18

Prob. 14–20

14–19. If the cord is subjected to a constant force of $F = 300$ N and the 15-kg smooth collar starts from rest at A, determine the velocity of the collar when it reaches point B. Neglect the size of the pulley.

14–21. The steel ingot has a mass of 1800 kg. It travels along the conveyor at a speed $v = 0.5$ m/s when it collides with the "nested" spring assembly. Determine the maximum deflection in each spring needed to stop the motion of the ingot. Take $k_A = 5$ kN/m, $k_B = 3$ kN/m.

Prob. 14–19

Prob. 14–21

14–22. The 25-lb block has an initial speed of $v_0 = 10$ ft/s when it is midway between springs A and B. After striking spring B, it rebounds and slides across the horizontal plane toward spring A, and continues to move back and forth. If the coefficient of kinetic friction between the plane and the block is $\mu_k = 0.4$, determine the total distance traveled by the block before it comes to rest.

Prob. 14–22

14–23. The force **F**, acting in a constant direction on the 20-kg block, has a magnitude which varies with position s of the block. Determine the speed of the block after it slides 3 m. When $s = 0$ the block is moving to the right at 2 m/s. The coefficient of kinetic friction between the block and surface is $\mu_k = 0.3$.

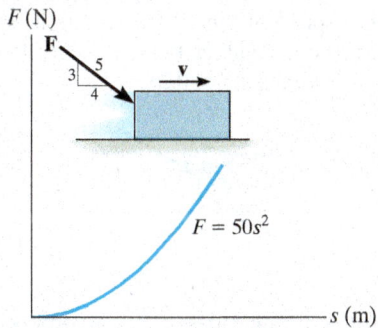

Prob. 14–23

*14–24. At a given instant the 10-lb block A is moving downward with a speed of 6 ft/s. Determine its speed 2 s later. Block B has a weight of 4 lb, and the coefficient of kinetic friction between it and the horizontal plane is $\mu_k = 0.2$. Neglect the mass of the cord and pulleys.

Prob. 14–24

14–25. The 5-lb cylinder is falling from A with a speed $v_A = 10$ ft/s onto the platform. Determine the maximum displacement of the platform, caused by the collision. The spring has an unstretched length of 1.75 ft and is originally kept in compression by the 1-ft-long cables attached to the platform. Neglect the mass of the platform and spring and any energy lost during the collision.

Prob. 14–25

14–26. The catapulting mechanism is used to propel the 10-kg slider A to the right along the smooth track. The propelling action is obtained by drawing the pulley attached to rod BC rapidly to the left by means of a piston P. If the piston applies a constant force $F = 20$ kN to rod BC such that it moves it 0.2 m, determine the speed attained by the slider if it was originally at rest. Neglect the mass of the pulleys, cable, piston, and rod BC.

Prob. 14–26

14–27. The "flying car" is a ride at an amusement park which consists of a car having wheels that roll along a track mounted inside a rotating drum. By design the car cannot fall off the track, however motion of the car is developed by applying the car's brake, thereby gripping the car to the track and allowing it to move with a constant speed of the track, $v_t = 3$ m/s. If the rider applies the brake when going from B to A and then releases it at the top of the drum, A, so that the car coasts freely down along the track to B ($\theta = \pi$ rad), determine the speed of the car at B and the normal reaction which the drum exerts on the car at B. Neglect friction during the motion from A to B. The rider and car have a total mass of 250 kg, and the center of mass of the car and rider moves along a circular path having a radius of 8 m.

Prob. 14–27

*14–28.** The 30-lb box A is released from rest and slides down along a *smooth* ramp and onto the surface of a cart. If the cart is *fixed from moving*, determine the distance s from the end of the cart to where the box stops. The coefficient of kinetic friction between the cart and the box is $\mu_k = 0.6$.

Prob. 14–28

14–29. The 2-lb block slides down the smooth parabolic surface, such that when it is at A it has a speed of 10 ft/s. Determine the magnitudes of the block's velocity and acceleration when it reaches point B, and the maximum height y_{max} reached by the block.

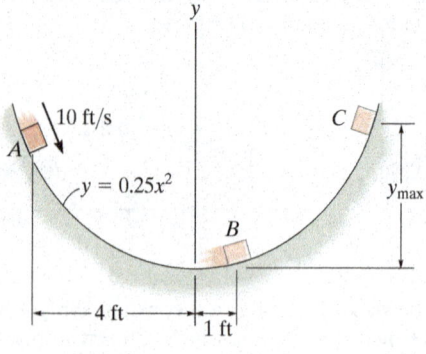

Prob. 14–29

14–30. The 2-kg block is subjected to a force having a constant direction and a magnitude $F = (300/(1 + s))$ N, where s is in meters. When $s = 4$ m, the block is moving to the left with a speed of 8 m/s. Determine its speed when $s = 12$ m. The coefficient of kinetic friction between the block and the ground is $\mu_k = 0.25$.

Prob. 14–30

14–31. Marbles having a mass of 5 g are dropped from rest at A through the smooth glass tube and accumulate in the can at C. Determine the placement R of the can from the end of the tube and the speed at which the marbles fall into the can. Neglect the size of the can.

Prob. 14–31

*14–32. The 8-kg cylinder A and 3-kg cylinder B are released from rest. Determine the speed of A after it has moved 2 m starting from rest. Neglect the mass of the cord and pulleys.

14–33. Cylinder A has a mass of 3 kg and cylinder B has a mass of 8 kg. Determine the speed of A after it has moved 2 m starting from rest. Neglect the mass of the cord and pulleys.

14–35. When the 150-lb skier is at point A he has a speed of 5 ft/s. Determine his speed when he reaches point B on the smooth slope. For this distance the slope follows the cosine curve shown. Also, what is the normal force on his skis at B, and his rate of increase in speed? Neglect friction and air resistance.

Prob. 14–35

Probs. 14–32/33

14-34 The 2-lb box slides on the smooth circular ramp. If the box has a velocity of 30 ft/s at A, determine the velocity of the box and normal force acting on the ramp when the box is located at B and C. Assume the radius of curvature of the path at C is still 5 ft.

*14–36. Roller coasters are designed so that riders will not experience a normal force that is more than 3.5 times their weight against the seat of the car. Determine the smallest radius of curvature ρ of the track at its lowest point if the car has a speed of 5 ft/s at the crest of the drop. Neglect friction.

Prob. 14–34

Prob. 14–36

14–37. If the force exerted by the motor M on the cable is 250 N, determine the speed of the 100-kg crate when it is hoisted to $s = 3$ m. The crate is at rest when $s = 0$.

14–39. If the 75-kg crate starts from rest at A, determine its speed when it reaches point B. The cable is subjected to a constant force of $F = 300$ N. Neglect friction and the size of the pulley.

*14–40. If the 75-kg crate starts from rest at A, and its speed is 6 m/s when it passes point B, determine the constant force $\mathbf{F}$ exerted on the cable. Neglect friction and the size of the pulley.

Prob. 14–37

Probs. 14–39/40

14–38. The 150-lb skater passes point A with a speed of 6 ft/s. Determine his speed when he reaches point B and the normal force exerted on him by the track at this point. Neglect friction.

14–41. The 10-lb block is pressed against the spring so as to compress it 2 ft when it is at A. If the plane is smooth, determine the distance d, measured from the wall, to where the block strikes the ground. Neglect the size of the block.

Prob. 14–38

Prob. 14–41

14.4 POWER AND EFFICIENCY

Power. The term "power" provides a useful basis for choosing the type of motor or machine which is required to do a certain amount of work in a given time. For example, two pumps may each be able to empty a reservoir if given enough time; however, the pump having the larger power will complete the job sooner.

The *instantaneous power* generated by a force that performs an amount of work dU within the time interval dt is therefore

$$P = \frac{dU}{dt}$$

If the work ΔU occurs over a finite time Δt, then the *average power* is determined from

$$P_{avg} = \frac{\Delta U}{\Delta t} \quad (14\text{--}9)$$

We can also obtain the instantaneous power of a force $\mathbf{F}$ by expressing the work as $dU = \mathbf{F} \cdot d\mathbf{r}$. Then

$$P = \frac{dU}{dt} = \frac{\mathbf{F} \cdot d\mathbf{r}}{dt} = \mathbf{F} \cdot \frac{d\mathbf{r}}{dt}$$

or

$$\boxed{P = \mathbf{F} \cdot \mathbf{v}} \quad (14\text{--}10)$$

Hence, power is a *scalar*, where in this formulation $\mathbf{v}$ represents the velocity of the force $\mathbf{F}$ which acts on the particle.

The basic units of power in the SI and FPS systems are the watt (W) and horsepower (hp), respectively. These units are defined as

$$1 \text{ W} = 1 \text{ J/s} = 1 \text{ N} \cdot \text{m/s}$$

$$1 \text{ hp} = 550 \text{ ft} \cdot \text{lb/s}$$

For conversion between the two systems of units, 1 hp = 746 W.

The power output of this locomotive comes from the driving frictional force developed at its wheels. It is this force that overcomes the frictional resistance of the cars in tow and is able to lift the weight of the train up a grade.

Efficiency. The *mechanical efficiency* of a machine is defined as the ratio of the output of useful power produced by the machine to the input of power supplied to the machine. Hence,

$$\boxed{\varepsilon = \frac{\text{power output}}{\text{power input}}} \quad (14\text{--}11)$$

If energy supplied to the machine occurs during the *same time interval* at which it is drawn, then the efficiency may also be expressed in terms of the ratio

$$\varepsilon = \frac{\text{energy output}}{\text{energy input}} \quad (14\text{--}12)$$

Since machines consist of a series of moving parts, frictional forces will always be developed within the machine, and as a result, extra energy or power is needed to overcome these forces. Consequently, power output will be less than power input, and so *the efficiency of a machine is always less than 1.*

PROCEDURE FOR ANALYSIS

To obtain the power developed by a force consider using the following procedure.

- Determine the magnitude of the force **F** acting on the body (or particle) which causes the motion. This force is usually developed by a machine or engine placed either within or external to the body.

- If the body (or particle) is accelerating, it may be necessary to draw its free-body diagram and apply the equation of motion ($\Sigma \mathbf{F} = m\mathbf{a}$) to determine **F**.

- Once **F** and its velocity **v** have been determined, then the power is obtained by multiplying the force magnitude with the component of velocity acting in the direction of **F**, i.e., $P = \mathbf{F} \cdot \mathbf{v} = Fv \cos\theta$.

- In some problems the average power may be found by calculating the work done by **F** during a time period Δt. It is $P_{avg} = \Delta U / \Delta t$.

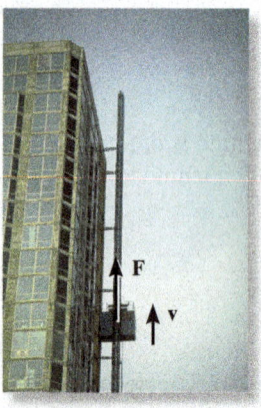

The power requirement of this hoist depends upon the vertical force **F** that acts on the elevator and causes it to move upward. If the velocity of the elevator is **v**, then the instantaneous power output is $P = \mathbf{F} \cdot \mathbf{v}$.

EXAMPLE 14.7

The man in Fig. 14–15a pushes on the 50-kg crate with a force of $F = 150$ N. Determine the instantaneous power supplied by the man when $t = 4$ s and the average power during the four seconds. The coefficient of kinetic friction between the floor and the crate is $\mu_k = 0.2$. Initially the crate is at rest.

SOLUTION

Instantaneous Power. To determine the power developed by the man, the velocity of the 150-N force must be obtained when $t = 4$ s. The free-body diagram of the crate is shown in Fig. 14–15b. Applying the equations of motion,

$$+\uparrow \Sigma F_y = ma_y; \quad N - (\tfrac{3}{5})150 \text{ N} - 50(9.81) \text{ N} = 0$$

$$N = 580.5 \text{ N}$$

$$\stackrel{+}{\rightarrow} \Sigma F_x = ma_x; \quad (\tfrac{4}{5})150 \text{ N} - 0.2(580.5 \text{ N}) = (50 \text{ kg})a$$

$$a = 0.078 \text{ m/s}^2$$

The velocity of the crate when $t = 4$ s is therefore

$$(\stackrel{+}{\rightarrow}) \quad v = v_0 + a_c t$$

$$v = 0 + (0.078 \text{ m/s}^2)(4 \text{ s}) = 0.312 \text{ m/s}$$

The power supplied to the crate by the man when $t = 4$ s is therefore

$$P = \mathbf{F} \cdot \mathbf{v} = F_x v = (\tfrac{4}{5})(150 \text{ N})(0.312 \text{ m/s})$$

$$= 37.4 \text{ W} \qquad \textit{Ans.}$$

Average Power. The displacement of the crate can be determined since the acceleration is known.

$$s = s_0 + v_0 t + \tfrac{1}{2} a_c t^2$$

$$s = 0 + 0 + \tfrac{1}{2}(0.078)(4)^2 = 0.624 \text{ m}$$

The work done by the 150-N force is therefore

$$U = 150(\tfrac{4}{5})(0.624) = 74.88 \text{ J}$$

Therefore, the average power is

$$P_{\text{avg}} = \frac{U}{\Delta t} = \frac{74.88}{4} = 18.7 \text{ W} \qquad \textit{Ans.}$$

Fig. 14–15

EXAMPLE 14.8

The motor M of the hoist in Fig. 14–16a lifts the 75-lb crate C so that the acceleration of point P is 4 ft/s^2. Determine the power that must be supplied to the motor at the instant P has a velocity of 2 ft/s. Neglect the mass of the pulley and cable and take $\varepsilon = 0.85$.

SOLUTION
In order to find the power output of the motor, it is first necessary to determine the tension in the cable since this force is developed by the motor.

From the free-body diagram, Fig. 14–16b, we have

$$+\downarrow \Sigma F_y = ma_y; \qquad -2T + 75 \text{ lb} = \frac{75 \text{ lb}}{32.2 \text{ ft/s}^2} a_C \qquad (1)$$

The acceleration of the crate can be obtained by using kinematics to relate it to the known acceleration of point P, Fig. 14–16a. Using the methods of absolute dependent motion, the coordinates s_C and s_P can be related to a constant portion of cable length l which is changing in the vertical and horizontal directions. We have $2s_C + s_P = l$. Taking the second time derivative of this equation yields

$$2a_C = -a_P \qquad (2)$$

Since $a_P = +4$ ft/s^2, then $a_C = -(4 \text{ ft/s}^2)/2 = -2$ ft/s^2. What does the negative sign indicate? Substituting this result into Eq. 1 and *retaining* the negative sign since the acceleration in *both* Eq. 1 and Eq. 2 was considered positive downward, we have

$$-2T + 75 \text{ lb} = \left(\frac{75 \text{ lb}}{32.2 \text{ ft/s}^2}\right)(-2 \text{ ft/s}^2)$$

$$T = 39.83 \text{ lb}$$

The power output, measured in units of horsepower, required to draw the cable in at a rate of 2 ft/s is therefore

$$P = \mathbf{T} \cdot \mathbf{v} = (39.83 \text{ lb})(2 \text{ ft/s})[1 \text{ hp}/(550 \text{ ft} \cdot \text{lb/s})]$$
$$= 0.1448 \text{ hp}$$

This *power output* requires that the motor provide a *power input* of

$$\text{power input} = \frac{1}{\varepsilon}(\text{power output})$$

$$= \frac{1}{0.85}(0.1448 \text{ hp}) = 0.170 \text{ hp} \qquad Ans.$$

NOTE: Since the velocity of the crate is constantly changing, the power requirement is *instantaneous*.

(a)

(b)

Fig. 14–16

FUNDAMENTAL PROBLEMS

F14–7. If the contact surface between the 20-kg crate and the ground is smooth, determine the power of force **F** when $t = 4$ s. Initially, the crate is at rest.

Prob. F14–7

F14–8. If $F = (10s)$ N, where s is in meters, and the contact surface between the 20-kg crate and the ground is smooth, determine the power of force **F** when $s = 5$ m. When $s = 0$, the crate is moving at $v = 1$ m/s.

Prob. F14–8

F14–9. If the motor winds in the cable with a constant speed of $v = 3$ ft/s, determine the power supplied to the motor at this instant. The load weighs 100 lb and the efficiency of the motor is $\varepsilon = 0.8$. Neglect the mass of the pulleys.

Prob. F14–9

F14–10. The coefficient of kinetic friction between the 20-kg crate and the inclined plane is $\mu_k = 0.2$. If the crate is traveling up the inclined plane with a constant velocity $v = 5$ m/s, determine the power of force **F**.

Prob. F14–10

F14–11. If the 50-kg load A is hoisted by motor M so that the load has a constant velocity of 1.5 m/s, determine the power input to the motor, which operates at an efficiency $\varepsilon = 0.8$.

Prob. F14–11

F14–12. At the instant shown, point P on the cable has a velocity $v_P = 12$ m/s, which is increasing at a rate of $a_P = 6$ m/s². Determine the power input to the motor M at this instant if the motor operates with an efficiency $\varepsilon = 0.8$. The mass of block A is 50 kg.

Prob. F14–12

PROBLEMS

All solutions must include a free-body diagram.

14–42. The jeep has a weight of 2500 lb and an engine which transmits a power of 100 hp to *all* the wheels. Assuming the wheels do not slip on the ground, determine the angle θ of the largest incline the jeep can climb at a constant speed $v = 30$ ft/s.

Prob. 14–42

14–43. Determine the power input for a motor necessary to lift 300 lb at a constant rate of 5 ft/s. The efficiency of the motor is $\varepsilon = 0.65$.

*14–44. An automobile having a mass of 2 Mg travels up a 7° slope at a constant speed of $v = 100$ km/h. If mechanical friction and wind resistance are neglected, determine the power developed by the engine if the automobile has an efficiency $\varepsilon = 0.65$.

Prob. 14–44

14–45. The Milkin Aircraft Co. manufactures a turbojet engine that is placed in a plane having a weight of 13000 lb. If the engine develops a constant thrust of 5200 lb, determine the power output of the plane when it is just ready to take off with a speed of 600 mi/h.

14–46. To dramatize the loss of energy from an automobile when it stops, consider a car having a weight of 5000 lb that is traveling at 35 mi/h. If the car is brought to a stop, determine how long a 100-W light bulb must burn to expend the same amount of energy. (1 mi = 5280 ft.)

14–47. An escalator is moving forward and upward with a speed of 2 ft/s. If the steps are 8 in. high and 15 in. in length, determine the required horsepower of a motor needed to lift an average load of 180 lb per step. The escalator rises to a height of 20 ft from the floor.

*14–48. The man having a weight of 150 lb is able to run up a 15-ft-high flight of stairs in 4 s. Determine the power generated. How long would a 100-W light bulb have to burn to expend the same amount of energy? *Conclusion*: Please turn off the lights when they are not in use!

Prob. 14–48

14–49. Using the biomechanical power curve shown, determine the maximum speed attained by the rider and his bicycle, which have a total mass of 92 kg, as the rider ascends the 20° slope starting from rest.

Prob. 14–49

14–50. A rocket having a total mass of 8 Mg is fired vertically from rest. If the engines provide a constant thrust of $T = 300$ kN, determine the power output of the engines as a function of time. Neglect the effect of drag resistance and the loss of fuel mass and weight.

Prob. 14–50

14–51. The 1000-lb elevator is hoisted by the pulley system and motor M. If the motor exerts a constant force of 500 lb on the cable, determine the power that must be supplied to the motor at the instant the load has been hoisted $s = 15$ ft starting from rest. The motor has an efficiency of $\varepsilon = 0.65$.

Prob. 14–51

*14–52. A truck has a weight of 25 000 lb and an engine which transmits a power of 350 hp to *all* the wheels. Assuming that the wheels do not slip on the ground, determine the angle θ of the largest incline the truck can climb at a constant speed of $v = 50$ ft/s.

Prob. 14–52

14–53. The sports car has a mass of 2.3 Mg, and while it is traveling at 28 m/s the driver causes it to accelerate at 5 m/s². If the drag resistance on the car due to the wind is $F_D = (0.3v^2)$ N, where v is the velocity in m/s, determine the power supplied to the engine at this instant. The engine has a running efficiency of $\varepsilon = 0.68$.

14–54. The sports car has a mass of 2.3 Mg and accelerates at 6 m/s², starting from rest. If the drag resistance on the car due to the wind is $F_D = (10v)$ N, where v is the velocity in m/s, determine the power supplied to the engine when $t = 5$ s. The engine has a running efficiency of $\varepsilon = 0.68$.

Probs. 14–53/54

14–55. The elevator E and its freight have a total mass of 400 kg. Hoisting is provided by the motor M and the 60-kg block C. If the motor has an efficiency of $\varepsilon = 0.6$, determine the power that must be supplied to the motor when the elevator is hoisted upward at a constant speed of $v_E = 4$ m/s.

Prob. 14–55

*14–56. The motor is used to lift the loaded 500-kg elevator with a constant velocity $v_E = 8$ m/s. If the motor draws 60 kW of electrical power, determine the motor's efficiency. Neglect the mass of the pulleys and cable.

14–57. The 500-kg elevator starts from rest and moves upward with a constant acceleration $a_c = 2$ m/s². Determine the power output of the motor M when $t = 3$ s. Neglect the mass of the pulleys and cable.

Probs. 14–56/57

14–58. The 50-lb block rests on the rough surface for which the coefficient of kinetic friction is $\mu_k = 0.2$. A force $F = (40 + s^2)$ lb, where s is in ft, acts on the block in the direction shown. If the spring is originally unstretched ($s = 0$) and the block is at rest, determine the power developed by the force the instant the block has moved $s = 1.5$ ft.

Prob. 14–58

14–59. The material hoist and the load have a total mass of 800 kg and the counterweight C has a mass of 150 kg. At a given instant, the hoist has an upward velocity of 2 m/s and an acceleration of 1.5 m/s². Determine the power generated by the motor M at this instant if it operates with an efficiency of $\varepsilon = 0.8$.

*14–60. The material hoist and the load have a total mass of 800 kg and the counterweight C has a mass of 150 kg. If the upward speed of the hoist increases uniformly from 0.5 m/s to 1.5 m/s in 1.5 s, determine the average power generated by the motor M during this time. The motor operates with an efficiency of $\varepsilon = 0.8$.

Probs. 14–59/60

14–61. The 50-lb crate is given a speed of 10 ft/s in $t=4$ s starting from rest. If the acceleration is constant, determine the power that must be supplied to the motor when $t=2$ s. The motor has an efficiency $\varepsilon=0.76$. Neglect the mass of the pulley and cable.

*14–64. The motor M is used to hoist the 500-kg elevator upward with a constant velocity $v_E = 8$ m/s. If the motor draws 60 kW of electrical power, determine the motor's efficiency. Neglect the mass of the pulleys and cable.

Prob. 14–61

Prob. 14–64

14–62. The 10-lb collar starts from rest at A and is lifted by applying a constant vertical force of $F = 25$ lb to the cord. If the rod is smooth, determine the power developed by the force at the instant $\theta = 60°$.

14–63. The 10-lb collar starts from rest at A and is lifted with a constant speed of 2 ft/s along the smooth rod. Determine the power developed by the force **F** at the instant shown.

14–65. Sand is being discharged from the silo at A to the conveyor and transported to the storage deck at the rate of 360 000 lb/h. An electric motor is attached to the conveyor to maintain the speed of the belt at 3 ft/s. Determine the average power generated by the motor.

Probs. 14–62/63

Prob. 14–65

14.5 CONSERVATIVE FORCES AND POTENTIAL ENERGY

Conservative Force. If the work of a force is independent of its path and depends only on the force's initial and final positions on the path, then the force is called a *conservative force*. Examples of conservative forces are the weight of a particle and the force developed by a spring. The work done by the weight depends only on the *vertical* displacement of the weight, and the work done by a spring force depends only on the spring's elongation or compression.

In contrast to a conservative force, consider the force of friction exerted on a sliding object by a fixed surface. The work done depends on the path—the longer the path, the greater the work. Consequently, frictional forces are *nonconservative*. The work is dissipated from the body in the form of heat.

Energy. *Energy* is defined as the capacity for doing work. There are essentially two types of energy, kinetic and potential. Kinetic energy is associated with the motion of the particle. For example, if a particle is originally at rest, then the principle of work and energy requires $\Sigma U_{1 \to 2} = T_2$. In other words, the work done on the particle is transferred into kinetic energy, which gives the particle a speed v. When the energy depends upon the position of the particle, measured from a fixed datum or reference plane, it is called *potential energy*. Thus, potential energy is a measure of the amount of work a conservative force must do to move a particle from a given position to the datum. In mechanics, the potential energy created by gravity (weight) and an elastic spring is important.

Gravitational Potential Energy. If a particle is located a distance y above an arbitrarily selected datum, as shown in Fig. 14–17, the particle's weight **W** has positive *gravitational potential energy*, V_g, since **W** has the capacity of doing positive work when the particle is moved back down to the datum. Likewise, if the particle is located a distance y below the datum, V_g is negative since the weight does negative work when the particle is moved back up to the datum. At the datum $V_g = 0$.*

In general then, if y is positive upward, the gravitational potential energy of the particle of weight W is

$$V_g = Wy \qquad (14\text{--}13)$$

Gravitational potential energy

Fig. 14–17

Gravitational potential energy of this weight is increased as it is hoisted upward.

---

*Here the weight is assumed to be *constant*. This assumption is suitable for small differences in elevation Δy. If the elevation change is significant, however, a variation of weight with elevation must be taken into account (see Prob. 14–82).

Elastic Potential Energy. When an elastic spring is elongated or compressed a distance s from its unstretched position, the force on the spring does work and thereby stores elastic potential energy V_e in the spring. This energy is

$$V_e = +\tfrac{1}{2}ks^2 \qquad (14\text{--}14)$$

Here V_e is always positive since, in the deformed position, the force of the spring has the *capacity* or "potential" for always doing positive work on the particle when the spring is returned to its unstretched position, Fig. 14–18.

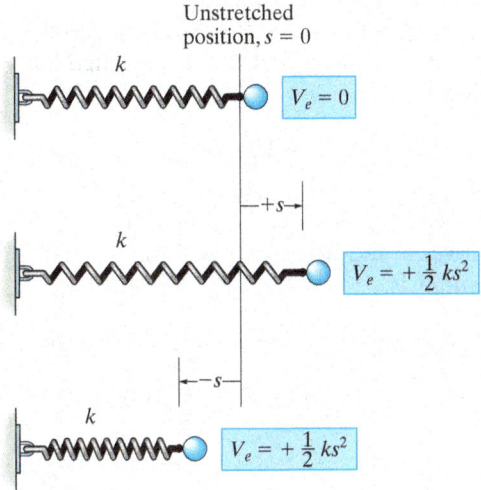

Elastic potential energy

Fig. 14–18

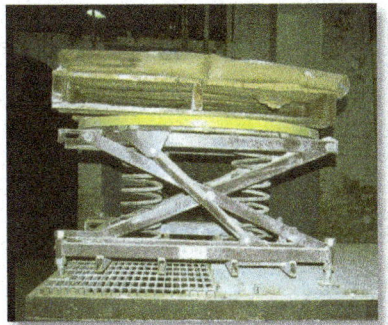

The weight of the sacks resting on this platform causes potential energy to be stored in the supporting springs. As each sack is removed, the platform will *rise* slightly since some of the potential energy within the springs will be transformed into an increase in gravitational potential energy of the remaining sacks. Such a device is useful for removing the sacks without having to bend over to pick them up as they are unloaded.

Potential Function. In the general case, if a particle is subjected to both gravitational and elastic forces, the particle's potential energy can be expressed as a *potential function*, which is the algebraic sum

$$V = V_g + V_e \qquad (14\text{-}15)$$

Measurement of V depends on the location of the particle with respect to a selected datum in accordance with Eqs. 14–13 and 14–14.

The work done by the gravitational and elastic forces in moving the particle from one point to another point is measured by the *difference* of this function, i.e.,

$$U_{1-2} = V_1 - V_2 \qquad (14\text{-}16)$$

For example, the potential function for a particle of weight W suspended from a spring can be expressed in terms of its position, s, measured from a datum located at the unstretched length of the spring, Fig. 14–19. We have

$$V = V_g + V_e$$
$$= -Ws + \tfrac{1}{2}ks^2$$

If the particle moves from s_1 to a lower position s_2, then applying Eq. 14–16 the work of $\mathbf{W}$ and $\mathbf{F}_s$ is

$$U_{1-2} = V_1 - V_2 = \left(-Ws_1 + \tfrac{1}{2}ks_1^2\right) - \left(-Ws_2 + \tfrac{1}{2}ks_2^2\right)$$
$$= W(s_2 - s_1) - \left(\tfrac{1}{2}ks_2^2 - \tfrac{1}{2}ks_1^2\right)$$

Fig. 14–19

14.5 CONSERVATIVE FORCES AND POTENTIAL ENERGY

When the displacement along the path is infinitesimal, i.e., from point (x, y, z) to $(x + dx, y + dy, z + dz)$, Eq. 14–16 becomes

$$dU = V(x, y, z) - V(x + dx, y + dy, z + dz)$$
$$= -dV(x, y, z) \qquad (14\text{--}17)$$

If we represent both the force and its displacement as Cartesian vectors, then the work can also be expressed as

$$dU = \mathbf{F} \cdot d\mathbf{r} = (F_x\mathbf{i} + F_y\mathbf{j} + F_z\mathbf{k}) \cdot (dx\mathbf{i} + dy\mathbf{j} + dz\mathbf{k})$$
$$= F_x\,dx + F_y\,dy + F_z\,dz$$

Substituting this result into Eq. 14–17 and expressing the differential $dV(x, y, z)$ in terms of its partial derivatives yields

$$F_x\,dx + F_y\,dy + F_z\,dz = -\left(\frac{\partial V}{\partial x}dx + \frac{\partial V}{\partial y}dy + \frac{\partial V}{\partial z}dz\right)$$

Since changes in x, y, and z are all independent of one another, this equation is satisfied provided

$$F_x = -\frac{\partial V}{\partial x}, \qquad F_y = -\frac{\partial V}{\partial y}, \qquad F_z = -\frac{\partial V}{\partial z} \qquad (14\text{--}18)$$

Thus,

$$\mathbf{F} = -\frac{\partial V}{\partial x}\mathbf{i} - \frac{\partial V}{\partial y}\mathbf{j} - \frac{\partial V}{\partial z}\mathbf{k} \qquad (14\text{--}19)$$

This equation relates a force $\mathbf{F}$ to its potential function V and thereby provides a mathematical criterion for proving that $\mathbf{F}$ is conservative. For example, the gravitational potential function for a weight located a distance y above a datum is $V_g = Wy$. To prove that $\mathbf{W}$ is conservative, it is necessary to show that it satisfies Eq. 14–18 (or Eq. 14–19), in which case

$$F_y = -\frac{\partial V}{\partial y}; \qquad F_y = -\frac{\partial}{\partial y}(Wy) = -W$$

The negative sign indicates that $\mathbf{W}$ acts downward, opposite to positive y, which is upward. Use Eq. 14–19 to show the spring force $F = kx$ is a conservative force.

14.6 CONSERVATION OF ENERGY

When a particle is acted upon by a system of *both* conservative and nonconservative forces, the portion of the work done by the *conservative forces* can be written in terms of the difference in their potential energies using Eq. 14–16, i.e., $(\Sigma U_{1-2})_{\text{cons.}} = V_1 - V_2$. As a result, the principle of work and energy can be written as

$$T_1 + V_1 + (\Sigma U_{1-2})_{\text{noncons.}} = T_2 + V_2 \qquad (14\text{–}20)$$

Here $(\Sigma U_{1-2})_{\text{noncons.}}$ represents the work of the nonconservative forces acting on the particle. If *only conservative forces* do work then we have

$$\boxed{T_1 + V_1 = T_2 + V_2} \qquad (14\text{–}21)$$

This equation is referred to as the **conservation of mechanical energy** or simply the **conservation of energy**. It states that during the motion the sum of the particle's kinetic and potential energies remains *constant*. For this to occur, kinetic energy must be transformed into potential energy, and vice versa.

For example, if a ball of weight **W** is dropped from a height h above the ground (datum), Fig. 14–20, the potential energy of the ball is maximum before it is dropped, at which time its kinetic energy is zero. The total mechanical energy of the ball in its initial position is thus

$$E = T_1 + V_1 = 0 + Wh = Wh$$

When the ball has fallen a distance $h/2$, its speed can be determined by using $v^2 = v_0^2 + 2a_c(y - y_0)$, which yields $v = \sqrt{2g(h/2)} = \sqrt{gh}$. The energy of the ball at the mid-height position is therefore

$$E = T_2 + V_2 = \frac{1}{2}\frac{W}{g}(\sqrt{gh})^2 + W\left(\frac{h}{2}\right) = Wh$$

Just before the ball strikes the ground, its potential energy is zero and its speed is $v = \sqrt{2gh}$. Here, again, the total energy of the ball is

$$E = T_3 + V_3 = \frac{1}{2}\frac{W}{g}(\sqrt{2gh})^2 + 0 = Wh$$

Note that when the ball comes in contact with the ground, it deforms somewhat, and provided the ground is hard enough, the ball will rebound off the surface, reaching a new height h', which will be *less* than the height h from which it was first released. Neglecting air friction, the difference in height accounts for an energy loss, $E_l = W(h - h')$, which occurs during the collision. Portions of this loss produce noise, localized deformation of the ball and ground, and heat.

Fig. 14–20

System of Particles. If a system of particles is *subjected only to conservative forces*, then an equation similar to Eq. 14–21 can be written for the particles. Applying the ideas of the preceding discussion, Eq. 14–8 ($\Sigma T_1 + \Sigma U_{1-2} = \Sigma T_2$) becomes

$$\Sigma T_1 + \Sigma V_1 = \Sigma T_2 + \Sigma V_2 \qquad (14\text{–}22)$$

Here, the sum of the system's initial kinetic and potential energies is equal to the sum of the system's final kinetic and potential energies. In other words, for the system $\Sigma T + \Sigma V =$ const.

PROCEDURE FOR ANALYSIS

The conservation of energy equation can be used to solve problems involving *velocity*, *displacement*, and *conservative force systems*. It is generally easier to apply than the principle of work and energy because this equation requires specifying the particle's or system's kinetic and potential energies at only *two points* along the path, rather than finding the work when the particle moves through a displacement. For application it is suggested that the following procedure be used.

Potential Energy.
- Draw two diagrams showing the particle or system in its initial and final position along the path.

- If the particle or system is subjected to a vertical displacement, establish the fixed horizontal datum from which to measure the particle's gravitational potential energy V_g.

- Determine the elevation y of the particle from the datum and the stretch or compression s of any connecting springs.

- Recall $V_g = Wy$, where y is positive upward from the datum and negative downward from the datum; also for a spring, $V_e = \frac{1}{2}ks^2$, which is *always positive*.

Conservation of Energy.
- Apply the equation $T_1 + V_1 = T_2 + V_2$.

- When determining the kinetic energy, $T = \frac{1}{2}mv^2$, remember that the particle's speed v must be measured from an inertial reference frame.

EXAMPLE 14.9

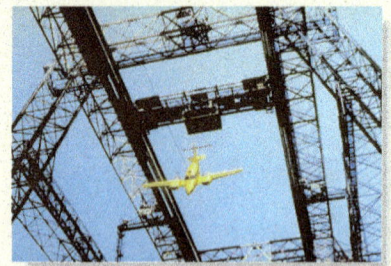

The gantry structure in the photo is used to test the response of an airplane during a crash. As shown in Fig. 14–21a, the plane, having a mass of 8 Mg, is hoisted back until $\theta = 60°$, and then the pull-back cable AC is released when the plane is at rest. Determine the speed of the plane just before it crashes into the ground, $\theta = 15°$. Also, what is the maximum tension developed in the supporting cable at this instant? Neglect the size of the airplane and the effect of lift caused by the wings during the motion.

SOLUTION
Since the force of the cable does *no work* on the plane, it must be obtained using the equation of motion. First, however, we must determine the plane's speed at B.

Potential Energy. For convenience, the datum has been established at the top of the gantry, Fig. 14–21a.

Conservation of Energy.

$$T_A + V_A = T_B + V_B$$
$$0 - 8000 \text{ kg } (9.81 \text{ m/s}^2)(20 \cos 60° \text{ m}) =$$
$$\tfrac{1}{2}(8000 \text{ kg})v_B^2 - 8000 \text{ kg } (9.81 \text{ m/s}^2)(20 \cos 15° \text{ m})$$
$$v_B = 13.52 \text{ m/s} = 13.5 \text{ m/s} \quad \textit{Ans.}$$

Equation of Motion. From the free-body diagram when the plane is at B, Fig. 14–21b, we have

$$+\nwarrow \Sigma F_n = ma_n;$$

$$T - (8000(9.81) \text{ N}) \cos 15° = (8000 \text{ kg}) \frac{(13.52 \text{ m/s})^2}{20 \text{ m}}$$

$$T = 149 \text{ kN} \quad \textit{Ans.}$$

Fig. 14–21

EXAMPLE 14.10

The ram R shown in Fig. 14-22a has a mass of 100 kg and is released from rest 0.75 m from the top of a spring, A, that has a stiffness $k_A = 12$ kN/m. If a second spring B, having a stiffness $k_B = 15$ kN/m, is "nested" in A, determine the maximum displacement of A needed to stop the downward motion of the ram. The unstretched length of each spring is indicated in the figure. Neglect the mass of the springs.

SOLUTION
Here we will consider the system to be the ram and the two springs.

Potential Energy. We will *assume* that the ram compresses *both* springs at the instant it comes to rest. The datum is located through the center of gravity of the ram at its initial position, Fig. 14-22b. When the kinetic energy is reduced to zero ($v_2 = 0$), A is compressed a distance s_A and B compresses $s_B = s_A - 0.1$ m.

Conservation of Energy.

$$T_1 + V_1 = T_2 + V_2$$

$$0 + 0 = 0 + [\tfrac{1}{2}k_A s_A^2 + \tfrac{1}{2}k_B(s_A - 0.1)^2 - Wh]$$

$$0 + 0 = 0 + [\tfrac{1}{2}(12\,000\text{ N/m})s_A^2 + \tfrac{1}{2}(15\,000\text{ N/m})(s_A - 0.1\text{ m})^2$$
$$- 981\text{ N}(0.75\text{ m} + s_A)]$$

Rearranging the terms,

$$13\,500 s_A^2 - 2481 s_A - 660.75 = 0$$

Using the quadratic formula and solving for the positive root, we have

$$s_A = 0.331 \text{ m} \qquad \textit{Ans.}$$

Since $s_B = 0.331$ m $-$ 0.1 m $=$ 0.231 m, which is positive, the assumption that *both* springs are compressed by the ram is correct.

NOTE: The second root, $s_A = -0.148$ m, does not represent the physical situation. Since positive s is measured downward, the negative sign indicates that spring A would have to be "extended" by an amount of 0.148 m to stop the ram.

(a)

(b)

Fig. 14-22

EXAMPLE 14.11

A smooth 2-kg collar, shown in Fig. 14–23a, fits loosely on the vertical shaft. If the spring is unstretched when the collar is in the position A, determine the speed at which the collar is moving when $y = 1$ m, if (a) it is released from rest at A, and (b) it is released at A with an *upward* velocity $v_A = 2$ m/s.

SOLUTION
The system here consists of the collar and spring.

Part (a) Potential Energy. For convenience, the datum is established through AB, Fig. 14–23b. When the collar is at C, the gravitational potential energy is $-(mg)y$, since the collar is *below* the datum. The elastic potential energy is $\frac{1}{2}ks_{CB}^2$, where $s_{CB} = 0.5$ m, which represents the *stretch* in the spring as shown in the figure.

Conservation of Energy.

$$T_A + V_A = T_C + V_C$$

$$0 + 0 = \tfrac{1}{2}mv_C^2 + [\tfrac{1}{2}ks_{CB}^2 - mgy]$$

$$0 + 0 = [\tfrac{1}{2}(2 \text{ kg})v_C^2] + [\tfrac{1}{2}(3 \text{ N/m})(0.5 \text{ m})^2 - 2(9.81) \text{ N}(1 \text{ m})]$$

$$v_C = 4.39 \text{ m/s} \downarrow \qquad \text{Ans.}$$

If this problem was solved using the equation of motion or the principle of work and energy, then for *both* of these methods, the variation of the magnitude and direction of the spring force must be taken into account (see Example 13.4). Here, however, the above solution is clearly advantageous since *only* the energy at the initial and final points of the path must be determined.

Fig. 14–23

Part (b) Conservation of Energy. If $v_A = 2$ m/s, using the data in Fig. 14–23b, we have

$$T_A + V_A = T_C + V_C$$

$$\tfrac{1}{2}mv_A^2 + 0 = \tfrac{1}{2}mv_C^2 + [\tfrac{1}{2}ks_{CB}^2 - mgy]$$

$$\tfrac{1}{2}(2 \text{ kg})(2 \text{ m/s})^2 + 0 = \tfrac{1}{2}(2 \text{ kg})v_C^2 + [\tfrac{1}{2}(3 \text{ N/m})(0.5 \text{ m})^2$$

$$- 2(9.81) \text{ N}(1 \text{ m})]$$

$$v_C = 4.82 \text{ m/s} \downarrow \qquad \text{Ans.}$$

NOTE: The kinetic energy of the collar depends only on the *magnitude* of velocity, and therefore it is immaterial if the collar is moving up or down at 2 m/s when released at A.

FUNDAMENTAL PROBLEMS

F14–13. The 2-kg pendulum bob is released from rest when it is at *A*. Determine the speed of the bob and the tension in the cord when the bob passes through its lowest position, *B*.

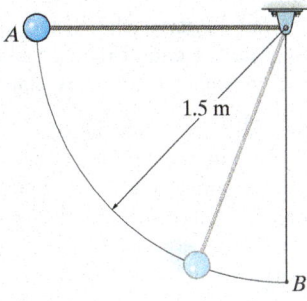

Prob. F14–13

F14–14. The 2-kg package leaves the conveyor belt at *A* with a speed of $v_A = 1$ m/s and slides down the smooth ramp. Determine the required speed of the conveyor belt at *B* so that the package can be delivered without slipping on the belt. Also, find the normal reaction the curved portion of the ramp exerts on the package at *B* if $\rho_B = 2$ m.

Prob. F14–14

F14–15. The 2-kg collar is given a downward velocity of 4 m/s when it is at *A*. If the spring has an unstretched length of 1 m and a stiffness of $k = 30$ N/m, determine the velocity of the collar at $s = 1$ m.

Prob. F14–15

F14–16. The 5-lb smooth collar is released from rest at *A*. Determine its speed when it strikes the stop *B*. The spring has an unstretched length of 0.5 ft.

Prob. F14–16

F14–17. The 75-lb block is released from rest 5 ft above the plate. Determine the compression of each spring when the block momentarily comes to rest after striking the plate. Neglect the mass of the plate. The springs are initially unstretched.

Prob. F14–17

F14–18. The 4-kg collar *C* has a velocity of $v_A = 2$ m/s when it is at *A*. If the guide rod is smooth, determine the speed of the collar when it is at *B*. The spring has an unstretched length of $l_0 = 0.2$ m.

Prob. F14–18

PROBLEMS

14–66. The girl has a mass of 40 kg and center of mass at G. If she is swinging to a maximum height defined by $\theta = 60°$, determine the force developed along each of the four supporting posts such as AB at the instant $\theta = 0°$. The swing is centrally located between the posts.

Prob. 14–66

14–67. The 30-lb block A is placed on top of two nested springs B and C and then pushed down to the position shown. If it is then released, determine the maximum height h to which it will rise.

Prob. 14–67

*__14–68.__ The 5-kg collar has a velocity of 5 m/s to the right when it is at A. It then travels down along the smooth guide. Determine the speed of the collar when it reaches point B, which is located just before the end of the curved portion of the rod. The spring has an unstretched length of 100 mm and B is located just before the end of the curved portion of the rod. Also, what is the normal force on the collar at this instant?

14–69. The 5-kg collar has a velocity of 5 m/s to the right when it is at A. It then travels along the smooth guide. Determine its speed when it reaches point B, which is located just before the end of the curved portion of the rod, and the normal force it exerts on the rod at this point. The spring has an unstretched length of 100 mm.

Probs. 14–68/69

14–70. The bob of the pendulum has a mass of 0.2 kg and is released from rest when it is in the horizontal position shown. Determine its speed and the tension in the cord at the instant the bob passes through its lowest position.

Prob. 14–70

14–71. The car C and its contents have a weight of 600 lb, whereas block B has a weight of 200 lb. If the car is released from rest, determine its speed when it travels 30 ft down the 20° incline. *Suggestion*: To measure the gravitational potential energy, establish separate datums at the initial elevations of B and C.

14–74. Determine the smallest amount the spring at B must be compressed against the 0.5-lb block so that when the block is released from B it slides along the smooth surface and reaches point A.

Prob. 14–71

Prob. 14–74

*****14–72.** The roller coaster car has a mass of 700 kg, including its passenger. If it starts from the top of the hill A with a speed $v_A = 3$ m/s, determine the minimum height h of the hill crest so that the car travels around the inside loops without leaving the track. Neglect friction, the mass of the wheels, and the size of the car. What is the normal reaction on the car when the car is at B and when it is at C? Take $\rho_B = 7.5$ m and $\rho_C = 5$ m.

14–73. The roller coaster car has a mass of 700 kg, including its passenger. If it is released from rest at the top of the hill A, determine the minimum height h of the hill crest so that the car travels around both inside loops without leaving the track. Neglect friction, the mass of the wheels, and the size of the car. What is the normal reaction on the car when the car is at B and when it is at C? Take $\rho_B = 7.5$ m and $\rho_C = 5$ m.

14–75. If the spring is compressed 3 in. against the 0.5-lb block and the block is released from rest, determine the normal force of the smooth surface on the block when it reaches point $x = 0.5$ ft.

Probs. 14–72/73

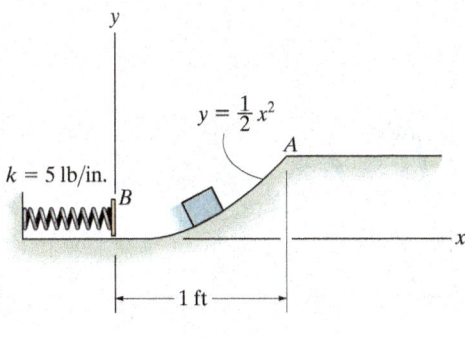

Prob. 14–75

*14–76. The spring has a stiffness $k = 50$ N/m and an unstretched length of 0.3 m. If it is attached to the 2-kg smooth collar and the collar is released from rest at A ($\theta = 0°$), determine the speed of the collar when $\theta = 60°$. The motion occurs in the horizontal plane. Neglect the size of the collar.

14–78. The spring has a stiffness $k = 200$ N/m and an unstretched length of 0.5 m. If it is attached to the 3-kg smooth collar and the collar is released from rest at A, determine the speed of the collar when it reaches B. Neglect the size of the collar.

Prob. 14–76

Prob. 14–78

14–77. The roller coaster car having a mass m is released from rest at point A. If the track is to be designed so that the car does not leave it at B, determine the required height h. Also, find the speed of the car when it reaches point C. Neglect friction.

14–79. A 2-lb block rests on the smooth semicylindrical surface at A. An elastic cord having a stiffness of $k = 2$ lb/ft is attached to the block at B and to the base of the semicylinder at C. If the block is released from rest at $\theta = 0$, A, determine the longest unstretched length of the cord so the block begins to leave the semicylinder at the instant $\theta = 45°$. Neglect the size of the block.

Prob. 14–77

Prob. 14–79

*14–80. When $s = 0$, the spring on the firing mechanism is unstretched. If the arm is pulled back such that $s = 100$ mm and released, determine the speed of the 0.3-kg ball and the normal reaction of the circular track on the ball when $\theta = 60°$. Assume all surfaces of contact to be smooth. Neglect the mass of the spring and the size of the ball.

Prob. 14–80

14–81. When $s = 0$, the spring on the firing mechanism is unstretched. If the arm is pulled back such that $s = 100$ mm and released, determine the maximum angle θ the 0.3-kg ball will travel without leaving the circular track. Assume all surfaces of contact to be smooth. Neglect the mass of the spring and the size of the ball.

Prob. 14–81

14–82. If the mass of the earth is M_e, show that the gravitational potential energy of a body of mass m located a distance r from the center of the earth is $V_g = -GM_em/r$. Recall that the gravitational force acting between the earth and the body is $F = G(M_em/r^2)$, Eq. 13–1. For the calculation, locate the datum at $r \to \infty$. Also, prove that F is a conservative force.

Prob. 14–82

14–83. A rocket of mass m is fired vertically from the surface of the earth, i.e., at $r = r_1$. Assuming that no mass is lost as it travels upward, determine the work it must do against gravity to reach a distance r_2. The force of gravity is $F = GM_em/r^2$ (Eq. 13–1), where M_e is the mass of the earth and r the distance between the rocket and the center of the earth.

Prob. 14–83

*14–84. The 4-kg smooth collar has a speed of 3 m/s when it is at $s = 0$. Determine the maximum distance s it travels before it stops momentarily. The spring has an unstretched length of 1 m.

Prob. 14–84

14–85. A 60-kg satellite travels in free flight along an elliptical orbit such that at A, where $r_A = 20$ Mm, it has a speed $v_A = 40$ Mm/h. What is the speed of the satellite when it reaches point B, where $r_B = 80$ Mm? Hint: See Prob. 14–82, where $M_e = 5.976(10^{24})$ kg and $G = 66.73(10^{-12})$ m^3/(kg·s^2).

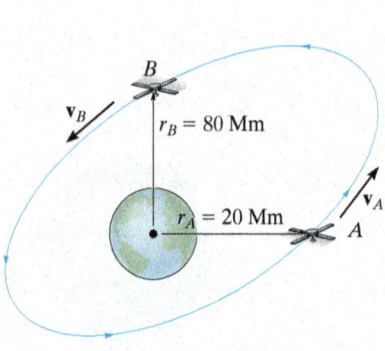

Prob. 14–85

14–86. The spring is unstretched when $s = 1$ m and the 15-kg block is released from rest at this position. Determine the speed of the block when $s = 3$ m. The spring remains horizontal during the motion, and the contact surfaces between the block and the inclined plane are smooth.

Prob. 14–86

14–87. The block has a mass of 20 kg and is released from rest when $s = 0.5$ m. If the mass of the bumpers A and B can be neglected, determine the maximum deformation of each spring due to the collision.

Prob. 14–87

*14–88. The 20-lb smooth collar is attached to the spring that has an unstretched length of 4 ft. If it is released from rest at position A, determine its speed when it reaches point B.

14–90. When the 5-kg box reaches point A it has a speed $v_A = 10$ m/s. Determine the normal force the box exerts on the surface when it reaches point B. Neglect friction and the size of the box.

Prob. 14–88

Prob. 14–90

14–91. The 0.75-kg bob of a pendulum is fired from rest at position A, by a spring which has a stiffness of $k = 6$ kN/m and is compressed 125 mm. Determine the speed of the bob and the tension in the cord when the bob is at positions B and C. Point B is located on the path where the radius of curvature is still 0.6 m, i.e., just before the cord becomes horizontal.

14–89. The 10-kg block A is released from rest and slides down the smooth plane. Determine the compression x of the spring when the block momentarily stops.

Prob. 14–89

Prob. 14–91

14–92. The roller coaster car has a speed of 15 ft/s when it is at the crest of a vertical parabolic track. Determine the car's velocity and the normal force it exerts on the track when it reaches point B. Neglect friction and the mass of the wheels. The total weight of the car and the passengers is 350 lb.

14–94. The 20-lb collar is constrained to move on the smooth rod. It is attached to the three springs which are unstretched when $s = 0$. If the collar is displaced $s = 0.5$ ft and released from rest, determine its speed when $s = 0$.

Prob. 14–92

Prob. 14–94

14–93. The spring has a stiffness $k = 3$ lb/ft and an unstretched length of 2 ft. If it is attached to the 5-lb smooth collar and the collar is released from rest at A, determine the speed of the collar just before it strikes the end of the rod at B. Neglect the size of the collar.

14–95. A tank car is stopped by two spring bumpers A and B, having a stiffness of $k_A = 15(10^3)$ lb/ft and $k_B = 20(10^3)$ lb/ft, respectively. Bumper A is attached to the car, whereas bumper B is attached to the wall. If the car has a weight of $25(10^3)$ lb and is freely coasting at 3 ft/s, determine the maximum deflection of each spring at the instant the bumpers stop the car.

Prob. 14–93

Prob. 14–95

***14–96.** The cylinder has a mass of 20 kg and is released from rest when $h = 0$. Determine its speed when $h = 3$ m. Each spring has a stiffness $k = 40$ N/m and an unstretched length of 2 m.

14–97. If the 20-kg cylinder is released from rest at $h = 0$, determine the required stiffness k of each spring so that its motion is arrested or stops when $h = 0.5$ m. Each spring has an unstretched length of 1 m.

Prob. 14–96

Prob. 14–97

CHAPTER REVIEW

Work of a Force

A force does work when it undergoes a displacement along its line of action. If the force varies with the displacement, then the work is $U = \int F \cos\theta \, ds$.

Graphically, this represents the area under the F–s diagram.

If the force is constant, then for a displacement Δs in the direction of the force, $U = F_c \Delta s$. A typical example of this case is the work of a weight, $U = -W(y_2 - y_1)$. Here, Δy is the vertical displacement.

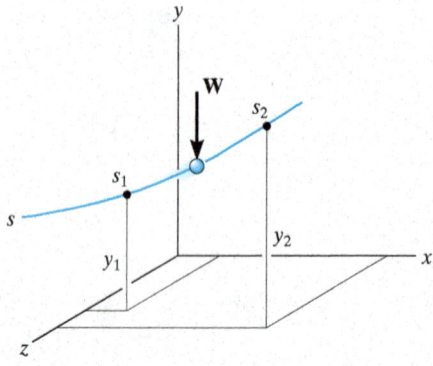

The work done by a spring force, $F = ks$, depends upon the stretch or compression s of the spring.

$$U = -(\tfrac{1}{2}ks_2^2 - \tfrac{1}{2}ks_1^2)$$

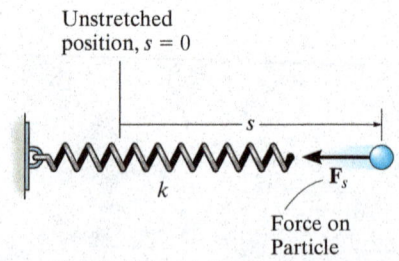

The Principle of Work and Energy

If the equation of motion in the tangential direction, $\Sigma F_t = ma_t$, is combined with the kinematic equation, $a_t \, ds = v \, dv$, we obtain the principle of work and energy. This equation states that the initial kinetic energy of a particle, plus the work done on the particle, is equal to the final kinetic energy of the particle.

$$T_1 + \Sigma U_{1\text{-}2} = T_2$$

The principle of work and energy is useful for solving problems that involve force, velocity, and displacement. For application, the free-body diagram of the particle should be drawn in order to identify the forces that do work.

Power and Efficiency

Power is the time rate of doing work. For application, the force **F** creating the power and its velocity **v** must be specified.

Average power is equal to the work done divided by the time.

Efficiency represents the ratio of power output to power input. Due to frictional losses, it is always less than one.

$$P = \mathbf{F} \cdot \mathbf{v}$$

$$P_{avg} = \frac{\Delta U}{\Delta t}$$

$$\varepsilon = \frac{\text{power output}}{\text{power input}}$$

Conservation of Energy

A conservative force does work that is independent of its path. Two examples are the weight of a particle and the spring force.

Friction is a nonconservative force since the work depends upon the length of the path. The longer the path, the more work.

The work of a conservative force depends upon its position relative to a datum. When this work is referenced from a datum, it is called potential energy. For a weight, it is $V_g = \pm Wy$, and for a spring it is $V_e = +\frac{1}{2}ks^2$.

Mechanical energy consists of kinetic energy T and gravitational and elastic potential energies V. According to the conservation of energy, this sum is constant for a system of particles and it has the same value at any position on the path.

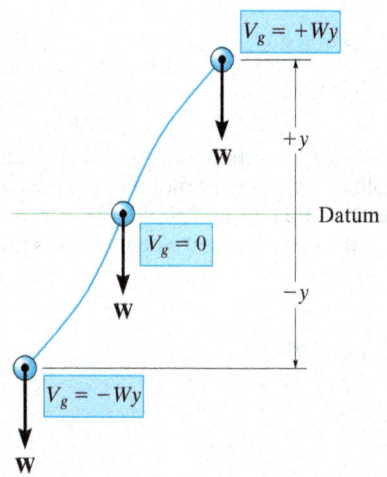

Gravitational potential energy

Elastic potential energy

$$T_1 + V_1 = T_2 + V_2$$

REVIEW PROBLEMS

R14–1. If a 150-lb crate is released from rest at A, determine its speed after it slides 30 ft down the plane. The coefficient of kinetic friction between the crate and plane is $\mu_k = 0.3$.

Prob. R14–1

R14–2. The small 2-lb collar starting from rest at A slides down along the smooth rod. During the motion, the collar is acted upon by a force $\mathbf{F} = \{10\mathbf{i} + 6y\mathbf{j} + 2z\mathbf{k}\}$ lb, where x, y, z are in feet. Determine the collar's speed when it strikes the wall at B.

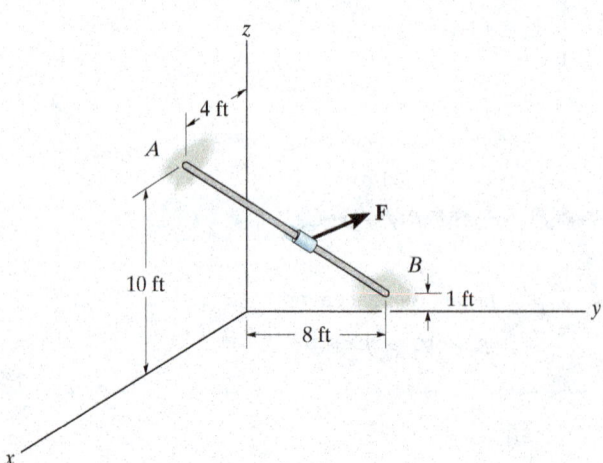

Prob. R14–2

R14–3. The block has a weight of 1.5 lb and slides along the smooth chute AB. It is released from rest at A, which has coordinates of $A(5\text{ ft}, 0, 10\text{ ft})$. Determine the speed at which it slides off at B, which has coordinates of $B(0, 8\text{ ft}, 0)$.

Prob. R14–3

R14–4. The block has a mass of 0.5 kg and moves within the smooth vertical slot. If the block starts from rest when the *attached* spring is in the unstretched position at A, determine the *constant* vertical force F which must be applied to the cord so that the block attains a speed $v_B = 2.5$ m/s when it reaches B; $s_B = 0.15$ m. Neglect the mass of the cord and pulley.

Prob. R14–4

R14–5. The crate, having a weight of 50 lb, is hoisted by the pulley system and motor M. If the crate starts from rest and, by constant acceleration, attains a speed of 12 ft/s after rising 10 ft, determine the power that must be supplied to the motor at the instant $s = 10$ ft. The motor has an efficiency $\varepsilon = 0.74$.

R14–7. The collar of negligible size has a mass of 0.25 kg and is attached to a spring having an unstretched length of 100 mm. If the collar is released from rest at A and travels along the smooth guide, determine its speed just before it strikes B.

Prob. R14–5

Prob. R14–7

R14–8. The blocks A and B weigh 10 and 30 lb, respectively. They are connected together by a light cord and ride in the frictionless grooves. Determine the speed of each block after block A moves 6 ft up along the plane. The blocks are released from rest.

R14–6. The 50-lb load is hoisted by the pulley system and motor M. If the motor exerts a constant force of 30 lb on the cable, determine the power that must be supplied to the motor if the load has been hoisted $s = 10$ ft starting from rest. The motor has an efficiency of $\varepsilon = 0.76$.

Prob. R14–6

Prob. R14–8

CHAPTER 15

Riders on this amusement park ride obey the conservation of angular momentum. As their motion changes, it can be determined as discussed in this chapter.

KINETICS OF A PARTICLE: IMPULSE AND MOMENTUM

CHAPTER OBJECTIVES

- To develop the principle of linear impulse and momentum for a particle and apply it to solve problems that involve force, velocity, and time.
- To discuss the conservation of linear momentum for particles.
- To analyze the mechanics of impact.
- To introduce the concept of angular impulse and momentum.
- To solve problems involving steady fluid streams and propulsion with variable mass.

15.1 PRINCIPLE OF LINEAR IMPULSE AND MOMENTUM

In this section we will integrate the equation of motion with respect to time and thereby obtain the principle of impulse and momentum. Since $a = dv/dt$, we have

$$\Sigma \mathbf{F} = m\mathbf{a} = m\frac{d\mathbf{v}}{dt} \qquad (15\text{--}1)$$

Rearranging the terms and integrating between the limits $\mathbf{v} = \mathbf{v}_1$ when $t = t_1$ and $\mathbf{v} = \mathbf{v}_2$ when $t = t_2$, we have

$$\Sigma \int_{t_1}^{t_2} \mathbf{F}\,dt = m\int_{\mathbf{v}_1}^{\mathbf{v}_2} d\mathbf{v}$$

or

$$\Sigma \int_{t_1}^{t_2} \mathbf{F} dt = m\mathbf{v}_2 - m\mathbf{v}_1 \qquad (15\text{–}2)$$

This equation is referred to as the **principle of linear impulse and momentum**. It provides a *direct means* of obtaining the particle's final velocity $\mathbf{v}_2$ after a specified time period when the particle's initial velocity is known and the forces acting on the particle are either constant or can be expressed as functions of time.

Linear Momentum. Each of the two vectors of the form $\mathbf{L} = m\mathbf{v}$ in Eq. 15–2 is referred to as the particle's *linear momentum*. Since the particle's mass m is a positive scalar, the linear-momentum vector has the same direction as $\mathbf{v}$, and its magnitude has units of mass times velocity, e.g., $kg \cdot m/s$, or $slug \cdot ft/s$.

Linear Impulse. The integral $\mathbf{I} = \int \mathbf{F} \, dt$ in Eq. 15–2 is referred to as the *linear impulse*. This term is a vector quantity which measures the effect of a force during the time the force acts. Since time is a positive scalar, the impulse acts in the same direction as the force, and its magnitude has units of force times time, e.g., $N \cdot s$ or $lb \cdot s$.*

If the force is expressed as a function of time, the impulse can be determined by direct evaluation of the integral. As a special case, if the force is constant in both magnitude and direction, then the impulse becomes

$$\mathbf{I} = \int_{t_1}^{t_2} \mathbf{F}_c dt = \mathbf{F}_c(t_2 - t_1)$$

Graphically the magnitude of the impulse can be represented by the shaded area under the curve of force versus time, Fig. 15–1. A constant force creates the shaded rectangular area shown in Fig. 15–2.

The impulse tool is used to remove the dent in the trailer fender. To do so its end is first screwed into a hole drilled in the fender, then the weight is gripped and jerked upwards, striking the stop ring. The impulse developed is transferred along the shaft of the tool and pulls suddenly on the dent.

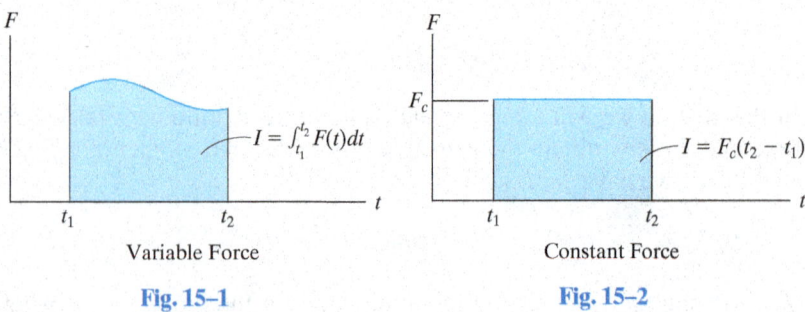

Variable Force

Fig. 15–1

Constant Force

Fig. 15–2

* Although the units for impulse and momentum are defined differently, it can be shown that Eq. 15–2 is dimensionally homogeneous.

15.1 Principle of Linear Impulse and Momentum

Principle of Linear Impulse and Momentum. For problem solving, Eq. 15–2 will be rewritten in the form

$$m\mathbf{v}_1 + \Sigma \int_{t_1}^{t_2} \mathbf{F}\, dt = m\mathbf{v}_2 \qquad (15\text{–}3)$$

which states that the initial momentum of the particle at time t_1 plus the sum of all the impulses applied to the particle from t_1 to t_2 is equivalent to the final momentum of the particle at time t_2. These three terms are illustrated graphically on the impulse and momentum diagrams shown in Fig. 15–3.

If each of the vectors in Eq. 15–3 is resolved into its x, y, z components, we can write the following three scalar equations of linear impulse and momentum.

$$m(v_x)_1 + \Sigma \int_{t_1}^{t_2} F_x\, dt = m(v_x)_2$$

$$m(v_y)_1 + \Sigma \int_{t_1}^{t_2} F_y\, dt = m(v_y)_2 \qquad (15\text{–}4)$$

$$m(v_z)_1 + \Sigma \int_{t_1}^{t_2} F_z\, dt = m(v_z)_2$$

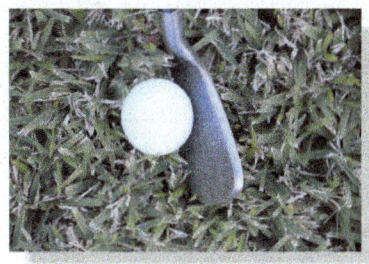

The dynamics of many types of sports, such as golf, requires application of the principle of linear impulse and momentum.

Initial momentum diagram + Impulse diagram = Final momentum diagram

Fig. 15–3

15.2 PRINCIPLE OF LINEAR IMPULSE AND MOMENTUM FOR A SYSTEM OF PARTICLES

The principle of linear impulse and momentum for a system of particles moving relative to an inertial reference, Fig. 15–4, is obtained from the equation of motion applied to all the particles in the system, i.e.,

$$\Sigma \mathbf{F}_i = \Sigma m_i \frac{d\mathbf{v}_i}{dt}$$

The term on the left side represents only the sum of the external forces $\mathbf{F}_i$ acting on the particles. The internal forces $\mathbf{f}_i$ acting between particles do not appear with this summation, since by Newton's third law they occur in equal but opposite collinear pairs and therefore cancel out. Multiplying both sides of the above equation by dt and integrating between the limits $t = t_1$, $\mathbf{v}_i = (\mathbf{v}_i)_1$ and $t = t_2$, $\mathbf{v}_i = (\mathbf{v}_i)_2$, we have

$$\boxed{\Sigma m_i(\mathbf{v}_i)_1 + \Sigma \int_{t_1}^{t_2} \mathbf{F}_i \, dt = \Sigma m_i(\mathbf{v}_i)_2} \quad (15\text{–}5)$$

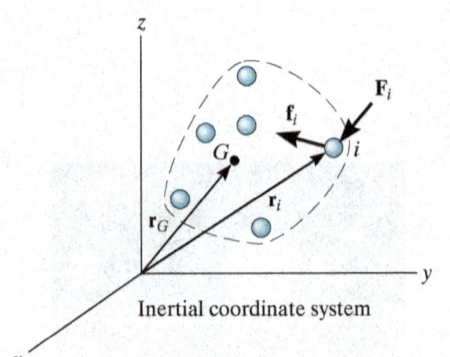

Fig. 15–4

Here the initial linear momentum of the system plus the impulses of all the external forces acting on the system from t_1 to t_2 is equal to the system's final linear momentum.

This equation can be simplified by noting that the location of the mass center G of the system is determined from $m\mathbf{r}_G = \Sigma m_i \mathbf{r}_i$, where $m = \Sigma m_i$ is the total mass of all the particles, Fig. 15–4. Taking the time derivative, we have

$$m\mathbf{v}_G = \Sigma m_i \mathbf{v}_i \quad (15\text{–}6)$$

In words, the total linear momentum of the system of particles, $\Sigma m_i \mathbf{v}_i$, is equivalent to the linear momentum of a "fictitious" aggregate particle of mass $m = \Sigma m_i$ moving with the velocity of the mass center of the system, $\mathbf{v}_G$. Substituting into Eq. 15–6 yields

$$m(\mathbf{v}_G)_1 + \Sigma \int_{t_1}^{t_2} \mathbf{F}_i \, dt = m(\mathbf{v}_G)_2 \quad (15\text{–}7)$$

If the system of particles collectively represents the particles composing a rigid body, then this result justifies application of the principle of linear impulse and momentum to a body of finite size.

PROCEDURE FOR ANALYSIS

The principle of linear impulse and momentum is used to solve problems involving *force*, *time*, and *velocity*, since these terms are involved in the formulation. For application it is suggested that the following procedure be used.*

Free-Body Diagram.

- Establish the *x*, *y*, *z* inertial frame of reference and draw the particle's free-body diagram in order to account for all the forces that produce impulses on the particle.

- The direction and sense of the particle's initial and final velocities should be established.

- If a vector is unknown, assume that the sense of its components is in the direction of the positive inertial coordinate(s).

- As an alternative procedure, draw the impulse and momentum diagrams for the particle as shown in Fig. 15–3.

Principle of Impulse and Momentum.

- Apply the principle of linear impulse and momentum, $m\mathbf{v}_1 + \Sigma \int_{t_1}^{t_2} \mathbf{F}\, dt = m\mathbf{v}_2$. If motion occurs in the *x*–*y* plane, the two scalar component equations can be formulated by either resolving the vector components of $\mathbf{F}$ from the free-body diagram, or by using the data on the impulse and momentum diagrams.

- Realize that every force acting on the particle's free-body diagram will create an impulse, even though some of these forces will do no work.

- Forces that are functions of time must be integrated to obtain the impulse. Graphically, the impulse is equal to the area under the force–time curve.

As the wheels of the pitching machine rotate, they apply frictional impulses to the ball, thereby giving it a linear momentum. These impulses are shown on the impulse diagram. Here both the frictional and normal impulses vary with time. By comparison, the weight impulse is constant and is very small since the time Δt the ball is in contact with the wheels will be very small.

*This procedure will be followed when developing the proofs and theory in the book.

EXAMPLE 15.1

The 100-kg crate shown in Fig. 15–5a is originally at rest on the smooth horizontal surface. If a towing force of 200 N, acting at an angle of 45°, is applied for 10 s, determine the final velocity of the crate and the normal force which the surface exerts on the crate during this time interval.

SOLUTION
This problem can be solved using the principle of impulse and momentum since it involves force, velocity, and time.

Free-Body Diagram. Since all the forces acting on the free-body diagram, Fig. 15–5b, are *constant*, the impulses are simply the product of the force magnitude and 10 s [$\mathbf{I} = \mathbf{F}_c(t_2 - t_1)$]. Note the alternative procedure of drawing the crate's impulse and momentum diagrams, Fig. 15–5c.

Principle of Impulse and Momentum. Applying Eqs. 15–4, we have

$$(\xrightarrow{+}) \qquad m(v_x)_1 + \Sigma \int_{t_1}^{t_2} F_x \, dt = m(v_x)_2$$

$$0 + 200 \text{ N} \cos 45°(10 \text{ s}) = (100 \text{ kg})v_2$$

$$v_2 = 14.1 \text{ m/s} \qquad \textit{Ans.}$$

$$(+\uparrow) \qquad m(v_y)_1 + \Sigma \int_{t_1}^{t_2} F_y \, dt = m(v_y)_2$$

$$0 + N_C(10 \text{ s}) - 981 \text{ N}(10 \text{ s}) + 200 \text{ N} \sin 45°(10 \text{ s}) = 0$$

$$N_C = 840 \text{ N} \qquad \textit{Ans.}$$

NOTE: Since no motion occurs in the y direction, direct application of the equilibrium equation $\Sigma F_y = 0$ gives the same result for N_C. Try to solve the problem by first applying $\Sigma F_x = ma_x$, then $v = v_0 + a_c t$.

Fig. 15–5

EXAMPLE 15.2

The 50-lb crate shown in Fig. 15–6a is acted upon by a force having a variable magnitude $P = (20t)$ lb, where t is in seconds. Determine the crate's velocity when $t = 2$ s if the initial velocity of the crate is $v_1 = 3$ ft/s down the plane. The coefficient of kinetic friction between the crate and the plane is $\mu_k = 0.3$.

(a)

SOLUTION

Free-Body Diagram. See Fig. 15–6b. Since the magnitude of force $P = 20t$ varies with time, the impulse it creates must be determined by integrating over the 2-s time interval.

Principle of Impulse and Momentum. Applying Eqs. 15–4 in the x direction, we have

$$(+\swarrow) \qquad m(v_x)_1 + \Sigma \int_{t_1}^{t_2} F_x \, dt = m(v_x)_2$$

$$\frac{50 \text{ lb}}{32.2 \text{ ft/s}^2}(3 \text{ ft/s}) + \int_0^{2\text{ s}} 20t \, dt - 0.3 N_C(2 \text{ s}) + (50 \text{ lb}) \sin 30°(2 \text{ s}) = \frac{50 \text{ lb}}{32.2 \text{ ft/s}^2} v_2$$

$$4.658 + 40 - 0.6 N_C + 50 = 1.553 v_2$$

The equation of equilibrium can be applied in the y direction. Why?

$$+\nwarrow \Sigma F_y = 0; \qquad N_C - 50 \cos 30° \text{ lb} = 0$$

Solving,

$$N_C = 43.30 \text{ lb}$$
$$v_2 = 44.2 \text{ ft/s} \qquad Ans.$$

NOTE: We can also solve this problem using the equation of motion. From Fig. 15–6b,

$$+\swarrow \Sigma F_x = ma_x; \qquad 20t - 0.3(43.30) + 50 \sin 30° = \frac{50}{32.2} a$$

$$a = 12.88t + 7.734$$

(b)

Fig. 15–6

Using kinematics

$$+\swarrow dv = a \, dt; \qquad \int_{3 \text{ ft/s}}^{v} dv = \int_0^{2\text{ s}} (12.88t + 7.734) \, dt$$

$$v = 44.2 \text{ ft/s} \qquad Ans.$$

By comparison, application of the principle of impulse and momentum eliminates the need for using kinematics ($a = dv/dt$) and thereby yields an easier method for solution.

EXAMPLE 15.3

(b)

Fig. 15–7

Blocks A and B shown in Fig. 15–7a have a mass of 3 kg and 5 kg, respectively. If the system is released from rest, determine the velocity of block B in 6 s. Neglect the mass of the pulleys and cord.

SOLUTION
Free-Body Diagram. See Fig. 15–7b. Since the weight of each block is constant, the cord tensions will also be constant. Furthermore, since the mass of pulley D is neglected, the cord tension $T_A = 2T_B$. Note that the blocks are both assumed to be moving downward in the positive coordinate directions, s_A and s_B.

Principle of Impulse and Momentum.
Block A:

$$(+\downarrow) \qquad m(v_A)_1 + \Sigma \int_{t_1}^{t_2} F_y\, dt = m(v_A)_2$$

$$0 - 2T_B(6\text{ s}) + 3(9.81)\text{ N}(6\text{ s}) = (3\text{ kg})(v_A)_2 \qquad (1)$$

Block B:

$$(+\downarrow) \qquad m(v_B)_1 + \Sigma \int_{t_1}^{t_2} F_y\, dt = m(v_B)_2$$

$$0 + 5(9.81)\text{ N}(6\text{ s}) - T_B(6\text{ s}) = (5\text{ kg})(v_B)_2 \qquad (2)$$

Kinematics. Since the blocks are subjected to dependent motion, the velocity of A can be related to that of B by using the kinematic analysis discussed in Sec. 12–9. A horizontal datum is established through the fixed point at C, Fig. 15–7a, and the position coordinates, s_A and s_B, are related to the constant total length l of the vertical segments of the cord by the equation

$$2s_A + s_B = l$$

Taking the time derivative yields

$$2v_A = -v_B \qquad (3)$$

As indicated by the negative sign, when B moves downward A moves upward. Substituting this result into Eq. 1 and solving Eqs. 1 and 2 yields

$$(v_B)_2 = 35.8 \text{ m/s} \downarrow$$
$$T_B = 19.2 \text{ N} \qquad\qquad Ans.$$

NOTE: Realize that the *positive* (downward) direction for $\mathbf{v}_A$ and $\mathbf{v}_B$ is consistent in Figs. 15–7a and 15–7b and in Eqs. 1 to 3. This is important since we are seeking a simultaneous solution of equations.

FUNDAMENTAL PROBLEMS

F15–1. The 0.5-kg ball strikes the rough ground and rebounds with the velocities shown. Determine the magnitude of the impulse the ground exerts on the ball. Assume that the ball does not slip when it strikes the ground, and neglect the size of the ball and the impulse produced by the weight of the ball.

Prob. F15–1

F15–2. If the coefficient of kinetic friction between the 150-lb crate and the ground is $\mu_k = 0.2$, determine the speed of the crate when $t = 4$ s. The crate starts from rest and is towed by the 100-lb force.

Prob. F15–2

F15–3. The motor exerts a force of $F = (20t^2)$ N on the cable, where t is in seconds. Determine the speed of the 25-kg crate when $t = 4$ s. The coefficients of static and kinetic friction between the crate and the plane are $\mu_s = 0.3$ and $\mu_k = 0.25$, respectively.

Prob. F15–3

F15–4. The wheels of the 1.5-Mg car generate the traction force **F** described by the graph. If the car starts from rest, determine its speed when $t = 6$ s.

Prob. F15–4

F15–5. The 2.5-Mg four-wheel-drive SUV tows the 1.5-Mg trailer. The traction force developed at the wheels is $F_D = 9$ kN. Determine the speed of the truck in 20 s, starting from rest. Also, determine the tension developed in the coupling, A, between the SUV and the trailer. Neglect the mass of the wheels.

Prob. F15–5

F15–6. The 10-lb block A attains a velocity of 1 ft/s in 5 seconds, starting from rest. Determine the tension in the cord and the coefficient of kinetic friction between block A and the horizontal plane. Neglect the weight of the pulley. Block B has a weight of 8 lb.

Prob. F15–6

PROBLEMS

All solutions must include a free-body diagram.

15–1. A hammer head H having a weight of 0.3 lb is moving downwards at 40 ft/s when it strikes the head of a nail of negligible mass and drives it into a block of wood. Find the impulse on the nail if it is assumed that the grip at A is loose, the handle has a negligible mass, and the hammer stays in contact with the nail while it comes to rest.

Prob. 15–1

15–2. A 20-lb block slides down a 30° inclined plane with an initial velocity of 2 ft/s. Determine the velocity of the block in 3 s if the coefficient of kinetic friction between the block and the plane is $\mu_k = 0.25$.

15–3. The uniform beam has a weight of 5000 lb. Determine the average tension in each of the two cables AB and AC if the beam is given an upward speed of 8 ft/s in 1.5 s starting from rest.

*****15–4.** Each of the cables can sustain a maximum tension of 5000 lb. If the uniform beam has a weight of 5000 lb, determine the shortest time possible to lift the beam with a speed of 10 ft/s starting from rest.

Probs. 15–3/4

15–5. The 60-lb hoisting block is operated by a motor M. Starting from rest, the supporting cable can be wound up by the motor so that it is wound into the motor with a relative velocity of 3 ft/s in 5 s. If the cable does not slip on the pulleys, determine the average force exerted on the cables during this time. Neglect the mass of each of the pulleys.

Prob. 15–5

15–6. The jet plane has a mass of 250 Mg and a horizontal velocity of 100 m/s when $t = 0$. If *both* engines provide a horizontal thrust which varies as shown in the graph, determine the plane's velocity in $t = 5$ s. Neglect air resistance and the loss of fuel during the motion.

Prob. 15–6

15–7. Crates A and B weigh 100 lb and 50 lb, respectively. If they start from rest, determine their speed when $t = 5$ s. Also, find the force exerted by crate A on crate B during the motion. The coefficient of kinetic friction between the crates and the ground is $\mu_k = 0.25$.

15–9. The 200-kg crate rests on the ground for which the coefficients of static and kinetic friction are $\mu_s = 0.5$ and $\mu_k = 0.4$, respectively. The winch delivers a horizontal towing force T to its cable at A which varies as shown in the graph. Determine the speed of the crate when $t = 4$ s. Originally the tension in the cable is zero. *Hint*: First determine the force needed to begin moving the crate.

Prob. 15–7

Prob. 15–9

*15–8. The automobile has a weight of 2700 lb and is traveling forward at 4 ft/s when it crashes into the wall. If the impact occurs in 0.06 s, determine the average impulsive force acting on the car. Assume the brakes are *not applied*. If the coefficient of kinetic friction between the wheels and the pavement is $\mu_k = 0.3$, calculate the impulsive force on the wall if the brakes *were applied* during the crash. The brakes are applied to all four wheels so that all the wheels slip.

15–10. The force acting on a projectile having a mass m as it passes horizontally through the barrel of the cannon is $F = C \sin(\pi t/t')$. Determine the projectile's velocity when $t = t'$. If the projectile reaches the end of the barrel at this instant, determine the length s.

Prob. 15–8

Prob. 15–10

15–11. During operation the jack hammer strikes the concrete surface with a force which is indicated in the graph. To achieve this the 2-kg spike S is fired into the surface at 90 m/s. Determine the speed of the spike just after rebounding.

15–14. A tankcar has a mass of 20 Mg and is freely rolling to the right with a speed of 0.75 m/s. If it strikes the barrier, determine the horizontal impulse needed to stop the car if the spring in the bumper B has a stiffness (a) $k \to \infty$ (bumper is rigid), and (b) $k = 15$ kN/m.

Prob. 15–11

Prob. 15–14

*15–12. For a short period of time, the frictional driving force acting on the wheels of the 2.5-Mg van is $F_D = (600t^2)$ N, where t is in seconds. If the van has a speed of 20 km/h when $t = 0$, determine its speed when $t = 5$ s.

15–15. The 10-lb slider block is moving to the right with a velocity of 10 ft/s when it is acted upon by the forces $\mathbf{F}_1$ and $\mathbf{F}_2$. If these loadings vary in the manner shown in the graph, determine the velocity of the block after 6 s. Neglect friction and the weight of the pulleys.

Prob. 15–12

15–13. The 2.5-Mg van is traveling with a speed of 100 km/h when the brakes are applied and all four wheels lock. If the speed decreases to 40 km/h in 5 s, determine the coefficient of kinetic friction between the tires and the road.

Prob. 15–13

Prob. 15–15

*15–16. The 180-lb iron worker is secured by a fall-arrest system consisting of a harness and lanyard AB, which is fixed to the beam. If the lanyard has a slack of 4 ft, determine the average impulsive force developed in the lanyard if he happens to fall 4 feet. Neglect his size in the calculation and assume the impulse takes place in 0.6 seconds.

Prob. 15–16

15–17. The towing force acting on the 400-kg safe varies as shown in the graph. Determine its speed, starting from rest, when $t = 8$ s. How far has it traveled during this time?

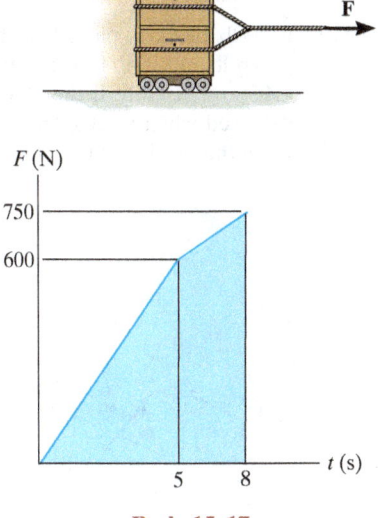

Prob. 15–17

15–18. The motor exerts a force F on the 40-kg crate as shown in the graph. Determine the speed of the crate when $t = 3$ s and when $t = 6$ s. When $t = 0$, the crate is moving downward at 10 m/s.

Prob. 15–18

15–19. The 30-kg slider block is moving to the left with a speed of 5 m/s when it is acted upon by the forces $\mathbf{F}_1$ and $\mathbf{F}_2$. If these loadings vary in the manner shown in the graph, determine the speed of the block at $t = 6$ s. Neglect friction and the mass of the pulleys and cords..

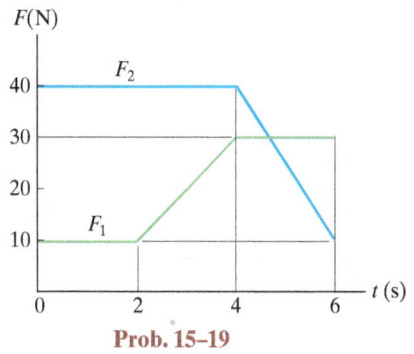

Prob. 15–19

*15–20. Determine the maximum speed attained by the 1.5-Mg rocket sled if the rockets provide the thrust shown in the graph. Initailly, the sled is at rest. Neglect friction and the loss of mass due to fuel consumption.

Prob. 15–20

15–21. If it takes 35 s for the 50-Mg tugboat to increase its speed uniformly to 25 km/h, starting from rest, determine the force of the rope on the tugboat. The propeller provides the propulsion force **F** on the tugboat which gives it forward motion, whereas the barge moves freely. Also, determine the magnitude of **F**. The barge has a mass of 75 Mg.

Prob. 15–21

15–22. The tanker has a mass of 130 Gg. If it is originally at rest, determine its speed when $t = 10$ s. The horizontal thrust provided by its propeller varies with time as shown in the graph. Neglect the effect of water resistance.

Prob. 15–22

15–23. The motor pulls on the cable at A with a force $F = (30 + t^2)$ lb, where t is in seconds. If the 34-lb crate is originally on the ground at $t = 0$, determine its speed in $t = 4$ s. Neglect the mass of the cable and pulleys. *Hint*: First find the time needed to begin lifting the crate.

Prob. 15–23

*15–24. The small 20-lb block is placed on the inclined plane and subjected to 6-lb and 15-lb forces that act parallel with edges AB and AC, respectively. If the block is initially at rest, determine its speed when $t = 3$ s. The coefficient of kinetic friction between the block and the plane is $\mu_k = 0.2$.

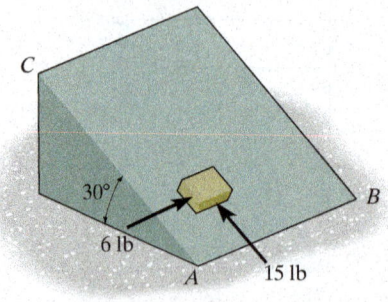

Prob. 15–24

15–25. The balloon has a total mass of 400 kg including the passengers and ballast. The balloon is rising at a constant velocity of 18 km/h when $h = 10$ m. If the man drops the 40-kg sand bag, determine the velocity of the balloon when the bag strikes the ground. Neglect air resistance.

15–27. The winch delivers a horizontal towing force **F** to its cable at A which varies as shown in the graph. Determine the speed of the 70-kg bucket when $t = 18$ s. Originally the bucket is moving upward at $v_1 = 3$ m/s.

*15–28. The winch delivers a horizontal towing force **F** to its cable at A which varies as shown in the graph. Determine the speed of the 80-kg bucket when $t = 24$ s. Originally the bucket is released from rest.

Prob. 15–25

Probs. 15–27/28

15–26. As indicated by the derivation, the principle of impulse and momentum is valid for observers in *any* inertial reference frame. Show that this is so, by considering the 10-kg block which slides along the smooth surface and is subjected to a horizontal force of 6 N. If observer A is in a *fixed* frame x, determine the final speed of the block in 4 s if it has an initial speed of 5 m/s measured from the fixed frame. Compare the result with that obtained by an observer B, attached to the x' axis that moves at a constant velocity of 2 m/s relative to A.

15–29. Determine the velocities of blocks A and B 2 s after they are released from rest. Neglect the mass of the pulleys and cables.

Prob. 15–26

Prob. 15–29

15–30. A jet plane having a mass of 7 Mg takes off from an aircraft carrier such that the engine thrust varies as shown by the graph. If the carrier is traveling forward with a speed of 40 km/h, determine the plane's airspeed after 5 s.

15–33. If the force T exerted on the cable by the motor M is indicated by the graph, determine the speed of the 500-lb crate when $t = 4$ s, starting from rest. The coefficients of static and kinetic friction are $\mu_s = 0.3$ and $\mu_k = 0.25$, respectively.

Prob. 15–30

Prob. 15–33

15–31. The 100-kg crate is hoisted by the motor M. If the velocity of the crate increases uniformly from 1.5 m/s to 4.5 m/s in 5 s, determine the tension developed in the cable during the motion.

*15–32. The 100-kg crate is hoisted by the motor M. The motor exerts a force on the cable of $T = (200t^{1/2} + 150)$ N, where t is in seconds. If the crate starts from rest at the ground, determine the speed of the crate when $t = 5$ s.

15–34. A cake and plate weighing 1.5 lb rest at the center of a circular table. Without touching the cake a boy attempts to remove the 2-ft-radius tablecloth by quickly pulling on it horizontally. If slipping between the cake plate and tablecloth occurs at all times, determine the longest time the tablecloth can remain in contact with the plate without having the cake wind up on the floor. The coefficient of kinetic friction between the cake plate and the tablecloth is $\mu_k = 0.3$ and between the cake plate and the table surface $\mu'_k = 0.4$. *Hint:* Apply the principles of impulse and momentum and work and energy to the cake for x ft, the distance the cake plate remains on the tablecloth, and for $(2 - x)$ ft when it remains on the table.

Probs. 15–31/32

Prob. 15–34

15.3 CONSERVATION OF LINEAR MOMENTUM FOR A SYSTEM OF PARTICLES

When the sum of the *external impulses* acting on a system of particles is *zero*, Eq. 15–5 reduces to a simplified form, namely,

$$\Sigma m_i(\mathbf{v}_i)_1 = \Sigma m_i(\mathbf{v}_i)_2 \qquad (15\text{–}8)$$

This equation is referred to as the ***conservation of linear momentum***. It states that the total linear momentum for a system of particles remains constant during the time period t_1 to t_2. From Eq. 15–7, we can also write

$$(\mathbf{v}_G)_1 = (\mathbf{v}_G)_2 \qquad (15\text{–}9)$$

which indicates that the velocity $\mathbf{v}_G$ of the mass center for the system of particles does not change if no external impulses are applied to the system.

The conservation of linear momentum is often applied when particles collide or interact. For application, a careful study of the free-body diagram for the *entire* system of particles should be made in order to identify the forces which create external impulses and thereby determine in what direction(s) linear momentum is conserved. If the time over which the motion is studied is *very short*, some of the impulses may also be neglected or considered approximately equal to zero. The forces causing these negligible impulses are called ***nonimpulsive forces***. By comparison, forces which are very large and act for a very short period of time produce a significant change in momentum and are called ***impulsive forces***. They, of course, cannot be neglected in the impulse–momentum analysis.

Impulsive forces normally occur due to an explosion or the striking of one body against another, whereas nonimpulsive forces may include the weight of a body, the force imparted by a slightly deformed spring having a relatively small stiffness, or for that matter, any force that is very small compared to other larger (impulsive) forces. When making this distinction between impulsive and nonimpulsive forces, it is important to realize that this only applies during the time t_1 to t_2. To illustrate, consider the effect of striking a tennis ball with a racket as shown in the photo on the next page. During the *very short* time of interaction, the force of the racket on the ball is impulsive since it changes the ball's momentum drastically. By

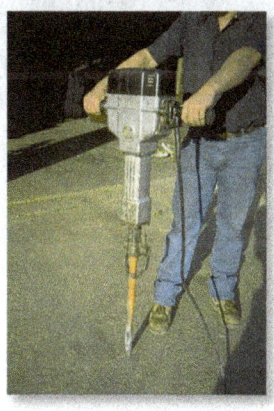

The hammer in the top photo applies an impulsive force to the stake. During this extremely short time of contact the weight of the stake can be considered nonimpulsive, and provided the stake is driven into soft ground, the impulse of the ground acting on the stake can also be considered nonimpulsive. By contrast, if the stake is used in a concrete chipper to break concrete, then two impulsive forces act on the stake: one at its top due to the chipper and the other on its bottom due to the rigidity of the concrete.

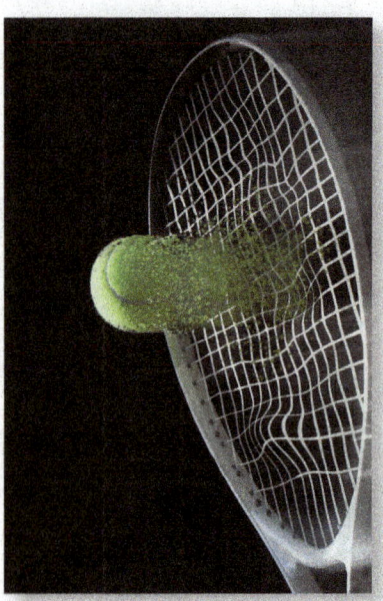

comparison, the ball's weight will have a negligible effect on the change in momentum, and therefore it is nonimpulsive. Consequently, it can be neglected from an impulse–momentum analysis during this time. If an impulse–momentum analysis is considered during the much longer time of flight after the racket–ball interaction, then the impulse of the ball's weight is important since it, along with air resistance, causes the change in the momentum of the ball.

PROCEDURE FOR ANALYSIS

Generally, the principle of linear impulse and momentum or the conservation of linear momentum are applied to a *system of particles* in order to determine the final velocities of the particles *just after* the time period considered. By applying these principles to the entire system, the internal impulses acting within the system are *eliminated* from the analysis. For application it is suggested that the following procedure be used.

Free-Body Diagram.
- Establish the x, y, z inertial frame of reference and draw the free-body diagram for each particle of the system in order to identify the internal and external forces.

- The conservation of linear momentum applies to the system in a direction which either has no external forces or the forces can be considered nonimpulsive.

- Establish the direction and sense of the particles' initial and final velocities. If the sense is unknown, assume it is in the positive coordinate direction.

Momentum Equations.
- Apply the principle of linear impulse and momentum or the conservation of linear momentum in the appropriate directions.

- If it is necessary to determine the internal impulse $\int F \, dt$ acting on only one particle of a system, then the particle must be *isolated* (free-body diagram), and the principle of linear impulse and momentum must be applied to this particle.

- After the impulse is calculated, and provided the time Δt for which the impulse acts is known, then the average impulsive force F_{avg} can be determined from $F_{avg} = \int F \, dt / \Delta t$.

EXAMPLE 15.4

The 15-Mg boxcar A is coasting at 1.5 m/s on the horizontal track when it encounters a 12-Mg tank car B coasting at 0.75 m/s toward it as shown in Fig. 15–8a. If the cars collide and couple together, determine (a) the speed of both cars just after the coupling, and (b) the average force between them if the coupling takes place in 0.8 s.

(a)

SOLUTION

Part (a) Free-Body Diagram.* As shown in Fig. 15–8b, we have considered *both* cars as a single system. By inspection, momentum is conserved in the x direction since the coupling force **F** is *internal* to the system and will therefore cancel out. It is assumed that when coupled, both cars move at $\mathbf{v}_2$ in the positive x direction.

Conservation of Linear Momentum.

$$(\stackrel{+}{\rightarrow}) \qquad m_A(v_A)_1 + m_B(v_B)_1 = (m_A + m_B)v_2$$

$$(15\,000 \text{ kg})(1.5 \text{ m/s}) - 12\,000 \text{ kg}(0.75 \text{ m/s}) = (27\,000 \text{ kg})v_2$$

$$v_2 = 0.5 \text{ m/s} \rightarrow \qquad \textit{Ans.}$$

Part (b). The average (impulsive) coupling force, $\mathbf{F}_{avg}$, can be determined by applying the principle of linear momentum to *either one* of the cars.

Free-Body Diagram. As shown in Fig. 15–8c, by isolating the boxcar the coupling force is *external* to the car.

Principle of Impulse and Momentum. Since $\int F \, dt = F_{avg} \Delta t = F_{avg}(0.8 \text{ s})$, we have

$$(\stackrel{+}{\rightarrow}) \qquad m_A(v_A)_1 + \Sigma \int F \, dt = m_A v_2$$

$$(15\,000 \text{ kg})(1.5 \text{ m/s}) - F_{avg}(0.8 \text{ s}) = (15\,000 \text{ kg})(0.5 \text{ m/s})$$

$$F_{avg} = 18.8 \text{ kN} \qquad \textit{Ans.}$$

NOTE: Try solving for F_{avg} by applying the principle of impulse and momentum to the tank car.

(b)

(c)

Fig. 15–8

---

*Only horizontal forces are shown on the free-body diagram.

EXAMPLE 15.5

(a)

(b)

Fig. 15–9

The bumper cars A and B in Fig. 15–9a each have a mass of 150 kg and are coasting with the velocities shown before they freely collide head on. If no energy is lost during the collision, determine their velocities after collision.

SOLUTION

Free-Body Diagram. The cars will be considered as a single system. The free-body diagram is shown in Fig. 15–9b.

Conservation of Momentum.

$$(\xrightarrow{+}) \qquad m_A(v_A)_1 + m_B(v_B)_1 = m_A(v_A)_2 + m_B(v_B)_2$$

$$(150 \text{ kg})(3 \text{ m/s}) + (150 \text{ kg})(-2 \text{ m/s}) = (150 \text{ kg})(v_A)_2 + (150 \text{ kg})(v_B)_2$$

$$(v_A)_2 = 1 - (v_B)_2 \qquad (1)$$

Conservation of Energy. Since no energy is lost, the conservation of energy theorem gives

$$T_1 + V_1 = T_2 + V_2$$

$$\tfrac{1}{2}m_A(v_A)_1^2 + \tfrac{1}{2}m_B(v_B)_1^2 + 0 = \tfrac{1}{2}m_A(v_A)_2^2 + \tfrac{1}{2}m_B(v_B)_2^2 + 0$$

$$\tfrac{1}{2}(150 \text{ kg})(3 \text{ m/s})^2 + \tfrac{1}{2}(150 \text{ kg})(2 \text{ m/s})^2 + 0 = \tfrac{1}{2}(150 \text{ kg})(v_A)_2^2$$

$$+ \tfrac{1}{2}(150 \text{ kg})(v_B)_2^2 + 0$$

$$(v_A)_2^2 + (v_B)_2^2 = 13 \qquad (2)$$

Substituting Eq. 1 into Eq. 2 and simplifying, we get

$$(v_B)_2^2 - (v_B)_2 - 6 = 0$$

Solving for the two roots,

$$(v_B)_2 = 3 \text{ m/s} \qquad \text{and} \qquad (v_B)_2 = -2 \text{ m/s}$$

Since $(v_B)_2 = -2$ m/s refers to the velocity of B just *before* collision, then the velocity of B just after the collision must be

$$(v_B)_2 = 3 \text{ m/s} \rightarrow \qquad \textit{Ans.}$$

Substituting this result into Eq. 1, we obtain

$$(v_A)_2 = 1 - 3 \text{ m/s} = -2 \text{ m/s} = 2 \text{ m/s} \leftarrow \qquad \textit{Ans.}$$

EXAMPLE 15.6

An 800-kg rigid pile P shown in Fig. 15–10a is driven into the ground using a 300-kg hammer. The hammer falls from rest at a height $y_0 = 0.5$ m and strikes the top of the pile. Determine the impulse which the pile exerts on the hammer if the pile is surrounded entirely by loose sand, so that after striking, the hammer does *not* rebound off the pile.

(a)

SOLUTION

Conservation of Energy. The velocity at which the hammer strikes the pile can be determined using the conservation of energy equation applied to the hammer. With the datum at the top of the pile, Fig. 15–10a, we have

$$T_0 + V_0 = T_1 + V_1$$

$$\tfrac{1}{2}m_H(v_H)_0^2 + W_H y_0 = \tfrac{1}{2}m_H(v_H)_1^2 + W_H y_1$$

$$0 + 300(9.81)\text{ N}(0.5\text{ m}) = \tfrac{1}{2}(300 \text{ kg})(v_H)_1^2 + 0$$

$$(v_H)_1 = 3.132 \text{ m/s}$$

Free-Body Diagram. From the physical aspects of the problem, the free-body diagram of the hammer and pile, Fig. 15–10b, indicates that during the *short time* from *just before* to *just after* the *collision*, the weights of the hammer and pile and the resistance force $\mathbf{F}_s$ of the sand are all *nonimpulsive*. The impulsive force $\mathbf{R}$ is internal to the system and therefore cancels. Consequently, momentum is conserved in the vertical direction during this short time.

(b)

Conservation of Momentum. Since the hammer does not rebound off the pile just after collision, then $(v_H)_2 = (v_P)_2 = v_2$.

$(+\downarrow)$
$$m_H(v_H)_1 + m_P(v_P)_1 = m_H v_2 + m_P v_2$$

$$(300 \text{ kg})(3.132 \text{ m/s}) + 0 = (300 \text{ kg})v_2 + (800 \text{ kg})v_2$$

$$v_2 = 0.8542 \text{ m/s}$$

Principle of Impulse and Momentum. The impulse which the pile imparts to the hammer can now be determined since v_2 is known. From the free-body diagram for the hammer, Fig. 15–10c, we have

$(+\downarrow)$
$$m_H(v_H)_1 + \Sigma \int_{t_1}^{t_2} F_y \, dt = m_H v_2$$

$$(300 \text{ kg})(3.132 \text{ m/s}) - \int R\, dt = (300 \text{ kg})(0.8542 \text{ m/s})$$

$$\int R\, dt = 683 \text{ N}\cdot\text{s} \qquad \textit{Ans.}$$

(c)

NOTE: The equal but opposite impulse acts on the pile. Try finding this impulse by applying the principle of impulse and momentum to the pile.

Fig. 15–10

EXAMPLE 15.7

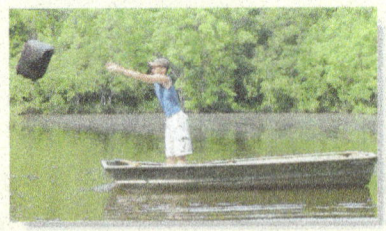

The 80-kg man can throw the 20-kg box horizontally at 4 m/s when standing on the ground. If instead he firmly stands in the 120-kg boat and throws the box, as shown in the photo, determine how far the boat will move in three seconds. Neglect water resistance.

SOLUTION

Free-Body Diagram.* If the man, boat, and box are considered as a single system, the horizontal forces between the man and the boat and the man and the box become internal to the system, Fig. 15–11a, and so linear momentum will be conserved along the x axis.

Conservation of Momentum. When writing the conservation of momentum equation, it is *important* that the velocities be measured from the same inertial coordinate system, assumed here to be fixed. From this coordinate system, we will assume that the boat and man go to the right while the box goes to the left, as shown in Fig. 15–11b.

Applying the conservation of linear momentum to the man, boat, box system,

(a)

(b)

Fig. 15–11

$$(\xrightarrow{+}) \quad 0 + 0 + 0 = (m_m + m_b)v_b - m_{box}v_{box}$$

$$0 = (80\,\text{kg} + 120\,\text{kg})v_b - (20\,\text{kg})v_{box}$$

$$v_{box} = 10v_b \quad (1)$$

Kinematics. Since the velocity of the box *relative* to the man (and boat), $v_{box/b}$, is known, then v_b can also be related to v_{box} using the relative velocity equation.

$$(\xrightarrow{+}) \quad v_{box} = v_b + v_{box/b}$$

$$-v_{box} = v_b - 4\,\text{m/s} \quad (2)$$

Solving Eqs. 1 and 2,

$$v_{box} = 3.64\,\text{m/s} \leftarrow$$

$$v_b = 0.3636\,\text{m/s} \rightarrow$$

The displacement of the boat in three seconds is therefore

$$s_b = v_b t = (0.3636\,\text{m/s})(3\,\text{s}) = 1.09\,\text{m} \qquad \textit{Ans.}$$

---

*Only horizontal forces are shown on the free-body diagram.

EXAMPLE 15.8

The 1200-lb cannon shown in Fig. 15–12a fires an 8-lb projectile with a muzzle velocity of 1500 ft/s measured relative to the cannon. If firing takes place in 0.03 s, determine the recoil velocity of the cannon just after firing. The cannon support is fixed to the ground, and the horizontal recoil of the cannon is absorbed by two springs.

SOLUTION

Part (a) Free-Body Diagram.* As shown in Fig. 15–12b, we will consider the projectile and cannon as a single system, since the impulsive forces, **F** and -**F**, between the cannon and projectile are *internal* to the system and will therefore cancel from the analysis. Furthermore, during the time $\Delta t = 0.03$ s, the two recoil springs which are attached to the support each exert a *nonimpulsive force* F_s on the cannon. This is because Δt is very short, so that during this time the cannon only moves through a *very small* distance x. Consequently, $F_s = kx \approx 0$, where k is the spring's stiffness, which is also considered to be relatively small. Hence it can be concluded that momentum for the system is conserved in the *horizontal direction*.

Conservation of Linear Momentum.

$(\stackrel{+}{\rightarrow})$ $\quad m_c(v_c)_1 + m_p(v_p)_1 = -m_c(v_c)_2 + m_p(v_p)_2$

$$0 + 0 = -\left(\frac{1200 \text{ lb}}{32.2 \text{ ft/s}^2}\right)(v_c)_2 + \left(\frac{8 \text{ lb}}{32.2 \text{ ft/s}^2}\right)(v_p)_2$$

$$(v_p)_2 = 150 (v_c)_2 \qquad (1)$$

Kinematics. These two unknown velocities are measured by a *fixed* observer. As in Example 15.7, they can also be related using the relative velocity equation.

$(\stackrel{+}{\rightarrow})$ $\quad (v_p)_2 = (v_c)_2 + (v_{p/c})_2$

$\qquad (v_p)_2 = -(v_c)_2 + 1500 \text{ ft/s} \qquad (2)$

Solving Eqs. 1 and 2 yields

$\qquad (v_c)_2 = 9.93 \text{ ft/s} \qquad$ *Ans.*

$\qquad (v_p)_2 = 1490 \text{ ft/s}$

Apply the principle of impulse and momentum to the projectile (or the cannon) and show that the average impulsive force on the projectile is 12.3 kip.

NOTE: If the cannon is firmly fixed to its support (no springs), the reactive force of the support on the cannon must be considered as an external impulse to the system, since the support would allow no movement of the cannon. In this case momentum is *not* conserved.

* Only horizontal forces are shown on the free-body diagram.

(b)

Fig. 15–12

FUNDAMENTAL PROBLEMS

F15–7. The freight cars A and B have a mass of 20 Mg and 15 Mg, respectively. Determine the velocity of A after collision if the cars collide and rebound, such that B moves to the right with a speed of 2 m/s. If A and B are in contact for 0.5 s, find the average impulsive force which acts between them.

Prob. F15–7

F15–8. The cart and package have a mass of 20 kg and 5 kg, respectively. If the cart has a smooth surface and it is initially at rest, while the velocity of the package is as shown, determine the final common velocity of the cart and package after the impact.

Prob. F15–8

F15–9. The 5-kg block A has an initial speed of 5 m/s as it slides down the smooth ramp, after which it collides with the stationary 8-kg block B. If the two blocks couple together after collision, determine their common velocity immediately after collision.

Prob. F15–9

F15–10. The spring is fixed to block A and block B is pressed against the spring. If the spring is compressed $s = 200$ mm and then the blocks are released, determine their velocity at the instant block B loses contact with the spring. The masses of blocks A and B are 10 kg and 15 kg, respectively.

Prob. F15–10

F15–11. Blocks A and B have a mass of 15 kg and 10 kg, respectively. If A is stationary and B has a velocity of 15 m/s just before collision, and the blocks couple together after impact, determine the maximum compression of the spring.

Prob. F15–11

F15–12. The cannon and support without a projectile have a mass of 250 kg. If a 20-kg projectile is fired from the cannon with a velocity of 400 m/s, measured *relative* to the cannon, determine the speed of the projectile as it leaves the cannon barrel. Neglect rolling resistance.

Prob. F15–12

PROBLEMS

15–35. The 5-Mg bus B is traveling to the right at 20 m/s. Meanwhile a 2-Mg car A is traveling at 15 m/s to the right. If the vehicles crash and become entangled, determine their common velocity just after the collision. Assume that the vehicles are free to roll during collision.

Prob. 15–35

*15–36. The 50-kg boy jumps on the 5-kg skateboard with a horizontal velocity of 5 m/s. Determine the distance s the boy reaches up the inclined plane before momentarily coming to rest. Neglect the skateboard's rolling resistance.

Prob. 15–36

15–37. The 2.5-Mg pickup truck is towing the 1.5-Mg car using a cable as shown. If the car is initially at rest and the truck is coasting with a velocity of 30 km/h when the cable is slack, determine the common velocity of the truck and the car just after the cable becomes taut. Also, find the loss of energy.

Prob. 15–37

15–38. A railroad car weighs 30,000 lb and is traveling horizontally at 30 ft/s. At the same time another car weighing 10,000 lb is traveling at 5 ft/s in the opposite direction. If the cars meet and couple together, determine the speed of both cars just after coupling. Determine the kinetic energy of both cars before and after coupling has occurred, and explain qualitatively what happened to the difference of these energies.

15–39. A ballistic pendulum consists of a 4-kg wooden block originally at rest, $\theta = 0°$. When a 2-g bullet strikes and becomes embedded in it, it is observed that the block swings upward to a maximum angle of $\theta = 6°$. Estimate the initial speed of the bullet.

Prob. 15–39

*15–40. A man wearing ice skates throws an 8-kg block with an initial velocity of 2 m/s, measured relative to himself, in the direction shown. If he is originally at rest and completes the throw in 1.5 s while keeping his legs rigid, determine the horizontal velocity of the man just after releasing the block. What is the vertical reaction of both his skates on the ice during the throw? The man has a mass of 70 kg. Neglect friction and the motion of his arms.

Prob. 15–40

15–41. The 5-kg spring-loaded gun rests on the smooth surface. It fires a ball having a mass of 1 kg with a velocity of $v' = 6$ m/s relative to the gun in the direction shown. If the gun is originally at rest, determine the horizontal distance d the ball is from the initial position of the gun at the instant the ball strikes the ground at D. Neglect the size of the gun.

15–42. The 5-kg spring-loaded gun rests on the smooth surface. It fires a ball having a mass of 1 kg with a velocity of $v' = 6$ m/s relative to the gun in the direction shown. If the gun is originally at rest, determine the distance the ball is from the gun at the instant the ball reaches its highest elevation C. Neglect the size of the gun.

*15–44.** A toboggan having a mass of 10 kg starts from rest at A and carries a girl and boy having a mass of 40 kg and 45 kg, respectively. When the toboggan reaches the bottom of the slope at B, the boy is pushed off from the back with a horizontal velocity of $v_{b/t} = 2$ m/s, measured relative to the toboggan. Determine the velocity of the toboggan afterwards. Neglect friction in the calculation.

Prob. 15–44

Probs. 15–41/42

15–45. The block of mass m travels at v_1 in the direction θ_1 shown at the top of the smooth slope. Determine its speed v_2 and its direction θ_2 when it reaches the bottom.

15–43. The 20-g bullet is traveling at 400 m/s when it becomes embedded in the 2-kg stationary block. Determine the distance the block will slide before it stops. The coefficient of kinetic friction between the block and the plane is $\mu_k = 0.2$.

Prob. 15–43

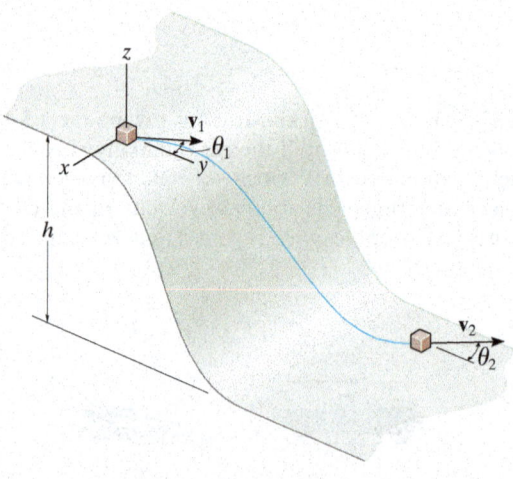

Prob. 15–45

15–46. The boy B jumps off the canoe at A with a velocity of 5 m/s relative to the canoe as shown. If he lands in the second canoe C, determine the final speed of both canoes after the motion. Each canoe has a mass of 40 kg. The boy's mass is 30 kg, and the girl D has a mass of 25 kg. Both canoes are originally at rest.

Prob. 15–46

15–47. The 30-Mg freight car A and 15-Mg freight car B are moving towards each other with the velocities shown. Determine the maximum compression of the spring mounted on car A. Neglect rolling resistance.

Prob. 15–47

*15–48. Blocks A and B have masses of 40 kg and 60 kg, respectively. They are placed on a smooth surface and the spring connected between them is stretched 2 m. If they are released from rest, determine the speeds of both blocks the instant the spring becomes unstretched.

Prob. 15–48

15–49. The crate has a mass m and rests on the barge which has a mass M and is initially at rest. The coefficient of kinetic friction between the crate and barge is μ. The rope is pulled (or jerked) such that it snaps and as a result the impulse gives the barge a sudden velocity $\mathbf{v}_0$. Determine the time the crate slides on the barge before coming to rest on the barge. Also, what is the final velocity of the barge?

15–50. Using the data in Prob. 15–49, determine the distance the crate slides on the barge before coming to rest.

Probs. 15–49/50

15–51. A 40-lb box slides from rest down the smooth ramp onto the surface of a 20-lb cart. Determine the speed of the box at the instant it stops sliding on the cart. If someone ties the cart to the ramp at B, determine the horizontal impulse the box will exert at C in order to stop its motion. Neglect friction and the size of the box.

Prob. 15–51

*15–52. Two cars A and B, each having a mass of 1.6 Mg, collide on the icy pavement of an intersection. The direction of motion of each car after collision is measured from snow tracks as shown. If the driver in car A states that he was going 50 km/h just before collision and that after collision he applied the brakes, so that his car skidded 4 m before stopping, determine the approximate speed of car B just before collision. Assume that the coefficient of kinetic friction between the car wheels and the pavement is $\mu_k = 0.15$. *Note:* The line of impact has not been defined; furthermore, this information is not needed for solution.

Prob. 15–52

15–53. Block A has a mass of 5 kg and is placed on the smooth triangular block B having a mass of 30 kg. If the system is released from rest, determine the distance B moves from point O when A reaches the bottom. Neglect the size of block A.

15–54. Solve Prob. 15–53 if the coefficient of kinetic friction between A and B is $\mu_k = 0.3$. Neglect friction between block B and the horizontal plane.

Probs. 15–53/54

15–55. A box A, having a mass of 20 kg, is released from rest at the position shown and slides freely down the smooth inclined ramp. When it reaches the bottom of the ramp, it slides horizontally onto the surface of a 10-kg cart for which the coefficient of kinetic friction between the cart and the box is $\mu_k = 0.6$. If $h = 0.2$ m, determine the final velocity of the cart once the block comes to rest on it. Also, determine the position s of the box on the cart after it comes to rest on the cart.

Prob. 15–55

*15–56. Two boxes A and B, each having a weight of 160 lb, sit on the 500-lb conveyor which is free to roll on the ground. If the belt starts from rest and begins to run with a speed of 3 ft/s, determine the final speed of the conveyor if (a) the boxes are not stacked and A falls off then B falls off, and (b) A is stacked on top of B and both fall off together.

Prob. 15–56

15–57. The two toy cars A and B have a weight of 0.4 lb and 0.6 lb, respectively. A spring having a stiffness of 30 lb/ft is attached to one of them and the cars are pressed together such that the spring is compressed 0.2 ft. Determine the speed of each car after they are released from rest.

Prob. 15–57

15.4 IMPACT

Impact occurs when two bodies collide with each other during a very short period of time, causing relatively large (impulsive) forces to be exerted between the bodies. The striking of a hammer on a nail, or a golf club on a ball, are common examples of impact loadings.

In general, there are two types of impact. ***Central impact*** occurs when the direction of motion of the mass centers of the two colliding particles is along the ***line of impact*** that passes through the mass centers of the particles. This line is perpendicular to the ***plane of contact***, as shown in Fig. 15–13*a*. When the motion of one or both of the particles makes an angle with the line of impact, Fig. 15–13*b*, the impact is said to be ***oblique impact***.

Central Impact. To illustrate the method for analyzing the mechanics of impact, consider the case involving the central impact of the two particles *A* and *B* shown in Fig. 15–14.

- The particles have the initial momenta shown in Fig. 15–14*a*. Provided $(v_A)_1 > (v_B)_1$, collision will eventually occur.

- During the collision the particles must be thought of as deformable or nonrigid. The particles will undergo a period of deformation such that they exert an equal but opposite ***deformation impulse*** $\int P \, dt$ on each other, Fig. 15–14*b*.

- Only at the instant of maximum deformation will both particles move with a common velocity **v**, since their relative motion is zero, Fig. 15–14*c*.

- Afterward a period of restitution occurs, in which case the particles will either return to their original shape or remain permanently deformed. The equal but opposite ***restitution impulse*** $\int R \, dt$ pushes the particles apart from one another, Fig. 15–14*d*. In reality, the physical properties of any two bodies are such that the deformation impulse will always be greater than the restitution impulse, i.e., $\int P \, dt > \int R \, dt$.

- Just after separation the particles will have the final momenta shown in Fig. 15–14*e*, where $(v_B)_2 > (v_A)_2$.

Central impact

(a)

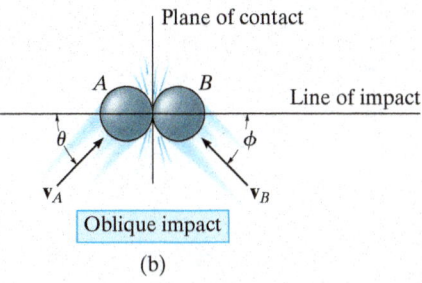

Oblique impact

(b)

Fig. 15–13

Before impact

(a)

Fig. 15–14

In most problems the initial velocities of the particles will be *known*, and it will be necessary to determine their final velocities $(v_A)_2$ and $(v_B)_2$. To do this we will need two equations that involve these unknowns. The first equation requires conservation of momentum for the two particle systems since during collision the internal impulses of deformation and restitution *cancel*. Hence, referring to Fig. 15–14a and Fig. 15–14e we require

$$(\pm) \qquad m_A(v_A)_1 + m_B(v_B)_1 = m_A(v_A)_2 + m_B(v_B)_2 \qquad (15\text{–}10)$$

To obtain a second equation, we will apply the principle of impulse and momentum to *each particle*. For example, during the deformation phase for particle *A*, Figs. 15–14a, 15–14b, and 15–14c, we have

$$(\pm) \qquad m_A(v_A)_1 - \int P \, dt = m_A v$$

For the restitution phase, Figs. 15–14c, 15–14d, and 15–14e,

$$(\pm) \qquad m_A v - \int R \, dt = m_A(v_A)_2$$

If we define the **coefficient of restitution**, *e*, as the ratio of the restitution impulse to the deformation impulse, then from the above equations, this value for particle *A* is

$$e = \frac{\int R \, dt}{\int P \, dt} = \frac{v - (v_A)_2}{(v_A)_1 - v}$$

In a similar manner, we can establish *e* by considering particle *B*, Fig. 15–14. This yields

$$e = \frac{\int R \, dt}{\int P \, dt} = \frac{(v_B)_2 - v}{v - (v_B)_1}$$

If the unknown *v* is eliminated from the above two equations, the coefficient of restitution can be expressed in terms of the particles' initial and final velocities as

$$(\pm) \qquad \boxed{e = \frac{(v_B)_2 - (v_A)_2}{(v_A)_1 - (v_B)_1}} \qquad (15\text{–}11)$$

Provided a value for e is specified, Eqs. 15–10 and 15–11 can be solved simultaneously to obtain $(v_A)_2$ and $(v_B)_2$. In doing so, however, it is important to carefully establish a sign convention for defining the positive direction for both $\mathbf{v}_A$ and $\mathbf{v}_B$ and then use it *consistently* when writing *both* equations. For the case shown in Fig. 15–14, we have defined the positive direction to the right when referring to the motions of both A and B. Consequently, if a negative value results from the solution of either $(v_A)_2$ or $(v_B)_2$, it indicates motion is to the left.

Coefficient of Restitution.
From Figs. 15–14a and 15–14e, it is seen that Eq. 15–11 states that e is equal to the ratio of the relative velocity of the particles' separation just after impact, $(v_B)_2 - (v_A)_2$, to the relative velocity of the particles' approach just before impact, $(v_A)_1 - (v_B)_1$, and so we can also write Eq. 15–11 as

$$e = \frac{\text{rel. vel. after impact}}{\text{rel. vel. before impact}} \qquad (15\text{–}12)$$

The quality of a manufactured tennis ball is measured by the height of its bounce, which can be related to its coefficient of restitution. Using the mechanics of oblique impact, engineers can design a separation device to remove substandard tennis balls from a production line.

By measuring these relative velocities experimentally, it has been found that e varies appreciably with impact velocity as well as with the size and shape of the colliding bodies. For these reasons the coefficient of restitution is reliable only when used with data which closely approximate the conditions which existed when measurements of e were made. In general e has a value between zero and one, and one should be aware of the physical meaning of these two limits.

Elastic Impact ($e = 1$).
If the collision between the two particles is perfectly elastic, the deformation impulse $\left(\int \mathbf{P} \, dt\right)$ is equal and opposite to the restitution impulse $\left(\int \mathbf{R} \, dt\right)$. Although in reality this can never be achieved, $e = 1$ for an elastic collision.

Plastic Impact ($e = 0$).
The impact is said to be inelastic or plastic when $e = 0$. In this case there is no restitution impulse $\left(\int \mathbf{R} \, dt = \mathbf{0}\right)$, so that after collision both particles couple or stick together and move with a common velocity.

From the above derivation it should be evident that the principle of work and energy cannot be used for the analysis of impact problems since it is not possible to know how the internal forces of deformation and restitution vary or displace during the collision. By knowing the particle's velocities before and after collision, however, the energy loss during collision can be calculated on the basis of the difference in the particle's kinetic energy. This energy loss, $\Sigma U_{1-2} = \Sigma T_2 - \Sigma T_1$, occurs because some of the initial kinetic energy of the particle is transformed into localized deformation of the material when the collision occurs. In particular, if the impact is perfectly elastic, no energy is lost in the collision; whereas if the collision is plastic, the energy lost during collision is a maximum.

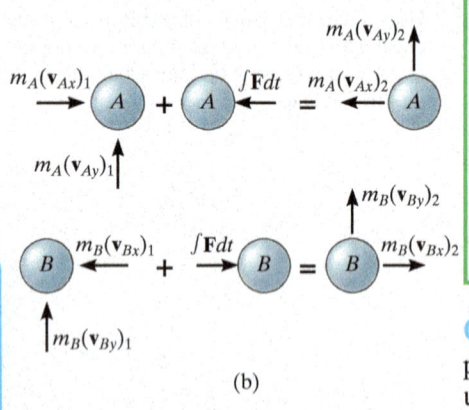

Fig. 15–15

PROCEDURE FOR ANALYSIS (CENTRAL IMPACT)

In most cases the final velocities of two smooth particles are to be determined just after they are subjected to direct central impact. Provided the coefficient of restitution, the mass of each particle, and each particle's initial velocity just before impact are known, the solution to this problem can be obtained using the following two equations.

- The conservation of momentum applies to the system of particles, $\Sigma m v_1 = \Sigma m v_2$.
- The coefficient of restitution, $e = [(v_B)_2 - (v_A)_2]/[(v_A)_1 - (v_B)_1]$, relates the relative velocities of the particles along the line of impact, just before and just after impact.

When applying these two equations, the sense of an unknown velocity can be assumed. If the solution yields a negative magnitude, the velocity acts in the opposite sense.

Oblique Impact. When oblique impact occurs between two smooth particles, the particles move away from each other with velocities having unknown directions as well as unknown magnitudes. Provided the initial velocities are known, then there are four unknowns in the problem. They are either $(v_A)_2, (v_B)_2, \theta_2$, and ϕ_2, Fig. 15–15a, or the x and y components of the final velocities, Fig. 15–15b.

PROCEDURE FOR ANALYSIS (OBLIQUE IMPACT)

If the y axis is established within the plane of contact and the x axis along the line of impact, then the impulsive forces of deformation and restitution act *only in the x direction*, Fig. 15–15. By resolving the velocity or momentum vectors into components along the x and y axes, it is then possible to write four independent scalar equations in order to determine $(v_{Ax})_2, (v_{Ay})_2, (v_{Bx})_2$, and $(v_{By})_2$.

- Momentum of the system is conserved *along the line of impact, x axis*, so that $\Sigma m(v_x)_1 = \Sigma m(v_x)_2$.
- The coefficient of restitution, $e = [(v_{Bx})_2 - (v_{Ax})_2]/[(v_{Ax})_1 - (v_{Bx})_1]$, relates the relative-velocity *components* of the particles *along the line of impact (x axis)*.
- If these two equations are solved simultaneously, we obtain $(v_{Ax})_2$ and $(v_{Bx})_2$.
- Momentum of particle A is conserved along the y axis, perpendicular to the line of impact, since no impulse acts on particle A in this direction. As a result $m_A(v_{Ay})_1 = m_A(v_{Ay})_2$ or $(v_{Ay})_1 = (v_{Ay})_2$.
- Momentum of particle B is conserved along the y axis, perpendicular to the line of impact, since no impulse acts on particle B in this direction. Consequently $(v_{By})_1 = (v_{By})_2$.

Application of these four equations is illustrated in Example 15.11.

EXAMPLE 15.9

The bag A, having a weight of 6 lb, is released from rest at the position $\theta = 0°$, as shown in Fig. 15–16a. After falling to $\theta = 90°$, it strikes an 18-lb box B. If the coefficient of restitution between the bag and box is $e = 0.5$, determine the velocities of the bag and box just after impact. What is the loss of energy during collision?

SOLUTION
This problem involves central impact. Why? Before analyzing the mechanics of the impact, however, it is first necessary to obtain the velocity of the bag *just before* it strikes the box.

Conservation of Energy. With the datum at $\theta = 0°$, Fig. 15–16b, we have

$$T_0 + V_0 = T_1 + V_1$$

$$0 + 0 = \frac{1}{2}\left(\frac{6 \text{ lb}}{32.2 \text{ ft/s}^2}\right)(v_A)_1^2 - 6 \text{ lb}(3 \text{ ft}); \quad (v_A)_1 = 13.90 \text{ ft/s}$$

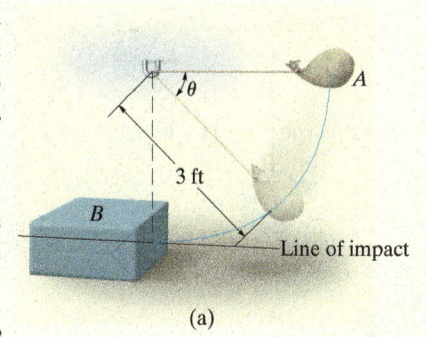

Conservation of Momentum. Just after impact we will assume both A and B travel to the left. Applying the conservation of momentum to the system, Fig. 15–16c, we have

$$(\xleftarrow{+}) \qquad m_B(v_B)_1 + m_A(v_A)_1 = m_B(v_B)_2 + m_A(v_A)_2$$

$$0 + \left(\frac{6 \text{ lb}}{32.2 \text{ ft/s}^2}\right)(13.90 \text{ ft/s}) = \left(\frac{18 \text{ lb}}{32.2 \text{ ft/s}^2}\right)(v_B)_2 + \left(\frac{6 \text{ lb}}{32.2 \text{ ft/s}^2}\right)(v_A)_2$$

$$(v_A)_2 = 13.90 - 3(v_B)_2 \qquad (1)$$

Coefficient of Restitution. Realizing that for separation to occur just after collision $(v_B)_2 > (v_A)_2$, Fig. 15–16c, we have

$$(\xleftarrow{+}) \qquad e = \frac{(v_B)_2 - (v_A)_2}{(v_A)_1 - (v_B)_1}; \qquad 0.5 = \frac{(v_B)_2 - (v_A)_2}{13.90 \text{ ft/s} - 0}$$

$$(v_A)_2 = (v_B)_2 - 6.950 \qquad (2)$$

Solving Eqs. 1 and 2 simultaneously yields

$$(v_A)_2 = -1.74 \text{ ft/s} = 1.74 \text{ ft/s} \rightarrow \quad \text{and} \quad (v_B)_2 = 5.21 \text{ ft/s} \leftarrow \quad Ans.$$

Loss of Energy. Applying the principle of work and energy to the bag and box just before and just after collision, we have

$$\Sigma U_{1-2} = T_2 - T_1;$$

$$\Sigma U_{1-2} = \left[\frac{1}{2}\left(\frac{18 \text{ lb}}{32.2 \text{ ft/s}^2}\right)(5.21 \text{ ft/s})^2 + \frac{1}{2}\left(\frac{6 \text{ lb}}{32.2 \text{ ft/s}^2}\right)(1.74 \text{ ft/s})^2\right]$$

$$- \left[\frac{1}{2}\left(\frac{6 \text{ lb}}{32.2 \text{ ft/s}^2}\right)(13.9 \text{ ft/s})^2\right]$$

$$\Sigma U_{1-2} = -10.1 \text{ ft} \cdot \text{lb} \qquad Ans.$$

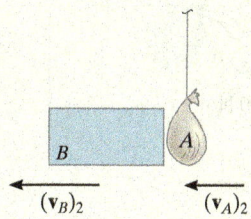

Fig. 15–16

NOTE: The energy loss occurs due to inelastic deformation during the collision.

EXAMPLE 15.10

Ball B shown in Fig. 15–17a has a mass of 1.5 kg and is suspended from the ceiling by a 1-m-long elastic cord. If the cord is *stretched* downward 0.25 m and the ball is released from rest, determine how far the cord stretches after the ball rebounds from the ceiling and falls downward. The stiffness of the cord is $k = 800$ N/m, and the coefficient of restitution between the ball and ceiling is $e = 0.8$. The ball makes a central impact with the ceiling.

SOLUTION
First we must obtain the velocity of the ball *just before* it strikes the ceiling using energy methods, then consider the impulse and momentum between the ball and ceiling, and finally again use energy methods to determine the stretch in the cord.

Conservation of Energy. With the datum located as shown in Fig. 15–17a, realizing that initially $y = y_0 = (1 + 0.25)$ m $= 1.25$ m, we have

$$T_0 + V_0 = T_1 + V_1$$

$$\tfrac{1}{2}m(v_B)_0^2 - W_B y_0 + \tfrac{1}{2}ks^2 = \tfrac{1}{2}m(v_B)_1^2 + 0$$

$$0 - 1.5(9.81)\text{N}(1.25 \text{ m}) + \tfrac{1}{2}(800 \text{ N/m})(0.25 \text{ m})^2 = \tfrac{1}{2}(1.5 \text{ kg})(v_B)_1^2$$

$$(v_B)_1 = 2.968 \text{ m/s} \uparrow$$

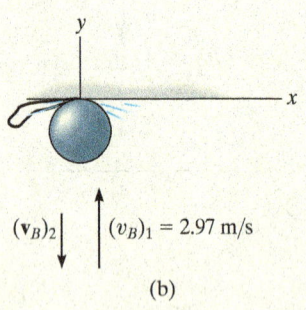

The interaction of the ball with the ceiling will now be considered using the principles of impact.* Since an unknown portion of the mass of the ceiling is involved in the impact, the conservation of momentum for the ball–ceiling system will not be written. The "velocity" of this portion of ceiling is zero since it (or the earth) are assumed to remain at rest *both* before and after impact.

Coefficient of Restitution. Fig. 15–17b.

$$(+\uparrow) \quad e = \frac{(v_B)_2 - (v_A)_2}{(v_A)_1 - (v_B)_1}; \qquad 0.8 = \frac{(v_B)_2 - 0}{0 - 2.968 \text{ m/s}}$$

$$(v_B)_2 = -2.374 \text{ m/s} = 2.374 \text{ m/s} \downarrow$$

Conservation of Energy. The maximum stretch s_3 in the cord can be determined by again applying the conservation of energy equation to the ball just after collision. Assuming that $y = y_3 = (1 + s_3)$ m, Fig. 15–17c, then

$$T_2 + V_2 = T_3 + V_3$$

$$\tfrac{1}{2}m(v_B)_2^2 + 0 = \tfrac{1}{2}m(v_B)_3^2 - W_B y_3 + \tfrac{1}{2}ks_3^2$$

$$\tfrac{1}{2}(1.5 \text{ kg})(2.37 \text{ m/s})^2 = 0 - 9.81(1.5)\text{ N}(1 \text{ m} + s_3) + \tfrac{1}{2}(800 \text{ N/m})s_3^2$$

$$400s_3^2 - 14.715s_3 - 18.94 = 0$$

Solving this quadratic equation for the positive root yields

Fig. 15–17

$$s_3 = 0.237 \text{ m} = 237 \text{ mm} \qquad \textit{Ans.}$$

*The weight of the ball is considered a nonimpulsive force.

EXAMPLE 15.11

Two smooth disks A and B, having a mass of 1 kg and 2 kg, respectively, slide on the horizontal plane and collide with the velocities shown in Fig. 15–18a. If the coefficient of restitution for the disks is $e = 0.75$, determine the x and y components of the final velocity of each disk just after collision.

SOLUTION
This problem involves *oblique impact*. Why? In order to solve it, we have established the x and y axes along the line of impact and the plane of contact, respectively, Fig. 15–18a.

Resolving each of the initial velocities into x and y components, we have

$(v_{Ax})_1 = 3 \cos 30° = 2.598$ m/s $\quad (v_{Ay})_1 = 3 \sin 30° = 1.50$ m/s
$(v_{Bx})_1 = -1 \cos 45° = -0.7071$ m/s $\quad (v_{By})_1 = -1 \sin 45° = -0.7071$ m/s

The four unknown velocity components after collision are *assumed to act in the positive directions*, Fig. 15–18b. Since the impact occurs in the x direction (line of impact), the conservation of momentum for *both* disks can be applied in this direction. Why?

Conservation of "x" Momentum. In reference to the momentum diagrams, we have

$(\xrightarrow{+}) \quad m_A(v_{Ax})_1 + m_B(v_{Bx})_1 = m_A(v_{Ax})_2 + m_B(v_{Bx})_2$
$1 \text{ kg}(2.598 \text{ m/s}) + 2 \text{ kg}(-0.707 \text{ m/s}) = 1 \text{ kg}(v_{Ax})_2 + 2 \text{ kg}(v_{Bx})_2$
$(v_{Ax})_2 + 2(v_{Bx})_2 = 1.184 \quad (1)$

Coefficient of Restitution (x).

$(\xrightarrow{+}) \quad e = \dfrac{(v_{Bx})_2 - (v_{Ax})_2}{(v_{Ax})_1 - (v_{Bx})_1}; \quad 0.75 = \dfrac{(v_{Bx})_2 - (v_{Ax})_2}{2.598 \text{ m/s} - (-0.7071 \text{ m/s})}$
$(v_{Bx})_2 - (v_{Ax})_2 = 2.482 \quad (2)$

Solving Eqs. 1 and 2 for $(v_{Ax})_2$ and $(v_{Bx})_2$ yields

$(v_{Ax})_2 = -1.26$ m/s $= 1.26$ m/s $\leftarrow \quad (v_{Bx})_2 = 1.22$ m/s $\rightarrow \quad$ *Ans.*

Conservation of "y" Momentum. The momentum of *each disk* is *conserved* in the y direction (plane of contact), since the disks are smooth and therefore no frictional impulse acts in this direction. From Fig. 15–18b,

$(+\uparrow) \; m_A(v_{Ay})_1 = m_A(v_{Ay})_2; \quad (v_{Ay})_2 = 1.50$ m/s $\uparrow \quad$ *Ans.*
$(+\uparrow) \; m_B(v_{By})_1 = m_B(v_{By})_2; \quad (v_{By})_2 = -0.707$ m/s $= 0.707$ m/s $\downarrow \quad$ *Ans.*

NOTE: Show that when the velocity components are summed vectorially, one obtains the results shown in Fig. 15–18c.

(a)

(b)

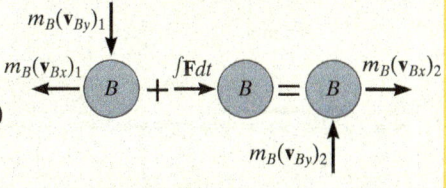

(c)

Fig. 15–18

FUNDAMENTAL PROBLEMS

F15–13. Determine the coefficient of restitution e between ball A and ball B. The velocities of A and B before and after the collision are shown.

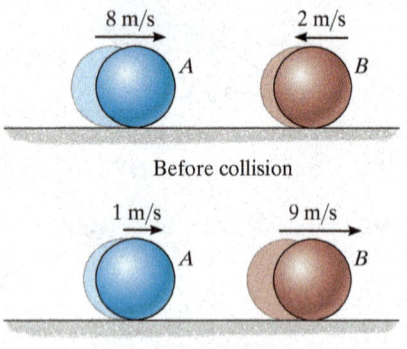

Prob. F15–13

F15–14. The 15-Mg tank car A and 25-Mg freight car B travel toward each other with the velocities shown. If the coefficient of restitution between the bumpers is $e = 0.6$, determine the velocity of each car just after the collision.

Prob. F15–14

F15–15. The 30-lb package A has a speed of 5 ft/s when it enters the smooth ramp. As it slides down the ramp, it strikes the 80-lb package B which is initially at rest. If the coefficient of restitution between A and B is $e = 0.6$, determine the velocity of B just after the impact.

Prob. F15–15

F15–16. The ball strikes the smooth wall with a velocity of $(v_b)_1 = 20$ m/s. If the coefficient of restitution between the ball and the wall is $e = 0.75$, determine the velocity of the ball just after the impact.

Prob. F15–16

F15–17. Disk A has a mass of 2 kg and slides on the smooth horizontal plane with a velocity of 3 m/s. Disk B has a mass of 11 kg and is initially at rest. If after impact A has a velocity of 1 m/s, parallel to the positive x axis, determine the speed of disk B after impact.

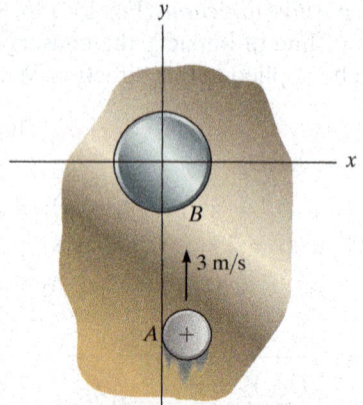

Prob. F15–17

F15–18. Two disks A and B each have a mass of 1 kg and the initial velocities shown just before they collide. If the coefficient of restitution is $e = 0.5$, determine their speeds just after impact.

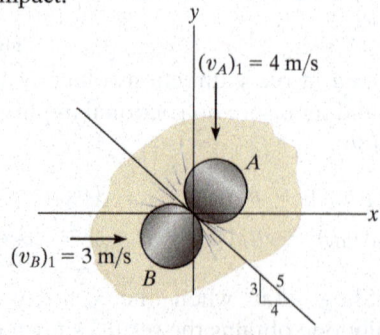

Prob F15–18

PROBLEMS

15–58. Disk A has a mass of 250 g and is sliding on a *smooth* horizontal surface with an initial velocity $(v_A)_1 = 2$ m/s. It makes a direct collision with disk B, which has a mass of 175 g and is originally at rest. If both disks are of the same size and the collision is perfectly elastic ($e = 1$), determine the velocity of each disk just after collision. Show that the kinetic energy of the disks before and after collision is the same.

15–59. The 5-Mg truck and 2-Mg car are traveling with the free-rolling velocities shown just before they collide. After the collision, the car moves with a velocity of 15 km/h to the right *relative* to the truck. Determine the coefficient of restitution between the truck and car and the loss of energy due to the collision.

Prob. 15–59

*15–60. The 1-lb ball A is thrown so that when it strikes the 10-lb block B it is traveling horizontally at 20 ft/s. Determine the averge normal force exerted between A and B if the impact occurs in 0.02 s. The coefficient of restitution between A and B is $e = 0.6$.

Prob. 15–60

15–61. The 15-kg block A has a velocity $v = 10$ m/s when it is $s = 4$ m from the 10-kg block B. If the unstretched spring has a stiffness $k = 1000$ N/m, determine the maximum compression of the spring after the collision. Take $e = 0.6$, and for block A, $\mu_k = 0.3$. Assume block B is smooth.

Prob. 15–61

15–62. The four smooth balls each have the same mass m. If A and B are rolling forward with a velocity $\mathbf{v}$ and strike C, explain why after collision C and D each move off with a velocity $\mathbf{v}$. Why doesn't D move off with a velocity $2\mathbf{v}$? The collision is elastic, $e = 1$. Neglect the size of each ball.

15–63. The four balls each have the same mass m. If A and B are rolling forward with a velocity $\mathbf{v}$ and strike C, determine the velocity of each ball after the first three collisions. Take $e = 0.5$ between each ball.

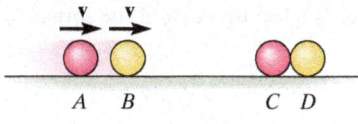

Probs. 15–62/63

*15–64. Ball A has a mass of 3 kg and is moving with a velocity of 8 m/s when it makes a direct collision with ball B, which has a mass of 2 kg and is moving with a velocity of 4 m/s. If $e = 0.7$, determine the velocity of each ball just after the collision. Neglect the size of the balls.

Prob. 15–64

15-65. The 1-lb ball A is thrown so that when it strikes the 10-lb block B it is traveling horizontally at 20 ft/s. If the coefficient of restitution between A and B is $e = 0.6$, and the coefficent of kinetic friction between the plane and the block is $\mu_k = 0.4$, determine the time before block B stops sliding.

15-66. The 1-lb ball A is thrown so that when it strikes the 10-lb block B it is traveling horizontally at 20 ft/s. If the coefficient of restitution between A and B is $e = 0.6$, and the coefficient of kinetic friction between the plane and the block is $\mu_k = 0.4$, determine the distance block B slides on the plane before stopping.

Probs. 15–65/66

15-67. The 5-lb box B is dropped from rest 5 ft from the top of the 10-lb plate P, which is supported by the spring having a stiffness of $k = 30$ lb/ft. If $e = 0.6$ between the box and plate, determine the maximum compression imparted to the spring. Neglect the mass of the spring.

Prob. 15–67

*15-68. Plates A and B each have a mass of 4 kg and are restricted to move along the smooth guides. If the coefficient of restitution between the plates is $e = 0.7$, determine (a) the speed of both plates just after collision and (b) the maximum compression of the spring. Plate A has a velocity of 4 m/s just before striking B. Plate B is originally at rest and the spring is unstretched.

Prob. 15–68

15-69. A 300-g ball is kicked with a velocity of $v_A = 25$ m/s at point A as shown. If the coefficient of restitution between the ball and the field is $e = 0.4$, determine the magnitude and direction θ of the velocity of the rebounding ball at B.

Prob. 15–69

15–70. Two identical balls A and B of mass m are suspended from cords of length $L/2$ and L, respectively. Ball A is released from rest when $\phi = 90°$ and swings down to $\phi = 0°$, where it strikes B. Determine the speed of each ball just after impact and the maximum angle θ through which B will swing. The coefficient of restitution between the balls is e.

15–73. Two smooth disks A and B each have a mass of 0.5 kg. If both disks are moving with the velocities shown when they collide, determine their final velocities just after collision. The coefficient of restitution is $e = 0.75$.

15–74. Two smooth disks A and B each have a mass of 0.5 kg. If both disks are moving with the velocities shown when they collide, determine the coefficient of restitution between the disks if after collision B travels along a line 30° counterclockwise from the y axis.

Prob. 15–70

15–71. The man A has a weight of 175 lb and jumps from rest $h = 8$ ft onto a platform P that has a weight of 60 lb. The platform is mounted on a spring, which has a stiffness $k = 200$ lb/ft. Determine (a) the velocities of A and P just after impact and (b) the maximum compression imparted to the spring by the impact. Assume the coefficient of restitution between the man and the platform is $e = 0.6$, and the man holds himself rigid during the motion.

*****15–72.** The man A has a weight of 100 lb and jumps from rest onto the platform P that has a weight of 60 lb. The platform is mounted on a spring, which has a stiffness $k = 200$ lb/ft. If the coefficient of restitution between the man and the platform is $e = 0.6$, and the man holds himself rigid during the motion, determine the required height h of the jump if the maximum compression of the spring becomes 2 ft.

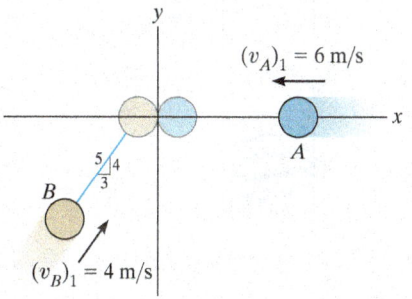

Probs. 15–73/74

15–75. The 0.5-kg ball is fired from the tube at A with a velocity of $v = 6$ m/s. If the coefficient of restitution between the ball and the surface is $e = 0.8$, determine the height h after it bounces off the surface.

Probs. 15–71/72

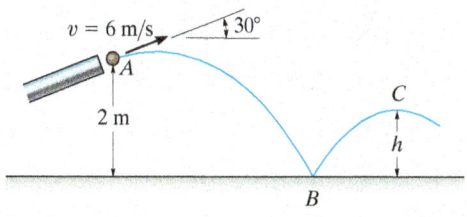

Prob. 15–75

*15–76. The 8-lb ball is released from rest 10 ft from the surface of a flat plate P which weighs 6 lb. Determine the maximum compression in the spring if the impact is perfectly elastic.

Prob. 15–76

15–77. The cue ball A is given an initial velocity $(v_A)_1 = 5$ m/s. If it makes a direct collision with ball B ($e = 0.8$), determine the velocity of B and the angle θ just after it rebounds from the cushion at C ($e' = 0.6$). Each ball has a mass of 0.4 kg. Neglect their size.

Prob. 15–77

15–78. Using a slingshot, the boy fires the 0.2-lb marble at the concrete wall, striking it at B. If the coefficient of restitution between the marble and the wall is $e = 0.5$, determine the speed of the marble after it rebounds from the wall.

Prob. 15–78

15–79. The two disks A and B have a mass of 3 kg and 5 kg, respectively. If they collide with the initial velocities shown, determine their velocities just after impact. The coefficient of restitution is $e = 0.65$.

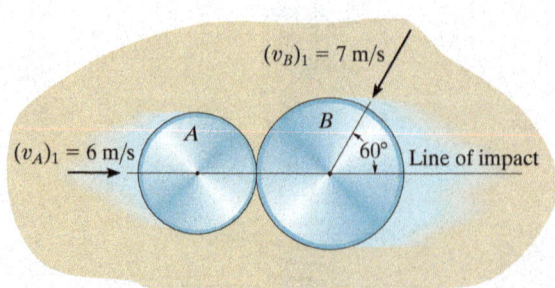

Prob. 15–79

*15–80. A ball of negligible size and mass m is given a velocity of $\mathbf{v}_0$ on the center of the cart which has a mass M and is originally at rest. If the coefficient of restitution between the ball and walls A and B is e, determine the velocity of the ball and the cart just after the ball strikes A. Also, determine the total time needed for the ball to strike A, rebound, then strike B, and rebound and then return to the center of the cart. Neglect friction.

15–82. The 20-lb box slides on the surface for which $\mu_k = 0.3$. The box has a velocity $v = 15$ ft/s when it is 2 ft from the plate. If it strikes the smooth plate, which has a weight of 10 lb and is held in position by an unstretched spring of stiffness $k = 400$ lb/ft, determine the maximum compression imparted to the spring. Take $e = 0.8$ between the box and the plate. Assume that the plate slides smoothly.

Prob. 15–80

Prob. 15–82

15–81. The girl throws the 0.5-kg ball toward the wall with an initial velocity $v_A = 10$ m/s. Determine (a) the velocity at which it strikes the wall at B, (b) the velocity at which it rebounds from the wall if the coefficient of restitution is $e = 0.5$, and (c) the distance s from the wall to where it strikes the ground at C.

15–83. The two hockey pucks A and B each have a mass of 250 g. If they collide at O and are deflected along the colored paths, determine their speeds just after impact. Assume that the icy surface over which they slide is smooth. *Hint:* Since the y' axis is *not* along the line of impact, apply the conservation of momentum along the x' and y' axes.

Prob. 15–81

Prob. 15–83

*15–84. A ball is thrown onto a rough floor at an angle θ. If it rebounds at an angle ϕ and the coefficient of kinetic friction is μ, determine the coefficient of restitution e. Neglect the size of the ball. *Hint:* Show that during impact, the average impulses in the x and y directions are related by $I_x = \mu I_y$. Since the time of impact is the same, $F_x \Delta t = \mu F_y \Delta t$ or $F_x = \mu F_y$.

15–85. A ball is thrown onto a rough floor at an angle of $\theta = 45°$. If it rebounds at the same angle $\phi = 45°$, determine the coefficient of kinetic friction between the floor and the ball. The coefficient of restitution is $e = 0.6$. *Hint:* Show that during impact, the average impulses in the x and y directions are related by $I_x = \mu I_y$. Since the time of impact is the same, $F_x \Delta t = \mu F_y \Delta t$ or $F_x = \mu F_y$.

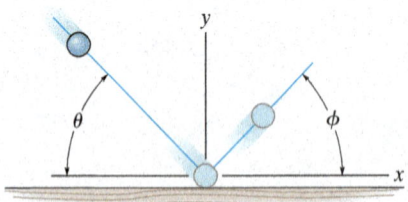

Probs. 15–84/85

15–86. Two disks A and B each have a weight of 2 lb and the initial velocities shown just before they collide. If the coefficient of restitution is $e = 0.5$, determine their speeds just after impact.

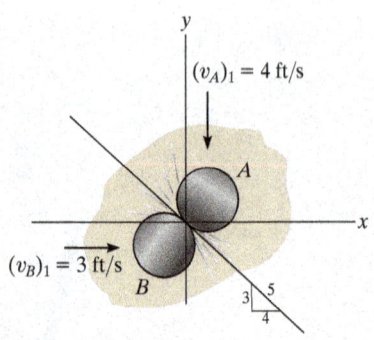

Prob. 15–86

15–87. It is observed that hailstones strike the surface of a frozen lake at an angle of 60° with the horizontal and rebound at an angle of 45°. Determine the coefficient of restitution.

*15–88. If the hailstones in Prob. 15–87 rise to a height of 0.5 m after rebounding, determine their velocity just before impact.

Probs. 15–87/88

15–89. The 0.2-lb ball bearing travels over the edge A with a velocity of $v_A = 3$ ft/s. Determine the speed at which it rebounds from the smooth inclined plane at B. Take $e = 0.8$.

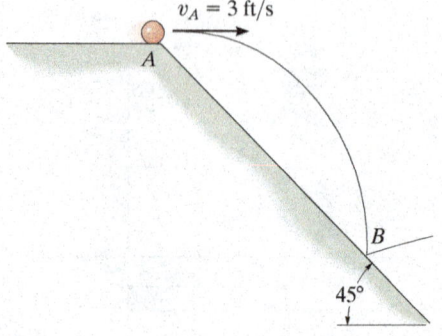

Prob. 15–89

15–90. Before a cranberry can make it to your dinner plate, it must pass a bouncing test which rates its quality. If cranberries having an $e \geq 0.8$ are to be accepted, determine the dimensions d and h for the barrier so that when a cranberry falls from rest at A it strikes the incline at B and bounces over the barrier at C.

***15–92.** The two billiard balls A and B are originally in contact with one another when a third ball C strikes each of them at the same time as shown. If ball C remains at rest after the collision, determine the coefficient of restitution. All the balls have the same mass. Neglect the size of each ball.

Prob. 15–92

Prob. 15–90

15–93. Disks A and B have a mass of 15 kg and 10 kg, respectively. If they are sliding on a smooth horizontal plane with the velocities shown, determine their speeds just after impact. The coefficient of restitution between them is $e = 0.8$.

15–91. The ping-pong ball has a mass of 2 g. If it is struck with the velocity shown, determine how high h it rises above the end of the smooth table after the rebound. Take $e = 0.8$.

Prob. 15–91

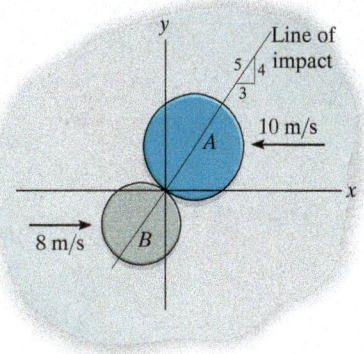

Prob. 15–93

15.5 ANGULAR MOMENTUM

The **angular momentum** of a particle about point O is defined as the "moment" of the particle's linear momentum about O. Since this concept is analogous to finding the moment of a force about a point, the angular momentum, $\mathbf{H}_O$, is sometimes referred to as the *moment of momentum*.

Fig. 15–19

Scalar Formulation. If a particle moves along a curve lying in the x–y plane, Fig. 15–19, the angular momentum at any instant can be determined about point O (actually the z axis) by using a scalar formulation. The *magnitude* of $\mathbf{H}_O$ is

$$H_O = (d)(mv) \quad (15\text{–}13)$$

Here d is the moment arm or perpendicular distance from O to the line of action of $m\mathbf{v}$. Common units for H_O are $kg \cdot m^2/s$ or $slug \cdot ft^2/s$. The *direction* of $\mathbf{H}_O$ is defined by the right-hand rule. As shown, the curl of the fingers of the right hand indicates the sense of rotation of $m\mathbf{v}$ about O, so that in this case the thumb (or $\mathbf{H}_O$) is directed perpendicular to the x–y plane along the $+z$ axis.

Vector Formulation. If the particle moves along a space curve, Fig. 15–20, the vector cross product can be used to determine the *angular momentum* about O. In this case

$$\mathbf{H}_O = \mathbf{r} \times m\mathbf{v} \quad (15\text{–}14)$$

Here $\mathbf{r}$ denotes a position vector drawn from point O to the particle. As shown in the figure, $\mathbf{H}_O$ is *perpendicular* to the shaded plane containing $\mathbf{r}$ and $m\mathbf{v}$.

If $\mathbf{r}$ and $m\mathbf{v}$ are expressed in terms of their Cartesian components, then the angular momentum can be determined by evaluating the determinant:

$$\mathbf{H}_O = \begin{vmatrix} \mathbf{i} & \mathbf{j} & \mathbf{k} \\ r_x & r_y & r_z \\ mv_x & mv_y & mv_z \end{vmatrix} \quad (15\text{–}15)$$

Fig. 15–20

15.6 RELATION BETWEEN THE MOMENT OF A FORCE AND ANGULAR MOMENTUM

The moments about point O of all the forces acting on the particle in Fig. 15–21a can be related to the particle's angular momentum by applying the equation of motion. Since the mass of the particle is constant, we may write

$$\Sigma \mathbf{F} = m\dot{\mathbf{v}}$$

The moments of the forces about point O can be obtained by performing a cross-product multiplication of each side of this equation by the position vector $\mathbf{r}$, which is directed from O to the particle. We have

$$\Sigma \mathbf{M}_O = \mathbf{r} \times \Sigma \mathbf{F} = \mathbf{r} \times m\dot{\mathbf{v}}$$

From Appendix B, the derivative of $\mathbf{r} \times m\mathbf{v}$ can be written as

$$\dot{\mathbf{H}}_O = \frac{d}{dt}(\mathbf{r} \times m\mathbf{v}) = \dot{\mathbf{r}} \times m\mathbf{v} + \mathbf{r} \times m\dot{\mathbf{v}}$$

The first term on the right side, $\dot{\mathbf{r}} \times m\mathbf{v} = m(\dot{\mathbf{r}} \times \dot{\mathbf{r}}) = \mathbf{0}$, since the cross product of a vector with itself is zero. Hence, the above equation becomes

$$\boxed{\Sigma \mathbf{M}_O = \dot{\mathbf{H}}_O} \qquad (15\text{–}16)$$

which states that *the resultant moment about point O of all the forces acting on the particle is equal to the time rate of change of the particle's angular momentum about point O.* This result is similar to Eq. 15–1, i.e.,

$$\boxed{\Sigma \mathbf{F} = \dot{\mathbf{L}}} \qquad (15\text{–}17)$$

Here *the resultant force acting on the particle is equal to the time rate of change of the particle's linear momentum*, $\mathbf{L} = m\mathbf{v}$.

From the derivations, it is seen that Eqs. 15–16 and 15–17 are actually another way of stating Newton's second law of motion. In other sections of this book it will be shown that these equations have many practical applications when extended and applied to problems involving either a system of particles or a rigid body.

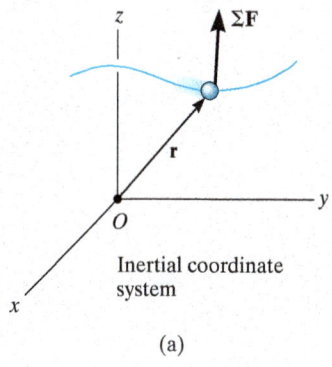

Inertial coordinate system

(a)

Fig. 15–21

System of Particles. An equation having the same form as Eq. 15–16 may be derived for the system of particles shown in Fig. 15–21b. The forces acting on the arbitrary *i*th particle of the system consist of a resultant *external force* $\mathbf{F}_i$ and a resultant *internal force* $\mathbf{f}_i$. Expressing the moments of these forces about point O, using the form of Eq. 15–16, we have

$$(\mathbf{r}_i \times \mathbf{F}_i) + (\mathbf{r}_i \times \mathbf{f}_i) = (\dot{\mathbf{H}}_i)_O$$

Here $(\dot{\mathbf{H}}_i)_O$ is the time rate of change of the angular momentum of the *i*th particle about O. Similar equations can be written for each of the other particles of the system. When the results are summed vectorially, the result is

$$\Sigma(\mathbf{r}_i \times \mathbf{F}_i) + \Sigma(\mathbf{r}_i \times \mathbf{f}_i) = \Sigma(\dot{\mathbf{H}}_i)_O$$

The second term is zero since the internal forces occur in equal but opposite collinear pairs, and hence the moment of each pair of forces about point O is zero. Dropping the index notation, the above equation can be written in a simplified form as

$$\Sigma \mathbf{M}_O = \dot{\mathbf{H}}_O \qquad (15\text{–}18)$$

which states that *the sum of the moments about point O of all the external forces acting on a system of particles is equal to the time rate of change of the total angular momentum of the system about point O.* Although O has been chosen here as the origin of coordinates, it actually can represent any *fixed point* in the inertial frame of reference.

(b)

Fig. 15–21(cont.)

EXAMPLE 15.12

The box shown in Fig. 15–22a has a mass m and travels down the smooth circular ramp such that when it is at the angle θ it has a speed v. Determine its angular momentum about point O at this instant and the rate of increase in its speed, i.e., a_t.

Fig. 15–22

SOLUTION
Since **v** is tangent to the path, applying Eq. 15–13 the angular momentum is

$$H_O = rmv \downarrow \qquad \text{Ans.}$$

The rate of increase in its speed (dv/dt) can be found by applying Eq. 15–16. From the free-body diagram of the box, Fig. 15–22b, it can be seen that only the weight $W = mg$ contributes a moment about point O. We have

$$\zeta + \Sigma M_O = \dot{H}_O; \qquad mg(r \sin \theta) = \frac{d}{dt}(rmv)$$

Since r and m are constant,

$$mgr \sin \theta = rm\frac{dv}{dt}$$

$$\frac{dv}{dt} = g \sin \theta \qquad \text{Ans.}$$

NOTE: This same result can, of course, be obtained from the equation of motion applied in the tangential direction, Fig. 15–22b, i.e.,

$$+\swarrow \Sigma F_t = ma_t; \qquad mg \sin \theta = m\left(\frac{dv}{dt}\right)$$

$$\frac{dv}{dt} = g \sin \theta \qquad \text{Ans.}$$

15.7 PRINCIPLE OF ANGULAR IMPULSE AND MOMENTUM

If Eq. 15–16 is rewritten in the form $\Sigma \mathbf{M}_O \, dt = d\mathbf{H}_O$ and integrated, assuming that at time $t = t_1$, $\mathbf{H}_O = (\mathbf{H}_O)_1$ and at time $t = t_2$, $\mathbf{H}_O = (\mathbf{H}_O)_2$, we have

$$\Sigma \int_{t_1}^{t_2} \mathbf{M}_O \, dt = (\mathbf{H}_O)_2 - (\mathbf{H}_O)_1$$

or

$$(\mathbf{H}_O)_1 + \Sigma \int_{t_1}^{t_2} \mathbf{M}_O \, dt = (\mathbf{H}_O)_2 \qquad (15\text{–}19)$$

This equation is referred to as the *principle of angular impulse and momentum*. The initial and final angular momenta $(\mathbf{H}_O)_1$ and $(\mathbf{H}_O)_2$ are defined as the moment of the linear momentum of the particle $(\mathbf{H}_O = \mathbf{r} \times m\mathbf{v})$ at the instants t_1 and t_2, respectively. The second term on the left side, $\Sigma \int \mathbf{M}_O \, dt$, is called the *angular impulse*. It is determined by integrating, with respect to time, the moments of all the forces acting on the particle over the time period t_1 to t_2. Since the moment of a force about point O is $\mathbf{M}_O = \mathbf{r} \times \mathbf{F}$, the angular impulse may be expressed in vector form as

$$\boxed{\text{angular impulse} = \int_{t_1}^{t_2} \mathbf{M}_O \, dt = \int_{t_1}^{t_2} (\mathbf{r} \times \mathbf{F}) \, dt} \qquad (15\text{–}20)$$

Here $\mathbf{r}$ is a position vector which extends from point O to any point on the line of action of $\mathbf{F}$.

In a similar manner, using Eq. 15–19, the principle of angular impulse and momentum for a system of particles may be written as

$$\Sigma(\mathbf{H}_O)_1 + \Sigma \int_{t_1}^{t_2} \mathbf{M}_O \, dt = \Sigma(\mathbf{H}_O)_2 \qquad (15\text{–}21)$$

Here the first and third terms represent the angular momenta of all the particles $[\Sigma \mathbf{H}_O = \Sigma(\mathbf{r}_i \times m\mathbf{v}_i)]$ at the instants t_1 and t_2. The second term is the sum of the angular impulses given to all the particles from t_1 to t_2. Recall that these impulses are created only by the moments of the external forces acting on the system where, for the *i*th particle, $\mathbf{M}_O = \mathbf{r}_i \times \mathbf{F}_i$.

Vector Formulation.
Using impulse and momentum principles, it is therefore possible to write two vector equations which define the particle's motion, namely, Eq. 15–3 and Eq. 15–19, restated as

$$m\mathbf{v}_1 + \Sigma \int_{t_1}^{t_2} \mathbf{F}\, dt = m\mathbf{v}_2$$

$$(\mathbf{H}_O)_1 + \Sigma \int_{t_1}^{t_2} \mathbf{M}_O\, dt = (\mathbf{H}_O)_2$$

(15–22)

Scalar Formulation.
In general, the above equations can be expressed in x, y, z component form. If the particle is confined to move in the x–y plane, then three scalar equations can be written to describe the motion, namely,

$$m(v_x)_1 + \Sigma \int_{t_1}^{t_2} F_x\, dt = m(v_x)_2$$

$$m(v_y)_1 + \Sigma \int_{t_1}^{t_2} F_y\, dt = m(v_y)_2$$

$$(H_O)_1 + \Sigma \int_{t_1}^{t_2} M_O\, dt = (H_O)_2$$

(15–23)

The first two of these equations represent the principle of linear impulse and momentum in the x and y directions, which has been discussed in Sec. 15.1, and the third equation represents the principle of angular impulse and momentum about the z axis.

280   CHAPTER 15   KINETICS OF A PARTICLE: IMPULSE AND MOMENTUM

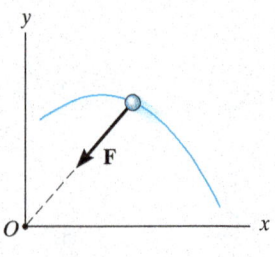

Fig. 15–23

Conservation of Angular Momentum. When the angular impulses acting on a particle are all zero during the time t_1 to t_2, Eq. 15–19 reduces to the following simplified form:

$$(\mathbf{H}_O)_1 = (\mathbf{H}_O)_2 \qquad (15\text{–}24)$$

This equation is known as the **conservation of angular momentum**. It states that from t_1 to t_2 the particle's angular momentum remains constant. Obviously, if no external impulse is applied to the particle, both linear and angular momentum will be conserved. In some cases, however, the particle's angular momentum will be conserved and linear momentum may not. An example of this occurs when the particle is subjected *only* to a *central force* (see Sec. 13.7). As shown in Fig. 15–23, the impulsive central force **F** is always directed toward point O as the particle moves along the path. Hence, the angular impulse (moment) created by **F** about the z axis is always zero, and therefore angular momentum of the particle is conserved about this axis.

From Eq. 15–21, we can also write the conservation of angular momentum for a system of particles as

$$\Sigma(\mathbf{H}_O)_1 = \Sigma(\mathbf{H}_O)_2 \qquad (15\text{–}25)$$

In this case the summation must include the angular momenta of all particles in the system.

Provided air resistance is neglected, the passengers on this amusement-park ride are subjected to a conservation of angular momentum about the z axis of rotation. As shown on the free-body diagram, the line of action of the normal force **N** of the seat on the passenger passes through this axis, and the passenger's weight **W** is parallel to it. Thus, no angular impulse acts around the z axis.

PROCEDURE FOR ANALYSIS

When applying the principles of angular impulse and momentum, or the conservation of angular momentum, it is suggested that the following procedure be used.

Free-Body Diagram.

- Draw the particle's free-body diagram in order to determine any axis about which angular momentum may be conserved. For this to occur, the moments of all the forces (or impulses) must either be parallel or pass through the axis so as to create zero moment throughout the time period t_1 to t_2.

- The direction and sense of the particle's initial and final velocities should also be established.

Momentum Equations.

- Apply the principle of angular impulse and momentum, $(\mathbf{H}_O)_1 + \Sigma \int_{t_1}^{t_2} \mathbf{M}_O dt = (\mathbf{H}_O)_2$, or if appropriate, the conservation of angular momentum, $(\mathbf{H}_O)_1 = (\mathbf{H}_O)_2$.

EXAMPLE 15.13

The 1.5-Mg car travels along the circular road as shown in Fig. 15–24a. If the traction force of the wheels on the road is $F = (150t^2)$ N, where t is in seconds, determine the speed of the car when $t = 5$ s. The car initially travels with a speed of 5 m/s. Neglect the size of the car.

(a)

Free-Body Diagram. The free-body diagram of the car is shown in Fig. 15–24b. If we apply the principle of angular impulse and momentum about the z axis, then the angular impulse created by the weight, normal force, and radial frictional force will be eliminated since they act parallel to the axis or pass through it.

Principle of Angular Impulse and Momentum.

$$(H_z)_1 + \Sigma \int_{t_1}^{t_2} M_z \, dt = (H_z)_2$$

$$r \, m_c(v_c)_1 + \int_{t_1}^{t_2} r F \, dt = r \, m_c(v_c)_2$$

$$(100 \text{ m})(1500 \text{ kg})(5 \text{ m/s}) + \int_0^{5 \text{ s}} (100 \text{ m})[(150t^2) \text{ N}] \, dt$$
$$= (100 \text{ m})(1500 \text{ kg})(v_c)_2$$

$$750(10^3) + 5000t^3 \Big|_0^{5 \text{ s}} = 150(10^3)(v_c)_2$$

$$(v_c)_2 = 9.17 \text{ m/s} \qquad Ans.$$

(b)

Fig. 15–24

EXAMPLE 15.14

(a)

The 0.8-lb ball B, shown in Fig. 15–25a, is attached to a cord which passes through a hole at A in a smooth table. When the ball is $r_1 = 1.75$ ft from the hole, it is rotating around in a circle such that its speed is $v_1 = 4$ ft/s. By applying the force $\mathbf{F}$ the cord is pulled downward through the hole with a constant speed $v_c = 6$ ft/s. Determine (a) the speed of the ball at the instant it is $r_2 = 0.6$ ft from the hole, and (b) the amount of work done by $\mathbf{F}$ in shortening the radial distance from r_1 to r_2. Neglect the size of the ball.

SOLUTION

Part (a) Free-Body Diagram. As the ball moves from r_1 to r_2, Fig. 15–25b, the cord force $\mathbf{F}$ on the ball always passes through the z axis, and the weight and $\mathbf{N}_B$ are parallel to it. Hence the moments, or angular impulses created by these forces, are all *zero* about this axis. Therefore, angular momentum is conserved about the z axis.

Conservation of Angular Momentum. The ball's velocity $\mathbf{v}_2$ is resolved into two components. The radial component, 6 ft/s, is known; however, it produces zero angular momentum about the z axis. Thus,

$$\mathbf{H}_1 = \mathbf{H}_2$$
$$r_1 m_B v_1 = r_2 m_B v_2'$$
$$1.75 \text{ ft}\left(\frac{0.8 \text{ lb}}{32.2 \text{ ft/s}^2}\right) 4 \text{ ft/s} = 0.6 \text{ ft}\left(\frac{0.8 \text{ lb}}{32.2 \text{ ft/s}^2}\right) v_2'$$
$$v_2' = 11.67 \text{ ft/s}$$

The speed of the ball is thus

$$v_2 = \sqrt{(11.67 \text{ ft/s})^2 + (6 \text{ ft/s})^2} = 13.1 \text{ ft/s} \qquad Ans.$$

Part (b). The only force that does work on the ball is $\mathbf{F}$. (The normal force and weight do not move vertically.) The initial and final kinetic energies of the ball can be determined so that from the principle of work and energy we have

$$T_1 + \Sigma U_{1-2} = T_2$$
$$\frac{1}{2}\left(\frac{0.8 \text{ lb}}{32.2 \text{ ft/s}^2}\right)(4 \text{ ft/s})^2 + U_F = \frac{1}{2}\left(\frac{0.8 \text{ lb}}{32.2 \text{ ft/s}^2}\right)(13.1 \text{ ft/s})^2$$
$$U_F = 1.94 \text{ ft} \cdot \text{lb} \qquad Ans.$$

NOTE: The force F is not constant because the normal component of acceleration, $a_n = v^2/r$, changes as r changes.

(b)

Fig. 15–25

EXAMPLE 15.15

The 2-kg disk shown in Fig. 15-26a rests on a smooth horizontal surface and is attached to an elastic cord that has a stiffness $k_c = 20$ N/m and is initially unstretched. If the disk is given a velocity $(v_D)_1 = 1.5$ m/s, perpendicular to the cord, determine the rate at which the cord is being stretched and the speed of the disk at the instant the cord is stretched 0.2 m.

SOLUTION

Free-Body Diagram. After the disk has been launched, it slides along the path shown in Fig. 15-26b. By inspection, angular momentum about point O (or the z axis) is *conserved*, since none of the forces produce an angular impulse about this axis. Also, when the distance is 0.7 m, only the transverse component $(v'_D)_2$ produces angular momentum of the disk about O.

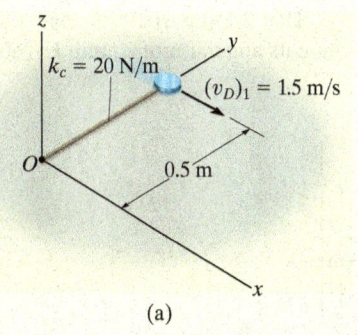

(a)

Conservation of Angular Momentum. To obtain $(v'_D)_2$ we have

$$(H_O)_1 = (H_O)_2$$

$$r_1 m_D (v_D)_1 = r_2 m_D (v'_D)_2$$

$$0.5 \text{ m } (2 \text{ kg})(1.5 \text{ m/s}) = 0.7 \text{ m}(2 \text{ kg})(v'_D)_2$$

$$(v'_D)_2 = 1.071 \text{ m/s}$$

Conservation of Energy. The speed of the disk can be obtained by applying the conservation of energy equation at the point where the disk was launched and at the point where the cord is stretched 0.2 m.

$$T_1 + V_1 = T_2 + V_2$$

$$\tfrac{1}{2} m_D (v_D)_1^2 + \tfrac{1}{2} k x_1^2 = \tfrac{1}{2} m_D (v_D)_2^2 + \tfrac{1}{2} k x_2^2$$

$$\tfrac{1}{2}(2 \text{ kg})(1.5 \text{ m/s})^2 + 0 = \tfrac{1}{2}(2 \text{ kg})(v_D)_2^2 + \tfrac{1}{2}(20 \text{ N/m})(0.2 \text{ m})^2$$

$$(v_D)_2 = 1.360 \text{ m/s} = 1.36 \text{ m/s} \qquad Ans.$$

(b)

Fig. 15-26

Having determined $(v_D)_2$ and its component $(v'_D)_2$, the rate of stretch of the cord, or radial component, $(v''_D)_2$ is determined from the Pythagorean theorem,

$$(v''_D)_2 = \sqrt{(v_D)_2^2 - (v'_D)_2^2}$$

$$= \sqrt{(1.360 \text{ m/s})^2 - (1.071 \text{ m/s})^2}$$

$$= 0.838 \text{ m/s} \qquad Ans.$$

FUNDAMENTAL PROBLEMS

F15–19. The 2-kg particle A has the velocity shown. Determine its angular momentum $\mathbf{H}_O$ about point O.

Prob. F15–19

F15–20. The 2-kg particle A has the velocity shown. Determine its angular momentum $\mathbf{H}_P$ about point P.

Prob. F15–20

F15–21. Initially the 5-kg block is moving with a constant speed of 2 m/s around the circular path centered at O on the smooth horizontal plane. If a constant tangential force $F = 5$ N is applied to the block, determine the block's speed when $t = 3$ s. Neglect the size of the block and the mass of the rod.

Prob. F15–21

F15–22. The 5-kg block is moving around the circular path centered at O on the smooth horizontal plane when it is subjected to the force $F = (10t)$ N, where t is in seconds. If the block starts from rest, determine its speed when $t = 4$ s. Neglect the size of the block and the mass of the rod. The force maintains the same constant angle tangent to the path.

Prob. F15–22

F15–23. The 2-kg sphere is attached to the rod, which rotates in the *horizontal plane* centered at O. If the system is subjected to a couple moment $M = (0.9t^2)$ N·m, where t is in seconds, determine the speed of the sphere at the instant $t = 5$ s starting from rest. Neglect the mass of the rod.

Prob. F15–23

F15–24. Two identical 10-kg spheres are attached to the rod, which rotates in the horizontal plane centered at pin O. If the spheres are subjected to tangential forces of $P = 10$ N, and the rod is subjected to a couple moment $M = (8t)$ N·m, where t is in seconds, determine the speed of the spheres at the instant $t = 4$ s. The system starts from rest. Neglect the size of the spheres and the mass of the rod.

Prob. F15–24

PROBLEMS

All solutions, where necessary, must include a free-body diagram.

15–94. Determine the angular momentum $\mathbf{H}_O$ of each of the two particles about point O. Use a scalar solution.

15–95. Determine the angular momentum $\mathbf{H}_P$ of each of the two particles about point P. Use a scalar solution.

15–98. Determine the angular momentum $\mathbf{H}_O$ of the particle about point O.

15–99. Determine the angular momentum $\mathbf{H}_P$ of the particle about point P.

Probs. 15–94/95

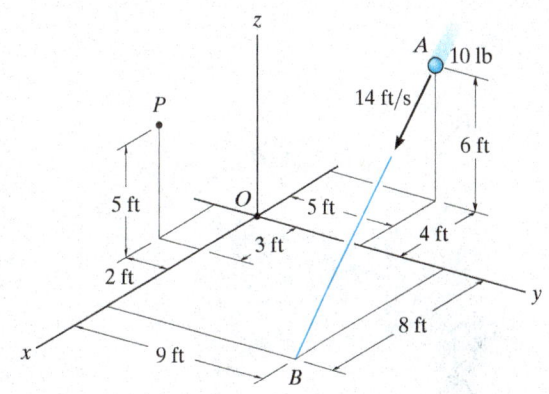

Probs. 15–98/99

*****15–96.** Determine the angular momentum $\mathbf{H}_o$ of each of the two particles about point O.

15–97. Determine the angular momentum $\mathbf{H}_p$ of each of the two particles about point P.

*****15–100.** The ball B has a mass of 10 kg and is attached to the end of a rod whose mass may be neglected. If the rod is subjected to a torque $M = (3t^2 + 5t + 2)$ N·m, where t is in seconds, determine the speed of the ball when $t = 2$ s. The ball has a speed $v = 2$ m/s when $t = 0$.

Probs. 15–96/97

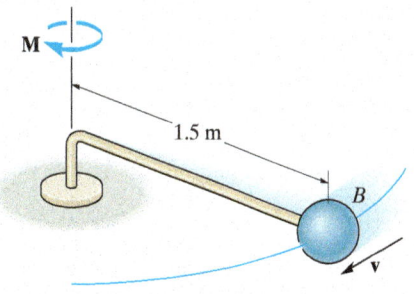

Prob. 15–100

15–101. The 800-lb roller-coaster car starts from rest on the track having the shape of a cylindrical helix. If the normal force component of the track on the car in the transverse direction is $N_\theta = 68$ lb, determine the time required for the car to obtain a velocity of 16 ft/s in the transverse direction. Neglect the size of the car.

15–102. The 800-lb roller-coaster car starts from rest on the track having the shape of a cylindrical helix. If the normal force of the tracks on the car has a transverse component of $N_\theta = 68$ lb, determine the transverse component of its velocity in $t = 4$ s. Also, what is the car's velocity when it descends 8 ft? Neglect the size of the car.

*15–104. At the instant $r = 1.5$ m, the 5-kg disk is given a speed of $v = 5$ m/s, perpendicular to the elastic cord. Determine the speed of the disk and the rate of shortening of the elastic cord at the instant $r = 1.2$ m. The disk slides on the smooth horizontal plane. Neglect its size. The cord has an unstretched length of 0.5 m.

Prob. 15–104

Probs. 15–101/102

15–105. Each of the 5-lb spheres is rigidly attached to the crossbar frame. The frame has a negligible weight and is free to rotate. If a couple moment $M = (5t + 2)$ lb·ft, where t is in seconds, is applied to the frame, determine the speed of each of the spheres in 4 s, starting from rest.

15–103. The ball B has a mass of 10 kg and is attached to the end of a rod whose mass can be neglected. If the shaft is subjected to a torque $M = (2t^2 + 4)$ N·m, where t is in seconds, determine the speed of the ball when $t = 2$ s. The ball has a speed $v = 2$ m/s when $t = 0$.

Prob. 15–103

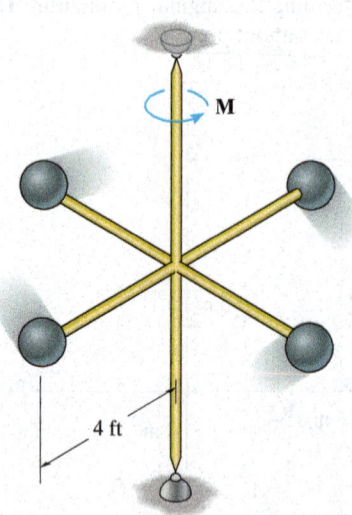

Prob. 15–105

15–106. The two spheres A and B each have a mass of 400 g. The spheres are fixed to the horizontal rods and their initial velocity is 2 m/s. If a couple moment of $M = 0.6 \text{ N} \cdot \text{m}$ is applied to the frame, determine the speed of the spheres in 3 s. The mass of the supporting frame is negligible and it is free to rotate. Neglect the size of the spheres.

*15–108. The projectile having a mass of 3 kg is fired from a cannon with a muzzle velocity of $v_0 = 500 \text{ m/s}$. Determine the projectile's angular momentum about point O at the instant it is at the maximum height of its trajectory.

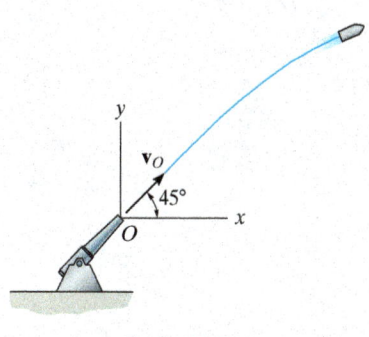

Prob. 15–108

Prob. 15–106

15–107. If the rod of negligible mass is subjected to a couple moment of $M = (30t^2) \text{ N} \cdot \text{m}$, and the engine of the car supplies a traction force of $F = (15t) \text{ N}$ to the wheels, where t is in seconds, determine the speed of the car at the instant $t = 5$ s. The car starts from rest. The total mass of the car and rider is 150 kg. Neglect the size of the car.

15–109. The 3-lb ball located at A is released from rest and travels down the curved path. If the ball exerts a normal force of 5 lb on the path when it reaches point B, determine the angular momentum of the ball about the center of curvature, point O. *Hint:* Neglect the size of the ball. The radius of curvature at point B must first be determined.

Prob. 15–107

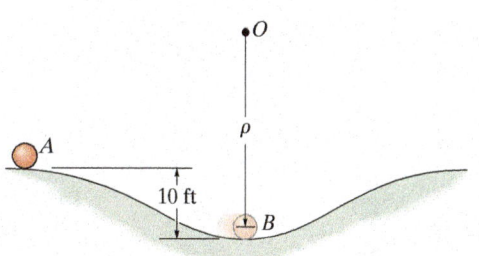

Prob. 15–109

15–110. The amusement park ride consists of a 200-kg car and passenger that are traveling at 3 m/s along a circular path having a radius of 8 m. If at $t = 0$, the cable OA is pulled in toward O at 0.5 m/s, determine the speed of the car when $t = 4$ s. Also, determine the work done to pull in the cable.

Prob. 15–110

15–111. A small block having a mass of 0.1 kg is given a horizontal velocity of $v_1 = 0.4$ m/s when $r_1 = 500$ mm. It slides along the smooth conical surface. Determine the distance h it must descend to reach a speed of $v_2 = 2$ m/s. Also, what is the angle of descent θ, that is, the angle measured from the horizontal to the tangent of the path?

Prob. 15–111

*15–112. The 2-kg ball rotates around a 0.5-m-diameter circular path with a constant speed. If the cord length is shortened from $l = 1$ m to $l' = 0.5$ m, by pulling the cord through the tube, determine the new diameter of the path d'. Also, what is the tension in the cord in each case?

Prob. 15–112

15–113. An earth satellite of mass 700 kg is launched into a free-flight trajectory about the earth with an initial speed of $v_A = 10$ km/s when the distance from the center of the earth is $r_A = 15$ Mm. If the launch angle at this position is $\phi_A = 70°$, determine the speed v_B of the satellite and its closest distance r_B from the center of the earth. The earth has a mass $M_e = 5.976(10^{24})$ kg. *Hint:* Under these conditions, the satellite is subjected only to the earth's gravitational force, $F = GM_e m_s / r^2$, Eq. 13–1. For part of the solution, use the conservation of energy.

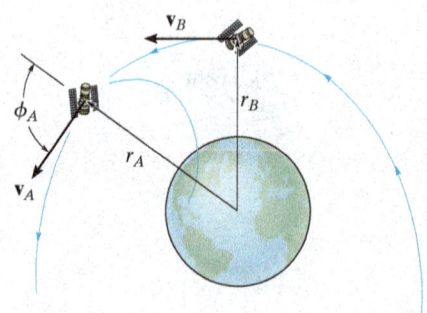

Prob. 15–113

15.8 BODIES SUBJECTED TO A MASS FLOW

In the previous sections we have considered application of impulse and momentum principles to a system of particles that compose a rigid body, having a closed boundary. In this section we will apply the principle of impulse and momentum to a system of particles that can flow either into or out of volume having an open boundary.

Normally the mass flow for these cases is analyzed using a control volume approach, and is discussed in the subject of fluid mechanics.* Here, we will use this method and consider two specific applications that have widespread application. Before we do this however, we will first derive a general equation that applies to both cases. The derivation can be developed by considering the flow of a system of particles, such as a fluid, that pass through a pipe, Fig. 15–27a. As shown in Fig. 15–27b, the initial momentum of the particles is $(m\mathbf{v})_1$. During the time Δt, impulsive forces $\Sigma \mathbf{F}$, such as caused by the wall of the pipe, or the pressure at the entrance and exit, act on the moving particles and change their velocity in both magnitude and direction, so the final momentum of the system is $(m\mathbf{v})_2$. If we apply the impulse and momentum principle to the particles, we have

$$(m\mathbf{v})_1 + \Sigma \mathbf{F} \Delta t = (m\mathbf{v})_2$$

$$\Sigma \mathbf{F} \Delta t = \Delta m \mathbf{v} \qquad (15\text{–}26)$$

Finding this time rate of change in momentum, $\Delta m\mathbf{v}$, for all the particles is rather difficult in this case because of their random motion within the system. A better way to determine $\Delta m\mathbf{v}$ is to consider this time rate of change relative to a specific region of space, called a **control volume**, Fig. 15–27a. For example, if the particles compose a fluid, then the control volume may define the size and shape of the solid boundaries and openings of a conduit or channel, a turbine, or a pump. A rocket or jet engine can also be selected as a control volume, and in this case we can consider it to be moving.

(a)

Fig. 15–27

(b)

Fig. 15–27

---

* See *Fluid Mechanics*, R. C. Hibbeler, Pearson Education.

290   CHAPTER 15   KINETICS OF A PARTICLE: IMPULSE AND MOMENTUM

(c)

Fig. 15–27

For our purposes we will take the control volume to be the region in the pipe shown in Fig. 15–27c. At time t all the particles in the *system* are contained within the control volume, and so the total momentum of the system is $(m\mathbf{v})_1 = (m\mathbf{v})_{1\text{cv}}$. But at $t + \Delta t$, some of the particles of the system will flow out of the control volume, while other particles, not part of the system, will flow into it. The momenta of these particles are $(\Delta m\mathbf{v})_{\text{out}}$ and $(\Delta m\mathbf{v})_{\text{in}}$, respectively. Within the control volume the remaining momentum of the system's particles has changed to $(m\mathbf{v})_{2\text{cv}}$. As a result the change in momentum becomes

$$\Delta m\mathbf{v} = (\Delta m\mathbf{v})_{\text{cv}} + (\Delta m\mathbf{v})_{\text{out}} - (\Delta m\mathbf{v})_{\text{in}}$$

To summarize, the change of momentum of the system as seen relative to the control volume consists of two parts, namely, the change that occurs *within* the control volume, $(\Delta m\mathbf{v})_{\text{cv}} = (m\mathbf{v})_{2\text{cv}} - (m\mathbf{v})_{1\text{cv}}$, and the *net change* in the momentum of the system that is flowing out of the control volume, $(\Delta m\mathbf{v})_{\text{out}} - (\Delta m\mathbf{v})_{\text{in}}$.

If we substitute the above equation into Eq. 15–26 and divide by Δt, allowing $\Delta t \to 0$, we get the result

$$\Sigma \mathbf{F} = \left(\frac{d}{dt}(m\mathbf{v})\right)_{\text{cv}} + \dot{m}_{\text{out}} \mathbf{v}_{\text{out}} - \dot{m}_{\text{in}} \mathbf{v}_{\text{in}} \qquad (15\text{–}27)$$

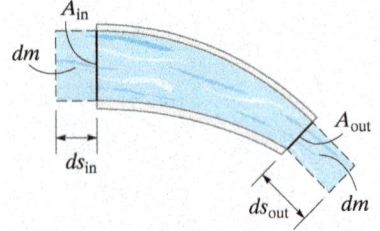

Fig. 15–28

The term $\dot{m} = dm/dt$ is called the **mass flow**. It indicates the number of particles that flow into or out of the control volume per unit time. If the particles compose a fluid, and the open control surfaces have cross-sectional areas A_{in} and A_{out}, Fig. 15–28, then the conservation of mass through these surfaces requires $dm = \rho\, dV = \rho_{\text{in}}(ds_{\text{in}} A_{\text{in}}) = \rho_{\text{out}}(ds_{\text{out}} A_{\text{out}})$, where ρ is the density of the system (fluid) and $dV = ds\, A$ is an element of volume, entering or leaving a control surface. Hence, during the time dt, since $v_{\text{in}} = ds_{\text{in}}/dt$ and $v_{\text{out}} = ds_{\text{out}}/dt$, we have $\dot{m} = \rho_{\text{in}} v_{\text{in}} A_{\text{in}} = \rho_{\text{out}} v_{\text{out}} A_{\text{out}}$, or in general

$$\dot{m} = \rho v A = \rho Q \qquad (15\text{–}28)$$

The term $Q = vA$ measures the volume of fluid flow per unit time, and is referred to as the **volumetric flow**. In the next two sections we will consider two important applications of Eq. 15–27.

15.9 STEADY FLOW OF A FLUID STREAM

Some problems in engineering require finding the forces acting on a fluid stream in order to divert it as the fluid passes through a conduit or over a blade. When this is the case, provided we consider the fluid to be *incompressible* and the *flow* to be *steady*, that is, independent of time, then no change in momentum will occur within the control volume. As a result, the first term on the right of Eq. 15–27 will be equal to zero. Also, because of the conservation of mass, $\dot{m} = \dot{m}_{in} = \dot{m}_{out}$. Therefore, our result becomes

$$\Sigma \mathbf{F} = \dot{m}(\mathbf{v}_{out} - \mathbf{v}_{in}) \tag{15-29}$$

The following procedure can be used whenever this equation is applied to determine the force acting on a device, such as a pipe or blade, that diverts a fluid stream having steady flow.

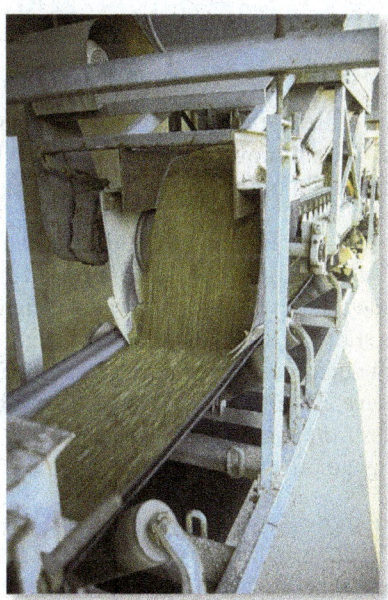

The conveyor belt must supply frictional forces to the gravel that falls upon it in order to change the momentum of the gravel stream, so that it begins to travel along the belt.

The air on one side of this fan is essentially at rest, and as it passes through the blades its momentum is increased. To change the momentum of the air flow in this manner, the blades must exert a horizontal thrust on the air stream. As the blades turn faster, the equal but opposite thrust of the air on the blades could overcome the rolling resistance of the wheels on the ground and begin to move the frame of the fan.

PROCEDURE FOR ANALYSIS

Control Volume. Select the control volume so its free-body diagram includes the unknown forces that are to be determined. This volume can contain both solid and fluid parts.

Free-Body Diagram. Draw the free-body diagram of the control volume in order to identify the forces $\Sigma \mathbf{F}$ that act on it. These forces may include the support reactions, the weight of the control volume, and the static gage pressure forces acting on the fluid at the open control surfaces, or on a closed surface or boundary.* This *gage pressure* is the pressure measured above or below atmospheric pressure. As a result, if an opening is exposed to the atmosphere, the gage pressure will be zero.

Momentum Equation. The mass flow is determined from Eq. 15–28. Establish an inertial x, y coordinate system and apply Eq. 15–29 along the axes using the components of velocity at the open control surfaces, and the components of the forces shown on the free-body diagram. If the control volume is moving with a velocity, then apply the equation *relative to the control volume* so that the steady flow will then occur relative to the open control surfaces.

---

* In the SI system, a unit of pressure is measured using the pascal (Pa), where 1 Pa = 1 N/m^2.

EXAMPLE 15.16

Determine the components of force reaction which the fixed pipe joint at A exerts on the elbow in Fig. 15–29a, if water flowing through the pipe is subjected to a static gage pressure of 100 kPa at A. The volumetric flow at B is $Q_B = 0.2$ m³/s. Water has a density $\rho_w = 1000$ kg/m³, and the water-filled elbow has a mass of 20 kg and center of mass at G.

SOLUTION

We will consider the control volume to be the elbow and the water within it. Using a *fixed* inertial coordinate system, the velocity of flow at A and B and the mass flow can be obtained from Eq. 15–28. Since the density of water is constant, $Q_B = Q_A = Q$. Hence,

$$\dot{m} = \rho_w Q = (1000 \text{ kg/m}^3)(0.2 \text{ m}^3/\text{s}) = 200 \text{ kg/s}$$

$$v_B = \frac{Q}{A_B} = \frac{0.2 \text{ m}^3/\text{s}}{\pi(0.05 \text{ m})^2} = 25.46 \text{ m/s} \downarrow$$

$$v_A = \frac{Q}{A_A} = \frac{0.2 \text{ m}^3/\text{s}}{\pi(0.1 \text{ m})^2} = 6.37 \text{ m/s} \rightarrow$$

Free-Body Diagram. As shown on the free-body diagram of the control volume (elbow) Fig. 15–29b, the *fixed* connection at A exerts a resultant couple moment $\mathbf{M}_O$ and force components $\mathbf{F}_x$ and $\mathbf{F}_y$ on the elbow. Due to the static pressure of water in the pipe, the pressure force acting on the open control surface at A is $F_A = p_A A_A$. Since 1 kPa = 1000 N/m²,

$$F_A = p_A A_A = [100(10^3) \text{ N/m}^2][\pi(0.1 \text{ m})^2] = 3141.6 \text{ N}$$

There is no static pressure acting at B, since the water is discharged at atmospheric pressure; i.e., the pressure measured by a gage at B is equal to zero, $p_B = 0$.

Fig. 15–29

Equations of Steady Flow.

$$\xrightarrow{+} \Sigma F_x = \dot{m}(v_{Bx} - v_{Ax}); \quad -F_x + 3141.6 \text{ N} = 200 \text{ kg/s}(0 - 6.37 \text{ m/s})$$

$$F_x = 4.41 \text{ kN} \qquad Ans.$$

$$+\uparrow \Sigma F_y = \dot{m}(v_{By} - v_{Ay}); \quad -F_y - 20(9.81) \text{ N} = 200 \text{ kg/s}(-25.46 \text{ m/s} - 0)$$

$$F_y = 4.90 \text{ kN} \qquad Ans.$$

NOTE: The couple moment $\mathbf{M}_O$ can be determined using an angular impulse and momentum equation, which is discussed in fluid mechanics.

EXAMPLE 15.17

A 2-in.-diameter water jet having a velocity of 25 ft/s impinges upon a single moving blade, Fig. 15–30a. If the blade moves with a constant velocity of 5 ft/s away from the jet, determine the horizontal and vertical components of force that the blade exerts on the water. What power does the water generate on the blade? Water has a specific weight of $\gamma_w = 62.4 \text{ lb/ft}^3$.

SOLUTION

Kinematic Diagram. Here the control volume will be the stream of water on the blade. The inertial coordinate system is moving at 5 ft/s, so that the flow onto the blade appears as steady flow, Fig. 15–30b. The rate at which water enters the control volume at A is, therefore

$$\mathbf{v}_w - \mathbf{v}_{cv} = 25\mathbf{i} - 5\mathbf{i} = \{20\mathbf{i}\} \text{ ft/s}$$

Also, the velocity of flow at B measured from the control volume is

$$\mathbf{v}_B = \{20\mathbf{j}\} \text{ ft/s}$$

The mass flow of water through the control volume is

$$\dot{m} = \rho_w(v_{w/cv})A_A = \left(\frac{62.4}{32.2}\right)(20)\left[\pi\left(\frac{1}{12}\right)^2\right] = 0.8456 \text{ slug/s}$$

Fig. 15–30

Free-Body Diagram. The free-body diagram of the control volume is shown in Fig. 15–30c. The weight of the water will be neglected in the calculation, since this force will be small compared to the reactive components $\mathbf{F}_x$ and $\mathbf{F}_y$.

Equations of Steady Flow.

$$\Sigma \mathbf{F} = \dot{m}(\mathbf{v}_B - \mathbf{v}_A)$$

$$-F_x\mathbf{i} + F_y\mathbf{j} = 0.8456(20\mathbf{j} - 20\mathbf{i})$$

Equating the respective $\mathbf{i}$ and $\mathbf{j}$ components gives

$$F_x = 0.8456(20) = 16.9 \text{ lb} \leftarrow \qquad \textit{Ans.}$$

$$F_y = 0.8456(20) = 16.9 \text{ lb} \uparrow \qquad \textit{Ans.}$$

The water exerts equal but opposite forces on the blade.

Since the water force that causes the blade to move forward horizontally with a velocity of 5 ft/s is $F_x = 16.9$ lb, then from Eq. 14–10 the power is

$$P = \mathbf{F} \cdot \mathbf{v}; \qquad P = \frac{16.9 \text{ lb}(5 \text{ ft/s})}{550 \text{ hp}/(\text{ft} \cdot \text{lb/s})} = 0.154 \text{ hp} \qquad \textit{Ans.}$$

15.10 BODIES THAT LOSE OR GAIN MASS

In this section we will apply Eq. 15–27 to cases where a body will lose or gain mass. Typical examples include an open ended balloon or a moving shovel. In both of these cases it is important to recognize that the loss or gain of mass will produce a force or impulse on the inner boundary of the control volume. Whatever the case, we will consider the control volume to be the body and its contents, which at a given instant is moving with a velocity $\mathbf{v}_{cv}$. Here the time rate of change of the momentum *within* the control volume must now account for the changes in both its mass and velocity, that is, $(d(mv)/dt)_{cv} = (\dot{m}\mathbf{v})_{cv} + (m\dot{\mathbf{v}})_{cv}$. Substituting into Eq. 15–27, we have

$$\Sigma \mathbf{F} = (\dot{m}\mathbf{v})_{cv} + (m\dot{\mathbf{v}})_{cv} + \dot{m}_{out}\mathbf{v}_{out} - \dot{m}_{in}\mathbf{v}_{in} \quad (15\text{–}30)$$

To illustrate an application, consider a balloon (control volume) which loses mass (air) as it moves upward with a velocity $\mathbf{v}$, Fig. 15–31a. The last term in the above equation is zero because the air is self contained within the balloon. If the rate at which the air is expelled is $\dot{m}_{out} = \dot{m}_e$, then there is a corresponding *loss* of mass within the control volume, so $\dot{m}_{cv} = -\dot{m}_e$. Also, if the velocity of the ejected air is $(\mathbf{v}_e)_{rel}$, as measured relative to the balloon, then the velocity of this air as seen by a fixed observer is $\mathbf{v}_{out} = \mathbf{v} - (\mathbf{v}_e)_{rel}$. Substituting these terms into the above equation and simplifying, we obtain

$$\Sigma \mathbf{F} = (m\dot{\mathbf{v}})_{cv} - \dot{m}_e(\mathbf{v}_e)_{rel} \quad (15\text{–}31)$$

As shown on the free-body diagram for the balloon, Fig. 15–31b, the only forces acting are its weight $\mathbf{W}$ and the atmospheric drag force $\mathbf{F}_D$. The last term on the right of Eq. 15–31 represents the thrust which the exhaust exerts on the balloon and propels it upwards, giving it an acceleration $\dot{\mathbf{v}}$.

In a similar manner, Eq. 15–30 can also be applied to the scoop in Fig. 15–32, which is moving to the right with a velocity $\mathbf{v}$ so that mass is being injected at the rate of $(\dot{m})_{cv} = \dot{m}_i$. Here, $\dot{m}_{out} = 0$ and $\dot{m}_{in} = \dot{m}_i$. If the relative velocity of the injected mass is $(\mathbf{v}_i)_{rel}$, then $\mathbf{v}_{in} = \mathbf{v} + (\mathbf{v}_i)_{rel}$, and Eq. 15–30 now becomes

$$\Sigma \mathbf{F} = (m\dot{\mathbf{v}})_{cv} - \dot{m}_i(\mathbf{v}_i)_{rel} \quad (15\text{–}32)$$

The last term represents the magnitude of the force that the injected mass exerts on the scoop.

As noted by these two examples, when applying Eq. 15–30, whatever the situation, it is important to first define the control volume and then draw its free-body diagram in order to identify the forces exerted on the body and identify the thrust or retardation caused by the fluid or particle stream.

(a)   (b)

Fig. 15–31

Fig. 15–32

The scraper box behind this tractor gains mass at the rate of $\dot{m}$. If the tractor maintains a constant velocity v, then $\dot{v} = 0$ and, because the soil is originally at rest, $(v_i)_{rel} = v$. Applying Eq. 15–32, the horizontal towing force on the scraper box is then $T = 0 + \dot{m}v$.

EXAMPLE 15.18

(© NASA)

Fig. 15–33

The initial combined mass of a rocket and its fuel is m_0. The mass m_f of fuel within the rocket is expelled at a constant rate of $\dot{m}_e = c$ and at a constant speed of u measured relative to the rocket. Determine the maximum velocity of the rocket, i.e., at the instant the fuel runs out. Neglect the change in the rocket's weight with altitude and the drag resistance of the air. The rocket is fired vertically from rest.

SOLUTION
Since the rocket loses mass as it moves upward, $\dot{m}_{out} = -c$. Also, $\dot{m}_{in} = 0$, and Eq. 15–31 applies. The only *external force* acting on the *control volume* (rocket) is its weight $\mathbf{W}$, Fig. 15–33. Hence,

$$+\uparrow \Sigma F = (m\dot{v})_{cv} - \dot{m}_e(v_e)_{rel}; \qquad -W = m\dot{v} - cu \qquad (1)$$

The rocket's velocity is obtained by integrating this equation.

At any given instant t during the flight, the mass of the rocket is $m = m_0 - (\dot{m}_e)t = m_0 - ct$. Since $W = mg$, Eq. 1 then becomes

$$-(m_0 - ct)g = (m_0 - ct)\frac{dv}{dt} - cu$$

Separating the variables and integrating, realizing that $v = 0$ when $t = 0$, we have

$$\int_0^v dv = \int_0^t \left(\frac{cu}{m_0 - ct} - g\right) dt$$

$$v = -u \ln(m_0 - ct) - gt \Big|_0^t = u \ln\left(\frac{m_0}{m_0 - ct}\right) - gt \qquad (2)$$

Note that liftoff requires the first term on the right to be greater than the second term during the initial phase of motion. To find the time t' needed to consume all the fuel, we have

$$m_f = ct'$$

or

$$t' = m_f/c$$

Substituting into Eq. 2 yields

$$v_{max} = u \ln\left(\frac{m_0}{m_0 - m_f}\right) - \frac{gm_f}{c} \qquad \textit{Ans.}$$

EXAMPLE 15.19

A chain of length l, Fig. 15-34a, has a mass m. Determine the magnitude of force **F** required to (a) raise the chain with a constant speed v_c, starting from rest when $y = 0$; and (b) lower the chain with a constant speed v_c, starting from rest when $y = l$.

SOLUTION

Part (a). As the chain is raised, all the suspended links are given a sudden downward impulse by each added link which is lifted off the ground. Thus, the *suspended portion* of the chain may be considered as a device which is *gaining mass*. The control volume to be considered is the length of chain y which is suspended by **F** at any instant, including the next link which is about to be added but is still at rest, Fig. 15-34b. The forces acting on the control volume *exclude* the internal forces **P** and −**P**, which act between the added link and the suspended portion of the chain. Hence, $\Sigma F = F - mg(y/l)$.

Since v_c is constant, $y = v_c t$. Thus, the mass of the control volume at any instant is $m = m(y/l) = m(v_c t/l)$, and therefore the *rate* at which mass is *added* to the control volume is

$$\dot{m}_i = m\left(\frac{v_c}{l}\right)$$

Applying Eq. 15-32, we have

$$+\uparrow \Sigma F = (m\dot{v})_{cv} + \dot{m}_i(v_i)_{rel}$$

$$F - mg\left(\frac{y}{l}\right) = 0 + m\left(\frac{v_c}{l}\right)v_c$$

or

$$F = (m/l)(gy + v_c^2) \qquad \textit{Ans.}$$

Part (b). When the chain is being lowered at v_c the (loose) links which are expelled (given zero velocity) *do not* impart an impulse or a force on the *remaining* suspended links, that is, they do not create an impulse or force on the control volume, instead, they simply "fall off." And so the control volume in Part (a) will not be considered. Rather, we can apply the equation of motion to obtain the solution. At time t the length of chain still off the floor is y. Using the free-body diagram for a suspended portion of the chain shown in Fig. 15-34c, we have

$$+\uparrow \Sigma F = ma; \qquad F - mg\left(\frac{y}{l}\right) = 0$$

$$F = mg\left(\frac{y}{l}\right) \qquad \textit{Ans.}$$

(a)

(b)

(c)

Fig. 15-34

PROBLEMS

15–114. Water is discharged at 16 m/s against the fixed cone diffuser. If the opening diameter of the nozzle is 40 mm, determine the horizontal force exerted by the water on the diffuser. $\rho_W = 1\,\text{Mg/m}^3$.

*15–116. The 200-kg boat is powered by the fan which develops a slipstream having a diameter of 0.75 m. If the fan ejects air with a speed of 14 m/s, measured relative to the boat, determine the initial acceleration of the boat if it is initially at rest. Assume that air has a constant density of $\rho_a = 1.22\,\text{kg/m}^3$ and that the entering air is essentially at rest. Neglect the drag resistance of the water.

Prob. 15–114

Prob. 15–116

15–115. The chute is used to divert the flow of water, $Q = 0.6\,\text{m}^3/\text{s}$. If the water has a cross-sectional area of $0.05\,\text{m}^2$, determine the force components at the pin D and roller C necessary for equilibrium. Neglect the weight of the chute and weight of the water on the chute. $\rho_w = 1\,\text{Mg/m}^3$.

15–117. The elbow for a 5-in.-diameter buried pipe is subjected to a static pressure of 10 lb/in². The speed of the water passing through it is $v = 8$ ft/s. Assuming the pipe connections at A and B do not offer any vertical force resistance on the elbow, determine the resultant vertical force F that the soil must then exert on the elbow in order to hold it in equilibrium. Neglect the weight of the elbow and the water within it. $\gamma_w = 62.4\,\text{lb/ft}^3$.

Prob. 15–115

Prob. 15–117

15–118. The car is used to scoop up water that is lying in a trough at the tracks. Determine the force needed to pull the car forward at constant velocity **v** for each of the three cases. The scoop has a cross-sectional area A and the density of water is ρ_w.

Prob. 15–118

15–119. The blade divides the jet of water having a diameter of 3 in. If one-fourth of the water flows downward while the other three-fourths flow upward, and the total flow is $Q = 0.5 \text{ ft}^3/\text{s}$, determine the horizontal and vertical components of force exerted on the blade by the jet, $\gamma_w = 62.4 \text{ lb/ft}^3$.

Prob. 15–119

*15–120. The gage pressure of water at A is 150.5 kPa. Water flows through the pipe at A with a velocity of 18 m/s, and out the pipe at B and C with the same velocity v. Determine the horizontal and vertical components of force exerted on the elbow necessary to hold the pipe assembly in equilibrium. Neglect the weight of water within the pipe and the weight of the pipe. The pipe has a diameter of 50 mm at A, and at B and C the diameter is 30 mm. $\rho_w = 1000 \text{ kg/m}^3$.

Prob. 15–120

15–121. The gage pressure of water at C is 40 lb/in². If water flows out of the pipe at A and B with velocities $v_A = 12$ ft/s and $v_B = 25$ ft/s, determine the horizontal and vertical components of force exerted on the elbow necessary to hold the pipe assembly in equilibrium. Neglect the weight of water within the pipe and the weight of the pipe. The pipe has a diameter of 0.75 in. at C, and at A and B the diameter is 0.5 in. $\gamma_w = 62.4 \text{ lb/ft}^3$.

Prob. 15–121

15–122. The fountain shoots water in the direction shown. If the water is discharged at 30° from the horizontal, and the cross-sectional area of the water stream is approximately 2 in², determine the force it exerts on the concrete wall at B. $\gamma_w = 62.4 \text{ lb/ft}^3$.

Prob. 15–122

15–123. A plow located on the front of a locomotive scoops up snow at the rate of 10 ft³/s and stores it in the train. If the locomotive is traveling at a constant speed of 12 ft/s, determine the resistance to motion caused by the shoveling. The specific weight of snow is $\gamma_s = 6 \text{ lb/ft}^3$.

*15–124. The boat has a mass of 180 kg and is traveling forward on a river with a constant velocity of 70 km/h, measured *relative* to the river. The river is flowing in the opposite direction at 5 km/h. If a tube is placed in the water, as shown, and it collects 40 kg of water in the boat in 80 s, determine the horizontal thrust T on the tube that is required to overcome the resistance due to the water collection and yet maintain the constant speed of the boat. $\rho_w = 1 \text{ Mg/m}^3$.

Prob. 15–124

15–125. Water is discharged from a nozzle with a velocity of 12 m/s and strikes the blade mounted on the 20-kg cart. Determine the tension developed in the cord, needed to hold the cart stationary, and the normal reaction of the wheels on the cart. The nozzle has a diameter of 50 mm and the density of water is $\rho_w = 1000 \text{ kg/m}^3$.

Prob. 15–125

15–126. The fan draws air through a vent with a speed of 12 ft/s. If the cross-sectional area of the vent is 2 ft², determine the horizontal thrust on the blade. The specific weight of the air is $\gamma_a = 0.076 \text{ lb/ft}^3$.

Prob. 15–126

15–127. The fan blows air at 6000 ft³/min. If the fan has a weight of 30 lb and a center of gravity at G, determine the smallest diameter d of its base so that it will not tip over. The specific weight of air is $\gamma_a = 0.076 \text{ lb/ft}^3$.

Prob. 15–127

*15–128. The nozzle has a diameter of 40 mm. If it discharges water uniformly with a downward velocity of 20 m/s against the fixed blade, determine the vertical force exerted by the water on the blade. $\rho_w = 1\,\text{Mg/m}^3$.

15–130. Sand drops onto the 2-Mg empty rail car at 50 kg/s from a conveyor belt. If the car is initially coasting at 4 m/s, determine the speed of the car as a function of time.

Prob. 15–128

Prob. 15–130

15–129. The toy sprinkler for children consists of a 0.2-kg cap and a hose that has a mass per length of 30 g/m. Determine the required rate of flow of water through the 5-mm-diameter tube so that the sprinkler will lift 1.5 m from the ground and hover from this position. Neglect the weight of the water in the tube. $\rho_w = 1\,\text{Mg/m}^3$.

15–131. The fireboat is equipped with two water cannons. The one at A has a nozzle diameter of 6 in. and can pump 3000 gal/min, and the one at B has a nozzle diameter of 3 in. and can pump 1000 gal/min. Determine the horizontal thrust T that the propeller must exert on the boat in order to hold the boat steady. Take $\gamma = 62.4\,\text{lb/ft}^3$, and there is 7.4805 gal/ft^3 of water.

Prob. 15–129

Prob. 15–131

*15–132. Aggregates fall off the conveyor belt at 500 kg/s and into the barge. Determine the force the rope at A exerts on the barge to hold it in place.

Prob. 15–132

15–133. The tractor together with the empty tank has a total mass of 4 Mg. The tank is filled with 2 Mg of water. The water is discharged at a constant rate of 50 kg/s with a constant velocity of 5 m/s, measured relative to the tractor. If the tractor starts from rest, and the rear wheels provide a resultant traction force of 250 N, determine the velocity and acceleration of the tractor at the instant the tank becomes empty.

Prob. 15–133

15–134. A rocket is in deep space, and on earth it has an empty weight of 500 lb and carries 300 lb of fuel. If the fuel is burned at the rate of 15 lb/s and ejected with a relative velocity of 4400 ft/s, determine the maximum speed attained by the rocket, starting from rest, while it is in deep space. Neglect the effect of gravitation on the rocket.

Prob. 15–134

15–135. A power lawn mower hovers very close over the ground. This is done by drawing air in at a speed of 6 m/s through an intake unit A, which has a cross-sectional area of $A_A = 0.25$ m^2, and then discharging it at the ground, B, where the cross-sectional area is $A_B = 0.25$ m^2. If air at A is subjected only to atmospheric pressure, determine the air pressure which the lawn mower exerts on the ground when the weight of the mower is freely supported and no load is placed on the handle. The mower has a mass of 15 kg with center of mass at G. Assume that air has a constant density of $\rho_a = 1.22$ kg/m^3.

Prob. 15–135

*15–136. The rocket car has an empty weight of 400 lb and is fired from rest along the horizontal tracks. There is 300 lb of fuel stored in the car, which is consumed at a constant rate of 15 lb/s and ejected with a relative velocity of 5000 ft/s. Determine the maximum speed attained by the car, starting from rest, if the frictional resistance due to the atmosphere is $F = (0.01v)$ lb, where v is the velocity in feet per second.

Prob. 15–136

15–137. The bin deposits gravel onto the conveyor belt at the rate of 1500 lb/min. If the speed of the belt is 5 ft/s, determine how much greater the tension in the top portion of the belt must be than that in the bottom portion in order to pull the belt forward.

Prob. 15–137

15–138. The second stage of a two-stage rocket weighs 2000 lb (empty) and is launched from the first stage with a velocity of 3000 mi/h. The fuel in the second stage weighs 1000 lb. If it is consumed at the rate of 50 lb/s and ejected with a relative velocity of 8000 ft/s, determine the acceleration of the second stage just after the engine is fired. What is the rocket's acceleration just before all the fuel is consumed? Neglect the effect of gravitation.

15–139. The missile weighs 40,000 lb. The constant thrust provided by the turbojet engine is $T = 15{,}000$ lb. Additional thrust is provided by *two* rocket boosters B. The propellant in each booster is burned at a constant rate of 150 lb/s, with a relative exhaust velocity of 3000 ft/s. If the mass of the propellant lost by the turbojet engine can be neglected, determine the velocity of the missile after the 4-s burn time of the boosters. The initial velocity of the missile is 300 mi/h.

Prob. 15–139

*15–140. The jet is traveling at a speed of 720 km/h. If the fuel is being spent at 0.8 kg/s, and the engine takes in air at 200 kg/s, whereas the exhaust gas (air and fuel) has a relative speed of 12,000 m/s, determine the acceleration of the plane at this instant. The drag resistance of the air is $F_D = (55\ v^2)$, where the speed is measured in m/s. The jet has a mass of 7 Mg.

Prob. 15–140

15–141. The 1.5-m-long smooth chain hangs over the edge of the table. If the chain is released from rest, determine its speed just after the last link clears the table. The chain has a mass per unit length of 6 kg/m.

Prob. 15–141

15–142. The hemispherical bowl of mass m is held in equilibrium by the vertical jet of water discharged through a nozzle of diameter d. If the discharge of the water through the nozzle is Q, determine the height h at which the bowl is suspended. The water density is ρ_w. Neglect the weight of the water jet.

Prob. 15–142

CHAPTER REVIEW

Impulse

An impulse is defined as the product of force and time. Graphically it represents the area under the F–t diagram. If the force is constant, then the impulse becomes $I = F_c(t_2 - t_1)$.

Principle of Impulse and Momentum

When the equation of motion, $\Sigma \mathbf{F} = m\mathbf{a}$, and the kinematic equation, $a = dv/dt$, are combined, we obtain the principle of impulse and momentum. This is a vector equation that can be resolved into rectangular components and used to solve problems that involve force, velocity, and time. For application, the free-body diagram should be drawn in order to account for all the impulses that act on the particle.

$$m\mathbf{v}_1 + \Sigma \int_{t_1}^{t_2} \mathbf{F}\, dt = m\mathbf{v}_2$$

Conservation of Linear Momentum

If the principle of impulse and momentum is applied to a *system of particles*, then the collisions between the particles produce internal impulses that are equal, opposite, and collinear, and therefore cancel from the equation. Furthermore, if an external impulse is small, that is, the force is small and the time is short, then the impulse can be classified as nonimpulsive and can be neglected. Consequently, momentum for the system of particles is conserved.

$$\Sigma m_i(\mathbf{v}_i)_1 = \Sigma m_i(\mathbf{v}_i)_2$$

The conservation-of-momentum equation is useful for finding the final velocity of a particle when internal impulses are exerted between two particles and the initial velocities of the particles are known. If the internal impulse is to be determined, then one of the particles is isolated and the principle of impulse and momentum is applied to this particle.

Impact

When two particles A and B have a direct impact, the internal impulse between them is equal, opposite, and collinear. Consequently the conservation of momentum for this system applies along the line of impact.

$$m_A(v_A)_1 + m_B(v_B)_1 = m_A(v_A)_2 + m_B(v_B)_2$$

If the final velocities are unknown, a second equation is needed for solution. We must use the coefficient of restitution, e. This experimentally determined coefficient depends upon the physical properties of the colliding particles. It can be expressed as the ratio of their relative velocity after collision to their relative velocity before collision. If the collision is elastic, no energy is lost and $e = 1$. For a plastic collision $e = 0$.	$$e = \frac{(v_B)_2 - (v_A)_2}{(v_A)_1 - (v_B)_1}$$
If the impact is oblique, then the conservation of momentum for the system and the coefficient-of-restitution equation apply along the line of impact. Also, conservation of momentum for each particle applies perpendicular to this line (plane of contact) because no impulse acts on the particles in this direction.	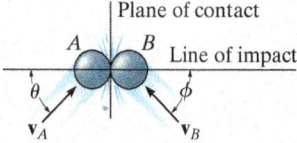

Principle of Angular Impulse and Momentum

The moment of the linear momentum about an axis (z) is called the angular momentum. The principle of angular impulse and momentum is often used to eliminate unknown impulses by summing the moments about an axis through which the lines of action of these impulses produce no moment. For this reason, a free-body diagram should accompany the solution.	$(H_O)_z = (d)(mv)$ $(H_O)_1 + \Sigma \int_{t_1}^{t_2} M_O \, dt = (H_O)_2$

Steady Fluid Streams

Impulse-and-momentum methods are often used to determine the forces that a device exerts on the mass flow of a fluid—liquid or gas. To do so, a free-body diagram of the fluid mass in contact with the device is drawn in order to identify these forces. Also, the velocity of the fluid as it flows into and out of a control volume for the device is calculated. The equations of steady flow involve summing the forces and the moments to determine these reactions.	$$\Sigma \mathbf{F} = \frac{dm}{dt}(\mathbf{v}_B - \mathbf{v}_A)$$ $$\Sigma \mathbf{M}_O = \frac{dm}{dt}(\mathbf{r}_B \times \mathbf{v}_B - \mathbf{r}_A \times \mathbf{v}_A)$$

Propulsion with Variable Mass

Some devices, such as a rocket, lose mass as they are propelled forward. Others gain mass, such as a shovel. We can account for this mass loss or gain by applying the principle of impulse and momentum to a control volume for the device. From this equation, the force exerted on the device by the mass flow can then be determined.	$\Sigma F_{cv} = m\dfrac{dv}{dt} - v_{D/e}\dfrac{dm_e}{dt}$ Loses Mass $\Sigma F_{cv} = m\dfrac{dv}{dt} + v_{D/i}\dfrac{dm_i}{dt}$ Gains Mass

FUNDAMENTAL REVIEW PROBLEMS

R15–1. Packages having a mass of 6 kg slide down a smooth chute and land horizontally with a speed of 3 m/s on the surface of a conveyor belt. If the coefficient of kinetic friction between the belt and a package is $\mu_k = 0.2$, determine the time needed to bring the package to rest on the belt if the belt is moving in the same direction as the package with a speed $v = 1$ m/s.

R15–3. A 20-kg block is originally at rest on a horizontal surface for which the coefficient of static friction is $\mu_s = 0.6$ and the coefficient of kinetic friction is $\mu_k = 0.5$. If a horizontal force F is applied such that it varies with time as shown, determine the speed of the block in 10 s. *Hint:* First determine the time needed to overcome friction and start the block moving.

Prob. R15–1

Prob. R15–3

R15–2. The 50-kg block is hoisted up the incline using the cable and motor arrangement shown. The coefficient of kinetic friction between the block and the surface is $\mu_k = 0.4$. If the block is initially moving up the plane at $v_0 = 2$ m/s, and at this instant ($t = 0$) the motor develops a tension in the cord of $T = (300 + 120\sqrt{t})$ N, where t is in seconds, determine the velocity of the block when $t = 2$ s.

R15–4. The three freight cars A, B, and C have masses of 10 Mg, 5 Mg, and 20 Mg, respectively. They are traveling along the track with the velocities shown. Car A collides with car B first, followed by car C. If the three cars couple together after collision, determine the common velocity of the cars after the two collisions have taken place.

Prob. R15–2

Prob. R15–4

R15–5. The 200-g projectile is fired with a velocity of 900 m/s towards the center of the 15-kg wooden block, which rests on a rough surface. If the projectile penetrates and emerges from the block with a velocity of 300 m/s, determine the velocity of the block just after the projectile emerges. How long does the block slide on the rough surface, after the projectile emerges, before it comes to rest again? The coefficient of kinetic friction between the surface and the block is $\mu_k = 0.2$.

R15–7. Two smooth billiard balls A and B have an equal mass of $m = 200$ g. If A strikes B with a velocity of $(v_A)_1 = 2$ m/s as shown, determine their final velocities just after collision. Ball B is originally at rest and the coefficient of restitution is $e = 0.75$.

Prob. R15–5

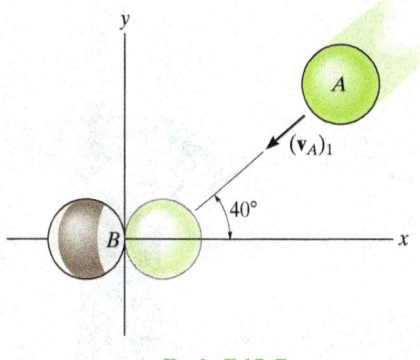

Prob. R15–7

R15–8. The small cylinder C has a mass of 10 kg and is attached to the end of a rod whose mass may be neglected. If the frame is subjected to a couple $M = (8t^2 + 5)$ N·m, where t is in seconds, and the cylinder is subjected to a force of 60 N, which is always directed as shown, determine the speed of the cylinder when $t = 2$ s. The cylinder has a speed $v_0 = 2$ m/s when $t = 0$.

R15–6. Block A has a mass of 3 kg and is sliding on a rough horizontal surface with a velocity $(v_A)_1 = 2$ m/s when it makes a direct collision with block B, which has a mass of 2 kg and is originally at rest. If the collision is perfectly elastic ($e = 1$), determine the velocity of each block just after collision and the distance between the blocks when they stop sliding. The coefficient of kinetic friction between the blocks and the plane is $\mu_k = 0.3$.

Prob. R15–6

Prob. R15–8

CHAPTER 16

Kinematics is important for the design of gear trains used in different mechanisms

PLANAR KINEMATICS OF A RIGID BODY

CHAPTER OBJECTIVES

- To classify the various types of rigid-body planar motion.
- To investigate rigid-body translation and angular motion about a fixed axis.
- To study planar motion using an absolute motion analysis.
- To provide a relative-motion analysis of velocity and acceleration using a translating frame of reference.
- To show how to find the instantaneous center of zero velocity and determine the velocity of a point on a body using this method.
- To provide a relative-motion analysis of velocity and acceleration using a rotating frame of reference.

16.1 PLANAR RIGID-BODY MOTION

In this chapter, the planar kinematics of a rigid body will be discussed. This study is important for the design of gears, cams, and mechanisms used for many structural and mechanical operations. Once the kinematics is thoroughly understood, then we can apply the equations of motion, which relate the forces on the body to the body's motion.

The *planar motion* of a body occurs when all the particles of a rigid body move along paths which are equidistant from a fixed plane. There are three types of rigid-body planar motion. In order of increasing complexity, they are

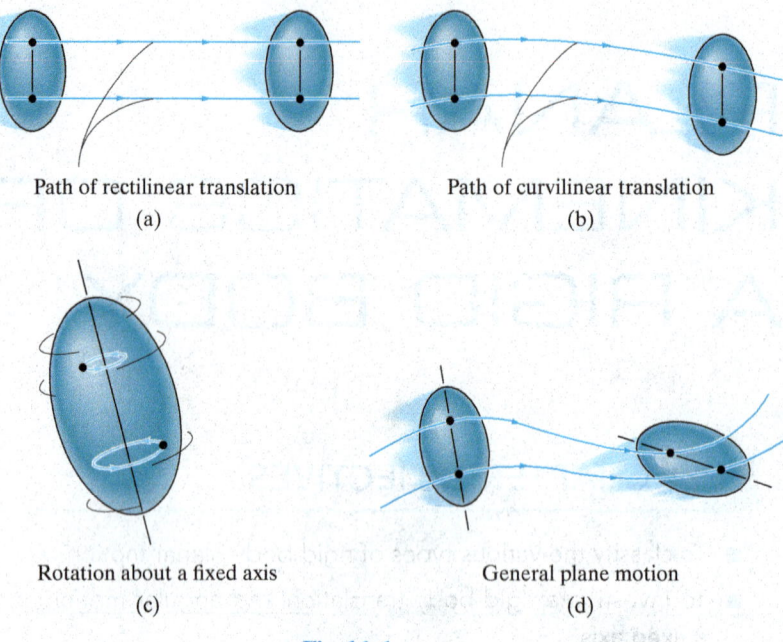

Path of rectilinear translation
(a)

Path of curvilinear translation
(b)

Rotation about a fixed axis
(c)

General plane motion
(d)

Fig. 16–1

- *Translation.* This type of motion occurs when a line in the body remains parallel to its original orientation throughout the motion. When the paths of motion for any two points on the body are parallel straight lines, the motion is called **rectilinear translation**, Fig. 16–1a. If the paths of motion are along curved lines, the motion is called **curvilinear translation**, Fig. 16–1b.

- *Rotation about a fixed axis.* When a rigid body rotates about a fixed axis, all the particles of the body, except those which lie on the axis of rotation, move along circular paths, Fig. 16–1c.

- *General plane motion.* When a body is subjected to general plane motion, it undergoes a simultaneous translation *and* rotation, Fig. 16–1d. The translation occurs within a reference plane, and the rotation occurs about an axis perpendicular to the reference plane.

In the following sections we will consider each of these motions in detail. Examples of bodies undergoing each type of motion are shown in Fig. 16–2.

Rectilinear translation

General plane motion

Curvilinear translation

Rotation about a fixed axis

Fig. 16–2

16.2 TRANSLATION

Consider a rigid body which is subjected to either rectilinear or curvilinear translation in the *x*–*y* plane, Fig. 16–3.

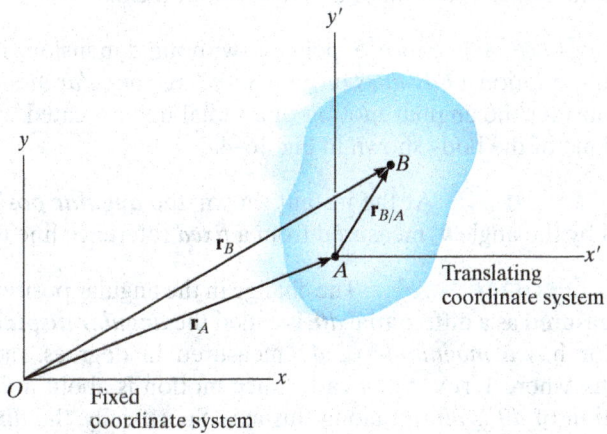

Fig. 16–3

Position. The locations of points *A* and *B* on the body are defined with respect to the fixed *x*, *y* reference frame using position vectors $\mathbf{r}_A$ and $\mathbf{r}_B$. The *translating* x', y' coordinate system is fixed in the body and has its origin at *A*, hereafter referred to as the **base point**. The position of *B* with respect to *A* is denoted by the relative-position vector $\mathbf{r}_{B/A}$ ("**r** of *B* with respect to *A*"). By vector addition,

$$\mathbf{r}_B = \mathbf{r}_A + \mathbf{r}_{B/A}$$

Velocity. A relation between the instantaneous velocities of *A* and *B* is obtained by taking the time derivative of this equation, which yields $\mathbf{v}_B = \mathbf{v}_A + d\mathbf{r}_{B/A}/dt$. Here $\mathbf{v}_A$ and $\mathbf{v}_B$ denote absolute velocities since they are measured with respect to the *x*, *y* axes. The term $d\mathbf{r}_{B/A}/dt = \mathbf{0}$, since the magnitude of $\mathbf{r}_{B/A}$ is constant by definition of a rigid body, and because the body is translating, the direction of $\mathbf{r}_{B/A}$ is also constant. Therefore,

$$\mathbf{v}_B = \mathbf{v}_A$$

Acceleration. Taking the time derivative of the velocity equation yields a similar relationship between the instantaneous accelerations of *A* and *B*:

$$\mathbf{a}_B = \mathbf{a}_A$$

The above two equations indicate that *all points in a rigid body subjected to either rectilinear or curvilinear translation move with the same velocity and acceleration*. As a result, the kinematics of points located in a translating rigid body can be studied using the kinematics of particle motion studied in Chapter 12.

Riders on this Ferris wheel are subjected to curvilinear translation, since the gondolas move in a circular path, yet they always remain in the upright position.

16.3 ROTATION ABOUT A FIXED AXIS

When a body rotates about a fixed axis, any point P located in the body travels along a circular path. To study the motion of P it is first necessary to discuss the angular motion of the body about the axis.

Angular Motion. Since a point is without dimension, it cannot have angular motion. *Only lines or bodies undergo angular motion.* Here we will consider the angular motion of a radial line r located within the shaded plane of the body shown in Fig. 16–4a.

Angular Position. At the instant shown, the **angular position** of r is defined by the angle θ, measured from a *fixed* reference line to r.

Angular Displacement. The change in the angular position, which can be measured as a differential $d\theta$, is called the **angular displacement**.* This vector has a *magnitude* of $d\theta$ measured in degrees, radians, or revolutions, where 1 rev = 2π rad. Since motion is about a *fixed axis*, the direction of $d\boldsymbol{\theta}$ is *always* along this axis. Specifically, the *direction* is determined by the right-hand rule; that is, the fingers of the right hand are curled with the sense of rotation, so that in this case the thumb, or $d\boldsymbol{\theta}$, points upward, Fig. 16–4a. In two dimensions, as shown by the top view of the shaded plane, Fig. 16–4b, both $\boldsymbol{\theta}$ and $d\boldsymbol{\theta}$ are counterclockwise, and so the thumb points outward from the page.

Angular Velocity. The time rate of change in the angular position is called the **angular velocity** ω (omega). Since $d\theta$ occurs during an instant of time dt, and maintains its fixed position, then

$$(\curvearrowleft+) \qquad \omega = \frac{d\theta}{dt} \qquad (16\text{–}1)$$

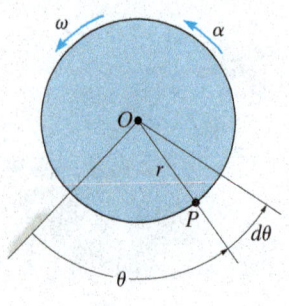

(a)

(b)

Fig. 16–4

The vector $\boldsymbol{\omega}$ has a *magnitude* which is often measured in rad/s. Its *direction* is always along the axis of rotation, Fig. 16–4a. When indicating the angular motion in the shaded plane, Fig. 16–4b, we can refer to the sense of rotation as clockwise or counterclockwise. Here we have *arbitrarily* chosen counterclockwise rotations as *positive* and indicated this by the curl shown in parentheses next to Eq. 16–1. Realize, however, that the directional sense of $\boldsymbol{\omega}$ is actually outward from the page.

*It is shown in Sec. 20.1 that finite rotations or finite angular displacements are *not* vector quantities, although differential rotations $d\boldsymbol{\theta}$ are vectors.

16.3 ROTATION ABOUT A FIXED AXIS

Angular Acceleration. The *angular acceleration* α (alpha) measures the time rate of change of the angular velocity. The *magnitude* of this vector is

($\downarrow +$) $$\alpha = \frac{d\omega}{dt} \quad (16\text{-}2)$$

Using Eq. 16–1, it is also possible to express α as

($\downarrow +$) $$\alpha = \frac{d^2\theta}{dt^2} \quad (16\text{-}3)$$

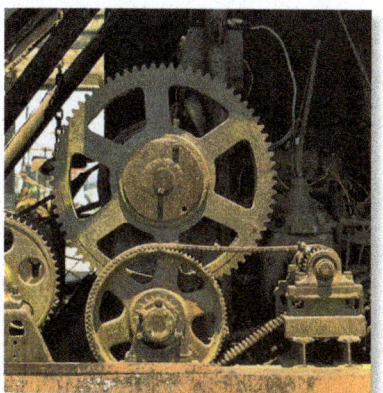

The gears used in the operation of a crane all rotate about fixed axes. Engineers must be able to relate their angular motions in order to properly design this gear system.

The line of action of $\boldsymbol{\alpha}$ is the same as that for $\boldsymbol{\omega}$, Fig. 16–4a; however, its sense of *direction* depends on whether $\boldsymbol{\omega}$ is increasing or decreasing. If $\boldsymbol{\omega}$ is decreasing, then $\boldsymbol{\alpha}$ is called an *angular deceleration*, and therefore it has a sense of direction which is opposite to $\boldsymbol{\omega}$.

By eliminating dt from Eqs. 16–1 and 16–2, we obtain a differential relation between the angular acceleration, angular velocity, and angular displacement, namely,

($\downarrow +$) $$\alpha \, d\theta = \omega \, d\omega \quad (16\text{-}4)$$

The similarity between the differential relations for angular motion and those developed for rectilinear motion of a particle ($v = ds/dt$, $a = dv/dt$, and $a \, ds = v \, dv$) should be apparent.

Constant Angular Acceleration. If the angular acceleration of the body is *constant*, $\boldsymbol{\alpha} = \boldsymbol{\alpha}_c$, then Eqs. 16–1, 16–2, and 16–4, when integrated, yield the following set of formulas.

($\downarrow +$) $$\omega = \omega_0 + \alpha_c t \quad (16\text{-}5)$$
($\downarrow +$) $$\theta = \theta_0 + \omega_0 t + \tfrac{1}{2}\alpha_c t^2 \quad (16\text{-}6)$$
($\downarrow +$) $$\omega^2 = \omega_0^2 + 2\alpha_c(\theta - \theta_0) \quad (16\text{-}7)$$
Constant Angular Acceleration

Here θ_0 and ω_0 are the initial values of the body's angular position and angular velocity, respectively. Note that these equations are similar to Eqs. 12–4 to 12–6 used for rectilinear motion.

(c)

Fig. 16–4 (cont.)

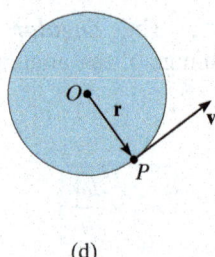

(d)

Motion of Point P. As the rigid body in Fig. 16–4c rotates, point P travels along a *circular path* of radius r with center at point O. This path is contained within the shaded plane shown in top view, Fig. 16–4d.

Position and Displacement. The position of P is defined by the position vector $\mathbf{r}$, which extends from O to P. If the body rotates $d\theta$ then P will displace $ds = r\,d\theta$.

Velocity. The velocity of P has a magnitude which can be found by dividing $ds = r\,d\theta$ by dt. This yields

$$v = \omega r \qquad (16\text{–}8)$$

As shown in Figs. 16–4c and 16–4d, the *direction* of $\mathbf{v}$ is *tangent* to the circular path.

Both the magnitude and direction of $\mathbf{v}$ can also be accounted for by using the cross product of $\boldsymbol{\omega}$ and $\mathbf{r}_P$ (see Appendix B). Here, $\mathbf{r}_P$ is directed from *any point* on the axis of rotation to point P, Fig. 16–4c. We have

$$\mathbf{v} = \boldsymbol{\omega} \times \mathbf{r}_P \qquad (16\text{–}9)$$

The order of the vectors in this formulation is important, since the cross product is not commutative, i.e., $\boldsymbol{\omega} \times \mathbf{r}_P \neq \mathbf{r}_P \times \boldsymbol{\omega}$. Notice in Fig. 16–4c how the correct direction of $\mathbf{v}$ is established by the right-hand rule. The fingers of the right hand are curled from $\boldsymbol{\omega}$ toward $\mathbf{r}_P$ ($\boldsymbol{\omega}$ "cross" $\mathbf{r}_P$). The thumb indicates the correct direction of $\mathbf{v}$, which is tangent to the path in the direction of motion. From Eq. B–8, the magnitude of $\mathbf{v}$ in Eq. 16–9 is $v = \omega r_P \sin\phi$, and since $r = r_P \sin\phi$, Fig. 16–4c, then $v = \omega r$, which agrees with Eq. 16–8. As a special case, the position vector $\mathbf{r}$ can be chosen for $\mathbf{r}_P$. Here $\mathbf{r}$ lies in the plane of motion and again the velocity of point P is

$$\mathbf{v} = \boldsymbol{\omega} \times \mathbf{r} \qquad (16\text{–}10)$$

Acceleration. The acceleration of P can be expressed in terms of its normal and tangential components. For the tangential component, applying Eq. 12–19 and Eq. 16–8, we have $a_t = dv/dt = d(\omega r)/dt = (d\omega/dt)r$, and since $\alpha = d\omega/dt$, we get

$$a_t = \alpha r \qquad (16\text{–}11)$$

The normal component is determined from Eq. 12–20 and Eq. 16–8. We have $a_n = v^2/\rho = (\omega r)^2/r$ or

$$a_n = \omega^2 r \qquad (16\text{–}12)$$

The *tangential component of acceleration*, Figs. 16–4e and 16–4f, represents the time rate of change in the velocity's magnitude. If the speed of P is increasing, then $\mathbf{a}_t$ acts in the same direction as $\mathbf{v}$; if the speed is decreasing, $\mathbf{a}_t$ acts in the opposite direction of $\mathbf{v}$; and finally, if the speed is constant, $\mathbf{a}_t$ is zero.

The *normal component of acceleration* represents the time rate of change in the velocity's direction. The *direction* of $\mathbf{a}_n$ is always toward O, the center of the circular path, Figs. 16–4e and 16–4f.

Like the velocity, the acceleration of point P can be expressed in terms of the vector cross product. Taking the time derivative of Eq. 16–9 we have

$$\mathbf{a} = \frac{d\mathbf{v}}{dt} = \frac{d\boldsymbol{\omega}}{dt} \times \mathbf{r}_P + \boldsymbol{\omega} \times \frac{d\mathbf{r}_P}{dt}$$

Since $\boldsymbol{\alpha} = d\boldsymbol{\omega}/dt$, and using Eq. 16–9 ($d\mathbf{r}_P/dt = \mathbf{v} = \boldsymbol{\omega} \times \mathbf{r}_P$), then

$$\mathbf{a} = \boldsymbol{\alpha} \times \mathbf{r}_P + \boldsymbol{\omega} \times (\boldsymbol{\omega} \times \mathbf{r}_P) \qquad (16\text{–}13)$$

From the definition of the cross product, the first term on the right has a magnitude $a_t = \alpha r_P \sin\phi = \alpha r$, and by the right-hand rule, $\boldsymbol{\alpha} \times \mathbf{r}_P$ is in the direction of $\mathbf{a}_t$, Fig. 16–4e. Likewise, the second term has a magnitude $a_n = \omega^2 r_P \sin\phi = \omega^2 r$, and applying the right-hand rule twice, first to determine the result $\mathbf{v}_P = \boldsymbol{\omega} \times \mathbf{r}_P$ then $\boldsymbol{\omega} \times \mathbf{v}_P$, it can be seen that this result is in the same direction as $\mathbf{a}_n$, shown in Fig. 16–4e. Noting that this is also the *same* direction as $-\mathbf{r}$, which lies in the plane of motion, we can express $\mathbf{a}_n$ in a much simpler form as $\mathbf{a}_n = -\omega^2 \mathbf{r}$. Hence, Eq. 16–13 can be identified by its two components as

$$\mathbf{a} = \mathbf{a}_t + \mathbf{a}_n$$
$$= \boldsymbol{\alpha} \times \mathbf{r} - \omega^2 \mathbf{r} \qquad (16\text{–}14)$$

Since $\mathbf{a}_t$ and $\mathbf{a}_n$ are perpendicular to one another, if needed the magnitude of acceleration can be determined from the Pythagorean theorem; namely, $a = \sqrt{a_n^2 + a_t^2}$, Fig. 16–4f.

(e)

(f)

If two rotating bodies are in contact, as in Fig. 16–5a, then the *points in contact A and A′ move along different circular paths*, and the velocity and *tangential components* of acceleration of the points will be the *same*, $\mathbf{v}_A = \mathbf{v}_{A'}$, Fig. 16–5b, and $(\mathbf{a}_A)_t = (\mathbf{a}_{A'})_t$, Fig. 16–5c. However, the *normal components* of acceleration will *not* be the same, $(\mathbf{a}_A)_n \neq (\mathbf{a}_{A'})_n$, and therefore, as shown in Fig. 16–5c, $\mathbf{a}_A \neq \mathbf{a}_{A'}$.

(a)

(b)

(c)

Fig. 16–5

IMPORTANT POINTS

- A body can undergo two types of translation. During rectilinear translation all points follow parallel straight-line paths, and during curvilinear translation the points follow curved paths that are the same shape.

- All the points on a translating body move with the same velocity and acceleration.

- Points located on a body that rotates about a fixed axis follow circular paths.

- The relation $\alpha\, d\theta = \omega\, d\omega$ is derived from $\alpha = d\omega/dt$ and $\omega = d\theta/dt$ by eliminating dt.

- Once the angular motions ω and α are known, the velocity and acceleration of any point on the body can be determined.

- The velocity always acts tangent to the path of motion.

- The acceleration has two components. The tangential acceleration measures the time rate of change in the magnitude of the velocity and can be determined from $a_t = \alpha r$. The normal acceleration measures the time rate of change in the direction of the velocity and can be determined from $a_n = \omega^2 r$.

16.3 ROTATION ABOUT A FIXED AXIS

PROCEDURE FOR ANALYSIS

The velocity and acceleration of a point located on a rigid body that is rotating about a fixed axis can be determined using the following procedure.

Angular Motion.

- Establish the positive sense of rotation about the axis of rotation and show it alongside each kinematic equation as it is applied.

- If a relation is known between any *two* of the four variables α, ω, θ, and t, then a third variable can be obtained by using one of the following kinematic equations which relates all three variables.

$$\omega = \frac{d\theta}{dt} \qquad \alpha = \frac{d\omega}{dt} \qquad \alpha\, d\theta = \omega\, d\omega$$

- If the body's angular acceleration is *constant*, then the following equations can be used:

$$\omega = \omega_0 + \alpha_c t$$
$$\theta = \theta_0 + \omega_0 t + \tfrac{1}{2}\alpha_c t^2$$
$$\omega^2 = \omega_0^2 + 2\alpha_c(\theta - \theta_0)$$

- Once the solution is obtained, the sense of θ, ω, and α is determined from the algebraic signs of their numerical quantities.

Motion of Point P.

- In most cases the velocity of P and its two components of acceleration can be determined from the scalar equations

$$v = \omega r$$
$$a_t = \alpha r$$
$$a_n = \omega^2 r$$

- If the geometry of the problem is difficult to visualize, the following vector equations should be used:

$$\mathbf{v} = \boldsymbol{\omega} \times \mathbf{r}_P = \boldsymbol{\omega} \times \mathbf{r}$$
$$\mathbf{a}_t = \boldsymbol{\alpha} \times \mathbf{r}_P = \boldsymbol{\alpha} \times \mathbf{r}$$
$$\mathbf{a}_n = \boldsymbol{\omega} \times (\boldsymbol{\omega} \times \mathbf{r}_P) = -\omega^2 \mathbf{r}$$

Here $\mathbf{r}_P$ is directed from any point on the axis of rotation to point P, whereas $\mathbf{r}$ lies in the plane of motion of P. Either of these vectors, along with $\boldsymbol{\omega}$ and $\boldsymbol{\alpha}$, should be expressed in terms of its $\mathbf{i}$, $\mathbf{j}$, $\mathbf{k}$ components, and, if necessary, the cross products determined using a determinant expansion (see Eq. B–12).

EXAMPLE 16.1

A cord is wrapped around the wheel in Fig. 16–6, which is initially at rest when $\theta = 0$. If a force is applied to the cord and gives it an acceleration $a = (4t)$ m/s^2, where t is in seconds, determine, as a function of time, (a) the angular velocity of the wheel, and (b) the angular position of line OP.

Fig. 16–6

SOLUTION
Part (a). The wheel is subjected to rotation about a fixed axis passing through point O. Thus, point P on the wheel has motion about a circular path, and the acceleration of this point has *both* tangential and normal components. The tangential component is $(a_P)_t = (4t)$ m/s^2, since the cord is wrapped around the wheel and moves *tangent* to it. Hence the angular acceleration of the wheel is

($\curvearrowright +$)
$$(a_P)_t = \alpha r$$
$$(4t) \text{ m/s}^2 = \alpha(0.2 \text{ m})$$
$$\alpha = (20t) \text{ rad/s}^2 \downarrow$$

Using this result, the wheel's angular velocity ω can now be determined from $\alpha = d\omega/dt$, since this equation relates α, t, and ω. Integrating, with the initial condition that $\omega = 0$ when $t = 0$, yields

($\curvearrowright +$)
$$\alpha = \frac{d\omega}{dt} = (20t) \text{ rad/s}^2$$
$$\int_0^\omega d\omega = \int_0^t 20t \, dt$$
$$\omega = 10t^2 \text{ rad/s} \downarrow \qquad \qquad Ans.$$

Part (b). Using this result, the angular position θ of OP can be found from $\omega = d\theta/dt$, since this equation relates θ, ω, and t. Integrating, with the initial condition $\theta = 0$ when $t = 0$, we have

($\curvearrowright +$)
$$\frac{d\theta}{dt} = \omega = (10t^2) \text{ rad/s}$$
$$\int_0^\theta d\theta = \int_0^t 10t^2 \, dt$$
$$\theta = 3.33t^3 \text{ rad} \qquad \qquad Ans.$$

NOTE: We cannot use the equation of constant angular acceleration, since α is a function of time.

EXAMPLE 16.2

The motor shown in the photo is used to turn the wheel. The details are shown in Fig. 16–7a. If the pulley A connected to the motor begins to rotate from rest with a constant angular acceleration of $\alpha_A = 2$ rad/s^2, determine the magnitudes of the velocity and acceleration of point P on the wheel, after the pulley has turned two revolutions. Assume the transmission belt does not slip on the pulley and wheel.

SOLUTION

Angular Motion. First we will convert the two revolutions to radians. Since there are 2π rad in one revolution, then

$$\theta_A = 2 \text{ rev}\left(\frac{2\pi \text{ rad}}{1 \text{ rev}}\right) = 12.57 \text{ rad}$$

Since α_A is constant, the angular velocity of pulley A is therefore

($\curvearrowright$+)
$$\omega^2 = \omega_0^2 + 2\alpha_c(\theta - \theta_0)$$
$$\omega_A^2 = 0 + 2(2 \text{ rad/s}^2)(12.57 \text{ rad} - 0)$$
$$\omega_A = 7.090 \text{ rad/s}$$

The belt has the same speed and tangential component of acceleration as it passes over the pulley and wheel. Thus,

$$v = \omega_A r_A = \omega_B r_B; \quad 7.090 \text{ rad/s }(0.15 \text{ m}) = \omega_B(0.4 \text{ m})$$
$$\omega_B = 2.659 \text{ rad/s}$$
$$a_t = \alpha_A r_A = \alpha_B r_B; \quad 2 \text{ rad/s}^2 \,(0.15 \text{ m}) = \alpha_B(0.4 \text{ m})$$
$$\alpha_B = 0.750 \text{ rad/s}^2$$

Motion of P. As shown on the kinematic diagram in Fig. 16–7b, we have

$$v_P = \omega_B r_B = 2.659 \text{ rad/s }(0.4 \text{ m}) = 1.06 \text{ m/s} \quad \textit{Ans.}$$
$$(a_P)_t = \alpha_B r_B = 0.750 \text{ rad/s}^2 \,(0.4 \text{ m}) = 0.3 \text{ m/s}^2$$
$$(a_P)_n = \omega_B^2 r_B = (2.659 \text{ rad/s})^2(0.4 \text{ m}) = 2.827 \text{ m/s}^2$$

Thus

$$a_P = \sqrt{(0.3 \text{ m/s}^2)^2 + (2.827 \text{ m/s}^2)^2} = 2.84 \text{ m/s}^2 \quad \textit{Ans.}$$

Fig. 16–7

FUNDAMENTAL PROBLEMS

F16–1. When the gear rotates 20 revolutions, it achieves an angular velocity of $\omega = 30$ rad/s, starting from rest. Determine its constant angular acceleration and the time required.

Prob. F16–1

F16–2. The flywheel rotates with an angular velocity of $\omega = (0.005\theta^2)$ rad/s, where θ is in radians. Determine the angular acceleration when the wheel has rotated 20 revolutions.

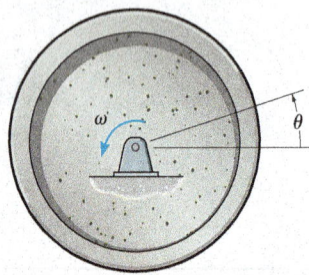

Prob. F16–2

F16–3. The flywheel rotates with an angular velocity of $\omega = (4\theta^{1/2})$ rad/s, where θ is in radians. Determine the time it takes to achieve an angular velocity of $\omega = 150$ rad/s. When $t = 0$, $\theta = 1$ rad.

Prob. F16–3

F16–4. The bucket is hoisted by the rope that wraps around a drum wheel. If the angular displacement of the wheel is $\theta = (0.5t^3 + 15t)$ rad, where t is in seconds, determine the velocity and acceleration of the bucket when $t = 3$ s.

Prob. F16–4

F16–5. A wheel has an angular acceleration of $\alpha = (0.5\,\theta)$ rad/s^2, where θ is in radians. Determine the magnitude of the velocity and acceleration of a point P located on its rim after the wheel has rotated 2 revolutions. The wheel has a radius of 0.2 m and starts when $\theta = 0$, $\omega_0 = 2$ rad/s.

F16–6. For a short period of time, the motor turns gear A with a constant angular acceleration of $\alpha_A = 4.5$ rad/s^2, starting from rest. Determine the velocity of the cylinder and the distance it travels in three seconds. The cord is wrapped around the drum D which is rigidly attached to gear B.

Prob. F16–6

PROBLEMS

16–1. The angular velocity of the disk is defined by $\omega = (5t^2 + 2)$ rad/s, where t is in seconds. Determine the magnitudes of the velocity and acceleration of point A on the disk when $t = 0.5$ s.

Prob. 16–1

16–2. The angular acceleration of the disk is defined by $\alpha = (3t^2 + 12)$ rad/s², where t is in seconds. If the disk is originally rotating at $\omega_0 = 12$ rad/s, determine the magnitude of the velocity and the n and t components of acceleration of point A on the disk when $t = 2$ s.

16–3. The disk is originally rotating at $\omega_0 = 8$ rad/s. If it is subjected to a constant angular acceleration of $\alpha = 20$ rad/s², determine the magnitudes of the velocity and the n and t components of acceleration of point A at the instant $t = 0.5$ s.

*16–4. The disk is originally rotating at $\omega_0 = 8$ rad/s. If it is subjected to a constant angular acceleration of $\alpha = 6$ rad/s², determine the magnitudes of the velocity and the n and t components of acceleration of point B just after the wheel undergoes 2 revolutions.

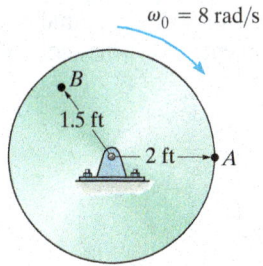

Probs. 16–3/4

16–5. The disk is driven by a motor such that the angular position of the disk is defined by $\theta = (20t + 4t^2)$ rad, where t is in seconds. Determine the number of revolutions, the angular velocity, and angular acceleration of the disk when $t = 90$ s.

Prob. 16–5

16–6. A wheel has an initial angular velocity of 10 rev/s and a constant angular acceleration of 3 rad/s². Determine the number of revolutions required for the wheel to have an angular velocity of 20 rev/s. What time is required?

16–7. For a short time, gear A of the automobile starter rotates with an angular acceleration of $\alpha_A = (450t^2 + 60)$ rad/s², where t is in seconds. Determine the angular velocity and angular displacement of gear B when $t = 2$ s, starting from rest. The radii of gears A and B are 10 mm and 25 mm, respectively.

*16–8. For a short time, gear A of the automobile starter rotates with an angular acceleration of $\alpha_A = (50\omega^{1/2})$ rad/s², where ω is in rad/s. Determine the angular velocity of gear B after gear A has rotated 50 rev, starting from rest. The radii of gears A and B are 10 mm and 25 mm, respectively.

Probs. 16–7/8

16–9. The 50-mm-radius pulley A of the clothes dryer rotates with an angular acceleration of $\alpha_A = (27\theta_A^{1/2})$ rad/s², where θ_A is in radians. Determine its angular acceleration when $t = 1$ s, starting from rest.

16–10. If the 50-mm-radius motor pulley A of the clothes dryer rotates with an angular acceleration of $\alpha_A = (10 + 50t)$ rad/s², where t is in seconds, determine its angular velocity when when $t = 3$ s, starting from rest.

*16–12.** The blade on the horizontal-axis windmill is turning with an angular velocity of $\omega_0 = 2$ rad/s. Determine the distance point P on the tip of the blade has traveled if the blade attains an angular velocity of $\omega = 5$ rad/s in 3 s. The angular acceleration is constant. Also, what is the magnitude of the acceleration of this point when $t = 3$ s?

16–13. The blade on the horizontal-axis windmill is turning with an angular velocity of $\omega_0 = 2$ rad/s. If is is given an angular acceleration of $\alpha_c = 0.6$ rad/s², determine the angular velocity and the magnitude of acceleration of point P on the tip of the blade when $t = 3$ s.

Probs. 16–12/13

Probs. 16–9/10

16–11. The disk is originally rotating at $\omega_0 = 8$ rad/s. If it is subjected to a constant angular acceleration $\alpha_c = 6$ rad/s², determine the magnitudes of the velocity and the n and t components of acceleration of point A at the instant $t = 3$ s.

16–14. The disk starts from rest and is given an angular acceleration $\alpha = (2t^2)$ rad/s², where t is in seconds. Determine the angular velocity of the disk and its angular displacement when $t = 4$ s.

16–15. The disk starts from rest and is given an angular acceleration $\alpha = (5\,t^{1/2})$ rad/s², where t is in seconds. Determine the magnitudes of the normal and tangential components of acceleration of a point P on the rim of the disk when $t = 2$ s.

*16–16.** The disk starts at $\omega_0 = 1$ rad/s when $\theta = 0$, and is given an angular acceleration $\alpha = (0.3\theta)$ rad/s², where θ is in radians. Determine the magnitudes of the normal and tangential components of acceleration of a point P on the rim of the disk when $\theta = 1$ rev.

Prob. 16–11

Probs. 16–14/15/16

16–17. A motor gives gear A an angular acceleration of $\alpha_A = (2 + 0.006\,\theta^2)$ rad/s², where θ is in radians. If this gear is initially turning at $\omega_A = 15$ rad/s, determine the angular velocity of gear B after A undergoes an angular displacement of 10 rev.

16–18. A motor gives gear A an angular acceleration of $\alpha_A = (2t^3)$ rad/s², where t is in seconds. If this gear is initially turning at $\omega_A = 15$ rad/s, determine the angular velocity of gear B when $t = 3$ s.

*16–20. A motor gives gear A an angular acceleration of $\alpha_A = (4t^3)$ rad/s², where t is in seconds. If this gear is initially turning at $(\omega_A)_0 = 20$ rad/s, determine the angular velocity of gear B when $t = 2$ s.

Probs. 16–17/18

Prob. 16–20

16–19. The vacuum cleaner's armature shaft S rotates with an angular acceleration of $\alpha = 4\omega^{3/4}$ rad/s², where ω is in rad/s. Determine the brush's angular velocity when $t = 4$ s, starting from $\omega_0 = 1$ rad/s, at $t = 0$. The radii of the shaft and the brush are 0.25 in. and 1 in., respectively. Neglect the thickness of the drive belt.

16–21. At the instant shown, the horizontal portion of the belt has an angular acceleration of 3 m/s², while points in contact with the outer edge of the pulleys have an acceleration magnitude of 5 m/s². If the belt does not slip on the pulleys, determine the belt's speed due to the motion.

Prob. 16–19

Prob. 16–21

16–22. If the motor turns gear A with an angular acceleration of $\alpha_A = 2$ rad/s² when the angular velocity is $\omega_A = 20$ rad/s, determine the angular acceleration and angular velocity of gear D.

*16–24. A pendulum has a swinging motion such that $\theta = [-0.101 \sin(4.95t) + 0.3 \cos(4.95t)]$ rad, where t is in seconds, and the arguments for the sine and cosine are in radians. Determine the magnitudes of the velocity and acceleration of the bob when $\theta = 0°$.

16–25. Determine the maximum angular displacement of the pendulum in Prob. 16–24.

Probs. 16–24/25

Prob. 16–22

16–23. If the motor turns gear A with an angular acceleration of $\alpha_A = 3$ rad/s² when the angular velocity is $\omega_A = 60$ rad/s, determine the angular acceleration and angular velocity of gear D.

16–26. The pinion gear A on the motor shaft is given a constant angular acceleration $\alpha = 3$ rad/s². If the gears A and B have the dimensions shown, determine the angular velocity and angular displacement of the output shaft C when $t = 2$ s, starting from rest. The shaft is fixed to B and turns with it.

Prob. 16–23

Prob. 16–26

16–27. The gear A on the drive shaft of the outboard motor has a radius $r_A = 0.7$ in. and the meshed pinion gear B on the propeller shaft has a radius $r_B = 1.4$ in. Determine the angular velocity of the propeller in $t = 1.3$ s if the drive shaft rotates with an angular acceleration $\alpha = (300\sqrt{t})$ rad/s², where t is in seconds. The propeller is originally at rest and the motor frame does not move.

16–29. Morse Industrial manufactures the speed reducer shown. If a motor drives the gear shaft S with an angular acceleration of $\alpha = (0.4e^t)$ rad/s², where t is in seconds, determine the angular velocity of shaft E when $t = 2$ s after starting from rest. The radius of each gear is listed in the figure. Note that gears B and C are fixed connected to the same shaft.

Prob. 16–27

$r_A = 20$ mm
$r_B = 80$ mm
$r_C = 30$ mm
$r_D = 120$ mm

Prob. 16–29

*16–28. The gear A on the drive shaft of the outboard motor has a radius $r_A = 0.7$ in. and the meshed pinion gear B on the propeller shaft has a radius $r_B = 1.4$ in. Determine the magnitudes of the velocity and acceleration of a point P located on the tip of the propeller at the instant $t = 0.75$ s. The drive shaft rotates with an angular acceleration $\alpha = (300\sqrt{t})$ rad/s², where t is in seconds. The propeller is originally at rest and the motor frame does not move.

16–30. Due to an increase in power, the motor M rotates the shaft A with an angular acceleration of $\alpha = (0.06\theta^2)$ rad/s², where θ is in radians. If the shaft is initially turning at $\omega_0 = 50$ rad/s, determine the angular velocity of gear B after the shaft undergoes an angular displacement $\Delta\theta = 10$ rev.

Prob. 16–28

Prob. 16–30

16–31. At the instant shown, gear A is rotating with a constant angular velocity of $\omega_A = 6$ rad/s. Determine the largest angular velocity of gear B and the maximum speed of point C.

16–33. The turntable T is driven by the frictional idler wheel A, which simultaneously bears against the inner rim of the turntable and the motor-shaft spindle B. Determine the required diameter d of the spindle if the motor turns it at 25 rad/s and it is required that the turntable rotate at 2 rad/s.

Prob. 16–31

Prob. 16–33

16–34. For a short time a motor of the random-orbit sander drives the gear A with an angular velocity of $\omega_A = 40(t^3 + 6t)$ rad/s, where t is in seconds. This gear is connected to gear B, which is fixed connected to the shaft CD. The end of this shaft is connected to the eccentric spindle EF and pad P, which causes the pad to orbit around shaft CD at a radius of 15 mm. Determine the magnitudes of the velocity and the tangential and normal components of acceleration of the spindle EF when $t = 2$ s after starting from rest.

*16–32. Determine the distance the load W is lifted in $t = 5$ s using the hoist. The shaft of the motor M turns with an angular velocity $\omega = 100(4 + t)$ rad/s, where t is in seconds.

Prob. 16–32

Prob. 16–34

16–35. If the shaft and plate rotates with a constant angular velocity of $\omega = 14$ rad/s, determine the velocity and acceleration of point C located on the corner of the plate at the instant shown. Express the result in Cartesian vector form.

Prob. 16–35

16–37. The rod assembly is supported by ball-and-socket joints at A and B. At the instant shown it is rotating about the y axis with an angular velocity $\omega = 5$ rad/s and has an angular acceleration $\alpha = 8$ rad/s^2. Determine the magnitudes of the velocity and acceleration of point C at this instant. Solve the problem using Cartesian vectors and Eqs. 16–9 and 16–13.

Prob. 16–37

*16–36. At the instant shown, the shaft and plate rotates with an angular velocity of $\omega = 14$ rad/s and angular acceleration of $\alpha = 7$ rad/s^2. Determine the velocity and acceleration of point D located on the corner of the plate at this instant. Express the result in Cartesian vector form.

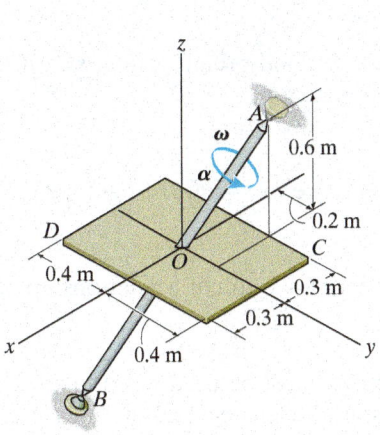

Prob. 16–36

16–38. Rotation of the robotic arm occurs due to linear movement of the hydraulic cylinders A and B. If this motion causes the gear at D to rotate clockwise at 5 rad/s, determine the magnitudes of velocity and acceleration of the part C held by the grips of the arm.

Prob. 16–38

*16.4 ABSOLUTE MOTION ANALYSIS

A body subjected to general plane motion undergoes a simultaneous translation and rotation. If the body is represented by a thin slab, the slab translates in the plane of the slab and rotates about an axis perpendicular to this plane. The motion can be completely specified by knowing *both* the angular rotation of a line fixed in the body and the motion of a point on the body. One way to relate these motions is to use a rectilinear position coordinate s to locate the point along its path and an angular position coordinate θ to specify the orientation of the line. The two coordinates are then related using the geometry of the problem. By direct application of the time-differential equations $v = ds/dt$, $a = dv/dt$, $\omega = d\theta/dt$, and $\alpha = d\omega/dt$, the motion of the point and the angular motion of the line can then be related. This procedure is similar to that used to solve dependent motion problems involving pulleys, Sec. 12.9. In some cases, this same procedure may be used to relate the motion of one body to the motion of a connected body.

The dumping bin on the truck rotates about a fixed axis passing through the pin at A. It is operated by the extension of the hydraulic cylinder BC. The angular position of the bin can be specified using the angular position coordinate θ, and the position of point C on the bin is specified using the rectilinear position coordinate s. Since a and b are fixed lengths, then the two coordinates can be related by the cosine law, $s = \sqrt{a^2 + b^2 - 2ab\cos\theta}$. The time derivative of this equation relates the speed at which the hydraulic cylinder extends to the angular velocity of the bin.

PROCEDURE FOR ANALYSIS

The velocity and acceleration of a point P undergoing rectilinear motion can be related to the angular velocity and angular acceleration of a line contained within a body using the following procedure.

Position Coordinate Equation.

- Locate point P on the body using a position coordinate s, which is measured from a *fixed origin* and is directed along the *straight-line path of motion* of point P.

- Measure from a fixed reference line the angular position θ of a line lying in the body.

- From the dimensions of the body, relate s to θ, $s = f(\theta)$, using geometry and/or trigonometry.

Time Derivatives.

- Take the first time derivative of $s = f(\theta)$ to get a relationship between v and ω.

- Take the second time derivative to get a relationship between a and α.

- In each case the chain rule must be used when taking the time derivatives of the position coordinate equation. See Appendix C.

EXAMPLE 16.3

The end of rod R shown in Fig. 16–8 maintains contact with the cam by means of a spring. If the cam rotates about an axis passing through point O with an angular acceleration α and angular velocity ω, determine the velocity and acceleration of the rod when the cam is in the arbitrary position θ.

Fig. 16–8

SOLUTION

Position Coordinate Equation. Coordinates θ and x are chosen in order to relate the *rotational motion* of the line segment OA on the cam to the *rectilinear translation* of the rod. These coordinates are measured from the *fixed point* O and can be related to each other using trigonometry. Since $OC = CB = r\cos\theta$, Fig. 16–8, then

$$x = 2r\cos\theta$$

Time Derivatives. Using the chain rule, we have

$$\frac{dx}{dt} = -2r(\sin\theta)\frac{d\theta}{dt}$$

$$v = -2r\omega\sin\theta \qquad \textit{Ans.}$$

$$\frac{dv}{dt} = -2r\left(\frac{d\omega}{dt}\right)\sin\theta - 2r\omega(\cos\theta)\frac{d\theta}{dt}$$

$$a = -2r(\alpha\sin\theta + \omega^2\cos\theta) \qquad \textit{Ans.}$$

NOTE: The negative signs indicate that v and a are opposite to the direction of positive x. This seems reasonable if you visualize the motion.

EXAMPLE 16.4

At a given instant, the cylinder of radius r, shown in Fig. 16–9, has an angular velocity ω and angular acceleration α. Determine the velocity and acceleration of its center G if the cylinder rolls without slipping.

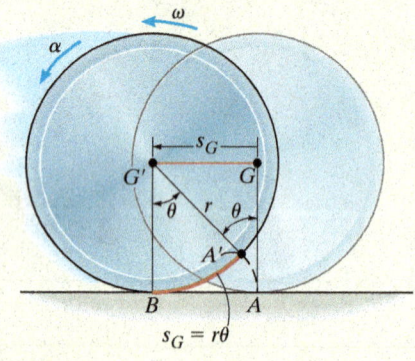

Fig. 16–9

SOLUTION

Position Coordinate Equation. The cylinder undergoes general plane motion since it simultaneously translates and rotates. By inspection, point G moves in a *straight line* to the left, from G to G', as the cylinder rolls, Fig. 16–9. Consequently its new position G' will be specified by the *horizontal* position coordinate s_G. Also, as the cylinder rolls (without slipping), the length of the arc $A'B$ on the rim, which was in contact with the ground from A to B, is equivalent to the length s_G. Consequently, the motion requires the radial line GA to rotate θ to the position $G'A'$. Since the arc $A'B = r\theta$, then G travels a distance

$$s_G = r\theta$$

Time Derivatives. Taking successive time derivatives of this equation, realizing that r is constant, $\omega = d\theta/dt$, and $\alpha = d\omega/dt$, gives the necessary relationships:

$$s_G = r\theta$$
$$v_G = r\omega \qquad \text{Ans.}$$
$$a_G = r\alpha \qquad \text{Ans.}$$

NOTE: Remember that these relationships are valid only if the cylinder (disk, wheel, ball, etc.) rolls *without* slipping.

EXAMPLE 16.5

The window in Fig. 16–10 is opened using a hydraulic cylinder AB. If the cylinder extends at a constant rate of 0.5 m/s, determine the angular velocity and angular acceleration of the window at the instant $\theta = 30°$.

SOLUTION
Position Coordinate Equation. The angular motion of the window can be obtained using the coordinate θ, whereas the extension or motion *along the hydraulic cylinder* is defined using the coordinate s. These coordinates can be related using the law of cosines, namely,

$$s^2 = (2 \text{ m})^2 + (1 \text{ m})^2 - 2(2 \text{ m})(1 \text{ m}) \cos \theta$$

$$s^2 = 5 - 4 \cos \theta \qquad (1)$$

When $\theta = 30°$,

$$s = 1.239 \text{ m}$$

Fig. 16–10

Time Derivatives. Taking the time derivatives of Eq. 1, we have

$$2s \frac{ds}{dt} = 0 - 4(-\sin \theta) \frac{d\theta}{dt}$$

$$s(v_s) = 2(\sin \theta)\omega \qquad (2)$$

Since $v_s = 0.5$ m/s, then at $\theta = 30°$,

$$(1.239 \text{ m})(0.5 \text{ m/s}) = 2 \sin 30° \omega$$

$$\omega = 0.6197 \text{ rad/s} = 0.620 \text{ rad/s} \qquad \textit{Ans.}$$

Taking the time derivative of Eq. 2 yields

$$\frac{ds}{dt} v_s + s \frac{dv_s}{dt} = 2(\cos \theta) \frac{d\theta}{dt} \omega + 2(\sin \theta) \frac{d\omega}{dt}$$

$$v_s^2 + s a_s = 2(\cos \theta)\omega^2 + 2(\sin \theta)\alpha$$

Since $a_s = dv_s/dt = 0$, then

$$(0.5 \text{ m/s})^2 + 0 = 2 \cos 30°(0.6197 \text{ rad/s})^2 + 2 \sin 30° \alpha$$

$$\alpha = -0.415 \text{ rad/s}^2 \qquad \textit{Ans.}$$

Because the result is negative, it indicates the window has an angular deceleration.

PROBLEMS

16–39. The bar DC rotates uniformly about the shaft at D with a constant angular velocity ω. Determine the velocity and acceleration of the bar AB, which is confined by the guides to move vertically.

16–41. Rod CD presses against AB, giving it an angular velocity. If the angular velocity of AB is maintained at $\omega = 5$ rad/s, determine the required magnitude of the velocity $\mathbf{v}$ of CD as a function of the angle θ of rod AB.

Prob. 16–39

Prob. 16–41

*16–40. The bar DC rotates uniformly about the shaft at D with an angular velocity ω and angular acceleration α. Determine the velocity and acceleration of the bar AB, which is confined by the guides to move vertically.

16–42. Determine the velocity and acceleration of the plate at the instant $\theta = 30°$, if at this instant the circular cam is rotating about the fixed point O with an angular velocity $\omega = 4$ rad/s and an angular acceleration $\alpha = 2$ rad/s^2.

Prob. 16–40

Prob. 16–42

16–43. Determine the velocity and acceleration of platform P as a function of the angle θ of cam C if the cam rotates with a constant angular velocity ω. The pin connection does not cause interference with the motion of P on C. The platform is constrained to move vertically by the smooth vertical guides.

16–45. The scaffold S is raised by moving the roller at A toward the pin at B. If A is approaching B with a speed of 1.5 ft/s, determine the speed at which the platform rises as a function of θ. The 4-ft links are pin connected at their midpoint.

Prob. 16–45

16–46. If the hydraulic cylinder AB is extending at a constant rate of 1 ft/s, determine the dumpster's angular velocity at the instant $\theta = 30°$.

Prob. 16–43

*16–44. Determine the velocity and acceleration of the follower rod CD as a function of θ when the contact between the cam and follower is along the straight region AB on the face of the cam. The cam rotates with a constant counterclockwise angular velocity ω.

Prob. 16–46

16–47. Determine the angular velocity of rod AB when $\theta = 30°$. The shaft and the center of the roller C move forward at a constant rate of $v = 5$ m/s.

Prob. 16–44

Prob. 16–47

*16–48. At the instant shown, $\theta = 60°$, and rod AB is subjected to a deceleration of 16 m/s² when the velocity is 10 m/s. Determine the angular velocity and angular acceleration of link CD at this instant.

16–50. End A of the bar moves to the left with a constant velocity v_A. Determine the angular velocity ω and angular acceleration α of the bar as a function of its position x.

Prob. 16–48

Prob. 16–50

16–49. Wheel A rolls without slipping over the surface of the *fixed* cylinder B. Determine the angular velocity of A if its center C has a speed of 2 m/s. How many revolutions will A make about its center after link DC completes one revolution?

16–51. The bar remains in contact with the floor and with point A. If point B moves to the right with a constant velocity v_B, determine the angular velocity and angular acceleration of the bar as a function of x.

Prob. 16–49

Prob. 16–51

*16–52. When the bar is at the angle θ, the rod is rotating clockwise with constant angular velocity ω. Determine the velocity of the weight A at this instant. The cord is 20 ft long.

16–54. The crate is transported on a platform which rests on rollers, each having a radius r. If the rollers do not slip, determine their angular velocity if the platform moves forward with a velocity $\mathbf{v}$.

Prob. 16–54

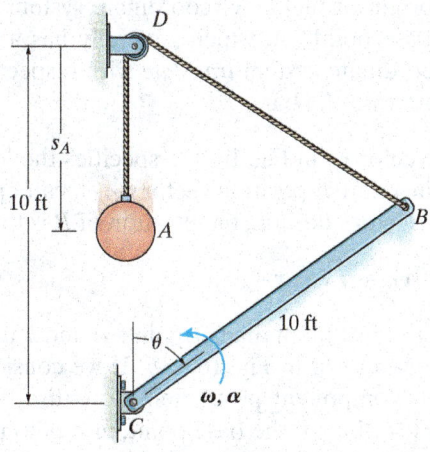

Prob. 16–52

16–55. Determine the angular velocity of rod CD at the instant $\theta = 30°$. Rod AB moves to the left at a constant rate $v_{AB} = 5 \text{ m/s}$.

16–53. The inclined plate moves to the left with a constant velocity $\mathbf{v}$. Determine the angular velocity and angular acceleration of the slender rod of length l. The rod pivots about the step at C as its slides on the plate.

*16–56. Determine the angular acceleration of rod CD at the instant $\theta = 30°$. Rod AB has zero velocity, i.e., $v_{AB} = 0$, and an acceleration of $a_{AB} = 2 \text{ m/s}^2$ to the right when $\theta = 30°$.

Prob. 16–53

Probs. 16–55/56

16.5 RELATIVE-MOTION ANALYSIS: VELOCITY

The general plane motion of a rigid body can be described as a combination of translation and rotation. To view these "component" motions separately we will use a relative-motion analysis involving two sets of coordinate axes. The x, y coordinate system is fixed and measures the *absolute* position of two points A and B on the body, here represented as a bar, Fig. 16–11a. The origin of the x', y' coordinate system will be attached to the selected "base point" A, which generally has a known motion. The axes of this coordinate system *translate* with respect to the fixed frame but do not rotate with the bar.

Position. The position vector $\mathbf{r}_A$ in Fig. 16–11a specifies the location of the "base point" A, and the relative-position vector $\mathbf{r}_{B/A}$ locates point B with respect to point A. By vector addition, the position of B is then

$$\mathbf{r}_B = \mathbf{r}_A + \mathbf{r}_{B/A}$$

Displacement. During an instant of time dt, points A and B undergo displacements $d\mathbf{r}_A$ and $d\mathbf{r}_B$ as shown in Fig. 16–11b. If we consider the general plane motion by its component parts then the entire bar first translates by an amount $d\mathbf{r}_A$ so that A, the base point, moves to its *final* position and point B moves to B', Fig. 16–11c. The bar is then rotated about A by an amount $d\theta$ so that B' undergoes the relative displacement $d\mathbf{r}_{B/A}$ and thus moves to its final position B. Due to the rotation about A, $dr_{B/A} = r_{B/A}\, d\theta$, and the displacement of B is

$$d\mathbf{r}_B = d\mathbf{r}_A + d\mathbf{r}_{B/A}$$

$\quad\quad\quad\quad\quad\ \ \ $ due to rotation about A
$\quad\quad\quad\ \ \ $ due to translation of A
$\quad\quad $ due to translation and rotation

Fig. 16–11

16.5 RELATIVE-MOTION ANALYSIS: VELOCITY

Velocity. To determine the relation between the velocities of points A and B, it is necessary to take the time derivative of the position equation, or simply divide the displacement equation by dt. This yields

$$\frac{d\mathbf{r}_B}{dt} = \frac{d\mathbf{r}_A}{dt} + \frac{d\mathbf{r}_{B/A}}{dt}$$

The terms $d\mathbf{r}_B/dt = \mathbf{v}_B$ and $d\mathbf{r}_A/dt = \mathbf{v}_A$ are measured with respect to the fixed x, y axes and represent the absolute velocities of points A and B, respectively. Since the relative displacement is caused by a rotation, the magnitude of the third term is $dr_{B/A}/dt = r_{B/A}\, d\theta/dt = r_{B/A}\dot\theta = r_{B/A}\omega$, where ω is the angular velocity of the body at the instant considered. We will denote this term as the relative velocity $\mathbf{v}_{B/A}$, since it represents the velocity of B with respect to A as measured by an observer fixed to the translating x', y' axes. In other words, *the bar appears to move as if it were rotating with an angular velocity $\boldsymbol{\omega}$ about the z' axis passing through A.* Consequently, $\mathbf{v}_{B/A}$ has a magnitude of $v_{B/A} = \omega r_{B/A}$ and a direction which is perpendicular to $\mathbf{r}_{B/A}$. We therefore have

$$\boxed{\mathbf{v}_B = \mathbf{v}_A + \mathbf{v}_{B/A}} \qquad (16\text{–}15)$$

where

$\mathbf{v}_B$ = velocity of point B
$\mathbf{v}_A$ = velocity of the base point A
$\mathbf{v}_{B/A}$ = velocity of B with respect to A

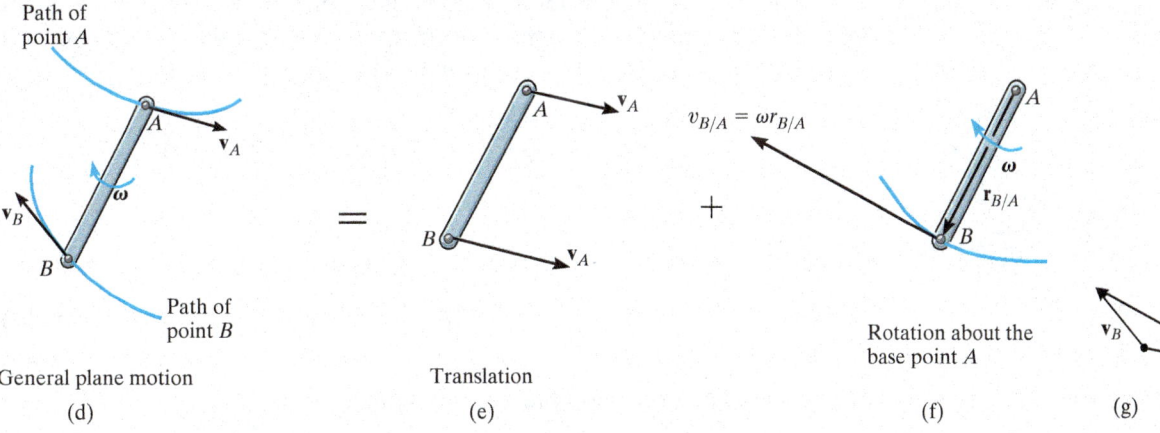

General plane motion
(d)

Translation
(e)

Rotation about the base point A
(f)

(g)

Fig. 16–11(cont.)

What the equation $\mathbf{v}_B = \mathbf{v}_A + \mathbf{v}_{B/A}$ states is that the velocity of B, Fig. 16–11d, is determined by considering the entire bar to translate with a velocity of $\mathbf{v}_A$, Fig. 16–11e, and rotate about A with an angular velocity $\boldsymbol{\omega}$, Fig. 16–11f. Vector addition of these two effects, applied to B, yields $\mathbf{v}_B$, as shown in Fig. 16–11g.

Since the relative velocity $\mathbf{v}_{B/A}$ represents the effect of *circular motion* about A, this term can be expressed by the cross product $\mathbf{v}_{B/A} = \boldsymbol{\omega} \times \mathbf{r}_{B/A}$, Eq. 16–9. Hence, for application using Cartesian vector analysis, we can also write Eq. 16–15 as

$$\mathbf{v}_B = \mathbf{v}_A + \boldsymbol{\omega} \times \mathbf{r}_{B/A} \tag{16-16}$$

where

$\mathbf{v}_B$ = velocity of B
$\mathbf{v}_A$ = velocity of the base point A
$\boldsymbol{\omega}$ = angular velocity of the body
$\mathbf{r}_{B/A}$ = position vector directed from A to B

When applying the velocity equation, points A and B should generally be selected as points on the body which are pin connected to other bodies, or as points in contact with adjacent bodies which have a *known motion*. For example, point A on link AB in Fig. 16–12a must move along a horizontal path, whereas point B moves on a circular path. The *directions* of $\mathbf{v}_A$ and $\mathbf{v}_B$ can therefore be established since they are always *tangent* to their paths of motion, Fig. 16–12b.

Fig. 16–12

Rolling Without Slipping. Consider the case of a disk (wheel, ball, etc.) which rolls *without slipping*, Fig. 16–13a. If we select point A at the ground, then it (momentarily) has zero velocity since the ground does not move. Furthermore, the center of the disk G moves along a horizontal path so that $\mathbf{v}_G$ is horizontal. To obtain this velocity we will apply Eq. 16–16. From the kinematic diagram, Fig. 16–13b, we have

$$\mathbf{v}_G = \mathbf{v}_A + \boldsymbol{\omega} \times \mathbf{r}_{G/A} = \mathbf{0} + (-\omega \mathbf{k}) \times (r\mathbf{j})$$

So that

$$v_G = \omega r \tag{16-17}$$

Fig. 16–13

This same result was obtained in Example 16–4.

PROCEDURE FOR ANALYSIS

The relative velocity equation can be applied either by using Cartesian vector analysis, or by writing the x and y scalar component equations directly. For application, it is suggested that the following procedure be used.

Vector Analysis

Kinematic Diagram.

- Establish the directions of the fixed x, y coordinates and draw a kinematic diagram of the body that shows the velocities $\mathbf{v}_A$, $\mathbf{v}_B$ of points A and B, the angular velocity $\boldsymbol{\omega}$, and the relative-position vector $\mathbf{r}_{B/A}$.

- If the magnitudes of $\mathbf{v}_A$, $\mathbf{v}_B$, or $\boldsymbol{\omega}$ are unknown, the sense of direction of these vectors can be assumed.

Velocity Equation.

- To apply $\mathbf{v}_B = \mathbf{v}_A + \boldsymbol{\omega} \times \mathbf{r}_{B/A}$, express the vectors in Cartesian vector form and substitute them into the equation. Evaluate the cross product and then equate the respective $\mathbf{i}$ and $\mathbf{j}$ components to obtain two scalar equations.

- If the solution yields a *negative* answer for an *unknown* magnitude, it indicates the sense of direction of the vector is opposite to that shown on the kinematic diagram.

Scalar Analysis

Kinematic Diagram.

- If the velocity equation is to be applied in scalar form, then the magnitude and direction of the relative velocity $\mathbf{v}_{B/A}$ must be established. Draw a kinematic diagram such as shown in Fig. 16–11*f*, which shows the relative motion. Since the body is considered to be "pinned" momentarily at the base point A, the magnitude of $\mathbf{v}_{B/A}$ is $v_{B/A} = \omega r_{B/A}$. The sense of direction of $\mathbf{v}_{B/A}$ is always perpendicular to $\mathbf{r}_{B/A}$ in accordance with the rotational motion $\boldsymbol{\omega}$ of the body.*

Velocity Equation.

- Write Eq. 16–15 in symbolic form, $\mathbf{v}_B = \mathbf{v}_A + \mathbf{v}_{B/A}$, and underneath each of the terms represent the vectors graphically by showing their magnitudes and directions. The scalar equations are determined from the x and y components of these vectors.

*The notation $\mathbf{v}_B = \mathbf{v}_A + \mathbf{v}_{B/A(\text{pin})}$ may be helpful in recalling that A is "pinned."

EXAMPLE 16.6

The link shown in Fig. 16–14a is guided by two blocks at A and B, which move in the fixed slots. If the velocity of A is 2 m/s downward, determine the velocity of B at the instant $\theta = 45°$.

SOLUTION I (VECTOR ANALYSIS)
Kinematic Diagram. Since points A and B are restricted to move along the fixed slots and $\mathbf{v}_A$ is directed downward, then $\mathbf{v}_B$ must be directed horizontally to the right, Fig. 16–14b. This motion causes the link to rotate counterclockwise; that is, by the right-hand rule the angular velocity $\boldsymbol{\omega}$ is directed outward, perpendicular to the plane of motion.

Velocity Equation. Expressing each of the vectors in Fig. 16–14b in terms of their $\mathbf{i}, \mathbf{j}, \mathbf{k}$ components and applying Eq. 16–16 to A, the base point, and B, we have

$$\mathbf{v}_B = \mathbf{v}_A + \boldsymbol{\omega} \times \mathbf{r}_{B/A}$$
$$v_B \mathbf{i} = -2\mathbf{j} + [\omega \mathbf{k} \times (0.2 \sin 45° \mathbf{i} - 0.2 \cos 45° \mathbf{j})]$$
$$v_B \mathbf{i} = -2\mathbf{j} + 0.2\omega \sin 45° \mathbf{j} + 0.2\omega \cos 45° \mathbf{i}$$

Equating the $\mathbf{i}$ and $\mathbf{j}$ components gives

$$v_B = 0.2\omega \cos 45° \qquad (1)$$
$$0 = -2 + 0.2\omega \sin 45° \qquad (2)$$

Thus,
$$\omega = 14.1 \text{ rad/s} \circlearrowleft \qquad v_B = 2 \text{ m/s} \rightarrow \qquad Ans.$$

SOLUTION II (SCALAR ANALYSIS)
The kinematic diagram of the relative "circular motion" which produces $\mathbf{v}_{B/A}$ is shown in Fig. 16–14c. Here $v_{B/A} = \omega(0.2 \text{ m})$.
Thus,
$$v_B = v_A + v_{B/A}$$

$$\begin{bmatrix} v_B \\ \rightarrow \end{bmatrix} = \begin{bmatrix} 2 \text{ m/s} \\ \downarrow \end{bmatrix} + \begin{bmatrix} \omega(0.2 \text{ m}) \\ \measuredangle 45° \end{bmatrix}$$

$(\xrightarrow{+}) \qquad v_B = 0 + \omega(0.2) \cos 45°$
$(+\uparrow) \qquad 0 = -2 + \omega(0.2) \sin 45°$

The solution produces the above Eqs. 1 and 2.

It should be emphasized that these results are *valid only* at the instant $\theta = 45°$. A recalculation for $\theta = 44°$ yields $v_B = 2.07$ m/s and $\omega = 14.4$ rad/s; whereas when $\theta = 46°$, $v_B = 1.93$ m/s and $\omega = 13.9$ rad/s, etc.

Relative motion
(c)

Fig. 16–14

NOTE: Since v_A and ω are *known*, the velocity of any other point on the link can be determined. As an exercise, see if you can apply Eq. 16–15 or Eq. 16–16 to points A and C or to points B and C and show that when $\theta = 45°$, $v_C = 3.16$ m/s, directed at an angle of 18.4° up from the horizontal.

EXAMPLE 16.7

The cylinder shown in Fig. 16–15a rolls without slipping on the surface of a conveyor belt which is moving at 2 ft/s. Determine the velocity of point A. The cylinder has a clockwise angular velocity $\omega = 15$ rad/s at the instant shown.

SOLUTION I (VECTOR ANALYSIS)
Kinematic Diagram. Since no slipping occurs, point B on the cylinder has the same velocity as the conveyor, Fig. 16–15b. Also, the angular velocity of the cylinder is known, so we can apply the velocity equation to B, the base point, and A to determine $\mathbf{v}_A$.

Velocity Equation.

$$\mathbf{v}_A = \mathbf{v}_B + \boldsymbol{\omega} \times \mathbf{r}_{A/B}$$
$$(v_A)_x \mathbf{i} + (v_A)_y \mathbf{j} = 2\mathbf{i} + (-15\mathbf{k}) \times (-0.5\mathbf{i} + 0.5\mathbf{j})$$
$$(v_A)_x \mathbf{i} + (v_A)_y \mathbf{j} = 2\mathbf{i} + 7.50\mathbf{j} + 7.50\mathbf{i}$$

so that

$$(v_A)_x = 2 + 7.50 = 9.50 \text{ ft/s}$$
$$(v_A)_y = 7.50 \text{ ft/s}$$

Thus,

$$v_A = \sqrt{(9.50)^2 + (7.50)^2} = 12.1 \text{ ft/s} \quad \textit{Ans.}$$

$$\theta = \tan^{-1}\frac{7.50}{9.50} = 38.3° \quad \textit{Ans.}$$

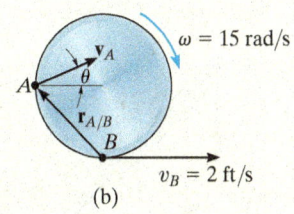

SOLUTION II (SCALAR ANALYSIS)
As an alternative procedure, the scalar components of $\mathbf{v}_A = \mathbf{v}_B + \mathbf{v}_{A/B}$ can be obtained directly. From the kinematic diagram showing the relative "circular" motion which produces $\mathbf{v}_{A/B}$, Fig. 16–15c, we have

$$v_{A/B} = \omega r_{A/B} = (15 \text{ rad/s})\left(\frac{0.5 \text{ ft}}{\cos 45°}\right) = 10.6 \text{ ft/s}$$

Relative motion
(c)

Fig. 16–15

Thus,

$$\mathbf{v}_A = \mathbf{v}_B + \mathbf{v}_{A/B}$$

$$\begin{bmatrix}(v_A)_x \\ \rightarrow\end{bmatrix} + \begin{bmatrix}(v_A)_y \\ \uparrow\end{bmatrix} = \begin{bmatrix}2 \text{ ft/s} \\ \rightarrow\end{bmatrix} + \begin{bmatrix}10.6 \text{ ft/s} \\ \measuredangle 45°\end{bmatrix}$$

Equating the x and y components gives the same results as before, namely,

$(\xrightarrow{+})$ $(v_A)_x = 2 + 10.6 \cos 45° = 9.50 \text{ ft/s}$
$(+\uparrow)$ $(v_A)_y = 0 + 10.6 \sin 45° = 7.50 \text{ ft/s}$

EXAMPLE 16.8

The collar C in Fig. 16–16a is moving downward with a velocity of 2 m/s. Determine the angular velocity of CB at this instant.

SOLUTION I (VECTOR ANALYSIS)

Kinematic Diagram. The downward motion of C causes B to move to the right along a curved path. Also, CB and AB rotate counterclockwise.

Velocity Equation. Link CB (general plane motion): See Fig. 16–16b.

$$\mathbf{v}_B = \mathbf{v}_C + \boldsymbol{\omega}_{CB} \times \mathbf{r}_{B/C}$$

$$v_B \mathbf{i} = -2\mathbf{j} + \omega_{CB}\mathbf{k} \times (0.2\mathbf{i} - 0.2\mathbf{j})$$

$$v_B \mathbf{i} = -2\mathbf{j} + 0.2\omega_{CB}\mathbf{j} + 0.2\omega_{CB}\mathbf{i}$$

$$v_B = 0.2\omega_{CB} \quad (1)$$

$$0 = -2 + 0.2\omega_{CB} \quad (2)$$

$$\omega_{CB} = 10 \text{ rad/s} \; \curvearrowleft \qquad \textit{Ans.}$$

$$v_B = 2 \text{ m/s} \rightarrow$$

SOLUTION II (SCALAR ANALYSIS)

The scalar component equations of $\mathbf{v}_B = \mathbf{v}_C + \mathbf{v}_{B/C}$ can be obtained directly. The kinematic diagram in Fig. 16–16c shows the relative "circular" motion which produces $\mathbf{v}_{B/C}$. We have

$$\mathbf{v}_B = \mathbf{v}_C + \mathbf{v}_{B/C}$$

$$\begin{bmatrix} v_B \\ \rightarrow \end{bmatrix} = \begin{bmatrix} 2 \text{ m/s} \\ \downarrow \end{bmatrix} + \begin{bmatrix} \omega_{CB}(0.2\sqrt{2} \text{ m}) \\ \angle 45° \end{bmatrix}$$

Resolving these vectors in the x and y directions yields

$(\xrightarrow{+}) \qquad v_B = 0 + \omega_{CB}(0.2\sqrt{2} \cos 45°)$

$(+\uparrow) \qquad 0 = -2 + \omega_{CB}(0.2\sqrt{2} \sin 45°)$

which is the same as Eqs. 1 and 2.

NOTE: Since link AB rotates about a fixed axis and v_B is known, Fig. 16–16d, its angular velocity is found from $v_B = \omega_{AB} r_{AB}$ or 2 m/s = $\omega_{AB}(0.2 \text{ m})$, $\omega_{AB} = 10$ rad/s.

(b)

Relative motion
(c)

(d)

Fig. 16–16

FUNDAMENTAL PROBLEMS

F16–7. If roller A moves to the right with a constant velocity of $v_A = 3$ m/s, determine the angular velocity of the link and the velocity of roller B at the instant $\theta = 30°$.

Prob. F16–7

F16–8. The wheel rolls without slipping with an angular velocity of $\omega = 10$ rad/s. Determine the magnitude of the velocity of point B at the instant shown.

Prob. F16–8

F16–9. Determine the angular velocity of the spool. The cable wraps around the inner core, and the spool does not slip on the platform P.

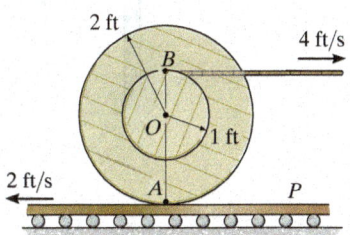

Prob. F16–9

F16–10. If crank OA rotates with an angular velocity of $\omega = 12$ rad/s, determine the velocity of piston B and the angular velocity of rod AB at the instant shown.

Prob. F16–10

F16–11. If rod AB slides along the horizontal collar with a velocity of 60 ft/s, determine the angular velocity of link BC and the velocity of point C at the instant shown.

Prob. F16–11

F16–12. End A of the link has a velocity of $v_A = 3$ m/s. Determine the velocity of the peg at B at this instant. The peg is constrained to move along the slot.

Prob. F16–12

PROBLEMS

16–57. At the instant shown the boomerang has an angular velocity $\omega = 4$ rad/s, and its mass center G has a velocity $v_G = 6$ in./s. Determine the velocity of point B at this instant.

16–59. A bowling ball is cast on the "alley" with a backspin of $\omega = 10$ rad/s while its center O has a forward velocity of $v_O = 8$ m/s. Determine the velocity of the contact point A in contact with the alley.

Prob. 16–57

Prob. 16–59

16–58. When the slider block C is in the position shown, the link AB has a clockwise angular velocity of 2 rad/s. Determine the velocity of block C at this instant.

*16–60. The slider block C moves at 8 m/s down the inclined groove. Determine the angular velocities of links AB and BC, at the instant shown.

Prob. 16–58

Prob. 16–60

16–61. Determine the angular velocity of links AB and BC at the instant $\theta = 30°$. Also, sketch the position of link BC when $\theta = 55°, 45°,$ and $30°$ to show its general plane motion.

16–63. If bar AB has an angular velocity $\omega_{AB} = 4$ rad/s, determine the velocity of the slider block C at the instant shown.

Prob. 16–61

Prob. 16–63

16–62. The center of the wheel is moving to the right with a speed of 2 ft/s. If no slipping occurs at the ground, A, determine the velocity of points B and C.

*16–64. The pinion gear A rolls on the fixed gear rack B with an angular velocity $\omega = 4$ rad/s. Determine the velocity of the gear rack C.

16–65. The pinion gear rolls on the gear racks. If B is moving to the right at 8 ft/s and C is moving to the left at 4 ft/s, determine the angular velocity of the pinion gear and the velocity of its center A.

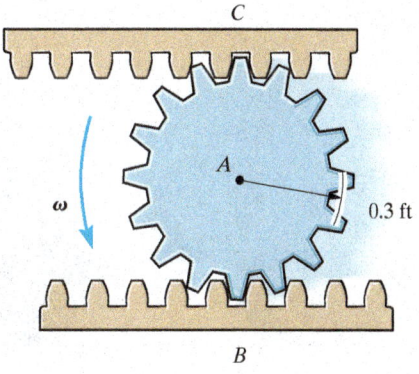

Prob. 16–62

Probs. 16–64/65

16–66. Knowing that the angular velocity of link AB is $\omega_{AB} = 4$ rad/s, determine the velocity of the collar at C and the angular velocity of link CB at the instant shown. Link CB is horizontal at this instant.

***16–68.** The inner hub of the roller bearing is rotating with an angular velocity of $\omega_i = 6$ rad/s, while the outer hub is rotating in the opposite direction at $\omega_o = 4$ rad/s. Determine the angular velocity of each of the rollers if they roll on the hubs without slipping.

Prob. 16–66

Prob. 16–68

16–69. The clockwise angular velocity of the link AB is 2 rad/s. Determine the angular velocity of the connecting links BC and CD at the instant shown.

16–67. At the instant shown, the truck travels to the right at 8 m/s. If the pipe does not slip at B, determine its angular velocity if its mass center G appears to remain stationary to an observer on the ground.

Prob. 16–67

Prob. 16–69

16–70. The wheel rolls on its hub without slipping on the horizontal surface. If the velocity of the center of the wheel is 2 ft/s to the right, determine the velocity of points A and B at the instant shown.

***16–72.** If the slider block A is moving downward at $v_A = 4$ m/s, determine the velocities of blocks B and C at the instant shown.

16–73. If the slider block A is moving downward at $v_A = 4$ m/s, determine the velocity of point E at the instant shown.

Probs. 16–72/73

Prob. 16–70

16–71. The rod CD has a downward velocity of 6 ft/s. The rod is pinned at C to gear B. Determine the velocity of the gear rack A at the instant shown.

16–74. The gage is used to indicate the safe load acting at the end of the boom, B, when it is in any angular position. It consists of a fixed dial plate D and an indicator arm ACE, which is pinned to the plate at C and to a short link EF. If the boom is pin connected to the trunk frame at G and is rotating downward at $\omega_B = 4$ rad/s, determine the velocity of the dial pointer A at the instant shown, i.e., when EF and AC are in the vertical position.

Prob. 16–71

Prob. 16–74

16–75. Piston P moves upward with a velocity of 300 in./s at the instant shown. Determine the angular velocity of the crankshaft AB at this instant.

*16–76. Determine the velocity of the center of gravity G of the connecting rod at the instant shown. Piston P is moving upward with a velocity of 300 in./s.

Probs. 16–75/76

16–77. The planetary gear system is used in an automatic transmission for an automobile. By locking or releasing certain gears, it has the advantage of operating the car at different speeds. Consider the case where the ring gear R is held fixed, $\omega_R = 0$, and the sun gear S is rotating at $\omega_S = 5$ rad/s. Determine the angular velocity of each of the planet gears P and shaft A.

Prob. 16–77

16–78. If the link AB has an angular velocity of $\omega = 4$ rad/s at the instant shown, determine the velocity of the slider block E at this instant.

Prob. 16–78

16–79. The pinion gear A rolls on the fixed gear rack B with an angular velocity $\omega = 4$ rad/s. Determine the velocity of the gear rack C.

Prob. 16–79

*16–80. If the slider block A is moving to the right at $v_A = 8$ ft/s, determine the velocities of blocks B and C at the instant shown. Member CD is pin connected to member ADB.

Prob. 16–80

16.6 INSTANTANEOUS CENTER OF ZERO VELOCITY

The velocity of any point B located on a rigid body can be obtained in a very direct way by choosing the base point A to be a point that has zero velocity at the instant considered. In this case, $\mathbf{v}_A = \mathbf{0}$, and therefore the velocity equation, $\mathbf{v}_B = \mathbf{v}_A + \boldsymbol{\omega} \times \mathbf{r}_{B/A}$, becomes $\mathbf{v}_B = \boldsymbol{\omega} \times \mathbf{r}_{B/A}$. For a body having general plane motion, point A so chosen is called the ***instantaneous center of zero velocity (IC)***, and it lies on the ***instantaneous axis of zero velocity***. This axis is always perpendicular to the plane of motion, and the intersection of the axis with this plane defines the location of the *IC*. Since point A coincides with the *IC*, then

$$\mathbf{v}_B = \boldsymbol{\omega} \times \mathbf{r}_{B/IC}$$

and so point B moves momentarily about the *IC* in a *circular path*. The *magnitude* of $\mathbf{v}_B$ is simply $v_B = \omega r_{B/IC}$, where ω is the angular velocity of the body. Due to the circular motion, the *direction* of $\mathbf{v}_B$ must always be *perpendicular* to $\mathbf{r}_{B/IC}$.

For example, the *IC* for the bicycle wheel in Fig. 16–17 is at the contact point with the ground since the wheel does not slip. If one imagines that the wheel is momentarily *pinned* at the *IC*, the velocities of various points on the wheel can be found using $v = \omega r$, where the radial distances shown in the photo must be determined from the geometry of the wheel.

Fig. 16–17

350   CHAPTER 16   PLANAR KINEMATICS OF A RIGID BODY

Location of IC
knowing $\mathbf{v}_A$ and $\boldsymbol{\omega}$
(a)

Location of IC
knowing the directions
of $\mathbf{v}_A$ and $\mathbf{v}_B$
(b)

(c)

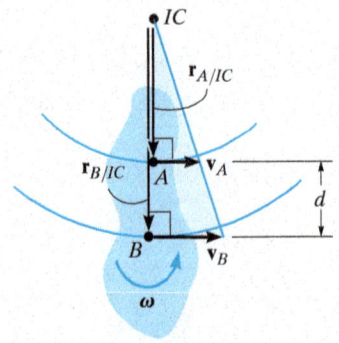

Location of IC
knowing $\mathbf{v}_A$ and $\mathbf{v}_B$
(d)

Fig. 16–18

Location of the IC. To locate the *IC* we can use the fact that the *velocity* of a point on the body is *always perpendicular* to the *relative-position vector* directed from the *IC* to the point. Several possibilities exist:

- The velocity $\mathbf{v}_A$ of a point A on the body and the angular velocity $\boldsymbol{\omega}$ of the body are known, Fig. 16–18a. In this case, the *IC* is located along the line drawn perpendicular to $\mathbf{v}_A$ at A, such that the distance from A to the *IC* is $r_{A/IC} = v_A/\omega$. Note that the *IC* lies up and to the right of A since $\mathbf{v}_A$ must cause a clockwise angular velocity $\boldsymbol{\omega}$ about the *IC*.

- The lines of action of two nonparallel velocities $\mathbf{v}_A$ and $\mathbf{v}_B$ are known, Fig. 16–18b. Construct at points A and B line segments that are perpendicular to $\mathbf{v}_A$ and $\mathbf{v}_B$. Extending these perpendiculars to their *point of intersection*, as shown, locates the *IC* at the instant considered.

- The magnitude and direction of two parallel velocities $\mathbf{v}_A$ and $\mathbf{v}_B$ are known. Here the location of the *IC* is determined by proportional triangles. Examples are shown in Figs. 16–18c and d. In both cases $r_{A/IC} = v_A/\omega$ and $r_{B/IC} = v_B/\omega$. If d is a known, distance between points A and B, then in Fig. 16–18c, $r_{A/IC} + r_{B/IC} = d$, and in Fig. 16–18d, $r_{B/IC} - r_{A/IC} = d$.

16.6 INSTANTANEOUS CENTER OF ZERO VELOCITY   351

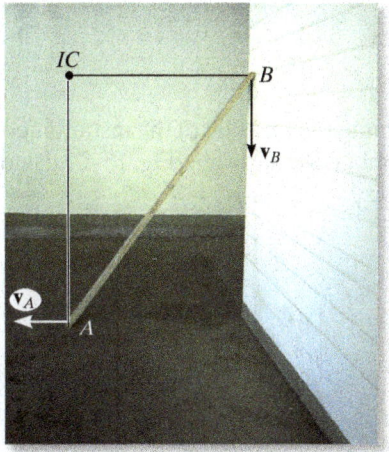

As the board slides downward to the left it is subjected to general plane motion. Since the directions of the velocities of its ends A and B are known, the IC is located as shown. At this instant the board will momentarily rotate about this point.

Realize that the point chosen as the instantaneous center of zero velocity for the body *can only be used at the instant considered* since the body changes its position from one instant to the next. The path described by points which define the location of the IC during the body's motion is called a *centrode*, Fig. 16–18a, and so each point on the centrode acts as the IC for the body only for an instant.

Although the IC may be conveniently used to determine the velocity of any point in a body, it generally *does not have zero acceleration* and therefore it *should not* be used for finding the accelerations of points in a body.

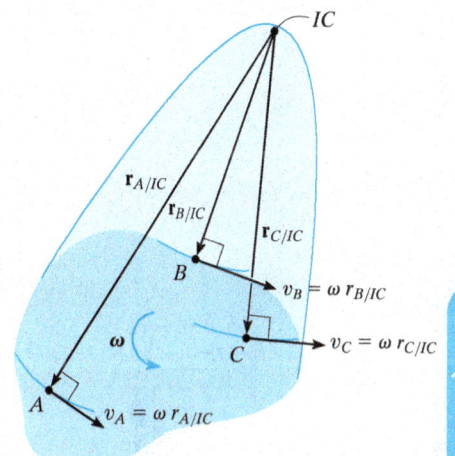

Fig. 16–19

PROCEDURE FOR ANALYSIS

The velocity of a point on a body which is subjected to general plane motion can be determined with reference to its instantaneous center of zero velocity provided the location of the IC is first established using one of the three methods described in Fig. 16–18.

- As shown on the kinematic diagram in Fig. 16–19, the body is imagined as "extended and pinned" at the IC so that, at the instant considered, it rotates about this pin with its angular velocity ω.

- The *magnitude* of velocity for each of the arbitrary points A, B, and C on the body can be determined by using the equation $v = \omega r$, where r is the radial distance from the IC to each point.

- The line of action of each velocity vector $\mathbf{v}$ is *perpendicular* to its associated radial line $\mathbf{r}$, and the velocity has a *sense of direction* which tends to move the point in a manner consistent with the angular rotation ω of the radial line, Fig. 16–19.

EXAMPLE 16.9

Show how to determine the location of the instantaneous center of zero velocity for (a) member *BC* shown in Fig. 16–20a; and (b) the link *CB* shown in Fig. 16–20c.

(a)

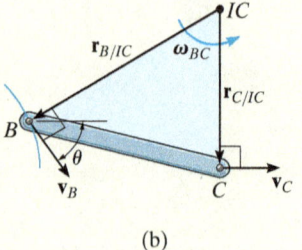

(b)

SOLUTION

Part (a). As shown in Fig. 16–20a, point *B* moves in a circular path such that $\mathbf{v}_B$ is perpendicular to *AB*. Therefore, it acts at an angle θ from the horizontal as shown in Fig. 16–20b. The motion of point *B* causes the piston to move forward *horizontally* with a velocity $\mathbf{v}_C$. When lines are drawn perpendicular to $\mathbf{v}_B$ and $\mathbf{v}_C$, Fig. 16–20b, they intersect at the *IC*.

Part (b). Points *B* and *C* follow circular paths of motion since links *AB* and *DC* are each subjected to rotation about a fixed axis, Fig. 16–20c. Since the velocity is always tangent to the path, at the instant considered, $\mathbf{v}_C$ on rod *DC* and $\mathbf{v}_B$ on rod *AB* are both directed vertically downward, along the axis of link *CB*, Fig. 16–20d. Radial lines drawn perpendicular to these two velocities form parallel lines which intersect at "infinity;" i.e., $r_{C/IC} \to \infty$ and $r_{B/IC} \to \infty$. Thus, $\omega_{CB} = (v_C/r_{C/IC}) \to 0$. As a result, link *CB* momentarily *translates*. An instant later, however, *CB* will move to a tilted position, causing the *IC* to move to some finite location.

(c)

(d)

Fig. 16–20

EXAMPLE 16.10

Block D shown in Fig. 16–21a moves with a speed of 3 m/s. Determine the angular velocities of links BD and AB at the instant shown.

(a)

SOLUTION
As D moves to the right, it causes AB to rotate clockwise about point A. Hence, $\mathbf{v}_B$ is directed perpendicular to AB, Fig. 16–21b. The instantaneous center of zero velocity for BD is located at the intersection of the line segments drawn perpendicular to $\mathbf{v}_B$ and $\mathbf{v}_D$. From the geometry,

$$r_{B/IC} = 0.4 \tan 45° \text{ m} = 0.4 \text{ m}$$

$$r_{D/IC} = \frac{0.4 \text{ m}}{\cos 45°} = 0.5657 \text{ m}$$

Since the magnitude of $\mathbf{v}_D$ is known, the angular velocity of link BD is

$$\omega_{BD} = \frac{v_D}{r_{D/IC}} = \frac{3 \text{ m/s}}{0.5657 \text{ m}} = 5.30 \text{ rad/s} \quad \textit{Ans.}$$

The velocity of B is therefore

$$v_B = \omega_{BD}(r_{B/IC}) = 5.30 \text{ rad/s } (0.4 \text{ m}) = 2.12 \text{ m/s } \searrow 45°$$

From Fig. 16–21c, the angular velocity of AB is

$$\omega_{AB} = \frac{v_B}{r_{B/A}} = \frac{2.12 \text{ m/s}}{0.4 \text{ m}} = 5.30 \text{ rad/s} \downarrow \quad \textit{Ans.}$$

NOTE: Try to solve this problem by applying $\mathbf{v}_D = \mathbf{v}_B + \mathbf{v}_{D/B}$ to member BD.

(b)

(c)

Fig. 16–21

EXAMPLE 16.11

The cylinder shown in Fig. 16–22a rolls without slipping between the two moving plates E and D. Determine the angular velocity of the cylinder and the velocity of its center C.

SOLUTION

Since no slipping occurs, the contact points A and B on the cylinder have the same velocities as the plates E and D, respectively. The radial lines drawn from these points are collinear, so that by the proportionality of right triangles the IC is located at a point on line AB, Fig. 16–22b. Assuming this point to be a distance x from B, we have

(b)

Fig. 16–22

$v_B = \omega x;$ $0.4 \text{ m/s} = \omega x$
$v_A = \omega(0.25 \text{ m} - x);$ $0.25 \text{ m/s} = \omega(0.25 \text{ m} - x)$

Dividing one equation into the other eliminates ω and yields

$$0.4(0.25 - x) = 0.25x$$

$$x = \frac{0.1}{0.65} = 0.1538 \text{ m}$$

Hence, the angular velocity of the cylinder is

$$\omega = \frac{v_B}{x} = \frac{0.4 \text{ m/s}}{0.1538 \text{ m}} = 2.60 \text{ rad/s} \downarrow \qquad \textit{Ans.}$$

The velocity of point C is therefore

$$v_C = \omega r_{C/IC} = 2.60 \text{ rad/s } (0.1538 \text{ m} - 0.125 \text{ m})$$

$$= 0.0750 \text{ m/s} \leftarrow \qquad \textit{Ans.}$$

EXAMPLE 16.12

The crankshaft AB turns with a clockwise angular velocity of 10 rad/s, Fig. 16–23a. Determine the velocity of the piston at the instant shown.

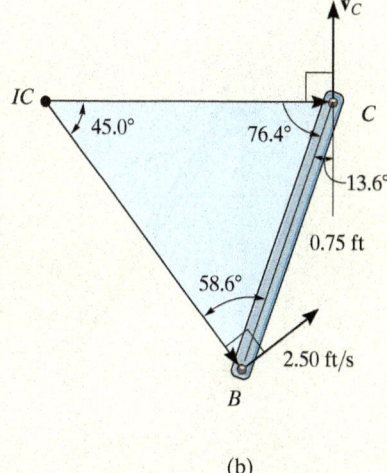

(a) (b)

Fig. 16–23

SOLUTION
The crankshaft rotates about a fixed axis, and so the velocity of point B is
$$v_B = 10 \text{ rad/s } (0.25 \text{ ft}) = 2.50 \text{ ft/s } \measuredangle 45°$$

Since the directions of the velocities of B and C are known, then the location of the IC for rod BC is at the intersection of the lines extended from these points, perpendicular to $\mathbf{v}_B$ and $\mathbf{v}_C$, Fig. 16–23b. The interior angles to the shaded triangle are determined, and the magnitudes of $\mathbf{r}_{B/IC}$ and $\mathbf{r}_{C/IC}$ can be obtained from the law of sines, i.e.,

$$\frac{0.75 \text{ ft}}{\sin 45°} = \frac{r_{B/IC}}{\sin 76.4°}$$

$$r_{B/IC} = 1.031 \text{ ft}$$

$$\frac{0.75 \text{ ft}}{\sin 45°} = \frac{r_{C/IC}}{\sin 58.6°}$$

$$r_{C/IC} = 0.9056 \text{ ft}$$

The rotational sense of ω_{BC} must be the same as the rotation caused by $\mathbf{v}_B$ about the IC, which is counterclockwise. Therefore,

$$\omega_{BC} = \frac{v_B}{r_{B/IC}} = \frac{2.50 \text{ ft/s}}{1.031 \text{ ft}} = 2.425 \text{ rad/s}$$

Using this result, the velocity of the piston is
$$v_C = \omega_{BC} r_{C/IC} = (2.425 \text{ rad/s})(0.9056 \text{ ft}) = 2.20 \text{ ft/s} \quad \textit{Ans.}$$

FUNDAMENTAL PROBLEMS

F16–13. Determine the angular velocity of the rod and the velocity of point C at the instant shown.

Prob. F16–13

F16–14. Determine the angular velocity of link BC and velocity of the piston C at the instant shown.

Prob. F16–14

F16–15. If the center O of the gear is moving with a speed of $v_O = 6$ m/s, determine the velocity of point A on the gear. The gear rack B is fixed.

Prob. F16–15

F16–16. If cable AB is unwound with a speed of 3 m/s, and the gear rack C has a speed of 1.5 m/s, determine the angular velocity of the gear and the velocity of its center O.

Prob. F16–16

F16–17. Determine the angular velocity of link BC and the velocity of the piston C at the instant shown.

Prob. F16–17

F16–18. Determine the angular velocity of links BC and CD at the instant shown.

Prob. F16–18

PROBLEMS

16–81. In each case show graphically how to locate the instantaneous center of zero velocity of link AB. Assume the geometry is known. The disk rolls without slipping.

*16–84. The conveyor belt is moving to the right at $v = 8$ ft/s, and at the same instant the cylinder is rolling counterclockwise at $\omega = 2$ rad/s without slipping. Determine the velocities of the cylinder's center C and point B at this instant.

16–85. The conveyor belt is moving to the right at $v = 12$ ft/s, and at the same instant the cylinder is rolling counterclockwise at $\omega = 6$ rad/s while its center has a velocity of 4 ft/s to the left. Determine the velocities of points A and B on the disk at this instant. Does the cylinder slip on the conveyor?

Prob. 16–81

16–82. Due to slipping, points A and B on the rim of the disk have the velocities shown. Determine the velocities of the center point C and point D at this instant.

16–83. Due to slipping, points A and B on the rim of the disk have the velocities shown. Determine the velocities of the center point C and point E at this instant.

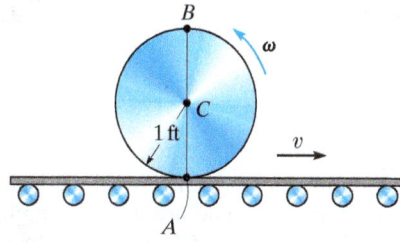

Probs. 16–84/85

16–86. The disk of radius r is confined to roll without slipping at A and B. If the plates have the velocities shown, determine the angular velocity of the disk.

Probs. 16–82/83

Prob. 16–86

16–87. The mechanism shown is used in a riveting machine. It consists of a driving piston A, three members, and a riveter which is attached to the slider block D. Determine the velocity of D at the instant shown, when the piston at A is traveling at $v_A = 30$ m/s.

16–89. The mechanism produces intermittent motion of link AB. If the sprocket S is turning with an angular velocity of $\omega_S = 6$ rad/s, determine the angular velocity of link AB at this instant. The sprocket S is mounted on a shaft which is separate from a collinear shaft attached to AB at A. The pin at C is attached to the vertical chain links.

Prob. 16–87

Prob. 16–89

*16–88. In an automobile transmission the planet pinions A and B rotate on shafts that are mounted on the planet-pinion carrier CD. As shown, CD is attached to a shaft at E which is aligned wih the center of the *fixed* sun gear S. This shaft is not attached to the sun gear. If CD is rotating at $\omega_{CD} = 8$ rad/s, determine the angular velocity of the ring gear R.

16–90. The transmission gear fixed onto the frame of an electric train turns with a constant rate of $\omega_t = 30$ rad/s. This gear is in mesh with the gear that is fixed to the axle of the engine. Determine the velocity of the train, assuming the wheels do not slip on the track.

16–91. The wheel of an electric train is driven by the transmission gear B, which is in mesh with the gear fixed to the wheel. If the wheel rolls without slipping, determine the angular velocity of the transmission gear needed to drive the train 40 ft/s. The transmission gear is fixed to the frame of the engine.

Prob. 16–88

Probs. 16–90/91

*16–92. Member AB is rotating at $\omega_{AB} = 6$ rad/s. Determine the velocity of point D and the angular velocity of members BPD and CD.

16–93. Member AB is rotating at $\omega_{AB} = 6$ rad/s. Determine the velocity of point P, and the angular velocity of member BPD.

16–95. As the car travels forward at 80 ft/s on a wet road, due to slipping, the rear wheels have an angular velocity $\omega = 100$ rad/s. Determine the speeds of points A, B, and C caused by the motion.

Prob. 16–95

Probs. 16–92/93

*16–96. If C has a velocity of $v_C = 3$ m/s, determine the angular velocity of the wheel at the instant shown.

16–94. If the center O of the gear is given a velocity of $v_O = 10$ m/s, determine the velocity of the slider block B at the instant shown.

Prob. 16–94

Prob. 16–96

16–97. The square plate is confined within the slots at A and B. When $\theta = 30°$, point A is moving at $v_A = 8$ m/s. Determine the velocity of point C at this instant.

16–98. The square plate is confined within the slots at A and B. When $\theta = 30°$, point A is moving at $v_A = 8$ m/s. Determine the velocity of point D at this instant.

*16–100.** The drums have the angular velocities at the instant shown. Determine the angular velocity of the pulley C and the velocity of the load D.

Prob. 16–100

Probs. 16–97/98

16–99. The crankshaft AB rotates at $\omega_{AB} = 50$ rad/s about the fixed axis through point A, and the disk at C is held fixed in its support at E. Determine the angular velocity of rod CD at the instant shown.

16–101. The mechanism used in a marine engine consists of a single crank AB and two connecting rods BC and BD. Determine the velocity of the piston at C the instant the crank is in the position shown and has an angular velocity of 5 rad/s.

16–102. The mechanism used in a marine engine consists of a single crank AB and two connecting rods BC and BD. Determine the velocity of the piston at D the instant the crank is in the position shown and has an angular velocity of 5 rad/s.

Prob. 16–99

Probs. 16–101/102

16.7 RELATIVE-MOTION ANALYSIS: ACCELERATION

An equation that relates the accelerations of two points A and B on the bar (rigid body) shown in Fig. 16–24a is determined by differentiating $\mathbf{v}_B = \mathbf{v}_A + \mathbf{v}_{B/A}$ with respect to time. This yields

$$\frac{d\mathbf{v}_B}{dt} = \frac{d\mathbf{v}_A}{dt} + \frac{d\mathbf{v}_{B/A}}{dt}$$

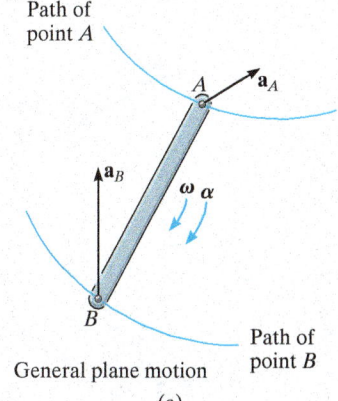

General plane motion
(a)

The terms $d\mathbf{v}_B/dt = \mathbf{a}_B$ and $d\mathbf{v}_A/dt = \mathbf{a}_A$ are measured with respect to a set of *fixed* x, y axes and represent the *absolute accelerations* of points B and A. The last term represents the acceleration of B with respect to A as measured by an observer fixed to translating x', y' axes which have their origin at the base point A. In Sec. 16.5 it was shown that to this observer point B appears to move along a *circular arc* that has a radius of curvature $r_{B/A}$. Consequently, $\mathbf{a}_{B/A}$ can be expressed in terms of its tangential and normal components; i.e., $\mathbf{a}_{B/A} = (\mathbf{a}_{B/A})_t + (\mathbf{a}_{B/A})_n$, where $(a_{B/A})_t = \alpha r_{B/A}$ and $(a_{B/A})_n = \omega^2 r_{B/A}$. Hence, the relative-acceleration equation can be written in the form

$$\mathbf{a}_B = \mathbf{a}_A + (\mathbf{a}_{B/A})_t + (\mathbf{a}_{B/A})_n \qquad (16\text{–}18)$$

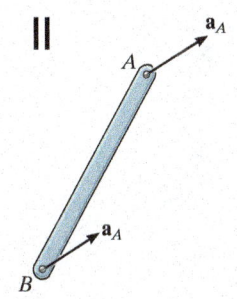

Translation
(b)

where

- $\mathbf{a}_B$ = acceleration of point B
- $\mathbf{a}_A$ = acceleration of point A
- $(\mathbf{a}_{B/A})_t$ = tangential acceleration component of B with respect to A. The *magnitude* is $(a_{B/A})_t = \alpha r_{B/A}$, and the *direction* is perpendicular to $\mathbf{r}_{B/A}$.
- $(\mathbf{a}_{B/A})_n$ = normal acceleration component of B with respect to A. The *magnitude* is $(a_{B/A})_n = \omega^2 r_{B/A}$, and the *direction* is always from B toward A.

Rotation about the base point A
(c)

These terms are represented graphically in Fig. 16–24. Here the acceleration of B, Fig. 16–24a, is determined by considering the bar to *translate* with an acceleration $\mathbf{a}_A$, Fig. 16–24b, and simultaneously *rotate* about the base point A with an instantaneous angular velocity $\boldsymbol{\omega}$ and angular acceleration $\boldsymbol{\alpha}$, Fig. 16–24c. Vector addition of these two effects, applied to B, yields $\mathbf{a}_B$, as shown in Fig. 16–24d.

(d)

Fig. 16–24

For a vector analysis, the relative-acceleration components can be expressed as $(\mathbf{a}_{B/A})_t = \boldsymbol{\alpha} \times \mathbf{r}_{B/A}$ and $(\mathbf{a}_{B/A})_n = -\omega^2 \mathbf{r}_{B/A}$, Eq. 16–14, and so, Eq. 16–18 then becomes

$$\mathbf{a}_B = \mathbf{a}_A + \boldsymbol{\alpha} \times \mathbf{r}_{B/A} - \omega^2 \mathbf{r}_{B/A} \qquad (16\text{–}19)$$

where

$\mathbf{a}_B$ = acceleration of point B
$\mathbf{a}_A$ = acceleration of the base point A
$\boldsymbol{\alpha}$ = angular acceleration of the body
ω = angular velocity of the body
$\mathbf{r}_{B/A}$ = position vector directed from A to B

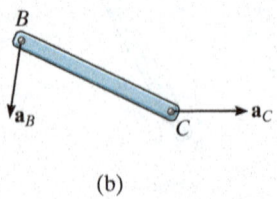

(b)

Fig. 16–25

When Eq. 16–18 or Eq. 16–19 is applied to study the accelerated motion of a rigid body which is *pin connected* to two other bodies, it should be realized that points which are *coincident at the pin* move with the *same acceleration*, since the path of motion over which they travel is the *same*. For example, point B lying on either rod BA or BC of the crank mechanism shown in Fig. 16–25a has the same acceleration, since the rods are pin connected at B. Here the motion of B is along a *circular path*, so that $\mathbf{a}_B$ can be expressed in terms of its tangential and normal components. At the other end of rod BC point C moves along a *straight-lined path*, which is defined by the piston. Hence, $\mathbf{a}_C$ is horizontal, Fig. 16–25b.

Rolling without slipping. Consider a wheel that rolls without slipping, Fig. 16–26. We have shown that the velocity of its center can be found using Eq. 16–17, namely

$$v_G = \omega r$$

Fig. 16–26

Since G moves along a *straight line*, its acceleration can be determined from the time derivative of its velocity.

$$\frac{dv_G}{dt} = \frac{d\omega}{dt} r$$

$$a_G = \alpha r \qquad (16\text{–}20)$$

This result was also obtained in Example 16–4. It applies as well to any circular object, such as a ball, gear, disk, etc., that *rolls without slipping*.

PROCEDURE FOR ANALYSIS

The relative-acceleration equation can be applied between any two points A and B on a body either by using a Cartesian vector analysis, or by writing the x and y scalar component equations directly.

Velocity Analysis.
- Determine the angular velocity $\boldsymbol{\omega}$ of the body by using a velocity analysis as discussed in Sec. 16.5 or 16.6. Also, determine the velocities $\mathbf{v}_A$ and $\mathbf{v}_B$ of points A and B if these points move along curved paths.

Vector Analysis
Kinematic Diagram.
- Establish the directions of the fixed x, y coordinates and draw a kinematic diagram of the body that shows $\mathbf{a}_A$, $\mathbf{a}_B$, $\boldsymbol{\omega}$, $\boldsymbol{\alpha}$, and $\mathbf{r}_{B/A}$.
- If points A and B move along *curved paths*, then their accelerations should be indicated in terms of their tangential and normal components, i.e., $\mathbf{a}_A = (\mathbf{a}_A)_t + (\mathbf{a}_A)_n$ and $\mathbf{a}_B = (\mathbf{a}_B)_t + (\mathbf{a}_B)_n$.

Acceleration Equation.
- To apply $\mathbf{a}_B = \mathbf{a}_A + \boldsymbol{\alpha} \times \mathbf{r}_{B/A} - \omega^2 \mathbf{r}_{B/A}$, express the vectors in Cartesian vector form and substitute them into the equation. Evaluate the cross product and then equate the respective $\mathbf{i}$ and $\mathbf{j}$ components to obtain two scalar equations.
- If the solution yields a *negative* answer for an *unknown* magnitude, it indicates that the sense of direction of the vector is opposite to that shown on the kinematic diagram.

Scalar Analysis
Kinematic Diagram.
- If the acceleration equation is applied in scalar form, then the magnitudes and directions of the relative-acceleration components $(\mathbf{a}_{B/A})_t$ and $(\mathbf{a}_{B/A})_n$ must be established. To do this draw a kinematic diagram such as shown in Fig. 16–27c. Since the body is considered to be momentarily "pinned" at the base point A, the *magnitudes* of these components are $(a_{B/A})_t = \alpha r_{B/A}$ and $(a_{B/A})_n = \omega^2 r_{B/A}$. Their *sense of direction* is established from the diagram such that $(\mathbf{a}_{B/A})_t$ acts perpendicular to $\mathbf{r}_{B/A}$, in accordance with the rotational motion $\boldsymbol{\alpha}$ of the body, and $(\mathbf{a}_{B/A})_n$ is directed from B toward A.*

Acceleration Equation.
- Represent the vectors $\mathbf{a}_B = \mathbf{a}_A + (\mathbf{a}_{B/A})_t + (\mathbf{a}_{B/A})_n$ graphically by showing their magnitudes and directions underneath each term. The scalar equations are determined from the x and y components of these vectors.

*The notation $\mathbf{a}_B = \mathbf{a}_A + (\mathbf{a}_{B/A(\text{pin})})_t + (\mathbf{a}_{B/A(\text{pin})})_n$ may be helpful in recalling that A is assumed to be pinned.

The mechanism for a window is shown. Here CA rotates about a fixed axis through C, and AB undergoes general plane motion. Since point A moves along a curved path it has two components of acceleration, whereas point B moves along a straight track and so the direction of its acceleration is specified.

EXAMPLE 16.13

(a)

(b)

The rod AB shown in Fig. 16–27a is confined to move along the inclined planes. If point A has an acceleration of 3 m/s² and a velocity of 2 m/s, both directed down the plane at the instant the rod is horizontal, determine the angular acceleration of the rod at this instant.

SOLUTION I (VECTOR ANALYSIS)

We will apply the acceleration equation to points A and B on the rod. To do so it is first necessary to determine the angular velocity of the rod. Show that it is $\omega = 0.283$ rad/s ↻ using either the velocity equation or the method of instantaneous centers.

Kinematic Diagram. Since points A and B both move along straight-line paths, they have *no* components of acceleration normal to the paths. There are two unknowns in Fig. 16–27b, namely, a_B and α.

Acceleration Equation.

$$\mathbf{a}_B = \mathbf{a}_A + \boldsymbol{\alpha} \times \mathbf{r}_{B/A} - \omega^2 \mathbf{r}_{B/A}$$

$$a_B \cos 45°\mathbf{i} + a_B \sin 45°\mathbf{j} = 3\cos 45°\mathbf{i} - 3\sin 45°\mathbf{j} + (\alpha\mathbf{k}) \times (10\mathbf{i}) - (0.283)^2(10\mathbf{i})$$

Carrying out the cross product and equating the $\mathbf{i}$ and $\mathbf{j}$ components yields

$$a_B \cos 45° = 3 \cos 45° - (0.283)^2(10) \quad (1)$$
$$a_B \sin 45° = -3 \sin 45° + \alpha(10) \quad (2)$$

Solving, we have

$$a_B = 1.87 \text{ m/s}^2 \; ⦨45°$$
$$\alpha = 0.344 \text{ rad/s}^2 \; ↻ \qquad \textit{Ans.}$$

(c)

Fig. 16–27

SOLUTION II (SCALAR ANALYSIS)

From the kinematic diagram, showing the relative-acceleration components $(\mathbf{a}_{B/A})_t$ and $(\mathbf{a}_{B/A})_n$, Fig. 16–27c, we have

$$\mathbf{a}_B = \mathbf{a}_A + (\mathbf{a}_{B/A})_t + (\mathbf{a}_{B/A})_n$$

$$\begin{bmatrix} a_B \\ ⦨45° \end{bmatrix} = \begin{bmatrix} 3 \text{ m/s}^2 \\ ⦩45° \end{bmatrix} + \begin{bmatrix} \alpha(10 \text{ m}) \\ \uparrow \end{bmatrix} + \begin{bmatrix} (0.283 \text{ rad/s})^2(10 \text{ m}) \\ \leftarrow \end{bmatrix}$$

Equating the x and y components of these four terms yields Eq. 1 and 2, and the solution proceeds as before.

EXAMPLE 16.14

The disk rolls without slipping and has the angular motion shown in Fig. 16–28a. Determine the acceleration of point A at this instant.

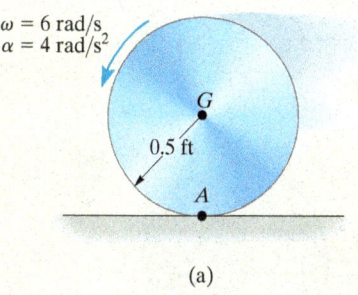

(a)

SOLUTION I (VECTOR ANALYSIS)
Kinematic Diagram. Since no slipping occurs, applying Eq. 16–20,

$$a_G = \alpha r = (4 \text{ rad/s}^2)(0.5 \text{ ft}) = 2 \text{ ft/s}^2$$

Acceleration Equation.
We will apply the acceleration equation to points G and A, Fig. 16–28b,

$$\mathbf{a}_A = \mathbf{a}_G + \boldsymbol{\alpha} \times \mathbf{r}_{A/G} - \omega^2 \mathbf{r}_{A/G}$$
$$\mathbf{a}_A = -2\mathbf{i} + (4\mathbf{k}) \times (-0.5\mathbf{j}) - (6)^2(-0.5\mathbf{j})$$
$$= \{18\mathbf{j}\} \text{ ft/s}^2$$

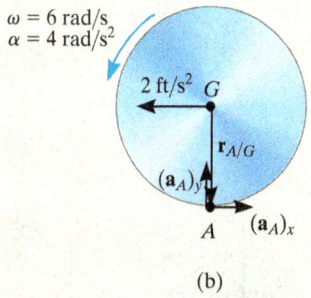

(b)

SOLUTION II (SCALAR ANALYSIS)
Using the result for $a_G = 2 \text{ ft/s}^2$ determined above, and from the kinematic diagram, showing the relative motion $\mathbf{a}_{A/G}$, Fig. 16–28c, we have

$$\mathbf{a}_A = \mathbf{a}_G + (\mathbf{a}_{A/G})_x + (\mathbf{a}_{A/G})_y$$

$$\begin{bmatrix}(a_A)_x \\ \rightarrow \end{bmatrix} + \begin{bmatrix}(a_A)_y \\ \uparrow \end{bmatrix} = \begin{bmatrix}2 \text{ ft/s}^2 \\ \leftarrow \end{bmatrix} + \begin{bmatrix}(4 \text{ rad/s}^2)(0.5 \text{ ft}) \\ \rightarrow \end{bmatrix} + \begin{bmatrix}(6 \text{ rad/s})^2(0.5 \text{ ft}) \\ \uparrow \end{bmatrix}$$

$$\xrightarrow{+} \quad (a_A)_x = -2 + 2 = 0$$

$$+\uparrow \quad (a_A)_y = 18 \text{ ft/s}^2$$

Therefore,

$$a_A = \sqrt{(0)^2 + (18 \text{ ft/s}^2)^2} = 18 \text{ ft/s}^2 \quad \textit{Ans.}$$

(c)

Fig. 16–28

NOTE: The fact that $a_A = 18 \text{ ft/s}^2$ indicates that the instantaneous center of zero velocity, point A, is *not* a point of zero acceleration.

EXAMPLE 16.15

The spool in Fig. 16–29a unravels from the cord, such that at the instant shown it has an angular velocity of 3 rad/s and an angular acceleration of 4 rad/s². Determine the magnitude of acceleration of point B.

SOLUTION I (VECTOR ANALYSIS)

The spool "appears" to be rolling downward without slipping at point A. Therefore, we can use the results of Eq. 16-20 to determine the acceleration of point G, i.e.,

$$a_G = \alpha r = (4 \text{ rad/s}^2)(0.5 \text{ ft}) = 2 \text{ ft/s}^2$$

We will apply the acceleration equation to points G and B.

Kinematic Diagram. Point B moves along a *curved path* having an *unknown* radius of curvature.* Its acceleration will be represented by its unknown x and y components as shown in Fig. 16–29b.

Acceleration Equation.

$$\mathbf{a}_B = \mathbf{a}_G + \boldsymbol{\alpha} \times \mathbf{r}_{B/G} - \omega^2 \mathbf{r}_{B/G}$$
$$(a_B)_x \mathbf{i} + (a_B)_y \mathbf{j} = -2\mathbf{j} + (-4\mathbf{k}) \times (0.75\mathbf{j}) - (3)^2(0.75\mathbf{j})$$

Equating the **i** and **j** terms, the component equations are

$$(a_B)_x = 4(0.75) = 3 \text{ ft/s}^2 \rightarrow \quad (1)$$
$$(a_B)_y = -2 - 6.75 = -8.75 \text{ ft/s}^2 = 8.75 \text{ ft/s}^2 \downarrow \quad (2)$$

The magnitude and direction of $\mathbf{a}_B$ are therefore

$$a_B = \sqrt{(3)^2 + (8.75)^2} = 9.25 \text{ ft/s}^2 \quad \textit{Ans.}$$

$$\theta = \tan^{-1}\frac{8.75}{3} = 71.1° \quad \textit{Ans.}$$

SOLUTION II (SCALAR ANALYSIS)

This problem may be solved by writing the scalar component equations directly. The kinematic diagram in Fig. 16–29c shows the relative-acceleration components $(\mathbf{a}_{B/G})_t$ and $(\mathbf{a}_{B/G})_n$. Thus,

$$\mathbf{a}_B = \mathbf{a}_G + (\mathbf{a}_{B/G})_t + (\mathbf{a}_{B/G})_n$$

$$\begin{bmatrix}(a_B)_x \\ \rightarrow\end{bmatrix} + \begin{bmatrix}(a_B)_y \\ \uparrow\end{bmatrix}$$

$$= \begin{bmatrix}2 \text{ ft/s}^2 \\ \downarrow\end{bmatrix} + \begin{bmatrix}4 \text{ rad/s}^2 (0.75 \text{ ft}) \\ \rightarrow\end{bmatrix} + \begin{bmatrix}(3 \text{ rad/s})^2(0.75 \text{ ft}) \\ \downarrow\end{bmatrix}$$

The x and y components yield Eqs. 1 and 2 above.

Fig. 16–29

*Realize that the path's radius of curvature ρ is not equal to the radius of the spool since the spool is not rotating about point G. Furthermore, ρ is not defined as the distance from A (IC) to B, since the location of the IC depends only on the velocity of a point and not the geometry of its path.

EXAMPLE 16.16

The collar C in Fig. 16–30a moves downward with an acceleration of 1 m/s². At the instant shown, it has a speed of 2 m/s which gives links CB and AB an angular velocity $\omega_{AB} = \omega_{CB} = 10$ rad/s. (See Example 16.8.) Determine the angular accelerations of CB and AB at this instant.

(a)

SOLUTION (VECTOR ANALYSIS)

Kinematic Diagram. The kinematic diagrams of both links AB and CB are shown in Fig. 16–30b. To solve, we will apply the appropriate kinematic equation to each link.

Acceleration Equation.

Link AB (rotation about a fixed axis):

$$\mathbf{a}_B = \boldsymbol{\alpha}_{AB} \times \mathbf{r}_B - \omega_{AB}^2 \mathbf{r}_B$$
$$\mathbf{a}_B = (\alpha_{AB}\mathbf{k}) \times (-0.2\mathbf{j}) - (10)^2(-0.2\mathbf{j})$$
$$\mathbf{a}_B = 0.2\alpha_{AB}\mathbf{i} + 20\mathbf{j}$$

Note that $\mathbf{a}_B$ has n and t components since it moves along a *circular path*.

Link BC (general plane motion): Using the result for $\mathbf{a}_B$ and applying Eq. 16–18, we have

$$\mathbf{a}_B = \mathbf{a}_C + \boldsymbol{\alpha}_{CB} \times \mathbf{r}_{B/C} - \omega_{CB}^2 \mathbf{r}_{B/C}$$
$$0.2\alpha_{AB}\mathbf{i} + 20\mathbf{j} = -1\mathbf{j} + (\alpha_{CB}\mathbf{k}) \times (0.2\mathbf{i} - 0.2\mathbf{j}) - (10)^2(0.2\mathbf{i} - 0.2\mathbf{j})$$
$$0.2\alpha_{AB}\mathbf{i} + 20\mathbf{j} = -1\mathbf{j} + 0.2\alpha_{CB}\mathbf{j} + 0.2\alpha_{CB}\mathbf{i} - 20\mathbf{i} + 20\mathbf{j}$$

Thus,

$$0.2\alpha_{AB} = 0.2\alpha_{CB} - 20$$
$$20 = -1 + 0.2\alpha_{CB} + 20$$

Solving,

$$\alpha_{CB} = 5 \text{ rad/s}^2 \;\circlearrowleft \qquad \qquad Ans.$$
$$\alpha_{AB} = -95 \text{ rad/s}^2 = 95 \text{ rad/s}^2 \;\circlearrowright \qquad Ans.$$

(b)

Fig. 16–30

EXAMPLE 16.17

The crankshaft AB turns with a clockwise angular acceleration of 20 rad/s^2, Fig. 16–31a. Determine the acceleration of the piston at the instant AB is in the position shown. At this instant $\omega_{AB} = 10 \text{ rad/s}$ and $\omega_{BC} = 2.43 \text{ rad/s}$. (See Example 16.12.)

SOLUTION (VECTOR ANALYSIS)

Kinematic Diagram. The kinematic diagrams for both AB and BC are shown in Fig. 16–31b. Here $\mathbf{a}_C$ is vertical since C moves along a straight-line path.

Acceleration Equation. Expressing each of the position vectors in Cartesian vector form

$$\mathbf{r}_B = \{-0.25 \sin 45°\mathbf{i} + 0.25 \cos 45°\mathbf{j}\} \text{ ft} = \{-0.177\mathbf{i} + 0.177\mathbf{j}\} \text{ ft}$$
$$\mathbf{r}_{C/B} = \{0.75 \sin 13.6°\mathbf{i} + 0.75 \cos 13.6°\mathbf{j}\} \text{ ft} = \{0.177\mathbf{i} + 0.729\mathbf{j}\} \text{ ft}$$

Crankshaft AB (rotation about a fixed axis):

$$\mathbf{a}_B = \boldsymbol{\alpha}_{AB} \times \mathbf{r}_B - \omega_{AB}^2 \mathbf{r}_B$$
$$= (-20\mathbf{k}) \times (-0.177\mathbf{i} + 0.177\mathbf{j}) - (10)^2(-0.177\mathbf{i} + 0.177\mathbf{j})$$
$$= \{21.21\mathbf{i} - 14.14\mathbf{j}\} \text{ ft/s}^2$$

Connecting Rod BC (general plane motion): Using the result for $\mathbf{a}_B$ and noting that $\mathbf{a}_C$ is in the vertical direction, we have

$$\mathbf{a}_C = \mathbf{a}_B + \boldsymbol{\alpha}_{BC} \times \mathbf{r}_{C/B} - \omega_{BC}^2 \mathbf{r}_{C/B}$$
$$a_C\mathbf{j} = 21.21\mathbf{i} - 14.14\mathbf{j} + (\alpha_{BC}\mathbf{k}) \times (0.177\mathbf{i} + 0.729\mathbf{j}) - (2.43)^2(0.177\mathbf{i} + 0.729\mathbf{j})$$
$$a_C\mathbf{j} = 21.21\mathbf{i} - 14.14\mathbf{j} + 0.177\alpha_{BC}\mathbf{j} - 0.729\alpha_{BC}\mathbf{i} - 1.04\mathbf{i} - 4.30\mathbf{j}$$
$$0 = 20.17 - 0.729\alpha_{BC}$$
$$a_C = 0.177\alpha_{BC} - 18.45$$

Solving yields

$$\alpha_{BC} = 27.7 \text{ rad/s}^2 \;\;\;\;\;$$
$$a_C = -13.5 \text{ ft/s}^2 \quad\quad\quad Ans.$$

NOTE: Since the piston is moving upward, the negative sign for a_C indicates that the piston is decelerating, i.e., $\mathbf{a}_C = \{-13.5\mathbf{j}\} \text{ ft/s}^2$. This causes the speed of the piston to decrease until AB becomes vertical, at which time the piston is momentarily at rest.

Fig. 16–31

FUNDAMENTAL PROBLEMS

F16–19. At the instant shown, end A of the rod has the velocity and acceleration shown. Determine the angular acceleration of the rod and acceleration of end B of the rod.

Prob. F16–19

F16–20. The gear rolls on the fixed rack with an angular velocity of $\omega = 12$ rad/s and angular acceleration of $\alpha = 6$ rad/s^2. Determine the horizontal and vertical components of acceleration of point A.

Prob. F16–20

F16–21. The gear rolls on the fixed rack B. At the instant shown, the center O of the gear moves with a velocity of $v_O = 6$ m/s and acceleration of $a_O = 3$ m/s^2. Determine the angular acceleration of the gear and the horizontal and vertical components of acceleration of point A at this instant.

Prob. F16–21

F16–22. At the instant shown, cable AB has a velocity of 3 m/s and acceleration of 1.5 m/s^2, while the gear rack has a velocity of 1.5 m/s and acceleration of 0.75 m/s^2. Determine the angular acceleration of the gear at this instant.

Prob. F16–22

F16-23 At the instant shown, the wheel rotates with an angular velocity of $\omega = 12$ rad/s and an angular acceleration of $\alpha = 6$ rad/s^2. Determine the angular acceleration of link BC at the instant shown.

Prob. F16–23

F16–24. At the instant shown, wheel A rotates with an angular velocity of $\omega = 6$ rad/s and an angular acceleration of $\alpha = 3$ rad/s^2. Determine the angular acceleration of link BC and the acceleration of piston C at this instant.

Prob. F16–24

PROBLEMS

16–103. The disk is rotating at a constant rate of 4 rad/s as it falls freely, its center having an acceleration of 32.2 ft/s². Determine the accelerations of points A and B on the rim of the disk at the instant shown.

16–106. At a given instant, the cables supporting the pipe have the motions shown. Determine the angular velocity and angular acceleration of the pipe, and the velocity and acceleration of point B located on the pipe.

Prob. 16–103

Prob. 16–106

*16–104. At a given instant A has the motion shown. Determine the acceleration of B and the angular acceleration of the bar at this instant.

16–105. At a given instant A has the motion shown. Determine the acceleration of point C at this instant.

16–107. At a given instant the roller A on the bar has the velocity and acceleration shown. Determine the velocity and acceleration of the roller B, and the bar's angular velocity and angular acceleration at this instant.

Probs. 16–104/105

Prob. 16–107

***16–108.** The rod is confined to move along the path due to the pins at its ends. At the instant shown, point A has the motion shown. Determine the velocity and acceleration of point B at this instant.

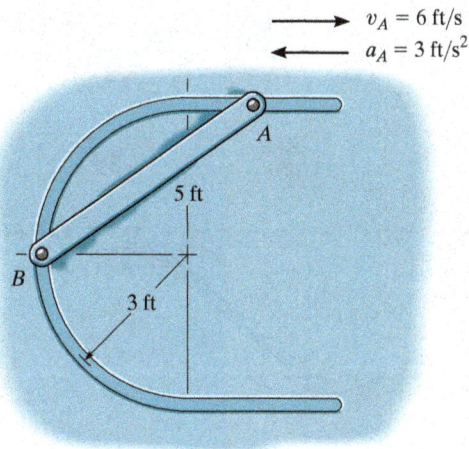

Prob. 16–108

16–109. The ends of bar AB are confined to move along the paths shown. At a given instant, A has a velocity of 8 ft/s and an acceleration of 3 ft/s². Determine the angular velocity and angular acceleration of AB at this instant.

Prob. 16–109

16–110. Determine the angular acceleration of link AB if link CD has the angular velocity and angular deceleration shown.

Prob. 16–110

16–111. At a given instant the slider block A is moving to the right with the motion shown. Determine the angular acceleration of link AB and the acceleration of point B at this instant.

Prob. 16–111

*16–112. Determine the angular acceleration of link CD if link AB has the angular velocity and angular acceleration shown.

Prob. 16–112

16–113. The hoop is cast on the rough surface such that it has an angular velocity $\omega = 4$ rad/s and an angular acceleration $\alpha = 5$ rad/s^2. Also, its center has a velocity $v_O = 5$ m/s and a deceleration $a_O = 2$ m/s^2. Determine the acceleration of point A at this instant.

16–114. The hoop is cast on the rough surface such that it has an angular velocity $\omega = 4$ rad/s and an angular acceleration $\alpha = 5$ rad/s^2. Also, its center has a velocity $v_O = 5$ m/s and a deceleration $a_O = 2$ m/s^2. Determine the acceleration of point B at this instant.

Probs. 16–113/114

16–115. At a given instant the slider block B is moving to the right with the motion shown. Determine the angular acceleration of link AB and the acceleration of point A at this instant.

Prob. 16–115

*16–116. The hydraulic cylinder D extends with a velocity of $v_B = 4$ ft/s and an acceleration of $a_B = 1.5$ ft/s^2. Determine the acceleration of A at the instant shown.

16–117. The hydraulic cylinder D extends with a velocity of $v_B = 4$ ft/s and an acceleration of $a_B = 1.5$ ft/s^2. Determine the acceleration of C at the instant shown.

Probs. 16–116/117

16–118. If member AB has the angular motion shown, determine the angular velocity and angular acceleration of member CD at the instant shown.

16–119. If member AB has the angular motion shown, determine the velocity and acceleration of point C at the instant shown.

Probs. 16–118/119

*16–120. Pulley A rotates with the angular velocity and angular acceleration shown. Determine the angular acceleration of pulley B at the instant shown.

16–121. Pulley A rotates with the angular velocity and angular acceleration shown. Determine the angular acceleration of block E at the instant shown.

Probs. 16–120/121

16–122. Gear A is held fixed, and arm DE rotates clockwise with an angular velocity of $\omega_{DE} = 6$ rad/s and an angular acceleration of $\alpha_{DE} = 3$ rad/s². Determine the angular acceleration of gear B at the instant shown.

16–123. Gear A rotates counterclockwise with a constant angular velocity of $\omega_A = 10$ rad/s, while arm DE rotates clockwise with an angular velocity of $\omega_{DE} = 6$ rad/s and an angular acceleration of $\alpha_{DE} = 3$ rad/s². Determine the angular acceleration of gear B at the instant shown.

Probs. 16–122/123

*16–124. The disk rolls without slipping such that it has an angular acceleration of $\alpha = 4$ rad/s² and angular velocity of $\omega = 2$ rad/s at the instant shown. Determine the acceleration of points A and B on the link and the link's angular acceleration at this instant. Assume point A lies on the periphery of the disk, 150 mm from C.

Prob. 16–124

16–125. The flywheel rotates with an angular velocity $\omega = 2$ rad/s and an angular acceleration $\alpha = 6$ rad/s². Determine the angular acceleration of links AB and BC at this instant.

16–127. The slider block moves with a velocity of $v_B = 5$ ft/s and an acceleration of $a_B = 3$ ft/s². Determine the angular acceleration of rod AB at the instant shown.

Prob. 16–125

Prob. 16–127

16–126. The hydraulic cylinder extends with a velocity of $v_A = 1.5$ m/s and an acceleration of $\alpha_A = 0.5$ m/s². Determine the angular acceleration of link ABC and the acceleration of end C at the instant shown. Point B is pin connected to the slider block.

*16–128. The slider block moves with a velocity of $v_B = 5$ ft/s and an acceleration of $a_B = 3$ ft/s². Determine the acceleration of A at the instant shown.

Prob. 16–126

Prob. 16–128

*16.8 RELATIVE-MOTION ANALYSIS USING ROTATING AXES

In the previous sections the relative-motion analysis for velocity and acceleration was described using a translating coordinate system. This type of analysis is useful for determining the motion of points on the *same* rigid body, or the motion of points located on several pin-connected bodies. In some problems, however, rigid bodies (mechanisms) are constructed such that *sliding* will occur at their connections. The kinematic analysis for such cases is best performed if the motion is analyzed using a coordinate system which both *translates* and *rotates*. Furthermore, this frame of reference is useful for analyzing the motions of two points on a mechanism which are *not* located in the *same* body and for specifying the kinematics of particle motion when the particle moves along a rotating path.

In the following analysis two equations will be developed which relate the velocity and acceleration of two points, one of which is the origin of a moving frame of reference subjected to both a translation and a rotation in the plane.*

Position. Consider the two points A and B shown in Fig. 16–32a. Their location is specified by the position vectors $\mathbf{r}_A$ and $\mathbf{r}_B$, which are measured with respect to the fixed X, Y, Z coordinate system. As shown in the figure, the "base point" A represents the origin of the x, y, z coordinate system, which is assumed to be both translating and rotating with respect to the X, Y, Z system. The position of B with respect to A is specified by the relative-position vector $\mathbf{r}_{B/A}$. The components of this vector may be expressed either in terms of unit vectors along the X, Y axes, i.e., $\mathbf{I}$ and $\mathbf{J}$, or by unit vectors along the x, y axes, i.e., $\mathbf{i}$ and $\mathbf{j}$. For the development which follows, $\mathbf{r}_{B/A}$ will be measured with respect to the moving x, y frame of reference. Thus, if B has coordinates (x_B, y_B), Fig. 16–32a, then

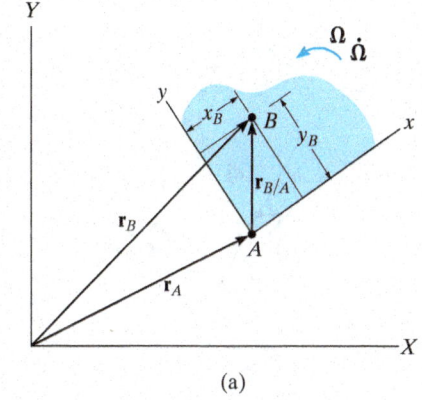

(a)

Fig. 16–32

$$\mathbf{r}_{B/A} = x_B\mathbf{i} + y_B\mathbf{j}$$

The three position vectors in Fig. 16–32a are related by vector addition.

$$\boxed{\mathbf{r}_B = \mathbf{r}_A + \mathbf{r}_{B/A}} \qquad (16\text{–}21)$$

At the instant considered, point A has a velocity $\mathbf{v}_A$ and an acceleration $\mathbf{a}_A$, while the angular velocity and angular acceleration of the x, y axes are $\boldsymbol{\Omega}$ (omega) and $\dot{\boldsymbol{\Omega}} = d\boldsymbol{\Omega}/dt$, respectively.

---
*The more general, three-dimensional motion of the points is developed in Sec. 20.4.

Velocity. The velocity of point B is determined by taking the time derivative of Eq. 16–21, which yields

$$\mathbf{v}_B = \mathbf{v}_A + \frac{d\mathbf{r}_{B/A}}{dt} \qquad (16\text{–}22)$$

The last term in this equation is evaluated as follows:

$$\frac{d\mathbf{r}_{B/A}}{dt} = \frac{d}{dt}(x_B\mathbf{i} + y_B\mathbf{j})$$

$$= \frac{dx_B}{dt}\mathbf{i} + x_B\frac{d\mathbf{i}}{dt} + \frac{dy_B}{dt}\mathbf{j} + y_B\frac{d\mathbf{j}}{dt}$$

$$= \left(\frac{dx_B}{dt}\mathbf{i} + \frac{dy_B}{dt}\mathbf{j}\right) + \left(x_B\frac{d\mathbf{i}}{dt} + y_B\frac{d\mathbf{j}}{dt}\right) \qquad (16\text{–}23)$$

The two terms in the first set of parentheses represent the components of velocity of point B as measured by an observer attached to the moving x, y, z coordinate system, $(\mathbf{v}_{B/A})_{xyz}$. In the second set of parentheses the time rate of change of the unit vectors $\mathbf{i}$ and $\mathbf{j}$ are measured by an observer located in the fixed X, Y, Z coordinate system. These changes, $d\mathbf{i}$ and $d\mathbf{j}$, are due *only* to the *rotation* $d\theta$ of the x, y, z axes, causing $\mathbf{i}$ to become $\mathbf{i}' = \mathbf{i} + d\mathbf{i}$ and $\mathbf{j}$ to become $\mathbf{j}' = \mathbf{j} + d\mathbf{j}$, Fig. 16–32b. As shown, the *magnitudes* of both $d\mathbf{i}$ and $d\mathbf{j}$ equal $1\, d\theta$, since $i = i' = j = j' = 1$. The *direction* of $d\mathbf{i}$ is defined by $+\mathbf{j}$, since $d\mathbf{i}$ is tangent to the path described by the arrowhead of $\mathbf{i}$ in the limit as $\Delta t \to dt$. Likewise, $d\mathbf{j}$ acts in the $-\mathbf{i}$ direction, Fig. 16–32b. Hence,

$$\frac{d\mathbf{i}}{dt} = \frac{d\theta}{dt}(\mathbf{j}) = \Omega\mathbf{j} \qquad \frac{d\mathbf{j}}{dt} = \frac{d\theta}{dt}(-\mathbf{i}) = -\Omega\mathbf{i}$$

(b)

Viewing the axes in three dimensions, Fig. 16–32c, and noting that $\Omega = \Omega \mathbf{k}$, we can express the above derivatives in terms of the cross product as

$$\frac{d\mathbf{i}}{dt} = \Omega \times \mathbf{i} \qquad \frac{d\mathbf{j}}{dt} = \Omega \times \mathbf{j} \qquad (16\text{–}24)$$

Substituting these results into Eq. 16–23 and using the distributive property of the vector cross product, we obtain

$$\frac{d\mathbf{r}_{B/A}}{dt} = (\mathbf{v}_{B/A})_{xyz} + \Omega \times (x_B\mathbf{i} + y_B\mathbf{j})$$

$$= (\mathbf{v}_{B/A})_{xyz} + \Omega \times \mathbf{r}_{B/A} \qquad (16\text{–}25)$$

(c)

Fig. 16–32 (cont.)

Hence, Eq. 16–22 becomes

$$\mathbf{v}_B = \mathbf{v}_A + \mathbf{\Omega} \times \mathbf{r}_{B/A} + (\mathbf{v}_{B/A})_{xyz} \qquad (16\text{–}26)$$

where

$\mathbf{v}_B$ = velocity of B, measured from the X, Y, Z reference
$\mathbf{v}_A$ = velocity of the origin A of the x, y, z reference, measured from the X, Y, Z reference
$(\mathbf{v}_{B/A})_{xyz}$ = velocity of "B with respect to A," as measured by an observer attached to the rotating x, y, z reference
$\mathbf{\Omega}$ = angular velocity of the x, y, z reference, measured from the X, Y, Z reference
$\mathbf{r}_{B/A}$ = position of B with respect to A

Comparing Eq. 16–26 with Eq. 16–16 ($\mathbf{v}_B = \mathbf{v}_A + \mathbf{\Omega} \times \mathbf{r}_{B/A}$), which is valid for a translating frame of reference, it can be seen that the only difference between these two equations is represented by the term $(\mathbf{v}_{B/A})_{xyz}$.

When applying Eq. 16–26 it is often useful to understand what each of the terms represents. In order of appearance, they are as follows:

$\mathbf{v}_B$ { absolute velocity of B } } motion of B observed from the X, Y, Z frame

(equals)

$\mathbf{v}_A$ { absolute velocity of the origin of x, y, z frame }

(plus)

$\mathbf{\Omega} \times \mathbf{r}_{B/A}$ { angular velocity effect caused by the rotation of x, y, z frame } } motion of x, y, z frame observed from the X, Y, Z frame

(plus)

$(\mathbf{v}_{B/A})_{xyz}$ { velocity of B with respect to A } } motion of B observed from the x, y, z frame

Acceleration. The acceleration of B, observed from the X, Y, Z coordinate system, may be expressed in terms of its motion measured with respect to the rotating system of coordinates by taking the time derivative of Eq. 16–26.

$$\frac{d\mathbf{v}_B}{dt} = \frac{d\mathbf{v}_A}{dt} + \frac{d\mathbf{\Omega}}{dt} \times \mathbf{r}_{B/A} + \mathbf{\Omega} \times \frac{d\mathbf{r}_{B/A}}{dt} + \frac{d(\mathbf{v}_{B/A})_{xyz}}{dt}$$

$$\mathbf{a}_B = \mathbf{a}_A + \dot{\mathbf{\Omega}} \times \mathbf{r}_{B/A} + \mathbf{\Omega} \times \frac{d\mathbf{r}_{B/A}}{dt} + \frac{d(\mathbf{v}_{B/A})_{xyz}}{dt} \quad (16\text{–}27)$$

Here $\dot{\mathbf{\Omega}} = d\mathbf{\Omega}/dt$ is the angular acceleration of the x, y, z coordinate system. Since $\mathbf{\Omega}$ is always perpendicular to the plane of motion, then $\dot{\mathbf{\Omega}}$ measures *only the change in magnitude* of $\mathbf{\Omega}$. The derivative $d\mathbf{r}_{B/A}/dt$ is defined by Eq. 16–25, so that

$$\mathbf{\Omega} \times \frac{d\mathbf{r}_{B/A}}{dt} = \mathbf{\Omega} \times (\mathbf{v}_{B/A})_{xyz} + \mathbf{\Omega} \times (\mathbf{\Omega} \times \mathbf{r}_{B/A}) \quad (16\text{–}28)$$

Finding the time derivative of $(\mathbf{v}_{B/A})_{xyz} = (v_{B/A})_x \mathbf{i} + (v_{B/A})_y \mathbf{j}$,

$$\frac{d(\mathbf{v}_{B/A})_{xyz}}{dt} = \left[\frac{d(v_{B/A})_x}{dt}\mathbf{i} + \frac{d(v_{B/A})_y}{dt}\mathbf{j}\right] + \left[(v_{B/A})_x \frac{d\mathbf{i}}{dt} + (v_{B/A})_y \frac{d\mathbf{j}}{dt}\right]$$

The two terms in the first set of brackets represent the components of acceleration of point B as measured by an observer attached to the rotating coordinate system. These terms will be denoted by $(\mathbf{a}_{B/A})_{xyz}$. The terms in the second set of brackets can be simplified using Eqs. 16–24.

$$\frac{d(\mathbf{v}_{B/A})_{xyz}}{dt} = (\mathbf{a}_{B/A})_{xyz} + \mathbf{\Omega} \times (\mathbf{v}_{B/A})_{xyz}$$

Substituting this and Eq. 16–28 into Eq. 16–27 and rearranging terms,

$$\mathbf{a}_B = \mathbf{a}_A + \dot{\mathbf{\Omega}} \times \mathbf{r}_{B/A} + \mathbf{\Omega} \times (\mathbf{\Omega} \times \mathbf{r}_{B/A}) + 2\mathbf{\Omega} \times (\mathbf{v}_{B/A})_{xyz} + (\mathbf{a}_{B/A})_{xyz}$$

$$(16\text{–}29)$$

where

$\mathbf{a}_B$ = acceleration of B, measured from the X, Y, Z reference
$\mathbf{a}_A$ = acceleration of the origin A of the x, y, z reference, measured from the X, Y, Z reference
$(\mathbf{a}_{B/A})_{xyz}, (\mathbf{v}_{B/A})_{xyz}$ = acceleration and velocity of B with respect to A, as measured by an observer attached to the rotating x, y, z reference
$\dot{\mathbf{\Omega}}, \mathbf{\Omega}$ = angular acceleration and angular velocity of the x, y, z reference, measured from the X, Y, Z reference
$\mathbf{r}_{B/A}$ = position of B with respect to A

16.8 RELATIVE-MOTION ANALYSIS USING ROTATING AXES

If Eq. 16–29 is compared with Eq. 16–18, $\mathbf{a}_B = \mathbf{a}_A + \dot{\mathbf{\Omega}} \times \mathbf{r}_{B/A} + \mathbf{\Omega} \times (\mathbf{\Omega} \times \mathbf{r}_{B/A})$, which is valid for a translating frame of reference, it can be seen that the difference between these two equations is represented by the terms $2\mathbf{\Omega} \times (\mathbf{v}_{B/A})_{xyz}$ and $(\mathbf{a}_{B/A})_{xyz}$. In particular, $2\mathbf{\Omega} \times (\mathbf{v}_{B/A})_{xyz}$ is called the **Coriolis acceleration**, named after the French engineer G. C. Coriolis, who was the first to determine it. This term represents the difference in the acceleration of B as measured from nonrotating and rotating x, y, z axes. As indicated by the vector cross product, the Coriolis acceleration will *always* be perpendicular to both $\mathbf{\Omega}$ and $(\mathbf{v}_{B/A})_{xyz}$. It is an important component of acceleration which must be considered whenever rotating reference frames are used. This often occurs, for example, when studying the accelerations and forces which act on rockets, long-range projectiles, or other bodies having motions whose measurements are significantly affected by the rotation of the earth.

The following interpretation of the terms in Eq. 16–29 may be useful when applying this equation to the solution of problems.

$\mathbf{a}_B$ $\quad\{$ absolute acceleration of B $\}\quad\{$ motion of B observed from the X, Y, Z frame $\}$

(equals)

$\mathbf{a}_A$ $\quad\{$ absolute acceleration of the origin of x, y, z frame $\}$

(plus)

$\dot{\mathbf{\Omega}} \times \mathbf{r}_{B/A}$ $\quad\{$ angular acceleration effect caused by rotation of x, y, z frame $\}\quad\{$ motion of x, y, z frame observed from the X, Y, Z frame $\}$

(plus)

$\mathbf{\Omega} \times (\mathbf{\Omega} \times \mathbf{r}_{B/A})$ $\quad\{$ angular velocity effect caused by rotation of x, y, z frame $\}$

(plus)

$2\mathbf{\Omega} \times (\mathbf{v}_{B/A})_{xyz}$ $\quad\{$ combined effect of B moving relative to x, y, z coordinates and rotation of x, y, z frame $\}\quad\{$ interacting motion $\}$

(plus)

$(\mathbf{a}_{B/A})_{xyz}$ $\quad\{$ acceleration of B with respect to A $\}\quad\{$ motion of B observed from the x, y, z frame $\}$

PROCEDURE FOR ANALYSIS

Equations 16–26 and 16–29 can be applied to the solution of problems involving the planar motion of particles or rigid bodies using the following procedure.

Coordinate Axes.

- Choose an appropriate location for the origin and proper orientation of the axes for both fixed X, Y, Z and moving x, y, z reference frames.
- Most often solutions are easily obtained if at the instant considered:
  1. the origins are coincident
  2. the corresponding axes are collinear
  3. the corresponding axes are parallel
- The moving frame should be selected fixed to the body or device along which the relative motion occurs.

Kinematic Equations.

- After defining the origin A of the moving reference and specifying the moving point B, Eqs. 16–26 and 16–29 should be written in symbolic form.

$$\mathbf{v}_B = \mathbf{v}_A + \mathbf{\Omega} \times \mathbf{r}_{B/A} + (\mathbf{v}_{B/A})_{xyz}$$
$$\mathbf{a}_B = \mathbf{a}_A + \dot{\mathbf{\Omega}} \times \mathbf{r}_{B/A} + \mathbf{\Omega} \times (\mathbf{\Omega} \times \mathbf{r}_{B/A}) + 2\mathbf{\Omega} \times (\mathbf{v}_{B/A})_{xyz} + (\mathbf{a}_{B/A})_{xyz}$$

- The Cartesian components of all these vectors may be expressed along either the X, Y, Z axes or the x, y, z axes. The choice is arbitrary provided a consistent set of unit vectors is used.

- Motion of the moving reference is expressed by $\mathbf{v}_A, \mathbf{a}_A, \mathbf{\Omega}$, and $\dot{\mathbf{\Omega}}$; and motion of B with respect to the moving reference is expressed by $\mathbf{r}_{B/A}, (\mathbf{v}_{B/A})_{xyz}$, and $(\mathbf{a}_{B/A})_{xyz}$.

The rotation of the dumping bin of the truck about point C is operated by the extension of the hydraulic cylinder AB. To determine the rotation of the bin due to this extension, we can use the equations of relative motion and fix the rotating x, y, z axes to the cylinder so that the relative motion of the cylinder's extension occurs along the y axis.

EXAMPLE 16.18

At the instant $\theta = 60°$, the rod in Fig. 16–33 has an angular velocity of 3 rad/s and an angular acceleration of 2 rad/s^2. At this same instant, collar C travels outward along the rod such that when $x = 0.2$ m its velocity is 2 m/s and its acceleration is 3 m/s^2, both measured relative to the rod. Determine the Coriolis acceleration and the velocity and acceleration of the collar at this instant.

SOLUTION
Coordinate Axes. The origin of both coordinate systems is located at point O, Fig. 16–33. Since motion of the collar is reported relative to the rod, the moving x, y, z frame of reference is attached to the rod.

Kinematic Equations.
$$\mathbf{v}_C = \mathbf{v}_O + \mathbf{\Omega} \times \mathbf{r}_{C/O} + (\mathbf{v}_{C/O})_{xyz} \qquad (1)$$
$$\mathbf{a}_C = \mathbf{a}_O + \dot{\mathbf{\Omega}} \times \mathbf{r}_{C/O} + \mathbf{\Omega} \times (\mathbf{\Omega} \times \mathbf{r}_{C/O}) + 2\mathbf{\Omega} \times (\mathbf{v}_{C/O})_{xyz} + (\mathbf{a}_{C/O})_{xyz} \qquad (2)$$

Fig. 16–33

It will be simpler to express all the vectors in terms of their $\mathbf{i}, \mathbf{j}, \mathbf{k}$ components rather than $\mathbf{I}, \mathbf{J}, \mathbf{K}$ components. Hence,

Motion of moving reference	Motion of C with respect to moving reference
$\mathbf{v}_O = 0$	$\mathbf{r}_{C/O} = \{0.2\mathbf{i}\}$ m
$\mathbf{a}_O = 0$	$(\mathbf{v}_{C/O})_{xyz} = \{2\mathbf{i}\}$ m/s
$\mathbf{\Omega} = \{-3\mathbf{k}\}$ rad/s	$(\mathbf{a}_{C/O})_{xyz} = \{3\mathbf{i}\}$ m/s^2
$\dot{\mathbf{\Omega}} = \{-2\mathbf{k}\}$ rad/s^2	

The Coriolis acceleration is defined as
$$\mathbf{a}_{Cor} = 2\mathbf{\Omega} \times (\mathbf{v}_{C/O})_{xyz} = 2(-3\mathbf{k}) \times (2\mathbf{i}) = \{-12\mathbf{j}\} \text{ m/s}^2 \qquad Ans.$$

This vector is shown dashed in Fig. 16–33. If desired, it may be resolved into $\mathbf{I}, \mathbf{J}$ components acting along the X and Y axes, respectively.
The velocity and acceleration of the collar are determined by substituting the data into Eqs. 1 and 2 and evaluating the cross products, which yields

$$\mathbf{v}_C = \mathbf{v}_O + \mathbf{\Omega} \times \mathbf{r}_{C/O} + (\mathbf{v}_{C/O})_{xyz}$$
$$= 0 + (-3\mathbf{k}) \times (0.2\mathbf{i}) + 2\mathbf{i}$$
$$= \{2\mathbf{i} - 0.6\mathbf{j}\} \text{ m/s} \qquad Ans.$$

$$\mathbf{a}_C = \mathbf{a}_O + \dot{\mathbf{\Omega}} \times \mathbf{r}_{C/O} + \mathbf{\Omega} \times (\mathbf{\Omega} \times \mathbf{r}_{C/O}) + 2\mathbf{\Omega} \times (\mathbf{v}_{C/O})_{xyz} + (\mathbf{a}_{C/O})_{xyz}$$
$$= 0 + (-2\mathbf{k}) \times (0.2\mathbf{i}) + (-3\mathbf{k}) \times [(-3\mathbf{k}) \times (0.2\mathbf{i})] + 2(-3\mathbf{k}) \times (2\mathbf{i}) + 3\mathbf{i}$$
$$= 0 - 0.4\mathbf{j} - 1.80\mathbf{i} - 12\mathbf{j} + 3\mathbf{i}$$
$$= \{1.20\mathbf{i} - 12.4\mathbf{j}\} \text{ m/s}^2 \qquad Ans.$$

EXAMPLE 16.19

Fig. 16–34

Rod AB, shown in Fig. 16–34, rotates clockwise such that it has an angular velocity $\omega_{AB} = 3$ rad/s and angular acceleration $\alpha_{AB} = 4$ rad/s^2 when $\theta = 45°$. Determine the angular motion of rod DE at this instant. The collar at C is pin connected to AB and slides over rod DE.

SOLUTION

Coordinate Axes. The origin of both the fixed and moving frames of reference is located at D, Fig. 16–34. Furthermore, the x, y, z reference is attached to and rotates with rod DE so that the relative motion of the collar is easy to follow.

Kinematic Equations.

$$\mathbf{v}_C = \mathbf{v}_D + \mathbf{\Omega} \times \mathbf{r}_{C/D} + (\mathbf{v}_{C/D})_{xyz} \tag{1}$$

$$\mathbf{a}_C = \mathbf{a}_D + \dot{\mathbf{\Omega}} \times \mathbf{r}_{C/D} + \mathbf{\Omega} \times (\mathbf{\Omega} \times \mathbf{r}_{C/D}) + 2\mathbf{\Omega} \times (\mathbf{v}_{C/D})_{xyz} + (\mathbf{a}_{C/D})_{xyz} \tag{2}$$

All vectors will be expressed in terms of $\mathbf{i}, \mathbf{j}, \mathbf{k}$ components.

Motion of moving reference	Motion of C with respect to moving reference
$\mathbf{v}_D = 0$	$\mathbf{r}_{C/D} = \{0.4\mathbf{i}\}$ m
$\mathbf{a}_D = 0$	$(\mathbf{v}_{C/D})_{xyz} = (v_{C/D})_{xyz}\mathbf{i}$
$\mathbf{\Omega} = -\omega_{DE}\mathbf{k}$	$(\mathbf{a}_{C/D})_{xyz} = (a_{C/D})_{xyz}\mathbf{i}$
$\dot{\mathbf{\Omega}} = -\alpha_{DE}\mathbf{k}$	

Motion of C: Since the collar moves along a *circular path* of radius AC, its velocity and acceleration can be determined using Eqs. 16–9 and 16–14.

$$\mathbf{v}_C = \boldsymbol{\omega}_{AB} \times \mathbf{r}_{C/A} = (-3\mathbf{k}) \times (0.4\mathbf{i} + 0.4\mathbf{j}) = \{1.2\mathbf{i} - 1.2\mathbf{j}\} \text{ m/s}$$

$$\mathbf{a}_C = \boldsymbol{\alpha}_{AB} \times \mathbf{r}_{C/A} - \omega_{AB}^2 \mathbf{r}_{C/A}$$
$$= (-4\mathbf{k}) \times (0.4\mathbf{i} + 0.4\mathbf{j}) - (3)^2(0.4\mathbf{i} + 0.4\mathbf{j}) = \{-2\mathbf{i} - 5.2\mathbf{j}\} \text{ m/s}^2$$

Substituting the data into Eqs. 1 and 2, we have

$$\mathbf{v}_C = \mathbf{v}_D + \mathbf{\Omega} \times \mathbf{r}_{C/D} + (\mathbf{v}_{C/D})_{xyz}$$
$$1.2\mathbf{i} - 1.2\mathbf{j} = 0 + (-\omega_{DE}\mathbf{k}) \times (0.4\mathbf{i}) + (v_{C/D})_{xyz}\mathbf{i}$$
$$1.2\mathbf{i} - 1.2\mathbf{j} = 0 - 0.4\omega_{DE}\mathbf{j} + (v_{C/D})_{xyz}\mathbf{i}$$
$$(v_{C/D})_{xyz} = 1.2 \text{ m/s}$$
$$\omega_{DE} = 3 \text{ rad/s} \downarrow \qquad Ans.$$

$$\mathbf{a}_C = \mathbf{a}_D + \dot{\mathbf{\Omega}} \times \mathbf{r}_{C/D} + \mathbf{\Omega} \times (\mathbf{\Omega} \times \mathbf{r}_{C/D}) + 2\mathbf{\Omega} \times (\mathbf{v}_{C/D})_{xyz} + (\mathbf{a}_{C/D})_{xyz}$$
$$-2\mathbf{i} - 5.2\mathbf{j} = 0 + (-\alpha_{DE}\mathbf{k}) \times (0.4\mathbf{i}) + (-3\mathbf{k}) \times [(-3\mathbf{k}) \times (0.4\mathbf{i})] + 2(-3\mathbf{k}) \times (1.2\mathbf{i}) + (a_{C/D})_{xyz}\mathbf{i}$$
$$-2\mathbf{i} - 5.2\mathbf{j} = -0.4\alpha_{DE}\mathbf{j} - 3.6\mathbf{i} - 7.2\mathbf{j} + (a_{C/D})_{xyz}\mathbf{i}$$
$$(a_{C/D})_{xyz} = 1.6 \text{ m/s}^2$$
$$\alpha_{DE} = -5 \text{ rad/s}^2 = 5 \text{ rad/s}^2 \nwarrow \qquad Ans.$$

EXAMPLE 16.20

Planes A and B fly at the same elevation and have the motions shown in Fig. 16-35. Determine the velocity and acceleration of A as measured by the pilot of B.

SOLUTION
Coordinate Axes. Since the relative motion of A with respect to the pilot in B is being sought, the x, y, z axes are attached to plane B, Fig. 16-35. At the *instant* considered, the origin B coincides with the origin of the fixed X, Y, Z frame.

Kinematic Equations.
$$\mathbf{v}_A = \mathbf{v}_B + \mathbf{\Omega} \times \mathbf{r}_{A/B} + (\mathbf{v}_{A/B})_{xyz} \qquad (1)$$
$$\mathbf{a}_A = \mathbf{a}_B + \dot{\mathbf{\Omega}} \times \mathbf{r}_{A/B} + \mathbf{\Omega} \times (\mathbf{\Omega} \times \mathbf{r}_{A/B}) + 2\mathbf{\Omega} \times (\mathbf{v}_{A/B})_{xyz} + (\mathbf{a}_{A/B})_{xyz} \qquad (2)$$

Motion of Moving Reference: Using $\mathbf{i}, \mathbf{j}, \mathbf{k}$ components,

$\mathbf{v}_B = \{600\mathbf{j}\}$ km/h

$(a_B)_n = \dfrac{v_B^2}{\rho} = \dfrac{(600)^2}{400} = 900$ km/h^2

$\mathbf{a}_B = (\mathbf{a}_B)_n + (\mathbf{a}_B)_t = \{900\mathbf{i} - 100\mathbf{j}\}$ km/h^2

$\Omega = \dfrac{v_B}{\rho} = \dfrac{600 \text{ km/h}}{400 \text{ km}} = 1.5$ rad/h ↓ $\quad \mathbf{\Omega} = \{-1.5\mathbf{k}\}$ rad/h

$\dot{\Omega} = \dfrac{(a_B)_t}{\rho} = \dfrac{100 \text{ km/h}^2}{400 \text{ km}} = 0.25$ rad/h^2 ↑ $\quad \dot{\mathbf{\Omega}} = \{0.25\mathbf{k}\}$ rad/h^2

Fig. 16-35

Motion of A with Respect to Moving Reference:

$\mathbf{r}_{A/B} = \{-4\mathbf{i}\}$ km $\quad (\mathbf{v}_{A/B})_{xyz} = ? \quad (\mathbf{a}_{A/B})_{xyz} = ?$

Substituting the data into Eqs. 1 and 2, realizing that $\mathbf{v}_A = \{700\mathbf{j}\}$ km/h and $\mathbf{a}_A = \{50\mathbf{j}\}$ km/h^2, we have

$$\mathbf{v}_A = \mathbf{v}_B + \mathbf{\Omega} \times \mathbf{r}_{A/B} + (\mathbf{v}_{A/B})_{xyz}$$
$$700\mathbf{j} = 600\mathbf{j} + (-1.5\mathbf{k}) \times (-4\mathbf{i}) + (\mathbf{v}_{A/B})_{xyz}$$
$$(\mathbf{v}_{A/B})_{xyz} = \{94\mathbf{j}\} \text{ km/h} \qquad \textit{Ans.}$$

$$\mathbf{a}_A = \mathbf{a}_B + \dot{\mathbf{\Omega}} \times \mathbf{r}_{A/B} + \mathbf{\Omega} \times (\mathbf{\Omega} \times \mathbf{r}_{A/B}) + 2\mathbf{\Omega} \times (\mathbf{v}_{A/B})_{xyz} + (\mathbf{a}_{A/B})_{xyz}$$
$$50\mathbf{j} = (900\mathbf{i} - 100\mathbf{j}) + (0.25\mathbf{k}) \times (-4\mathbf{i})$$
$$\quad + (-1.5\mathbf{k}) \times [(-1.5\mathbf{k}) \times (-4\mathbf{i})] + 2(-1.5\mathbf{k}) \times (94\mathbf{j}) + (\mathbf{a}_{A/B})_{xyz}$$
$$(\mathbf{a}_{A/B})_{xyz} = \{-1191\mathbf{i} + 151\mathbf{j}\} \text{ km/h}^2 \qquad \textit{Ans.}$$

NOTE: The solution of this problem should be compared with that of Example 12.26, where it is seen that $(v_{B/A})_{xyz} \neq (v_{A/B})_{xyz}$ and $(a_{B/A})_{xyz} \neq (a_{A/B})_{xyz}$.

PROBLEMS

16–129. The man stands on the platform at O and runs out towards the edge such that when he is at A, $y = 5$ ft, his mass center has a velocity of 2 ft/s and an acceleration of 3 ft/s², both measured relative to the platform and directed along the positive y axis. If the platform has the angular motions shown, determine the velocity and acceleration of his mass center at this instant.

16–131. The slider block B, which is attached to a cord, moves along the slot of the horizontal circular disk. If the cord is pulled down through the central hole A in the disk at a constant rate of $\dot{x} = -3$ m/s, measured relative to the disk, determine the acceleration of the block at the instant $x = 0.1$ m. The disk has a constant angular velocity of $\omega_D = 2$ rad/s.

*__**16–132.**__ Solve Prob. 16–131 assuming that at the instant $x = 0.1$ m, $\dot{x} = -3$ m/s, $\ddot{x} = 1.25$ m/s², $\omega_D = 2$ rad/s, and the disk has an angular deceleration $\alpha_D = -4$ rad/s².

Prob. 16–129

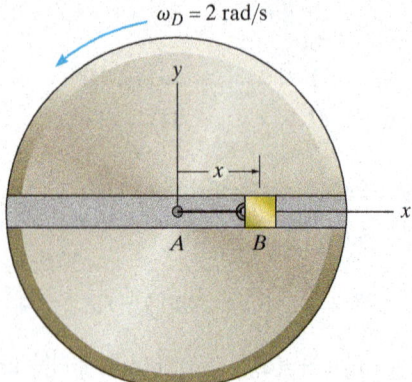

Probs. 16–131/132

16–130. The ball B of negligible size rolls through the tube such that at the instant shown it has a velocity of 5 ft/s and an acceleration of 3 ft/s², measured relative to the tube. If the tube has an angular velocity of $\omega = 3$ rad/s and an angular acceleration of $\alpha = 5$ rad/s² at this same instant, determine the velocity and acceleration of the ball.

16–133. Water leaves the impeller of the centrifugal pump with a velocity of 25 m/s and acceleration of 30 m/s², both measured relative to the impeller along the blade line AB. Determine the velocity and acceleration of a water particle at A as it leaves the impeller at the instant shown. The impeller rotates with a constant angular velocity of $\omega = 15$ rad/s.

Prob. 16–130

Prob. 16–133

16–134. Block A, which is attached to a cord, moves along the slot of a horizontal forked rod. At the instant shown, the cord is pulled down through the hole at O with an acceleration of 4 m/s² and its velocity is 2 m/s. Determine the acceleration of the block at this instant. The rod rotates about O with a constant angular velocity ω = 4 rad/s.

16–137. At the instant shown, rod AB has an angular velocity ω_{AB} = 3 rad/s and an angular acceleration α_{AB} = 5 rad/s². Determine the angular velocity and angular acceleration of rod CD at this instant. The collar at C is pin connected to CD and slides over AB.

Prob. 16–134

Prob. 16–137

16–135. Rod AB rotates counterclockwise with a constant angular velocity ω = 3 rad/s. Determine the velocity of point C located on the double collar when θ = 30°. The collar consists of two pin-connected slider blocks which are constrained to move along the circular path and the rod AB.

*16–136. Rod AB rotates counterclockwise with a constant angular velocity ω = 3 rad/s. Determine the velocity and acceleration of point C located on the double collar when θ = 45°. The collar consists of two pin-connected slider blocks which are constrained to move along the circular path and the rod AB.

16–138. At the instant θ = 60°, link CD has an angular velocity ω_{CD} = 4 rad/s and an angular acceleration α_{CD} = 2 rad/s². Determine the angular velocity and angular acceleration of rod AB at this instant. The collar at C is pin connected to DC and slides over AB.

Probs. 16–135/136

Prob. 16–138

16–139. The block B of the "quick-return" mechanism is confined to move within the slot in member CD. If AB is rotating at a constant rate of $\omega_{AB} = 3$ rad/s, determine the angular velocity and angular acceleration of member CD at the instant shown.

16–141. The disk rotates with the angular motion shown. Determine the angular velocity and angular acceleration of the slottted link AC at this instant. The peg at B is fixed to the disk.

Prob. 16–141

Prob. 16–139

*16–140. At the instant shown rod AB has an angular velocity $\omega_{AB} = 4$ rad/s and an angular acceleration $\alpha_{AB} = 2$ rad/s². Determine the angular velocity and angular acceleration of rod CD at this instant. The collar at C is pin connected to CD and slides freely along AB.

16–142. The wheel is rotating with the angular velocity and angular acceleration at the instant shown. Determine the angular velocity and angular acceleration of the rod at this instant. The rod slides freely through the smooth collar.

Prob. 16–140

Prob. 16–142

16–143. Peg B on the gear slides freely along the slot in link AB. If the gear's center O moves with the velocity and acceleration shown, determine the angular velocity and angular acceleration of the link at this instant.

Prob. 16–143

*16–144. A ride in an amusement park consists of a rotating platform P, having a constant angular velocity $\omega_P = 1.5$ rad/s, and four cars, C, mounted on the platform, which have constant angular velocities $\omega_{C/P} = 2$ rad/s measured relative to the platform. Determine the velocity and acceleration of the passenger at B at the instant shown.

Prob. 16–144

16–145. A ride in an amusement park consists of a rotating arm AB having a constant angular velocity $\omega_{AB} = 2$ rad/s. The car mounted at the end of the arm has a constant angular velocity $\omega' = \{-0.5\mathbf{k}\}$ rad/s, measured relative to the arm. At the instant shown, determine the velocity and acceleration of the passenger at C.

16–146. A ride in an amusement park consists of a rotating arm AB that has an angular acceleration of $\alpha_{AB} = \{1\mathbf{k}\}$ rad/s² when $\omega_{AB} = \{2\mathbf{k}\}$ rad/s at the instant shown. Also at this instant the car mounted at the end of the arm has an angular acceleration of $\alpha = \{-0.6\mathbf{k}\}$ rad/s² and angular velocity of $\omega' = \{-0.5\mathbf{k}\}$ rad/s, measured relative to the arm. Determine the velocity and acceleration of the passenger at C at this instant.

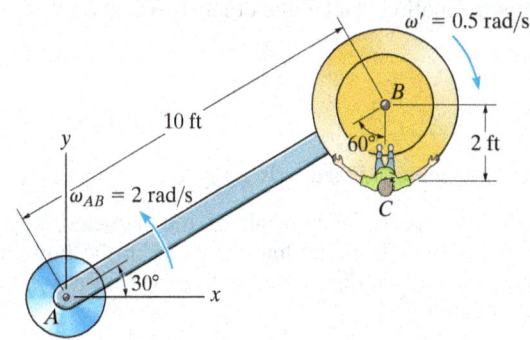

Probs. 16–145/146

16–147. If the slider block C is fixed to the disk that has a constant counterclockwise angular velocity of 4 rad/s, determine the angular velocity and angular acceleration of the slotted arm AB at the instant shown.

Prob. 16–147

CHAPTER REVIEW

Rigid-Body Planar Motion

A rigid body undergoes three types of planar motion: translation, rotation about a fixed axis, and general plane motion.

Translation

When a body has rectilinear translation, all the particles of the body travel along parallel straight-line paths. If the paths have the same radius of curvature, then curvilinear translation occurs. Provided we know the motion of one of the particles, then the motion of all of the others is also known.

Path of rectilinear translation

Path of curvilinear translation

Rotation about a Fixed Axis

For this type of motion, all of the particles move along circular paths. Here, all line segments in the body undergo the same angular displacement, angular velocity, and angular acceleration.

Once the angular motion of the body is known, then the velocity of any particle a distance r from the axis can be obtained.

The acceleration of any particle has two components. The tangential component accounts for the change in the magnitude of the velocity, and the normal component accounts for the change in the velocity's direction.

Rotation about a fixed axis

$\omega = d\theta/dt$
$\alpha = d\omega/dt$ or
$\alpha\, d\theta = \omega\, d\omega$

$v = \omega r$

$\omega = \omega_0 + \alpha_c t$
$\theta = \theta_0 + \omega_0 t + \tfrac{1}{2}\alpha_c t^2$
$\omega^2 = \omega_0^2 + 2\alpha_c(\theta - \theta_0)$

Constant α_c

$a_t = \alpha r \qquad a_n = \omega^2 r$

General Plane Motion

When a body undergoes general plane motion, it simultaneously translates and rotates. There are several methods for analyzing this motion.

Absolute Motion Analysis
If the motion of a point on a body or the angular motion of a line is known, then it may be possible to relate this motion to that of another point or line using an absolute motion analysis. To do so, linear position coordinates s or angular position coordinates θ are established (measured from a fixed point or line). These position coordinates are then related using the geometry of the body. The time derivative of this equation gives the relationship between the velocities and/or the angular velocities. A second time derivative relates the accelerations and/or the angular accelerations.

General plane motion

Relative Motion Using Translating Axes

General plane motion can also be analyzed using a relative-motion analysis between two points A and B located on the body. This method considers the motion in parts: first a translation of the selected base point A, then a relative "rotation" of the body about point A, which is measured from a translating axis. Since the relative motion is viewed as circular motion about the base point, point B will have a velocity $\mathbf{v}_{B/A}$ that is tangent to the circle. It also has two components of acceleration, $(\mathbf{a}_{B/A})_t$ and $(\mathbf{a}_{B/A})_n$. It is also important to realize that $\mathbf{a}_A$ and $\mathbf{a}_B$ will have tangential and normal components if these points move along curved paths.

$$\mathbf{v}_B = \mathbf{v}_A + \boldsymbol{\omega} \times \mathbf{r}_{B/A}$$

$$\mathbf{a}_B = \mathbf{a}_A + \boldsymbol{\alpha} \times \mathbf{r}_{B/A} - \omega^2 \mathbf{r}_{B/A}$$

Instantaneous Center of Zero Velocity

If the base point A is selected as having zero velocity, then the relative velocity equation becomes $\mathbf{v}_B = \boldsymbol{\omega} \times \mathbf{r}_{B/A}$. In this case, motion appears as if the body rotates about an instantaneous axis passing through A.

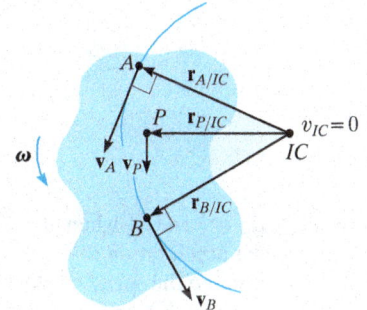

The instantaneous center of rotation (IC) can be established provided the directions of the velocities of any two points on the body are known, or the velocity of a point and the angular velocity are known. Since a radial line r will always be perpendicular to each velocity, then the IC is at the point of intersection of these two radial lines. Its measured location is determined from the geometry of the body. Once it is established, then the velocity of any point P on the body can be determined from $v = \omega r$, where r extends from the IC to point P.

Relative Motion Using Rotating Axes

Problems that involve connected members that slide relative to one another or points not located on the same body can be analyzed using a relative-motion analysis referenced from a rotating frame. This gives rise to the term $2\boldsymbol{\Omega} \times (\mathbf{v}_{B/A})_{xyz}$ that is called the Coriolis acceleration.

$$\mathbf{v}_B = \mathbf{v}_A + \boldsymbol{\Omega} \times \mathbf{r}_{B/A} + (\mathbf{v}_{B/A})_{xyz}$$

$$\mathbf{a}_B = \mathbf{a}_A + \dot{\boldsymbol{\Omega}} \times \mathbf{r}_{B/A} + \boldsymbol{\Omega} \times (\boldsymbol{\Omega} \times \mathbf{r}_{B/A}) + 2\boldsymbol{\Omega} \times (\mathbf{v}_{B/A})_{xyz} + (\mathbf{a}_{B/A})_{xyz}$$

REVIEW PROBLEMS

R16–1. The hoisting gear A has an initial angular velocity of 60 rad/s and a constant deceleration of 1 rad/s^2. Determine the velocity and deceleration of the block which is being hoisted by the hub on gear B when $t = 3$ s.

Prob. R16–1

R16–2. Starting at $(\omega_A)_0 = 3$ rad/s, when $\theta = 0$, $s = 0$, pulley A is given an angular acceleration $\alpha = (0.6\theta)$ rad/s^2, where θ is in radians. Determine the speed of block B when it has risen $s = 0.5$ m. The pulley has an inner hub D which is fixed to C and turns with it.

Prob. R16–2

R16–3. At the instant shown, the board has an acceleration of 0.5 m/s^2 to the right, while at the same instant points on the outer rim of each drum have an acceleration with a magnitude of 3 m/s^2. If the board does not slip on the drums, determine its speed due to the motion.

Prob. R16–3

R16–4. If bar AB has an angular velocity $\omega_{AB} = 6$ rad/s, determine the velocity of the slider block C at the instant shown.

Prob. R16–4

R16–5. The center of the pulley is being lifted vertically with an acceleration of 4 m/s² at the instant it has a velocity of 2 m/s. If the cable does not slip on the pulley's surface, determine the accelerations of the cylinder B and point C on the pulley.

Prob. R16–5

R16–6. At the instant shown, link AB has an angular velocity $\omega_{AB} = 2$ rad/s and an angular acceleration $\alpha_{AB} = 6$ rad/s². Determine the acceleration of the pin at C and the angular acceleration of link CB at this instant, when $\theta = 60°$.

Prob. R16–6

R16–7. The disk is moving to the left such that it has an angular acceleration $\alpha = 8$ rad/s² and angular velocity $\omega = 3$ rad/s at the instant shown. If it does not slip at A, determine the acceleration of point B.

Prob. R16–7

R16–8. At the given instant member AB has the angular motions shown. Determine the velocity and acceleration of the slider block C at this instant.

Prob. R16–8

CHAPTER 17

Motorcycles and other vehicles can be subjected to severe dynamic loadings. In this chapter we will show how to determine these loadings for planar motion.

PLANAR KINETICS OF A RIGID BODY: FORCE AND ACCELERATION

CHAPTER OBJECTIVES

- To introduce the methods used to determine the mass moment of inertia of a body.
- To develop the planar kinetic equations of motion for a symmetric rigid body.
- To discuss applications of these equations to bodies undergoing translation, rotation about a fixed axis, and general plane motion.

17.1 MASS MOMENT OF INERTIA

Since a body has a definite size and shape, an applied nonconcurrent force system can cause the body to both translate and rotate. The translational aspects of the motion were studied in Chapter 13 and are governed by the equation $\mathbf{F} = m\mathbf{a}$. It will be shown in the next section that the rotational aspects, caused by a moment $\mathbf{M}$, are governed by an equation of the form $\mathbf{M} = I\boldsymbol{\alpha}$. The symbol I in this equation is termed the mass moment of inertia. This property is a measure of the resistance of a body to angular acceleration ($\mathbf{M} = I\boldsymbol{\alpha}$) in the same way that mass is a measure of the body's resistance to acceleration ($\mathbf{F} = m\mathbf{a}$).

The flywheel on the engine of this tractor has a large moment of inertia about its axis of rotation. Once it is set into motion, it will be difficult to stop, and this in turn will prevent the engine from stalling and instead will allow it to maintain a constant power.

We define the ***moment of inertia*** as the integral of the "second moment" about an axis of all the elements of mass dm which compose the body.* For example, the body's moment of inertia about the z axis in Fig. 17–1 is

$$I = \int_m r^2 \, dm \qquad (17\text{–}1)$$

Here the "moment arm" r is the perpendicular distance from the z axis to the arbitrary element dm. Since the formulation involves r, the value of I will be different for each axis about which it is calculated. In the study of planar kinetics, the axis most often chosen passes through the body's mass center G and is always perpendicular to the plane of motion. The moment of inertia about this axis will be denoted as I_G. Since r is squared in Eq. 17–1, the mass moment of inertia is always a positive quantity. Common units used for its measurement are $kg \cdot m^2$ or $slug \cdot ft^2$.

If the body consists of material having a variable density, $\rho = \rho(x,y,z)$, the elemental mass dm of the body can be expressed in terms of its density and volume as $dm = \rho \, dV$, Fig. 17–2a. Substituting dm into Eq. 17–1, the body's moment of inertia is then calculated using volume elements for integration, i.e.,

Fig. 17–1

$$I = \int_V r^2 \rho \, dV \qquad (17\text{–}2)$$

---

*Another property of the body, which measures the symmetry of the body's mass with respect to a coordinate system, is the product of inertia. This property applies to the three-dimensional motion of a body and will be discussed in Chapter 21.

17.1 MASS MOMENT OF INERTIA

If ρ is *constant*, then this term may be factored out of the integral, and the integration is then simply a function of geometry,

$$I = \rho \int_V r^2 \, dV \qquad (17\text{–}3)$$

When the volume element chosen for integration has infinitesimal dimensions in all three directions, Fig. 17–2a, then the moment of inertia of the body must be determined using "triple integration." However, for bodies that have symmetry about an axis, as in Fig. 17–2, the integration process can be simplified to a *single integration*. Here a shell element, Fig. 17–2b, or disk element, Fig. 17–2c, are often used for this purpose.

(a)

PROCEDURE FOR ANALYSIS

To obtain the moment of inertia by integration, we will consider only symmetric bodies having volumes which are generated by revolving a curve about an axis, as in Fig. 17–2a. Two types of differential elements can be chosen.

Shell Element.

- If a *shell element* having a height z, radius $r = y$, and thickness dy is chosen, Fig. 17–2b, then the volume is $dV = (2\pi y)(z)dy$.

- This element may be used in Eq. 17–2 or 17–3 for determining the moment of inertia I_z of the body about the z axis, since the *entire element*, due to its "thinness," lies at the *same* perpendicular distance $r = y$ from the z axis (see Example 17.1).

Disk Element.

- If a disk element having a radius y and a thickness dz is chosen, Fig. 17–2c, then the volume is $dV = (\pi y^2)dz$.

- This element is *finite* in the radial direction, and consequently its parts *do not* all lie at the *same radial distance* r from the z axis. As a result, Eq. 17–2 or 17–3 *cannot* be used to determine I_z directly. Instead, to perform the integration it is first necessary to determine the moment of inertia *of the element* about the z axis and then integrate this result (see Example 17.2).

(b)

(c)

Fig. 17–2

EXAMPLE 17.1

Determine the moment of inertia of the cylinder shown in Fig. 17–3a about the z axis. The density of the material, ρ, is constant.

(a)   (b)

Fig. 17–3

SOLUTION

Shell Element. This problem will be solved using the shell element in Fig. 17–3b and a single integration. The volume of the element is $dV = (2\pi r)(h)\, dr$, so that its mass is $dm = \rho\, dV = \rho(2\pi h r\, dr)$. Since the entire element lies at the same distance r from the z axis, the moment of inertia of the element is

$$dI_z = r^2 dm = \rho 2\pi h r^3\, dr$$

Integrating over the entire region of the cylinder, we have

$$I_z = \int_m r^2\, dm = \rho 2\pi h \int_0^R r^3\, dr = \frac{\rho \pi}{2} R^4 h$$

The mass of the cylinder is

$$m = \int_m dm = \rho 2\pi h \int_0^R r\, dr = \rho \pi h R^2$$

And so,

$$I_z = \frac{1}{2} m R^2 \qquad \text{Ans.}$$

EXAMPLE 17.2

If the density of the material is 5 slug/ft³, determine the moment of inertia of the solid in Fig. 17-4a about the y axis.

Fig. 17–4

SOLUTION

Disk Element. The moment of inertia will be found using a disk element, shown in Fig. 17–4b. Here the element intersects the curve at the arbitrary point (x,y) and has a mass

$$dm = \rho \, dV = \rho(\pi x^2) \, dy$$

Although all portions of the element are *not* located at the same distance from the y axis, it is still possible to determine the moment of inertia dI_y of the element about the y axis. In the preceding example it was shown that the moment of inertia of a cylinder about its longitudinal axis is $I = \frac{1}{2}mR^2$, where m and R are the mass and radius of the cylinder. Since the height is not involved in this formula, the disk can be thought of as a cylinder. Thus, for the disk element in Fig. 17–4b, we have

$$dI_y = \tfrac{1}{2}(dm)x^2 = \tfrac{1}{2}[\rho(\pi x^2) \, dy]x^2$$

Substituting $x = y^2$, $\rho = 5$ slug/ft³, and integrating with respect to y, from $y = 0$ to $y = 1$ ft, we get

$$I_y = \frac{\pi(5 \text{ slug/ft}^3)}{2} \int_0^{1 \text{ ft}} x^4 \, dy = \frac{\pi(5)}{2} \int_0^{1 \text{ ft}} y^8 \, dy = 0.873 \text{ slug} \cdot \text{ft}^2 \quad Ans.$$

Fig. 17–5

Parallel-Axis Theorem. If the moment of inertia of the body about an axis passing through the body's mass center is known, then the moment of inertia about any other parallel axis can be determined by using the ***parallel-axis theorem***. This theorem can be derived by considering the body shown in Fig. 17–5. Here the z' axis passes through the mass center G, whereas the corresponding parallel z axis lies at a constant distance d away. Selecting the differential element of mass dm, which is located at point (x', y'), and using the Pythagorean theorem, $r^2 = (d + x')^2 + y'^2$, we can express the moment of inertia of the body about the z axis as

$$I = \int_m r^2 \, dm = \int_m [(d + x')^2 + y'^2] \, dm$$

$$= \int_m (x'^2 + y'^2) \, dm + 2d \int_m x' \, dm + d^2 \int_m dm$$

Since $r'^2 = x'^2 + y'^2$, the first integral represents I_G. The second integral equals *zero*, since the z' axis passes through the body's mass center, i.e., $\int x' \, dm = \bar{x}' m = 0$ since $\bar{x}' = 0$. Finally, the third integral represents the total mass m of the body. Hence, the moment of inertia about the z axis can be written as

$$I = I_G + md^2 \qquad (17\text{–}4)$$

where

I_G = moment of inertia about the z' axis passing through the mass center G
m = mass of the body
d = perpendicular distance between the parallel z and z' axes

Radius of Gyration. Occasionally, the moment of inertia of a body about a specified axis is reported in handbooks using the *radius of gyration*, k. This is a geometrical property which has units of length. When it and the body's mass m are known, the body's moment of inertia is determined from the equation

$$I = mk^2 \quad \text{or} \quad k = \sqrt{\frac{I}{m}} \qquad (17\text{–}5)$$

Note the similarity between the definition of k in this formula and r in the equation $dI = r^2\,dm$, which defines the moment of inertia of an elemental mass dm of the body about an axis.

Composite Bodies. If a body consists of a number of simple shapes such as disks, spheres, and rods, the moment of inertia of the body about any axis can be determined by adding algebraically the moments of inertia of all the composite shapes calculated about the axis. Algebraic addition is necessary since a composite part must be considered as a negative quantity if it has already been counted as a piece of another part—for example, a "hole" subtracted from a solid plate. The parallel-axis theorem is needed for the calculations if the center of mass of each composite part does not lie on the axis about which the moment of inertia is to be determined. For the calculation, then, $I = \Sigma(I_G + md^2)$. Here I_G for each of the composite parts is determined by integration, or for simple shapes, such as rods and disks, it can be found from a table, such as the one given on the inside back cover of this book.

EXAMPLE 17.3

If the plate shown in Fig. 17–6a has a density of 8000 kg/m³ and a thickness of 10 mm, determine its moment of inertia about an axis directed perpendicular to the page and passing through point O.

Fig. 17–6

SOLUTION
The plate consists of two composite parts, the 250-mm-radius disk *minus* a 125-mm-radius disk, Fig. 17–6b. The moment of inertia about O can be determined by calculating the moment of inertia of each of these parts about O and then adding the results *algebraically*. The calculations are performed by using the parallel-axis theorem in conjunction with the data listed in the table on the inside back cover.

Disk. The moment of inertia of a disk about the centroidal axis perpendicular to the plane of the disk is $I_G = \frac{1}{2}mr^2$. The mass center of the disk is located at a distance of 0.25 m from point O. Thus,

$$m_d = \rho_d V_d = 8000 \text{ kg/m}^3 \, [\pi(0.25 \text{ m})^2(0.01 \text{ m})] = 15.71 \text{ kg}$$

$$(I_d)_O = \tfrac{1}{2}m_d r_d^2 + m_d d^2$$

$$= \tfrac{1}{2}(15.71 \text{ kg})(0.25 \text{ m})^2 + (15.71 \text{ kg})(0.25 \text{ m})^2$$

$$= 1.473 \text{ kg} \cdot \text{m}^2$$

Hole. For the 125-mm-radius disk (hole), we have

$$m_h = \rho_h V_h = 8000 \text{ kg/m}^3 \, [\pi(0.125 \text{ m})^2(0.01 \text{ m})] = 3.927 \text{ kg}$$

$$(I_h)_O = \tfrac{1}{2}m_h r_h^2 + m_h d^2$$

$$= \tfrac{1}{2}(3.927 \text{ kg})(0.125 \text{ m})^2 + (3.927 \text{ kg})(0.25 \text{ m})^2$$

$$= 0.276 \text{ kg} \cdot \text{m}^2$$

The moment of inertia of the plate about point O is therefore

$$I_O = (I_d)_O - (I_h)_O$$

$$= 1.473 \text{ kg} \cdot \text{m}^2 - 0.276 \text{ kg} \cdot \text{m}^2$$

$$= 1.20 \text{ kg} \cdot \text{m}^2 \qquad \textit{Ans.}$$

EXAMPLE 17.4

The pendulum in Fig. 17–7 is suspended from the pin at O and consists of two thin rods. Rod OA weighs 10 lb, and BC weighs 8 lb. Determine the moment of inertia of the pendulum about an axis passing through (a) point O, and (b) the mass center G of the pendulum.

SOLUTION

Part (a). Using the table on the inside back cover, the moment of inertia of rod OA about an axis perpendicular to the page and passing through point O of the rod is $I_O = \frac{1}{3}ml^2$. Hence,

$$(I_{OA})_O = \tfrac{1}{3}ml^2 = \frac{1}{3}\left(\frac{10 \text{ lb}}{32.2 \text{ ft/s}^2}\right)(2 \text{ ft})^2 = 0.414 \text{ slug} \cdot \text{ft}^2$$

This same value can be obtained using $I_G = \frac{1}{12}ml^2$ and the parallel-axis theorem.

Fig. 17–7

$$(I_{OA})_O = \tfrac{1}{12}ml^2 + md^2 = \frac{1}{12}\left(\frac{10 \text{ lb}}{32.2 \text{ ft/s}^2}\right)(2 \text{ ft})^2 + \left(\frac{10 \text{ lb}}{32.2 \text{ ft/s}^2}\right)(1 \text{ ft})^2$$

$$= 0.414 \text{ slug} \cdot \text{ft}^2$$

For rod BC we have

$$(I_{BC})_O = \tfrac{1}{12}ml^2 + md^2 = \frac{1}{12}\left(\frac{8 \text{ lb}}{32.2 \text{ ft/s}^2}\right)(1.5 \text{ ft})^2 + \left(\frac{8 \text{ lb}}{32.2 \text{ ft/s}^2}\right)(2 \text{ ft})^2$$

$$= 1.040 \text{ slug} \cdot \text{ft}^2$$

The moment of inertia of the pendulum about O is therefore

$$I_O = 0.414 + 1.040 = 1.454 = 1.45 \text{ slug} \cdot \text{ft}^2 \qquad \textit{Ans.}$$

Part (b). The mass center G will be located relative to point O. Assuming this distance to be $\bar{y}$, Fig. 17–7, and using the formula for determining the mass center, we have

$$\bar{y} = \frac{\Sigma \tilde{y} m}{\Sigma m} = \frac{1(10/32.2) + 2(8/32.2)}{(10/32.2) + (8/32.2)} = 1.444 \text{ ft}$$

The moment of inertia I_G may be found in the same manner as I_O, which requires successive applications of the parallel-axis theorem to transfer the moments of inertia of rods OA and BC to G. A more direct solution, however, involves using the result for I_O, i.e.,

$$I_O = I_G + md^2; \quad 1.454 \text{ slug} \cdot \text{ft}^2 = I_G + \left(\frac{18 \text{ lb}}{32.2 \text{ ft/s}^2}\right)(1.444 \text{ ft})^2$$

$$I_G = 0.288 \text{ slug} \cdot \text{ft}^2 \qquad \textit{Ans.}$$

PROBLEMS

17–1. Determine the moment of inertia I_y for the slender rod. The rod's density ρ and cross-sectional area A are constant. Express the result in terms of the rod's total mass m.

Prob. 17–1

17–2. The solid cylinder has an outer radius R, height h, and is made from a material having a density that varies from its center as $\rho = k + ar^2$, where k and a are constants. Determine the mass of the cylinder and its moment of inertia about the z axis.

Prob. 17–2

17–3. Determine the moment of inertia of the thin ring about the z axis. The ring has a mass m.

Prob. 17–3

*17–4. The right circular cone is formed by revolving the shaded area around the x axis. Determine the moment of inertia I_x and express the result in terms of the total mass m of the cone. The cone has a constant density ρ.

Prob. 17–4

17–5. Determine the mass moment of inertia I_z of the cone formed by revolving the shaded area around the z axis. The total density of the material is ρ. Express the results in terms of the mass m of the cone.

17–7. Determine the moment of inertia of the semi ellipsoid with respect to the x axis and express the result in terms of the mass m of the semi ellipsoid. The material has a constant density ρ.

Prob. 17–5

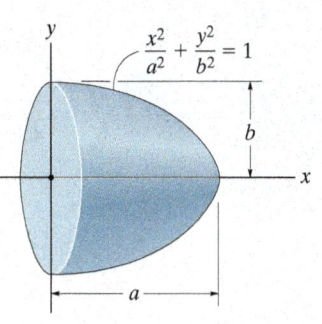

Prob. 17–7

17–6. Determine the moment of inertia of the assembly about an axis which is perpendicular to the page and passes through point O. The material has a specific weight of $\gamma = 90 \text{ lb/ft}^3$.

***17–8** Determine the moment of inertia I_z of the torus. The mass of the torus is m and the density ρ is constant. *Suggestion*: Use a shell element.

Prob. 17–6

Prob. 17–8

17–9. Determine the moment of inertia I_z of the frustum of the cone, which has a conical depression. The material has a density of 200 kg/m³.

Prob. 17–9

17–10. The pendulum consists of a plate having a weight of 12 lb and a slender rod having a weight of 4 lb. Determine the radius of gyration of the pendulum about an axis perpendicular to the page and passing through point O.

Prob. 17–10

17–11 The pendulum consists of a 2-kg disk and slender rods AB and DC which have a mass per unit length of 2 kg/m. Determine the length L of DC so that the center of mass is at the bearing O. What is the moment of inertia of this assembly about an axis perpendicular to the page and passing through point O?

Prob. 17–11

*17–12. Determine the moment of inertia of the solid steel assembly about the x axis. Steel has a specific weight of $\gamma_{st} = 490$ lb/ft³.

Prob. 17–12

17–13. The wheel consists of a thin ring having a mass of 10 kg and four spokes made from slender rods and each having a mass of 2 kg. Determine the wheel's moment of inertia about an axis perpendicular to the page and passing through point A.

Prob. 17–13

17–14. If the large ring, small ring, and each of the spokes weigh 100 lb, 15 lb, and 20 lb, respectively, determine the mass moment of inertia of the wheel about an axis perpendicular to the page and passing through point A.

Prob. 17–14

17–15. Determine the moment of inertia of the center crank about the x axis. The material is steel having a specific weight of $\gamma_{st} = 490 \text{ lb/ft}^3$.

17–17. Determine the location $\bar{y}$ of the center of mass G of the assembly, and then calculate the moment of inertia about an axis perpendicular to the page and passing through G. The block has a mass of 3 kg and the semicylinder has a mass of 5 kg.

17–18. Determine the moment of inertia of the assembly about an axis perpendicular to the page and passing through point O. The block has a mass of 3 kg and the semicylinder has a mass of 5 kg.

Prob. 17–15

Probs. 17–17/18

*__17–16.__ Determine the mass moment of inertia of the thin plate about an axis perpendicular to the page and passing through point O. The material has a mass per unit area of 20 kg/m^2.

17–19. Determine the moment of inertia of the wheel about an axis which is perpendicular to the page and passes through the center of mass G. The material has a specific weight $\gamma = 90 \text{ lb/ft}^3$.

Prob. 17–16

Prob. 17–19

*17–20. Determine the moment of inertia of the wheel about an axis which is perpendicular to the page and passes through point O. The material has a specific weight $\gamma = 90 \text{ lb/ft}^3$.

17–22. Determine the moment of inertia of the overhung crank about the x axis. The material is steel having a density of $\rho = 7.85 \text{ Mg/m}^3$.

Prob. 17–20

Prob. 17–22

17–21. The slender rods have a weight of 3 lb/ft. Determine the moment of inertia of the assembly about an axis perpendicular to the page and passing through the pin at A.

17–23 Determine the moment of inertia of the overhung crank about the x' axis. The material is steel having a density of $\rho = 7.85 \text{ Mg/m}^3$.

Prob. 17–21

Prob. 17–23

17.2 PLANAR KINETIC EQUATIONS OF MOTION

In the following analysis we will limit our study of planar kinetics to rigid bodies which, along with their loadings, are considered to be *symmetrical* with respect to a fixed reference plane.* Since the motion of the body can be viewed within the reference plane, all the forces acting on the body can then be projected onto the plane. An example of an arbitrary body of this type is shown in Fig. 17–8a. Here the ***inertial frame of reference*** x, y, z has its origin *coincident* with the arbitrary point P in the body. By definition, these axes do not rotate and either are fixed or translate with constant velocity.

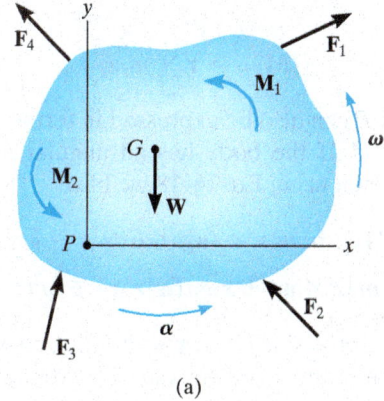

(a)

Fig. 17–8

Equation of Translational Motion. The external forces acting on the body in Fig. 17–8a represent the effect of gravitational, electrical, magnetic, or contact forces between adjacent bodies. Since this force system has been considered previously in Sec. 13.3 for the analysis of a system of particles, the resulting Eq. 13–6 can be used here for a rigid body, in which case,

$$\Sigma \mathbf{F} = m\mathbf{a}_G$$

This equation is referred to as the ***translational equation of motion*** for the mass center of a rigid body. It states that *the sum of all the external forces acting on the body is equal to the body's mass times the acceleration of its mass center G.*

For motion of the body in the x–y plane, the translational equation of motion may be written in the form of two independent scalar equations, namely,

$$\Sigma F_x = m(a_G)_x$$
$$\Sigma F_y = m(a_G)_y$$

---

*By doing this, the rotational equation of motion reduces to a rather simplified form. The more general case of body shape and loading is considered in Chapter 21.

CHAPTER 17 PLANAR KINETICS OF A RIGID BODY: FORCE AND ACCELERATION

Particle free-body diagram

(b)

Particle kinetic diagram

(c)

Fig. 17–8 (cont.)

Equation of Rotational Motion. We will now determine the effects caused by the moments of the external force system calculated about an axis perpendicular to the plane of motion (the z axis) and passing through point P. As shown on the free-body diagram of the ith particle, Fig. 17–8b, $\mathbf{F}_i$ represents the *resultant external force* acting on the particle, and $\mathbf{f}_i$ is the *resultant of the internal forces* caused by interactions with adjacent particles. If the particle has a mass m_i and its acceleration is $\mathbf{a}_i$, then its kinetic diagram is shown in Fig. 17–8c. Summing moments about point P, we require

$$\mathbf{r} \times \mathbf{F}_i + \mathbf{r} \times \mathbf{f}_i = \mathbf{r} \times m_i \mathbf{a}_i$$

or

$$(\mathbf{M}_P)_i = \mathbf{r} \times m_i \mathbf{a}_i$$

The moments about P can also be expressed in terms of the acceleration of point P, Fig. 17–8d. If the body has an angular acceleration $\boldsymbol{\alpha}$ and angular velocity $\boldsymbol{\omega}$, then using Eq. 16–18 we have

$$(\mathbf{M}_P)_i = m_i \mathbf{r} \times (\mathbf{a}_P + \boldsymbol{\alpha} \times \mathbf{r} - \omega^2 \mathbf{r})$$
$$= m_i[\mathbf{r} \times \mathbf{a}_P + \mathbf{r} \times (\boldsymbol{\alpha} \times \mathbf{r}) - \omega^2(\mathbf{r} \times \mathbf{r})]$$

The last term is zero, since $\mathbf{r} \times \mathbf{r} = \mathbf{0}$. Expressing the vectors as Cartesian components and carrying out the cross-product operations yields

$$(M_P)_i \mathbf{k} = m_i\{(x\mathbf{i} + y\mathbf{j}) \times [(a_P)_x \mathbf{i} + (a_P)_y \mathbf{j}]$$
$$+ (x\mathbf{i} + y\mathbf{j}) \times [\alpha \mathbf{k} \times (x\mathbf{i} + y\mathbf{j})]\}$$
$$(M_P)_i \mathbf{k} = m_i[-y(a_P)_x + x(a_P)_y + \alpha x^2 + \alpha y^2]\mathbf{k}$$
$$\zeta (M_P)_i = m_i[-y(a_P)_x + x(a_P)_y + \alpha r^2]$$

Letting $m_i \rightarrow dm$ and integrating over the entire mass m of the body, we obtain

$$\zeta \Sigma M_P = -\left(\int_m y\, dm\right)(a_P)_x + \left(\int_m x\, dm\right)(a_P)_y + \left(\int_m r^2 dm\right)\alpha$$

Here ΣM_P represents only the moment of the *external forces* acting on the body about point P. The resultant moment of the internal forces is zero, since for the entire body these forces occur in equal and opposite collinear pairs and thus the moment of each pair of forces about P cancels. The integrals in the first and second terms on the right are used to locate the body's center of mass G with respect to P, since $\bar{y}m = \int y\, dm$ and $\bar{x}m = \int x\, dm$. See Fig. 17–8d. The last integral represents the body's moment of inertia about the z axis, i.e., $I_P = \int r^2 dm$. Thus,

$$\zeta \Sigma M_P = -\bar{y}m(a_P)_x + \bar{x}m(a_P)_y + I_P\alpha \qquad (17\text{–}6)$$

It is possible to reduce this equation to a simpler form if point P coincides with the mass center G for the body. If this is the case, then $\bar{x} = \bar{y} = 0$, and therefore*

$$\boxed{\Sigma M_G = I_G \alpha} \qquad (17\text{–}7)$$

This **rotational equation of motion** states that *the sum of the moments of all the external forces about the body's mass center G is equal to the product of the moment of inertia of the body about an axis passing through G and the body's angular acceleration.*

Equation 17–6 can also be rewritten in terms of the x and y components of $\mathbf{a}_G$ and the body's moment of inertia I_G. If point G is located at $(\bar{x}, \bar{y})$, Fig. 17–8d, then by the parallel-axis theorem, $I_P = I_G + m(\bar{x}^2 + \bar{y}^2)$. Substituting into Eq. 17–6 and rearranging terms, we get

$$\zeta \Sigma M_P = \bar{y}m[-(a_P)_x + \bar{y}\alpha] + \bar{x}m[(a_P)_y + \bar{x}\alpha] + I_G\alpha \qquad (17\text{–}8)$$

From the kinematic diagram of Fig. 17–8d, $\mathbf{a}_P$ can be expressed in terms of $\mathbf{a}_G$ as

$$\mathbf{a}_G = \mathbf{a}_P + \boldsymbol{\alpha} \times \bar{\mathbf{r}} - \omega^2 \bar{\mathbf{r}}$$
$$(a_G)_x \mathbf{i} + (a_G)_y \mathbf{j} = (a_P)_x \mathbf{i} + (a_P)_y \mathbf{j} + \alpha \mathbf{k} \times (\bar{x}\mathbf{i} + \bar{y}\mathbf{j}) - \omega^2(\bar{x}\mathbf{i} + \bar{y}\mathbf{j})$$

Carrying out the cross product and equating the respective $\mathbf{i}$ and $\mathbf{j}$ components yields the two scalar equations

$$(a_G)_x = (a_P)_x - \bar{y}\alpha - \bar{x}\omega^2$$
$$(a_G)_y = (a_P)_y + \bar{x}\alpha - \bar{y}\omega^2$$

From these equations, $[-(a_P)_x + \bar{y}\alpha] = [-(a_G)_x - \bar{x}\omega^2]$ and $[(a_P)_y + \bar{x}\alpha] = [(a_G)_y + \bar{y}\omega^2]$. Substituting these results into Eq. 17–8 and simplifying gives

$$\zeta \Sigma M_P = -\bar{y}m(a_G)_x + \bar{x}m(a_G)_y + I_G\alpha \qquad (17\text{–}9)$$

This important result indicates that *when moments of the external forces shown on the free-body diagram are summed about point P, Fig. 17–8e, they are equivalent to the sum of the "kinetic moments" of the components of $m\mathbf{a}_G$ about P plus the "kinetic moment" of $I_G\boldsymbol{\alpha}$*, Fig. 17–8f. Note that $I_G\boldsymbol{\alpha}$ can be treated as a free vector and therefore act at any point. It is important to remember, however, that $m\mathbf{a}_G$ and $I_G\boldsymbol{\alpha}$ are not the same as a force or a couple moment. Instead, they are caused by the external effects of forces and couple moments acting on the body. As a result, we can therefore write Eq. 17–9 in a more general form as

$$\boxed{\Sigma M_P = \Sigma(\mathcal{M}_k)_P} \qquad (17\text{–}10)$$

Free-body diagram

(e)

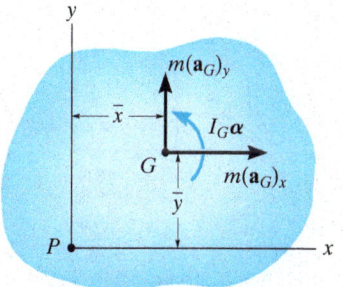

Kinetic diagram

(f)

Fig. 17–8 (cont.)

---

*It also reduces to a simpler form $\Sigma M_P = I_P\alpha$ if point P is a *fixed point* (see Eq. 17–16) or if the acceleration of point P is directed along the line PG.

410   CHAPTER 17   PLANAR KINETICS OF A RIGID BODY: FORCE AND ACCELERATION

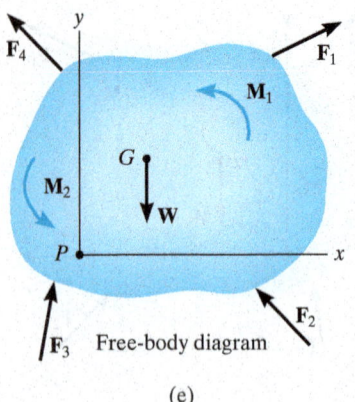

Free-body diagram
(e)

General Application of the Equations of Motion. To summarize this analysis, three independent scalar equations can be written to describe the general plane motion of a symmetrical rigid body. They are

$$\Sigma F_x = m(a_G)_x$$
$$\Sigma F_y = m(a_G)_y$$
$$\Sigma M_G = I_G \alpha$$

or

$$\Sigma M_P = \Sigma (\mathcal{M}_k)_P \qquad (17\text{–}11)$$

When applying these equations, one should always draw a free-body diagram, Fig. 17–8e, in order to account for the terms involved in ΣF_x, ΣF_y, ΣM_G, or ΣM_P. In some problems it may also be helpful to draw the kinetic diagram for the body, Fig. 17–8f. This diagram graphically accounts for the terms $m(\mathbf{a}_G)_x$, $m(\mathbf{a}_G)_y$, and $I_G \alpha$. It is especially convenient when used to determine the moment of these three components when applying $\Sigma (\mathcal{M}_k)_P$.

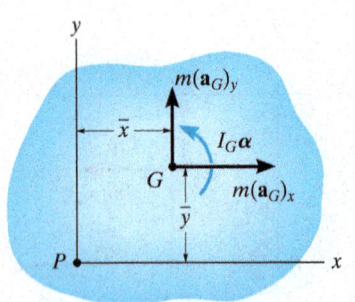

Kinetic diagram
(f)

Fig. 17–8 (repeated)

17.3 EQUATIONS OF MOTION: TRANSLATION

When the rigid body in Fig. 17–9a undergoes a translation, all the particles of the body have the same acceleration. Furthermore, $\alpha = \mathbf{0}$, in which case the rotational equation of motion applied at point G reduces to a simplified form, namely, $\Sigma M_G = 0$. Application of this and the force equations of motion will now be discussed for each of the two types of translation.

Rectilinear Translation. When a body is subjected to rectilinear translation, all the particles of the body (slab) travel along parallel straight-line paths. The free-body and kinetic diagrams are shown in Fig. 17–9b. Since $I_G \alpha = \mathbf{0}$, only $m\mathbf{a}_G$ is shown on the kinetic diagram. Hence, the equations of motion which apply in this case become

$$\Sigma F_x = m(a_G)_x$$
$$\Sigma F_y = m(a_G)_y \qquad (17\text{–}12)$$
$$\Sigma M_G = 0$$

(a)

Fig. 17–9

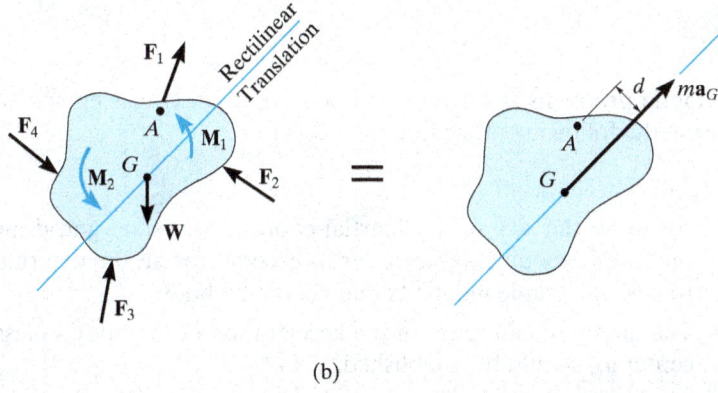

(b)

It is also possible to sum moments about some other point on or off the body, in which case the moment of $m\mathbf{a}_G$ must be taken into account. For example, if point A is chosen, which lies at a perpendicular distance d from the line of action of $m\mathbf{a}_G$, the following moment equation applies:

$$\zeta + \Sigma M_A = \Sigma(\mathcal{M}_k)_A; \qquad \Sigma M_A = (ma_G)d$$

Here the sum of moments of the external forces and couple moments about A (ΣM_A, free-body diagram) equals the moment of $m\mathbf{a}_G$ about A ($\Sigma(\mathcal{M}_k)_A$, kinetic diagram).

Curvilinear Translation. When a rigid body is subjected to curvilinear translation, then all the particles of the body have the same accelerations as they travel along curved paths, as noted in Sec. 16.1. For analysis, it is often convenient to use an inertial coordinate system having an origin which coincides with the body's mass center at the instant considered, and axes which are oriented in the normal and tangential directions to the path of motion, Fig. 17–9c. The three scalar equations of motion are then

$$\boxed{\begin{aligned} \Sigma F_n &= m(a_G)_n \\ \Sigma F_t &= m(a_G)_t \\ \Sigma M_G &= 0 \end{aligned}} \qquad (17\text{–}13)$$

If moments are summed about another point such as B, Fig. 17–9c, then it is necessary to account for the moments, $\Sigma(\mathcal{M}_k)_B$, of the two components $m(\mathbf{a}_G)_n$ and $m(\mathbf{a}_G)_t$ about this point. From the kinetic diagram, h and e represent the perpendicular distances (or "moment arms") from B to the lines of action of the components. The required moment equation therefore becomes

$$\zeta + \Sigma M_B = \Sigma(\mathcal{M}_k)_B; \qquad \Sigma M_B = e[m(a_G)_t] - h[m(a_G)_n]$$

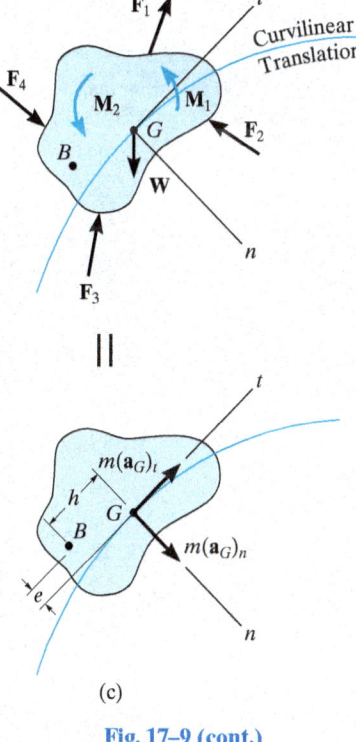

(c)

Fig. 17–9 (cont.)

The free-body and kinetic diagrams are shown below. If moments are summed about the mass center, G, then $\Sigma M_G = 0$. However, if moments are summed about point B then $\curvearrowleft + \Sigma M_B = m a_G(d)$.

PROCEDURE FOR ANALYSIS

Kinetic problems involving rigid-body *translation* can be solved using the following procedure.

Free-Body Diagram.

- Establish the x, y or n, t inertial coordinate system and draw the free-body diagram in order to account for all the external forces and couple moments that act on the body.
- The direction and sense of the acceleration of the body's mass center $\mathbf{a}_G$ should be established.
- Identify the unknowns in the problem.
- If it is decided that the rotational equation of motion $\Sigma M_P = \Sigma(\mathcal{M}_k)_P$ is to be used in the solution, then consider drawing the kinetic diagram, since it graphically accounts for the components $m(\mathbf{a}_G)_x$, $m(\mathbf{a}_G)_y$ or $m(\mathbf{a}_G)_t$, $m(\mathbf{a}_G)_n$ and is therefore convenient for "visualizing" the terms needed in the moment sum $\Sigma(\mathcal{M}_k)_P$.

Equations of Motion.

- Apply the three equations of motion in accordance with the established sign convention.
- To simplify the analysis, the moment equation $\Sigma M_G = 0$ can be replaced by the more general equation $\Sigma M_P = \Sigma(\mathcal{M}_k)_P$, where point P is usually located at the intersection of the lines of action of as many unknown forces as possible.
- If the body is in contact with a *rough surface* and slipping occurs, use the friction equation $F = \mu_k N$. Remember, $\mathbf{F}$ always acts on the body so as to oppose the motion of the body relative to the surface it contacts.

Kinematics.

- Use kinematics to determine the velocity and position of the body.
- For rectilinear translation with *variable acceleration*

  $a_G = dv_G/dt \qquad a_G ds_G = v_G dv_G$

- For rectilinear translation with *constant acceleration*

  $v_G = (v_G)_0 + a_G t \qquad v_G^2 = (v_G)_0^2 + 2a_G[s_G - (s_G)_0]$
  $s_G = (s_G)_0 + (v_G)_0 t + \tfrac{1}{2} a_G t^2$

- For curvilinear translation

  $(a_G)_n = v_G^2/\rho$
  $(a_G)_t = dv_G/dt \qquad (a_G)_t ds_G = v_G dv_G$

EXAMPLE 17.5

The car shown in Fig. 17–10a has a mass of 2 Mg and a center of mass at G. Determine the acceleration if the rear "driving" wheels are always slipping, whereas the front wheels are free to rotate. Neglect the mass of the wheels. The coefficient of kinetic friction between the wheels and the road is $\mu_k = 0.25$.

(a)

SOLUTION I

Free-Body Diagram. As shown in Fig. 17–10b, the rear-wheel frictional force $\mathbf{F}_B$ pushes the car forward, and since slipping occurs, $F_B = 0.25 N_B$. The frictional forces acting on the front wheels are zero, since these wheels have negligible mass.* There are three unknowns in this problem, N_A, N_B, and a_G.

Equations of Motion. Here we will sum moments about the mass center. The car (point G) accelerates to the left, i.e., in the negative x direction, Fig. 17–10b.

$\xrightarrow{+} \Sigma F_x = m(a_G)_x;$ $\quad -0.25 N_B = -(2000 \text{ kg}) a_G$ (1)

$+\uparrow \Sigma F_y = m(a_G)_y;$ $\quad N_A + N_B - 2000(9.81) \text{ N} = 0$ (2)

$\zeta + \Sigma M_G = 0;$ $\quad -N_A(1.25 \text{ m}) - 0.25 N_B(0.3 \text{ m}) + N_B(0.75 \text{ m}) = 0$ (3)

Solving,

$$a_G = 1.59 \text{ m/s}^2 \leftarrow \quad \textit{Ans.}$$
$$N_A = 6.88 \text{ kN}$$
$$N_B = 12.7 \text{ kN}$$

(b)

SOLUTION II

Free-Body and Kinetic Diagrams. If the "moment" equation is applied about point A, then the unknown N_A will be eliminated from the equation. To "visualize" the moment of $m\mathbf{a}_G$ about A, we will include the kinetic diagram as part of the analysis, Fig. 17–10c.

Equation of Motion.

$\zeta + \Sigma M_A = \Sigma (\mathcal{M}_k)_A;$ $\quad N_B(2 \text{ m}) - [2000(9.81) \text{ N}](1.25 \text{ m}) = (2000 \text{ kg}) a_G(0.3 \text{ m})$

Solving this and Eq. 1 for a_G leads to a simpler solution than that obtained from Eqs. 1 to 3.

(c)

Fig. 17–10

---

*With negligible wheel mass, $I\alpha = 0$ and the frictional force at A required to turn the wheel is zero. If the wheels' mass were included, then the solution would be more involved, since a general-plane-motion analysis of the wheels would have to be considered (see Sec. 17.5).

EXAMPLE 17.6

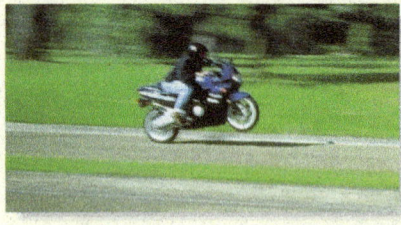

The motorcycle shown in Fig. 17–11a has a mass of 125 kg and a center of mass at G_1, while the rider has a mass of 75 kg and a center of mass at G_2. Determine the minimum coefficient of static friction between the wheels and the pavement in order for the rider to do a "wheely," i.e., lift the front wheel off the ground as shown in the photo. What acceleration is necessary to do this? Neglect the mass of the wheels and assume that the front wheel is free to roll.

Fig. 17–11

SOLUTION

Free-Body and Kinetic Diagrams. In this problem we will consider both the motorcycle and the rider as a single *system*. It is possible first to determine the location of the center of mass for this "system" by using the equations $\bar{x} = \Sigma \tilde{x} m / \Sigma m$ and $\bar{y} = \Sigma \tilde{y} m / \Sigma m$. Here, however, we will consider the weight and mass of the motorcycle and rider separately as shown on the free-body and kinetic diagrams, Fig. 17–11b. Both of these parts move with the *same* acceleration. We have assumed that the front wheel is *about* to leave the ground, so that the normal reaction $N_A \approx 0$. The three unknowns in the problem are N_B, F_B, and a_G.

Equations of Motion.

$\xrightarrow{+} \Sigma F_x = m(a_G)_x;$ $\quad F_B = (75 \text{ kg} + 125 \text{ kg}) a_G$

$+ \uparrow \Sigma F_y = m(a_G)_y;$ $\quad N_B - 735.75 \text{ N} - 1226.25 \text{ N} = 0$

$\zeta + \Sigma M_B = \Sigma (\mathcal{M}_k)_B;$ $\quad -(735.75 \text{ N})(0.4 \text{ m}) - (1226.25 \text{ N})(0.8 \text{ m}) =$
$\quad -(75 \text{ kg } a_G)(0.9 \text{ m}) - (125 \text{ kg } a_G)(0.6 \text{ m})$

Solving,

$$a_G = 8.95 \text{ m/s}^2 \rightarrow \quad \textit{Ans.}$$
$$N_B = 1962 \text{ N}$$
$$F_B = 1790 \text{ N}$$

Thus the minimum coefficient of static friction is

$$(\mu_s)_{min} = \frac{F_B}{N_B} = \frac{1790 \text{ N}}{1962 \text{ N}} = 0.912 \quad \textit{Ans.}$$

EXAMPLE 17.7

The 100-kg beam BD shown in Fig. 17-12a is suspended from two rods having negligible mass. Determine the force developed in each rod if at the instant $\theta = 30°$, $\omega = 6$ rad/s.

SOLUTION
Free-Body and Kinetic Diagrams. The beam moves with curvilinear translation since all points on the beam move along circular paths, each path having the same radius of 0.5 m, but different centers of curvature. Using normal and tangential coordinates, the free-body and kinetic diagrams for the beam are shown in Fig. 17–12b. Because of the translation, G has the same motion as the pin at B, which is connected to both the rod and the beam. Note that the tangential component of acceleration acts downward to the left due to the clockwise direction of α, Fig. 17–12c. Furthermore, the normal component of acceleration is always directed toward the center of curvature (toward point A for rod AB). Since the angular velocity of AB is 6 rad/s when $\theta = 30°$, then

$$(a_G)_n = \omega^2 r = (6 \text{ rad/s})^2 (0.5 \text{ m}) = 18 \text{ m/s}^2$$

The three unknowns are T_B, T_D, and $(a_G)_t$.

Fig. 17–12

Equations of Motion.

$+\nwarrow \Sigma F_n = m(a_G)_n;\quad T_B + T_D - 981 \cos 30° \text{ N} = 100 \text{ kg}(18 \text{ m/s}^2) \quad (1)$

$+\swarrow \Sigma F_t = m(a_G)_t;\quad 981 \sin 30° = 100 \text{ kg}(a_G)_t \quad (2)$

$\zeta + \Sigma M_G = 0;\quad -(T_B \cos 30°)(0.4 \text{ m}) + (T_D \cos 30°)(0.4 \text{ m}) = 0 \quad (3)$

Simultaneous solution of these three equations gives

$$T_B = T_D = 1.32 \text{ kN} \qquad \textit{Ans.}$$

$$(a_G)_t = 4.905 \text{ m/s}^2$$

NOTE: It is also possible to apply the equations of motion along horizontal and vertical x, y axes, but the solution becomes more involved.

FUNDAMENTAL PROBLEMS

F17–1. The cart and its load have a total mass of 100 kg. Determine the acceleration of the cart and the normal reactions on the pair of wheels at A and B. Neglect the mass of the wheels.

Prob. F17–1

F17–2. If the 80-kg cabinet is allowed to roll down the inclined plane, determine the acceleration of the cabinet and the normal reactions on the pair of rollers at A and B that have negligible mass.

Prob. F17–2

F17–3. The 20-lb link AB is pinned to a moving frame at A and held in a vertical position by means of a string BC which can support a maximum tension of 10 lb. Determine the maximum acceleration of the frame without breaking the string. What are the corresponding components of reaction at the pin A?

Prob. F17–3

F17–4. Determine the maximum acceleration of the truck without causing the 100-kg assembly to move relative to the truck. Also what is the corresponding normal reaction on legs A and B? The assembly has a mass center at G and the coefficient of static friction between its legs and the truck bed is $\mu_s = 0.2$.

Prob. F17–4

F17–5. At the instant shown both rods of negligible mass swing with a counterclockwise angular velocity of $\omega = 5$ rad/s, while the 50-kg bar is subjected to the 100-N horizontal force. Determine the tension developed in the rods and the angular acceleration of the rods at this instant.

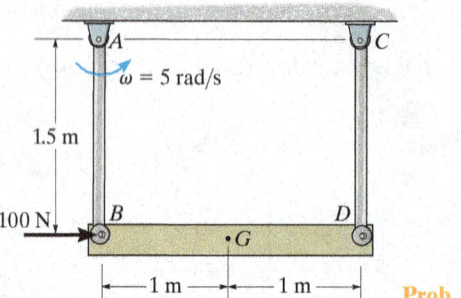

Prob. F17–5

F17–6. At the instant shown, rod CD rotates with an angular velocity of $\omega = 6$ rad/s. If it is subjected to a couple moment $M = 450$ N·m, determine the force developed in rod AB, the horizontal and vertical component of reaction on pin D, and the angular acceleration of rod CD at this instant. The bar has a mass of 50 kg and center of mass at G. Neglect the mass of rods AB and CD.

Prob. F17–6

PROBLEMS

All solutions must include a free-body diagram and a kinetic diagram.

***17–24.** The door has a weight of 200 lb and a center of gravity at G. Determine how far the door moves in 2 s, starting from rest, if a man pushes on it at C with a horizontal force $F = 30$ lb. Also, find the vertical reactions at the rollers A and B.

17–25. The door has a weight of 200 lb and a center of gravity at G. Determine the constant force F that must be applied to the door to push it open 12 ft to the right in 5 s, starting from rest. Also, find the vertical reactions at the rollers A and B.

Probs. 17–24/25

17–26. The jet aircraft has a total mass of 22 Mg and a center of mass at G. Initially at takeoff the engines provide a thrust $2T = 4$ kN and $T' = 1.5$ kN. Determine the acceleration of the plane and the normal reactions on the nose wheel at A and each of the *two* wing wheels located at B. Neglect the mass of the wheels and, due to low velocity, neglect any lift caused by the wings.

Prob. 17–26

17–27. The sports car has a weight of 4500 lb and center of gravity at G. If it starts from rest it causes the rear wheels to slip as it accelerates. Determine how long it takes for it to reach a speed of 10 ft/s. Also, what are the normal reactions at each of the four wheels on the road? The coefficients of static and kinetic friction at the road are $\mu_s = 0.5$ and $\mu_k = 0.3$, respectively. Neglect the mass of the wheels.

Prob. 17–27

***17–28.** The bicycle and rider have a mass of 80 kg with center of mass located at G. If the coefficient of kinetic friction at the rear tire is $\mu_B = 0.8$, determine the normal reactions at the tires A and B, and the deceleration of the rider, when the rear wheel locks for braking. What is the normal reaction at the rear wheel when the bicycle is traveling at constant velocity and the brakes are not applied? Neglect the mass of the wheels.

17–29. The bicycle and rider have a mass of 80 kg with center of mass located at G. Determine the minimum coefficient of kinetic friction between the road and the wheels so that the rear wheel B starts to lift off the ground when the rider applies the brakes to the front wheel. Neglect the mass of the wheels.

Probs. 17–28/29

17–30. At the *start* of takeoff, the propeller on the 2-Mg plane exerts a horizontal thrust of 600 N on the plane. Determine the plane's acceleration and the vertical reactions at the nose wheel A and each of the *two* wing wheels B. Neglect the lifting force of the wings since the plane is originally at rest. The mass center is at G.

Prob. 17–30

17–31. Block A weighs 50 lb and the platform weighs 10 lb. If $P = 100$ lb, determine the normal force exerted by block A on B. Neglect the weight of the pulleys and bars of the triangular frame.

Prob. 17–31

*****17–32.** The motorcycle and rider have a total mass of 900 kg and mass center at G. If it is traveling up the slope with a constant acceleration of 3 m/s^2, determine the normal reactions on the front and rear wheels. Also, what is the required friction force developed on the rear wheel? The front wheel is free to roll.

17–33. The motorcycle and rider have a total mass of 900 kg and mass center at G. Determine the minimum coefficient of static friction between the rear wheel and the slope in order for the rider to perform a "wheely," that is, to begin to lift the front wheel off the road. Also, what is the corresponding acceleration? The front wheel is free to roll.

Probs. 17–32/33

17–34. The sports car has been designed so that the 260-kg engine E and 165-kg transmission T have been placed over the front and rear wheels, respectively. Their mass centers are located at G_E and G_T. The mass center of the remaining 512-kg body and frame is located at G_b. If power is supplied to the rear wheels only, determine the shortest time it takes for the car to reach a speed of 80 km/h, starting from rest. The front wheels are free rolling, and the coefficient of static friction between the powered wheels and the road is $\mu_s = 0.4$. Neglect the mass of the wheels and driver.

Prob. 17–34

17–35. The dresser has a weight of 80 lb and is pushed along the floor. If the coefficient of static friction at A and B is $\mu_s = 0.3$ and the coefficient of kinetic friction is $\mu_k = 0.2$, determine the smallest horizontal force P needed to cause motion. If this force is increased slightly, determine the acceleration of the dresser. Also, what are the normal reactions at A and B when it begins to move?

*17–36. The dresser has a weight of 80 lb and is pushed along the floor. If the coefficient of static friction at A and B is $\mu_s = 0.3$ and the coefficient of kinetic friction is $\mu_k = 0.2$, determine the maximum horizontal force P that can be applied without causing the dresser to tip over.

Probs. 17–35/36

17–37. The crate of mass m is supported on a cart of negligible mass. Determine the maximum force P that can be applied a distance d from the cart bottom without causing the crate to tip on the cart.

Prob. 17–37

17–38. A 75-kg man and 40-kg girl sit on the seesaw, which has negligible mass. At the instant the man lifts his feet from the ground, determine their accelerations if each sits upright, i.e., they do not rotate. The centers of mass of the man and girl are at G_m and G_g, respectively.

Prob. 17–38

17–39. The bar has a weight per length w and is supported by the smooth collar. If it is released from rest, determine the internal normal force, shear force, and bending moment in the bar as a function of x.

Prob. 17–39

*17–40. The smooth 180-lb pipe has a length of 20 ft and a negligible diameter. It is carried on a truck as shown. Determine the maximum acceleration which the truck can have without causing the normal reaction at A to be zero. Also determine the horizontal and vertical components of force which the truck exerts on the pipe at B.

17–41. The smooth 180-lb pipe has a length of 20 ft and a negligible diameter. It is carried on a truck as shown. If the truck accelerates at $a = 5$ ft/s^2, determine the normal reaction at A and the horizontal and vertical components of force which the truck exerts on the pipe at B.

17–43. A man stands in the aisle of a moving train. If the coefficient of static friction between his shoes and the floor is $\mu_s = 0.5$, and he has a mass of 85 kg, with center of mass at G, determine the smallest train acceleration to the right to cause the man to either slide or tip. Assume the man holds himself rigid.

Prob. 17–43

Probs. 17–40/41

17–42. The truck carries the 1200-lb safe, which has a center of mass at G. Determine the largest acceleration of the truck so that the safe will not slip or tip on the truck bed. The coefficient of static friction between the safe and the truck is $\mu_s = 0.5$.

*17–44. The uniform bar of mass m is pin connected to the collar, which slides along the smooth horizontal rod. If the collar is given a constant acceleration of $\mathbf{a}$, determine the bar's inclination angle θ. Neglect the collar's mass.

Prob. 17–44

Prob. 17–42

17–45. Determine the greatest possible acceleration of the 975-kg race car so that its front tires do not leave the ground and none of the tires slip on the track. The coefficients of static and kinetic friction are $\mu_s = 0.8$ and $\mu_k = 0.6$, respectively. Neglect the mass of the tires. The car has rear-wheel drive and the front tires are free to roll.

17–46. Determine the greatest possible acceleration of the 975-kg race car so that its front wheels do not leave the ground and none of the tires slip on the track. The coefficients of static and kinetic friction are $\mu_s = 0.8$ and $\mu_k = 0.6$, respectively. Neglect the mass of the tires. The car has four-wheel drive.

Probs. 17–45/46

17–47. The four-wheeler has a weight of 335 lb and a center of gravity at G_1, whereas the rider has a weight of 150 lb and a center of gravity at G_2. If the engine can develop enough torque to cause the rear wheels to slip, determine the largest coefficient of static friction between the rear wheels and the ground so that the vehicle will accelerate without tipping over. What is this maximum acceleration? In order to *increase* the acceleration, should the rider crouch down or sit up straight from the position shown? Explain. The front wheels are free to roll. Neglect the mass of the wheels in the calculation.

*17–48. The four-wheeler has a weight of 335 lb and a center of gravity at G_1, whereas the rider has a weight of 150 lb and a center of gravity at G_2. If the coefficient of static friction between the rear wheels and the ground is $\mu_s = 0.3$, determine the greatest acceleration the vehicle can have without tipping over. The front wheels are free to roll. Neglect the mass of the wheels in the calculation.

Probs. 17–47/48

17–49. If the cart's mass is 30 kg and it is subjected to a horizontal force of $P = 90$ N, determine the tension in cord AB and the horizontal and vertical components of reaction on end C of the uniform 15-kg rod BC.

17–50. If the cart's mass is 30 kg, determine the horizontal force P that should be applied to the cart so that the cord AB just becomes slack. The uniform rod BC has a mass of 15 kg.

Probs. 17–49/50

17–51. The forklift travels forward with a constant speed of 9 ft/s. Determine the shortest stopping distance without causing any of the wheels to leave the ground. The forklift has a weight of 2000 lb with center of gravity at G_1, and the load weighs 900 lb with center of gravity at G_2. Neglect the weight of the wheels.

*17–52. If the forklift's rear wheels supply a combined traction force of $F_A = 300$ lb, determine its acceleration and the normal reactions on the pairs of rear wheels and front wheels. The forklift has a weight of 2000 lb, with center of gravity at G_1, and the load weighs 900 lb, with center of gravity at G_2. The front wheels are free to roll. Neglect the weight of the wheels.

Probs. 17–51/52

17–53. The crate C has a weight of 150 lb and rests on the truck elevator for which the coefficient of static friction is $\mu_s = 0.4$. Determine the largest initial angular acceleration α, starting from rest, which the parallel links AB and DE can have without causing the crate to slip. No tipping occurs.

Prob. 17–53

17–54. The crate C has a weight of 150 lb and rests on the truck elevator. Determine the initial friction and normal force of the elevator on the crate if the parallel links are given an angular acceleration $\alpha = 2$ rad/s² starting from rest.

Prob. 17–54

17–55. A uniform plate has a weight of 50 lb. Link AB is subjected to a couple moment of $M = 10$ lb · ft and has a clockwise angular velocity of 2 rad/s at the instant $\theta = 30°$. Determine the force developed in link CD and the tangential component of the acceleration of the plate's mass center at this instant. Neglect the mass of links AB and CD.

Prob. 17–55

*****17–56.** If the hydraulic cylinder BE exerts a vertical force of $F = 1.5$ kN on the platform, determine the force developed in links AB and CD at the instant $\theta = 90°$. The platform is at rest when $\theta = 45°$. Neglect the mass of the links and the platform. The 200-kg crate does not slip on the platform.

Prob. 17–56

17.4 EQUATIONS OF MOTION: ROTATION ABOUT A FIXED AXIS

Consider the rigid body (or slab) shown in Fig. 17–13a, which is constrained to rotate in the vertical plane about a fixed axis perpendicular to the page and passing through the pin at O. The angular velocity and angular acceleration are caused by the external force and couple moment system acting on the body. Because the body's center of mass G moves around a circular path, the acceleration of this point is best represented by its tangential and normal components. The tangential component of acceleration has a magnitude of $(a_G)_t = \alpha r_G$ and must act in a direction which is consistent with the body's angular acceleration α. The magnitude of the normal component of acceleration is $(a_G)_n = \omega^2 r_G$. This component is always directed from point G towards O, regardless of the rotational sense of ω.

The free-body and kinetic diagrams for the body are shown in Fig. 17–13b. The two components $m(\mathbf{a}_G)_t$ and $m(\mathbf{a}_G)_n$, shown on the kinetic diagram, are associated with the tangential and normal components of acceleration of the body's mass center. The $I_G\alpha$ vector acts in the same direction as α and has a magnitude of $I_G\alpha$. From the derivation given in Sec. 17.2, the equations of motion which apply to the body can be written in the form

$$\Sigma F_n = m(a_G)_n = m\omega^2 r_G$$
$$\Sigma F_t = m(a_G)_t = m\alpha r_G \qquad (17\text{–}14)$$
$$\Sigma M_G = I_G\alpha$$

The moment equation can be replaced by a moment summation about any arbitrary point P on or off the body provided one accounts for the moments $\Sigma(\mathcal{M}_k)_P$ produced by $I_G\alpha$, $m(\mathbf{a}_G)_t$, and $m(\mathbf{a}_G)_n$ about the point.

Moment Equation About Point O. Often it is convenient to sum moments about the pin at O in order to eliminate the unknown force $\mathbf{F}_O$. From the kinetic diagram, Fig. 17–13b, this requires

$$\zeta+\Sigma M_O = \Sigma(\mathcal{M}_k)_O; \qquad \Sigma M_O = r_G m(a_G)_t + I_G\alpha \qquad (17\text{–}15)$$

The moment of $m(\mathbf{a}_G)_n$ is not included here since the line of action of this vector passes through O. Substituting $(a_G)_t = r_G\alpha$, we may rewrite the above equation as $\zeta+\Sigma M_O = (I_G + mr_G^2)\alpha$. From the parallel-axis theorem, $I_O = I_G + md^2$, and therefore the term in parentheses represents the moment of inertia of the body about the fixed axis of rotation passing through O.* Consequently, we can also write the three equations of motion for the body as

$$\Sigma F_n = m(a_G)_n = m\omega^2 r_G$$
$$\Sigma F_t = m(a_G)_t = m\alpha r_G \qquad (17\text{–}16)$$
$$\Sigma M_O = I_O\alpha$$

When using these equations, remember that "$I_O\alpha$" accounts for the "moment" of both $m(\mathbf{a}_G)_t$ and $I_G\alpha$ about point O, Fig. 17–13b. In other words, $\Sigma M_O = \Sigma(\mathcal{M}_k)_O = I_O\alpha$.

*The result $\Sigma M_O = I_O\alpha$ can also be obtained directly from Eq. 17–6 by selecting point P to coincide with O, realizing that $(a_P)_x = (a_P)_y = 0$.

(a)

(b)

Fig. 17–13

PROCEDURE FOR ANALYSIS

Kinetic problems which involve the rotation of a body about a fixed axis can be solved using the following procedure.

Free-Body Diagram.

- Establish the inertial n, t coordinate system and specify the direction and sense of the accelerations $(\mathbf{a}_G)_n$ and $(\mathbf{a}_G)_t$ and the angular acceleration $\boldsymbol{\alpha}$ of the body. Recall that $(\mathbf{a}_G)_t$ must act in a direction which is in accordance with the rotational sense of $\boldsymbol{\alpha}$, whereas $(\mathbf{a}_G)_n$ always acts toward the axis of rotation, point O.

- Draw the free-body diagram to account for all the external forces and couple moments that act on the body.

- Determine the moment of inertia I_G or I_O.

- Identify the unknowns in the problem.

- If it is decided that the rotational equation of motion $\Sigma M_P = \Sigma(\mathcal{M}_k)_P$ is to be used, i.e., P is a point other than G or O, then consider drawing the kinetic diagram in order to help "visualize" the "moments" developed by the components $m(\mathbf{a}_G)_n$, $m(\mathbf{a}_G)_t$, and $I_G\boldsymbol{\alpha}$ when writing the terms for the moment sum $\Sigma(\mathcal{M}_k)_P$.

Equations of Motion.

- Apply the three equations of motion in accordance with the established sign convention.

- If moments are summed about the body's mass center, G, then $\Sigma M_G = I_G\alpha$, since $(m\mathbf{a}_G)_t$ and $(m\mathbf{a}_G)_n$ create no moment about G.

- If moments are summed about the pin support O on the axis of rotation, then $(m\mathbf{a}_G)_n$ creates no moment about O, and it can be shown that $\Sigma M_O = I_O\alpha$.

Kinematics.

- Use kinematics if a complete solution cannot be obtained strictly from the equations of motion.

- If the angular acceleration is variable, use

$$\omega = \frac{d\theta}{dt} \qquad \alpha = \frac{d\omega}{dt} \qquad \alpha\, d\theta = \omega\, d\omega$$

- If the angular acceleration is constant, use

$$\omega = \omega_0 + \alpha_c t$$
$$\theta = \theta_0 + \omega_0 t + \tfrac{1}{2}\alpha_c t^2$$
$$\omega^2 = \omega_0^2 + 2\alpha_c(\theta - \theta_0)$$

The crank on the oil-pumping rig undergoes rotation about a fixed axis which is caused by a driving torque **M** of the motor. The loadings shown on the free-body diagram cause the effects shown on the kinetic diagram. If moments are summed about the mass center, G, then $\Sigma M_G = I_G\alpha$. However, if moments are summed about point O, noting that $(a_G)_t = \alpha d$, then $\zeta + \Sigma M_O = I_G\alpha + m(a_G)_t d + m(a_G)_n(0) = (I_G + md^2)\alpha = I_O\alpha$.

EXAMPLE 17.8

The unbalanced 50-lb flywheel shown in Fig. 17–14a has a radius of gyration of $k_G = 0.6$ ft about an axis passing through its mass center G. If it is released from rest, determine the horizontal and vertical components of reaction at the pin O at this instant.

SOLUTION

Free-Body and Kinetic Diagrams. Since G moves in a circular path, it will have both normal and tangential components of acceleration. Also, since α, which is caused by the flywheel's weight, acts clockwise, the tangential component of acceleration must act downward. Why? Since $\omega = 0$, only $m(a_G)_t = m\alpha r_G$ and $I_G\alpha$ are shown on the kinetic diagram in Fig. 17–14b. Here, the moment of inertia about G is

$$I_G = mk_G^2 = (50 \text{ lb}/32.2 \text{ ft/s}^2)(0.6 \text{ ft})^2 = 0.559 \text{ slug} \cdot \text{ft}^2$$

The three unknowns are O_n, O_t, and α.

(a)

Equations of Motion.

$\overset{+}{\leftarrow} \Sigma F_n = m\omega^2 r_G;$ $\qquad O_n = 0$ $\qquad$ Ans.

$+\downarrow \Sigma F_t = m\alpha r_G;$ $\qquad -O_t + 50 \text{ lb} = \left(\dfrac{50 \text{ lb}}{32.2 \text{ ft/s}^2}\right)(\alpha)(0.5 \text{ ft})$ $\qquad$ (1)

$\zeta + \Sigma M_G = I_G\alpha;$ $\qquad O_t(0.5 \text{ ft}) = (0.5590 \text{ slug} \cdot \text{ft}^2)\alpha$

Solving,

$$\alpha = 26.4 \text{ rad/s}^2 \quad O_t = 29.5 \text{ lb} \quad \text{Ans.}$$

Moments can also be summed about point O in order to eliminate $\mathbf{O}_n$ and $\mathbf{O}_t$ and thereby obtain a direct solution for α, Fig. 17–14b. This can be done in one of two ways.

$\zeta + \Sigma M_O = \Sigma(\mathcal{M}_k)_O;$

$(50 \text{ lb})(0.5 \text{ ft}) = (0.5590 \text{ slug} \cdot \text{ft}^2)\alpha + \left[\left(\dfrac{50 \text{ lb}}{32.2 \text{ ft/s}^2}\right)\alpha(0.5 \text{ ft})\right](0.5 \text{ ft})$

$\qquad\qquad 50 \text{ lb}(0.5 \text{ ft}) = 0.9472\alpha$ $\qquad$ (2)

If $\Sigma M_O = I_O\alpha$ is applied, then by the parallel-axis theorem the moment of inertia of the flywheel about O is

$$I_O = I_G + mr_G^2 = 0.559 + \left(\dfrac{50}{32.2}\right)(0.5)^2 = 0.9472 \text{ slug} \cdot \text{ft}^2$$

Hence,

$\zeta + \Sigma M_O = I_O\alpha;$ $\quad (50 \text{ lb})(0.5 \text{ ft}) = (0.9472 \text{ slug} \cdot \text{ft}^2)\alpha$

which is the same as Eq. 2. Solving for α and substituting into Eq. 1 yields the answer for O_t obtained previously.

(b)

Fig. 17–14

EXAMPLE 17.9

At the instant shown in Fig. 17–15a, the 20-kg slender rod has an angular velocity of $\omega = 5$ rad/s. Determine the angular acceleration and the horizontal and vertical components of reaction of the pin on the rod at this instant.

(a)

SOLUTION

Free-Body and Kinetic Diagrams. Fig. 17–15b. As shown on the kinetic diagram, point G moves around a circular path and so it has two components of acceleration. It is important that the tangential component $a_t = \alpha r_G$ act downward since it must be in accordance with the rotational sense of α. The three unknowns are O_n, O_t, and α.

Equation of Motion.

$\xrightarrow{+} \Sigma F_n = m\omega^2 r_G;\quad O_n = (20\text{ kg})(5\text{ rad/s})^2(1.5\text{ m})$

$+\downarrow \Sigma F_t = m\alpha r_G;\quad -O_t + 20(9.81)\text{ N} = (20\text{ kg})(\alpha)(1.5\text{ m})$

$\zeta+\Sigma M_G = I_G \alpha;\quad O_t(1.5\text{ m}) + 60\text{ N}\cdot\text{m} = \left[\frac{1}{12}(20\text{ kg})(3\text{ m})^2\right]\alpha$

Solving

$O_n = 750\text{ N}\quad O_t = 19.05\text{ N}\quad \alpha = 5.90\text{ rad/s}^2 \quad Ans.$

Fig. 17–15

A more direct solution to this problem would be to sum moments about point O to eliminate O_n and O_t and obtain a *direct solution* for α. Here,

$\zeta+\Sigma M_O = \Sigma(\mathcal{M}_k)_O;\quad 60\text{ N}\cdot\text{m} + 20(9.81)\text{ N}(1.5\text{ m}) =$
$\left[\frac{1}{12}(20\text{ kg})(3\text{ m})^2\right]\alpha + [20\text{ kg}(\alpha)(1.5\text{ m})](1.5\text{ m})$

$\alpha = 5.90\text{ rad/s}^2 \quad Ans.$

Also, since $I_O = \frac{1}{3}ml^2$ for a slender rod, we can apply

$\zeta+\Sigma M_O = I_O \alpha;\quad 60\text{ N}\cdot\text{m} + 20(9.81)\text{ N}(1.5\text{ m}) = \left[\frac{1}{3}(20\text{ kg})(3\text{ m})^2\right]\alpha$

$\alpha = 5.90\text{ rad/s}^2 \quad Ans.$

NOTE: By comparison, the last equation provides the simplest solution for α and *does not* require use of the kinetic diagram.

EXAMPLE 17.10

The drum shown in Fig. 17–16a has a mass of 60 kg and a radius of gyration $k_O = 0.25$ m. A cord of negligible mass is wrapped around the periphery of the drum and attached to a block having a mass of 20 kg. If the block is released, determine the drum's angular acceleration.

SOLUTION I
Free-Body Diagram. Here we will consider the drum and block separately, Fig. 17–16b. Assuming the block accelerates downward at **a**, then it creates a counterclockwise angular acceleration α of the drum. The moment of inertia of the drum is

$$I_O = mk_O^2 = (60 \text{ kg})(0.25 \text{ m})^2 = 3.75 \text{ kg} \cdot \text{m}^2$$

There are five unknowns, namely $O_x, O_y, T, a,$ and α.

Equations of Motion. Moments will be summed about O to eliminate O_x and O_y. For the drum and block, we have

$$\zeta + \Sigma M_O = I_O \alpha; \quad T(0.4 \text{ m}) = (3.75 \text{ kg} \cdot \text{m}^2)\alpha \quad (1)$$
$$+\uparrow \Sigma F_y = m(a_G)_y; \quad -20(9.81) \text{ N} + T = -(20 \text{ kg})a \quad (2)$$

Kinematics. Since the point of contact A between the cord and drum has a tangential component of acceleration **a**, then

$$\zeta + a = \alpha r; \quad a = \alpha(0.4 \text{ m}) \quad (3)$$

Solving the above equations,

$$T = 106 \text{ N} \quad a = 4.52 \text{ m/s}^2$$
$$\alpha = 11.3 \text{ rad/s}^2 \zeta \qquad Ans.$$

SOLUTION II
Free-Body and Kinetic Diagrams. The cable tension T can be eliminated from the analysis by considering the drum and block as a single system, Fig. 17–16c. The kinetic diagram is shown since moments will be summed about point O.

Equations of Motion. Using Eq. 3 and applying the moment equation about O to eliminate the unknowns O_x and O_y, we have

$$\zeta + \Sigma M_O = \Sigma(\mathcal{M}_k)_O; \quad [20(9.81) \text{ N}](0.4 \text{ m}) = $$
$$(3.75 \text{ kg} \cdot \text{m}^2)\alpha + [20 \text{ kg}(\alpha\, 0.4 \text{ m})](0.4 \text{ m})$$
$$\alpha = 11.3 \text{ rad/s}^2 \zeta \qquad Ans.$$

NOTE: If the block were removed and a force of 20(9.81) N were applied to the cord, show that $\alpha = 20.9$ rad/s². This value is larger than our result since the block has an inertia, or resistance to acceleration.

(a)

(b)

(c)

Fig. 17–16

EXAMPLE 17.11

(a)

(b)

Fig. 17–17

The slender rod shown in Fig. 17–17a has a mass m and length l and is released from rest when $\theta = 0°$. Determine the horizontal and vertical components of force which the pin at A exerts on the rod at the instant $\theta = 90°$.

SOLUTION
Free-Body and Kinetic Diagrams. The free-body diagram for the rod in the general position θ is shown in Fig. 17–17b. For convenience, the force components at A are shown acting in the n and t directions. Note that α acts clockwise and so $(\mathbf{a}_G)_t$ acts in the $+t$ direction.

The moment of inertia of the rod about point A is $I_A = \frac{1}{3}ml^2$. For a given angle θ there are four unknowns, namely, A_n, A_t, ω, and α.

Equations of Motion. Moments will be summed about A in order to eliminate A_n and A_t.

$$+\nwarrow \Sigma F_n = m\omega^2 r_G; \quad A_n - mg\sin\theta = m\omega^2(l/2) \quad (1)$$
$$+\swarrow \Sigma F_t = m\alpha r_G; \quad A_t + mg\cos\theta = m\alpha(l/2) \quad (2)$$
$$\zeta + \Sigma M_A = I_A\alpha; \quad mg\cos\theta(l/2) = \left(\tfrac{1}{3}ml^2\right)\alpha \quad (3)$$

Kinematics. As shown by Eq. 3, α is not constant; rather, it depends on the position θ of the rod. The necessary fourth equation is obtained using kinematics, where α and ω can be related to θ by the equation

$$(\zeta+) \qquad \omega\, d\omega = \alpha\, d\theta \quad (4)$$

Note that the positive clockwise direction for this equation agrees with that of Eq. 3. This is important since we are seeking a simultaneous solution.

In order to solve for ω at $\theta = 90°$, eliminate α from Eqs. 3 and 4, which yields

$$\omega\, d\omega = (1.5g/l)\cos\theta\, d\theta$$

Since $\omega = 0$ at $\theta = 0°$, we have

$$\int_0^\omega \omega\, d\omega = (1.5g/l) \int_{0°}^{90°} \cos\theta\, d\theta$$

$$\omega^2 = 3g/l$$

Substituting this value into Eq. 1 with $\theta = 90°$ and solving Eqs. 1 to 3 yields

$$\alpha = 0$$
$$A_t = 0 \qquad A_n = 2.5mg \qquad \textit{Ans.}$$

NOTE: If $\Sigma M_A = \Sigma(\mathcal{M}_k)_A$ is used, one must account for the moments of $I_G\alpha$ and $m(\mathbf{a}_G)_t$ about A.

FUNDAMENTAL PROBLEMS

F17–7. The 100-kg wheel has a radius of gyration about its center O of $k_O = 500$ mm. If the wheel starts from rest, determine its angular velocity in $t = 3$ s.

Prob. F17–7

F17–8. The 50-kg disk is subjected to the couple moment of $M = (9t)$ N·m, where t is in seconds. Determine the angular velocity of the disk when $t = 4$ s starting from rest.

Prob. F17–8

F17–9. At the instant shown, the uniform 30-kg slender rod has a counterclockwise angular velocity of $\omega = 6$ rad/s. Determine the tangential and normal components of reaction of pin O on the rod and the angular acceleration of the rod at this instant.

Prob. F17–9

F17–10. At the instant shown, the 30-kg disk has a counterclockwise angular velocity of $\omega = 10$ rad/s. Determine the tangential and normal components of reaction of the pin O on the disk and the angular acceleration of the disk at this instant.

Prob. F17–10

F17–11. The uniform slender rod has a mass of 15 kg. Determine the horizontal and vertical components of reaction at the pin O, and the angular acceleration of the rod just after the cord is cut.

Prob. F17–11

F17–12. The uniform 30-kg slender rod is being lifted by the cord that passes over the small smooth peg at A. If the rod has a counterclockwise angular velocity of $\omega = 6$ rad/s at the instant shown, determine the tangential and normal components of reaction at the pin O and the angular acceleration of the rod.

Prob. F17–12

PROBLEMS

All solutions must include a free-body diagram.

17–57. The 10-kg wheel has a radius of gyration $k_A = 200$ mm. If the wheel is subjected to a moment $M = (5t)$ N·m, where t is in seconds, determine its angular velocity when $t = 3$ s starting from rest. Also, calculate the reactions which the fixed pin A exerts on the wheel during the motion.

Prob. 17–57

17–58. The uniform 24-kg plate is released from rest at the position shown. Determine its initial angular acceleration and the horizontal and vertical reactions at the pin A.

Prob. 17–58

17–59. The uniform slender rod has a mass m. If it is released from rest when $\theta = 0°$, determine the magnitude of the reactive force exerted on it by pin B when $\theta = 90°$.

Prob. 17–59

*17–60. The bent rod has a mass of 2 kg/m. If it is released from rest in the position shown, determine its initial angular acceleration and the horizontal and vertical components of reaction at A.

Prob. 17–60

17–61. The uniform slender rod has a mass of 5 kg. If the cord at A is cut, determine the reaction at the pin O, (a) when the rod is still in the horizontal position, and (b) when the rod swings to the vertical position.

Prob. 17–61

17-62. A motor supplies a constant torque $M = 2$ N·m to a 50-mm-diameter shaft O connected to the center of the 30-kg flywheel. The resultant bearing friction **F**, which the bearing exerts on the shaft, acts tangent to the shaft and has a magnitude of 50 N. Determine how long the torque must be applied to the shaft to increase the flywheel's angular velocity from 4 rad/s to 15 rad/s. The flywheel has radius of gyration $k_O = 0.15$ m about its center O.

17-63. If the motor of Prob. 17-62 is disengaged from the shaft once the flywheel is rotating at 15 rad/s, so that $M = 0$, determine how long it will take before the resultant bearing friction $F = 50$ N stops the flywheel from rotating.

Probs. 17-62/63

*17-64. In order to experimentally determine the moment of inertia I_G of a 4-kg connecting rod, the rod is suspended horizontally at A by a cord and at B by a bearing and piezoelectric sensor, an instrument used for measuring force. Under these equilibrium conditions, the force at B is measured as 14.6 N. If at the instant the cord is released the reaction at B is measured as 9.3 N, determine the value of I_G. The support at B does not move when the measurement is taken. For the calculation, the horizontal location of G must be determined.

Prob. 17-64

17-65. Disk A has a weight of 5 lb and disk B has a weight of 10 lb. If no slipping occurs between them, determine the couple moment M which must be applied to disk A to give it an angular acceleration of 4 rad/s^2.

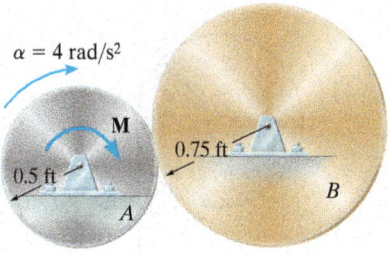

Prob. 17-65

17-66. The kinetic diagram representing the general rotational motion of a rigid body about a fixed axis passing through O is shown in the figure. Show that $I_G \alpha$ may be eliminated by moving the vectors $m(\mathbf{a}_G)_t$ and $m(\mathbf{a}_G)_n$ to point P, located a distance $r_{GP} = k_G^2/r_{OG}$ from the center of mass G of the body. Here k_G represents the radius of gyration of the body about an axis passing through G. The point P is called the *center of percussion* of the body.

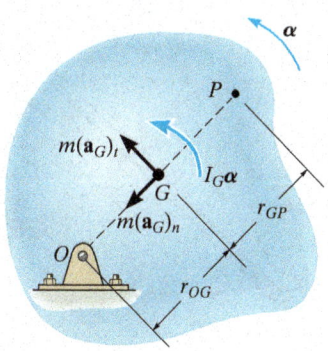

Prob. 17-66

17–67. The disk has a mass M and a radius R. If the block of mass m is attached to the cord, determine the angular velocity of the disk when the block is released from rest and falls a distance h.

Prob. 17–67

*17–68. The rod has a length L and mass m. If it is released from rest when $\theta \approx 0°$, determine its angular velocity as a function of θ. Also express the horizontal and vertical components of reaction at the pin O as a function of θ.

Prob. 17–68

17–69. The passengers, the gondola, and its swing frame have a total mass of 50 Mg, a mass center at G, and a radius of gyration $k_B = 3.5$ m. Additionally, the 3-Mg steel block at A can be considered as a point of concentrated mass. Determine the horizontal and vertical components of reaction at pin B if the gondola swings freely at $\omega = 1$ rad/s when it reaches its lowest point as shown. Also, what is the gondola's angular acceleration at this instant?

17–70. The passengers, the gondola, and its swing frame have a total mass of 50 Mg, a mass center at G, and a radius of gyration $k_B = 3.5$ m. Additionally, the 3-Mg steel block at A can be considered as a point of concentrated mass. Determine the angle θ to which the gondola will swing before it stops momentarily, if it has an angular velocity of $\omega = 1$ rad/s at its lowest point.

Probs. 17–69/70

17–71. The reel of cable has a mass of 400 kg and a radius of gyration of $k_A = 0.75$ m. Determine its angular velocity when $t = 2$ s, starting from rest, if the force $P = (20t^2 + 80)$ N is applied, where t is in seconds. Neglect the mass of the unwound cable, and assume it is always at a radius of 0.5 m.

Prob. 17–71

*17–72. The 30-kg disk is originally spinning at $\omega = 125$ rad/s. If it is placed on the ground, for which the coefficient of kinetic friction is $\mu_C = 0.5$, determine the time required for the motion to stop. What are the horizontal and vertical components of force which the member AB exerts on the pin at A during this time? Neglect the mass of AB.

17–74. The 5-kg cylinder is initially at rest when it is placed in contact with the wall B and the rotor at A. If the rotor always maintains a constant clockwise angular velocity $\omega = 6$ rad/s, determine the initial angular acceleration of the cylinder. The coefficient of kinetic friction at the contacting surfaces B and C is $\mu_k = 0.2$.

Prob. 17–72

Prob. 17–74

17–73. Cable is unwound from a spool supported on small rollers at A and B by exerting a force $T = 300$ N on the cable. Determine the time needed to unravel 5 m of cable from the spool if the spool and cable have a total mass of 600 kg and a radius of gyration of $k_O = 1.2$ m. For the calculation, neglect the mass of the cable being unwound and the mass of the rollers at A and B. The rollers turn with no friction.

17–75. The man pulls up on the cord with a force of 80 N. Determine the angular acceleration of the spool if there is no slipping at A. The spool has a mass of 60 kg and a radius of gyration about its mass center G of $k_G = 0.35$ m.

Prob. 17–73

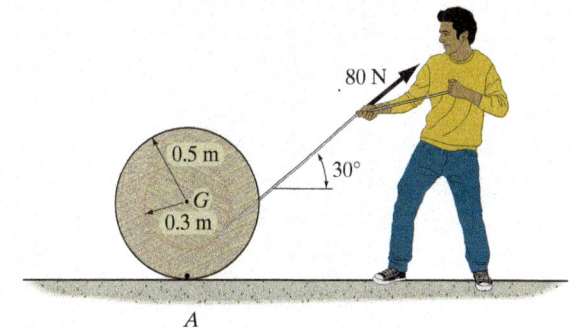

Prob. 17–75

*17–76. The 20-kg roll of paper has a radius of gyration $k_A = 120$ mm about an axis passing through point A. It is pin supported at both ends by two brackets AB. The roll rests on the floor, for which the coefficient of kinetic friction is $\mu_k = 0.2$. If a horizontal force $F = 60$ N is applied to the end of the paper, determine the initial angular acceleration of the roll as the paper unrolls.

17–78. Two cylinders A and B, having a weight of 10 lb and 5 lb, respectively, are attached to the ends of a cord which passes over a 3-lb pulley (disk). If the cylinders are released from rest, determine their speed in $t = 0.5$ s. The cord does not slip on the pulley. Neglect the mass of the cord. *Suggestion*: Analyze the "system" consisting of both the cylinders and the pulley.

Prob. 17–78

Prob. 17–76

17–77. Disk D turns with a constant clockwise angular velocity of 30 rad/s. Disk E has a weight of 60 lb and is initially at rest when it is brought into contact with D. Determine the time required for disk E to attain the same angular velocity as disk D. The coefficient of kinetic friction between the two disks is $\mu_k = 0.3$. Neglect the weight of bar BC.

17–79. The 50-kg flywheel has a radius of gyration about its center of the mass of $k_O = 250$ mm. It rotates with a constant angular velocity of 1200 rev/min before the brake is applied. If the coefficient of kinetic friction between the brake pad B and the wheel's rim is $\mu_k = 0.5$, and a force of $P = 300$ N is applied to the braking mechanism's handle, determine the time required to stop the wheel.

*17–80. The 50-kg flywheel has a radius of gyration about its center of mass of $k_O = 250$ mm. It rotates with a constant angular velocity of 1200 rev/min before the brake is applied. If the coefficient of kinetic friction between the brake pad B and the wheel's rim is $\mu_k = 0.5$, determine the constant force P that must be applied to the braking mechanism's handle in order to stop the wheel in 100 revolutions.

Prob. 17–77

Probs. 17–79/80

17–81. Determine the angular acceleration of the 25-kg diving board, and the horizontal and vertical components of reaction at the pin A, the instant the man dives off. Assume that the board is uniform and rigid, and that at the instant he dives off, the spring is compressed a maximum amount of 200 mm, $\omega = 0$, and the board is horizontal. Take $k = 7$ kN/m.

17–83. If the support at B is suddenly removed, determine the initial reactions at the pin A. The plate has a weight of 30 lb.

Prob. 17–83

Prob. 17–81

*17–84. The furnace cover has a mass of 20 kg and a radius of gyration $k_G = 0.25$ m about its mass center G. If an operator applies a force $F = 120$ N to the handle in order to open the cover, determine the cover's initial angular acceleration and the horizontal and vertical components of reaction which the pin at A exerts on the cover at the instant the cover begins to open. Neglect the mass of the handle BAC in the calculation.

17–82. The turbine consists of a rotor which is powered from a torque applied at its center. At the instant the rotor is horizontal it has an angular velocity of 15 rad/s and a clockwise angular acceleration of 8 rad/s². Determine the internal normal force, shear force, and moment at a section through A. Assume the rotor is a 50-m-long slender rod, having a mass of 3 kg/m.

Prob. 17–82

Prob. 17–84

17–85. The bar has a weight per length of w. If it is rotating in the vertical plane at a constant rate ω about point O, determine the internal normal force, shear force, and moment as a function of x and θ.

Prob. 17–85

17–86. The uniform rod of length L has a mass of m. If the couple moment $\mathbf{M}_0$ is applied to its end, determine the internal normal force and shear force, and the bending moment in the rod as a function of x just after $\mathbf{M}_0$ is applied. The bearing at A allows free rotation.

Prob. 17–86

17–87. The 4-kg slender rod is supported horizontally by a spring at A and a cord at B. Determine the angular acceleration of the rod and the acceleration of the rod's mass center at the instant the cord at B is cut. *Hint*: The stiffness of the spring is not needed for the calculation.

Prob. 17–87

*17–88. The 4-kg slender rod is initially supported horizontally by a spring at B and pin at A. Determine the angular acceleration of the rod and the acceleration of the rod's mass center at the instant the 100-N force is applied.

Prob. 17–88

17–89. The "Catherine wheel" is a firework that consists of a coiled tube of powder which is pinned at its center. If the powder burns at a constant rate of 20 g/s such that the exhaust gases always exert a force having a constant magnitude of 0.3 N, directed tangent to the wheel, determine the angular velocity of the wheel when 75% of the mass is burned off. Initially, the wheel is at rest and has a mass of 100 g and a radius of $r = 75$ mm. For the calculation, consider the wheel to always be a thin disk.

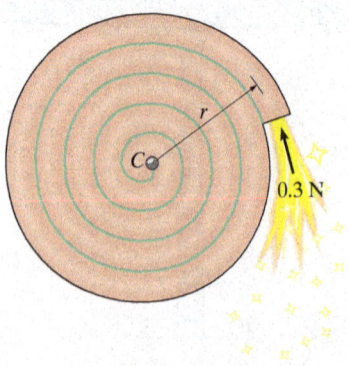

Prob. 17–89

17.5 EQUATIONS OF MOTION: GENERAL PLANE MOTION

The rigid body (or slab) shown in Fig. 17–18a is subjected to general plane motion caused by the externally applied force and couple-moment system. The free-body and kinetic diagrams for the body are shown in Fig. 17–18b. If an x and y inertial coordinate system is established as shown, the three equations of motion are

$$\Sigma F_x = m(a_G)_x$$
$$\Sigma F_y = m(a_G)_y \quad (17\text{--}17)$$
$$\Sigma M_G = I_G \alpha$$

In some problems it may be convenient to sum moments about a point P other than G in order to eliminate as many unknown forces as possible from the moment summation. When used for this more general case, the three equations of motion are

$$\Sigma F_x = m(a_G)_x$$
$$\Sigma F_y = m(a_G)_y \quad (17\text{--}18)$$
$$\Sigma M_P = \Sigma(\mathcal{M}_k)_P$$

Here $\Sigma(\mathcal{M}_k)_P$ represents the moment sum of $I_G \alpha$ and $m\mathbf{a}_G$ (or its components) about P.

Moment Equation About the IC.
Some types of problems involve a uniform disk, or body of circular shape, that rolls on a rough surface *without slipping*, Fig. 17–19. If we sum the moments about the instantaneous center of zero velocity, then $\Sigma(\mathcal{M}_k)_{IC}$ becomes $I_{IC}\alpha$, so that

$$\Sigma M_{IC} = I_{IC}\alpha \quad (17\text{--}19)$$

This result compares with $\Sigma M_O = I_O \alpha$, which is used for a body pinned at point O, Eq. 17–16. See Prob. 17–90.

Fig. 17–18

Fig. 17–19

As the soil compactor, or "sheep's foot roller" moves forward, the roller has general plane motion. The forces shown on its free-body diagram cause the effects shown on the kinetic diagram. If moments are summed about the mass center, G, then $\Sigma M_G = I_G \alpha$. However, if moments are summed about point A (the IC) then $\zeta + \Sigma M_A = I_G \alpha + (m a_G)d = I_A \alpha$ since $a_G = \alpha d$.

PROCEDURE FOR ANALYSIS

Kinetic problems involving general plane motion of a rigid body can be solved using the following procedure.

Free-Body Diagram.

- Establish the x, y inertial coordinate system and draw the free-body diagram for the body.
- Specify the direction and sense of the acceleration of the mass center, $\mathbf{a}_G$, and the angular acceleration $\boldsymbol{\alpha}$ of the body.
- Determine the moment of inertia I_G.
- Identify the unknowns in the problem.
- If it is decided that the rotational equation of motion $\Sigma M_P = \Sigma(\mathcal{M}_k)_P$ is to be used, then consider drawing the kinetic diagram in order to help "visualize" the "moments" developed by the components $m(\mathbf{a}_G)_x$, $m(\mathbf{a}_G)_y$, and $I_G \boldsymbol{\alpha}$ when writing the terms in the moment sum $\Sigma(\mathcal{M}_k)_P$.

Equations of Motion.

- Apply the three equations of motion in accordance with the established sign convention.

Kinematics.

- Use kinematics if a complete solution cannot be obtained strictly from the equations of motion.
- If the body's motion is constrained due to its supports, additional equations may be obtained by using $\mathbf{a}_B = \mathbf{a}_A + \mathbf{a}_{B/A}$, which relates the accelerations of any two points A and B on the body.
- When a wheel, disk, cylinder, or ball *rolls without slipping*, then $a_G = \alpha r$.

EXAMPLE 17.12

Determine the angular acceleration of the spool in Fig. 17–20a. The spool has a mass of 8 kg and a radius of gyration of $k_G = 0.35$ m. The cords of negligible mass are wrapped around its inner hub and outer rim.

SOLUTION I
Free-Body and Kinetic Diagrams. Fig. 17–20b. The 100-N force causes $\mathbf{a}_G$ to act upward. Also, $\boldsymbol{\alpha}$ acts clockwise, since the spool winds around the cord at A.
The moment of inertia of the spool about its mass center is

$$I_G = mk_G^2 = 8 \text{ kg}(0.35 \text{ m})^2 = 0.980 \text{ kg} \cdot \text{m}^2$$

There are three unknowns T, a_G, and α.

(a)

Equations of Motion.

$+\uparrow \Sigma F_y = m(a_G)_y; \quad T + 100 \text{ N} - 78.48 \text{ N} = (8 \text{ kg})a_G \quad (1)$

$\zeta + \Sigma M_G = I_G \alpha; \quad 100 \text{ N}(0.2 \text{ m}) - T(0.5 \text{ m}) = (0.980 \text{ kg} \cdot \text{m}^2)\alpha \quad (2)$

Kinematics. A complete solution is obtained if kinematics is used to relate a_G to α. In this case the spool "rolls without slipping" on the cord at A. Hence, we can use the results of Example 16.4 or 16.15, so that

$(\zeta +) \, a_G = \alpha r; \qquad a_G = \alpha (0.5 \text{ m}) \quad (3)$

Solving Eqs. 1 to 3, we have

$$\alpha = 10.3 \text{ rad/s}^2 \quad Ans.$$
$$a_G = 5.16 \text{ m/s}^2$$
$$T = 19.8 \text{ N}$$

SOLUTION II
Equations of Motion. We can eliminate the unknown T by summing moments about point A. From the free-body and kinetic diagrams, Figs. 17–20b, we have

$\zeta + \Sigma M_A = \Sigma (\mathcal{M}_k)_A; \quad 100 \text{ N}(0.7 \text{ m}) - 78.48 \text{ N}(0.5 \text{ m})$
$\qquad = (0.980 \text{ kg} \cdot \text{m}^2)\alpha + [(8 \text{ kg})a_G](0.5 \text{ m})$

Using Eq. 3,

$$\alpha = 10.3 \text{ rad/s}^2 \quad Ans.$$

SOLUTION III
Equations of Motion. The simplest way to solve this problem is to realize that point A is the IC for the spool. Then Eq. 17–19 applies.

$\zeta + \Sigma M_A = I_A \alpha; \quad (100 \text{ N})(0.7 \text{ m}) - (78.48 \text{ N})(0.5 \text{ m})$
$\qquad = [0.980 \text{ kg} \cdot \text{m}^2 + (8 \text{ kg})(0.5 \text{ m})^2]\alpha$
$\qquad \alpha = 10.3 \text{ rad/s}^2 \quad Ans.$

(b)

Fig. 17–20

EXAMPLE 17.13

(a)

(b)

Fig. 17–21

The 50-lb wheel shown in Fig. 17–21 has a radius of gyration $k_G = 0.70$ ft. If a 35-lb·ft couple moment is applied to the wheel, determine the acceleration of its mass center G. The coefficients of static and kinetic friction between the wheel and the plane at A are $\mu_s = 0.3$ and $\mu_k = 0.25$, respectively.

SOLUTION
Free-Body and Kinetic Diagrams. By inspection of Fig. 17–21b, it is seen that the couple moment causes the wheel to have a clockwise angular acceleration of α. As a result, the acceleration of the mass center, $\mathbf{a}_G$, is directed to the right. The moment of inertia is

$$I_G = mk_G^2 = \frac{50 \text{ lb}}{32.2 \text{ ft/s}^2}(0.70 \text{ ft})^2 = 0.7609 \text{ slug} \cdot \text{ft}^2$$

The unknowns are $N_A, F_A, a_G,$ and α.

Equations of Motion.

$$\xrightarrow{+} \Sigma F_x = m(a_G)_x; \quad F_A = \left(\frac{50 \text{ lb}}{32.2 \text{ ft/s}^2}\right)a_G \quad (1)$$

$$+\uparrow \Sigma F_y = m(a_G)_y; \quad N_A - 50 \text{ lb} = 0 \quad (2)$$

$$\zeta + \Sigma M_G = I_G \alpha; \quad 35 \text{ lb} \cdot \text{ft} - 1.25 \text{ ft}(F_A) = (0.7609 \text{ slug} \cdot \text{ft}^2)\alpha \quad (3)$$

A fourth equation is needed for a complete solution.

Kinematics (No Slipping). If no slipping occurs, then

$$(\zeta +) \qquad a_G = (1.25 \text{ ft})\alpha \quad (4)$$

Solving Eqs. 1 to 4,

$$N_A = 50.0 \text{ lb} \qquad F_A = 21.3 \text{ lb}$$
$$\alpha = 11.0 \text{ rad/s}^2 \qquad a_G = 13.7 \text{ ft/s}^2$$

This solution requires $F_A \leq \mu_s N_A$. However, since 21.3 lb $>$ 0.3(50 lb) = 15 lb, the wheel slips as it rolls.

(Slipping). Equation 4 is not valid, and so $F_A = \mu_k N_A$, or

$$F_A = 0.25 N_A \quad (5)$$

Solving Eqs. 1 to 3 and 5 yields

$$N_A = 50.0 \text{ lb} \qquad F_A = 12.5 \text{ lb}$$
$$\alpha = 25.5 \text{ rad/s}^2$$
$$a_G = 8.05 \text{ ft/s}^2 \rightarrow \qquad \textit{Ans.}$$

EXAMPLE 17.14

The uniform slender pole shown in Fig. 17–22a has a mass of 100 kg. If the coefficients of static and kinetic friction between the end of the pole and the surface are $\mu_s = 0.3$ and $\mu_k = 0.25$, respectively, determine the pole's angular acceleration at the instant the 400-N horizontal force is applied. The pole is originally at rest.

SOLUTION
Free-Body and Kinetic Diagrams. Figure 17–22b. The path of motion of the mass center G will be along an unknown curved path having a radius of curvature ρ, which is initially on a vertical line. However, there is no normal or y component of acceleration since the pole is originally at rest, i.e., $\mathbf{v}_G = \mathbf{0}$, so that $(a_G)_y = v_G^2/\rho = 0$. We will assume the mass center accelerates to the right and that the pole has a clockwise angular acceleration of α. The unknowns are N_A, F_A, a_G, and α.

Equation of Motion.

$\xrightarrow{+} \Sigma F_x = m(a_G)_x;\qquad 400\text{ N} - F_A = (100\text{ kg})a_G \qquad (1)$

$+\uparrow \Sigma F_y = m(a_G)_y;\qquad N_A - 981\text{ N} = 0 \qquad (2)$

$\zeta + \Sigma M_G = I_G\alpha;\ F_A(1.5\text{ m}) - (400\text{ N})(1\text{ m}) = [\tfrac{1}{12}(100\text{ kg})(3\text{ m})^2]\alpha \quad (3)$

A fourth equation is needed for a complete solution.

Kinematics (No Slipping). With this assumption, point A acts as a "pivot" so that α is clockwise, then a_G is directed to the right.

$a_G = \alpha r_{AG};\qquad a_G = (1.5\text{ m})\alpha \qquad (4)$

Solving Eqs. 1 to 4 yields

$$N_A = 981\text{ N}\qquad F_A = 300\text{ N}$$
$$a_G = 1\text{ m/s}^2\qquad \alpha = 0.667\text{ rad/s}^2$$

The assumption of no slipping requires $F_A \le \mu_s N_A$. However, $300\text{ N} > 0.3(981\text{ N}) = 294\text{ N}$ and so the pole slips at A.

(Slipping). For this case Eq. 4 does *not* apply. Instead the frictional equation $F_A = \mu_k N_A$ must be used. Hence,

$$F_A = 0.25 N_A \qquad (5)$$

Solving Eqs. 1 to 3 and 5 simultaneously yields

$$N_A = 981\text{ N}\quad F_A = 245\text{ N}\quad a_G = 1.55\text{ m/s}^2$$
$$\alpha = -0.428\text{ rad/s}^2 = 0.428\text{ rad/s}^2\ \zeta \qquad Ans.$$

(a)

(b)

Fig. 17–22

EXAMPLE 17.15

The uniform 50-kg bar in Fig. 17–23a is held in the equilibrium position by cords AC and BD. Determine the tension in BD and the angular acceleration of the bar immediately after AC is cut.

SOLUTION
Free-Body and Kinetic Diagrams. Fig. 17–23b. There are four unknowns, T_B, $(a_G)_x$, $(a_G)_y$, and α.

Equations of Motion.

$\xrightarrow{+} \Sigma F_x = m(a_G)_x;$ $\quad 0 = 50 \text{ kg }(a_G)_x$
$\quad (a_G)_x = 0$

$+\uparrow \Sigma F_y = m(a_G)_y;$ $\quad T_B - 50(9.81) \text{ N} = -50 \text{ kg }(a_G)_y \quad (1)$

$\zeta + \Sigma M_G = I_G \alpha;$ $\quad T_B(1.5 \text{ m}) = \left[\dfrac{1}{12}(50 \text{ kg})(3 \text{ m})^2\right]\alpha \quad (2)$

Kinematics. Since the bar is at rest just after the cable is cut, then its angular velocity and the velocity of point B at this instant are equal to zero. Thus $(a_B)_n = v_B^2/\rho_{BD} = 0$. Therefore, $\mathbf{a}_B$ only has a tangential component, which is directed along the x axis, Fig. 17–23c. Applying the relative acceleration equation to points G and B,

$$\mathbf{a}_G = \mathbf{a}_B + \boldsymbol{\alpha} \times \mathbf{r}_{G/B} - \omega^2 \mathbf{r}_{G/B}$$

$$-(a_G)_y \mathbf{j} = a_B \mathbf{i} + (\alpha \mathbf{k}) \times (-1.5\mathbf{i}) - \mathbf{0}$$

$$-(a_G)_y \mathbf{j} = a_B \mathbf{i} - 1.5\alpha \mathbf{j}$$

Equating the **i** and **j** components of both sides of this equation,

$0 = a_B$

$(a_G)_y = 1.5\alpha \quad (3)$

Solving Eqs. 1 through 3 yields

$\alpha = 4.905 \text{ rad/s}^2 \quad$ *Ans.*
$T_B = 123 \text{ N} \quad$ *Ans.*
$(a_G)_y = 7.36 \text{ m/s}^2$

Fig. 17–23

FUNDAMENTAL PROBLEMS

F17–13. The uniform 60-kg slender bar is initially at rest on a smooth *horizontal plane* when the forces are applied. Determine the acceleration of the bar's mass center and the angular acceleration of the bar at this instant.

Prob. F17–13

F17–14. The 100-kg cylinder rolls without slipping on the horizontal plane. Determine the acceleration of its mass center and its angular acceleration.

Prob. F17–14

F17–15. The 20-kg wheel has a radius of gyration about its center G of $k_G = 300$ mm. When the wheel is subjected to the couple moment, it slips as it rolls. Determine the angular acceleration of the wheel and the acceleration of G. The coefficient of kinetic friction between the wheel and the plane is $\mu_k = 0.5$.

Prob. F17–15

F17–16. The 20-kg sphere rolls down the inclined plane without slipping. Determine the angular acceleration of the sphere and the acceleration of its mass center.

Prob. F17–16

F17–17. The 200-kg spool has a radius of gyration about its mass center of $k_G = 300$ mm. If the couple moment is applied to the spool and the coefficient of kinetic friction between the spool and the ground is $\mu_k = 0.2$, determine the angular acceleration of the spool, the acceleration of G, and the tension in the cable.

Prob. F17–17

F17–18. The 12-kg slender rod is pinned to a small roller A that slides freely along the slot. If the rod is released from rest at $\theta = 0°$, determine the angular acceleration of the rod and the acceleration of the roller immediately after the release.

Prob. F17–18

PROBLEMS

All solutions must include a free-body diagram and a kinetic diagram.

17–90. If the disk in Fig. 17–19 *rolls without slipping*, show that when moments are summed about the instantaneous center of zero velocity, IC, it is possible to use the moment equation $\Sigma M_{IC} = I_{IC}\alpha$, where I_{IC} represents the moment of inertia of the disk calculated about the instantaneous axis of zero velocity.

17–91. The 75-kg wheel has a radius of gyration about the z axis of $k_z = 150$ mm. If the belt of negligible mass is subjected to a force of $P = 150$ N, determine the acceleration of the mass center and the angular acceleration of the wheel. The surface is smooth and the wheel is free to slide.

Prob. 17–91

*17–92. The uniform 150-lb beam is initially at rest when the forces are applied to the cables. Determine the magnitude of the acceleration of the mass center and the angular acceleration of the beam at this instant.

Prob. 17–92

17–93. The ship has a weight of $4(10^6)$ lb and center of gravity at G. Two tugboats of negligible weight are used to turn it. If each tugboat pushes on it with a force of $T = 2000$ lb, determine the initial acceleration of its center of gravity G and its angular acceleration. Its radius of gyration about its center of gravity is $k_G = 125$ ft. Neglect water resistance.

Prob. 17–93

17–94. The 75-kg wheel has a radius of gyration about its mass center of $k_G = 375$ mm. If it is subjected to a torque of $M = 100$ N·m, determine its angular acceleration. The coefficients of static and kinetic friction between the wheel and the ground are $\mu_s = 0.2$ and $\mu_k = 0.15$, respectively.

17–95. The 75-kg wheel has a radius of gyration about its mass center of $k_G = 375$ mm. If it is subjected to a torque of $M = 150$ N·m, determine its angular acceleration. The coefficients of static and kinetic friction between the wheel and the ground are $\mu_s = 0.2$ and $\mu_k = 0.15$, respectively.

Probs. 17–94/95

*17–96. The spool has a mass of 100 kg and radius of gyration about its mass center of $k_G = 0.3$ m. Determine its angular acceleration when the force of 250 N is applied to the cable. The spool rolls without slipping.

17–97. The spool has a mass of 100 kg and radius of gyration about its mass center of $k_G = 0.3$ m. Determine the smallest coefficient of static friction that will prevent slipping on the rail at A. What is the angular acceleration of the spool?

Probs. 17–96/97

17–98. The spool has a mass of 100 kg and a radius of gyration $k_G = 0.3$ m. If the coefficients of static and kinetic friction at A are $\mu_s = 0.2$ and $\mu_k = 0.15$, respectively, and $P = 600$ N, determine the angular acceleration of the spool.

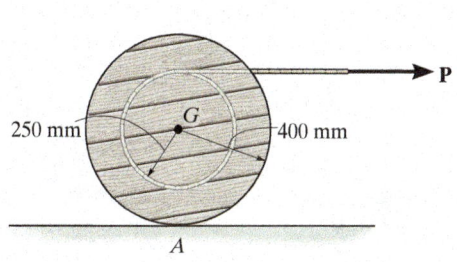

Prob. 17–98

17–99. The 15-lb circular plate is suspended from a pin at A. If the pin is connected to a track which is given an acceleration $a_A = 3$ ft/s², determine the initial horizontal and vertical components of reaction at A and the acceleration of the plate's mass center G. The plate is originally at rest.

Prob. 17–99

*17–100. A force of $F = 10$ N is applied to the 10-kg ring as shown. If slipping does not occur, determine the ring's initial angular acceleration, and the acceleration of its mass center, G. Neglect the thickness of the ring.

Prob. 17–100

17–101. If the coefficient of static friction at C is $\mu_s = 0.3$, determine the largest force F that can be applied to the 5-kg ring without causing it to slip. Neglect the thickness of the ring.

Prob. 17–101

17–102. The slender, 200-kg beam is suspended by a cable at its end as shown. If a man pushes on its other end with a horizontal force of 30 N, determine the initial acceleration of its mass center G, the beam's angular acceleration, the tension in the cable AB, and the initial acceleration of the end A.

***17–104.** If $P = 30$ lb, determine the angular acceleration of the 50-lb roller. Assume the roller to be a uniform cylinder and that no slipping occurs.

17–105. If the coefficient of static friction between the 50-lb roller and the ground is $\mu_s = 0.25$, determine the maximum force P that can be applied to the handle so that the roller rolls on the ground without slipping. Also, find the angular acceleration of the roller. Assume the roller to be a uniform cylinder.

Prob. 17–102

Probs. 17–104/105

17–103. The spool has a mass of 75 kg and a radius of gyration $k_G = 0.380$ m. It rests on the inclined surface for which the coefficient of kinetic friction is $\mu_k = 0.15$. If the spool is released from rest and slips at A, determine the initial tension in the cord and the angular acceleration of the spool.

17–106. The truck carries the spool which has a weight of 500 lb and a radius of gyration of $k_G = 2$ ft. Determine the angular acceleration of the spool if it is not tied down on the truck and the truck begins to accelerate at 3 ft/s². Assume the spool does not slip on the bed of the truck.

17–107. The truck carries the spool which has a weight of 200 lb and a radius of gyration of $k_G = 2$ ft. Determine the angular acceleration of the spool if it is not tied down on the truck and the truck begins to accelerate at 5 ft/s². The coefficients of static and kinetic friction between the spool and the truck bed are $\mu_s = 0.15$ and $\mu_k = 0.1$, respectively.

Prob. 17–103

Probs. 17–106/107

*17–108. The semicircular disk having a mass of 10 kg is rotating at $\omega = 4$ rad/s at the instant $\theta = 60°$. If the coefficient of static friction at A is $\mu_s = 0.5$, determine if the disk slips at this instant.

Prob. 17–108

17–109. The 10-lb hoop or thin ring is given an initial angular velocity of 6 rad/s when it is placed on the surface. If the coefficient of kinetic friction between the hoop and the surface is $\mu_k = 0.3$, determine the distance the hoop moves before it stops slipping.

Prob. 17–109

17–110. Wheel C has a mass of 60 kg and a radius of gyration of 0.4 m, whereas wheel D has a mass of 40 kg and a radius of gyration of 0.35 m. Determine the angular acceleration of each wheel at the instant shown. Neglect the mass of the link and assume the assembly does not slip on the plane.

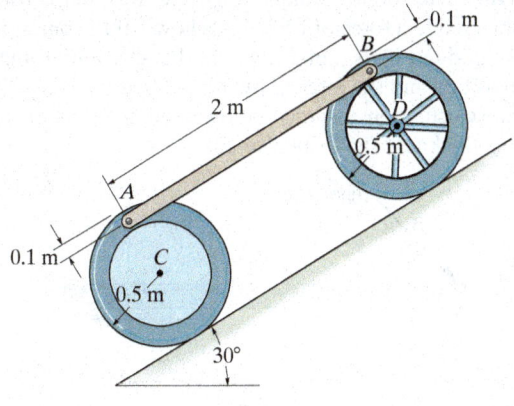

Prob. 17–110

17–111. The uniform bar of mass m and length L is balanced in the vertical position when the horizontal force P is applied to the roller at A. Determine the bar's initial angular acceleration and the acceleration of its top point B.

Prob. 17–111

*17–112. The circular concrete culvert rolls with an angular velocity of $\omega = 0.5$ rad/s when the man is at the position shown. At this instant the center of gravity of the culvert and the man is located at point G, and the radius of gyration about G is $k_G = 3.5$ ft. Determine the angular acceleration of the culvert. The combined weight of the culvert and the man is 500 lb. Assume that the culvert rolls without slipping, and the man does not move within the culvert.

Prob. 17–112

17–113. The uniform disk of mass m is rotating with an angular velocity of ω_0 when it is placed on the floor. Determine the initial angular acceleration of the disk and the acceleration of its mass center. The coefficient of kinetic friction between the disk and the floor is μ_k.

17–114. The uniform disk of mass m is rotating with an angular velocity of ω_0 when it is placed on the floor. Determine the time before it starts to roll without slipping. What is the angular velocity of the disk at this instant? The coefficient of kinetic friction between the disk and the floor is μ_k.

Probs. 17–113/114

17–115. The uniform plate weighs 40 lb and is supported by a roller at A. If a horizontal force of $F = 70$ lb is applied to the roller, determine the acceleration of the center of the roller at the instant the force is applied. The plate has a moment of inertia about its center of mass of $I_G = 0.414$ slug · ft². Neglect the weight and the size d of the roller.

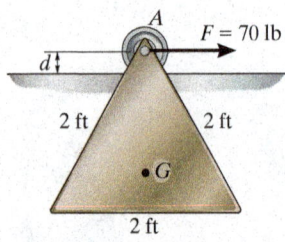

Prob. 17–115

*****17–116.** The cylinder A has a weight of 15 lb and is attached to the 20-lb block B using the cord and pulley system shown. Determine the acceleration of the block when it is released. The coefficients of static friction are indicated in the figure.

Prob. 17–116

17–117. The uniform beam has a weight W. If it is originally at rest while being supported at A and B by cables, determine the tension in cable A if cable B suddenly fails. Assume the beam is a slender rod.

Prob. 17–117

17–118. The 500-lb beam is supported at A and B when it is subjected to a force of 1000 lb as shown. If the pin support at A suddenly fails, determine the beam's initial angular acceleration and the force of the roller support on the beam. For the calculation, assume that the beam is a slender rod so that its thickness can be neglected.

Prob. 17–118

CHAPTER REVIEW

Moment of Inertia

The moment of inertia is a measure of the resistance of a body to a change in its angular velocity. It is defined by $I = \int r^2 dm$ and will be different for each axis about which it is calculated.

Many bodies are composed of simple shapes. If this is the case, then tabular values of I can be used, such as the ones given on the inside back cover of this book. To obtain the moment of inertia of a composite body about any specified axis, the moment of inertia of each part is determined about the axis and the results are added together. Doing this often requires use of the parallel-axis theorem.

$$I = I_G + md^2$$

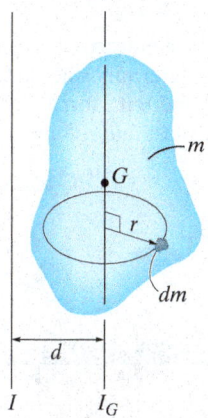

Planar Equations of Motion

The equations of motion define the translational and rotational motion of a rigid body. In order to account for all of the terms in these equations, a free-body diagram should always accompany their application, and for some problems, it may also be convenient to draw the kinetic diagram which shows $m\mathbf{a}_G$ and $I_G\alpha$.

$\Sigma F_x = m(a_G)_x$
$\Sigma F_y = m(a_G)_y$
$\Sigma M_G = 0$

Rectilinear translation

$\Sigma F_n = m(a_G)_n$
$\Sigma F_t = m(a_G)_t$
$\Sigma M_G = 0$

Curvilinear translation

$\Sigma F_n = m(a_G)_n = m\omega^2 r_G$
$\Sigma F_t = m(a_G)_t = m\alpha r_G$
$\Sigma M_G = I_G\alpha$ or $\Sigma M_O = I_O\alpha$

Rotation About a Fixed Axis

$\Sigma F_x = m(a_G)_x$
$\Sigma F_y = m(a_G)_y$
$\Sigma M_G = I_G\alpha$ or $\Sigma M_P = \Sigma(\mathcal{M}_k)_P$

General Plane Motion

REVIEW PROBLEMS

R17–1. The handcart has a mass of 200 kg and center of mass at G. Determine the normal reactions at *each* of the wheels at A and B if a force $P=50$ N is applied to the handle. Neglect the mass and rolling resistance of the wheels.

R17–3. The car has a mass of 1.50 Mg and a mass center at G. Determine the maximum acceleration it can have if power is supplied only to the rear wheels. Neglect the mass of the wheels in the calculation, and assume that the wheels that do not receive power are free to roll. Also, assume that slipping of the powered wheels occurs, where the coefficient of kinetic friction is $\mu_k = 0.3$.

Prob. R17–1

Prob. R17–3

R17–2. The two 3-lb rods EF and HI are fixed (welded) to the link AC at E. Determine the internal axial force E_x, shear force E_y, and moment M_E, which the bar AC exerts on FE at E if at the instant $\theta = 30°$ link AB has an angular velocity $\omega = 5$ rad/s and an angular acceleration $\alpha = 8$ rad/s² as shown.

R17–4. A 20-kg roll of paper, originally at rest, is pin supported at its ends to bracket AB. The roll rests against a wall for which the coefficient of kinetic friction at C is $\mu_C = 0.3$. If a force of 40 N is applied uniformly to the end of the sheet, determine the initial angular acceleration of the roll and the tension in the bracket as the paper unwraps. For the calculation, treat the roll as a cylinder.

Prob. R17–2

Prob. R17–4

REVIEW PROBLEMS   451

R17–5. At the instant shown, two forces act on the 30-lb slender rod which is pinned at O. Determine the magnitude of force **F** and the initial angular acceleration of the rod so that the horizontal reaction which the *pin exerts on the rod* is 5 lb directed to the right.

R17–7. The spool and wire wrapped around its core have a mass of 20 kg and a centroidal radius of gyration $k_G = 250$ mm. If the coefficient of kinetic friction at the ground is $\mu_B = 0.1$, determine the angular acceleration of the spool when the 30-N·m couple moment is applied.

Prob. R17–5

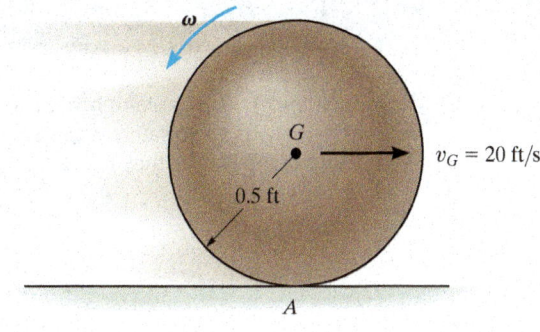

Prob. R17–7

R17–6. The pendulum consists of a 30-lb sphere and a 10-lb slender rod. Compute the reaction at the pin O just after the cord AB is cut.

R17–8. Determine the backspin ω which should be given to the 20-lb ball so that when its center is given an initial horizontal velocity $v_G = 20$ ft/s it stops spinning and translating at the same instant. The coefficient of kinetic friction is $\mu_A = 0.3$.

Prob. R17–6

Prob. R17–8

CHAPTER 18

Roller coasters must be able to coast over loops and through turns, and have enough energy to do so safely. Accurate calculation of this energy must account for the size of the cars as they move along the track.

PLANAR KINETICS OF A RIGID BODY: WORK AND ENERGY

CHAPTER OBJECTIVES

■ To develop formulations for the kinetic energy of a body, and define the various ways a force and couple do work.

■ To apply the principle of work and energy to solve rigid-body planar kinetic problems.

■ To show how the conservation of energy can be used to solve rigid-body planar kinetic problems.

18.1 KINETIC ENERGY

In this chapter we will apply work and energy methods to solve planar motion problems involving force, velocity, and displacement. But first it will be necessary to develop a means of obtaining the body's kinetic energy when the body is subjected to translation, rotation about a fixed axis, or general plane motion.

To do this we will consider the rigid body shown in Fig. 18–1, which is represented here by a *slab* moving in the inertial *x–y* reference plane. An arbitrary *i*th particle of the body, having a mass dm, is located a distance r from the arbitrary point P. If at the *instant* shown the particle has a velocity $\mathbf{v}_i$, then the particle's kinetic energy is $T_i = \frac{1}{2} dm\, v_i^2$.

Fig. 18–1

Fig. 18–1 (repeated)

The kinetic energy of the entire body is determined by writing similar expressions for each particle of the body and integrating the results, i.e.,

$$T = \frac{1}{2}\int_m dm\, v_i^2$$

This equation can also be expressed in terms of the velocity of a specific point P rather than the velocity v_i. If the body has an angular velocity $\boldsymbol{\omega}$, then from Fig. 18–1, we have

$$\mathbf{v}_i = \mathbf{v}_P + \mathbf{v}_{i/P}$$
$$= (v_P)_x\mathbf{i} + (v_P)_y\mathbf{j} + \omega\mathbf{k} \times (x\mathbf{i} + y\mathbf{j})$$
$$= [(v_P)_x - \omega y]\mathbf{i} + [(v_P)_y + \omega x]\mathbf{j}$$

The square of the magnitude of $\mathbf{v}_i$ is thus

$$\mathbf{v}_i \cdot \mathbf{v}_i = v_i^2 = [(v_P)_x - \omega y]^2 + [(v_P)_y + \omega x]^2$$
$$= (v_P)_x^2 - 2(v_P)_x\omega y + \omega^2 y^2 + (v_P)_y^2 + 2(v_P)_y\omega x + \omega^2 x^2$$
$$= v_P^2 - 2(v_P)_x\omega y + 2(v_P)_y\omega x + \omega^2 r^2$$

Substituting this into the equation of kinetic energy yields

$$T = \frac{1}{2}\left(\int_m dm\right)v_P^2 - (v_P)_x\omega\left(\int_m y\,dm\right) + (v_P)_y\omega\left(\int_m x\,dm\right) + \frac{1}{2}\omega^2\left(\int_m r^2\,dm\right)$$

The first integral on the right represents the entire mass m of the body. Since $\bar{y}m = \int y\,dm$ and $\bar{x}m = \int x\,dm$, the second and third integrals locate the body's center of mass G with respect to P. The last integral represents the body's moment of inertia I_P, calculated about the z axis passing through point P. Thus,

$$T = \tfrac{1}{2}mv_P^2 - (v_P)_x\omega\bar{y}m + (v_P)_y\omega\bar{x}m + \tfrac{1}{2}I_P\omega^2 \qquad (18\text{–}1)$$

As a special case, if point P coincides with the mass center G of the body, then $\bar{y} = \bar{x} = 0$, and therefore

$$T = \tfrac{1}{2}mv_G^2 + \tfrac{1}{2}I_G\omega^2 \qquad (18\text{–}2)$$

Both terms on the right side are *always positive*, since v_G and ω are squared. The first term represents the translational kinetic energy, referenced from the mass center, and the second term represents the body's rotational kinetic energy about the mass center.

18.1 KINETIC ENERGY

Translation. When a rigid body of mass m is subjected to either rectilinear or curvilinear translation, Fig. 18–2, the kinetic energy due to rotation is zero, since $\omega = 0$. The kinetic energy of the body is therefore

$$T = \tfrac{1}{2} m v_G^2 \qquad (18\text{–}3)$$

Rotation about a Fixed Axis. When a rigid body rotates about a fixed axis passing through point O, Fig. 18–3, the body has both translational and rotational kinetic energy, so that

$$T = \tfrac{1}{2} m v_G^2 + \tfrac{1}{2} I_G \omega^2 \qquad (18\text{–}4)$$

The body's kinetic energy can also be formulated from its motion about point O. Since $v_G = r_G \omega$, then $T = \tfrac{1}{2}(I_G + m r_G^2)\omega^2$. By the parallel-axis theorem, the terms inside the parentheses represent the moment of inertia I_O of the body about an axis perpendicular to the plane of motion and passing through point O. Hence,*

$$T = \tfrac{1}{2} I_O \omega^2 \qquad (18\text{–}5)$$

General Plane Motion. When a rigid body is subjected to general plane motion, Fig. 18–4, it has an angular velocity ω and its mass center has a velocity $\mathbf{v}_G$. Therefore, the kinetic energy is

$$T = \tfrac{1}{2} m v_G^2 + \tfrac{1}{2} I_G \omega^2 \qquad (18\text{–}6)$$

We can also calculate this kinetic energy in terms of the body's motion about its instantaneous center of zero velocity, i.e.,

$$T = \tfrac{1}{2} I_{IC} \omega^2 \qquad (18\text{–}7)$$

where I_{IC} is the moment of inertia of the body about its instantaneous center. The proof is similar to that of Eq. 18–5. (See Prob. 18–1.)

Translation

Fig. 18–2

Rotation About a Fixed Axis

Fig. 18–3

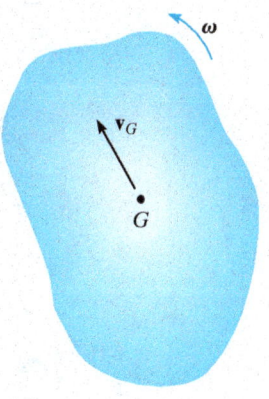

General Plane Motion

Fig. 18–4

---

*The similarity between this derivation and that of $\Sigma M_O = I_O \alpha$ should be noted. Also the same result can be obtained directly from Eq. 18–1 by selecting point P at O, realizing that $v_O = 0$.

System of Bodies. Because energy is a scalar quantity, the total kinetic energy for a system of connected rigid bodies is the sum of the kinetic energies of all its moving parts. Depending on the type of motion, the kinetic energy of each body is found by applying Eq. 18–2 or one of its alternative forms.

18.2 THE WORK OF A FORCE

Several types of forces are often encountered in planar kinetics problems involving a rigid body. The work of each of these forces has been presented in Sec. 14.1 and is listed below as a summary.

Work of a Variable Force. If an external force $\mathbf{F}$ acts on a body, the work done by the force when the body moves along the path s, Fig. 18–5, is

$$U_F = \int \mathbf{F} \cdot d\mathbf{r} = \int_s F \cos\theta \, ds \qquad (18\text{–}8)$$

Here θ is the angle between the "tails" of the force and the differential displacement. The integration must account for the variation of the force's direction and magnitude.

Fig. 18–5

Work of a Constant Force. If an external force $\mathbf{F}_c$ acts on a body, Fig. 18–6, and maintains a constant magnitude F_c and constant direction θ, while the body undergoes a translation s, then the above equation can be integrated, so that the work becomes

$$U_{F_c} = (F_c \cos\theta)s \qquad (18\text{–}9)$$

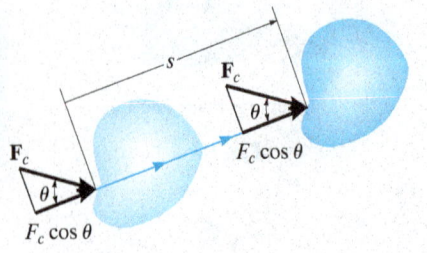

Fig. 18–6

18.2 THE WORK OF A FORCE

Work of a Weight. The weight of a body does work only when the body's center of mass G undergoes a vertical displacement y. If this displacement is *downward*, Fig. 18–7, the work is positive, since the weight is in the same direction as the displacement.

$$U_W = Wy \quad (18\text{–}10)$$

Fig. 18–7

Likewise, if the displacement is *upward* $(-y)$ the work becomes *negative*. In both cases the elevation change is considered to be small so that **W**, which is caused by gravitation, is constant.

Work of a Spring Force. If a linear elastic spring is attached to a body, the spring force $F_s = ks$ acting on the body does work when the spring either stretches or compresses from s_1 to a farther position s_2. In both cases the work will be negative since the displacement of the body is in the opposite direction to the force, Fig. 18–8. The work is

$$U_s = -\left(\tfrac{1}{2}ks_2^2 - \tfrac{1}{2}ks_1^2\right) \quad (18\text{–}11)$$

where $|s_2| > |s_1|$.

Fig. 18–8

Forces That Do No Work. There are some external forces that do no work when the body is displaced. These forces act either at fixed points on the body, or they have a direction that is perpendicular to their displacement. Examples include the reactions at a pin support about which a body rotates, the normal reaction acting on a body that moves along a fixed surface, and the weight of a body when the center of gravity of the body moves in a horizontal plane, Fig. 18–9. A frictional force $\mathbf{F}_f$ acting on a round body as it *rolls without slipping* over a rough surface also does no work.* This is because, during any instant of time dt, $\mathbf{F}_f$ acts at a point on the body which has zero velocity (instantaneous center, *IC*) and so the work done by the force on the point is zero. In other words, the point is not displaced in the direction of the force during this instant. Since $\mathbf{F}_f$ contacts successive points for only an instant, the work of $\mathbf{F}_f$ will be zero.

Fig. 18–9

---

*The work done by a frictional force *when the body slips* is discussed in Sec. 14.3.

18.3 THE WORK OF A COUPLE MOMENT

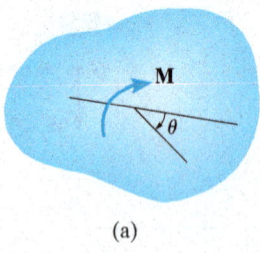

(a)

Consider the body in Fig. 18–10a, which is subjected to a couple moment $M = Fr$. If the body undergoes a differential displacement, then the work done by the couple forces can be found by considering the displacement as the sum of a separate translation plus rotation. When the body translates, the work of each force is produced only by the component of displacement along the line of action of the forces ds_t, Fig. 18–10b. Clearly the "positive" work of one force cancels the "negative" work of the other. When the body undergoes a differential rotation $d\theta$ about the arbitrary point O, Fig. 18–10c, then each force undergoes a displacement $ds_\theta = (r/2)\, d\theta$ in the direction of the force. Hence, the total work done is

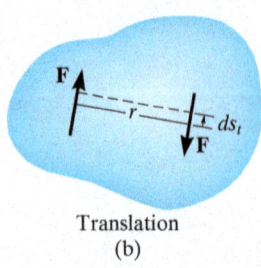

Translation
(b)

$$dU_M = F\left(\frac{r}{2} d\theta\right) + F\left(\frac{r}{2} d\theta\right) = (Fr)\, d\theta$$

$$= M\, d\theta$$

The work is *positive* when **M** and $d\boldsymbol{\theta}$ have the same sense of direction and negative if these vectors are in the opposite sense.

When the body rotates in the plane through a finite angle θ measured in radians, from θ_1 to θ_2, the work of a couple moment is therefore

Rotation
(c)

Fig. 18–10

$$U_M = \int_{\theta_1}^{\theta_2} M\, d\theta \qquad (18\text{–}12)$$

If the couple moment **M** has a constant magnitude, then

$$U_M = M(\theta_2 - \theta_1) \qquad (18\text{–}13)$$

EXAMPLE 18.1

The bar shown in Fig. 18–11a has a mass of 10 kg and is subjected to a couple moment of $M = 50$ N·m, and a force of $P = 80$ N, which is always applied perpendicular to the end of the bar. Also, the spring has an unstretched length of 0.5 m and remains in the vertical position due to the roller guide at B. Determine the total work done by all the forces acting on the bar when it has rotated downward from $\theta = 0°$ to $\theta = 90°$.

SOLUTION
First the free-body diagram of the bar is drawn in order to account for all the forces that act on it, Fig. 18–11b.

Weight W. Since the weight $10(9.81)$ N $= 98.1$ N is displaced downward 1.5 m, the work is

$$U_W = 98.1 \text{ N}(1.5 \text{ m}) = 147.2 \text{ J}$$

Why is the work positive?

Couple Moment M. The couple moment rotates through an angle of $\theta = \pi/2$ rad. Hence,

$$U_M = 50 \text{ N·m}(\pi/2) = 78.5 \text{ J}$$

Spring Force F_s. When $\theta = 0°$, the spring is stretched $(0.75 \text{ m} - 0.5 \text{ m}) = 0.25$ m, and when $\theta = 90°$, the stretch is $(2 \text{ m} + 0.75 \text{ m}) - 0.5 \text{ m} = 2.25$ m. Thus,

$$U_s = -\left[\tfrac{1}{2}(30 \text{ N/m})(2.25 \text{ m})^2 - \tfrac{1}{2}(30 \text{ N/m})(0.25 \text{ m})^2\right] = -75.0 \text{ J}$$

By inspection the spring does negative work on the bar since $\mathbf{F}_s$ acts in the opposite direction to displacement. This checks with the result.

Force P. As the bar moves downward, the force is displaced through a distance of $(\pi/2)(3 \text{ m}) = 4.712$ m. The work is positive. Why?

$$U_P = 80 \text{ N}(4.712 \text{ m}) = 377.0 \text{ J}$$

Pin Reactions. Forces $\mathbf{A}_x$ and $\mathbf{A}_y$ do no work since they are not displaced.

Total Work. The work of all the forces when the bar is displaced is thus

$$U = 147.2 \text{ J} + 78.5 \text{ J} - 75.0 \text{ J} + 377.0 \text{ J} = 528 \text{ J} \quad \textit{Ans.}$$

(a)

(b)

Fig. 18–11

18.4 PRINCIPLE OF WORK AND ENERGY

By applying the principle of work and energy, developed in Sec. 14.2, to each of the particles of a rigid body and adding the results algebraically, since energy is a scalar, we obtain a similar result for a rigid body.

$$T_1 + \Sigma U_{1-2} = T_2 \qquad (18\text{–}14)$$

This equation states that the body's initial translational and rotational kinetic energy, plus the work done by all the external forces and couple moments acting on the body as the body moves from its initial to its final position, is equal to the body's final translational and rotational kinetic energy. The work of the body's internal forces does not have to be considered, since these forces occur in equal but opposite collinear pairs, so that when the body moves, the work of one force cancels that of its counterpart. Furthermore, since the body is rigid, no relative movement between these forces occurs, so that no internal work is done.

When several rigid bodies are pin connected, connected by inextensible cables, or in mesh with one another, Eq. 18–14 can also be applied to the entire system of connected bodies. In all these cases the internal forces, which hold the various members together, do no work and hence are eliminated from the analysis.

The counterweight A on this bascule bridge does positive work as the bridge is lifted, and thereby cancels the negative work done by the weight of the bridge.

The work of the torque or moment developed by the motors is transformed into kinetic energy of rotation of the drum.

PROCEDURE FOR ANALYSIS

The principle of work and energy is used to solve kinetic problems that involve *velocity, force*, and *displacement*, since these terms are involved in the formulation. For application, it is suggested that the following procedure be used.

Kinetic Energy (Kinematic Diagrams).

- The kinetic energy of a body is made up of two parts. Kinetic energy of translation is referenced to the velocity of the mass center, $T = \frac{1}{2}mv_G^2$, and kinetic energy of rotation is determined using the moment of inertia of the body about the mass center, $T = \frac{1}{2}I_G\omega^2$. In the special case of rotation about a fixed axis (or rotation about the *IC*), the kinetic energy can be determined relative to motion about the axis of rotation, $T = \frac{1}{2}I_O\omega^2$.
- *Kinematic diagrams* for velocity may be useful for determining v_G and ω or for establishing a relationship between v_G and ω.

Work (Free-Body Diagram).

- Draw a free-body diagram of the body when it is located at an intermediate point along the path in order to account for all the forces and couple moments which do work on the body as it moves along the path.
- A force does work when it moves through a displacement in the direction of the force.
- Forces that are functions of displacement must be integrated to obtain the work. Graphically, the work is equal to the area under the force–displacement curve.
- The work of a weight is the product of its magnitude and the vertical displacement, $U_W = Wy$. It is positive when the weight moves downwards.
- The work of a spring is of the form $U_s = \frac{1}{2}ks^2$, where k is the spring stiffness and s is the stretch or compression of the spring.
- The work of a couple is the product of the couple moment and the angle in radians through which it rotates, $U_M = M\theta$.
- Since *algebraic addition* of the work terms is required, it is important that the proper sign of each term be specified. Specifically, work is *positive* when the force (couple moment) is in the *same direction* as its displacement (rotation); otherwise, it is negative.

Principle of Work and Energy.

- Apply the principle of work and energy, $T_1 + \Sigma U_{1-2} = T_2$. Since this is a scalar equation, it can be used to solve for only one unknown when it is applied to a single rigid body.

EXAMPLE 18.2

The 30-kg disk shown in Fig. 18–12a is pin supported at its center. Determine the angle through which it must rotate to attain an angular velocity of 2 rad/s starting from rest. It is acted upon by a constant couple moment $M = 5 \text{ N} \cdot \text{m}$. The spring is originally unstretched and its cord wraps around the rim of the disk.

(a)

SOLUTION
Kinetic Energy. Since the disk rotates about a fixed axis, and it is initially at rest, then

$$T_1 = 0$$
$$T_2 = \tfrac{1}{2} I_O \omega_2^2 = \tfrac{1}{2}[\tfrac{1}{2}(30 \text{ kg})(0.2 \text{ m})^2](2 \text{ rad/s})^2 = 1.2 \text{ J}$$

Work (Free-Body Diagram). As shown in Fig. 18–12b, the pin reactions $\mathbf{O}_x$ and $\mathbf{O}_y$ and the weight (294.3 N) do no work, since they are not displaced. The *couple moment*, having a constant magnitude, does positive work $U_M = M\theta$ as the disk *rotates* through a clockwise angle of θ rad, and the spring does negative work $U_s = -\tfrac{1}{2} k s^2$.

Principle of Work and Energy.

$$\{T_1\} + \{\Sigma U_{1-2}\} = \{T_2\}$$

$$\{T_1\} + \left\{M\theta - \tfrac{1}{2} k s^2\right\} = \{T_2\}$$

$$\{0\} + \left\{(5 \text{ N} \cdot \text{m})\theta - \tfrac{1}{2}(10 \text{ N/m})[\theta(0.2 \text{ m})]^2\right\} = \{1.2 \text{ J}\}$$

$$-0.2\theta^2 + 5\theta - 1.2 = 0$$

Solving this quadratic equation for the smallest positive root,

$$\theta = 0.2423 \text{ rad} = 0.2423 \text{ rad}\left(\frac{180°}{\pi \text{ rad}}\right) = 13.9° \qquad Ans.$$

(b)

Fig. 18–12

EXAMPLE 18.3

The wheel shown in Fig. 18–13a weighs 40 lb and has a radius of gyration $k_G = 0.6$ ft about its mass center G. If it is subjected to a clockwise couple moment of 15 lb·ft and rolls from rest without slipping, determine its angular velocity after its center G moves 0.5 ft. The spring has a stiffness $k = 10$ lb/ft and is initially unstretched when the couple moment is applied.

(a)

SOLUTION
Kinetic Energy (Kinematic Diagram). Since the wheel is initially at rest,

$$T_1 = 0$$

The kinematic diagram of the wheel when it is in the final position is shown in Fig. 18–13b. The final kinetic energy is determined from

$$T_2 = \tfrac{1}{2} I_{IC} \omega_2^2$$

$$= \frac{1}{2}\left[\frac{40\text{ lb}}{32.2\text{ ft/s}^2}(0.6\text{ ft})^2 + \left(\frac{40\text{ lb}}{32.2\text{ ft/s}^2}\right)(0.8\text{ ft})^2\right]\omega_2^2$$

$$T_2 = 0.6211\, \omega_2^2$$

(b)

Work (Free-Body Diagram). As shown in Fig. 18–13c, only the spring force $\mathbf{F}_s$ and the couple moment do work. The normal force does not move along its line of action and the frictional force does *no work*, since the wheel does not slip as it rolls.

The work of $\mathbf{F}_s$ is found using $U_s = -\tfrac{1}{2}ks^2$. Here the work is negative since $\mathbf{F}_s$ is in the opposite direction to displacement. Since the wheel does not slip when the center G moves 0.5 ft, then the wheel rotates $\theta = s_G/r_{G/IC} = 0.5\text{ ft}/0.8\text{ ft} = 0.625$ rad,* Fig. 18–13b. Hence, the spring stretches $s = \theta r_{A/IC} = (0.625\text{ rad})(1.6\text{ ft}) = 1$ ft.

Principle of Work and Energy.

$$\{T_1\} + \{\Sigma U_{1-2}\} = \{T_2\}$$

$$\{T_1\} + \{M\theta - \tfrac{1}{2}ks^2\} = \{T_2\}$$

$$\{0\} + \left\{15\text{ lb}\cdot\text{ft}(0.625\text{ rad}) - \frac{1}{2}(10\text{ lb/ft})(1\text{ ft})^2\right\} = \{0.6211\,\omega_2^2\text{ ft}\cdot\text{lb}\}$$

$$\omega_2 = 2.65\text{ rad/s} \quad\quad Ans.$$

(c)

Fig. 18–13

*See Example 16.4.

EXAMPLE 18.4

The 700-kg pipe is equally suspended from the two tines of the fork lift shown in the photo. It is undergoing a swinging motion such that when $\theta = 30°$ it is momentarily at rest. Determine the normal and frictional forces acting on each tine which are needed to support the pipe at the instant $\theta = 0°$. Measurements of the pipe and the suspender are shown in Fig. 18–14a. Neglect the mass of the suspender and the thickness of the pipe.

(a)

Fig. 18–14

SOLUTION

We must use the equations of motion to find the forces on the tines since these forces do no work. Before doing this, however, we will apply the principle of work and energy to determine the angular velocity of the pipe when $\theta = 0°$.

Kinetic Energy (Kinematic Diagram). Since the pipe is originally at rest, then

$$T_1 = 0$$

The final kinetic energy may be calculated with reference to either the fixed point O or the center of mass G. For the calculation we will consider the pipe to be a thin ring so that $I_G = mr^2$. (See the table on the inside back cover.) If point G is considered, we have

$$T_2 = \tfrac{1}{2}m(v_G)_2^2 + \tfrac{1}{2}I_G\omega_2^2$$
$$= \tfrac{1}{2}(700 \text{ kg})[(0.4 \text{ m})\omega_2]^2 + \tfrac{1}{2}[700 \text{ kg}(0.15 \text{ m})^2]\omega_2^2$$
$$= 63.875\omega_2^2$$

If point O is considered then the parallel-axis theorem must be used to determine I_O. Hence,

$$T_2 = \tfrac{1}{2}I_O\omega_2^2 = \tfrac{1}{2}[700 \text{ kg}(0.15 \text{ m})^2 + 700 \text{ kg}(0.4 \text{ m})^2]\omega_2^2$$
$$= 63.875\omega_2^2$$

Work (Free-Body Diagram). Fig. 18–14b. The normal and frictional forces on the tines do no work since they do not move as the pipe swings. The weight does positive work since the weight moves downward through a vertical distance $y = 0.4$ m $- 0.4 \cos 30°$ m $= 0.05359$ m.

Principle of Work and Energy.

$$\{T_1\} + \{\Sigma U_{1-2}\} = \{T_2\}$$

$$\{0\} + \{700(9.81) \text{ N}(0.05359 \text{ m})\} = \{63.875\omega_2^2\}$$

$$\omega_2 = 2.400 \text{ rad/s}$$

Equations of Motion. Referring to the free-body and kinetic diagrams shown in Fig. 18–14c, and using the result for ω_2, we have

$\xleftarrow{+} \Sigma F_t = m(a_G)_t;$ $F_T = (700 \text{ kg})(a_G)_t$

$+\uparrow \Sigma F_n = m(a_G)_n;$ $N_T - 700(9.81) \text{ N} = (700 \text{ kg})(2.400 \text{ rad/s})^2(0.4 \text{ m})$

$\zeta + \Sigma M_O = I_O \alpha;$ $0 = [(700 \text{ kg})(0.15 \text{ m})^2 + (700 \text{ kg})(0.4 \text{ m})^2]\alpha$

Since $(a_G)_t = (0.4 \text{ m})\alpha$, then

$$\alpha = 0 \quad (a_G)_t = 0$$

$$F_T = 0$$

$$N_T = 8.480 \text{ kN}$$

There are two tines used to support the load, therefore

$$F'_T = 0 \qquad \qquad Ans.$$

$$N'_T = \frac{8.480 \text{ kN}}{2} = 4.24 \text{ kN} \qquad Ans.$$

NOTE: Due to the swinging motion the tines are subjected to a *greater* normal force than would be the case if the load were static, in which case $N'_T = 700(9.81) \text{ N}/2 = 3.43$ kN.

(b)

Fig. 18–14 (cont.)

EXAMPLE 18.5

The 10-kg rod shown in Fig. 18–15a is constrained so that its ends move along the grooved slots. The rod is initially at rest when $\theta = 0°$. If the slider block at B is subjected to the force $P = 50$ N, determine the angular velocity of the rod at the instant $\theta = 45°$. Neglect friction and the mass of blocks A and B.

SOLUTION
Why can the principle of work and energy be used to solve this problem?

Kinetic Energy (Kinematic Diagrams). Two kinematic diagrams of the rod, when it is in the initial position 1 and final position 2, are shown in Fig. 18–15b. When the rod is in position 1, $T_1 = 0$ since $(v_G)_1 = \omega_1 = 0$. In position 2 the angular velocity is ω_2 and the velocity of the mass center is $(v_G)_2$. Hence, the kinetic energy is

$$T_2 = \tfrac{1}{2}m(v_G)_2^2 + \tfrac{1}{2}I_G\omega_2^2$$
$$= \tfrac{1}{2}(10\text{ kg})(v_G)_2^2 + \tfrac{1}{2}[\tfrac{1}{12}(10\text{ kg})(0.8\text{ m})^2]\omega_2^2$$
$$= 5(v_G)_2^2 + 0.2667(\omega_2)^2$$

The two unknowns $(v_G)_2$ and ω_2 can be related from the instantaneous center of zero velocity for the rod. Fig. 18–15b. It is seen that as A moves downward with a velocity $(v_A)_2$, B moves horizontally to the left with a velocity $(v_B)_2$. Knowing these directions, the IC is located as shown in the figure. Hence,

$$(v_G)_2 = r_{G/IC}\omega_2 = (0.4 \tan 45°\text{ m})\omega_2$$
$$= 0.4\omega_2$$

Therefore,

$$T_2 = 0.8\omega_2^2 + 0.2667\omega_2^2 = 1.0667\omega_2^2$$

Of course, we can also determine this result using $T_2 = \tfrac{1}{2}I_{IC}\omega_2^2$.

Work (Free-Body Diagram). Fig. 18–15c. The normal forces $\mathbf{N}_A$ and $\mathbf{N}_B$ do no work as the rod is displaced. Why? The 98.1-N weight is displaced a vertical distance of $\Delta y = (0.4 - 0.4 \cos 45°)$ m; whereas the 50-N force moves a horizontal distance of $s = (0.8 \sin 45°)$ m. Both of these forces do positive work. Why?

Principle of Work and Energy.

$$\{T_1\} + \{\Sigma U_{1-2}\} = \{T_2\}$$
$$\{T_1\} + \{W\Delta y + Ps\} = \{T_2\}$$
$$\{0\} + \{98.1\text{ N}(0.4\text{ m} - 0.4\cos 45°\text{ m}) + 50\text{ N}(0.8\sin 45°\text{ m})\}$$
$$= \{1.0667\omega_2^2\text{ J}\}$$

Solving for ω_2 gives

$$\omega_2 = 6.11\text{ rad/s} \qquad Ans.$$

Fig. 18–15

FUNDAMENTAL PROBLEMS

F18–1. The 80-kg wheel has a radius of gyration about its mass center O of $k_O = 400$ mm. Determine its angular velocity after it has rotated 20 revolutions starting from rest.

Prob. F18–1

F18–2. The uniform 50-lb slender rod is subjected to a couple moment of $M = 100$ lb·ft. If the rod is at rest when $\theta = 0°$, determine its angular velocity when $\theta = 90°$.

Prob. F18–2

F18–3. The uniform 50-kg slender rod is at rest in the position shown when $P = 600$ N is applied. Determine the angular velocity of the rod when the rod reaches the vertical position.

Prob. F18–3

F18–4. The 50-kg wheel is subjected to a force of 50 N. If the wheel starts from rest and rolls without slipping, determine its angular velocity after it has rotated 10 revolutions. The radius of gyration of the wheel about its mass center G is $k_G = 0.3$ m.

Prob. F18–4

F18–5. If the uniform 30-kg slender rod starts from rest at the position shown, determine its angular velocity after it has rotated 4 revolutions. The forces remain perpendicular to the rod.

Prob. F18–5

F18–6. The 20-kg wheel has a radius of gyration about its center G of $k_G = 300$ mm. When it is subjected to a couple moment of $M = 50$ N·m, it rolls without slipping. Determine the angular velocity of the wheel after its mass center G has traveled through a distance of $s_G = 20$ m, starting from rest.

Prob. F18–6

PROBLEMS

All solutions must include a free-body diagram.

18–1. At a given instant the body of mass m has an angular velocity ω and its mass center has a velocity $\mathbf{v}_G$. Show that its kinetic energy can be represented as $T = \frac{1}{2}I_{IC}\omega^2$, where I_{IC} is the moment of inertia of the body determined about the instantaneous axis of zero velocity, located a distance $r_{G/IC}$ from the mass center as shown.

Prob. 18–1

18–2. The spool has a mass of 60 kg and a radius of gyration $k_G = 0.3$ m. If it is released from rest, determine how far its center descends down the smooth plane before it attains an angular velocity of $\omega = 6$ rad/s. Neglect friction and the mass of the cord which is wound around the central core.

18–3. Solve Prob. 18–2 if the coefficient of kinetic friction between the spool and plane at A is $\mu_k = 0.2$.

Probs. 18–2/3

*18–4. The spool weighs 20 lb and has a radius of gyration of $k_G = 3$ in. If it is released from rest, determine how far it must drop before the velocity of its center becomes 20 ft/s. The cable is wrapped around the inner hub of the spool and unravels as the spool rotates. Neglect the mass of the cable.

Prob. 18–4

18–5. A force of $P = 20$ N is applied to the cable, which causes the 175-kg reel to turn since it is resting on the two rollers A and B of the dispenser. Determine the angular velocity of the reel after it has made two revolutions starting from rest. Neglect the mass of the rollers and the mass of the cable. The radius of gyration of the reel about its center axis is $k_G = 0.42$ m.

18–6. A force of $P = 20$ N is applied to the cable, which causes the 175-kg reel to turn without slipping on the two rollers A and B of the dispenser. Determine the angular velocity of the reel after it has made two revolutions starting from rest. Neglect the mass of the cable. Each roller can be considered as an 18-kg cylinder, having a radius of 0.1 m. The radius of gyration of the reel about its center axis is $k_G = 0.42$ m.

Probs. 18–5/6

18–7. The double pulley consists of two parts that are attached to one another. It has a weight of 50 lb and a centroidal radius of gyration of $k_O = 0.6$ ft and is turning with an angular velocity of 20 rad/s clockwise. Determine the kinetic energy of the system. Assume that neither cable slips on the pulley.

*18–8. The double pulley consists of two parts that are attached to one another. It has a weight of 50 lb and a centroidal radius of gyration of $k_O = 0.6$ ft and is turning with an angular velocity of 20 rad/s clockwise. Determine the angular velocity of the pulley at the instant the 20-lb weight moves 2 ft downward.

Probs. 18–7/8

18–9. The disk, which has a mass of 20 kg, is subjected to the couple moment of $M = (2\theta + 4)$ N · m, where θ is in radians. If it starts from rest, determine its angular velocity when it has made two revolutions.

Prob. 18–9

18–10. A motor delivers 50 lb · ft of torque to the pinion gear D, which is in mesh with the larger gear A and the attached drum over which the cable winds. If A has a weight of 120 lb and a radius of gyration $k_A = 0.6$ ft, determine the velocity of the 50-lb crate C when it has risen 5 ft starting from rest. Neglect the mass of the pinion gear and the cable and pulley at B.

Prob. 18–10

18–11. The revolving door consists of four doors which are attached to an axle AB. Each door can be assumed to be a 50-lb thin plate. Friction at the axle contributes a moment of 2 lb · ft which resists the rotation of the doors. If a woman passes through one door by always pushing with a force $P = 15$ lb perpendicular to the plane of the door as shown, determine the door's angular velocity after it has rotated 90°. The doors are originally at rest.

Prob. 18–11

*18–12. Each gear has mass of 2 kg and a radius of gyration about its pinned mass center A or B, respectively, of $k_g = 40$ mm. Each link has a mass of 2 kg and a radius of gyration about its pinned end, A or B, of $k_l = 50$ mm. If originally the spring is unstretched when the couple moment $M = 20$ N·m is applied to link AC, determine the angular velocity of the links at the instant link AC rotates $\theta = 45°$. Each gear and link are connected together and rotate about the fixed pins A and B.

Prob. 18–12

18–13. The 10-kg uniform slender rod is suspended at rest from the pin at O when the force of $F = 150$ N is applied to its end. Determine the angular velocity of the rod when it has rotated 90° clockwise from the position shown. The force is always perpendicular to the rod.

18–14. The 10-kg uniform slender rod is suspended at rest from the pin at O when the force of $F = 150$ N is applied to its end. Determine the angular velocity of the rod when it has rotated 180° clockwise from the position shown. The force is always perpendicular to the rod.

Probs. 18–13/14

18–15. The pendulum consists of a 10-kg uniform disk and a 3-kg uniform slender rod. If it is released from rest in the position shown, determine its angular velocity when it rotates clockwise 90°.

Prob. 18–15

*18–16. The system is released from rest at $\theta = 0°$ when a constant couple moment $M = 100$ N·m is applied. If the mass of the cable and links AB and BC can be neglected, and each pulley can be treated as a disk having a mass of 6 kg, determine the speed of the 10-kg block at the instant link AB has rotated $\theta = 90°$. Note that point C moves along the vertical guide. Also, the cable does not slip on the pulleys.

Prob. 18–16

18–17. The center O of the thin ring of mass m is given an angular velocity of ω_0. If the ring rolls without slipping, determine its angular velocity after it has traveled a distance of s down the plane. Neglect its thickness.

18–19. The rotary screen S is used to wash limestone. When empty it has a mass of 800 kg and a radius of gyration of $k_G = 1.75$ m. Rotation is achieved by applying a torque of $M = 280$ N·m about the drive wheel at A. If no slipping occurs at A and the supporting wheel at B is free to roll, determine the angular velocity of the screen after it has rotated 5 revolutions. Neglect the mass of A and B.

Prob. 18–17

Prob. 18–19

*18–20.** A man having a weight of 150 lb crouches down on the end of a diving board as shown. In this position the radius of gyration about his center of gravity is $k_G = 1.2$ ft. While holding this position at $\theta = 0°$, he rotates about his toes at A until he loses contact with the board when $\theta = 90°$. If he remains rigid, determine approximately how many revolutions he makes before striking the water after falling 30 ft.

18–18. The wheel has a mass of 100 kg and a radius of gyration of $k_O = 0.2$ m. A motor supplies a torque $M = (40\theta + 900)$ N·m, where θ is in radians, about the drive shaft at O. Determine the speed of the loading car, which has a mass of 300 kg, after it travels $s = 4$ m. Initially the car is at rest when $s = 0$ and $\theta = 0°$. Neglect the mass of the attached cable and the mass of the car's wheels.

Prob. 18–18

Prob. 18–20

18–21. If corner A of the 60-kg plate is subjected to a vertical force of $P = 500$ N, and the plate is released from rest when $\theta = 0°$, determine the angular velocity of the plate when $\theta = 45°$.

Prob. 18–21

18–22. If the 50-lb bucket C is released from rest, determine its velocity after it has fallen a distance of 10 ft. The windlass A can be considered as a 30-lb cylinder, while the spokes are slender rods, each having a weight of 2 lb. Neglect the pulley's weight.

Prob. 18–22

18–23. The coefficient of kinetic friction between the 100-lb disk and the surface of the conveyor belt is $\mu_A = 0.2$. If the conveyor belt is moving with a speed of $v_C = 6$ ft/s when the disk is placed in contact with it, determine the number of revolutions the disk makes before it reaches a constant angular velocity.

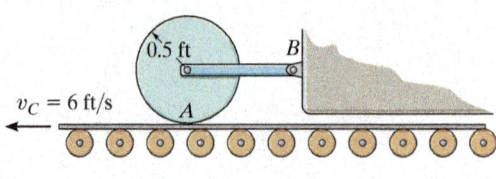

Prob. 18–23

*18–24. The 50-lb cylinder A is descending with a speed of 20 ft/s when the brake is applied. If wheel B must be brought to a stop after it has rotated 5 revolutions, determine the constant force P that must be applied to the brake arm. The coefficient of kinetic friction between the brake pad C and the wheel is $\mu_k = 0.5$. The wheel's weight is 25 lb, and the radius of gyration about its center of mass is $k = 0.6$ ft.

18–25. When a force of $P = 30$ lb is applied to the brake arm, the 50-lb cylinder A is descending with a speed of 20 ft/s. Determine the number of revolutions wheel B will rotate before it is brought to a stop. The coefficient of kinetic friction between the brake pad C and the wheel is $\mu_k = 0.5$. The wheel's weight is 25 lb, and the radius of gyration about its center of mass is $k = 0.6$ ft.

Probs. 18–24/25

18–26. A uniform ladder having a weight of 30 lb is released from rest when it is in the vertical position. If it is allowed to fall freely, determine the angle θ at which the bottom end A starts to lift off the ground. For the calculation, assume the ladder to be a slender rod and neglect friction at A.

Prob. 18–26

18–27. The link AB is subjected to a couple moment of $M = 40$ N·m. If the ring gear C is fixed, determine the angular velocity of the 15-kg inner gear when the link has made two revolutions starting from rest. Neglect the mass of the link and assume the inner gear is a disk. Motion occurs in the vertical plane.

Prob. 18–27

*18–28. The 10-kg rod AB is pin connected at A and subjected to a couple moment of $M = 15$ N·m. If the rod is released from rest when the spring is unstretched at $\theta = 30°$, determine the rod's angular velocity at the instant $\theta = 60°$. As the rod rotates, the spring always remains horizontal, because of the roller support at C.

Prob. 18–28

18–29. Gears C and H each have a weight of 0.4 lb and a radius of gyration about their mass center of $k_A = k_B = 2$ in. The uniform link AB has a weight of 0.2 lb and a radius of gyration of $k_A = 3$ in., whereas link DE has a weight of 0.15 lb and a radius of gyration of $k_B = 4.5$ in. If a couple moment $M = 3$ lb·ft is applied to link AB and the assembly is originally at rest, determine the angular velocity of link DE when link AB has rotated 360°. Gear C is fixed from rotating and motion occurs in the horizontal plane. Also, gear H and link DE rotate together on the same shaft at B.

Prob. 18–29

18–30. Motor M exerts a constant force of $P = 750$ N on the rope. If the 100-kg post is at rest when $\theta = 0°$, determine the angular velocity of the post at the instant $\theta = 60°$. Neglect the mass of the pulley and its size, and consider the post as a slender rod.

Prob. 18–30

18–31. A ball of mass m and radius r is cast onto the horizontal surface such that it rolls without slipping. Determine its angular velocity at the instant $\theta = 90°$, if it has an initial speed of v_G as shown.

*18–32. A ball of mass m and radius r is cast onto the horizontal surface such that it rolls without slipping. Determine the minimum speed v_G of its mass center G so that it rolls completely around the loop of radius $R + r$ without leaving the track.

Probs. 18–31/32

18–33. The two 2-kg gears A and B are attached to the ends of a 3-kg slender bar. The gears roll within the fixed ring gear C, which lies in the horizontal plane. If a 10-N·m torque is applied to the center of the bar as shown, determine the number of revolutions the bar must rotate starting from rest in order for it to have an angular velocity of $\omega_{AB} = 20$ rad/s. For the calculation, assume the gears can be approximated by thin disks. What is the result if the gears lie in the vertical plane?

Prob. 18–33

18–34. The spool has a mass of 100 kg and a radius of gyration of 400 mm about its center of mass O. If it is released from rest, determine its angular velocity after its center O has moved down the plane a distance of 2 m. The contact surface between the spool and the inclined plane is smooth.

18–35. The spool has a mass of 100 kg and a radius of gyration of 400 mm about its center of mass O. If it is released from rest, determine its angular velocity after its center O has moved down the plane a distance of 2 m. The coefficient of kinetic friction between the spool and the inclined plane is $\mu_k = 0.15$.

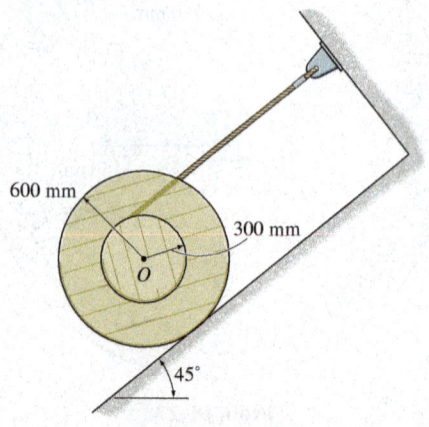

Probs. 18–34/35

18.5 CONSERVATION OF ENERGY

When a force system acting on a rigid body consists only of *conservative forces*, then the conservation of energy theorem can be used to solve a problem that otherwise would be solved using the principle of work and energy. This theorem is often easier to apply since the work of a conservative force is independent of the path and depends only on the initial and final positions of the body. It was shown in Sec. 14.5 that the work of a conservative force can be expressed as the difference in the body's potential energy measured from an arbitrarily selected reference or datum.

Gravitational Potential Energy. Since the total weight of a body can be considered concentrated at its center of gravity, the gravitational potential energy of the body is determined by knowing the height of the body's center of gravity above or below a horizontal datum.

$$V_g = Wy_G \qquad (18\text{--}15)$$

Here the potential energy is positive when G is above the datum, since the weight has the ability to do positive work when the body moves back to the datum, Fig. 18–16. Likewise, if G is located below the datum ($-y_G$), the gravitational potential energy is negative, since the weight does negative work when the body returns to the datum.

Gravitational potential energy

Fig. 18–16

Elastic Potential Energy. The force developed by an elastic spring is also a conservative force. The elastic potential energy which a spring imparts to an attached body when the spring is stretched or compressed from an initial undeformed position ($s = 0$) to a final position s, Fig. 18–17, is

$$V_e = \tfrac{1}{2}ks^2 \qquad (18\text{--}16)$$

In the deformed position, the spring force acting on the body always has the ability for doing positive work when the spring returns back to its original undeformed position (see Sec. 14.5).

Elastic potential energy

Fig. 18–17

Conservation of Energy. In general, if a body is subjected to both gravitational and elastic forces, the total potential energy can be expressed as a potential function represented as the algebraic sum

$$V = V_g + V_e \qquad (18\text{--}17)$$

Realizing that the work of conservative forces can be written as a difference in their potential energies, i.e., $(\Sigma U_{1-2})_{\text{cons}} = V_1 - V_2$, Eq. 14–16, we can write the principle of work and energy for a rigid body as

$$T_1 + V_1 + (\Sigma U_{1-2})_{\text{noncons}} = T_2 + V_2 \qquad (18\text{--}18)$$

Here $(\Sigma U_{1-2})_{\text{noncons}}$ represents the work of the nonconservative forces such as friction. If this term is zero, then

$$T_1 + V_1 = T_2 + V_2 \qquad (18\text{--}19)$$

This equation is referred to as the *conservation of energy*. It states that the sum of the potential and kinetic energies of the body remains constant when the body moves from one position to another. It also applies to a *system* of smooth, pin-connected rigid bodies, bodies connected by inextensible cords, and bodies in mesh with other bodies. In all these cases the forces acting at the points of contact are eliminated from the analysis, since they occur in equal but opposite collinear pairs and each pair of forces moves through an equal distance when the system undergoes a displacement.

It is important to remember that only problems involving conservative force systems can be solved by using Eq. 18–19. As stated in Sec. 14.5, dry friction or fluid drag-resistant forces, which depend on velocity or acceleration, are nonconservative. The work of such forces is transformed into thermal energy used to heat up the surfaces of contact, and consequently this energy is dissipated into the surroundings and may not be recovered. Therefore, problems involving frictional forces can be solved by using either the principle of work and energy written in the form of Eq. 18–18, if it applies, or the equations of motion.

The torsional springs located at the top of the garage door wind up as the door is lowered. When the door is raised, the potential energy stored in the springs is then transferred into gravitational potential energy of the door's weight, thereby making it easy to open.

PROCEDURE FOR ANALYSIS

The conservation of energy equation is used to solve problems involving *velocity, displacement*, and *conservative force systems*. For application it is suggested that the following procedure be used.

Potential Energy.
- Draw two diagrams showing the body or system located at its initial and final positions on the path.

- If the center of gravity, G, is subjected to a vertical displacement, establish a fixed horizontal datum from which to measure the body's gravitational potential energy V_g.

- Data pertaining to the elevation y_G of the body's center of gravity from the datum and the extension or compression of any connecting springs can be determined from the problem geometry and listed on the two diagrams.

- The potential energy is determined from $V = V_g + V_e$. Here $V_g = Wy_G$, which can be positive or negative, and $V_e = \frac{1}{2}ks^2$, which is always positive.

Kinetic Energy.
- The kinetic energy of the body consists of two parts, namely, translational kinetic energy referenced from the mass center, $T = \frac{1}{2}mv_G^2$, and rotational kinetic energy about the mass center, $T = \frac{1}{2}I_G\omega^2$.

- Kinematic diagrams for velocity may be useful for establishing a relationship between v_G and ω.

Conservation of Energy.
- Apply the conservation of energy equation $T_1 + V_1 = T_2 + V_2$.

EXAMPLE 18.6

The 10-kg rod AB shown in Fig. 18–18a is confined so that its ends move in the horizontal and vertical slots. The spring has a stiffness of $k = 800$ N/m and is unstretched when $\theta = 0°$. Determine the angular velocity of AB when $\theta = 0°$, if the rod is released from rest when $\theta = 30°$. Neglect the mass of the slider blocks.

SOLUTION
Here the system consists of the rod and spring.

Potential Energy. The two diagrams of the rod, when it is located at its initial and final positions, are shown in Fig. 18–18b. The datum, used to measure the gravitational potential energy, is placed in line with the rod when $\theta = 0°$.

When the rod is in position 1, the center of gravity G is located below the datum so its gravitational potential energy is negative. Positive elastic potential energy is stored in the spring, since it is stretched a distance of $s_1 = (0.4 \sin 30°)$ m. Thus,

$$V_1 = -Wy_1 + \tfrac{1}{2}ks_1^2$$
$$= -(98.1 \text{ N})(0.2 \sin 30° \text{ m}) + \tfrac{1}{2}(800 \text{ N/m})(0.4 \sin 30° \text{ m})^2 = 6.19 \text{ J}$$

When the rod is in position 2, the potential energy of the rod is zero, since the center of gravity G is located at the datum, and the spring is unstretched, $s_2 = 0$. Thus,

$$V_2 = 0$$

Kinetic Energy. The rod is released from rest from position 1, thus $(\mathbf{v}_G)_1 = \boldsymbol{\omega}_1 = \mathbf{0}$, and so

$$T_1 = 0$$

In position 2, the angular velocity is ω_2 and the rod's mass center has a velocity of $(\mathbf{v}_G)_2$. Thus,

$$T_2 = \tfrac{1}{2}m(v_G)_2^2 + \tfrac{1}{2}I_G\omega_2^2$$
$$= \tfrac{1}{2}(10 \text{ kg})(v_G)_2^2 + \tfrac{1}{2}\left[\tfrac{1}{12}(10 \text{ kg})(0.4 \text{ m})^2\right]\omega_2^2$$

Using kinematics, $(\mathbf{v}_G)_2$ can be related to ω_2 as shown in Fig. 18–18c. At the instant considered, the instantaneous center of zero velocity (IC) for the rod is at point A; hence, $(v_G)_2 = (r_{G/IC})\omega_2 = (0.2 \text{ m})\omega_2$. Substituting into the above expression and simplifying (or using $\tfrac{1}{2}I_{IC}\omega_2^2$), we get

$$T_2 = 0.2667\omega_2^2$$

Conservation of Energy.

$$\{T_1\} + \{V_1\} = \{T_2\} + \{V_2\}$$
$$\{0\} + \{6.19 \text{ J}\} = \{0.2667\omega_2^2\} + \{0\}$$
$$\omega_2 = 4.82 \text{ rad/s} \;\;\;\; \textit{Ans.}$$

Fig. 18–18

EXAMPLE 18.7

The wheel shown in Fig. 18–19a has a weight of 30 lb and a radius of gyration of $k_G = 0.6$ ft. It is attached to a spring which has a stiffness $k = 2$ lb/ft and an unstretched length of 1 ft. If the wheel is released from rest in the position shown and rolls without slipping, determine its angular velocity at the instant G moves 3 ft to the left.

SOLUTION
Here the system consists of the wheel and spring.

Potential Energy. Two diagrams of the wheel, when it is at the initial and final positions, are shown in Fig. 18–19b. A gravitational datum is not needed here since the weight is not displaced vertically. From the problem geometry the spring is stretched $s_1 = (\sqrt{3^2 + 4^2} - 1) = 4$ ft in the initial position, and spring $s_2 = (4 - 1) = 3$ ft in the final position. Hence, the positive spring potential energy is

$$V_1 = \tfrac{1}{2}ks_1^2 = \tfrac{1}{2}(2 \text{ lb/ft})(4 \text{ ft})^2 = 16 \text{ ft} \cdot \text{lb}$$
$$V_2 = \tfrac{1}{2}ks_2^2 = \tfrac{1}{2}(2 \text{ lb/ft})(3 \text{ ft})^2 = 9 \text{ ft} \cdot \text{lb}$$

Kinetic Energy. The wheel is released from rest and so $(\mathbf{v}_G)_1 = \mathbf{0}$, $\boldsymbol{\omega}_1 = \mathbf{0}$. Therefore,

$$T_1 = 0$$

Since the instantaneous center of zero velocity is at the ground, Fig. 18–19c, we have

$$T_2 = \tfrac{1}{2}I_{IC}\omega_2^2$$
$$= \tfrac{1}{2}\left[\left(\dfrac{30 \text{ lb}}{32.2 \text{ ft/s}^2}\right)(0.6 \text{ ft})^2 + \left(\dfrac{30 \text{ lb}}{32.2 \text{ ft/s}^2}\right)(0.75 \text{ ft})^2\right]\omega_2^2$$
$$= 0.4297\omega_2^2$$

Conservation of Energy.

$$\{T_1\} + \{V_1\} = \{T_2\} + \{V_2\}$$
$$\{0\} + \{16 \text{ ft} \cdot \text{lb}\} = \{0.4297\omega_2^2\} + \{9 \text{ ft} \cdot \text{lb}\}$$
$$\omega_2 = 4.04 \text{ rad/s} \quad\quad\quad \textit{Ans.}$$

NOTE: If the principle of work and energy were used to solve this problem, then the work of the spring would have to be determined by considering both the change in magnitude and direction of the spring force.

Fig. 18–19

EXAMPLE 18.8

The 10-kg disk shown in Fig. 18–20a is attached to a uniform 5-kg rod AB. If the assembly is released from rest when $\theta = 60°$, determine the angular velocity of the rod when $\theta = 0°$. Assume that the disk rolls without slipping. Neglect friction along the guide and the mass of the collar at B.

SOLUTION
The system consists of the disk and rod.

Potential Energy. Two diagrams for the rod and disk, when they are located at their initial and final positions, are shown in Fig. 18–20b. For convenience the datum passes through point A.
When the system is in position 1, only the rod's weight has positive potential energy. Thus,

$$V_1 = W_r y_1 = (49.05 \text{ N})(0.3 \sin 60° \text{ m}) = 12.74 \text{ J}$$

When the system is in position 2, both the weight of the rod and the weight of the disk have zero potential energy. Why? Thus,

$$V_2 = 0$$

Kinetic Energy. Since the entire system is at rest at the initial position,

$$T_1 = 0$$

In the final position the rod has an angular velocity $(\omega_r)_2$ and its mass center has a velocity $(v_G)_2$, Fig. 18–20c. Since the rod is *fully extended* in this position, the disk is momentarily at rest, so $(\omega_d)_2 = 0$ and $(v_A)_2 = 0$. For the rod $(v_G)_2$ can be related to $(\omega_r)_2$ from the instantaneous center of zero velocity, which is located at point A, Fig. 18–20c. Hence, $(v_G)_2 = r_{G/IC}(\omega_r)_2$ or $(v_G)_2 = 0.3(\omega_r)_2$. Thus,

$$T_2 = \frac{1}{2}m_r(v_G)_2^2 + \frac{1}{2}I_G(\omega_r)_2^2 + \frac{1}{2}m_d(v_A)_2^2 + \frac{1}{2}I_A(\omega_d)_2^2$$

$$= \frac{1}{2}(5 \text{ kg})[(0.3 \text{ m})(\omega_r)_2]^2 + \frac{1}{2}\left[\frac{1}{12}(5 \text{ kg})(0.6 \text{ m})^2\right](\omega_r)_2^2 + 0 + 0$$

$$= 0.3(\omega_r)_2^2$$

Conservation of Energy.

$$\{T_1\} + \{V_1\} = \{T_2\} + \{V_2\}$$
$$\{0\} + \{12.74 \text{ J}\} = \{0.3(\omega_r)_2^2\} + \{0\}$$
$$(\omega_r)_2 = 6.52 \text{ rad/s} \,\downarrow \qquad Ans.$$

Fig. 18–20

NOTE: We can also determine the final kinetic energy of the rod using $T_2 = \frac{1}{2}I_{IC}\omega_2^2$.

FUNDAMENTAL PROBLEMS

F18–7. If the 30-kg disk is released from rest when $\theta = 0°$, determine its angular velocity when $\theta = 90°$.

Prob. F18–7

F18–8. The 50-kg reel has a radius of gyration about its center G of $k_G = 300$ mm. If it is released from rest, determine its angular velocity when its center G has traveled 6 m down the smooth inclined plane.

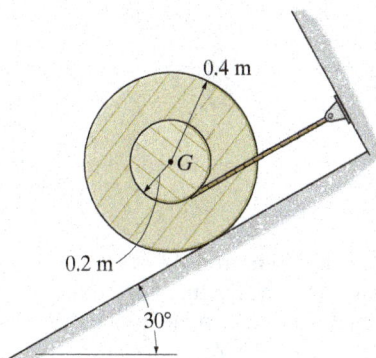

Prob. F18–8

F18–9. The 60-kg rod OA is released from rest when $\theta = 0°$. Determine its angular velocity when $\theta = 45°$. The spring remains vertical during the motion and is unstretched when $\theta = 0°$.

Prob. F18–9

F18–10. The 30-kg rod is released from rest when $\theta = 0°$. Determine the angular velocity of the rod when $\theta = 90°$. The spring is unstretched when $\theta = 0°$.

Prob. F18–10

F18–11. The 30-kg rod is released from rest when $\theta = 45°$. Determine the angular velocity of the rod when $\theta = 0°$. The spring is unstretched when $\theta = 45°$.

Prob. F18–11

F18–12. The 20-kg rod is released from rest when $\theta = 0°$. Determine its angular velocity when $\theta = 90°$. The spring has an unstretched length of 0.5 m.

Prob. F18–12

PROBLEMS

*18–36. The assembly consists of a 3-kg pulley A and 10-kg pulley B. If a 2-kg block is suspended from the cord, determine the block's speed after it descends 0.5 m starting from rest. Neglect the mass of the cord and treat the pulleys as thin disks. No slipping occurs.

18–37. The assembly consists of a 3-kg pulley A and 10-kg pulley B. If a 2-kg block is suspended from the cord, determine the distance the block must descend, starting from rest, in order to cause B to have an angular velocity of 6 rad/s. Neglect the mass of the cord and treat the pulleys as thin disks. No slipping occurs.

Probs. 18–36/37

18–38. A chain that has a negligible mass is draped over the sprocket which has a mass of 2 kg and a radius of gyration of $k_O = 50$ mm. If the 4-kg block A is released from rest in the position $s = 1$ m, determine the angular velocity of the sprocket at the instant $s = 2$ m.

18–39. Solve Prob. 18–38 if the chain has a mass of 0.8 kg/m. For the calculation, neglect the portion of the chain that wraps over the sprocket.

Probs. 18–38/39

*18–40. The 6-kg rod ABC is connected to the 3-kg rod CD. If the system is released from rest when $\theta = 0°$, determine the angular velocity of rod ABC at the instant it becomes horizontal.

Prob. 18–40

18–41. The 25-lb slender rod AB is attached to spring BC which has an unstretched length of 4 ft. If the rod is released from rest when $\theta = 30°$, determine its angular velocity at the instant $\theta = 90°$.

18–42. The 25-lb slender rod AB is attached to spring BC which has an unstretched length of 4 ft. If the rod is released from rest when $\theta = 30°$, determine the angular velocity of the rod the instant the spring becomes unstretched.

Probs. 18–41/42

18–43. The tire has a mass of 8 kg, a radius of 0.3 m, and a radius of gyration about its center of gravity G of $k_G = 0.25$ m. If it is released from rest at the top of the hill, determine the velocity of its center at the bottom of the hill if it rolls without slipping.

Prob. 18–43

*18–44. The uniform 150-lb stone (rectangular block) is being turned over on its side by pulling the vertical cable *slowly* upward until the stone begins to tip. If it then falls freely ($T = 0$) from an essentially balanced at-rest position, determine the speed at which the corner A strikes the pad at B. The stone does not slip at its corner C as it falls.

Prob. 18–44

18–45. The two bars are released from rest at the position θ. Determine their angular velocities at the instant they become horizontal. Neglect the mass of the roller at C. Each bar has a mass m and length L.

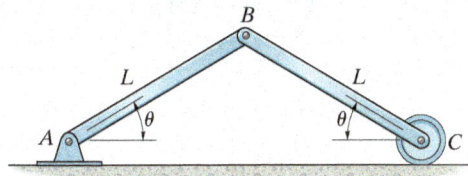

Prob. 18–45

18–46. The two bars are released from rest at the position $\theta = 90°$. Determine their angular velocities at the instant they become horizontal. Neglect the mass of the roller at C. Each bar has a mass m and length L.

Prob. 18–46

18–47. The 40-kg wheel has a radius of gyration about its center of gravity G of $k_G = 250$ mm. If it rolls without slipping, determine its angular velocity when it has rotated clockwise 90° from the position shown. The spring AB has a stiffness $k = 100$ N/m and an unstretched length of 500 mm. The wheel is released from rest.

Prob. 18–47

*18–48. The assembly consists of two 8-lb bars which are pin connected to the two 10-lb disks. If the bars are released from rest when $\theta = 60°$, determine their angular velocities at the instant $\theta = 0°$. Assume the disks roll without slipping.

18–49. The assembly consists of two 8-lb bars which are pin connected to the two 10-lb disks. If the bars are released from rest when $\theta = 60°$, determine their angular velocities at the instant $\theta = 30°$. Assume the disks roll without slipping.

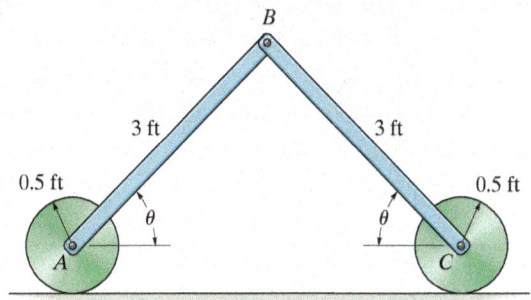

Probs. 18–48/49

18–50. The pavement compactor is traveling down the incline at $v_G = 5$ ft/s when the motor is disengaged. Determine the angular velocity of the roller B when the compactor has traveled 20 ft down the plane. The body of the compactor, excluding the rollers, has a weight of 8000 lb and a center of gravity at G. Each of the two rear rollers weighs 400 lb and has a radius of gyration of $k_A = 3.3$ ft. The front roller has a weight of 800 lb and a radius of gyration of $k_B = 1.8$ ft. The rollers do not slip as they rotate.

Prob. 18–50

18–51. The uniform garage door has a mass of 150 kg and is guided along smooth tracks at its ends. Lifting is done using the two springs, which are attached to the anchor bracket at A and to the counterbalance shaft at B and C. As the door is raised, the springs begin to unwind from the shaft, thereby assisting the lift. If each spring provides a torsional moment of $M = (0.7\theta)$ N·m, where θ is in radians, determine the angle θ_0 at which both the left-wound and right-wound spring should be attached at A so that the door is completely balanced by the springs, i.e., when the door is in the vertical position and is given a slight force upward, the springs will lift the door along the side tracks to the horizontal plane with no final angular velocity. *Note:* The elastic potential energy of a torsional spring is $V_e = \frac{1}{2}k\theta^2$, where $M = k\theta$ and in this case $k = 0.7$ N·m/rad.

***18–52.** The two 12-kg slender rods are pin connected and released from rest at the position $\theta = 60°$. If the spring has an unstretched length of 1.5 m, determine the angular velocity of rod BC when the system is at the position $\theta = 0°$. Neglect the mass of the roller at C.

Prob. 18–52

18–53. A large roll of paper having a mass of 20 kg and a radius $r = 150$ mm is resting over the edge of a corner, such that the end of the paper on the roll is attached to the horizontal surface. If the roll is disturbed slightly from its equilibrium position, determine the angle θ at which it begins to leap off the corner A as it falls. The radius of gyration of the roll is $k_G = 75$ mm.

Prob. 18–51

Prob. 18–53

18–54. If the 250-lb block is released from rest when the spring is unstretched, determine the velocity of the block after it has descended 5 ft. The drum has a weight of 50 lb and a radius of gyration of $k_O = 0.5$ ft about its center of mass O.

Prob. 18–54

18–55. The slender 15-kg bar is initially at rest and standing in the vertical position when the bottom end A is displaced slightly to the right. If the track in which it moves is smooth, determine the speed at which end A strikes the corner D. The bar is constrained to move in the vertical plane. Neglect the mass of the cord BC.

Prob. 18–55

***18–56.** The small gear has a mass m and may be treated as a uniform disk. If it is released from rest at $\theta = 0°$, and rolls along the fixed circular gear rack, determine the angular velocity of the radial line AB at the instant $\theta = 90°$.

Prob. 18–56

18–57. The slender rod has a mass of 10 kg and is released from rest when it is in the vertical position. Determine its angular velocity at the instant it becomes horizontal. The attached block at D has a mass of 4 kg. The mass of the cable and pulley at C can be neglected.

Prob. 18–57

18–58. The uniform rectangular door panel has a mass of 25 kg and is held in equilibrium above the horizontal at the position $\theta = 60°$ by rod BC. Determine the required stiffness of the torsional spring at A, so that the door's angular velocity becomes zero when the door reaches the closed position ($\theta = 0°$) once the supporting rod BC is removed. The spring is undeformed when $\theta = 60°$.

Prob. 18–58

18–59. The arm and seat of the amusement-park ride have a mass of 1.5 Mg, with the center of mass located at point G_1. The passenger seated at A has a mass of 125 kg, with the center of mass located at G_2. If the arm is raised to a position where $\theta = 150°$ and released from rest, determine the speed of the passenger at the instant $\theta = 0°$. The arm has a radius of gyration of $k_{G1} = 12$ m about its center of mass G_1. Neglect the size of the passenger.

Prob. 18–59

*18–60. The uniform bar AB has a mass of 12 kg and is pin connected at A. If the support at B is removed ($\theta = 90°$), determine the velocity of the 5-kg block C at the instant the bar rotates downward to $\theta = 150°$. Neglect the size and mass of the pulley at D.

Prob. 18–60

18–61. Pulley A has a weight of 30 lb and a centroidal radius of gyration $k_B = 0.6$ ft. Determine the speed of the 20-lb crate C at the instant $s = 10$ ft. Initially, the crate is released from rest when $s = 5$ ft. The pulley at P "rolls" downward on the cord without slipping. For the calculation, neglect the mass of this pulley and the cord as it unwinds from the inner and outer hubs of pulley A.

Prob. 18–61

18–62. The 50-lb wheel has a radius of gyration about its center of gravity G of $k_G = 0.7$ ft. If it rolls without slipping, determine its angular velocity when it has rotated clockwise 90° from the position shown. The spring AB has a stiffness $k = 1.20$ lb/ft and an unstretched length of 0.5 ft. The wheel is released from rest.

*18–64.** Gear A has a mass of 0.5 kg and a radius of gyration of $k_A = 40$ mm, and gear B has a mass of 0.8 kg and a radius of gyration of $k_B = 55$ mm. The link is pinned at C and has a mass of 0.35 kg. If the link can be treated as a slender rod, determine the angular velocity of the rod after the assembly is released from rest when $\theta = 0°$ and falls to $\theta = 90°$.

Prob. 18–62

Prob. 18–64

18–63. The system consists of 60-lb and 20-lb blocks A and B, respectively, and 5-lb pulleys C and D that can be treated as thin disks. Determine the speed of block A after block B has risen 5 ft, starting from rest. Assume that the cord does not slip on the pulleys, and neglect the mass of the cord.

18–65. The gear is pinned at A and has a mass of 0.3 kg and a radius of gyration of $k_A = 25$ mm. The gear rack has a mass of 0.5 kg. If the spring is stretched 50 mm and released from rest, determine the angular velocity of the gear at the instant the spring compresses 20 mm. Neglect friction.

Prob. 18–63

Prob. 18–65

CHAPTER REVIEW

Kinetic Energy

The kinetic energy of a rigid body that undergoes planar motion can be calculated using the motion of its mass center or its motion about a fixed axis or instantaneous axis of rotation. It includes a scalar sum of its translational and rotational kinetic energies.

Translation

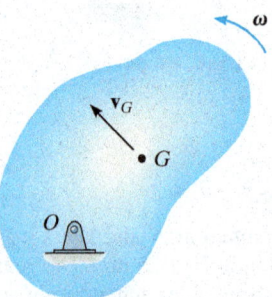

Rotation About a Fixed Axis

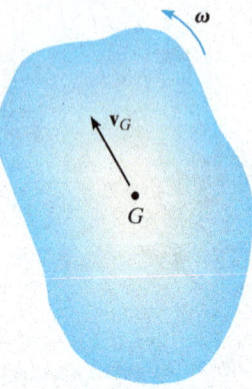

General Plane Motion

Translation

$$T = \tfrac{1}{2}mv_G^2$$

Rotation About a Fixed Axis

$$T = \tfrac{1}{2}mv_G^2 + \tfrac{1}{2}I_G\omega^2$$

or

$$T = \tfrac{1}{2}I_O\omega^2$$

General Plane Motion

$$T = \tfrac{1}{2}mv_G^2 + \tfrac{1}{2}I_G\omega^2$$

or

$$T = \tfrac{1}{2}I_{IC}\omega^2$$

Work of a Force and a Couple Moment

A force does work when it undergoes a displacement ds in the direction of the force, and a couple moment does work when it undergoes a rotation.

$$U_F = \int F \cos \theta \, ds$$

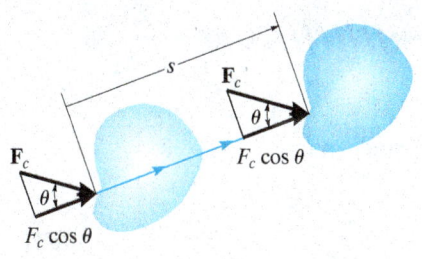

$$U_{F_c} = (F_c \cos \theta) s$$

Constant Force

$$U_W = Wy$$

Weight

$$U = -\frac{1}{2} k s^2$$

Spring

$$U_M = \int_{\theta_1}^{\theta_2} M \, d\theta$$

$$U_M = M(\theta_2 - \theta_1)$$

Constant Magnitude

Principle of Work and Energy

Problems that involve velocity, force, and displacement can be solved using the principle of work and energy. For application, a free-body diagram should be drawn in order to account for the work of all of the forces and couple moments that act on the body as it moves along the path.

$$T_1 + \Sigma U_{1-2} = T_2$$

Conservation of Energy

If a rigid body is subjected only to conservative forces, then the conservation-of-energy equation can be used to solve the problem. This equation requires that the sum of the potential and kinetic energies of the body remain the same at any two points along the path.

$$T_1 + V_1 = T_2 + V_2$$

where $V = V_g + V_e$

The potential energy is the sum of the body's gravitational and elastic potential energies. The gravitational potential energy will be positive if the body's center of gravity is located above a datum. If it is below the datum, then it will be negative. The elastic potential energy is always positive, regardless if the spring is stretched or compressed.

Gravitational potential energy

$$V_e = \tfrac{1}{2} k s^2$$

Elastic potential energy

REVIEW PROBLEMS

R18–1. The pendulum of the Charpy impact machine has a mass of 50 kg and a radius of gyration of $k_A = 1.75$ m. If it is released from rest when $\theta = 0°$, determine its angular velocity just before it strikes the specimen S, $\theta = 90°$.

R18–3. The drum has a mass of 50 kg and a radius of gyration about the pin at O of $k_O = 0.23$ m. Starting from rest, the suspended 15-kg block B is allowed to fall 3 m without applying the brake ACD. Determine the speed of the block at this instant. If the coefficient of kinetic friction at the brake pad C is $\mu_k = 0.5$, determine the force P that must be applied at the brake handle which will then stop the block after it descends *another* 3 m. Neglect the thickness of the handle.

Prob. R18–1

Prob. R18–3

R18–2. The 50-kg flywheel has a radius of gyration of $k_O = 200$ mm about its center of mass. If it is subjected to a torque of $M = (9\theta^{1/2} + 1)$ N·m, where θ is in radians, determine its angular velocity when it has rotated 5 revolutions, starting from rest.

R18–4. The spool has a mass of 40 kg and a radius of gyration of $k_O = 0.3$ m. If the 10-kg block is released from rest, determine the distance the block must fall in order for the spool to have an angular velocity $\omega = 15$ rad/s. Also, what is the tension in the cord while the block is in motion? Neglect the mass of the cord.

Prob. R18–2

Prob. R18–4

R18–5. The gear rack has a mass of 6 kg, and the gears each have a mass of 4 kg and a radius of gyration of $k = 30$ mm at their centers. If the rack is originally moving downward at 2 m/s, when $s = 0$, determine the speed of the rack when $s = 600$ mm. The gears are free to turn about their centers A and B.

R18–7. The system consists of a 20-lb disk A, 4-lb slender rod BC, and a 1-lb smooth collar C. If the disk rolls without slipping, determine the velocity of the collar at the instant the rod becomes horizontal, i.e., $\theta = 0°$. The system is released from rest when $\theta = 45°$.

Prob. R18–5

Prob. R18–7

R18–6. The compound disk pulley consists of a hub and attached outer rim. If it has a mass of 3 kg and a radius of gyration of $k_G = 45$ mm, determine the speed of block A after A descends 0.2 m from rest. Blocks A and B each have a mass of 2 kg. Neglect the mass of the cords.

R18–8. At the instant the spring becomes undeformed, the center of the 40-kg disk has a speed of 4 m/s. From this point determine the distance d the disk moves down the plane before momentarily stopping. The disk rolls without slipping.

Prob. R18–6

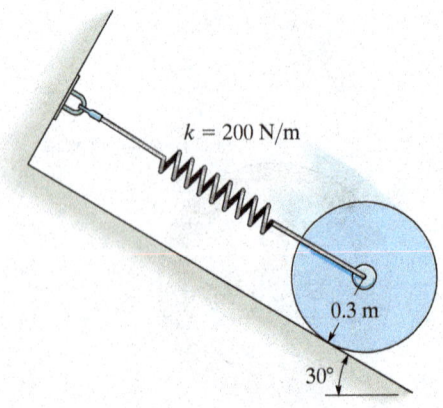

Prob. R18–8

CHAPTER 19

The astronaut maneuvers in space by applying short impulses using the jetpack. The results of the motion can be determined using impulse and momentum methods.

PLANAR KINETICS OF A RIGID BODY: IMPULSE AND MOMENTUM

CHAPTER OBJECTIVES

- To develop formulations for the linear and angular momentum of a body.
- To apply the principles of linear and angular impulse and momentum to solve rigid-body planar kinetic problems.
- To discuss application of the conservation of momentum.
- To analyze the mechanics of eccentric impact.

19.1 LINEAR AND ANGULAR MOMENTUM

In this chapter we will use the principles of linear and angular impulse and momentum to solve problems involving force, velocity, and time as it relates to the planar motion of a rigid body. Before doing this, we will first formalize the methods for obtaining a body's linear and angular momentum, assuming the body is symmetric with respect to an inertial x–y reference plane.

Linear Momentum. The linear momentum of a rigid body is determined by summing vectorially the linear momenta of all the particles of the body, i.e., $\mathbf{L} = \Sigma m_i \mathbf{v}_i$. Since $\Sigma m_i \mathbf{v}_i = m \mathbf{v}_G$ (see Sec. 15.2) we can then reference the linear momentum from the mass center, that is,

$$\mathbf{L} = m\mathbf{v}_G \qquad (19\text{–}1)$$

This equation states that the body's linear momentum is a vector quantity having a magnitude mv_G, which is commonly measured in units of kg · m/s or slug · ft/s, and a direction defined by $\mathbf{v}_G$, the velocity of the body's mass center.

496   CHAPTER 19   PLANAR KINETICS OF A RIGID BODY: IMPULSE AND MOMENTUM

Angular Momentum. Consider the body in Fig. 19–1a, which is subjected to general plane motion. At the instant shown, the arbitrary *i*th particle of the body has a mass m_i and a velocity $\mathbf{v}_i$. The angular momentum of this particle about the point P is equal to the "moment" of the particle's linear momentum about P, that is,

$$(\mathbf{H}_P)_i = \mathbf{r} \times m_i \mathbf{v}_i$$

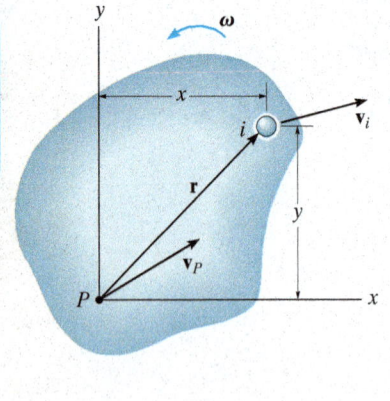

This momentum can also be expressed in terms of the velocity of point P rather than the velocity $\mathbf{v}_i$. If the body has an angular velocity $\boldsymbol{\omega}$, then from Fig. 19–1a, we have

$$\mathbf{v}_i = \mathbf{v}_P + \mathbf{v}_{i/P} = \mathbf{v}_P + \boldsymbol{\omega} \times \mathbf{r}$$

(a)

Letting $m_i \to dm_i$, expressing $\mathbf{v}_i$ in terms of $\mathbf{v}_P$ and using Cartesian vectors, we have

$$(H_P)_i \mathbf{k} = m_i (x\mathbf{i} + y\mathbf{j}) \times [(v_P)_x \mathbf{i} + (v_P)_y \mathbf{j} + \omega \mathbf{k} \times (x\mathbf{i} + y\mathbf{j})]$$
$$(H_P)_i = -dm_i y(v_P)_x + dm_i x(v_P)_y + dm_i \omega r^2$$

Integrating over the entire mass m of the body, we obtain

$$H_P = -\left(\int_m y\, dm\right)(v_P)_x + \left(\int_m x\, dm\right)(v_P)_y + \left(\int_m r^2\, dm\right)\omega$$

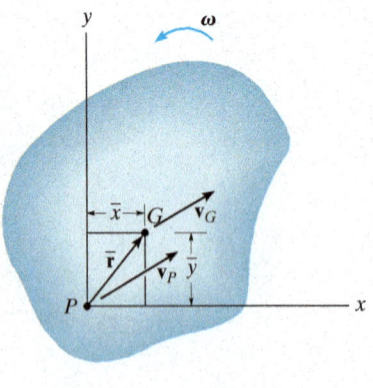

(b)

Fig. 19–1

Here H_P represents the angular momentum of the body about an axis (the z axis) perpendicular to the plane of motion that passes through point P. Since $\bar{y}m = \int y\, dm$ and $\bar{x}m = \int x\, dm$, the integrals for the first and second terms on the right are used to locate the body's center of mass G with respect to P, Fig. 19–1b. The last integral represents the body's moment of inertia about point P. Thus,

$$H_P = -\bar{y}m(v_P)_x + \bar{x}m(v_P)_y + I_P \omega \qquad (19\text{--}2)$$

This equation reduces to a simpler form if P coincides with the mass center G for the body,* in which case $\bar{x} = \bar{y} = 0$. Hence,

$$\boxed{H_G = I_G \omega} \qquad (19\text{--}3)$$

---

* It also reduces to a simpler form, $H_P = I_P \omega$, if point P is a *fixed point* (see Eq. 19–9) or if the velocity of P is directed along the line PG.

Here the angular momentum of the body about G is equal to the product of the moment of inertia of the body about an axis passing through G and the body's angular velocity. Realize that $\mathbf{H}_G$ is a vector quantity having a magnitude $I_G\omega$, which is commonly measured in units of $kg \cdot m^2/s$ or $slug \cdot ft^2/s$, and a direction defined by $\boldsymbol{\omega}$, which is always perpendicular to the plane of motion.

Equation 19–2 can also be rewritten in terms of the x and y components of the velocity of the body's mass center, $(\mathbf{v}_G)_x$ and $(\mathbf{v}_G)_y$, and the body's moment of inertia I_G. Since G is located at coordinates $(\bar{x}, \bar{y})$, then by the parallel-axis theorem, $I_P = I_G + m(\bar{x}^2 + \bar{y}^2)$. Substituting into Eq. 19–2 and rearranging terms, we have

$$H_P = \bar{y}m[-(v_P)_x + \bar{y}\omega] + \bar{x}m[(v_P)_y + \bar{x}\omega] + I_G\omega \quad (19\text{–}4)$$

From the kinematic diagram of Fig. 19–1b, $\mathbf{v}_G$ can be expressed in terms of $\mathbf{v}_P$ as

$$\mathbf{v}_G = \mathbf{v}_P + \boldsymbol{\omega} \times \bar{\mathbf{r}}$$
$$(v_G)_x \mathbf{i} + (v_G)_y \mathbf{j} = (v_P)_x \mathbf{i} + (v_P)_y \mathbf{j} + \omega \mathbf{k} \times (\bar{x}\mathbf{i} + \bar{y}\mathbf{j})$$

Carrying out the cross product and equating the respective $\mathbf{i}$ and $\mathbf{j}$ components yields the two scalar equations

$$(v_G)_x = (v_P)_x - \bar{y}\omega$$
$$(v_G)_y = (v_P)_y + \bar{x}\omega$$

Substituting these results into Eq. 19–4 yields

$$(\zeta +) H_P = -\bar{y}m(v_G)_x + \bar{x}m(v_G)_y + I_G\omega \quad (19\text{–}5)$$

As shown in Fig. 19–1c, this result indicates that when the angular momentum of the body is calculated about point P, it is equivalent to the moment of the linear momentum $m\mathbf{v}_G$, or its components $m(\mathbf{v}_G)_x$ and $m(\mathbf{v}_G)_y$, about P plus the angular momentum $I_G\boldsymbol{\omega}$. Using these results, we will now consider three types of motion.

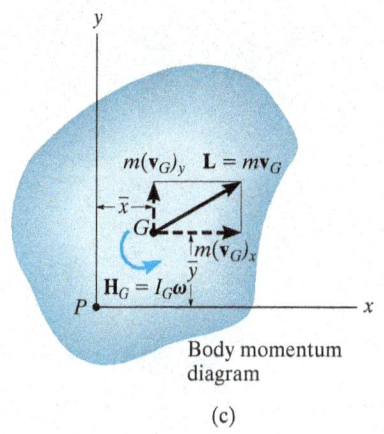

Body momentum diagram

(c)

Fig. 19–1

Translation
(a)

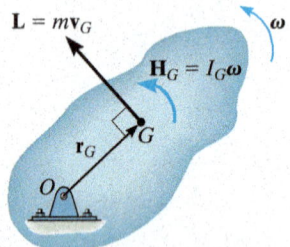

Rotation about a fixed axis
(b)

Fig. 19–2

Translation. When a rigid body is subjected to either rectilinear or curvilinear translation, Fig. 19–2a, then $\boldsymbol{\omega} = \mathbf{0}$ and its mass center has a velocity of $\mathbf{v}_G = \mathbf{v}$. Hence, the linear momentum, and the angular momentum about G, become

$$\boxed{\begin{array}{c} L = mv_G \\ H_G = 0 \end{array}} \qquad (19\text{–}6)$$

If the angular momentum is calculated about some other point A, the "moment" of the linear momentum $\mathbf{L}$ must be found about the point. Since d is the "moment arm," as shown in Fig. 19–2a, then in accordance with Eq. 19–5, $H_A = (d)(mv_G)\,\uparrow$.

Rotation About a Fixed Axis. When a rigid body is rotating about a fixed axis, Fig. 19–2b, the linear momentum, and the angular momentum about G, are

$$\boxed{\begin{array}{c} L = mv_G \\ H_G = I_G \omega \end{array}} \qquad (19\text{–}7)$$

It is sometimes convenient to calculate the angular momentum about the pin at O. Noting that $\mathbf{L}$ (or $\mathbf{v}_G$) is always *perpendicular* to $\mathbf{r}_G$, we have

$$(\zeta +) \quad H_O = I_G \omega + r_G (m v_G) \qquad (19\text{–}8)$$

Since $v_G = r_G \omega$, this equation can be written as $H_O = (I_G + m r_G^2)\omega$. Using the parallel-axis theorem,*

$$\boxed{H_O = I_O \omega} \qquad (19\text{–}9)$$

For the calculation, then, either Eq. 19–8 or 19–9 can be used.

---

*The similarity between this derivation and that of Eq. 17–16 ($\Sigma M_O = I_O \alpha$) and Eq. 18–5 ($T = \frac{1}{2} I_O \omega^2$) should be noted. Also, this same result can be obtained from Eq. 19–2 by selecting point P at O, realizing that $(v_O)_x = (v_O)_y = 0$.

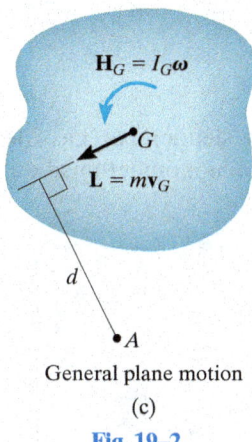

General plane motion
(c)

Fig. 19–2

General Plane Motion.
When a rigid body is subjected to general plane motion, Fig. 19–2c, the linear momentum, and the angular momentum about G, become

$$\boxed{\begin{aligned} L &= mv_G \\ H_G &= I_G\omega \end{aligned}} \qquad (19\text{–}10)$$

If the angular momentum is calculated about point A, Fig. 19–2c, it is necessary to include the moment of $\mathbf{L}$ and $\mathbf{H}_G$ about this point. In this case,

$$(\downarrow +) \quad H_A = I_G\omega + (d)(mv_G)$$

Here d is the moment arm, as shown in the figure.

As a special case, if point A is the instantaneous center of zero velocity, then, like Eq. 19–9, we can write the above equation in simplified form as

$$\boxed{H_{IC} = I_{IC}\omega} \qquad (19\text{–}11)$$

where I_{IC} is the moment of inertia of the body about the IC. (See Prob. 19–2.)

EXAMPLE 19.1

At a given instant the 5-kg slender bar has the motion shown in Fig. 19–3a. Determine its angular momentum about point G and about the IC at this instant.

(a)

SOLUTION

Bar. The bar undergoes *general plane motion*. The IC is established in Fig. 19–3b, so that

$$\omega = \frac{2 \text{ m/s}}{4 \text{ m} \cos 30°} = 0.5774 \text{ rad/s}$$

$$v_G = (0.5774 \text{ rad/s})(2 \text{ m}) = 1.155 \text{ m/s}$$

Thus,

$$(\zeta+)H_G = I_G\omega = \left[\tfrac{1}{12}(5 \text{ kg})(4 \text{ m})^2\right](0.5774 \text{ rad/s}) = 3.85 \text{ kg} \cdot \text{m}^2/\text{s} \downarrow$$
Ans.

The angular momentum about the IC is

$$(\zeta+)H_{IC} = I_G\omega + d(mv_G)$$

$$= \left[\tfrac{1}{12}(5 \text{ kg})(4 \text{ m})^2\right](0.5774 \text{ rad/s}) + (2 \text{ m})(5 \text{ kg})(1.155 \text{ m/s})$$

$$= 15.4 \text{ kg} \cdot \text{m}^2/\text{s} \downarrow$$
Ans.

We can also use

$$(\zeta+)H_{IC} = I_{IC}\omega$$

$$= \left[\tfrac{1}{12}(5 \text{ kg})(4 \text{ m})^2 + (5 \text{ kg})(2 \text{ m})^2\right](0.5774 \text{ rad/s})$$

$$= 15.4 \text{ kg} \cdot \text{m}^2/\text{s} \downarrow$$
Ans.

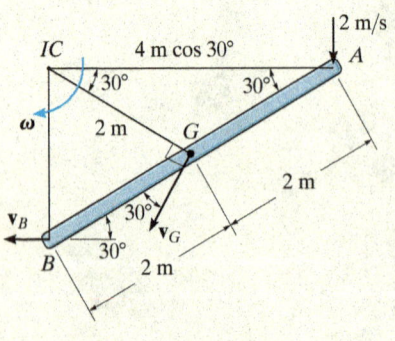

(b)

Fig. 19–3

19.2 PRINCIPLE OF IMPULSE AND MOMENTUM

Like the case for particle motion, the principle of impulse and momentum for a rigid body can be developed by combining the equation of motion with kinematics.

Principle of Linear Impulse and Momentum. The equation of translational motion for a rigid body can be written as $\Sigma \mathbf{F} = m\mathbf{a}_G = m(d\mathbf{v}_G/dt)$. Since the mass of the body is constant,

$$\Sigma \mathbf{F} = \frac{d}{dt}(m\mathbf{v}_G)$$

Multiplying both sides by dt and integrating from $t = t_1$, $\mathbf{v}_G = (\mathbf{v}_G)_1$ to $t = t_2$, $\mathbf{v}_G = (\mathbf{v}_G)_2$ yields

$$\Sigma \int_{t_1}^{t_2} \mathbf{F}\, dt = m(\mathbf{v}_G)_2 - m(\mathbf{v}_G)_1$$

This equation is referred to as the *principle of linear impulse and momentum*. It states that the sum of all the impulses created by the external force system which acts on the body during the time interval t_1 to t_2 is equal to the change in the linear momentum of the body during this time interval.

Principle of Angular Impulse and Momentum. If the body has general plane motion then $\Sigma M_G = I_G \alpha = I_G(d\omega/dt)$. Since the moment of inertia is constant,

$$\Sigma M_G = \frac{d}{dt}(I_G \omega)$$

Multiplying both sides by dt and integrating from $t = t_1$, $\omega = \omega_1$ to $t = t_2$, $\omega = \omega_2$ gives

$$\Sigma \int_{t_1}^{t_2} M_G\, dt = I_G \omega_2 - I_G \omega_1 \qquad (19\text{--}12)$$

In a similar manner, for rotation about a fixed axis passing through point O, Eq. 17–16 ($\Sigma M_O = I_O (d\omega/dt)$) when integrated becomes

$$\Sigma \int_{t_1}^{t_2} M_O\, dt = I_O \omega_2 - I_O \omega_1 \qquad (19\text{--}13)$$

Equations 19–12 and 19–13 are referred to as the *principle of angular impulse and momentum*. Both equations state that the sum of the angular impulses acting on the body during the time interval t_1 to t_2 is equal to the change in the body's angular momentum during this time interval.

To summarize these concepts, if motion occurs in the x–y plane, the following three scalar equations can be written to describe the planar motion of the body.

$$m(v_{Gx})_1 + \Sigma \int_{t_1}^{t_2} F_x \, dt = m(v_{Gx})_2$$

$$m(v_{Gy})_1 + \Sigma \int_{t_1}^{t_2} F_y \, dt = m(v_{Gy})_2 \qquad (19\text{–}14)$$

$$I_G \omega_1 + \Sigma \int_{t_1}^{t_2} M_G \, dt = I_G \omega_2$$

The terms in these equations can be shown graphically on the impulse and momentum diagrams for the body, Fig. 19–4. The linear momentum $m\mathbf{v}_G$ is applied at the body's mass center, Figs. 19–4a and 19–4c; whereas the angular momentum $I_G \boldsymbol{\omega}$ is a free vector, and therefore, like a couple moment, it can be applied at any point on the body. When the impulse diagram is constructed, Fig. 19–4b, the forces $\mathbf{F}$ and moment $\mathbf{M}$ vary with time, and are indicated by the integrals. However, if $\mathbf{F}$ and $\mathbf{M}$ are constant, then integration of the impulses yields $\mathbf{F}(t_2 - t_1)$ and $\mathbf{M}(t_2 - t_1)$. Such is the case for the body's weight $\mathbf{W}$, Fig. 19–4b.

Equations 19–14 can also be applied to an entire system of connected bodies, rather than to each body separately. This eliminates the need to include interaction impulses which occur at the connections since they are internal to the system and cancel out. The resultant equations may be written in symbolic form as

$$\left(\Sigma \begin{array}{c}\text{syst. linear}\\ \text{momentum}\end{array}\right)_{x1} + \left(\Sigma \begin{array}{c}\text{syst. linear}\\ \text{impulse}\end{array}\right)_{x(1-2)} = \left(\Sigma \begin{array}{c}\text{syst. linear}\\ \text{momentum}\end{array}\right)_{x2}$$

$$\left(\Sigma \begin{array}{c}\text{syst. linear}\\ \text{momentum}\end{array}\right)_{y1} + \left(\Sigma \begin{array}{c}\text{syst. linear}\\ \text{impulse}\end{array}\right)_{y(1-2)} = \left(\Sigma \begin{array}{c}\text{syst. linear}\\ \text{momentum}\end{array}\right)_{y2}$$

$$\left(\Sigma \begin{array}{c}\text{syst. angular}\\ \text{momentum}\end{array}\right)_{O1} + \left(\Sigma \begin{array}{c}\text{syst. angular}\\ \text{impulse}\end{array}\right)_{O(1-2)} = \left(\Sigma \begin{array}{c}\text{syst. angular}\\ \text{momentum}\end{array}\right)_{O2}$$

$$(19\text{–}15)$$

As indicated by the third equation, the system's angular momentum and angular impulse must be calculated with respect to the same reference point O for all the bodies of the system.

Initial momentum diagram

(a)

Impulse diagram

(b)

Final momentum diagram

(c)

Fig. 19–4

PROCEDURE FOR ANALYSIS

Impulse and momentum principles are used to solve kinetic problems that involve *velocity, force*, and *time* since these terms are involved in the formulation.

Free-Body Diagram.
- Establish the x, y, z inertial frame of reference and draw the free-body diagram in order to account for all the forces and couple moments that produce impulses on the body.
- The direction and sense of the initial and final velocity of the body's mass center, $\mathbf{v}_G$, and the body's angular velocity ω should be established. If any of these motions is unknown, assume that the sense of its components is in the direction of the positive inertial coordinates.
- Calculate the moment of inertia I_G or I_O.
- As an alternative procedure, draw the impulse and momentum diagrams for the body or system of bodies as shown in Fig. 19–4. These diagrams are particularly helpful in order to visualize the "moment" terms used in the principle of angular impulse and momentum, if application is about a fixed point O, the *IC*, or another point other than the body's mass center G.

Principle of Impulse and Momentum.
- Apply the three scalar equations of impulse and momentum.
- The angular momentum of a rigid body rotating about a fixed axis is the moment of $m\mathbf{v}_G$ plus $I_G\boldsymbol{\omega}$ about the axis. This is equal to $H_O = I_O\omega$, where I_O is the moment of inertia of the body about the axis.
- Forces that are functions of time must be integrated to obtain the impulse.
- The principle of angular impulse and momentum is often used to eliminate unknown impulsive forces that are parallel to or pass through a common axis, since the moment of these forces is zero about this axis.

Kinematics.
- If more than three equations are needed for a complete solution, it may be possible to relate the velocity of the body's mass center to the body's angular velocity using kinematics.

EXAMPLE 19.2

The 20-lb disk shown in Fig. 19–5a is acted upon by a couple moment of 4 lb·ft and a force of 10 lb which is applied to the cord wrapped around its periphery. Determine the angular velocity of the disk two seconds after starting from rest. Also, what are the force components of reaction at the pin?

SOLUTION

Since angular velocity, force, and time are involved in the problem, we will apply the principle of impulse and momentum to the solution.

Free-Body Diagram. Fig. 19–5b. The disk's mass center does not move; however, the loading causes the disk to rotate clockwise.
The moment of inertia of the disk about its fixed axis of rotation is

$$I_A = \frac{1}{2}mr^2 = \frac{1}{2}\left(\frac{20 \text{ lb}}{32.2 \text{ ft/s}^2}\right)(0.75 \text{ ft})^2 = 0.1747 \text{ slug}\cdot\text{ft}^2$$

Principle of Impulse and Momentum.

$(\xrightarrow{+})$
$$m(v_{Ax})_1 + \Sigma\int_{t_1}^{t_2} F_x\, dt = m(v_{Ax})_2$$
$$0 + A_x(2 \text{ s}) = 0$$

$(+\uparrow)$
$$m(v_{Ay})_1 + \Sigma\int_{t_1}^{t_2} F_y\, dt = m(v_{Ay})_2$$
$$0 + A_y(2 \text{ s}) - 20 \text{ lb}(2 \text{ s}) - 10 \text{ lb}(2 \text{ s}) = 0$$

$(\zeta+)$
$$I_A\omega_1 + \Sigma\int_{t_1}^{t_2} M_A\, dt = I_A\omega_2$$
$$0 + 4 \text{ lb}\cdot\text{ft}(2 \text{ s}) + [10 \text{ lb}(2 \text{ s})](0.75 \text{ ft}) = 0.1747\omega_2$$

Solving these equations yields

$$A_x = 0 \qquad \text{Ans.}$$
$$A_y = 30 \text{ lb} \qquad \text{Ans.}$$
$$\omega_2 = 132 \text{ rad/s} \downarrow \qquad \text{Ans.}$$

(a)

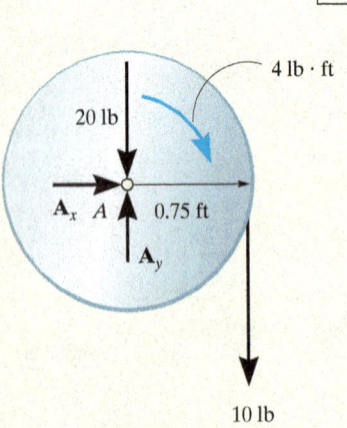

(b)

Fig. 19–5

EXAMPLE 19.3

The 100-kg spool shown in Fig. 19–6a has a radius of gyration $k_G = 0.35$ m. A cable is wrapped around the central hub of the spool, and a horizontal force having a variable magnitude of $P = (t + 10)$ N is applied, where t is in seconds. If the spool is initially at rest, determine its angular velocity in 5 s. Assume that the spool rolls without slipping at A.

Fig. 19–6

SOLUTION

Free-Body Diagram. From the free-body diagram, Fig. 19–6b, the *variable* force **P** will cause the friction force $\mathbf{F}_A$ to be variable, and thus the impulses created by both **P** and $\mathbf{F}_A$ must be determined by integration. Force **P** causes the mass center to have a velocity $\mathbf{v}_G$ to the right, and so the spool has a clockwise angular velocity ω.

Principle of Impulse and Momentum. A direct solution for ω can be obtained by applying the principle of angular impulse and momentum about point A, the IC, in order to eliminate the unknown friction impulse.

$(\zeta+)$
$$I_A \omega_1 + \Sigma \int M_A \, dt = I_A \omega_2$$

$$0 + \left[\int_0^{5\,\text{s}} (t + 10) \, \text{N} \, dt \right] (0.75 \text{ m} + 0.4 \text{ m}) = [100 \text{ kg} (0.35 \text{ m})^2 + (100 \text{ kg})(0.75 \text{ m})^2]\omega_2$$

$$62.5(1.15) = 68.5 \omega_2$$

$$\omega_2 = 1.05 \text{ rad/s} \downarrow \qquad\qquad Ans.$$

NOTE: Try solving this problem by applying the principle of impulse and momentum about G and using the principle of linear impulse and momentum in the x direction.

EXAMPLE 19.4

The cylinder B, shown in Fig. 19–7a has a mass of 6 kg. It is attached to a cord which is wrapped around the periphery of a 20-kg disk that has a moment of inertia $I_A = 0.40$ kg·m². If the cylinder is initially moving downward with a speed of 2 m/s, determine its speed in 3 s. Neglect the mass of the cord in the calculation.

(a)

SOLUTION I

Free-Body Diagram. The free-body diagrams of the cylinder and disk are shown in Fig. 19–7b. All the forces are constant since the weight of the cylinder causes the motion. The downward motion of the cylinder, $\mathbf{v}_B$, causes $\boldsymbol{\omega}$ of the disk to be clockwise.

Principle of Impulse and Momentum. We can eliminate $\mathbf{A}_x$ and $\mathbf{A}_y$ from the analysis by applying the principle of angular impulse and momentum about point A. Hence

Disk

$(\curvearrowright +)$
$$I_A \omega_1 + \Sigma \int M_A \, dt = I_A \omega_2$$

$$0.40 \text{ kg} \cdot \text{m}^2(\omega_1) + T(3 \text{ s})(0.2 \text{ m}) = (0.40 \text{ kg} \cdot \text{m}^2)\omega_2$$

Cylinder

$(+\uparrow)$
$$m_B(v_B)_1 + \Sigma \int F_y \, dt = m_B(v_B)_2$$

$$-6 \text{ kg}(2 \text{ m/s}) + T(3 \text{ s}) - 58.86 \text{ N}(3 \text{ s}) = -6 \text{ kg}(v_B)_2$$

Kinematics. Since $\omega = v_B/r$, then $\omega_1 = (2 \text{ m/s})/(0.2 \text{ m}) = 10$ rad/s and $\omega_2 = (v_B)_2/0.2 \text{ m} = 5(v_B)_2$. Substituting and solving the equations simultaneously for $(v_B)_2$ yields

$$(v_B)_2 = 13.0 \text{ m/s} \downarrow \qquad Ans.$$

(b)

Fig. 19–7

SOLUTION II

Impulse and Momentum Diagrams. We can obtain $(v_B)_2$ directly by considering the *system* consisting of the cylinder, the cord, and the disk. The impulse and momentum diagrams have been drawn to clarify application of the principle of angular impulse and momentum about point A, Fig. 19–7c.

Principle of Angular Impulse and Momentum. Realizing that $\omega_1 = 10$ rad/s and $\omega_2 = 5(v_B)_2$, we have

$$(\curvearrowleft +)\left(\sum \substack{\text{syst. angular}\\ \text{momentum}}\right)_{A1} + \left(\sum \substack{\text{syst. angular}\\ \text{impulse}}\right)_{A(1-2)} = \left(\sum \substack{\text{syst. angular}\\ \text{momentum}}\right)_{A2}$$

$$(6 \text{ kg})(2 \text{ m/s})(0.2 \text{ m}) + (0.40 \text{ kg} \cdot \text{m}^2)(10 \text{ rad/s}) + (58.86 \text{ N})(3 \text{ s})(0.2 \text{ m})$$
$$= (6 \text{ kg})(v_B)_2(0.2 \text{ m}) + (0.40 \text{ kg} \cdot \text{m}^2)[5(v_B)_2]$$
$$(v_B)_2 = 13.0 \text{ m/s} \downarrow \qquad\qquad Ans.$$

(c)

Fig. 19–7

EXAMPLE 19.5

A Charpy impact test is used to test a material to determine its energy absorption characteristics during impact. The test is performed using the pendulum shown in Fig. 19–8a, which has a mass m, mass center at G, and a radius of gyration k_G about G. Determine the distance r_P from the pin at A to the point P where the impact with the specimen S should occur so that the horizontal force at the pin A is essentially zero during the impact. For the calculation, assume the specimen absorbs all the pendulum's kinetic energy during the time it falls and thereby stops the pendulum from swinging when $\theta = 0°$.

SOLUTION

Free-Body Diagram. As shown on the free-body diagram, Fig. 19–8b, the conditions of the problem require the horizontal force at A to be zero. Just before impact, the pendulum has a clockwise angular velocity ω_1, and the mass center of the pendulum is moving to the left at $(v_G)_1 = \bar{r}\omega_1$.

Principle of Impulse and Momentum. We will apply the principle of angular impulse and momentum about point A. Thus,

$$I_A\omega_1 + \Sigma \int M_A \, dt = I_A\omega_2$$

$(\curvearrowright +)$
$$I_A\omega_1 - \left(\int F \, dt\right)r_P = 0$$

$$m(v_G)_1 + \Sigma \int F \, dt = m(v_G)_2$$

$(\xrightarrow{+})$
$$-m(\bar{r}\omega_1) + \int F \, dt = 0$$

Fig. 19–8

Eliminating the impulse $\int F \, dt$ and substituting $I_A = mk_G^2 + m\bar{r}^2$ yields

$$[mk_G^2 + m\bar{r}^2]\omega_1 - m(\bar{r}\omega_1)r_P = 0$$

Factoring out $m\omega_1$ and solving for r_P, we obtain

$$r_P = \bar{r} + \frac{k_G^2}{\bar{r}} \qquad \qquad Ans.$$

NOTE: Point P, so defined, is called the *center of percussion*. By placing the striking point at P, the force developed at the pin will be minimized. Many sports rackets, clubs, etc., are designed so that collision with the object being struck occurs at the center of percussion. As a consequence, no "sting" or little sensation occurs in the hand of the player. (Also see Probs. 17–66 and 19–1.)

FUNDAMENTAL PROBLEMS

F19–1. The 60-kg wheel has a radius of gyration about its center O of $k_O = 300$ mm. If it is subjected to a couple moment of $M = (3t^2)$ N·m, where t is in seconds, determine the angular velocity of the wheel when $t = 4$ s, starting from rest.

Prob. F19–1

F19–2. The 300-kg wheel has a radius of gyration about its mass center G of $k_G = 400$ mm. If the wheel is subjected to a couple moment of $M = 300$ N·m and no slipping occurs, determine its angular velocity 6 s after it starts from rest. Also, determine the friction force that the ground applies to the wheel.

Prob. F19–2

F19–3. If rod OA of negligible mass is subjected to the couple moment $M = 9$ N·m, determine the angular velocity of the 10-kg inner gear $t = 5$ s after it starts from rest. The gear has a radius of gyration about its mass center of $k_A = 100$ mm, and it rolls on the fixed outer gear, B. Motion occurs in the horizontal plane.

Prob. F19–3

F19–4. Gears A and B of mass 10 kg and 50 kg have radii of gyration about their respective mass centers of $k_A = 80$ mm and $k_B = 150$ mm. If gear A is subjected to the couple moment $M = 10$ N·m when it is at rest, determine the angular velocity of gear B when $t = 5$ s.

Prob. F19–4

F19–5. The 50-kg spool is subjected to a horizontal force of $P = 150$ N. If the spool rolls without slipping, determine its angular velocity 3 s after it starts from rest. The radius of gyration of the spool about its center of mass is $k_G = 175$ mm.

Prob. F19–5

F19–6. The reel has a weight of 150 lb and a radius of gyration about its center of gravity of $k_G = 1.25$ ft. If it is subjected to a torque of $M = 25$ lb·ft, and starts from rest when the torque is applied, determine its angular velocity in 3 seconds. The coefficient of kinetic friction between the reel and the horizontal plane is $\mu_k = 0.15$.

Prob. F19–6

PROBLEMS

All solutions must include a free-body diagram.

19–1. The rigid body (slab) has a mass m and rotates with an angular velocity ω about an axis passing through the fixed point O. Show that the momenta of all the particles composing the body can be represented by a single vector having a magnitude $m v_G$ and acting through point P, called the *center of percussion*, which lies at a distance $r_{P/G} = k_G^2/r_{G/O}$ from the mass center G. Here k_G is the radius of gyration of the body, calculated about an axis perpendicular to the plane of motion and passing through G.

Prob. 19–1

19–2. At a given instant, the body has a linear momentum $\mathbf{L} = m\mathbf{v}_G$ and an angular momentum $\mathbf{H}_G = I_G\boldsymbol{\omega}$ calculated about its mass center. Show that the angular momentum of the body calculated about the instantaneous center of zero velocity IC can be expressed as $\mathbf{H}_{IC} = I_{IC}\boldsymbol{\omega}$, where I_{IC} represents the body's moment of inertia calculated about the instantaneous axis of zero velocity. As shown, the IC is located at a distance $r_{G/IC}$ away from the mass center G.

Prob. 19–2

19–3. Show that if a slab is rotating about a fixed axis perpendicular to the slab and passing through its mass center G, the angular momentum is the same when calculated about any other point P.

Prob. 19–3

*19–4. The two tugboats each exert a constant force $\mathbf{F}$ on the boat such that these forces are always directed perpendicular to the boat's centerline. If the boat has a mass m and a radius of gyration k_G about its center of mass G, determine the angular velocity of the boat as a function of time t. The boat is originally at rest.

Prob. 19–4

19–5. A sphere and cylinder are released from rest on the ramp at $t = 0$. If each has a mass m and a radius r, determine their angular velocities at time t. Assume no slipping occurs.

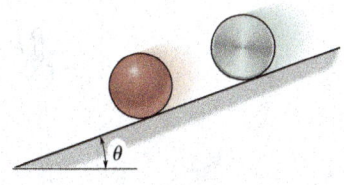

Prob. 19–5

19–6. Determine the height h at which a bullet can strike the disk that has a weight of 15 lb and cause it to roll without slipping at A. *Note:* This requires that the frictional force at A be essentially zero.

Prob. 19–6

19–7. The 10-lb rectangular plate is at rest on a smooth *horizontal* floor. If it is given the horizontal impulses shown, determine its angular velocity and the velocity of the mass center.

Prob. 19–7

*19–8. The spool has a mass of 30 kg and a radius of gyration $k_O = 0.25$ m. Block A has a mass of 25 kg, and block B has a mass of 10 kg. If they are released from rest, determine the time required for block A to attain a speed of 2 m/s. Neglect the mass of the ropes.

Prob. 19–8

19–9. The 50-kg cylinder has an angular velocity of 30 rad/s when it is brought into contact with the surface at C. If the coefficient of kinetic friction is $\mu_C = 0.2$, determine how long it takes for the cylinder to stop spinning. What force is developed in the link AB during this time? The axis of the cylinder is connected to *two* symmetrical links. (Only AB is shown.) Neglect the weight of the links.

Prob. 19–9

19–10. The frame of the roller has a mass of 5.5 Mg and a center of mass at G. The roller has a mass of 2 Mg and a radius of gyration about its mass center of $k_A = 0.45$ m. If a torque of $M = 600$ N·m is applied to the rear wheels, determine the speed of the compactor in $t = 4$ s, starting from rest. No slipping occurs. Neglect the mass of the driving wheels.

Prob. 19–10

19–11. A wire of negligible mass is wrapped around the outer surface of the 2-kg disk. If the disk is released from rest, determine its angular velocity in 3 s.

19–13. The smooth rod assembly shown is at rest when it is struck by a hammer at A with an impulse of $10\ \text{N}\cdot\text{s}$. Determine the angular velocity of the assembly and the magnitude of velocity of its mass center immediately after it has been struck. The rods have a mass per unit length of 6 kg/m.

Prob. 19–13

Prob. 19–11

19–14. The car strikes the side of a light pole, which is designed to break away from its base with negligible resistance. From a video taken of the collision it is observed that the pole was given an angular velocity of 60 rad/s when AC was vertical. The pole has a mass of 175 kg, a center of mass at G, and a radius of gyration about an axis perpendicular to the plane of the pole assembly and passing through G of $k_G = 2.25$ m. Determine the horizontal impulse which the car exerts on the pole at the instant AC is essentially vertical.

*19–12. The drum of mass m, radius r, and radius of gyration k_G rolls along an inclined plane for which the coefficient of static friction is μ_s. If the drum is released from rest, determine the maximum angle θ for the incline so that it rolls without slipping.

Prob. 19–12

Prob. 19–14

19–15. If the boxer hits the 75-kg punching bag with an impulse of $I = 20\text{N} \cdot \text{s}$, determine the angular velocity of the bag immediately after it has been hit. Also, find the location d of point B, about which the bag appears to rotate. Treat the bag as a uniform cylinder.

19–17. A cord of negligible mass is wrapped around the outer surface of the 50-lb cylinder and its end is subjected to a constant horizontal force of $P = 2$ lb. If the cylinder rolls without slipping at A, determine its angular velocity in 4 s starting from rest. Neglect the thickness of the cord.

Prob. 19–15

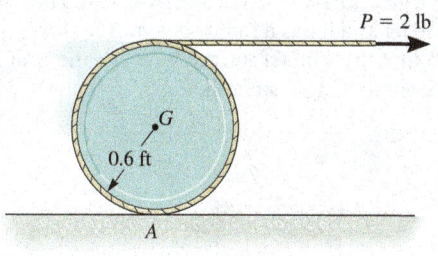

Prob. 19–17

*19–16.** Gear A rotates along the inside of the circular gear rack R having a radius 2.25 ft. If A has a weight of 4 lb and a radius of gyration of $k_B = 0.5$ ft, determine its angular momentum about point C when $\omega_{CB} = 30$ rad/s and (a) $\omega_R = 0$, (b) $\omega_R = 20$ rad/s.

19–18. The 200-kg satellite has a radius of gyration about the centroidal z axis of $k_z = 1.25$ m. Initially it is rotating with a constant angular velocity of $\omega_0 = \{1500\mathbf{k}\}$ rev/min. If the two jets A and B, both in the x–y plane, are fired simultaneously and produce a thrust of $T = (5e^{-0.1t})$ kN, where t is in seconds, determine the angular velocity of the satellite, five seconds after firing.

Prob. 19–16

Prob. 19–18

19–19. If the cord is subjected to a horizontal force of $P = 150$ N, and the gear rack is fixed to the horizontal plane, determine the angular velocity of the gear in 4 s, starting from rest. The mass of the gear is 50 kg, and it has a radius of gyration about its center of mass O of $k_O = 125$ mm.

*19–20. If the cord is subjected to a horizontal force of $P = 150$ N, and the gear is supported by a fixed pin at O, determine the angular velocity of the gear and the velocity of the 20-kg gear rack in 4 s, starting from rest. The mass of the gear is 50 kg and it has a radius of gyration of $k_O = 125$ mm. Assume that the contact surface between the gear rack and the horizontal plane is smooth.

Probs. 19–19/20

19–21. The spool has a weight of 30 lb and a radius of gyration $k_O = 0.45$ ft. A cord is wrapped around its inner hub and the end subjected to a horizontal force $P = 5$ lb. Determine the spool's angular velocity in 4 s starting from rest. Assume the spool rolls without slipping.

Prob. 19–21

19–22. The two gears A and B have weights and radii of gyration of $W_A = 15$ lb, $k_A = 0.5$ ft and $W_B = 10$ lb, $k_B = 0.35$ ft, respectively. If a motor transmits a couple moment to gear B of $M = 2(1 - e^{-0.5t})$ lb·ft, where t is in seconds, determine the angular velocity of gear A in $t = 5$ s, starting from rest.

Prob. 19–22

19–23. The 12-kg disk has an angular velocity of $\omega = 20$ rad/s. If the brake ABC is applied such that the magnitude of force $\mathbf{P}$ varies with time as shown, determine the time needed to stop the disk. The coefficient of kinetic friction at B is $\mu_k = 0.4$. Neglect the thickness of the brake.

Prob. 19–23

19–24. The 30-kg gear is subjected to a force of $P = (20t)$ N, where t is in seconds. Determine the angular velocity of the gear at $t = 4$ s, starting from rest. Gear rack B is fixed to the horizontal plane, and the gear's radius of gyration about its mass center O is $k_O = 125$ mm.

Prob. 19–24

19–25. The body and bucket of a skid steer loader have a weight of 2000 lb, and its center of gravity is located at G. Each of the four wheels has a weight of 100 lb and a radius of gyration about its center of gravity of 1 ft. If the engine supplies a torque of $M = 100$ lb · ft to each of the rear-drive wheels, determine the speed of the loader in $t = 10$ s, starting from rest. The wheels roll without slipping.

19–26. The body and bucket of a skid steer loader have a weight of 2000 lb, and its center of gravity is located at G. Each of the four wheels has a weight of 100 lb and a radius of gyration about its center of gravity of 1 ft. If the loader attains a speed of 20 ft/s in 10 s, starting from rest, determine the torque M supplied to each of the rear-drive wheels. The wheels roll without slipping.

Probs. 19–25/26

19–27. The double pulley consists of two wheels which are attached to one another and turn at the same rate. The pulley has a mass of 15 kg and a radius of gyration of $k_O = 110$ mm. If the block at A has a mass of 40 kg and the container at B has a mass of 85 kg, including its contents, determine the speed of the container when $t = 3$ s after it is released from rest.

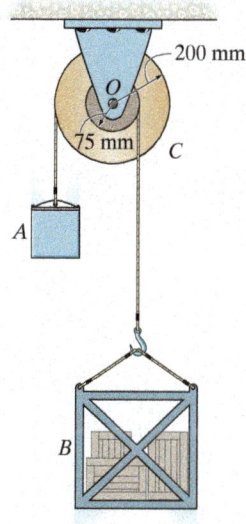

Prob. 19–27

19–28. A thin rod having a mass of 4 kg is balanced vertically as shown. Determine the height h at which it can be struck with a horizontal force **F** and not slip on the floor. This requires that the frictional force at A be essentially zero.

Prob. 19–28

19.3 CONSERVATION OF MOMENTUM

Conservation of Linear Momentum. If the sum of all the *linear impulses* acting on a system of connected rigid bodies is *zero* in a specific direction, then the linear momentum of the system is constant, or conserved in this direction, that is,

$$\left(\sum \begin{array}{c}\text{syst. linear}\\\text{momentum}\end{array}\right)_1 = \left(\sum \begin{array}{c}\text{syst. linear}\\\text{momentum}\end{array}\right)_2 \quad (19\text{–}16)$$

This equation is referred to as the *conservation of linear momentum*.

Without introducing appreciable errors in the calculations, it may also be possible to apply Eq. 19–16 in a specified direction for which the linear impulses are small or *nonimpulsive*. Specifically, nonimpulsive forces occur when small forces act over very short periods of time. Typical examples include the force of a slightly deformed spring, the initial contact force with soft ground, and in some cases the weight of the body.

Conservation of Angular Momentum. The angular momentum of a system of connected rigid bodies is conserved about the system's center of mass G, or a fixed point O, when the sum of all the angular impulses about these points is zero or appreciably small (nonimpulsive). The third of Eqs. 19–15 then becomes

$$\left(\sum \begin{array}{c}\text{syst. angular}\\\text{momentum}\end{array}\right)_{O1} = \left(\sum \begin{array}{c}\text{syst. angular}\\\text{momentum}\end{array}\right)_{O2} \quad (19\text{–}17)$$

This equation is referred to as the *conservation of angular momentum*.

In the case of a single rigid body, Eq. 19–17 applied to point G becomes $(I_G \omega)_1 = (I_G \omega)_2$. For example, consider a swimmer who executes a somersault after jumping off a diving board. Angular momentum is conserved about his center of mass G since his resultant weight passes through this point. By tucking his arms and legs in close to his chest, he decreases his body's moment of inertia and thus increases his angular velocity ($I_G \omega$ must be constant). If he straightens out just before entering the water, his body's moment of inertia is increased, and so his angular velocity decreases. Since his weight creates a linear impulse during the motion, this example also illustrates how the angular momentum of a body can be conserved and yet the linear momentum is not. Such cases occur whenever the external forces creating the linear impulse pass through either the center of mass of the body or a fixed axis of rotation.

PROCEDURE FOR ANALYSIS

The conservation of linear or angular momentum should be applied using the following procedure.

Free-Body Diagram.

- Establish the x, y inertial frame of reference and draw the free-body diagram for the body or system of bodies during the time of impact. From this diagram classify each of the applied forces as being either "impulsive" or "nonimpulsive."

- By inspection of the free-body diagram, the *conservation of linear momentum* applies in a given direction when *no* external impulsive forces act on the body or system in that direction; whereas the *conservation of angular momentum* applies about a fixed point O or at the mass center G of a body or system of bodies when all the external impulsive forces acting on the body or system create zero moment (or zero angular impulse) about O or G.

- As an alternative procedure, draw the impulse and momentum diagrams for the body or system of bodies. These diagrams are particularly helpful in order to visualize the "moment" terms used in the conservation of angular momentum equation, if it is decided that angular momenta are to be calculated about a point other than the body's mass center G.

Conservation of Momentum.

- Apply the conservation of linear or angular momentum in the appropriate directions.

Kinematics.

- If the motion appears to be complicated, kinematic (velocity) diagrams may be helpful in obtaining the necessary kinematic relations.

EXAMPLE 19.6

Fig. 19–9

The 10-kg wheel shown in Fig. 19–9a has a moment of inertia $I_G = 0.156 \text{ kg} \cdot \text{m}^2$. Assuming that the wheel does not slip or rebound, determine the minimum velocity $\mathbf{v}_G$ it must have to just roll over the obstruction at A.

SOLUTION

Impulse and Momentum Diagrams. Since no slipping or rebounding occurs, the wheel essentially *pivots* about point A during contact. This condition is shown in Fig. 19–9b, which indicates, respectively, the momentum of the wheel *just before impact*, the impulses given to the wheel *during impact*, and the momentum of the wheel *just after impact*. Only two impulses (forces) act on the wheel. By comparison, the force at A is much greater than that of the weight, and since the time of impact is very short, the weight can be considered nonimpulsive. The impulsive force $\mathbf{F}$ at A has both an unknown magnitude and an unknown direction θ. We can eliminate this force from the analysis by noting that angular momentum about A is essentially *conserved* since $(98.1 \Delta t)d \approx 0$.

Conservation of Angular Momentum. With reference to Fig. 19–9b,

$(\zeta+)$
$$(H_A)_1 = (H_A)_2$$
$$r'm(v_G)_1 + I_G\omega_1 = rm(v_G)_2 + I_G\omega_2$$
$$(0.2 \text{ m} - 0.03 \text{ m})(10 \text{ kg})(v_G)_1 + (0.156 \text{ kg} \cdot \text{m}^2)(\omega_1) =$$
$$(0.2 \text{ m})(10 \text{ kg})(v_G)_2 + (0.156 \text{ kg} \cdot \text{m}^2)(\omega_2)$$

Kinematics. Since no slipping occurs, in general $\omega = v_G/r = v_G/0.2 \text{ m} = 5v_G$. Substituting this into the above equation and simplifying yields

$$(v_G)_2 = 0.8921(v_G)_1 \quad (1)$$

Conservation of Energy.* In order to roll over the obstruction, the wheel must pass position 3 shown in Fig. 19–9c. Hence, if $(v_G)_2$ [or $(v_G)_1$] is to be a minimum, it is necessary that the kinetic energy of the wheel at position 2 be equal to the potential energy at position 3. Placing the datum through the center of gravity, as shown in the figure, and applying the conservation of energy equation, we have

$$\{T_2\} + \{V_2\} = \{T_3\} + \{V_3\}$$
$$\{\tfrac{1}{2}(10 \text{ kg})(v_G)_2^2 + \tfrac{1}{2}(0.156 \text{ kg} \cdot \text{m}^2)\omega_2^2\} + \{0\} =$$
$$\{0\} + \{(98.1 \text{ N})(0.03 \text{ m})\}$$

Substituting $\omega_2 = 5(v_G)_2$ and Eq. 1 into this equation, and solving,

$$(v_G)_1 = 0.729 \text{ m/s} \rightarrow \qquad Ans.$$

*This principle *does not apply during impact*, since energy is *lost* during the collision. However, just after impact, as in Fig. 19–9c, it can be used.

EXAMPLE 19.7

The 5-kg slender rod shown in Fig. 19–10a is pinned at O and is initially at rest. If a 4-g bullet is fired into the rod with a velocity of 400 m/s, as shown in the figure, determine the angular velocity of the rod just after the bullet becomes embedded in it.

SOLUTION

Impulse and Momentum Diagrams. The impulse which the bullet exerts on the rod can be eliminated from the analysis by considering the bullet and rod as a single system. To clarify the principles involved, the impulse and momentum diagrams are shown in Fig. 19–10b. The momentum diagrams are drawn *just before and just after impact*. As shown on the impulse diagram, the external impulses are caused by the reactions at O and the weights of the bullet and rod. Since the time of impact, Δt, is very short, the rod moves only a slight amount, and so the "moments" of the weight impulses about point O are essentially zero. Therefore angular momentum is conserved about this point.

(a)

(b)

Conservation of Angular Momentum. From Fig. 19–10b, we have

$(\downarrow +)$
$$\Sigma(H_O)_1 = \Sigma(H_O)_2$$

$m_B(v_B)_1 \cos 30°(0.75 \text{ m}) = m_B(v_B)_2(0.75 \text{ m}) + m_R(v_G)_2(0.5 \text{ m}) + I_G \omega_2$

$(0.004 \text{ kg})(400 \cos 30° \text{ m/s})(0.75 \text{ m}) =$

$(0.004 \text{ kg})(v_B)_2(0.75 \text{ m}) + (5 \text{ kg})(v_G)_2(0.5 \text{ m}) + [\frac{1}{12}(5 \text{ kg})(1 \text{ m})^2]\omega_2$

or

$$1.039 = 0.003(v_B)_2 + 2.50(v_G)_2 + 0.4167\omega_2$$

Kinematics. Since the rod is pinned at O, from Fig. 19–10c we have

$$(v_G)_2 = (0.5 \text{ m})\omega_2 \quad (v_B)_2 = (0.75 \text{ m})\omega_2$$

Substituting into the above equation and solving yields

$$\omega_2 = 0.623 \text{ rad/s} \uparrow \qquad Ans.$$

(c)

Fig. 19–10

*19.4 ECCENTRIC IMPACT

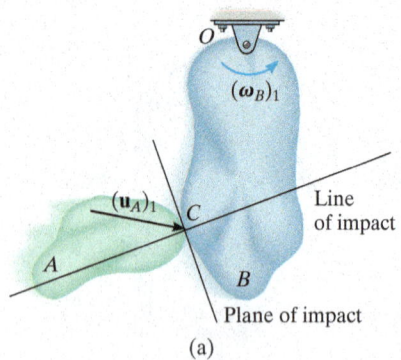

Fig. 19–11

The concepts involving central and oblique impact of particles were presented in Sec. 15.4. We will now expand this treatment and discuss the eccentric impact of two bodies. **Eccentric impact** occurs when the line connecting the mass centers of the two bodies does not coincide with the line of impact.* This type of impact often occurs when one or both of the bodies are constrained to rotate about a fixed axis. For example, consider the collision at C between the two bodies A and B, shown in Fig. 19–11a. It is assumed that just before collision B is rotating counterclockwise with an angular velocity $(\omega_B)_1$, and the velocity of the contact point C located on A is $(\mathbf{u}_A)_1$. Kinematic diagrams for both bodies just before collision are shown in Fig. 19–11b. Provided the bodies are smooth, the impulsive forces they exert on each other are directed along the line of impact. Hence, the component of velocity of point C on body B, which is directed along the line of impact, is $(v_B)_1 = (\omega_B)_1 r$, Fig. 19–11b. Likewise, on body A the component of velocity $(\mathbf{u}_A)_1$ along the line of impact is $(\mathbf{v}_A)_1$. In order for a collision to occur, $(v_A)_1 > (v_B)_1$.

During the impact an equal but opposite impulsive force $\mathbf{P}$ is exerted between the bodies which *deforms* their shapes at the point of contact. The resulting deformation impulse is shown on the impulse diagrams for both bodies, Fig. 19–11c. In addition, the impulsive force at point C on the rotating body creates impulsive pin reactions at O. On these diagrams it is assumed that these impulses are much larger than the nonimpulsive weights of the bodies, which are not shown. When the deformation at point C is a maximum, point C on both bodies moves with a common velocity $\mathbf{v}$ along the line of impact, Fig. 19–11d. A period of *restitution* then occurs in which the bodies tend to regain their original shapes. The restitution phase creates an equal but opposite restitution impulse between the bodies as shown on the impulse diagram, Fig. 19–11e. After restitution the bodies move apart such that point C on body B has a velocity $(\mathbf{v}_B)_2$ and point C on body A has a velocity $(\mathbf{u}_A)_2$, Fig. 19–11f, where $(v_B)_2 > (v_A)_2$.

In general, a problem involving the impact of two bodies requires determining the *two unknowns* $(v_A)_2$ and $(v_B)_2$, assuming $(v_A)_1$ and $(v_B)_1$ are known. To solve such problems, two equations must be written. The *first equation* generally involves application of *the conservation of angular momentum*. For example, in the case of bodies A and B, angular momentum is conserved about point O since the impulses at C are internal to the system and the impulses at O create zero moment (or zero angular impulse) about O. The *second equation* can be obtained using the definition of the **coefficient of restitution**, e, which is a ratio of the restitution impulse to the deformation impulse.

Here is an example of eccentric impact occurring between this bowling ball and pin.

* When these lines coincide, central impact occurs and the problem can be analyzed as discussed in Sec. 15.4.

19.4 ECCENTRIC IMPACT

To establish a useful form of the coefficient of restitution equation we will first apply the principle of angular impulse and momentum about point O to body B. Proceeding from the time just before the collision to the instant of maximum deformation, Figs. 19–11b, 19–11c, and 19–11d, we have

$$(\downarrow+) \qquad I_O(\omega_B)_1 + r\int P\, dt = I_O\omega \qquad (19\text{–}18)$$

Here I_O is the moment of inertia of B about point O. Similarly, applying the principle of angular impulse and momentum from the instant of maximum deformation to the time just after the impact, Figs. 19-11d, 19-11e, and 19-11f, yields

$$(\downarrow+) \qquad I_O\omega + r\int R\, dt = I_O(\omega_B)_2 \qquad (19\text{–}19)$$

Solving Eqs. 19–18 and 19–19 for $\int P\, dt$ and $\int R\, dt$, respectively, and formulating e, we have

$$e = \frac{\int R\, dt}{\int P\, dt} = \frac{r(\omega_B)_2 - r\omega}{r\omega - r(\omega_B)_1} = \frac{(v_B)_2 - v}{v - (v_B)_1}$$

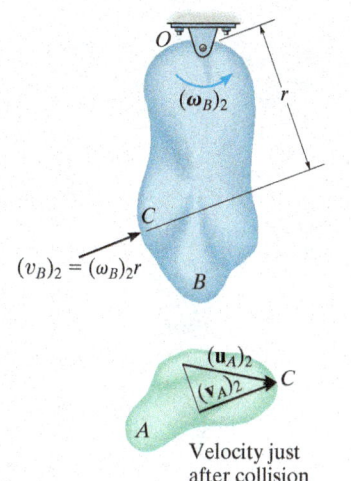

Fig. 19–11

In the same manner, we can write an equation which relates the magnitudes of velocity $(v_A)_1$ and $(v_A)_2$ of body A. The result is

$$e = \frac{v - (v_A)_2}{(v_A)_1 - v}$$

Combining the above two equations by eliminating the common velocity v yields the desired result, i.e.,

$(+\nearrow)$
$$e = \frac{(v_B)_2 - (v_A)_2}{(v_A)_1 - (v_B)_1} \qquad (19\text{–}20)$$

This equation is identical to Eq. 15–11, which was derived for the central impact between two particles. It states that the coefficient of restitution is equal to the ratio of the relative velocity of *separation* of the points of contact (C) *just after impact* to the relative velocity at which the points *approach* one another *just* before impact. In deriving this equation, we assumed that the points of contact for both bodies move up and to the right *both* before and after impact. If motion of any one of the contacting points occurs down and to the left, the velocity of this point should be considered a negative quantity in Eq. 19–20.

It is important to realize, however, that this analysis has only a very limited application in engineering, because values of e for eccentric impact have been found to be highly sensitive to the material, geometry, and the velocity of each of the colliding bodies.

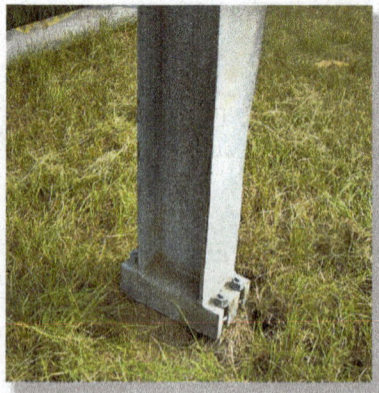

During impact the columns of many highway signs are intended to break out of their supports and easily collapse at their joints. This is shown by the slotted connections at their base and the clean breaks at the column's midsection.

EXAMPLE 19.8

The 10-lb slender rod is suspended from the pin at A, Fig. 19–12a. If a 2-lb ball B is thrown at the rod and strikes its center with a velocity of 30 ft/s, determine the angular velocity of the rod just after impact. The coefficient of restitution is $e = 0.4$.

(a)

SOLUTION

Conservation of Angular Momentum. If we consider the ball and rod as a single system, Fig. 19–12b, then angular momentum is conserved about point A since the impulsive force between the rod and ball is *internal*. Also, the *weights* of the ball and rod are *nonimpulsive*. Assuming the directions of the velocities of the ball and rod just after impact are as shown on the kinematic diagram, Fig. 19–12c, we require

$(\downarrow+)$ $(H_A)_1 = (H_A)_2$

$m_B(v_B)_1(1.5 \text{ ft}) = m_B(v_B)_2(1.5 \text{ ft}) + m_R(v_G)_2(1.5 \text{ ft}) + I_G \omega_2$

$\left(\dfrac{2 \text{ lb}}{32.2 \text{ ft/s}^2}\right)(30 \text{ ft/s})(1.5 \text{ ft}) = \left(\dfrac{2 \text{ lb}}{32.2 \text{ ft/s}^2}\right)(v_B)_2(1.5 \text{ ft}) +$

$\left(\dfrac{10 \text{ lb}}{32.2 \text{ ft/s}^2}\right)(v_G)_2(1.5 \text{ ft}) + \left[\dfrac{1}{12}\left(\dfrac{10 \text{ lb}}{32.2 \text{ ft/s}^2}\right)(3 \text{ ft})^2\right]\omega_2$

(b)

Since $(v_G)_2 = 1.5\omega_2$ then

$2.795 = 0.09317(v_B)_2 + 0.9317\omega_2$ (1)

Coefficient of Restitution. With reference to Fig. 19–12c, we have

$(\xrightarrow{+})$ $e = \dfrac{(v_G)_2 - (v_B)_2}{(v_B)_1 - (v_G)_1}$ $0.4 = \dfrac{(1.5 \text{ ft})\omega_2 - (v_B)_2}{30 \text{ ft/s} - 0}$

$12.0 = 1.5\omega_2 - (v_B)_2$ (2)

Solving Eqs. 1 and 2, yields

$(v_B)_2 = -6.52 \text{ ft/s} = 6.52 \text{ ft/s} \leftarrow$

$\omega_2 = 3.65 \text{ rad/s} \curvearrowleft$ *Ans.*

(c)

Fig. 19–12

PROBLEMS

19–29. The turntable T of a record player has a mass of 0.75 kg and a radius of gyration $k_z = 125$ mm. It is *turning freely* at $\omega_T = 2$ rad/s when a 50-g record (thin disk) falls on it. Determine the final angular velocity of the turntable just after the record stops slipping on the turntable.

Prob. 19–29

19–30. The 10-g bullet having a velocity of 800 m/s is fired into the 5-kg disk as shown. Determine the angular velocity of the disk just after the bullet becomes embedded. Also, calculate the angle θ the disk will swing when it stops. The disk is originally at rest. Neglect the mass of the rod AB and the final angular momentum of the embedded bullet.

19–31. The 10-g bullet having a velocity of 800 m/s is fired into the 5-kg disk as shown. Determine the angular velocity of the disk just after the bullet becomes embedded. Also, calculate the angle θ the disk will swing when it stops. The disk is originally at rest. The rod AB has a mass of 3 kg. Neglect the final angular momentum of the embedded bullet.

Probs. 19–30/31

*19–32. The man sits on the swivel chair holding two 5-lb weights with his arms outstretched. If he is rotating at 3 rad/s in this position, determine his angular velocity when the weights are drawn in and held 0.3 ft from the axis of rotation. Assume he weighs 160 lb and has a radius of gyration $k_z = 0.55$ ft about the z axis. Neglect the mass of his arms and the size of the weights for the calculation.

Prob. 19–32

19–33. The platform swing consists of a 200-lb flat plate suspended by four rods of negligible weight. When the swing is at rest, the 150-lb man jumps off the platform when his center of gravity G is 10 ft from the pin at A. This is done with a horizontal velocity of 5 ft/s, measured relative to the swing at the level of G. Determine the angular velocity he imparts to the swing just after jumping off.

Prob. 19–33

19–34. Two children A and B, each having a mass of 30 kg, sit at the edge of the merry-go-round which is rotating at $\omega = 2$ rad/s. Excluding the children, the merry-go-round has a mass of 180 kg and a radius of gyration $k_z = 0.6$ m. Determine the angular velocity of the merry-go-round if A jumps off horizontally in the $-n$ direction with a speed of 2 m/s, measured with respect to the merry-go-round. What is the merry-go-round's angular velocity if B then jumps off horizontally in the $+t$ direction with a speed of 2 m/s, measured with respect to the merry-go-round? Neglect friction and the size of each child.

Prob. 19–34

19–35. The Hubble Space Telescope is powered by two solar panels as shown. The body of the telescope has a mass of 11 Mg and radii of gyration $k_x = 1.64$ m and $k_y = 3.85$ m, whereas the solar panels can be considered as thin plates, each having a mass of 54 kg. Due to an internal drive, the panels are given an angular velocity of $\{0.6\mathbf{j}\}$ rad/s, measured relative to the telescope. Determine the angular velocity of the telescope due to the rotation of the panels. Prior to rotating the panels, the telescope was originally traveling at $\mathbf{v}_G = \{-400\mathbf{i} + 250\mathbf{j} + 175\mathbf{k}\}$ m/s. Neglect its orbital rotation.

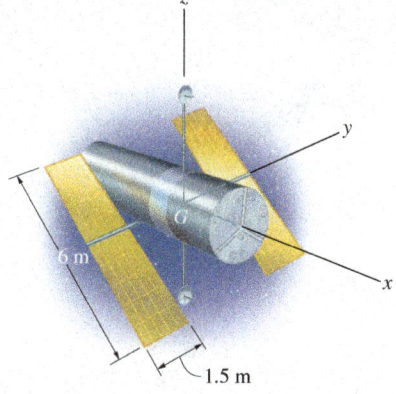

Prob. 19–35

*19–36. Each of the two slender rods and the disk have the same mass m. Also, the length of each rod is equal to the diameter d of the disk. If the assembly is rotating with an angular velocity ω_1 when the rods are directed outward, determine the angular velocity of the assembly if by internal means the rods are brought to an upright vertical position.

Prob. 19–36

19–37. Disk A has a weight of 20 lb. An inextensible cable is attached to the 10-lb weight and wrapped around the disk. The weight is dropped 2 ft before the slack is taken up. If the impact is perfectly elastic, i.e., $e = 1$, determine the angular velocity of the disk just after impact.

Prob. 19–37

19–38. The rod has a length L and mass m. A smooth collar having a negligible size and one-fourth the mass of the rod is placed on the rod at its midpoint. If the rod is freely rotating at ω about its end and the collar is released, determine the rod's angular velocity just before the collar flies off the rod. Also, what is the speed of the collar as it leaves the rod?

Prob. 19–38

19–39. The 12-kg rod AB is pinned to the 40-kg disk. If the disk is given an angular velocity $\omega_D = 100$ rad/s while the rod is held stationary, and the assembly is then released, determine the angular velocity of the rod after the disk has stopped spinning relative to the rod due to frictional resistance at the bearing B. Motion is in the *horizontal plane*. Neglect friction at the pin A.

Prob. 19–39

*19–40. A thin rod of mass m has an angular velocity ω_0 while rotating on a smooth surface. Determine its new angular velocity just after its end strikes and hooks onto the peg and the rod starts to rotate about P without rebounding. Solve the problem (a) using the parameters given, (b) setting $m = 2$ kg, $\omega_0 = 4$ rad/s, $l = 1.5$ m.

Prob. 19–40

19–41. Tests of impact on the fixed crash dummy are conducted using the 300-lb ram that is released from rest at $\theta = 30°$, and allowed to fall and strike the dummy at $\theta = 90°$. If the coefficient of restitution between the dummy and the ram is $e = 0.4$, determine the angle θ to which the ram will rebound before momentarily coming to rest.

Prob. 19–41

19–42. The vertical shaft is rotating with an angular velocity of 3 rad/s when $\theta = 0°$. If a force $\mathbf{F}$ is applied to the collar so that $\theta = 90°$, determine the angular velocity of the shaft. Also, find the work done by force $\mathbf{F}$. Neglect the mass of rods GH and EF and the collars I and J. The rods AB and CD each have a mass of 10 kg.

Prob. 19–42

19–43. The rigid 30-lb plank is struck by the 15-lb hammer head H. Just before the impact the hammer is gripped loosely and has a vertical velocity of 75 ft/s. If the coefficient of restitution between the hammer head and the plank is $e = 0.5$, determine the maximum height attained by the 50-lb block D. The block can slide freely along the two vertical guide rods. The plank is initially in a horizontal position.

Prob. 19–43

*19–44.** The pendulum consists of a 5-lb slender rod AB and a 10-lb wooden block. A projectile weighing 0.2 lb is fired into the center of the block with a velocity of 1000 ft/s. If the pendulum is initially at rest, and the projectile embeds itself into the block, determine the angular velocity of the pendulum just after the impact. Neglect the final angular momentum of the embedded projectile.

Prob. 19–44

19–45. The 10-lb block is sliding on the smooth surface when the corner D hits a stop block S. Determine the minimum velocity $\mathbf{v}$ the block should have which would allow it to tip over on its side and land in the position shown. Neglect the size of S. *Hint:* During impact consider the weight of the block to be nonimpulsive.

Prob. 19–45

19–46. Determine the height h of the bumper of the pool table, so that when the pool ball of mass m strikes it, no frictional force will be developed between the ball and the table at A. Assume the bumper exerts only a horizontal force on the ball.

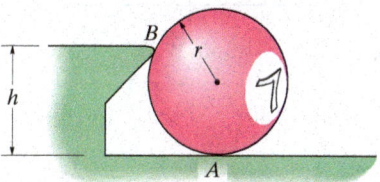

Prob. 19–46

19–47. The pendulum consists of a 15-kg solid ball and 6-kg rod. If it is released from rest when $\theta_1 = 90°$, determine the angle θ_2 after the ball strikes the wall, rebounds, and the pendulum swings up to the point of momentary rest. Take $e = 0.6$.

Prob. 19–47

*19–48. The 4-lb rod AB is hanging in the vertical position. A 2-lb block, sliding on a smooth horizontal surface with a velocity of 12 ft/s, strikes the rod at its end B. Determine the velocity of the block immediately after the collision. The coefficient of restitution between the block and the rod at B is $e = 0.8$.

Prob. 19–48

19–49. The hammer consists of a 10-kg solid cylinder C and 6-kg uniform slender rod AB. If the hammer is released from rest when $\theta = 90°$ and strikes the 30-kg block D when $\theta = 0°$, determine the velocity of block D and the angular velocity of the hammer immediately after the impact. The coefficient of restitution between the hammer and the block is $e = 0.6$.

Prob. 19–49

19–50. The 20-kg disk strikes the step without rebounding. Determine the largest angular velocity ω_1 the disk can have without losing contact with the step A.

Prob. 19–50

19–51. The solid ball of mass m is dropped with a velocity $\mathbf{v}_1$ onto the edge of the rough step. If it rebounds horizontally off the step with a velocity $\mathbf{v}_2$, determine the angle θ at which contact occurs. Assume no slipping when the ball strikes the step. The coefficient of restitution is e.

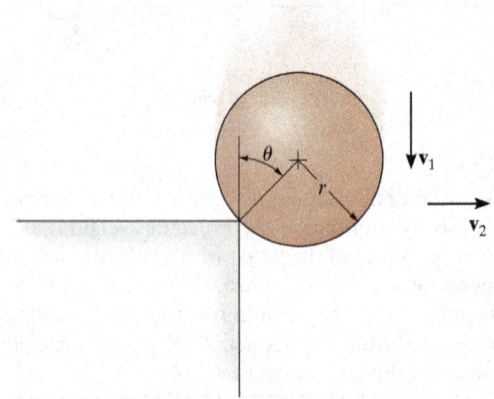

Prob. 19–51

*19–52. The uniform rod assembly rotates with an angular velocity of ω_0 on the smooth horizontal plane just before the hook strikes the peg P without rebound. Determine the angular velocity of the assembly immediately after the impact. Each rod has a mass of m.

Prob. 19–52

19–53. A 50-g pencil (uniform slender rod) falls down onto the table with a velocity of 0.2 m/s just before impact. If it is at an angle of 60° with the horizontal, and it does not slip during the impact, determine its angular velocity ω and the velocity v_G of its center of mass just after rebounding. The coefficient of restitution is $e = 0.8$.

Prob. 19–53

19–54. The rod of mass m and length L is released from rest without rotating. When it falls a distance L, the end A strikes the hook S, which provides a permanent connection. Determine the angular velocity ω of the rod after it has rotated 90°. Treat the rod's weight during impact as a nonimpulsive force.

Prob. 19–54

19–55. The 15-lb rod AB is released from rest in the vertical position. If the coefficient of restitution between the floor and the cushion at B is $e = 0.7$, determine how high the end of the rod rebounds after impact with the floor.

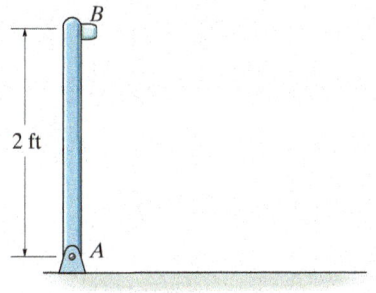

Prob. 19–55

*19–56. A ball having a mass of 8 kg and initial speed of $v_1 = 0.2$ m/s rolls over a 30-mm-long depression. Assuming that the ball rolls off the edges of contact, first A then B, without slipping, determine its final velocity v_2 when it reaches the other side.

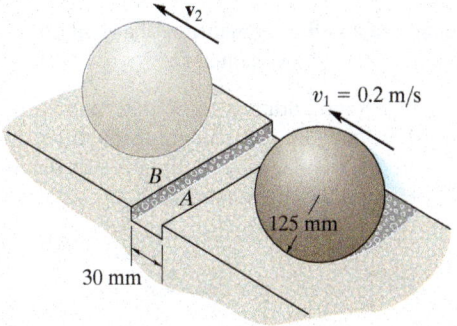

Prob. 19–56

19–57. The 2-kg disk is thrown down onto the rough surface with the velocity and angular velocity shown. If there is no slipping, and the coefficient of restitution is $e = 0.5$, determine the velocity of the disk and its angular velocity just after rebounding.

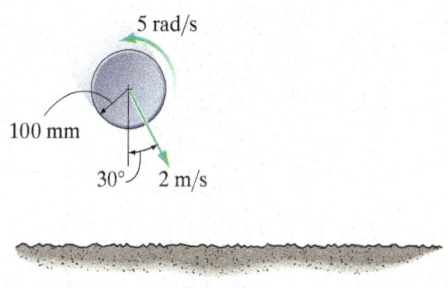

Prob. 19–57

19–58. The pendulum consists of a 10-lb solid ball and 4-lb rod. If it is released from rest when $\theta = 0°$, determine the angle θ_1 of rebound after the ball strikes the wall and the pendulum swings up to the point of momentary rest. Take $e = 0.6$.

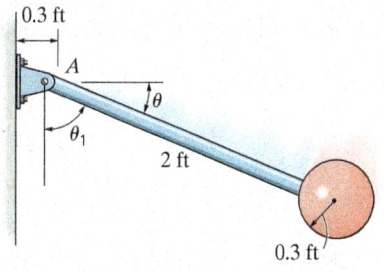

Prob. 19–58

CHAPTER REVIEW

Linear and Angular Momentum

The linear and angular momentum of a rigid body can be referenced to its mass center G.

If the angular momentum is to be determined about an axis other than the one passing through the mass center, then the angular momentum is determined by summing $\mathbf{H}_G$ and the moment of $\mathbf{L}$ about this axis.

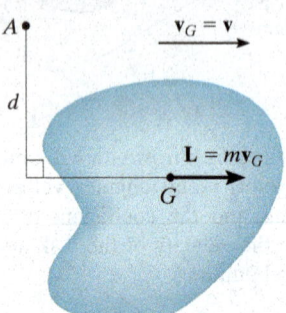

Translation

$$L = mv_G$$
$$H_G = 0$$
$$H_A = (mv_G)d$$

Rotation about a fixed axis

$$L = mv_G$$
$$H_G = I_G\omega$$
$$H_O = I_O\omega$$

General plane motion

$$L = mv_G$$
$$H_G = I_G\omega$$
$$H_A = I_G\omega + (mv_G)d$$

Principle of Impulse and Momentum

The principles of linear and angular impulse and momentum are used to solve problems that involve force, velocity, and time. Before applying these equations, it is important to establish the x, y, z inertial coordinate system. The free-body diagram for the body should also be drawn in order to account for all of the forces and couple moments that produce impulses on the body.

$$m(v_{Gx})_1 + \Sigma \int_{t_1}^{t_2} F_x \, dt = m(v_{Gx})_2$$

$$m(v_{Gy})_1 + \Sigma \int_{t_1}^{t_2} F_y \, dt = m(v_{Gy})_2$$

$$I_G\omega_1 + \Sigma \int_{t_1}^{t_2} M_G \, dt = I_G\omega_2$$

Conservation of Momentum

Provided the sum of the linear impulses acting on a system of connected rigid bodies is zero in a particular direction, then the linear momentum for the system is conserved in this direction. Conservation of angular momentum occurs if the impulses pass through an axis or are parallel to it. Momentum is also conserved if the external forces are small and thereby create nonimpulsive forces on the system. A free-body diagram should accompany any application in order to classify the forces as impulsive or nonimpulsive and to determine an axis about which the angular momentum may be conserved.

$$\left(\Sigma \begin{matrix}\text{syst. linear}\\\text{momentum}\end{matrix}\right)_1 = \left(\Sigma \begin{matrix}\text{syst. linear}\\\text{momentum}\end{matrix}\right)_2$$

$$\left(\Sigma \begin{matrix}\text{syst. angular}\\\text{momentum}\end{matrix}\right)_{O1} = \left(\Sigma \begin{matrix}\text{syst. angular}\\\text{momentum}\end{matrix}\right)_{O2}$$

Eccentric Impact

If the line of impact does not coincide with the line connecting the mass centers of two colliding bodies, then eccentric impact will occur. If the motion of the bodies just after the impact is to be determined, then it is necessary to consider a conservation of momentum equation for the system and use the coefficient of restitution equation.

$$e = \frac{(v_B)_2 - (v_A)_2}{(v_A)_1 - (v_B)_1}$$

REVIEW PROBLEMS

R19–1. The cable is subjected to a force of $P = (10t^2)$ lb, where t is in seconds. Determine the angular velocity of the spool 3 s after **P** is applied, starting from rest. The spool has a weight of 150 lb and a radius of gyration of 1.25 ft about its center, O.

R19–3. The tire has a mass of 9 kg and a radius of gyration $k_O = 225$ mm. If it is released from rest and rolls down the plane without slipping, determine the speed of its center O when $t = 3$ s.

Prob. R19–1

Prob. R19–3

R19–2. The space capsule has a mass of 1200 kg and a moment of inertia $I_G = 900$ kg·m² about an axis passing through G and directed perpendicular to the page. If it is traveling forward with a speed $v_G = 800$ m/s and executes a turn by means of two jets, which provide a constant thrust of 400 N for 0.3 s, determine the capsule's angular velocity just after the jets are turned off.

R19–4. The wheel having a mass of 100 kg and a radius of gyration about the z axis of $k_z = 300$ mm rests on the smooth horizontal plane. If the belt is subjected to a force of $P = 200$ N, determine the angular velocity of the wheel and the speed of its center of mass O, three seconds after the force is applied.

Prob. R19–2

Prob. R19–4

R19–5. The spool has a weight of 30 lb and a radius of gyration $k_O = 0.65$ ft. If a force of 40 lb is applied to the cord at A, determine the angular velocity of the spool in $t = 3$ s starting from rest. Neglect the mass of the pulley and cord.

R19–7. A thin disk of mass m has an angular velocity ω_1 while rotating on a smooth surface. Determine its new angular velocity just after the hook at its edge strikes the peg P and the disk starts to rotate about P without rebounding.

Prob. R19–7

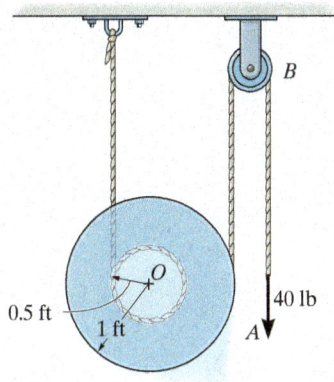

Prob. R19–5

R19–8. The space satellite has a mass of 125 kg and a moment of inertia $I_z = 0.940$ kg·m², excluding the four solar panels $A, B, C,$ and D. Each solar panel has a mass of 20 kg and can be approximated as a thin plate. If the satellite is originally spinning about the z axis at a constant rate $\omega_z = 0.5$ rad/s when $\theta = 90°$, determine the rate of spin if all the panels are raised and reach the upward position, $\theta = 0°$, at the same instant.

R19–6. Spool B is at rest and spool A is rotating at 6 rad/s when the slack in the cord connecting them is taken up. If the cord does not stretch, determine the angular velocity of each spool immediately after the cord is jerked tight. The spools A and B have weights and radii of gyration $W_A = 30$ lb, $k_A = 0.8$ ft, $W_B = 15$ lb, $k_B = 0.6$ ft, respectively.

Prob. R19–6

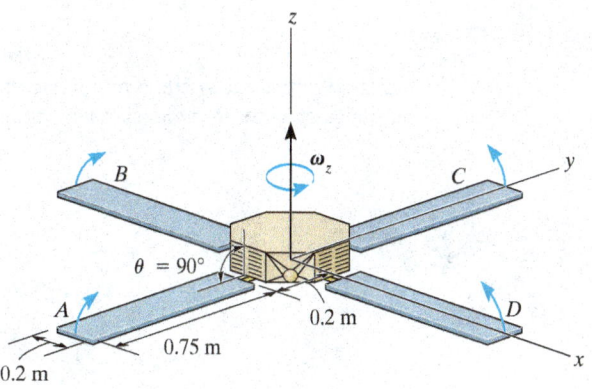

Prob. R19–8

CHAPTER 20

Operation of this tower crane requires controlling the kinematics of its three-dimensional motion.

THREE-DIMENSIONAL KINEMATICS OF A RIGID BODY

CHAPTER OBJECTIVES

- To analyze the kinematics of a body subjected to rotation about a fixed point and to general plane motion.
- To provide a relative-motion analysis of a rigid body using translating and rotating axes.

20.1 ROTATION ABOUT A FIXED POINT

When a rigid body rotates about a fixed point, the path of motion for a particle on the body lies on the surface of a sphere, having a radius r and centered at the fixed point. Since motion along this path is formed from a series of rotations, we will first develop a familiarity with some of the properties of rotational displacements.

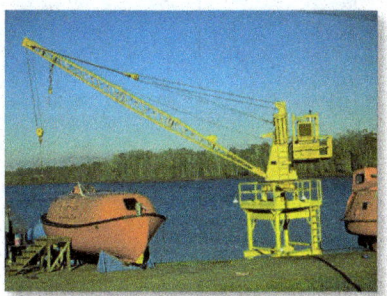

The boom is subjected to rotation about a fixed point because it can rotate up and down, and the frame can turn about a vertical axis.

536   CHAPTER 20   THREE-DIMENSIONAL KINEMATICS OF A RIGID BODY

Euler's Theorem. Euler's theorem states that two "component" rotations about different axes passing through a point are equivalent to a single resultant rotation about an axis passing through the same point. If more than two rotations are applied, they can all be combined into pairs, and each pair can be further reduced and combined into a single rotation.

Finite Rotations. If the component rotations used in Euler's theorem are finite, then it is important that the order in which they are applied be maintained. For example, consider the two finite rotations $\boldsymbol{\theta}_1 + \boldsymbol{\theta}_2$ applied to the block in Fig. 20–1a. Each rotation has a magnitude of 90° and a direction defined by the right-hand rule, as indicated by the arrow. The final position of the block is shown at the right. If these two rotations are applied in the reverse order, $\boldsymbol{\theta}_2 + \boldsymbol{\theta}_1$, as shown in Fig. 20–1b, then the final position of the block will not be the same as it is in Fig. 20–1a. Because finite rotations do not obey the commutative law of addition ($\boldsymbol{\theta}_1 + \boldsymbol{\theta}_2 \neq \boldsymbol{\theta}_2 + \boldsymbol{\theta}_1$), they cannot be classified as vectors. Realize that if smaller, yet finite, rotations had been used to illustrate this point, e.g., 10° instead of 90°, the final position of the block would still be different; however, in this case, the difference would only be a small amount.

Fig. 20–1

20.1 ROTATION ABOUT A FIXED POINT

Infinitesimal Rotations. When defining the angular motions of a body subjected to three-dimensional motion, only rotations which are infinitesimally small will be considered. Such rotations can be classified as vectors, since they can be added vectorially in any manner. To show this, for purposes of simplicity let us consider the rigid body itself to be a sphere which is allowed to rotate about its central fixed point O, Fig. 20–2a. If we impose two infinitesimal rotations $d\boldsymbol{\theta}_1 + d\boldsymbol{\theta}_2$ on the body, it is seen that point P moves along the path $d\boldsymbol{\theta}_1 \times \mathbf{r} + d\boldsymbol{\theta}_2 \times \mathbf{r}$ and ends up at P'. Had the two successive rotations occurred in the order $d\boldsymbol{\theta}_2 + d\boldsymbol{\theta}_1$, then the resultant displacements of P would have been $d\boldsymbol{\theta}_2 \times \mathbf{r} + d\boldsymbol{\theta}_1 \times \mathbf{r}$ and again P would move to P'. This occurs because the vector cross product obeys the distributive law, i.e., $(d\boldsymbol{\theta}_1 + d\boldsymbol{\theta}_2) \times \mathbf{r} = (d\boldsymbol{\theta}_2 + d\boldsymbol{\theta}_1) \times \mathbf{r}$, and so infinitesimal rotations are vectors, since these quantities have both a magnitude and direction for which the order of (vector) addition is not important, i.e., $d\boldsymbol{\theta}_1 + d\boldsymbol{\theta}_2 = d\boldsymbol{\theta}_2 + d\boldsymbol{\theta}_1$. The two "component" rotations $d\boldsymbol{\theta}_1$ and $d\boldsymbol{\theta}_2$ in Fig. 20–2a are therefore equivalent to a single resultant rotation $d\boldsymbol{\theta} = d\boldsymbol{\theta}_1 + d\boldsymbol{\theta}_2$, a consequence of Euler's theorem.

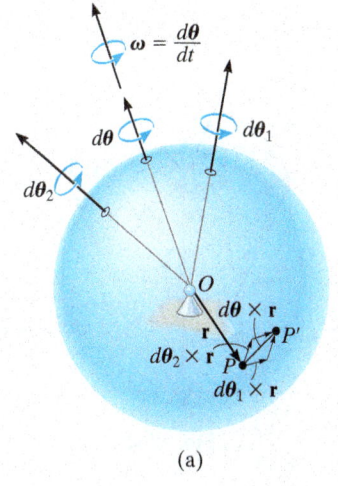

(a)

Angular Velocity. If the body is subjected to an angular rotation $d\boldsymbol{\theta}$ about a fixed point, the angular velocity of the body is defined by its time derivative,

$$\boldsymbol{\omega} = \dot{\boldsymbol{\theta}} \qquad (20\text{--}1)$$

The line specifying the direction of $\boldsymbol{\omega}$, which is collinear with $d\boldsymbol{\theta}$, is referred to as the *instantaneous axis of rotation*, Fig. 20–2b. In general, this axis changes direction during each instant of time. Since $d\boldsymbol{\theta}$ is a vector quantity, so too is $\boldsymbol{\omega}$, and it follows from vector addition that if the body is subjected to two component angular motions, $\boldsymbol{\omega}_1 = \dot{\boldsymbol{\theta}}_1$ and $\boldsymbol{\omega}_2 = \dot{\boldsymbol{\theta}}_2$, the resultant angular velocity is $\boldsymbol{\omega} = \boldsymbol{\omega}_1 + \boldsymbol{\omega}_2$.

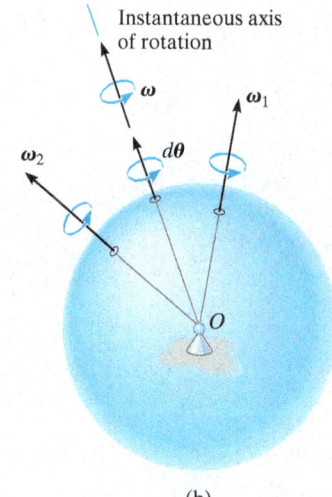

(b)

Fig. 20–2

Angular Acceleration. The body's angular acceleration is determined from the time derivative of its angular velocity, i.e.,

$$\boldsymbol{\alpha} = \dot{\boldsymbol{\omega}} \qquad (20\text{--}2)$$

For motion about a fixed point, $\boldsymbol{\alpha}$ must account for a change in *both* the magnitude and direction of $\boldsymbol{\omega}$, so that, in general, $\boldsymbol{\alpha}$ will not be directed along the instantaneous axis of rotation, Fig. 20–3.

As the direction of the instantaneous axis of rotation (or the line of action of $\boldsymbol{\omega}$) changes in space, the locus of the axis generates a fixed *space cone*, Fig. 20–4. If the change in the direction of this axis is viewed with respect to the rotating body, the locus of the axis generates a *body cone*.

Fig. 20–3

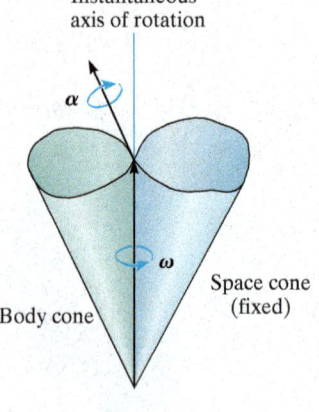

Fig. 20–4

At any given instant, these cones meet along the instantaneous axis of rotation, and when the body is in motion, the body cone appears to roll either on the inside or the outside surface of the space cone. Provided the paths defined by the open ends of the cones are described by the head of the $\boldsymbol{\omega}$ vector, then $\boldsymbol{\alpha}$ must act tangent to these paths at any given instant, since the time rate of change of $\boldsymbol{\omega}$ is equal to $\boldsymbol{\alpha}$, Fig. 20–4.

To illustrate this concept, consider the disk in Fig. 20–5a that *spins* about the rod at $\boldsymbol{\omega}_s$, while the rod and disk *precess* about the vertical axis at $\boldsymbol{\omega}_p$. The resultant angular velocity of the disk is therefore $\boldsymbol{\omega} = \boldsymbol{\omega}_s + \boldsymbol{\omega}_p$. Since both point O and the contact point P have zero velocity, then all points on a line between these points must have zero velocity. Thus, both $\boldsymbol{\omega}$ and the instantaneous axis of rotation are along OP. Therefore, as the disk rotates, this axis appears to move along the surface of the fixed space cone shown in Fig. 20–5b. If the axis is observed from the rotating disk, the axis then appears to move on the surface of the body cone. If $\boldsymbol{\omega}$ has a constant magnitude, then $\boldsymbol{\alpha}$ indicates only the change in the direction of $\boldsymbol{\omega}$, which is tangent to the cones at the tip of $\boldsymbol{\omega}$ as shown in Fig. 20–5b.

Velocity. Once $\boldsymbol{\omega}$ is specified, the velocity of any point on a body rotating about a fixed point can be determined using the same methods as for a body rotating about a fixed axis. Hence, by the cross product,

$$\mathbf{v} = \boldsymbol{\omega} \times \mathbf{r} \qquad (20\text{–}3)$$

where $\mathbf{r}$ defines the position of the point measured from the fixed point O, Fig. 20–3.

Acceleration. If $\boldsymbol{\omega}$ and $\boldsymbol{\alpha}$ are known at a given instant, the acceleration of a point can then be obtained from the time derivative of Eq. 20–3. This yields

$$\mathbf{a} = \boldsymbol{\alpha} \times \mathbf{r} + \boldsymbol{\omega} \times (\boldsymbol{\omega} \times \mathbf{r}) \qquad (20\text{–}4)$$

*20.2 THE TIME DERIVATIVE OF A VECTOR MEASURED FROM A FIXED OR TRANSLATING-ROTATING SYSTEM

(a)

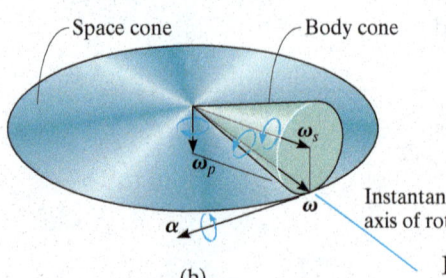

(b)

Fig. 20–5

In many types of problems involving the motion of a body about a fixed point, the angular velocity $\boldsymbol{\omega}$ is specified in terms of its components. Then, if the angular acceleration $\boldsymbol{\alpha}$ of the body is to be determined, it is often easier to find the time derivative of $\boldsymbol{\omega}$ using a coordinate system that has a rotation defined by one or more of the components of $\boldsymbol{\omega}$. For example, in the case of the disk in Fig. 20–5a, where $\boldsymbol{\omega} = \boldsymbol{\omega}_s + \boldsymbol{\omega}_p$, the x, y, z axes can be given an angular velocity of $\boldsymbol{\omega}_p$, and then $\boldsymbol{\alpha}$ can be determined relative to these axes. For this reason, and for other uses later, an equation will now be derived, which relates the time derivative of any vector $\mathbf{A}$ defined from a translating-rotating reference to its time derivative defined from a fixed reference.

*20.2 THE TIME DERIVATIVE OF A VECTOR MEASURED FROM A FIXED OR TRANSLATING-ROTATING SYSTEM

Consider the x, y, z axes of the moving frame of reference to be rotating with an angular velocity $\boldsymbol{\Omega}$, which is measured from the fixed X, Y, Z axes, Fig. 20–6a. In the following discussion, it will be convenient to express vector $\mathbf{A}$ in terms of its $\mathbf{i}, \mathbf{j}, \mathbf{k}$ components, which define the directions of the moving axes. Hence,

$$\mathbf{A} = A_x\mathbf{i} + A_y\mathbf{j} + A_z\mathbf{k}$$

If the time derivative is taken with respect to the moving frame of reference, only the change in the magnitudes of the components of $\mathbf{A}$ must be accounted for, since the directions of the components do not change with respect to the moving reference. Hence,

$$(\dot{\mathbf{A}})_{xyz} = \dot{A}_x\mathbf{i} + \dot{A}_y\mathbf{j} + \dot{A}_z\mathbf{k} \qquad (20\text{–}5)$$

When the time derivative of $\mathbf{A}$ is taken with respect to the fixed frame of reference, then we have

$$\dot{\mathbf{A}} = \dot{A}_x\mathbf{i} + \dot{A}_y\mathbf{j} + \dot{A}_z\mathbf{k} + A_x\dot{\mathbf{i}} + A_y\dot{\mathbf{j}} + A_z\dot{\mathbf{k}}$$

The time derivatives of the unit vectors will now be considered. For example, $\dot{\mathbf{i}} = d\mathbf{i}/dt$ represents only the change in the direction of $\mathbf{i}$ with respect to time, since $\mathbf{i}$ always has a magnitude of 1 unit. As shown in Fig. 20–6b, this change, $d\mathbf{i}$, is tangent to the path described by the arrowhead of $\mathbf{i}$, as $\mathbf{i}$ swings due to the rotation $\boldsymbol{\Omega}$. The rate of change in the magnitude of $d\mathbf{i}$ is $1(d\theta/dt) = 1\Omega$, and so we can account for both the rate of change in the magnitude and direction of $d\mathbf{i}$ using the cross product, $\dot{\mathbf{i}} = \boldsymbol{\Omega} \times \mathbf{i}$. In general, then*

$$\dot{\mathbf{i}} = \boldsymbol{\Omega} \times \mathbf{i} \qquad \dot{\mathbf{j}} = \boldsymbol{\Omega} \times \mathbf{j} \qquad \dot{\mathbf{k}} = \boldsymbol{\Omega} \times \mathbf{k}$$

Substituting these results into the above equation and using Eq. 20–5, we get

$$\boxed{\dot{\mathbf{A}} = (\dot{\mathbf{A}})_{xyz} + \boldsymbol{\Omega} \times \mathbf{A}} \qquad (20\text{–}6)$$

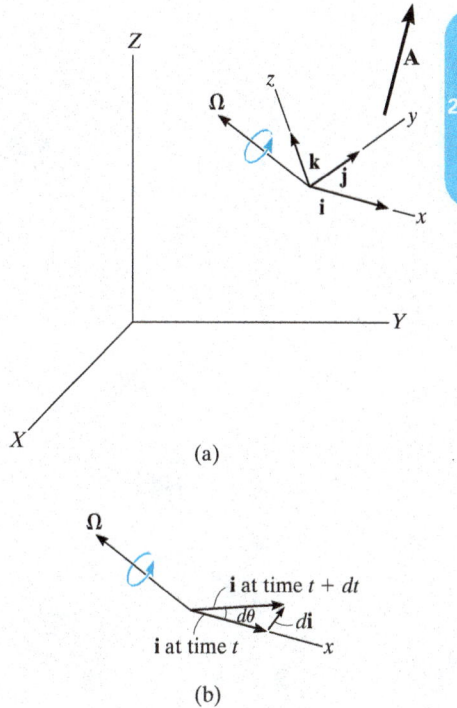

Fig. 20–6

This important result will be used throughout Sec. 20.4 and Chapter 21. It states that the time derivative of any vector $\mathbf{A}$ as observed from the fixed X, Y, Z frame of reference is equal to the time rate of change of $\mathbf{A}$ as observed from the x, y, z translating-rotating frame of reference, Eq. 20–5, plus $\boldsymbol{\Omega} \times \mathbf{A}$, the change of $\mathbf{A}$ caused by the rotation of the x, y, z frame. As a result, Eq. 20–6 should always be used whenever $\boldsymbol{\Omega}$ produces a change in the direction of $\mathbf{A}$ as seen from the X, Y, Z reference. Notice that if the x, y, z axes are translating, then $\boldsymbol{\Omega} = \mathbf{0}$, and so $\dot{\mathbf{A}} = (\dot{\mathbf{A}})_{xyz}$. In other words, the time rate of change of $\mathbf{A}$ as observed from both coordinate systems will be the *same*.

---

*These formulations were also developed in Sec. 16.8, regarding planar motion of the axes.

EXAMPLE 20.1

The disk shown in Fig. 20–7 spins about its axle with a constant angular velocity $\omega_s = 3$ rad/s, while the horizontal platform on which the disk is mounted rotates about the vertical axis at a constant rate $\omega_p = 1$ rad/s. Determine the angular acceleration of the disk and the velocity and acceleration of point A on the disk when it is in the position shown.

Fig. 20–7

SOLUTION

Point O represents a fixed point of rotation for the disk if one imagines an extension of the disk in the shape of a cone to this point. To determine the velocity and acceleration of point A, it is first necessary to determine the angular velocity $\boldsymbol{\omega}$ and angular acceleration $\boldsymbol{\alpha}$ of the disk, since these vectors are used in Eqs. 20–3 and 20–4.

Angular Velocity. The angular velocity, which is measured from X, Y, Z, is simply the vector addition of its two component motions. Thus,

$$\boldsymbol{\omega} = \boldsymbol{\omega}_s + \boldsymbol{\omega}_p = \{3\mathbf{j} - 1\mathbf{k}\} \text{ rad/s}$$

Angular Acceleration. Since the magnitude of $\boldsymbol{\omega}$ is constant, only a change in its direction, as seen from the fixed reference, creates the angular acceleration $\boldsymbol{\alpha}$ of the disk. One way to obtain $\boldsymbol{\alpha}$ is to calculate the time derivative of *each of the two components* of $\boldsymbol{\omega}$ using Eq. 20–6. At the instant shown in Fig. 20–7, imagine the fixed X, Y, Z and a rotating x, y, z frame to be coincident. If the rotating x, y, z frame is chosen to have an angular velocity of $\boldsymbol{\Omega} = \boldsymbol{\omega}_p = \{-1\mathbf{k}\}$ rad/s, then $\boldsymbol{\omega}_s$ will always be directed along the y (not Y) axis, and the time rate of change of $\boldsymbol{\omega}_s$ as seen from x, y, z is zero; i.e., $(\dot{\boldsymbol{\omega}}_s)_{xyz} = \mathbf{0}$ (the magnitude and direction of $\boldsymbol{\omega}_s$ is constant). Thus,

$$\dot{\boldsymbol{\omega}}_s = (\dot{\boldsymbol{\omega}}_s)_{xyz} + \boldsymbol{\omega}_p \times \boldsymbol{\omega}_s = \mathbf{0} + (-1\mathbf{k}) \times (3\mathbf{j}) = \{3\mathbf{i}\} \text{ rad/s}^2$$

By the same choice of axes rotation, $\boldsymbol{\Omega} = \boldsymbol{\omega}_p$, or even with $\boldsymbol{\Omega} = \mathbf{0}$, the time derivative $(\dot{\boldsymbol{\omega}}_p)_{xyz} = \mathbf{0}$, since $\boldsymbol{\omega}_p$ has a constant magnitude and direction with respect to x, y, z. Hence,

$$\dot{\boldsymbol{\omega}}_p = (\dot{\boldsymbol{\omega}}_p)_{xyz} + \boldsymbol{\omega}_p \times \boldsymbol{\omega}_p = \mathbf{0} + \mathbf{0} = \mathbf{0}$$

The angular acceleration of the disk is therefore

$$\boldsymbol{\alpha} = \dot{\boldsymbol{\omega}} = \dot{\boldsymbol{\omega}}_s + \dot{\boldsymbol{\omega}}_p = \{3\mathbf{i}\} \text{ rad/s}^2 \qquad \textit{Ans.}$$

Velocity and Acceleration. The velocity and acceleration of point A can be found using Eqs. 20–3 and 20–4. Realizing that $\mathbf{r}_A = \{1\mathbf{j} + 0.25\mathbf{k}\}$ m, Fig. 20–7, we have

$$\mathbf{v}_A = \boldsymbol{\omega} \times \mathbf{r}_A = (3\mathbf{j} - 1\mathbf{k}) \times (1\mathbf{j} + 0.25\mathbf{k}) = \{1.75\mathbf{i}\} \text{ m/s} \qquad \textit{Ans.}$$

$$\mathbf{a}_A = \boldsymbol{\alpha} \times \mathbf{r}_A + \boldsymbol{\omega} \times (\boldsymbol{\omega} \times \mathbf{r}_A)$$
$$= (3\mathbf{i}) \times (1\mathbf{j} + 0.25\mathbf{k}) + (3\mathbf{j} - 1\mathbf{k}) \times [(3\mathbf{j} - 1\mathbf{k}) \times (1\mathbf{j} + 0.25\mathbf{k})]$$
$$= \{-2.50\mathbf{j} - 2.25\mathbf{k}\} \text{ m/s}^2 \qquad \textit{Ans.}$$

EXAMPLE 20.2

At the instant $\theta = 60°$, the gyrotop in Fig. 20–8 has three components of angular motion directed as shown and having magnitudes defined as:

Spin: $\omega_s = 10$ rad/s, increasing at the rate of 6 rad/s^2
Nutation: $\omega_n = 3$ rad/s, increasing at the rate of 2 rad/s^2
Precession: $\omega_p = 5$ rad/s, increasing at the rate of 4 rad/s^2

Determine the angular velocity and angular acceleration of the top.

SOLUTION

Angular Velocity. The top rotates about the fixed point O. If the fixed and rotating frames are coincident at the instant shown, then the angular velocity can be expressed in terms of $\mathbf{i}, \mathbf{j}, \mathbf{k}$ components, referenced from the x, y, z frame; i.e.,

$$\boldsymbol{\omega} = -\omega_n \mathbf{i} + \omega_s \sin\theta \mathbf{j} + (\omega_p + \omega_s \cos\theta)\mathbf{k}$$
$$= -3\mathbf{i} + 10\sin 60°\mathbf{j} + (5 + 10\cos 60°)\mathbf{k}$$
$$= \{-3\mathbf{i} + 8.66\mathbf{j} + 10\mathbf{k}\} \text{ rad/s} \qquad Ans.$$

Angular Acceleration. As in the solution of Example 20.1, the angular acceleration $\boldsymbol{\alpha}$ will be determined by investigating separately the time rate of change of each of the angular velocity components. We will choose an $\boldsymbol{\Omega}$ for the x, y, z reference so that the component of $\boldsymbol{\omega}$ being considered is viewed as having a constant direction when observed from x, y, z.

Careful examination of the motion of the top reveals that $\boldsymbol{\omega}_s$ has a constant direction relative to x, y, z if these axes rotate at $\boldsymbol{\Omega} = \boldsymbol{\omega}_n + \boldsymbol{\omega}_p$. Thus,

$$\dot{\boldsymbol{\omega}}_s = (\dot{\boldsymbol{\omega}}_s)_{xyz} + (\boldsymbol{\omega}_n + \boldsymbol{\omega}_p) \times \boldsymbol{\omega}_s$$
$$= (6\sin 60°\mathbf{j} + 6\cos 60°\mathbf{k}) + (-3\mathbf{i} + 5\mathbf{k}) \times (10\sin 60°\mathbf{j} + 10\cos 60°\mathbf{k})$$
$$= \{-43.30\mathbf{i} + 20.20\mathbf{j} - 22.98\mathbf{k}\} \text{ rad/s}^2$$

Since $\boldsymbol{\omega}_n$ always lies in the fixed X–Y plane, this vector has a constant direction if the motion is viewed from axes x, y, z having a rotation of $\boldsymbol{\Omega} = \boldsymbol{\omega}_p$ (not $\boldsymbol{\Omega} = \boldsymbol{\omega}_s + \boldsymbol{\omega}_p$). Thus,

$$\dot{\boldsymbol{\omega}}_n = (\dot{\boldsymbol{\omega}}_n)_{xyz} + \boldsymbol{\omega}_p \times \boldsymbol{\omega}_n = -2\mathbf{i} + (5\mathbf{k}) \times (-3\mathbf{i}) = \{-2\mathbf{i} - 15\mathbf{j}\} \text{ rad/s}^2$$

Finally, the component $\boldsymbol{\omega}_p$ is always directed along the Z axis so that here it is not necessary to think of x, y, z as rotating, i.e., $\boldsymbol{\Omega} = 0$. We therefore have

$$\dot{\boldsymbol{\omega}}_p = (\dot{\boldsymbol{\omega}}_p)_{xyz} + 0 \times \boldsymbol{\omega}_p = \{4\mathbf{k}\} \text{ rad/s}^2$$

Thus, the angular acceleration of the top is

$$\boldsymbol{\alpha} = \dot{\boldsymbol{\omega}}_s + \dot{\boldsymbol{\omega}}_n + \dot{\boldsymbol{\omega}}_p = \{-45.3\mathbf{i} + 5.20\mathbf{j} - 19.0\mathbf{k}\} \text{ rad/s}^2 \qquad Ans.$$

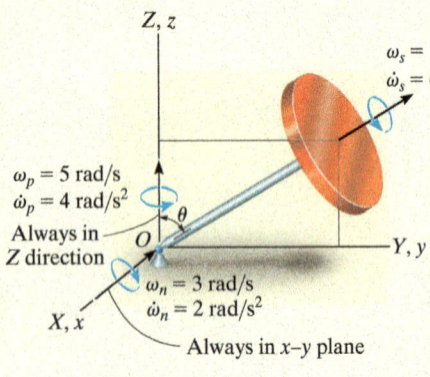

Fig. 20–8

20.3 GENERAL MOTION

Shown in Fig. 20–9 is a rigid body that has general motion in three dimensions for which the angular velocity is $\boldsymbol{\omega}$ and the angular acceleration is $\boldsymbol{\alpha}$. If point A has a known motion of $\mathbf{v}_A$ and $\mathbf{a}_A$, the motion of any other point B can be determined by using a relative-motion analysis. In this section a translating coordinate system will be used to define the relative motion, and in the next section a reference that is both rotating and translating will be considered.

If the origin of the translating system x, y, z ($\boldsymbol{\Omega} = \mathbf{0}$) is located at the "base point" A, then, at the instant shown, the motion of the body can be regarded as the sum of an instantaneous translation of the body having a motion of $\mathbf{v}_A$, and $\mathbf{a}_A$, and a rotation of the body about an instantaneous axis passing through point A. Since the body is rigid, the motion of point B measured by an observer located at A is therefore the same as the rotation of the body about a fixed point. This relative motion occurs about the instantaneous axis of rotation and is defined by $\mathbf{v}_{B/A} = \boldsymbol{\omega} \times \mathbf{r}_{B/A}$, Eq. 20–3, and $\mathbf{a}_{B/A} = \boldsymbol{\alpha} \times \mathbf{r}_{B/A} + \boldsymbol{\omega} \times (\boldsymbol{\omega} \times \mathbf{r}_{B/A})$, Eq. 20–4. For translating axes, the relative motions are related to absolute motions by $\mathbf{v}_B = \mathbf{v}_A + \mathbf{v}_{B/A}$ and $\mathbf{a}_B = \mathbf{a}_A + \mathbf{a}_{B/A}$, Eqs. 16–15 and 16–17, so that the absolute velocity and acceleration of point B can be determined from the equations

Fig. 20–9

$$\mathbf{v}_B = \mathbf{v}_A + \boldsymbol{\omega} \times \mathbf{r}_{B/A} \tag{20–7}$$

and

$$\mathbf{a}_B = \mathbf{a}_A + \boldsymbol{\alpha} \times \mathbf{r}_{B/A} + \boldsymbol{\omega} \times (\boldsymbol{\omega} \times \mathbf{r}_{B/A}) \tag{20–8}$$

These two equations are essentially the same as those describing the general plane motion of a rigid body, Eqs. 16–16 and 16–18. However, difficulty in application arises for three-dimensional motion, because $\boldsymbol{\alpha}$ must measure the change in *both* the magnitude and direction of $\boldsymbol{\omega}$.

For many applications, an easy way to obtain $\mathbf{v}_B$ and $\mathbf{a}_B$ is to note that $\mathbf{v}_{B/A} = \mathbf{v}_B - \mathbf{v}_A$, and so Eq. 20–7 becomes $\mathbf{v}_{B/A} = \boldsymbol{\omega} \times \mathbf{r}_{B/A}$. The cross product indicates that $\mathbf{v}_{B/A}$ is *perpendicular* to both $\mathbf{r}_{B/A}$ and $\boldsymbol{\omega}$, and so, by Eq. B–14 of Appendix B, we require

$$\mathbf{r}_{B/A} \cdot \mathbf{v}_{B/A} = 0 \tag{20–9}$$

Taking the time derivative to obtain $\mathbf{a}_{B/A}$, we have

$$\mathbf{v}_{B/A} \cdot \mathbf{v}_{B/A} + \mathbf{r}_{B/A} \cdot \mathbf{a}_{B/A} = 0 \tag{20–10}$$

Then $\mathbf{a}_B = \mathbf{a}_A + \mathbf{a}_{B/A}$.

Solution II of the following example illustrates application of this idea.

EXAMPLE 20.3

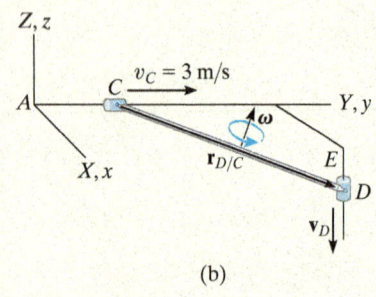

Fig. 20–10

If the collar at C in Fig. 20–10a moves toward B with a speed of 3 m/s, determine the velocity of the collar at D and the angular velocity of the bar at the instant shown. The bar is connected to the collars at its end points by ball-and-socket joints.

SOLUTION I
Bar CD is subjected to general motion. Why? The velocity of point D on the bar can be related to the velocity of point C by the equation

$$\mathbf{v}_D = \mathbf{v}_C + \boldsymbol{\omega} \times \mathbf{r}_{D/C}$$

The fixed and translating frames of reference are assumed to coincide at the instant considered, Fig. 20–10b. We have

$$\mathbf{v}_D = -v_D \mathbf{k} \qquad \mathbf{v}_C = \{3\mathbf{j}\} \text{ m/s}$$

$$\mathbf{r}_{D/C} = \{1\mathbf{i} + 2\mathbf{j} - 0.5\mathbf{k}\} \text{ m} \qquad \boldsymbol{\omega} = \omega_x \mathbf{i} + \omega_y \mathbf{j} + \omega_z \mathbf{k}$$

Substituting into the above equation we get

$$-v_D \mathbf{k} = 3\mathbf{j} + \begin{vmatrix} \mathbf{i} & \mathbf{j} & \mathbf{k} \\ \omega_x & \omega_y & \omega_z \\ 1 & 2 & -0.5 \end{vmatrix}$$

Expanding and equating the respective **i**, **j**, **k** components yields

$$-0.5\omega_y - 2\omega_z = 0 \qquad (1)$$

$$0.5\omega_x + 1\omega_z + 3 = 0 \qquad (2)$$

$$2\omega_x - 1\omega_y + v_D = 0 \qquad (3)$$

These equations contain four unknowns.* A fourth equation can be written if the direction of $\boldsymbol{\omega}$ is specified. In this regard, any component of $\boldsymbol{\omega}$ acting along the bar's axis has no effect on moving the collars, rather this component only spins the bar about its axis. Therefore, if $\boldsymbol{\omega}$ is specified as acting perpendicular to the axis of the bar, then $\boldsymbol{\omega}$ must have a unique magnitude to satisfy the above equations. Hence,

$$\boldsymbol{\omega} \cdot \mathbf{r}_{D/C} = (\omega_x \mathbf{i} + \omega_y \mathbf{j} + \omega_z \mathbf{k}) \cdot (1\mathbf{i} + 2\mathbf{j} - 0.5\mathbf{k}) = 0$$

$$1\omega_x + 2\omega_y - 0.5\omega_z = 0 \qquad (4)$$

*Although this is the case, the magnitude of v_D can be obtained. For example, solve Eqs. 1 and 2 for ω_y and ω_x in terms of ω_z and substitute this into Eq. 3. You will find ω_z will cancel out, which will allow a solution for v_D.

Solving Eqs. 1 through 4 simultaneously yields

$$\omega_x = -4.86 \text{ rad/s} \quad \omega_y = 2.29 \text{ rad/s} \quad \omega_z = -0.571 \text{ rad/s}$$

$$v_D = 12.0 \text{ m/s, so that} \quad \omega = 5.40 \text{ rad/s} \quad \textit{Ans.}$$

SOLUTION II

Applying Eq. 20–9, $\mathbf{v}_{D/C} = \mathbf{v}_D - \mathbf{v}_C = -v_D \mathbf{k} - 3\mathbf{j}$, so that

$$\mathbf{r}_{D/C} \cdot \mathbf{v}_{D/C} = (1\mathbf{i} + 2\mathbf{j} - 0.5\mathbf{k}) \cdot (-v_D \mathbf{k} - 3\mathbf{j}) = 0$$

$$(1)(0) + (2)(-3) + (-0.5)(-v_D) = 0$$

$$v_D = 12 \text{ m/s} \quad \textit{Ans.}$$

Since $\boldsymbol{\omega}$ is *perpendicular* to $\mathbf{r}_{D/C}$, then from Eq. 20–7, $\mathbf{v}_{D/C} = \boldsymbol{\omega} \times \mathbf{r}_{D/C}$. We have

$$v_{D/C} = \omega \, r_{D/C}$$

$$\sqrt{(-12)^2 + (-3)^2} = \omega \sqrt{(1)^2 + (2)^2 + (-0.5)^2}$$

$$\omega = 5.40 \text{ rad/s} \quad \textit{Ans.}$$

PROBLEMS

20–1. The propeller of an airplane is rotating at a constant speed $\omega_x \mathbf{i}$, while the plane is undergoing a turn at a constant rate ω_t. Determine the angular acceleration of the propeller if (a) the turn is horizontal, i.e., $\omega_t \mathbf{k}$, and (b) the turn is vertical, downward, i.e., $\omega_t \mathbf{j}$.

20–2. The disk rotates about the z axis at a constant rate $\omega_z = 0.5$ rad/s without slipping on the horizontal plane. Determine the velocity and the acceleration of point A on the disk.

Prob. 20–1

Prob. 20–2

20–3. The ladder of the fire truck rotates around the z axis with an angular velocity $\omega_1 = 0.15$ rad/s, which is increasing at 0.8 rad/s². At the same instant it is rotating upward at a constant rate $\omega_2 = 0.6$ rad/s. Determine the velocity and acceleration of point A located at the top of the ladder at this instant.

*20–4. The ladder of the fire truck rotates around the z axis with an angular velocity of $\omega_1 = 0.15$ rad/s, which is increasing at 0.2 rad/s². At the same instant it is rotating upward at $\omega_2 = 0.6$ rad/s while increasing at 0.4 rad/s². Determine the velocity and acceleration of point A located at the top of the ladder at this instant.

Probs. 20–3/4

20–5. The cone rolls in a circle and rotates about the z axis at a constant rate $\omega_z = 8$ rad/s. Determine the angular velocity and angular acceleration of the cone if it rolls without slipping. Also, what are the velocity and acceleration of point A?

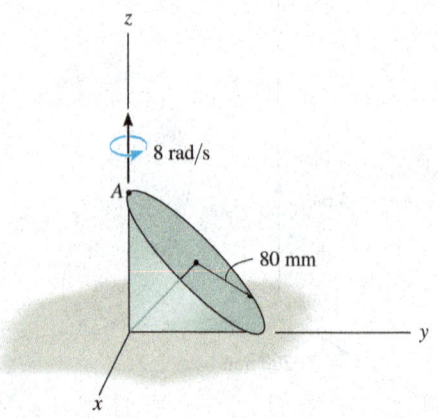

Prob. 20–5

20–6. The antenna is following the motion of a jet plane. At the instant $\theta = 25°$ and $\phi = 75°$, the constant angular rates of change are $\dot{\theta} = 0.4$ rad/s and $\dot{\phi} = 0.6$ rad/s. Determine the velocity and acceleration of the signal horn A at this instant. The distance OA is 0.8 m.

Prob. 20–6

20–7. At a given instant, the antenna has an angular motion $\omega_1 = 3$ rad/s and $\dot{\omega}_1 = 2$ rad/s² about the z axis. At this same instant, $\theta = 30°$, the angular motion about the x axis is $\omega_2 = 1.5$ rad/s and $\dot{\omega}_2 = 4$ rad/s². Determine the velocity and acceleration of the signal horn A at this instant. The distance from O to A is $d = 3$ ft.

Prob. 20–7

*20–8. The disk rotates about the shaft S, while the shaft is turning about the z axis at a rate of $\omega_z = 4$ rad/s, which is increasing at 2 rad/s². Determine the velocity and acceleration of point A on the disk at the instant shown. No slipping occurs.

20–9. The disk rotates about the shaft S, while the shaft is turning about the z axis at a rate of $\omega_z = 4$ rad/s, which is increasing at 2 rad/s². Determine the velocity and acceleration of point B on the disk at the instant shown. No slipping occurs.

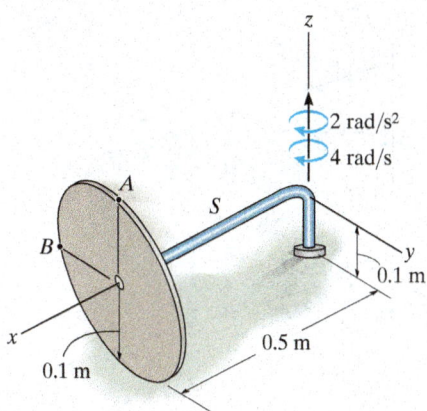

Probs. 20–8/9

20–10. The electric fan is mounted on a swivel support such that the fan rotates about the z axis at a constant rate of $\omega_z = 1$ rad/s and the fan blade is spinning at a constant rate $\omega_s = 60$ rad/s. If $\phi = 45°$ for the motion, determine the angular velocity and the angular acceleration of the blade.

20–11. The electric fan is mounted on a swivel support such that the fan rotates about the z axis at a constant rate of $\omega_z = 1$ rad/s and the fan blade is spinning at a constant rate $\omega_s = 60$ rad/s. If at the instant $\phi = 45°$, $\dot{\phi} = 2$ rad/s for the motion, determine the angular velocity and the angular acceleration of the blade.

Probs. 20–10/11

*20–12. The motion of the top is such that at the instant shown it rotates about the z axis at $\omega_1 = 0.6$ rad/s, while it spins at $\omega_2 = 8$ rad/s. Determine the angular velocity and angular acceleration of the top at this instant. Express the result as a Cartesian vector.

Prob. 20–12

20–13. The bevel gear A rolls on the fixed gear B. If at the instant shown the shaft to which A is attached is rotating at 2 rad/s and has an angular acceleration of 4 rad/s², determine the angular velocity and angular acceleration of gear A.

Prob. 20–13

20–14. If the frame rotates with a constant angular velocity of $\omega_p = \{-10\mathbf{k}\}$ rad/s and the horizontal gear B rotates with a constant angular velocity of $\omega_B = \{5\mathbf{k}\}$ rad/s, determine the angular velocity and angular acceleration of the bevel gear A.

Prob. 20–14

20–15. The telescope is mounted on the frame F that allows it to be directed to any point in the sky. At the instant $\theta = 30°$, the frame has an angular acceleration of $\alpha_{y'} = 0.2$ rad/s² and an angular velocity of $\omega_{y'} = 0.3$ rad/s about the y' axis, and $\ddot{\theta} = 0.5$ rad/s² while $\dot{\theta} = 0.4$ rad/s. Determine the velocity and acceleration of the instrument capsule at C at this instant.

Prob. 20–15

*20–16. The truncated cone rotates about the z axis at a constant rate $\omega_z = 0.4$ rad/s without slipping on the horizontal plane. Determine the velocity and acceleration of point A on the cone.

20–17. The truncated cone rotates about the z axis at $\omega_z = 0.4$ rad/s without slipping on the horizontal plane. If at this same instant ω_z is increasing at $\dot{\omega}_z = 0.5$ rad/s², determine the velocity and acceleration of point A on the cone.

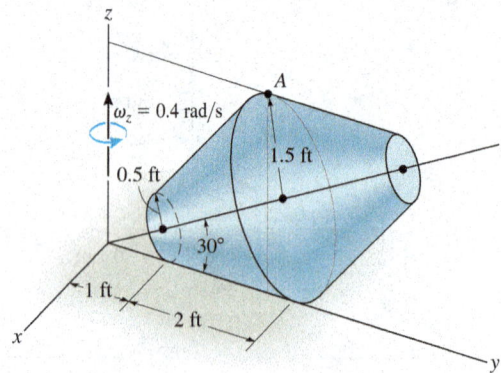

Probs. 20–16/17

20–18. Gear C is driven by shaft DE, while gear B spins freely about its axle GF, which precesses freely about shaft DE. If gear A is held fixed ($\omega_A = 0$), and shaft DE rotates with a constant angular velocity of $\omega_{DE} = 10$ rad/s, determine the angular velocity of gear B.

Prob. 20–18

20–19. Shaft BD is connected to a ball-and-socket joint at B, and a beveled gear A is attached to its other end. The gear is in mesh with a fixed gear C. If the shaft and gear A are *spinning* with a constant angular velocity $\omega_1 = 8$ rad/s, determine the angular velocity and angular acceleration of gear A.

Prob. 20–19

*20–20. Gear B is driven by a motor mounted on turntable C. If gear A is held fixed, and the motor shaft rotates with a constant angular velocity of $\omega_y = 30$ rad/s, determine the angular velocity and angular acceleration of gear B.

20–21. Gear B is driven by a motor mounted on turntable C. If gear A and the motor shaft rotate with constant angular speeds of $\omega_A = \{10\mathbf{k}\}$ rad/s and $\omega_y = \{30\mathbf{j}\}$ rad/s, respectively, determine the angular velocity and angular acceleration of gear B.

Probs. 20–20/21

20–22. The crane boom OA rotates about the z axis with a constant angular velocity of $\omega_1 = 0.15$ rad/s, while it is rotating downward with a constant angular velocity of $\omega_2 = 0.2$ rad/s. Determine the velocity and acceleration of point A located at the end of the boom at the instant shown.

Prob. 20–22

20–23. The differential of an automobile allows the two rear wheels to rotate at different speeds when the automobile travels along a curve. For operation, the rear axles are attached to the wheels at one end and have beveled gears A and B on their other ends. The differential case D is placed over the left axle but can rotate about C independent of the axle. The case supports a pinion gear E on a shaft, which meshes with gears A and B. Finally, a ring gear G is *fixed* to the differential case so that the case rotates with the ring gear when the latter is driven by the drive pinion H. This gear, like the differential case, is free to rotate about the left wheel axle. If the drive pinion is turning at $\omega_H = 100$ rad/s and the pinion gear E is spinning about its shaft at $\omega_E = 30$ rad/s, determine the angular velocity, ω_A and ω_B, of each axle.

Prob. 20–23

*20–24. The rod AB is attached to collars at its ends by ball-and-socket joints. If collar A has a velocity $v_A = 15$ ft/s at the instant shown, determine the velocity of collar B.

20–25. The rod AB is attached to collars at its ends by ball-and-socket joints. If collar A has an acceleration of $a_A = 2$ ft/s^2 at the instant shown, determine the acceleration of collar B.

*20–28. Rod AB is attached to the rotating arm using ball-and-socket joints. If AC is rotating with a constant angular velocity of 8 rad/s about the pin at C, determine the angular velocity of link BD at the instant shown.

20–29. Rod AB is attached to the rotating arm using ball-and-socket joints. If AC is rotating about point C with an angular velocity of 8 rad/s and has an angular acceleration of 6 rad/s^2 at the instant shown, determine the angular velocity and angular acceleration of link BD at this instant.

Probs. 20–24/25

20–26. Rod AB is attached to collars at its ends by using ball-and-socket joints. If collar A moves along the fixed rod at $v_A = 5$ m/s, determine the angular velocity of the rod and the velocity of collar B at the instant shown. Assume that the rod's angular velocity is directed perpendicular to the axis of the rod.

20–27. Rod AB is attached to collars at its ends by using ball-and-socket joints. If collar A moves along the fixed rod with a velocity of $v_A = 5$ m/s and has an acceleration $a_A = 2$ m/s^2 at the instant shown, determine the angular acceleration of the rod and the acceleration of collar B at this instant. Assume that the rod's angular velocity and angular acceleration are directed perpendicular to the axis of the rod.

Probs. 20–28/29

20–30. Rod AB is attached to collars at its ends by ball-and-socket joints. If collar A has a speed $v_A = 4$ m/s, determine the speed of collar B at the instant $z = 2$ m. Assume the angular velocity of the rod is directed perpendicular to the rod.

Probs. 20–26/27

Prob. 20–30

20–31. Rod CD is attached to the rotating arms using ball-and-socket joints. If AC has the motion shown, determine the angular velocity of link BD at the instant shown.

20–32. Rod CD is attached to the rotating arms using ball-and-socket joints. If AC has the motion shown, determine the angular acceleration of link BD at this instant.

20–35. Disk A rotates at a constant angular velocity of 10 rad/s. If rod BC is joined to the disk and a collar by ball-and-socket joints, determine the velocity of collar B at the instant shown. Also, what is the rod's angular velocity ω_{BC} if it is directed perpendicular to the axis of the rod?

Probs. 20–31/32

20–33. If the rod is attached with ball-and-socket joints to smooth collars A and B at its end points, determine the velocity of B at the instant shown if A is moving upward at a constant speed of $v_A = 5$ ft/s. Also, determine the angular velocity of the rod if it is directed perpendicular to the axis of the rod.

*20–34. If the collar A in Prob. 20–33 is moving upward with an acceleration of $\mathbf{a}_A = \{-2\mathbf{k}\}$ ft/s² at the instant its speed is $v_A = 5$ ft/s, determine the acceleration of collar B at this instant.

Prob. 20–35

*20–36. Solve Prob. 20–33 if the connection at B consists of a pin as shown in the figure below, rather than a ball-and-socket joint. *Hint:* The constraint allows rotation of the rod both along the bar (**j** direction) and along the axis of the pin (**n** direction). Since there is no rotational component in the **u** direction, i.e., perpendicular to **n** and **j** where $\mathbf{u} = \mathbf{j} \times \mathbf{n}$, an additional equation for solution can be obtained from $\boldsymbol{\omega} \cdot \mathbf{u} = 0$. The vector **n** is in the same direction as $\mathbf{r}_{D/B} \times \mathbf{r}_{C/B}$.

Probs. 20–33/34

Prob. 20–36

*20.4 RELATIVE-MOTION ANALYSIS USING TRANSLATING AND ROTATING AXES

The most general way to analyze the three-dimensional motion of a rigid body requires the use of x, y, z axes that both translate and rotate relative to a second frame X, Y, Z. This analysis also provides a means to determine the motions of two points A and B located on separate members of a mechanism, and the relative motion of one particle with respect to another when one or both particles are moving along *curved paths.*

As shown in Fig. 20–11, the locations of points A and B are specified relative to the X, Y, Z frame of reference by position vectors $\mathbf{r}_A$ and $\mathbf{r}_B$. The base point A represents the origin of the x, y, z coordinate system, which is translating and rotating with respect to X, Y, Z. At the instant considered, the velocity and acceleration of point A are $\mathbf{v}_A$ and $\mathbf{a}_A$, and the angular velocity and angular acceleration of the x, y, z axes are $\boldsymbol{\Omega}$ and $\dot{\boldsymbol{\Omega}} = d\boldsymbol{\Omega}/dt$. All these vectors are *measured* with respect to the X, Y, Z frame of reference, although they can be expressed in Cartesian component form along either set of axes.

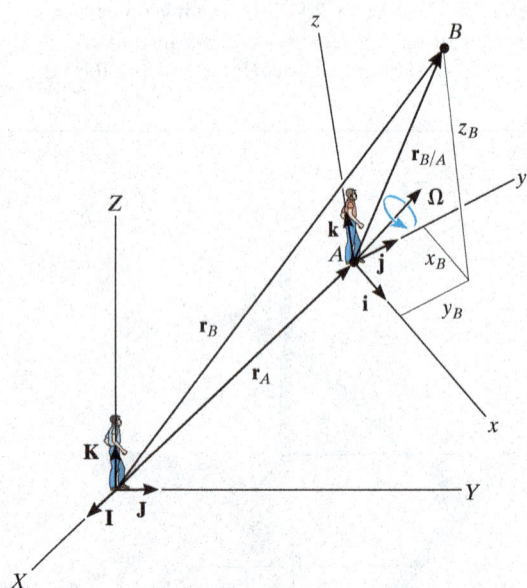

Fig. 20–11

*20.4 Relative-Motion Analysis Using Translating and Rotating Axes

Position. If the position of "B with respect to A" is specified by the *relative-position vector* $\mathbf{r}_{B/A}$, Fig. 20–11, then, by vector addition,

$$\mathbf{r}_B = \mathbf{r}_A + \mathbf{r}_{B/A} \qquad (20\text{–}11)$$

where

$\mathbf{r}_B$ = position of B
$\mathbf{r}_A$ = position of the origin A
$\mathbf{r}_{B/A}$ = position of "B with respect to A"

Velocity. The velocity of point B measured from X, Y, Z can be determined by taking the time derivative of Eq. 20–11,

$$\dot{\mathbf{r}}_B = \dot{\mathbf{r}}_A + \dot{\mathbf{r}}_{B/A}$$

The first two terms represent $\mathbf{v}_B$ and $\mathbf{v}_A$. The last term must be evaluated by applying Eq. 20–6, since $\mathbf{r}_{B/A}$ is measured from the rotating reference. Hence,

$$\dot{\mathbf{r}}_{B/A} = (\dot{\mathbf{r}}_{B/A})_{xyz} + \mathbf{\Omega} \times \mathbf{r}_{B/A} = (\mathbf{v}_{B/A})_{xyz} + \mathbf{\Omega} \times \mathbf{r}_{B/A} \quad (20\text{–}12)$$

Therefore,

$$\mathbf{v}_B = \mathbf{v}_A + \mathbf{\Omega} \times \mathbf{r}_{B/A} + (\mathbf{v}_{B/A})_{xyz} \qquad (20\text{–}13)$$

where

$\mathbf{v}_B$ = velocity of B
$\mathbf{v}_A$ = velocity of the origin A of the x, y, z frame of reference
$(\mathbf{v}_{B/A})_{xyz}$ = velocity of "B with respect to A" as measured by an observer attached to the rotating x, y, z frame of reference
$\mathbf{\Omega}$ = angular velocity of the x, y, z frame of reference
$\mathbf{r}_{B/A}$ = position of "B with respect to A"

Acceleration. The acceleration of point B measured from X, Y, Z is determined by taking the time derivative of Eq. 20–13.

$$\dot{\mathbf{v}}_B = \dot{\mathbf{v}}_A + \dot{\boldsymbol{\Omega}} \times \mathbf{r}_{B/A} + \boldsymbol{\Omega} \times \dot{\mathbf{r}}_{B/A} + \frac{d}{dt}(\mathbf{v}_{B/A})_{xyz}$$

The term on the left and the first term on the right represent $\mathbf{a}_B$ and $\mathbf{a}_A$, respectively. The last term can be evaluated using Eq. 20–12, and the last term is evaluated by applying Eq. 20–6, which yields

$$\frac{d}{dt}(\mathbf{v}_{B/A})_{xyz} = (\dot{\mathbf{v}}_{B/A})_{xyz} + \boldsymbol{\Omega} \times (\mathbf{v}_{B/A})_{xyz} = (\mathbf{a}_{B/A})_{xyz} + \boldsymbol{\Omega} \times (\mathbf{v}_{B/A})_{xyz}$$

Here $(\mathbf{a}_{B/A})_{xyz}$ is the acceleration of B with respect to A measured from x, y, z. Substituting this result and Eq. 20–12 into the above equation and simplifying, we have

$$\boxed{\mathbf{a}_B = \mathbf{a}_A + \dot{\boldsymbol{\Omega}} \times \mathbf{r}_{B/A} + \boldsymbol{\Omega} \times (\boldsymbol{\Omega} \times \mathbf{r}_{B/A}) + 2\boldsymbol{\Omega} \times (\mathbf{v}_{B/A})_{xyz} + (\mathbf{a}_{B/A})_{xyz}}$$

$$(20\text{–}14)$$

where

$\mathbf{a}_B$ = acceleration of B
$\mathbf{a}_A$ = acceleration of the origin A of the x, y, z frame of reference
$(\mathbf{a}_{B/A})_{xyz}, (\mathbf{v}_{B/A})_{xyz}$ = relative acceleration and relative velocity of "B with respect to A" as measured by an observer attached to the rotating x, y, z frame of reference
$\dot{\boldsymbol{\Omega}}, \boldsymbol{\Omega}$ = angular acceleration and angular velocity of the x, y, z frame of reference
$\mathbf{r}_{B/A}$ = position of "B with respect to A"

Complicated spatial motion of the concrete bucket B occurs due to the rotation of the boom about the Z axis, motion of the carriage A along the boom, and extension and swinging of the cable AB. A translating-rotating x, y, z coordinate system can be established on the carriage, and a relative-motion analysis can then be applied to study this motion.

Equations 20–13 and 20–14 are identical to those used in Sec. 16.8 for analyzing relative plane motion.* In that case, however, application is simplified since $\boldsymbol{\Omega}$ and $\dot{\boldsymbol{\Omega}}$ have a *constant direction* which is always perpendicular to the plane of motion. For three-dimensional motion, $\dot{\boldsymbol{\Omega}}$ must be determined by using Eq. 20–6, since $\dot{\boldsymbol{\Omega}}$ depends on the change in *both* the magnitude and direction of $\boldsymbol{\Omega}$.

*Refer to Sec. 16.8 for an interpretation of the terms.

PROCEDURE FOR ANALYSIS

Three-dimensional motion of particles or rigid bodies can be analyzed with Eqs. 20–13 and 20–14 by using the following procedure.

Coordinate Axes.
- Select the location and orientation of the X, Y, Z and x, y, z coordinate axes. Most often solutions can be easily obtained if at the instant considered:
  (1) the origins are coincident
  (2) the axes are collinear
  (3) the axes are parallel

- If several components of angular velocity are involved in a problem, the calculations will be reduced if the x, y, z axes are selected such that only one component of angular velocity is observed with respect to this frame ($\mathbf{\Omega}_{xyz}$) and the frame rotates with $\mathbf{\Omega}$ defined by the other components of angular velocity.

Kinematic Equations.
- After the origin of the moving reference, A, is defined and the moving point B is specified, Eqs. 20–13 and 20–14 should then be written in symbolic form as

$$\mathbf{v}_B = \mathbf{v}_A + \mathbf{\Omega} \times \mathbf{r}_{B/A} + (\mathbf{v}_{B/A})_{xyz}$$

$$\mathbf{a}_B = \mathbf{a}_A + \dot{\mathbf{\Omega}} \times \mathbf{r}_{B/A} + \mathbf{\Omega} \times (\mathbf{\Omega} \times \mathbf{r}_{B/A}) + 2\mathbf{\Omega} \times (\mathbf{v}_{B/A})_{xyz} + (\mathbf{a}_{B/A})_{xyz}$$

- If $\mathbf{r}_A$ and $\mathbf{\Omega}$ appear to *change direction* when observed from the fixed X, Y, Z reference, then use a set of primed reference axes, x', y', z', having a rotation $\mathbf{\Omega}' = \mathbf{\Omega}$. Equation 20–6 is then used to determine $\dot{\mathbf{\Omega}}$ and the motion $\mathbf{v}_A$ and $\mathbf{a}_A$ of the origin of the moving x, y, z axes.

- If $\mathbf{r}_{B/A}$ and $\mathbf{\Omega}_{xyz}$ appear to change direction as observed from x, y, z, then use a set of double-primed reference axes x'', y'', z'', having a rotation $\mathbf{\Omega}'' = \mathbf{\Omega}_{xyz}$, and apply Eq. 20–6 to determine $\dot{\mathbf{\Omega}}_{xyz}$ and the relative motion $(\mathbf{v}_{B/A})_{xyz}$ and $(\mathbf{a}_{B/A})_{xyz}$.

- After the final forms of $\dot{\mathbf{\Omega}}$, $\mathbf{v}_A$, $\mathbf{a}_A$, $\dot{\mathbf{\Omega}}_{xyz}$, $(\mathbf{v}_{B/A})_{xyz}$, and $(\mathbf{a}_{B/A})_{xyz}$ are obtained, numerical problem data can be substituted and the terms evaluated. The components of all these vectors can be selected either along the X, Y, Z or along the x, y, z axes. The choice is arbitrary, provided a consistent set of unit vectors is used.

EXAMPLE 20.4

A motor and attached rod AB have the angular motions shown in Fig. 20–12. A collar C on the rod is located 0.25 m from A and is moving downward along the rod with a velocity of 3 m/s and an acceleration of 2 m/s². Determine the velocity and acceleration of C at this instant.

SOLUTION

Coordinate Axes. The origin of the fixed X, Y, Z reference is chosen at the center of the platform, and the origin of the moving x, y, z frame is at point A, Fig. 20–12. Since the collar is subjected to two components of angular motion, $\boldsymbol{\omega}_p$ and $\boldsymbol{\omega}_M$, it will be viewed as having an angular velocity of $\boldsymbol{\Omega}_{xyz} = \boldsymbol{\omega}_M$ in x, y, z. Therefore, the x, y, z axes will be attached to the platform so that $\boldsymbol{\Omega} = \boldsymbol{\omega}_p$.

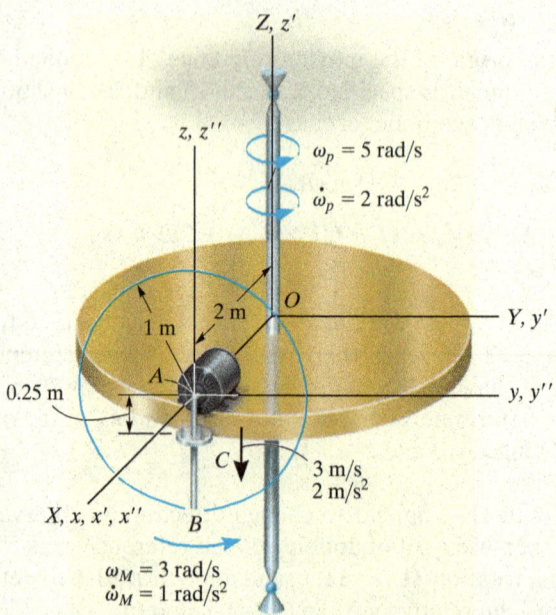

Fig. 20–12

Kinematic Equations. Equations 20–13 and 20–14, applied to points C and A, become

$$\mathbf{v}_C = \mathbf{v}_A + \boldsymbol{\Omega} \times \mathbf{r}_{C/A} + (\mathbf{v}_{C/A})_{xyz}$$
$$\mathbf{a}_C = \mathbf{a}_A + \dot{\boldsymbol{\Omega}} \times \mathbf{r}_{C/A} + \boldsymbol{\Omega} \times (\boldsymbol{\Omega} \times \mathbf{r}_{C/A}) + 2\boldsymbol{\Omega} \times (\mathbf{v}_{C/A})_{xyz} + (\mathbf{a}_{C/A})_{xyz}$$

Motion of A. Here $\mathbf{r}_A$ changes direction relative to X, Y, Z. To find the time derivatives of $\mathbf{r}_A$ we will use a set of x', y', z' axes coincident with the X, Y, Z axes that rotate at $\boldsymbol{\Omega}' = \boldsymbol{\omega}_p$. Thus,

$$\boldsymbol{\Omega} = \boldsymbol{\omega}_p = \{5\mathbf{k}\} \text{ rad/s } (\boldsymbol{\Omega} \text{ does not change direction relative to } X, Y, Z.)$$
$$\dot{\boldsymbol{\Omega}} = \dot{\boldsymbol{\omega}}_p = \{2\mathbf{k}\} \text{ rad/s}^2$$
$$\mathbf{r}_A = \{2\mathbf{i}\} \text{ m}$$
$$\mathbf{v}_A = \dot{\mathbf{r}}_A = (\dot{\mathbf{r}}_A)_{x'y'z'} + \boldsymbol{\omega}_p \times \mathbf{r}_A = 0 + 5\mathbf{k} \times 2\mathbf{i} = \{10\mathbf{j}\} \text{ m/s}$$
$$\mathbf{a}_A = \ddot{\mathbf{r}}_A = [(\ddot{\mathbf{r}}_A)_{x'y'z'} + \boldsymbol{\omega}_p \times (\dot{\mathbf{r}}_A)_{x'y'z'}] + \dot{\boldsymbol{\omega}}_p \times \mathbf{r}_A + \boldsymbol{\omega}_p \times \dot{\mathbf{r}}_A$$
$$= [0 + 0] + 2\mathbf{k} \times 2\mathbf{i} + 5\mathbf{k} \times 10\mathbf{j} = \{-50\mathbf{i} + 4\mathbf{j}\} \text{ m/s}^2$$

Motion of C with Respect to A. Here $\mathbf{r}_{C/A}$ changes direction relative to x, y, z, and so to find its time derivatives use a set of x'', y'', z'' axes that rotate at $\boldsymbol{\Omega}'' = \boldsymbol{\Omega}_{xyz} = \boldsymbol{\omega}_M$. Thus,

$$\boldsymbol{\Omega}_{xyz} = \boldsymbol{\omega}_M = \{3\mathbf{i}\} \text{ rad/s } (\boldsymbol{\Omega}_{xyz} \text{ does not change direction relative to } x, y, z.)$$
$$\dot{\boldsymbol{\Omega}}_{xyz} = \dot{\boldsymbol{\omega}}_M = \{1\mathbf{i}\} \text{ rad/s}^2$$
$$\mathbf{r}_{C/A} = \{-0.25\mathbf{k}\} \text{ m}$$
$$(\mathbf{v}_{C/A})_{xyz} = (\dot{\mathbf{r}}_{C/A})_{xyz} = (\dot{\mathbf{r}}_{C/A})_{x''y''z''} + \boldsymbol{\omega}_M \times \mathbf{r}_{C/A}$$
$$= -3\mathbf{k} + [3\mathbf{i} \times (-0.25\mathbf{k})] = \{0.75\mathbf{j} - 3\mathbf{k}\} \text{ m/s}$$
$$(\mathbf{a}_{C/A})_{xyz} = (\ddot{\mathbf{r}}_{C/A})_{xyz} = [(\ddot{\mathbf{r}}_{C/A})_{x''y''z''} + \boldsymbol{\omega}_M \times (\dot{\mathbf{r}}_{C/A})_{x''y''z''}] + \dot{\boldsymbol{\omega}}_M \times \mathbf{r}_{C/A} + \boldsymbol{\omega}_M \times (\dot{\mathbf{r}}_{C/A})_{xyz}$$
$$= [-2\mathbf{k} + 3\mathbf{i} \times (-3\mathbf{k})] + (1\mathbf{i}) \times (-0.25\mathbf{k}) + (3\mathbf{i}) \times (0.75\mathbf{j} - 3\mathbf{k})$$
$$= \{18.25\mathbf{j} + 0.25\mathbf{k}\} \text{ m/s}^2$$

Motion of C.
$$\mathbf{v}_C = \mathbf{v}_A + \boldsymbol{\Omega} \times \mathbf{r}_{C/A} + (\mathbf{v}_{C/A})_{xyz}$$
$$= 10\mathbf{j} + [5\mathbf{k} \times (-0.25\mathbf{k})] + (0.75\mathbf{j} - 3\mathbf{k})$$
$$= \{10.75\mathbf{j} - 3\mathbf{k}\} \text{ m/s} \qquad \textit{Ans.}$$
$$\mathbf{a}_C = \mathbf{a}_A + \dot{\boldsymbol{\Omega}} \times \mathbf{r}_{C/A} + \boldsymbol{\Omega} \times (\boldsymbol{\Omega} \times \mathbf{r}_{C/A}) + 2\boldsymbol{\Omega} \times (\mathbf{v}_{C/A})_{xyz} + (\mathbf{a}_{C/A})_{xyz}$$
$$= (-50\mathbf{i} + 4\mathbf{j}) + [2\mathbf{k} \times (-0.25\mathbf{k})] + 5\mathbf{k} \times [5\mathbf{k} \times (-0.25\mathbf{k})]$$
$$\quad + 2[5\mathbf{k} \times (0.75\mathbf{j} - 3\mathbf{k})] + (18.25\mathbf{j} + 0.25\mathbf{k})$$
$$= \{-57.5\mathbf{i} + 22.25\mathbf{j} + 0.25\mathbf{k}\} \text{ m/s}^2 \qquad \textit{Ans.}$$

EXAMPLE 20.5

Fig. 20-13

The pendulum shown in Fig. 20–13 consists of two rods; AB is pin supported at A and swings only in the Y–Z plane, whereas a bearing at B allows the attached rod BD to spin about rod AB. At a given instant, the rods have the angular motions shown. Also, a collar C, located 0.2 m from B, has a velocity of 3 m/s and an acceleration of 2 m/s^2 along the rod. Determine the velocity and acceleration of the collar at this instant.

SOLUTION I

Coordinate Axes. The origin of the fixed X, Y, Z frame will be placed at A. Motion of the collar is conveniently observed from B, so the origin of the x, y, z frame is located at this point. We will choose $\Omega = \omega_1$ and $\Omega_{xyz} = \omega_2$.

Kinematic Equations.
$$\mathbf{v}_C = \mathbf{v}_B + \Omega \times \mathbf{r}_{C/B} + (\mathbf{v}_{C/B})_{xyz}$$
$$\mathbf{a}_C = \mathbf{a}_B + \dot{\Omega} \times \mathbf{r}_{C/B} + \Omega \times (\Omega \times \mathbf{r}_{C/B}) + 2\Omega \times (\mathbf{v}_{C/B})_{xyz} + (\mathbf{a}_{C/B})_{xyz}$$

Motion of B. To find the time derivatives of $\mathbf{r}_B$ let the x', y', z' axes rotate with $\Omega' = \omega_1$. Then
$$\Omega' = \omega_1 = \{4\mathbf{i}\} \text{ rad/s} \quad \dot{\Omega}' = \dot{\omega}_1 = \{1.5\mathbf{i}\} \text{ rad/s}^2$$
$$\mathbf{r}_B = \{-0.5\mathbf{k}\} \text{ m}$$
$$\mathbf{v}_B = \dot{\mathbf{r}}_B = (\dot{\mathbf{r}}_B)_{x'y'z'} + \omega_1 \times \mathbf{r}_B = 0 + 4\mathbf{i} \times (-0.5\mathbf{k}) = \{2\mathbf{j}\} \text{ m/s}$$
$$\mathbf{a}_B = \ddot{\mathbf{r}}_B = [(\ddot{\mathbf{r}}_B)_{x'y'z'} + \omega_1 \times (\dot{\mathbf{r}}_B)_{x'y'z'}] + \dot{\omega}_1 \times \mathbf{r}_B + \omega_1 \times \dot{\mathbf{r}}_B$$
$$= [0 + 0] + 1.5\mathbf{i} \times (-0.5\mathbf{k}) + 4\mathbf{i} \times 2\mathbf{j} = \{0.75\mathbf{j} + 8\mathbf{k}\} \text{ m/s}^2$$

Motion of C with Respect to B. To find the time derivatives of $\mathbf{r}_{C/B}$ relative to x, y, z, let the x'', y'', z'' axes rotate with $\Omega_{xyz} = \omega_2$. Then
$$\Omega_{xyz} = \omega_2 = \{5\mathbf{k}\} \text{ rad/s} \quad \dot{\Omega}_{xyz} = \dot{\omega}_2 = \{-6\mathbf{k}\} \text{ rad/s}^2$$
$$\mathbf{r}_{C/B} = \{0.2\mathbf{j}\} \text{ m}$$
$$(\mathbf{v}_{C/B})_{xyz} = (\dot{\mathbf{r}}_{C/B})_{xyz} = (\dot{\mathbf{r}}_{C/B})_{x''y''z''} + \omega_2 \times \mathbf{r}_{C/B} = 3\mathbf{j} + 5\mathbf{k} \times 0.2\mathbf{j} = \{-1\mathbf{i} + 3\mathbf{j}\} \text{ m/s}$$
$$(\mathbf{a}_{C/B})_{xyz} = (\ddot{\mathbf{r}}_{C/B})_{xyz} = [(\ddot{\mathbf{r}}_{C/B})_{x''y''z''} + \omega_2 \times (\dot{\mathbf{r}}_{C/B})_{x''y''z''}] + \dot{\omega}_2 \times \mathbf{r}_{C/B} + \omega_2 \times (\dot{\mathbf{r}}_{C/B})_{xyz}$$
$$= (2\mathbf{j} + 5\mathbf{k} \times 3\mathbf{j}) + (-6\mathbf{k} \times 0.2\mathbf{j}) + [5\mathbf{k} \times (-1\mathbf{i} + 3\mathbf{j})]$$
$$= \{-28.8\mathbf{i} - 3\mathbf{j}\} \text{ m/s}^2$$

Motion of C.
$$\mathbf{v}_C = \mathbf{v}_B + \Omega \times \mathbf{r}_{C/B} + (\mathbf{v}_{C/B})_{xyz} = 2\mathbf{j} + 4\mathbf{i} \times 0.2\mathbf{j} + (-1\mathbf{i} + 3\mathbf{j})$$
$$= \{-1\mathbf{i} + 5\mathbf{j} + 0.8\mathbf{k}\} \text{ m/s} \hspace{2em} \textit{Ans.}$$
$$\mathbf{a}_C = \mathbf{a}_B + \dot{\Omega} \times \mathbf{r}_{C/B} + \Omega \times (\Omega \times \mathbf{r}_{C/B}) + 2\Omega \times (\mathbf{v}_{C/B})_{xyz} + (\mathbf{a}_{C/B})_{xyz}$$
$$= (0.75\mathbf{j} + 8\mathbf{k}) + (1.5\mathbf{i} \times 0.2\mathbf{j}) + [4\mathbf{i} \times (4\mathbf{i} \times 0.2\mathbf{j})]$$
$$+ 2[4\mathbf{i} \times (-1\mathbf{i} + 3\mathbf{j})] + (-28.8\mathbf{i} - 3\mathbf{j})$$
$$= \{-28.8\mathbf{i} - 5.45\mathbf{j} + 32.3\mathbf{k}\} \text{ m/s}^2 \hspace{2em} \textit{Ans.}$$

SOLUTION II
Coordinate Axes. Here we will let the x, y, z axes rotate at
$$\Omega = \omega_1 + \omega_2 = \{4\mathbf{i} + 5\mathbf{k}\} \text{ rad/s}$$
Then $\Omega_{xyz} = 0$.

Motion of B. From the constraints of the problem ω_1 does not change direction relative to X, Y, Z; however, the direction of ω_2 is changed by ω_1. Thus, to obtain $\dot{\Omega}$ consider x', y', z' axes coincident with the X, Y, Z axes at A, so that $\Omega' = \omega_1$. Then taking the derivative of the components of Ω,

$$\dot{\Omega} = \dot{\omega}_1 + \dot{\omega}_2 = [(\dot{\omega}_1)_{x'y'z'} + \omega_1 \times \omega_1] + [(\dot{\omega}_2)_{x'y'z'} + \omega_1 \times \omega_2]$$
$$= [1.5\mathbf{i} + 0] + [-6\mathbf{k} + 4\mathbf{i} \times 5\mathbf{k}] = \{1.5\mathbf{i} - 20\mathbf{j} - 6\mathbf{k}\} \text{ rad/s}^2$$

Also, ω_1 changes the direction of $\mathbf{r}_B$ so that the time derivatives of $\mathbf{r}_B$ can be found using the primed axes defined above. Hence,

$$\mathbf{v}_B = \dot{\mathbf{r}}_B = (\dot{\mathbf{r}}_B)_{x'y'z'} + \omega_1 \times \mathbf{r}_B$$
$$= 0 + 4\mathbf{i} \times (-0.5\mathbf{k}) = \{2\mathbf{j}\} \text{ m/s}$$
$$\mathbf{a}_B = \ddot{\mathbf{r}}_B = [(\ddot{\mathbf{r}}_B)_{x'y'z'} + \omega_1 \times (\dot{\mathbf{r}}_B)_{x'y'z'}] + \dot{\omega}_1 \times \mathbf{r}_B + \omega_1 \times \dot{\mathbf{r}}_B$$
$$= [0 + 0] + 1.5\mathbf{i} \times (-0.5\mathbf{k}) + 4\mathbf{i} \times 2\mathbf{j} = \{0.75\mathbf{j} + 8\mathbf{k}\} \text{ m/s}^2$$

Motion of C with Respect to B.
$$\Omega_{xyz} = 0$$
$$\dot{\Omega}_{xyz} = 0$$
$$\mathbf{r}_{C/B} = \{0.2\mathbf{j}\} \text{ m}$$
$$(\mathbf{v}_{C/B})_{xyz} = \{3\mathbf{j}\} \text{ m/s}$$
$$(\mathbf{a}_{C/B})_{xyz} = \{2\mathbf{j}\} \text{ m/s}^2$$

Motion of C.
$$\mathbf{v}_C = \mathbf{v}_B + \Omega \times \mathbf{r}_{C/B} + (\mathbf{v}_{C/B})_{xyz}$$
$$= 2\mathbf{j} + [(4\mathbf{i} + 5\mathbf{k}) \times (0.2\mathbf{j})] + 3\mathbf{j}$$
$$= \{-1\mathbf{i} + 5\mathbf{j} + 0.8\mathbf{k}\} \text{ m/s} \qquad \text{Ans.}$$
$$\mathbf{a}_C = \mathbf{a}_B + \dot{\Omega} \times \mathbf{r}_{C/B} + \Omega \times (\Omega \times \mathbf{r}_{C/B}) + 2\Omega \times (\mathbf{v}_{C/B})_{xyz} + (\mathbf{a}_{C/B})_{xyz}$$
$$= (0.75\mathbf{j} + 8\mathbf{k}) + [(1.5\mathbf{i} - 20\mathbf{j} - 6\mathbf{k}) \times (0.2\mathbf{j})]$$
$$+ (4\mathbf{i} + 5\mathbf{k}) \times [(4\mathbf{i} + 5\mathbf{k}) \times 0.2\mathbf{j}] + 2[(4\mathbf{i} + 5\mathbf{k}) \times 3\mathbf{j}] + 2\mathbf{j}$$
$$= \{-28.8\mathbf{i} - 5.45\mathbf{j} + 32.3\mathbf{k}\} \text{ m/s}^2 \qquad \text{Ans.}$$

PROBLEMS

20–37. Solve Example 20.5 such that the x, y, z axes move with curvilinear translation, $\Omega = 0$, in which case the collar appears to have both an angular velocity $\Omega_{xyz} = \omega_1 + \omega_2$ and radial motion.

20–38. Solve Example 20.5 by fixing x, y, z axes to rod BD so that $\Omega = \omega_1 + \omega_2$. In this case the collar appears only to move radially outward along BD; hence $\Omega_{xyz} = 0$.

20–39. At the instant $\theta = 60°$, the telescopic boom AB of the construction lift is rotating with a constant angular velocity about the z axis of $\omega_1 = 0.5$ rad/s and about the pin at A with a constant angular speed of $\omega_2 = 0.25$ rad/s. Simultaneously, the boom is extending with a velocity of 1.5 ft/s, and it has an acceleration of 0.5 ft/s², both measured relative to the construction lift. Determine the velocity and acceleration of point B located at the end of the boom at this instant.

*20–40. At the instant $\theta = 60°$, the construction lift is rotating about the z axis with an angular velocity of $\omega_1 = 0.5$ rad/s, and an angular acceleration of $\dot{\omega}_1 = 0.25$ rad/s² while the telescopic boom AB rotates about the pin at A with an angular velocity of $\omega_2 = 0.25$ rad/s and angular acceleration of $\dot{\omega}_2 = 0.1$ rad/s². Simultaneously, the boom is extending with a velocity of 1.5 ft/s, and it has an acceleration of 0.5 ft/s², both measured relative to the frame. Determine the velocity and acceleration of point B located at the end of the boom at this instant.

Probs. 20–39/40

20–41. At a given instant the boom AB of the tower crane rotates about the z axis with the motion shown. At this same instant, $\theta = 60°$ and the boom is rotating downward such that $\dot{\theta} = 0.4$ rad/s and $\ddot{\theta} = 0.6$ rad/s². Determine the velocity and acceleration of the end of the boom A at this instant. The boom has a length of $l_{AB} = 40$ m.

Prob. 20–41

20–42. At the instant $\theta = 30°$, the frame of the crane and the boom AB rotate with a constant angular velocity of $\omega_1 = 1.5$ rad/s and $\omega_2 = 0.5$ rad/s, respectively. Determine the velocity and acceleration of point B at this instant.

20–43. At the instant $\theta = 30°$, the frame of the crane is rotating with an angular velocity of $\omega_1 = 1.5$ rad/s and angular acceleration of $\dot{\omega}_1 = 0.5$ rad/s², while the boom AB rotates with an angular velocity of $\omega_2 = 0.5$ rad/s and angular acceleration of $\dot{\omega}_2 = 0.25$ rad/s². Determine the velocity and acceleration of point B at this instant.

Probs. 20–42/43

*20–44. At the instant shown, the arm AB is rotating about the fixed pin A with an angular velocity $\omega_1 = 4$ rad/s and an angular acceleration $\dot{\omega}_1 = 3$ rad/s². At the same instant, rod BD is rotating relative to rod AB with an angular velocity $\omega_2 = 5$ rad/s which is increasing at $\dot{\omega}_2 = 7$ rad/s². Also, the collar C is moving along rod BD with a velocity of 3 m/s and an acceleration of 2 m/s², both measured relative to the rod. Determine the velocity and acceleration of the collar at this instant.

20–46. At the instant shown, the industrial manipulator is rotating about the z axis at $\omega_1 = 5$ rad/s, and about joint B at $\omega_2 = 2$ rad/s. Determine the velocity and acceleration of the grip A at this instant, when $\phi = 30°$, $\theta = 45°$, and $r = 1.6$ m.

20–47. At the instant shown, the industrial manipulator is rotating about the z axis at $\omega_1 = 5$ rad/s, and $\dot{\omega}_1 = 2$ rad/s²; and about joint B at $\omega_2 = 2$ rad/s and $\dot{\omega}_2 = 3$ rad/s². Determine the velocity and acceleration of the grip A at this instant, when $\phi = 30°$, $\theta = 45°$, and $r = 1.6$ m.

Prob. 20–44

Probs. 20–46/47

20–45. At a given instant, rod BD is rotating about the x axis with an angular velocity $\omega_{BD} = 2$ rad/s and an angular acceleration $\alpha_{BD} = 5$ rad/s². Also, when $\theta = 45°$ link AC is rotating at $\dot{\theta} = 4$ rad/s and $\ddot{\theta} = 2$ rad/s². Determine the velocity and acceleration of point A on the link at this instant.

*20–48. At a given instant, the rod has the angular motions shown, while the collar C is moving down *relative* to the rod with a velocity of 6 ft/s and an acceleration of 2 ft/s². Determine the collar's velocity and acceleration at this instant.

Prob. 20–45

Prob. 20–48

20–49. At a given instant, the crane is moving along the track with a velocity $v_{CD} = 8$ m/s and acceleration of 9 m/s². Simultaneously, it has the angular motions shown. If the trolley T is moving outwards along the boom AB with a relative speed of 3 m/s and relative acceleration of 5 m/s², determine the velocity and acceleration of the trolley.

Prob. 20–49

20–50. At the instant shown, the arm OA of the conveyor belt is rotating about the z axis with a constant angular velocity $\omega_1 = 6$ rad/s, while at the same instant the arm is rotating upward at a constant rate $\omega_2 = 4$ rad/s. If the conveyor is running at a constant rate $\dot{r} = 5$ ft/s, determine the velocity and acceleration of the package P at the instant shown. Neglect the size of the package.

20–51. At the instant shown, the arm OA of the conveyor belt is rotating about the z axis with a constant angular velocity $\omega_1 = 6$ rad/s, while at the same instant the arm is rotating upward at a constant rate $\omega_2 = 4$ rad/s. If the conveyor is running at a rate $\dot{r} = 5$ ft/s, which is increasing at $\ddot{r} = 8$ ft/s², determine the velocity and acceleration of the package P at the instant shown. Neglect the size of the package.

Probs. 20–50/51

*20–52. At the instant shown, the frame of the brush cutter is traveling forward in the x direction with a constant velocity of 1 m/s, and the cab is rotating about the vertical axis with an angular velocity of $\omega_1 = 0.5$ rad/s. At the same instant the boom AB has an angular velocity of $\dot{\theta} = 0.8$ rad/s, in the direction shown. Determine the velocity and acceleration of point B at the connection to the mower at this instant.

20–53. At the instant shown, the frame of the brush cutter is traveling forward in the x direction with a constant velocity of 1 m/s, and the cab is rotating about the vertical axis with an angular velocity of $\omega_1 = 0.5$ rad/s, which is increasing at $\dot{\omega}_1 = 0.4$ rad/s². At the same instant the boom AB has an angular velocity of $\dot{\theta} = 0.8$ rad/s, which is increasing at $\ddot{\theta} = 0.9$ rad/s². Determine the velocity and acceleration of point B at the connection to the mower at this instant.

Probs. 20–52/53

20–54. At the given instant, the rod is spinning about the z axis with an angular velocity $\omega_1 = 8$ rad/s and angular acceleration $\dot{\omega}_1 = 2$ rad/s². At this same instant, the disk is spinning at a constant rate $\omega_2 = 4$ rad/s, measured *relative* to the rod. Determine the velocity and acceleration of point P on the disk at this instant.

Prob. 20–54

CHAPTER REVIEW

Rotation About a Fixed Point

When a body rotates about a fixed point O, then points on the body follow a path that lies on the surface of a sphere centered at O.

Since the angular acceleration is a time rate of change in the angular velocity, then it is necessary to determine both the magnitude and directional changes of $\boldsymbol{\omega}$ when finding its time derivative. To do this, the angular velocity is often specified in terms of its component motions, such that the direction of some of these components will remain constant relative to rotating x, y, z axes which are rotating at $\boldsymbol{\Omega}$. If this is the case, then the time derivative relative to the fixed axis can be determined using $\dot{\mathbf{A}} = (\dot{\mathbf{A}})_{xyz} + \boldsymbol{\Omega} \times \mathbf{A}$.

Once $\boldsymbol{\omega}$ and $\boldsymbol{\alpha}$ are known, the velocity and acceleration of any point P on the body can then be determined.

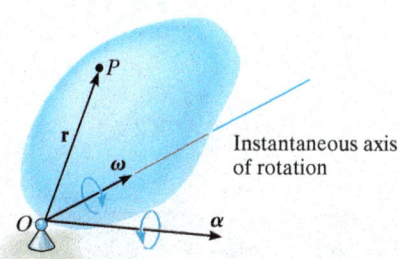

$$\mathbf{v}_P = \boldsymbol{\omega} \times \mathbf{r}$$

$$\mathbf{a}_P = \boldsymbol{\alpha} \times \mathbf{r} + \boldsymbol{\omega} \times (\boldsymbol{\omega} \times \mathbf{r})$$

General Motion

If the body undergoes general motion, then the motion of a point B on the body can be related to the motion of another point A using a relative motion analysis, with translating axes attached to A.

$$\mathbf{v}_B = \mathbf{v}_A + \boldsymbol{\omega} \times \mathbf{r}_{B/A}$$

$$\mathbf{a}_B = \mathbf{a}_A + \boldsymbol{\alpha} \times \mathbf{r}_{B/A} + \boldsymbol{\omega} \times (\boldsymbol{\omega} \times \mathbf{r}_{B/A})$$

Relative-Motion Analysis Using Translating and Rotating Axes

The motion of two points A and B on a body, on a series of connected bodies, or where each point is located on two different paths, can be related using a relative-motion analysis with rotating and translating axes at A.

When applying the equations to find $\mathbf{v}_B$ and $\mathbf{a}_B$, it is important to account for both the magnitude and directional changes of $\mathbf{r}_A$, $\mathbf{r}_{B/A}$, $\boldsymbol{\Omega}$, and $\boldsymbol{\Omega}_{xyz}$ when taking their time derivatives to find $\mathbf{v}_A$, $\mathbf{a}_A$, $(\mathbf{v}_{B/A})_{xyz}$, $(\mathbf{a}_{B/A})_{xyz}$, $\dot{\boldsymbol{\Omega}}$, and $\dot{\boldsymbol{\Omega}}_{xyz}$. To do this properly, one must use Eq. 20–6.

$$\mathbf{v}_B = \mathbf{v}_A + \boldsymbol{\Omega} \times \mathbf{r}_{B/A} + (\mathbf{v}_{B/A})_{xyz}$$

$$\mathbf{a}_B = \mathbf{a}_A + \dot{\boldsymbol{\Omega}} \times \mathbf{r}_{B/A} + \boldsymbol{\Omega} \times (\boldsymbol{\Omega} \times \mathbf{r}_{B/A}) + 2\boldsymbol{\Omega} \times (\mathbf{v}_{B/A})_{xyz} + (\mathbf{a}_{B/A})_{xyz}$$

CHAPTER 21

The forces acting on this motorcycle can be determined using the equations of motion as discussed in this chapter.

THREE-DIMENSIONAL KINETICS OF A RIGID BODY

CHAPTER OBJECTIVES

- To introduce the methods for finding the moments of inertia and products of inertia of a body about various axes.
- To show how to apply the principles of work and energy and linear and angular impulse and momentum to a rigid body having three-dimensional motion.
- To develop and apply the equations of motion in three dimensions.
- To study gyroscopic and torque-free motion.

*21.1 MOMENTS AND PRODUCTS OF INERTIA

When studying the planar motion of a body, it was necessary to introduce the moment of inertia I_G, which was determined about an axis perpendicular to the plane of motion and passing through the body's mass center G. For the analysis of three-dimensional motion it will sometimes be necessary to calculate six inertial quantities. These terms, called the moments and products of inertia, describe in a particular way the distribution of mass for a body relative to a given coordinate system that has a specified orientation and point of origin.

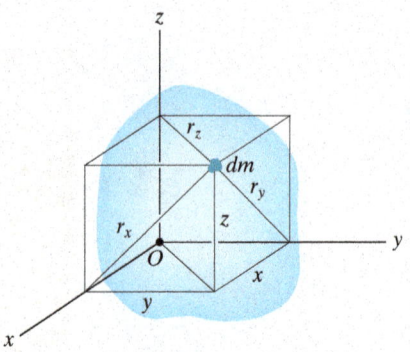

Fig. 21–1

Moment of Inertia. The *moment of inertia* for a differential element dm of a body about any one of the three coordinate axes is defined as the product of the mass of the element and the square of the shortest distance from the axis to the element. For example, as noted in Fig. 21–1, $r_x = \sqrt{y^2 + z^2}$, so that the mass moment of inertia of the element about the x axis is

$$dI_{xx} = r_x^2\, dm = (y^2 + z^2)\, dm$$

The moment of inertia I_{xx} for the body can be determined by integrating this expression over the entire mass of the body. Hence, for each of the axes, we can write

$$\begin{aligned} I_{xx} &= \int_m r_x^2\, dm = \int_m (y^2 + z^2)\, dm \\ I_{yy} &= \int_m r_y^2\, dm = \int_m (x^2 + z^2)\, dm \\ I_{zz} &= \int_m r_z^2\, dm = \int_m (x^2 + y^2)\, dm \end{aligned} \qquad (21\text{–}1)$$

In each of these cases the moment of inertia is always a positive quantity, since it is the summation of the product of the mass dm, which is always positive, and the distances squared.

Product of Inertia. The *product of inertia* for a differential element dm with respect to a set of two orthogonal planes is defined as the product of the mass of the element and the perpendicular (or shortest) distances from each plane to the element. For example, this distance is x to the y–z plane and it is y to the x–z plane, Fig. 21–1. The product of inertia dI_{xy} for the element is therefore

$$dI_{xy} = xy\, dm$$

Note also that $dI_{yx} = dI_{xy}$. Integrating over the entire mass, the products of inertia of the body with respect to each combination of planes then become

$$\begin{aligned} I_{xy} = I_{yx} &= \int_m xy\, dm \\ I_{yz} = I_{zy} &= \int_m yz\, dm \\ I_{zx} = I_{xz} &= \int_m xz\, dm \end{aligned} \qquad (21\text{–}2)$$

$I_{xy} = I_{zx} = 0$

(a)

$I_{xy} = I_{yz} = I_{zx} = 0$

(b)

Fig. 21–2

Unlike the moment of inertia, which is always positive, the product of inertia may be positive, negative, or zero. The result depends on the algebraic signs of the two defining coordinates, which vary independently from one another. If either one or both of the orthogonal planes are planes of symmetry for the mass, the product of inertia with respect to these planes will be zero. In such cases, elements of mass will occur in pairs located on each side of the plane of symmetry. On one side of the plane the product of inertia for the element will be positive, while on the other side the product of inertia of the corresponding element will be negative, the sum therefore yielding zero. For example, in Fig. 21–2a, the y–z plane is a plane of symmetry, and so $I_{xy} = I_{zx} = 0$. Calculation of I_{yz} will yield a positive result, since all elements of mass are located using only positive y and z coordinates. For the cylinder, with the coordinate axes located as shown in Fig. 21–2b, the x–z and y–z planes are both planes of symmetry. Thus, $I_{xy} = I_{yz} = I_{zx} = 0$.

Parallel-Axis and Parallel-Plane Theorems.

The techniques of integration used to determine the moment of inertia of a body were described in Sec. 17.1. Also discussed were methods to determine the moment of inertia of a composite body, i.e., a body composed of simpler segments, as tabulated on the inside back cover. In both of these cases the **parallel-axis theorem** was often used for the calculations. This theorem transfers the moment of inertia of a body from an axis passing through its mass center G to a parallel axis passing through some other point. If G has coordinates x_G, y_G, z_G, Fig. 21–3, then the parallel-axis equations used to calculate the moments of inertia about the x, y, z axes are

$$\begin{aligned} I_{xx} &= (I_{x'x'})_G + m(y_G^2 + z_G^2) \\ I_{yy} &= (I_{y'y'})_G + m(x_G^2 + z_G^2) \\ I_{zz} &= (I_{z'z'})_G + m(x_G^2 + y_G^2) \end{aligned} \quad (21\text{–}3)$$

Fig. 21–3

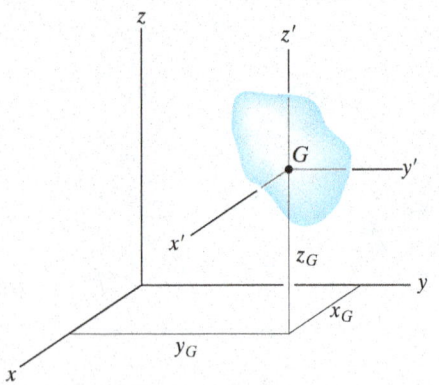

Fig. 21–3 (repeated)

The products of inertia of a composite body are determined in the same manner as the body's moments of inertia. Here, however, the ***parallel-plane theorem*** is important. This theorem is used to transfer the products of inertia of the body with respect to a set of three orthogonal planes passing through the body's mass center to a corresponding set of three parallel planes passing through some other point O. Defining the perpendicular distances between the planes as x_G, y_G, z_G, Fig. 21–3, the parallel-plane theorem for each product of inertia can be written as

$$I_{xy} = (I_{x'y'})_G + m x_G y_G$$
$$I_{yz} = (I_{y'z'})_G + m y_G z_G \qquad (21\text{–}4)$$
$$I_{zx} = (I_{z'x'})_G + m z_G x_G$$

The derivation of these formulas is similar to that given for the parallel-axis equation, Sec. 17.1.

Inertia Tensor. The inertial properties of a body are therefore completely characterized by nine terms, six of which are independent of one another. This set of terms is defined using Eqs. 21–1 and 21–2 and can be written as

$$\begin{pmatrix} I_{xx} & -I_{xy} & -I_{xz} \\ -I_{yx} & I_{yy} & -I_{yz} \\ -I_{zx} & -I_{zy} & I_{zz} \end{pmatrix}$$

This array is called an ***inertia tensor.*** * It has a unique set of values for a body when it is determined for each location of the origin O and orientation of the coordinate axes.

In general, for point O we can always specify a unique axes inclination for which the products of inertia for the body are zero. When this is done, the inertia tensor is said to be "diagonalized" and may be written in the simplified form

$$\begin{pmatrix} I_x & 0 & 0 \\ 0 & I_y & 0 \\ 0 & 0 & I_z \end{pmatrix}$$

Here $I_x = I_{xx}$, $I_y = I_{yy}$, and $I_z = I_{zz}$ are termed the ***principal moments of inertia*** for the body, which are calculated with respect to the ***principal axes of inertia***. Of these three principal moments of inertia, one will be a maximum and another a minimum of the body's moment of inertia.

The dynamics of the space shuttle while it orbits the earth can be predicted only if its moments and products of inertia are known relative to its mass center.

---

* Negative signs are included here in order to later describe the body's angular momentum. See Eqs. 21–10.

The mathematical determination of the directions of principal axes of inertia will not be discussed here (see Prob. 21–22). However, there are many cases in which the principal axes can be determined by inspection. From the previous discussion it was noted that if the coordinate axes are oriented such that two of the three orthogonal planes containing the axes are planes of symmetry for the body, then all the products of inertia for the body are zero with respect to these coordinate planes, and hence these coordinate axes are principal axes of inertia. For example, the x, y, z axes shown in Fig. 21–2b represent the principal axes of inertia for the cylinder at point O.

Moment of Inertia About an Arbitrary Axis.
Consider the body shown in Fig. 21–4, where the nine elements of the inertia tensor have been determined with respect to the x, y, z axes having an origin at O. Here we wish to determine the moment of inertia of the body about the Oa axis, defined by the unit vector $\mathbf{u}_a$. By definition, $I_{Oa} = \int b^2 \, dm$, where b is the perpendicular distance from the element dm to Oa. If the position of dm is located using $\mathbf{r}$, then $b = r \sin \theta$, which is the magnitude of the cross product $\mathbf{u}_a \times \mathbf{r}$. The moment of inertia can therefore be expressed as

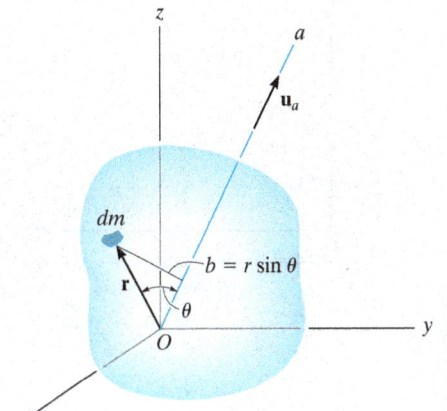

Fig. 21–4

$$I_{Oa} = \int_m |(\mathbf{u}_a \times \mathbf{r})|^2 dm = \int_m (\mathbf{u}_a \times \mathbf{r}) \cdot (\mathbf{u}_a \times \mathbf{r}) dm$$

In general, $\mathbf{u}_a = u_x \mathbf{i} + u_y \mathbf{j} + u_z \mathbf{k}$ and $\mathbf{r} = x\mathbf{i} + y\mathbf{j} + z\mathbf{k}$, and so $\mathbf{u}_a \times \mathbf{r} = (u_y z - u_z y)\mathbf{i} + (u_z x - u_x z)\mathbf{j} + (u_x y - u_y x)\mathbf{k}$. And so, after substituting and performing the dot-product operation, we get

$$I_{Oa} = \int_m [(u_y z - u_z y)^2 + (u_z x - u_x z)^2 + (u_x y - u_y x)^2] dm$$

$$= u_x^2 \int_m (y^2 + z^2) dm + u_y^2 \int_m (z^2 + x^2) dm + u_z^2 \int_m (x^2 + y^2) dm$$

$$- 2u_x u_y \int_m xy \, dm - 2u_y u_z \int_m yz \, dm - 2u_z u_x \int_m zx \, dm$$

Recognizing the integrals to be the moments and products of inertia of the body, Eqs. 21–1 and 21–2, we have

$$\boxed{I_{Oa} = I_{xx} u_x^2 + I_{yy} u_y^2 + I_{zz} u_z^2 - 2I_{xy} u_x u_y - 2I_{yz} u_y u_z - 2I_{zx} u_z u_x} \quad (21\text{–}5)$$

Thus, if the inertia tensor is specified for the x, y, z axes, the moment of inertia of the body about the inclined Oa axis can be found. For the calculation, the direction cosines u_x, u_y, u_z of the axes must be determined. These terms specify the cosines of the coordinate direction angles α, β, γ made between the positive Oa axis and the positive x, y, z axes, respectively (see Appendix B).

EXAMPLE 21.1

Determine the moment of inertia of the bent rod shown in Fig. 21–5a about the Oa axis. The mass of each of the three segments is given in the figure.

SOLUTION
Before applying Eq. 21–5, it is first necessary to determine the moments and products of inertia of the rod with respect to the x, y, z axes. This is done using the formula for the moment of inertia of a slender rod, $I = \frac{1}{12}ml^2$, and the parallel-axis and parallel-plane theorems, Eqs. 21–3 and 21–4. Dividing the rod into three parts and locating the mass center of each segment, Fig. 21–5b, we have

$$I_{xx} = \left[\tfrac{1}{12}(2)(0.2)^2 + 2(0.1)^2\right] + [0 + 2(0.2)^2]$$
$$+ \left[\tfrac{1}{12}(4)(0.4)^2 + 4((0.2)^2 + (0.2)^2)\right] = 0.480 \text{ kg} \cdot \text{m}^2$$

$$I_{yy} = \left[\tfrac{1}{12}(2)(0.2)^2 + 2(0.1)^2\right] + \left[\tfrac{1}{12}(2)(0.2)^2 + 2((-0.1)^2 + (0.2)^2)\right]$$
$$+ [0 + 4((-0.2)^2 + (0.2)^2)] = 0.453 \text{ kg} \cdot \text{m}^2$$

$$I_{zz} = [0 + 0] + \left[\tfrac{1}{12}(2)(0.2)^2 + 2(-0.1)^2\right] + \left[\tfrac{1}{12}(4)(0.4)^2 +\right.$$
$$\left. 4((-0.2)^2 + (0.2)^2)\right] = 0.400 \text{ kg} \cdot \text{m}^2$$

$$I_{xy} = [0 + 0] + [0 + 0] + [0 + 4(-0.2)(0.2)] = -0.160 \text{ kg} \cdot \text{m}^2$$

$$I_{yz} = [0 + 0] + [0 + 0] + [0 + 4(0.2)(0.2)] = 0.160 \text{ kg} \cdot \text{m}^2$$

$$I_{zx} = [0 + 0] + [0 + 2(0.2)(-0.1)] +$$
$$[0 + 4(0.2)(-0.2)] = -0.200 \text{ kg} \cdot \text{m}^2$$

Fig. 21–5

The Oa axis is defined by the unit vector

$$\mathbf{u}_{Oa} = \frac{\mathbf{r}_A}{r_A} = \frac{-0.2\mathbf{i} + 0.4\mathbf{j} + 0.2\mathbf{k}}{\sqrt{(-0.2)^2 + (0.4)^2 + (0.2)^2}} = -0.408\mathbf{i} + 0.816\mathbf{j} + 0.408\mathbf{k}$$

Thus,

$$u_x = -0.408 \quad u_y = 0.816 \quad u_z = 0.408$$

Substituting these results into Eq. 21–5 yields

$$I_{Oa} = I_{xx}u_x^2 + I_{yy}u_y^2 + I_{zz}u_z^2 - 2I_{xy}u_x u_y - 2I_{yz}u_y u_z - 2I_{zx}u_z u_x$$
$$= 0.480(-0.408)^2 + (0.453)(0.816)^2 + 0.400(0.408)^2$$
$$- 2(-0.160)(-0.408)(0.816) - 2(0.160)(0.816)(0.408)$$
$$- 2(-0.200)(0.408)(-0.408)$$
$$= 0.169 \text{ kg} \cdot \text{m}^2 \qquad \textit{Ans.}$$

PROBLEMS

21–1. Show that the sum of the moments of inertia of a body, $I_{xx} + I_{yy} + I_{zz}$, is independent of the orientation of the x, y, z axes and thus depends only on the location of the origin.

21–2. Determine the moment of inertia of the cone with respect to a vertical $\bar{y}$ axis passing through the cone's center of mass. What is the moment of inertia about a parallel axis y' that passes through the diameter of the base of the cone? The cone has a mass m.

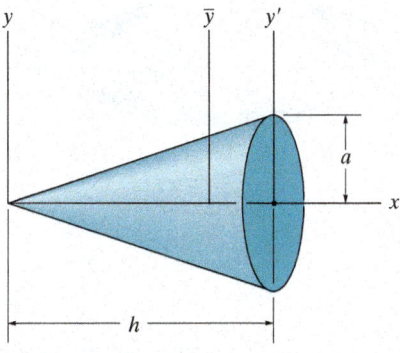

Prob. 21–2

21–3. Determine the product of inertia I_{xy} for the homogeneous tetrahedron. The density of the material is ρ. Express the result in terms of the total mass m of the solid. *Suggestion*: Use a triangular element of thickness dz and then express dI_{xy} in terms of the size and mass of the element using the result of Prob. 21–6.

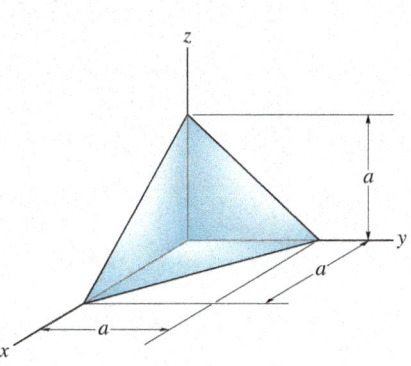

Prob. 21–3

*__**21–4.**__ Determine the moments of inertia for the homogeneous cylinder of mass m about the x', y', z' axes.

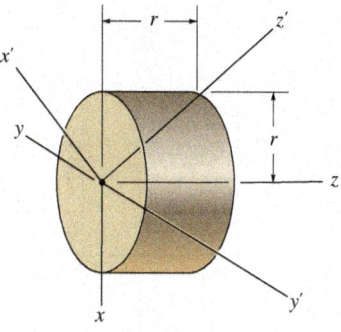

Prob. 21–4

21–5. Determine by direct integration the product of inertia I_{yz} for the homogeneous prism. The density of the material is ρ. Express the result in terms of the total mass m of the prism.

21–6. Determine by direct integration the product of inertia I_{xy} for the homogeneous prism. The density of the material is ρ. Express the result in terms of the total mass m of the prism.

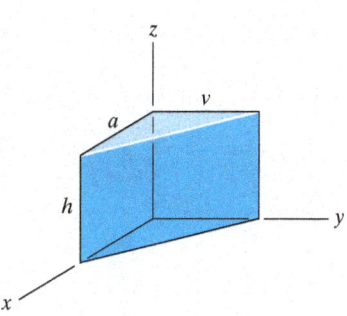

Probs. 21–5/6

21–7. Determine the product of inertia I_{xy} for the bent rod. The rod has a mass per unit length of 2 kg/m.

*21–8. Determine the moments of inertia I_{xy}, I_{yy}, I_{zz} for the bent rod. The rod has a mass per unit length of 2 kg/m.

21–10. Determine the radii of gyration k_x and k_y for the solid formed by revolving the shaded area about the y axis. The density of the material is ρ.

Prob. 21–10

Probs. 21–7/8

21–11. Determine the moment of inertia about the z axis of the assembly which consists of the 1.5-kg rod CD and the 7-kg disk.

21–9. Determine the moment of inertia of the cone about the z' axis. The weight of the cone is 15 lb, the *height* is $h = 1.5$ ft, and the radius is $r = 0.5$ ft.

Prob. 21–9

Prob. 21–11

*21–12. Determine the moment of inertia I_{xx} of the composite plate assembly. The plates have a specific weight of 6 lb/ft².

21–13. Determine the product of inertia I_{yz} of the composite plate assembly. The plates have a weight of 6 lb/ft².

Probs. 21–12/13

21–14. Determine the products of inertia I_{xy}, I_{yz}, and I_{xz} of the thin plate. The material has a density per unit area of 50 kg/m².

Prob. 21–14

21–15. Determine the moment of inertia of both the 1.5-kg rod and 4-kg disk about the z' axis.

Prob. 21–15

*21–16. Determine the moments of inertia about the x, y, z axes of the rod assembly. The rods have a mass of 0.75 kg/m.

Prob. 21–16

21–17. The bent rod has a weight of 1.5 lb/ft. Locate the center of gravity $G(\bar{x}, \bar{y})$ and determine the principal moments of inertia $I_{x'}$, $I_{y'}$, and $I_{z'}$ of the rod with respect to the x', y', z' axes.

Prob. 21–17

21–18. Determine the products of inertia I_{xy}, I_{yz}, and I_{xz} of the thin plate. The material has a mass per unit area of 50 kg/m².

*21–20. The bent rod has a mass of 4 kg/m. Determine the moment of inertia of the rod about the OA axis.

Prob. 21–20

Prob. 21–18

21–19. Determine the moment of inertia of the rod-and-thin-ring assembly about the z axis. The rods and ring have a mass per unit length of 2 kg/m.

21–21. Determine the products of inertia I_{xy}, I_{yz}, and I_{xz} of the solid. The material is steel, which has a specific weight of 490 lb/ft³.

Prob. 21–19

Prob. 21–21

21.2 ANGULAR MOMENTUM

In this section we will develop the necessary equations used to determine the angular momentum of a rigid body about an arbitrary point. These equations will provide a means for developing both the principle of impulse and momentum and the equations of rotational motion for a rigid body.

Consider the rigid body in Fig. 21–6, which has a mass m and center of mass at G. The X, Y, Z coordinate system represents an inertial frame of reference, and hence, its axes are fixed or translate with a constant velocity. The angular momentum as measured from this reference will be determined relative to the arbitrary point A. The position vectors $\mathbf{r}_A$ and $\boldsymbol{\rho}_A$ are drawn from the origin of coordinates to point A and from A to the ith particle of the body. If this particle's mass is m_i, then the angular momentum about point A is

$$(\mathbf{H}_A)_i = \boldsymbol{\rho}_A \times m_i \mathbf{v}_i$$

where $\mathbf{v}_i$ represents the particle's velocity measured from the X, Y, Z coordinate system. If the body has an angular velocity $\boldsymbol{\omega}$ at the instant considered, $\mathbf{v}_i$ may be related to the velocity of A by applying Eq. 20–7, i.e.,

$$\mathbf{v}_i = \mathbf{v}_A + \boldsymbol{\omega} \times \boldsymbol{\rho}_A$$

Thus,

$$(\mathbf{H}_A)_i = \boldsymbol{\rho}_A \times m_i(\mathbf{v}_A + \boldsymbol{\omega} \times \boldsymbol{\rho}_A)$$
$$= (\boldsymbol{\rho}_A m_i) \times \mathbf{v}_A + \boldsymbol{\rho}_A \times (\boldsymbol{\omega} \times \boldsymbol{\rho}_A) m_i$$

Summing the moments of all the particles of the body requires an integration. Since $m_i \to dm$, we have

$$\mathbf{H}_A = \left(\int_m \boldsymbol{\rho}_A \, dm \right) \times \mathbf{v}_A + \int_m \boldsymbol{\rho}_A \times (\boldsymbol{\omega} \times \boldsymbol{\rho}_A) \, dm \qquad (21\text{–}6)$$

Inertial coordinate system

Fig. 21–6

Fixed Point
(a)

Center of Mass
(b)

Fig. 21–7

Fixed Point O.
If A becomes a *fixed point* O in the body, Fig. 21–7a, then $\mathbf{v}_A = \mathbf{0}$ and Eq. 21–6 reduces to

$$\mathbf{H}_O = \int_m \boldsymbol{\rho}_O \times (\boldsymbol{\omega} \times \boldsymbol{\rho}_O) \, dm \qquad (21\text{–}7)$$

Center of Mass G.
If A is located at the *center of mass G* of the body, Fig. 21–7b, then $\int_m \boldsymbol{\rho}_A \, dm = \mathbf{0}$ and

$$\mathbf{H}_G = \int_m \boldsymbol{\rho}_G \times (\boldsymbol{\omega} \times \boldsymbol{\rho}_G) \, dm \qquad (21\text{–}8)$$

Arbitrary Point A.
In general, A can be a point other than O or G, Fig. 21–7c, in which case Eq. 21–6 may nevertheless be simplified to the following form (see Prob. 21–23).

$$\mathbf{H}_A = \boldsymbol{\rho}_{G/A} \times m\mathbf{v}_G + \mathbf{H}_G \qquad (21\text{–}9)$$

Here the angular momentum consists of two parts—the moment of the linear momentum $m\mathbf{v}_G$ of the body about point A added (vectorially) to the angular momentum $\mathbf{H}_G$. Equation 21–9 can also be used to determine the angular momentum of the body about a fixed point O. The results, of course, will be the same as those found using the more convenient Eq. 21–7.

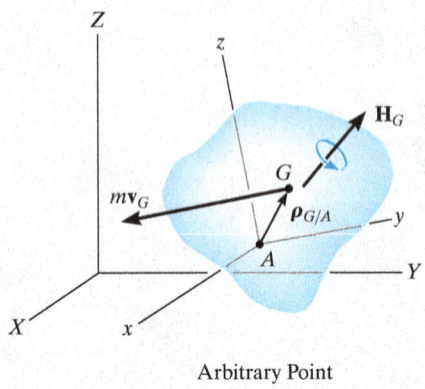

Arbitrary Point
(c)

Rectangular Components of H. To make practical use of Eqs. 21–7 through 21–9, the angular momentum must be expressed in terms of its scalar components. For a general formulation, note that Eqs. 21–7 and 21–8 are both of the form

$$\mathbf{H} = \int_m \boldsymbol{\rho} \times (\boldsymbol{\omega} \times \boldsymbol{\rho}) dm$$

Expressing $\mathbf{H}$, $\boldsymbol{\rho}$, and $\boldsymbol{\omega}$ in terms of their x, y, z components, we have

$$H_x \mathbf{i} + H_y \mathbf{j} + H_z \mathbf{k} = \int_m (x\mathbf{i} + y\mathbf{j} + z\mathbf{k}) \times [(\omega_x \mathbf{i} + \omega_y \mathbf{j} + \omega_z \mathbf{k}) \\ \times (x\mathbf{i} + y\mathbf{j} + z\mathbf{k})] dm$$

Expanding the cross products and combining terms yields

$$H_x \mathbf{i} + H_y \mathbf{j} + H_z \mathbf{k} = \left[\omega_x \int_m (y^2 + z^2) dm - \omega_y \int_m xy\, dm - \omega_z \int_m xz\, dm \right] \mathbf{i}$$
$$+ \left[-\omega_x \int_m xy\, dm + \omega_y \int_m (x^2 + z^2) dm - \omega_z \int_m yz\, dm \right] \mathbf{j}$$
$$+ \left[-\omega_x \int_m zx\, dm - \omega_y \int_m yz\, dm + \omega_z \int_m (x^2 + y^2) dm \right] \mathbf{k}$$

Equating the respective $\mathbf{i}$, $\mathbf{j}$, $\mathbf{k}$ components and recognizing that the integrals represent the moments and products of inertia, we obtain

$$\boxed{\begin{aligned} H_x &= I_{xx}\omega_x - I_{xy}\omega_y - I_{xz}\omega_z \\ H_y &= -I_{yx}\omega_x + I_{yy}\omega_y - I_{yz}\omega_z \\ H_z &= -I_{zx}\omega_x - I_{zy}\omega_y + I_{zz}\omega_z \end{aligned}} \qquad (21\text{–}10)$$

These equations can be simplified further if the x, y, z coordinate axes are oriented such that they become *principal axes of inertia* for the body at the point. When these axes are used, the products of inertia $I_{xy} = I_{yz} = I_{zx} = 0$, and if the principal moments of inertia about the x, y, z axes are represented as $I_x = I_{xx}$, $I_y = I_{yy}$, $I_z = I_{zz}$, then the three components of angular momentum become

$$\boxed{H_x = I_x \omega_x \quad H_y = I_y \omega_y \quad H_z = I_z \omega_z} \qquad (21\text{–}11)$$

578   CHAPTER 21   THREE-DIMENSIONAL KINETICS OF A RIGID BODY

The motion of the astronaut is controlled by use of small directional jets attached to his or her space suit. The impulses these jets provide must be carefully specified in order to prevent tumbling and loss of orientation. (© NASA)

Principle of Impulse and Momentum. Now that the formulation of the angular momentum for a body has been developed, the *principle of impulse and momentum*, as discussed in Sec. 19.2, can be used to solve kinetic problems which involve *force, velocity, and time*. For this case, the following two vector equations are available:

$$m(\mathbf{v}_G)_1 + \Sigma \int_{t_1}^{t_2} \mathbf{F}\, dt = m(\mathbf{v}_G)_2 \qquad (21\text{–}12)$$

$$(\mathbf{H}_O)_1 + \Sigma \int_{t_1}^{t_2} \mathbf{M}_O\, dt = (\mathbf{H}_O)_2 \qquad (21\text{–}13)$$

In three dimensions each vector term can be represented by its three scalar components, and therefore a total of six scalar equations can be written. Three equations relate the linear impulse and momentum in the x, y, z directions, and the other three equations relate the body's angular impulse and momentum about the x, y, z axes. Before applying Eqs. 21–12 and 21–13 to the solution of problems, the material in Secs. 19.2 and 19.3 should be reviewed.

21.3 KINETIC ENERGY

In order to apply the principle of work and energy to solve problems involving general rigid body motion, it is first necessary to formulate expressions for the kinetic energy of the body. If the *i*th particle of the body in Fig. 21–8 has a mass m_i and velocity $\mathbf{v}_i$, measured relative to the inertial X, Y, Z frame of reference, then its kinetic energy is

$$T_i = \tfrac{1}{2} m_i v_i^2 = \tfrac{1}{2} m_i (\mathbf{v}_i \cdot \mathbf{v}_i)$$

Provided the velocity of an arbitrary point A in the body is known, $\mathbf{v}_i$ can be related to $\mathbf{v}_A$ by the equation $\mathbf{v}_i = \mathbf{v}_A + \boldsymbol{\omega} \times \boldsymbol{\rho}_A$. Using this expression, the kinetic energy for the particle becomes

$$T_i = \tfrac{1}{2} m_i (\mathbf{v}_A + \boldsymbol{\omega} \times \boldsymbol{\rho}_A) \cdot (\mathbf{v}_A + \boldsymbol{\omega} \times \boldsymbol{\rho}_A)$$
$$= \tfrac{1}{2} (\mathbf{v}_A \cdot \mathbf{v}_A) m_i + \mathbf{v}_A \cdot (\boldsymbol{\omega} \times \boldsymbol{\rho}_A) m_i + \tfrac{1}{2} (\boldsymbol{\omega} \times \boldsymbol{\rho}_A) \cdot (\boldsymbol{\omega} \times \boldsymbol{\rho}_A) m_i$$

The kinetic energy for the entire body is obtained by summing the kinetic energies of all the particles of the body. This requires an integration. Since $m_i \rightarrow dm$, we get

$$T = \tfrac{1}{2} m (\mathbf{v}_A \cdot \mathbf{v}_A) + \mathbf{v}_A \cdot \left(\boldsymbol{\omega} \times \int_m \boldsymbol{\rho}_A\, dm \right) + \tfrac{1}{2} \int_m (\boldsymbol{\omega} \times \boldsymbol{\rho}_A) \cdot (\boldsymbol{\omega} \times \boldsymbol{\rho}_A)\, dm$$

Inertial coordinate system

Fig. 21–8

The last term on the right can be rewritten using the vector identity $\mathbf{a} \times \mathbf{b} \cdot \mathbf{c} = \mathbf{a} \cdot \mathbf{b} \times \mathbf{c}$, where $\mathbf{a} = \boldsymbol{\omega}$, $\mathbf{b} = \boldsymbol{\rho}_A$, and $\mathbf{c} = \boldsymbol{\omega} \times \boldsymbol{\rho}_A$. The final result is

$$T = \tfrac{1}{2}m(\mathbf{v}_A \cdot \mathbf{v}_A) + \mathbf{v}_A \cdot \left(\boldsymbol{\omega} \times \int_m \boldsymbol{\rho}_A\, dm\right)$$
$$+ \tfrac{1}{2}\boldsymbol{\omega} \cdot \int_m \boldsymbol{\rho}_A \times (\boldsymbol{\omega} \times \boldsymbol{\rho}_A)\, dm \qquad (21\text{--}14)$$

Simplification occurs if the reference point A is either a fixed point or the center of mass.

Fixed Point O. If A is a *fixed point O* in the body, Fig. 21–7a, then $\mathbf{v}_A = \mathbf{0}$, and using Eq. 21–7, we can express Eq. 21–14 as

$$T = \tfrac{1}{2}\boldsymbol{\omega} \cdot \mathbf{H}_O$$

If the x, y, z axes represent the principal axes of inertia for the body, then $\boldsymbol{\omega} = \omega_x \mathbf{i} + \omega_y \mathbf{j} + \omega_z \mathbf{k}$ and $\mathbf{H}_O = I_x \omega_x \mathbf{i} + I_y \omega_y \mathbf{j} + I_z \omega_z \mathbf{k}$. Substituting and performing the dot-product operations yields

$$\boxed{T = \tfrac{1}{2}I_x \omega_x^2 + \tfrac{1}{2}I_y \omega_y^2 + \tfrac{1}{2}I_z \omega_z^2} \qquad (21\text{--}15)$$

Center of Mass G. If A is located at the *center of mass G* of the body, Fig. 21-7b, then $\int \boldsymbol{\rho}_A\, dm = \mathbf{0}$ and, using Eq. 21–8, we can write Eq. 21–14 as

$$T = \tfrac{1}{2}m v_G^2 + \tfrac{1}{2}\boldsymbol{\omega} \cdot \mathbf{H}_G$$

In a manner similar to that for a fixed point, the last term on the right side can be represented in scalar form, in which case

$$\boxed{T = \tfrac{1}{2}m v_G^2 + \tfrac{1}{2}I_x \omega_x^2 + \tfrac{1}{2}I_y \omega_y^2 + \tfrac{1}{2}I_z \omega_z^2} \qquad (21\text{--}16)$$

Here it is seen that the kinetic energy consists of two parts; namely, the translational kinetic energy of the mass center, $\tfrac{1}{2}m v_G^2$, and the body's rotational kinetic energy.

Principle of Work and Energy. Having formulated the kinetic energy for a body, the **principle of work and energy** can be applied to solve kinetic problems which involve *force, velocity, and displacement*. For this case only one scalar equation can be written for each body, namely,

$$\boxed{T_1 + \Sigma U_{1\text{-}2} = T_2} \qquad (21\text{--}17)$$

Before applying this equation, the material in Secs. 18.1 through 18.4 should be reviewed.

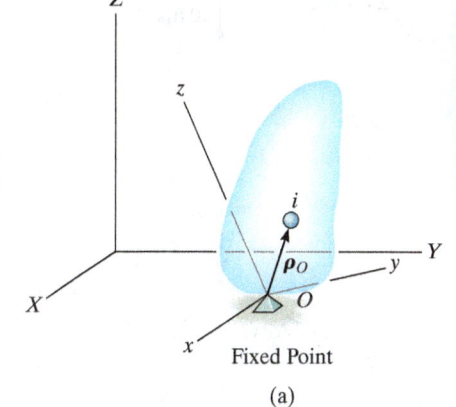

Fixed Point
(a)

Fig. 21–7 (repeated)

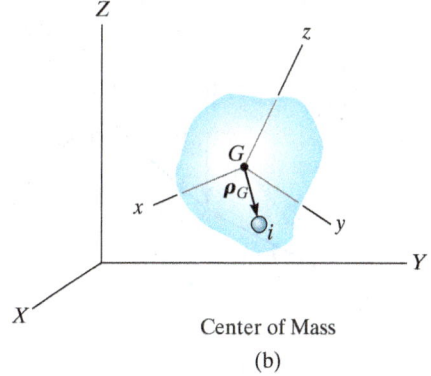

Center of Mass
(b)

Fig. 21–7 (repeated)

EXAMPLE 21.2

The rod in Fig. 21–9a has a weight per unit length of 1.5 lb/ft. Determine its angular velocity just after the end A falls onto the hook at E. The hook provides a permanent connection for the rod due to the spring-lock mechanism S. Just before striking the hook the rod is falling downward with a speed $(v_G)_1 = 10$ ft/s.

SOLUTION
The principle of impulse and momentum will be used since impact occurs.

Impulse and Momentum Diagrams. Fig. 21–9b. During the short time Δt, the impulsive force **F** acting at A changes the momentum of the rod. (The impulse created by the rod's weight **W** during this time is small compared to $\int \mathbf{F}\, dt$, so that it can be neglected, i.e., the weight is a nonimpulsive force.) Hence, the angular momentum of the rod is *conserved* about point A since the moment of $\int \mathbf{F}\, dt$ about A is zero.

Conservation of Angular Momentum. Equation 21–9 must be used to find the angular momentum of the rod, since A does not become a *fixed point* until *after* the impulsive interaction with the hook. With reference to Fig. 21–9b, $(\mathbf{H}_A)_1 = (\mathbf{H}_A)_2$, or

$$\mathbf{r}_{G/A} \times m(\mathbf{v}_G)_1 = \mathbf{r}_{G/A} \times m(\mathbf{v}_G)_2 + (\mathbf{H}_G)_2 \qquad (1)$$

From Fig. 21–9a, $\mathbf{r}_{G/A} = \{-0.667\mathbf{i} + 0.5\mathbf{j}\}$ ft. Furthermore, the primed axes are principal axes of inertia for the rod because $I_{x'y'} = I_{x'z'} = I_{z'y'} = 0$. Hence, from Eqs. 21–11, $(\mathbf{H}_G)_2 = I_{x'}\omega_x\mathbf{i} + I_{y'}\omega_y\mathbf{j} + I_{z'}\omega_z\mathbf{k}$. The principal moments of inertia are $I_{x'} = 0.0272$ slug·ft², $I_{y'} = 0.0155$ slug·ft², $I_{z'} = 0.0427$ slug·ft² (see Prob. 21–17). Substituting into Eq. 1, we have

$$(-0.667\mathbf{i} + 0.5\mathbf{j}) \times \left[\left(\frac{4.5}{32.2}\right)(-10\mathbf{k})\right] = (-0.667\mathbf{i} + 0.5\mathbf{j}) \times \left[\left(\frac{4.5}{32.2}\right)(-v_G)_2\mathbf{k}\right]$$
$$+ 0.0272\omega_x\mathbf{i} + 0.0155\omega_y\mathbf{j} + 0.0427\omega_z\mathbf{k}$$

Expanding and equating the respective **i, j, k** components yields

$$-0.699 = -0.0699(v_G)_2 + 0.0272\omega_x \qquad (2)$$
$$-0.932 = -0.0932(v_G)_2 + 0.0155\omega_y \qquad (3)$$
$$0 = 0.0427\omega_z \qquad (4)$$

Kinematics. There are four unknowns in the above equations; however, another equation may be obtained by relating $\boldsymbol{\omega}$ to $(\mathbf{v}_G)_2$ using kinematics. Since $\omega_z = 0$ (Eq. 4) and after impact the rod rotates about the fixed point A, Eq. 20–3 can be applied, in which case $(\mathbf{v}_G)_2 = \boldsymbol{\omega} \times \mathbf{r}_{G/A}$, or

$$-(v_G)_2\mathbf{k} = (\omega_x\mathbf{i} + \omega_y\mathbf{j}) \times (-0.667\mathbf{i} + 0.5\mathbf{j})$$
$$-(v_G)_2 = 0.5\omega_x + 0.667\omega_y \qquad (5)$$

Solving Eqs. 2, 3, and 5 simultaneously yields

$$(\mathbf{v}_G)_2 = \{-8.41\mathbf{k}\}\ \text{ft/s} \qquad \boldsymbol{\omega} = \{-4.09\mathbf{i} - 9.55\mathbf{j}\}\ \text{rad/s} \qquad Ans.$$

Fig. 21–9

EXAMPLE 21.3

A 5-N·m torque is applied to the vertical shaft *CD* shown in Fig. 21–10a, which causes the 10-kg gear *A* to turn freely about the axle *CE*. Assuming that *A* starts from rest, determine the angular velocity of *CD* after it has turned two revolutions. Neglect the mass of *CD* and *CE* and assume that *A* can be approximated by a thin disk. Gear *B* is fixed.

SOLUTION
The principle of work and energy can be used for the solution. Why?

Work. If *CD*, *CE*, and *A* are considered as a system of connected bodies, only the applied torque **M** does work. For two revolutions of *CD*, this work is $\Sigma U_{1-2} = (5 \text{ N} \cdot \text{m})(4\pi \text{ rad}) = 62.83$ J.

Kinetic Energy. Since the gear is initially at rest, its initial kinetic energy is zero. A kinematic diagram for the gear when it is moving is shown in Fig. 21–10b. If the angular velocity of *CD* is taken as ω_{CD}, then the angular velocity of *A* is $\boldsymbol{\omega}_A = \boldsymbol{\omega}_{CD} + \boldsymbol{\omega}_{CE}$. The gear may be imagined as a portion of a massless extended body which is rotating about the *fixed point C*. The instantaneous axis of rotation for this body is along line *CH*, because both points *C* and *H* on the body (gear) have zero velocity and must therefore lie on this axis. This requires that the components ω_{CD} and ω_{CE} be related by the equation $\omega_{CD}/0.1 \text{ m} = \omega_{CE}/0.3 \text{ m}$ or $\omega_{CE} = 3\omega_{CD}$. Thus,

$$\boldsymbol{\omega}_A = -\omega_{CE}\mathbf{i} + \omega_{CD}\mathbf{k} = -3\omega_{CD}\mathbf{i} + \omega_{CD}\mathbf{k} \qquad (1)$$

The *x*, *y*, *z* axes in Fig. 21–10a represent *principal axes of inertia* at *C* for the gear. Since point *C* is a fixed point of rotation, Eq. 21–15 may be applied to determine the kinetic energy, i.e.,

$$T = \tfrac{1}{2}I_x\omega_x^2 + \tfrac{1}{2}I_y\omega_y^2 + \tfrac{1}{2}I_z\omega_z^2 \qquad (2)$$

Using the parallel-axis theorem, the moments of inertia of the gear about point *C* are as follows:

$$I_x = \tfrac{1}{2}(10 \text{ kg})(0.1 \text{ m})^2 = 0.05 \text{ kg} \cdot \text{m}^2$$
$$I_y = I_z = \tfrac{1}{4}(10 \text{ kg})(0.1 \text{ m})^2 + 10 \text{ kg}(0.3 \text{ m})^2 = 0.925 \text{ kg} \cdot \text{m}^2$$

Since $\omega_x = -3\omega_{CD}$, $\omega_y = 0$, $\omega_z = \omega_{CD}$, Eq. 2 becomes

$$T_A = \tfrac{1}{2}(0.05)(-3\omega_{CD})^2 + 0 + \tfrac{1}{2}(0.925)(\omega_{CD})^2 = 0.6875\omega_{CD}^2$$

Principle of Work and Energy. Applying the principle of work and energy, we obtain

$$T_1 + \Sigma U_{1-2} = T_2$$
$$0 + 62.83 = 0.6875\omega_{CD}^2$$
$$\omega_{CD} = 9.56 \text{ rad/s} \qquad \qquad Ans.$$

(a)

(b)

Fig. 21–10

PROBLEMS

21–22. If a body contains *no planes of symmetry*, the principal moments of inertia can be determined mathematically. To show how this is done, consider the rigid body which is spinning with an angular velocity $\boldsymbol{\omega}$, directed along one of its principal axes of inertia. If the principal moment of inertia about this axis is I, the angular momentum can be expressed as $\mathbf{H} = I\boldsymbol{\omega} = I\omega_x \mathbf{i} + I\omega_y \mathbf{j} + I\omega_z \mathbf{k}$. The components of $\mathbf{H}$ may also be expressed by Eqs. 21-10, where the inertia tensor is assumed to be known. Equate the $\mathbf{i}, \mathbf{j}$, and $\mathbf{k}$ components of both expressions for $\mathbf{H}$ and consider ω_x, ω_y, and ω_z to be unknown. The solution of these three equations is obtained provided the determinant of the coefficients is zero. Show that this determinant, when expanded, yields the cubic equation

$$I^3 - (I_{xx} + I_{yy} + I_{zz})I^2$$
$$+ (I_{xx}I_{yy} + I_{yy}I_{zz} + I_{zz}I_{xx} - I_{xy}^2 - I_{yz}^2 - I_{zx}^2)I$$
$$- (I_{xx}I_{yy}I_{zz} - 2I_{xy}I_{yz}I_{zx} - I_{xx}I_{yz}^2 - I_{yy}I_{zx}^2 - I_{zz}I_{xy}^2) = 0$$

The three positive roots of I, obtained from the solution of this equation, represent the principal moments of inertia I_x, I_y, and I_z.

21–23. Show that if the angular momentum of a body is determined with respect to an arbitrary point A, then $\mathbf{H}_A$ can be expressed by Eq. 21-9. This requires substituting $\boldsymbol{\rho}_A = \boldsymbol{\rho}_G + \boldsymbol{\rho}_{G/A}$ into Eq. 21-6 and expanding, noting that $\int \boldsymbol{\rho}_G \, dm = \mathbf{0}$ by definition of the mass center and $\mathbf{v}_G = \mathbf{v}_A + \boldsymbol{\omega} \times \boldsymbol{\rho}_{G/A}$.

Prob. 21–23

*21–24. A thin plate, having a mass of 4 kg is suspended from one of its corners by a ball-and-socket joint O. If a stone strikes the plate perpendicular to its surface at an adjacent corner A with an impulse of $\mathbf{I}_s = \{-60\mathbf{i}\} \, \mathrm{N \cdot s}$, determine the instantaneous axis of rotation for the plate and the impulse created at O.

Prob. 21–22

Prob. 21–24

21–25. The large gear has a mass of 5 kg and a radius of gyration of $k_z = 75$ mm. Gears B and C each have a mass of 200 g and a radius of gyration about the axis of their connecting shaft of 15 mm. If the gears are in mesh and C has an angular velocity of $\omega_C = \{15\mathbf{j}\}$ rad/s, determine the total angular momentum for the system of three gears about point A.

*21–28. The rod assembly is supported at G by a ball-and-socket joint. Each segment has a mass of 0.5 kg/m. If the assembly is originally at rest and an impulse of $\mathbf{I} = \{-8\mathbf{k}\}$ N·s is applied at D, determine the angular velocity of the assembly just after the impact.

Prob. 21–25

Prob. 21–28

21–26. The 10-kg circular disk spins about its axle with a constant angular velocity of $\omega_1 = 15$ rad/s. Simultaneously, arm OB and shaft OA rotate about their axes with constant angular velocities of $\omega_2 = 0$ and $\omega_3 = 6$ rad/s, respectively. Determine the angular momentum of the disk about point O, and its kinetic energy.

21–27. The 10-kg circular disk spins about its axle with a constant angular velocity of $\omega_1 = 15$ rad/s. Simultaneously, arm OB and shaft OA rotate about their axes with constant angular velocities of $\omega_2 = 10$ rad/s and $\omega_3 = 6$ rad/s, respectively. Determine the angular momentum of the disk about point O, and its kinetic energy.

21–29. The 4-kg rod AB is attached to the 1-kg collar at A and a 2-kg link BC using ball-and-socket joints. If the rod is released from rest in the position shown, determine the angular velocity of the link after it has rotated 180°.

Probs. 21–26/27

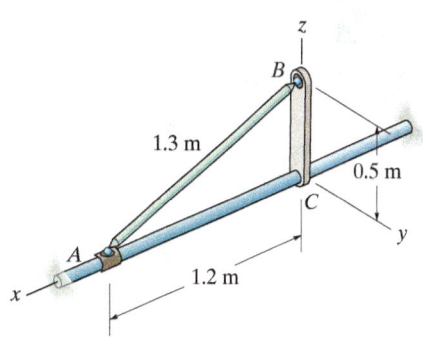

Prob. 21–29

21–30. The rod weighs 3 lb/ft and is suspended from parallel cords at A and B. If the rod has an angular velocity of 2 rad/s about the z axis at the instant shown, determine how high the center of the rod rises at the instant the rod momentarily stops swinging.

*21–32. The 2-kg thin disk is connected to the slender rod which is fixed to the ball-and-socket joint at A. If it is released from rest in the position shown, determine the spin of the disk about the rod when the disk reaches its lowest position. Neglect the mass of the rod. The disk rolls without slipping.

Prob. 21–30

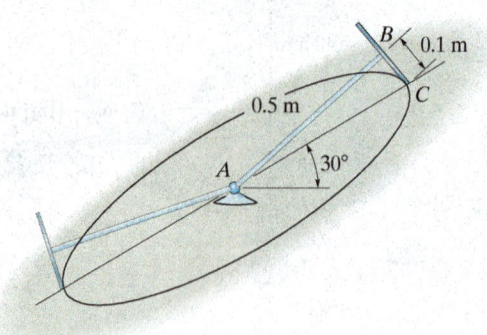

Prob. 21–32

21–31. The rod assembly has a mass of 2.5 kg/m and is rotating with a constant angular velocity of $\omega = \{2\mathbf{k}\}$ rad/s when the looped end at C encounters a hook at S, which provides a permanent connection. Determine the angular velocity of the assembly immediately after impact.

21–33. The 20-kg sphere rotates about the axle with a constant angular velocity of $\omega_s = 60$ rad/s. If shaft AB is subjected to a torque of $M = 50$ N·m, causing it to rotate, determine the value of ω_p after the shaft has turned 90° from the position shown. Initially, $\omega_p = 0$. Neglect the mass of arm CDE.

Prob. 21–31

Prob. 21–33

21–34. The 25-lb thin plate is suspended from a ball-and-socket joint at O. A 0.2-lb projectile is fired with a velocity of $\mathbf{v} = \{-300\mathbf{i} - 250\mathbf{j} + 300\mathbf{k}\}$ ft/s into the plate and becomes embedded in the plate at point A. Determine the angular velocity of the plate just after impact and the axis about which it begins to rotate. Neglect the mass of the projectile after it embeds into the plate.

21–35. Solve Prob. 21–34 if the projectile emerges from the plate with a velocity of 275 ft/s in the same direction.

Probs. 21–34/35

*21–36. The 15-kg rectangular plate is free to rotate about the y axis because of the bearing supports at A and B. When the plate is balanced in the vertical plane, a 3-g bullet is fired into it, perpendicular to its surface, with a velocity $\mathbf{v} = \{-2000\mathbf{i}\}$ m/s. Determine the angular velocity of the plate at the instant it has rotated 180°. If the bullet strikes corner D with the same velocity $\mathbf{v}$, instead of at C, does the angular velocity remain the same? Why or why not?

Prob. 21–36

21–37. The 5-kg thin plate is suspended at O using a ball-and-socket joint. It is rotating with a constant angular velocity $\boldsymbol{\omega} = \{2\mathbf{k}\}$ rad/s when the corner A strikes the hook at S, which provides a permanent connection. Determine the angular velocity of the plate immediately after impact.

Prob. 21–37

21–38. Rod AB has a weight of 6 lb and is attached to two smooth collars at its end points by ball-and-socket joints. If collar A is moving downward at a speed of 8 ft/s, determine the kinetic energy of the rod at the instant shown. *Hint:* This is determined from the angular velocity component of the rod that is directed perpendicular to the rod's axis.

21–39. At the instant shown the collar at A on the 6-lb rod AB has a velocity of $v_A = 8$ ft/s. Determine the kinetic energy of the rod after the collar A has descended 3 ft. Each collar is attached to the rod using a ball-and-socket joint. Neglect friction, the thickness of the rod, and the mass of the collars. *Hint:* Use the angular velocity component of the rod that is directed perpendicular to the rod's axis.

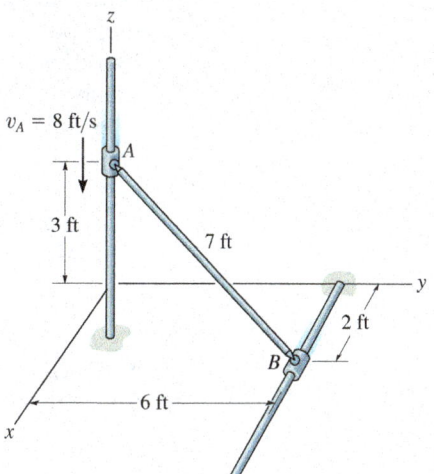

Probs. 21–38/39

*21.4 EQUATIONS OF MOTION

Having become familiar with the techniques used to describe both the inertial properties and the angular momentum of a body, we can now write the equations which describe the motion of the body in their most useful forms.

Equations of Translational Motion.
The *translational motion* of a body is defined in terms of the acceleration of the body's mass center, which is measured from an inertial X, Y, Z reference. This equation can be written in vector form as

$$\Sigma \mathbf{F} = m\mathbf{a}_G \qquad (21\text{–}18)$$

or by the three scalar equations

$$\begin{aligned} \Sigma F_x &= m(a_G)_x \\ \Sigma F_y &= m(a_G)_y \\ \Sigma F_z &= m(a_G)_z \end{aligned} \qquad (21\text{–}19)$$

Here, $\Sigma \mathbf{F} = \Sigma F_x \mathbf{i} + \Sigma F_y \mathbf{j} + \Sigma F_z \mathbf{k}$ represents the sum of all the external forces acting on the body.

Equations of Rotational Motion.
In Sec. 15.6, we developed Eq. 15–17, namely,

$$\Sigma \mathbf{M}_O = \dot{\mathbf{H}}_O \qquad (21\text{–}20)$$

which states that the sum of the moments of all the external forces acting on a system of particles contained in a rigid body about a *fixed point O* is equal to the time rate of change of the total angular momentum of the particles about point O. When moments of the external forces are summed about the system's *mass center G*, one again obtains the same simple form of Eq. 21–20, relating the moment summation $\Sigma \mathbf{M}_G$ to the angular momentum $\mathbf{H}_G$.

To show this, consider the system of particles in Fig. 21–11, where the x, y, z axes have their origin at G, and translate relative to the X, Y, Z inertial coordinate system. In general, G is *accelerating*, so by definition the translating frame is *not* an inertial reference. The angular momentum of the ith particle with respect to this frame is, however,

$$(\mathbf{H}_i)_G = \mathbf{r}_{i/G} \times m_i \mathbf{v}_{i/G}$$

where $\mathbf{r}_{i/G}$ and $\mathbf{v}_{i/G}$ represent the position and velocity of the ith particle with respect to G. Taking the time derivative we have

$$(\dot{\mathbf{H}}_i)_G = \dot{\mathbf{r}}_{i/G} \times m_i \mathbf{v}_{i/G} + \mathbf{r}_{i/G} \times m_i \dot{\mathbf{v}}_{i/G}$$

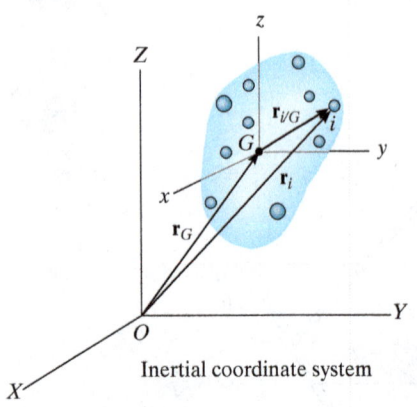

Inertial coordinate system

Fig. 21–11

21.4 EQUATIONS OF MOTION

By definition, $\mathbf{v}_{i/G} = \dot{\mathbf{r}}_{i/G}$. Thus, the first term on the right side is zero since the cross product of the same vectors is zero. Also, $\mathbf{a}_{i/G} = \dot{\mathbf{v}}_{i/G}$, so that

$$(\dot{\mathbf{H}}_i)_G = (\mathbf{r}_{i/G} \times m_i \mathbf{a}_{i/G})$$

Similar expressions can be written for the other particles of the body. When the results are summed, we get

$$\dot{\mathbf{H}}_G = \Sigma(\mathbf{r}_{i/G} \times m_i \mathbf{a}_{i/G})$$

Here $\dot{\mathbf{H}}_G$ is the time rate of change of the total angular momentum of the body about G.

The relative acceleration for the ith particle is defined by the equation $\mathbf{a}_{i/G} = \mathbf{a}_i - \mathbf{a}_G$, where $\mathbf{a}_i$ and $\mathbf{a}_G$ represent, respectively, the accelerations of the ith particle and point G measured with respect to the inertial frame of reference. Substituting and expanding, using the distributive property of the vector cross product, yields

$$\dot{\mathbf{H}}_G = \Sigma(\mathbf{r}_{i/G} \times m_i \mathbf{a}_i) - (\Sigma m_i \mathbf{r}_{i/G}) \times \mathbf{a}_G$$

By definition of the mass center, the sum $(\Sigma m_i \mathbf{r}_{i/G}) = (\Sigma m_i)\bar{\mathbf{r}}$ is equal to zero, since the position vector $\bar{\mathbf{r}}$ relative to G is zero. Hence, the last term in the above equation is zero. Using the equation of motion, the product $m_i \mathbf{a}_i$ can be replaced by the resultant external force $\mathbf{F}_i$ acting on the ith particle. Denoting $\Sigma \mathbf{M}_G = \Sigma(\mathbf{r}_{i/G} \times \mathbf{F}_i)$, the final result can be written as

$$\Sigma \mathbf{M}_G = \dot{\mathbf{H}}_G \qquad (21\text{-}21)$$

For some problems it is convenient to allow the x, y, z axes to *rotate* rather then just translate. If these axes have an angular velocity $\boldsymbol{\Omega}$, then the time derivative $\dot{\mathbf{H}} = d\mathbf{H}/dt$, as used in Eqs. 21–20 and 21–21, must account for this rotation as measured from the inertial X, Y, Z axes. This requires application of Eq. 20–6, in which case Eqs. 21–20 and 21–21 become

$$\Sigma \mathbf{M}_O = (\dot{\mathbf{H}}_O)_{xyz} + \boldsymbol{\Omega} \times \mathbf{H}_O$$
$$\Sigma \mathbf{M}_G = (\dot{\mathbf{H}}_G)_{xyz} + \boldsymbol{\Omega} \times \mathbf{H}_G \qquad (21\text{-}22)$$

Here $(\dot{\mathbf{H}})_{xyz}$ is the time rate of change of $\mathbf{H}$ measured from the x, y, z reference.

There are three ways in which one can define the motion of the x, y, z axes. Obviously, motion of this reference should be chosen so that it will yield the simplest set of moment equations for the solution of a particular problem.

x, y, z Axes Having Motion $\Omega = 0$. The x, y, z axes can be chosen such that the axes only *translate* relative to the inertial X, Y, Z frame of reference. Doing this simplifies Eq. 21–22, since $\Omega = 0$. However, the body may have a rotation ω about these axes, and if this is the case then the moments and products of inertia of the body will have to be expressed as *functions of time*. In most cases this would be a difficult task, so that such a choice of axes has restricted application.

x, y, z Axes Having Motion $\Omega = \omega$. The x, y, z axes can be chosen such that they are *fixed in and move with the body*. The moments and products of inertia of the body relative to these axes will then be *constant* during the motion. Since $\Omega = \omega$, Eqs. 21–22 become

$$\Sigma \mathbf{M}_O = (\dot{\mathbf{H}}_O)_{xyz} + \boldsymbol{\omega} \times \mathbf{H}_O$$
$$\Sigma \mathbf{M}_G = (\dot{\mathbf{H}}_G)_{xyz} + \boldsymbol{\omega} \times \mathbf{H}_G \tag{21-23}$$

We can express each of these vector equations as three scalar equations using Eqs. 21–10. Neglecting the subscripts O and G, we have

$$\begin{aligned}
\Sigma M_x &= I_{xx}\dot{\omega}_x - (I_{yy} - I_{zz})\omega_y\omega_z - I_{xy}(\dot{\omega}_y - \omega_z\omega_x) \\
&\quad - I_{yz}(\omega_y^2 - \omega_z^2) - I_{zx}(\dot{\omega}_z + \omega_x\omega_y) \\
\Sigma M_y &= I_{yy}\dot{\omega}_y - (I_{zz} - I_{xx})\omega_z\omega_x - I_{yz}(\dot{\omega}_z - \omega_x\omega_y) \\
&\quad - I_{zx}(\omega_z^2 - \omega_x^2) - I_{xy}(\dot{\omega}_x + \omega_y\omega_z) \\
\Sigma M_z &= I_{zz}\dot{\omega}_z - (I_{xx} - I_{yy})\omega_x\omega_y - I_{zx}(\dot{\omega}_x - \omega_y\omega_z) \\
&\quad - I_{xy}(\omega_x^2 - \omega_y^2) - I_{yz}(\dot{\omega}_y + \omega_z\omega_x)
\end{aligned} \tag{21-24}$$

If the x, y, z axes are chosen as *principal axes of inertia*, the products of inertia are zero, $I_{xx} = I_x$, etc., and the above equations become

$$\begin{aligned}
\Sigma M_x &= I_x\dot{\omega}_x - (I_y - I_z)\omega_y\omega_z \\
\Sigma M_y &= I_y\dot{\omega}_y - (I_z - I_x)\omega_z\omega_x \\
\Sigma M_z &= I_z\dot{\omega}_z - (I_x - I_y)\omega_x\omega_y
\end{aligned} \tag{21-25}$$

This set of equations is known historically as the **Euler equations of motion**, named after the Swiss mathematician Leonhard Euler, who first developed them. They apply *only* for moments summed about either point O or G.

When applying these equations it should be realized that $\dot{\omega}_x$, $\dot{\omega}_y$, $\dot{\omega}_z$ represent the time derivatives of the magnitudes of the x, y, z components of $\boldsymbol{\omega}$ as observed from x, y, z. To determine them, it is first necessary to find $\omega_x, \omega_y, \omega_z$ when the x, y, z axes are oriented in a *general position* and *then* take the time derivative, i.e., $(\dot{\boldsymbol{\omega}})_{xyz}$. However, for this case the x, y, z axes are rotating at $\boldsymbol{\Omega} = \boldsymbol{\omega}$, and so Eq. 20–6 becomes $\dot{\boldsymbol{\omega}} = (\dot{\boldsymbol{\omega}})_{xyz} + \boldsymbol{\omega} \times \boldsymbol{\omega}$. Since $\boldsymbol{\omega} \times \boldsymbol{\omega} = \mathbf{0}$, then $\dot{\boldsymbol{\omega}} = (\dot{\boldsymbol{\omega}})_{xyz}$. This important result indicates that the time derivative of $\boldsymbol{\omega}$ with respect to the X, Y, Z axes, that is $\dot{\boldsymbol{\omega}}$, can also be used to obtain $(\dot{\boldsymbol{\omega}})_{xyz}$. Generally this is the easiest way to determine the result, as shown in Example 21.5.

x, y, z Axes Having Motion $\Omega \neq \omega$.

To simplify the calculations for finding $\dot{\boldsymbol{\omega}}$, it is often convenient to choose the x, y, z axes having an angular velocity $\boldsymbol{\Omega}$ which is different from the angular velocity $\boldsymbol{\omega}$ of the body. This is particularly suitable for the analysis of spinning tops and gyroscopes which are *symmetrical* about their spinning axes.* When this is the case, the moments and products of inertia will remain constant about the axis of spin.

Equations 21–22 are applicable for such a set of axes. Each of these two vector equations can be reduced to a set of three scalar equations which are derived in a manner similar to Eqs. 21–25,† i.e.,

$$\Sigma M_x = I_x \dot{\omega}_x - I_y \Omega_z \omega_y + I_z \Omega_y \omega_z$$
$$\Sigma M_y = I_y \dot{\omega}_y - I_z \Omega_x \omega_z + I_x \Omega_z \omega_x \qquad (21\text{--}26)$$
$$\Sigma M_z = I_z \dot{\omega}_z - I_x \Omega_y \omega_x + I_y \Omega_x \omega_y$$

Here Ω_x, Ω_y, Ω_z represent the x, y, z components of $\boldsymbol{\Omega}$, measured from the inertial frame of reference, and $\dot{\omega}_x$, $\dot{\omega}_y$, $\dot{\omega}_z$ must be determined relative to the x, y, z axes that have the rotation $\boldsymbol{\Omega}$. See Example 21.6.

Any one of these sets of moment equations, Eqs. 21–24, 21–25, or 21–26, represents a series of three first-order nonlinear differential equations that are "coupled," since the angular-velocity components are present in all the terms. Success in determining the solution for a particular problem therefore depends upon what is unknown in these equations. Difficulty certainly arises when one attempts to solve for the unknown components of $\boldsymbol{\omega}$ when the external moments are functions of time. Further complications can arise if the moment equations are coupled to the three scalar equations of translational motion, Eqs. 21–19. For example, this can happen when the rotation of the body is related to the translation of its mass center, as in the case of a hoop which rolls without slipping. Problems such as this, that require the simultaneous

---

* A detailed discussion of such devices is given in Sec. 21.5.
† See Prob. 21–42.

solution of differential equations, are generally solved using numerical methods with the aid of a computer. In many engineering problems, however, we are given information about the motion of the body and are required to determine the applied moments acting on the body. Most of these problems have direct solutions, so that there is no need to resort to computer techniques.

PROCEDURE FOR ANALYSIS

Problems involving the three-dimensional motion of a rigid body can be solved using the following procedure.

Free-Body Diagram.

- Draw a *free-body diagram* of the body at the instant considered and specify the x, y, z coordinate system. The origin of this reference must be located either at the body's mass center G, or at point O, considered fixed in an inertial reference frame and located either in the body or on a massless extension of the body.

- Unknown reactive force components can be shown having a positive sense of direction.

- Depending on the nature of the problem, decide what type of rotational motion Ω the x, y, z coordinate system should have, i.e., $\Omega = 0$, $\Omega = \omega$, or $\Omega \neq \omega$. When choosing, keep in mind that the moment equations are simplified when the axes move in such a manner that they represent principal axes of inertia for the body at all times.

- Calculate the necessary moments and products of inertia for the body relative to the x, y, z axes.

Kinematics.

- Determine the x, y, z components of the body's angular velocity and find the time derivatives of $\boldsymbol{\omega}$.

- Note that if $\Omega = \omega$, then $\dot{\boldsymbol{\omega}} = (\dot{\boldsymbol{\omega}})_{xyz}$. Therefore we can either find the time derivative of $\boldsymbol{\omega}$ with respect to the X, Y, Z axes, $\dot{\boldsymbol{\omega}}$, and then determine its components $\dot{\omega}_x, \dot{\omega}_y, \dot{\omega}_z$, or we can find the components of $\boldsymbol{\omega}$ along the x, y, z axes, when the axes are oriented in a general position, and then take the time derivative of the magnitudes of these components, $(\dot{\boldsymbol{\omega}})_{xyz}$.

Equations of Motion.

- Apply either the two vector equations 21–18 and 21–22 or the six scalar component equations appropriate for the x, y, z coordinate axes chosen for the problem.

EXAMPLE 21.4

The gear shown in Fig. 21–12a has a mass of 10 kg and is mounted at an angle of 10° with the rotating shaft having negligible mass. If $I_z = 0.1$ kg·m², $I_x = I_y = 0.05$ kg·m², and the shaft is rotating with a constant angular velocity of $\omega = 30$ rad/s, determine the components of reaction that the thrust bearing A and journal bearing B exert on the shaft at the instant shown.

SOLUTION
Free-Body Diagram. Fig. 21–12b. The origin of the x, y, z coordinate system is located at the gear's center of mass G, which is also a fixed point. The axes are fixed in and rotate with the gear so that these axes always represent the principal axes of inertia for the gear. Hence $\Omega = \omega$.

Kinematics. As shown in Fig. 21–12c, the angular velocity ω of the gear is constant in magnitude and is always directed along the axis of the shaft AB. Since this vector is measured from the X, Y, Z inertial frame of reference, for any position of the x, y, z axes,

$$\omega_x = 0 \quad \omega_y = -30 \sin 10° \quad \omega_z = 30 \cos 10°$$

These components remain constant for any general orientation of the x, y, z axes, and so relative to the x, y, z axes $\dot\omega_x = \dot\omega_y = \dot\omega_z = 0$. Also note that since $\Omega = \omega$, then $\dot\omega = (\dot\omega)_{xyz}$. Therefore, we can also find these time derivatives of ω relative to the X, Y, Z axes. Since ω has a constant magnitude and direction $(+Z)$ then $\dot\omega = 0$, and so again $\dot\omega_x = \dot\omega_y = \dot\omega_z = 0$. Furthermore, since G is a fixed point, $(a_G)_x = (a_G)_y = (a_G)_z = 0$.

Equations of Motion. Applying Eqs. 21–25 ($\Omega = \omega$) yields

$$\Sigma M_x = I_x \dot\omega_x - (I_y - I_z)\omega_y \omega_z$$
$$-(A_Y)(0.2) + (B_Y)(0.25) = 0 - (0.05 - 0.1)(-30 \sin 10°)(30 \cos 10°)$$
$$-0.2 A_Y + 0.25 B_Y = -7.70 \quad (1)$$

$$\Sigma M_y = I_y \dot\omega_y - (I_z - I_x)\omega_z \omega_x$$
$$A_X(0.2) \cos 10° - B_X(0.25) \cos 10° = 0 - 0$$
$$A_X = 1.25 B_X \quad (2)$$

$$\Sigma M_z = I_z \dot\omega_z - (I_x - I_y)\omega_x \omega_y$$
$$A_X(0.2) \sin 10° - B_X(0.25) \sin 10° = 0 - 0$$
$$A_X = 1.25 B_X \text{(check)}$$

Fig. 21–12

Applying Eqs. 21–19, we have

$\Sigma F_X = m(a_G)_X;$ $\quad A_X + B_X = 0 \quad (3)$
$\Sigma F_Y = m(a_G)_Y;$ $\quad A_Y + B_Y - 98.1 = 0 \quad (4)$
$\Sigma F_Z = m(a_G)_Z;$ $\quad A_Z = 0 \quad$ **Ans.**

Solving Eqs. 1 through 4 simultaneously gives

$\quad A_X = B_X = 0 \quad A_Y = 71.6$ N $\quad B_Y = 26.5$ N $\quad$ **Ans.**

592   CHAPTER 21   THREE-DIMENSIONAL KINETICS OF A RIGID BODY

EXAMPLE 21.5

The airplane shown in Fig. 21–13a is in the process of making a steady *horizontal* turn at the rate of ω_p. During this motion, the propeller is spinning at the rate of ω_s. If the propeller has two blades, determine the moments which the propeller shaft exerts on the propeller at the instant the blades are in the vertical position. For simplicity, assume the blades to be a uniform slender bar having a moment of inertia I about an axis perpendicular to the blades passing through the center of the bar, and having zero moment of inertia about a longitudinal axis.

(a)

SOLUTION

Free-Body Diagram. Fig. 21–13b. The reactions of the connecting shaft on the propeller are indicated by the resultants $\mathbf{F}_R$ and $\mathbf{M}_R$. (The propeller's weight is assumed to be negligible.) The x, y, z axes will be taken fixed to the propeller, such that these axes always represent the principal axes of inertia for the propeller. The moments of inertia I_x and I_y are equal ($I_x = I_y = I$) and $I_z = 0$.

Kinematics. The angular velocity of the propeller observed from the X, Y, Z axes, coincident with the x, y, z axes, Fig. 21–13c, is $\boldsymbol{\omega} = \boldsymbol{\omega}_s + \boldsymbol{\omega}_p = \omega_s\mathbf{i} + \omega_p\mathbf{k}$, so that the x, y, z components of $\boldsymbol{\omega}$ are

$$\omega_x = \omega_s \qquad \omega_y = 0 \qquad \omega_z = \omega_p$$

Since $\boldsymbol{\Omega} = \boldsymbol{\omega}$, then $\dot{\boldsymbol{\omega}} = (\dot{\boldsymbol{\omega}})_{xyz}$. To find $\dot{\boldsymbol{\omega}}$, which is the time derivative with respect to the fixed X, Y, Z axes, we must use Eq. 20–6 since $\boldsymbol{\omega}$ changes direction relative to X, Y, Z. The time rate of change of each component $\dot{\boldsymbol{\omega}} = \dot{\boldsymbol{\omega}}_s + \dot{\boldsymbol{\omega}}_p$ relative to the X, Y, Z axes can be obtained by introducing a third coordinate system x', y', z', which has an angular velocity $\boldsymbol{\Omega}' = \boldsymbol{\omega}_p$ and is coincident with the X, Y, Z axes at the instant shown. Thus

Fig. 21–13

$$\begin{aligned}
\dot{\boldsymbol{\omega}} &= (\dot{\boldsymbol{\omega}})_{x'y'z'} + \boldsymbol{\omega}_p \times \boldsymbol{\omega} \\
&= (\dot{\boldsymbol{\omega}}_s)_{x'y'z'} + (\dot{\boldsymbol{\omega}}_p)_{x'y'z'} + \boldsymbol{\omega}_p \times (\boldsymbol{\omega}_s + \boldsymbol{\omega}_p) \\
&= \mathbf{0} + \mathbf{0} + \boldsymbol{\omega}_p \times \boldsymbol{\omega}_s + \boldsymbol{\omega}_p \times \boldsymbol{\omega}_p \\
&= \mathbf{0} + \mathbf{0} + \omega_p \mathbf{k} \times \omega_s \mathbf{i} + \mathbf{0} = \omega_p \omega_s \mathbf{j}
\end{aligned}$$

Since the X, Y, Z axes are coincident with the x, y, z axes at the instant shown, the components of $\dot{\boldsymbol{\omega}}$ along x, y, z are therefore

$$\dot{\omega}_x = 0 \quad \dot{\omega}_y = \omega_p \omega_s \quad \dot{\omega}_z = 0$$

These same results can also be determined by direct calculation of $(\dot{\boldsymbol{\omega}})_{xyz}$; however, this will involve a bit more work. To do this, it will be necessary to view the propeller (or the x, y, z axes) in some *general position* such as shown in Fig. 21–13d. Here the plane has turned through an angle ϕ (phi) and the propeller has turned through an angle ψ (psi) relative to the plane. Notice that $\boldsymbol{\omega}_p$ is always directed along the fixed Z axis and $\boldsymbol{\omega}_s$ follows the x axis. Thus the general components of $\boldsymbol{\omega}$ are

$$\omega_x = \omega_s \quad \omega_y = \omega_p \sin \psi \quad \omega_z = \omega_p \cos \psi$$

Since ω_s and ω_p are constant, the time derivatives of these components become

$$\dot{\omega}_x = 0 \quad \dot{\omega}_y = \omega_p \cos \psi \, \dot{\psi} \quad \dot{\omega}_z = -\omega_p \sin \psi \, \dot{\psi}$$

But $\phi = \psi = 0°$ and $\dot{\psi} = \omega_s$ at the instant considered. Thus,

$$\omega_x = \omega_s \quad \omega_y = 0 \quad \omega_z = \omega_p$$
$$\dot{\omega}_x = 0 \quad \dot{\omega}_y = \omega_p \omega_s \quad \dot{\omega}_z = 0$$

which are the same results as those obtained previously.

Fig. 21–13

Equations of Motion.
Using Eqs. 21–25, we have

$$\Sigma M_x = I_x \dot{\omega}_x - (I_y - I_z)\omega_y \omega_z = I(0) - (I - 0)(0)\omega_p$$
$$M_x = 0 \quad\quad Ans.$$

$$\Sigma M_y = I_y \dot{\omega}_y - (I_z - I_x)\omega_z \omega_x = I(\omega_p \omega_s) - (0 - I)\omega_p \omega_s$$
$$M_y = 2I\omega_p \omega_s \quad\quad Ans.$$

$$\Sigma M_z = I_z \dot{\omega}_z - (I_x - I_y)\omega_x \omega_y = 0(0) - (I - I)\omega_s(0)$$
$$M_z = 0 \quad\quad Ans.$$

EXAMPLE 21.6

The 10-kg flywheel (or thin disk) shown in Fig. 21–14a rotates (spins) about the shaft at a constant angular velocity of $\omega_s = 6$ rad/s. At the same time, the shaft rotates (precesses) about the bearing at A with an angular velocity of $\omega_p = 3$ rad/s. If A is a thrust bearing and B is a journal bearing, determine the components of force reaction at each of these supports due to the motion.

SOLUTION I

Free-Body Diagram. Fig. 21–14b. The origin of the x, y, z coordinate system is located at the center of mass G of the flywheel. Here we will let these coordinates have an angular velocity of $\mathbf{\Omega} = \boldsymbol{\omega}_p = \{3\mathbf{k}\}$ rad/s. Although the flywheel spins relative to these axes, the moments of inertia *remain constant,*[*] i.e.,

$$I_x = I_z = \tfrac{1}{4}(10 \text{ kg})(0.2 \text{ m})^2 = 0.1 \text{ kg} \cdot \text{m}^2$$

$$I_y = \tfrac{1}{2}(10 \text{ kg})(0.2 \text{ m})^2 = 0.2 \text{ kg} \cdot \text{m}^2$$

Fig. 21–14

Kinematics. From the coincident inertial X, Y, Z frame of reference, Fig. 21–14c, the flywheel has an angular velocity of $\boldsymbol{\omega} = \{6\mathbf{j} + 3\mathbf{k}\}$ rad/s, so that

$$\omega_x = 0 \quad \omega_y = 6 \text{ rad/s} \quad \omega_z = 3 \text{ rad/s}$$

The time derivative of $\boldsymbol{\omega}$ must be determined relative to the x, y, z axes. In this case both $\boldsymbol{\omega}_p$ and $\boldsymbol{\omega}_s$ do not change their magnitude or direction, and so

$$\dot{\omega}_x = 0 \quad \dot{\omega}_y = 0 \quad \dot{\omega}_z = 0$$

Equations of Motion. Applying Eqs. 21–26 ($\mathbf{\Omega} \neq \boldsymbol{\omega}$) yields

$$\Sigma M_x = I_x \dot{\omega}_x - I_y \Omega_z \omega_y + I_z \Omega_y \omega_z$$

$$-A_z(0.5) + B_z(0.5) = 0 - (0.2)(3)(6) + 0 = -3.6$$

$$\Sigma M_y = I_y \dot{\omega}_y - I_z \Omega_x \omega_z + I_x \Omega_z \omega_x$$

$$0 = 0 - 0 + 0$$

$$\Sigma M_z = I_z \dot{\omega}_z - I_x \Omega_y \omega_x + I_y \Omega_x \omega_y$$

$$A_x(0.5) - B_x(0.5) = 0 - 0 + 0$$

---

[*]This would not be true for the propeller in Example 21.5.

Applying Eqs. 21–19, we have

$\Sigma F_X = m(a_G)_X;$ $A_x + B_x = 0$

$\Sigma F_Y = m(a_G)_Y;$ $A_y = -10(0.5)(3)^2$

$\Sigma F_Z = m(a_G)_Z;$ $A_z + B_z - 10(9.81) = 0$

Solving these equations, we obtain

$A_x = 0$ $A_y = -45.0$ N $A_z = 52.6$ N   *Ans.*
$B_x = 0$ $B_z = 45.4$ N   *Ans.*

NOTE: If the precession ω_p had not occurred, the z component of force at A and B would be equal to 49.05 N. In this case, however, the difference in these components is caused by the "gyroscopic moment" created whenever a spinning body precesses about another axis. We will study this effect in detail in the next section.

SOLUTION II
This example can also be solved using Euler's equations of motion, Eqs. 21–25. In this case $\Omega = \omega = \{6\mathbf{j} + 3\mathbf{k}\}$ rad/s, and the time derivative $(\dot{\omega})_{xyz}$ can also be obtained with reference to the fixed X, Y, Z axes since $\dot{\omega} = (\dot{\omega})_{xyz}$. This calculation can be performed by choosing x', y', z' axes to have an angular velocity of $\Omega' = \omega_p$, Fig. 21–14c. Applying Eq. 20–6,

$\dot{\omega} = (\dot{\omega})_{x'y'z'} + \omega_p \times \omega = 0 + 3\mathbf{k} \times (6\mathbf{j} + 3\mathbf{k}) = \{-18\mathbf{i}\}$ rad/s²

$\dot{\omega}_x = -18$ rad/s $\dot{\omega}_y = 0$ $\dot{\omega}_z = 0$

The moment equations then become

$\Sigma M_x = I_x \dot{\omega}_x - (I_y - I_z)\omega_y \omega_z$

$-A_z(0.5) + B_z(0.5) = 0.1(-18) - (0.2 - 0.1)(6)(3) = -3.6$

$\Sigma M_y = I_y \dot{\omega}_y - (I_z - I_x)\omega_z \omega_x$

$0 = 0 - 0$

$\Sigma M_z = I_z \dot{\omega}_z - (I_x - I_y)\omega_x \omega_y$

$A_x(0.5) - B_x(0.5) = 0 - 0$

The solution then proceeds as before.

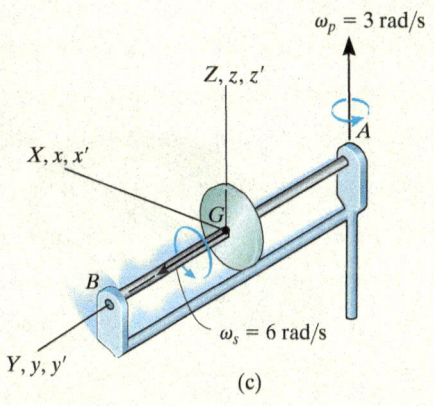

Fig. 21–14

PROBLEMS

*21–40. Derive the scalar form of the rotational equation of motion about the x axis if $\Omega \neq \omega$ and the moments and products of inertia of the body are *not constant* with respect to time.

21–41. Derive the scalar form of the rotational equation of motion about the x axis if $\Omega \neq \omega$ and the moments and products of inertia of the body are *constant* with respect to time.

21–42. Derive the Euler equations of motion for $\Omega \neq \omega$, i.e., Eqs. 21–26.

21–43. Rod CD of mass m and length L is rotating with a constant angular rate of ω_1 about axle AB, while shaft EF rotates with a constant angular rate of ω_2. Determine the X, Y, and Z components of reaction at thrust bearing E and journal bearing F at the instant shown. Neglect the mass of the other members.

Prob. 21–43

*21–44. The 40-kg flywheel (disk) is mounted 20 mm off its true center at G. If the shaft is rotating at a constant speed $\omega = 8 \, \text{rad/s}$, determine the minimum reactions exerted on the journal bearings at A and B during the motion.

Prob. 21–44

21–45. The 40-kg flywheel (disk) is mounted 20 mm off its true center at G. If the shaft is rotating at a constant speed $\omega = 8 \, \text{rad/s}$, determine the maximum reactions exerted on the journal bearings at A and B.

Prob. 21–45

21–46. The 4-lb bar rests along the smooth corners of an open box. At the instant shown, the box has a velocity $\mathbf{v} = \{5\mathbf{k}\}$ ft/s and an acceleration $\mathbf{a} = \{2\mathbf{k}\}$ ft/s^2. Determine the x, y, z components of force which the corners exert on the bar.

21–47. The 4-lb bar rests along the smooth corners of an open box. At the instant shown, the box has a velocity $\mathbf{v} = \{3\mathbf{j}\}$ ft/s and an acceleration $\mathbf{a} = \{-6\,\mathbf{j}\}$ ft/s^2. Determine the x, y, z components of force which the corners exert on the bar.

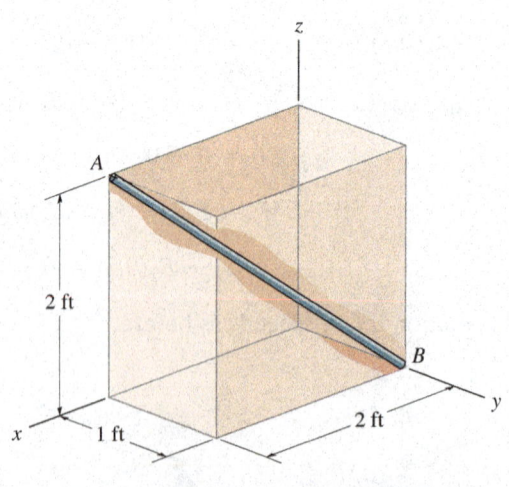

Probs. 21–46/47

*21–48. The man sits on a swivel chair which is rotating with a constant angular velocity of 3 rad/s. He holds the uniform 5-lb rod AB horizontal. He suddenly gives it an angular acceleration of 2 rad/s², measured relative to him, as shown. Determine the required force and moment components at the grip, A, necessary to do this. Establish axes at the rod's center of mass G, with $+z$ upward, and $+y$ directed along the axis of the rod toward A.

21–50. The 20-lb plate is mounted on the shaft AB so that the plane of the plate makes an angle $\theta = 30°$ with the vertical. If the shaft is turning in the direction shown with an angular velocity of 25 rad/s, determine the vertical reactions at the bearing supports A and B when the plate is in the position shown.

Prob. 21–48

Prob. 21–50

21–49. The shaft is constructed from a rod which has a mass per unit of 2 kg/m. Determine the x, y, z components of reaction at the bearings A and B if at the instant shown the shaft spins freely and has an angular velocity of $\omega = 30$ rad/s. What is the angular acceleration of the shaft at this instant? Bearing A can support a component of force in the y direction, whereas bearing B cannot.

21–51. The uniform hatch door, having a mass of 15 kg and a mass center at G, is supported in the horizontal plane by bearings at A and B. If a vertical force $F = 300$ N is applied to the door as shown, determine the components of reaction at the bearings and the angular acceleration of the door. The bearing at A will resist a component of force in the y direction, whereas the bearing at B will not. For the calculation, assume the door to be a thin plate and neglect the size of each bearing. The door is originally at rest.

Prob. 21–49

Prob. 21–51

*21–52. The 20-lb disk is mounted on the horizontal shaft AB such that its plane forms an angle of 10° with the vertical. If the shaft rotates with an angular velocity of 3 rad/s, determine the vertical reactions developed at the bearings when the disk is in the position shown.

21–54. The 10-kg disk turns around the shaft AB, while the shaft rotates about BC at a constant rate of $\omega_x = 5$ rad/s. If the disk does not slip, determine the normal and frictional force it exerts on the ground. Neglect the mass of shaft AB.

Prob. 21–52

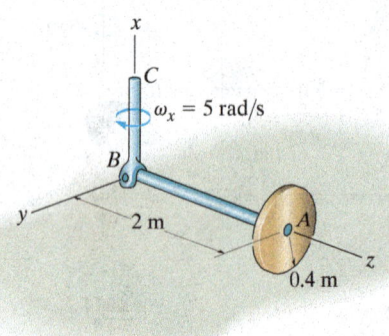

Prob. 21–54

21–53. The car travels around the curved road of radius ρ such that its mass center has a constant speed v_G. Write the equations of rotational motion with respect to the x, y, z axes. Assume that the car's six moments and products of inertia with respect to these axes are known.

21–55. The cylinder has a mass of 30 kg and is mounted on an axle which is supported by bearings at A and B. If the axle is subjected to a couple moment $\mathbf{M} = \{-30\mathbf{j}\}$ N·m, and at the instant shown has an angular velocity $\boldsymbol{\omega} = \{-40\mathbf{j}\}$ rad/s, determine the vertical components of force acting at the bearings at this instant.

Prob. 21–53

Prob. 21–55

*21–56. The rod assembly has a weight of 5 lb/ft. It is supported at B by a smooth journal bearing, which develops x and y force reactions, and at A by a smooth thrust bearing, which develops x, y, and z force reactions. If a 50-lb·ft torque is applied along rod AB, determine the components of reaction at the bearings when the assembly has an angular velocity $\omega = 10$ rad/s at the instant shown.

21–58. The 50-lb disk spins with a constant angular rate of $\omega_1 = 50$ rad/s about its axle. Simultaneously, the shaft rotates with a constant angular rate of $\omega_2 = 10$ rad/s. Determine the x, y, z components of the moment developed in the arm at A at the instant shown. Neglect the weight of arm AB.

Prob. 21–56

Prob. 21–58

21–57. The slender rod AB has a mass m and it is connected to the bracket by a smooth pin at A. The bracket is rigidly attached to the shaft. Determine the required constant angular velocity ω of the shaft, in order for the rod to make an angle θ with the vertical.

21–59. Four spheres are connected to shaft AB. If $m_C = 1$ kg and $m_E = 2$ kg, determine the mass of spheres D and F and the angles of the rods, θ_D and θ_F, so that the shaft is dynamically balanced, that is, so that the bearings at A and B exert only vertical reactions on the shaft as it rotates. Neglect the mass of the rods.

Prob. 21–57

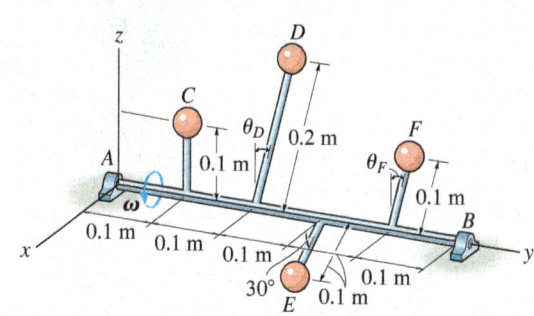

Prob. 21–59

600   CHAPTER 21   THREE-DIMENSIONAL KINETICS OF A RIGID BODY

*21.5 GYROSCOPIC MOTION

In this section we will develop the equations defining the motion of a body (top) which is symmetrical with respect to an axis and rotating about a fixed point. These equations also apply to the motion of a particularly interesting device, the gyroscope.

The body's motion will be analyzed using **Euler angles** ϕ, θ, ψ (phi, theta, psi). To show how they define the final position of a body, we will consider the top shown in Fig. 21–15a, which is in the final position shown in Fig. 21–15d. We will use stationary X, Y, Z axes and a second set of x, y, z axes fixed in the top. Starting with the X, Y, Z and x, y, z axes in coincidence, Fig. 21–15a, the final position of the top is determined using the following three steps:

1. Rotate the top about the Z (or z) axis through an angle ϕ ($0 \leq \phi < 2\pi$), Fig. 21–15b.

2. Rotate the top about the x axis through an angle θ ($0 \leq \theta \leq \pi$), Fig. 21–15c.

3. Rotate the top about the z axis through an angle ψ ($0 \leq \psi < 2\pi$) to obtain the final position, Fig. 21–15d.

The sequence of these three angles, ϕ, θ, then ψ, must be maintained, since finite rotations are *not vectors* (see Fig. 20–1). Although this is the case, the differential rotations $d\phi$, $d\theta$, $d\psi$ are vectors, and so the angular velocity $\boldsymbol{\omega}$ of the top can be expressed in terms of the time derivatives of the Euler angles. The angular-velocity components $\dot{\phi}$, $\dot{\theta}$, and $\dot{\psi}$ are known as the **precession**, **nutation**, and **spin**, respectively.

(a)

Precession $\dot{\phi}$
(b)

Nutation $\dot{\theta}$
(c)

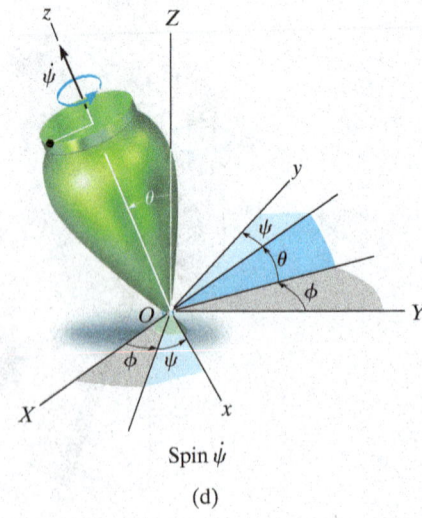
Spin $\dot{\psi}$
(d)

Fig. 21–15

Their positive directions are shown in Fig. 21–16. It is seen that these vectors are not all perpendicular to one another; however, $\boldsymbol{\omega}$ of the top can still be expressed in terms of these three components.

Since the body (top) is symmetric with respect to the z or spin axis, there is no need to attach the x, y, z axes to the top since the inertial properties of the top will remain constant with respect to this frame during the motion. Therefore $\boldsymbol{\Omega} = \boldsymbol{\omega}_p + \boldsymbol{\omega}_n$, Fig. 21–16, and so the angular velocity of the top is

$$\boldsymbol{\omega} = \omega_x \mathbf{i} + \omega_y \mathbf{j} + \omega_z \mathbf{k}$$
$$= \dot{\theta}\mathbf{i} + (\dot{\phi}\sin\theta)\mathbf{j} + (\dot{\phi}\cos\theta + \dot{\psi})\mathbf{k} \quad (21\text{–}27)$$

And the angular velocity of the axes is

$$\boldsymbol{\Omega} = \Omega_x \mathbf{i} + \Omega_y \mathbf{j} + \Omega_z \mathbf{k}$$
$$= \dot{\theta}\mathbf{i} + (\dot{\phi}\sin\theta)\mathbf{j} + (\dot{\phi}\cos\theta)\mathbf{k} \quad (21\text{–}28)$$

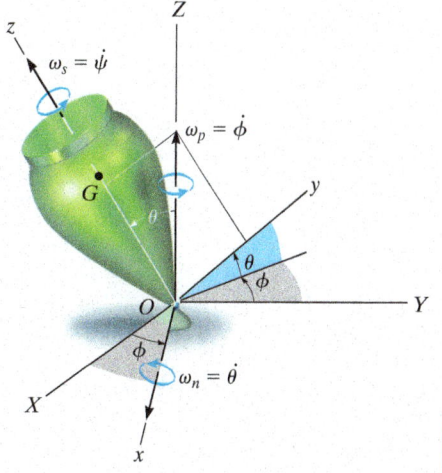

Fig. 21–16

The x, y, z axes represent principal axes of inertia for the top, and so the moments of inertia will be represented as $I_{xx} = I_{yy} = I$ and $I_{zz} = I_z$. Since $\boldsymbol{\Omega} \neq \boldsymbol{\omega}$, Eqs. 21–26 are used to establish the rotational equations of motion. Substituting into these equations the respective angular-velocity components defined by Eqs. 21–27 and 21–28, their corresponding time derivatives, and the moment of inertia components, yields

$$\Sigma M_x = I(\ddot{\theta} - \dot{\phi}^2 \sin\theta \cos\theta) + I_z\dot{\phi}\sin\theta(\dot{\phi}\cos\theta + \dot{\psi})$$
$$\Sigma M_y = I(\ddot{\phi}\sin\theta + 2\dot{\phi}\dot{\theta}\cos\theta) - I_z\dot{\theta}(\dot{\phi}\cos\theta + \dot{\psi}) \quad (21\text{–}29)$$
$$\Sigma M_z = I_z(\ddot{\psi} + \ddot{\phi}\cos\theta - \dot{\phi}\dot{\theta}\sin\theta)$$

Each moment summation applies only at the fixed point O or the center of mass G. Since the equations represent a coupled set of nonlinear second-order differential equations, in general a closed-form solution can not be obtained. Instead, the Euler angles ϕ, θ, and ψ may be obtained graphically as functions of time using numerical analysis and computer techniques.

A special case, however, does exist for which simplification of Eqs. 21–29 is possible. Commonly referred to as **steady precession**, it occurs when the nutation angle θ, precession $\dot{\phi}$, and spin $\dot{\psi}$ all remain *constant*. Equations 21–29 then reduce to the form

$$\boxed{\Sigma M_x = -I\dot{\phi}^2 \sin\theta \cos\theta + I_z\dot{\phi}\sin\theta(\dot{\phi}\cos\theta + \dot{\psi})} \quad (21\text{–}30)$$

$$\Sigma M_y = 0$$
$$\Sigma M_z = 0$$

Equation 21–30 can be further simplified by noting that, from Eq. 21–27, $\omega_z = \dot{\phi}\cos\theta + \dot{\psi}$, so that

$$\Sigma M_x = -I\dot{\phi}^2 \sin\theta \cos\theta + I_z\dot{\phi}(\sin\theta)\omega_z$$

or

$$\Sigma M_x = \dot{\phi}\sin\theta(I_z\omega_z - I\dot{\phi}\cos\theta) \qquad (21\text{–}31)$$

It is interesting to note what effects the spin $\dot{\psi}$ has on the moment about the x axis. To show this, consider the spinning rotor in Fig. 21–17. Here $\theta = 90°$, in which case Eq. 21–30 reduces to the form

$$\Sigma M_x = I_z\dot{\phi}\dot{\psi}$$

or

$$\Sigma M_x = I_z\Omega_y\omega_z \qquad (21\text{–}32)$$

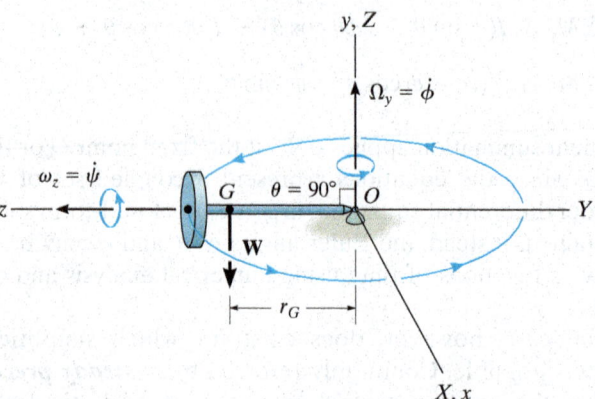

Fig. 21–17

Instinctively, one would expect the rotor to fall down under the influence of gravity! However, this is not the case at all, provided the product $I_z\Omega_y\omega_z$ is correctly chosen to counterbalance the moment $\Sigma M_x = Wr_G$ of the rotor's weight about O. This unusual phenomenon of rigid-body motion is often referred to as the **gyroscopic effect**.

21.5 GYROSCOPIC MOTION

Perhaps a more intriguing demonstration of the gyroscopic effect comes from studying the action of a **gyroscope**, frequently referred to as a **gyro**. A gyro is a rotor which spins at a very high rate about its axis of symmetry. This spin is considerably greater than its precession about the vertical axis. Hence, for all practical purposes, the angular momentum of the gyro can be assumed directed along its axis of spin. Thus, for the gyro rotor shown in Fig. 21–18, $\omega_z \gg \Omega_y$, and the magnitude of the angular momentum about point O, as determined from Eqs. 21–11, reduces to the form $H_O = I_z \omega_z$. Since both the magnitude and direction of $\mathbf{H}_O$ are constant as observed from x, y, z, direct application of Eq. 21–22 yields

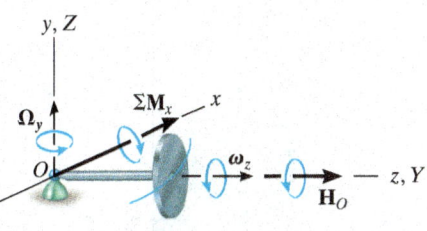

Fig. 21–18

$$\Sigma \mathbf{M}_x = \boldsymbol{\Omega}_y \times \mathbf{H}_O \qquad (21\text{–}33)$$

Using the right-hand rule applied to the cross product, it can be seen that $\boldsymbol{\Omega}_y$ always swings $\mathbf{H}_O$ (or $\boldsymbol{\omega}_z$) toward the sense of $\Sigma \mathbf{M}_x$. In effect, the *change in direction* of the gyro's angular momentum, $d\mathbf{H}_O$, is equivalent to the angular impulse caused by the gyro's weight about O, i.e., $d\mathbf{H}_O = \Sigma \mathbf{M}_x\, dt$, Eq. 21–20. Also, since $H_O = I_z \omega_z$ and $\Sigma \mathbf{M}_x$, $\boldsymbol{\Omega}_y$, and $\mathbf{H}_O$ are mutually perpendicular, Eq. 21–33 reduces to Eq. 21–32.

When a gyro is mounted in gimbal rings, Fig. 21–19, it becomes *free* of external moments applied to its base. Thus, in theory, its angular momentum **H** will never precess but, instead, maintain its same fixed orientation along the axis of spin when the base is rotated. This type of gyroscope is called a *free gyro* and is useful as a gyrocompass when the spin axis of the gyro is directed north. In reality, however, the gimbal mechanism is never completely free of friction, so such a device is useful only for the local navigation of ships and aircraft. The gyroscopic effect is also useful as a means of stabilizing both the rolling motion of ships at sea and the trajectories of missiles and projectiles. Furthermore, this effect is of significant importance in the design of shafts and bearings for rotors which are subjected to forced precessions, as discussed in Example 21.6.

Fig. 21–19

The spinning of the gyro within the frame of this toy gyroscope produces angular momentum $\mathbf{H}_O$, which is changing direction as the frame precesses $\boldsymbol{\omega}_p$ about the vertical axis. The gyroscope will not fall down since the moment of its weight **W** about the support is balanced by the change in direction of $\mathbf{H}_O$.

EXAMPLE 21.7

Fig. 21–20

The top shown in Fig. 21–20a has a mass of 0.5 kg and is precessing about the vertical axis at a constant angle of $\theta = 60°$. If it spins with an angular velocity $\omega_s = 100$ rad/s, determine its precession ω_p. Assume that the axial and transverse moments of inertia of the top are $0.45(10^{-3})$ kg·m² and $1.20(10^{-3})$ kg·m², respectively, measured with respect to the fixed point O.

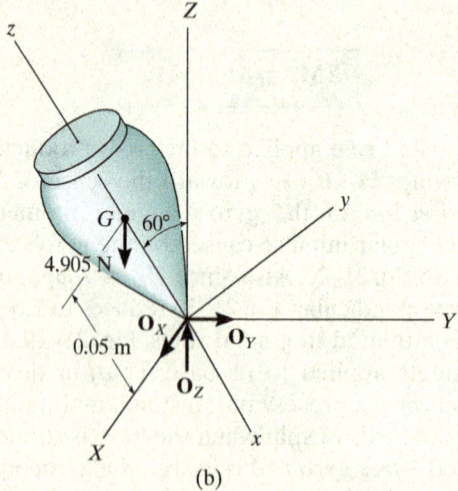

SOLUTION

Equation 21–30 will be used for the solution since the motion is *steady precession*. As shown on the free-body diagram, Fig. 21–20b, the coordinate axes are established in the usual manner, that is, with the positive z axis in the direction of spin, the positive Z axis in the direction of precession, and the positive x axis in the direction of the moment ΣM_x (refer to Fig. 21–16). Thus,

$$\Sigma M_x = -I\dot{\phi}^2 \sin\theta \cos\theta + I_z\dot{\phi}\sin\theta(\dot{\phi}\cos\theta + \dot{\psi})$$

4.905 N$(0.05$ m$)\sin 60° = -[1.20(10^{-3})$ kg·m² $\dot{\phi}^2]\sin 60°\cos 60°$
$\quad + [0.45(10^{-3})$ kg·m²$]\dot{\phi}\sin 60°(\dot{\phi}\cos 60° + 100$ rad/s$)$

or

$$\dot{\phi}^2 - 120.0\dot{\phi} + 654.0 = 0 \qquad (1)$$

Solving this quadratic equation for the precession gives

$$\dot{\phi} = 114 \text{ rad/s} \quad \text{(high precession)} \qquad \textit{Ans.}$$

and

$$\dot{\phi} = 5.72 \text{ rad/s} \quad \text{(low precession)} \qquad \textit{Ans.}$$

NOTE: In reality, low precession of the top would generally be observed, since high precession would require a larger kinetic energy.

EXAMPLE 21.8

The 1-kg disk shown in Fig. 21–21a spins about its axis with a constant angular velocity $\omega_D = 70$ rad/s. The block at B has a mass of 2 kg, and by adjusting its position s one can change the precession of the disk about its supporting pivot at O while the shaft remains horizontal. Determine the position s that will enable the disk to have a constant precession $\omega_p = 0.5$ rad/s about the pivot. Neglect the weight of the shaft.

Fig. 21–21

SOLUTION

The free-body diagram of the assembly is shown in Fig. 21–21b. The origin for both the x, y, z and X, Y, Z coordinate systems is located at the fixed point O. In the conventional sense, the Z axis is chosen along the axis of precession, and the z axis is along the axis of spin, so that $\theta = 90°$. Since the precession is *steady*, Eq. 21–32 can be used for the solution.

$$\Sigma M_x = I_z \Omega_y \omega_z$$

Substituting the required data gives

$$(9.81 \text{ N})(0.2 \text{ m}) - (19.62 \text{ N})s = \left[\tfrac{1}{2}(1 \text{ kg})(0.05 \text{ m})^2\right]0.5 \text{ rad/s}(-70 \text{ rad/s})$$

$$s = 0.102 \text{ m} = 102 \text{ mm} \qquad \textit{Ans.}$$

21.6 TORQUE-FREE MOTION

When the only external force acting on a body is caused by gravity, the general motion of the body is referred to as *torque-free motion*. This type of motion is characteristic of planets, artificial satellites, and projectiles—provided air friction is neglected.

In order to describe the characteristics of this motion, the distribution of the body's mass will be assumed *axisymmetric*. The satellite shown in Fig. 21–22 is an example of such a body, where the z axis represents an axis of symmetry. The origin of the x, y, z coordinates is located at the mass center G, such that $I_{zz} = I_z$ and $I_{xx} = I_{yy} = I$. Since gravity is the only external force present, the summation of moments about the mass center is zero. From Eq. 21–21, this requires the angular momentum of the body to be constant, i.e.,

$$\mathbf{H}_G = \text{constant}$$

At the instant considered, it will be assumed that the inertial frame of reference is oriented so that the positive Z axis is directed along $\mathbf{H}_G$ and the y axis lies in the plane formed by the z and Z axes, Fig. 21–22. The Euler angle formed between Z and z is θ, and therefore, with this choice of axes the angular momentum can be expressed as

$$\mathbf{H}_G = H_G \sin\theta\, \mathbf{j} + H_G \cos\theta\, \mathbf{k}$$

Furthermore, using Eqs. 21–11, we have

$$\mathbf{H}_G = I\omega_x \mathbf{i} + I\omega_y \mathbf{j} + I_z \omega_z \mathbf{k}$$

Equating the respective $\mathbf{i}, \mathbf{j}$, and $\mathbf{k}$ components of the above two equations yields

Fig. 21–22

$$\omega_x = 0 \quad \omega_y = \frac{H_G \sin\theta}{I} \quad \omega_z = \frac{H_G \cos\theta}{I_z} \quad (21\text{--}34)$$

or

$$\boldsymbol{\omega} = \frac{H_G \sin\theta}{I}\mathbf{j} + \frac{H_G \cos\theta}{I_z}\mathbf{k} \quad (21\text{--}35)$$

In a similar manner, equating the respective **i, j, k** components of Eq. 21–27 to those of Eq. 21–34, we obtain

$$\dot\theta = 0$$
$$\dot\phi \sin\theta = \frac{H_G \sin\theta}{I}$$
$$\dot\phi \cos\theta + \dot\psi = \frac{H_G \cos\theta}{I_z}$$

Solving, we get

$$\boxed{\begin{aligned} \theta &= \text{constant} \\ \dot\phi &= \frac{H_G}{I} \\ \dot\psi &= \frac{I - I_z}{I I_z} H_G \cos\theta \end{aligned}} \quad (21\text{--}36)$$

Thus, for torque-free motion of an axisymmetrical body, the angle θ formed between the angular-momentum vector and the spin of the body remains constant. Furthermore, the angular momentum $\mathbf{H}_G$, precession $\dot\phi$, and spin $\dot\psi$ for the body remain constant at all times during the motion.

608   CHAPTER 21   THREE-DIMENSIONAL KINETICS OF A RIGID BODY

$I > I_z$
(a)

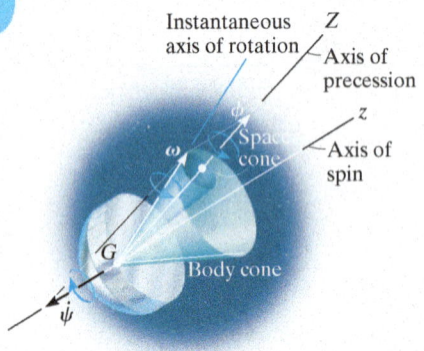

$I < I_z$
(b)

Fig. 21–23

Eliminating H_G from the second and third of Eqs. 21–36 yields the following relation between the spin and precession:

$$\dot{\psi} = \frac{I - I_z}{I_z}\dot{\phi}\cos\theta \qquad (21\text{–}37)$$

These two components of angular motion can be studied by using the body and space cone models introduced in Sec. 20.1. The *space cone*, which is oriented along the axis of precession, is fixed from rotating, since the precession has a fixed direction, while the surface of the *body cone* rolls on the space cone's surface. Try to imagine this satellite motion in Fig. 21–23a. The interior angle of each cone is chosen such that the angular velocity $\boldsymbol{\omega}$ of the satellite is directed along the line of contact of the two cones. This line represents the instantaneous axis of rotation for the body cone, and hence the angular velocity of both the body cone and the body must be directed along this line. Since the spin is a function of the moments of inertia I and I_z of the body, Eq. 21–36, the cone model in Fig. 21–23a is satisfactory for describing the motion, provided $I > I_z$. Torque-free motion which meets these requirements is called **regular precession**. If $I < I_z$, the spin is negative and the precession positive. This motion is represented by the satellite motion shown in Fig. 21–23b ($I < I_z$). The cone model can again be used to represent the motion; however, to preserve the correct vector addition of spin and precession to obtain the angular velocity $\boldsymbol{\omega}$, the inside surface of the body cone must roll on the outside surface of the (fixed) space cone. This motion is referred to as **retrograde precession**.

Satellites are often given a spin before they are launched. If their angular momentum is not collinear with the axis of spin, they will exhibit precession. For the satellite on the left, regular precession will occur since $I > I_z$, and for the one on the right, retrograde precession will occur since $I < I_z$.

EXAMPLE 21.9

The motion of a football is observed using a slow-motion projector. From the film, the spin of the football is seen to be directed 30° from the horizontal, as shown in Fig. 21–24a. Also, the football is precessing about the vertical axis at a rate $\dot{\phi} = 3$ rad/s. If the ratio of the axial to transverse moments of inertia of the football is $\frac{1}{3}$, measured with respect to the center of mass, determine the magnitude of the football's spin and its angular velocity. Neglect the effect of air resistance.

Fig. 21–24

SOLUTION
Since the weight of the football is the only force acting, the motion is torque free. In the conventional sense, if the z axis is established along the axis of spin and the Z axis along the precession axis, as shown in Fig. 21–24b, then the angle $\theta = 60°$. Applying Eq. 21–37, the spin is

$$\dot{\psi} = \frac{I - I_z}{I_z}\dot{\phi}\cos\theta = \frac{I - \frac{1}{3}I}{\frac{1}{3}I}(3)\cos 60°$$

$$= 3 \text{ rad/s} \qquad \textit{Ans.}$$

Using Eqs. 21–34, where $H_G = \dot{\phi}I$ (Eq. 21–36), we have

$$\omega_x = 0$$

$$\omega_y = \frac{H_G \sin\theta}{I} = \frac{3I \sin 60°}{I} = 2.60 \text{ rad/s}$$

$$\omega_z = \frac{H_G \cos\theta}{I_z} = \frac{3I \cos 60°}{\frac{1}{3}I} = 4.50 \text{ rad/s}$$

Thus,

$$\omega = \sqrt{(\omega_x)^2 + (\omega_y)^2 + (\omega_z)^2}$$
$$= \sqrt{(0)^2 + (2.60)^2 + (4.50)^2}$$
$$= 5.20 \text{ rad/s} \qquad \textit{Ans.}$$

PROBLEMS

*21–60. Show that the angular velocity of a body, in terms of Euler angles ϕ, θ, and ψ, can be expressed as $\boldsymbol{\omega} = (\dot{\phi} \sin\theta \sin\psi + \dot{\theta}\cos\psi)\mathbf{i} + (\dot{\phi}\sin\theta\cos\psi - \dot{\theta}\sin\psi)\mathbf{j} + (\dot{\phi}\cos\theta + \dot{\psi})\mathbf{k}$, where $\mathbf{i}$, $\mathbf{j}$, and $\mathbf{k}$ are directed along the x, y, z axes as shown in Fig. 21–15d.

21–61. A thin rod is initially coincident with the Z axis when it is given three rotations defined by the Euler angles $\phi = 30°$, $\theta = 45°$, and $\psi = 60°$. If these rotations are given in the order stated, determine the coordinate direction angles α, β, γ of the axis of the rod with respect to the X, Y, and Z axes. Are these directions the same for any order of the rotations? Why?

21–62. The motor weighs 50 lb and has a radius of gyration of 0.2 ft about the z axis. The shaft of the motor is supported by bearings at A and B, and spins at a constant rate of $\boldsymbol{\omega}_s = \{100\mathbf{k}\}$ rad/s, while the frame has an angular velocity of $\boldsymbol{\omega}_y = \{2\mathbf{j}\}$ rad/s. Determine the moment which the bearing forces at A and B exert on the shaft due to this motion.

Prob. 21–62

21–63. At the moment of takeoff, the landing gear of an airplane is retracted with a constant angular velocity of $\omega_p = 2$ rad/s, while the wheel continues to spin. If the plane takes off with a speed of $v = 320$ km/h, determine the torque at A due to the gyroscopic effect. The wheel has a mass of 50 kg, and the radius of gyration about its spinning axis is $k = 300$ mm.

Prob. 21–63

*21–64. The top consists of a thin disk that has a weight of 8 lb and a radius of 0.3 ft. The rod has a negligible mass and a length of 0.5 ft. If the top is spinning with an angular velocity $\omega_s = 300$ rad/s, determine the steady-state precessional angular velocity ω_p of the rod when $\theta = 40°$.

Prob. 21–64

21–65. The top consists of a thin disk that has a weight of 8 lb and a radius of 0.3 ft. The rod has a negligible mass and a length of 0.5 ft. If the top is spinning with an angular velocity $\omega_s = 300$ rad/s, determine the steady-state precessional angular velocity ω_p of the rod when $\theta = 90°$.

Prob. 21–65

21–66. The propeller on a single-engine airplane has a mass of 15 kg and a centroidal radius of gyration of 0.3 m calculated about the axis of spin. When viewed from the front of the airplane, the propeller is turning clockwise at 350 rad/s about the spin axis. If the airplane enters a vertical curve having a radius of 80 m and is traveling at 200 km/h, determine the gyroscopic bending moment which the propeller exerts on the bearings of the engine when the airplane is in its lowest position.

*21–68.** The car travels at a constant speed of $v_C = 100$ km/h around the horizontal curve having a radius of 80 m. If each wheel has a mass of 16 kg, a radius of gyration $k_G = 300$ mm about its spinning axis, and a radius of 400 mm, determine the difference between the normal forces of the rear wheels, caused by the gyroscopic effect. The distance between the wheels is 1.30 m.

Prob. 21–66

Prob. 21–68

21–67. The turbine on a ship has a mass of 400 kg and is mounted on bearings A and B as shown. Its center of mass is at G, its radius of gyration is $k_z = 0.3$ m, and $k_x = k_y = 0.5$ m. If it is spinning at 200 rad/s, determine the vertical reactions at the bearings when the ship undergoes each of the following motions: (a) rolling, $\omega_1 = 0.2$ rad/s, (b) turning, $\omega_2 = 0.8$ rad/s, (c) pitching, $\omega_3 = 1.4$ rad/s.

21–69. The top has a mass of 90 g, a center of mass at G, and a radius of gyration $k = 18$ mm about its axis of symmetry. About any transverse axis acting through point O the radius of gyration is $k_t = 35$ mm. If the top is connected to a ball-and-socket joint at O and the precession is $\omega_p = 0.5$ rad/s, determine the spin ω_s.

Prob. 21–67

Prob. 21–69

21–70. The 1-lb top has a center of gravity at point G. If it spins about its axis of symmetry and precesses about the vertical axis at constant rates of $\omega_s = 60$ rad/s and $\omega_p = 10$ rad/s, respectively, determine the steady state angle θ. The radius of gyration of the top about the z axis is $k_z = 1$ in., and about the x and y axes it is $k_x = k_y = 4$ in.

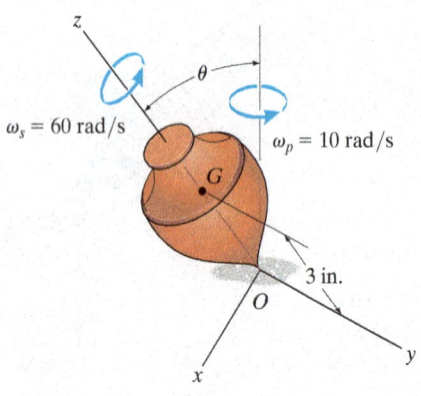

Prob. 21–70

21–71. The 20-kg disk is spinning about its center at $\omega_s = 20$ rad/s while the supporting axle is rotating at $\omega_y = 6$ rad/s. Determine the gyroscopic moment caused by the force reactions which the pin A exerts on the disk due to the motion.

Prob. 21–71

*21–72. The radius of gyration about an axis passing through the axis of symmetry of the 2.5-Mg satellite is $k_z = 2.3$ m, and about any transverse axis passing through the center of mass G, $k_t = 3.4$ m. If the satellite has a steady-state precession of two revolutions per hour about the Z axis, determine the rate of spin about the z axis.

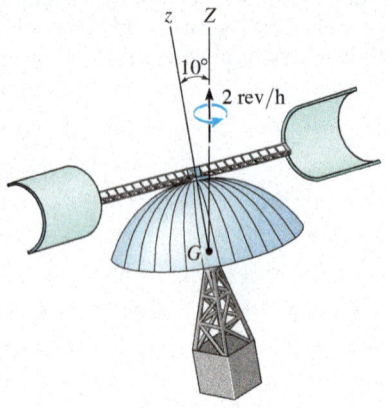

Prob. 21–72

21–73. The projectile shown is subjected to torque-free motion. The transverse and axial moments of inertia are I and I_z, respectively. If θ represents the angle between the precessional axis Z and the axis of symmetry z, and β is the angle between the angular velocity ω and the z axis, show that β and θ are related by the equation $\tan \theta = (I/I_z) \tan \beta$.

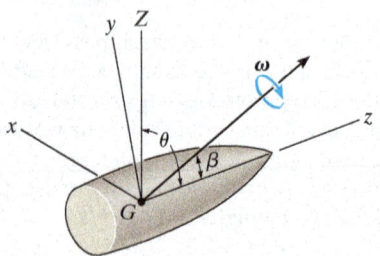

Prob. 21–73

21–74. The radius of gyration about an axis passing through the axis of symmetry of the 1.6-Mg space capsule is $k_z = 1.2$ m and about any transverse axis passing through the center of mass G, $k_t = 1.8$ m. If the capsule has a known steady-state precession of two revolutions per hour about the Z axis, determine the rate of spin about the z axis.

Prob. 21–74

21–75. The projectile has a mass of 0.9 kg and axial and transverse radii of gyration of $k_z = 20$ mm and $k_t = 25$ mm, respectively. If it is spinning at $\omega_s = 6$ rad/s when it leaves the barrel of a gun, determine its angular momentum. Precession occurs about the Z axis.

21–77. The satellite has a mass of 1.8 Mg, and about axes passing through the mass center G the axial and transverse radii of gyration are $k_z = 0.8$ m and $k_t = 1.2$ m, respectively. If it is spinning at $\omega_s = 6$ rad/s when it is launched, determine its angular momentum. Precession occurs about the Z axis.

Prob. 21–77

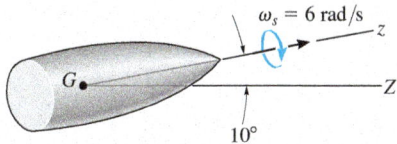

Prob. 21–75

*__21–76.__ The satellite has a mass of 100 kg and radii of gyration about its axis of symmetry (z axis) and its transverse axes (x or y axis) of $k_z = 300$ mm and $k_x = k_y = 900$ mm, respectively. If the satellite spins about the z axis at a constant rate of $\dot{\psi} = 200$ rad/s, and precesses about the Z axis, determine the precession $\dot{\phi}$ and the magnitude of its angular momentum $\mathbf{H}_G$.

21–78. A wheel of mass m and radius r rolls with constant spin ω about a circular path having a radius a. If the angle of inclination is θ, determine the rate of precession. Treat the wheel as a thin ring. No slipping occurs.

Prob. 21–76

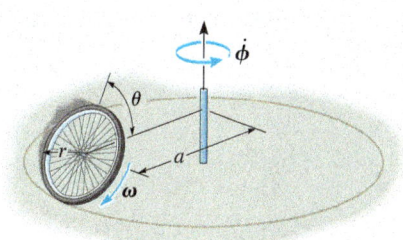

Prob. 21–78

CHAPTER REVIEW

Moments and Products of Inertia

A body has six components of inertia for any specified x, y, z axes. Three of these are moments of inertia about each of the axes, I_{xx}, I_{yy}, I_{zz}, and three are products of inertia, each defined from two orthogonal planes, I_{xy}, I_{yz}, I_{xz}. If either one or both of these planes are planes of symmetry, then the product of inertia with respect to these planes will be zero.

$$I_{xx} = \int_m r_x^2 \, dm = \int_m (y^2 + z^2) \, dm \qquad I_{xy} = I_{yx} = \int_m xy \, dm$$

$$I_{yy} = \int_m r_y^2 \, dm = \int_m (x^2 + z^2) \, dm \qquad I_{yz} = I_{zy} = \int_m yz \, dm$$

$$I_{zz} = \int_m r_z^2 \, dm = \int_m (x^2 + y^2) \, dm \qquad I_{xz} = I_{zx} = \int_m xz \, dm$$

The moments and products of inertia can be determined by direct integration or by using tabulated values. If these quantities are to be determined with respect to axes or planes that do not pass through the mass center, then the parallel-axis and parallel-plane theorems must be used to calculate their values.

Provided the six components of inertia are known, then the moment of inertia about any axis can be determined using the inertia transformation equation.

$$I_{Oa} = I_{xx} u_x^2 + I_{yy} u_y^2 + I_{zz} u_z^2 - 2 I_{xy} u_x u_y - 2 I_{yz} u_y u_z - 2 I_{zx} u_z u_x$$

Principal Moments of Inertia

At any point on or off the body, the x, y, z axes can be oriented so that the products of inertia will be zero. The resulting moments of inertia are called the principal moments of inertia. In general, one will be a maximum and the other a minimum.

$$\begin{pmatrix} I_x & 0 & 0 \\ 0 & I_y & 0 \\ 0 & 0 & I_z \end{pmatrix}$$

Principle of Impulse and Momentum

The angular momentum for a body can be determined about any arbitrary point A.

Once the linear and angular momentum for the body have been formulated, then the principle of impulse and momentum can be used to solve problems that involve force, velocity, and time.

$$m(\mathbf{v}_G)_1 + \Sigma \int_{t_1}^{t_2} \mathbf{F} \, dt = m(\mathbf{v}_G)_2$$

$$(\mathbf{H}_O)_1 + \Sigma \int_{t_1}^{t_2} \mathbf{M}_O \, dt = (\mathbf{H}_O)_2$$

where

$$H_x = I_{xx} \omega_x - I_{xy} \omega_y - I_{xz} \omega_z$$
$$H_y = -I_{yx} \omega_x + I_{yy} \omega_y - I_{yz} \omega_z$$
$$H_z = -I_{zx} \omega_x - I_{zy} \omega_y + I_{zz} \omega_z$$

$$\mathbf{H}_O = \int_m \boldsymbol{\rho}_O \times (\boldsymbol{\omega} \times \boldsymbol{\rho}_O) \, dm$$
Fixed Point O

$$\mathbf{H}_G = \int_m \boldsymbol{\rho}_G \times (\boldsymbol{\omega} \times \boldsymbol{\rho}_G) \, dm$$
Center of Mass

$$\mathbf{H}_A = \boldsymbol{\rho}_{G/A} \times m \mathbf{v}_G + \mathbf{H}_G$$
Arbitrary Point

Principle of Work and Energy

The kinetic energy for a body is usually determined relative to a fixed point or the body's mass center.

These formulations can be used with the principle of work and energy to solve problems that involve force, velocity, and displacement.

$$T = \tfrac{1}{2}I_x\omega_x^2 + \tfrac{1}{2}I_y\omega_y^2 + \tfrac{1}{2}I_z\omega_z^2$$
Fixed Point

$$T = \tfrac{1}{2}mv_G^2 + \tfrac{1}{2}I_x\omega_x^2 + \tfrac{1}{2}I_y\omega_y^2 + \tfrac{1}{2}I_z\omega_z^2$$
Center of Mass

$$T_1 + \Sigma U_{1-2} = T_2$$

Equations of Motion

There are three scalar equations of translational motion for a rigid body that moves in three dimensions.

The three scalar equations of rotational motion depend upon the motion of the x, y, z reference. Most often, these axes are oriented so that they are principal axes of inertia. If the axes are fixed in and move with the body so that $\Omega = \omega$, then the equations are referred to as the Euler equations of motion.

A free-body diagram should always accompany the application of the equations of motion.

$$\Sigma F_x = m(a_G)_x$$
$$\Sigma F_y = m(a_G)_y$$
$$\Sigma F_z = m(a_G)_z$$
$$\Sigma M_x = I_x\dot\omega_x - (I_y - I_z)\omega_y\omega_z$$
$$\Sigma M_y = I_y\dot\omega_y - (I_z - I_x)\omega_z\omega_x$$
$$\Sigma M_z = I_z\dot\omega_z - (I_x - I_y)\omega_x\omega_y$$
$$\Omega = \omega$$

$$\Sigma M_x = I_x\dot\omega_x - I_y\Omega_z\omega_y + I_z\Omega_y\omega_z$$
$$\Sigma M_y = I_y\dot\omega_y - I_z\Omega_x\omega_z + I_x\Omega_z\omega_x$$
$$\Sigma M_z = I_z\dot\omega_z - I_x\Omega_y\omega_x + I_y\Omega_x\omega_y$$
$$\Omega \ne \omega$$

Gyroscopic Motion

The angular motion of a gyroscope is best described using the three Euler angles ϕ, θ, and ψ. The angular velocity components are called the precession $\dot\phi$, the nutation $\dot\theta$, and the spin $\dot\psi$.

If $\dot\theta = 0$ and $\dot\phi$ and $\dot\psi$ are constant, then the motion is referred to as steady precession.

It is the spin of a gyro rotor that is responsible for holding a rotor from falling downward, and instead causing it to precess about a vertical axis. This phenomenon is called the gyroscopic effect.

$$\Sigma M_x = -I\dot\phi^2 \sin\theta \cos\theta + I_z\dot\phi \sin\theta(\dot\phi \cos\theta + \dot\psi)$$
$$\Sigma M_y = 0, \ \Sigma M_z = 0$$

Torque-Free Motion

A body that is only subjected to a gravitational force will have no moments on it about its mass center, and so the motion is referred to as torque-free motion. The angular momentum for the body about its mass center will remain constant and so the body can have both a spin and a precession. For a symmetric body the motion depends upon the magnitude of the moment of inertia about the spin axis, I_z, versus that about a perpendicular axis, I.

$$\theta = \text{constant}$$
$$\dot\phi = \frac{H_G}{I}$$
$$\dot\psi = \frac{I - I_z}{I\,I_z} H_G \cos\theta$$

CHAPTER 22

The springs supporting this rail car truck and wheels absorb the impact loads of the track and the induced vibration of the car. Both effects should be considered in design.

VIBRATIONS

CHAPTER OBJECTIVES

- To discuss undamped one-degree-of-freedom vibration of a rigid body using the equation of motion and energy methods.
- To study the analysis of undamped forced vibration and viscous damped forced vibration.

22.1 UNDAMPED FREE VIBRATION

A *vibration* is the oscillating motion of a body or system of connected bodies displaced from a position of equilibrium. In general, there are two types of vibration, free and forced. *Free vibration* occurs when the motion is maintained by gravitational or elastic restoring forces, such as the swinging motion of a pendulum or the vibration of an elastic rod. *Forced vibration* is caused by an external periodic or intermittent force applied to the system. Both of these types of vibration can either be damped or undamped. *Undamped* vibrations exclude frictional effects in the analysis. Since in reality both internal and external frictional forces are present, the motion of all vibrating bodies is actually *damped*.

618   Chapter 22   Vibrations

(a)

(b)

Fig. 22–1

The simplest type of vibrating motion is undamped free vibration, which can be represented by the block and spring model shown in Fig. 22–1a. Provided the supporting surface is smooth, vibrating motion occurs when the block is released from a displaced position x. The spring then pulls on the block, causing it to attain a velocity, and then move out of equilibrium when $x = 0$. Compression of the spring will arrest the motion, at position $-x$, and the back and forth oscillation is repeated.

The time-dependent path of motion of the block can be determined by applying the equation of motion to the block when it is in the displaced position x. The free-body diagram is shown in Fig. 22–1b. The elastic restoring force $F = kx$ is always directed toward the equilibrium position, whereas the acceleration $\mathbf{a}$ is assumed to act in the direction of positive displacement. Since $a = d^2x/dt^2 = \ddot{x}$, we have

$$\overset{+}{\rightarrow} \Sigma F_x = ma_x; \qquad -kx = m\ddot{x}$$

Because the acceleration is proportional to the block's displacement, the motion is called **simple harmonic motion**. Rearranging the terms into a "standard form" gives

$$\ddot{x} + \omega_n^2 x = 0 \qquad (22\text{--}1)$$

The constant ω_n, generally reported in rad/s, is called the **natural frequency**, and in this case

$$\boxed{\omega_n = \sqrt{\frac{k}{m}}} \qquad (22\text{--}2)$$

(a)

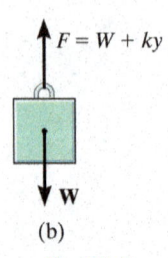

(b)

Fig. 22–2

Equation 22–1 can also be obtained by considering the block to be suspended so that the displacement y is measured from the block's *equilibrium position*, Fig. 22–2a. When the block is in equilibrium, the spring exerts an upward force of $F = W = mg$ on the block. Hence, when the block is displaced a distance y downward from this position, the magnitude of the spring force is $F = W + ky$, Fig. 22–2b. Applying the equation of motion gives

$$+\downarrow \Sigma F_y = ma_y; \qquad -W - ky + W = m\ddot{y}$$

or

$$\ddot{y} + \omega_n^2 y = 0$$

which is the same form as Eq. 22–1 and ω_n is defined by Eq. 22–2.

22.1 UNDAMPED FREE VIBRATION

Equation 22–1 is a homogeneous, second-order, linear differential equation with constant coefficients. It can be shown that the general solution is

$$x = A \sin \omega_n t + B \cos \omega_n t \qquad (22\text{–}3)$$

where A and B represent two constants of integration. The block's velocity and acceleration are determined by taking successive time derivatives, which yields

$$v = \dot{x} = A\omega_n \cos \omega_n t - B\omega_n \sin \omega_n t \qquad (22\text{–}4)$$

$$a = \ddot{x} = -A\omega_n^2 \sin \omega_n t - B\omega_n^2 \cos \omega_n t \qquad (22\text{–}5)$$

When Eqs. 22–3 and 22–5 are substituted into Eq. 22–1, the differential equation will be satisfied, showing that Eq. 22–3 is indeed the solution to Eq. 22–1.

The integration constants in Eq. 22–3 are generally determined from the initial conditions of the problem. For example, suppose that the block in Fig. 22–1a has been displaced a distance x_1 to the right from its equilibrium position and given an initial (positive) velocity $\mathbf{v}_1$ directed to the right. Substituting $x = x_1$ when $t = 0$ into Eq. 22–3 yields $B = x_1$. And since $v = v_1$ when $t = 0$, using Eq. 22–4 we obtain $A = v_1/\omega_n$. If these values are substituted into Eq. 22–3, the equation describing the motion becomes

$$x = \frac{v_1}{\omega_n} \sin \omega_n t + x_1 \cos \omega_n t \qquad (22\text{–}6)$$

Equation 22–3 may also be expressed in terms of simple sinusoidal motion. To show this, let

$$A = C \cos \phi \qquad (22\text{–}7)$$

and

$$B = C \sin \phi \qquad (22\text{–}8)$$

where C and ϕ are new constants to be determined in place of A and B. Substituting into Eq. 22–3 yields

$$x = C \cos \phi \sin \omega_n t + C \sin \phi \cos \omega_n t$$

And since $\sin(\theta + \phi) = \sin \theta \cos \phi + \cos \theta \sin \phi$, then

$$x = C \sin(\omega_n t + \phi) \qquad (22\text{–}9)$$

If this equation is plotted on an x versus $\omega_n t$ axis, the graph shown in Fig. 22–3 is obtained. The maximum displacement of the block from its equilibrium position is the **amplitude** of vibration. From either the figure or Eq. 22–9 the amplitude is C. The angle ϕ is called the **phase angle** since it represents the amount by which the curve is displaced from the origin when $t = 0$. We can relate these two constants to A and B using Eqs. 22–7 and 22–8. Squaring and adding these equations, the amplitude becomes

$$C = \sqrt{A^2 + B^2} \qquad (22\text{–}10)$$

If Eq. 22–8 is divided by Eq. 22–7, the phase angle is then

$$\phi = \tan^{-1}\frac{B}{A} \qquad (22\text{–}11)$$

Note that the sine curve, Eq. 22–9, completes one cycle in time $t = \tau$ (tau) when $\omega_n \tau = 2\pi$, or

$$\boxed{\tau = \frac{2\pi}{\omega_n}} \qquad (22\text{–}12)$$

This time interval is called the **period**, Fig. 22–3. Using Eq. 22–2, the period can also be represented as

$$\tau = 2\pi\sqrt{\frac{m}{k}} \qquad (22\text{–}13)$$

Finally, the **frequency** f is defined as the number of cycles completed per unit of time, which is the reciprocal of the period; that is,

$$\boxed{f = \frac{1}{\tau} = \frac{\omega_n}{2\pi}} \qquad (22\text{–}14)$$

or

$$f = \frac{1}{2\pi}\sqrt{\frac{k}{m}} \qquad (22\text{–}15)$$

The unit for the frequency is the **hertz** (Hz), where 1 Hz = 1 cycle/s = 2π rad/s.

When a body or system of connected bodies is given an initial displacement from its equilibrium position and released, it will vibrate with the **natural frequency**, ω_n. Provided the system has a single degree of freedom, that is, it requires only one coordinate to specify completely the position of the system at any time, then the vibrating motion will have the same characteristics as the simple harmonic motion of the block and spring model. Consequently, the motion is described by a differential equation of the same "standard form" as Eq. 22–1, i.e.,

$$\boxed{\ddot{x} + \omega_n^2 x = 0} \qquad (22\text{–}16)$$

Hence, if the natural frequency ω_n is known, the period of vibration τ, frequency f, and other vibrating characteristics can be established using Eqs. 22–3 through 22–15.

Fig. 22–3

IMPORTANT POINTS

- Free vibration occurs when the motion is maintained by gravitational or elastic restoring forces.
- The amplitude is the maximum displacement of the body.
- The period is the time required to complete one cycle.
- The frequency is the number of cycles completed per unit of time, where 1 Hz = 1 cycle/s.
- Only one position coordinate is needed to describe the location of a one-degree-of-freedom system.

PROCEDURE FOR ANALYSIS

As in the case of the block and spring, the natural frequency ω_n of a body or system of connected bodies having a single degree of freedom can be determined using the following procedure.

Free-Body Diagram.

- Draw the free-body diagram of the body when the body is displaced a *small amount* from its equilibrium position.

- Locate the body with respect to its equilibrium position by using an appropriate position coordinate q. The acceleration of the body's mass center $\mathbf{a}_G$ or the body's angular acceleration α should have an assumed sense of direction which is in the *positive direction* of the position coordinate.

- If the rotational equation of motion $\Sigma M_P = \Sigma(\mathcal{M}_k)_P$ is to be used, then consider drawing the kinetic diagram since it graphically accounts for the components $m(\mathbf{a}_G)_x$, $m(\mathbf{a}_G)_y$, and $I_G\alpha$, and thereby makes it convenient for visualizing the terms needed in the moment sum $\Sigma(\mathcal{M}_k)_P$.

Equation of Motion.

- Apply the equation of motion to relate the elastic or gravitational *restoring* forces and couple moments acting on the body to the body's accelerated motion.

Kinematics.

- Using kinematics, express the body's accelerated motion in terms of the second time derivative of the position coordinate, $\ddot{q}$.

- Substitute the result into the equation of motion and determine ω_n by rearranging the terms so that the resulting equation is in the "standard form," $\ddot{q} + \omega_n^2 q = 0$.

EXAMPLE 22.1

(a)

(b)

Fig. 22-4

Determine the period of oscillation for the simple pendulum shown in Fig. 22–4a. The bob has a mass m and is attached to a cord of length l. Neglect the size of the bob.

SOLUTION

Free-Body Diagram. Motion of the system will be described in terms of the position coordinate ($q =$) θ, Fig. 22–4b. When the bob is displaced by a small angle θ, the *restoring force* acting on the bob is caused by the tangential component of its weight, $mg \sin \theta$. We assume $\mathbf{a}_t$ acts in the direction of *increasing s* (or θ).

Equation of Motion. Applying the equation of motion in the *tangential direction*, since it involves the restoring force, yields

$$+\nearrow \Sigma F_t = ma_t; \qquad -mg \sin \theta = ma_t \qquad (1)$$

Kinematics. $a_t = d^2s/dt^2 = \ddot{s}$. Also, s can be related to θ by the equation $s = l\theta$, so that $a_t = l\ddot{\theta}$. Hence, Eq. 1 becomes,

$$\ddot{\theta} + \frac{g}{l} \sin \theta = 0 \qquad (2)$$

The solution of this equation involves an elliptic integral; however, for *small displacements*, $\sin \theta \approx \theta$, and so

$$\ddot{\theta} + \frac{g}{l} \theta = 0 \qquad (3)$$

Comparing this equation with Eq. 22–16 ($\ddot{x} + \omega_n^2 x = 0$), it is seen that $\omega_n = \sqrt{g/l}$. From Eq. 22–12, the period of time required for the bob to make one complete swing is therefore

$$\tau = \frac{2\pi}{\omega_n} = 2\pi \sqrt{\frac{l}{g}} \qquad Ans.$$

This interesting result, originally discovered by Galileo Galilei through experiment, indicates that the period depends only on the length of the cord and not on the mass of the pendulum bob or the angle θ.

NOTE: The solution of Eq. 3 is given by Eq. 22–3, where $\omega_n = \sqrt{g/l}$ and θ is substituted for x. Like the block and spring, the constants A and B in this problem can be determined if, for example, one knows the displacement and velocity of the bob at a given instant.

EXAMPLE 22.2

The 10-kg rectangular plate shown in Fig. 22–5a is suspended at its center from a rod having a torsional stiffness $k = 1.5 \text{ N} \cdot \text{m/rad}$. Determine the natural period of vibration of the plate when it is given a small angular displacement θ in the plane of the plate.

(a)

SOLUTION

Free-Body Diagram. Fig. 22–5b. Since the plate is displaced in its own plane, the torsional *restoring* moment created by the rod is $M = k\theta$. This moment acts in the direction opposite to the angular displacement θ. The angular acceleration $\ddot{\theta}$ acts in the direction of *positive* θ.

Equation of Motion.

$$\Sigma M_O = I_O \alpha; \qquad -k\theta = I_O \ddot{\theta}$$

or

$$\ddot{\theta} + \frac{k}{I_O}\theta = 0$$

Since this equation is in the "standard form," the natural frequency is $\omega_n = \sqrt{k/I_O}$.

From the table on the inside back cover, the moment of inertia of the plate about an axis coincident with the rod is $I_O = \frac{1}{12}m(a^2 + b^2)$. Hence,

$$I_O = \frac{1}{12}(10 \text{ kg})\left[(0.2 \text{ m})^2 + (0.3 \text{ m})^2\right] = 0.1083 \text{ kg} \cdot \text{m}^2$$

And so the natural period of vibration is

$$\tau = \frac{2\pi}{\omega_n} = 2\pi\sqrt{\frac{I_O}{k}} = 2\pi\sqrt{\frac{0.1083}{1.5}} = 1.69 \text{ s} \qquad Ans.$$

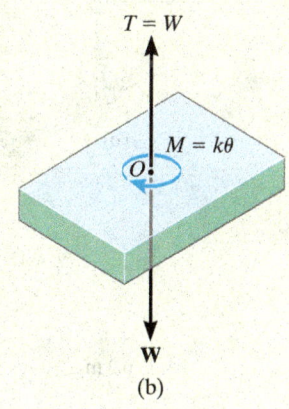

(b)

Fig. 22–5

EXAMPLE 22.3

(a)

(b)

(c)

Fig. 22–6

The bent rod shown in Fig. 22–6a has a negligible mass and supports a 5-kg collar at its end. If the rod is in the equilibrium position shown, determine the natural period of vibration for the system.

SOLUTION

Free-Body and Kinetic Diagrams. Fig. 22–6b. Here the rod is displaced by a small angle θ from the equilibrium position. Since the spring is subjected to an initial compression of x_{st} for equilibrium, then when the displacement $x > x_{st}$ the spring exerts a force of $F_s = kx - kx_{st}$ on the rod. To obtain the "standard form," Eq. 22–16, $5a_y$ must act *upward*, which is in accordance with positive θ displacement.

Equation of Motion. Moments will be summed about point B to eliminate the reactions $\mathbf{B}_x$ and $\mathbf{B}_y$. Since θ is small,

$$\zeta + \Sigma M_B = \Sigma (\mathcal{M}_k)_B;$$

$$kx(0.1 \text{ m}) - kx_{st}(0.1 \text{ m}) + 49.05 \text{ N}(0.2 \text{ m}) = -(5 \text{ kg})a_y(0.2 \text{ m})$$

The second term on the left side, $-kx_{st}(0.1 \text{ m})$, represents the moment created by the spring force which is necessary to hold the collar in *equilibrium*, i.e., at $x = 0$. Since this moment is equal and opposite to the moment 49.05 N(0.2 m) created by the weight of the collar, these two terms cancel, and so

$$kx(0.1) = -5a_y(0.2) \qquad (1)$$

Kinematics. The deformation of the spring and the position of the collar can be related to the angle θ, Fig. 22–6c. Since θ is small, $x = (0.1 \text{ m})\theta$ and $y = (0.2 \text{ m})\theta$. Therefore, $a_y = \ddot{y} = 0.2\ddot{\theta}$. Substituting into Eq. 1 yields

$$400(0.1\theta)0.1 = -5(0.2\ddot{\theta})0.2$$

Rewriting this equation in the "standard form" gives

$$\ddot{\theta} + 20\theta = 0$$

Compared with $\ddot{x} + \omega_n^2 x = 0$ (Eq. 22–16), we have

$$\omega_n^2 = 20 \qquad \omega_n = 4.47 \text{ rad/s}$$

The natural period of vibration is therefore

$$\tau = \frac{2\pi}{\omega_n} = \frac{2\pi}{4.47} = 1.40 \text{ s} \qquad \textit{Ans.}$$

EXAMPLE 22.4

A 10-lb block is suspended from a cord that passes over a 15-lb disk, as shown in Fig. 22–7a. The spring has a stiffness $k = 200$ lb/ft. Determine the natural period of vibration for the system.

SOLUTION

Free-Body and Kinetic Diagrams. Fig. 22–7b. The *system* consists of the disk, which undergoes a rotation defined by the angle θ, and the block, which translates by an amount s. The vector $I_O\ddot{\theta}$ acts in the direction of *positive* θ, and consequently $m_B \mathbf{a}_b$ acts downward in the direction of *positive s*.

Equation of Motion. Summing moments about point O to eliminate the reactions $\mathbf{O}_x$ and $\mathbf{O}_y$, realizing that $I_O = \frac{1}{2}mr^2$, yields

$\zeta + \Sigma M_O = \Sigma(\mathcal{M}_k)_O;$

$$10\text{ lb}(0.75\text{ ft}) - F_s(0.75\text{ ft}) = \frac{1}{2}\left(\frac{15\text{ lb}}{32.2\text{ ft/s}^2}\right)(0.75\text{ ft})^2\ddot{\theta} + \left(\frac{10\text{ lb}}{32.2\text{ ft/s}^2}\right)a_b(0.75\text{ ft})$$

(1)

Kinematics. As shown on the kinematic diagram in Fig. 22–7c, a small positive displacement θ of the disk causes the block to lower by an amount $s = 0.75\theta$; hence, $a_b = \ddot{s} = 0.75\ddot{\theta}$. When $\theta = 0°$, the spring force required for *equilibrium* of the disk is 10 lb, acting to the right. For position θ, the spring force is $F_s = (200\text{ lb/ft})(0.75\theta\text{ ft}) + 10\text{ lb}$. Substituting these results into Eq. 1 and simplifying yields

$$\ddot{\theta} + 368\theta = 0$$

Hence,

$$\omega_n^2 = 368 \qquad \omega_n = 19.18 \text{ rad/s}$$

Therefore, the natural period of vibration is

$$\tau = \frac{2\pi}{\omega_n} = \frac{2\pi}{19.18} = 0.328 \text{ s} \qquad Ans.$$

Fig. 22–7

PROBLEMS

22–1. A spring is stretched 175 mm by an 8-kg block. If the block is displaced 100 mm downward from its equilibrium position and given a downward velocity of 1.50 m/s, determine the differential equation which describes the motion. Assume that positive displacement is downward. Also, determine the position of the block when $t = 0.22$ s.

22–2. A spring has a stiffness of 800 N/m. If a 2-kg block is attached to the spring, pushed 50 mm above its equilibrium position, and released from rest, determine the equation that describes the block's motion. Assume that positive displacement is downward.

22–3. A spring is stretched 200 mm by a 15-kg block. If the block is displaced 100 mm downward from its equilibrium position and given a downward velocity of 0.75 m/s, determine the equation which describes the motion. What is the phase angle? Assume that positive displacement is downward.

*22–4. When a 20-lb weight is suspended from a spring, the spring is stretched a distance of 4 in. Determine the natural frequency and the period of vibration for a 10-lb weight attached to the same spring.

22–5. When a 3-kg block is suspended from a spring, the spring is stretched a distance of 60 mm. Determine the natural frequency and the period of vibration if a 0.2-kg block is attached to the same spring rather than a 3-kg block.

22–6. An 8-kg block is suspended from a spring having a stiffness $k = 80$ N/m. If the block is given an upward velocity of 0.4 m/s when it is 90 mm above its equilibrium position, determine the equation which describes the motion and the maximum upward displacement of the block measured from the equilibrium position. Assume that positive displacement is downward.

22–7. A 2-lb weight is suspended from a spring having a stiffness $k = 2$ lb/in. If the weight is pushed 1 in. upward from its equilibrium position and then released from rest, determine the equation which describes the motion. What is the amplitude and the natural frequency of the vibration? Assume that positive displacement is downward.

*22–8. A pendulum has a 0.4-m-long cord and is given a tangential velocity of 0.2 m/s toward the vertical from a position $\theta = 0.3$ rad. Determine the equation which describes the angular motion.

22–9. A 3-kg block is suspended from a spring having a stiffness of $k = 200$ N/m. If the block is pushed 50 mm upward from its equilibrium position and then released from rest, determine the equation that describes the motion. What are the amplitude and the natural frequency of the vibration? Assume that positive displacement is downward.

22–10. The uniform rod of mass m is supported by a pin at A and a spring at B. If B is given a small sideward displacement and released, determine the natural period of vibration.

Prob. 22–10

22–11. The two identical gears each have a mass of m and a radius of gyration about their center of mass of k_O. They are in mesh with the gear rack, which has a mass of M and is attached to a spring having a stiffness k. If the gear rack is displaced slightly horizontally, determine the natural period of oscillation.

Prob. 22–11

*22–12. The 3-kg target slides freely along the smooth horizontal guides *BC* and *DE*, which are "nested" in springs that each have a stiffness of $k = 9$ kN/m. If a 60-g bullet is fired with a velocity of 900 m/s and embeds into the target, determine the amplitude and frequency of oscillation of the target.

22–14. The 20-lb rectangular plate has a natural period of vibration $\tau = 0.3$ s, as it oscillates around the axis of rod *AB*. Determine the torsional stiffness k, measured in lb · ft/rad, of the rod. Neglect the mass of the rod.

Prob. 22–12

Prob. 22–14

22–13. The body of arbitrary shape has a mass *m*, mass center at *G*, and a radius of gyration about *G* of k_G. If it is displaced a slight amount θ from its equilibrium position and released, determine the natural period of vibration.

22–15. The rod of mass *m* is supported by two cords, each having a length *l*. If the rod is given a slight rotation about a vertical axis through its center and released, determine the period of oscillation.

Prob. 22–13

Prob. 22–15

*22–16. A block of mass m is suspended from two springs having a stiffness of k_1 and k_2, arranged a) parallel to each other, and b) as a series. Determine the equivalent stiffness of a single spring with the same oscillation characteristics and the period of oscillation for each case.

22–17. The 15-kg block is suspended from two springs having a different stiffness and arranged a) parallel to each other, and b) as a series. If the natural periods of oscillation of the parallel system and series system are observed to be 0.5 s and 1.5 s, respectively, determine the spring stiffnesses k_1 and k_2.

Probs. 22–16/17

22–18. The 50-lb wheel has a radius of gyration about its mass center G of $k_G = 0.7$ ft. Determine the frequency of vibration if it is displaced slightly from the equilibrium position and released. Assume no slipping.

Prob. 22–18

22–19. Determine the natural frequency for small oscillations of the 10-lb sphere when the rod is displaced a slight distance and released. Neglect the size of the sphere and the mass of the rod. The spring has an unstretched length of 1 ft.

Prob. 22–19

*22–20. A uniform board is supported on two wheels which rotate in opposite directions at a constant angular speed. If the coefficient of kinetic friction between the wheels and board is μ, determine the frequency of vibration of the board if it is displaced slightly, a distance x from the midpoint between the wheels, and released.

Prob. 22–20

22–21. The disk has a weight of 10 lb and rolls without slipping on the horizontal surface as it oscillates about its equilibrium position. If the disk is displaced, by rolling it counterclockwise 0.4 rad, determine the equation which describes its oscillatory motion when it is released. Initially, the disk is at rest.

Prob. 22–21

22–22. A platform, having an unknown mass, is supported by *four* springs, each having the same stiffness k. When nothing is on the platform, the period of vertical vibration is measured as 2.35 s; whereas if a 3-kg block is supported on the platform, the period of vertical vibration is 5.23 s. Determine the mass of a block placed on the (empty) platform which causes the platform to vibrate vertically with a period of 5.62 s. What is the stiffness k of each of the springs?

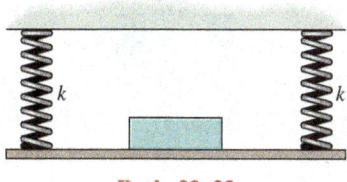

Prob. 22–22

22–23. The bell has a mass of 375 kg, a center of mass at G, and a radius of gyration about point D of $k_D = 0.4$ m. The tongue consists of a slender rod attached to the inside of the bell at C. If an 8-kg mass is attached to the end of the rod, determine the length l of the rod so that the bell will "ring silent," i.e., so that the natural period of vibration of the tongue is the same as that of the bell. For the calculation, neglect the small distance between C and D and neglect the mass of the rod.

Prob. 22–23

*22–24. The 10-kg disk is pin connected at its mass center. Determine the natural period of vibration of the disk if the springs have sufficient tension in them to prevent the cord from slipping on the disk as it oscillates. *Hint*: Assume that the initial stretch in each spring is δ_O.

22–25. If the springs in Prob. 22–24 are originally unstretched, determine the frequency of vibration.

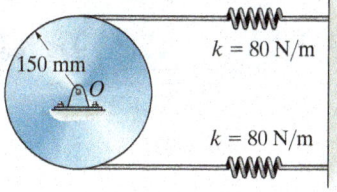

Probs. 22–24/25

22–26. The block has a mass m and is supported by a rigid bar of negligible mass. If the spring has a stiffness k, determine the natural period of vibration for the block.

Prob. 22–26

22–27. Determine the frequency of vibration for the block. The springs are originally compressed Δ.

Prob. 22–27

*22–28. The cylinder of radius r and mass m is displaced a small amount on the curved surface. If it rolls without slipping, determine the frequency of oscillation when it is released.

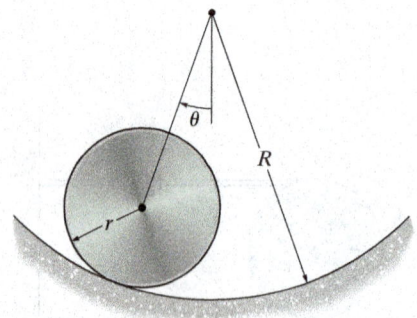

Prob. 22–28

22–29. The 25-lb weight is fixed to the end of the rod assembly. If both springs are unstretched when the assembly is in the position shown, determine the natural period of vibration for the weight when it is displaced slightly and released. Neglect the size of the block and the mass of the rods.

Prob. 22–29

*22.2 ENERGY METHODS

The simple harmonic motion of a body, discussed in the previous section, is due only to gravitational and elastic restoring forces acting on the body. Since these forces are *conservative*, it is also possible to use the conservation of energy equation to obtain the body's natural frequency or period of vibration. To show how to do this, consider again the block and spring model in Fig. 22–8. When the block is in a general displaced position x from the equilibrium position, the kinetic energy is $T = \frac{1}{2}mv^2 = \frac{1}{2}m\dot{x}^2$ and the potential energy is $V = \frac{1}{2}kx^2$. Since energy is conserved, it is necessary that

$$T + V = \text{constant}$$

$$\frac{1}{2}m\dot{x}^2 + \frac{1}{2}kx^2 = \text{constant} \tag{22–17}$$

The equation describing the *accelerated motion* of the block can be obtained by *differentiating* this equation with respect to time; i.e.,

$$m\dot{x}\ddot{x} + kx\dot{x} = 0$$

$$\dot{x}(m\ddot{x} + kx) = 0$$

Since the velocity $\dot{x}$ is not *always* zero in a vibrating system, then

$$\ddot{x} + \omega_n^2 x = 0 \qquad \omega_n = \sqrt{k/m}$$

which is the same as Eq. 22–1.

If the conservation of energy equation is written for a *system of connected bodies*, the natural frequency or the equation of motion can also be determined by time differentiation. It is *not necessary* to dismember the system to account for the internal forces acting between the bodies because they do no work.

Fig. 22–8

The suspension of a railroad car consists of a set of springs which are mounted between the frame of the car and the wheel truck. This will give the car a natural frequency of vibration.

PROCEDURE FOR ANALYSIS

The natural frequency ω_n of a body or system of connected bodies can be determined by applying the conservation of energy equation using the following procedure.

Energy Equation.

- Draw the body when it is displaced by a *small amount* from its equilibrium position and define the location of the body from its equilibrium position by an appropriate position coordinate q.

- Formulate the conservation of energy for the body, $T + V = $ constant, in terms of the position coordinate.

- In general, the kinetic energy must account for both the body's translational and rotational motion, $T = \frac{1}{2}mv_G^2 + \frac{1}{2}I_G\omega^2$, Eq. 18–2.

- The potential energy is the sum of the gravitational and elastic potential energies of the body, $V = V_g + V_e$, Eq. 18–17. In particular, V_g should be measured from a datum for which $q = 0$ (equilibrium position).

Time Derivative.

- Take the time derivative of the energy equation using the chain rule and factor out the common terms. The resulting differential equation represents the equation of motion for the system. The natural frequency ω_n is obtained after rearranging the terms in the "standard form," $\ddot{q} + \omega_n^2 q = 0$.

EXAMPLE 22.5

The thin hoop shown in Fig. 22–9a is supported by the peg at O. Determine the natural period of oscillation for small amplitudes of swing. The hoop has a mass m.

SOLUTION

Energy Equation. A diagram of the hoop when it is displaced a small amount $(q =)\ \theta$ from the equilibrium position is shown in Fig. 22–9b. Using the table on the inside back cover and the parallel-axis theorem to determine I_O, the kinetic energy is

$$T = \tfrac{1}{2}I_O\omega_n^2 = \tfrac{1}{2}[mr^2 + mr^2]\dot\theta^2 = mr^2\dot\theta^2$$

If a horizontal datum is placed through point O, then in the displaced position, the potential energy is

$$V = -mg(r\cos\theta)$$

The total energy is

$$T + V = mr^2\dot\theta^2 - mgr\cos\theta$$

(a)

(b)

Fig. 22–9

Time Derivative.

$$mr^2(2\dot\theta)\ddot\theta + mgr(\sin\theta)\dot\theta = 0$$

$$mr\dot\theta(2r\ddot\theta + g\sin\theta) = 0$$

Since $\dot\theta$ is not always equal to zero, then the terms in parentheses give

$$\ddot\theta + \frac{g}{2r}\sin\theta = 0$$

For small angle θ, $\sin\theta \approx \theta$.

$$\ddot\theta + \frac{g}{2r}\theta = 0$$

$$\omega_n = \sqrt{\frac{g}{2r}}$$

so that

$$\tau = \frac{2\pi}{\omega_n} = 2\pi\sqrt{\frac{2r}{g}} \qquad \textit{Ans.}$$

EXAMPLE 22.6

A 10-kg block is suspended from a cord wrapped around a 5-kg disk, as shown in Fig. 22–10a. If the spring has a stiffness $k = 200$ N/m, determine the natural period of vibration for the system.

SOLUTION

Energy Equation. A diagram of the block and disk when they are displaced by respective amounts s and θ from the equilibrium position is shown in Fig. 22–10b. Since $s = (0.15 \text{ m})\theta$, then $v_b \approx \dot{s} = (0.15 \text{ m})\dot{\theta}$. Thus, the kinetic energy of the system is

$$T = \tfrac{1}{2}m_b v_b^2 + \tfrac{1}{2}I_O \omega_d^2$$

$$= \tfrac{1}{2}(10 \text{ kg})[(0.15 \text{ m})\dot{\theta}]^2 + \tfrac{1}{2}\left[\tfrac{1}{2}(5 \text{ kg})(0.15 \text{ m})^2\right](\dot{\theta})^2$$

$$= 0.1406(\dot{\theta})^2$$

Establishing the datum at the equilibrium position of the block and realizing that the spring stretches s_{st} for equilibrium, the potential energy is

$$V = \tfrac{1}{2}k(s_{st} + s)^2 - Ws$$

$$= \tfrac{1}{2}(200 \text{ N/m})[s_{st} + (0.15 \text{ m})\theta]^2 - 98.1 \text{ N}[(0.15 \text{ m})\theta]$$

The total energy for the system is therefore,

$$T + V = 0.1406(\dot{\theta})^2 + 100(s_{st} + 0.15\theta)^2 - 14.715\theta$$

Time Derivative.

$$0.28125(\dot{\theta})\ddot{\theta} + 200(s_{st} + 0.15\theta)0.15\dot{\theta} - 14.72\dot{\theta} = 0$$

Since $s_{st} = 98.1/200 = 0.4905$ m, the above equation reduces to the "standard form"

$$\ddot{\theta} + 16\theta = 0$$

so that

$$\omega_n = \sqrt{16} = 4 \text{ rad/s}$$

Thus,

$$\tau = \frac{2\pi}{\omega_n} = \frac{2\pi}{4} = 1.57 \text{ s} \qquad \textit{Ans.}$$

(a)

(b)

Fig. 22–10

PROBLEMS

22–30. Determine the frequency of oscillation of the cylinder of mass m when it is pulled down slightly and released. Neglect the mass of the small pulley.

Prob. 22–30

22–31. Determine the differential equation of motion of the block of mass m when it is displaced slightly and released. Motion occurs in the vertical plane. The springs are attached to the block.

Prob. 22–31

*22–32. Determine the natural period of vibration of the 10-lb semicircular disk.

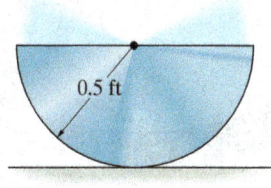

Prob. 22–32

22–33. A torsional spring of stiffness k is attached to a wheel that has a mass of M. If the wheel is given a small angular displacement of θ about the z axis, determine the natural period of oscillation. The wheel has a radius of gyration about the z axis of k_z.

Prob. 22–33

22–34. Determine the differential equation of motion of the 15-kg spool. Assume that it does not slip at the surface of contact as it oscillates. The radius of gyration of the spool about its center of mass is $k_G = 125$ mm. The springs are originally unstretched.

Prob. 22–34

22–35. The 5-lb sphere is attached to a rod of negligible mass and rests in the horizontal position. Determine the natural frequency of vibration. Neglect the size of the sphere.

Prob. 22–35

*22–36. If the lower end of the 6-kg slender rod is displaced a small amount and released from rest, determine the natural frequency of vibration. Each spring has a stiffness of $k = 200$ N/m and is unstretched when the rod is hanging vertically.

Prob. 22–36

22–37. Determine the natural period of vibration of the disk having a mass m and radius r. Assume the disk does not slip on the surface of contact as it oscillates.

Prob. 22–37

22–38. Without an adjustable screw, A, the 1.5-lb pendulum has a center of gravity at G. If it is required that it oscillates with a period of 1 s, determine the distance a from pin O to the screw. The pendulum's radius of gyration about O is $k_O = 8.5$ in. and the screw has a weight of 0.05 lb.

Prob. 22–38

22–39. The bar has a mass of 8 kg and is suspended from two springs such that when it is in equilibrium, the springs make an angle of 45° with the horizontal as shown. Determine the natural period of vibration if the bar is pulled down a short distance and released. Each spring has a stiffness of $k = 40$ N/m.

Prob. 22–39

*22–40. If the wheel is given a small angular displacement of θ and released from rest, it is observed that it oscillates with a natural period of τ. Determine the wheel's radius of gyration about its center of mass G. The wheel has a mass of m and rolls on the rails without slipping.

Prob. 22–40

*22.3 UNDAMPED FORCED VIBRATION

Undamped forced vibration is considered to be one of the most important types of vibrating motion in engineering. Its principles can be used to describe the motion of many types of machines and structures.

Periodic Force. The block and spring shown in Fig. 22–11a provide a convenient model which represents the vibrational characteristics of a system subjected to a periodic force $F = F_0 \sin \omega_0 t$. This force has an amplitude of F_0 and a *forcing frequency* ω_0. The free-body diagram for the block when it is displaced a distance x is shown in Fig. 22–11b. Applying the equation of motion, we have

$$\stackrel{+}{\rightarrow} \Sigma F_x = ma_x; \qquad F_0 \sin \omega_0 t - kx = m\ddot{x}$$

or

$$\ddot{x} + \frac{k}{m}x = \frac{F_0}{m}\sin \omega_0 t \qquad (22\text{–}18)$$

This equation is a nonhomogeneous second-order differential equation. The general solution consists of a complementary solution, x_c, plus a particular solution, x_p.

The **complementary solution** is determined by setting the term on the right side of Eq. 22–18 equal to zero and solving the resulting homogeneous equation. The solution is defined by Eq. 22–9, i.e.,

$$x_c = C \sin(\omega_n t + \phi) \qquad (22\text{–}19)$$

where ω_n is the natural frequency, $\omega_n = \sqrt{k/m}$, Eq. 22–2.

Since the motion is periodic, the **particular solution** of Eq. 22–18 can be determined by assuming a solution of the form

$$x_p = X \sin \omega_0 t \qquad (22\text{–}20)$$

where X is a constant. Taking the second time derivative and substituting into Eq. 22–18 yields

$$-X\omega_0^2 \sin \omega_0 t + \frac{k}{m}(X \sin \omega_0 t) = \frac{F_0}{m}\sin \omega_0 t$$

Factoring out $\sin \omega_0 t$ and solving for X gives

$$X = \frac{F_0/m}{(k/m) - \omega_0^2} = \frac{F_0/k}{1 - (\omega_0/\omega_n)^2} \qquad (22\text{–}21)$$

Substituting into Eq. 22–20, we obtain

$$\boxed{x_p = \frac{F_0/k}{1 - (\omega_0/\omega_n)^2}\sin \omega_0 t} \qquad (22\text{–}22)$$

Fig. 22–11

Shaker tables provide forced vibration and are used to separate out granular materials.

The *general solution* is therefore the sum of two sine functions having different frequencies.

$$x = x_c + x_p = C \sin(\omega_n t + \phi) + \frac{F_0/k}{1 - (\omega_0/\omega_n)^2} \sin \omega_0 t \quad (22\text{--}23)$$

The *complementary solution* x_c defines the *free vibration*, which depends on the natural frequency $\omega_n = \sqrt{k/m}$ and the constants C and ϕ. The *particular solution* x_p describes the *forced vibration* of the block caused by the applied force $F = F_0 \sin \omega_0 t$. Since all vibrating systems are subject to *friction*, the free vibration, x_c, will in time dampen out. For this reason the free vibration is referred to as **transient**, and the forced vibration is called **steady-state**, since it is the only vibration that remains.

From Eq. 22–21 it is seen that the amplitude of forced or steady-state vibration depends on the **frequency ratio** ω_0/ω_n. If the **magnification factor** MF is defined as the ratio of the amplitude of steady-state vibration, X, to the static deflection, F_0/k, which would be produced by the amplitude of the periodic force F_0, then, from Eq. 22–21,

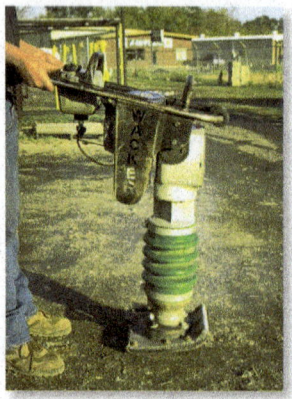

The soil compactor operates by forced vibration developed by an internal motor. It is important that the forcing frequency not be close to the natural frequency of vibration of the compactor, which can be determined when the motor is turned off; otherwise resonance will occur and the machine will become uncontrollable.

$$\text{MF} = \frac{X}{F_0/k} = \frac{1}{1 - (\omega_0/\omega_n)^2} \qquad (22\text{–}24)$$

This equation is graphed in Fig. 22–12. Note that if the force or displacement is applied with a frequency close to the natural frequency of the system, i.e., $\omega_0/\omega_n \approx 1$, the amplitude of vibration of the block becomes extremely large. This occurs because the force **F** is applied to the block so that it always follows the motion of the block. This condition is called **resonance**, and in practice, resonating vibrations can cause tremendous stress and rapid failure of parts.*

Periodic Support Displacement. Forced vibrations can also arise from the periodic excitation of the support of a system. For example, the model shown in Fig. 22–13a represents the periodic vibration of a block which is caused by harmonic movement $\delta = \delta_0 \sin \omega_0 t$ of the support. The free-body diagram for the block in this case is shown in Fig. 22–13b. If the displacement δ of the support is measured from the point of zero displacement, i.e., when the radial line OA coincides with OB, then the general deformation of the spring is $(x - \delta_0 \sin \omega_0 t)$. Applying the equation of motion,

$$\xrightarrow{+} F_x = ma_x; \qquad -k(x - \delta_0 \sin \omega_0 t) = m\ddot{x}$$

or

$$\ddot{x} + \frac{k}{m}x = \frac{k\delta_0}{m} \sin \omega_0 t \qquad (22\text{–}25)$$

By comparison, this equation is identical to the form of Eq. 22–18, provided F_0 is replaced by $k\delta_0$. If this substitution is also made into Eqs. 22–21 to 22–23, the results are appropriate for describing the motion of the block when subjected to the support displacement $\delta = \delta_0 \sin \omega_0 t$.

Fig. 22–12

(a)

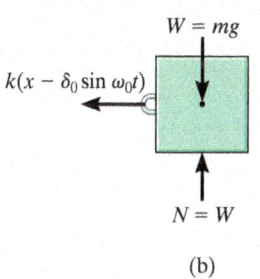

(b)

Fig. 22–13

*A swing has a natural period of vibration, as determined in Example 22.1. If someone pushes on the swing only when it reaches its highest point, neglecting drag or wind resistance, resonance will occur since the natural and forcing frequencies are the same.

EXAMPLE 22.7

The instrument shown in Fig. 22–14 is rigidly attached to a platform P, which in turn is supported by *four* springs, each having a stiffness $k = 800 \text{ N/m}$. If the floor is subjected to a vertical displacement $\delta = 10 \sin(8t)$ mm, where t is in seconds, determine the amplitude of steady-state vibration. What is the frequency of the floor vibration required to cause resonance? The instrument and platform have a total mass of 20 kg.

Fig. 22–14

SOLUTION
The natural frequency is

$$\omega_n = \sqrt{\frac{k}{m}} = \sqrt{\frac{4(800 \text{ N/m})}{20 \text{ kg}}} = 12.65 \text{ rad/s}$$

The amplitude of steady-state vibration is found using Eq. 22–21, with $k\delta_0$ replacing F_0.

$$X = \frac{\delta_0}{1 - (\omega_0/\omega_n)^2} = \frac{10}{1 - [(8 \text{ rad/s})/(12.65 \text{ rad/s})]^2} = 16.7 \text{ mm} \quad \textit{Ans.}$$

Resonance will occur when the amplitude of vibration X caused by the floor displacement approaches infinity. This requires

$$\omega_0 = \omega_n = 12.6 \text{ rad/s} \quad \textit{Ans.}$$

*22.4 VISCOUS DAMPED FREE VIBRATION

The vibration analysis considered thus far has not included the effects of friction or damping in the system, and as a result, the solutions obtained are only in close agreement with the actual motion. Since all vibrations die out in time, the presence of damping forces should be included in the analysis.

In many cases damping is attributed to the resistance created by the fluid, such as water, oil, or air, in which the system vibrates. Provided the body moves slowly through this fluid, then the resistance to motion is directly proportional to the body's speed. The type of force developed under these conditions is called a *viscous damping force*. The magnitude of this force is expressed by an equation of the form

$$F = c\dot{x} \quad (22\text{--}26)$$

where the constant c is called the *coefficient of viscous damping* and has units of $N \cdot s/m$ or $lb \cdot s/ft$.

The vibrating motion of a body or system having viscous damping can be characterized by the block and spring shown in Fig. 22–15a. The effect of damping is provided by the *dashpot* connected to the block on the right side. Damping occurs when the piston P moves to the right or left within the enclosed cylinder. The cylinder contains a fluid, and the motion of the piston is retarded since the fluid must flow around or through a small hole in the piston. The dashpot is assumed to have a coefficient of viscous damping c.

If the block is displaced a distance x from its equilibrium position, the resulting free-body diagram is shown in Fig. 22–15b. Both the spring and damping force oppose the forward motion of the block, and so the equation of motion becomes

$$\xrightarrow{+} \Sigma F_x = ma_x; \qquad -kx - c\dot{x} = m\ddot{x}$$

or

$$m\ddot{x} + c\dot{x} + kx = 0 \quad (22\text{--}27)$$

This linear, second-order, homogeneous, differential equation has a solution of the form

$$x = e^{\lambda t}$$

where e is the base of the natural logarithm and λ (lambda) is a constant. The value of λ can be obtained by substituting this solution and its time derivatives into Eq. 22–27, which yields

$$m\lambda^2 e^{\lambda t} + c\lambda e^{\lambda t} + k e^{\lambda t} = 0$$

or

$$e^{\lambda t}(m\lambda^2 + c\lambda + k) = 0$$

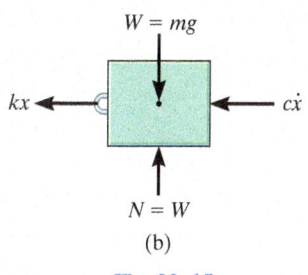

Fig. 22–15

Since $e^{\lambda t}$ can never be zero, a solution is possible provided

$$m\lambda^2 + c\lambda + k = 0$$

Hence, by the quadratic formula, the two values of λ are

$$\lambda_1 = -\frac{c}{2m} + \sqrt{\left(\frac{c}{2m}\right)^2 - \frac{k}{m}}$$
$$\lambda_2 = -\frac{c}{2m} - \sqrt{\left(\frac{c}{2m}\right)^2 - \frac{k}{m}}$$
(22–28)

The general solution of Eq. 22–27 is therefore a combination of exponentials which involves both of these roots. There are three possible combinations of λ_1 and λ_2 which must be considered. Before discussing these combinations, however, we will first define the **critical damping coefficient** c_c as the value of c which makes the radical in Eqs. 22–28 equal to zero; i.e.,

$$\left(\frac{c_c}{2m}\right)^2 - \frac{k}{m} = 0$$

or

$$c_c = 2m\sqrt{\frac{k}{m}} = 2m\omega_n \qquad (22\text{–}29)$$

Overdamped System. When $c > c_c$, the roots λ_1 and λ_2 are both real. The general solution of Eq. 22–27 can then be written as

$$x = Ae^{\lambda_1 t} + Be^{\lambda_2 t} \qquad (22\text{–}30)$$

Motion corresponding to this solution is nonvibrating. The effect of damping is so strong that when the block is displaced and released, it simply creeps back to its original position without oscillating. The system is said to be **overdamped**.

Critically Damped System. If $c = c_c$, then $\lambda_1 = \lambda_2 = -c_c/2m = -\omega_n$. This situation is known as **critical damping**, since it represents a condition where c has the smallest value necessary to cause the system to be nonvibrating. Using the methods of differential equations, it can be shown that the solution to Eq. 22–27 for critical damping is

$$x = (A + Bt)e^{-\omega_n t} \qquad (22\text{–}31)$$

Underdamped System. Most often $c < c_c$, in which case the system is referred to as **underdamped**. In this case the roots λ_1 and λ_2 are complex numbers, and it can be shown that the general solution of Eq. 22–27 can be written as

$$x = D[e^{-(c/2m)t} \sin(\omega_d t + \phi)] \qquad (22\text{–}32)$$

where D and ϕ are constants generally determined from the initial conditions of the problem. The constant ω_d is called the **damped natural frequency** of the system. It has a value of

$$\omega_d = \sqrt{\frac{k}{m} - \left(\frac{c}{2m}\right)^2} = \omega_n \sqrt{1 - \left(\frac{c}{c_c}\right)^2} \qquad (22\text{–}33)$$

where the ratio c/c_c is called the **damping factor**.

The graph of Eq. 22–32 is shown in Fig. 22–16. The initial limit of motion, D, diminishes with each cycle of vibration, since motion is confined within the bounds of the exponential curve. Using the damped natural frequency ω_d, the period of damped vibration can be written as

$$\tau_d = \frac{2\pi}{\omega_d} \qquad (22\text{–}34)$$

Since $\omega_d < \omega_n$, Eq. 22–33, the period of damped vibration, τ_d, will be greater than that of free vibration, $\tau = 2\pi/\omega_n$.

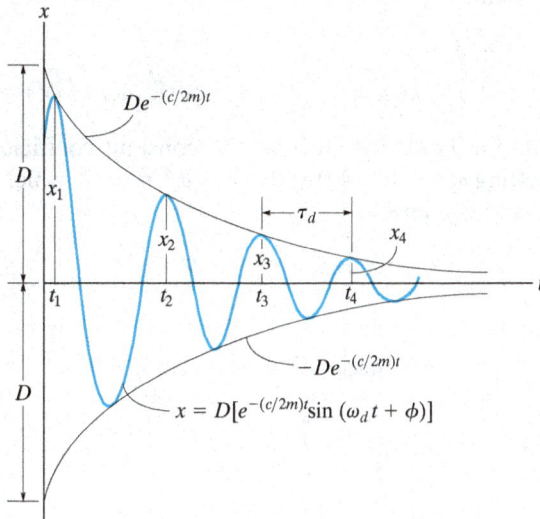

Fig. 22–16

*22.5 VISCOUS DAMPED FORCED VIBRATION

The most general case of single-degree-of-freedom vibrating motion occurs when the system includes the effects of forced motion and induced damping. The analysis of this particular type of vibration is of practical value when applied to systems having significant damping characteristics.

If a dashpot is attached to the block and spring shown in Fig. 22–11a, the differential equation which describes the motion becomes

$$m\ddot{x} + c\dot{x} + kx = F_0 \sin \omega_0 t \qquad (22\text{–}35)$$

A similar equation can be written for a block and spring having a periodic support displacement, Fig. 22–13a, which includes the effects of damping. In that case, however, F_0 is replaced by $k\delta_0$. Since Eq. 22–35 is nonhomogeneous, the general solution is the sum of a complementary solution, x_c, and a particular solution, x_p. The complementary solution is determined by setting the right side of Eq. 22–35 equal to zero and solving the homogeneous equation, which is equivalent to Eq. 22–27. The solution is therefore given by Eq. 22–30, 22–31, or 22–32, depending on the values of λ_1 and λ_2. Because all systems are subjected to friction, this solution will dampen out with time. Only the particular solution, which describes the *steady-state vibration* of the system, will remain. Since the applied forcing function is harmonic, the steady-state motion will also be harmonic. Consequently, the particular solution will be of the form

$$x_P = X' \sin(\omega_0 t - \phi') \qquad (22\text{–}36)$$

The constants X' and ϕ' are determined by taking the first and second time derivatives and substituting them into Eq. 22–35, which after simplification yields

$$-X'm\omega_0^2 \sin(\omega_0 t - \phi') + X'c\omega_0 \cos(\omega_0 t - \phi') + X'k \sin(\omega_0 t - \phi') = F_0 \sin \omega_0 t$$

Since this equation holds for all time, the constant coefficients can be obtained by setting $\omega_0 t - \phi' = 0$ and $\omega_0 t - \phi' = \pi/2$, which causes the above equation to become

$$X'c\omega_0 = F_0 \sin \phi'$$

and

$$-X'm\omega_0^2 + X'k = F_0 \cos \phi'$$

The amplitude is obtained by squaring these equations, adding the results, and using the identity $\sin^2\phi' + \cos^2\phi' = 1$, which gives

$$X' = \frac{F_0}{\sqrt{(k - m\omega_0^2)^2 + c^2\omega_0^2}} \qquad (22\text{–}37)$$

Dividing the first equation by the second gives

$$\phi' = \tan^{-1}\left[\frac{c\omega_0}{k - m\omega_0^2}\right] \qquad (22\text{–}38)$$

Since $\omega_n = \sqrt{k/m}$ and $c_c = 2m\omega_n$, then the above equations can also be written as

$$X' = \frac{F_0/k}{\sqrt{[1 - (\omega_0/\omega_n)^2]^2 + [2(c/c_c)(\omega_0/\omega_n)]^2}}$$

$$\phi' = \tan^{-1}\left[\frac{2(c/c_c)(\omega_0/\omega_n)}{1 - (\omega_0/\omega_n)^2}\right] \qquad (22\text{–}39)$$

The angle ϕ' represents the phase difference between the applied force and the resulting steady-state vibration of the damped system.

The *magnification factor* MF has been defined in Sec. 22.3 as the ratio of the amplitude of deflection caused by the forced vibration to the deflection caused by a static force F_0. Thus,

$$\text{MF} = \frac{X'}{F_0/k} = \frac{1}{\sqrt{[1 - (\omega_0/\omega_n)^2]^2 + [2(c/c_c)(\omega_0/\omega_n)]^2}} \qquad (22\text{–}40)$$

The MF is plotted in Fig. 22–17 versus the frequency ratio ω_0/ω_n for various values of the damping factor c/c_c. It can be seen from this graph that the magnification of the amplitude increases as the damping factor decreases. Resonance obviously occurs only when the damping factor is zero and the frequency ratio equals 1.

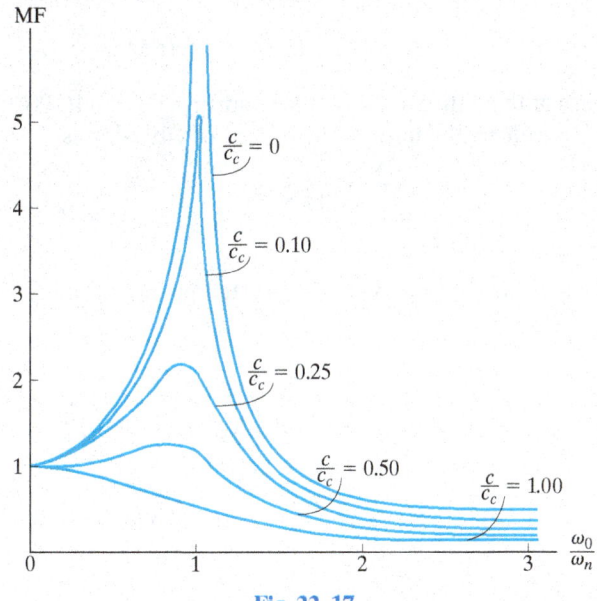

Fig. 22–17

EXAMPLE 22.8

The 30-kg electric motor shown in Fig. 22–18 is confined to move vertically, and is supported by *four* springs, each spring having a stiffness of 200 N/m. If the rotor is unbalanced such that its effect is equivalent to a 4-kg mass located 60 mm from the axis of rotation, determine the amplitude of vibration when the rotor is turning at $\omega_0 = 10$ rad/s. The damping factor is $c/c_c = 0.15$.

Fig. 22–18

SOLUTION
The periodic force which causes the motor to vibrate is the centrifugal force due to the unbalanced rotor. This force has a constant magnitude of

$$F_0 = ma_n = mr\omega_0^2 = 4 \text{ kg}(0.06 \text{ m})(10 \text{ rad/s})^2 = 24 \text{ N}$$

The stiffness of the entire system of four springs is $k = 4(200 \text{ N/m}) = 800 \text{ N/m}$. Therefore, the natural frequency of vibration is

$$\omega_n = \sqrt{\frac{k}{m}} = \sqrt{\frac{800 \text{ N/m}}{30 \text{ kg}}} = 5.164 \text{ rad/s}$$

Since the damping factor is known, the steady-state amplitude can be determined from the first of Eqs. 22–39, i.e.,

$$X' = \frac{F_0/k}{\sqrt{[1 - (\omega_0/\omega_n)^2]^2 + [2(c/c_c)(\omega_0/\omega_n)]^2}}$$

$$= \frac{24/800}{\sqrt{[1 - (10/5.164)^2]^2 + [2(0.15)(10/5.164)]^2}}$$

$$= 0.0107 \text{ m} = 10.7 \text{ mm} \qquad Ans.$$

*22.6 ELECTRICAL CIRCUIT ANALOGS

The characteristics of a vibrating mechanical system can be represented by an electric circuit. Consider the circuit shown in Fig. 22–19a, which consists of an inductor L, a resistor R, and a capacitor C. When a voltage $E(t)$ is applied, it causes a current of magnitude i to flow through the circuit. As the current flows past the inductor the voltage drop is $L(di/dt)$, when it flows across the resistor the drop is Ri, and when it arrives at the capacitor the drop is $(1/C)\int i\,dt$. Since current cannot flow past a capacitor, it is only possible to measure the charge q acting on the capacitor. The charge can, however, be related to the current by the equation $i = dq/dt$. Thus, the voltage drops which occur across the inductor, resistor, and capacitor become $L\,d^2q/dt^2$, $R\,dq/dt$, and q/C, respectively. According to Kirchhoff's voltage law, the applied voltage balances the sum of the voltage drops around the circuit. Therefore,

(a)

$$L\frac{d^2q}{dt^2} + R\frac{dq}{dt} + \frac{1}{C}q = E(t) \qquad (22\text{–}41)$$

Consider now the model of a single-degree-of-freedom mechanical system, Fig. 22–19b, which is subjected to both a general forcing function $F(t)$ and damping. The equation of motion for this system was established in the previous section and can be written as

(b)

Fig. 22–19

$$m\frac{d^2x}{dt^2} + c\frac{dx}{dt} + kx = F(t) \qquad (22\text{–}42)$$

By comparison, it is seen that Eqs. 22–41 and 22–42 have the same form, and therefore mathematically the procedure of analyzing an electric circuit is the same as that of analyzing a vibrating mechanical system. The analogs between the two equations are given in Table 22–1.

This analogy has important application to experimental work, for it is much easier to simulate the vibration of a complex mechanical system using an electric circuit, which can be constructed on an analog computer, than to make an equivalent mechanical spring-and-dashpot model.

TABLE 22–1
Electrical–Mechanical Analogs

Electrical		Mechanical	
Electric charge	q	Displacement	x
Electric current	i	Velocity	dx/dt
Voltage	$E(t)$	Applied force	$F(t)$
Inductance	L	Mass	m
Resistance	R	Viscous damping coefficient	c
Reciprocal of capacitance	$1/C$	Spring stiffness	k

PROBLEMS

22–41. If the block-and-spring model is subjected to the periodic force $F = F_0 \cos \omega t$, show that the differential equation of motion is $\ddot{x} + (k/m)x = (F_0/m) \cos \omega t$, where x is measured from the equilibrium position of the block. What is the general solution of this equation?

22–46. A 5-kg block is suspended from a spring having a stiffness of 300 N/m. If the block is acted upon by a vertical force $F = (7 \sin 8t)$ N, where t is in seconds, determine the equation which describes the motion of the block when it is pulled down 100 mm from the equilibrium position and released from rest at $t = 0$. Assume that positive displacement is downward.

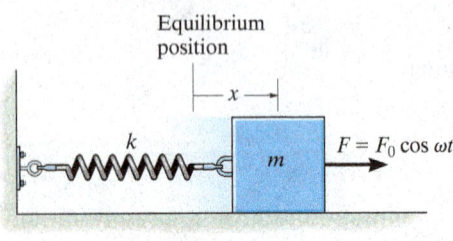

Prob. 22–41

22–42. A block which has a mass m is suspended from a spring having a stiffness k. If an impressed downward vertical force $F = F_O$ acts on the weight, determine the equation which describes the position of the block as a function of time.

22–43. A 4-lb weight is attached to a spring having a stiffness $k = 10$ lb/ft. The weight is drawn downward a distance of 4 in. and released from rest. If the support moves with a vertical displacement $\delta = (0.5 \sin 4t)$ in., where t is in seconds, determine the equation which describes the position of the weight as a function of time.

*__22–44.__ A 4-kg block is suspended from a spring that has a stiffness of $k = 600$ N/m. The block is drawn downward 50 mm from the equilibrium position and released from rest when $t = 0$. If the support moves with an impressed displacement of $\delta = (10 \sin 4t)$ mm, where t is in seconds, determine the equation that describes the vertical motion of the block. Assume positive displacement is downward.

22–45. The barrel of a cannon has a mass of 700 kg, and after firing it recoils a distance of 0.64 m. If it returns to its original position by means of a single recuperator having a damping coefficient of 2 kN·s/m, determine the required stiffness of each of the two springs fixed to the base and attached to the barrel so that the barrel recuperates without vibration.

Prob. 22–46

22–47. If the dashpot has a damping coefficient of $c = 50$ N·s/m, and the spring has a stiffness of $k = 600$ N/m, show that the system is underdamped, and then find the pendulum's period of oscillation. The uniform rods have a mass per unit length of 10 kg/m.

Prob. 22–47

*22–48. The 30-lb block is attached to two springs having a stiffness of 10 lb/ft. A periodic force $F = (8 \cos 3t)$ lb, where t is in seconds, is applied to the block. Determine the maximum speed of the block after frictional forces cause the free vibrations to dampen out.

Prob. 22–48

22–49. The light elastic rod supports a 4-kg sphere. When an 18-N vertical force is applied to the sphere, the rod deflects 14 mm. If the wall oscillates with harmonic frequency of 2 Hz and has an amplitude of 15 mm, determine the amplitude of vibration for the sphere.

Prob. 22–49

22–50. Find the differential equation for small oscillations in terms of θ for the uniform rod of mass m. Also show that if $c < \sqrt{mk}/2$, then the system remains underdamped. The rod is in a horizontal position when it is in equilibrium.

Prob. 22–50

22–51. If the 30-kg block is subjected to a periodic force of $P = (300 \sin 5t)$ N, $k = 1500$ N/m, and $c = 300$ N · s/m, determine the equation that describes the steady-state vibration as a function of time.

Prob. 22–51

*22–52. Using a block-and-spring model, like that shown in Fig. 22–13a, but suspended from a vertical position and subjected to a periodic support displacement of $\delta = \delta_0 \cos \omega_0 t$, determine the equation of motion for the system, and obtain its general solution. Define the displacement y measured from the static equilibrium position of the block when $t = 0$.

22–53. The fan has a mass of 25 kg and is fixed to the end of a horizontal beam that has a negligible mass. The fan blade is mounted eccentrically on the shaft such that it is equivalent to an unbalanced 3.5-kg mass located 100 mm from the axis of rotation. If the static deflection of the beam is 50 mm as a result of the weight of the fan, determine the angular velocity of the fan blade at which resonance will occur. *Hint*: See the first part of Example 22.8.

22–54. In Prob. 22–53, determine the amplitude of steady-state vibration of the fan if its angular velocity is 10 rad/s.

22–55. What will be the amplitude of steady-state vibration of the fan in Prob. 22–53 if the angular velocity of the fan blade is 18 rad/s? *Hint*: See the first part of Example 22.8.

Probs. 22–53/54/55

***22–56.** The engine is mounted on a foundation block which is spring supported. Describe the steady-state vibration if the block and engine have a total weight of 1500 lb, and the engine, when running, creates an impressed force $F = (50 \sin 2t)$ lb, where t is in seconds. Assume that the system vibrates only in the vertical direction, with the positive displacement measured downward, and that the total stiffness of the springs is $k = 2000$ lb/ft.

Prob. 22–56

22–57. The electric motor turns an eccentric flywheel which is equivalent to an unbalanced 0.25-lb weight located 10 in. from the axis of rotation. If the static deflection of the beam is 1 in. because of the weight of the motor, determine the angular velocity of the flywheel at which resonance will occur. The motor weighs 150 lb. Neglect the mass of the beam.

22–58. What will be the amplitude of steady-state vibration of the motor in Prob. 22–57 if the angular velocity of the flywheel is 20 rad/s?

22–59. Determine the angular velocity of the flywheel in Prob. 22–57 which will produce an amplitude of vibration of 0.25 in.

Probs. 22–57/58/59

***22–60.** The 450-kg trailer is pulled with a constant speed over the surface of a bumpy road, which may be approximated by a cosine curve having an amplitude of 50 mm and wave length of 4 m. If the two springs s which support the trailer each have a stiffness of 800 N/m, determine the speed v which will cause the greatest vibration (resonance) of the trailer. Neglect the weight of the wheels.

22–61. Determine the amplitude of vibration of the trailer in Prob. 22–60 if the speed $v = 15$ km/h.

Probs. 22–60/61

22–62. The motor of mass M is supported by a simply supported beam of negligible mass. If block A of mass m is clipped onto the rotor, which is turning at constant angular velocity of ω, determine the amplitude of the steady-state vibration. *Hint*: When the beam is subjected to a concentrated force of P at its mid-span, it deflects $\delta = PL^3/48EI$ at this point. Here E is Young's modulus of elasticity, a property of the material, and I is the moment of inertia of the beam's cross-sectional area.

Prob. 22–62

22–63. A block having a mass of 0.8 kg is suspended from a spring having a stiffness of 120 N/m. If a dashpot provides a damping force of 2.5 N when the speed of the block is 0.2 m/s, determine the period of free vibration.

*22–64. The block, having a weight of 15 lb, is immersed in a liquid such that the damping force acting on the block has a magnitude of $F = (0.8|v|)$ lb, where v is the velocity of the block in ft/s. If the block is pulled down 0.8 ft and released from rest, determine the position of the block as a function of time. The spring has a stiffness of $k = 40$ lb/ft. Consider positive displacement to be downward.

Probs. 22–64

22–65. A 7-lb block is suspended from a spring having a stiffness of $k = 75$ lb/ft. The support to which the spring is attached is given simple harmonic motion which may be expressed as $\delta = (0.15 \sin 2t)$ ft, where t is in seconds. If the damping factor is $c/c_c = 0.8$, determine the phase angle ϕ of steady-state vibration.

22–66. Determine the magnification factor of the block, spring, and dashpot combination in Prob. 22–65.

22–67. A block having a mass of 7 kg is suspended from a spring that has a stiffness $k = 600$ N/m. If the block is given an upward velocity of 0.6 m/s from its equilibrium position at $t = 0$, determine its position as a function of time. Assume that positive displacement of the block is downward and that motion takes place in a medium which furnishes a damping force $F = (50|v|)$ N, where v is in m/s.

*22–68. The 200-lb electric motor is fastened to the midpoint of the simply supported beam. It is found that the beam deflects 2 in. when the motor is not running. The motor turns an eccentric flywheel which is equivalent to an unbalanced weight of 1 lb located 5 in. from the axis of rotation. If the motor is turning at 100 rpm, determine the amplitude of steady-state vibration. The damping factor is $c/c_c = 0.20$. Neglect the mass of the beam.

Prob. 22–68

22–69. If the 12-kg rod is subjected to a periodic force of $F = (30 \sin 6t)$ N, where t is in seconds, determine the steady-state vibration amplitude θ_{max} of the rod about the pin B. Assume θ is small.

Prob. 22–69

22–70. The damping factor, c/c_c, may be determined experimentally by measuring the successive amplitudes of vibrating motion of a system. If two of these maximum displacements can be approximated by x_1 and x_2, as shown in Fig. 22–16, show that $\ln(x_1/x_2) = 2\pi(c/c_c)/\sqrt{1-(c/c_c)^2}$. The quantity $\ln(x_1/x_2)$ is called the *logarithmic decrement*.

22–71. The 10-kg block-spring-damper system is damped. If the block is displaced to $x = 50$ mm and released from rest, determine the time required for it to return to the position $x = 2$ mm.

Prob. 22–71

*22–72. The block, having a weight of 12 lb, is immersed in a liquid such that the damping force acting on the block has a magnitude of $F = (0.7|v|)$ lb, where v is in ft/s. If the block is pulled down 0.62 ft and released from rest, determine the position of the block as a function of time. The spring has a stiffness of $k = 53$ lb/ft. Assume that positive displacement is downward.

Prob. 22–72

22–73. The bar has a weight of 6 lb. If the stiffness of the spring is $k = 8$ lb/ft and the dashpot has a damping coefficient $c = 60$ lb·s/ft, determine the differential equation which describes the motion in terms of the angle θ of the bar's rotation. Also, what should be the damping coefficient of the dashpot if the bar is to be critically damped?

Prob. 22–73

22–74. A bullet of mass m has a velocity $\mathbf{v}_0$ just before it strikes the target of mass M. If the bullet embeds in the target, and the vibration is to be critically damped, determine the dashpot's damping coefficient, and the springs' maximum compression. The target is free to move along the two horizontal guides that are "nested" in the springs.

22–75. A bullet of mass m has a velocity $\mathbf{v}_0$ just before it strikes the target of mass M. If the bullet embeds in the target, and the dashpot's damping coefficient is $0 < c \ll c_c$, determine the springs' maximum compression. The target is free to move along the two horizontal guides that are "nested" in the springs.

Probs. 22–74/75

*22–76. Determine the differential equation of motion for the damped vibratory system shown. What type of motion occurs? Take $k = 100$ N/m, $c = 200$ N·s/m, $m = 25$ kg.

Prob. 22–76

22–77. The 20-kg block is subjected to the action of the harmonic force $F = (90 \cos 6t)$ N, where t is in seconds. Write the equation which describes the steady-state motion.

Prob. 22–77

22–78. Determine the differential equation of motion for the damped vibratory system shown. What type of motion occurs?

Prob. 22–78

22–79. Draw the electrical circuit that is equivalent to the mechanical system shown. What is the differential equation which describes the charge q in the circuit?

Prob. 22–79

CHAPTER REVIEW

Undamped Free Vibration

A body has free vibration when gravitational or elastic restoring forces cause the motion. This motion is undamped when friction forces are neglected. The periodic motion of an undamped, freely vibrating body can be studied by displacing the body from the equilibrium position and then applying the equation of motion along the path.

For a one-degree-of-freedom system, the resulting differential equation can be written in terms of its natural frequency ω_n.

$$\ddot{x} + \omega_n^2 x = 0 \qquad \tau = \frac{2\pi}{\omega_n} \qquad f = \frac{1}{\tau} = \frac{\omega_n}{2\pi}$$

Energy Methods

Provided the restoring forces acting on the body are gravitational and elastic, then the conservation of energy theorem can also be used to determine the simple harmonic motion. To do this, the body is displaced a small amount from its equilibrium position, and an expression for its kinetic and potential energy is written. The time derivative of this equation can then be rearranged in the standard form $\ddot{x} + \omega_n^2 x = 0$.

Undamped Forced Vibration

When the equation of motion is applied to a body which is subjected to a periodic force, or the support has a displacement with a frequency ω_0, then the solution of the differential equation consists of a complementary solution and a particular solution. The complementary solution is caused by the free vibration and can be neglected. The particular solution is caused by the forced vibration.

Resonance will occur if the natural frequency of vibration ω_n is equal to the forcing frequency ω_0. This should be avoided, since the motion will tend to become unbounded.

$$x_p = \frac{F_0/k}{1 - (\omega_0/\omega_n)^2} \sin \omega_0 t$$

Viscous Damped Free Vibration

A viscous damping force is caused by fluid drag on the system as it vibrates. If the motion is slow, this drag force will be proportional to the velocity, that is, $F = c\dot{x}$, where c is the coefficient of viscous damping. By comparing its value to the critical damping coefficient $c_c = 2m\omega_n$, we can specify the type of vibration that occurs. If $c > c_c$, it is an overdamped system; if $c = c_c$, it is a critically damped system; and if $c < c_c$, it is an underdamped system.

Viscous Damped Forced Vibration

The most general type of vibration for a one-degree-of-freedom system occurs when the system is damped and subjected to periodic forced motion. The solution provides insight as to how the damping factor, c/c_c, and the frequency ratio, ω_0/ω_n, influence the vibration.

Resonance is avoided provided $c/c_c \neq 0$ and $\omega_0/\omega_n \neq 1$.

Electrical Circuit Analogs

The vibrating motion of a complex mechanical system can be studied by modeling it as an electrical circuit. This is possible since the differential equations that govern the behavior of each system are the same.

APPENDIX A

MATHEMATICAL EXPRESSIONS

Quadratic Formula

If $ax^2 + bx + c = 0$, then $x = \dfrac{-b \pm \sqrt{b^2 - 4ac}}{2a}$

Hyperbolic Functions

$\sinh x = \dfrac{e^x - e^{-x}}{2}$, $\cosh x = \dfrac{e^x + e^{-x}}{2}$,

$\tanh x = \dfrac{\sinh x}{\cosh x}$

Trigonometric Identities

$\sin \theta = \dfrac{A}{C}$, $\csc \theta = \dfrac{C}{A}$

$\cos \theta = \dfrac{B}{C}$, $\sec \theta = \dfrac{C}{B}$

$\tan \theta = \dfrac{A}{B}$, $\cot \theta = \dfrac{B}{A}$

$\sin^2 \theta + \cos^2 \theta = 1$

$\sin(\theta \pm \phi) = \sin \theta \cos \phi \pm \cos \theta \sin \phi$

$\sin 2\theta = 2 \sin \theta \cos \theta$

$\cos(\theta \pm \phi) = \cos \theta \cos \phi \mp \sin \theta \sin \phi$

$\cos 2\theta = \cos^2 \theta - \sin^2 \theta$

$\cos \theta = \pm \sqrt{\dfrac{1 + \cos 2\theta}{2}}$, $\sin \theta = \pm \sqrt{\dfrac{1 - \cos 2\theta}{2}}$

$\tan \theta = \dfrac{\sin \theta}{\cos \theta}$

$1 + \tan^2 \theta = \sec^2 \theta$ $1 + \cot^2 \theta = \csc^2 \theta$

Power-Series Expansions

$\sin x = x - \dfrac{x^3}{3!} + \cdots$ $\sinh x = x + \dfrac{x^3}{3!} + \cdots$

$\cos x = 1 - \dfrac{x^2}{2!} + \cdots$ $\cosh x = 1 + \dfrac{x^2}{2!} + \cdots$

Derivatives

$\dfrac{d}{dx}(u^n) = n u^{n-1} \dfrac{du}{dx}$

$\dfrac{d}{dx}(uv) = u \dfrac{dv}{dx} + v \dfrac{du}{dx}$

$\dfrac{d}{dx}\left(\dfrac{u}{v}\right) = \dfrac{v \dfrac{du}{dx} - u \dfrac{dv}{dx}}{v^2}$

Appendix A Mathematical Expressions

$\dfrac{d}{dx}(\cot u) = -\csc^2 u \dfrac{du}{dx}$

$\dfrac{d}{dx}(\sec u) = \tan u \sec u \dfrac{du}{dx}$

$\dfrac{d}{dx}(\csc u) = -\csc u \cot u \dfrac{du}{dx}$

$\dfrac{d}{dx}(\sin u) = \cos u \dfrac{du}{dx}$

$\dfrac{d}{dx}(\cos u) = -\sin u \dfrac{du}{dx}$

$\dfrac{d}{dx}(\tan u) = \sec^2 u \dfrac{du}{dx}$

$\dfrac{d}{dx}(\sinh u) = \cosh u \dfrac{du}{dx}$

$\dfrac{d}{dx}(\cosh u) = \sinh u \dfrac{du}{dx}$

Integrals

$\displaystyle\int x^n\, dx = \dfrac{x^{n+1}}{n+1} + C,\ n \neq -1$

$\displaystyle\int \dfrac{dx}{a+bx} = \dfrac{1}{b}\ln(a+bx) + C$

$\displaystyle\int \dfrac{dx}{a+bx^2} = \dfrac{1}{2\sqrt{-ba}}\ln\left[\dfrac{a + x\sqrt{-ab}}{a - x\sqrt{-ab}}\right] + C,\ ab < 0$

$\displaystyle\int \dfrac{x\, dx}{a+bx^2} = \dfrac{1}{2b}\ln(bx^2 + a) + C$

$\displaystyle\int \dfrac{x^2\, dx}{a+bx^2} = \dfrac{x}{b} - \dfrac{a}{b\sqrt{ab}}\tan^{-1}\dfrac{x\sqrt{ab}}{a} + C,\ ab > 0$

$\displaystyle\int \dfrac{dx}{a^2 - x^2} = \dfrac{1}{2a}\ln\left[\dfrac{a+x}{a-x}\right] + C,\ a^2 > x^2$

$\displaystyle\int \sqrt{a+bx}\, dx = \dfrac{2}{3b}\sqrt{(a+bx)^3} + C$

$\displaystyle\int x\sqrt{a+bx}\, dx = \dfrac{-2(2a - 3bx)\sqrt{(a+bx)^3}}{15b^2} + C$

$\displaystyle\int x^2\sqrt{a+bx}\, dx = \dfrac{2(8a^2 - 12abx + 15b^2x^2)\sqrt{(a+bx)^3}}{105b^3} + C$

$\displaystyle\int \sqrt{a^2 - x^2}\, dx = \dfrac{1}{2}\left[x\sqrt{a^2 - x^2} + a^2 \sin^{-1}\dfrac{x}{a}\right] + C,\ a > 0$

$\displaystyle\int x\sqrt{x^2 \pm a^2}\, dx = \dfrac{1}{3}\sqrt{(x^2 \pm a^2)^3} + C$

$\displaystyle\int x^2\sqrt{a^2 - x^2}\, dx = -\dfrac{x}{4}\sqrt{(a^2 - x^2)^3}$
$\qquad + \dfrac{a^2}{8}\left(x\sqrt{a^2 - x^2} + a^2 \sin^{-1}\dfrac{x}{a}\right) + C,\ a > 0$

$\displaystyle\int \sqrt{x^2 \pm a^2}\, dx = \dfrac{1}{2}\Big[x\sqrt{x^2 \pm a^2}$
$\qquad \pm a^2 \ln\left(x + \sqrt{x^2 \pm a^2}\right)\Big] + C$

$\displaystyle\int x\sqrt{a^2 - x^2}\, dx = -\dfrac{1}{3}\sqrt{(a^2 - x^2)^3} + C$

$\displaystyle\int x^2\sqrt{x^2 \pm a^2}\, dx = \dfrac{x}{4}\sqrt{(x^2 \pm a^2)^3} \mp \dfrac{a^2}{8}x\sqrt{x^2 \pm a^2}$
$\qquad - \dfrac{a^4}{8}\ln\left(x + \sqrt{x^2 \pm a^2}\right) + C$

$\displaystyle\int \dfrac{dx}{\sqrt{a+bx}} = \dfrac{2\sqrt{a+bx}}{b} + C$

$\displaystyle\int \dfrac{x\, dx}{\sqrt{x^2 \pm a^2}} = \sqrt{x^2 \pm a^2} + C$

$\displaystyle\int \dfrac{dx}{\sqrt{a + bx + cx^2}} = \dfrac{1}{\sqrt{c}}\ln\left[\sqrt{a + bx + cx^2}\right.$
$\qquad \left. + x\sqrt{c} + \dfrac{b}{2\sqrt{c}}\right] + C,\ c > 0$

$\qquad = \dfrac{1}{\sqrt{-c}}\sin^{-1}\left(\dfrac{-2cx - b}{\sqrt{b^2 - 4ac}}\right) + C,\ c < 0$

$\displaystyle\int \sin x\, dx = -\cos x + C$

$$\int \cos x \, dx = \sin x + C$$

$$\int x \cos(ax) \, dx = \frac{1}{a^2}\cos(ax) + \frac{x}{a}\sin(ax) + C$$

$$\int x^2 \cos(ax) \, dx = \frac{2x}{a^2}\cos(ax) + \frac{a^2 x^2 - 2}{a^3}\sin(ax) + C$$

$$\int e^{ax} \, dx = \frac{1}{a}e^{ax} + C$$

$$\int x e^{ax} \, dx = \frac{e^{ax}}{a^2}(ax - 1) + C$$

$$\int \sinh x \, dx = \cosh x + C$$

$$\int \cosh x \, dx = \sinh x + C$$

APPENDIX B

VECTOR ANALYSIS

The following discussion provides a brief review of vector analysis. A more detailed treatment of these topics is given in *Engineering Mechanics: Statics*.

Vector. A vector, **A**, is a quantity which has magnitude and direction, and adds according to the parallelogram law. As shown in Fig. B–1, **A** = **B** + **C**, where **A** is the *resultant vector* and **B** and **C** are *component vectors*.

Unit Vector. A unit vector, $\mathbf{u}_A$, has a magnitude of one "dimensionless" unit and acts in the same direction as **A**. It is determined by dividing **A** by its magnitude A, i.e.,

$$\mathbf{u}_A = \frac{\mathbf{A}}{A} \qquad (B\text{–}1)$$

Fig. B–1

Fig. B–2

Cartesian Vector Notation. The directions of the positive x, y, z axes are defined by the Cartesian unit vectors $\mathbf{i}, \mathbf{j}, \mathbf{k}$, respectively.

As shown in Fig. B–2, vector $\mathbf{A}$ is formulated by the addition of its x, y, z components

$$\mathbf{A} = A_x\mathbf{i} + A_y\mathbf{j} + A_z\mathbf{k} \tag{B–2}$$

The *magnitude* of $\mathbf{A}$ is determined from

$$A = \sqrt{A_x^2 + A_y^2 + A_z^2} \tag{B–3}$$

The *direction* of $\mathbf{A}$ is defined in terms of its *coordinate direction angles*, α, β, γ, measured from the *tail* of $\mathbf{A}$ to the *positive* x, y, z axes, Fig. B–3. These angles are determined from *direction cosines* which represent the $\mathbf{i}, \mathbf{j}, \mathbf{k}$ components of the unit vector $\mathbf{u}_A$; i.e., from Eqs. B–1 and B–2

$$\mathbf{u}_A = \frac{A_x}{A}\mathbf{i} + \frac{A_y}{A}\mathbf{j} + \frac{A_z}{A}\mathbf{k} \tag{B–4}$$

so that the direction cosines are

$$\cos\alpha = \frac{A_x}{A} \quad \cos\beta = \frac{A_y}{A} \quad \cos\gamma = \frac{A_z}{A} \tag{B–5}$$

Hence, $\mathbf{u}_A = \cos\alpha\,\mathbf{i} + \cos\beta\,\mathbf{j} + \cos\gamma\,\mathbf{k}$, and using Eq. B–3, it is seen that

$$\cos^2\alpha + \cos^2\beta + \cos^2\gamma = 1 \tag{B–6}$$

Fig. B–3

Fig. B–4

The Cross Product.

The cross product of two vectors **A** and **B**, which yields the resultant vector **C**, is written as

$$\mathbf{C} = \mathbf{A} \times \mathbf{B} \qquad (B\text{–}7)$$

and reads **C** equals **A** "cross" **B**. The *magnitude* of **C** is

$$C = AB \sin \theta \qquad (B\text{–}8)$$

where θ is the angle between the tails of **A** and **B** ($0° \leq \theta \leq 180°$). The *direction* of **C** is determined by the right-hand rule, whereby the fingers of the right hand are curled *from* **A** *to* **B** and the thumb points in the direction of **C**, Fig. B–4. This vector is perpendicular to the plane containing vectors **A** and **B**.

The vector cross product is *not* commutative, i.e., $\mathbf{A} \times \mathbf{B} \neq \mathbf{B} \times \mathbf{A}$. Rather,

$$\mathbf{A} \times \mathbf{B} = -\mathbf{B} \times \mathbf{A} \qquad (B\text{–}9)$$

The distributive law is valid; i.e.,

$$\mathbf{A} \times (\mathbf{B} + \mathbf{D}) = \mathbf{A} \times \mathbf{B} + \mathbf{A} \times \mathbf{D} \qquad (B\text{–}10)$$

And the cross product may be multiplied by a scalar m in any manner; i.e.,

$$m(\mathbf{A} \times \mathbf{B}) = (m\mathbf{A}) \times \mathbf{B} = \mathbf{A} \times (m\mathbf{B}) = (\mathbf{A} \times \mathbf{B})m \qquad (B\text{–}11)$$

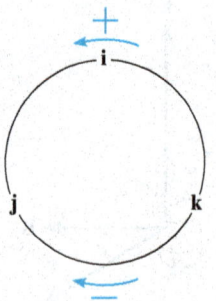

Fig. B–5

Equation B–7 can be used to find the cross product of any pair of Cartesian unit vectors. For example, to find $\mathbf{i} \times \mathbf{j}$, the magnitude is $(i)(j) \sin 90° = (1)(1)(1) = 1$, and its direction $+\mathbf{k}$ is determined from the right-hand rule, applied to $\mathbf{i} \times \mathbf{j}$, Fig. B–2. A simple scheme shown in Fig. B–5 may be helpful in obtaining this and other results when the need arises. If the circle is constructed as shown, then "crossing" two of the unit vectors in a *counterclockwise* fashion around the circle yields a *positive* third unit vector, e.g., $\mathbf{k} \times \mathbf{i} = \mathbf{j}$. Moving *clockwise*, a *negative* unit vector is obtained, e.g., $\mathbf{i} \times \mathbf{k} = -\mathbf{j}$.

If $\mathbf{A}$ and $\mathbf{B}$ are expressed in Cartesian component form, then the cross product, Eq. B–7, may be evaluated by expanding the determinant

$$\mathbf{C} = \mathbf{A} \times \mathbf{B} = \begin{vmatrix} \mathbf{i} & \mathbf{j} & \mathbf{k} \\ A_x & A_y & A_z \\ B_x & B_y & B_z \end{vmatrix} \qquad (B\text{–}12)$$

which yields

$$\mathbf{C} = (A_y B_z - A_z B_y)\mathbf{i} - (A_x B_z - A_z B_x)\mathbf{j} + (A_x B_y - A_y B_x)\mathbf{k}$$

Recall that the cross product is used in statics to define the moment of a force $\mathbf{F}$ about point O, in which case

$$\mathbf{M}_O = \mathbf{r} \times \mathbf{F} \qquad (B\text{–}13)$$

where $\mathbf{r}$ is a position vector directed from point O to *any point* on the line of action of $\mathbf{F}$.

The Dot Product. The dot product of two vectors **A** and **B**, which yields a scalar, is defined as

$$\mathbf{A} \cdot \mathbf{B} = AB \cos \theta \qquad \text{(B–14)}$$

and is read as **A** "dot" **B**. The angle θ is formed between the *tails* of **A** and **B** ($0° \leq \theta \leq 180°$).

The dot product is commutative; i.e.,

$$\mathbf{A} \cdot \mathbf{B} = \mathbf{B} \cdot \mathbf{A} \qquad \text{(B–15)}$$

The distributive law is valid; i.e.,

$$\mathbf{A} \cdot (\mathbf{B} + \mathbf{D}) = \mathbf{A} \cdot \mathbf{B} + \mathbf{A} \cdot \mathbf{D} \qquad \text{(B–16)}$$

And scalar multiplication can be performed in any manner, i.e.,

$$m(\mathbf{A} \cdot \mathbf{B}) = (m\mathbf{A}) \cdot \mathbf{B} = \mathbf{A} \cdot (m\mathbf{B}) = (\mathbf{A} \cdot \mathbf{B})m \qquad \text{(B–17)}$$

Using Eq. B–14, the dot product between any two Cartesian vectors can be determined. For example, $\mathbf{i} \cdot \mathbf{i} = (1)(1) \cos 0° = 1$ and $\mathbf{i} \cdot \mathbf{j} = (1)(1) \cos 90° = 0$.

If **A** and **B** are expressed in Cartesian component form, then the dot product, Eq. C–14, can be determined from

$$\boxed{\mathbf{A} \cdot \mathbf{B} = A_x B_x + A_y B_y + A_z B_z} \qquad \text{(B–18)}$$

The dot product may be used to determine the angle θ formed between two vectors. From Eq. B–14,

$$\theta = \cos^{-1}\left(\frac{\mathbf{A} \cdot \mathbf{B}}{AB}\right) \qquad \text{(B–19)}$$

It is also possible to find the component of a vector in a given direction using the dot product. For example, the magnitude of the component (or projection) of vector **A** in the direction of **B**, Fig. B–6, is defined by $A \cos \theta$. From Eq. B–14, this magnitude is

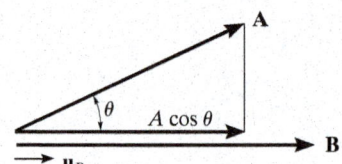

$$A \cos \theta = \mathbf{A} \cdot \frac{\mathbf{B}}{B} = \mathbf{A} \cdot \mathbf{u}_B \qquad (B\text{–}20)$$

Fig. B–6

where $\mathbf{u}_B$ represents a unit vector acting in the direction of **B**, Fig. B–6.

Differentiation and Integration of Vector Functions. The rules for differentiation and integration of the sums and products of scalar functions also apply to vector functions. Consider, for example, the two vector functions $\mathbf{A}(s)$ and $\mathbf{B}(s)$. Provided these functions are smooth and continuous for all s, then

$$\frac{d}{ds}(\mathbf{A} + \mathbf{B}) = \frac{d\mathbf{A}}{ds} + \frac{d\mathbf{B}}{ds} \qquad (B\text{–}21)$$

$$\int (\mathbf{A} + \mathbf{B}) \, ds = \int \mathbf{A} \, ds + \int \mathbf{B} \, ds \qquad (B\text{–}22)$$

For the cross product,

$$\frac{d}{ds}(\mathbf{A} \times \mathbf{B}) = \left(\frac{d\mathbf{A}}{ds} \times \mathbf{B}\right) + \left(\mathbf{A} \times \frac{d\mathbf{B}}{ds}\right) \qquad (B\text{–}23)$$

Similarly, for the dot product,

$$\frac{d}{ds}(\mathbf{A} \cdot \mathbf{B}) = \frac{d\mathbf{A}}{ds} \cdot \mathbf{B} + \mathbf{A} \cdot \frac{d\mathbf{B}}{ds} \qquad (B\text{–}24)$$

APPENDIX C

THE CHAIN RULE

The chain rule of calculus is used to determine the time derivative of a composite function. For example, if y is a function of x and x is a function of t, then we can find the derivative of y with respect to t as follows

$$\dot{y} = \frac{dy}{dt} = \frac{dy}{dx}\frac{dx}{dt} = \frac{dy}{dx}\dot{x}$$

In other words, to find $\dot{y}$ we find the change in y with respect to x, and then the change in x with respect to time. And so,

$$\boxed{\dot{y} = \frac{dy}{dx}\dot{x}} \qquad \text{(C–1)}$$

If a second time derivative of this equation is to be determined, then the product rule $d(uv) = du\,v + u\,dv$ must be used along with the chain rule when taking the time derivatives because both dy/dx and $\dot{x}$ are functions of time. And so letting $u = dy/dx$ and $v = \dot{x}$, we have

$$\ddot{y} = \frac{d}{dt}\left(\frac{dy}{dx}\right)\dot{x} + \frac{dy}{dx}\frac{d}{dt}(\dot{x})$$

Applying the chain rule to the first term on the right, we first find the change in (dy/dx) with respect to x, and then the change in x with respect to time.

$$\ddot{y} = \left(\frac{d^2y}{dx^2}\frac{dx}{dt}\right)\dot{x} + \left(\frac{dy}{dx}\right)\ddot{x}$$

or

$$\ddot{y} = \frac{d^2y}{dx^2}\dot{x}^2 + \frac{dy}{dx}\ddot{x} \qquad \text{(C–2)}$$

The following examples illustrate these ideas.

EXAMPLE C–1

If $y = x^3$ and $x = t^4$, find $\ddot{y}$, the second derivative of y with respect to time.

SOLUTION
Using the chain rule, Eq. C–1,

$$\dot{y} = 3x^2\dot{x}$$

To obtain the second time derivative we must use the product rule since x and $\dot{x}$ are both functions of time, and also, for $3x^2$ the chain rule must be applied. Thus, with $u = 3x^2$ and $v = \dot{x}$, we have

$$d(uv) = du\, v + u\, dv$$
$$\ddot{y} = [6x\dot{x}]\dot{x} + 3x^2[\ddot{x}]$$
$$= 3x[2\dot{x}^2 + x\ddot{x}]$$

Since $x = t^4$, then $\dot{x} = 4t^3$ and $\ddot{x} = 12t^2$ so that

$$\ddot{y} = 3(t^4)[2(4t^3)^2 + t^4(12t^2)]$$
$$= 132t^{10} \qquad \text{Ans.}$$

We can also apply Eq. C–2

$$\ddot{y} = (6x)(4t^3)^2 + (3x^2)(12t^2)$$

Setting $x = t^4$, we get the result

$$y = 132t^{10} \qquad \text{Ans.}$$

Note that this result can also be obtained by combining the functions, then taking the time derivatives, that is,

$$y = x^3 = (t^4)^3 = t^{12}$$
$$\dot{y} = 12t^{11}$$
$$\ddot{y} = 132t^{10} \qquad \text{Ans.}$$

EXAMPLE C–2

If $y = xe^x$, and $x = t^2$, find $\ddot{y}$.

SOLUTION
Since x and e^x are both functions of time and so the product and chain rules must be applied. Setting $u = x$ and $v = e^x$.

$$d(uv) = du\, v + u\, dv$$
$$\dot{y} = [\dot{x}]e^x + x[e^x \dot{x}]$$

The second time derivative also requires application of the product and chain rules. In this case the product rule applies to the three time variables in the last term, i.e., x, e^x, and $\dot{x}$. We have

$$\ddot{y} = \{[\ddot{x}]e^x + \dot{x}[e^x \dot{x}]\} + \{[\dot{x}]e^x \dot{x} + x[e^x \dot{x}]\dot{x} + xe^x[\ddot{x}]\}$$
$$= e^x[\ddot{x}(1 + x) + \dot{x}^2(2 + x)]$$

Since $x = t^2$ then $\dot{x} = 2t$, $\ddot{x} = 2$ so that in terms in t, we have

$$\ddot{y} = e^{t^2}[2(1 + t^2) + 4t^2(2 + t^2)]$$

EXAMPLE C–3

If the path in radial coordinates is given as $r = 5\theta^2$, where θ is a known function of time, find $\ddot{r}$.

SOLUTION
First, using the chain rule then the chain and product rules where $u = 10\theta$ and $v = \dot{\theta}$, we have

$$r = 5\theta^2$$
$$\dot{r} = 10\theta\dot{\theta}$$
$$\dot{r} = \frac{dr}{d\theta}\dot{\theta};$$
$$d(uv) = du\, v + u\, dv$$
$$\ddot{r} = 10[(\dot{\theta})\dot{\theta} + \theta(\ddot{\theta})]$$
$$= 10\dot{\theta}^2 + 10\theta\ddot{\theta}$$

Apply Eq. C–2 and show that you get the same result.

Fundamental Problems Solutions And Answers

Chapter 12

F12–1. $v = v_0 + a_c t$
$10 = 35 + a_c(15)$
$a_c = -1.67 \text{ m/s}^2 = 1.67 \text{ m/s}^2 \leftarrow$ Ans.

F12–2. $s = s_0 + v_0 t + \frac{1}{2} a_c t^2$
$0 = 0 + 15t + \frac{1}{2}(-9.81)t^2$
$t = 3.06 \text{ s}$ Ans.

F12–3. $\int_0^s ds = \int_0^t (4t - 3t^2) dt$
$s = (2t^2 - t^3) \text{ m}$
$s = 2(4^2) - 4^3$
$= -32 \text{ m} = 32 \text{ m} \leftarrow$ Ans.

F12–4. $a = \frac{dv}{dt} = \frac{d}{dt}(0.5t^3 - 8t)$
$a = (1.5t^2 - 8) \text{ m/s}^2$
When $t = 2$ s,
$a = 1.5(2^2) - 8 = -2 \text{ m/s}^2 = 2 \text{ m/s}^2 \leftarrow$ Ans.

F12–5. $v = \frac{ds}{dt} = \frac{d}{dt}(2t^2 - 8t + 6) = (4t - 8) \text{ m/s}$
$v = 0 = (4t - 8)$
$t = 2 \text{ s}$ Ans.
$s|_{t=0} = 2(0^2) - 8(0) + 6 = 6 \text{ m}$
$s|_{t=2} = 2(2^2) - 8(2) + 6 = -2 \text{ m}$
$s|_{t=3} = 2(3^2) - 8(3) + 6 = 0 \text{ m}$
Plot these positions, then
$(\Delta s)_{\text{Tot}} = 8 \text{ m} + 2 \text{ m} = 10 \text{ m}$ Ans.

F12–6. $v \, dv = a \, ds$
$\int_{5 \text{ m/s}}^v v \, dv = \int_0^s (10 - 0.2s) ds$
$v = \left(\sqrt{20s - 0.2s^2 + 25}\right) \text{ m/s}$
At $s = 10$ m,
$v = \sqrt{20(10) - 0.2(10^2) + 25}$
$= 14.3 \text{ m/s} \rightarrow$ Ans.

F12–7. $v = \int (4t^2 - 2) dt$
$v = \frac{4}{3}t^3 - 2t + C_1$
$s = \int \left(\frac{4}{3}t^3 - 2t + C_1\right) dt$
$s = \frac{1}{3}t^4 - t^2 + C_1 t + C_2$
$t = 0, s = -2, C_2 = -2$
$t = 2, s = -20, C_1 = -9.67$
$t = 4, s = 28.7 \text{ m}$ Ans.

F12–8. $a = v \frac{dv}{ds}$
$= (20 - 0.05s^2)(-0.1s)$
At $s = 15$ m,
$a = -13.1 \text{ m/s}^2 = 13.1 \text{ m/s}^2 \leftarrow$ Ans.

F12–9. $v = \frac{ds}{dt} = \frac{d}{dt}(0.5t^3) = 1.5t^2$
$v = \frac{ds}{dt} = \frac{d}{dt}(108) = 0$ Ans.

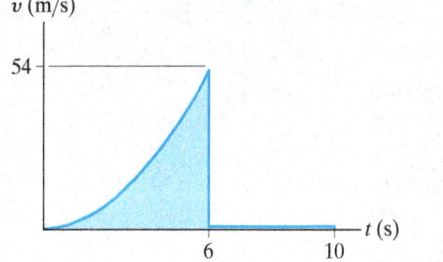

F12–10. $ds = v \, dt$
$\int_0^s ds = \int_0^t (-4t + 80) dt$
$s = -2t^2 + 80t$
$a = \frac{dv}{dt} = \frac{d}{dt}(-4t + 80) = -4 \text{ ft/s}^2 = 4 \text{ ft/s}^2 \leftarrow$
Also,
$a = \frac{\Delta v}{\Delta t} = \frac{0 - 80 \text{ ft/s}}{20 \text{ s} - 0} = -4 \text{ ft/s}^2$

F12–11. $a\,ds = v\,dv$

$a = v\dfrac{dv}{ds} = 0.25s\dfrac{d}{ds}(0.25s) = 0.0625s$

$a|_{s=40\,\text{m}} = 0.0625(40\,\text{m}) = 2.5\,\text{m/s}^2 \rightarrow$

F12–12. For $0 \le s \le 10\,\text{m}$

$a = s$

$\displaystyle\int_0^v v\,dv = \int_0^s s\,ds$

$v = s$

at $s = 10\,\text{m}$, $v = 10\,\text{m/s}$

For $10\,\text{m} \le s \le 15$

$a = 10$

$\displaystyle\int_{10}^v v\,dv = \int_{10}^s 10\,ds$

$\dfrac{1}{2}v^2 - 50 = 10s - 100$

$v = \sqrt{20s - 100}$

at $s = 15\,\text{m}$

$v = 14.1\,\text{m/s}$ Ans.

F12–13. $0 \le t < 5\,\text{s}$,

$dv = a\,dt \quad \displaystyle\int_0^v dv = \int_0^t 20\,dt$

$v = (20t)\,\text{m/s}$

$5\,\text{s} < t \le t'$,

$(\overset{+}{\rightarrow})\quad dv = a\,dt \quad \displaystyle\int_{100\,\text{m/s}}^v dv = \int_{5\,\text{s}}^t -10\,dt$

$v - 100 = (50 - 10t)\,\text{m/s}$,

$0 = 150 - 10t'$

$t' = 15\,\text{s}$

Also,

$\Delta v = 0 =$ Area under the a–t graph

$0 = (20\,\text{m/s}^2)(5\,\text{s}) + [-(10\,\text{m/s})(t' - 5)\,\text{s}]$

$t' = 15\,\text{s}$

[v (m/s) graph: triangle peaking at 100 at t=5, returning to 0 at t=15]

F12–14. $0 \le t \le 5\,\text{s}$,

$ds = v\,dt \quad \displaystyle\int_0^s ds = \int_0^t 30t\,dt$

$s\big|_0^s = 15t^2\big|_0^t$

$s = (15t^2)\,\text{m}$

$5\,\text{s} < t \le 15\,\text{s}$,

$(\overset{+}{\rightarrow})\, ds = v\,dt;\quad \displaystyle\int_{375\,\text{m}}^s ds = \int_{5\,\text{s}}^t (-15t + 225)dt$

$s = (-7.5t^2 + 225t - 562.5)\,\text{m}$

$s = (-7.5)(15)^2 + 225(15) - 562.5\,\text{m}$

$= 1125\,\text{m}$ Ans.

Also,

$\Delta s =$ Area under the v–t graph

$= \tfrac{1}{2}(150\,\text{m/s})(15\,\text{s})$

$= 1125\,\text{m}$ Ans.

[s (m) graph: curve rising to 1125 at t=15, passing through 375 at t=5]

F12–15. $\displaystyle\int_0^x dx = \int_0^t 32t\,dt$

$x = (16t^2)\,\text{m}$ (1)

$\displaystyle\int_0^y dy = \int_0^t 8\,dt$

$t = \dfrac{y}{8}$ (2)

Substituting Eq. (2) into Eq. (1), get

$y = 2\sqrt{x}$ Ans.

F12–16. $y = 0.75(8t) = 6t$

$v_x = \dot{x} = \frac{dx}{dt} = \frac{d}{dt}(8t) = 8 \text{ m/s} \rightarrow$

$v_y = \dot{y} = \frac{dy}{dt} = \frac{d}{dt}(6t) = 6 \text{ m/s} \uparrow$

$v = \sqrt{v_x^2 + v_y^2} = \sqrt{(8 \text{ m/s})^2 + (6 \text{ m/s})^2}$

$= 10 \text{ m/s}$ *Ans.*

F12–17. $y = (4t^2) \text{ m}$

$v_x = \dot{x} = \frac{d}{dt}(4t^4) = (16t^3) \text{ m/s} \rightarrow$

$v_y = \dot{y} = \frac{d}{dt}(4t^2) = (8t) \text{ m/s} \uparrow$

When $t = 0.5 \text{ s}$,

$v = \sqrt{v_x^2 + v_y^2} = \sqrt{(2 \text{ m/s})^2 + (4 \text{ m/s})^2}$

$= 4.47 \text{ m/s}$ *Ans.*

$a_x = \dot{v}_x = \frac{d}{dt}(16t^3) = (48t^2) \text{ m/s}^2$

$a_y = \dot{v}_y = \frac{d}{dt}(8t) = 8 \text{ m/s}^2$

When $t = 0.5 \text{ s}$,

$a = \sqrt{a_x^2 + a_y^2} = \sqrt{(12 \text{ m/s}^2)^2 + (8 \text{ m/s}^2)^2}$

$= 14.4 \text{ m/s}^2$ *Ans.*

F12–18. $y = 0.5x$

$\dot{y} = 0.5\dot{x}$

$v_y = t^2$

When $t = 4 \text{ s}$,

$v_x = 32 \text{ m/s} \quad v_y = 16 \text{ m/s}$

$v = \sqrt{v_x^2 + v_y^2} = 35.8 \text{ m/s}$ *Ans.*

$a_x = \dot{v}_x = 4t$

$a_y = \dot{v}_y = 2t$

When $t = 4 \text{ s}$,

$a_x = 16 \text{ m/s}^2 \quad a_y = 8 \text{ m/s}^2$

$a = \sqrt{a_x^2 + a_y^2} = \sqrt{16^2 + 8^2} = 17.9 \text{ m/s}^2$ *Ans.*

F12–19. $v_y = \dot{y} = 0.5 \times \dot{x} = 0.5(8)(8) = 32 \text{ m/s}$

Thus,

$v = \sqrt{v_x^2 + v_y^2} = 33.0 \text{ m/s}$ *Ans.*

$a_y = \dot{v}_y = 0.5 \quad \dot{x}^2 + 0.5x\ddot{x}$

$= 0.5(8)^2 + 0.5(8)(4)$

$= 48 \text{ m/s}^2$

$a_x = 4 \text{ m/s}^2$

Thus,

$a = \sqrt{a_x^2 + a_y^2} = \sqrt{4^2 + 48^2} = 48.2 \text{ m/s}^2$ *Ans.*

F12–20. $\dot{y} = 0.1x\dot{x}$

$v_y = 0.1(5)(-3) = -1.5 \text{ m/s} = 1.5 \text{ m/s} \downarrow$ *Ans.*

$\ddot{y} = 0.1[\dot{x}\dot{x} + x\ddot{x}]$

$a_y = 0.1[(-3)^2 + 5(-1.5)] = 0.15 \text{ m/s}^2 \uparrow$ *Ans.*

F12–21. $(v_B)_y^2 = (v_A)_y^2 + 2a_y(y_B - y_A)$

$0^2 = (5 \text{ m/s})^2 + 2(-9.81 \text{ m/s}^2)(h - 0)$

$h = 1.27 \text{ m}$ *Ans.*

F12–22. $y_C = y_A + (v_A)_y t_{AC} + \frac{1}{2} a_y t_{AC}^2$

$0 = 0 + (5 \text{ m/s})t_{AC} + \frac{1}{2}(-9.81 \text{ m/s}^2)t_{AC}^2$

$t_{AC} = 1.0194 \text{ s}$

$(v_C)_y = (v_A)_y + a_y t_{AC}$

$(v_C)_y = 5 \text{ m/s} + (-9.81 \text{ m/s}^2)(1.0194 \text{ s})$

$= -5 \text{ m/s} = 5 \text{ m/s} \downarrow$

$v_C = \sqrt{(v_C)_x^2 + (v_C)_y^2}$

$= \sqrt{(8.660 \text{ m/s})^2 + (5 \text{ m/s})^2} = 10 \text{ m/s}$ *Ans.*

$R = x_A + (v_A)_x t_{AC} = 0 + (8.660 \text{ m/s})(1.0194 \text{ s})$

$= 8.83 \text{ m}$ *Ans.*

F12–23. $s = s_0 + v_0 t$

$10 = 0 + v_A \cos 30° t$

$s = s_0 + v_0 t + \frac{1}{2} a_c t^2$

$3 = 1.5 + v_A \sin 30° t + \frac{1}{2}(-9.81)t^2$

$t = 0.9334 \text{ s}, \quad v_A = 12.4 \text{ m/s}$ *Ans.*

F12–24. $s = s_0 + v_0 t$

$R\left(\frac{4}{5}\right) = 0 + 20\left(\frac{3}{5}\right)t$

$s = s_0 + v_0 t + \frac{1}{2} a_c t^2$

$-R\left(\frac{3}{5}\right) = 0 + 20\left(\frac{4}{5}\right)t + \frac{1}{2}(-9.81)t^2$

$t = 5.10 \text{ s}$

$R = 76.5 \text{ m}$ *Ans.*

F12–25. $x_B = x_A + (v_A)_x t_{AB}$

$12 \text{ ft} = 0 + (0.8660 \, v_A)t_{AB}$

$v_A t_{AB} = 13.856$ (1)

$y_B = y_A + (v_A)_y t_{AB} + \frac{1}{2} a_y t_{AB}^2$

$(8 - 3)\text{ ft} = 0 + 0.5v_A t_{AB} + \frac{1}{2}(-32.2 \text{ ft/s}^2)t_{AB}^2$

Using Eq. (1),

$5 = 0.5(13.856) - 16.1\, t_{AB}^2$

$t_{AB} = 0.3461$ s

$v_A = 40.0$ ft/s   Ans.

F12–26. $y_B = y_A + (v_A)_y t_{AB} + \frac{1}{2} a_y t_{AB}^2$

$-150\text{ m} = 0 + (90\text{ m/s})t_{AB} + \frac{1}{2}(-9.81\text{ m/s}^2)t_{AB}^2$

$t_{AB} = 19.89$ s

$x_B = x_A + (v_A)_x t_{AB}$

$R = 0 + 120\text{ m/s}(19.89\text{ s}) = 2386.37$ m

$= 2.39$ km   Ans.

F12–27. $a_t = \dot{v} = \frac{dv}{dt} = \frac{d}{dt}(0.0625t^2) = (0.125t)\text{ m/s}^2 \big|_{t=10\text{ s}}$

$= 1.25$ m/s^2

$a_n = \frac{v^2}{\rho} = \frac{(0.0625t^2)^2}{40\text{ m}} = [97.656(10^{-6})t^4]\text{ m/s}^2\big|_{t=10\text{ s}}$

$= 0.9766$ m/s^2

$a = \sqrt{a_t^2 + a_n^2} = \sqrt{(1.25\text{ m/s}^2)^2 + (0.9766\text{ m/s}^2)^2}$

$= 1.59$ m/s^2 Ans.

F12–28. $v = 2s\big|_{s=10} = 20$ m/s

$a_n = \frac{v^2}{\rho} = \frac{(20\text{ m/s})^2}{50\text{ m}} = 8$ m/s^2

$a_t = v\frac{dv}{ds} = 4s\big|_{s=10} = 40$ m/s^2

$a = \sqrt{a_t^2 + a_n^2} = \sqrt{(40\text{ m/s}^2)^2 + (8\text{ m/s}^2)^2}$

$= 40.8$ m/s^2 Ans.

F12–29. $v_C^2 = v_A^2 + 2a_t(s_C - s_A)$

$(15\text{ m/s})^2 = (25\text{ m/s})^2 + 2a_t(300\text{ m} - 0)$

$a_t = -0.6667$ m/s^2

$v_B^2 = v_A^2 + 2a_t(s_B - s_A)$

$v_B^2 = (25\text{ m/s})^2 + 2(-0.6667\text{ m/s}^2)(250\text{ m} - 0)$

$v_B = 17.08$ m/s

$(a_B)_n = \frac{v_B^2}{\rho} = \frac{(17.08\text{ m/s})^2}{300\text{ m}} = 0.9722$ m/s^2

$a_B = \sqrt{(a_B)_t^2 + (a_B)_n^2}$

$= \sqrt{(-0.6667\text{ m/s}^2)^2 + (0.9722\text{ m/s}^2)^2}$

$= 1.18$ m/s^2 Ans.

F12–30. $\tan\theta = \frac{dy}{dx} = \frac{d}{dx}\left(\frac{1}{24}x^2\right) = \frac{1}{12}x$

$\theta = \tan^{-1}\left(\frac{1}{12}x\right)\bigg|_{x=10\text{ ft}}$

$= \tan^{-1}\left(\frac{10}{12}\right) = 39.81° = 39.8°$ ⤴   Ans.

$\rho = \frac{[1 + (dy/dx)^2]^{3/2}}{|d^2y/dx^2|} = \frac{[1 + (\frac{1}{12}x)^2]^{3/2}}{|\frac{1}{12}|}\bigg|_{x=10\text{ ft}}$

$= 26.468$ ft

$a_n = \frac{v^2}{\rho} = \frac{(20\text{ ft/s})^2}{26.468\text{ ft}} = 15.11$ ft/s^2

$a = \sqrt{(a_t)^2 + (a_n)^2} = \sqrt{(6\text{ ft/s}^2)^2 + (15.11\text{ ft/s}^2)^2}$

$= 16.3$ ft/s^2 Ans.

F12–31. $(a_B)_t = -0.001s = (-0.001)(300\text{ m})\left(\frac{\pi}{2}\text{ rad}\right)$ m/s^2

$= -0.4712$ m/s^2

$v\, dv = a_t\, ds$

$\int_{25\text{ m/s}}^{v_B} v\, dv = \int_0^{150\pi\text{ m}} -0.001s\, ds$

$v_B = 20.07$ m/s

$(a_B)_n = \frac{v_B^2}{\rho} = \frac{(20.07\text{ m/s})^2}{300\text{ m}} = 1.343$ m/s^2

$a_B = \sqrt{(a_B)_t^2 + (a_B)_n^2}$

$= \sqrt{(-0.4712\text{ m/s}^2)^2 + (1.343\text{ m/s}^2)^2}$

$= 1.42$ m/s^2 Ans.

F12–32. $a_t\, ds = v\, dv$

$a_t = v\frac{dv}{ds} = (0.2s)(0.2) = (0.04s)$ m/s^2

$a_t = 0.04(50\text{ m}) = 2$ m/s^2

$v = 0.2(50\text{ m}) = 10$ m/s

$$a_n = \frac{v^2}{\rho} = \frac{(10 \text{ m/s})^2}{500 \text{ m}} = 0.2 \text{ m/s}^2$$
$$a = \sqrt{a_t^2 + a_n^2} = \sqrt{(2 \text{ m/s}^2)^2 + (0.2 \text{ m/s}^2)^2}$$
$$= 2.01 \text{ m/s}^2 \qquad \text{Ans.}$$

F12–33. $v_r = \dot{r} = 0$
$v_\theta = r\dot{\theta} = (400\dot{\theta}) \text{ ft/s}$
$v = \sqrt{v_r^2 + v_\theta^2}$
$55 \text{ ft/s} = \sqrt{0^2 + [(400\dot{\theta}) \text{ ft/s}]^2}$
$\dot{\theta} = 0.1375 \text{ rad/s}$ \qquad Ans.

F12–34. $r = 0.1t^3\big|_{t=1.5\text{s}} = 0.3375 \text{ m}$
$\dot{r} = 0.3t^2\big|_{t=1.5\text{s}} = 0.675 \text{ m/s}$
$\ddot{r} = 0.6t\big|_{t=1.5\text{s}} = 0.900 \text{ m/s}^2$
$\theta = 4t^{3/2}\big|_{t=1.5\text{s}} = 7.348 \text{ rad}$
$\dot{\theta} = 6t^{1/2}\big|_{t=1.5\text{s}} = 7.348 \text{ rad/s}$
$\ddot{\theta} = 3t^{-1/2}\big|_{t=1.5\text{s}} = 2.449 \text{ rad/s}^2$
$v_r = \dot{r} = 0.675 \text{ m/s}$
$v_\theta = r\dot{\theta} = (0.3375 \text{ m})(7.348 \text{ rad/s}) = 2.480 \text{ m/s}$
$a_r = \ddot{r} - r\dot{\theta}^2$
$= (0.900 \text{ m/s}^2) - (0.3375 \text{ m})(7.348 \text{ rad/s})^2$
$= -17.325 \text{ m/s}^2$
$a_\theta = r\ddot{\theta} + 2\dot{r}\dot{\theta} = (0.3375 \text{ m})(2.449 \text{ rad/s}^2)$
$\qquad + 2(0.675 \text{ m/s})(7.348 \text{ rad/s}) = 10.747 \text{ m/s}^2$
$v = \sqrt{v_r^2 + v_\theta^2}$
$= \sqrt{(0.675 \text{ m/s})^2 + (2.480 \text{ m/s})^2}$
$= 2.57 \text{ m/s}$ \qquad Ans.
$a = \sqrt{a_r^2 + a_\theta^2}$
$= \sqrt{(-17.325 \text{ m/s}^2)^2 + (10.747 \text{ m/s}^2)^2}$
$= 20.4 \text{ m/s}^2$ \qquad Ans.

F12–35. $r = 2\theta$
$\dot{r} = 2\dot{\theta}$
$\ddot{r} = 2\ddot{\theta}$
At $\theta = \pi/4$ rad,
$r = 2\left(\frac{\pi}{4}\right) = \frac{\pi}{2} \text{ ft}$
$\dot{r} = 2(3 \text{ rad/s}) = 6 \text{ ft/s}$
$\ddot{r} = 2(1 \text{ rad/s}) = 2 \text{ ft/s}^2$
$a_r = \ddot{r} - r\dot{\theta}^2 = 2 \text{ ft/s}^2 - \left(\frac{\pi}{2} \text{ ft}\right)(3 \text{ rad/s})^2$
$= -12.14 \text{ ft/s}^2$

$a_\theta = r\ddot{\theta} + 2\dot{r}\dot{\theta}$
$= \left(\frac{\pi}{2} \text{ ft}\right)(1 \text{ rad/s}^2) + 2(6 \text{ ft/s})(3 \text{ rad/s})$
$= 37.57 \text{ ft/s}^2$
$a = \sqrt{a_r^2 + a_\theta^2}$
$= \sqrt{(-12.14 \text{ ft/s}^2)^2 + (37.57 \text{ ft/s}^2)^2}$
$= 39.5 \text{ ft/s}^2$ \qquad Ans.

F12–36. $r = e^\theta$
$\dot{r} = e^\theta \dot{\theta}$
$\ddot{r} = e^\theta \ddot{\theta} + e^\theta \dot{\theta}^2$
$a_r = \ddot{r} - r\dot{\theta}^2 = (e^\theta \ddot{\theta} + e^\theta \dot{\theta}^2) - e^\theta \dot{\theta}^2 = e^{\pi/4}(4)$
$= 8.77 \text{ m/s}^2$ \qquad Ans.
$a_\theta = r\ddot{\theta} + 2\dot{r}\dot{\theta} = (e^\theta \ddot{\theta}) + (2(e^\theta \dot{\theta})\dot{\theta}) = e^\theta(\ddot{\theta} + 2\dot{\theta}^2)$
$= e^{\pi/4}(4 + 2(2)^2)$
$= 26.3 \text{ m/s}^2$ \qquad Ans.

F12–37. $r = [0.2(1 + \cos\theta)] \text{ m}\big|_{\theta=30°} = 0.3732 \text{ m}$
$\dot{r} = [-0.2(\sin\theta)\dot{\theta}] \text{ m/s}\big|_{\theta=30°}$
$= -0.2 \sin 30°(3 \text{ rad/s})$
$= -0.3 \text{ m/s}$
$v_r = \dot{r} = -0.3 \text{ m/s}$
$v_\theta = r\dot{\theta} = (0.3732 \text{ m})(3 \text{ rad/s}) = 1.120 \text{ m/s}$
$v = \sqrt{v_r^2 + v_\theta^2} = \sqrt{(-0.3 \text{ m/s})^2 + (1.120 \text{ m/s})^2}$
$= 1.16 \text{ m/s}$ \qquad Ans.

F12–38. $30 \text{ m} = r\sin\theta$
$r = \left(\frac{30 \text{ m}}{\sin\theta}\right) = (30 \csc\theta) \text{ m}$
$r = (30 \csc\theta)\big|_{\theta=45°} = 42.426 \text{ m}$
$\dot{r} = -30 \csc\theta \cot\theta \, \dot{\theta}\big|_{\theta=45°} = -(42.426\dot{\theta}) \text{ m/s}$
$v_r = \dot{r} = -(42.426\dot{\theta}) \text{ m/s}$
$v_\theta = r\dot{\theta} = (42.426\dot{\theta}) \text{ m/s}$
$v = \sqrt{v_r^2 + v_\theta^2}$
$2 = \sqrt{(-42.426\dot{\theta})^2 + (42.426\dot{\theta})^2}$
$\dot{\theta} = 0.0333 \text{ rad/s}$ \qquad Ans.

F12–39. $l_T = 3s_D + s_A$
$0 = 3v_D + v_A$
$0 = 3v_D + 3 \text{ m/s}$
$v_D = -1 \text{ m/s} = 1 \text{ m/s} \uparrow$ \qquad Ans.

F12–40. $s_B + 2s_A + 2h = l$
$v_B + 2v_A = 0$
$6 + 2v_A = 0 \quad v_A = -3 \text{ m/s} = 3 \text{ m/s} \uparrow$ *Ans.*

F12–41. $3s_A + s_B = l$
$3v_A + v_B = 0$
$3v_A + 1.5 = 0 \quad v_A = -0.5 \text{ m/s} = 0.5 \text{ m/s} \uparrow$ *Ans.*

F12–42. $l_T = 4s_A + s_F$
$0 = 4v_A + v_F$
$0 = 4v_A + 3 \text{ m/s}$
$v_A = -0.75 \text{ m/s} = 0.75 \text{ m/s} \uparrow$ *Ans.*

F12–43. $s_A + 2(s_A - a) + (s_A - s_P) = l$
$4s_A - s_P = l + 2a$
$4v_A - v_P = 0$
$4v_A - (-4) = 0$
$4v_A + 4 = 0 \quad v_A = -1 \text{ m/s} = 1 \text{ m/s} \nearrow$ *Ans.*

F12–44. $s_C + s_B = l_{CED}$ (1)
$(s_A - s_C) + (s_B - s_C) + s_B = l_{ACDF}$
$s_A + 2s_B - 2s_C = l_{ACDF}$ (2)
Thus
$v_C + v_B = 0$
$v_A + 2v_B - 2v_C = 0$
Eliminating v_C,
$v_A + 4v_B = 0$
Thus,
$4 \text{ ft/s} + 4v_B = 0$
$v_B = -1 \text{ ft/s} = 1 \text{ ft/s} \uparrow$ *Ans.*

F12–45. $\mathbf{v}_B = \mathbf{v}_A + \mathbf{v}_{B/A}$
$100\mathbf{i} = 80\mathbf{j} + \mathbf{v}_{B/A}$
$\mathbf{v}_{B/A} = 100\mathbf{i} - 80\mathbf{j}$
$v_{B/A} = \sqrt{(v_{B/A})_x^2 + (v_{B/A})_y^2}$
$= \sqrt{(100 \text{ km/h})^2 + (-80 \text{ km/h})^2}$
$= 128 \text{ km/h}$ *Ans.*
$\theta = \tan^{-1}\left[\frac{(v_{B/A})_y}{(v_{B/A})_x}\right] = \tan^{-1}\left(\frac{80 \text{ km/h}}{100 \text{ km/h}}\right) = 38.7° \searrow$ *Ans.*

F12–46. $\mathbf{v}_B = \mathbf{v}_A + \mathbf{v}_{B/A}$
$(-400\mathbf{i} - 692.82\mathbf{j}) = (650\mathbf{i}) + \mathbf{v}_{B/A}$
$\mathbf{v}_{B/A} = \{-1050\mathbf{i} - 692.82\mathbf{j}\} \text{ km/h}$
$v_{B/A} = \sqrt{(v_{B/A})_x^2 + (v_{B/A})_y^2}$
$= \sqrt{(1050 \text{ km/h})^2 + (692.82 \text{ km/h})^2}$
$= 1258 \text{ km/h}$ *Ans.*
$\theta = \tan^{-1}\left[\frac{(v_{B/A})_y}{(v_{B/A})_x}\right] = \tan^{-1}\left(\frac{692.82 \text{ km/h}}{1050 \text{ km/h}}\right) = 33.4° \swarrow$ *Ans.*

F12–47. $\mathbf{v}_B = \mathbf{v}_A + \mathbf{v}_{B/A}$
$(5\mathbf{i} + 8.660\mathbf{j}) = (12.99\mathbf{i} + 7.5\mathbf{j}) + \mathbf{v}_{B/A}$
$\mathbf{v}_{B/A} = \{-7.990\mathbf{i} + 1.160\mathbf{j}\} \text{ m/s}$
$v_{B/A} = \sqrt{(-7.990 \text{ m/s})^2 + (1.160 \text{ m/s})^2}$
$= 8.074 \text{ m/s}$
$d_{AB} = v_{B/A} t = (8.074 \text{ m/s})(4 \text{ s}) = 32.3 \text{ m}$ *Ans.*

F12–48. $\mathbf{v}_A = \mathbf{v}_B + \mathbf{v}_{A/B}$
$-20 \cos 45° \mathbf{i} + 20 \sin 45° \mathbf{j} = 65\mathbf{i} + \mathbf{v}_{A/B}$
$\mathbf{v}_{A/B} = -79.14\mathbf{i} + 14.14\mathbf{j}$
$v_{A/B} = \sqrt{(-79.14)^2 + (14.14)^2}$
$= 80.4 \text{ km/h}$ *Ans.*
$\mathbf{a}_A = \mathbf{a}_B + \mathbf{a}_{A/B}$
$\frac{(20)^2}{0.1} \cos 45° \mathbf{i} + \frac{(20)^2}{0.1} \sin 45° \mathbf{j} = 1200\mathbf{i} + \mathbf{a}_{A/B}$
$\mathbf{a}_{A/B} = 1628\mathbf{i} + 2828\mathbf{j}$
$a_{A/B} = \sqrt{(1628)^2 + (2828)^2}$
$= 3.26(10^3) \text{ km/h}^2$ *Ans.*

Chapter 13

F13–1. $s = s_0 + v_0 t + \frac{1}{2} a_c t^2$
$6 \text{ m} = 0 + 0 + \frac{1}{2} a (3 \text{ s})^2$
$a = 1.333 \text{ m/s}^2$
Draw FBD.
$\Sigma F_y = ma_y; \quad N_A - 20(9.81) \text{ N} \cos 30° = 0$
$N_A = 169.91 \text{ N}$
$\Sigma F_x = ma_x; \quad T - 20(9.81) \text{ N} \sin 30°$
$- 0.3(169.91 \text{ N}) = (20 \text{ kg})(1.333 \text{ m/s}^2)$
$T = 176 \text{ N}$ *Ans.*

F13–2. $(F_f)_{\max} = \mu_s N_A = 0.3(245.25 \text{ N}) = 73.575 \text{ N}$
Since $F = 100 \text{ N} > (F_f)_{\max}$ when $t = 0$, the crate will start to move immediately after $\mathbf{F}$ is applied.
Draw FBD.
$+\uparrow \Sigma F_y = ma_y; \quad N_A - 25(9.81) \text{ N} = 0$
$N_A = 245.25 \text{ N}$
$\xrightarrow{+} \Sigma F_x = ma_x;$
$10t^2 + 100 - 0.25(245.25 \text{ N}) = (25 \text{ kg})a$
$a = (0.4t^2 + 1.5475) \text{ m/s}^2$
$dv = a \, dt$
$\int_0^v dv = \int_0^{4 \text{ s}} (0.4t^2 + 1.5475) \, dt$
$v = 14.7 \text{ m/s} \rightarrow$ *Ans.*

F13–3. Draw FBD.

$\xrightarrow{+} \Sigma F_x = ma_x;$

$\left(\frac{4}{5}\right)500 \text{ N} - (500s) \text{ N} = (10 \text{ kg})a$

$a = (40 - 50s) \text{ m/s}^2$

$v \, dv = a \, ds$

$\int_0^v v \, dv = \int_0^{0.5 \text{ m}} (40 - 50s) \, ds$

$\left.\frac{v^2}{2}\right|_0^v = \left.(40s - 25s^2)\right|_0^{0.5 \text{ m}}$

$v = 5.24 \text{ m/s}$ *Ans.*

F13–4. Draw FBD.

$\xrightarrow{+} \Sigma F_x = ma_x; \quad 100(s + 1) \text{ N} = (2000 \text{ kg})a$

$a = (0.05(s + 1)) \text{ m/s}^2$

$v \, dv = a \, ds$

$\int_0^v v \, dv = \int_0^{10 \text{ m}} 0.05(s + 1) \, ds$

$v = 2.45 \text{ m/s}$

F13–5. $F_{sp} = k(l - l_0) = (200 \text{ N/m})(0.5 \text{ m} - 0.3 \text{ m})$
$= 40 \text{ N}$

$\theta = \tan^{-1}\left(\frac{0.3 \text{ m}}{0.4 \text{ m}}\right) = 36.86°$

Draw FBD.

$\xrightarrow{+} \Sigma F_x = ma_x;$

$100 \text{ N} - (40 \text{ N})\cos 36.86° = (25 \text{ kg})a$

$a = 2.72 \text{ m/s}^2$

F13–6. Consider no slipping between A and B.

Draw FBD.

$\xrightarrow{+} \Sigma F_x = ma_x; \quad 6 = \frac{70}{32.2} a; \quad a = 2.76 \text{ ft/s}^2$

Check if slipping occurs between A and B.

$\xrightarrow{+} \Sigma F_x = ma_x; \quad 6 - F = \frac{20}{32.2}(2.76);$ No slipping

$F = 4.29 \text{ lb} < 0.4(20) = 8 \text{ lb}$

$a_A = a_B = 2.76 \text{ ft/s}^2$ *Ans.*

F13–7. Draw FBD.

$\Sigma F_n = m\frac{v^2}{\rho}; \quad (0.3)m(9.81) = m\frac{v^2}{2}$

$v = 2.43 \text{ m/s}$ *Ans.*

F13–8. Draw FBD.

$+\downarrow \Sigma F_n = ma_n; \quad m(32.2) = m\left(\frac{v^2}{250}\right)$

$v = 89.7 \text{ ft/s}$ *Ans.*

F13–9. Draw FBD.

$+\downarrow \Sigma F_n = ma_n; \quad 150 + N_p = \frac{150}{32.2}\left(\frac{(120)^2}{400}\right)$

$N_p = 17.7 \text{ lb}$ *Ans.*

F13–10. Draw FBD.

$\xrightarrow{+} \Sigma F_n = ma_n;$

$N_c \sin 30° + 0.2 N_c \cos 30° = m\frac{v^2}{500}$

$+\uparrow \Sigma F_b = 0;$

$N_c \cos 30° - 0.2 N_c \sin 30° - m(32.2) = 0$

$v = 119 \text{ ft/s}$ *Ans.*

F13–11. Draw FBD.

$\Sigma F_t = ma_t; \quad 10(9.81) \text{ N} \cos 45° = (10 \text{ kg})a_t$

$a_t = 6.94 \text{ m/s}^2$ *Ans.*

$\Sigma F_n = ma_n;$

$T - 10(9.81) \text{ N} \sin 45° = (10 \text{ kg})\frac{(3 \text{ m/s})^2}{2 \text{ m}}$

$T = 114 \text{ N}$ *Ans.*

F13–12. Draw FBD.

$\Sigma F_n = ma_n;$

$F_n = (500 \text{ kg})\frac{(15 \text{ m/s})^2}{200 \text{ m}} = 562.5 \text{ N}$

$\Sigma F_t = ma_t;$

$F_t = (500 \text{ kg})(1.5 \text{ m/s}^2) = 750 \text{ N}$

$F = \sqrt{F_n^2 + F_t^2} = \sqrt{(562.5 \text{ N})^2 + (750 \text{ N})^2}$

$= 938 \text{ N}$ *Ans.*

F13–13. $a_r = \ddot{r} - r\dot{\theta}^2 = 0 - (1.5 \text{ m} + (8 \text{ m})\sin 45°)\dot{\theta}^2$
$= (-7.157 \, \dot{\theta}^2) \text{ m/s}^2$

Draw FBD.

$\Sigma F_z = ma_z;$

$T \cos 45° - m(9.81) = m(0) \quad T = 13.87 \, m$

$\Sigma F_r = ma_r;$

$-(13.87 m) \sin 45° = m(-7.157 \, \dot{\theta}^2)$

$\dot{\theta} = 1.17 \text{ rad/s}$ *Ans.*

F13–14. $\theta = \pi t^2 \big|_{t=0.5\,s} = (\pi/4)$ rad
$\dot{\theta} = 2\pi t \big|_{t=0.5\,s} = \pi$ rad/s
$\ddot{\theta} = 2\pi$ rad/s^2
$r = 0.6 \sin\theta \big|_{\theta=\pi/4\,\text{rad}} = 0.4243$ m
$\dot{r} = 0.6 (\cos\theta)\dot{\theta} \big|_{\theta=\pi/4\,\text{rad}} = 1.3329$ m/s
$\ddot{r} = 0.6 [(\cos\theta)\ddot{\theta} - (\sin\theta)\dot{\theta}^2] \big|_{\theta=\pi/4\,\text{rad}} = -1.5216$ m/s^2
$a_r = \ddot{r} - r\dot{\theta}^2 = -1.5216$ m/s^2 − (0.4243 m)(π rad/s)^2
$\quad = -5.7089$ m/s^2
$a_\theta = r\ddot{\theta} + 2\dot{r}\dot{\theta} = 0.4243$ m(2π rad/s^2)
$\quad + 2(1.3329\text{ m/s})(\pi\text{ rad/s})$
$\quad = 11.0404$ m/s^2
Draw FBD.
$\Sigma F_r = ma_r;$
$F\cos 45° - N\cos 45° - 0.2(9.81)\cos 45°$
$\quad = 0.2(-5.7089)$
$\Sigma F_\theta = ma_\theta;$
$F\sin 45° + N\sin 45° - 0.2(9.81)\sin 45°$
$\quad = 0.2(11.0404)$
$N = 2.37$ N $F = 2.72$ N   Ans.

F13–15. $r = 50e^{2\theta} \big|_{\theta=\pi/6\,\text{rad}} = [50e^{2(\pi/6)}]$ m = 142.48 m
$\dot{r} = 50(2e^{2\theta}\dot{\theta}) = 100e^{2\theta}\dot{\theta} \big|_{\theta=\pi/6\,\text{rad}}$
$\quad = [100e^{2(\pi/6)}(0.05)] = 14.248$ m/s
$\ddot{r} = 100((2e^{2\theta}\dot{\theta})\dot{\theta} + e^{2\theta}(\ddot{\theta})) \big|_{\theta=\pi/6\,\text{rad}}$
$\quad = 100[2e^{2(\pi/6)}(0.05^2) + e^{2(\pi/6)}(0.01)]$
$\quad = 4.274$ m/s^2
$a_r = \ddot{r} - r\dot{\theta}^2 = 4.274$ m/s^2 − 142.48 m(0.05 rad/s)^2
$\quad = 3.918$ m/s^2
$a_\theta = r\ddot{\theta} + 2\dot{r}\dot{\theta} = 142.48$ m(0.01 rad/s^2)
$\quad + 2(14.248\text{ m/s})(0.05\text{ rad/s})$
$\quad = 2.850$ m/s^2
Draw FBD.
$\Sigma F_r = ma_r;$
$F_r = (2000\text{ kg})(3.918\text{ m/s}^2) = 7836.55$ N
$\Sigma F_\theta = ma_\theta;$
$F_\theta = (2000\text{ kg})(2.850\text{ m/s}^2) = 5699.31$ N
$F = \sqrt{F_r^2 + F_\theta^2}$
$\quad = \sqrt{(7836.55\text{ N})^2 + (5699.31\text{ N})^2}$
$\quad = 9689.87$ N = 9.69 kN

F13–16.
$r = (0.6\cos 2\theta)$ m $\big|_{\theta=0°} = [0.6\cos 2(0°)]$ m = 0.6 m
$\dot{r} = (-1.2\sin 2\theta\dot{\theta})$ m/s $\big|_{\theta=0°}$
$\quad = [-1.2\sin 2(0°)(-3)]$ m/s = 0
$\ddot{r} = -1.2(\sin 2\theta\ddot{\theta} + 2\cos 2\theta\dot{\theta}^2)$ m/s^2 $\big|_{\theta=0°}$
$\quad = -21.6$ m/s^2
Thus,
$a_r = \ddot{r} - r\dot{\theta}^2 = -21.6$ m/s^2 − 0.6 m(−3 rad/s)^2
$\quad = -27$ m/s^2
$a_\theta = r\ddot{\theta} + 2\dot{r}\dot{\theta} = 0.6$ m(0) + 2(0)(−3 rad/s) = 0
Draw FBD.
$\Sigma F_\theta = ma_\theta;$ $F - 0.2(9.81)$ N = 0.2 kg(0)
$\quad\quad\quad\quad\quad F = 1.96$ N ↑   Ans.

Chapter 14

F14–1. Draw FBD.
$T_1 + \Sigma U_{1-2} = T_2$
$0 + (\frac{4}{5})(500\text{ N})(0.5\text{ m}) - \frac{1}{2}(500\text{ N/m})(0.5\text{ m})^2$
$\quad\quad\quad\quad\quad = \frac{1}{2}(10\text{ kg})v^2$
$v = 5.24$ m/s   Ans.

F14–2. Draw FBD.
$\Sigma F_y = ma_y;$ $N_A - 20(9.81)$ N $\cos 30° = 0$
$N_A = 169.91$ N
$T_1 + \Sigma U_{1-2} = T_2$
$0 + 300$ N(10 m) − 0.3(169.91 N)(10 m)
$\quad - 20(9.81)$ N (10 m) $\sin 30°$
$\quad = \frac{1}{2}(20\text{ kg})v^2$
$v = 12.3$ m/s   Ans.

F14–3. Draw FBD.
$T_1 + \Sigma U_{1-2} = T_2$
$0 + 2\left[\int_0^{15\,\text{m}}(600 + 2s^2)\text{ N }ds\right] - 100(9.81)$ N(15 m)
$\quad\quad\quad\quad\quad = \frac{1}{2}(100\text{ kg})v^2$
$v = 12.5$ m/s   Ans.

FUNDAMENTAL PROBLEMS

F14–4. Draw FBD.
$$T_1 + \Sigma U_{1-2} = T_2$$
$$\tfrac{1}{2}(1800 \text{ kg})(125 \text{ m/s})^2 - \left[\tfrac{(50\,000 \text{ N} + 20\,000 \text{ N})}{2}(400 \text{ m})\right]$$
$$= \tfrac{1}{2}(1800 \text{ kg})v^2$$
$$v = 8.33 \text{ m/s} \qquad \qquad Ans.$$

F14–5. Draw FBD.
$$T_1 + \Sigma U_{1-2} = T_2$$
$$\tfrac{1}{2}(10 \text{ kg})(5 \text{ m/s})^2 + 100 \text{ N}s' + [10(9.81) \text{ N}]\, s' \sin 30°$$
$$-\tfrac{1}{2}(200 \text{ N/m})(s')^2 = 0$$
$$s' = 2.09 \text{ m}$$
$$s = 0.6 \text{ m} + 2.09 \text{ m} = 2.69 \text{ m} \qquad Ans.$$

F14–6. Draw FBD.
$$T_A + \Sigma U_{A-B} = T_B$$
Consider difference in cord length $AC - BC$, which is distance F moves.
$$0 + 10 \text{ lb}(\sqrt{(3 \text{ ft})^2 + (4 \text{ ft})^2} - 3 \text{ ft})$$
$$= \tfrac{1}{2}\left(\tfrac{5}{32.2} \text{ slug}\right)v_B^2$$
$$v_B = 16.0 \text{ ft/s} \qquad \qquad Ans.$$

F14–7. Draw FBD.
$$\xrightarrow{+} \Sigma F_x = ma_x;$$
$$30\left(\tfrac{4}{5}\right) = 20a \quad a = 1.2 \text{ m/s}^2 \rightarrow$$
$$v = v_0 + a_c t$$
$$v = 0 + 1.2(4) = 4.8 \text{ m/s}$$
$$P = \mathbf{F} \cdot \mathbf{v} = F(\cos\theta)v$$
$$= 30\left(\tfrac{4}{5}\right)(4.8)$$
$$= 115 \text{ W} \qquad \qquad Ans.$$

F14–8. Draw FBD.
$$\xrightarrow{+} \Sigma F_x = ma_x;$$
$$10s = 20a \quad a = 0.5s \text{ m/s}^2 \rightarrow$$
$$v\, dv = a\, ds$$
$$\int_1^v v\, dv = \int_0^{5 \text{ m}} 0.5\, s\, ds$$
$$v = 3.674 \text{ m/s}$$
$$P = \mathbf{F} \cdot \mathbf{v} = [10(5)](3.674) = 184 \text{ W} \qquad Ans.$$

F14–9. Draw FBD.
$$+\uparrow \Sigma F_y = 0;$$
$$T_1 - 100 \text{ lb} = 0 \qquad T_1 = 100 \text{ lb}$$
$$+\uparrow \Sigma F_y = 0;$$
$$100 \text{ lb} + 100 \text{ lb} - T_2 = 0 \quad T_2 = 200 \text{ lb}$$
$$P_{\text{out}} = \mathbf{T}_B \cdot \mathbf{v}_B = (200 \text{ lb})(3 \text{ ft/s}) = 1.091 \text{ hp}$$
$$P_{\text{in}} = \frac{P_{\text{out}}}{\varepsilon} = \frac{1.091 \text{ hp}}{0.8} = 1.36 \text{ hp} \qquad Ans.$$

F14–10. Draw FBD.
$$\Sigma F_{y'} = ma_{y'}; \quad N - 20(9.81)\cos 30° = 20(0)$$
$$N = 169.91 \text{ N}$$
$$\Sigma F_{x'} = ma_{x'};$$
$$F - 20(9.81)\sin 30° - 0.2(169.91) = 0$$
$$F = 132.08 \text{ N}$$
$$P = \mathbf{F} \cdot \mathbf{v} = 132.08(5) = 660 \text{ W} \qquad Ans.$$

F14–11. Draw FBD.
$$+\uparrow \Sigma F_y = ma_y;$$
$$T - 50(9.81) = 50(0) \quad T = 490.5 \text{ N}$$
$$P_{\text{out}} = \mathbf{T} \cdot \mathbf{v} = 490.5(1.5) = 735.75 \text{ W}$$
Also, for a point on the other cable
$$P_{\text{out}} = \left(\frac{490.5}{2}\right)(1.5)(2) = 735.75 \text{ W}$$
$$P_{\text{in}} = \frac{P_{\text{out}}}{\varepsilon} = \frac{735.75}{0.8} = 920 \text{ W} \qquad Ans.$$

F14–12. $2s_A + s_P = l$
$$2a_A + a_P = 0$$
$$2a_A + 6 = 0$$
$$a_A = -3 \text{ m/s}^2 = 3 \text{ m/s}^2 \uparrow$$
Draw FBD.
$$\Sigma F_y = ma_y; \quad T_A - 490.5 \text{ N} = (50 \text{ kg})(3 \text{ m/s}^2)$$
$$T_A = 640.5 \text{ N}$$
$$P_{\text{out}} = \mathbf{T} \cdot \mathbf{v} = (640.5 \text{ N}/2)(12) = 3843 \text{ W}$$
$$P_{\text{in}} = \frac{P_{\text{out}}}{\varepsilon} = \frac{3843}{0.8} = 4803.75 \text{ W} = 4.80 \text{ kW} \quad Ans.$$

F14–13. Datum at B.
$$T_A + V_A = T_B + V_B$$
$$0 + 2(9.81)(1.5) = \tfrac{1}{2}(2)(v_B)^2 + 0$$
$$v_B = 5.42 \text{ m/s} \qquad \qquad Ans.$$
$$+\uparrow \Sigma F_n = ma_n; \quad T - 2(9.81) = 2\left(\frac{(5.42)^2}{1.5}\right)$$
$$T = 58.9 \text{ N} \qquad \qquad Ans.$$

F14–14. Datum at B.

$T_A + V_A = T_B + V_B$

$\frac{1}{2} m_A v_A^2 + mgh_A = \frac{1}{2} m_B v_B^2 + mgh_B$

$[\frac{1}{2}(2 \text{ kg})(1 \text{ m/s})^2] + [2(9.81) \text{ N}(4 \text{ m})]$

$= [\frac{1}{2}(2 \text{ kg})v_B^2] + [0]$

$v_B = 8.915 \text{ m/s} = 8.92 \text{ m/s}$ Ans.

$+\uparrow \Sigma F_n = ma_n;\quad N_B - 2(9.81) \text{ N}$

$= (2 \text{ kg})\left(\dfrac{(8.915 \text{ m/s})^2}{2 \text{ m}}\right)$

$N_B = 99.1 \text{ N}$ Ans.

F14–15. Datum at A.

$T_1 + V_1 = T_2 + V_2$

$\frac{1}{2}(2)(4)^2 + 0 + \frac{1}{2}(30)(2-1)^2$

$= \frac{1}{2}(2)(v)^2 - 2(9.81)(1) + \frac{1}{2}(30)(\sqrt{5}-1)^2$

$v = 5.26 \text{ m/s}$ Ans.

F14–16. Datum at B.

$T_A + V_A = T_B + V_B$

$0 + \frac{1}{2}(4)(2.5 - 0.5)^2 + 5(2.5)$

$= \frac{1}{2}\left(\frac{5}{32.2}\right)v_B^2 + 0 + \frac{1}{2}(4)(1-0.5)^2$

$v_B = 16.0 \text{ ft/s}$ Ans.

F14–17. Datum at initial position.

$T_1 + V_1 = T_2 + V_2$

$\frac{1}{2}mv_1^2 + mgy_1 + \frac{1}{2}ks_1^2$

$= \frac{1}{2}mv_2^2 + mgy_2 + \frac{1}{2}ks_2^2$

$[0] + [0] + [0] = [0] +$

$[-75 \text{ lb}(5 \text{ ft} + s)] + [2(\frac{1}{2}(1000 \text{ lb/ft})s^2)$

$+ \frac{1}{2}(1500 \text{ lb/ft})(s - 0.25 \text{ ft})^2]$

$s = s_A = s_C = 0.580 \text{ ft}$ Ans.

Also,

$s_B = 0.5803 \text{ ft} - 0.25 \text{ ft} = 0.330 \text{ ft}$ Ans.

F14–18. Datum at A.

$T_A + V_A = T_B + V_B$

$\frac{1}{2}mv_A^2 + (\frac{1}{2}ks_A^2 + mgy_A)$

$= \frac{1}{2}mv_B^2 + (\frac{1}{2}ks_B^2 + mgy_B)$

$\frac{1}{2}(4 \text{ kg})(2 \text{ m/s})^2 + \frac{1}{2}(400 \text{ N/m})(0.1 \text{ m} - 0.2 \text{ m})^2 + 0$

$= \frac{1}{2}(4 \text{ kg})v_B^2 + \frac{1}{2}(400 \text{ N/m})(\sqrt{(0.4 \text{ m})^2 + (0.3 \text{ m})^2}$

$- 0.2 \text{ m})^2 + [4(9.81) \text{ N}](-(0.1 \text{ m} + 0.3 \text{ m}))$

$v_B = 1.962 \text{ m/s} = 1.96 \text{ m/s}$ Ans.

Chapter 15

F15–1. Draw FBD.

$(\xrightarrow{+})\quad m(v_1)_x + \Sigma \int_{t_1}^{t_2} F_x \, dt = m(v_2)_x$

$(0.5 \text{ kg})(25 \text{ m/s})\cos 45° - \int F_x \, dt$

$= (0.5 \text{ kg})(10 \text{ m/s})\cos 30°$

$I_x = \int F_x \, dt = 4.509 \text{ N} \cdot \text{s}$

$(+\uparrow)\quad m(v_1)_y + \Sigma \int_{t_1}^{t_2} F_y \, dt = m(v_2)_y$

$-(0.5 \text{ kg})(25 \text{ m/s})\sin 45° + \int F_y \, dt$

$= (0.5 \text{ kg})(10 \text{ m/s})\sin 30°$

$I_y = \int F_y \, dt = 11.339 \text{ N} \cdot \text{s}$

$I = \int F \, dt = \sqrt{(4.509 \text{ N} \cdot \text{s})^2 + (11.339 \text{ N} \cdot \text{s})^2}$

$= 12.2 \text{ N} \cdot \text{s}$ Ans.

F15–2. Draw FBD.

$(+\uparrow)\quad m(v_1)_y + \Sigma \int_{t_1}^{t_2} F_y \, dt = m(v_2)_y$

$0 + N(4 \text{ s}) + (100 \text{ lb})(4 \text{ s})\sin 30°$

$-(150 \text{ lb})(4 \text{ s}) = 0$

$N = 100 \text{ lb}$

$(\xrightarrow{+})\quad m(v_1)_x + \Sigma \int_{t_1}^{t_2} F_x \, dt = m(v_2)_x$

$0 + (100 \text{ lb})(4 \text{ s})\cos 30° - 0.2(100 \text{ lb})(4 \text{ s})$

$= \left(\dfrac{150}{32.2} \text{ slug}\right)v$

$v = 57.2 \text{ ft/s}$ Ans.

F15–3. Draw FBD.

Time to start motion,

$+\uparrow \Sigma F_y = 0;\quad N - 25(9.81) \text{ N} = 0\quad N = 245.25 \text{ N}$

$\xrightarrow{+} \Sigma F_x = 0;\quad 20t^2 - 0.3(245.25 \text{ N}) = 0\quad t = 1.918 \text{ s}$

$(\xrightarrow{+})\quad m(v_1)_x + \Sigma \int_{t_1}^{t_2} F_x \, dt = m(v_2)_x$

$0 + \int_{1.918 \text{ s}}^{4 \text{ s}} 20t^2 \, dt - (0.25(245.25 \text{ N}))(4 \text{ s} - 1.918 \text{ s})$

$= (25 \text{ kg})v$

$v = 10.1 \text{ m/s}$ Ans.

F15–4. Draw FBD.

$$(\xrightarrow{+}) \quad m(v_1)_x + \Sigma \int_{t_1}^{t_2} F_x \, dt = m(v_2)_x$$

$$(1500 \text{ kg})(0) + \left[\tfrac{1}{2}(6000 \text{ N})(2 \text{ s}) + (6000 \text{ N})(6 \text{ s} - 2 \text{ s})\right]$$

$$= (1500 \text{ kg}) \, v$$

$$v = 20 \text{ m/s} \qquad \text{Ans.}$$

F15–5. Draw FBD.
SUV and trailer,

$$m(v_1)_x + \Sigma \int_{t_1}^{t_2} F_x \, dt = m(v_2)_x$$

$$0 + (9000 \text{ N})(20 \text{ s}) = (1500 \text{ kg} + 2500 \text{ kg})v$$

$$v = 45.0 \text{ m/s} \qquad \text{Ans.}$$

Trailer,

$$m(v_1)_x + \Sigma \int_{t_1}^{t_2} F_x \, dt = m(v_2)_x$$

$$0 + T(20 \text{ s}) = (1500 \text{ kg})(45.0 \text{ m/s})$$

$$T = 3375 \text{ N} = 3.375 \text{ kN} \qquad \text{Ans.}$$

F15–6. Draw FBD.
Block B:

$$(+\downarrow) \, mv_1 + \int F \, dt = mv_2$$

$$0 + 8(5) - T(5) = \tfrac{8}{32.2}(1)$$

$$T = 7.95 \text{ lb} \qquad \text{Ans.}$$

Block A:

$$(\xrightarrow{+}) \, mv_1 + \int F \, dt = mv_2$$

$$0 + 7.95(5) - \mu_k(10)(5) = \tfrac{10}{32.2}(1)$$

$$\mu_k = 0.789 \qquad \text{Ans.}$$

F15–7. $(\xrightarrow{+}) \, m_A(v_A)_1 + m_B(v_B)_1 = m_A(v_A)_2 + m_B(v_B)_2$

$$(20(10^3) \text{ kg})(3 \text{ m/s}) + (15(10^3) \text{ kg})(-1.5 \text{ m/s})$$

$$= (20(10^3) \text{ kg})(v_A)_2 + (15(10^3) \text{ kg})(2 \text{ m/s})$$

$$(v_A)_2 = 0.375 \text{ m/s} \rightarrow \qquad \text{Ans.}$$

$$(\xrightarrow{+}) \, m(v_B)_1 + \Sigma \int_{t_1}^{t_2} F \, dt = m(v_B)_2$$

$$(15(10^3) \text{ kg})(-1.5 \text{ m/s}) + F_{\text{avg}}(0.5 \text{ s})$$

$$= (15(10^3) \text{ kg})(2 \text{ m/s})$$

$$F_{\text{avg}} = 105(10^3) \text{ N} = 105 \text{ kN} \qquad \text{Ans.}$$

F15–8. $(\xrightarrow{+}) \, m_p[(v_p)_1]_x + m_c[(v)_1]_x = (m_p + m_c)v_2$

$$5[10(\tfrac{4}{5})] + 0 = (5 + 20)v_2$$

$$v_2 = 1.6 \text{ m/s} \qquad \text{Ans.}$$

F15–9. Datum at B.

$$T_1 + V_1 = T_2 + V_2$$

$$\tfrac{1}{2} m_A (v_A)_1^2 + (V_g)_1 = \tfrac{1}{2} m_A (v_A)_2^2 + (V_g)_2$$

$$\tfrac{1}{2}(5)(5)^2 + 5(9.81)(1.5) = \tfrac{1}{2}(5)(v_A)_2^2 + 0$$

$$(v_A)_2 = 7.378 \text{ m/s}$$

$$(\xrightarrow{+}) \, m_A(v_A)_2 + m_B(v_B)_2 = (m_A + m_B)v$$

$$5(7.378) + 0 = (5 + 8)v$$

$$v = 2.84 \text{ m/s} \qquad \text{Ans.}$$

F15–10. $(\xrightarrow{+}) \, m_A(v_A)_1 + m_B(v_B)_1 = m_A(v_A)_2 + m_B(v_B)_2$

$$0 + 0 = 10(v_A)_2 + 15(v_B)_2 \qquad (1)$$

$$T_1 + V_1 = T_2 + V_2$$

$$\tfrac{1}{2} m_A(v_A)_1^2 + \tfrac{1}{2} m_B(v_B)_1^2 + (V_e)_1$$

$$= \tfrac{1}{2} m_A(v_A)_2^2 + \tfrac{1}{2} m_B(v_B)_2^2 + (V_e)_2$$

$$0 + 0 + \tfrac{1}{2}[5(10^3)](0.2^2)$$

$$= \tfrac{1}{2}(10)(v_A)_2^2 + \tfrac{1}{2}(15)(v_B)_2^2 + 0$$

$$5(v_A)_2^2 + 7.5 \, (v_B)_2^2 = 100 \qquad (2)$$

Solving Eqs. (1) and (2),

$$(v_B)_2 = 2.31 \text{ m/s} \rightarrow \qquad \text{Ans.}$$

$$(v_A)_2 = -3.464 \text{ m/s} = 3.46 \text{ m/s} \leftarrow \qquad \text{Ans.}$$

F15–11. $(\xrightarrow{+}) \, m_A(v_A)_1 + m_B(v_B)_1 = (m_A + m_B)v_2$

$$0 + 10(15) = (15 + 10)v_2$$

$$v_2 = 6 \text{ m/s}$$

$$T_1 + V_1 = T_2 + V_2$$

$$\tfrac{1}{2}(m_A + m_B)v_2^2 + (V_e)_2 = \tfrac{1}{2}(m_A + m_B)v_3^2 + (V_e)_3$$

$$\tfrac{1}{2}(15 + 10)(6^2) + 0 = 0 + \tfrac{1}{2}[10(10^3)]s_{\max}^2$$

$$s_{\max} = 0.3 \text{ m} = 300 \text{ mm} \qquad \text{Ans.}$$

F15–12. $(\xrightarrow{+}) \, 0 + 0 = m_p(v_p)_x - m_c v_c$

$$0 = (20 \text{ kg})(v_p)_x - (250 \text{ kg})v_c$$

$$(v_p)_x = 12.5 \, v_c \qquad (1)$$

$$\mathbf{v}_p = \mathbf{v}_c + \mathbf{v}_{p/c}$$

$$(v_p)_x \mathbf{i} + (v_p)_y \mathbf{j} = -v_c \mathbf{i} + [(400 \text{ m/s}) \cos 30°\mathbf{i}$$

$$+ (400 \text{ m/s}) \sin 30°\mathbf{j}]$$

$$(v_p)_x \mathbf{i} + (v_p)_y \mathbf{j} = (346.41 - v_c)\mathbf{i} + 200\mathbf{j}$$

$$(v_p)_x = 346.41 - v_c$$

$$(v_p)_y = 200 \text{ m/s}$$

$$(v_p)_x = 320.75 \text{ m/s} \quad v_c = 25.66 \text{ m/s}$$

$$v_p = \sqrt{(v_p)_x^2 + (v_p)_y^2}$$

$$= \sqrt{(320.75 \text{ m/s})^2 + (200 \text{ m/s})^2}$$

$$= 378 \text{ m/s} \qquad \text{Ans.}$$

F15–13. $(\xrightarrow{+})$ $e = \dfrac{(v_B)_2 - (v_A)_2}{(v_A)_1 - (v_B)_1}$

$= \dfrac{(9 \text{ m/s}) - (1 \text{ m/s})}{(8 \text{ m/s}) - (-2 \text{ m/s})} = 0.8$

F15–14. $(\xrightarrow{+})$ $m_A(v_A)_1 + m_B(v_B)_1 = m_A(v_A)_2 + m_B(v_B)_2$
$[15(10^3) \text{ kg}](5 \text{ m/s}) + [25(10^3)](-7 \text{ m/s})$
$= [15(10^3) \text{ kg}](v_A)_2 + [25(10^3)](v_B)_2$
$15(v_A)_2 + 25(v_B)_2 = -100$ (1)

Using the coefficient of restitution equation,

$(\xrightarrow{+})$ $e = \dfrac{(v_B)_2 - (v_A)_2}{(v_A)_1 - (v_B)_1}$

$0.6 = \dfrac{(v_B)_2 - (v_A)_2}{5 \text{ m/s} - (-7 \text{ m/s})}$

$(v_B)_2 - (v_A)_2 = 7.2$ (2)

Solving,

$(v_B)_2 = 0.2 \text{ m/s} \rightarrow$ Ans.
$(v_A)_2 = -7 \text{ m/s} = 7 \text{ m/s} \leftarrow$ Ans.

F15–15. Datum at B.
$T_1 + V_1 = T_2 + V_2$
$\tfrac{1}{2} m(v_A)_1^2 + mg(h_A)_1 = \tfrac{1}{2} m(v_A)_2^2 + mg(h_A)_2$
$\tfrac{1}{2}(\tfrac{30}{32.2} \text{ slug})(5 \text{ ft/s})^2 + (30 \text{ lb})(10\text{ft})$
$= \tfrac{1}{2}(\tfrac{30}{32.2} \text{ slug})(v_A)_2^2 + 0$
$(v_A)_2 = 25.87 \text{ ft/s} \leftarrow$

$(\xleftrightarrow{+})$ $m_A(v_A)_2 + m_B(v_B)_2 = m_A(v_A)_3 + m_B(v_B)_3$
$(\tfrac{30}{32.2} \text{ slug})(25.87 \text{ ft/s}) + 0$
$= (\tfrac{30}{32.2} \text{ slug})(v_A)_3 + (\tfrac{80}{32.2} \text{ slug})(v_B)_3$
$30(v_A)_3 + 80(v_B)_3 = 775.95$ (1)

$(\xleftrightarrow{+})$ $e = \dfrac{(v_B)_3 - (v_A)_3}{(v_A)_2 - (v_B)_2}$

$0.6 = \dfrac{(v_B)_3 - (v_A)_3}{25.87 \text{ ft/s} - 0}$

$(v_B)_3 - (v_A)_3 = 15.52$ (2)

Solving Eqs. (1) and (2) yields
$(v_B)_3 = 11.3 \text{ ft/s} \leftarrow$
$(v_A)_3 = -4.23 \text{ ft/s} = 4.23 \text{ ft/s} \rightarrow$ Ans.

F15–16. $(+\uparrow)$ $m[(v_b)_1]_y = m[(v_b)_2]_y$
$[(v_b)_2]_y = [(v_b)_1]_y = (20 \text{ m/s}) \sin 30° = 10 \text{ m/s} \uparrow$

$(\xrightarrow{+})$ $e = \dfrac{(v_w)_2 - [(v_b)_2]_x}{[(v_b)_1]_x - (v_w)_1}$

$0.75 = \dfrac{0 - [(v_b)_2]_x}{(20 \text{ m/s}) \cos 30° - 0}$

$[(v_b)_2]_x = -12.99 \text{ m/s} = 12.99 \text{ m/s} \leftarrow$
$(v_b)_2 = \sqrt{[(v_b)_2]_x^2 + [(v_b)_2]_y^2}$
$= \sqrt{(12.99 \text{ m/s})^2 + (10 \text{ m/s})^2}$
$= 16.4 \text{ m/s}$ Ans.

$\theta = \tan^{-1}\left(\dfrac{[(v_b)_2]_y}{[(v_b)_2]_x}\right) = \tan^{-1}\left(\dfrac{10 \text{ m/s}}{12.99 \text{ m/s}}\right)$
$= 37.6°$ Ans.

F15–17. $\Sigma m(v_x)_1 = \Sigma m(v_x)_2$
$0 + 0 = 2(1) + 11 (v_{Bx})_2$
$(v_{Bx})_2 = -0.1818 \text{ m/s}$
$\Sigma m(v_y)_1 = \Sigma m(v_y)_2$
$2(3) + 0 = 0 + 11 (v_{By})_2$
$(v_{By})_2 = 0.545 \text{ m/s}$
$(v_B)_2 = \sqrt{(-0.1818)^2 + (0.545)^2}$
$= 0.575 \text{ m/s}$ Ans.

F15–18. $+\nearrow$ $1(3)(\tfrac{3}{5}) - 1(4)(\tfrac{4}{5})$
$= 1(v_B)_{2x} + 1(v_A)_{2x}$
$+\nearrow$ $0.5 = [(v_A)_{2x} - (v_B)_{2x}]/[(3)(\tfrac{3}{5}) - (-4)(\tfrac{4}{5})]$
Solving,
$(v_A)_{2x} = 0.550 \text{ m/s}, (v_B)_{2x} = -1.95 \text{ m/s}$
Disc A,
$+\nwarrow$ $-1(4)(\tfrac{3}{5}) = 1(v_A)_{2y}$
$(v_A)_{2y} = -2.40 \text{ m/s}$
Disc B,
$-1(3)(\tfrac{4}{5}) = 1(v_B)_{2y}$
$(v_B)_{2y} = -2.40 \text{ m/s}$
$(v_A)_2 = \sqrt{(0.550)^2 + (2.40)^2} = 2.46 \text{ m/s}$ Ans.
$(v_B)_2 = \sqrt{(1.95)^2 + (2.40)^2} = 3.09 \text{ m/s}$ Ans.

F15–19. $H_O = \Sigma mvd$;
$H_O = [2(10)(\frac{4}{5})](4) - [2(10)(\frac{3}{5})](3)$
$= 28 \text{ kg} \cdot \text{m}^2/\text{s}$

F15–20. $H_P = \Sigma mvd$;
$H_P = [2(15) \sin 30°](2) - [2(15) \cos 30°](5)$
$= -99.9 \text{ kg} \cdot \text{m}^2/\text{s} = 99.9 \text{ kg} \cdot \text{m}^2/\text{s} \downarrow$

F15–21. Draw an FBD of the rod and block.
$(H_z)_1 + \Sigma \int M_z \, dt = (H_z)_2$
$5(2)(1.5) + 5(1.5)(3) = 5v(1.5)$
$v = 5 \text{ m/s}$ *Ans.*

F15–22. Draw an FBD of the rod and block.
$(H_z)_1 + \Sigma \int M_z \, dt = (H_z)_2$
$0 + \int_0^{4\text{ s}} (10t)(\frac{4}{5})(1.5) dt = 5v(1.5)$
$v = 12.8 \text{ m/s}$ *Ans.*

F15–23. Draw an FBD of the rod and sphere.
$(H_z)_1 + \Sigma \int M_z \, dt = (H_z)_2$
$0 + \int_0^{5\text{ s}} 0.9t^2 \, dt = 2v(0.6)$
$v = 31.2 \text{ m/s}$ *Ans.*

F15–24. Draw an FBD of the rod and spheres.
$(H_z)_1 + \Sigma \int M_z \, dt = (H_z)_2$
$0 + \int_0^{4\text{ s}} 8t \, dt + 2(10)(0.5)(4) = 2[10v(0.5)]$
$v = 10.4 \text{ m/s}$ *Ans.*

Chapter 16

F16–1. $\theta = (20 \text{ rev})(\frac{2\pi \text{ rad}}{1 \text{ rev}}) = 40\pi \text{ rad}$
$\omega^2 = \omega_0^2 + 2\alpha_c(\theta - \theta_0)$
$(30 \text{ rad/s})^2 = 0^2 + 2\alpha_c[(40\pi \text{ rad}) - 0]$
$\alpha_c = 3.581 \text{ rad/s}^2 = 3.58 \text{ rad/s}^2$ *Ans.*
$\omega = \omega_0 + \alpha_c t$
$30 \text{ rad/s} = 0 + (3.581 \text{ rad/s}^2)t$
$t = 8.38 \text{ s}$ *Ans.*

F16–2. $\frac{d\omega}{d\theta} = 2(0.005\theta) = (0.01\theta)$
$\alpha = \omega \frac{d\omega}{d\theta} = (0.005 \, \theta^2)(0.01\theta) = 50(10^{-6})\theta^3 \text{ rad/s}^2$
When $\theta = 20 \text{ rev}(2\pi \text{ rad}/1 \text{ rev}) = 40\pi \text{ rad}$,
$\alpha = [50(10^{-6})(40\pi)^3] \text{ rad/s}^2$
$= 99.22 \text{ rad/s}^2 = 99.2 \text{ rad/s}^2$ *Ans.*

F16–3. $\omega = 4\theta^{1/2}$
$150 \text{ rad/s} = 4\,\theta^{1/2}$
$\theta = 1406.25 \text{ rad}$
$dt = \frac{d\theta}{\omega}$
$\int_0^t dt = \int_{1 \text{ rad}}^\theta \frac{d\theta}{4\theta^{1/2}}$
$t\big|_0^t = \frac{1}{2}\theta^{1/2}\big|_{1 \text{ rad}}^\theta$
$t = \frac{1}{2}\theta^{1/2} - \frac{1}{2}$
$t = \frac{1}{2}(1406.25)^{1/2} - \frac{1}{2} = 18.25 \text{ s}$ *Ans.*

F16–4. $\omega = \frac{d\theta}{dt} = (1.5t^2 + 15) \text{ rad/s}$
$\alpha = \frac{d\omega}{dt} = (3t) \text{ rad/s}^2$
$\omega = [1.5(3^2) + 15] \text{ rad/s} = 28.5 \text{ rad/s}$
$\alpha = 3(3) \text{ rad/s}^2 = 9 \text{ rad/s}^2$
$v = \omega r = (28.5 \text{ rad/s})(0.75 \text{ ft}) = 21.4 \text{ ft/s}$ *Ans.*
$a = \alpha r = (9 \text{ rad/s}^2)(0.75 \text{ ft}) = 6.75 \text{ ft/s}^2$ *Ans.*

F16–5. $\omega \, d\omega = \alpha \, d\theta$
$\int_{2 \text{ rad/s}}^\omega \omega \, d\omega = \int_0^\theta 0.5\theta \, d\theta$
$\frac{\omega^2}{2}\bigg|_{2 \text{ rad/s}}^\omega = 0.25\theta^2 \bigg|_0^\theta$
$\omega = (0.5\theta^2 + 4)^{1/2} \text{ rad/s}$
When $\theta = 2 \text{ rev} = 4\pi \text{ rad}$,
$\omega = [0.5(4\pi)^2 + 4]^{1/2} \text{ rad/s} = 9.108 \text{ rad/s}$
$v_P = \omega r = (9.108 \text{ rad/s})(0.2 \text{ m}) = 1.82 \text{ m/s}$ *Ans.*
$(a_P)_t = \alpha r = (0.5\theta \text{ rad/s}^2)(0.2 \text{ m})\big|_{\theta=4\pi \text{ rad}}$
$= 1.257 \text{ m/s}^2$
$(a_P)_n = \omega^2 r = (9.108 \text{ rad/s})^2(0.2 \text{ m}) = 16.59 \text{ m/s}^2$
$a_P = \sqrt{(a_P)_t^2 + (a_P)_n^2}$
$= \sqrt{(1.257 \text{ m/s}^2)^2 + (16.59 \text{ m/s}^2)^2}$
$= 16.6 \text{ m/s}^2$ *Ans.*

F16–6. The tangential acceleration of P is the same for A and B.

$$\alpha_B = \alpha_A \left(\frac{r_A}{r_B}\right)$$
$$= (4.5 \text{ rad/s}^2)\left(\frac{0.075 \text{ m}}{0.225 \text{ m}}\right) = 1.5 \text{ rad/s}^2$$
$$\omega_B = (\omega_B)_0 + \alpha_B t$$
$$\omega_B = 0 + (1.5 \text{ rad/s}^2)(3 \text{ s}) = 4.5 \text{ rad/s}$$
$$\theta_B = (\theta_B)_0 + (\omega_B)_0 t + \tfrac{1}{2}\alpha_B t^2$$
$$\theta_B = 0 + 0 + \tfrac{1}{2}(1.5 \text{ rad/s}^2)(3 \text{ s})^2$$
$$\theta_B = 6.75 \text{ rad}$$
$$v_C = \omega_B r_D = (4.5 \text{ rad/s})(0.125 \text{ m})$$
$$= 0.5625 \text{ m/s} \qquad \text{Ans.}$$
$$s_C = \theta_B r_D = (6.75 \text{ rad})(0.125 \text{ m}) = 0.84375 \text{ m}$$
$$= 844 \text{ mm} \qquad \text{Ans.}$$

F16–7. Vector Analysis
$$\mathbf{v}_B = \mathbf{v}_A + \boldsymbol{\omega} \times \mathbf{r}_{B/A}$$
$$-v_B \mathbf{j} = (3\mathbf{i}) \text{ m/s}$$
$$+ (\omega \mathbf{k}) \times (-1.5\cos 30°\mathbf{i} + 1.5 \sin 30°\mathbf{j})$$
$$-v_B \mathbf{j} = [3 - \omega_{AB}(1.5 \sin 30°)]\mathbf{i} - \omega(1.5 \cos 30°)\mathbf{j}$$
$$0 = 3 - \omega(1.5 \sin 30°) \qquad (1)$$
$$-v_B = 0 - \omega(1.5 \cos 30°) \qquad (2)$$
$$\omega = 4 \text{ rad/s} \qquad v_B = 5.20 \text{ m/s} \qquad \text{Ans.}$$
Scalar Analysis
$$\mathbf{v}_B = \mathbf{v}_A + \mathbf{v}_{B/A}$$
$$\left[\downarrow v_B\right] = \left[\underset{\rightarrow}{3}\right] + \left[\omega(1.5) \nearrow e30°\right]$$
This yields Eqs. (1) and (2).

F16–8. Vector Analysis
$$\mathbf{v}_B = \mathbf{v}_A + \boldsymbol{\omega} \times \mathbf{r}_{B/A}$$
$$(v_B)_x \mathbf{i} + (v_B)_y \mathbf{j} = 0 + (-10\mathbf{k}) \times (-0.6\mathbf{i} + 0.6\mathbf{j})$$
$$(v_B)_x \mathbf{i} + (v_B)_y \mathbf{j} = 6\mathbf{i} + 6\mathbf{j}$$
$$(v_B)_x = 6 \text{ m/s} \text{ and } (v_B)_y = 6 \text{ m/s}$$
$$v_B = \sqrt{(v_B)_x^2 + (v_B)_y^2}$$
$$= \sqrt{(6 \text{ m/s})^2 + (6 \text{ m/s})^2}$$
$$= 8.49 \text{ m/s} \qquad \text{Ans.}$$
Scalar Analysis
$$\mathbf{v}_B = \mathbf{v}_A + \mathbf{v}_{B/A}$$

$$\left[(v_B)_x \underset{\rightarrow}{}\right] + \left[(v_B)_y \uparrow\right] = \left[0\right] + \left[\nearrow 45°\ 10\left(\frac{0.6}{\cos 45°}\right)\right]$$
$$\overset{+}{\rightarrow}(v_B)_x = 0 + 10(0.6/\cos 45°)\cos 45° = 6 \text{ m/s} \rightarrow$$
$$+\uparrow (v_B)_y = 0 + 10(0.6/\cos 45°)\sin 45° = 6 \text{ m/s} \uparrow$$

F16–9. Vector Analysis
$$\mathbf{v}_B = \mathbf{v}_A + \boldsymbol{\omega} \times \mathbf{r}_{B/A}$$
$$(4 \text{ ft/s})\mathbf{i} = (-2 \text{ ft/s})\mathbf{i} + (-\omega\mathbf{k}) \times (3 \text{ ft})\mathbf{j}$$
$$4\mathbf{i} = (-2 + 3\omega)\mathbf{i}$$
$$\omega = 2 \text{ rad/s} \qquad \text{Ans.}$$
Scalar Analysis
$$\mathbf{v}_B = \mathbf{v}_A + \mathbf{v}_{B/A}$$
$$\left[\underset{\rightarrow}{4}\right] = \left[\underset{\leftarrow}{2}\right] + \left[\underset{\rightarrow}{\omega(3)}\right]$$
$$\overset{+}{\rightarrow} 4 = -2 + \omega(3); \qquad \omega = 2 \text{ rad/s}$$

F16–10. Vector Analysis
$$\mathbf{v}_A = \boldsymbol{\omega}_{OA} \times \mathbf{r}_A$$
$$= (12 \text{ rad/s})\mathbf{k} \times (0.3 \text{ m})\mathbf{j}$$
$$= [-3.6\mathbf{i}] \text{ m/s}$$
$$\mathbf{v}_B = \mathbf{v}_A + \boldsymbol{\omega}_{AB} \times \mathbf{r}_{B/A}$$
$$v_B \mathbf{j} = (-3.6 \text{ m/s})\mathbf{i}$$
$$+ (\omega_{AB} \mathbf{k}) \times (0.6\cos 30°\mathbf{i} - 0.6 \sin 30°\mathbf{j}) \text{ m}$$
$$v_B \mathbf{j} = [\omega_{AB}(0.6\sin 30°) - 3.6]\mathbf{i} + \omega_{AB}(0.6\cos 30°)\mathbf{j}$$
$$0 = \omega_{AB}(0.6 \sin 30°) - 3.6 \qquad (1)$$
$$v_B = \omega_{AB}(0.6 \cos 30°) \qquad (2)$$
$$\omega_{AB} = 12 \text{ rad/s} \quad v_B = 6.24 \text{ m/s} \uparrow \qquad \text{Ans.}$$
Scalar Analysis
$$\mathbf{v}_B = \mathbf{v}_A + \mathbf{v}_{B/A}$$
$$\left[v_B \uparrow\right] = \left[\underset{\leftarrow}{12(0.3)}\right] + \left[\nwarrow 30° \omega(0.6)\right]$$
This yields Eqs. (1) and (2).

F16–11. Vector Analysis
$$\mathbf{v}_C = \mathbf{v}_B + \boldsymbol{\omega}_{BC} \times \mathbf{r}_{C/B}$$
$$v_C \mathbf{j} = (-60\mathbf{i}) \text{ ft/s}$$
$$+ (-\omega_{BC}\mathbf{k}) \times (-2.5\cos 30°\mathbf{i} + 2.5\sin 30°\mathbf{j}) \text{ ft}$$
$$v_C \mathbf{j} = (-60)\mathbf{i} + 2.165\omega_{BC}\mathbf{j} + 1.25\omega_{BC}\mathbf{i}$$
$$0 = -60 + 1.25\omega_{BC} \qquad (1)$$
$$v_C = 2.165 \omega_{BC} \qquad (2)$$
$$\omega_{BC} = 48 \text{ rad/s} \qquad \text{Ans.}$$
$$v_C = 104 \text{ ft/s} \qquad \text{Ans.}$$

Scalar Analysis

$$\mathbf{v}_C = \mathbf{v}_B + \mathbf{v}_{C/B}$$

$$\left[v_C \uparrow \right] = \left[\begin{matrix} v_B \\ \leftarrow \end{matrix} \right] + \left[\diagdown 30° \; \omega(2.5) \right]$$

This yields Eqs. (1) and (2).

F16–12. Vector Analysis

$$\mathbf{v}_B = \mathbf{v}_A + \boldsymbol{\omega} \times \mathbf{r}_{B/A}$$
$$-v_B \cos 30° \mathbf{i} + v_B \sin 30° \mathbf{j} = (-3 \text{ m/s})\mathbf{j} +$$
$$(-\omega \mathbf{k}) \times (-2 \sin 45° \mathbf{i} - 2 \cos 45° \mathbf{j}) \text{ m}$$
$$-0.8660 v_B \mathbf{i} + 0.5 v_B \mathbf{j}$$
$$= -1.4142 \omega \mathbf{i} + (1.4142 \omega - 3)\mathbf{j}$$
$$-0.8660 v_B = -1.4142 \omega \quad (1)$$
$$0.5 v_B = 1.4142 \omega - 3 \quad (2)$$
$$\omega = 5.02 \text{ rad/s} \quad v_B = 8.20 \text{ m/s} \quad Ans.$$

Scalar Analysis

$$\mathbf{v}_B = \mathbf{v}_A + \mathbf{v}_{B/A}$$

$$\left[\diagdown 30° \; v_B \right] = \left[\downarrow 3 \right] + \left[\diagdown 45° \; \omega(2) \right]$$

This yields Eqs. (1) and (2).

F16–13. $\omega_{AB} = \dfrac{v_A}{r_{A/IC}} = \dfrac{6}{3} = 2 \text{ rad/s} \quad$ Ans.

$r_{C/IC} = \sqrt{(3)^2 + (2.5)^2 - 2(3)(2.5) \cos 53.13°} = 2.5 \text{ m}$

$v_C = \omega_{AB} r_{C/IC} = 2(2.5) = 5 \text{ m/s} \quad$ Ans.

$\phi = \tan^{-1}\left(\dfrac{2}{1.5}\right) = 53.13°$

$\theta = 90° - \phi = 90° - 53.13° = 36.9° \diagdown \quad$ Ans.

F16–14. $v_B = \omega_{AB} r_{B/A} = 12(0.6) = 7.2 \text{ m/s} \downarrow$

$v_C = 0 \quad$ Ans.

$\omega_{BC} = \dfrac{v_B}{r_{B/IC}} = \dfrac{7.2}{1.2} = 6 \text{ rad/s} \quad$ Ans.

F16–15. $\omega = \dfrac{v_O}{r_{O/IC}} = \dfrac{6}{0.3} = 20 \text{ rad/s} \quad$ Ans.

$r_{A/IC} = \sqrt{0.3^2 + 0.6^2} = 0.6708 \text{ m}$

$\phi = \tan^{-1}\left(\dfrac{0.3}{0.6}\right) = 26.57°$

$v_A = \omega r_{A/IC} = 20(0.6708) = 13.4 \text{ m/s} \quad$ Ans.

$\theta = 90° - \phi = 90° - 26.57° = 63.4° \measuredangle \quad$ Ans.

F16–16. The location of IC can be determined using similar triangles.

$$\dfrac{0.5 - r_{C/IC}}{3} = \dfrac{r_{C/IC}}{1.5} \quad r_{C/IC} = 0.1667 \text{ m}$$

$$\omega = \dfrac{v_C}{r_{C/IC}} = \dfrac{1.5}{0.1667} = 9 \text{ rad/s} \quad Ans.$$

Also, $r_{O/IC} = 0.3 - r_{C/IC} = 0.3 - 0.1667$
$= 0.1333 \text{ m}$

$v_O = \omega r_{O/IC} = 9(0.1333) = 1.20 \text{ m/s} \quad$ Ans.

F16–17. $v_B = \omega r_{B/A} = 6(0.2) = 1.2 \text{ m/s}$

$r_{B/IC} = 0.8 \tan 60° = 1.3856 \text{ m}$

$r_{C/IC} = \dfrac{0.8}{\cos 60°} = 1.6 \text{ m}$

$\omega_{BC} = \dfrac{v_B}{r_{B/IC}} = \dfrac{1.2}{1.3856} = 0.8660 \text{ rad/s}$

$= 0.866 \text{ rad/s} \quad$ Ans.

Then,

$v_C = \omega_{BC} r_{C/IC} = 0.8660(1.6) = 1.39 \text{ m/s} \quad$ Ans.

F16–18. $v_B = \omega_{AB} r_{B/A} = 10(0.2) = 2 \text{ m/s}$

$v_C = \omega_{CD} r_{C/D} = \omega_{CD}(0.2) \rightarrow$

$r_{B/IC} = \dfrac{0.4}{\cos 30°} = 0.4619 \text{ m}$

$r_{C/IC} = 0.4 \tan 30° = 0.2309 \text{ m}$

$\omega_{BC} = \dfrac{v_B}{r_{B/IC}} = \dfrac{2}{0.4619} = 4.330 \text{ rad/s}$

$= 4.33 \text{ rad/s} \quad$ Ans.

$v_C = \omega_{BC} r_{C/IC}$

$\omega_{CD}(0.2) = 4.330(0.2309)$

$\omega_{CD} = 5 \text{ rad/s} \quad$ Ans.

F16–19. $\omega = \dfrac{v_A}{r_{A/IC}} = \dfrac{6}{3} = 2 \text{ rad/s}$

Vector Analysis

$\mathbf{a}_B = \mathbf{a}_A + \boldsymbol{\alpha} \times \mathbf{r}_{B/A} - \omega^2 \mathbf{r}_{B/A}$

$a_B \mathbf{i} = -5\mathbf{j} + (\alpha \mathbf{k}) \times (3\mathbf{i} - 4\mathbf{j}) - 2^2(3\mathbf{i} - 4\mathbf{j})$

$a_B \mathbf{i} = (4\alpha - 12)\mathbf{i} + (3\alpha + 11)\mathbf{j}$

$a_B = 4\alpha - 12 \quad (1)$

$0 = 3\alpha + 11 \quad (2)$

$\alpha = -3.67 \text{ rad/s}^2 \quad$ Ans.

$a_B = -26.7 \text{ m/s}^2 \quad$ Ans.

The negative signs indicate decelerations.

Scalar Analysis

$a_B = a_A + a_{B/A}$

$\begin{bmatrix} a_B \\ \rightarrow \end{bmatrix} = \begin{bmatrix} \downarrow 5 \end{bmatrix} + \begin{bmatrix} \alpha(5) \underset{4}{\overset{5\,3}{\nearrow}} \end{bmatrix} + \begin{bmatrix} 4\underset{3}{\overset{5}{\nearrow}}(2)^2(5) \end{bmatrix}$

This yields Eqs. (1) and (2).

F16–20. Vector Analysis

$a_A = a_O + \alpha \times r_{A/O} - \omega^2 r_{A/O}$
$= 1.8i + (-6k) \times (0.3j) - 12^2(0.3j)$
$= \{3.6i - 43.2j\}$ m/s²   Ans.

Scalar Analysis

$a_A = a_O + a_{A/O}$

$\begin{bmatrix} (a_A)_x \\ \rightarrow \end{bmatrix} + \begin{bmatrix} (a_A)_y \uparrow \end{bmatrix} = \begin{bmatrix} (6)(0.3) \\ \rightarrow \end{bmatrix} + \begin{bmatrix} (6)(0.3) \\ \rightarrow \end{bmatrix}$
$\qquad + \begin{bmatrix} \downarrow (12)^2(0.3) \end{bmatrix}$

$\overset{+}{\rightarrow}\quad (a_A)_x = 1.8 + 1.8 = 3.6$ m/s² $\rightarrow$
$+\uparrow \quad (a_A)_y = -43.2$ m/s²

F16–21. Using

$v_O = \omega r; \quad 6 = \omega(0.3)$
$\qquad\qquad \omega = 20$ rad/s
$a_O = \alpha r; \quad 3 = \alpha(0.3)$
$\qquad\qquad \alpha = 10$ rad/s²   Ans.

Vector Analysis

$a_A = a_O + \alpha \times r_{A/O} - \omega^2 r_{A/O}$
$= 3i + (-10k) \times (-0.6i) - 20^2(-0.6i)$
$= \{243i + 6j\}$ m/s²   Ans.

Scalar Analysis

$a_A = a_O + a_{A/O}$

$\begin{bmatrix} (a_A)_x \\ \rightarrow \end{bmatrix} + \begin{bmatrix} (a_A)_y \\ \uparrow \end{bmatrix} = \begin{bmatrix} 3 \\ \rightarrow \end{bmatrix} + \begin{bmatrix} 10(0.6) \\ \uparrow \end{bmatrix} + \begin{bmatrix} (20)^2(0.6) \\ \rightarrow \end{bmatrix}$

$\overset{+}{\rightarrow}\quad (a_A)_x = 3 + 240 = 243$ m/s²
$+\uparrow \quad (a_A)_y = 10(0.6) = 6$ m/s² $\uparrow$

F16–22. $\dfrac{r_{A/IC}}{3} = \dfrac{0.5 - r_{A/IC}}{1.5}$; $\quad r_{A/IC} = 0.3333$ m

$\omega = \dfrac{v_A}{r_{A/IC}} = \dfrac{3}{0.3333} = 9$ rad/s

Vector Analysis

$a_A = a_C + \alpha \times r_{A/C} - \omega^2 r_{A/C}$
$1.5i - (a_A)_n j = -0.75i + (a_C)_n j$
$\qquad\qquad + (-\alpha k) \times 0.5j - 9^2(0.5j)$
$1.5i - (a_A)_n j = (0.5\alpha - 0.75)i + [(a_C)_n - 40.5]j$
$1.5 = 0.5\alpha - 0.75$
$\alpha = 4.5$ rad/s²   Ans.

Scalar Analysis

$a_A = a_C + a_{A/C}$

$\begin{bmatrix} 1.5 \\ \rightarrow \end{bmatrix} + \begin{bmatrix} (a_A)_n \\ \downarrow \end{bmatrix} = \begin{bmatrix} 0.75 \\ \leftarrow \end{bmatrix} + \begin{bmatrix} (a_C)_n \\ \uparrow \end{bmatrix} + \begin{bmatrix} \alpha(0.5) \\ \rightarrow \end{bmatrix}$
$\qquad\qquad\qquad\qquad\qquad + \begin{bmatrix} (9)^2(0.5) \\ \downarrow \end{bmatrix}$

$\overset{+}{\rightarrow}\quad 1.5 = -0.75 + \alpha(0.5)$
$\alpha = 4.5$ rad/s²

F16–23. $v_B = \omega r_{B/A} = 12(0.3) = 3.6$ m/s

The IC is at C.

$\omega_{BC} = \dfrac{v_B}{r_{B/IC}} = \dfrac{3.6}{1.2} = 3$ rad/s, $\omega_{CD} = 0$

Vector Analysis

$a_B = \alpha \times r_{B/A} - \omega^2 r_{B/A}$
$= (-6k) \times (0.3i) - 12^2(0.3i)$
$= \{-43.2i - 1.8j\}$ m/s²

$a_C = a_B + \alpha_{BC} \times r_{C/B} - \omega_{BC}^2 r_{C/B}$
$a_C i = (-43.2i - 1.8j)$
$\qquad + (\alpha_{BC} k) \times (1.2i) - 3^2(1.2i)$
$a_C i = -54i + (1.2\alpha_{BC} - 1.8)j$
$a_C = -54$ m/s² = 54 m/s² $\leftarrow$ Ans.
$0 = 1.2\alpha_{BC} - 1.8 \quad \alpha_{BC} = 1.5$ rad/s²   Ans.

Scalar Analysis

$a_C = a_B + a_{C/B}$

$\begin{bmatrix} a_C \\ \leftarrow \end{bmatrix} = \begin{bmatrix} 6(0.3) \\ \downarrow \end{bmatrix} + \begin{bmatrix} (12)^2(0.3) \\ \leftarrow \end{bmatrix} + \begin{bmatrix} \alpha_{BC}(1.2) \\ \uparrow \end{bmatrix} + \begin{bmatrix} (3)^2(1.2) \\ \leftarrow \end{bmatrix}$

$\overset{+}{\leftarrow}\quad a_C = 43.2 + 10.8 = 54$ m/s² $\leftarrow$
$+\uparrow \quad 0 = -6(0.3) + 1.2\alpha_{BC}$
$\alpha_{BC} = 1.5$ rad/s²

F16–24. $v_B = \omega\, r_{B/A} = 6(0.2) = 1.2$ m/s $\rightarrow$

$r_{B/IC} = 0.8 \tan 60° = 1.3856$ m

$$\omega_{BC} = \frac{v_B}{r_{B/IC}} = \frac{1.2}{1.3856} = 0.8660 \text{ rad/s}$$

Vector Analysis

$\mathbf{a}_B = \boldsymbol{\alpha} \times \mathbf{r}_{B/A} - \omega^2 \mathbf{r}_{B/A}$

$\quad = (-3\mathbf{k}) \times (0.2\mathbf{j}) - 6^2(0.2\mathbf{j})$

$\quad = (0.6\mathbf{i} - 7.2\mathbf{j})$ m/s

$\mathbf{a}_C = \mathbf{a}_B + \boldsymbol{\alpha}_{BC} \times \mathbf{r}_{C/B} - \omega^2 \mathbf{r}_{C/B}$

$a_C \cos 30°\mathbf{i} + a_C \sin 30°\mathbf{j}$

$\quad = (0.6\mathbf{i} - 7.2\mathbf{j}) + (\alpha_{BC}\mathbf{k} \times 0.8\mathbf{i}) - 0.8660^2(0.8\mathbf{i})$

$0.8660 a_C\, \mathbf{i} + 0.5 a_C\, \mathbf{j} = (0.8\alpha_{BC} - 7.2)\mathbf{j}$

$0.8660 a_C = 0$ (1)

$0.5 a_C = 0.8\alpha_{BC} - 7.2$ (2)

$a_C = 0 \qquad \alpha_{BC} = 9 \text{ rad/s}^2$ *Ans.*

Scalar Analysis

$\mathbf{a}_C = \mathbf{a}_B + \mathbf{a}_{C/B}$

$\begin{bmatrix} a_C \\ \angle 30° \end{bmatrix} = \begin{bmatrix} 3(0.2) \\ \rightarrow \end{bmatrix} + \begin{bmatrix} (6)^2(0.2) \\ \downarrow \end{bmatrix} + \begin{bmatrix} \alpha_{BC}(0.8) \\ \uparrow \end{bmatrix}$

$\qquad\qquad + \begin{bmatrix} (0.8660)^2(0.8) \\ \leftarrow \end{bmatrix}$

This yields Eqs. (1) and (2).

Chapter 17

F17–1. $\xrightarrow{+} \Sigma F_x = m(a_G)_x; \quad 100\left(\frac{4}{5}\right) = 100a$

$a = 0.8$ m/s^2 $\rightarrow$ *Ans.*

$+\uparrow \Sigma F_y = m(a_G)_y;$

$N_A + N_B - 100\left(\frac{3}{5}\right) - 100(9.81) = 0$ (1)

$\zeta + \Sigma M_G = 0;$

$N_A(0.6) + 100\left(\frac{3}{5}\right)(0.7)$

$\qquad - N_B(0.4) - 100\left(\frac{4}{5}\right)(0.7) = 0$ (2)

$N_A = 430.4$ N $= 430$ N *Ans.*

$N_B = 610.6$ N $= 611$ N *Ans.*

F17–2. $+\swarrow \Sigma F_{x'} = m(a_G)_{x'}; \quad 80(9.81) \sin 15° = 80a$

$a = 2.54$ m/s^2 *Ans.*

$+\nwarrow \Sigma F_{y'} = m(a_G)_{y'};$

$N_A + N_B - 80(9.81) \cos 15° = 0$ (1)

$\zeta + \Sigma M_G = 0;$

$N_A(0.5) - N_B(0.5) = 0$ (2)

$N_A = N_B = 379$ N *Ans.*

F17–3. $\zeta + \Sigma M_A = \Sigma(\mathcal{M}_k)_A; \quad 10\left(\frac{3}{5}\right)(7) = \frac{20}{32.2}a(3.5)$

$a = 19.3$ ft/s^2 *Ans.*

$\xrightarrow{+} \Sigma F_x = m(a_G)_x; \quad A_x + 10\left(\frac{3}{5}\right) = \frac{20}{32.2}(19.32)$

$A_x = 6$ lb *Ans.*

$+\uparrow \Sigma F_y = m(a_G)_y; \quad A_y - 20 + 10\left(\frac{4}{5}\right) = 0$

$A_y = 12$ lb *Ans.*

F17–4. $F_A = \mu_s N_A = 0.2 N_A \quad F_B = \mu_s N_B = 0.2 N_B$

$\xrightarrow{+} \Sigma F_x = m(a_G)_x;$

$0.2 N_A + 0.2 N_B = 100a$ (1)

$+\uparrow \Sigma F_y = m(a_G)_y;$

$N_A + N_B - 100(9.81) = 0$ (2)

$\zeta + \Sigma M_G = 0;$

$0.2 N_A(0.75) + N_A(0.9) + 0.2 N_B(0.75)$

$\qquad - N_B(0.6) = 0$ (3)

Solving Eqs. (1), (2), and (3),

$N_A = 294.3$ N $= 294$ N *Ans.*

$N_B = 686.7$ N $= 687$ N *Ans.*

$a = 1.96$ m/s^2 *Ans.*

Since N_A is positive, the table will indeed slide before it tips.

F17–5. $(a_G)_t = \alpha r = \alpha(1.5 \text{ m})$

$(a_G)_n = \omega^2 r = (5 \text{ rad/s})^2(1.5 \text{ m}) = 37.5 \text{ m/s}^2$

$\Sigma F_t = m(a_G)_t; \quad 100 \text{ N} = 50 \text{ kg}[\alpha(1.5 \text{ m})]$

$\qquad\qquad \alpha = 1.33 \text{ rad/s}^2$ *Ans.*

$\Sigma F_n = m(a_G)_n; \quad T_{AB} + T_{CD} - 50(9.81)$ N

$\qquad\qquad = 50 \text{ kg}(37.5 \text{ m/s}^2)$

$T_{AB} + T_{CD} = 2365.5$

$\zeta + \Sigma M_G = 0; \quad T_{CD}(1 \text{ m}) - T_{AB}(1 \text{ m}) = 0$

$T_{AB} = T_{CD} = 1182.75$ N $= 1.18$ kN *Ans.*

F17–6. Member CD,

$\zeta + \Sigma M_C = 0;$

$D_y(0.6) - 450 = 0 \quad D_y = 750$ N *Ans.*

$\mathbf{a}_G = \mathbf{a}_D = \mathbf{a}_B$

$(a_G)_n = \omega^2 r = 6^2(0.6) = 21.6 \text{ m/s}^2$

$(a_G)_t = \alpha r = \alpha(0.6)$

Member DB

$+\uparrow \Sigma F_t = m(a_G)_t;$

$750 - 50(9.81) = 50[\alpha(0.6)]$

$\alpha = 8.65 \text{ rad/s}^2$ Ans.

$\xrightarrow{+} \Sigma F_n = m(a_G)_n;$

$F_{AB} + D_x = 50(21.6)$ (1)

$\zeta + \Sigma M_G = 0;$

$D_x(0.4) + 750(0.1) - F_{AB}(0.4) = 0$ (2)

$D_x = 446.25 \text{ N} = 446 \text{ N}$ Ans.

$F_{AB} = 633.75 \text{ N} = 634 \text{ N}$ Ans.

F17–7. $I_O = mk_O^2 = 100(0.5^2) = 25 \text{ kg} \cdot \text{m}^2$

$\zeta + \Sigma M_O = I_O \alpha;$ $-100(0.6) = -25\alpha$

$\alpha = 2.4 \text{ rad/s}^2$

$\omega = \omega_0 + \alpha_c t$

$\omega = 0 + 2.4(3) = 7.2 \text{ rad/s}$ Ans.

F17–8. $I_O = \frac{1}{2} mr^2 = \frac{1}{2}(50)(0.3^2) = 2.25 \text{ kg} \cdot \text{m}^2$

$\zeta + \Sigma M_O = I_O \alpha;$

$-9t = -2.25\alpha$ $\alpha = (4t) \text{ rad/s}^2$

$d\omega = \alpha \, dt$

$\int_0^\omega d\omega = \int_0^t 4t \, dt$

$\omega = (2t^2) \text{ rad/s}$

$\omega = 2(4^2) = 32 \text{ rad/s}$ Ans.

F17–9. $(a_G)_t = \alpha r_G = \alpha(0.15)$

$(a_G)_n = \omega^2 r_G = 6^2(0.15) = 5.4 \text{ m/s}^2$

$I_O = I_G + md^2 = \frac{1}{12}(30)(0.9^2) + 30(0.15^2)$

$= 2.7 \text{ kg} \cdot \text{m}^2$

$\zeta + \Sigma M_O = I_O \alpha;$ $60 - 30(9.81)(0.15) = 2.7\alpha$

$\alpha = 5.872 \text{ rad/s}^2 = 5.87 \text{ rad/s}^2$ Ans.

$\xrightarrow{+} \Sigma F_n = m(a_G)_n;$ $O_n = 30(5.4) = 162 \text{ N}$ Ans.

$+\uparrow \Sigma F_t = m(a_G)_t;$

$O_t - 30(9.81) = 30[5.872(0.15)]$

$O_t = 320.725 \text{ N} = 321 \text{ N}$ Ans.

F17–10. $(a_G)_t = \alpha r_G = \alpha(0.3)$

$(a_G)_n = \omega^2 r_G = 10^2(0.3) = 30 \text{ m/s}^2$

$I_O = I_G + md^2 = \frac{1}{2}(30)(0.3^2) + 30(0.3^2)$

$= 4.05 \text{ kg} \cdot \text{m}^2$

$\zeta + \Sigma M_O = I_O \alpha;$

$50(\frac{3}{5})(0.3) + 50(\frac{4}{5})(0.3) = 4.05\alpha$

$\alpha = 5.185 \text{ rad/s}^2 = 5.19 \text{ rad/s}^2$ Ans.

$+\uparrow \Sigma F_n = m(a_G)_n;$

$O_n + 50(\frac{3}{5}) - 30(9.81) = 30(30)$

$O_n = 1164.3 \text{ N} = 1.16 \text{ kN}$ Ans.

$\xrightarrow{+} \Sigma F_t = m(a_G)_t;$

$O_t + 50(\frac{4}{5}) = 30[5.185(0.3)]$

$O_t = 6.67 \text{ N}$ Ans.

F17–11. $I_G = \frac{1}{12} ml^2 = \frac{1}{12}(15 \text{ kg})(0.9 \text{ m})^2 = 1.0125 \text{ kg} \cdot \text{m}^2$

$(a_G)_n = \omega^2 r_G = 0$

$(a_G)_t = \alpha(0.15 \text{ m})$

$I_O = I_G + md_{OG}^2$

$= 1.0125 \text{ kg} \cdot \text{m}^2 + 15 \text{ kg}(0.15 \text{ m})^2$

$= 1.35 \text{ kg} \cdot \text{m}^2$

$\zeta + \Sigma M_O = I_O \alpha;$

$[15(9.81) \text{ N}](0.15 \text{ m}) = (1.35 \text{ kg} \cdot \text{m}^2)\alpha$

$\alpha = 16.35 \text{ rad/s}^2$ Ans.

$+\downarrow \Sigma F_t = m(a_G)_t;$ $-O_t + 15(9.81) \text{ N}$

$= (15 \text{ kg})[16.35 \text{ rad/s}^2(0.15 \text{ m})]$

$O_t = 110.36 \text{ N} = 110 \text{ N}$ Ans.

$\xrightarrow{+} \Sigma F_n = m(a_G)_n;$ $O_n = 0$ Ans.

F17–12. $(a_G)_t = \alpha r_G = \alpha(0.45)$

$(a_G)_n = \omega^2 r_G = 6^2(0.45) = 16.2 \text{ m/s}^2$

$I_O = \frac{1}{3} ml^2 = \frac{1}{3}(30)(0.9^2) = 8.1 \text{ kg} \cdot \text{m}^2$

$\zeta + \Sigma M_O = I_O \alpha;$

$300(\frac{4}{5})(0.6) - 30(9.81)(0.45) = 8.1\alpha$

$\alpha = 1.428 \text{ rad/s}^2 = 1.43 \text{ rad/s}^2$ Ans.

$\xrightarrow{+} \Sigma F_n = m(a_G)_n;$ $O_n + 300(\frac{3}{5}) = 30(16.2)$

$O_n = 306 \text{ N}$ Ans.

$+\uparrow \Sigma F_t = m(a_G)_t;$ $O_t + 300(\frac{4}{5}) - 30(9.81)$

$= 30[1.428(0.45)]$

$O_t = 73.58 \text{ N} = 73.6 \text{ N}$ Ans.

F17–13. $I_G = \frac{1}{12}ml^2 = \frac{1}{12}(60)(3^2) = 45 \text{ kg} \cdot \text{m}^2$
$+\uparrow \Sigma F_y = m(a_G)_y;\quad 80 - 20 = 60a_G$
$a_G = 1 \text{ m/s}^2 \uparrow$
$\zeta + \Sigma M_G = I_G \alpha;\quad 80(1) + 20(0.75) = 45\alpha$
$\alpha = 2.11 \text{ rad/s}^2$ *Ans.*

F17–14. $\zeta + \Sigma M_{IC} = (\mathcal{M}_k)_{IC};$
$-200(0.3) = -100a_G(0.3) - \frac{1}{2}(100)(0.3)^2\alpha$
$30a_G + 4.5\alpha = 60$ (1)
$a_G = \alpha r = \alpha(0.3)$ (2)
$\alpha = 4.44 \text{ rad/s}^2 \quad a_G = 1.33 \text{ m/s}^2 \rightarrow$ *Ans.*
$I_G = 20(0.3)^2 = 1.8 \text{ kg} \cdot \text{m}^2$

F17–15. $I_G = 20(0.3)^2 = 1.8 \text{ kg} \cdot \text{m}^2$
$+\uparrow \Sigma F_y = m(a_G)_y;$
$N - 20(9.81) = 0 \quad N = 196.2 \text{ N}$
$\xrightarrow{+} \Sigma F_x = m(a_G)_x;\quad 0.5(196.2) = 20a_O$
$a_O = 4.905 \text{ m/s}^2 \rightarrow$ *Ans.*
$\zeta + \Sigma M_O = I_O\alpha;$
$0.5(196.2)(0.4) - 100 = -1.8\alpha$
$\alpha = 33.8 \text{ rad/s}^2$ *Ans.*

F17–16. Sphere $I_G = \frac{2}{5}(20)(0.15)^2 = 0.18 \text{ kg} \cdot \text{m}^2$
$\zeta + \Sigma M_{IC} = (\mathcal{M}_k)_{IC};$
$20(9.81)\sin 30°(0.15) = 0.18\alpha + (20a_G)(0.15)$
$0.18\alpha + 3a_G = 14.715$
$a_G = \alpha r = \alpha(0.15)$
$\alpha = 23.36 \text{ rad/s}^2 = 23.4 \text{ rad/s}^2$ *Ans.*
$a_G = 3.504 \text{ m/s}^2 = 3.50 \text{ m/s}^2$ *Ans.*

F17–17. $I_G = 200(0.3)^2 = 18 \text{ kg} \cdot \text{m}^2$
$+\uparrow \Sigma F_y = m(a_G)_y;$
$N - 200(9.81) = 0 \quad N = 1962 \text{ N}$
$\xrightarrow{+} \Sigma F_x = m(a_G)_x;$
$T - 0.2(1962) = 200a_G$ (1)
The IC is at A.
$\zeta + \Sigma M_A = (\mathcal{M}_k)_A;\quad 450 - 0.2(1962)(1)$
$= 18\alpha + 200a_G(0.4)$ (2)
$(a_A)_t = 0 \quad a_A = (a_A)_n$
$\mathbf{a}_G = \mathbf{a}_A + \boldsymbol{\alpha} \times \mathbf{r}_{G/A} - \omega^2 \mathbf{r}_{G/A}$

$a_G \mathbf{i} = -a_A \mathbf{j} + \alpha \mathbf{k} \times (-0.4\mathbf{j}) - \omega^2(-0.4\mathbf{j})$
$a_G \mathbf{i} = 0.4\alpha \mathbf{i} + (0.4\omega^2 - a_A)\mathbf{j}$
$a_G = 0.4\alpha$ (3)
Solving Eqs. (1), (2), and (3),
$\alpha = 1.15 \text{ rad/s}^2 \quad a_G = 0.461 \text{ m/s}^2$
$T = 485 \text{ N}$ *Ans.*

F17–18. $\xrightarrow{+} \Sigma F_x = m(a_G)_x;\quad 0 = 12(a_G)_x \quad (a_G)_x = 0$
$\zeta + \Sigma M_A = (\mathcal{M}_k)_A;$
$-12(9.81)(0.3) = 12(a_G)_y(0.3) - \frac{1}{12}(12)(0.6)^2\alpha$
$0.36\alpha - 3.6(a_G)_y = 35.316$ (1)
$\omega = 0$
$\mathbf{a}_G = \mathbf{a}_A + \boldsymbol{\alpha} \times \mathbf{r}_{G/A} - \omega^2 \mathbf{r}_{G/A}$
$(a_G)_y \mathbf{j} = a_A \mathbf{i} + (-\alpha \mathbf{k}) \times (0.3\mathbf{i}) - 0$
$(a_G)_y \mathbf{j} = (a_A)\mathbf{i} - 0.3\alpha \mathbf{j}$
$a_A = 0$ *Ans.*
$(a_G)_y = -0.3\alpha$ (2)
Solving Eqs. (1) and (2)
$\alpha = 24.5 \text{ rad/s}^2$
$(a_G)_y = -7.36 \text{ m/s}^2 = 7.36 \text{ m/s}^2 \downarrow$ *Ans.*

Chapter 18

F18–1. $I_O = mk_O^2 = 80(0.4^2) = 12.8 \text{ kg} \cdot \text{m}^2$
$T_1 = 0$
$T_2 = \frac{1}{2}I_O\omega^2 = \frac{1}{2}(12.8)\omega^2 = 6.4\omega^2$
$s = \theta r = 20(2\pi)(0.6) = 24\pi \text{ m}$
$T_1 + \Sigma U_{1-2} = T_2$
$0 + 50(24\pi) = 6.4\omega^2$
$\omega = 24.3 \text{ rad/s}$ *Ans.*

F18–2. $T_1 = 0$
$T_2 = \frac{1}{2}m(v_G)_2^2 + \frac{1}{2}I_G\omega_2^2$
$= \frac{1}{2}\left(\frac{50}{32.2} \text{ slug}\right)(2.5\omega_2)^2$
$+ \frac{1}{2}\left[\frac{1}{12}\left(\frac{50}{32.2} \text{ slug}\right)(5 \text{ ft})^2\right]\omega_2^2$
$T_2 = 6.4700\omega_2^2$
Or,
$I_O = \frac{1}{3}ml^2 = \frac{1}{3}\left(\frac{50}{32.2} \text{ slug}\right)(5 \text{ ft})^2$
$= 12.9400 \text{ slug} \cdot \text{ft}^2$

688   SOLUTIONS AND ANSWERS

So that
$$T_2 = \tfrac{1}{2} I_O \omega_2^2 = \tfrac{1}{2}(12.9400 \text{ slug} \cdot \text{ft}^2)\omega_2^2$$
$$= 6.4700\omega_2^2$$
$$T_1 + \Sigma U_{1-2} = T_2$$
$$T_1 + [-Wy_G + M\theta] = T_2$$
$$0 + [-(50 \text{ lb})(2.5 \text{ ft}) + (100 \text{ lb} \cdot \text{ft})(\tfrac{\pi}{2})]$$
$$= 6.4700\omega_2^2$$
$$\omega_2 = 2.23 \text{ rad/s} \qquad \textit{Ans.}$$

F18–3. $(v_G)_2 = \omega_2 r_{G/IC} = \omega_2(2.5)$
$$I_G = \tfrac{1}{12} m l^2 = \tfrac{1}{12}(50)(5^2) = 104.17 \text{ kg} \cdot \text{m}^2$$
$$T_1 = 0$$
$$T_2 = \tfrac{1}{2} m(v_G)_2^2 + \tfrac{1}{2} I_G \omega_2^2$$
$$= \tfrac{1}{2}(50)[\omega_2(2.5)]^2 + \tfrac{1}{2}(104.17)\omega_2^2 = 208.33\omega_2^2$$
$$U_P = P s_P = 600(3) = 1800 \text{ J}$$
$$U_W = -Wh = -50(9.81)(2.5 - 2) = -245.25 \text{ J}$$
$$T_1 + \Sigma U_{1-2} = T_2$$
$$0 + 1800 + (-245.25) = 208.33\omega_2^2$$
$$\omega_2 = 2.732 \text{ rad/s} = 2.73 \text{ rad/s} \qquad \textit{Ans.}$$

F18–4. $T = \tfrac{1}{2} m v_G^2 + \tfrac{1}{2} I_G \omega^2$
$$= \tfrac{1}{2}(50 \text{ kg})(0.4\omega)^2 + \tfrac{1}{2}[50 \text{ kg}(0.3 \text{ m})^2]\omega^2$$
$$= 6.25\omega^2 \text{ J}$$
Or,
$$T = \tfrac{1}{2} I_{IC} \omega^2$$
$$= \tfrac{1}{2}[50 \text{ kg}(0.3 \text{ m})^2 + 50 \text{ kg}(0.4 \text{ m})^2]\omega^2$$
$$= 6.25\omega^2 \text{ J}$$
$$s_G = \theta r = 10(2\pi \text{ rad})(0.4 \text{ m}) = 8\pi \text{ m}$$
$$T_1 + \Sigma U_{1-2} = T_2$$
$$T_1 + P \cos 30° s_G = T_2$$
$$0 + (50 \text{ N}) \cos 30°(8\pi \text{ m}) = 6.25\omega^2 \text{ J}$$
$$\omega = 13.2 \text{ rad/s} \qquad \textit{Ans.}$$

F18–5. $I_G = \tfrac{1}{12} m l^2 = \tfrac{1}{12}(30)(3^2) = 22.5 \text{ kg} \cdot \text{m}^2$
$$T_1 = 0$$
$$T_2 = \tfrac{1}{2} m v_G^2 + \tfrac{1}{2} I_G \omega^2$$
$$= \tfrac{1}{2}(30)[\omega(0.5)]^2 + \tfrac{1}{2}(22.5)\omega^2 = 15\omega^2$$
Or,
$$I_O = I_G + md^2 = \tfrac{1}{12}(30)(3^2) + 30(0.5^2)$$
$$= 30 \text{ kg} \cdot \text{m}^2$$
$$T_2 = \tfrac{1}{2} I_O \omega^2 = \tfrac{1}{2}(30)\omega^2 = 15\omega^2$$

$$s_1 = \theta r_1 = 8\pi(0.5) = 4\pi \text{ m}$$
$$s_2 = \theta r_2 = 8\pi(1.5) = 12\pi \text{ m}$$
$$U_{P_1} = P_1 s_1 = 30(4\pi) = 120\pi \text{ J}$$
$$U_{P_2} = P_2 s_2 = 20(12\pi) = 240\pi \text{ J}$$
$$U_M = M\theta = 20[4(2\pi)] = 160\pi \text{ J}$$
$$U_W = 0 \text{ since bar returns to the same position}$$
$$T_1 + \Sigma U_{1-2} = T_2$$
$$0 + 120\pi + 240\pi + 160\pi = 15\omega^2$$
$$\omega = 10.44 \text{ rad/s} = 10.4 \text{ rad/s} \qquad \textit{Ans.}$$

F18–6. $v_G = \omega r = \omega(0.4)$
$$I_G = m k_G^2 = 20(0.3^2) = 1.8 \text{ kg} \cdot \text{m}^2$$
$$T_1 = 0$$
$$T_2 = \tfrac{1}{2} m v_G^2 + \tfrac{1}{2} I_G \omega^2$$
$$= \tfrac{1}{2}(20)[\omega(0.4)]^2 + \tfrac{1}{2}(1.8)\omega^2$$
$$= 2.5\omega^2$$
$$U_M = M\theta = M\left(\tfrac{s_O}{r}\right) = 50\left(\tfrac{20}{0.4}\right) = 2500 \text{ J}$$
$$T_1 + \Sigma U_{1-2} = T_2$$
$$0 + 2500 = 2.5\omega^2$$
$$\omega = 31.62 \text{ rad/s} = 31.6 \text{ rad/s} \qquad \textit{Ans.}$$

F18–7. $v_G = \omega r = \omega(0.3)$
$$I_G = \tfrac{1}{2} m r^2 = \tfrac{1}{2}(30)(0.3^2) = 1.35 \text{ kg} \cdot \text{m}^2$$
$$T_1 = 0$$
$$T_2 = \tfrac{1}{2} m(v_G)_2^2 + \tfrac{1}{2} I_G \omega_2^2$$
$$= \tfrac{1}{2}(30)[\omega_2(0.3)]^2 + \tfrac{1}{2}(1.35)\omega_2^2 = 2.025\omega_2^2$$
Datum at O.
$$(V_g)_1 = W y_1 = 0$$
$$(V_g)_2 = -W y_2 = -30(9.81)(0.3) = -88.29 \text{ J}$$
$$T_1 + V_1 = T_2 + V_2$$
$$0 + 0 = 2.025\omega_2^2 + (-88.29)$$
$$\omega_2 = 6.603 \text{ rad/s} = 6.60 \text{ rad/s} \qquad \textit{Ans.}$$

F18–8. $v_O = \omega r_{O/IC} = \omega(0.2)$
$$I_O = m k_O^2 = 50(0.3^2) = 4.5 \text{ kg} \cdot \text{m}^2$$
$$T_1 = 0$$
$$T_2 = \tfrac{1}{2} m(v_O)_2^2 + \tfrac{1}{2} I_O \omega_2^2$$
$$= \tfrac{1}{2}(50)[\omega_2(0.2)]^2 + \tfrac{1}{2}(4.5)\omega_2^2$$
$$= 3.25\omega_2^2$$
Datum at initial position.
$$(V_g)_1 = W y_1 = 0$$
$$(V_g)_2 = -W y_2 = -50(9.81)(6 \sin 30°)$$
$$= -1471.5 \text{ J}$$

$T_1 + V_1 = T_2 + V_2$

$0 + 0 = 3.25\omega_2^2 + (-1471.5)$

$\omega_2 = 21.28 \text{ rad/s} = 21.3 \text{ rad/s}$ **Ans.**

F18–9. $v_G = \omega r_G = \omega(1.5)$

$I_G = \frac{1}{12}(60)(3^2) = 45 \text{ kg} \cdot \text{m}^2$

$T_1 = 0$

$T_2 = \frac{1}{2}m(v_G)_2^2 + \frac{1}{2}I_G\omega_2^2$

$= \frac{1}{2}(60)[\omega_2(1.5)]^2 + \frac{1}{2}(45)\omega_2^2$

$= 90\omega_2^2$

Or,

$T_2 = \frac{1}{2}I_O\omega_2^2 = \frac{1}{2}[45 + 60(1.5^2)]\omega_2^2 = 90\omega_2^2$

$(V_g)_1 = Wy_1 = 0$

$(V_g)_2 = -Wy_2 = -60(9.81)(1.5 \sin 45°)$

$= -624.30 \text{ J}$

$(V_e)_1 = \frac{1}{2}ks_1^2 = 0$

$(V_e)_2 = \frac{1}{2}ks_2^2 = \frac{1}{2}(150)(3 \sin 45°)^2 = 337.5 \text{ J}$

$T_1 + V_1 = T_2 + V_2$

$0 + 0 = 90\omega_2^2 + [-624.30 + 337.5]$

$\omega_2 = 1.785 \text{ rad/s} = 1.79 \text{ rad/s}$ **Ans.**

F18–10. $v_G = \omega r_G = \omega(0.75)$

$I_G = \frac{1}{12}(30)(1.5^2) = 5.625 \text{ kg} \cdot \text{m}^2$

$T_1 = 0$

$T_2 = \frac{1}{2}m(v_G)_2^2 + \frac{1}{2}I_G\omega_2^2$

$= \frac{1}{2}(30)[\omega(0.75)]^2 + \frac{1}{2}(5.625)\omega_2^2 = 11.25\omega_2^2$

Or,

$T_2 = \frac{1}{2}I_O\omega_2^2 = \frac{1}{2}[5.625 + 30(0.75^2)]\omega_2^2$

$= 11.25\omega_2^2$

Datum at O.

$(V_g)_1 = Wy_1 = 0$

$(V_g)_2 = -Wy_2 = -30(9.81)(0.75)$

$= -220.725 \text{ J}$

$(V_e)_1 = \frac{1}{2}ks_1^2 = 0$

$(V_e)_2 = \frac{1}{2}ks_2^2 = \frac{1}{2}(80)(\sqrt{2^2 + 1.5^2} - 0.5)^2 = 160 \text{ J}$

$T_1 + V_1 = T_2 + V_2$

$0 + 0 = 11.25\omega_2^2 + (-220.725 + 160)$

$\omega_2 = 2.323 \text{ rad/s} = 2.32 \text{ rad/s}$ **Ans.**

F18–11. $(v_G)_2 = \omega_2 r_{G/IC} = \omega_2(0.75)$

$I_G = \frac{1}{12}(30)(1.5^2) = 5.625 \text{ kg} \cdot \text{m}^2$

$T_1 = 0$

$T_2 = \frac{1}{2}m(v_G)_2^2 + \frac{1}{2}I_G\omega_2^2$

$= \frac{1}{2}(30)[\omega_2(0.75)]^2 + \frac{1}{2}(5.625)\omega_2^2 = 11.25\omega_2^2$

Datum at B.

$(V_g)_1 = Wy_1 = 30(9.81)(0.75 \sin 45°) = 156.08 \text{ J}$

$(V_g)_2 = -Wy_2 = 0$

$(V_e)_1 = \frac{1}{2}ks_1^2 = 0$

$(V_e)_2 = \frac{1}{2}ks_2^2 = \frac{1}{2}(300)(1.5 - 1.5 \cos 45°)^2$

$= 28.95 \text{ J}$

$T_1 + V_1 = T_2 + V_2$

$0 + (156.08 + 0) = 11.25\omega_2^2 + (0 + 28.95)$

$\omega_2 = 3.362 \text{ rad/s} = 3.36 \text{ rad/s}$ **Ans.**

F18–12. $T_1 = 0$

$T_2 = \frac{1}{2}I_A\omega^2 = \frac{1}{2}\left[\frac{1}{3}(20 \text{ kg})(2 \text{ m})^2\right]\omega^2$

$= 13.3333\omega^2$

Datum at A.

$(V_g)_1 = -Wy_1 = -[20(9.81) \text{ N}](1 \text{ m}) = -196.2 \text{ J}$

$(V_g)_2 = 0$

$(V_e)_1 = \frac{1}{2}ks_1^2$

$= \frac{1}{2}(100 \text{ N/m})\left(\sqrt{(3 \text{ m})^2 + (2 \text{ m})^2} - 0.5 \text{ m}\right)^2$

$= 482.22 \text{ J}$

$(V_e)_2 = \frac{1}{2}ks_2^2 = \frac{1}{2}(100 \text{ N/m})(1 \text{ m} - 0.5 \text{ m})^2$

$= 12.5 \text{ J}$

$T_1 + V_1 = T_2 + V_2$

$0 + [-196.2 \text{ J} + 482.22 \text{ J}]$

$= 13.3333\omega_2^2 + [0 + 12.5 \text{ J}]$

$\omega_2 = 4.53 \text{ rad/s}$ **Ans.**

Chapter 19

F19–1. $\zeta + I_O\omega_1 + \Sigma \int_{t_1}^{t_2} M_O \, dt = I_O\omega_2$

$0 + \int_0^{4 \text{ s}} 3t^2 \, dt = [60(0.3)^2]\omega_2$

$\omega_2 = 11.85 \text{ rad/s} = 11.9 \text{ rad/s}$ **Ans.**

F19–2. $\zeta + (H_A)_1 + \Sigma \int_{t_1}^{t_2} M_A dt = (H_A)_2$

$0 + 300(6) = 300(0.4^2)\omega_2 + 300[\omega(0.6)](0.6)$

$\omega_2 = 11.54 \text{ rad/s} = 11.5 \text{ rad/s}$ **Ans.**

$\xrightarrow{+} \quad m(v_1)_x + \Sigma \int_{t_1}^{t_2} F_x dt = m(v_2)_x$

$$0 + F_f(6) = 300[11.54(0.6)]$$
$$F_f = 346 \text{ N} \qquad \text{Ans.}$$

F19–3. $v_A = \omega_A r_{A/IC} = \omega_A(0.15)$

$\zeta + \Sigma M_O = 0;\quad 9 - A_t(0.45) = 0 \quad A_t = 20 \text{ N}$

$$\zeta + (H_C)_1 + \Sigma \int_{t_1}^{t_2} M_C \, dt = (H_C)_2$$
$$0 + [20(5)](0.15)$$
$$= 10[\omega_A(0.15)](0.15)$$
$$+ \left[10(0.1^2)\right]\omega_A$$
$$\omega_A = 46.2 \text{ rad/s} \qquad \text{Ans.}$$

F19–4. $I_A = mk_A^2 = 10(0.08^2) = 0.064 \text{ kg} \cdot \text{m}^2$
$I_B = mk_B^2 = 50(0.15^2) = 1.125 \text{ kg} \cdot \text{m}^2$

$\omega_A r_A = \omega_B r_B$

$\omega_A = \left(\dfrac{r_B}{r_A}\right)\omega_B = \left(\dfrac{0.2}{0.1}\right)\omega_B = 2\omega_B$

F is the force between the gears.

$$\zeta + I_A(\omega_A)_1 + \Sigma \int_{t_1}^{t_2} M_A \, dt = I_A(\omega_A)_2$$

$$0 + 10(5) - \int_0^{5s} F(0.1) \, dt = 0.064[2(\omega_B)_2]$$

$$\int_0^{5s} F \, dt = 500 - 1.28(\omega_B)_2 \qquad (1)$$

$$\zeta + I_B(\omega_B)_1 + \Sigma \int_{t_1}^{t_2} M_B \, dt = I_B(\omega_B)_2$$

$$0 + \int_0^{5s} F(0.2) \, dt = 1.125(\omega_B)_2$$

$$\int_0^{5s} F \, dt = 5.625(\omega_B)_2 \qquad (2)$$

Equating Eqs. (1) and (2),
$$500 - 1.28(\omega_B)_2 = 5.625(\omega_B)_2$$
$$(\omega_B)_2 = 72.41 \text{ rad/s} = 72.4 \text{ rad/s} \qquad \text{Ans.}$$

F19–5. $(\xrightarrow{+})\; m[(v_O)_x]_1 + \Sigma \int F_x \, dt = m[(v_O)_x]_2$

$$0 + (150 \text{ N})(3 \text{ s}) + F_A(3 \text{ s})$$
$$= (50 \text{ kg})(0.3\omega_2)$$

$$\zeta + I_G\omega_1 + \Sigma \int M_G \, dt = I_G\omega_2$$

$$0 + (150 \text{ N})(0.2 \text{ m})(3 \text{ s}) - F_A(0.3 \text{ m})(3 \text{ s})$$
$$= [(50 \text{ kg})(0.175 \text{ m})^2]\omega_2$$
$$\omega_2 = 37.3 \text{ rad/s} \qquad \text{Ans.}$$
$$F_A = 36.53 \text{ N}$$

Also,
$$I_{IC}\omega_1 + \Sigma \int M_{IC} \, dt = I_{IC}\omega_2$$
$$0 + [(150 \text{ N})(0.2 + 0.3) \text{ m}](3 \text{ s})$$
$$= [(50 \text{ kg})(0.175 \text{ m})^2 + (50 \text{ kg})(0.3 \text{ m})^2]\omega_2$$
$$\omega_2 = 37.3 \text{ rad/s} \qquad \text{Ans.}$$

F19–6. $(+\uparrow)\; m[(v_G)_1]_y + \Sigma \int F_y \, dt = m[(v_G)_2]_y$

$$0 + N_A(3 \text{ s}) - (150 \text{ lb})(3 \text{ s}) = 0$$
$$N_A = 150 \text{ lb}$$

$$\zeta + (H_{IC})_1 + \Sigma \int M_{IC} \, dt = (H_{IC})_2$$

$$0 + (25 \text{ lb} \cdot \text{ft})(3 \text{ s}) - [0.15(150 \text{ lb})(3 \text{ s})](0.5 \text{ ft})$$
$$+ \left[\tfrac{150}{32.2} \text{ slug}(1.25 \text{ ft})^2\right]\omega_2 + \left(\tfrac{150}{32.2} \text{ slug}\right)[\omega_2(1 \text{ ft})](1 \text{ ft})$$
$$\omega_2 = 3.46 \text{ rad/s} \qquad \text{Ans.}$$

Review Problem Answers

Chapter 12

R12–1. $a_{max} = 42 \text{ ft/s}^2$ Ans.
$v_{max} = 135 \text{ ft/s}$ Ans.

R12–2. $s = 20.0 \text{ ft}$ Ans.

R12–3. $s = 1.84 \text{ m}$ Ans.

R12–4. $0 \le t \le 5 \quad a = 4 \text{ m/s}^2$ Ans.
$5 \le t \le 20 \quad a = 0 \text{ m/s}^2$ Ans.
$20 \le t \le 30 \quad a = -2 \text{ m/s}^2$ Ans.
$s_1 = 50 \text{ m}$ Ans.
$s_2 = 350 \text{ m}$ Ans.
$s_3 = 450 \text{ m}$ Ans.

R12–5. $\mathbf{a}_{AB} = \{0.404\mathbf{i} + 7.07\mathbf{j}\} \text{ m/s}^2$ Ans.
$\mathbf{a}_{AC} = \{2.50\mathbf{i}\} \text{ m/s}^2$ Ans.

R12–6. $v_0 = 67.7 \text{ ft/s}$ Ans.
$\theta = 58.9°$ Ans.

R12–7. $v = 4.58 \text{ m/s}$ Ans.
$a = 0.653 \text{ m/s}^2$ Ans.

R12–8. $v = 3.19 \text{ m/s}$ Ans.
$a = 4.22 \text{ m/s}^2$ Ans.

R12–9. $a = 29.0 \text{ m/s}^2$ Ans.

R12–10. $t = 160 \text{ s}$ Ans.

R12–11. $v_{B/A} = 875 \text{ km/h}$ Ans.
$\theta = 41.5°$ Ans.

Chapter 13

R13–1. $F = 85.7 \text{ N}$ Ans.

R13–2. $T = 158 \text{ N}$ Ans.

R13–3. $a_A = 9.66 \text{ ft/s}^2$ Ans.
$a_B = 3.86 \text{ ft/s}^2$ Ans.

R13–4. $T = 46.7 \text{ N}$ Ans.

R13–5. $a_t = 3.36 \text{ m/s}^2$ Ans.
$T = 361 \text{ N}$ Ans.

R13–6. $v = 5.38 \text{ ft/s}$ Ans.

R13–7. $N = 5.88 \text{ lb}$ Ans.
$a_t = 23.0 \text{ ft/s}^2$ Ans.

R13–8. $F_r = 68 \text{ N}$ Ans.
$N = 153 \text{ N}$ Ans.

Chapter 14

R14–1. $v_2 = 21.5 \text{ ft/s}$ Ans.

R14–2. $v_B = 47.8 \text{ ft/s}$ Ans.

R14–3. $v_B = 25.4 \text{ ft/s}$ Ans.

R14–4. $F = 38.5 \text{ N}$ Ans.

R14–5. $P_i = 1.80 \text{ hp}$ Ans.

R14–6. $P_i = 1.63 \text{ hp}$ Ans.

R14–7. $v_B = 10.4 \text{ m/s}$ Ans.

R14–8. $v_2 = 13.8 \text{ ft/s}$ Ans.

Chapter 15

R15–1. $t = 1.02 \text{ s}$ Ans.

R15–2. $v_2 = 1.92 \text{ m/s}$ Ans.

R15–3. $v_2 = 31.7 \text{ m/s}$ Ans.

R15–4. $v_{ABC} = 2.18 \text{ m/s} \leftarrow$ Ans.

R15–5. $(v_B)_2 = 8 \text{ m/s} \rightarrow$
$t = 4.077 \text{ s} = 4.08 \text{ s}$ Ans.

R15–6. $(v_A)_2 = 0.400 \text{ m/s} \rightarrow$ Ans.
$(v_B)_2 = 2.40 \text{ m/s} \rightarrow$ Ans.
$d = 0.951 \text{ m}$ Ans.

R15–7. $(v_B)_2 = 1.34 \text{ m/s} \leftarrow$ Ans.
$(v_A)_2 = 1.30 \text{ m/s}$ Ans.
$(\theta_A)_2 = 8.47°$ Ans.

R15–8. $v = 13.4 \text{ m/s}$ Ans.

Chapter 16

R16–1. $v_W = 14.2$ ft/s　　Ans.
$a_W = 0.25$ ft/s^2　　Ans.

R16–2. $v_B = 0.394$ m/s　　Ans.

R16–3. $v_B = 0.860$ m/s　　Ans.

R16–4. $v_C = 1.34$ m/s　　Ans.

R16–5. $a_C = 50.6$ m/s^2　　Ans.
$\theta = 9.09°\ \measuredangle\ \theta$　　Ans.
The cylinder moves up with an acceleration
$a_B = 8.00$ m/s$^2 \uparrow$　　Ans.

R16–6. $\alpha_{CB} = 7.52$ rad/s^2　　Ans.
$a_C = 4.22$ m/s^2　　Ans.
$\theta = 60.8°\ \measuredangle$　　Ans.

R16–7. $a_B = 2.25$ m/s^2　　Ans.
$\theta = 32.6°\ \measuredangle$　　Ans.

R16–8. $v_C = 9.38$ in./s　　Ans.
$a_C = 54.7$ in./s^2　　Ans.
$\alpha = 3.00$ rad/s^2　　Ans.

Chapter 17

R17–1. At each wheel
$N'_A = 383$ N　　Ans.
$N'_B = 620$ N　　Ans.

R17–2. $E_x = 9.87$ lb　　Ans.
$E_y = 4.86$ lb　　Ans.
$M_E = 7.29$ lb·ft　　Ans.

R17–3. $a_G = 1.38$ m/s^2　　Ans.

R17–4. $T = 218$ N　　Ans.
$\alpha = 21.0$ rad/s^2　　Ans.

R17–5. $\alpha = 12.1$ rad/s^2　　Ans.
$F = 30.0$ lb　　Ans.

R17–6. $F_o = 6.14$ lb $\rightarrow$　　Ans.

R17–7. $\alpha = 8.89$ rad/s^2　　Ans.

R17–8. $\omega = 100$ rad/s　　Ans.

Chapter 18

R18–1. $\omega_2 = 2.83$ rad/s　　Ans.

R18–2. $\omega = 33.0$ rad/s　　Ans.

R18–3. $v_B = 2.58$ m/s　　Ans.
$P = 141$ N　　Ans.

R18–4. $s_G = 0.859$ m　　Ans.

R18–5. $v_2 = 3.46$ m/s　　Ans.

R18–6. $\omega = 2.30$ rad/s　　Ans.

R18–7. $v_C = 13.3$ ft/s　　Ans.

R18–8. $d = 3.38$ m　　Ans.

Chapter 19

R19–1. $\omega_2 = 12.4$ rad/s　　Ans.

R19–3. $(v_G)_2 = 9.42$ m/s　　Ans.

R19–4. $(v_O)_2 = 6$ m/s　　Ans.
$\omega_2 = 26.7$ rad/s　　Ans.

R19–5. $\omega = 215$ rad/s　　Ans.

R19–6. $\omega_A = 1.70$ rad/s　　Ans.
$\omega_B = 5.10$ rad/s　　Ans.

R19–7. $\omega_2 = \dfrac{1}{3}\omega_1$　　Ans.

R19–8. $\omega = 3.56$ rad/s　　Ans.

Answers to Selected Problems

Chapter 12

12-1 $v = 0, s = 80.7$ m

12-2 $s = 20$ ft

12-3 $a = -24$ m/s^2, $\Delta s = -880$ m, $s_T = 912$ m

12-5 $a = -12$ m/s^2, $s = 0$, $s_T = 8$ m, $(v_{sp})_{avg} = 2.67$ m/s

12-6 $s = 11.9$ m, $v = 0.250$ m/s

12-7 $v_{avg} = 15$ m/s, $(v_{sp})_{avg} = 19.7$ m/s

12-9 $v = 2.63$ m/s

12-10 $v = 1.29$ m/s

12-11 $v_{avg} = 0.222$ m/s, $(v_{sp})_{avg} = 2.22$ m/s

12-13 Normal: $d = 517$ ft, Drunk: $d = 616$ ft

12-14 $t = 2.45s, t = 6.67s$

12-15 $h = 127$ ft, $v = 90.6$ ft/s $\downarrow$

12-17 $d = 16.9$ ft

12-18 $s_A = 3200$ ft

12-19 $k = -4.80$ ft/s^3

12-21 $s = 123$ ft, $a = 2.99$ ft/s^2

12-22 $v = 22.4$ m/s

12-23 $v = (20e^{-2t})$ m/s, $a = (-40e^{-2t})$ m/s^2, $s = 10(1 - e^{-2t})$ m

12-25 (a) $v = 45.5$ m/s, (b) $v_{max} = 100$ m/s

12-26 $h = 4.54$ m

12-27 $t = 0.549\left(\dfrac{v_f}{g}\right)$

12-29 $s_{AB} = 35.7$ m

12-30 $s = 54.0$ m

12-31 $s_{tot} = 13.0$ ft, $v_{avg} = -1.00$ ft/s, $(v_{sp})_{avg} = 2.60$ ft/s

12-33 $v = 11.2$ km/s

12-34 $v = -R\sqrt{\dfrac{2g_0(y_0 - y)}{(R + y)(R + y_0)}}$, $v_{imp} = 3.02$ km/s

12-35 $t' = 27.3$ s. When $t = 27.3$ s, $v = 13.7$ ft/s.

12-37 $\Delta s = 1.11$ km

12-38 $a|_{t=0} = -4$ m/s^2, $a|_{t=2s} = 0$, $a|_{t=4s} = 4$ m/s^2, $v|_{t=0} = 3$ m/s, $v|_{t=2s} = -1$ m/s, $v|_{t=4s} = 3$ m/s

12-39 $s = 2\sin\left(\dfrac{\pi}{5}t\right) + 4$, $v = \dfrac{2\pi}{5}\cos\left(\dfrac{\pi}{5}t\right)$, $a = -\dfrac{2\pi^2}{25}\sin\left(\dfrac{\pi}{5}t\right)$

12-41 $t = 7.48$ s. When $t = 2.14$ s, $v = v_{max} = 10.7$ ft/s, $h = 11.4$ ft

12-42 $t = 11.25$ s

12-43 $t' = 35$ s For $0 \le t < 10$ s, $s = \{300t\}$ ft, $v = 300$ ft/s

For 10 s $< t < 20$ s,
$s = \left\{\dfrac{1}{6}t^3 - 15t^2 + 550t - 1167\right\}$ ft
$v = \left\{\dfrac{1}{2}t^2 - 30t + 550\right\}$ ft/s

For 20 s $< t \le 35$ s, $s = \{-5t^2 + 350t + 167\}$ ft
$v = (-10t + 350)$ ft/s

12-45 When $t = 0.1$ s, $s = 0.5$ m and a changes from 100 m/s^2 to -100 m/s^2. When $t = 0.2$ s, $s = 1$ m.

12-46 $v|_{s=75\,ft} = 27.4$ ft/s, $v|_{s=125\,ft} = 37.4$ ft/s

12-47 For $0 \le t < 9$ s
$v = (4t^{3/2})$ m/s $\quad v|_{t=9s} = 108$ m/s
$s = \left(\dfrac{8}{5}t^{5/2}\right)$ m $\quad s|_{t=9s} = 389$ m
For 9 s $< t \le 14$ s
$v = (2t^2 - 18t + 108)$ m/s $\quad v|_{t=14s} = 248$ m/s
$s = \left(\dfrac{2}{3}t^3 - 9t^2 + 108t - 340\right)$ m
$s|_{t=14s} = 1237$ m

12-49 $v_{max} = 100$ m/s, $t' = 40$ s

12-50 For $0 \le t < 10$ s
$a = 8$ m/s^2
$s = (4t^2)$ m $\quad s|_{t=10s} = 400$ m
For 10 s $< t \le 40$ s
$a = 0$
$s = (80t - 400)$ m
$s|_{t=40s} = 2800$ m

12-51 For $0 \le t < 60$ s, $s = \left\{\dfrac{1}{20}t^2\right\}$ m, $a = 0.1$ m/s^2.
For 60 s $< t < 120$ s, $s = \{6t - 180\}$ m, $a = 0$.
For 120 s $< t \le 180$ s, $s = \left\{\dfrac{1}{30}t^2 - 2t + 300\right\}$ m, $a = 0.0667$ m/s^2.

12-53 At $t = 8$ s, $a = 0$ and $s = 30$ m. At $t = 12$ s, $a = -1$ m/s^2 and $s = 48$ m.

12-54 For $0 \le t < 5$ s, $s = \{0.2t^3\}$ m, $a = \{1.2t\}$ m/s^2
For 5 s $< t \le 15$ s, $s = \left\{\dfrac{1}{4}(90t - 3t^2 - 275)\right\}$ m

Answers to Selected Problems

12-55
$a = -1.5 \text{ m/s}^2$,
At $t = 15 \text{ s}$, $s = 100 \text{ m}$, $v_{avg} = 6.67 \text{ m/s}$
$t' = 33.3 \text{ s}$, $s|_{t=5\text{s}} = 550 \text{ ft}$, $s|_{t=15\text{s}} = 1500 \text{ ft}$,
$s|_{t=20\text{s}} = 1800 \text{ ft}$, $s|_{t=33.3\text{s}} = 2067 \text{ ft}$

12-57 For $0 \le s < 100 \text{ ft}$, $v = \left\{\sqrt{\dfrac{1}{50}(800s - s^2)}\right\} \text{ ft/s}$,

For $100 \text{ ft} < s \le 150 \text{ ft}$,
$v = \left\{\dfrac{1}{5}\sqrt{-3s^2 + 900s - 25000}\right\} \text{ ft/s}$

12-58 For $0 \le t < 6 \text{ s}$
$v = \left(\dfrac{1}{18}t^3\right) \text{ m/s}$
$s = \left(\dfrac{1}{72}t^4\right) \text{ m}$
For $6 \text{ s} < t \le 10 \text{ s}$
$v = (6t - 24) \text{ m/s}$
$s = (3t^2 - 24t + 54) \text{ m}$
$s = 114 \text{ m}$

12-59 $s_T = 980 \text{ m}$

12-61 When $t = 5 \text{ s}$, $s_B = 62.5 \text{ m}$.
When $t = 10 \text{ s}$, $v_A = (v_A)_{max} = 40 \text{ m/s}$ and $s_A = 200 \text{ m}$.
When $t = 15 \text{ s}$, $s_A = 400 \text{ m}$ and $s_B = 312.5 \text{ m}$.
$\Delta s = s_A - s_B = 87.5 \text{ m}$

12-62 $s|_{t=0.8335} = 2.08 \text{ ft}$
$s|_{t=1.675} = 0$
$s|_{t=2.5\text{s}} = -6.25 \text{ ft}$.
$v = \{5 - 6t\} \text{ ft/s}$
$v|_{t=0} = 5 \text{ ft/s}$,
$v = 0$ at $t = 0.833 \text{ s}$,
$v|_{t=2.5\text{s}} = -10 \text{ m/s}$.
$a = -6 \text{ ft/s}^2$ for all t.

12-63 For $0 \le s < 225 \text{ m}$
$a = 12.5 \text{ m/s}^2$
For $225 \text{ m} < s \le 525 \text{ m}$
$a = \{0.04s - 24\} \text{ m/s}^2$
$a|_{s=225\text{ m}} = -15 \text{ m/s}^2$ $a|_{s=525\text{ m}} = -3 \text{ m/s}^2$

12-65 $v = 354 \text{ ft/s}$, $t = 5.32 \text{ s}$

12-66 For $0 \le s < 100 \text{ m}$
$v = \{0.1415\} \text{ m/s}$ $v|_{s=100\text{ m}} = 14.1 \text{ m/s}$,
For $100 \text{ m} < s \le 200 \text{ m}$
$v = \left\{\sqrt{4s - 200}\right\} \text{ m/s}$ $v|_{s=200\text{ m}} = 24.5 \text{ m/s}$

12-67 For $0 \le s < 100 \text{ ft}$
$a = \{0.015 + 0.53\} \text{ ft/s}$ $a|_{s=0} = 0.5 \text{ ft/s}^2$
$a|_{s=100\text{ ft}} = 1.5 \text{ ft/s}^2$
For $100 \text{ ft} < s < 350 \text{ ft}$
$a = \{0.00165 - 0.76\} \text{ ft/s}^2$
$a|_{s=100\text{ ft}} = -0.6 \text{ ft/s}^2$
$a|_{s=350\text{ ft}} = -0.2 \text{ ft/s}^2$

12-69 $a = 5.31 \text{ m/s}^2$, $\alpha = 53.0°$ $\beta = 37.0°$, $\gamma = 90.0°$
12-70 $v_x = 0.398 \text{ ft/s}$, $v_y = 3.98 \text{ ft/s}$
12-71 $a_x = \dfrac{3}{4}\sqrt{\dfrac{c}{3b}}\dfrac{1}{\sqrt{t}}$, $a_y = 2ct$
12-73 (5.15 ft, 1.33 ft)
12-74 $v = 2.69 \text{ ft/s}$, $a = 0.0200 \text{ ft/s}^2$
12-75 $(v_{sp})_{avg} = 4.28 \text{ m/s}$
12-77 $\theta = 45°$
12-78 $v = 1003 \text{ m/s}$, $a = 103 \text{ m/s}^2$
12-79 $d = 4.00 \text{ ft}$, $a = 37.8 \text{ ft/s}^2$
12-81 $(\mathbf{a}_{AB})_{avg} = \{1.2\mathbf{i}+1.6\mathbf{j}\} \text{ m/s}^2$,
$(\mathbf{a}_{AC})_{avg} = \{-1.2\mathbf{i}\} \text{ m/s}^2$
12-82 $v = \sqrt{c^2 k^2 + b^2}$, $a = ck^2$
12-83 $v = 10.4 \text{ m/s}$, $a = 38.5 \text{ m/s}^2$
12-85 $d = 204 \text{ m}$, $v = 41.8 \text{ m/s}$ $a = 4.66 \text{ m/s}^2$
12-86 $\theta = 58.3°$, $(v_0)_{min} = 9.76 \text{ m/s}$
12-87 $\theta = 76.0°$, $v_A = 49.8 \text{ ft/s}$, $h = 39.7 \text{ ft}$
12-89 $R_{max} = 10.2 \text{ m}$, $\theta = 45°$
12-90 $R = 8.83 \text{ m}$
12-91 $x = 13.2 \text{ ft}$, $y = -6.95 \text{ ft}$
12-93 $d = 166 \text{ ft}$
12-94 $t = 3.57 \text{ s}$, $v_B = 67.4 \text{ ft/s}$
12-95 $v_A = 36.7 \text{ ft/s}$, $h = 11.5 \text{ ft}$
12-97 $R = 8.83 \text{ ft}$
12-98 $v_A = 6.49 \text{ m/s}$, $t = 0.890 \text{ s}$
12-99 $v_A = 28.0 \text{ m/s}$
12-101 $v_{min} = 0.838 \text{ m/s}$, $v_{max} = 1.76 \text{ m/s}$
12-102 $s = 8.68 \text{ ft}$, $s = 34.4 \text{ ft}$
12-103 $\theta = 23.8°$ and $77.5°$
12-105 $t_A = 0.553 \text{ s}$, $x = 3.46 \text{ m}$
12-106 $R = 19.0 \text{ m}$, $t = 2.48 \text{ s}$
12-107 $\theta_1 = 24.9°$ ⦨, $\theta_2 = 85.2°$ ⦨
12-109 $\theta = 76.0°$, $v_A = 49.8 \text{ ft/s}$, $h = 39.7 \text{ ft}$
12-110 $a = 0.0719 \text{ ft/s}^2$
12-111 $a_n = 8.00 \text{ ft/s}^2$, $a_t = 24 \text{ ft/s}^2$
12-113 $v = 4.40 \text{ m/s}$, $a_t = 5.04 \text{ m/s}^2$, $a_n = 1.39 \text{ m/s}^2$
12-114 $a_t = 8.66 \text{ ft/s}^2$, $\rho = 1280 \text{ ft}$
12-115 $a = 1.30 \text{ m/s}^2$
12-117 $v = 5.66 \text{ m/s}$, $a = 1.02 \text{ m/s}^2$
12-118 $v = 1.80 \text{ m/s}$, $a = 1.20 \text{ m/s}^2$
12-119 $h = 5.99 \text{ Mm}$
12-121 $a = 2.75 \text{ m/s}^2$
12-122 $a = 1.68 \text{ m/s}^2$
12-123 $a = \dfrac{9}{20\left[1 + 4(10^{-6})x^2\right]^{3/2}}$
12-125 $a_A = 4.44 \text{ m/s}^2$
12-126 $a_B = 0.556 \text{ m/s}^2$
12-127 $a_t = 3.62 \text{ m/s}^2$, $\rho = 29.6 \text{ m}$
12-129 $t = 7.00 \text{ s}$, $s = 98.0 \text{ m}$
12-130 $a = 7.42 \text{ ft/s}^2$
12-131 $a = 2.36 \text{ m/s}^2$

Answers to Selected Problems

12-133 $a = 3.05 \text{ m/s}^2$
12-134 $a = 0.763 \text{ m/s}^2$
12-135 $a = 0.952 \text{ m/s}^2$
12-137 $y = -0.0766x^2$, $v = 8.37 \text{ m/s}$, $\theta_v = 17.0°$ ⤦
         $a_n = 9.38 \text{ m/s}^2$, $a_t = 2.88 \text{ m/s}^2$
12-138 $d = 106 \text{ ft}$, $a_A = 9.88 \text{ ft/s}^2$, $a_B = 1.28 \text{ ft/s}^2$
12-139 $a = 0.824 \text{ m/s}^2$
12-141 $a = 1.81 \text{ ft/s}^2$
12-142 $t = 1.21 \text{ s}$
12-143 $a = 7.85 \text{ ft/s}^2$
12-145 $d = 11.0 \text{ m}$, $a_A = 19.0 \text{ m/s}^2$, $a_B = 12.8 \text{ m/s}^2$
12-146 $t = 2.51 \text{ s}$, $a_A = 22.2 \text{ m/s}^2$, $a_B = 65.1 \text{ m/s}^2$
12-147 $v = 51.1 \text{ ft/s}$, $a = 10.3 \text{ ft/s}^2$
12-149 $a = 0.511 \text{ m/s}^2$
12-150 $a = 0.309 \text{ m/s}^2$
12-151 $a = 322 \text{ mm/s}^2$, $\theta = 26.6°$ ⤦
12-153 $v = 43.0 \text{ m/s}$, $a = 6.52 \text{ m/s}^2$
12-154 $a = 7.48 \text{ ft/s}^2$
12-155 $v_r = -1.66 \text{ m/s}$, $v_\theta = -2.07 \text{ m/s}$, $a_r = -4.20 \text{ m/s}^2$,
         $a_\theta = 2.97 \text{ m/s}^2$
12-157 $v_r = a \sin\theta \dot{\theta}$, $v_\theta = (b - a\cos\theta)\dot{\theta}$,
         $a_r = (2a\cos\theta - b)\dot{\theta}^2 + a\sin\theta\ddot{\theta}$,
         $a_\theta = (b - a\cos\theta)\ddot{\theta} + 2a\dot{\theta}^2 \sin\theta$
12-158 $v = 464 \text{ ft/s}$, $a = 43.2(10^3) \text{ ft/s}^2$
12-159 $\mathbf{v} = \{-14.4\mathbf{u}_r - 24.0\mathbf{u}_z\} \text{ m/s}$
         $\mathbf{a} = \{-3.61\mathbf{u}_r - 6.00\mathbf{u}_z\} \text{ m/s}^2$
12-161 $v_r = -2\sin t$, $v_\theta = \cos t$, $a_r = -\dfrac{5}{2}\cos t$,
         $a_\theta = -2\sin t$
12-162 $v_r = -2.33 \text{ m/s}$, $v_\theta = 7.91 \text{ m/s}$, $a_r = -158 \text{ m/s}^2$,
         $a_\theta = -18.6 \text{ m/s}^2$
12-163 $\dot{\theta} = 0.302 \text{ rad/s}$
12-165 $\dot{\mathbf{a}} = (\dddot{r} - 3\dot{r}\dot{\theta}^2 - 3r\dot{\theta}\ddot{\theta})\mathbf{u}_r$
         $+ (3\ddot{r}\dot{\theta} + \dot{r}\ddot{\theta} + 3\dot{r}\ddot{\theta} - r\dot{\theta}^3)\mathbf{u}_\theta + (\dddot{z})\mathbf{u}_z$
12-166 $v = 146 \text{ mm/s}$, $a = 90.0 \text{ mm/s}^2$
12-167 $v = 16.8 \text{ ft/s}$, $a = 199 \text{ ft/s}^2$
12-169 $v_r = -\dfrac{20}{\sqrt{1+\theta^2}}$, $v_\theta = \dfrac{20\theta}{\sqrt{1+\theta^2}}$, $v_r = -14.1 \text{ ft/s}$
         $v_\theta = 14.1 \text{ ft/s}$
12-170 $v_r = 16.0 \text{ ft/s}$, $v_\theta = 1.94 \text{ ft/s}$, $a_r = 7.76 \text{ ft/s}^2$,
         $a_\theta = 1.94 \text{ ft/s}^2$
12-171 $v = 0.242 \text{ m/s}$, $a = 0.169 \text{ m/s}^2$,
         $(v_A)_x = 0.162 \text{ m/s} \leftarrow$, $(v_A)_y = 0.180 \text{ m/s} \uparrow$
12-173 $a = 27.8 \text{ m/s}^2$
12-174 $v = lc \sec^2 ct$, $a = 2lc^2 \sec^2 ct \tan ct$
12-175 $v = 242 \text{ mm/s}$, $a = 169 \text{ mm/s}^2$,
         $(v_A)_x = 162 \text{ mm/s} \leftarrow$,
         $(v_A)_y = 180 \text{ mm/s} \uparrow$
12-177 $v = 4.16 \text{ m/s}$, $a = 33.1 \text{ m/s}^2$
12-178 $v = 447 \text{ mm/s}$, $a = 3.66 \text{ m/s}^2$
12-179 $v_{AB} = 25.9 \text{ mm/s}$, $a_{AB} = -9.28 \text{ mm/s}^2$
12-181 $v = 8.49 \text{ m/s}$, $a = 88.2 \text{ m/s}^2$

12-182 $v = 5.16 \text{ m/s}$, $a = 39.1 \text{ m/s}^2$
12-183 $v_r = -24.2 \text{ ft/s}$, $v_\theta = 25.3 \text{ ft/s}$
12-185 $v = 1.32 \text{ m/s}$
12-186 $a = 8.66 \text{ m/s}^2$
12-187 $v_{AB} = -250 \text{ mm/s}$, $a_{AB} = -2.17 \text{ m/s}^2$
12-189 $v_r = 32.0 \text{ ft/s}$, $v_\theta = 50.3 \text{ ft/s}$, $a_r = -201 \text{ ft/s}^2$,
         $a_\theta = 256 \text{ ft/s}^2$
12-190 $v_r = 32.0 \text{ ft/s}$, $v_\theta = 50.3 \text{ ft/s}$, $a_r = -161 \text{ ft/s}^2$,
         $a_\theta = 319 \text{ ft/s}^2$
12-191 $v = 5.95 \text{ ft/s}$, $a = 3.44 \text{ ft/s}^2$
12-193 $v = 2a\dot{\theta}$, $a = 4a\dot{\theta}^2$
12-194 $v = 2a\dot{\theta}$, $a = 2a\sqrt{4\dot{\theta}^4 + \ddot{\theta}^2}$
12-195 $v_B = 0.5 \text{ m/s}$
12-197 $\Delta s_B = 2 \text{ ft} \uparrow$
12-198 $v_{A/B} = 2.90 \text{ m/s} \uparrow$
12-199 $\Delta s_B = 1.33 \text{ ft} \rightarrow$
12-201 $t = 3.83 \text{ s}$
12-202 $v_B = 0.75 \text{ m/s} \leftarrow$
12-203 $t = 5.00 \text{ s}$
12-205 $v_B = 1 \text{ m/s} \uparrow$
12-206 $v_B = 1.50 \text{ m/s} \leftarrow$
12-207 $v_A = 12 \text{ ft/s} \rightarrow$
12-209 $v_C = 0.667 \text{ m/s} \uparrow$
12-210 $v_A = 14.7 \text{ m/s}$
12-211 $v_B = 2.40 \text{ m/s} \uparrow$, $a_B = 3.25 \text{ m/s}^2 \uparrow$
12-213 $v_A = 4 \text{ ft/s} \leftarrow$
12-215 $v_A = 10.0 \text{ m/s} \leftarrow$, $a_A = 46.0 \text{ m/s}^2 \leftarrow$
12-217 $v_C = 0.175 \text{ m/s} \uparrow$
12-218 $v_{B/A} = 1044 \text{ km/h}$, $\theta = 54.5°$ ⤢
12-219 $v_{B/A} = 28.5 \text{ mi/h}$, $\theta_v = 44.5°$ ⤢,
         $a_{B/A} = 3418 \text{ mi/h}^2$, $\theta_a = 80.6°$ ⤢
12-221 $v_{A/B} = 13.5 \text{ ft/s}$, $t = 1.85 \text{ min}$
12-222 $v_{A/C} = 21.5 \text{ m/s}$, $\theta_v = 34.9°$ ⤥,
         $a_{A/C} = 4.20 \text{ m/s}^2$, $\theta_a = 75.4°$ ⤦
12-223 $v_{B/C} = 18.6 \text{ m/s}$, $\theta_v = 66.2°$ ⤢, $a_{B/C} = 0.959 \text{ m/s}^2$,
         $\theta_a = 8.57°$ ⤦
12-225 $v_{A/B} = 98.4 \text{ ft/s}$, $\theta_v = 67.6°$ ⤦, $a_{A/B} = 19.8 \text{ ft/s}^2$,
         $\theta_a = 57.4°$ ⤢
12-226 $v_{A/B} = 6.69 \text{ m/s}$, $\theta = 53.3°$ ⤦, $a_{A/B} = 1.52 \text{ m/s}^2$,
         $\theta = 41.9°$ ⤦
12-227 $v_{B/A} = 20.5 \text{ m/s}$, $\theta_v = 43.1°$ ⤦, $a_{B/A} = 4.92 \text{ m/s}^2$,
         $\theta_a = 6.04°$ ⤦
12-229 $v_b = 6.21 \text{ m/s}$, $t = 11.4 \text{ s}$
12-230 $v_m = 4.87 \text{ ft/s}$, $t = 10.3 \text{ s}$
12-231 $v_{B/A} = 26.0 \text{ m/s}$, $\theta = 30°$ ⤥, $a_{B/A} = 1.98 \text{ m/s}^2$,
         $\phi = 35.2°$ ⤦

Chapter 13

13-1 $s = 97.4 \text{ ft}$
13-2 $v = 3.36 \text{ m/s}$, $s = 5.04 \text{ m}$
13-3 $P = 224 \text{ N}$
13-5 $F = 6.37 \text{ N}$
13-6 $v = 3.29 \text{ m/s}$

Answers to Selected Problems

13-7 $P = 392$ N
13-9 $v = 59.8$ ft/s
13-10 $v = 60.7$ ft/s
13-11 $t = 0.249$ s
13-13 $a_A = 9.66$ ft/s² ←, $a_B = 15.0$ ft/s² →
13-14 $T = 11.25$ kN, $F = 33.75$ kN
13-15 $A_x = 685$ N, $A_y = 1.19$ kN, $M_A = 4.74$ kN·m
13-17 $P = 574$ N
13-18 $R = 5.30$ ft, $t = 1.82$ s
13-19 $R = 5.08$ ft, Total time $= 1.48$ s
13-21 $s = 5.43$ m
13-22 $v_B = 5.70$ m/s ↑
13-23 $a_A = -\frac{5}{11}g = \frac{5}{11}g\uparrow, a_B = \frac{3}{11}g\downarrow, a_C = \frac{7}{11}g\downarrow$
13-25 $\theta = 22.6°$
13-26 $v_A = 17.4$ ft/s ↓
13-27 $t = 0.815$ s
13-29 $v = 2.01$ ft/s
13-30 $F = 23.5$ N
13-31 $T = 1.63$ kN
13-33 $a_B = 4.89$ ft/s², $N_B = 35.60$ lb, $a_A = 1.31$ ft/s²
13-34 $P = 31.9$ N, $t = 3.07$ s
13-35 $t = 5.66$ s
13-37 $a_B = 2.30$ ft/s²↑
13-38 $a = 6.92$ m/s² (The initial velocity of block B will not affect its acceleration)
13-39 $v = \frac{1}{m}\sqrt{1.09F_0^2 t^2 + 2F_0 tmv_0 + m^2v_0^2}$,

$x = \frac{y}{0.3} + v_0\left(\sqrt{\frac{2m}{0.3F_0}}\right)y^{1/2}$

13-41   a) $a_C = 6.94$ m/s² ✓   b) $a_C = 6.94$ m/s² ✓
c) $a_{C/A} = 5.225$ m/s² ↗
$a_C = 7.08$ m/s², $\theta = 56.5°$ ↗
13-42 $a_C = 7.49$ m/s² $\theta = 22.8°$ ↗
13-43 $v = \sqrt{\frac{mg}{k}}\left[\frac{e^{2t\sqrt{mg/k}} - 1}{e^{2t\sqrt{mg/k}} + 1}\right]$, $v_t = \sqrt{\frac{mg}{k}}$
13-45 $x = \frac{mv_0}{k}\cos\theta_0(1 - e^{-(k/m)t})$, $y = -\frac{mgt}{k} +$
$\frac{m}{k}\left(v_0 \sin\theta_0 + \frac{mg}{k}\right)(1 - e^{-(k/m)t})$,
$x_{\max} = \frac{mv_0 \cos\theta_0}{k}$
13-46 $P = 2mg \tan\theta$
13-47 $P = 2mg\left(\frac{\sin\theta + \mu_s \cos\theta}{\cos\theta - \mu_s \sin\theta}\right)$
13-49 $A_x = 35.6$ lb, $A_y = 236$ lb, $M_A = 678$ lb·ft
13-50 $v = \frac{mg}{c}[1 - e^{-(c/m)t}]$,
$s = \frac{mg}{c}\left\{t + \frac{m}{c}[e^{-(c/m)t} - 1]\right\}$

13-51 $d = \frac{(m_A + m_B)g}{k}$
13-53 $r = 1.36$ m
13-54 $v = 10.5$ m/s
13-55 $v_{\max} = 251$ m/s, $N = 6.18$ kN
13-57 $v = 9.37$ m/s, $N_z = 589$ N, $N_n = 340$ N, $N_t = 0$
13-58 $v = 0.969$ m/s
13-59 $v = 1.48$ m/s
13-61 $v = 9.29$ ft/s, $T = 38.0$ lb
13-62 $v = 2.10$ m/s
13-63 $F_n = 323$ lb, $F_t = 0$
13-65 $\theta = 15.0°, \phi = 12.6°$
13-66 $v = 22.1$ m/s
13-67 $N = 279$ lb, $N = 0$
13-69 $d = 383$ mm
13-70 $N_A = -11.8$ lb, $N_B = -29.5$ lb
13-71 $N = 277$ lb, $F = 13.4$ lb
13-73 $N = 466$ N, $a_t = 0.986$ m/s²
13-74 $a_t = 29.4$ ft/s², $N = 69.0$ lb
13-75 $l_0 = 1.5$ ft
13-77 $v_G = 2.83$ m/s
13-78 $a = 7.32$ m/s², $N = 59.2$ N
13-79 $N_P = 711$ N, $\rho = 951$ m
13-81 $\theta = 37.7°$
13-82 $N = 4.16$ kN, $a_t = 1.92$ m/s²
13-83 $N = 20.216$, $v = 10.4$ ft/s
13-85 $(F_z)_{\min} = 15.6$ N, $(F_z)_{\max} = 23.6$ N
13-86 $F = 210$ N
13-87 $F = 1.60$ lb
13-89 $F_r = -29.4$ N, $F_\theta = 0$, $F_z = 392$ N
13-90 $F_r = 102$ N, $N_z = 375$ N, $N_\theta = 79.7$ N
13-91 $N = 4.90$ N, $F = 4.17$ N
13-93 $N = 4.22$ lb, $P = -126$ lb
13-94 $N = 12.1$ N, $F = 7.67$ N
13-95 $\psi = 84.3°$, $a_t = 12$ m/s²
13-97 $\dot{\theta} = 7.00$ rad/s
13-98 $N = 4.77$ lb, $P = 1.68$ lb
13-99 $\dot{\theta} = 5.70$ rad/s
13-101 $F_r = -100$ N, $F_\theta = 0$, $F_z = 2.45$ kN
13-102 $a_t = 12$ m/s²; $\theta = 5.71°$ ↘
13-103 $F_r = -131$ N, $F_\theta = -38.4$ N, $F_z = 215$ N
13-105 $N = 5.76$ N, $F = 6.48$ N
13-106 $N = 0.328$ N, $F = 10.5$ N
13-107 $F = 0.163$ lb
13-109 $F_r = 25.6$ N, $F_{OA} = 0$
13-110 $F_r = 20.7$ N, $F_{OA} = 0$
13-111 $r = 0.198$ m
13-113 $v_o = 30.4$ km/s,
$\frac{1}{r} = 0.348\,(10^{-12})\cos\theta + 6.74\,(10^{-12})$
13-114 $h = 35.9$ Mm, $v_s = 3.07$ km/s
13-115 $v_0 = 7.45$ km/s

Answers to Selected Problems

13-117 $T^2 = \left(\dfrac{4\pi^2}{GM_s}\right)a^3$
13-118 $v_p = 7.76$ km/s, $v_A = 4.52$ km/s, $T = 3.35$ hr
13-119 $v_A = 6.67(10^3)$ m/s, $v_B = 2.77(10^3)$ m/s
13-121 $v = 7.31$ km/s
13-122 $v_A = 6.33$ km/s, $t = 2.9$ hr, $\Delta v = 578$ m/s
13-123 $(v_A)_C = 5.27(10^3)$ m/s, $\Delta v = 684$ m/s
13-125 $v_0 = 23.9(10^3)$ ft/s
13-126 $v_{A'} = 7.30(10^3)$ ft/s, $t = 1.69$ h
13-127 $\Delta v = \sqrt{\dfrac{GM_e}{r_0}}\left(\sqrt{2} - \sqrt{1+e}\right)$ occur at perigee
13-129 $v_A = 2.01(10^3)$ m/s
13-130 $v_{A'} = 521$ m/s, $t = 21.8$ h
13-131 $v_A = 7.01(10^3)$ m/s

Chapter 14
14-1 $s = 59.7$ m
14-2 $x_{max} = 3.24$ ft
14-3 $v = 6.89$ m/s
14-5 $h = 39.3$ m, $\rho = 26.2$ m
14-6 $d = 120$ m
14-7 Observer A: $v_2 = 6.08$ m/s, Observer B: $v_2 = 4.08$ m/s
14-9 $P = 207$ N
14-10 $s = 20.5$ m
14-11 $v = 2.67$ ft/s
14-13 $v_B = 31.5$ ft/s, $d = 22.6$ ft, $v_C = 54.1$ ft/s
14-14 $x = 5.44$ ft
14-15 $v_A = 3.52$ ft/s
14-17 $X_{max} = \dfrac{mg}{k} + \sqrt{\left(\dfrac{mg}{k}\right)^2 + \dfrac{m}{k}(v_0^2 + 2gh)}$
14-18 $y = 3.81$ ft
14-19 $v_B = 3.34$ m/s
14-21 $s_A = 0.255$ m, $s_B = 0.205$ m
14-22 $s_{Tot} = 3.88$ ft
14-23 $v = 3.77$ m/s
14-25 $s = 0.0735$ ft
14-26 $v_A = 28.3$ m/s
14-27 $v_B = 18.0$ m/s, $N_B = 12.5$ kN
14-29 $v_B = 18.5$ ft/s, $a_B = 123$ ft/s^2, $y_{max} = 5.55$ ft
14-30 $v_2 = 15.4$ m/s
14-31 $R = 2.83$ m, $v_C = 7.67$ m/s
14-33 $v_A = 3.82$ m/s
14-34 $v_B = 24.0$ ft/s, $N_B = 7.18$ lb, $v_C = 16.0$ ft/s, $N_C = 1.18$ lb
14-35 $v_B = 42.2$ ft/s, $N = 50.6$ lb, $a_t = 26.2$ ft/s^2
14-37 $v = 1.07$ m/s
14-38 $v_B = 26.1$ ft/s, $N_B = 135$ lb
14-39 $v_B = 5.42$ m/s
14-41 $d = 36.2$ ft
14-42 $\theta = 47.2°$
14-43 $P_i = 4.20$ hp
14-45 $P = 8.32\,(10^3)$ hp
14-46 $t = 46.2$ min
14-47 $P = 9.24$ hp
14-49 $v = 4.86$ m/s
14-50 $P = 8.31t$ MW
14-51 $P = 92.2$ hp
14-53 $P_i = 483$ kW
14-54 $P_i = 622$ kW
14-55 $P_i = 22.2$ kW
14-57 $P_{out} = 35.4$ kW
14-58 $P = 0.231$ hp
14-59 $P_{in} = 19.5$ kW
14-61 $P_{in} = 0.645$ hp
14-62 $P = 0.229$ hp
14-63 $P = 0.0364$ hp
14-65 $P = 3.64$ hp
14-66 $F = 227$ N
14-67 $h = 113$ in.
14-69 $v_B = 5.33$ m/s, $N = 694$ N
14-70 $v = 3.84$ m/s; $T = -5.89$ N
14-71 $v_C = 17.7$ ft/s
14-73 $N_B = 0$, $h = 18.75$ m, $N_C = 17.2$ kN
14-74 $x - 1.10$ in.
14-75 $N = 3.04$ lb
14-77 $h = 23.75$ m, $v_C = 21.6$ m/s
14-78 $v_B = 15.5$ m/s
14-79 $l = 2.77$ ft
14-81 $\theta = 118°$
14-83 $U_{1-2} = GM_em\left(\dfrac{1}{r_1} - \dfrac{1}{r_2}\right)$
14-85 $v_B = 34.8$ Mm/h
14-86 $v_2 = 2.15$ m/s
14-87 $s_B = 0.638$ m, $s_A = 1.02$ m
14-89 $x = 453$ mm
14-90 $N = 78.6$ N
14-91 $v_B = 10.6$ m/s, $T_B = 142$ N, $v_C = 9.47$ m/s, $T_C = 48.7$ N
14-93 $v_B = 27.2$ ft/s
14-94 $v = 4.49$ ft/s
14-95 $x_A = 0.516$ ft, $x_B = 0.387$ ft
14-97 $k = 773$ N/m

Chapter 15
15-1 $I = 0.373$ lb·s
15-2 $v = 29.4$ ft/s
15-3 $T = 3643$ lb
15-5 $T = 20.1$ lb

Answers to Selected Problems

15-6 $v = 104$ m/s

15-7 $F_{AB} = 16.7$ lb, $v = 13.4$ ft/s

15-9 $v = 6.62$ m/s

15-10 $v_2 = \dfrac{2Ct'}{\pi m}, s = \dfrac{Ct'^2}{\pi m}$

15-11 $v = 60.0$ m/s ↑

15-13 $\mu_k = 0.340$

15-15 $v = 23.5$ ft/s

15-17 $v = 8.81$ m/s, $s = 24.8$ m

15-18 $v|_{t=3\,s} = 5.68$ m/s ↓, $v|_{t=6\,s} = 21.1$ m/s ↑

15-21 $T = 14.9$ kN, $F = 24.8$ kN

15-22 $v = 0.849$ m/s

15-23 $v = 10.1$ ft/s

15-25 $v = 7.21$ m/s ↑

15-26   Observer A: $v = 7.40$ m/s, Observer B: $v = 5.40$ m/s

15-27 $v_2 = 21.8$ m/s

15-29 $v_B = 21.5$ ft/s ↓, $v_A = 21.5$ ft/s ↑

15-30 $v = 16.1$ m/s

15-31 $T = 520.5$ N

15-33 $v = 8.80$ ft/s

15-34 $t = 0.486$ s

15-35 $v = 18.6$ m/s →

15-37 $v = 5.21$ m/s ←, $\Delta T = 32.6$ kJ

15-38 $v = 21.2$ ft/s, $T_1 = 423$ ft·kip, $T_2 = 280$ ft·kip

15-39 $v = 733$ m/s

15-41 $d = 3.18$ m

15-42 $d = 1.40$ m

15-43 $s = 4.00$ m

15-45 $v_2 = \sqrt{v_1^2 + 2gh}, \theta_2 = \sin^{-1}\left(\dfrac{v_1 \sin\theta}{\sqrt{v_1^2 + 2gh}}\right)$

15-46 $v_A = 1.86$ m/s ←, $v_C = 0.781$ m/s →

15-47 $s_{max} = 481$ mm

15-49 $v = \dfrac{Mv_0}{m + M}, t = \dfrac{v_0}{\mu g \left(1 + \dfrac{m}{M}\right)}$

15-50 $x_{c/b} = \left|\dfrac{v_0^2}{2\mu g \left(1 + \dfrac{m}{M}\right)}\right|$

15-51 $v_2 = 20.7$ ft/s, $I = 38.6$ lb·s

15-53 $s_B = 71.4$ mm →

15-54 $s_B = 71.4$ mm →

15-55 $v = 1.32$ m/s, $s = 2.39$ m

15-57 $v_B = 5.08$ ft/s, $v_A = 7.61$ ft/s

15-58 $(v_A)_2 = 0.353$ m/s, $(v_B)_2 = 2.35$ m/s

15-59 $e = 0.75$, $\Delta T = -9.65$ kJ

15-61 $x_{max} = 0.839$ m

15-63 $v_C = 0.1875v →, v_D = 0.5625v →, v_B = 0.8125v →, v_A = 0.4375v →$

15-65 $t = 0.226$ s

15-66 $d = 0.329$ ft

15-67 $x_{max} = 1.31$ ft

15-69 $v'_B = 22.2$ m/s, $\theta = 13.0°$

15-70 $(v_A)_3 = \left(\dfrac{1-e}{2}\right)\sqrt{gL}, (v_B)_3 = \left(\dfrac{1+e}{2}\right)\sqrt{gL},$

$\theta = \cos^{-1}\left[1 - \dfrac{(1+e)^2}{8}\right]$

15-71 $v_{p2} = 27.0$ ft/s, $v_{m2} = 13.4$ ft/s, $x = 2.61$ ft

15-73 $v_A = 1.35$ m/s →, $v_B = 5.89$ m/s, $\theta = 32.9°$

15-75 $h = 1.57$ m

15-77 $(v_B)_3 = 3.24$ m/s, $\theta = 43.9°$

15-78 $v'_B = 31.8$ ft/s

15-79 $(v_A)_2 = 3.80$ m/s ←, $(v_B)_2 = 6.51$ m/s, $(\theta_B)_2 = 68.6°$

15-81   (a) $(v_B)_1 = 8.81$ m/s, $\theta = 10.5°$,
(b) $(v_B)_2 = 4.62$ m/s, $\phi = 20.3°$,
(c) $s = 3.96$ m

15-82 $s = 0.456$ ft

15-83 $(v_A)_2 = 6.90$ m/s, $(v_B)_2 = 75.7$ m/s

15-85 $\mu_k = 0.25$

15-86 $(v_A)_2 = 2.46$ ft/s, $(v_B)_2 = 3.09$ ft/s

15-87 $e = 0.577$

15-89 $(v_B)_2 = 6.59$ ft/s

15-90 $d = 1.15$ ft, $h = 0.770$ ft

15-91 $h = 1.57$ m

15-93 $(v_A)_2 = 8.19$ m/s, $(v_B)_2 = 9.38$ m/s

15-94 $(H_A)_O = 72.0$ kg·m²/s, $(H_B)_O = 59.5$ kg·m²/s

15-95 $(H_A)_P = 66.0$ kg·m²/s, $(H_B)_P = 73.9$ kg·m²/s

15-97 $(\mathbf{H}_A)_P = \{-52.8\mathbf{k}\}$ kg·m²/s, $(\mathbf{H}_B)_P = \{-118\mathbf{k}\}$ kg·m²/s

15-98 $\mathbf{H}_O = \{-16.8\mathbf{i} + 14.9\mathbf{j} - 23.6\mathbf{k}\}$ slug·ft²/s

15-99 $\mathbf{H}_P = \{-14.3\mathbf{i} - 9.32\mathbf{j} - 34.8\mathbf{k}\}$ slug·ft²/s

15-101 $t = 5.85$ s

15-102 $v_\theta = 10.9$ ft/s, $v = 22.7$ ft/s

15-103 $v = 4.67$ m/s

15-105 $v = 19.3$ ft/s

15-106 $v = 9.50$ m/s

15-107 $v = 3.33$ m/s

15-109 $H_O = 70.9$ slug·ft²/s

15-110 $v_2 = 4.03$ m/s, $\Sigma U_{1-2} = 725$ J

15-111 $h = 196$ mm, $\theta = 75.0°$

15-113 $v_B = 10.2$ km/s, $r_B = 13.8$ Mm

15-114 $F_x = 11.0$ N

15-115 $C_x = 4.97$ kN, $D_x = 2.23$ kN, $D_y = 7.20$ kN

15-117 $F = 302$ lb

15-118 $F = v^2 \rho A$ for all cases

15-119 $F_x = 9.87$ lb, $F_y = 4.93$ lb

15-121 $F_x = 19.5$ lb, $F_y = 1.96$ lb

15-122 $F = 20.0$ lb

15-123 $F = 22.4$ lb
15-125 $T = 82.8$ N, $N = 396$ N
15-126 $T = 0.680$ lb
15-127 $d = 2.56$ ft
15-129 $Q = 0.217(10^{-3})$ m³/s
15-130 $v = \left\{\dfrac{8000}{2000 + 50t}\right\}$ m/s
15-131 $F = 482$ lb
15-133 $a = 0.125$ m/s², $v = 4.05$ m/s
15-134 $v_{max} = 2.07\,(10^3)$ ft/s
15-135   452 Pa
15-137 $\Delta T = 3.88$ lb
15-138 $a_i = 133$ ft/s², $a_f = 200$ ft/s²
15-139 $v = 580$ ft/s
15-141 $v = 3.12$ m/s
15-142 $h = \dfrac{8Q^2}{\pi^2 d^4 g} - \dfrac{m^2 g}{8\rho_w^2 Q^2}$

Chapter 16
16-1 $v_A = 2.60$ m/s, $a_A = 9.35$ m/s²
16-5 $\theta = 5443$ rev, $\omega = 740$ rad/s, $\alpha = 8$ rad/s²
16-6 $\theta = 314$ rev, $t = 20.9$ s
16-7 $\omega_B = 528$ rad/s, $\theta_B = 288$ rad
16-9 $\alpha_A = 60.8$ rad/s²
16-10 $\omega_A = 225$ rad/s
16-11 $v = 52.0$ ft/s, $a_t = 12.0$ ft/s², $a_n = 1.35(10^3)$ ft/s²
16-13 $\omega = 3.80$ rad/s, $a_p = 217$ ft/s²
16-14 $\omega = 42.7$ rad/s, $\theta = 42.7$ rad
16-15 $a_t = 2.83$ m/s², $a_n = 35.6$ m/s²
16-19 $\omega_B = 156$ rad/s
16-21 $v = 1$ m/s
16-22 $\omega_D = 4.00$ rad/s, $\alpha_D = 0.400$ rad/s²
16-23 $\omega_D = 12.0$ rad/s, $\alpha_D = 0.600$ rad/s²
16-25 $\theta_{max} = 18.1°$
16-26 $\omega_C = 1.68$ rad/s, $\theta_C = 1.68$ rad
16-29 $\omega_E = 0.160$ rad/s
16-30 $\omega_B = 22.3$ rad/s
16-33 $d = 1.44$ in.
16-34 $v_E = 3$ m/s, $(a_E)_t = 2.70$ m/s², $(a_E)_n = 600$ m/s²
16-35 $\mathbf{v}_C = \{-4.8\mathbf{i} - 3.6\mathbf{j} - 1.2\mathbf{k}\}$ m/s,
         $\mathbf{a}_C = \{38.4\mathbf{i} - 64.8\mathbf{j} + 40.8\mathbf{k}\}$ m/s²
16-37 $v_C = 2.50$ m/s, $a_C = 13.1$ m/s²
16-38 $v_C = 21.2$ ft/s, $a_C = 106$ ft/s²
16-39 $v_{AB} = \omega l \cos\theta$, $a = -\omega^2 l \sin\theta$
16-41 $v = (-10 \csc^2\theta)$ ft/s
16-42 $v = 0.416$ m/s, $a = -0.752$ m/s²
16-43 $v = \omega r \cos\theta$, $a = -\omega^2 r \sin\theta$
16-45 $\dot{y} = 1.5 \cot\theta$
16-46 $\dot{\theta} = 0.0841$ rad/s ↻
16-47 $\omega = 14.4$ rad/s ↻
16-49 $\omega_A = 13.3$ rad/s, $\theta = 2$ rev

16-50 $\omega = -\left(\dfrac{r}{x\sqrt{x^2 - r^2}}\right)v_A$, $\alpha = \left[\dfrac{r(2x^2 - r^2)}{x^2(x^2 - r^2)^{3/2}}\right]v_A^2$
16-51 $\omega = \dfrac{h}{h^2 + x^2}v_B$, $\alpha = \dfrac{-2xh}{(h^2 + x^2)^2}v_B^2$
16-53 $\omega = \dfrac{v \sin\phi}{l \cos(\phi - \theta)}$, $\alpha = \dfrac{-v^2 \sin^2\phi \sin(\phi - \theta)}{l^2 \cos^3(\phi - \theta)}$
16-54 $\omega = \dfrac{v}{2r}$ ↻
16-55 $\omega_{CD} = 4.17$ rad/s ↻
16-57 $v_B = 12.6$ in./s, 65.7° ↙
16-58 $v_C = 3.54$ ft/s
16-59 $v_A = 9.20$ m/s →
16-61 $\omega_{BC} = 2.31$ rad/s ↻, $\omega_{AB} = 3.46$ rad/s ↻
16-62 $v_C = 4$ ft/s →, $v_B = 3.46$ ft/s, $\theta = 30°$ ↖
16-63 $v_C = 1.04$ m/s →
16-65 $\omega = 20$ rad/s ↻, $v_A = 2$ ft/s →
16-66 $v_C = 2.45$ m/s ↘, $\omega_{BC} = 7.81$ rad/s ↻
16-67 $\omega = 5.33$ rad/s ↻
16-69 $\omega_{BC} = 1.5$ rad/s ↻, $\omega_{CD} = 1.70$ rad/s ↻
16-70 $v_B = 7.33$ ft/s →, $v_A = 2.83$ ft/s, $\theta = 45°$ ↖
16-71 $v_A = 8$ ft/s →
16-73 $v_E = 4.00$ m/s, $\theta = 52.7°$ ↙
16-74 $v_A = 1.36$ m/s →
16-75 $\omega_{AB} = 330$ rad/s ↻
16-77 $\omega_P = 5$ rad/s ↺, $\omega_A = 1.67$ rad/s ↻
16-78 $v_E = 4$ ft/s ←
16-79 $v_C = 2.40$ ft/s ↑
16-85 $v_A = 2$ ft/s →, $v_B = 10$ ft/s ←.
         The cylinder slips.
16-86 $\omega = 1.5\dfrac{v}{r}$ ↻
16-87 $v_D = 10.6$ m/s ↓
16-89 $\omega_{AB} = 7.17$ rad/s ↺
16-90 $v_A = 45$ ft/s →
16-91 $\omega_t = 26.7$ rad/s ↻
16-93 $\omega_{BPD} = 3.00$ rad/s ↺, $v_P = 1.79$ m/s ←
16-94 $v_B = 29.7$ m/s ↗
16-95 $v_A = 60.0$ ft/s →, $v_C = 220$ ft/s ←,
         $v_B = 161$ ft/s, $\theta = 60.3°$ ↙
16-97 $v_C = 8.69$ m/s, $\theta = 22.9°$ ↗
16-98 $v_D = 5.72$ m/s, $\theta = 36.2°$ ↙
16-99 $\omega_{CD} = 57.7$ rad/s ↻
16-101 $v_C = 0.897$ m/s ↗
16-102 $v_D = 0.518$ m/s ↖
16-103 $a_A = 56.2$ ft/s² ↓, $a_B = 40.2$ ft/s², $\theta = 53.3°$ ↘
16-105 $a_C = 1.80$ ft/s², 33.7° ↖
16-106 $\omega = 0.25$ rad/s ↺, $v_B = 5.00$ ft/s ↓,
         $\alpha = 0.875$ rad/s² ↻, $a_B = 1.51$ ft/s², $\theta = 85.2°$ ↙
16-107 $\omega = 6.67$ rad/s ↻, $v_B = 4.00$ m/s ↘
         $\alpha = 15.7$ rad/s² ↺, $a_B = 24.8$ m/s² ↖

Answers to Selected Problems

16-109 $\omega = 2$ rad/s $\circlearrowright$, $\alpha = 7.68$ rad/s^2 $\circlearrowright$
16-110 $\alpha_{AB} = 36$ rad/s^2 $\circlearrowright$
16-111 $\alpha_{AB} = 4.62$ rad/s^2 $\circlearrowright$, $a_B = 13.3$ m/s^2, $\theta = 37.0°$
16-113 $a_A = 5.94$ m/s^2, $\theta = 53.9°$
16-114 $a_B = 6.21$ m/s^2, $\theta = 45.8°$
16-115 $\alpha_{AB} = 3$ rad/s^2 $\circlearrowright$, $a_A = 13.9$ ft/s^2, $\theta = 59.7°$
16-117 $a_c = 6.96$ ft/s^2, $\theta = 18.8°$
16-118 $\omega_{CD} = 7.79$ rad/s $\circlearrowright$, $\omega_{CD} = 136$ rad/s^2 $\circlearrowright$
16-119 $v_c = 1.56$ m/s $\leftarrow$, $a_C = 29.7$ m/s^2, $\theta = 24.1°$
16-121 $a_E = 0.0714$ m/s^2 $\uparrow$
16-122 $\alpha_B = 7.50$ rad/s^2 $\circlearrowright$
16-123 $\alpha_B = 7.50$ rad/s^2 $\circlearrowright$
16-125 $\alpha_{AB} = 0.75$ rad/s^2 $\circlearrowright$, $\alpha_{BC} = 3.94$ rad/s^2 $\circlearrowright$
16-126 $\alpha_{ABC} = 41.6$ rad/s^2 $\circlearrowright$, $a_C = 38.2$ m/s^2, $\theta = 39.4°$
16-127 $\alpha_{AB} = 3.70$ rad/s^2 $\circlearrowright$
16-129 $\mathbf{v}_A = \{-2.50\mathbf{i} + 2.00\mathbf{j}\}$ ft/s,
$\mathbf{a}_A = \{-3.00\mathbf{i} + 1.75\mathbf{j}\}$ ft/s^2
16-130 $\mathbf{v}_B = \{5\mathbf{i} + 6\mathbf{j}\}$ ft/s, $\mathbf{a}_B = \{-15\mathbf{i} + 40\mathbf{j}\}$ ft/s^2
16-131 $\mathbf{a}_B = \{-0.4\mathbf{i} - 12\mathbf{j}\}$ m/s^2
16-133 $\mathbf{v}_A = \{-17.2\mathbf{i} + 12.5\mathbf{j}\}$ m/s,
$\mathbf{a}_A = \{349\mathbf{i} + 597\mathbf{j}\}$ m/s^2
16-134 $\mathbf{a}_A = \{-5.60\mathbf{i} - 16\mathbf{j}\}$ m/s^2
16-135 $v_C = 2.40$ m/s, $\theta = 30°$
16-137 $\omega_{CD} = 9.00$ rad/s $\circlearrowright$, $\alpha_{CD} = 249$ rad/s^2 $\circlearrowright$
16-138 $\omega_{AB} = 2.00$ rad/s $\circlearrowright$, $\alpha_{AB} = 1.00$ rad/s^2 $\circlearrowright$
16-139 $\omega_{CD} = 0.375$ rad/s $\circlearrowright$, $\alpha_{CD} = 2.44$ rad/s^2 $\circlearrowright$
16-141 $\omega_{AC} = 0$, $\alpha_{AC} = 14.4$ rad/s^2 $\circlearrowright$
16-142 $\omega_{AB} = 1.18$ rad/s $\circlearrowright$, $\alpha_{AB} = 15.4$ rad/s^2 $\circlearrowright$
16-143 $\omega_{AB} = 5$ rad/s $\circlearrowright$, $\alpha_{AB} = 2.50$ rad/s^2 $\circlearrowright$
16-145 $\mathbf{v}_C = \{-7.00\mathbf{i} + 17.3\mathbf{j}\}$ ft/s,
$\mathbf{a}_C = \{-34.6\mathbf{i} - 15.5\mathbf{j}\}$ ft/s^2
16-146 $\mathbf{v}_C = \{-7.00\mathbf{i} + 17.3\mathbf{j}\}$ ft/s,
$\mathbf{a}_C = \{-38.8\mathbf{i} - 6.84\mathbf{j}\}$ ft/s^2
16-147 $\omega_{AB} = 0.667$ rad/s $\circlearrowright$, $\alpha_{AB} = 3.08$ rad/s^2 $\circlearrowright$

Chapter 17

17-1 $I_y = \frac{1}{3}ml^2$
17-2 $m = \pi h R^2 \left(k + \frac{aR^2}{2}\right)$, $I_z = \frac{\pi h R^4}{2}\left[k + \frac{2aR^2}{3}\right]$
17-3 $I_z = mR^2$
17-5 $I_z = \frac{3}{10}mr^2$
17-6 $I_O = 282$ slug · ft^2
17-7 $I_x = \frac{2}{5}mb^2$
17-9 $I_z = 34.2$ kg · m^2
17-10 $k_O = 3.15$ ft
17-11 $L = 2.39$ m, $I_O = 5.93$ kg · m^2
17-13 $I_A = 7.67$ kg · m^2
17-14 $I_A = 222$ slug · ft^2
17-15 $I_x = 0.402$ slug · in^2
17-17 $I_G = 0.230$ kg · m^2
17-18 $I_O = 0.560$ kg · m^2
17-19 $I_G = 118$ slug · ft^2
17-21 $I_A = 2.17$ slug · ft^2
17-22 $I_x = 0.00325$ kg · m^2
17-23 $I_{x'} = 0.00719$ kg · m^2
17-25 $F = 5.96$ lb, $N_B = 99.0$ lb, $N_A = 101$ lb
17-26 $A_y = 72.6$ kN, $B_y = 71.6$ kN, $a_G = 0.250$ m/s^2
17-27 $N_A = 1393$ lb, $N_B = 857$ lb, $t = 2.72$ s
17-29 $\mu_k = 0.458$
17-30 $a_G = 0.3$ m/s^2, $N_A = 5.43$ kN, $N_B = 7.09$ kN
17-31 $R = 33.3$ lb
17-33 $a_G = 4.33$ m/s^2, $(\mu_s)_{min} = 0.833$
17-34 $t = 11.2$ s
17-35 $P = 24$ lb, $a_G = 3.22$ ft/s^2, $N_B = 14.7$ lb, $N_A = 65.3$ lb
17-37 $P_{max} = \dfrac{mgb}{2(d - \frac{h}{2})}$
17-38 $a_m = 1.45$ m/s^2, $a_b = 1.94$ m/s^2
17-39 $N = 0.433wx$, $V = 0.25wx$, $M = 0.125wx^2$
17-41 $B_x = 73.9$ lb, $B_y = 69.7$ lb, $N_A = 120$ lb
17-42 $a_G = 16.1$ ft/s^2
17-43 $a_G = 2.94$ m/s^2
17-45 $a_G = 6.78$ m/s^2
17-46 $a_G = 6.78$ m/s^2
17-47 $a = 53.5$ ft/s^2, $\mu = 1.66$
17-49 $F_{AB} = 112$ N, $C_x = 26.2$ N, $C_y = 49.8$ N
17-50 $P = 765$ N
17-51 $s = 2.74$ ft
17-53 $\alpha = 9.67$ rad/s^2
17-54 $F_C = 16.1$ lb, $N_C = 159$ lb
17-55 $F_{CD} = 9.17$ lb, $(a_G)_t = 32.2$ ft/s^2
17-57 $\omega = 56.2$ rad/s, $A_x = 0$, $A_y = 98.1$ N
17-58 $\alpha = 14.7$ rad/s^2, $A_x = 88.3$ N, $A_y = 147$ N
17-59 $F_A = \dfrac{3}{2}mg$
17-61 $F_O = 91.1$ N
17-62 $t = 9.90$ s
17-63 $t = 8.10$ s
17-65 $M = 0.233$ lb · ft
17-67 $\omega = \dfrac{2}{R}\sqrt{\dfrac{mgh}{M + 2m}}$
17-69 $\alpha = 0$, $B_x = 0$, $B_y = 761$ kN
17-70 $\theta = 30.1°$
17-71 $\omega = 0.474$ rad/s
17-73 $t = 6.71$ s
17-74 $\alpha = 14.2$ rad/s^2
17-75 $\alpha = 0.476$ rad/s^2
17-77 $t = 1.09$ s
17-78 $v = 4.88$ ft/s

17-79 $t = 6.40$ s
17-81 $A_x = 0, A_y = 289$ N, $\alpha = 23.1$ rad/s^2
17-82 $N_A = 177$ kN, $V_A = 5.86$ kN, $M_A = 50.7$ kN·m
17-83 $F_A = 21.9$ lb
17-85 $N = \dfrac{wx}{2g}(2\omega^2 L - \omega^2 x + 2g\cos\theta)$, $V = wx\sin\theta$, $M = \dfrac{1}{2}wx^2 \sin\theta$
17-86 $N = 0$, $V_t = \dfrac{3M_0}{2L^3}(2Lx - x^2)$, $V_z = \dfrac{mg}{L}x$, $T = 0$, $M_t = \dfrac{mg}{2L}x^2$, $M_z = \dfrac{M_0}{2L^3}(3Lx^2 - x^3)$
17-87 $\alpha = 14.7$ rad/s^2, $a_G = 4.90$ m/s^2
17-89 $\omega = 800$ rad/s
17-91 $a_G = 2$ m/s^2, $\alpha = 22.22$ rad/s^2
17-93 $a = 0, \alpha = 0.309(10^{-3})$ rad/s^2
17-94 $\alpha = 3.89$ rad/s^2
17-95 $\alpha = 9.51$ rad/s^2
17-97 $\alpha = 12.3$ rad/s^2, $(\mu_s)_{min} = 0.0585$
17-98 $\alpha = 15.6$ rad/s^2
17-99 $A_y = 15.0$ lb, $A_x = 0.466$ lb, $a_G = 1.00$ ft/s$^2 \rightarrow$
17-101 $F = 42.3$ N
17-102 $a_G = 0.15$ m/s$^2 \rightarrow$, $T_{AB} = 1.96$ kN$\uparrow$, $\alpha = 0.225$ rad/s$^2 \circlearrowright$, $a_A = 0.3$ m/s$^2 \leftarrow$
17-103 $\alpha = 4.65$ rad/s^2, $T = 359$ N
17-105 $\alpha = 18.9$ rad/s^2, $P = 76.4$ lb
17-106 $\alpha = 0.692$ rad/s^2
17-107 $\alpha = 1.15$ rad/s^2
17-109 $s = 1.40$ in.
17-110 $a = 6.21$ rad/s^2
17-111 $\alpha = \dfrac{6P}{mL}\circlearrowright, a_B = \dfrac{2P}{m}\rightarrow$
17-113 $a_G = \mu_k g \leftarrow, \alpha = \dfrac{2\mu_k g}{r}\circlearrowright$
17-114 $\omega = \dfrac{1}{3}\omega_0, t = \dfrac{\omega_0 r}{3\mu_k g}$
17-115 $a_A = 282$ ft/s$^2 \rightarrow$
17-117 $T_A = \dfrac{4}{7}W$
17-118 $\alpha = 23.4$ rad/s$^2 \circlearrowright$, $B_y = 9.62$ lb

Chapter 18
18-2 $s = 0.661$ m
18-3 $s = 0.859$ m
18-5 $\omega = 2.02$ rad/s
18-6 $\omega = 1.88$ rad/s
18-7 $T = 283$ ft·lb
18-9 $\omega = 21.5$ rad/s
18-10 $v_C = 10.4$ ft/s
18-11 $\omega = 2.45$ rad/s
18-13 $\omega = 6.11$ rad/s
18-14 $\omega = 8.64$ rad/s
18-15 $\omega = 3.16$ rad/s
18-17 $\omega = \sqrt{\omega_0^2 + \dfrac{g}{r^2}s\sin\theta}$
18-18 $v_C = 7.49$ m/s
18-19 $\omega = 6.92$ rad/s
18-21 $\omega_2 = 2.06$ rad/s
18-22 $v_C = 19.6$ ft/s
18-23 $\theta = 0.445$ rev
18-25 $\theta = 5.18$ rev
18-26 $\theta = 70.5°$
18-27 $\omega = 44.6$ rad/s
18-29 $\omega_{DE} = 132$ rad/s
18-30 $\omega = 2.50$ rad/s
18-31 $\omega_2 = \dfrac{\sqrt{7}}{7r}\sqrt{7v_G^2 - 10gR}$
18-33 $\theta = 0.891$ rev, regardless of orientation
18-34 $\omega = 10.5$ rad/s
18-35 $\omega = 7.81$ rad/s
18-37 $s_C = 78.0$ mm
18-38 $\omega = 41.8$ rad/s
18-39 $\omega = 39.3$ rad/s
18-41 $\omega = 1.18$ rad/s
18-42 $\omega = 2.82$ rad/s
18-43 $\omega = 6.81$ m/s
18-45 $\omega_{AB} = \sqrt{\dfrac{3g}{L}\sin\theta}$
18-46 $\omega_{AB} = \sqrt{\dfrac{3g}{L}}$
18-47 $\omega = 3.28$ rad/s
18-49 $\omega_{AB} = 2.21$ rad/s
18-50 $\omega_B = 11.1$ rad/s
18-51 $\theta_0 = 8.94$ rev
18-53 $\theta = 52.0°$
18-54 $v_B = 15.5$ ft/s
18-55 $v_A = 4.00$ m/s
18-57 $\omega_{AB} = 5.48$ rad/s
18-58 $k = 232$ N·m/rad
18-59 $v = 20.7$ m/s
18-61 $v_C = 11.3$ ft/s
18-62 $\omega = 1.80$ rad/s
18-63 $v_A = 21.0$ ft/s
18-65 $v = 8.55$ rad/s

Chapter 19
19-5 $\omega_s = \left(\dfrac{5g\sin\theta}{7r}\right)t, \omega_C = \left(\dfrac{2g\sin\theta}{3r}\right)t$
19-6 $h = 0.375$ ft
19-7 $\omega = 119$ rad/s, $\theta = 22.5° \measuredangle$
19-9 $F_{AB} = 49.0$ N, $v_G = 64.5$ ft/s, $t = 1.53$ s

ANSWERS TO SELECTED PROBLEMS

19-10	$v_G = 0.557$ m/s			$\mathbf{a}_A = \{-7.24\mathbf{j} - 7.24\mathbf{k}\}$ m/s^2
19-11	$\omega_2 = 245$ rad/s	20-6		$\mathbf{v}_A = \{0.0828\mathbf{i} - 0.464\mathbf{j} + 0.124\mathbf{k}\}$ m/s,
19-13	$v_G = 1.39$ m/s, $\omega = 9.49$ rad/s			$\mathbf{a}_A = \{-0.371\mathbf{i} - 0.108\mathbf{j} - 0.278\mathbf{k}\}$ m/s^2
19-14	$\int F\, dt = 15.2$ kN·s	20-7		$\mathbf{v}_A = \{-7.79\mathbf{i} - 2.25\mathbf{j} + 3.90\mathbf{k}\}$ ft/s,
19-15	$\omega = 0.328$ rad/s, $d = 0.0625$ m			$\mathbf{a}_A = \{8.30\mathbf{i} - 35.2\mathbf{j} + 7.02\mathbf{k}\}$ ft/s^2
19-17	$\omega = 11.4$ rad/s	20-9		$\mathbf{v}_B = \{-0.4\mathbf{i} - 2\mathbf{j} - 2\mathbf{k}\}$ m/s,
19-18	$\omega_2 = \{-31.8\mathbf{k}\}$ rad/s			$\mathbf{a}_B = \{-8.20\mathbf{i} + 40.6\mathbf{j} - \mathbf{k}\}$ rad/s^2
19-19	$\omega = 70.8$ rad/s	20-10		$\omega = \{42.4\mathbf{j} + 43.4\mathbf{k}\}$ rad/s, $\alpha = \{-42.4\mathbf{i}\}$ rad/s^2
19-21	$\omega = 12.7$ rad/s $\circlearrowright$	20-11		$\omega = \{2\mathbf{i} + 42.4\mathbf{j} + 42.4\mathbf{k}\}$ rad/s,
19-22	$\omega_A = 47.3$ rad/s $\circlearrowleft$			$\alpha = \{-42.4\mathbf{i} - 82.9\mathbf{j} + 84.9\mathbf{k}\}$ rad/s^2
19-23	$t = 5.08$ s	20-13		$\omega = \{-3.46\mathbf{j}\}$ rad/s, $\dot{\omega} = \{6.93\mathbf{i} - 6.93\mathbf{j}\}$ rad/s^2
19-25	$v = 19.4$ ft/s	20-14		$\omega = \{26.0\mathbf{i} + 5\mathbf{k}\}$ rad/s, $\alpha = [260\mathbf{i}]$ rad/s^2
19-26	$M = 103$ lb·ft	20-15		$\mathbf{v}_c = [2.60\mathbf{i} + 4.00\mathbf{j}]$ m/s,
19-27	$v_B = 1.59$ m/s			$\mathbf{a}_c = [1.73\mathbf{i} + 5.39\mathbf{j} - 2.275\mathbf{k}]$ m/s^2
19-29	$\omega = 1.91$ rad/s	20-17		$\mathbf{v}_A = \{-1.80\mathbf{i}\}$ ft/s,
19-30	$\omega_2 = 0.658$ rad/s, $\theta = 18.8°$			$\mathbf{a}_A = \{-0.750\mathbf{i} - 0.720\mathbf{j} - 0.831\mathbf{k}\}$ ft/s^2
19-31	$\omega_2 = 0.578$ rad/s, $\theta = 15.8°$	20-18		$\omega_B = \{5\mathbf{j} + 5\mathbf{k}\}$ rad/s
19-33	$\omega = 0.190$ rad/s	20-19		$\omega = \{4.35\mathbf{i} + 12.7\mathbf{j}\}$ rad/s,
19-34	$\omega_1 = 2.41$ rad/s, $\omega_2 = 1.86$ rad/s			$\alpha = \{-26.1\mathbf{k}\}$ rad/s^2
19-35	$\omega_T = 1.19(10^{-3})$ rad/s	20-21		$\omega = \{30\mathbf{j} - 5\mathbf{k}\}$ rad/s, $\alpha = \{150\mathbf{i}\}$ rad/s^2
19-37	$\omega = 22.7$ rad/s	20-22		$\mathbf{v}_A = \{10\mathbf{i} + 14.7\mathbf{j} - 19.6\mathbf{k}\}$ ft/s,
19-38	$\omega' = \dfrac{19}{28}\omega$, $v'' = 0.985\omega L$			$\mathbf{a}_A = \{-6.12\mathbf{i} + 3\mathbf{j} - 2\mathbf{k}\}$ ft/s^2
		20-23		$\omega_A = 47.8$ rad/s, $\omega_B = 7.78$ rad/s
19-39	$\omega_2 = 1.01$ rad/s	20-25		$\mathbf{a}_B = \{110\mathbf{k}\}$ ft/s^2
19-41	$\theta = 66.9°$	20-26		$\omega_{AB} = \{-1.00\mathbf{i} - 0.500\mathbf{j} + 2.50\mathbf{k}\}$ rad/s,
19-42	$\omega_2 = 57$ rad/s, $U_F = 367$ J			$\mathbf{v}_B = \{-2.50\mathbf{j} - 2.50\mathbf{k}\}$ m/s
19-43	$h = 4.99$ ft	20-27		$\alpha_{AB} = \{-7.90\mathbf{i} - 3.95\mathbf{j} + 4.75\mathbf{k}\}$ rad/s^2,
19-45	$v = 5.96$ ft/s			$\mathbf{a}_B = \{-19.75\mathbf{j} - 19.75\mathbf{k}\}$ m/s^2
19-46	$h = \dfrac{7}{5}r$	20-29		$\omega_{BD} = \{-2.00\mathbf{i}\}$ rad/s
				$\alpha_{BD} = \{34.5\mathbf{i}\}$ rad/s^2
19-47	$\theta = 50.2°$	20-30		$\mathbf{v}_B = 3.20$ m/s
19-49	$(v_D)_3 = 1.54$ m/s, $\omega_3 = 0.934$ rad/s	20-31		$\omega_{BD} = \{-1.20\mathbf{j}\}$ rad/s
19-50	$\omega_1 = 7.17$ rad/s	20-33		$\omega_{AB} = \{-0.663\mathbf{i} + 0.612\mathbf{j} - 0.918\mathbf{k}\}$ rad/s,
19-51	$\theta = \tan^{-1}\left(\sqrt{\dfrac{7}{5}}e\right)$			$\mathbf{v}_B = \{-7.50\mathbf{j}\}$ ft/s
		20-34		$\alpha_B = \{-37.6\mathbf{j}\}$ ft/s^2
19-53	$\omega = 2.25$ rad/s $\circlearrowright$, $(v_G)_x = 0.149$ m/s,	20-35		$\omega_{BC} = \{0.204\mathbf{i} - 0.612\mathbf{j} + 1.36\mathbf{k}\}$ rad/s,
	$\theta = 38.4°$ $\measuredangle$			$\mathbf{v}_B = [-0.333\mathbf{j}]$ m/s
19-54	$\omega = \sqrt{7.5\dfrac{g}{L}}$	20-37		$\mathbf{v}_C = \{-1.00\mathbf{i} + 5.00\mathbf{j} + 0.800\mathbf{k}\}$ m/s,
				$\mathbf{a}_C = \{-28.8\mathbf{i} - 5.45\mathbf{j} + 32.3\mathbf{k}\}$ m/s^2
19-55	$h_B = 0.980$ ft	20-38		$\mathbf{v}_C = \{-1\mathbf{i} + 5\mathbf{j} + 0.8\mathbf{k}\}$ m/s,
19-57	$v_G = 1.00$ m/s, $\theta = 60°$ $\measuredangle$, $\omega = 5.00$ rad/s $\circlearrowright$			$\mathbf{a}_C = \{-28.8\mathbf{i} - 5.45\mathbf{j} + 32.3\mathbf{k}\}$ m/s^2
19-58	$\theta_1 = 39.8°$	20-39		$\mathbf{v}_B = \{-2.75\mathbf{i} - 2.50\mathbf{j} + 3.17\mathbf{k}\}$ m/s,
				$\mathbf{a}_B = \{2.50\mathbf{i} - 2.24\mathbf{j} - 0.00389\mathbf{k}\}$ ft/s^2
		20-41		$\mathbf{v}_A = \{13.9\mathbf{i} + 40.0\mathbf{j} - 8.00\mathbf{k}\}$ m/s,
Chapter 20				$\mathbf{a}_A = \{-62.4\mathbf{i} + 115\mathbf{j} - 17.5\mathbf{k}\}$ m/s^2
20-1	(a) $\alpha = \omega_x\omega_t\mathbf{j}$, (b) $\alpha = -\omega_x\omega_t\mathbf{k}$	20-42		$\mathbf{v}_B = \{-17.8\mathbf{i} - 3\mathbf{j} + 5.20\mathbf{k}\}$ m/s,
20-2	$\mathbf{v}_A = \{-0.225\mathbf{i}\}$ m/s,			$\mathbf{a}_B = \{9\mathbf{i} - 29.4\mathbf{j} - 1.5\mathbf{k}\}$ m/s^2
	$\mathbf{a}_A = \{-0.1125\mathbf{j} - 0.130\mathbf{k}\}$ m/s^2	20-43		$\mathbf{v}_B = \{-17.8\mathbf{i} - 3\mathbf{j} + 5.20\mathbf{k}\}$ m/s,
20-3	$\mathbf{v}_A = \{-5.20\mathbf{i} - 12\mathbf{j} + 20.8\mathbf{k}\}$ ft/s,			$\mathbf{a}_B = \{3.05\mathbf{i} - 30.9\mathbf{j} + 1.10\mathbf{k}\}$ m/s^2
	$\mathbf{a}_A = \{-24.1\mathbf{i} - 13.3\mathbf{j} - 7.20\mathbf{k}\}$ ft/s^2	20-45		$\mathbf{v}_A = \{4.24\mathbf{i} + 8.49\mathbf{j} + 8.49\mathbf{k}\}$ ft/s,
20-5	$\omega = \{-8.00\mathbf{j}\}$ rad/s, $\alpha = \{64.0\mathbf{i}\}$ rad/s^2,			$\mathbf{a}_A = \{-23.3\mathbf{i} - 29.7\mathbf{j} + 46.7\mathbf{k}\}$ ft/s^2
	$\mathbf{v}_A = \{-0.905\mathbf{i}\}$ m/s,			

Answers to Selected Problems

20-46 $\mathbf{v}_A = \{-8.66\mathbf{i} + 2.26\mathbf{j} + 2.26\mathbf{k}\}$ m/s,
$\mathbf{a}_A = \{-22.6\mathbf{i} - 47.8\mathbf{j} + 45.3\mathbf{k}\}$ m/s^2

20-47 $\mathbf{v}_A = \{-8.66\mathbf{i} + 2.26\mathbf{j} + 2.26\mathbf{k}\}$ m/s,
$\mathbf{a}_A = \{-26.1\mathbf{i} - 44.4\mathbf{j} + 7.92\mathbf{k}\}$ m/s^2

20-49 $\mathbf{v}_T = \{-9.50\mathbf{i} + 3\mathbf{j} + 1.20\mathbf{k}\}$ m/s,
$\mathbf{a}_T = \{-14.4\mathbf{i} + 3.77\mathbf{j} + 4.20\mathbf{k}\}$ m/s^2

20-50 $\mathbf{v}_P = \{-25.5\mathbf{i} - 13.4\mathbf{j} + 20.5\mathbf{k}\}$ ft/s,
$\mathbf{a}_P = \{161\mathbf{i} - 249\mathbf{j} - 39.6\mathbf{k}\}$ ft/s^2

20-51 $\mathbf{v}_P = \{-25.5\mathbf{i} - 13.4\mathbf{j} + 20.5\mathbf{k}\}$ ft/s,
$\mathbf{a}_P = \{161\mathbf{i} - 243\mathbf{j} - 33.9\mathbf{k}\}$ ft/s^2

20-53 $\mathbf{v}_B = \{5\mathbf{i} - 0.5\mathbf{j} + 6.4\mathbf{k}\}$ m/s,
$\mathbf{a}_B = \{2.95\mathbf{i} - 7.52\mathbf{j} + 7.20\mathbf{k}\}$ m/s^2

20-54 $\mathbf{v}_P = \{8.00\mathbf{i} + 32.0\mathbf{j} - 4.00\mathbf{k}\}$ ft/s,
$\mathbf{a}_P = \{-270\mathbf{i} + 72.0\mathbf{j}\}$ ft/s^2

Chapter 21

21-2 $I_{\bar{y}} = \dfrac{3m}{80}(h^2 + 4a^2)$, $I_{y'} = \dfrac{m}{20}(2h^2 + 3a^2)$

21-3 $I_{xy} = \dfrac{ma^2}{20}$

21-5 $I_{yz} = \dfrac{m}{6}ah$

21-6 $I_{xy} = \dfrac{m}{12}a^2$

21-7 $I_{xy} = 0.330$ kg·m^2

21-9 $I_{z'z'} = 0.0961$ slug·ft^2

21-10 $k_y = 2.35$ ft, $k_x = 1.80$ ft

21-11 $I_z = 0.0915$ kg·m^2

21-13 $I_{yz} = 0$

21-14 $I_{xy} = 0.32$ kg·m^2, $I_{yz} = 0.08$ kg·m^2, $I_{xz} = 0$

21-15 $I_{z'} = 0.0595$ kg·m^2

21-17 $\bar{y} = 0.5$ ft, $\bar{x} = -0.667$ ft, $I_{x'} = 0.0272$ slug·ft^2,
$I_{y'} = 0.0155$ slug·ft^2, $I_{z'} = 0.0427$ slug·ft^2

21-18 $I_{xy} = 0.32$ kg·m^2, $I_{yz} = 0.257$ kg·m^2,
$I_{xz} = 0$ kg·m^2

21-19 $I_z = 0.429$ kg·m^2

21-21 $I_{xy} = 0.0966$ slug·ft^2 $I_{yz} = 0.0483$ slug·ft^2
$I_{xz} = 0.0372$ slug·ft^2

21-25 $\mathbf{H} = \{-477(10^{-6})\mathbf{i} + 198(10^{-6})\mathbf{j} + 0.169\mathbf{k}\}$ kg·m^2/s

21-26 $\mathbf{H}_O = \{21.9\mathbf{i} + 1.69\mathbf{k}\}$ kg·m^2/s, $T = 78.5$ J

21-27 $\mathbf{H}_O = \{21.9\mathbf{i} + 0.5625\mathbf{j} + 1.69\mathbf{k}\}$ kg·m^2/s,
$T = 81.3$ J

21-29 $\omega_{BC} = 10.8$ rad/s

21-30 $h = 2.24$ in.

21-31 $\boldsymbol{\omega} = \{-0.444\mathbf{j} + 0.444\mathbf{k}\}$ rad/s

21-33 $\omega_p = 4.82$ rad/s

21-34 $\boldsymbol{\omega} = \{-2.16\mathbf{i} + 5.40\mathbf{j} + 7.20\mathbf{k}\}$ rad/s,
$\mathbf{u}_o = -0.233\mathbf{i} + 0.583\mathbf{j} + 0.778\mathbf{k}$

21-35 $\boldsymbol{\omega} = \{-0.954\mathbf{i} + 2.38\mathbf{j} + 3.18\mathbf{k}\}$ rad/s,
$\mathbf{u}_A = -0.233\mathbf{i} + 0.583\mathbf{j} + 0.778\mathbf{k}$

21-37 $\boldsymbol{\omega} = \{-0.750\mathbf{j} + 1.00\mathbf{k}\}$ rad/s

21-38 $T = 6.46$ ft·lb

21-39 $T_2 = 15.5$ ft·lb

21-41 $\Sigma M_x = (I_x\dot{\omega}_x - I_{xy}\dot{\omega}_y - I_{xz}\dot{\omega}_z)$,
$-\Omega_z(I_y\omega_y - I_{yz}\omega_z - I_{yx}\omega_x)$,
$+\Omega_y(I_z\omega_z - I_{zx}\omega_x - I_{zy}\omega_y)$ Similarly for ΣM_y and ΣM_z.

21-42 $\Sigma M_x = I_x\dot{\omega}_x - I_y\Omega_z\omega_y + I_z\Omega_y\omega_z$

21-43 $E_X = 0$, $F_Y = \dfrac{mL^2\omega_1\omega_2}{12a}$, $E_Y = \dfrac{mL^2\omega_1\omega_2}{12a}$,
$E_Z = F_Z = \dfrac{mg}{2}$

21-45 $F_A = 277$ N, $F_B = 166$ N

21-46 $A_x = -2.12$ lb, $A_y = 1.06$ lb, $B_x = 2.12$ lb,
$B_y = -1.06$ lb $B_z = 4.25$ lb

21-47 $A_x = -2.00$ lb, $A_y = 0.627$ lb, $B_x = 2.00$ lb,
$B_y = -1.37$ lb $B_z = 4$ lb

21-49 $\dot{\omega}_y = 58.9$ rad/s^2, $A_y = 0$, $A_x = B_x = 72.0$ N,
$A_z = B_z = 12.9$ N

21-50 $F_A = 8.83$ lb, $F_B = 11.2$ lb

21-51 $\dot{\omega}_y = -102$ rad/s^2, $A_x = B_x = 0$, $A_y = 0$,
$A_z = 297$ N, $B_z = -143$ N

21-53 $M_z = 0$, $A_x = 0$, $M_y = 0$, $\theta = 64.1°$,
$A_y = 1.30$ lb, $A_z = 20.2$ lb

21-54 $N = 148$ N, $F_f = 0$

21-55 $\Sigma M_x = \dfrac{I_{yz}}{\rho^2}v_G^2$

$\Sigma M_y = -\dfrac{I_{zx}}{\rho^2}v_G^2$

$\Sigma M_z = 0$

21-57 $\omega = \sqrt{\dfrac{3g\tan\theta}{L(2\sin\theta + 1)}}$

21-58 $M_Y = -218$ lb·ft, $M_Z = 0$, $M_X = 100$ lb·ft

21-59 $\theta_D = 139°$, $m_D = 0.661$ kg, $\theta_F = 40.9°$,
$m_F = 1.32$ kg

21-61 $\alpha = 69.3°$, $\beta = 128°$, $\gamma = 45°$. No, the orientation will not be the same for any order. Finite rotations are not vectors.

21-62 $M_x = 12.4$ lb·ft, $M_y = 0$, $M_z = 0$

21-63 $M_x = 2$ kN·m

21-65 $\omega_P = 1.19$ rad/s

21-66 $M_x = 328$ N·m

21-67 a) $A_y = 1.49$ kN, $B_y = 2.43$ kN,
b) $A_y = -1.24$ kN, $B_y = 5.17$ kN,
c) $A_y = 1.49$ kN, $B_y = 2.43$ kN

21-69 $\omega_s = 3.63(10^3)$ rad/s

21-70 $\theta = 68.1°$

21-71 $M_x = 27.0$ N·m

21-74 $\dot{\psi} = 2.35$ rev/h

21-75 $H_G = 6.09(10^{-3})$ kg·m^2/s

Chapter 22

21-77	$H_G = 12.5 \text{ Mg} \cdot \text{m}^2/\text{s}$
21-78	$\dot{\phi} = \left(\dfrac{2g \cot\theta}{a + r\cos\theta}\right)^{1/2}$

Chapter 22

22-1	$\ddot{y} + 56.1\, y = 0,\ y\vert_{t=0.22\text{ s}} = 0.192$ m
22-2	$y = -0.05 \cos(20t)$
22-3	$y = 0.107 \sin(7.00t) + 0.100 \cos(7.00t),\ \phi = 43.0°$
22-5	$\omega_n = 49.5$ rad/s, $\tau = 0.127$ s
22-6	$y = \{-0.126 \sin(3.16t) - 0.09 \cos(3.16t)\}$ m $C = 0.155$ m
22-7	$\omega_n = 19.7$ rad/s, $C = 1$ in. $y = (-0.0833 \cos 19.7t)$ ft
22-9	$f = 1.30$ Hz, $y = -0.05 \cos(8.16t),\ C = 50$ mm
22-10	$\tau = 2\pi\sqrt{\dfrac{2mL}{3mg + 6kL}}$
22-11	$\tau = 2\pi\sqrt{\dfrac{Mr^2 + 2mk_O{}^2}{kr^2}}$
22-13	$\tau = 2\pi\sqrt{\dfrac{k_G^2 + d^2}{gd}}$
22-14	$k = 90.8$ lb·ft/rad
22-15	$\tau = 2\pi\sqrt{\dfrac{l}{3g}}$
22-17	$k_1 = 2067$ N/m, $k_2 = 302$ N/m, or vice versa
22-18	$f = 0.624$ Hz
22-19	$f = 0.597$ Hz
22-21	$\theta = 0.4 \cos(20.7t)$
22-22	$k = 1.36$ N/m, $m_B = 3.58$ kg
22-23	$l = 0.457$ m
22-25	$f = 0.900$ Hz
22-26	$\tau = 2\pi\left(\dfrac{a}{b}\right)\sqrt{\dfrac{m}{k}}$
22-27	$f = \dfrac{1}{\pi}\sqrt{\dfrac{k}{m}}$
22-29	$\tau = 0.910$ s
22-30	$f = \dfrac{1}{\pi}\sqrt{\dfrac{k}{m}}$
22-31	$\ddot{y} + \dfrac{(k_1 + k_2)}{m} y = 0$
22-33	$\tau = 2\pi\sqrt{\dfrac{Mk_z^2}{k}}$
22-34	$\ddot{\theta} + 26.0\theta = 0$
22-35	$f = 0.851$ Hz
22-37	$\tau = 3.85\sqrt{\dfrac{m}{k}}$
22-38	$a = 1.05$ ft
22-39	$\tau = 2.81$ s
22-41	$x = A \sin \omega_n t + B \cos \omega_n t + \dfrac{F_0/k}{1 - \left(\dfrac{\omega}{\omega_n}\right)^2} \cos \omega t$
22-42	$y = A \sin \omega_n t + B \cos \omega_n t + \dfrac{F_0}{k}$
22-43	$y = \{-0.0232 \sin 8.97\, t + 0.333 \cos 8.97 t$ $\quad + 0.0520 \sin 4t\}$ ft
22-45	$k \leq 714$ N/m
22-46	$y = (361 \sin 7.75t + 100 \cos 7.75t,$ $\quad - 350 \sin 8t)$ mm
22-47	$\tau_d = 0.734$ s
22-49	$(x_p)_{\max} = 29.5$ mm
22-50	$\ddot{\theta} + \dfrac{4c}{m}\dot{\theta} + \dfrac{k}{m}\theta = 0$
22-51	$y_P = 0.111 \sin(5t - 0.588)$ m
22-53	$\omega = 14.0$ rad/s
22-54	$(x_p)_{\max} = 14.6$ mm
22-55	$(x_p)_{\max} = 35.5$ mm
22-57	$\omega = 19.7$ rad/s
22-58	$C = 0.490$ in.
22-59	$\omega = 19.0$ rad/s
22-61	$(x_p)_{\max} = 4.53$ mm
22-62	$Y = \dfrac{mr\omega^2 L^3}{48EI - M\omega^2 L^3}$
22-63	$\tau_d = 0.666$ s
22-65	$\phi = 9.89°$
22-66	MF $= 0.997$
22-67	$y = [-0.0702\, e^{-3.57t} \sin(8.54t)]$ m
22-69	$\theta_{\max} = 0.106$ rad
22-71	$t = 3.99$ s
22-73	$1.55\ddot{\theta} + 540\dot{\theta} + 200\theta = 0,$ $(c_{dp})_c = 3.92$ lb·s/ft
22-74	$c = \sqrt{8(m + M)k},\ x_{\max} = \left[\dfrac{m}{e}\sqrt{\dfrac{1}{2k(m + M)}}\right] v_0$
22-75	$x_{\max} = \dfrac{2mv_0}{\sqrt{8k(m + M) - c^2}} e^{-\pi c/(2\sqrt{8k(m+M)-c^2})}$
22-77	$x = 0.119 \cos(6t - 1.46°)$ m
22-78	$\ddot{y} + 16\dot{y} + 4y = 0$; System is overdamped.
22-79	$L\ddot{q} + R\dot{q} + \dfrac{1}{C}q = 0$

Index

A

a–s (acceleration–position) graphs, 21
a–t (acceleration–time) graphs, 19–20
Absolute acceleration, 94, 361
Absolute dependent motion analysis, *see* Dependent motion analysis
Absolute motion analysis, 328–331, 388
Absolute position, 93
Absolute velocity, 93, 337
Acceleration (a), 7–8, 19–21, 33, 35, 39, 54–55, 69, 94, 102, 104, 108–173, 311, 313, 315–317, 361–368, 378–379, 388–389, 392–451, 537–542, 554
   absolute, 94, 361
   angular (α), 69, 313, 317, 393, 537–542
   average, 7, 33
   centripetal, 55
   circular motion and, 315–316, 361–363
   constant, 8, 39, 117, 313
   continuous motion and, 7–8
   coordinating fixed and translating reference frames, 361–368, 389
   Coriolis, 379, 389
   curvilinear motion and, 33, 35, 54–55, 69, 104
   cylindrical components and, 69, 148–152, 171
   deceleration and, 7
   displacement and, 117
   equations of motion for, 110, 112–122, 134–139, 148–152, 171, 407–415, 423–428, 437–442, 449
   erratic motion and, 19–21
   fixed-axis rotation and, 313, 315–317, 388, 423–428, 449
   fixed-point rotation and, 537–542
   force (**F**) and, 108–173, 392–451
   general plane motion and, 361–368, 378–379, 388–389, 437–442, 449
   graphs of variables, 19–21, 102
   gravitational (g), 111
   hodographs and, 33
   inertia and, 109, 112–113
   instantaneous, 7, 33
   kinematics of particles and, 7–8, 33, 35, 39, 54–55, 69, 94, 102
   kinetics of a particle, 108–173
   magnitude of, 35, 54–55, 69, 361, 378
   mass (m) and, 109–122
   moment of inertia (I) and, 393, 423, 437–442, 449
   normal (n) components of, 54–55, 134–139, 171, 315–316
   normal force (N) and, 148–149
   planar kinetics, equations of motion for, 407–415, 449
   planar kinetics of rigid bodies, 392–451
   planar kinematics of rigid bodies, 311, 313, 315–317, 361–368, 378–379, 388–389
   procedure for analysis of, 317, 363
   projectile motion and, 39
   rectangular coordinates and, 35, 104, 116–122, 171
   rectilinear kinematics and, 7–8, 19–21, 102
   relative, 94
   relative-motion analysis and, 94, 361–368, 378–379, 389, 554
   resistance of body to, 393
   rigid-body kinematics for, 537–542
   rotating axes, 378–379, 389, 554
   rotation and, 313, 315–317, 361–368, 388–389
   rotational equations of motion, 408–409, 423–428, 449
   sign convention for, 7, 8
   tangential (t) components of, 54–55, 134–139, 171, 315–316
   tangential force (tan) and, 148–149
   three-dimensional rigid-body motion, 537–542
   time and, 117
   time derivative for, 537–538
   translating axes, 94, 361–368, 389, 554
   translation and, 311, 361–368, 389, 407
   translation and rotation, 378–379
   translational equations of motion, 407, 410–415, 449
   unbalanced force and, 109–110, 115
   velocity (v) and, 7–8, 33
Amplitude of vibration, 619–620, 630
Angular acceleration (α), 69, 313, 388, 393, 537–542
   cylindrical components, 69
   fixed-axis rotation, 313, 388
   fixed-point rotation, 537–542
   time derivative for, 537–538
Angular deceleration, 313
Angular displacement ($d\theta$), 312
Angular impulse and momentum, principle of, 278–283
Angular momentum (**H**), 274–283, 305, 496–503, 516–519, 520–523, 530–531, 575–578, 603, 606–607, 614
   angular impulse and, 278–283, 305
   arbitrary point A for, 576
   center of mass (G) for, 496–497, 576
   conservation of, 280, 516–519, 520–523, 531
   direction of, 274
   eccentric impact and, 520–523, 531
   fixed-axis rotation and, 498, 530
   fixed point O for, 576
   free-body diagrams for, 274, 280–283
   general plane motion and, 499, 530
   gyroscopic motion and, 603
   kinetics of a particle, 274–283
   magnitude of, 274
   moment of a force relations with, 275–277
   moment of momentum, 274
   principle axes of inertia, 676
   principle of impulse and, 278–283, 305, 501–503, 578, 614
   procedures for analysis of, 280, 517
   rectangular components of momentum, 577
   right-hand rule for, 274
   rigid-body planar motion, 496–503, 516–519, 520–523, 530–531
   scalar formulation, 274, 279
   system of particles, 276, 278–279
   three-dimensional rigid bodies, 575–578, 603, 606–607, 614
   torque-free motion and, 606–607
   translation and, 498, 530
   units of, 274
   vector formulation, 274, 279
Angular motion, 312–313, 317, 388, 537–542, 603–605, 615
   fixed-axis rotation, 312–313, 317, 388
   fixed-point rotation, 537–542
   gyroscopic, 603–605, 615
Angular position (θ), 312
Angular velocity (ω), 68, 312, 388, 537–542, 600–603
   cylindrical components, 68
   Euler angles for, 600
   fixed-axis rotation, 313, 388

705

706   INDEX

Angular velocity (ω) (*Continued*)
   fixed-point rotation, 537–542
   gyroscopic motion and, 600–603
   nutation angle (θ), 600–602
   precession (φ), 600–602
   spin (ψ), 600–602
   time derivative for, 537–538
Apogee, 165
Areal velocity, 160
Average acceleration, 7, 33
Average power, 199
Average speed, 6
Average velocity, 6, 32
Axes, 53–59, 93–97, 105, 310, 312–319, 336–342, 361–368, 375–383, 388–389, 423–428, 449, 455, 488, 498, 530, 552–559, 568–569, 586–595
   acceleration (*a*) of, 94, 361–368, 389, 554
   angular motion and, 312–313, 317
   arbitrary, moment of inertia about, 569
   circular motion and, 314–317, 361–363
   constant motion, 588
   coordinating fixed and translating reference frames, 336–342, 361–368, 388–389
   curvilinear motion, 53–59
   equations of motion for, 423–428, 449, 586–595
   Euler's equations for, 588–589
   fixed, 588–590
   fixed, rotation about, 310, 312–319, 388, 423–428, 449, 455, 488
   fixed reference frame, 53–59
   impulse and momentum of, 498, 530
   inertia (*I*), principle axes of, 568–569
   kinematics of a particle, 53–59, 93–97, 105
   kinematics of rigid bodies, 310, 312–319, 336–342, 361–368, 375–383, 388–389, 552–559
   kinetic energy and, 455, 488
   moments of inertia (*I*) about, 423, 568–569
   pinned-end members, 336–342, 361–368
   planar (normal and tangential) motion, 53–55
   planes of symmetry, 569
   position coordinates for, 93–97, 105
   position vector (**r**) for, 553
   procedure for analysis of, 56, 94, 339
   relative-motion analysis of, 93–97, 105, 336–342, 361–368, 375–383, 389, 552–559
   rigid-body planar motion, 310, 312–319, 336–342, 361–368, 375–383, 423–428, 449, 455, 488, 498, 530
   rotating, 375–383, 389, 552–559
   rotation about, 310, 312–319, 361–368, 388, 488, 498, 530
   rotational equations of motion for, 423–428, 449, 455, 488, 586–587
   slipping, 375–383
   symmetrical spinning axes, 589–590
   three-dimensional particle motion, 55
   three-dimensional rigid-body motion, 552–559, 568–569, 586–595
   translating, 93–97, 105, 389, 552–559, 588
   translating frames of reference, 93–97, 105
   translation and rotation, 336–342, 361–368, 375–383
   translational equations of motion for, 449, 488, 586
   velocity (*v*), 93, 336–342, 389, 553
Axis of rotation, 537–538, 543
Axisymmetric characteristics, 606

B
Base point, 311
Binormal coordinates, 55, 134, 171
Body cone, 537–538

C
Cartesian vector notation, 660
Center of curvature, 53
Center of mass (*G*), 115, 496–497, 565, 567–568, 576, 579, 586–587
   angular momentum (**H**) and, 496–497, 576
   kinetic energy and, 579
   moments of inertia and, 565, 567–568
   parallel-axis theorem and, 567–568
   rigid-body planar motion, 496–497
   rotational equations of motion and, 586–587
   systems of particles and, 115
   three-dimensional rigid bodies, 567–568, 576, 579, 586–587
Center of percussion, 508
Central-force motion, 160–167, 171, 280
   areal velocity, 160
   circular orbit, 164
   conservation of angular momentum, 280
   directrix, 162
   eccentricity (*e*), 162–163, 171
   elliptical orbit, 164–165
   equations of motion, 160–163
   escape velocity, 164
   focus, 162
   gravitational attraction (*G*) and, 161–163
   Kepler's laws, 166
   kinetics of a particle, 160–167
   parabolic path, 164
   path of motion, 160–161
   space mechanics and, 160–167
   trajectories, 161–167, 171
   velocity (*v*) and, 160–161
Central impact, 259–261, 262, 304–305
   coefficient of restitution (*e*), 260, 305
   conservation of momentum for, 260, 304
   deformation impulse, 259
   kinetics of a particle, 259–261, 262
   principle of impulse and momentum for, 260
   procedure for analysis of, 262
   restitution impulse, 259
Centripetal acceleration, 55
Centripetal force, 134
Centrode, 350–351
Chain Rule, 664–666
Circular motion, 314–317, 338–342, 349–355, 361–363, 389
   acceleration (*a*), 315–316, 361–363
   instantaneous center (IC) of zero velocity, 349–355, 389
   planar rigid-body motion, 314–317, 338–342, 349–355, 361–363
   position and displacement from, 314
   procedures for analysis of, 317, 351
   relative-motion analysis of, 338–342, 349–355, 361–363
   relative velocity and, 338
   right-hand rule for, 314
   rolling without slipping, 338, 362
   rotation about a fixed axis, 314–317
   velocity (*v*), 314, 338–342, 349–355, 389

Circular orbit, 164
Circular path of a point, 314–316, 349, 361–362
Coefficient of restitution, 260–265, 305, 520–523, 531
Coefficient of viscous damping, 641
Complimentary solution, vibration, 637–638
Composite bodies, moment of inertia for, 399
Conservation of energy, 212–218, 226, 475–480, 490, 631–634, 654
　conservative forces and, 212–218, 226, 475–480, 631–634
　displacement and, 631, 654
　elastic potential energy, 475, 490
　gravitational potential energy, 475, 490
　kinetic energy and, 212–213
　kinetics of a particle, 212–218, 226
　natural frequency (ω_n) from, 631–632, 654
　nonconservative forces and, 476
　potential energy (V) and, 212–218, 226, 475–480, 490
　procedures for analysis using, 213, 477, 632
　rigid-body planar motion, 475–480, 490
　system of connected bodies, 631
　systems of particles, 213
　time derivative for, 631–632, 654
　vibration and, 631–634, 654
　weight (W), displacement of, 212, 226
　work (W) and, 212–218, 226, 475–480, 490
Conservation of mass, 291
Conservation of momentum, 247–253, 260, 262–265, 280, 304, 516–519, 520–523, 531
　angular, 280, 516–519, 531
　eccentric impact and, 520–523
　impact and, 260, 262–265, 304
　impulsive forces and, 247–248
　kinetics of a particle, 247–253, 260, 262–265, 280
　linear, 247–253, 260, 262–265, 304, 516–519, 531
　nonimpulsive forces and, 247–248
　procedures for analysis of, 248, 262, 280, 517

rigid-body planar motion, 516–519, 520–523, 531
systems of particles, 247–253, 280, 304
Conservative force, 208–218, 226, 475–480, 490, 631–634
　conservation of energy, 212–218, 226, 475–480, 490, 631–634
　elastic potential energy, 209, 475, 490
　friction force compared to, 208, 226
　gravitational potential energy, 208, 210, 226, 475, 490
　potential energy (V) and, 208–211, 226, 475–480, 490
　potential function for, 210–211
　procedure for analysis of, 213
　spring force as, 208–211, 226, 475, 490
　vibration and, 631–634
　weight (W), displacement of, 208–211, 226
　work (U) and, 208–218, 226, 475–480, 490
Constant acceleration, 8, 39, 117, 313
Constant force, work of, 177, 208, 225, 456, 489
Constant velocity, 39
Continuous motion, 5–14, 102
　acceleration (a), 7–8
　displacement (Δ), 5
　kinematics of particles, 5–14, 102
　position (s), 5, 8
　procedure for analysis of, 9
　rectilinear kinematics of, 5–14, 102
　time (t), functions of, 8
　velocity (v), 6–8
Control volume, 289–297
　fluid flow, 289–297
　fluid streams, 291–294
　kinematics of a particle, 289–297
　mass flow, 289–290, 295
　mass gain and loss (propulsion), 295–297
　procedure for analysis of, 292
　steady flow, 291–294
　volumetric flow (discharge), 290
Coordinates, 5, 34–38, 53–59, 67–74, 81–86, 93–97, 103–105, 116–122, 134–139, 148–152, 171, 312, 314–316, 336, 388, 407–410, 543–545, 552, 575–579, 588–589

acceleration (**a**) and, 35, 54–55, 69, 94, 104, 116–122, 134–139, 148–152, 171, 315–316
angular motion, 312
angular momentum (**H**) and, 575–577
binormal, 55, 134, 171
centripetal force, 134
circular motion, 314–316
continuous motion, 5
coordinating fixed and translating reference frames, 336
curvilinear motion, 34–38, 53–59, 67–74, 103–105
cylindrical (r, θ, z), 70, 148–152, 171
dependent motion analysis and, 81–86, 105
directional angle (ψ), 148–149
dot notation for, 34
equations of motion and, 116–122, 134–139, 148–152, 171, 407–410, 588–589
fixed origin (O), 5
fixed reference frame, 34, 53–59, 67–74, 336, 588–589
force (F) and, 116–122, 134–139
frictional forces (**F**) and, 148
inertial, 575, 578–579
kinematics of a particle, 5, 34–38, 53–59, 67–74, 81–86, 93–97, 103–105
kinetic energy, 578–579
kinetics of a particle, 116–122, 134–139, 148–152, 171
normal (n), 53–55, 104, 134–139, 171, 315–316
normal forces (**N**) and, 148–149
planar motion, 53–55
polar, 67–70
position (s), 5
position-coordinate equations, 81–86, 105
position vector (**r**), 34, 68, 93–97, 314
procedures for analysis using, 36, 56, 70, 82, 94, 116–117, 135, 149
radial (r), 67–70
rectangular (x, y, z), 34–38, 103, 116–122, 171, 407–410, 577
relative-motion analysis and, 93–97, 105, 336, 552
rigid-body planar motion, 312, 314–316, 336, 407–410, 575–577

708 INDEX

Coordinates (*Continued*)
   rotating reference frame for, 552
   rotation about a fixed axis, 312, 314–317
   rotational motion, 408–409
   symmetrical bodies, 407–410
   tangential (t), 53–55, 104, 134–139, 171, 315–316
   tangential forces (tan) and, 148–149
   three-dimensional motion, 55, 543–545, 552, 575–579, 588–589,
   translating axes and, 93–97, 105
   translating reference frame, 93–97, 105, 336, 407
   translating systems, 543–545, 552
   transverse (θ), 67–70
   velocity (**v**) and, 34–35, 53, 68, 93, 314
Coriolis acceleration, 379, 389
Couple moment (M), work (W) of a, 458–459, 489
Critical damping coefficient, 642
Critically damped vibration systems, 642
Cross product, 660–661
Curvilinear motion, 32–38, 53–59, 67–74, 103–104
   acceleration (a), 33, 35, 54–55, 69
   center of curvature (O'), 53
   coordinates for, 34–38, 53–59, 67–74, 103–104
   cylindrical components, 67–74, 104
   cylindrical (r, θ, z) coordinates, 70
   displacement (Δ), 32
   fixed reference frames for, 34, 53–59, 67–74
   general, 32–33
   normal (n) axes, 53–59, 104
   kinematics of a particle, 32–38, 53–59, 67–74, 103–104
   planar motion, 53–55
   polar coordinates, 67–70, 104
   position (s), 32, 34, 68
   procedures for analysis of, 36, 56, 70
   radial coordinate (r), 67–70
   radius of curvature (ρ), 53
   rectangular (x, y, z) coordinates, 34–38, 103
   tangential (t) axes, 53–59, 103
   time derivatives of, 34–35, 70
   three-dimensional motion, 55
   transverse coordinate (θ), 67–70

   velocity (*v*), 32–35, 53, 68
Curvilinear translation, 310–311, 388, 411, 449
Cycle, vibration frequency, 620
Cylindrical components, 67–74, 104, 148–152, 171
   acceleration (a) and, 69, 148–152, 171
   curvilinear motion, 67–74, 104
   directional angle (ψ), 148–149
   cylindrical (r, θ, z) coordinates, 70, 148–152, 171
   equations of motion and, 148–152, 171
   friction (**F**) force, 148
   normal force (N) and, 148–149
   polar coordinates for, 67–70, 104
   position vector (**r**) for, 68
   procedures for analysis using, 70, 149
   radial coordinate (r), 67–70
   tangential force and, 148–149
   time derivatives of, 70
   transverse coordinate (θ), 67–70
   velocity (v) and, 68

D

D' Alembert principle, 112
Damped vibrations, 617, 641–646, 654–655
   critically damped systems, 642
   motion of, 617
   overdamped systems, 642
   resonance from, 645, 655
   underdamped systems, 643
   viscous forced, 644–646, 655
   viscous free, 641–643, 654
Damping factor, 643
Dashpot, 641
Deceleration, 7, 313
Deformation, 182–183, 259–265, 520–523
   central impact and, 259–261, 262
   coefficient of restitution (e), 260–265, 520–523
   conservation of momentum and, 520–523
   displacement and, 182–183
   eccentric impact and, 520–523
   elastic impact and, 261
   friction force and, 183
   impact and, 259–265, 520–523

   impulse, 259
   kinetics of a particle, 182–183, 259–265
   localized, 183
   maximum, 259
   oblique impact and, 259, 262
   period of, 259
   plastic impact and, 261
   principles of work and energy and, 182–183
   restitution phase, 259–262, 520
   rigid-body planar motion, 520–523
   separation of contact points, 523
   sliding and, 183
   systems of particles, 182–183
Dependent motion analysis, 81–86, 105
   particle kinematics, 81–86, 105
   position coordinates for, 81–82, 105
   procedure for, 82
   time derivatives for, 81–82, 105
   time-differential equations, 81–82, 105
Derivative equations, 656–657
Diagrams for impulse and momentum, 233
Direction, 6, 32, 34–35, 53, 175–176, 274, 312–314, 350, 361, 554
   acceleration (a) and, 361
   angular acceleration (α), 313
   angular displacement ($d\theta$), 312
   angular momentum, 274
   constant, 554
   force displacement and, 175–176
   instantaneous center (IC) of zero velocity from, 350
   relative-motion analysis, 350, 361, 554
   right-hand rule for, 274, 312, 314
   rotation about a fixed axis, 312–314
   three-dimensional rigid bodies, 554
   translation and rotation, 350, 361
   velocity (v) and, 6, 32, 34–35, 53, 250, 314
Directional angle (ψ), 148–149
Directrix, 162
Disk elements, moment of inertia of, 395
Displacement (Δ or d), 5, 20, 32, 117, 175–177, 182–183, 226, 312, 314, 336, 456–458, 489, 535–538, 617–625, 639, 654
   acceleration (a) as a function of, 117

amplitude, 619–620
angular ($d\theta$), 312
  circular motion and, 314
  conservation of energy and, 631, 654
  continuous motion and, 5
  couple moment (M) and, 458, 489
  curvilinear motion, 32
  deformation from, 182–183
  erratic motion, 20
  graphing determination of, 20
  kinematics of a particle, 5, 20, 32
  kinetics of a particle, 117, 175–177, 182–183
  periodic support, 639
  phase angle (ϕ), 620
  position change as, 5, 312, 314
  principle of work and energy, 182–183, 226
  relative-motion analysis and, 336
  resonance and, 639, 645
  right-hand rule for direction of, 312, 314
  rigid-body planar motion, 312, 314, 336, 458, 489
  rotation about a fixed point, 312, 314, 535–538
  simple harmonic motion, 618
  sliding, 183
  spring force and, 618, 620–625
  system of particles, 182–183
  three-dimensional rigid bodies, 535–538
  translation and rotation causing, 336
  vertical, 177, 457
  vibration and, 617–625, 639, 654
  weight (W) and, 177, 447–448
  work of a force and, 175–177, 456–458, 489
  work of a weight and, 457
Distance, 5. *See also* Displacement
Dot notation, 34
Dot product, 176, 662
Drag force, 295
Dynamic equilibrium, 112
Dynamics, 3–4
  principles of, 3–4
  procedure for problem solving, 4
  study of, 3

E

Eccentric impact, 520–523, 531
Eccentricity (e), 162–163, 171

Efficiency (ε), 199–202, 226
  energy (E) and, 199–202, 226
  mechanical, 199–200
  power (P) and, 199–202, 226
  procedure for analysis of, 200
Elastic impact, 261
Elastic potential energy, 209, 226, 475, 490. *See also* Spring force
Electrical circuit analogs, vibrations and, 647, 655
Elliptical orbit, 164–165
Energy (E), 174–228, 225–226, 452–492, 578–581, 615, 631–634, 654
  conservative forces and, 208–218, 226, 475–480, 490
  conservation of, 212–218, 226, 475–480, 490, 631–634, 654
  efficiency (ε) and, 199–202, 226
  elastic potential, 209, 475, 490
  gravitational potential, 208, 475, 490
  heat generation, 183
  internal, 183
  kinetic, 180–182, 208, 212–213, 225, 453–456, 460–466, 476–477, 488, 578–581, 615
  kinetics of a particle, 174–228
  mechanical, 212–218, 226
  natural frequency (ω_n) and, 631–632, 654
  potential (V), 208–218, 226, 475–480, 490
  power (P) and, 199–202, 226
  principle of work and, 180–188, 225–226, 460–466, 490, 615
  procedures for analysis of, 181, 200, 213, 461, 477
  rigid-body planar motion and, 452–492
  system of connected bodies, 631
  systems of particles, 182–188, 213
  three-dimensional rigid bodies, 578–581, 615
  time derivative for, 631–632, 654
  work (U) and, 174–228, 452–492
  vibration and, 631–634, 654
Equations of motion, 110, 112–122, 134–139, 148–152, 160–163, 171, 181–182, 233–234, 407–415, 423–428, 437–442, 449, 586–595, 615
  acceleration (a) and, 110, 112–122, 134–139, 148–152, 407–415, 423–428, 449

central-force motion, 160–163
centripetal force, 134
cylindrical (r, θ, z) coordinates, 148–152, 171
dynamic equilibrium and, 112
external force, 114–115, 407–409
fixed-axis motion, 588–591
fixed-axis rotation, 423–428, 449
force (**F**) and, 110, 112–122, 134–139, 148–149, 407–415, 423–428, 449
free-body diagrams for, 112–113, 171, 407–410
friction (**F**) force, 117, 148
general plane motion, 410, 437–442, 449
gravitational attraction (G), 110–111, 161–163
inertial reference frame for, 112–113, 171, 407–410, 586–587
instantaneous center (IC) of zero velocity and, 437
internal force, 114–115, 408–409
kinetic diagrams for, 112, 171, 408–410
kinetics of a particle, 110, 112–122, 134–139, 148–152, 160–163, 171, 181–182, 233–234
linear impulse and momentum, 233–234
mass (m) and, 110, 112–122
moment equation about instantaneous center (IC), 437
moment equation about point O, 423
moments of inertia (I) and, 423, 437–442
Newton's second law, 110, 171
normal (n) coordinates, 134–139, 171
normal (**N**) force, 148–149, 171
planar kinetics, 407–415, 449
principle of work and energy, 180–182
procedures for analysis using, 116–117, 135, 149, 412, 424, 438, 590
rectangular (x, y, z) coordinates, 116–122, 171, 407–410
rigid-body planar motion, 407–415, 423–428, 437–442, 449
rotational equations of motion, 408–409, 423–428, 449, 586–587, 615
slipping and, 437
spring force, 117

710 INDEX

Equations of motion (*Continued*)
  symmetrical spinning axes, 589–590
  symmetry of reference frames for, 407–410
  systems of particles, 114–115
  tangential (*t*) coordinates, 134–139, 171
  tangential force, 148–149, 171
  three-dimensional rigid bodies, 586–595, 615
  trajectories, 161–167
  translational equations of motion, 407, 410–415, 449, 586, 615
Equilibrium, equations of motion and, 112
Equilibrium position, vibrations, 618–620, 631, 637, 654
Erratic motion, 19–24, 102
  a–s (acceleration–position), 21
  a–t (acceleration–time), 19–20
  integration of equations for, 20
  particle rectilinear kinematics for, 19–24, 102
  s–t (position–time), 19–20
  v–s (velocity–position), 21
  v–t (velocity–time), 19–20
Escape velocity, 164
Euler angles, 600
Euler's equations, 589–590
Euler's theorem, 536
External force, 114–115, 180, 234, 276, 407–409
External impulses, 247–248
External work, 183

F

Finite rotation, 536
Fixed-axis motion, 586–590, 600–605, 615
  equations of motion for, 586–590, 615
  Euler's equations of motion for, 588–589
  gyroscopic motion, 589–590, 600–605, 615
  three-dimensional rigid bodies, 586–590, 600–605, 615
  symmetrical spinning, 589–590
Fixed-axis rotation, 310, 312–319, 388, 423–428, 449, 455, 488, 498, 530
  acceleration (*a*) of, 313, 315–317, 388, 423–428, 449
  angular acceleration (*α*), 313

angular displacement (*dθ*), 312
angular motion and, 312–313, 317, 388
angular position (*θ*), 312
angular velocity (*ω*), 312
circular motion, 314–317
circular path of a point, 314–316
displacement from, 312, 314
equations of motion for, 423–428, 449, 455
force (*F*) of, 423–428, 449
impulse and momentum for, 498, 530
kinetic energy and, 455, 488
kinetics, 423–428, 449, 455, 488, 498, 530
moment equation about point *O*, 423
normal (*n*) coordinates, 315–316
path of, 310
position and, 312, 314
procedure for analysis of, 317, 424
right-hand rule for, 312, 314
rigid-body planar motion, 310, 312–319, 388, 423–428, 449, 455, 488, 498, 530
tangential (*t*) coordinates 315–316
velocity (*v*) of, 312, 314, 388
Fixed origin (*O*), 5
Fixed-point motion, 576, 579, 615
  angular momentum (**H**) for, 576
  kinetic energy of, 579, 615
  three-dimensional rigid bodies, 576, 579, 615
Fixed-point rotation, 535–542, 563, 600–605
  acceleration (*a*) and, 538
  angular acceleration (*α*) of, 537–542
  angular velocity components of, 537–542, 600–602
  displacement from, 535–538
  Euler's angles for, 600
  Euler's theorem for, 536
  finite rotation, 536
  fixed-axis and, 600–605
  infinitesimal rotation, 537
  sphere as representation of, 537, 563
  three-dimensional rigid bodies, 535–542, 563, 600–605
  time derivatives for, 537–538
  velocity (*v*) and, 538
Fixed reference frame, 34, 53–59, 67–74, 93, 336

Fluid flow, 289–297, 305
  control volume for, 289–297
  fluid stream, 291–294, 305
  kinetics of a particle, 289–297
  mass flow, 289–290
  mass gain and loss (propulsion), 295–297, 305
  procedure for analysis of, 292
  steady, 291–294, 305
  system of particles, 289–290
  volumetric flow (discharge), 290
Focus, 162
Force (**F**), 108–173, 175–188, 208–218, 225–226, 231–238, 247–248, 275–277, 295–297, 305, 392–451, 456–459, 475–480, 489, 618, 620–625, 631–634, 637–646. *See also* Central-force motion
  acceleration (*a*) and, 108–173, 392–451
  angular momentum relations with, 275–277
  central-force motion and, 160–167, 171
  centripetal, 134
  conservation of energy and, 212–218, 226, 475–480, 631–634
  conservation of linear momentum and, 247–248
  conservative, 208–218, 226, 475–480, 631–634
  constant, 177, 225, 456, 489
  couple moment (*M*) and, 458–459, 489
  damping, 641–646
  displacement (*d*) from, 175–176, 456–458, 489
  elastic potential energy from, 209, 226
  energy and, 180–188, 208–218
  equations of motion for, 110, 112–122, 134–139, 148–152, 407–415, 423–428, 437–442, 449
  external, 114–115, 180, 234, 276, 407–409
  fixed-axis rotation and, 423–428, 449
  free-body diagrams for, 112–113, 171, 407–410
  friction (**F**), 117, 148, 183, 208, 226
  general plane motion and, 437–442
  gravitational attraction (*G*) as, 110–111, 161–163

impulsive, 247–248
inertia force vector, 112
internal, 114–115, 276, 408–409
kinetics of a particle, 108–173,
    175–188, 208–218, 225–226,
    231–238, 275–277
linear impulse and momentum of,
    231–238
planar kinetics, equations of motion
    for, 407–415, 449
potential energy (V) and, 208–218,
    226, 475–480
principle of work and energy and,
    180–188
procedure for analysis of, 213
mass (m) and, 109–122
moments of a, 275–277
moments of inertia (I) and, 393, 423,
    437–442, 449
Newton's laws and, 109–111, 171
nonconservative, 208, 476
normal (**N**), 148–149
normal coordinates for, 134–139
periodic, 637–640
planar motion and, 392–451,
    456–459, 489
potential energy (V) and, 208–218,
    226
potential function for, 210–211
propulsion (mass gain and loss),
    295–297, 305
resultant, 112, 183, 275–276, 408
rigid-body kinetics, 392–451,
    456–459, 489
rolling without slipping, 437
rotational equations of motion,
    408–409, 423–428, 449
slipping (no work), 457
spring, 117, 178–179, 208–211,
    225–226, 457, 489, 618, 620–625
straight-line path of, 177
system of particles, 114–115,
    182–188, 213, 247–248
tangential, 148–149
tangential coordinates for,
    134–139
trajectories, 161–167, 171
translational equations of motion,
    407, 410–415, 449
unbalanced, 119–110, 115
units of, 176
variable, 176, 456

vector, 112
vibrations and, 618, 620–625,
    631–634, 637–646
viscous damping, 641–646
weight (W), 111, 177, 208–211,
    225–226, 457, 489
work (U) of, 175–188, 208–211,
    225–226, 456–459, 489
Forced vibrations, 617, 637–640,
    644–646, 654–655
damped, 644–646, 655
equilibrium position of, 637, 654
forcing frequency (ω_0) for, 637–640,
    654
magnification factor (MF) for,
    638–639, 645
motion of, 617
periodic force and, 637–640
periodic support displacement of,
    639
resonance from, 639, 645
steady-state of, 638–639, 644–645
undamped, 637–640, 654
viscous damped, 644–646, 655
Free-body diagrams, 112–113, 171, 235,
    274, 280–283, 407–410
angular momentum, 274, 280–283
equations of motion and, 112–113,
    407–410
inertial reference frames, 112–113,
    407–410
kinetics of particles using, 112–113,
    171, 274, 280–283
linear impulse and momentum, 235
rigid-body planar motion, 407–410
rotational motion, 408–409
translational motion, 407
Free-flight trajectory, 161, 163
Free vibrations, 617–625, 638, 641–643,
    654
motion of, 617
transient state of, 638
undamped, 617–625, 654
viscous damped, 641–643, 654
Frequency (f), 618, 620–621, 631–632,
    637–640, 643, 654
cycles of, 620
damped natural (ω_d), 643
energy conservation and, 631–632
forcing (ω_0), 637–640, 654
natural (ω_n), 618, 620–621, 631–632,
    654

procedure for analysis of, 621, 632
unit of, 620
vibration and, 618, 620–621,
    631–632, 637–640, 654
Frequency ratio, 638
Friction force (**F**), 117, 148, 183, 208, 226
conservative forces compared to,
    208, 226
equations of motion for, 117, 148
principle of work and energy for,
    183
work of caused by sliding, 183

G

Gage pressure, 291
General plane motion, 310, 336–342,
    349–355, 361–368, 375–383,
    388–389, 437–442, 449, 455, 488,
    499, 530. *See also* Planar motion
absolute motion analysis for,
    328–331, 388
acceleration (a), 361–368, 361–368,
    378–379, 389, 437–442, 449
angular momentum and, 499
displacement (d) from, 336
equations of motion for, 437–442,
    449, 455
force (**F**) and, 437–442, 449
impulse and momentum for, 498,
    530
instantaneous center (IC) of zero
    velocity, 349–355, 389, 453
kinetic energy and, 455, 488
linear momentum and, 499
moment equation about the
    instantaneous center (IC), 453
path of, 310
position (**r**) and, 336
procedure for analysis of, 339, 363,
    380, 438
relative-motion analysis for, 336–
    332, 361–368, 375–383, 388–389
rigid-body kinematics, 310, 336–342,
    349–355, 361–368, 388–389
rigid-body kinetics, 437–442, 449,
    455, 488, 498, 530
rolling without slipping, 338, 437
rotating axes, 375–383, 389
rotation and translation, 328–331,
    361–368
velocity (v), 336–342, 349–355,
    376–377, 389

General three-dimensional motion, 543–545, 563
Graphs, 19–24, 102, 232, 304
  constant and variable force representation, 232
  erratic motion represented by, 19–24, 102
  impulse represented by, 232, 304
  rectilinear kinematic solutions using, 19–24, 102
Gravitational acceleration (g), 111
Gravitational attraction (G), 110–111, 161–163
  central-force motion and, 161–163
  Newton's law of, 110–111
Gravitational potential energy, 208, 210, 226, 475, 490
Gyroscopic motion, 589–590, 600–605, 615
  angular momentum (**H**) and, 603
  angular motion of, 603–605, 615
  angular velocity components for, 600–603
  equations of motion for, 589–590
  Euler angles for, 600
  gyro, 603
  gyroscopic effect, 602–603, 615
  nutation angle (θ), 600–602
  precession (ϕ), 600–602, 615
  spin (ψ), 600–602
  symmetrical spinning axes, 589–590

H

Heat, friction forces from sliding and, 183
Hertz (Hz), unit of, 620
Hodographs, particle acceleration and, 33
Horizontal projectile motion, 39–40
Horsepower (hp), unit of, 199
Hyperbolic functions, 656
Hyperbolic trajectory, 164

I

Impact, 259–265, 304–305, 520–523, 531
  central, 259–261, 262, 304–305
  coefficient of restitution (e), 260–265, 305, 520–523, 531
  conservation of momentum, 260, 262–265, 304–305
  deformation and, 259–265, 520–523
  eccentric, 520–523, 531
  elastic, 261, 305
  energy loss from, 261
  kinetics of a particle, 259–265, 304–305
  line of, 259, 262, 304–305, 520
  oblique, 259, 262, 305
  plane of contact, 259, 262, 304
  plastic (inelastic), 261, 305
  principle of impulse and momentum for, 260
  procedures for analysis of, 262
  restitution phase, 259–262, 520
  rigid-body planar motion, 520–523, 531
  separation points of contact (C), 522
Impulse, 230–307, 494–533, 578, 614
  angular, 278–283, 305, 501–503, 578, 614
  conservation of angular momentum and, 280
  conservation of linear momentum and, 247–253, 304
  control volumes, 289–297
  diagrams, 233, 235
  equations of motion for, 233–234
  external, 247–248
  graphical representation of, 232, 304
  impact and, 259–265, 304–305, 520–523
  internal, 247–248
  kinetics, 230–307, 494–533
  linear, 231–238, 304, 501–503
  magnitude of, 232
  mass flow, 289–290
  mass gain and loss (propulsion), 295–297, 305
  momentum and, 230–307, 494–533
  particles, 230–307
  principle of momentum and, 231–238, 278–283, 304, 501–508, 530, 578, 614
  procedures for analysis of, 235, 248, 280, 503
  restitution, 259, 521
  rigid-body planar motion, 494–533
  steady flow and, 291–294, 305
  three-dimensional rigid bodies, 578, 614
  time (t), as functions of, 231–233
Impulsive forces, 247–248
Inertia (I), 393–401, 423, 437, 449, 565–570, 577, 614
  acceleration (**a**) and, 393, 423, 437, 449
  angular acceleration (α) and, 393
  angular momentum (**H**) and, 577
  arbitrary axis, moment of about, 569
  composite bodies, 399
  equations of motion and, 423, 437
  integration of, 394–395, 566–567
  mass center (G) and, 565, 567–568
  mass moments of, 393–401
  moment equation about the instantaneous center (IC), 437
  moment equation about point O, 423
  moment of, 393–401, 423, 437, 449, 566, 568–570, 577, 614
  parallel-axis theorem, 398–399, 567
  parallel-plane theorem, 568
  principle axes of, 568–569, 577
  principle moments of, 568, 614
  procedure for analysis of, 395
  products of, 566–567, 614
  radius of gyration, 399
  resistance of body to acceleration, 393
  rigid-body planar motion and, 393–401, 423, 437, 449
  slipping and, 437
  three-dimensional rigid-body motion, 565–570, 577, 614
  volume elements for integration of, 394–395
Inertia tensor, 568–569
Inertial reference frames, 112–113, 171, 407–410, 453–454, 575–576, 578–579, 586–587
  angular momentum (**H**), 575–576
  equations of motion, 112–113, 171, 407–410, 586–587
  force vector, 113
  kinetic energy, 453–454, 578–579
  kinetics of a particle, 112–113, 171
  rigid-body planar motion, 407–410, 453–454
  rotational motion, 408–409, 586–587
  slab in, 453–454
  symmetry of, 407–410
  three-dimensional rigid-body motion, 575–576, 578–579, 586–587
  translational motion, 407, 586

Infinitesimal rotation, 537
Instantaneous acceleration, 7, 33
Instantaneous axis of rotation, 537–538, 563
Instantaneous center (IC), 349–355, 389, 437
  centrode, 351
  circular motion and, 349–355, 389
  general plane motion, 349–355, 437
  location of, 350–355
  moment equation about, 437
  procedure for analysis of, 351
  relative-position vector for, 349–350
  slipping and, 437
  zero velocity, 349–355, 389, 437
Instantaneous power, 199
Instantaneous velocity, 6, 32
Integral equations, 657–658
Integration of equations, 20, 394–395, 566–567, 578–579
  erratic motion, 20
  kinetic energy, 578–579
  moments of inertia, 394–395, 566
  products of inertia, 566–567
  three-dimensional rigid bodies, 566–567, 578–579
Internal energy, 183
Internal force, 114–115, 180, 234, 276, 408–409
Internal impulses, 247–248

J
Joule (J), unit of, 176

K
Kepler's laws, 166
Kinematics, 2–107, 308–391, 534–563. *See also* Planar motion
  absolute motion analysis, 328–331, 388
  continuous motion, 5–14
  coordinates for, 34–38, 53–59, 67–74, 103–105, 543–545
  curvilinear motion, 32–38, 53–59, 67–74, 103–104, 310–311
  cylindrical components, 67–74
  cylindrical (r, θ, z) coordinates, 70
  dependent motion analysis, 81–86, 105
  erratic motion, 19–24, 102
  fixed-axis rotation, 310, 312–319, 388
  fixed-point rotation, 535–542, 563
  graphs for solution of, 19–24, 102
  instantaneous center (IC) of zero velocity, 349–355, 389
  motion and, 3
  normal (n) axes, 53–59, 104
  particles and, 2–107
  planar, 308–391
  polar coordinates, 67–70
  position (**r**), 311, 336
  position coordinates, 5, 81–86, 104
  principle of, 3
  procedures for analysis of, 9, 36, 40, 56, 82, 94, 317, 328, 339, 351, 363, 380, 555
  projectile motion, 39–43, 103
  radial (r) coordinate, 67–70
  rectangular (x, y, z) coordinates, 34–38, 103
  rectilinear, 5–14, 19–24, 102, 310–311
  relative-motion analysis, 93–97, 105, 336–342, 361–368, 375–383, 389, 552–559, 563
  rigid bodies, 308–391, 534–563
  rotating axes, 375–383, 389, 535–542, 552–559, 563
  rotation and, 310, 312–319, 328–331, 336–342, 375–383, 388–389
  sign conventions for, 5–7
  tangential (t) axes, 53–59, 104
  three-dimensional motion, 55, 534–563
  time derivatives, 34–35, 70, 81–82, 537–542
  translating axes, 93–97, 105, 375–383, 389, 543–545, 552–559, 563
  translating-rotating systems, 538–542
  translation and, 310–311, 328–331, 336–342, 375–383, 388–389
  transverse (θ) coordinate, 67–70
Kinetic diagram, 112, 171, 408–410
Kinetic energy, 180–182, 208, 212–213, 225, 453–456, 460–466, 488, 578–581, 615
  conservation of energy and, 212–213, 476–477
  center of mass (G) for, 579
  fixed-point O for, 579
  general plane motion and, 455, 488
  inertial coordinate system for, 578–579
  integration for, 453–456, 578–579
  kinetics of particles, 180–182, 208, 212–213, 225
  potential energy and, 208, 212–213
  principle of work and energy, 180–182, 225, 460–466, 579, 615
  procedure for analysis of, 213, 461, 477
  rigid-body planar motion and, 453–456, 460–466, 476–477, 488
  rotation about a fixed axis and, 455, 488
  slab in inertial reference for, 453–454
  system of bodies, 456
  three-dimensional rigid-body motion, 578–581, 615
  translation for, 455, 488
Kinetic moments, 409
Kinetics, 3, 108–173, 174–228, 230–307, 392–451, 452–492, 494–533, 564–615. *See also* Planar motion; Space mechanics
  acceleration (a) and, 108–173, 392–451
  angular momentum (**H**), 274–283, 305, 496–503, 516–519, 530–531, 575–578, 614
  central-force motion, 160–167, 171
  conservation of energy, 212–218, 226
  conservation of momentum, 247–253, 280, 304, 516–519, 531
  conservative forces and, 208–218, 226
  control volumes, 289–297
  cylindrical (r, θ, z) coordinates, 148–152, 171
  eccentric impact, 520–523, 531
  efficiency (ε) and, 199–202, 226
  energy (E) and, 174–228, 452–492, 578–581
  equations of motion, 110, 112–122, 134–139, 148–152, 407–415, 423–428, 437–442, 449, 586–595, 615
  fixed-axis rotation, 423–428, 449, 455, 488, 498, 530
  fluid flow, 289–297, 305
  force (F) and, 108–173, 175–179, 208–218, 225–226, 392–451
  free-body diagrams for, 112–113, 171, 407–410
  gyroscopic motion, 589–590, 600–605, 615

Kinetics (*Continued*)
 impact, 259–265, 304–305, 520–523, 531
 impulse, 230–307, 494–533, 614
 inertia (I), 109, 112–113, 393–401, 437, 449, 565–570, 614
 inertial reference frame for 112–113, 171
 linear momentum, 247–253, 495, 498–501, 503, 516–519
 mass (m), 109–122
 mass flow, 289–290
 mass gain and loss (propulsion), 295–297, 305
 mass moments of inertia, 393–401, 449
 momentum, 230–307, 494–533, 614
 Newton's laws and, 109–111, 171
 normal (n) coordinates, 134–139, 171
 particles, 108–173, 174–228, 230–307
 planar motion, 392–451, 452–492, 494–533
 potential energy and, 208–218, 226
 power (P), 199–202, 226
 principle of, 3
 principle of impulse and momentum, 231–238, 278–283, 304–305, 501–508, 578, 614
 principle of work and energy, 180–188, 225–226, 579, 615
 procedures for analysis of, 116–117, 135, 149, 181, 200, 213, 235, 248, 262, 280, 292, 395, 412, 424, 438, 461, 503, 590
 propulsion (mass gain and loss), 295–297, 305
 rectangular (x, y, z) coordinates, 116–122, 171, 577
 rigid-bodies, 182, 392–451, 452–492, 494–533
 rotation and, 498, 530
 rotational equations of motion, 408–409, 423–428, 449, 615
 spring force, 178–179, 209–211, 225–226
 steady flow, 291–294, 305
 systems of particles, 247–253, 289–290
 tangential (t) coordinates, 134–139, 171
 three-dimensional rigid bodies, 564–615
 torque-free motion, 606–609, 615

 trajectories, 161–167, 171
 translation and, 498, 530
 translational equations of motion, 407, 410–415, 449, 615
 work (U) and, 174–228, 452–492, 579, 615

L
Line of action, 350
Line of impact, 259, 262, 304, 520
Linear impulse and momentum, 231–238, 247–253, 304, 495, 498–503, 516–519, 530–531
 conservation of momentum, 247–253, 304, 516–519, 531
 diagrams for, 233, 235
 equations of motion for, 233–234
 fixed-axis rotation and, 498, 530
 force (F) and, 231–238
 impulsive forces and, 247–248
 general plane motion and, 499, 530
 kinetics of a particle, 231–238, 247–253, 304
 principle of, 231–238, 501–503, 530
 procedures for analysis of, 235, 248, 517
 rigid-body planar motion, 495, 498–503, 516–519
 systems of particles, 234–238, 247–253, 304
 time (t), as functions of, 231–233
 translation and, 498, 530
 vectors, 232

M
Magnification factor (MF), 638–639, 645
Magnitude, 5–7, 32, 34–35, 53–55, 68–69, 104, 232, 274, 312, 313, 350, 361, 376, 378, 458, 489
 acceleration (a), 7, 35, 55, 69, 361, 378
 angular acceleration (α), 313
 angular displacement ($d\theta$) and, 312
 angular velocity (ω) and, 312
 angular momentum (**H**), 274
 average speed, 6
 constant, 458, 489
 couple moment (M), work of and, 458, 489

 curvilinear motion and, 32, 34–35, 53–55, 68–69, 104
 distance as, 5
 fixed-axis rotation and, 312
 graphical representation of, 232
 instantaneous center (IC) of zero velocity from, 350
 kinematics of a particle, 5–7, 32, 34–35, 53–55, 68–69, 104, 232, 274
 linear impulse, 232
 planar kinematics, 312
 position vector (**r**) and, 34
 rectilinear kinematics and, 5–7
 relative-motion analysis and, 350, 361, 376, 378
 rigid-body planar motion, 312, 458, 489
 rotating axes, changes in motion from, 376, 378
 rotation, changes in motion from, 312
 speed as, 6, 32, 55, 68, 104
 time rate of change of, 55
 velocity (v), 6, 32, 34–35, 53, 68, 350, 376
Mass (m), 109–122, 289–297, 295–297, 305, 393–401, 495–508. *See also* Center of mass (G)
 acceleration (a) and, 109–122, 393
 conservation of, 291
 control volumes for, 289–297
 equations of motion and, 110, 112–122
 fluid flow, 289–297
 gravitational attraction and, 110–111
 linear momentum and, 495–500
 loss and gain of (propulsion), 295–297, 305
 moments (M) of inertia (I), 393–401
 Newton's laws and, 109–111
 particle, 109–111
 principle of impulse and momentum, 501–508
 propulsion (gain and loss of), 295–297, 305
 rigid-body planar motion, 393–401, 495–508
 steady flow of fluid systems and, 291–294, 305
 system of particles and, 114–115, 289–290
 weight and, 111

Mass flow, 289–290
Mathematical expressions, 656–658
Maximum deformation, 259
Mechanical efficiency, 199–200
Mechanical energy, 212–218, 226. *See also* Conservation of energy
Mechanics, study of, 3
Moment arm, 394
Moment of a force, 275–277
   angular momentum and, 275–277
   external force, 276
   internal force, 276
   kinetics of a particle, 275–277
   resultant force and, 275–276
   systems of particles, 276
Moment of inertia, 393–401, 423, 437–442, 449, 565–570, 614
   acceleration (a) and, 393, 423, 437–442, 449
   arbitrary axis, about, 569
   body resistance to acceleration, 393
   composite bodies, 399
   disk elements, 395
   equation about the instantaneous center (IC), 437
   equation about point O, 423
   equations of motion and, 423, 437–442
   fixed-axis rotation, 423
   force (F) and, 393, 437–442, 449
   general plane motion, 437–442, 449
   inertia tensor for, 568–569
   integration of, 394–395, 566
   mass, 393–401
   mass center (G), 565, 567–568
   parallel-axis theorem for, 398–399, 449, 567
   parallel-plane theorem for, 568
   principal, 568, 614
   procedure for analysis of, 395
   products of inertia and, 565–570
   radius of gyration for, 399
   rigid-body planar motion, 393–401, 423, 437–442, 449
   rotational equation of motion and, 423
   shell elements, 395
   single integration of, 395
   slipping and, 437
   three-dimensional rigid-body motion, 565–570, 614
   volume elements for integration of, 394–395
Moment of momentum, 274. *See also* Angular momentum
Moments, work of a couple, 458–459, 489
Momentum, 230–307, 494–533, 575–578, 614
   angular (H), 274–283, 305, 496–500, 501–503, 516–519, 530–531, 575–578, 614
   conservation of, 247–253, 260, 262–265, 280, 304, 516–519, 531
   control volumes, 289–297
   diagrams for, 233
   equations of motion for, 233–234
   fixed-axes rotation and, 498
   general plane motion and, 499, 530
   impact and, 259–265, 304–305, 520–523, 531
   impulse and, 230–307, 494–533
   kinetics, 230–307, 494–533
   linear (L), 231–238, 247–253, 304, 495, 498–503, 516–519, 530–531
   mass flow, 289–290
   mass gain and loss (propulsion), 295–297, 305
   moments of force and, 275–277
   particle, 230–307
   principle axes of inertia and, 577
   principle of impulse and, 231–238, 260, 278–283, 304–305, 501–508, 530, 578, 614
   procedures for analysis of, 235, 248, 262, 280, 503, 517
   rectangular components for, 577
   rigid-body planar motion, 494–533
   steady flow and, 291–294, 305
   systems of particles, 234–238, 247–253, 276, 304
   three-dimensional rigid bodies, 575–578, 614
   time (t), as functions of, 231–233
   translation and, 498, 530
   vector form, 232

N
Natural frequency (ω_n), 618, 620–621, 631–632, 643, 654
   damped (ω_d), 643
   energy conservation and, 631–632
   procedures for analysis of, 621, 632
   undamped free vibration, 618, 620–621, 654
   underdamped systems, 643
   vibration and, 618, 620–621, 631–632, 643, 654
Newton's laws, 109–113, 171
   body mass and weight from, 111
   equation of motion, 110, 112–113, 171
   gravitational attraction, 110–111
   inertia and, 109, 111–113
   kinetics of particles and, 109–111, 171
   second law of motion, 109–111, 171
   unbalanced force and, 109–110
Newtonian inertial reference frame, 112
Nonconservative force, 208, 476
Nonimpulsive forces, 247–248
Nonrigid bodies, principle of work and energy for, 182
Normal (n) coordinates, 53–59, 134–139, 171, 315–316
   acceleration (a) and, 54–55, 134–139, 315–316
   circular motion components, 315–316
   curvilinear motion components, 53–59
   equations of motion and, 134–139, 171
   kinematics, 53–59, 315–316
   kinetics, 134–139, 171
   particles, 53–59, 134–139
   planar motion and, 53, 315–316
   procedure for analysis of, 56
   rigid-bodies, 315–316
   rotation about a fixed axis, 315–316
   three-dimensional motion, 55
   velocity (v) and, 53
Normal (**N**) force, 148–149
Nutation angle (θ), 600–602

O
Oblique impact, 259, 262, 305
Orbit, central-force motion of, 164–167. *See also* Trajectories
Orbital revolution, 166
Osculating plane, 53
Overdamped vibration systems, 642

P

Parabolic path, 164
Parallel-axis theorem, 398–399, 449, 567
Parallel-plane theorem, 568
Particles, 2–107, 108–173, 174–228, 230–307
  acceleration (a), 7–8, 33, 35, 54–55, 69, 94, 102, 108–173
  angular impulse of, 278–279
  angular momentum ($\mathbf{H}$) of, 274–283, 305
  central-force motion of, 160–167, 171
  conservation of angular momentum, 280, 305
  conservation of energy, 212–218, 226
  conservation of linear momentum, 247–253, 260, 262–265, 304
  conservative forces and, 208–218, 226
  continuous motion of, 5–14
  control volume, 289–297
  coordinates for, 34–38, 53–59, 67–74, 103–105, 116–122, 134–139, 148–152, 171
  curvilinear motion of, 32–38, 53–59, 67–74, 103–104
  deformation of, 182–183, 259–265
  dependent motion analysis, 81–86, 105
  displacement (Δ), 5, 32, 182–183
  efficiency (ε) and, 199–202, 226
  energy (E) and, 174–228
  equations of motion, 110, 112–122, 134–139, 148–152, 160–163, 171
  erratic motion of, 19–24, 102
  force ($\mathbf{F}$) and, 108–173, 175–179, 208–218, 225–226
  free-body diagrams, 112–113, 171
  gravitational attraction (G), 110–111, 161–163
  hodographs, 33
  impact, 259–265, 304–305
  impulse, 230–307
  impulsive forces, 247–248
  inertial reference frame, 112–113, 171
  kinematics of, 2–107
  kinetic diagrams, 112, 171
  kinetic energy of, 180–182, 208, 212–213
  kinetics of, 108–173, 174–228, 230–307
  mass (m), 109–122
  mass flow, 289–290
  mass gain and loss (propulsion), 295–297, 305
  momentum, 230–307
  Newton's second law of motion, 109–111, 171
  normal coordinates (n) for, 134–139
  planar motion of, 53–55
  position (s), 5, 8, 32, 102
  position vector ($\mathbf{r}$), 32, 34, 68, 93
  position-coordinate equations, 81–86
  potential energy of, 208–218, 226
  power (P) and, 199–202, 226
  principle of work and energy for, 180–188, 225–226
  principles of impulse and momentum, 231–238, 260, 278–283, 304
  procedures for analysis of, 9, 36, 40, 56, 70, 82, 94, 116–117, 135, 149, 181, 200, 213, 235, 248, 262, 280, 292
  projectile motion of, 39–43, 103
  rectilinear kinematics of, 5–14, 19–24, 102
  relative motion analysis, 93–97, 103
  resultant force on, 112, 183, 275–276
  speed (magnitude), 6, 7, 32, 34, 35, 68, 102
  spring force, 178–179, 208–211, 225–226
  straight-line path of, 177
  system of, 114–115, 182–188, 213, 234–238, 247–253, 276, 278–279, 304
  tangential coordinates (t) for, 134–139
  three-dimensional motion of, 55
  time (t), functions of, 8
  time derivatives, 34–35, 70, 81–82, 104–105
  trajectories, 161–167, 171
  translating axes, two particles on, 93–97, 105
  velocity (v), 6–8, 32–35, 53, 68, 93, 102
  work (U) and, 174–228
Particular solution, vibration, 637–638
Path of motion, 160–161
Perigee, 165
Period of deformation, 259
Period of time, vibration, 620
Periodic force, 637–640
Periodic support displacement, 639
Phase angle (ϕ), 620
Pinned-end members, 336–342, 361–368, 389
  acceleration (a) and, 361–368
  coordinating fixed and translating reference frames, 336–342, 389
  relative-motion analysis of, 336–342, 361–368, 389
  velocity (v) and, 336–342
Planar motion, 53–55, 308–391, 392–451, 452–492, 494–533. *See also* General plane motion
  absolute motion analysis, 328–331, 388
  acceleration (a) and, 54–55, 311, 313, 315–316, 361–368, 378–379, 388–389, 392–451
  angular momentum, 496–500, 501–503, 516–519
  angular motion and, 312–313, 388
  conservation of energy, 475–480, 490
  conservation of momentum, 516–519, 531
  couple moment (M) in, 458–459, 489
  curvilinear, 53–55
  displacement (d), 312, 314, 336
  eccentric impact, 520–523, 531
  energy (E) and, 452–492
  equations of motion for, 407–415, 423–428, 437–442, 449
  fixed-axis rotation, 310, 312–319, 388, 423–428, 449, 498, 530
  force (F) and, 392–451, 456–459, 489
  free-body diagrams for, 407–410
  impact (eccentric), 520–523, 531
  impulse, 494–533
  instantaneous center (IC) of zero velocity, 349–355, 389
  kinematics, 53–55, 308–391, 361–368
  kinetic diagrams for, 408–410
  kinetic energy and, 453–456, 460–466, 488
  kinetics, 392–451, 452–492, 494–533
  linear momentum, 498–503, 516–519
  moment of inertia (I) for, 393–401, 423, 423–442, 449
  momentum, 494–533
  normal component (n) coordinates, 53–55
  osculating plane and, 53
  particles, 53–55

paths of, 310
position (**r**) and, 311, 312, 314, 336–337, 375
potential energy (*V*) of, 475–480, 490
principle of work and energy, 460–466, 490
principles of impulse and momentum, 501–508, 530
procedures for analysis of, 317, 328, 339, 351, 363, 380, 395, 412, 424, 438, 461, 477, 503, 517
relative-motion analysis, 336–342, 361–368, 375–383, 389
rigid bodies, 308–391, 392–451, 452–492, 494–533
rotating axes, 375–383, 389
rotation and, 310, 312–319, 388–389, 498, 530
rotational equation of motion, 408–409, 423–428, 449
tangential component (*t*) coordinates, 53–55
time derivatives for, 328
translation, 310–311, 388–389, 407, 498, 530
translation and rotation, 336–342, 361–368, 375–383
translational equation of motion, 407, 410–415, 449
velocity (*v*) and, 53, 311, 312, 314, 336–342, 349–355, 376–377, 388–389
work (*U*) and, 452–492
Plane of contact, 259, 262
Plastic (inelastic) impact, 261
Polar coordinates, 67–70, 104. *See also* Cylindrical coordinates
Position, 5, 8, 19–21, 32, 34, 68, 93–97, 102–103, 105, 311, 312, 314, 336–337, 349–350, 375, 553
angular (θ), 312
base point, 311
continuous motion and, 5, 8
curvilinear motion and, 32, 34, 68
displacement (*d*) from changes of, 5, 312, 314, 336
erratic motion and, 19–21
fixed reference frame, 34, 93
graphs of variables, 19–21, 102
instantaneous center (IC) of zero velocity, 349–350
kinematics of particles and, 5, 8, 32, 34, 68, 93–97, 12
magnitude and, 34

planar kinematics of rigid bodies and, 311, 312, 314, 336–337, 375
rectangular components, 34, 103
rectilinear kinematics and, 5, 8, 19–21, 102
relative-motion analysis and, 93–97, 105, 336–337, 375, 553
relative vector, 93, 349–350
rotating axes, 375, 553
rotation about fixed axis, 312, 314
three-dimensional rigid-body motion, 553
time (*t*), as a function of, 8
translating axes, 93–97, 553
translating reference frame, 93–97, 105
translation and, 311, 336
vector (**r**), 32, 34, 68, 93–97, 311, 336–337, 375
velocity (*v*) as a function of, 8, 93, 337
Position coordinates (*s*), 5, 81–86, 102, 105
dependent-motion analysis using, 81–86, 105
equations for, 81–86, 105
kinematics of particles and, 5, 81–86, 102, 105
origin (*O*), 5
procedure for analysis using, 82
rectilinear kinematics, 5, 102
time derivatives and, 81–82, 105
Potential energy (*V*), 208–218, 226, 475–480, 490
conservation of energy and, 212–218, 226, 475–480, 490
conservative forces and, 208–218, 226, 475–480, 490
elastic, 209, 226, 475, 490
equations for conservation of, 476
gravitational, 208, 210, 226, 475, 490
kinetic energy and, 208, 212–213
kinetics of a particle, 208–218, 226
potential function for, 210–211
procedure for analysis of, 213, 477
rigid-body planar motion, 475–480, 490
spring force and, 208–211, 226, 475, 490
weight (*W*), displacement of, 208–211, 226, 475
work (*U*) and, 208–218, 226, 475–480, 490
Power (*P*), 199–202, 226

average, 199
efficiency (ε) and, 199–202, 226
energy (*E*) and, 199–202, 226
instantaneous, 199
procedure for analysis of, 200
units of, 199
Power-flight trajectory, 161, 163
Power-series expansions, 656
Precession (φ), 600–602, 607–608
Principal moments of inertia (*I*), 568, 614
Principal normal axis, 55
Principle axes of inertia (*I*), 568–569, 588, 615
Principle of impulse and momentum, 231–238, 260, 278–283, 304–305, 501–508, 530, 578, 614
angular impulse, 278–279
angular impulse and momentum, 501–503, 530
angular momentum (**H**), 278–283, 305, 508
conservation of angular momentum, 280
diagrams for, 233
external and internal forces, 234
graphs for, 232
impact and, 260
kinetics of a particle, 231–238, 278–283, 304–305
linear, 231–238, 304, 501–503, 530
procedures for analysis using, 235, 280, 503
rigid-body planar motion, 501–508, 530
scalar formulation, 279
systems of particles, 234–238
three-dimensional rigid-body motion, 578, 614
time (*t*) and, 231–233
vector formulation, 279
Principle of work and energy, 180–188, 225–226, 460–466, 490, 579, 615
deformation and, 182–183
displacement and, 182–183, 226
equation for, 180, 225
kinetic energy and, 180–182, 225, 460–466, 579, 615
kinetics of a particle, 180–188, 225–226
procedures for analysis using, 181, 461
rigid-body planar motion, 460–466, 490

718   INDEX

Principle of work and energy (*Continued*)
  three-dimensional rigid bodies, 579, 615
  systems of particles, 182–188
  units of, 180
  work of friction caused by sliding, 183
Problem solving procedure, 4
Products of inertia, 565–570, 614
Projectile motion, 39–43, 103
  horizontal, 39–40
  particle kinematics and, 39–43, 103
  procedure for analysis of, 40
  vertical, 39–40
Propulsion (mass gain and loss), 295–297, 305. *See also* Control volume

Q
Quadratic formula, 656

R
Radial component (v_r), 68
Radial coordinate (r), 67–70
Radius of curvature (ρ), 53
Radius of gyration, 399
Rectangular (x, y, z) coordinates, 34–38, 103, 116–122, 171, 577
  acceleration (a), 35, 116–122
  angular momentum (**H**) and, 577
  curvilinear motion, 34–38, 103
  dot notation for, 34
  equations of motion and, 116–122, 171
  kinematics of a particle, 34–38, 103
  kinetics of a particle, 116–122, 171
  position vector, 34
  procedures for analysis using, 36, 116–117
  three-dimensional rigid-plane motion and, 577
  velocity (v), 34–35
Rectilinear kinematics, 5–14, 19–24, 102
  acceleration (a), 7–8, 19–21, 102
  continuous motion, 5–14
  displacement (Δ), 5, 20
  erratic motion, 19–24
  graphs for solution of, 19–24, 102
  particles and, 5–15, 19–24, 102
  position (s), 5, 8, 19–21, 102
  procedure for analysis of, 9

sign conventions for, 5–7
time (t) and, 8, 19–20, 102
velocity (v), 6–8, 19–21, 102
Rectilinear motion, 116
Rectilinear translation, 310–311, 388, 410–411, 449
Reference frames, 34, 53–59, 67–74, 93–97, 112–113, 171, 314–316, 336–342, 361–362, 388–389, 407–410, 453–454, 537–542, 563, 575–576
  acceleration (a), 361–362
  angular momentum (**H**) and, 575–576
  circular path, 314–316, 338, 361–362
  curvilinear motion, 34, 53–59, 67–74
  coordinating fixed and translating axes, 336–342, 389
  equations of motion and, 112–113, 171, 407–410
  fixed, 34, 53–59, 67–74, 93, 336, 388, 537–542
  inertial, 112–113, 171, 407, 453–454, 575–576
  instantaneous axis of rotation, 537–538, 563
  kinematics of particles, 34, 53–59, 67–74, 93–97
  kinetic energy, 453–454
  kinetics of particles, 112–113, 171
  Newtonian inertial, 112–113
  relative-motion analysis, 93–97, 336–342, 361–362
  rigid-body planar motion, 314–136, 336–342, 361–362, 407–410, 453–454
  rotation about a fixed axis, 314–316
  rotational motion, 408–409
  three-dimensional rigid-body motion, 537–542, 563, 575–576
  translation and rotation, 336–342, 361–362
  translational motion, 407
  translating, 93–97, 105, 336
  translating-rotating systems, 538–542
  symmetry of, 407–410
  velocity (v), 336–342
Relative acceleration, 94
Relative-motion analysis, 93–97, 105, 336–342, 349–355, 361–368, 375–383, 389, 552–559, 563
  acceleration (a) and, 94, 361–368, 378–379, 389, 554
  circular motion, 338–342, 349–355, 361–362, 389

circular path, 338
coordinate systems for, 93–97, 105, 336, 552
coordinating fixed and translating reference frames, 336–342, 361–368, 375–383, 389
displacement and, 336
instantaneous center (IC) of zero velocity, 349–355, 389
kinematics of a particle, 93–97, 105
pinned-end members, 336–342, 361–368, 375–383
position vectors (**r**) and, 93, 105, 336–337, 375, 553
procedures for analysis using, 94, 339, 351, 363, 380, 555
rigid-body planar motion, 336–342, 349–355, 361–368, 375–383, 389
rolling without slipping, 338, 362
rotating axes, 375–383, 389, 552–559, 563
rotation and, 336–342, 361–368, 389
three-dimensional rigid-body motion, 552–559, 563
translating axes, 93–97, 105, 336–342, 361–368, 389, 552–559, 563
translating reference frames, 93–97, 105, 336
translation and rotation, 336–342, 361–368, 552–559
velocity (v) and, 93, 336–342, 349–355, 376–377, 389, 553
Relative-position vector, 93, 349–350, 553
Relative velocity, 93, 337–338, 389
Resonance, 639, 645, 654–655
Restitution, 259–265, 520–523
  central impact and, 259–261
  coefficient (e) of, 260–265, 520–523
  conservation of angular momentum for, 520–521
  deformation and, 259–262, 520–523
  eccentric impact and, 520–523
  elastic impact and, 261
  impact and, 259–265, 520–523
  impulse, 259, 521
  oblique impact and, 262
  period of, 259, 520
  plastic impact and, 261
  rigid-body planar motion, 520–523
Resultant force, 112, 183, 275–276, 408
Resultant vector, 659

Retrograde precession, 608
Right-hand rule, 274, 312, 314
Rigid bodies, 182, 308–391, 392–451, 452–492, 494–533, 534–563, 564–615
  absolute motion analysis, 328–331, 388
  acceleration (a) and, 311, 312, 315–316, 361–368, 378–379, 388–389, 392–451, 537–542, 554
  angular momentum, 496–500, 501–503, 516–519, 575–578, 603, 606–607, 614
  angular motion, 312–313, 317, 537–542
  circular motion, 314–316, 337–338, 349–355, 361–362, 388–389
  circular path, 314–316, 338, 361
  conservation of energy, 475–480, 490
  conservation of momentum, 516–519, 531
  coordinating fixed and translating reference frames, 336–342, 361–368, 375–383, 389
  couple moment (M) in, 458–459, 489
  displacement (d) of, 312, 314, 456–458, 489
  eccentric impact of, 520–523, 531
  energy (E) and, 452–492
  equations of motion for, 407–415, 423–428, 437–442, 449, 488, 586–595, 615
  fixed-axis rotation, 310, 312–319, 388, 423–428, 449, 455, 488, 498, 530
  fixed-point rotation, 535–542, 563, 579
  force (F) and, 392–451, 456–459, 489
  free-body diagrams for, 407–410
  general plane motion, 310, 336–342, 437–442, 449, 455, 488, 499, 530
  general three-dimensional motion, 543–545, 563
  gyroscopic motion, 589–590, 615
  impact of, 520–523, 531
  impulse, 494–533, 575–578, 614
  inertia and, 565–570, 614
  instantaneous center (IC) of zero velocity, 349–355, 389
  kinematics of, 308–391, 534–563
  kinetic energy and, 453–456, 488, 578–581, 615
  kinetics of, 182, 392–451, 452–492, 494–533, 564–615
  linear momentum, 498–503, 516–519
  moments of inertia (I) for, 393–401, 423, 437–442, 449, 565–570
  momentum, 494–533, 575–578, 614
  pinned-end members, 336–342, 361–368, 375–383
  planar motion, 308–391, 392–451, 452–492
  position ($\mathbf{r}$), 311, 312, 314, 336–337, 375, 553
  potential energy (V) of, 475–480, 490
  products of inertia (I) of, 565–570
  principle of impulse and momentum, 501–508, 578, 614
  principle of work and energy, 182, 460–466, 490, 579, 615
  procedures for analysis of, 317, 328, 339, 351, 363, 380, 412, 424, 438, 461, 477, 503, 517, 555, 590
  relative-motion analysis, 336–342, 349–355, 361–368, 375–383, 389, 552–559, 563
  rolling without slipping, 338, 362
  rotating axes, 375–383, 389, 552–559, 563
  rotation and translation, 336–342
  rotation of, 310, 312–319, 328–331, 361–368, 388–389, 455, 498, 530
  rotational equations of motion, 408–409, 423–428, 449, 488, 586–587, 615
  systems of particles and, 182
  systems of bodies and, 456
  three-dimensional, 534–563, 564–615
  time derivatives for, 537–542
  torque-free motion, 606–609, 615
  translating axes, 375–383, 389, 552–559, 563
  translation of, 310–311, 328–331, 361–368, 388–389, 455, 498, 530
  translational equations of motion, 407, 410–415, 449, 488, 586, 615
  velocity (v), 311, 312, 314, 336–342, 349–355, 376–377, 388–389, 537–542, 553
  work (U) and, 452–492
  zero velocity, 349–355, 389
Rolling without slipping, 338, 362, 437
Rotating axes, 375–383, 389, 535–542, 552–559, 563
  acceleration (a) of, 378–379, 537–542, 554
  instantaneous axis of rotation, 537–538, 563
  Coriolis acceleration of, 379, 389
  direction of, 376
  fixed reference frame, 535–542
  magnitude, change of, 376, 378
  position vectors ($\mathbf{r}$) for, 375, 553
  procedure for analysis of, 380, 555
  relative-motion analysis for, 375–383, 389, 552–559, 563
  rigid-body planar motion, 375–383, 389
  three-dimensional motion and, 535–542, 552–559, 563
  time derivatives for, 537–542
  translating-rotating systems, 538–542
  velocity (v) of, 376–377, 537–542, 553
Rotation, 310, 312–319, 328–331, 336–342, 349–355, 388–389, 488, 498, 530, 535–545, 563
  absolute motion analysis, 328–331, 388
  acceleration (a) and, 313, 315–316, 388
  angular momentum and, 498
  angular motion and, 312–313, 317, 388, 537–542
  circular motion and, 314–317, 349–355, 388–389
  circular path, 314–319, 338
  coordinating fixed and translating reference frames, 336–342, 389
  deceleration, 313
  displacement (d) and, 312, 314, 336, 535–538
  Euler's theorem for, 536
  finite, 536
  fixed-axis, 310, 312–319, 388, 488, 498, 530
  fixed-point, 535–542, 563
  general planar motion for, 336–342
  general three-dimensional motion, 543–545
  impulse and momentum of, 498, 530
  infinitesimal, 537
  instantaneous axis of, 537–538, 563
  instantaneous center (IC) of zero velocity, 349–355, 389
  kinetic energy and, 455, 488
  linear momentum and, 498
  line of action, 350
  paths of, 310
  position ($\mathbf{r}$) and, 312, 314, 336–337

Rotation (*Continued*)
  procedures for analysis of, 317, 328, 339, 351
  relative-motion analysis, 336–342, 349–355, 389
  right-hand rule for, 312, 314
  rigid-body planar motion and, 310, 312–319, 328–331, 336–342, 349–355, 388–389, 488, 498, 530
  rotating and translating axes, 538–542
  three-dimensional rigid bodies, 535–545, 563
  time derivatives for, 328, 537–542
  translation and, 328–331, 336–342
  velocity (v) and, 312, 314, 336–342, 349–355
Rotational equations of motion, 408–409, 423–428, 449, 455, 488, 586–587, 615
  acceleration (a) and, 408–409, 423–428, 449
  center of mass (G) for, 586–587
  fixed axes, 423–428, 449, 455, 488
  fixed point, 586–587
  force (F) and, 408–409, 423–428, 449
  inertial reference frame for, 408–409, 586–587
  kinetic energy and, 455, 488
  kinetic moments, 409
  moment equation about point O, 423
  moment of inertia, 423
  procedure for analysis of, 424
  rigid-body planar motion, 408–409, 423–428, 449, 455, 488
  three-dimensional rigid-body motion, 586–587, 615
  time derivative for, 587
  symmetry of reference frames for, 408–409

S

s–t (position–time) graphs, 19–20
Scalar formulation of angular momentum, 274, 279
Separation points of contact after impact, 523
Shell elements, moment of inertia of, 395
Simple harmonic motion, 618, 654
Slab in inertial reference frame, 453–454
Sliding, 183, 375–383
  acceleration (a) and, 378–379
  position and, 375
  principle of work and energy for, 183
  procedure for analysis of, 380
  relative-motion analysis for, 375–383
  velocity (v) and, 376–377
  work of friction by, 183
Slipping, 338, 362, 437, 457
  circular motion and, 338, 362
  equations of motion and, 437
  forces that do no work, 457
  general plane motion, 437
  moment of inertia and, 437
  relative-motion analysis and, 338, 362
  rigid-body planar motion, 338, 362, 437, 457
  rolling without, 338, 362
  zero velocity and, 338, 457
Space cone, 537–538, 608
Space mechanics, 160–167, 171, 289–290, 295–297, 305, 567–568, 578, 606–609, 615
  areal velocity, 160
  central-force motion and, 160–167, 171
  circular orbit, 164
  control volume of particles, 289–290
  drag force, 295
  eccentricity (e) of, 162–163, 171
  elliptical orbit, 164–165
  free-flight trajectory, 161, 163
  gravitational attraction (G) and, 161–163
  inertia (I) and, 567–568
  Kepler's laws, 166
  kinetics of a particle, 160–167, 171
  mass flow, 289–290
  orbital revolution, 166
  parabolic path, 164
  parallel-axis and parallel-plane theorems for, 567–568
  power-flight trajectory, 161, 163
  precession (ϕ), 607–608
  principle of impulse and momentum for, 578
  propulsion (mass gain and loss), 295–297, 305
  spin (ψ), 607–608
  three-dimensional rigid-body motion and, 567–568, 578, 606–609, 615
  thrust, 295
  torque-free motion, 606–609, 615
  trajectories, 161–167
Speed, 6–7, 32, 55, 68, 102. *See also* Acceleration; Magnitude
Spheres, fixed-point rotation and, 537, 563
Spin (ψ), 600–602, 607–608
Spinning axes, equations of motion for, 589–590
Spring force, 117, 178–179, 208–211, 225–226, 457, 475, 489–490, 618, 620–625
  conservation of energy and, 475, 490
  conservative force of, 208–211, 226
  displacement by, 457
  elastic potential energy and, 209, 226, 475, 490
  equations of motion for, 117
  kinetics of a particle, 117, 178–179, 208–211, 225–226
  potential function for, 210–211
  rigid-body planar motion, 457, 475, 489–490
  vibrations and, 618, 620–625
  weight and, 209–211
  work (U) of, 178–179, 208–211, 225, 457, 475, 489
Statics, study of, 3
Steady flow, 291–294, 305
  control volume, 291–294
  fluid streams, 291–294, 305
  linear impulse and momentum, 291–294
  procedure for analysis of, 292
Steady-state vibration, 638–639, 644–645
Symmetrical spinning axes, *see* Gyroscopic motion
Systems, 114–115, 182–188, 213, 234–238, 247–253, 276, 278–279, 280, 289–290, 295–297, 456, 538–542, 631, 642–643
  angular impulse of, 278–279
  angular momentum of, 276, 278–279
  angular motion of, 538–542
  center of mass (G), 115
  connected bodies, 631

conservation of angular momentum, 280
conservation of energy, 213, 631
conservation of linear momentum, 247–253
conservative forces and, 213
critically damped, 642
deformation in bodies, 182–183
equations of motion for, 114–115, 456
external and internal forces of, 114–115, 234, 276
fixed, 538–542
kinetic energy and, 456
kinetics of a particle, 114–115, 182–188, 218, 234–238, 247–253, 276, 278–279, 289–290
mass flow, 289–290, 295
mass gain and loss (propulsion), 295–297, 305
moment of a force, 276
nonrigid bodies, 182–183
overdamped, 642
potential energy (V) and, 213
principles of impulse and momentum for, 234–238, 278–279
principle of work and energy for, 182–188
procedure for analysis of, 235, 248
rigid bodies, 182, 456, 538–542
sliding and, 183
time derivatives for, 538–542
translating-rotating, 538–542
underdamped, 643
vibration, 631, 642–643
work of friction and, 183

T

Tangential (t) coordinates, 53–59, 134–139, 171, 315–316
acceleration (a) and, 54–55, 134–139, 315–316
circular motion components, 315–316
curvilinear motion components, 53–59
equations of motion and, 134–139, 171
kinematics, 53–59, 315–316
kinetics, 134–139, 171
particles, 53–59, 134–139, 171
planar motion and, 53, 315–316
procedure for analysis of, 56, 135
rigid-body planar motion, 315–316
rotation about a fixed axis, 315–316
three-dimensional motion, 55
velocity (v) and, 53
Tangential force, 148–149, 171
Three-dimensional motion, 55, 534–563, 564–615
angular, 537–542
angular momentum of, 575–578, 614
angular velocity of, 600–603
binormal axis, 55
curvilinear, 55
displacement from, 535–538
equations of motion for, 586–595, 615
Euler's equations for, 588–589
Euler's theorem for, 536
fixed-axis motion, 588–595
fixed-point rotation, 535–542, 563
frames of reference for, 537–542, 575–576
general, 543–545, 563
gyroscopic motion, 589–590, 600–605, 615
inertia, moments and products of, 565–570, 614
inertial frame of reference for, 575–576
kinematics of, 55, 534–563
kinetic energy of, 578–581, 615
kinetics of, 564–615
particles, 55
principle normal axis, 55
principle of impulse and momentum, 578, 614
principle of work and energy of, 579, 615
procedures for analysis of, 555, 590
rectangular (x, y, z) coordinates, 577
relative-motion analysis of, 552–559, 563
rigid bodies, 534–563, 564–615
rotating axes, 535–542, 552–559, 563
symmetrical spinning axes, 589–590
time derivatives for, 537–542
torque-free motion, 606–609, 615
translating axes, 552–559
translating coordinate systems for, 543–545
translating-rotating systems, 538–542, 563
Thrust, 295

Time (t), 8, 19–20, 32, 33, 102, 117, 166, 231–233, 620
acceleration (a) as a function of, 117
continuous motion and, 8
curvilinear motion and, 32, 33
cycle, 620
erratic motion and, 19–20
graphs of variables, 19–20, 102
linear impulse and momentum as a functions of, 231–233
orbital revolution, 166
period, 620
position (s) as a function of, 8
principle of impulse and momentum and, 231–233
rectilinear kinematics and, 8, 19–20, 102
velocity (v) as a function of, 8, 32, 33
vibration and, 620
Time derivatives, 34–35, 70, 81–82, 104–105, 328, 537–542, 587, 631–632, 654
absolute dependent motion analysis using, 81–82, 105
angular motion, 537–538
conservation of energy and, 631–632, 654
curvilinear motion, 34–35, 70, 104
fixed-point rotation, 537–542
rigid-body planar motion, 328
rotational equations of motion using, 587
three-dimensional rigid-body motion, 537–542, 587
time-differential equations, 81–82, 328
translating-rotating systems, 538–542
vibration, 631–632, 654
Torque-free motion, 606–609, 615
Trajectories, 161–167, 171
central-force motion of, 161–167, 171
circular orbit, 164
eccentricity (e) of, 162–163, 171
elliptical orbit, 164–165
escape velocity of, 164
free-flight, 161, 163
gravitational attraction (G) and, 161–163
hyperbolic, 164
orbital, 164–167
parabolic path, 164
power-flight, 161, 163
space mechanics, 161–167, 171

Transient state of vibration, 638
Translating axes, 93–97, 105, 336–342, 361–368, 389, 538–542, 552–559, 563
 acceleration (*a*), 94, 361–368, 554
 angular motion and, 538–542
 kinematics of particles, 93–97, 105
 observers (fixed and translating), 93, 105
 position coordinates for, 93
 position vectors (**r**) for, 93, 336–337, 553
 procedures for analysis of, 94, 339, 363, 555
 relative-motion analysis of, 93–97, 105, 336–342, 361–368, 389, 552–559, 563
 rigid-body planar motion, 336–342, 361–368, 389
 rotation and, 336–342, 361–368, 388
 three-dimensional rigid bodies, 538–542, 552–559, 563
 time derivatives for systems, 538–542
 translating reference frames, 93–97
 translating-rotating systems, 538–542
 velocity (*v*) of, 93, 336–342, 389, 553
Translating coordinate systems, 543–545, 552
Translating observer, 93
Translating reference frames, 93–97, 105, 336
Translation, 310–311, 328–331, 336–342, 361–368, 375–383, 388–389, 455, 488, 498, 530
 absolute motion analysis, 328–331, 388
 acceleration (*a*) and, 311, 361–368, 378–379, 388
 angular momentum and, 498
 circular motion and, 337–338
 coordinate system axes, 311, 336–342, 388
 coordinating fixed and translating reference frames, 336–342, 389
 curvilinear, 310–311, 388
 displacement (**r**) and, 336–337
 impulse and momentum, 498, 530
 kinetic energy and, 455, 488
 linear momentum and, 498
 paths of, 310
 position vectors (**r**), 311, 375
 procedures for analysis of, 328, 363, 380

rectilinear, 310–311, 388
relative-motion analysis, 336–342, 361–368, 375–383, 389
rigid-body planar motion, 310–311, 328–331, 375–383, 388–389, 455, 488, 498, 530
 rotating axes with, 375–383, 389
 rotation and, 328–331, 336–342, 361–368
 time derivatives for, 328
 velocity (*v*) and, 311, 336–342, 376–377, 388
Translational equations of motion, 407, 410–415, 449, 488, 586, 615
 acceleration (*a*) for, 407, 410–415, 449
 curvilinear translation, 310–311, 388, 411, 449
 force (*F*) for, 407, 410–415, 449
 kinetic energy and, 455, 488
 procedure for analysis using, 412, 615
 rectilinear translation, 410–411, 449
 rigid-body planar motion, 407, 410–415, 449, 455, 488
 symmetry of reference frames for, 407
 three-dimensional rigid-body motion, 586, 615
Transverse component (**v**$_\theta$), 69
Transverse coordinate (θ), 67–70
Trigonometric identities, 656

U

Unbalanced force, 109–110, 115
Undamped vibrations, 617–625, 637–640, 654
 amplitude of, 619–620, 638
 displacement and, 617–625
 equilibrium position for, 618–620, 654
 forcing frequency (ω_0) for, 637–640, 654
 forced, 617, 637–640, 654
 free, 617–625, 654
 frequency (*f*) of, 620
 frequency ratio, 638
 natural frequency (ω_n) for, 618, 620–621, 654
 period of, 620
 periodic force and, 637–640
 periodic support displacement of, 639
 phase angle (ϕ), 620

 procedure for analysis of, 621
 resonance, 639, 654
 simple harmonic motion of, 618
 spring force and, 618, 620–625
Underdamped vibration systems, 643
Unit vectors, 659

V

v–s (velocity–position) graphs, 21
v–t (velocity–time) graphs, 19–20
Variable force, work of, 176, 456
Vector analysis, 659–663
Vector formulation of angular momentum, 274, 279
Vector functions, 663
Vector of forces, 112
Vector quantity, particle position and displacement as, 5, 34
Velocity (*v*), 6–8, 19–21, 32–35, 53, 68, 93, 102, 160–161, 164, 311, 312, 314, 317, 336–342, 349–355, 376–377, 388–389, 457, 537–542, 553, 600–603
 absolute, 93, 337
 acceleration (*a*) and, 7–8, 33
 angular (ω), 68, 312, 317, 537–542, 600–603
 areal, 160
 average, 6, 32
 central-force motion and, 160–161
 circular motion and, 314, 336–342
 continuous motion and, 6–8
 coordinating fixed and translating reference frames, 336–342, 389
 curvilinear motion and, 32–35, 53, 68
 cylindrical components and, 68
 direction and, 6, 32, 34–35, 53, 68, 376
 erratic motion and, 19–21
 escape, 164
 fixed-axis rotation and, 312, 314, 388
 fixed-point rotation and, 537–542
 forces doing no work, 457
 graphs of variables, 19–21, 102
 gyroscopic motion and, 600–603
 instantaneous, 6, 32
 instantaneous center (IC) of zero, 349–355, 389
 kinematics of particles and, 6–8, 19–21, 32–35, 53, 68, 93, 102
 magnitude of, 6, 32, 34–35, 53, 68, 376, 388

normal component (*n*) coordinates, 53, 388
position (**r**) and, 312, 314, 336–337
position (*s*), as a function of, 8
procedures for analysis of, 317, 339, 351
radial component ($\mathbf{v}_r$), 68
rectangular components and, 34–35
rectilinear kinematics and, 6–8, 19–21, 102
relative, 93, 337–338
relative-motion analysis and, 93, 336–342, 376–377, 389, 553
rigid-body planar motion, 311, 312, 314, 317, 336–342, 349–355, 376–377, 388–389
rolling without slipping, 338
rotating axis, 376–377, 389, 553
rotation and, 312, 314, 317, 388–389
sign convention for, 6
slipping and, 338, 376–377, 457
space mechanics, 160–161
speed (magnitude), 6, 32, 68
tangential component (*t*) coordinates, 53, 68, 388
three-dimensional rigid-body motion, 537–542, 553, 600–603
time (*t*), as a function of, 8
time derivative for, 537
translating axes and, 336–342, 389, 553
translating observer of, 93
translation and, 311, 388
translation and rotation, 336–342, 376–377
transverse component ($\mathbf{v}_\theta$), 68
zero, 338, 349–355, 389, 457
Vertical displacement (Δ), 177, 457
Vertical projectile motion, 39–40
Vibrations, 616–655
amplitude of, 619–620, 638
complementary solution for, 637–638
critically damped systems, 642
cycle, 620
damped, 617, 641–646, 654–655
displacement and, 617–625
electrical circuit analogs and, 647, 655
energy methods for conservation of, 631–634, 654
equilibrium position for, 618–620, 631, 637, 654
forced, 617, 637–640, 644–646, 654–655
forcing frequency (ω_0), 637–640, 654
free, 617–625, 641–643, 654
frequency (*f*), 618, 620–621, 637–640, 643
frequency ratio, 638
magnification factor (MF) for, 638–639, 645
natural frequency (ω_n), 618, 620–621, 631–632, 643, 654
overdamped systems, 642
particular solution for, 637–638
period of time, 620
periodic force and, 637–640
periodic support displacement of, 639
phase angle (ϕ), 620
procedures for analysis of, 621, 632
resonance, 639, 645, 654–655
simple harmonic motion of, 618
spring force and, 618, 620–625
steady-state of, 638–639
transient state of, 638
undamped forced, 637–640, 654
undamped free, 617–625, 654
underdamped systems, 643
viscous damped, 641–646, 654–655
Viscous damped vibration, 641–646, 654–655
coefficient of damping, 641
critically damped systems, 642
damping force, 641–646, 654
forced, 644–646, 655
free, 641–643, 654
overdamped systems, 642
steady-state, 644–646
underdamped systems, 643
Viscous damping force, 641–646, 654
Volume elements, integration of moments of inertia using, 394–395
Volumetric flow (discharge), 290

W

Watt (W), unit of, 199
Weight (*W*), 111, 177, 208–212, 225–226, 295, 457, 475, 489
conservation of energy and, 212, 226, 475
conservative forces and, 208–212, 226
constant, 208
displacement of, 177, 208–211, 225
gravitational attraction and, 111
gravitational potential energy of, 208, 475
mass gain and loss, 295
mass of a body and, 111
potential energy (*V*) and, 208–212, 475
potential function for, 210–211
spring force and, 209–211
vertical displacement of, 457
work (*U*) of a, 177, 208–212, 225, 457, 475, 489
Work (*U*), 174–228, 452–492, 579, 615
conservation of energy and, 212–218, 226, 475–480, 490
conservative forces and, 208–218, 226, 475–480
constant force, 177, 225, 456, 489
couple moment (*M*), of a, 458–459, 489
deformation and, 182–183
displacement and (*d*), 175–177, 182–183, 225, 457–458, 489
energy (*E*) and, 174–228, 452–492, 579
external, 183
force (**F**) as, 175–179, 182–188, 225–226, 456–459, 489
friction caused by sliding, 183
internal, 183
kinetic energy and, 212–213, 579
kinetics of a particle, 174–228
nonconservative forces and, 476
potential energy (*V*) and, 208–211, 226, 475–480, 490
potential function for, 210–211
principle of energy and, 180–188, 225–226, 460–466, 490, 579, 615
procedures for analysis of, 181, 461, 477
rigid-body planar motion, 452–492
slipping (no work), 457
spring force as, 178–179, 208–211, 225, 457, 489–490
system of particles, 182–188
three-dimensional rigid body motion, 579, 615
units of, 176
variable force, of a, 176, 456
weight (*W*) as, 177, 208–212, 225, 457, 489
zero velocity and (no work), 457

Z

Zero velocity, 338, 349–355, 389, 437, 457
  general plane motion, 437
  instantaneous axis of, 349
  instantaneous center (IC) of, 349–355, 389, 437
  procedure for analysis of, 351
  relative-motion analysis, 338, 389
  rigid-body planar motion, 349–355, 389
  rolling without slipping, 338
  slipping (no work) and, 457

Geometric Properties of Line and Area Elements

Centroid Location	Centroid Location	Area Moment of Inertia
Circular arc segment $L = 2\theta r$, $\dfrac{r \sin \theta}{\theta}$	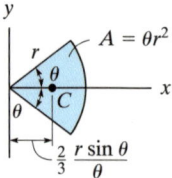 Circular sector area $A = \theta r^2$, $\dfrac{2}{3}\dfrac{r \sin \theta}{\theta}$	$I_x = \dfrac{1}{4} r^4 \left(\theta - \dfrac{1}{2}\sin 2\theta\right)$ $I_y = \dfrac{1}{4} r^4 \left(\theta + \dfrac{1}{2}\sin 2\theta\right)$
Quarter and semicircle arcs $L = \dfrac{\pi}{2} r$, $\dfrac{2r}{\pi}$; $L = \pi r$	Quarter circle area $A = \dfrac{1}{4}\pi r^2$, $\dfrac{4r}{3\pi}$	$I_x = \dfrac{1}{16}\pi r^4$ $I_y = \dfrac{1}{16}\pi r^4$ $I_{x'} = \left(\dfrac{\pi}{16} - \dfrac{4}{9\pi}\right) r^4$ $I_{y'} = \left(\dfrac{\pi}{16} - \dfrac{4}{9\pi}\right) r^4$
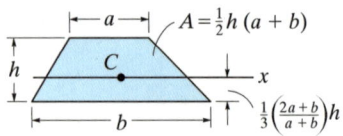 Trapezoidal area $A = \dfrac{1}{2} h (a+b)$, $\dfrac{1}{3}\left(\dfrac{2a+b}{a+b}\right) h$	Semicircular area $A = \dfrac{\pi r^2}{2}$, $\dfrac{4r}{3\pi}$	$I_x = \dfrac{1}{8}\pi r^4$ $I_y = \dfrac{1}{8}\pi r^4$
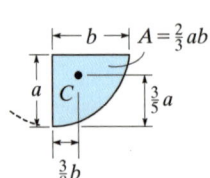 Semiparabolic area $A = \dfrac{2}{3} ab$, $\dfrac{3}{5} a$, $\dfrac{3}{8} b$	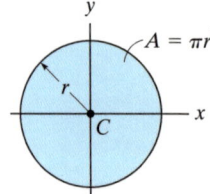 Circular area $A = \pi r^2$	$I_x = \dfrac{1}{4}\pi r^4$ $I_y = \dfrac{1}{4}\pi r^4$
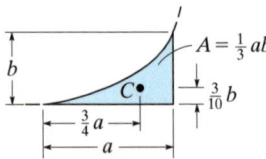 Exparabolic area $A = \dfrac{1}{3} ab$, $\dfrac{3}{4} a$, $\dfrac{3}{10} b$	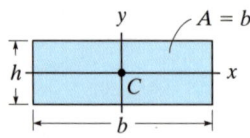 Rectangular area $A = bh$	$I_x = \dfrac{1}{12} bh^3$ $I_y = \dfrac{1}{12} hb^3$
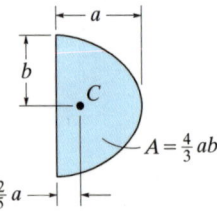 Parabolic area $A = \dfrac{4}{3} ab$, $\dfrac{2}{5} a$	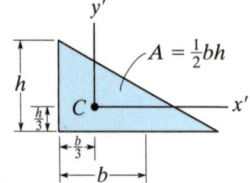 Triangular area $A = \dfrac{1}{2} bh$	$I_{x'} = \dfrac{1}{36} bh^3$ $I_{y'} = \dfrac{1}{36} hb^3$